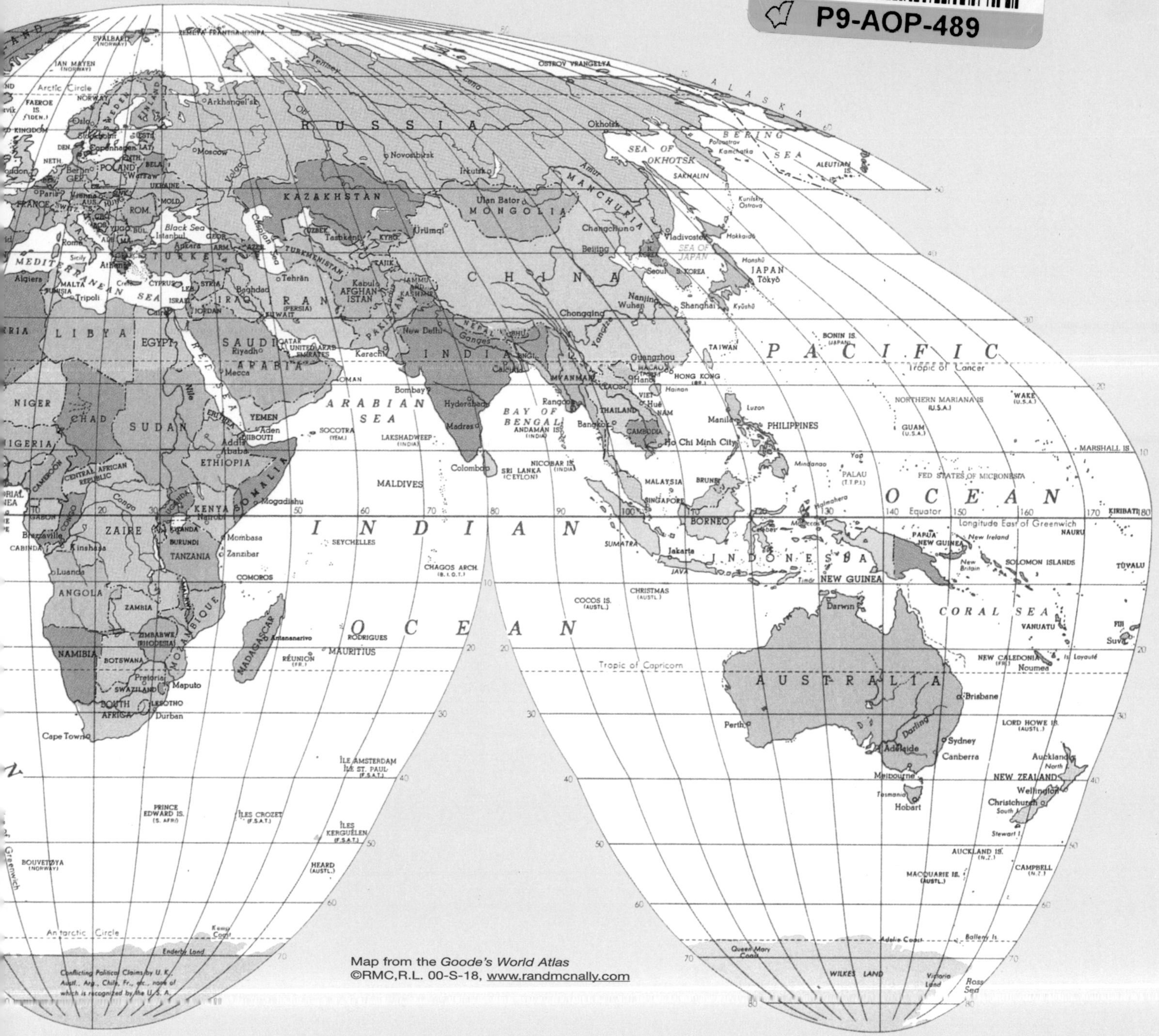

Goode's Homolosine Equal Area Projection

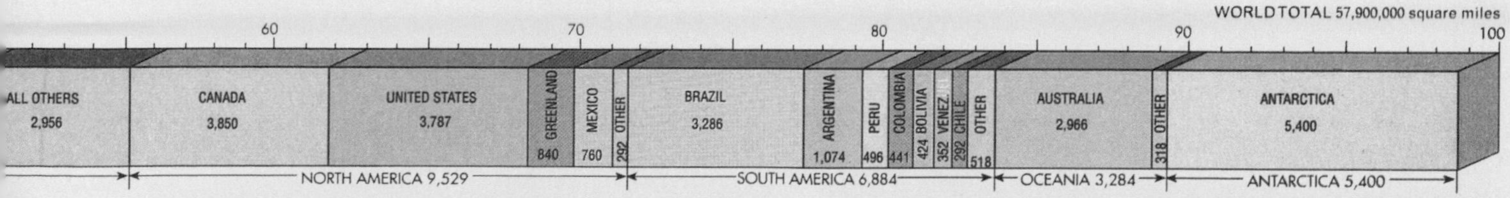

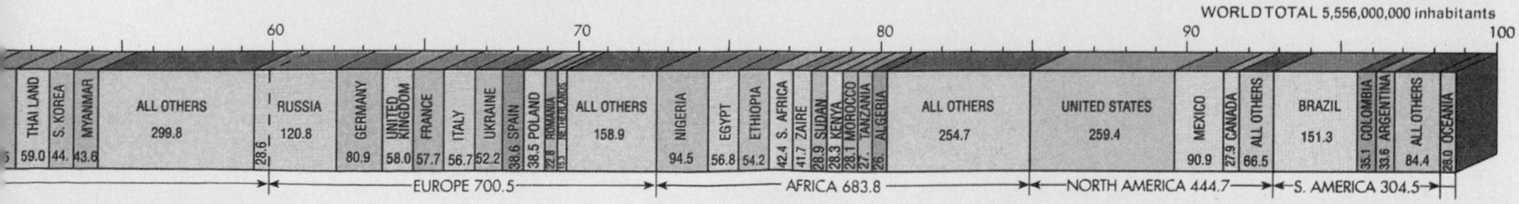

Environmental Science

SIXTH EDITION

Environmental Science

A GLOBAL CONCERN

William P. Cunningham
University of Minnesota

Barbara Woodworth Saigo
Saiwood Biology Resources
Saiwood Publications

McGraw Hill

Boston Burr Ridge, IL Dubuque, IA Madison, WI New York San Francisco St. Louis
Bangkok Bogotá Caracas Lisbon London Madrid
Mexico City Milan New Delhi Seoul Singapore Sydney Taipei Toronto

McGraw-Hill Higher Education
A Division of The McGraw-Hill Companies

ENVIRONMENTAL SCIENCE: A GLOBAL CONCERN, SIXTH EDITION

Published by McGraw-Hill, an imprint of The McGraw-Hill Companies, Inc., 1221 Avenue of the Americas, New York, NY 10020.

This book is printed on recycled, acid-free paper containing 10% postconsumer waste.

2 3 4 5 6 7 8 9 0 QPH/QPH 0 9 8 7 6 5 4 3 2 1

ISBN 0-07-290932-3
ISBN 0-07-118072-9 (ISE)

Vice president and editor-in-chief: *Kevin T. Kane*
Publisher: *Michael D. Lange*
Senior sponsoring editor: *Margaret J. Kemp*
Senior developmental editor: *Kathleen R. Loewenberg*
Editorial assistant: *Dianne Berning*
Marketing managers: *Michelle Watnick/Heather K. Wagner*
Project manager: *Mary Lee Harms*
Production supervisor: *Laura Fuller*
Design manager: *Stuart D. Paterson*
Cover/interior designer: *Jamie O'Neal*
Cover image: *Bryan and Cherry Alexander Photography*
Senior photo research coordinator: *Lori Hancock*
Photo research: *Connie Mueller*
Supplement coordinator: *Stacy A. Patch*
Compositor: *Precision Graphics*
Typeface: *10/12 Times Roman*
Printer: *Quebecor Printing Book Group/Hawkins, TN*

The credits section for this book begins on page 613 and is considered an extension of the copyright page.

In the cover photo, Kavavow Kiguktak, an Inuit hunter from the village of Grise Fiord on Ellesmere Island high in the Canadian arctic, jumps across ice floes as he hunts seals on the Arctic Ocean. He depends on experience, courage, and traditional knowledge to find his way through this dangerous, constantly shifting environment. As remote and exotic as Kav's life may seem to most of us, he also is interconnected to the modern world. In 1999, the government of Canada transferred two million square kilometers of land back to the Inuit people. Commercial air routes and modern telecommunications now connect the widely dispersed residents of the vast new territory of Nunavut.

But links to the outside world also bring problems to these northern people. Alcoholism, AIDS, and suicide are the main killers of Inuit people between 15 and 35 years old. Air pollutants from cities and industries in Europe, Russia, China, and North America drift northward into the Arctic. During the winter, the industrial smog in remote areas of Nunavut can be higher than in London or New York. Furthermore, as contaminants accumulate and are concentrated in arctic food webs, they reach toxic levels in the fish, seals, and sea birds eaten by the Inuit.

This photo provides a good metaphor for our global environmental situation. Like Kav, we face dangerous, rapidly changing environmental conditions. It will take courage, knowledge, and effort to find our way out of our current environmental dilemmas. I hope you'll find information and inspiration in this book to help in your journey toward a better, safe, more equitable world.

Library of Congress Cataloging-in-Publication Data

Cunningham, William P.
Environmental science : a global concern / William P. Cunningham, Barbara Woodworth Saigo. —6th ed.
p. cm.
Includes bibliographical references and index.
ISBN 0-07-290932-3
1. Environmental sciences. I. Saigo, Barbara Woodworth. II. Title.

GE105 .C86 2001
363.7—dc21 00-024459
CIP

INTERNATIONAL EDITION ISBN 0-07-118072-9

www.mhhe.com

Contents in Brief

A GLOBAL CONCERN

Contents
A GLOBAL CONCERN

Preface

A Global Concern

危机

The Chinese ideogram for "crisis" combines the character *wei,* meaning danger, with the character *ji,* denoting opportunity. This is a good description of our global environment situation. We are dangerously close to pushing biological communities and biophysical processes beyond the point from which they can recover. At the same time, our greatly increased understanding of the natural world and how it works gives us an opportunity to repair the damage we have caused and to find new, more efficient, and more environmentally friendly ways of providing the goods and services we need. British ecologist Norman Myers points out that the current generation of students has the advantage of being the first in history to have the information, resources, and motivation to do something to solve our environmental crisis. Unfortunately, he adds, if solutions to some of our most pressing problems aren't found quickly, this generation also may be the last to have a chance to do so.

I hope that you will find this book a valuable source of information about our global environment, as well as an inspiration for solutions to the dilemmas we face. Everyone has a role to play in this endeavor. Whether as students, educators, researchers, activists, or consumers, each of us can find ways to contribute in solving our common problems.

WHO MAKES UP THE AUDIENCE FOR THIS TEXT?

This book is intended for use in a one- or two-semester course in environmental science, human ecology, or environmental studies at the college or advanced placement high school level. Because most students who will use this book are freshman or sophomore nonscience majors, I have tried to make the text readable and accessible without technical jargon or a presumption of prior science background. At the same time, enough data and depth are presented to make this book suitable for many upper-division classes and a valuable resource for students who will keep it in their personal libraries after their formal studies are completed.

WHY DID I WRITE THIS BOOK?

I have taught aspects of environmental science in a variety of settings for about 40 years. Although the earliest of these classes focused primarily on natural history and conservation, I found my interests and concerns changing in the 1970s. Two broad areas of environmental science that seemed important to me weren't covered in the existing textbooks. One of these is global concerns. We live in an highly interconnected world; the coal burned in China, or the nuclear waste dumped in the ocean by Russia, or the pesticides used on farm crops in Central America affect all of us. The other area is environmental justice and the human dimensions of environmental issues. Although my original interests in the environment were primarily wilderness and wildlife issues, the Civil Rights Movement of the 1960s opened my eyes to the conditions in inner cities and the role of economics, health, policy, and law in environmental concerns. Ecology remains the heart of environmental science, but students also need to know something about the roles of human institutions and social sciences to be educated environmental citizens. It's gratifying to see that since the first edition of this book was published in 1990, concerns about environmental ethics and social justice are appearing in other textbooks as well.

WHY THIS COVER PHOTO?

I'm especially pleased with the cover photograph of this edition of *Environmental Science.* It shows Kavavow Kiguktak, an Inuk hunter jumping across ice floes as he looks for seals off the coast of Ellesmere Island in Canada's high arctic. In spite of the extremely harsh conditions they face, the Inuit people have used traditional knowledge and a strong ethic of cooperation and sharing to live in harmony with their environment for thousands of years. Cultural traditions inform Kav about how many seals he can take from the common pool resource, the respect he must show his prey, and how he shares his catch with other members of his community. He depends on experience and courage to find his way across the constantly shifting ice floes.

In several ways, this image also serves as a good metaphor for our current environmental issues. As he jumps from one ice block to the next in the ever-shifting ice pack, Kav is constantly looking ahead, evaluating risks and opportunities, and making reasoned judgments. He faces great danger—a fall into the frigid ocean could be fatal—but the ice also provides an opportunity to move far out from the shore as he hunts. Simply standing still just isn't an option in this situation; Kav must keep moving to stay afloat. In much the same way, it takes courage, foresight, and nimble footwork for us to find a way out of the complex, constantly changing environmental problems we now face. We can't simply remain where we are, but must use every skill at our disposal to find a safe path to the future.

Although he lives in a remote and exotic part of the world, Kav is affected by events occurring far away. As a resident of the new territory of Nunavut, he is participating in a bold new experiment in self-rule and self-determination. New job opportunities are opening up for First Nations people as they take control of their ancestral territory. During the brief arctic summer, Kav, for instance, is a seasonal ranger at the remote Ellesmere National Park. In the summer of 1998, I made a two-week backpacking trip to this harsh, but beautiful landscape (see figure 15.9). One of the main attractions of hiking there is the opportunity to see arctic animals such as the white wolves, musk ox, Perry caribou, narwhal, polar bears, and arctic hares, all of which have had so little contact with humans that they are generally unafraid of us. Unfortunately, there seems to be a widespread decline in many of these arctic species. The cause isn't known for certain, but one suspicion is that global climate change coupled with pollution carried by wind currents from industrial centers far to the south may be playing a role. While a disappointment to visitors, this decline could be devastating for residents like Kav who try to maintain traditional ways of life.

WHAT'S NEW IN THIS EDITION?

The most difficult thing about writing a broad, introductory text like this is incorporating all the interesting and important material, keeping it current, and still resisting the temptation to let it grow to encyclopedic size. By careful pruning, I've managed to add significant new features to each edition of this book while still maintaining a reasonable length. I have had the good fortune in working on this revision to have more than one hundred reviews from colleagues at colleges and universities across North America. They've offered many useful criticisms and suggestions for improving the book. If you are one of those reviewers, you have my most sincere thanks.

Two New Chapters

- A new introductory chapter presents suggestions to students about why environmental science is interesting and useful, how to study, how to prepare for tests, critical thinking, and concept maps. These topics are presented in the beginning of the book so students can begin to use them immediately. This is the kind of information that most of us cover in the first lecture of a class. No other book goes into the *fundamentals of critical thinking theory and application* found here.
- The other new chapter, "Environmental Policy, Planning and Law," begins with a discussion of the policy cycle of agenda setting, problem definition, implementation, and evaluation by which public policy is established. The environmental law section defines statutory, case, and administrative law, a level of sophistication never before presented in an environmental science textbook. Also included are current concepts such as alternative dispute resolution, wicked problems, resilience in ecosystems and institutions, the precautionary principle, arbitration and mediation, and collaborative approaches to community-based planning. If our students are going to be educated environmental citizens, they need to know how these processes work.

New Case Studies

Every chapter in this book begins with a case study designed to introduce the main topic and pique student interest. Three-quarters of these case studies are new to this edition and most are based on *very recent news stories* to emphasize the currency of environmental issues. All of them have been expanded from previous versions to be more substantive and meaningful. Overall, the total number of boxed readings have been reduced in this edition so that the remaining ones can be more substantial.

More Environmental Ethics

Because critical thinking has been moved from chapter 2, "Environmental Ethics and Philosophy," to the new introductory chapter, space is now available to present environmental ethics in greater detail. Notice that this discussion takes a pluralist approach. There is no prescribed "earthmanship ethics" that divide the world into bipolar camps. A variety of worldviews and ethical perspectives are presented and students are invited to think for themselves. Similarly, the discussion of science isn't limited to positivist, reductionist approaches, but recognizes the validity of descriptive and interpretive sciences.

New Information on Ecological Economics

Chapter 8, "Ecological Economics," has been revised to include a major new section on green business, eco-efficient economy, the Natural Step movement, and "design for the environment." These topics are both very current and are also positive examples of what we can do to improve environmental quality.

Other Significant Changes (A complete list of changes to this edition can be found on the accompanying website.)

- Data in tables, graphs, figures, and the text copy have been thoroughly updated.
- Chapter 9, "Environmental Health and Toxicology," has a new treatment of infectious diseases that emphasizes emergent diseases and how drug resistance is selected for different microorganisms.
- Major revisions and corrections have been incorporated in chapter 16, "Environmental Geology," including new information on earthquakes and flooding, and a new opening case study, "Earthquake in Turkey."
- Chapter 17, "Air, Climate, and Weather," has a helpful new illustration on how tornadoes form, and a new diagram and discussion on El Niño/Southern Oscillations.
- Chapters 19, "Water Use and Management," and 20, "Water Pollution," have been updated with new data and current information.
- Chapter 21, "Conventional Energy," opens with a new case study about the potential for a vast new oil supply under and around the Caspian Sea and how that affects regional politics, including the war in Chechnya.
- An important section has been added to chapter 22, "Sustainable Energy," on the current topics of fuel cells and hybrid gas/electric automobiles.
- Much new and/or corrected material appears in chapter 23, "Solid, Toxic, and Hazardous Waste," such as international toxic shipping, and in chapter 25, "What Then Shall We Do?." Chapter 25 now focuses much more succinctly than in the past on ways we can work individually and cooperatively to build a better world.

HOW IMPORTANT IS SUSTAINABILITY AND ENVIRONMENTAL CITIZENSHIP?

Ultimately the aims of this book are to foster attitudes of stewardship and environmental citizenship, and to encourage the goals of economic, ecological, and social sustainability. In the preamble to the United Nations Earth Charter, the authors declare, "In an increasingly interdependent world, it is imperative that we, the citizens of Earth, declare our responsibility to one another, the greater community of life, and future generations." Among the principles proposed by the Earth Charter are:

1. Respect Earth and all life, recognize the interdependence and intrinsic value of all beings.
2. Care for the community of life in all its diversity as a responsibility shared by everyone.
3. Strive to build free, just, participatory, and sustainable communities.
4. Secure peace and Earth's abundance and beauty for present and future generations.

I hope the readers of this book will come to share those goals and to understand the reasons they are so important.

—William P. Cunningham

RELATED TITLES OF INTEREST

1. *Field and Laboratory Activities Manual* (0-07-290913-7) by Enger and Smith.
2. *Annual Editions: Environment 99/00* (0-07-228498-6) Editor: John L. Allen.
3. *Taking Sides: Clashing Views on Controversial Environmental Issues* (0-07-303184-4) Editor: Theodore D. Goldfarb.
4. *Sources: Notable Selections in Environmental Studies* (0-07-303186-0) Editor: Theodore D. Goldfarb.
5. *The Dushkin Student Atlas of Environmental Issues* (0-697-36520-4) Editor: John Allen. University of Connecticut.
6. *Life Science Living Lexicon* (CD = 0-697-37993-0; Print = 0-697-12133-X) by William Marchuk.
7. *You Can Make a Difference: Be Environmentally Responsible* (0-07-292416-0) by Judy Getis.
8. *Environmental Ethics: Divergence and Convergence* (0-07-006180-7) by Botzler and Armstrong.
9. *Environmental Problem-Solving: A Case Study Approach* (0-07-027686-2) by Isobel W. Heathcote.
10. *Eyewitness World Atlas CD-ROM* (0-07-233220-4), published by Dorling-Kindersley.

ACKNOWLEDGMENTS

Many people have contributed in a variety of ways to each edition of this book. Mary Ann Cunningham provided advice and helpful suggestions throughout the writing process. She also has written a superb series of essays on current global environmental issues that appears on the web page world map. Laura Ragan wrote the initial draft of the case study about Chattanooga for chapter 24. Joel Burken of the University of Missouri provided information for the case study about arsenic in drinking water in India for chapter 20. I am particularly grateful to Amanda Woods McConney of Western Washington University, Jim Oris of Miami University of Ohio, Eric Anderson of the University of Wisconsin—Stevens Point, Jerry Hinckley of the College of Lake County, and Darby Nelson of Anoka Ramsey Community College, all of whom attended a workshop that provided some extremely valuable insights for this edition. In addition, Amanda has written an excellent class activities and assessment guide that will be a great help for instructors teaching with this book. Jim has created a set of web-page lecture notes that make an outstanding template for environmental science courses. And Darby has written a thoughtful student study guide that will be exceptionally helpful in assisting students to study in an effective and efficient way. I remain grateful to Karen Warren of Macalester College from whom I learned most of what I know about critical thinking and environmental philosophy. Caroline Getty, a hiking companion in Ellesmere National Park, led me to the photography of Cherry and Brian Alexander that appears on the book cover.

I'm also indebted to all the students and teachers who have sent helpful suggestions, corrections, and recommendations for improving this book. Unfortunately, space didn't permit inclusion of all the excellent ideas that were provided. All have been saved, however, and will be helpful in future editions. I hope that those who read this edition will offer their advice and insights as well. Little of the vast range of material in this book represents my own personal research. All of us owe a great debt to the many scholars whose work forms the basis of our understanding of environmental science. We stand on the shoulders of giants. If errors persist in spite of my best efforts to root them out, I accept responsibility and ask for your indulgence.

I want to express my appreciation to the entire McGraw-Hill book team for their wonderful work in putting together this edition. Kathy Loewenberg oversaw the developmental stages and has made many creative contributions to this book. Mary Lee Harms was production project manager and kept everything running smoothly. Cathy Conroy did an excellent job of copyediting and spotting errors and inconsistencies. Connie Mueller found superb photographs. The folks at Precision Graphics did an excellent job of composition and page layout. Michelle Watnick and Marge Kemp have supported this project with their enthusiasm and creative ideas.

I gratefully acknowledge the constructive criticism of the many colleagues from across North America who provided reviews of individual chapters for this revision. They include:

Lewis E. Allen
SUNY-Empire State College

Martha S. Andersen
Erie Community College

Max Anderson
University of Wisconsin—Platteville

Donald F. Anthrop
San Jose State University

Marilyn Barger
FAMU-FSC College of Engineering

Sharmistha Basu-Dutt
State University of West Georgia

Robert G. Benson
Adams State College

Joel G. Burken
University of Missouri-Rolla

Joseph G. Burleigh
University of Arkansas—Pine Bluff

Judith Byrnes-Enoch
SUNY-Empire State College

Ted T. Cable
Kansas State University

Sharelle Maree Campbell
Louisiana State University-Shreveport

Kenneth J. Camplese
Skagit Valley College

Hugh Canham
SUNY College of Environmental Science & Forestry

W.B. Clapham, Jr.
Cleveland State University

Terence H. Cooper
University of Minnesota

William C. Culver
St. Petersburg Junior College

William Dritschilo
Green Mountain College

John C. England
Central Carolina Technical College

David Engle
Oklahoma State University

David L. Evans
Pennsylvania College of Technology

Henriette K. Evans
Pennsylvania College of Technology

Jess W. Everett
Rowan University

Harry Field
Oklahoma State University

John T. Finn
University of Massachusetts

D. E. Fisher
University of Miami

Jianbang Gan
Tuskegee University

Dalton R. Gossett
Louisiana State University—Shreveport

Gian Gupta
University of Maryland-Eastern Shore

Oliver J. Hao
University of Maryland

Robert M. Hordon
Rutgers University

Sam Huffman
University of Wisconsin—River Falls

Donald W. Humphreys
Temple University

Huey-Min Hwang
Jackson State University

Jeffery Jack
Western Kentucky University

Gina Johnston
California State University-Chico

Michael E. Karian
St. Cloud State University

Carol A. Kimmons
University of Tennessee—Chattanooga

Robert W. Kingsolver
Kentucky Wesleyan College

John C. Kinworthy
Concordia University

Ned J. Knight
Linfield College

Penelope M. Koines
University of Maryland

Mark Kozubowski
Bethany College

Jayant S. Lal
Philadelphia College of Textiles & Science

James W. Langdon
University of South Alabama

Danita Maynard LaSage
University of Kentucky, Lexington Community College

Michael Levandowsky
School of Visual Arts, New York

L. Yu Lin
Christian Brothers University

John H. Litcher
Wake Forest University

Paul E. Lutz
Lenoir-Rhyne College

Thomas R. MacDonald
University of San Francisco

Michael J. Manetas
Humboldt State University

Michael Howard Marcovitz
Midland Lutheran College

Harry A. Meyer
McNeese State University

L. Ray Miller
York College

Patricia Mosto
Rowan University

Muthena Naseri
Moorpark College

Udeme J. Ndon
San Jose State University

Lisa H. Newton
Fairfield University

Thomas G. O'Brien
Nassau Community College

Chuks A. Ogbonnaya
Mountain Empire Community College

William P. Patterson
Syracuse University

John Pichtel
Ball State University

Mark Plunkett
Bellevue Community College

Rosann Poltrone
Arapahoe Community College

Paul H. Reitan
University at Buffalo-SUNY

Penny ReVelle
Essex Community College

Lynette Rushton
South Puget Sound Community College

John M. Rybczyk
California University of Pennsylvania

Robert Sanford
University of Southern Maine

Panagiotis D. Scarlatos
Florida Atlantic University

R. Wane Schneiter
Virginia Military Institute

Jeffrey L. Seibert
Baker College

Janet Anne Sherman
Pennsylvania College of Technology

Kathryn Springsteen
Colby-Sawyer College

Dion C. Stewart
Adams State College

R. Bruce Sundrud
Harrisburg Area Community College

Max R. Terman
Tabor College

Francesco Tubiello
Columbia University

S. Kant Vajpayee
University of Southern Mississippi

Stephen J. Ventura
University of Wisconsin—Madison

Warren Viessman, Jr.
University of Florida

John A. Wiggins
New Jersey Institute of Technology

Richard J. Wright
Valencia Community College

LEARNING AIDS

This book is designed to be useful as a self-education tool for students. To facilitate studying and encourage higher-level thinking, each chapter begins with a set of **Objectives** based on major concepts that students should master, and a listing of **ESP CD-ROM Topics** that are chapter-related. A short **Opening Story,** taken from recent news events, follows, setting the subject in context and illuminating the importance of the material to be discussed. **Key Terms,** indicated by boldface type, are defined in context where they are first used and are also listed in the **Glossary** for quick reference.

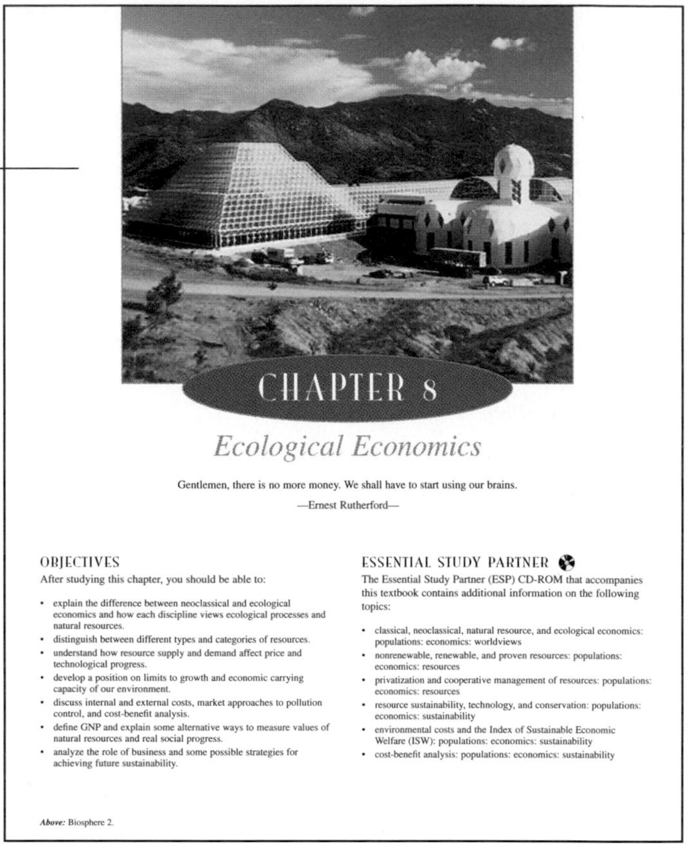

CHAPTER 8

Ecological Economics

Gentlemen, there is no more money. We shall have to start using our brains.

—Ernest Rutherford—

OBJECTIVES

After studying this chapter, you should be able to:

- explain the difference between neoclassical and ecological economics and how each discipline views ecological processes and natural resources.
- distinguish between different types and categories of resources.
- understand how resource supply and demand affect price and technological progress.
- develop a position on limits to growth and economic carrying capacity of our environment.
- discuss internal and external costs, market approaches to pollution control, and cost-benefit analysis.
- define GNP and explain some alternative ways to measure values of natural resources and real social progress.
- analyze the role of business and some possible strategies for achieving future sustainability.

ESSENTIAL STUDY PARTNER

The Essential Study Partner (ESP) CD-ROM that accompanies this textbook contains additional information on the following topics:

- classical, neoclassical, natural resource, and ecological economics: populations: economics: worldviews
- nonrenewable, renewable, and proven resources: populations: economics: resources
- privatization and cooperative management of resources: populations: economics: resources
- resource sustainability, technology, and conservation: populations: economics: sustainability
- environmental costs and the Index of Sustainable Economic Welfare (ISW): populations: economics: sustainability
- cost-benefit analysis: populations: economics: sustainability

Above: Biosphere 2.

IN DEPTH. . . .

Arsenic in Drinking Water

When we think of water pollution, we usually visualize sewage or industrial effluents pouring out of a discharge pipe, but there are natural toxins that threaten us as well. One of these is arsenic, a common contaminate in drinking water that may be poisoning millions of people around the world. Arsenic has been known since the fourth century B.C. to be a potent poison. It has been used for centuries as a rodenticide, insecticide, and weed killer, as well as a way of assassinating enemies. Because it isn't metabolized or excreted from the body, arsenic accumulates in hair and fingernails, where it can be detected long after death. Napoleon Bonaparte was found recently to have high enough levels of arsenic in his body to suggest he was poisoned.

Perhaps the largest population to be threatened by naturally occurring groundwater contamination by arsenic is in West Bengal, India, and adjacent areas of Bangladesh. Arsenic, in the form of insoluble salts, occurs naturally in the bedrock that underlies much of this region. Under normal conditions, the groundwater stays relatively free of arsenic in a soluble form. Rapid population growth, industrialization, and intensification of agricultural irrigation, however, have put increasing stresses on the limited surface water supplies of this region (see Chapter 19). Groundwater has all but replaced other water sources for most people in West Bengal, especially in the dry season.

In the 1960s, thousands of deep tube wells were sunk throughout the region to improve water supplies. Much of this humanitarian effort was financed by loans from the world bank in the name of human development. At first, villagers were suspicious of well water, regarding it as unnatural and possibly evil. But as surface water supplies diminished, dusty Bengali villages became more and more dependent on this new source of supposedly fresh, clean water. By the late 1980s, health workers

West Bengal and adjoining areas of Bangladesh have hundreds of millions of people who may be exposed to dangerous arsenic levels in well water.

Sediment

Rivers have always carried sediment to the oceans, but erosion rates in many areas have been greatly accelerated by human activities. As chapters 11 and 19 describe, some rivers carry astounding loads of sediment. Erosion and runoff from croplands contribute about 25 billion metric tons of soil, sediment, and suspended solids to world surface waters each year. Forests, grazing lands, urban construction sites, and other sources of erosion and runoff add at least 50 billion additional tons. This sediment fills lakes and reservoirs, obstructs shipping channels, clogs hydroelectr... and makes purifi-
cation of drinking water more cost...
beds in which insects take refuge a...
is blocked so that plants cannot ca...
gen levels decline. Murky, cloudy...
swimming, boating, fishing, and o...
Sediment also can be benef...
ishes floodplain farm fields. Se...
river mouths creates valuable del...
for instance, builds up islands in...
colonized by land-hungry peop...
of sediment in the Mississippi...
upstream) is causing biologic...
away. Sediment also can be ha...

454

what do you think?

THE WORLD AROUND YOU

Cultural Whaling in the Pacific Northwest

On May 17, 1999, for the first time in more than 70 years, members of the Makah tribe from Washington's Olympic peninsula struck and killed a gray whale. Using a mixture of modern technology and ancient traditions, nine men paddling a handmade cedar canoe drew close enough to harpoon the 20-ton whale, after which high-speed motorboats closed in to kill the wounded animal with a .50 caliber anti-tank rifle.

For months before the hunt, native people, animal rights groups, sea conservation societies, and members of the whale-watching industry had engaged in shouting matches, racial slurs, political maneuvering, media campaigns, and dangerous motorboat confrontations as each side tried to promote its own views. At issue is a complex mixture of international environmental policy, aboriginal culture and hunting rights, ethical values, commercial interests, and politics that is not easy to resolve. This case study presents a good opportunity for you to practice critical thinking skills.

For at least 1500 years the Makah people collected fish and shellfish and hunted for seals, whales, and sea otters along this section of fog-shrouded coast. The 1855 Treaty of Neah Bay recognized the right of the Makah people to hunt whales, the only such covenant in United States history. By the beginning of the twentieth century, however, commercial exploitation had driven nearly into extinction the whales, seals, otters, and fish on which the Makah once depended. Fewer than 2000 Pacific gray whales survived out of a historic population of about 30,000. By 1920, most Makah had given up their traditional subsistence culture and turned to farming, logging, or factory jobs.

Since their protection under the Endangered Species Act and the International Whaling Ban, gray whales have made a remarkable and encouraging recovery. The population today is thought to be around 26,000 animals and is growing at about 2.5 percent per year. It may well have reached the carrying capacity of its habitat. In 1999 scientists estimated that about 800 whales died of starvation and disease during their 4000-mile migration from Baja California to the Arctic. A sustainable harvest might actually be beneficial for the whale population. When the species was removed from the U.S. endangered species list in 1994, the Makah announced their intention to resume whaling under their treaty rights. They consider this an important step in recovering their traditional culture and combating the poverty, drugs, alcoholism, violence, and despair that have afflicted their community. Reviving the rituals, discipline, and pride of whaling, they believe, could lead to a cultural renaissance for their people.

To anti-hunting groups, killing magnificent, intelligent, social animals like whales is tantamount to murder. In their view, calling for a sustainable harvest of whales is equivalent to establishing a sustainable level of genocide or torture. This isn't subsistence hunting, they claim, because the Makah haven't depended on whale meat for many years and don't need it now. To the Makah and their supporters from whaling nations such as Japan and Norway, on the other hand, it is hypocritical to condemn hunting marine mammals when most Americans happily eat the millions of cattle, pigs, sheep, and other animals slaughtered every year in commercial packing plants.

How would you approach this complex question? What information would you need to determine whether Indian claims for cultural traditions should take precedence over ethical rights of wild animals? Does the fact that native people have suffered economically and socially give them special standing? Should subsistence hunting be limited to traditional tools and people who have a demonstrated nutritional need? What information sources would you regard as reliable and authoritative in this situation? How would you decide whom or what to believe?

As you study environmental science, you will find that many of the dilemmas we face, like this question of hunting whales by native people, require a combination of scientific data and contextual sensitivity and empathy. Because humans are involved, questions of history, culture, ethics, and politics also come into play. How would you resolve this clash in world views? Can you suggest a compromise that might defuse this tense situation?

Members of a Makah whaling crew set out in their handmade cedar canoe on a practice run. Whale hunting is central in the Makah culture, but is opposed by animal rights groups.

Case Studies, "In Depth" Boxes, and **"What Do You Think?" Essays,** many with **"Ethical Considerations"** attached, give students real-life examples to evaluate and a chance to practice critical thinking skills and formulate reasoned opinions. These boxes are carefully planned to build upon chapter content and are long enough to present important, real information.

The **"What Can You Do?" Listings** help students to learn that small, individual steps can make a real difference in affecting our environment.

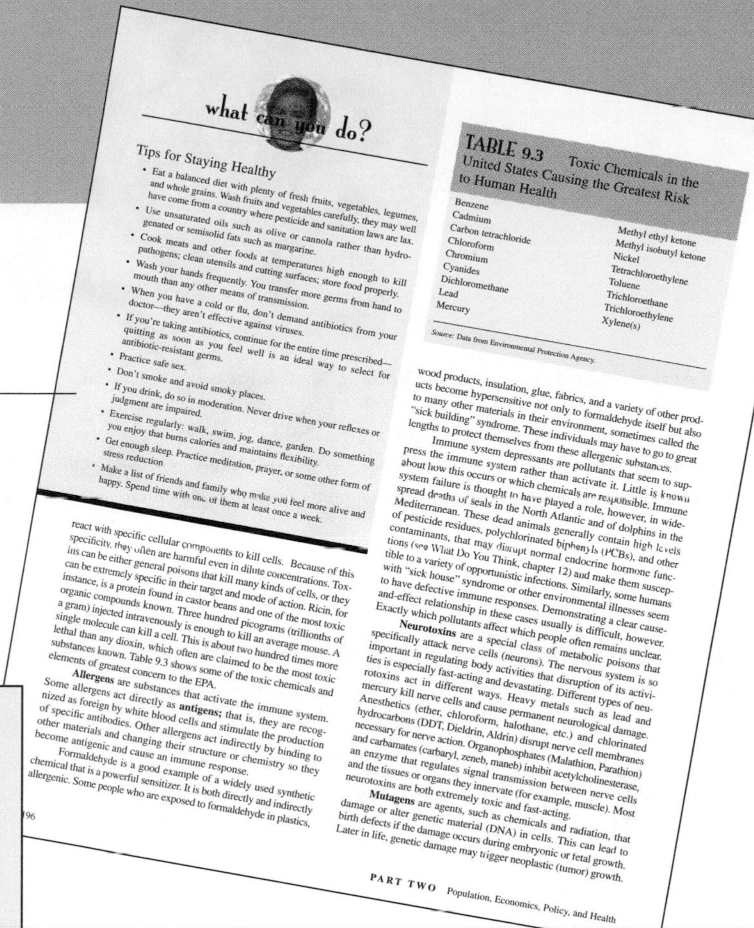

what can you do?

Tips for Staying Healthy

- Eat a balanced diet with plenty of fresh fruits, vegetables, legumes, and whole grains. Wash fruits and vegetables carefully, they may well have come from a country where pesticide and sanitation laws are lax.
- Use unsaturated oils such as olive or canola rather than hydrogenated or semisolid fats such as margarine.
- Cook meats and other foods at temperatures high enough to kill pathogens; clean utensils and cutting surfaces; store food properly.
- Wash your hands frequently. You transfer more germs from hand to mouth than any other means of transmission.
- When you have a cold or flu, don't demand antibiotics from your doctor—they aren't effective against viruses.
- If you're taking antibiotics, continue for the entire time prescribed—quitting as soon as you feel well is an ideal way to select for antibiotic-resistant germs.
- Practice safe sex.
- Don't smoke and avoid smoky places.
- If you drink, do so in moderation. Never drive when your reflexes or judgment are impaired.
- Exercise regularly: walk, swim, jog, dance, garden. Do something you enjoy that burns calories and maintains flexibility.
- Get enough sleep. Practice meditation, prayer, or some other form of stress reduction.
- Make a list of friends and family who make you feel more alive and happy. Spend time with one of them at least once a week.

TABLE 9.3 Toxic Chemicals in the United States Causing the Greatest Risk to Human Health

Benzene	Methyl ethyl ketone
Cadmium	Methyl isobutyl ketone
Carbon tetrachloride	Nickel
Chloroform	Tetrachloroethylene
Chromium	Toluene
Cyanides	Trichloroethane
Dichloromethane	Trichloroethylene
Lead	Xylene(s)
Mercury	

Source: Data from Environmental Protection Agency.

PART TWO Population, Economics, Policy, and Health

Profiles of ordinary people in environmental careers are scattered throughout the text. Giving students models with whom they can identify will help them consider how they might pursue a career in an environmental area.

PROFILE

Fisheries Biologist

Robyn Angliss

Would you like a job that includes field surveys of whales and porpoises in Alaska and developing marine policy? Robyn Angliss has done both in her job at the United States National Marine Fisheries Service (NMFS). Although her title is fisheries biologist, Robyn specializes in mammals, especially marine mammals.

How can you get work that takes you to beautiful places to study charismatic animals? Robyn was in her junior year of college and was thinking about a career in marine intertidal ecology when a friend invited her to be part of a summer field study in Prince William Sound, Alaska, to photograph whales and examine the effects of the *Exxon Valdez* oil spill on killer whales and humpback whales. ...

372

At the end of each chapter, a **Summary** and a set of **Questions for Review** provide an opportunity for students to test their understanding of the material just covered, while **Questions for Critical Thinking** are designed to stimulate creative, analytical thinking and to serve as a springboard for class discussions. **Additional Information on the Internet** lists important chapter topics for which there are hyperlinks, available on the accompanying website, to help students with study and research.

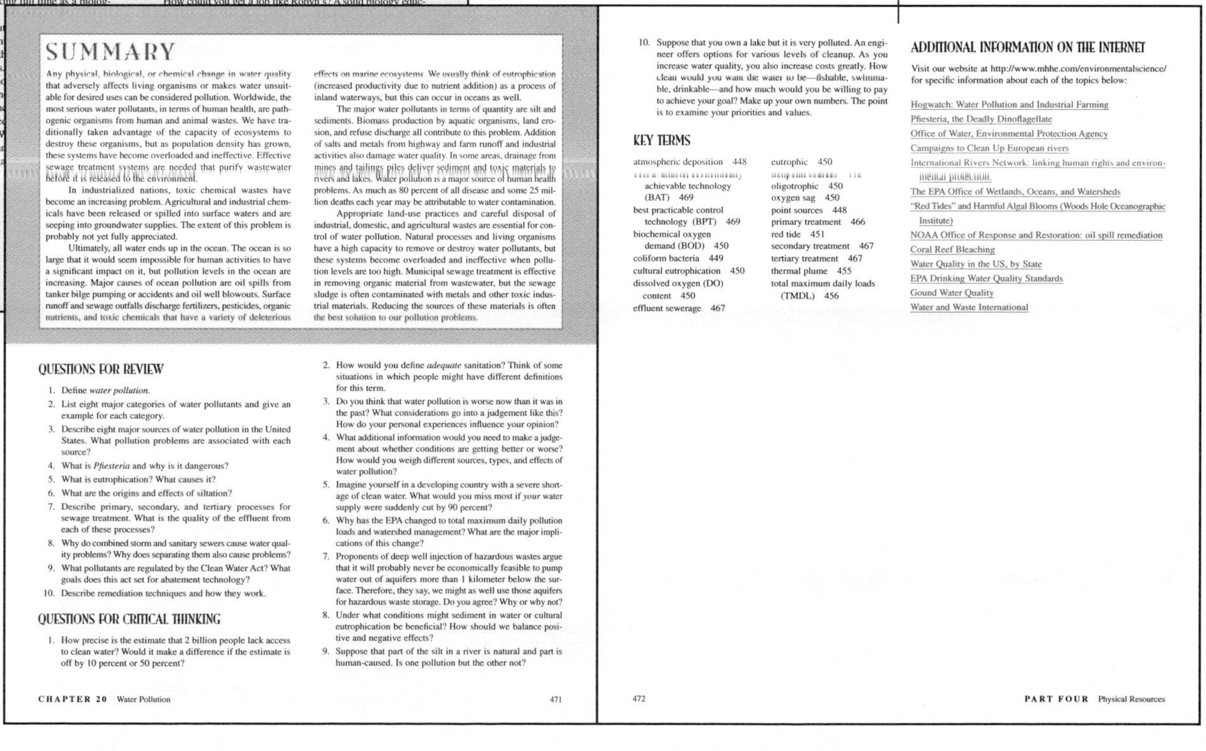

SUMMARY

QUESTIONS FOR REVIEW

1. Define *water pollution.*
2. List eight major categories of water pollutants and give an example for each category.
3. Describe eight major sources of water pollution in the United States. What pollution problems are associated with each source?
4. What is *Pfiesteria* and why is it dangerous?
5. What is eutrophication? What causes it?
6. What are the origins and effects of siltation?
7. Describe primary, secondary, and tertiary processes for sewage treatment. What is the quality of the effluent from each of these processes?
8. Why do combined storm and sanitary sewers cause water quality problems? Why does separating them also cause problems?
9. What pollutants are regulated by the Clean Water Act? What goals does this act set for abatement technology?
10. Describe remediation techniques and how they work.

QUESTIONS FOR CRITICAL THINKING

1. How precise is the estimate that 2 billion people lack access to clean water? Would it make a difference if the estimate is off by 10 percent or 50 percent?

CHAPTER 20 Water Pollution 471

KEY TERMS

ADDITIONAL INFORMATION ON THE INTERNET

Visit our website at http://www.mhhe.com/environmentalscience/ for specific information about each of the topics below:

472 PART FOUR Physical Resources

SUPPLEMENTS

1. **Class Activities and Assessment Guide (CAAG).** Proper assessment is critical to your grading system, and this brand new (free) ancillary for instructors reflects the most recent studies of assessment and ways that students learn. Available in print, or on the website, key components of this invaluable text include:
   - a "Resource Locator" that pulls together appropriate material from numerous sources to help instructors get the most out of every chapter
   - a variety of classroom activities, all cross-referenced in the Resource Locator and complete with assessment guides
   - guidelines for constructing a good test
   - Bloom's taxonomy for identifying levels of questions
   - objective test questions with answers
   - short answer/essay questions with assessment guides
2. **Computerized Testing Software** offers the objective test questions that are in the CAAG in electronic format for ease in class testing and grading.
3. A set of **100 Transparencies** is available to users of the text. These acetates include key figures from the text, including new art from this edition.
4. **Environmental Science and Ecology Visual Resource Library (VRL) CD-ROM.** This classroom presentation CD includes images from an ecology text and two environmental science textbooks, plus hundreds of photographs.
5. The **Student Study Guide** is completely reworked for this edition, linked to the ESP and to the Online Learning Center, and includes activities based on both of these resources. The student study guide is available in print and on the website.

The **Essential Study Partner** (ESP) CD-ROM is a student tutorial CD containing high-quality 3-D animations, interactive study activities, illustrated overviews of key topics in environmental science, and self-quizzes and exams for each important unit.

A comprehensive **Website** (http://www.mhhe.com/environmentalscience/) offers numerous resources for both students and instructors in the form of our Online Learning Center. Features include chapter-related hyperlinks, interactive lecture notes, answers to critical thinking questions, study questions, case studies, practice quizzing, key term flashcards, PageOut—our web page design tool, animations, current global environmental issues, and much more.

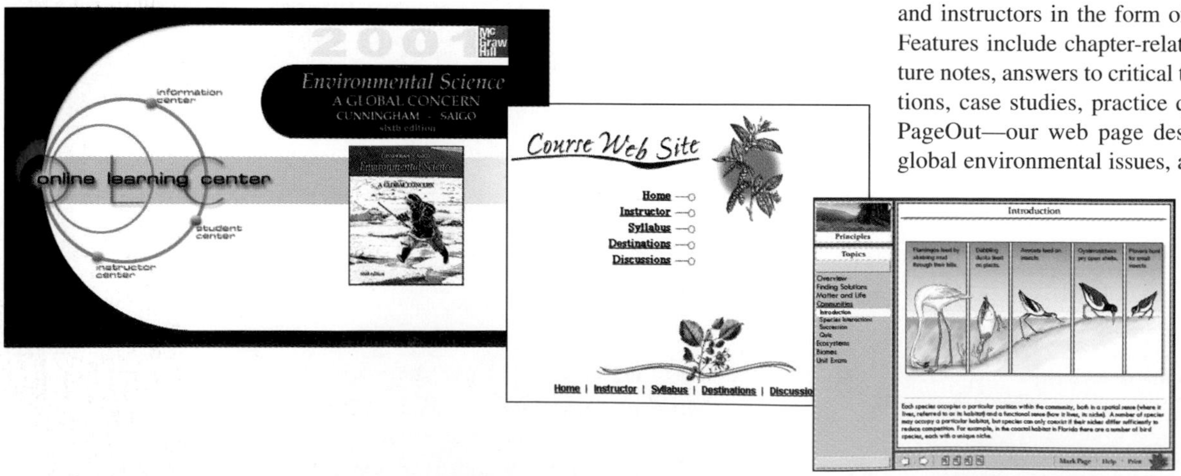

The Online Learning Center

Your Password to Success

www.mhhe.com/environmentalscience (click on cover)

This text-specific website allows students and instructors from all over the world to communicate. Instructors can create a more interactive course with the integration of this site, and students will find tools such as practice quizzing, key term flashcards, case studies, global environmental issues, and chapter-related hyperlinks, that will help them improve their grades.

Student Resources

Study questions
Practice quizzing
Hyperlinks to chapter topics
Case studies
Global environmental issues
Key term flashcards

Instructor Resources

Class activities and assignment guide
Links to related websites to expand on particular topics
Interactive lecture outlines
Case studies
Test item file
Global environmental issues
Sample syllabi

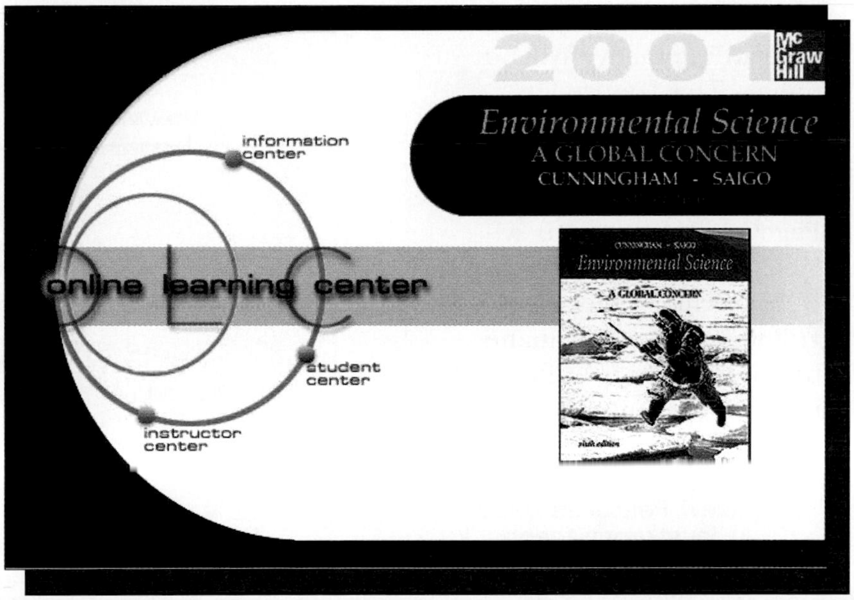

Imagine the advantages of having so many learning and teaching tools all in one place—all at your fingertips—FREE.

Contact your McGraw-Hill sales representative for more information or visit *www.mhhe.com*.

Essential Study Partner CD-ROM

A free study partner that engages, investigates, and reinforces what you are learning from your textbook. You'll find the **Essential Study Partner** for *Environmental Science: A Global Concern* to be a complete, interactive student study tool packed with hundreds of animations and learning activities. From quizzes to interactive diagrams, you'll find that there has never been a better study partner to ensure the mastery of core concepts. Best of all, it's **FREE** with your new textbook purchase.

The unit pop-up menu is accessible at any time within the program. Clicking on the current unit will bring up a menu of other units available in the program.

The topic menu contains an interactive list of the available topics. Clicking on any of the listings within this menu will open your selection and will show the specific concepts presented within this topic. Clicking any of the concepts will move you to your selection. You can use the UP and DOWN arrow keys to move through the topics.

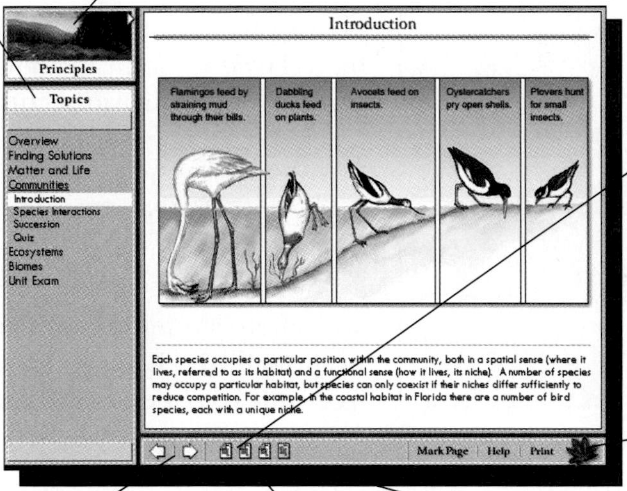

To the right of the arrows is a row of icons that represent the number of screens in a concept. There are three different icons, each representing different functions that a screen in that section will serve. The screen that is currently displayed will highlight yellow and visited ones will be checked.

The activity icon represents an interactive learning activity.

Along the bottom of the screen you will find various navigational aids. At the left are arrows that allow you to page forward and backward through text screens or interactive exercise screens. You can also use the LEFT and RIGHT arrows on your keyboard to perform the same function.

The page icon represents a page of informational text.

The film icon represents an animation screen.

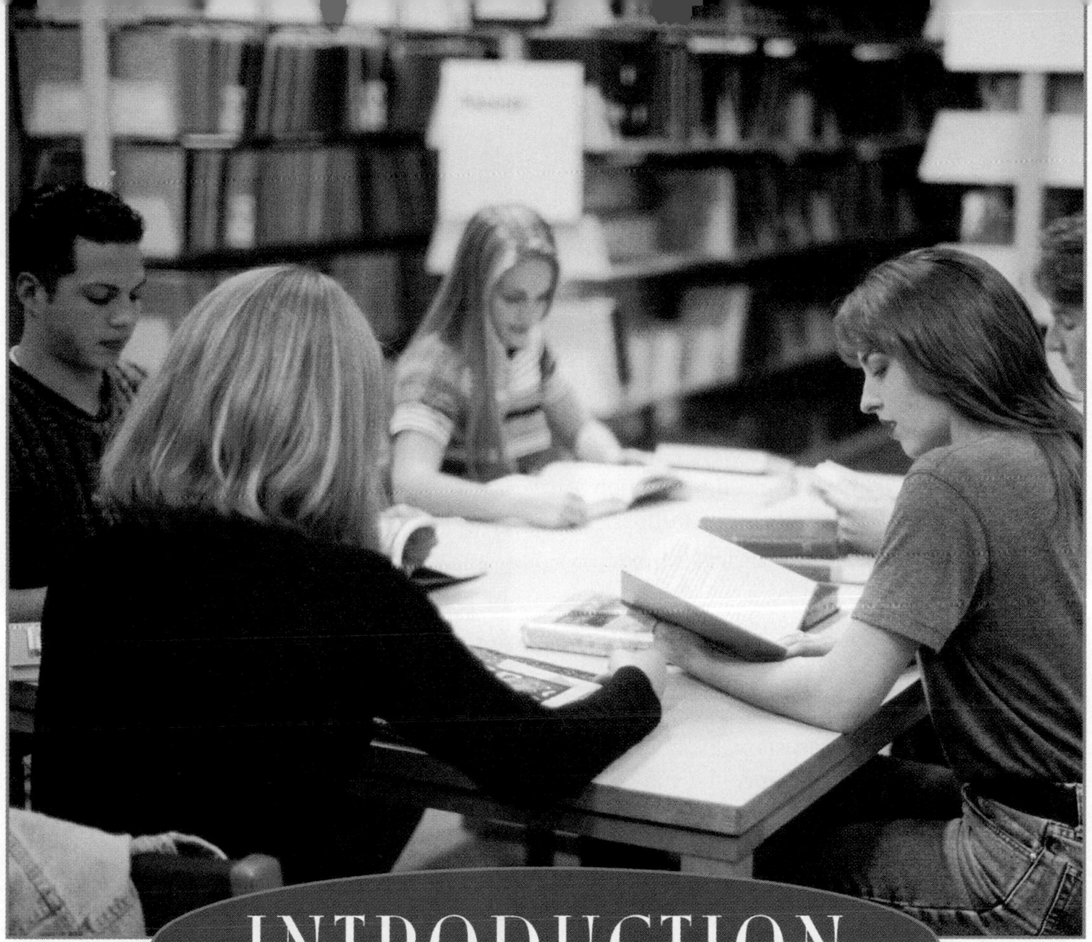

INTRODUCTION

Learning to Learn

What kind of world do you want to live in?
Demand that your teachers teach you
what you need to know to build it.

—Peter Kropotkin—

OBJECTIVES

After studying this introduction, you should be able to:

- form a plan to organize your efforts and become a more effective and efficient student.
- make an honest assessment of the strengths and weaknesses of your current study skills.
- assess what you need to do to get the grade you want in this class.
- set goals, schedule your time, and evaluate your study space.
- use this textbook effectively, practice active reading, and prepare for exams.
- be prepared to apply critical and reflective thinking in environmental science.
- understand the advantages and uses of concept mapping in your studying.

Above: Many different learning styles are helpful in studying environmental science. Reviewing material with a partner or study group can increase your understanding and retention of the wide variety of information in this book.

WHY STUDY ENVIRONMENTAL SCIENCE?

Welcome to environmental science. I hope you'll enjoy learning about the material presented in this book, and that you'll find it both engaging and useful. There should be something here for just about everyone, whether your interests are in basic ecology, natural resources, or the broader human condition. You'll see, as you go through the book, that it covers a wide range of topics. It defines our environment, not only the natural world, but also the built world of technology, cities, and machines, as well as human social or cultural institutions. All of these interrelated aspects of our life affect us, and, in turn, are affected by what we do.

You'll find that many issues discussed here are part of current news stories on television or in newspapers. Becoming an educated environmental citizen will give you a toolkit of skills and attitudes that will help you understand current events and be a more interesting person. Because this book contains information from so many different disciplines, you will find connections here with many of your other classes. Seeing material in an environmental context may assist you in mastering subject matter in many courses, as well as in life after you leave school.

One of the most useful skills you can learn in any of your classes is critical thinking—a principal topic of this chapter. Much of the most important information in environmental science is highly contested. Facts vary depending on when and by whom they were gathered. For every opinion there is an equal and opposite opinion. How can you make sense out of this welter of ever-changing information? The answer is that you need to develop a capacity to think independently, systematically, and skillfully to form your own opinions (fig. I.1). These qualities and abilities can help you in many aspects of life. Throughout this book you will find "What Do You Think?" boxes that invite you to practice your critical and reflective thinking skills.

There is much to be worried about in our global environment. Evidence is growing relentlessly that we are degrading our environment and consuming resources at unsustainable rates. Biodiversity is disappearing at a pace unequaled since the end of the age of dinosaurs 65 million years ago. Irreplaceable topsoil washes off farm fields, threatening global food supplies. Ancient forests are being destroyed to make newsprint and toilet paper. Rivers and lakes are polluted with untreated sewage, while soot and smoke obscure our skies. Even our global climate seems to be changing to a new regime that could have catastrophic consequences.

At the same time, we have better tools and knowledge than any previous generation to do something about these crises. Worldwide public awareness of—and support for—environmental protection is at an all-time high. Over the past 50 years, human ingenuity and enterprise have brought about a breathtaking pace of technological innovations and scientific breakthroughs. We have learned to produce more goods and services with less material. The breathtaking spread of communication technology makes it possible to share information worldwide nearly instantaneously. Since World War II, the average real income in developing countries has doubled; malnutrition has declined by almost one-third; child death rates have been halved; average life expectancy has increased by 30 percent; and the percentage of rural families with access to safe drinking water has risen from less than 10 percent to almost 75 percent.

The world's gross domestic product has increased more than tenfold over the past five decades, but the gap between the rich and poor has grown ever wider. More than a billion people now live in abject poverty without access to adequate food, shelter, medical care, education, and other resources required for a healthy, secure life. The challenge for us is to spread the benefits of our technological and economic progress more equably and to find ways to live sustainably over the long run without diminishing the natural resources and vital ecological services on which all life depends. I've tried to strike a balance in this book between enough doom and gloom to give you a realistic view of our problems, and enough positive examples to give hope that we can discover workable solutions.

FIGURE I.1 What does it all mean? Studying environmental science gives you an opportunity to develop creative, reflective, and critical thinking skills.

What would it mean to become a responsible environmental citizen? What rights and privileges do you enjoy as a member of the global community? What duties and responsibilities go with that citizenship? In many chapters of this book you will find practical advice on things you can do to conserve resources and decrease adverse environmental impacts. Ethical perspectives are an important part of our relationship to the environment and the other people with whom we share it. The discussion of ethical principles and worldviews in chapter 2 is a key section of this book. I hope you'll think about the ethics of how we treat our common environment.

Clearly, to become responsible and productive environmental citizens each of us needs a basis in scientific principles, as well as some insights into the social, political, and economic systems that impact our global environment. I hope this book and the class you're taking will give you the information you need to reach those goals. As the noted Senegalese conservationist and educator, Baba Dioum, once said, "in the end, we will conserve only what we love, we will love only what we understand, and we will understand only what we are taught."

HOW CAN I GET AN A IN THIS CLASS?

"What have I gotten myself into?" you are probably wondering as you began to read this book. "Will environmental science be worth my while? Do I have a chance to get a good grade?" The answers to these questions depend, to a large extent, on you and how you decide to apply yourself. Expecting to be interested and to do either well or poorly in your classes often turns out to be a self-fulfilling prophecy. As Henry Ford once said, "If you think you can do a thing, or think you can't do a thing, you're right." Cultivating good study skills can help you to reach your goals and make your experience in environmental science a satisfying and rewarding one. The purpose of this introduction is to give you some tips to help you get off to a good start in studying. You'll find that many of these techniques are also useful in other courses and after you graduate, as well.

Environmental science, as you can see by skimming through the table of contents of this book, is a complex, transdisciplinary field that draws from many academic specialties. It is loaded with facts, ideas, theories, and confusing data. It is also a dynamic, highly contested subject. Topics such as environmental contributions to cancer rates, potential dangers of pesticides, or when and how much global warming may be caused by human activities are widely disputed. Often you will find distinguished and persuasive experts who take completely opposite positions on any particular question. It will take an active, organized approach on your part to make sense of the vast amount of information you'll encounter here. And it will take critical, thoughtful reasoning to formulate your own position on the many controversial theories and ideas in environmental science. Learning to learn will help you keep up-to-date on important issues after you leave this course. Becoming educated voters and consumers is essential for a sustainable future.

TABLE I.1 Assessing Your Study Skills

Rate yourself on each of the following study skills and habits on a scale of 1 (excellent) to 5 (needs improvement). If you rate yourself below 3 on any item, think about an action plan to improve that competence or behavior.

_______	How strong is your commitment to be successful in this class?
_______	How well do you manage your time (e.g., do you always run late or do you complete assignments on time)?
_______	Do you have a regular study environment that reduces distraction and encourages concentration?
_______	How effective are you at reading and note-taking (e.g., do you remember what you've read; can you decipher your notes after you've made them)?
_______	Do you attend class regularly and listen for instructions and important ideas? Do you participate actively in class discussions and ask meaningful questions?
_______	Do you generally read assigned chapters in the textbook before attending class or do you wait until the night before the exam?
_______	Are you usually prepared before class with questions about material that needs clarification or that expresses your own interest in the subject matter?
_______	How do you handle test anxiety (e.g., do you usually feel prepared for exams and quizzes or are you terrified of them? Do you have techniques to reduce anxiety or turn it into positive energy)?
_______	Do you actively evaluate how you are doing in a course based on feedback from your instructor and then make corrections to improve your effectiveness?
_______	Do you seek out advice and assistance outside of class from your instructors or teaching assistants?

Developing Good Study Habits

What are your current study skills and habits? Making a frank and honest assessment of your strengths and weaknesses will help you set goals and make plans for achieving them during this class. Answer the questions in table I.1 as a way of assessing where you are as you begin to study environmental science and where you need to work to improve your study habits.

One of the first requirements for success is to set clear, honest, attainable goals for yourself. Are you willing to commit the time and effort necessary to do well in this class? Make goals for yourself in terms that you can measure and in time frames within which you can see progress and adjust your approach if it isn't taking you where you want to go. Be positive but realistic. It's more effective to try to accomplish a positive action than to avoid a negative one. When you set your goals, use proactive language that states what you want rather than negative language about what you're trying to avoid. It's good to be optimistic, but setting impossibly high standards will only lead to disappointment. Be objective about the obstacles you face and be willing to modify your goals if necessary. As you gain more experience and information, you may need to adjust your expectations either up or down. Take stock from time to time to see whether you are on track to accomplish what you expect from your studies. In environmental planning, this is called adaptive management.

One of the most common mistakes many of us make is to procrastinate and waste time. Be honest, are you habitually late for meetings or in getting assignments done? Do you routinely leave your studying until the last minute and then frantically cram the night before your exams? If so, you need to organize your schedule so that you can get your work done and still have a life. Make a study schedule for yourself and stick to it. Allow enough time for sleep, regular meals, exercise, and recreation so that you will be rested, healthy, and efficient when you do study. Schedule regular study times between your classes and work. Plan some study times during the day when you are fresh; don't leave all your work until late night hours when you don't get much done. Divide your work into reasonable sized segments that you can accomplish on a daily basis. Plan to have all your reading and assignments completed several days before your exams so you will have adequate time to review and process information. Carry a calendar so you will remember appointments and assignments.

Establish a regular study space in which you can be effective and productive. It might be a desk in your room, a carrel in the library, or some other quiet, private environment. Find a place that works for you and be disciplined about sticking to what you need to do. If you get in the habit of studying in a particular place and time, you will find it easier to get started and to stick to your tasks. Many students make the mistake of thinking that they can study while talking to their friends or watching TV. They may put in many hours but not really accomplish much. On the other hand, some people think most clearly in the anonymity of a crowd. The famous philosopher, Immanuel Kant, found that he could think best while wandering through the noisy, crowded streets of Königsberg, his home town.

How you behave in class and interact with your instructor can have a big impact on how much you learn and what grade you get. Make an effort to get to know your instructor. She or he is probably not nearly as formidable as you might think. Sit near the front of the room where you can see and be seen. Pay attention and ask questions that show your interest in the subject matter. Attend every class and arrive on time. Don't fold up your papers and prepare to leave until after the class period is over. Arriving late and leaving early says to your instructor that you don't care much about either the class or your grade. If you think of yourself as a good student and act like one, you may well get the benefit of the doubt when your grade is assigned.

Practice active, purposeful learning. It isn't enough to passively absorb knowledge provided by your instructor and this textbook. You need to actively engage the material in order to really understand it. The more you invest yourself in the material, the easier it will be to comprehend and remember. It is very helpful to have a study buddy or study group with whom you can compare notes and try out ideas. You will get a valuable perspective on whether you're getting the main points and understanding an adequate amount by comparing. It's an old adage that the best way to learn something is to teach it to someone else. Take turns with your study buddy explaining the material you're studying. You may think you've mastered a topic by quickly skimming the text but you're likely to find that you have to struggle to give a clear description in your own words. Anticipating possible exam questions and taking turns quizzing each other can be a very good way to prepare for tests.

Learning Styles

Each of us has ways that we learn most effectively. Discovering techniques that work for you and fit the material you need to learn is an important step in reaching your goals. Do any of the following fit your preferred ways of learning?

- **Visual learner:** understands and remembers best by reading, looking at photographs, figures, and diagrams. Good with maps and picture puzzles. Visualizes image or spatial location for recall. Uses flash cards for memorization.
- **Verbal Learner:** understands and remembers best by listening to lectures, reading out loud, and talking things through with a study partner. May like poetry and word games. Memorizes by repeating item verbally.
- **Logical Learner:** understands and remembers best by thinking through a subject and finding reasons that make sense. Good at logical puzzles and mysteries. May prefer to find patterns and logical connections between items rather than memorize.
- **Active Learner:** understands and remembers best those ideas and skills linked to physical activity. Takes notes, makes lists, uses cognitive maps. Good at working with hands and learning by doing. Remembers best by writing, drawing, or physically manipulating items.

The list above represents only a few of the learning styles identified by educational psychologists. How can you determine which approaches are right for you? Think about the one thing in life that you most enjoy and in which you have the greatest skills. What hobbies or special interests do you have? How do you learn new material in that area? Can you read about a procedure in a book and then do it, or do you throw away the manual and just mess around until you've figured out how things work? Do you need to see a diagram or a picture before things make sense, or are spoken directions most memorable and meaningful for you? Some people like to learn by themselves in a quiet place where there are no distractions, while others need to discuss ideas with another person to feel really comfortable about what they're learning.

Sometimes you have to adjust your preferred learning style to the specific material you're trying to master. You may be primarily an oral learner, but if what you need to remember for a particular subject is spatial or structural, you may need to try some visual learning techniques. Memorizing vocabulary items might be best accomplished by oral repetition, while developing your ability to work quantitative problems should be approached by practicing analytical or logical skills.

Using This Textbook Effectively

An important part of productive learning is to read assigned material in a purposeful, deliberate manner. Ask yourself questions as you read. What is the main point being made here? Does the evidence presented adequately support the assertions being made? What personal experience have you had or what prior knowledge can you bring to bear on this question? Can you suggest alternative explanations for the phenomena being discussed? What additional information would you need in order to make an informed judgement about this subject and how might you go about obtaining that information or making that judgement?

A study technique developed by Frances Robinson more than 40 years ago and called the SQ3R method (table I.2) can be a valuable aid in improving your reading comprehension. Start your study session with a *survey* of the entire chapter or section you are about to read so you'll have an idea of how the whole thing fits together. What are the major headings and subdivisions? Notice that there is usually a hierarchical organization that gives you clues about the relationship between the various parts. This survey will help you plan your strategy for approaching the material. Next, *question,* what the main points are likely to be in each of the sections. Which parts look most important or interesting? Ask yourself where you

TABLE I.2 The SQ3R Method for Studying Texts

Survey:	Preview the information to be studied before reading.
Question:	Ask yourself critical questions about the content of what you are reading.
Read:	Conduct the actual reading in small segments.
Recite:	Stop periodically to recite to yourself what you have just read.
Review:	Once you have completed the section, review the main points to make sure you remember them clearly.

should invest the most time and effort. Is one section or topic likely to be more relevant to your particular class? Has your instructor emphasized any of the topics you see? Being alert for important material can help you plan the most efficient way to study.

After developing a general plan, begin *active reading* of the text. Read in small segments and stop frequently for reflection and to make notes. Don't fall into a trance in which the words swim by without leaving any impression. Highlight or underline the main points but be careful that you don't just paint the whole page yellow. If you highlight too much, nothing will stand out. Try to distinguish what is truly central to the argument being presented. Make brief notes in the margins that identify main points. This can be very helpful in finding important sections or ideas when you are reviewing. Check your comprehension at the end of each major section. Ask yourself: Did I understand what I just read? What are the main points being made here? Does this relate to my own personal experiences or previous knowledge? Are there details or ideas that need clarification or elaboration?

As you read, stop periodically to *recite* the information you've just acquired. Summarize the information in your own words to be sure that you really understand and are not just depending on rote memory. This is a good time to have a study buddy. Taking turns to summarize and explain material really helps you internalize it. If you don't have a study partner and you feel awkward talking to yourself, you can try writing your summary. Finally, *review* the section. Did you miss any important points? Do you understand things differently the second time through? This is a chance to think critically about the material. Do you agree with the conclusions suggested by the authors? Can you think of alternative explanations for the same evidence? As you review each section, think about how this may be covered on the test. Put yourself in the position of the instructor. What would be some good questions based on this material? Don't try to memorize everything but try to anticipate what might be the most important points.

After class, compare your lecture notes with your study notes. Do they agree? If not, where are the discrepancies? Is it possible that you misunderstood what was said in class, or does your instructor differ with what's printed in the textbook? Are there things that your instructor emphasized in lecture that you missed in your pre-class reading? This is a good time to go back over the readings to reinforce your understanding and memory of the material.

Will This Be on the Test?

Students often complain that test results don't adequately reflect what they know and how much they've learned in studying. It may well be that test questions won't cover what you think is important or use a style that appeals to you, but you'll probably be more successful if you adapt yourself to the realities of your instructor's test methods rather than trying to force your instructor to accommodate to your preferences. One of your first priorities in studying, therefore, should be to learn your instructor's test style. Are you likely to have short-answer objective questions (multiple choice, true or false, fill in the blank) or does your instructor prefer essay questions? If you have an essay test, will the questions be broad and general or more analytical? You should develop a very different study strategy depending on whether you are expected to remember and choose between a multitude of facts and details, or whether you will be asked to write a paragraph summarizing some broad topic.

Organize the ideas you're reading and hearing in lecture. This course will probably include a great deal of information. Unless you have a photographic memory, you won't be able to remember every detail. What's most important? What's the big picture? If you see how pieces of the course fit together, it will all make more sense and be easier to remember. As you read and review, ask yourself what might be some possible test questions in each section. If you're likely to have factual questions, what are the most significant facts in the material you've read? Memorize some benchmark figures. Just a few will help a lot. Pay special attention to tables, graphs, and diagrams. They were chosen because they illustrate important points.

You probably won't be expected to remember all the specific numbers in this book but you probably should know orders of magnitude. The world population is about six *billion* people, not thousands, millions, or trillions. Highlight facts and figures in your lecture notes about which your instructor seemed especially interested. There is a good chance you'll see those topics again on a test. It often helps to remember facts and figures if you can relate them to some other familiar example. The United States, for instance, has about 275 million residents. The European Union is slightly larger, India is almost four times and China is nearly five times as large. Be sure you're familiar with the bold-face key terms in the textbook. Vocabulary terms make good objective questions. The questions for review at the end of each chapter generally cover objective material that makes good short-answer questions.

A number of strategies can help you be successful in test-taking (fig. I.2). Look over the whole test at the beginning and answer the questions you know well first, then tackle the harder ones. On multiple choice tests, find out whether there is a penalty for guessing. Use the process of elimination to narrow down the possible choices and improve the odds for guessing. Often you can get hints from the context of the question or from other similar questions. Notice that the longest or most specific answer often is right while those that are vague or general are more likely wrong. Be alert for absolutes (such as always, never, all) which could indicate wrong choices. Qualifiers (such as sometimes, may, or could)

FIGURE I.2 Cooperative learning is an important part of mastering environmental science. You never grasp material as clearly as when you explain it to someone else.

on the other hand, often point to correct answers. Exactly opposite answers may indicate that one of them is correct.

If you anticipate essay questions, practice writing one- or two-paragraph summaries of major points in each chapter. Develop your ability to generalize and to make connections between important facts and ideas. Notice that questions for review at the end of each chapter are open-ended topics that can work well either for discussion groups or as questions for an essay test. You'll have a big advantage on a test if you have some carefully thought out arguments for and against the major ideas presented in each chapter. If you don't have any idea what a particular essay question means, you often can make a transition to something you do understand. Look for a handle that links the question to a topic you are prepared to answer. Even if you have no idea what the question means, make an educated guess. You might get some credit. Anything is better than a zero. Sometimes if you explain your answer, you'll get at least some points. "If the question means such and such, then the answer would be ______" may get you partial credit.

Does your instructor like thought questions? Does she/he expect you to be able to interpret graphs or to draw inferences from a data table? Might you be asked to read a paragraph and describe what it means or relate it to other cases you've covered in the class? If so, you should practice these skills. Making up and sharing these types of questions with your study group can greatly increase your understanding of the material as well as improve your performance on exams. Notice that the end of every chapter in this book includes a list of questions for critical and reflective thinking that makes excellent essay questions. Writing a paragraph answer for each of these questions could be a very good way to study for an essay test.

Concentrate on positive attitudes and building confidence before your tests. If you have fears and test anxiety, practice relaxation techniques and visualize success. Be sure you are rested and well prepared. You certainly won't do well if you're sleep-deprived and a bundle of nerves. Often the worst thing you can do is to stay up all night to cram your brain with a jumble of data. Being able to think clearly and express yourself well may count much more than knowing a pile of unrelated facts. Review your test when it is returned to learn what you did well and where you need to improve. Ask your instructor for pointers on how you might have answered the questions better. Carefully add your score to be sure you got all the points you deserve. Sometimes graders make simple mathematical errors in adding up points.

THINKING ABOUT THINKING

Perhaps the most valuable skill you can learn in any of your classes is the ability to think clearly, creatively, and purposefully. In a rapidly moving field such as environmental science, facts and explanations change constantly. It's often said that in six years approximately half the information you learn from this class will be obsolete. During your lifetime you will probably change careers four to six times. Unfortunately, we don't know which of the ideas we now hold will be outdated or what qualifications you will need for those future jobs. Developing the ability to learn new skills, examine new facts, evaluate new theories, and formulate your own interpretations is essential to keep up in a changing world. In other words, you need to learn how to learn on your own.

Even in our everyday lives most of us are inundated by a flood of information and misinformation. Competing claims and contradictory ideas battle for our attention. The rapidly growing complexity of our world and our lives intensifies the difficulties in knowing what to believe or how to act. Consider how the communications revolution has brought us computers, e-mail, cell phones, mobile faxes, pagers, the World Wide Web, hundreds of channels of satellite TV, and direct mail or electronic marketing that overwhelm us with conflicting information. We have more choices than we can possibly manage, and know more about the world around us than ever before but, perhaps, understand less. How can we deal with the barrage of often contradictory news and advice that inundates us?

To complicate our difficulty in knowing what to believe, distinguished authorities disagree vehemently about many important topics. A law of environmental science might be that for any expert there is always an equal and opposite expert. How can you decide what is true and meaningful in such a welter of confusing information? Is it simply a matter of what feels good at the moment or supports our preconceived notions? Or are there ways to use logical, orderly, creative thinking procedures to reach decisions?

FIGURE I.3 "There is absolutely no cause for alarm at the nuclear plant!"

By now, most of us know not to believe everything we read or hear (fig. I.3). "Tastes great . . . Low, low sale price . . . Vote for me . . . . Lose 30 pounds in 3 weeks . . . You may already be a winner . . . Causes no environmental harm . . . I'll never lie to you . . . Two out of three doctors recommend . . ." More and more of the information we use to buy, elect, advise, judge, or heal has been created not to expand our knowledge but to sell a product or advance a cause. It would be unfortunate if we become cynical and apathetic due to information overload. It does make a difference what we think and how we act.

Approaches to Truth and Knowledge

A number of skills, attitudes, and approaches can help us evaluate information and make decisions. **Analytical thinking** asks, "How can I break this problem down into its constituent parts?" **Creative thinking** asks, "How might I approach this problem in new and inventive ways?" **Logical thinking** asks, "How can orderly, deductive reasoning help me think clearly?" **Critical thinking** asks, "What am I trying to accomplish here and how will I know when I've succeeded?" **Reflective thinking** asks, "What does it all mean?" In this section, we'll look more closely at critical and reflective thinking as a foundation for your study of environmental science. I hope you will apply these ideas consistently as you read this book.

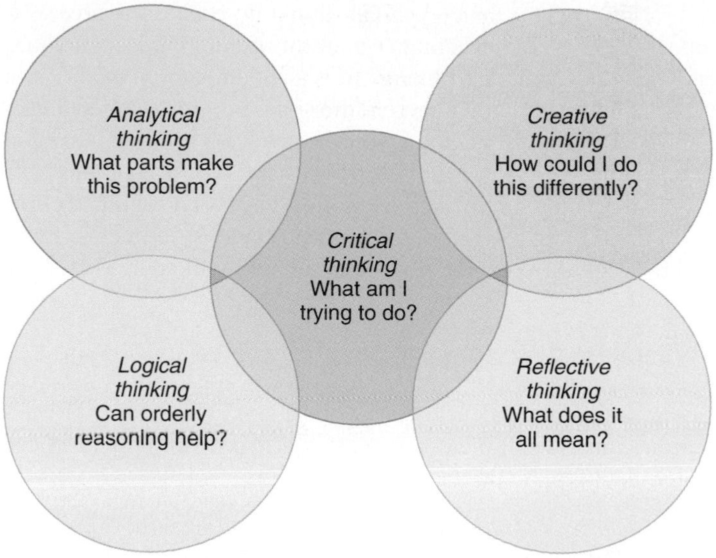

FIGURE I.4 Different approaches to thinking are used to solve different kinds of problems or to study alternate aspects of a single issue.

As figure I.4 suggests, critical thinking is central in the constellation of thinking skills. It challenges us to examine theories, facts, and options in a systematic, purposeful, and responsible manner. It shares many methods and approaches with other methods of reasoning but adds some important contextual skills, attitudes, and dispositions. Furthermore, it challenges us to plan methodically and to assess the process of thinking as well as the implications of our decisions. Thinking critically can help us discover hidden ideas and means, develop strategies for evaluating reasons and conclusions in arguments, recognize the differences between facts and values, and avoid jumping to conclusions. Richard Paul, chair of the National Council for Critical Thinking, identifies ten steps in critical thinking (table I.3).

TABLE I.3 Steps in Critical Thinking

1. What is the *purpose* of my thinking?
2. What precise *question* am I trying to answer?
3. Within what *point of view* am I thinking?
4. What *information* am I using?
5. How am I *interpreting* that information?
6. What *concepts* or ideas are central to my thinking?
7. What *conclusions* am I aiming towards?
8. What am I taking for granted; what *assumptions* am I making?
9. If I accept the conclusions, what are the *implications*?
10. What would the *consequences* be, if I put my thoughts into action?

Notice that many critical thinking processes are self-reflective and self-correcting. This form of thinking is sometimes called "thinking about thinking." It is an attempt to plan rationally how to analyze a problem, to monitor your progress while you are doing it, and to evaluate how your strategy worked and what you have learned when you are finished. It is not critical in the sense of finding fault, but it makes a conscious, active, disciplined effort to be aware of hidden motives and assumptions, to uncover bias, and to recognize the reliability or unreliability of sources.

What Do I Need To Think Critically?

Certain attitudes, tendencies, and dispositions are essential for well-reasoned analysis. Professor Karen J. Warren of Macalester College suggests the following list:

- *Skepticism and independence.* Question authority. Don't believe everything you hear or read—including this book; even experts sometimes are wrong.
- *Open-mindedness and flexibility.* Be willing to consider differing points of view and to entertain alternative explanations. Try arguing from a viewpoint different from your own. It will help you identify weaknesses and limitations in your own position.
- *Accuracy and orderliness.* Strive for as much precision as the subject permits or warrants. Deal systematically with parts of a complex whole. Be disciplined in the standards you apply.
- *Persistence and relevance.* Stick to the main point. Don't allow diversions or personal biases to lead you astray. Information may be interesting or even true, but is it relevant?
- *Contextual sensitivity and empathy.* Consider the total situation, relevant context, feelings, level of knowledge, and sophistication of others as you evaluate information. Imagine being in someone else's place to try to understand how they feel.
- *Decisiveness and courage.* Draw conclusions and take a stand when the evidence warrants doing so. Although we often wish for more definitive information, sometimes a well-reasoned but conditional position has to be the basis for action.
- *Humility.* Realize that you may be wrong and that reconsideration may be called for in the future. Be careful about making absolute declarations; you may need to change your mind someday.

While critical thinking shares many of the orderly, systematic approaches of formal logic, it also invokes traits like empathy, sensitivity, courage, and humility. Formulating intelligent opinions about some of the complex issues you'll encounter in environmental science requires more than simple logic. Developing these attitudes and skills is not easy or simple. It takes practice. You have to develop your mental faculties just as you need to train for a sport. Traits such as intellectual integrity, modesty, fairness, compassion, and fortitude are not things you can use only occasionally. They must be cultivated until they become your normal way of thinking.

Applying Critical Thinking

We all use critical or reflective thinking at times. Suppose a television commercial tells you that a new breakfast cereal is tasty and good for you. You may be suspicious and ask yourself a few questions. What do they mean by good? Good for whom or what? Does "tasty" simply mean more sugar and salt? Might the sources of this information have other motives in mind besides your health and happiness? Although you may not have been aware of it, you already have been using some of the techniques of critical analysis. Working to expand these skills helps you recognize the ways information and analysis can be distorted, misleading, prejudiced, superficial, unfair, or otherwise defective.

Here are some steps in critical thinking:

1. *Identify and evaluate premises and conclusions in an argument.* What is the basis for the claims made here? What evidence is presented to support these claims and what conclusions are drawn from this evidence? If the premises and evidence are correct, does it follow that the conclusions are necessarily true?
2. *Acknowledge and clarify uncertainties, vagueness, equivocation, and contradictions.* Do the terms used have more than one meaning? If so, are all participants in the argument using the same meanings? Are ambiguity or equivocation deliberate? Can all the claims be true simultaneously?
3. *Distinguish between facts and values.* Are claims made that can be tested? (If so, these are statements of fact and should be able to be verified by gathering evidence.) Are claims made about the worth or lack of worth of something? (If so, these are value statements or opinions and probably cannot be verified objectively.) For example, claims of what we *ought* to do to be moral or righteous or to respect nature are generally value statements.
4. *Recognize and assess assumptions.* Given the backgrounds and views of the protagonists in this argument, what underlying reasons might there be for the premises, evidence, or conclusions presented? Does anyone have an "axe to grind" or a personal agenda in this issue? What do they think you know, need, want, or believe? Is there a subtext based on race, gender, ethnicity, economics, or some belief system that distorts this discussion?
5. *Distinguish the reliability or unreliability of a source.* What makes the experts qualified in this issue? What special knowledge or information do they have? What evidence do they present? How can we determine whether the information offered is accurate, true, or even plausible?
6. *Recognize and understand conceptual frameworks.* What are the basic beliefs, attitudes, and values that this person, group, or society holds? What dominating philosophy or

ethics control their outlook and actions? How do these beliefs and values affect the way people view themselves and the world around them? If there are conflicting or contradictory beliefs and values, how can these differences be resolved?

Some Clues for Unpacking an Argument

In logic, an argument is made up of one or more introductory statements (called **premises**), and a **conclusion** that supposedly follows logically from the premises. Often in ordinary conversation, different kinds of statements are mixed together, so it is difficult to distinguish between them or to decipher hidden or implied meanings. Social theorists call the process of separating and analyzing textual components *unpacking*. Applying this type of analysis to an argument can be useful.

An argument's premises are usually claimed to be based on facts; conclusions are usually opinions and values drawn from or used to interpret, those facts. Words that often introduce a premise include: *as, because, assume that, given that, since, whereas,* and *we all know that* . . . Words that generally indicate a conclusion or statement of opinion or values include: *and, so, thus, therefore, it follows that, consequently, the evidence shows,* and *we can conclude that.*

For instance, in the example we used earlier, the television ad might have said: "*Since* we all need vitamins, and *since* this cereal contains vitamins, *consequently* the cereal must be good for you." Which are the premises and which is the conclusion? Does one necessarily follow from the other? Remember that even if the facts in a premise are correct, the conclusions drawn from them may not be. Information may be withheld from the argument such as the fact that the cereal is also loaded with unhealthy amounts of sugar.

Avoiding Logical Errors and Fallacies

Formal logic catalogs a large number of fallacies and errors that invalidate arguments. Although we don't have room here to include all of these fallacies and errors, it may be helpful to review a few of the more common ones.

- *Red herring:* Introducing extraneous information to divert attention from the important point.
- *Ad hominem attacks:* Criticizing the opponent rather than the logic of the argument.
- *Hasty generalization:* Drawing conclusions about all members of a group based on evidence that pertains only to a selected sample.
- *False cause:* Drawing a link between premises and conclusions that depends on some imagined causal connection that does not, in fact, exist.
- *Appeal to ignorance:* Because some facts are in doubt, therefore a conclusion is impossible.
- *Appeal to authority:* It's true because ____ says so.
- *Begging the question:* Using some trick to make a premise seem true when it is not.
- *Equivocation:* Using words with double meanings to mislead the listener.
- *Slippery slope:* A claim that some event or action will cause some subsequent action.
- *False dichotomy:* Giving either/or alternatives as if they are the only choices.

Avoiding these fallacies yourself or being aware of them in another's argument can help you be more logical and have more logical and reasonable discussions.

Using Critical Thinking in Environmental Science

As you go through this book, you will have many opportunities to practice critical thinking skills. Every chapter includes many facts, figures, opinions, and theories. Are all of them true? No, probably not. They were the best information available when this text was written, but much in environmental science is in a state of flux. Data change constantly as does our interpretation of them. Do the ideas presented here give a complete picture of the state of our environment? Unfortunately, they probably don't. No matter how comprehensive our discussion is of this complex, diverse subject, it can never capture everything worth knowing, nor can it reveal all possible points of view.

When reading this text, try to distinguish between statements of fact and opinion. Ask yourself if the premises support the conclusions drawn from them. Although I have tried to be fair and even-handed in presenting controversies, I, like everyone, have biases and values—some of which I may not even recognize—that affect how I see issues and present arguments. Watch for cases in which you need to think for yourself and utilize your critical and reflective thinking skills to find the truth.

CONCEPT MAPS

Concept mapping is a learning strategy that many students find useful in understanding complex ideas and clarifying ambiguous relationships. Creating a graphic representation of a topic often can help you visualize key concepts and organize your knowledge more clearly than will other methods of study. You may find that the physical process of drawing a map of a topic engages a different part of your brain than does ordinary reading or taking notes. Taking time to think carefully about what is most important about a particular topic will help you remember it better at a later date. It can also point out weaknesses in your understanding as well as areas in which you need more study. Practicing this technique as you think about environmental science can help you comprehend difficult material and prepare for exams.

From ancient times, people have made maps to help them understand and remember important aspects of the world around them. Maps integrate and summarize knowledge. They suggest

what do you think?

THE WORLD AROUND YOU

Don't Believe Everything You See on the Internet

"On the Internet, nobody knows you're a dog."
Drawing by P. Steiner; ©1993, The New Yorker Magazine, Inc.

The Internet (the World Wide Web, e-mail, news groups, etc.) provides a wonderful resource for students of environmental science. You can find a vast amount of information to update or supplement topics in this textbook. An Internet search that takes just seconds often will provide you with hundreds, even thousands, of sources for almost any subject you can imagine. The chances are that you've already gotten material for a term paper or class project using the Internet. But you have to be very careful about evaluating the reliability of information that you obtain from these sources. Skepticism and critical thinking skills should be part of every Internet session.

One of the best things about the Internet is also its greatest problem: Anyone with access to a server can put up anything they want. There is no editorial control. On the positive side, alternative groups that ordinarily wouldn't have access to mainstream media can have a voice. People with unusual interests can find each other to share ideas and information. It also means, however, that anyone can post rumors, myths, speculation, or outright lies anonymously and inexpensively. You can find conspiracy theories, vicious racist diatribes, recipes for building bombs, groups convinced they've seen Elvis at the local mall, or people who've been abducted by little green men in space ships. Just about any crazy idea that you can imagine is intermixed with real news and valuable information. How can you know what, or whom, to believe in an information free-for-all?

One of the first questions you should ask is what information you have about the source of a particular story. Is it a person or group whose history and interests you know anything about? What motivations or interest would they be likely to have? Think about why this person or group might be taking this particular stand. What special interest might they be promoting? A good way to evaluate the reliability of a source on the Internet is to look at the kind of language they use. Is it moderate, balanced reporting, or is it a shrill, strident, call-to-action? Does it use broad generalizations and stock characterizations to denounce its opponents, or is it a well-reasoned analysis? Is the purpose of the piece to inform or outrage you? Sometimes outrage is appropriate, but be aware if your emotions are being manipulated.

What sources are quoted or what links to other pages are referenced in the information you have found? Be aware that someone may claim to have obtained data from a well-known source but it may not be true. Check the original, claimed source to see if they have the same story. Knowing the affiliations of a group can help you assess what their agenda and funding sources may be. An important check on any information you find on the Internet is to ask whether it is corroborated by other independent sources. It may well be that the mainstream media have chosen to ignore or suppress a particular story. The Internet may be the only way to find out what's happening. But be wary of claims that aren't supported by any other evidence.

Beware of deceptive names that portray groups as something they aren't. Anti-environmental groups often hide behind innocuous-sounding names. Take the Information Council for the Environment, for example. Run by a public relations firm, it represents coal companies and mining firms trying to convince the government and the public that global warming is a myth. Similarly, Citizens for the Environment claims to be a grassroots citizen group but has no citizen members. Instead, it represents transnational corporations in supporting unfettered free enterprise. And the Environmental Conservation Organization, a "wise use" group organized by real estate developers, primarily opposes wetland regulations and land-use planning. You can find more about these and many other similar organizations at the Clearinghouse on Environmental Advocacy and Research.

Finally, don't take any single source or study as the last word on a subject. Think of it as one rung on a ladder leading to the truth. Continuing to search for an ever closer approximation to the truth takes effort, but it's important.

linkages that we may not have seen before, and they suggest routes for further exploration. But maps can never record every possible detail of the world or the ideas they represent. Only the most useful and meaningful information is put onto the map so that important points can easily be seen and remembered. No map is ever complete and finished. As we learn more, we revise our maps, correcting errors, adding new information, removing unnecessary features, and refining the presentation. The act of drawing a map exposes doubtful knowledge and shows us where we need more data and a clearer understanding.

If you ask different people to construct a map of the same area of a city, you would probably get very different results. A young child, for example, might draw in only the locations of her school, home, and playground. A commuter from the suburbs, on the other hand, might show the locations of the major office buildings, filling stations, freeway ramps, and the cheapest parking lots in that same city. Neither of these representations is wrong, they just emphasize the aspects of greatest importance to each person.

When we think about maps, we usually visualize physical features such as mountains, lakes, roads, buildings, and so forth. But we can also create maps of biodiversity, magnetic fields, economic flows, cultures, language families, or anything else that can be presented in graphical form. A **concept map** is a two-dimensional representation of the relationship between key ideas. It shows us how we think and suggests affinities and associations that might not otherwise be obvious. At first glance, a concept map looks like a flow chart in which key terms are placed in boxes connected by directional arrows. Based on educational psychology theories of how we organize information, concept maps are hierarchical, with broader, more general items at the top and more specific topics arranged in a cascade below them. They are metacognitive tools that empower the learner to take charge of his/her own learning in a highly organized and meaningful manner.

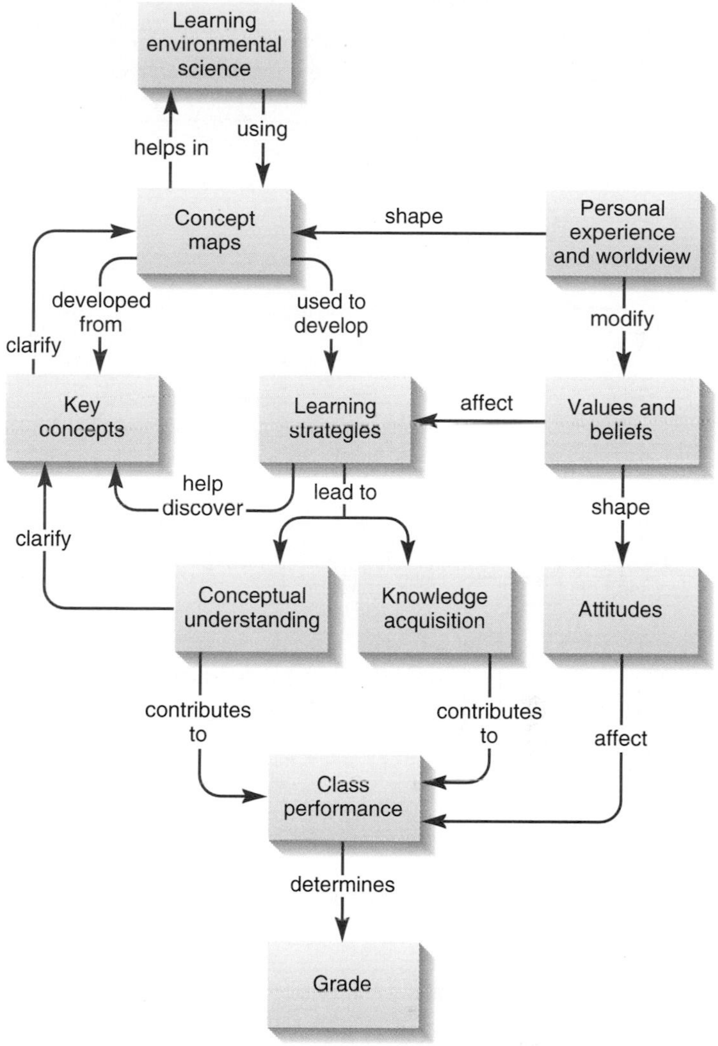

FIGURE I.5 How concept map use can affect your grade.

How do I Create a Concept Map?

To create a concept map, start with what you already know. Build from what's familiar. What are the key components or ideas in the topic you're trying to understand? Place each concept in its own individual circle, box, or other geometrical shape. You might want to use different shapes to indicate relative levels or types of ideas. Connect concept boxes with directional arrows to show relationships. Label each arrow with descriptive terms so that your diagram can be read as a statement or proposition by following interconnections from the top down. In figure I.5, for example, you can read the proposition that "concept maps are used to develop learning strategies that lead to knowledge acquisition that contributes to class performance and determines your grade" as one set of associations. Following another pathway, you can see that "values and beliefs affect learning strategies which help discover key concepts that clarify concept maps."

As you can see in this example, branches to one side or the other of the key concepts show related ideas. Where appropriate, cross links or bridges can connect branches of your map. Linking arrows can be bidirectional to indicate mutual interactions, but be careful not to make everything connected to everything else. Focus on the most important concepts and the most significant relationships. View this as an exercise in discrimination. Don't try to make your map perfect. Sometimes working briskly will help you cut away the superfluous while fostering creativity and synthesis. The point is not to create a work of art but to organize, discover, and understand central meanings. The map helps you learn how to learn.

A concept map can show just a small part or a subset of a broader field of knowledge. The top or key concept in one map may be subsumed into a lower position in a map with a different focus. A small branch in a general map could be expanded into a much more specific map of its own. Figure I.6, for instance, shows some key concepts about ecosystems, a topic that many students find intimidating. Notice that just a few lower elements of this map—the roles played by decomposers, consumers, and producers, for example—could be expanded and made into a map of their own to clarify the ways that energy and materials move through ecosystems.

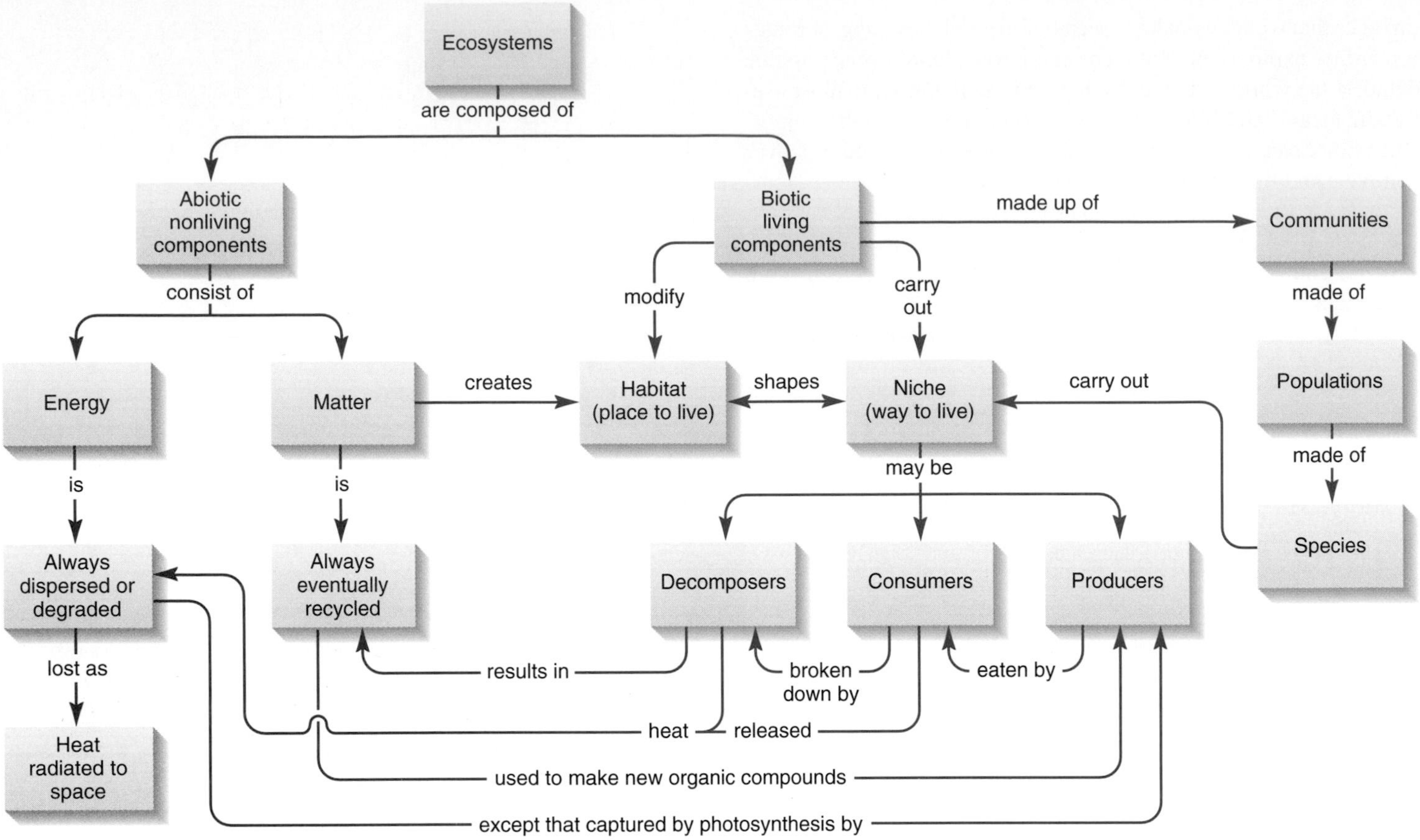

FIGURE 1.6 A concept map of ecosystem components and relationships.

Remember that concept maps are works in progress. There are no right or wrong maps. Each one represents one possible way of understanding a particular set of ideas by an individual at the time the map was made. Expect to do several iterations, right from the outset. There are probably as many ways of representing a collection of concepts as there are concepts in the collection, but some—typically those discovered by a process of trial and error—are more elegant and easier to understand than others.

The benefits of mapping are mainly to the individual making the map. The process of simplifying concepts and arranging them on a page forces you to think about what's most important. It helps you clarify your thoughts and understandings and makes learning more meaningful. A concept map can be a heuristic device: i.e., a process in which you make discoveries and uncover meanings through trial and error. Although you benefit most directly from making your own map, it can be instructive to compare your map to that of a fellow student to see how her or his take on a topic compares to yours.

SUMMARY

Whether you find environmental science interesting and useful depends to a large extent on your own attitudes and efforts. Developing good study habits, setting realistic goals for yourself, taking the initiative to look for interesting topics, finding an appropriate study space and working with a study partner can both make your study time more efficient and improve your final grade. Each of us has his/her own learning style. You may understand and remember things best if you see them in writing, hear them spoken by someone else, reason them out for yourself, or learn by doing. By determining your preferred style, you can study in the way that is most comfortable and effective for you. Many students find the SQ3R method a helpful study technique. The acronym stands for study, question, read, recite, and review.

Although grades shouldn't be the only reason to study, they are a important fact in every student's life. Knowing what kinds of test questions you will be asked can aid you in knowing how to study. Some simple analytic keys can help you choose the best answer in objective tests. Practicing writing short (a paragraph or two) essays such as those that answer the critical thinking questions at the end of every chapter in this book can be a great help if you are likely to have this kind of test. Practicing drawing the concept maps described at the end of this chapter may help you analyze and organize complex topics.

Perhaps the most useful skill that you could learn (or practice if you already know it) in this course is critical thinking. This is a conscious effort to "think about thinking." It challenges you to think about what you are trying to accomplish in a systematic, purposeful, and responsible manner. It also asks how you will know when you have reached an adequate conclusion. While critical thinking shares many of the skills and techniques of formal logic (as well as drawing on analytical, creative, and reflective thinking), it introduces attitudes such as open-mindedness, flexibility, skepticism, independence, persistence, relevance, contextual sensitivity, empathy, decisiveness, courage, and humility that will help you reach understanding in the complex and uncertain issues you will encounter in this course. One of the best places to practice critical thinking skills is in evaluating the reliability of information you see on the Internet. Although this new medium is a wonderful source of information, everything you see there should be approached with a sense of skepticism and caution.

QUESTIONS FOR REVIEW

1. Which study skills in table I.1 do you need to improve?
2. Describe some ways you can avoid procrastination and keep on schedule.
3. List four learning styles. Which fits you best?
4. Describe the SQ3R study techniques.
5. What are ten steps in critical thinking?
6. Name (and describe) seven attributes or dispositions essential for critical thinking.
7. List some adverbs or adverbial phrases that introduce premises and conclusions.
8. Distinguish between an *ad hominem* attack and an appeal to ignorance.
9. Describe three questions you'd ask to evaluate the reliability of Internet information.
10. Define concept mapping and show how it can be used in environmental science.

QUESTIONS FOR CRITICAL THINKING

1. What is critical thinking? Why is it important?
2. Suppose your preferred learning style doesn't match the teaching style of your instructor. How might you find a common ground?
3. You may find that you fit two or more of the learning profiles described in this chapter. How would you decide which is most effective or appropriate for you?
4. Why is critical thinking important in environmental science?
5. What is the difference between critical and reflective thinking?
6. Suppose you see a claim on television or the Internet, how would you evaluate its reliability?
7. Why are empathy and contextual sensitivity important in critical thinking?
8. If some facts in this book are vague or doubtful, why mention them at all? How would you decide which "facts" to use and which to ignore?
9. Why are concept maps usually hierarchical and linked by active verbs?

KEY TERMS

active learner 4
analytical thinking 7
concept maps 11
conclusion 9
creative thinking 7
critical thinking 7
logical learner 4
logical thinking 7
premises 9
reflective thinking 7
verbal learner 4
visual learner 4

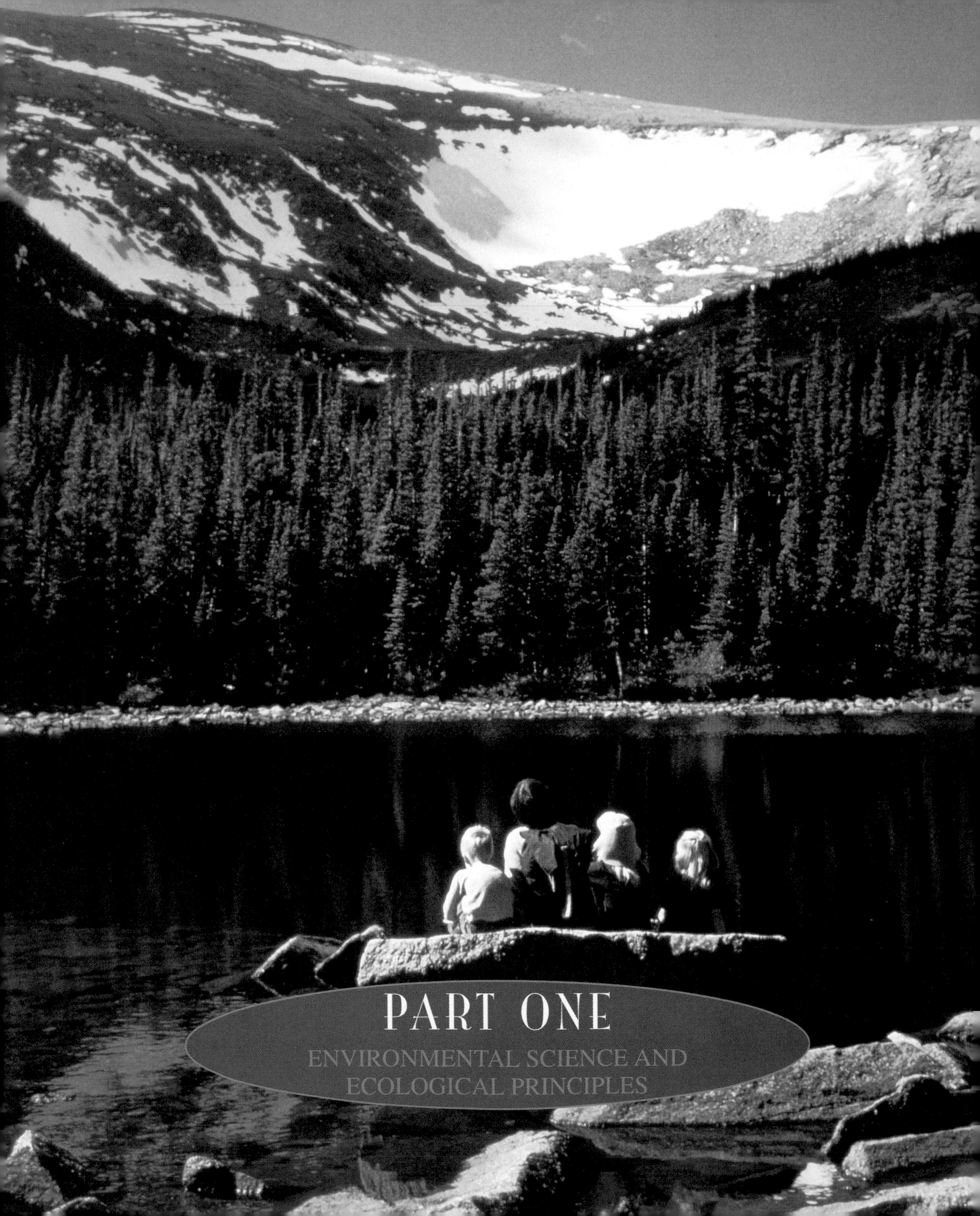

PART ONE

ENVIRONMENTAL SCIENCE AND ECOLOGICAL PRINCIPLES

CHAPTER ONE

Understanding Our Environment

We travel together, passengers on a little space ship, dependent upon its vulnerable reserves of air and soil, all committed for our safety to its security and peace; preserved from annihilation only by the care, the work, and I will say, the love that we give to our fragile craft.

—Adlai Stevenson—

OBJECTIVES

After studying this chapter, you should be able to:

- define the term, environment, and identify some important environmental concerns that we face today.
- discuss the history of conservation and the different attitudes toward nature revealed by utilitarian conservation and biocentric preservation.
- briefly describe some major environmental dilemmas and issues that shape our current environmental agenda.
- understand the connection between poverty and environmental degradation, as well as the division between the wealthy, industrialized countries and the poorer, developing countries of the world.
- recognize some of the reasons for feeling both optimistic and pessimistic about our environmental future.

ESSENTIAL STUDY PARTNER

The Essential Study partner (ESP) CD-ROM that accompanies this textbook contains additional information on the following topics:

- environmental science history and activism: principles: overview: history
- roots of public pollution awareness: principles: overview: history
- needs and environmental impact of growing world population: principles: overview: problems
- land degradation, resource consumption, and pollution production of the rich and poor: principles: overview: problems
- sustainable development: populations: economics: sustainability

Opposite: We live in a world of beauty and wonder. Will it remain so for our children and our children's children?

Above: A girl from the San Blas Islands in Panama holds a pet parrot. The San Blas people have been very successful in protecting biodiversity in their traditional lands.

Deformed Frogs

In 1995, school children on a summer field trip to a marsh in southern Minnesota discovered dozens of frogs with malformed, missing, or extra legs (fig. 1.1). The students' questions about these deformities led to studies that seem to show a sudden, widespread increase in developmental abnormalities in amphibians. Subsequent investigations in California, Iowa, Kansas, Missouri, New York, Oregon, Texas, Vermont, and Quebec have found as many as 60 percent of some frog or salamander species in certain areas have abnormal limbs, digits, or eyes. When dissected, many of these animals were found to have internal problems as well, including nonfunctioning digestive systems, grossly distended bladders, and abnormal reproductive organs.

What could be causing these problems? The widespread use of pesticides and herbicides in agriculture, together with industrial toxins released into the air and water, make toxic synthetic chemicals likely suspects. Since frogs spend most of their lives in water, they easily absorb toxins through their thin skin. And there seems to be more deformities in ponds near agricultural areas, compared to relatively "pesticide-free" areas, but these studies are not yet conclusive. We *do* know, however, that a number of synthetic chemicals disrupt endocrine hormone functions critical in regulating normal growth and development of many species.

Other factors could also play a role in these deformities. Retnoids (such as vitamin A) are known to regulate key processes in embryo development. They have been found in ponds with deformed frogs, perhaps from pesticide residues. Metals such as arsenic, mercury, selenium, and cadmium that are released by industry or carried in agricultural runoff also may be involved. Ultraviolet radiation from the sun is increasing in many areas because of damage to the stratospheric ozone shield and frog eggs and young are known to be sensitive to UV radiation. Parasites such as trematodes (flukes) are known to burrow into tadpoles during early development, forming cysts that interfere with normal growth and development. Unfortunately, there may be no single, simple cause for all the diverse problems observed at different places and times. Varying combinations of some or all of these factors may be at work.

What should we make of this apparently sudden increase in deformed animals? We don't have good historic data against which we can compare recent surveys; but scientists with many years of field experience report they have never seen the numbers of aberrant frogs found recently. Some people see this as an ominous warning that human activities are disrupting nature in ways that threaten not only amphibians but us as well. They see frogs as an indicator species, much like the canaries that early miners took into the mines to warn of life-threatening conditions.

Skeptics argue that frog abnormalities have probably always existed. Perhaps we're finding more now simply because we are looking for them. Just because the deformed frogs look pathetic, they contend, we shouldn't let our emotions run wild. Their condition may not mean anything for us. Clearly we need more careful research and reliable data before we know how serious the problem may be.

FIGURE 1.1 An alarming number of deformed frogs have been found in recent years. What causes these abnormalities is not yet known. Are they a warning of pollution or other serious environmental problems?

Deformed limbs are not the only threatening condition facing frogs and their amphibian kin today. Over the past 20 or 30 years, populations of frogs, toads, salamanders, and newts have diminished or even vanished entirely from wetlands around the world where they once were common. Wetlands are drained for agriculture and urban expansion, forests are cleared by lumber operations, and erosion fills streams and ponds with silt. Water pollution, acid rain, pesticide runoff, and other sources of toxic chemicals are known to be highly damaging to aquatic species. Naturally occurring diseases appear to be more virulent because hormone-disrupting chemicals damage immune system protection. Weather changes, perhaps caused by global warming, may be altering habitat.

These situations introduce a number of important themes in environmental science. What are we doing to other species and what threats do our activities expose *us* to as well? What can we do to reduce our negative impact and live in harmony with our environment? In this book we will survey a variety of global environmental issues and offer some suggestions for what we might do about them. In many cases—such as the example of deformed frogs—the problems are complex and only partially understood. Environmental science is a new and evolving field. It will require active participation on your part to think critically about these topics and to work out what you personally believe is correct. You are undertaking a journey of discovery that will continue, we hope, long after you finish the course for which you bought this book.

WHAT IS ENVIRONMENTAL SCIENCE?

Humans have always inhabited two worlds. One is the natural world of plants, animals, soils, air, and water that preceded us by billions of years and of which we are a part. The other is the world of social institutions and artifacts that we create for ourselves using science, technology, and political organization. Both worlds are essential to our lives, but integrating them successfully causes enduring tensions.

Where earlier people had limited ability to alter their surroundings, we now have power to extract and consume resources, produce wastes, and modify our world in ways that threaten both our continued existence and that of many organisms with which we share the planet. To ensure a sustainable future for ourselves and future generations, we need to understand something about how our world works, what we are doing to it, and what we can do to protect and improve it.

Environment (from the French *environner:* to encircle or surround) can be defined as (1) the circumstances or conditions that surround an organism or group of organisms, or (2) the complex of social or cultural conditions that affect an individual or community. Since humans inhabit the natural world as well as the "built" or technological, social, and cultural world, all constitute important parts of our environment (fig. 1.2).

Environmental science, then, is the systematic study of our environment and our proper place in it. A relatively new field, environmental science is highly interdisciplinary, integrating natural sciences, social sciences, and humanities in a broad, holistic study of the world around us. In contrast to more theoretical disciplines, environmental science is mission-oriented. That is, it seeks new, valid, contextual knowledge about the natural world and our impacts on it, but obtaining this information creates a responsibility to get involved in trying to do something about the problems we have created.

As distinguished economist Barbara Ward pointed out, for an increasing number of environmental issues, the difficulty is not to identify remedies. Remedies are now well understood. The problem is to make them socially, economically, and politically acceptable. Foresters know how to plant trees, but not how to establish conditions under which villagers in developing countries can manage plantations for themselves. Engineers know how to control pollution, but not how to persuade factories to install the necessary equipment. City planners know how to build housing and design safe drinking water systems, but not how to make them affordable for the poorest members of society. The solutions to these problems increasingly involve human social systems as well as natural science.

Criteria for environmental literacy suggested by the National Environmental Education Advancement Project in Wisconsin include: awareness and appreciation of the natural and built environment; knowledge of natural systems and ecological concepts; understanding of current environmental issues; and the ability to use critical-thinking and problem-solving skills on environmental issues. These are good overall goals to keep in mind as you study this book.

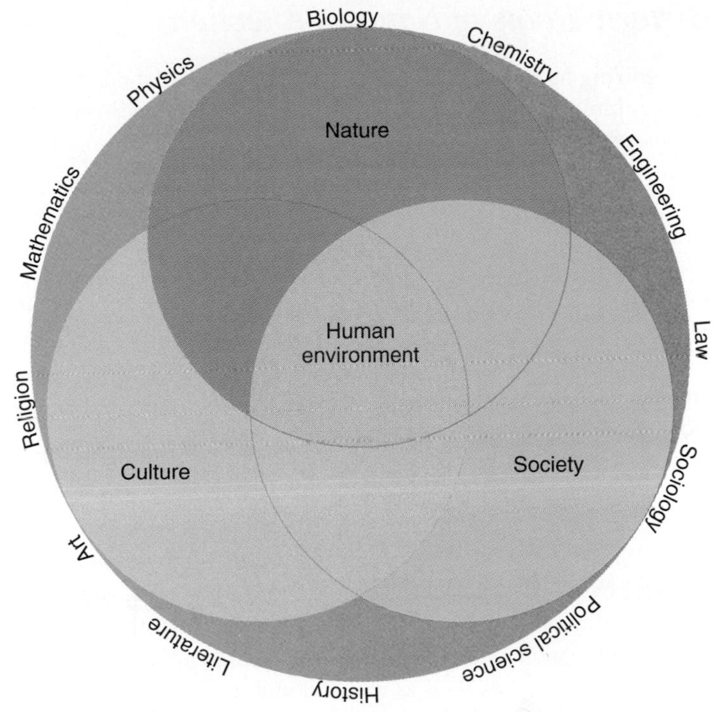

FIGURE 1.2 The intersections of the natural world with the social and cultural worlds encompass our environment. Many disciplines contribute to environmental science and help us understand how our worlds intertwine as well as our proper role in them.

Chapter 2 looks more closely at science as a way of knowing, environmental ethics, and other tools that help us analyze and understand the world around us. For the remainder of this chapter, we'll complete our overview with a short history of environmental thought and a survey of some important current issues that face us.

A BRIEF HISTORY OF CONSERVATION AND ENVIRONMENTALISM

Although many early societies had negative impacts on their surroundings, others lived in relative harmony with nature. In modern times, however, growing human populations and the power of our technology have heightened our concern about what we are doing to our environment. We can divide conservation history and environmental activism into at least four distinct stages: (1) pragmatic resource conservation, (2) moral and aesthetic nature preservation, (3) a growing concern about health and ecological damage caused by pollution, and (4) global environmental citizenship. Each era focused on different problems and each suggested a distinctive set of solutions. These stages are not necessarily mutually exclusive, however; parts of each persist today in the environmental movement and one person may embrace them all simultaneously.

Historic Roots of Nature Protection

Recognizing human misuse of nature is not unique to modern times. Plato complained in the fourth century B.C. that Greece once was blessed with fertile soil and clothed with abundant forests of fine trees. After the trees were cut to build houses and ships, however, heavy rains washed the soil into the sea, leaving only a rocky "skeleton of a body wasted by disease" (fig. 1.3). Springs and rivers dried up while farming became all but impossible. Many classical authors regarded Earth as a living being, vulnerable to aging, illness, and even mortality. Periodic threats about the impending death of nature as a result of human misuse have persisted into our own time. Many of these dire warnings have proven to be premature or greatly exaggerated, but others remain relevant to our own times. As Mostafa K. Tolba, former Executive Director of the United Nations Environment Programme has said, "The problems that overwhelm us today are precisely those we failed to solve decades ago."

Some of the earliest scientific studies of environmental damage were carried out in the eighteenth century by French and British colonial administrators who often were trained scientists and who considered responsible environmental stewardship as an aesthetic and moral priority, as well as an economic necessity. These early conservationists observed and understood the connection between deforestation, soil erosion, and local climate change. The pioneering British plant physiologist, Stephen Hales, for instance, suggested that conserving green plants preserved rainfall. His ideas were put into practice in 1764 on the Caribbean island of Tobago, where about 20 percent of the land was marked as "reserved in wood for rains."

Pierre Poivre, an early French governor of Mauritius, an island in the Indian Ocean, was appalled at the environmental and social devastation caused by destruction of wildlife (such as the flightless dodo) and the felling of ebony forests on the island by early European settlers. In 1769, Poivre ordered that one-quarter of the island was to be preserved in forests, particularly on steep mountain slopes and along waterways. Mauritius remains a model for balancing nature and human needs. Its forest reserves shelter a larger percentage of its original flora and fauna than most other human-occupied islands.

Pragmatic Resource Conservation

Many historians consider the publication of *Man and Nature* in 1864 by geographer George Perkins Marsh as the wellspring of environmental protection in North America. Marsh, who also was a lawyer, politician, and diplomat, traveled widely around the Mediterranean as part of his diplomatic duties in Turkey and Italy. He read widely in the classics (including Plato) and personally observed the damage caused by the excessive grazing by goats and sheep and by the deforesting of steep hillsides. Alarmed by the wanton destruction and profligate waste of resources still occurring on the American frontier in his lifetime, he warned of its ecological consequences. Largely as a result of his book, national forest reserves were established in the United States in 1873 to protect dwindling timber supplies and endangered watersheds.

FIGURE 1.3 Nearly 2500 years ago, Plato lamented land degradation that denuded the hills of Greece. Have we learned from history's lessons?

Among those influenced by Marsh's warnings were President Theodore Roosevelt and his chief conservation advisor, Gifford Pinchot. In 1905, Roosevelt, who was the leader of the populist, progressive movement, moved the Forest Service out of the corruption-filled Interior Department into the Department of Agriculture. Pinchot, who was the first native-born professional forester in North America, became the founding head of this new agency. He put resource management on an honest, rational, and scientific basis for the first time in our history. Together with naturalists and activists such as John Muir, William Brewster, and George Bird Grinnell, Roosevelt and Pinchot established the framework of our national forest, park, and wildlife refuge systems, passed game protection laws, and tried to stop some of the most flagrant abuses of the public domain. In 1908, Pinchot organized and chaired the White House Conference on Natural Resources, perhaps the most prestigious and influential environmental meeting ever held in the United States.

The basis of Roosevelt's and Pinchot's policies was pragmatic **utilitarian conservation.** They argued that the forests should be saved "not because they are beautiful or because they shelter wild creatures of the wilderness, but only to provide homes and jobs for people." Resources should be used "for the greatest good, for the greatest number for the longest time." "There has been a fundamental misconception," Pinchot said, "that conservation means nothing but husbanding of resources for future generations. Nothing could be further from the truth. The first principle of conservation is development and use of the natural resources now existing on this continent for the benefit of the people who live here now. There may be just as much waste in neglecting the development and use of certain natural resources as there is in their destruction." This pragmatic approach still can be seen in the multiple use policies of the Forest Service.

FIGURE 1.4 President Teddy Roosevelt (*left*) and naturalist John Muir pose at Glacier Point in Yosemite National Park during a historical meeting in 1903. How many modern presidents would take time for a camping trip to enjoy and celebrate the natural world?

FIGURE 1.5 The National Park Service has generally attempted to preserve pristine wilderness—such as this area in Redwood National Park—along the altruistic preservation principles proposed by John Muir.

Moral and Aesthetic Nature Preservation

John Muir (fig. 1.4), geologist, author, and first president of the Sierra Club, strenuously opposed Pinchot's influence and policies. Muir argued that nature deserves to exist for its own sake, regardless of its usefulness to us. Aesthetic and spiritual values formed the core of his philosophy of nature protection. This outlook has been called **biocentric preservation** because it emphasizes the fundamental right of other organisms to exist and to pursue their own interests. Muir wrote: "The world, we are told, was made for man. A presumption that is totally unsupported by the facts . . . Nature's object in making animals and plants might possibly be first of all the happiness of each one of them. . . . Why ought man to value himself as more than an infinitely small unit of the one great unit of creation?"

Muir, who was an early explorer and interpreter of the Sierra Nevada Mountains in California, fought long and hard for establishment of Yosemite and King's Canyon National Parks. The National Park Service, established in 1916, was first headed by Muir's disciple, Stephen Mather, and has always been oriented toward preservation of nature in its purest state (fig. 1.5). It has often been at odds with Pinchot's utilitarian Forest Service. Environmental ethics is discussed further in chapter 2.

Modern Environmentalism

The undesirable effects of pollution probably have been recognized at least as long as those of forest destruction. In 1273, King Edward I of England threatened to hang anyone burning coal in London because of the acrid smoke it produced. In 1661, the English diarist John Evelyn complained about the noxious air pollution caused by coal fires and factories and suggested that sweet-smelling trees be planted to purify city air. Increasingly dangerous smog attacks in Britain led, in 1880, to formation of a national Fog and Smoke Committee to combat this problem.

The tremendous industrial expansion during and after the Second World War added a new set of concerns to the environmental agenda. *Silent Spring,* written by Rachel Carson (fig. 1.6) and published in 1962, awakened the public to the threats of pollution and toxic chemicals to humans as well as other species. The movement she engendered might be called **environmentalism** because its concerns are extended to include both environmental resources and pollution. Among the pioneers of this movement are activist David Brower and scientist Barry Commoner. Brower, while executive director of the Sierra Club, Friends of the Earth, and, more recently, Earth Island Institute, introduced many of the techniques of modern environmentalism, including litigation, intervention in regulatory hearings, book and calendar publishing, and using mass media for publicity campaigns. Commoner, who was trained as a molecular biologist, has been a leader in analyzing the links between science, technology, and society. Both activism and research remain hallmarks of the modern environmental movement.

Under the leadership of a number of other brilliant and dedicated activists and scientists, the environmental agenda was expanded

FIGURE 1.6 Rachel Carson's book *Silent Spring* was a landmark in modern environmental history. She alerted readers to the dangers of indiscriminate pesticide use.

FIGURE 1.7 We live in a bountiful and beautiful world. Ours is a unique and irreplaceable planet on whose life-sustaining systems we are totally dependent.

in the 1960s and 1970s to include issues such as human population growth, atomic weapons testing and atomic power, fossil fuel extraction and use, recycling, air and water pollution, wilderness protection, and a host of other pressing problems that are addressed in this textbook. Environmentalism has become well established on the public agenda since the first national Earth Day in 1970. A majority of Americans now consider themselves environmentalists, although there is considerable variation in what that term means.

Global Concerns

Increased opportunities to travel, as well as greatly expanded international communications, now enable us to know about daily events in places unknown to our parents or grandparents. We have become, as Marshal McLuhan announced in the 1960s, a global village (In Depth, p. 32). As in a village, we are all interconnected in various ways. Events that occur on the other side of the globe have profound and immediate effects on our lives.

Photographs of the earth from space (fig. 1.7) provide a powerful icon for the fourth wave of ecological concern that might be called **global environmentalism.** These photos remind us how small, fragile, beautiful, and rare our home planet is. We all share a common environment at this global scale. As our attention shifts from questions of preserving particular landscapes or preventing pollution of a specific watershed or airshed, we begin to worry about the life-support systems of the whole planet.

Minnesota geologist Roger Hooke estimates that current human earth-moving activities now rival those of natural geological forces. In addition, we are changing planetary weather systems and atmospheric chemistry, reducing the natural variety of organisms, and degrading ecosystems in ways that could have devastating effects, both on humans and on all other life-forms. Protecting our environment has become an international cause and it will take international cooperation to bring about many necessary changes.

Among the leaders of this worldwide environmental movement have been British economist Barbara Ward, French/American scientist René Dubos, Norwegian Prime Minister Gro Harlem Brundtland, and Canadian diplomat Maurice Strong. All have been central in major international environmental conventions, such as the 1972 U.N. Conference on the Human Environment in Stockholm or the 1992 U.N. "Earth Summit" on Environment and Development in Rio de Janeiro. Once again, new issues have become part of the agenda as our field of vision widens. We have begun to appreciate the links between poverty, injustice, oppression, and exploitation of humans and our environment. We will discuss human development in more detail later in this chapter.

CURRENT CONDITIONS

As you probably already know, many environmental problems now face us. Before surveying them in the following section, we should pause for a moment to consider the extraordinary natural world that we inherited and that we hope to pass on to future generations in as good—perhaps even better—a condition than when we arrived.

FIGURE 1.8 We are fortunate to live in a beautiful, prolific, agreeable world. It will take care and hard work to keep it this way.

A Marvelous Planet

Imagine that you are an astronaut returning to Earth after a long trip to the moon or Mars. What a relief it would be to come back to this beautiful, bountiful planet (fig. 1.8) after experiencing the hostile, desolate environment of outer space. Although there are dangers and difficulties here, we live in a remarkably prolific and hospitable world that is, as far as we know, unique in the universe. Compared to the conditions on other planets in our solar system, temperatures on the earth are mild and relatively constant. Plentiful supplies of clean air, fresh water, and fertile soil are regenerated endlessly and spontaneously by geological and biological cycles (discussed in chapters 3 and 4).

Perhaps the most amazing feature of our planet is the rich diversity of life that exists here. Millions of beautiful and intriguing species populate the earth and help sustain a habitable environment. This vast multitude of life creates complex, interrelated communities where towering trees and huge animals live together with, and depend upon, tiny life-forms such as viruses, bacteria, and fungi. Together, all these organisms make up delightfully diverse, self-sustaining communities, including dense, moist forests, vast sunny savannas, and richly colorful coral reefs. From time to time, we should pause to remember that, in spite of the challenges and complications of life on Earth, we are incredibly lucky to be here. We should ask ourselves: what is our proper place in nature? What *ought* we do and what *can* we do to protect the irreplaceable habitat that produced and supports us? These are some of the central questions of environmental science.

Environmental Dilemmas

While there are many things to appreciate and celebrate about the world in which we live, many pressing environmental problems cry out for our attention. Human populations have grown at alarming rates in this century. More than 6 billion people now occupy the earth and we are adding about 85 million more each year. In the next decade, our numbers will increase by almost as many people as are now alive in India. Most of that growth will be in the poorer countries where resources and services are already strained by present populations.

Some demographers believe that this unprecedented growth rate will slow in the next century and that the population might eventually drop back below its present size. Others warn that the number of humans a century from now could be four or five times our present population if we don't act quickly to bring birthrates into balance with death rates. Whether there are sufficient resources to support 6 billion humans—let alone 25 billion—on a sustainable basis is one of the most important questions we face. How we might stabilize population and what level of resource consumption we and future generations can afford are equally difficult parts of this challenging equation.

Food shortages and famines already are too familiar in many places and may increase in frequency and severity if population growth, soil erosion, and nutrient depletion continue at the same rate in the future as they have in the past (fig. 1.9). We are coming to realize, however, that food security often has more to do with poverty, democracy, and equitable distribution than it does with the amount of food available. Water deficits and contamination of existing water supplies threaten to be critical environmental issues in the future for agricultural production as well as for domestic and industrial uses. Many countries already have serious water shortages and more than one billion people lack access to clean water or adequate sanitation. Violent conflicts over control of natural resources may flare up in many places if we don't learn to live within nature's budget.

How we obtain and use energy is likely to play a crucial role in our environmental future. Fossil fuels (oil, coal, and natural gas) presently supply about 80 percent of the energy used in industrialized countries (fig. 1.10). Supplies of these fuels are diminishing at an alarming rate and problems associated with their acquisition and use—air and water pollution, mining damage, shipping accidents, and political insecurity—may limit where and how we use remaining reserves. Cleaner renewable energy resources—solar power, wind, and biomass—together with conservation, may replace environmentally destructive energy sources if we invest in appropriate technology in the next few years.

As we burn fossil fuels, we release carbon dioxide and other heat-absorbing gases that cause global warming and may bring about sea-level rises and catastrophic climate changes. Acids formed in the air as a result of fossil fuel combustion

FIGURE 1.9 Three-quarters of the world's poorest nations are in Africa. Millions of people lack adequate food, housing, medical care, clean water, and safety. The human suffering engendered by this poverty is tragic.

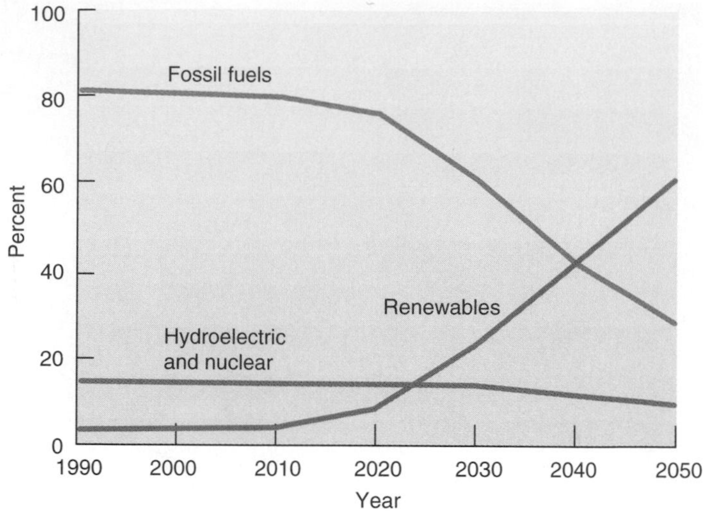

FIGURE 1.10 Fossil fuel supplies are dwindling at an alarming rate. They could be replaced by renewables such as solar power, wind, and biomass energy that could be more equitably distributed and would be less polluting than our current energy sources.

Sources: Data from World Bank estimates; D. Anderson and C. D. Bird, "Carbon Accumulations and Technical Progress—A Simulation Study of Costs," *Oxford Bulletin of Economics & Statistics* 54(1):1–29.

already have caused extensive damage to building materials and sensitive ecosystems in many places (fig. 1.11). Continued fossil fuel use without pollution control measures could cause even more extensive damage. Chlorinated compounds, such as the chlorofluorocarbons used in refrigeration and air conditioning, also contribute to global warming, as well as damaging the stratospheric ozone that protects us from cancer-causing ultraviolet radiation in sunlight.

Destruction of tropical forests, coral reefs, wetlands, and other biologically rich landscapes is causing an alarming loss of species and a reduction of biological variety and abundance that could severely limit our future options. Many rare and endangered species are threatened directly or indirectly by human activities (fig. 1.12). In addition to practical values, aesthetic and ethical considerations suggest that we should protect these species and the habitat necessary for their survival. On the other hand, sometimes cultural or economic interests compete with biodiversity conservation (What Do You Think? p. 23).

Toxic air and water pollutants, along with mountains of solid and hazardous wastes, are becoming overwhelming problems in industrialized countries. We produce hundreds of millions of tons of these dangerous materials annually, and much of it is disposed of in dangerous and irresponsible ways. No one wants this noxious stuff dumped in their own backyard, but too often the solution is to export it to someone else's. We may come to a political impasse where our failure to decide where to put our wastes or how to dispose of them safely will close down industries and result in wastes being spread everywhere. The health effects of pollution, toxic wastes, stress, and the other environmental ills of modern society have become a greater threat than infectious diseases for many of us in industrialized countries.

These and other similarly serious problems illustrate the importance of environmental science and environmental education for everyone. What we are doing to our world, and what that may mean for our future and that of our children is of paramount concern as we enter the twenty-first century. We trust that's why you're reading this book.

Signs of Hope

The dismal litany of problems we have just reviewed seems pretty overwhelming, doesn't it? But is there hope that we may find solutions to these problems? We think so. As you will see in subsequent chapters, progress has been made in many areas in controlling air and water pollution and reducing wasteful resource uses. Many

what do you think?

THE WORLD AROUND YOU

Cultural Whaling in the Pacific Northwest

On May 17, 1999, for the first time in more than 70 years, members of the Makah tribe from Washington's Olympic peninsula struck and killed a gray whale. Using a mixture of modern technology and ancient traditions, nine men paddling a handmade cedar canoe drew close enough to harpoon the 20-ton whale, after which high-speed motorboats closed in to kill the wounded animal with a .50 caliber anti-tank rifle.

For months before the hunt, native people, animal rights groups, sea conservation societies, and members of the whale-watching industry had engaged in shouting matches, racial slurs, political maneuvering, media campaigns, and dangerous motorboat confrontations as each side tried to promote its own views. At issue is a complex mixture of international environmental policy, aboriginal culture and hunting rights, ethical values, commercial interests, and politics that is not easy to resolve. This case study presents a good opportunity for you to practice critical thinking skills.

For at least 1500 years the Makah people collected fish and shellfish and hunted for seals, whales, and sea otters along this section of fog-shrouded coast. The 1855 Treaty of Neah Bay recognized the right of the Makah people to hunt whales, the only such covenant in United States history. By the beginning of the twentieth century, however, commercial exploitation had driven nearly into extinction the whales, seals, otters, and fish on which the Makah once depended. Fewer than 2000 Pacific gray whales survived out of a historic population of about 30,000. By 1920, most Makah had given up their traditional subsistence culture and turned to farming, logging, or factory jobs.

Members of a Makah whaling crew set out in their handmade cedar canoe on a practice run. Whale hunting is central in the Makah culture, but is opposed by animal rights groups.

Since their protection under the Endangered Species Act and the International Whaling Ban, gray whales have made a remarkable and encouraging recovery. The population today is thought to be around 26,000 animals and is growing at about 2.5 percent per year. It may well have reached the carrying capacity of its habitat. In 1999 scientists estimated that about 800 whales died of starvation and disease during their 4000-mile migration from Baja California to the Arctic. A sustainable harvest might actually be beneficial for the whale population. When the species was removed from the U.S. endangered species list in 1994, the Makah announced their intention to resume whaling under their treaty rights. They consider this an important step in recovering their traditional culture and combating the poverty, drugs, alcoholism, violence, and despair that have afflicted their community. Reviving the rituals, discipline, and pride of whaling, they believe, could lead to a cultural renaissance for their people.

To anti-hunting groups, killing magnificent, intelligent, social animals like whales is tantamount to murder. In their view, calling for a sustainable harvest of whales is equivalent to establishing a sustainable level of genocide or torture. This isn't subsistence hunting, they claim, because the Makah haven't depended on whale meat for many years and don't need it now. To the Makah and their supporters from whaling nations such as Japan and Norway, on the other hand, it is hypocritical to condemn hunting marine mammals when most Americans happily eat the millions of cattle, pigs, sheep, and other animals slaughtered every year in commercial packing plants.

How would you approach this complex question? What information would you need to determine whether Indian claims for cultural traditions should take precedence over ethical rights of wild animals? Does the fact that native people have suffered economically and socially give them special standing? Should subsistence hunting be limited to traditional tools and people who have a demonstrated nutritional need? What information sources would you regard as reliable and authoritative in this situation? How would you decide whom or what to believe?

As you study environmental science, you will find that many of the dilemmas we face, like this question of hunting whales by native people, require a combination of scientific data and contextual sensitivity and empathy. Because humans are involved, questions of history, culture, ethics, and politics also come into play. How would you resolve this clash in world views? Can you suggest a compromise that might defuse this tense situation?

cities in North America and Europe are cleaner and less polluted than they were a generation or so ago. Population has stabilized in most industrialized countries and even some very poor countries where social security and democracy have been established. Over the last twenty years, the average number of children born per woman worldwide has decreased from 6.1 to 3.4. This is still above the zero population growth rate of 2.1 children per couple, but it is an encouraging improvement. If this rate of progress continues in the next twenty years as it has in the past twenty, the world population could stabilize early in the twenty-first century.

The incidence of life-threatening infectious diseases has been reduced sharply in most countries during the past century, while the average life-expectancy has nearly doubled. Many new resources have been discovered and more efficient ways of using existing supplies have been invented that allow us to enjoy luxuries and conveniences that would have seemed miraculous only a

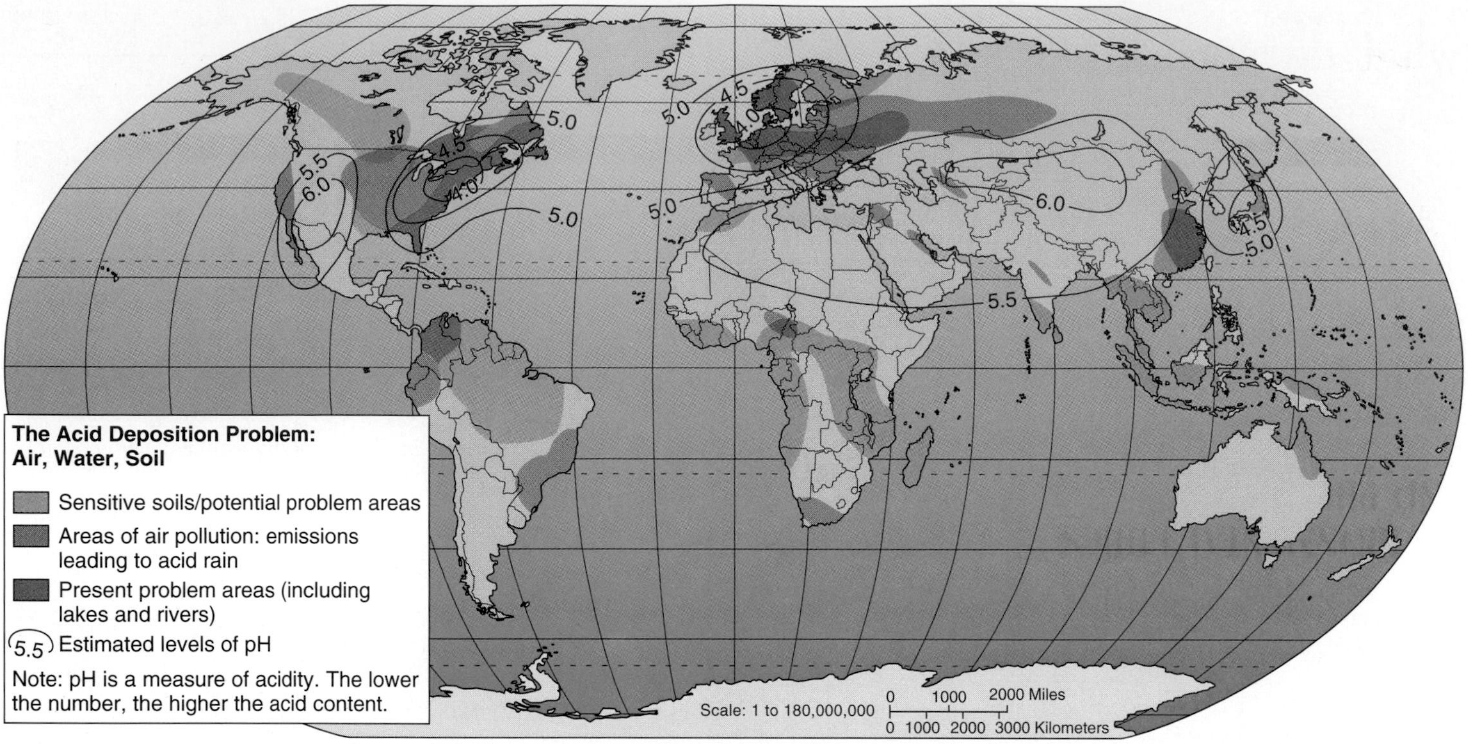

FIGURE 1.11 Sulfuric and nitric acids formed in the air by burning fossil fuels and other industrial activities threaten crops, water quality, building materials, and living organisms, especially in areas where soils have little buffering capacity.
Source: John Allen, *Student Atlas of Environmental Issues,* Dushkin/McGraw-Hill, 1997.

few generations ago. Although modern life has many stresses and strains, few people would willingly return to conditions that existed 10,000, 1000, or even 100 years ago. Would you?

Still, we can do much more, both individually and collectively, to protect and restore our environment. Being aware of the problems we face is the first step toward finding their solutions. Increased media coverage has brought environmental issues to public attention. More than 80 percent of the Americans polled in public opinion surveys agree that "protecting the environment is so important that requirements and standards cannot be too high and continuing environmental improvements must be made regardless of cost." This growing understanding and concern are themselves hopeful signs. Young people today may be in a unique position to address these issues because, for the first time in history, we now have resources, motivation, and knowledge to do something about our environmental problems. Unfortunately, if we don't act now, we may not have another chance to do so.

NORTH/SOUTH: A DIVIDED WORLD

We live in a world of haves and have-nots; a few of us live in increasing luxury while many others lack the basic necessities for a decent, healthy, productive life. The World Bank estimates that more than 1.3 billion people—about one-fifth of the world—live in **acute poverty** with an income of less than $1 (U.S.) per day. These poorest of the poor generally lack access to an adequate diet, decent housing, basic sanitation, clean water, education, medical care, and other essentials for a humane existence. Seventy percent of those people are women and children. In fact, four out of five people in the world live in what would be considered poverty in the United States or Canada. The plight of these poor people is not just a humanitarian concern. Policymakers are becoming aware that eliminating poverty and protecting our common environment are inextricably interlinked.

The world's poorest people have become both the victims and the agents of environmental degradation. The poorest people are too often forced to meet short-term survival needs at the cost of long-term sustainability. Desperate for croplands to feed themselves and their families, many move into virgin forests or cultivate steep, erosion-prone hillsides where soil nutrients are exhausted after only a few years. Others migrate to the grimy, crowded slums and ramshackle shantytowns that now surround most major cities in the developing world. With no way to dispose of wastes, the residents often foul their environment further and contaminate the air they breathe and the water on which they depend for washing and drinking (fig. 1.13).

The cycle of poverty, illness, and limited opportunities can become a self-sustaining process that passes from one generation to another. People who are malnourished and ill can't work productively to obtain food, shelter, or medicine for themselves or their children, who also are malnourished and ill. About 250 million

FIGURE 1.12 This emperor tamarin is a marmoset from South America. Although we hear a great deal about destruction of tropical *rainforests,* the *dry* tropical forests that once lined the Atlantic coast of South America and the Pacific coast of Central America have been destroyed to a much greater extent. More than 99% of the once-extensive dry tropical forests of Latin America have been destroyed, threatening a host of interesting and important species with extinction.

FIGURE 1.13 While many of us live in luxury, more than a billion people lack access to food, housing, clean water, sanitation, education, medical care, and other essentials for a healthy, productive life. Often the poorest people are both the victims and agents of environmental degradation as they struggle to survive. Helping them meet their needs is not only humane, it is essential to protect our mutual environment.

children—mostly in Asia and Africa and some as young as 4 years old—are forced to work under appalling conditions weaving carpets, making ceramics and jewelry, or in the sex trade. Growing up in these conditions leads to mental and developmental deficits that condemn these children to perpetuate this cycle.

Faced with immediate survival needs and few options, these unfortunate people often have no choice but to overharvest resources; in doing so, however, they diminish not only their own options, but also those of future generations. And in an increasingly interconnected world, the environments and resource bases damaged by poverty and ignorance are directly linked to those on which we depend. It is in our own self-interest to help everyone find better ways to live.

Rich and Poor Countries

Where do the rich and poor live? About one-fifth of the world's population lives in the twenty richest countries where the average per capita income is above $25,000 (U.S.) per year (fig. 1.14). Most of these countries are in North America or Western Europe, but Japan, Singapore, Australia, New Zealand, the United Arab Emirates, and Israel also fall into this group. Almost every country, however, even the richest, such as the United States and Canada, has poor people. No doubt everyone reading this book knows about homeless people or other individuals who lack resources for a safe, productive life. The Journal of Public Health reported in 1999 that 10 million Americans—including 4 million children—do not have enough to eat.

The other four-fifths of the world live in middle- or low-income countries where nearly everyone is poor by our standards. More than 3 billion people live in the poorest nations where the average per capita income is below $620 (U.S.) per year. China and India are the largest of these countries, with a combined population of about 2 billion people. Among the 41 other nations in this category, 33 are in sub-Saharan Africa. All the other lowest-income nations, except Haiti, are in Asia. Although poverty levels in countries such as China and Indonesia have fallen in recent years, most countries in sub-Saharan Africa and much of Latin America have made little progress. The destabilizing and impoverishing effects of earlier colonialism continue to play important roles in the ongoing problems of these unfortunate countries. Meanwhile, the relative gap between rich and poor has increased dramatically (fig 1.15).

The ten poorest countries in the world, in 1999, were (in ascending order from very poorest): Mozambique, Ethiopia, Tanzania, Congo, Malawi, Rwanda, Chad, Sierra Leone, Nepal, and Niger. Each of these countries has annual per capita Gross National Product (GNP) of less than $200 (U.S.) per year. They also have low levels of food security, social welfare, and quality of life as indicated by table 1.1.

By contrast, each of the ten richest countries in the world—Switzerland, Japan, Norway, Denmark, Germany, the United States,

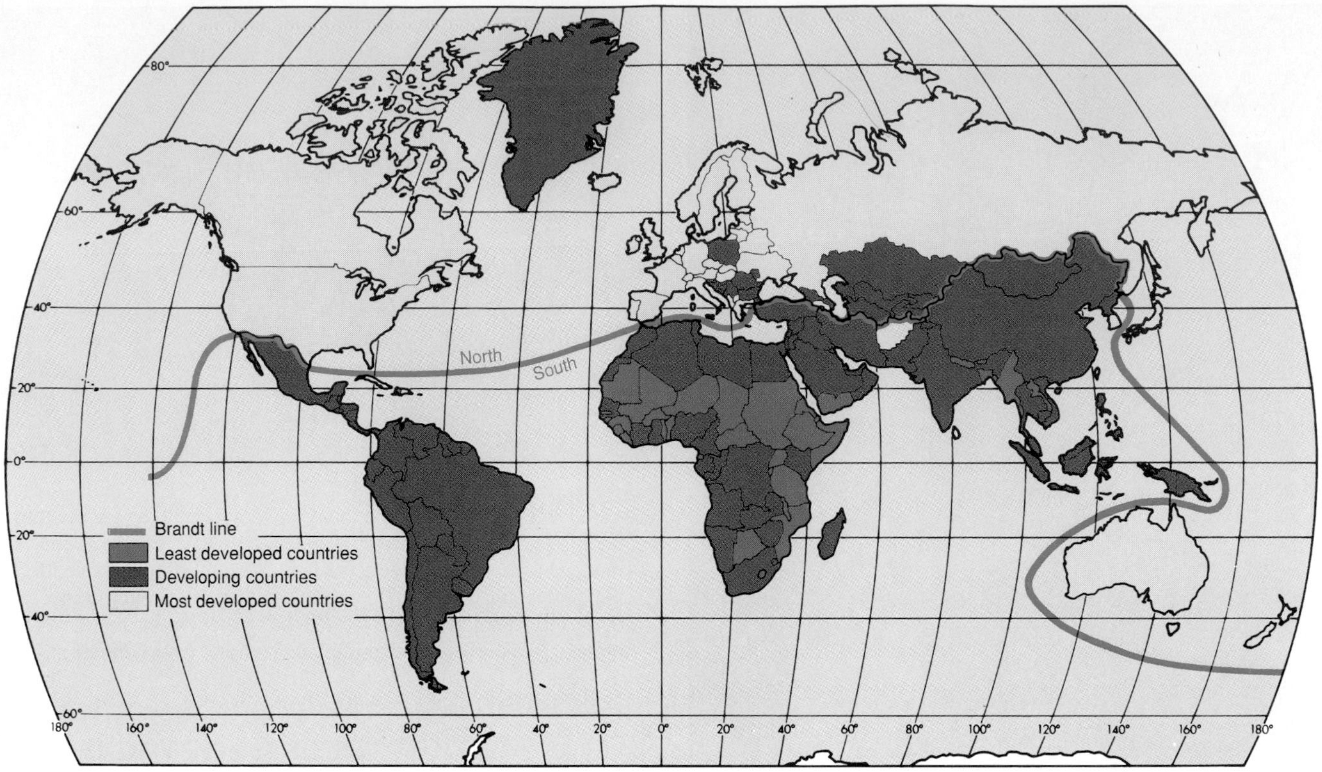

FIGURE 1.14 The world of underdevelopment. The Brandt line, denoted by the Independent Commission on International Development Issues, separates the richer, industrialized nations of Europe, North America, Japan, Australia, and New Zealand from their southern neighbors in the "developing" world. The 42 "least developed" countries, according to the United Nations Development Programme, are mostly in Africa and Asia. Note that since this map was drawn, conditions have worsened in parts of the former USSR and Africa but have often improved elsewhere.

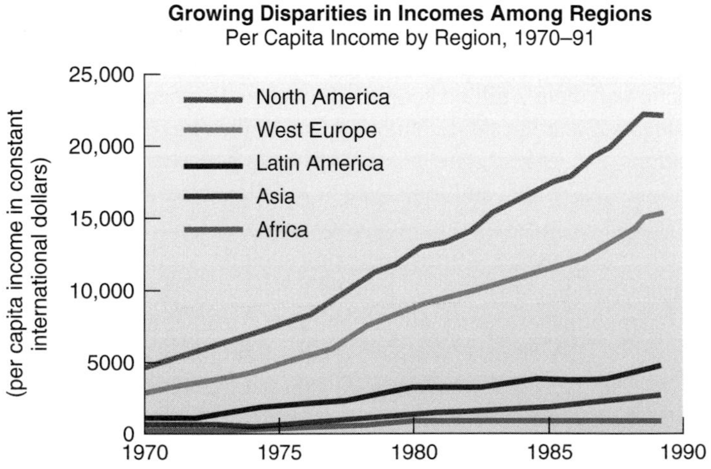

FIGURE 1.15 Although the percentage of the world's population living in poverty has decreased slightly over the past 30 years, the relative gap between rich and poor nations has increased sharply.
Source: United Nations (U.N.), *Critical Trends: Global Change and Sustainable Development* (U.N., New York, 1997), p. 58.

TABLE 1.1 Average Indicators of Quality of Life for the Ten Richest and Poorest Countries[1]

INDICATOR	POOR COUNTRIES	RICH COUNTRIES
GNP/capita	$170	$29,946
Life expectancy	47.4 years	77.9 years
Infant mortality[2]	114	5.7
Child deaths[3]	194	7.5
Safe drinking water	42%	NA[4]
Female literacy	38%	97%
Birth rate[5]	45.2	11.4

[1]Averaged as a group
[2]per1000 live births
[3]per 1000 children before age 5
[4]not available, but close to 100 percent
[5]per 1000 people
Source: World Resources Institute 1998–99 Oxford University Press: New York.

Austria, Singapore, France, and Iceland (in descending order from richest)—has an annual per capita GNP more than a hundred times that of the poorest countries. As you can see in table 1.1, other conditions in the rich countries reflect this wide disparity in wealth.

The gulf between rich and poor is even greater at the individual level. The richest 200 people in the world have a combined wealth of $1 trillion. This is more than the total wealth of the poorest half—some 3 billion people—of the world's population.

A Fair Share of Resources?

The affluent lifestyle that many of us in the richer countries enjoy consumes an inordinate share of the world's natural resources and produces a shockingly high proportion of pollutants and wastes. The United States, for instance, with less than 5 percent of the total population, consumes about one-quarter of most commercially traded commodities and produces a quarter to half of most industrial wastes (table 1.2).

To get an average American through the day takes about 450 kg (nearly 1000 lbs) of raw materials, including 18 kg (40 lbs) of fossil fuels, 13 kg (29 lbs) of other minerals, 12 kg (26 lbs) of farm products, 10 kg (22 lbs) of wood and paper, and 450 liters (119 gal) of water. Every year we throw away some 160 million tons of garbage, including 50 million tons of paper, 67 billion cans and bottles, 25 billion styrofoam cups, 18 billion disposable diapers, and 2 billion disposable razors (fig. 1.16).

This profligate resource consumption and waste disposal strains the life-support systems of the Earth on which we all depend. If everyone in the world tried to live at consumption levels approaching ours, the results would be disastrous. Unless we find ways to curb our desires and produce the things we truly need in less destructive ways, the sustainability of human life on our planet is questionable.

North/South Division

Notice in the lists of rich and poor countries discussed earlier that the wealthiest countries tend to be in the north, while the poorest countries tend to be located closer to the equator. It is common to speak of a **North/South division** of wealth and power, even though many poorer nations such as India and China are in the Northern Hemisphere and some relatively rich nations like Australia and New Zealand are in the Southern. Another way of describing the rich nations is that they tend to be highly industrialized and generally are members of the Organization for Economic Cooperation and Development (OECD), made up of Western Europe, North America, Japan, Korea, Turkey, Australia, and New Zealand. This group makes up about 20 percent of the world's population but consumes more than half of most resources. The other, poorer four-fifths of humanity are sometimes called the majority world because they make up the majority of the population.

FIGURE 1.16 "And may we continue to be worthy of consuming a disproportionate share of this planet's resources."
Drawing by Lorenz; © 1992 The New Yorker Magazine, Inc.

TABLE 1.2 The United States, with 4.5 Percent of the World's Population

CONSUMES	PRODUCES
26 percent of all oil	50 percent of all toxic wastes
24 percent of aluminum	26 percent of nitrogen oxides
20 percent of copper	25 percent of sulfur oxides
19 percent of nickel	22 percent of chlorofluorocarbons
13 percent of steel	26 percent of carbon dioxide

Source: Data from World Resources Institute 1998–99.

Since poverty and social welfare are closely linked with environmental degradation, there will be many instances in this book in which we will want to distinguish between conditions in groups of countries. In fact, income, social services, human development, and other indicators of quality of life don't always fit neatly in just two categories. Nations actually spread across a spectrum—or perhaps a three-dimensional matrix—of these indicators, making it difficult to establish inclusive categories. The following categories and corresponding acronyms are among those we will refer to throughout the rest of this book.

Political Economies

Countries are sometimes classified according to their economic system. **First World** describes the industrialized, market-oriented, democracies of Western Europe, North America, Japan, Australia, New Zealand, and their allies. **Second World** originally described the centrally planned, socialist countries, such as the former Soviet Union and its Eastern European allies. Several Asian socialist countries, such as China, Mongolia, North Korea, and Vietnam, also once belonged in this category but most are rapidly changing to market economies.

In the 1960s, the nonaligned, nonindustrial, ex-colonial nations such as India, Indonesia, Malaysia, Iran, Syria, and many African countries labeled themselves the **Third World** to show their independence from either of the superpower groups. This category was not intended to suggest that its members are third rate or third class. It has come to be used to describe developing countries in general. Some economists suggest that another category, the **Fourth World,** is necessary to classify the poorest nations with neither market economies nor central planning, as well as the indigenous communities within wealthy nations.

The amazing political and economic upheavals of the past decade have thrown all of these categories into a turmoil. Few countries any more are either purely socialist or capitalist. Nearly every government plans centrally and intervenes in its economy to some extent, and nearly every nation has at least some market-orientation. These categories have decreasing significance as old political alliances break down, forcing us to look for other ways to describe nations and peoples.

HUMAN DEVELOPMENT

Every year, the United Nations releases a report ranking countries by a **human development index** based primarily on average life expectancy, percentage of literate adults, mean years of schooling, and annual income per capita. Some other indicators factored into this index include infant mortality rates, daily calorie supply, child malnutrition, and access to clean water. The highest possible human development index (HDI) is 1.0; the lowest is 0.0. In 1999, Canada was highest in the world with an HDI of 0.96. As you might expect, there is a close correlation between wealth and human development. All the top twenty nations (all above 0.92) were in North America or Western Europe, except Japan and Singapore. The lowest HDI was Sierra Leone at 0.19. In fact, all of the twenty lowest rankings (all below 0.35) were in Africa, except Haiti and Bhutan.

Developmental Discrepancies

Of course aggregate numbers such as these hide many important issues. One of these is gender inequities. Men generally fare better than women on almost every socioeconomic indicator. For the few countries that keep such data, Japan had the lowest female-to-male wage ratio (51 percent), while Sweden had the highest (90 percent). Similarly, the female-to-male ratio in nonagricultural work varies from a low of 22 percent in Bahrain to 89 percent in Finland. If the HDI is weighed for these gender discrepancies, egalitarian Scandinavian countries like Sweden, Finland, and Norway move into top place.

Race is another variable that determines socioeconomic status in many countries. If white South Africa were a separate country, it would rank 24th in the world in human development (just after Spain). Black South Africa, on the other hand, would rank 123rd in the world (just above the Congo). In some countries, regional or ethnic differences create disparities. In Nigeria, the state of Bendel, with a HDI of 0.66 is equivalent to Cuba, while the poor state of Borno, with a HDI of 0.156 is lower than any country in the world.

The United Nations states that these vast discrepancies and the grinding poverty experienced by the poorest of the poor are the greatest threats to political stability, social cohesion, and environmental health on our planet. It warns that a number of nations such as Egypt, South Africa, Nigeria, and Brazil are in danger of joining the list of failed states because of these great disparities. Some suggested strategies for poverty reduction and social justice include:

- basic social services, especially basic education and primary health care;
- agrarian reform for more equitable land distribution;
- credit to tide those without resources through tough times;
- employment to ensure a secure livelihood for all;
- civil rights that enable everyone to participate in planning and management decisions;
- a social safety net to catch those whom markets exclude;
- economic growth that benefits the poor; and
- sustainable resource use that reduces material-intensive and energy-intensive lifestyles and turns, instead, to renewables.

Good News and Bad News

Over the past fifty years, human ingenuity and enterprise have brought about a breathtaking pace of technological innovations and scientific breakthroughs. The world's gross domestic product increased more than tenfold during that period, from $2 trillion to $22 trillion per year. While not all that increased wealth was applied to human development, there has been significant progress in increasing general standard of living nearly everywhere. In 1960, for instance, nearly three-quarters of the world's population lived in abysmal conditions (HDI below 0.5). Now, less than one-third are still at this low level of development.

Since World War II, average real income in developing countries has doubled; malnutrition declined by almost one-third; child death rates have been reduced by two-thirds; average life expectancy increased by 30 percent; and the percentage of rural families with access to safe drinking water has risen from less than 10 percent to almost 75 percent. Overall, poverty rates have decreased more in the last fifty years than in the previous five hundred. Nonetheless, while general welfare has increased, so has the gap between rich and poor worldwide. In 1960, the income ratio between the richest 20 percent of the world and the poorest 20 percent was 30 to 1. In 1999, this ratio was 100 to 1. Because perceptions of poverty are relative, people may feel worse off compared to their rich neighbors than development indices suggest they are.

Sustainable Development

By now, it is clear that security and living standards for the world's poorest people are inextricably linked to environmental protection. One of the most important questions in environmental science is how we can continue improvements in human welfare within the limits of the earth's natural resources. A possible solution to this dilemma is **sustainable development,** a term popularized by *Our Common Future,* the 1987 report of the World Commission on Environment and Development, chaired by Norwegian Prime Minister Gro Harlem Brundtland (and consequently called the Brundtland Commission). In the words of this report, sustainable development means "meeting the needs of the present without compromising the ability of future generations to meet their own needs."

Another way of saying this is that development means improving people's lives. Sustainable development, then, means progress in human well-being that can be extended or prolonged over many generations rather than just a few years. To be truly enduring, the benefits of sustainable development must be available to all humans rather than to just the members of a privileged group.

To many economists, it seems obvious that economic growth is the only way to bring about a long-range transformation to more advanced and productive societies and to provide resources to improve the lot of all people. As former President John F. Kennedy said, "A rising tide lifts all boats." But economic growth is not sufficient in itself to meet all essential needs. As the Brundtland Commission pointed out, political stability, democracy, and equitable economic distribution are needed to ensure that the poor will get a fair share of the benefits of greater wealth in a society.

Can Development Be Truly Sustainable?

Many ecologists regard "sustainable" growth of any sort as impossible in the long run because of the limits imposed by nonrenewable resources and the capacity of the biosphere to absorb our wastes. Using ever-increasing amounts of goods and services to make human life more comfortable, pleasant, or agreeable must inevitably interfere with the survival of other species and, eventually, of humans themselves in a world of fixed resources. But, supporters of sustainable development assure us, both technology and social organization can be managed in ways that meet essential needs and provide long-term—but not infinite—growth within natural limits, if we use ecological knowledge in our planning.

While economic growth makes possible a more comfortable lifestyle, it doesn't automatically result in a cleaner environment. As figure 1.17 shows, people will purchase clean water and sanitation if they can afford to do so. For low-income people, however, more money tends to result in higher air pollution because they can afford to burn more fuel for transportation and heating. Given enough money, people will be able to afford both convenience *and* clean air. Some environmental problems, such as waste generation and carbon dioxide emissions, continue to rise sharply with increasing wealth because their effects are diffuse and delayed. If we are able to sustain economic growth, we will need to develop personal restraint or social institutions to deal with these problems.

Some projects intended to foster development have been environmental, economic, and social disasters. Large-scale hydropower projects, like that in the James Bay region of Quebec or the Brazilian Amazon that were intended to generate valuable electrical power, also displaced indigenous people, destroyed

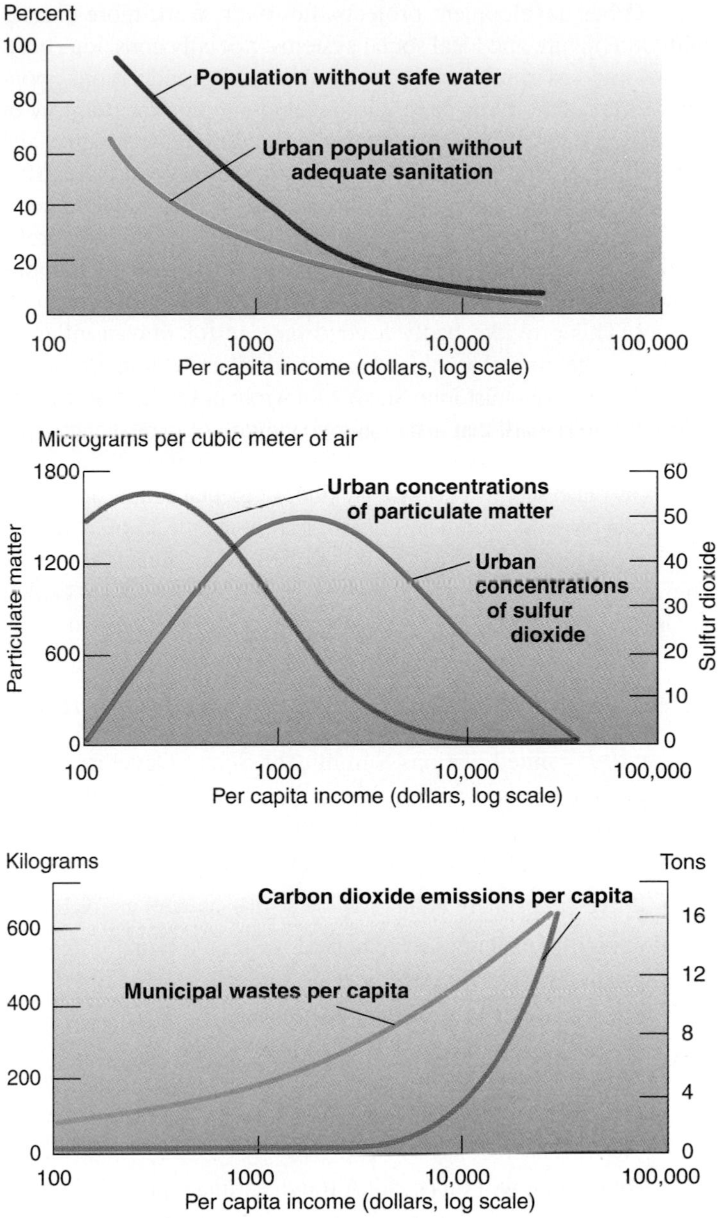

FIGURE 1.17 Environmental indicators show different patterns as average national income increases. When people have more money they invariably will purchase clean water and better sanitation. Rising income may temporarily produce increased urban air pollution (for example, particulates and sulfur dioxide) as people burn more fuel; eventually, however, people can afford both clean air and the benefits of technology. Some environmental problems such as waste generation and carbon dioxide emissions rise sharply with increasing wealth because of increased demands for goods and services without recognition of adverse environmental consequences. *Source:* Data from World Bank, *World Development Report* 1996.

wildlife, and poisoned local ecosystems with acids from decaying vegetation and heavy metals leached out of flooded soils. Similarly, introduction of "miracle" crop varieties in Asia and huge grazing projects in Africa financed by international lending agencies crowded out wildlife, diminished the diversity of traditional crops, and destroyed markets for small-scale farmers.

Other development projects, however, work more closely with both nature and local social systems. Socially conscious businesses and environmental, nongovernmental organizations sponsor ventures that allow people in developing countries to grow or make high-value products—often using traditional techniques and designs—that can be sold on world markets for good prices (fig. 1.18). Appleseed Fundraising, for example, a company started in 1989 by two Minnesota college students, buys textiles and crafts directly from artisans in Guatemala and sells them to schools, churches, and camps throughout the United States to use for fundraising. This provides badly needed income for Mayan villagers while also helping charitable organizations here at home.

As the economist John Stuart Mill wrote in 1857, "It is scarcely necessary to remark that a stationary condition of capital and population implies no stationary state of human improvement. There would be just as much scope as ever for all kinds of mental culture and moral and social progress; as much room for improving the art of living and much more likelihood of its being improved when minds cease to be engrossed by the art of getting on." Somehow, in our rush to exploit nature and consume resources, we have forgotten this sage advice.

FIGURE 1.18 A Mayan woman from Guatemala weaves on a backstrap loom. A member of a women's weaving cooperative, she sells her work to nonprofit organizations in the United States at much higher prices than she would get at the local market.

The 20:20 Compact for Human Development

At the 1995 United Nations Summit for Social Development in Copenhagen, a world social charter was passed that calls on all nations to ensure basic human needs for everyone. Among the goals in this world action plan to vanquish poverty and injustice are:

- universal primary education—for girls as well as for boys;
- adult illiteracy rates to be halved—with the female rate to be no higher than the male one;
- elimination of severe malnutrition;
- family planning services for all who wish them;
- safe drinking water and sanitation for all;
- credit for all—to ensure self-employment opportunities.

How much will this cost? A rough estimate provided by the United Nations Development Agency is that an additional $30 to $40 billion a year is needed to meet these targets. This is a large amount but not an impossible one. Developing countries now devote, on average, only $57 billion per year (13 percent of their national budgets) to basic human needs. Military spending by these countries, in contrast, averages $125 billion per year. If weapons purchases were cut in half, human development could be doubled.

An innovative suggestion called the 20:20 Compact was offered by the World Social Summit. Donor countries now allocate, on average, only 7 percent of their aid to humanitarian concerns. If they were to shift 20 percent of aid to social development, it would provide an additional $12 billion a year. Similarly, if developing countries would earmark 20 percent of their budgets to human priority concerns, $88 billion per year would be available. The $100 billion raised by this compact is dwarfed by the one trillion dollar annual total world military spending, but it would be three times the minimum needed for the human development agenda (fig. 1.19).

Indigenous People

Often at the absolute bottom of the social strata, whether in rich or poor countries, are the indigenous or native peoples who are generally the least powerful, most neglected groups in the world. Typically descendants of the original inhabitants of an area taken over by more powerful outsiders, they are distinct from their country's dominant language, culture, religion, and racial communities. Of the world's nearly 6000 recognized cultures, 5000 are indigenous ones that account for only about 10 percent of the total world population. In many countries, these indigenous people are repressed by traditional caste systems, discriminatory laws, economics, or prejudice. Unique cultures are disappearing along with biological diversity as natural habitats are destroyed to satisfy industrialized world appetites for resources. Traditional ways of life are disrupted further by dominant Western culture sweeping around the globe.

At least half of the world's 6000 distinct languages are dying because they are no longer taught to children. When the last few elders who still speak the language die, so will the culture that was

FIGURE 1.19 Every year, military spending equals the total income of half the world's people. The cost of a single large aircraft carrier equals ten years of human development aid given by all the world's industrialized countries. In this post-Cold War era, we ought to be able to declare a "peace dividend" for human development and environmental protection.

its origin. Lost with those cultures will be a rich repertoire of knowledge about nature and a keen understanding about a particular environment and a way of life (fig. 1.20).

Nonetheless, in many places, the 500 million indigenous people who remain in traditional homelands still possess valuable ecological wisdom and remain the guardians of little-disturbed habitats that are the refuge for rare and endangered species and undamaged ecosystems. Author Alan Durning estimates that indigenous homelands harbor more biodiversity than all the world's nature reserves and that greater understanding of nature is encoded in the languages, customs, and practices of native people than is stored in all the libraries of modern science. Interestingly, just 12 countries account for 60 percent of all human languages (fig. 1.21). Seven of those are also among the "megadiversity" countries that contain more than half of all unique plant and animal species. Conditions that support evolution of many unique species seem to favor development of equally diverse human cultures as well.

Recognizing native land rights and promoting political pluralism is often one of the best ways to safeguard ecological processes and endangered species. As the Kuna Indians of Panama say, "Where there are forests, there are native people, and where there are native people, there are forests." A few countries, such as Papua New Guinea, Fiji, Ecuador, Canada, and Australia acknowledge indigenous title to extensive land areas.

In other countries, unfortunately, the rights of native people are ignored. Indonesia, for instance, claims ownership of nearly three-quarters of its forest lands and all waters and offshore fishing rights, ignoring the interests of indigenous people who have lived in these areas for millennia. Similarly, the Philippine government claims possession of all uncultivated land in its territory, while Cameroon and Tanzania recognize no rights at all for forest-dwelling pygmies who represent one of the world's oldest cultures.

FIGURE 1.20 Indigenous and tribal cultures often have environmental wisdom that may help us understand our proper relationship with nature. Unfortunately, indigenous cultures are disappearing as rapidly as the endangered species and habitats of the world.

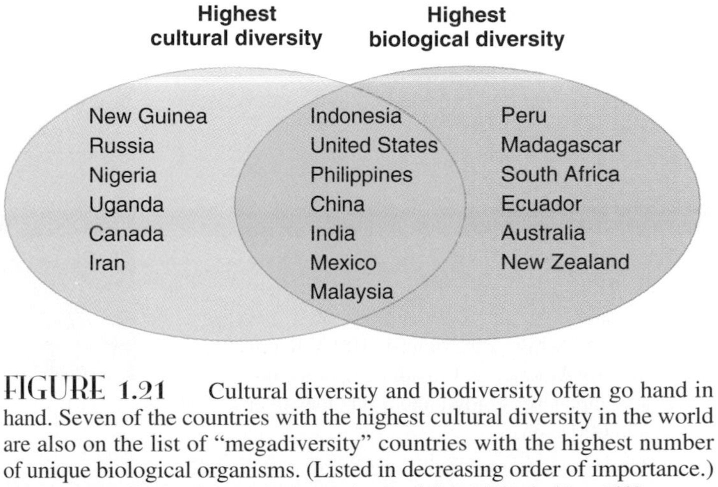

FIGURE 1.21 Cultural diversity and biodiversity often go hand in hand. Seven of the countries with the highest cultural diversity in the world are also on the list of "megadiversity" countries with the highest number of unique biological organisms. (Listed in decreasing order of importance.) *Source:* Norman Myers, Conservation International and Cultural Survival Inc., 1999.

IN DEPTH. . . . . .

Getting to Know Our Neighbors

Imagine the world as a village of 1000 people: who are our neighbors?

- Ethnically, this global village has 385 East Asians, 209 South Asians, 129 Africans, 91 Europeans, 78 Latin Americans, 52 individuals from the former USSR, 51 from the United States and Canada, and 5 from Oceania (Australia, New Zealand, and the Pacific Islands).
- About half the residents speak one or more of the six major world languages (Mandarin, English, Hindi/Urdu, Spanish, Russian, and Arabic—in that order), while the rest speak some of the other 6000 known languages.
- Children under age 15 make up 32 percent of the village, while those over age 65 make up 6 percent. The number of senior citizens is increasing rapidly while the number of children is falling slowly. At some point in the future, there may be more retirees than workers in the village, bankrupting the social security system.
- Every year, 26 children are born in the village and 9 people die. One-third of yearly deaths are children under age 5, mostly from infectious diseases aggravated by malnutrition. One person dies from heart disease, one from cancer, two from chronic or acute lung diseases, and two from violence, accidents, and various infectious diseases.
- The average life expectancy is 70 years for women and 65.6 for men, but there is considerable variation in longevity. In the richest families, women live to an average age of 80 and men to an average of 78 years, while in the poorest families the life expectancy is less than 48 years for both men and women.
- The average annual income for the village is $4880 (U.S.), but this average also obscures great discrepancies. More than half the village lives in households where the annual income is less than $620 per person, while the richest 100 citizens (mostly Americans, Canadians, Europeans, and Japanese) enjoy annual incomes over $25,000 each.
- As a consequence of this income gap, the richest 200 villagers own or control 80 percent of the resources and consume 80 percent of all products sold in the marketplace, while the other 800 people make do with 20 percent of the goods and merchandise available for sale.
- Half of the 620 adults in the village are illiterate. Lacking an education, most of these people work as day laborers or seasonal farm workers. Among the poorer families, girls are half as likely as boys to attend school.
- Women and girls make up slightly more than half the village population. They do two-thirds of all manual labor, receive one-tenth of the wages, and own less than one-hundredth of the property. Seventy percent of the poorest members of the population are women and children.
- About 400 villagers suffer from ill health at any given time. Much of that illness is related to lack of clean water, sanitation, and food. Some 250 of our neighbors don't have clean water to drink or adequate sanitation. About 150 are chronically hungry, lacking the calories and nutrients needed for normal growth and development in children or a healthy, productive life for adults.
- Only 452 people actually live in the village itself; the other 548 live in the surrounding countryside. Sixty percent of the rural families are landless or have too little land to subsist. They make up a majority of those who lack clean water, sanitation, food, housing, and health care.
- Generally the worst pollution problems are borne by those who live in the poorest parts of the village, where air and water pollution, noise, congestion, and toxic wastes are most common. Those who live in the better parts of town actually enjoy a cleaner, safer environment than their parents did only a few decades ago.
- In the past, people didn't often travel from one part of the village to another. Today, travel is easier and cheaper than ever before. Furthermore, 90 percent of the population has access to television so that the lower class is exposed to both consumer pressures and news about how the upper class lives.
- How long do you suppose these great discrepancies in opportunity and quality of life will persist? What might we do to reduce them?

Nine out of ten people in the world now have access to television. There may be only one TV per village, as in this scene from Nigeria, but it opens a window to the wider world. Most of what everyone watches is made in the United States and glorifies consumerism and commidification.

ENVIRONMENTAL PERSPECTIVES

In the *Christmas Carol* by Charles Dickens, Scrooge questions the Ghost of Christmas Future after seeing the disparity between the rich and poor in London. "Answer me one question," says Scrooge, "are these the shadows of the things that will be or are they shadows of the things that may be only?" We could ask something similar today. Are the problems and dilemmas outlined in this book warnings of what *will* be, or only what *may* be if we fail to take heed and adjust our course of action?

What will be our environmental future and what can we do to shape it? These are, perhaps, the most important questions in environmental science. There are many interpretations of what the data show and how we should interpret them. Think about the following worldviews and tactical positions as you read the subsequent chapters in this book.

Pessimism and Outrage

You will find much in this book that justifies pessimism. A number of very serious environmental problems threaten us. Many environmental scientists see our world as one of scarcity and competition in which too many people fight for too few resources. This viewpoint is often called **neo-Malthusian** after Thomas Malthus, who predicted a dismal cycle of overpopulation, misery, vice, and starvation as a result of human fallibility. We will discuss Malthus and his predictions in subsequent chapters as part of our discussion of population growth and resource economics.

This grim view of human nature and resources persuades many environmentalists to issue dire warnings of impending doom unless we make immediate, drastic changes in our way of life. While it is easy to feel moral outrage about the excesses and abuses that have occurred, shock, shame, and fear are generally poor motivation for positive action. How helpful is it in getting you to abandon your bad habits to have someone harp about how guilty and vile you are? Pioneering ecologist Aldo Leopold said: "I have no hope for a conservation based on fear." Furthermore, when the predicted disasters fail to occur, the public may assume that nothing was wrong and that there is no reason to change anything. How many times can we cry wolf before people stop paying attention?

Optimism

Science and technology have provided many benefits to humanity; they also have caused many difficulties, as you will see in the course of this book. **Technological optimists** believe that human ingenuity and enterprise will find cures for all our problems. They see the world as one of abundance and opportunity. Geographer Martin Lewis calls this **"Promethean environmentalism"** after the Greek Titan who stole fire from the Gods and gave it to humans. He fails to mention, however, that Prometheus suffered eternal torment for this sin of pride. Critics describe this optimistic outlook as **"cornucopian fallacy"** (after the mythical horn of plenty) and see it as either wishful thinking or deliberate denial. Optimists argue, however, that they merely expect historic patterns of progress to continue in the future as they have in the past.

While emphasizing only good news is popular and comfortable, it also can mislead the public. Blind faith in technology can be merely an excuse for business as usual. By telling us that everything is just fine, optimists may lull us into complacency and apathy. What do you think is the most appropriate and useful balance between Malthusian and cornucopian worldviews? How can we use environmental information to bring about positive change?

Lessons from the Past

You will see as you read this book that we have tried to be honest about the problems we face but cautiously optimistic about our chances for success. We believe that obstacles can be overcome if we face problems candidly, reasonably, and creatively. Furthermore, dwelling excessively on problems and failures has a way of producing self-fulfilling prophecies. If we expect the worst and believe that all efforts to avert doom are useless, the disasters we fear will probably occur. If, on the other hand, we hope for the best and work to build a better world, perhaps we can find ways to improve our lives and protect the environment at the same time.

You will have to decide for yourself as you study environmental science how much hope the future holds. If you feel overwhelmed and discouraged as you read about the many problems in the world, it might help to recall the disasters faced by the people who preceded us. Surely, to many who have experienced the horrors of war, oppression, and terror, the world must have seemed a dreadful place. But they have found the strength to persevere and survive even in horrendous circumstances.

Consider the fourteenth century, when Europe was wracked by a terrible century-long war in which innocent civilians were slaughtered by both sides (fig. 1.22). The climate changed inexplicably in the "little ice age," so that crops failed across Europe and famine compounded the horrors of war. Claims that nature was dying because of human sins were not unlike some of the apocalyptic warnings we hear today.

Adding to the inexplicable misery of that time, bubonic plague, the "Black Death," swept repeatedly across Europe, killing at least one-third of the population in just its first pass; as many as 70 percent died in some cities. Civilization crumbled as bands of brigands roamed the countryside and people fled in panic. Parents abandoned their children; everything seemed lost. Many interpreted this series of calamities as God's punishment for their sins and fully expected the end of the world to be imminent. Life went on, however, and things slowly got better. We haven't stopped killing each other yet, but at least those who kill and torture now generally try to hide their deeds. Perhaps progress, although halting and slow, is possible after all.

FIGURE 1.22 This painting of warfare in fourteenth-century Europe captures some of the cruelty and destructiveness that led people to fear the end of the world. Are we now at a turning point in history or can we learn to live in harmony with each other and with our environment?

SUMMARY

Humans always have inhabited two worlds: one of nature and another of human society and technology. Environmental science is the systematic study of the intersection of these worlds. An interdisciplinary field, environmental science draws from many areas of inquiry to help us understand the worlds in which we live and our proper role in them.

The most amazing features of our planet may be the self-sustaining ecological systems that make life possible and the rich diversity of life that is part of, and dependent upon, those ecological processes. In spite of the many problems that beset us, the earth is wonderfully bountiful and beautiful.

Concerns about pollution and land degradation date back at least 2500 years. Clearly, we have pragmatic interests in conserving resources and preserving a habitable environment. There also are ethical reasons to believe that nature has a right to continue to exist for its own sake. Unprecedented population growth, food shortages, scarce energy supplies, air and water pollution, and destruction of habitats and biological resources are all serious threats to our environment and our way of life.

As international travel and communication become easier, we realize that these problems encompass our whole planet and require global cooperation to find solutions. Still, there is good news. Pollution has been reduced and population growth has slowed in many places. Perhaps we can extend these advances to other areas as well.

The 20 percent of us in the world's richest countries consume an inordinate amount of resources and produce a shocking amount of waste and pollution. More than 1.3 billion people live in acute poverty and lack access to an adequate diet, decent housing, basic sanitation, clean water, education, medical care, and other essentials for a humane existence.

Concern for the poor is more than a humanitarian issue. Faced with immediate survival needs, these desperate people often have no choice but to overharvest resources and reduce long-term sustainability for themselves and their children. Since we share the same environment it is in our own self-interest to help them find better options than they currently have.

There are several ways of describing the economic and developmental status of different countries. The First World is generally industrialized and more highly developed. Many Third World countries have made encouraging progress in improving the quality of life for residents, but much remains to be done.

Indigenous or native peoples are generally among the poorest and most oppressed of any group. Nevertheless, they possess valuable ecological knowledge and remain the guardians of nature in many places. Recognizing the rights of indigenous people and minority communities is an important way to protect natural resources and environmental quality.

There are valid reasons to be pessimistic about our environmental conditions, but we must be careful that dire predictions don't overwhelm us and become self-fulfilling prophecies. Many people find an optimistic outlook a better motivation than fear, but blind faith in technological progress can be simply an excuse for business as usual. Although we still have far to go in protecting our environment, some heartening progress already has been made toward building a just and sustainable world.

QUESTIONS FOR REVIEW

1. Define environment and environmental science.
2. List six environmental dilemmas that we now face and describe how each concerns us.
3. Describe the differences between the North/South or rich/poor or more developed/less developed nations. What do we mean by First, Second, and Third World?
4. Compare some indicators of quality of life between the richest and poorest nations.
5. Why should we be concerned about the plight of the poor? How do they affect us?
6. What benefit to us would there be in protecting the rights of indigenous people?
7. Give some reasons for pessimism and optimism about our environmental future and summarize how you feel personally about the major environmental problems that we face.
8. Do you think that environmental conditions are better now or worse than they were 20 or 100 or 1000 years ago? Why?

QUESTIONS FOR CRITICAL THINKING

1. How could we determine whether the deformity of frogs has some grand, global significance, or is merely a random, local event?
2. What are the fundamental differences between utilitarian conservation and altruistic preservation? Which do you favor? Why?
3. Do the issues discussed in this chapter as global environmentalism belong in an environmental *science* text? Why would anyone ask this question?
4. Some people argue that we can't afford to be generous, tolerant, fair, or patient. There isn't enough to go around as it is, they say. What questions would you ask such a person?
5. Others claim that we live in a world of bounty. They believe there would be plenty for all if we just shared equitably. What questions would you ask such a person?
6. Around 200 million children are forced into dangerous, degrading slave labor each year. Is it our business what goes on in other countries?
7. What would it take for human development to be really sustainable? What does sustainable mean to you?
8. Are there enough resources in the world for 8 or 10 billion people to live decent, secure, happy, fulfilling lives? What do those terms mean to you? Try to imagine what they mean to others in our global village.
9. What responsibilities do we have to future generations? What have they done for us? Why not use whatever resources we want right now?
10. Do you see any similarities between current conditions and those of the fourteenth century? Have we made any real progress or do things just stay the same?

KEY TERMS

acute poverty 24
biocentric preservation 19
cornucopian fallacy 33
environment 17
environmentalism 19
environmental science 17
First World 27
Fourth World 28
global environmentalism 20
human development index 28
neo-Malthusian 33
North/South division 27
Promethean environmentalism 33
Second World 27
sustainable development 29
technological optimists 33
Third World 28
utilitarian conservation 18

ADDITIONAL INFORMATION ON THE INTERNET

Visit our website at http://www.mhhe.com/environmentalscience/ for specific information on each of the topics below:

Deformed Frogs: information from the Minnesota Pollution Control Agency

Frogweb

Major Figures in Conservation History: John Muir, Rachel Carson, Gifford Pinchot

Indigenous Peoples: Center for World Indigenous Studies

Indigenous Environmental Network

First Nations homepage

World Resources Institute

Earthwatch

United Nations Environment Programme information

Environmental Organization Web Directory Search Engine

CHAPTER 2

Environmental Ethics and Philosophy

A thing is right when it tends to preserve the integrity, stability, and beauty of the biotic community. It is wrong when it tends otherwise.

—Aldo Leopold—

OBJECTIVES

After studying this chapter, you should be able to:

- understand some principles of environmental ethics and philosophy.
- compare and contrast how different ethical perspectives shape our view of nature and our role in it.
- realize how your own worldview and core values shape your perceptions of nature.
- explain anthropocentrism, biocentrism, ecocentrism, utilitarianism, and ecofeminism, and what each says about human/nature relationships.
- summarize the methods, applications, and limitations of the scientific method.
- discuss the role of technology in *causing* environmental problems as well as helping us solve them.

ESSENTIAL STUDY PARTNER

The Essential Study Partner (ESP) CD-ROM that accompanies this textbook contains additional information on the following topics:

- attitudes and worldviews toward nature: principles: finding solutions: worldviews
- environmental justice: principles: finding solutions: worldviews
- hypothesis testing and the scientific method: principles: finding solutions: critical thinking
- critical thinking and policy decisions: principles: finding solutions: critical thinking

Above: A forest service scientist measures annual growth of a young fir in a reforested area of the Olympic National Forest, Washington.

People or Wildlife in South Africa?

South Africa has some of the most successful wildlife refuges in Africa. Although big game populations have been declining elsewhere on the continent, South African parks and refuges have a surplus of rhinos, elephants, and other endangered species. Game wardens have to cull herds periodically to protect their habitat.

Under apartheid, villagers were forcibly removed from national parks and game preserves. Boundaries were patrolled by heavily armed rangers drawn from the ranks of the South African Defense Force. Trespassers were shot on sight, even if they were only gathering firewood or hunting small animals to feed their families. "We had poachers in 1982 and 1983," boasted Anthony Hall-Martin, chief research officer for Pretoria's National Parks Board, "and we killed them off."

More than 560 South African national parks and nature preserves now enclose 3 million hectares (7.4 million acres) of forest, grasslands, and open woodlands. They clearly are responsible for saving many rare and endangered species that might otherwise have been lost. Just outside some splendid natural areas, however, former occupants—many never compensated for seizure of their lands—live in squalid shantytowns of mud and tin shacks (fig. 2.1). As apartheid is being dismantled, rural communities are beginning to demand a return of ancestral lands. They claim a rightful place in the landscape and argue they could live in harmony with wildlife.

The new government formed by the African National Congress is sympathetic to their claims, but ecotourism and big game safaris bring much-needed cash to this struggling country. And hunters and tourists like to believe they are in pristine wilderness. They don't like to shoot elephants or watch birds in the midst of cattle herds or domestic fields. Safaris are big business: minimum cost for a three-week hunting trip is $50,000 per person. Depending on how many animals you want to kill, you can easily spend upwards of $200,000 per person. Little of that money goes to local communities, however.

This case raises some interesting ethical questions. Should protecting wildlife take precedence over human needs? Is it fair to forbid natives to kill animals for food so that rich foreigners can kill them for fun? Is killing other animals justified under any circumstances? Does establishing a national park require removal of all local people and any sign of their occupancy, or can a landscape that includes traditional human habitation be an acceptable representation of nature?

FIGURE 2.1 Kilptown squatters camp in South Africa. Many former residents of lands seized for national parks now live in squalid shantytowns such as this.

To some conservationists it seems that wildlife species such as rhinos are so rare that they should have precedence. To social justice advocates, however, the plight of poor people in shantytowns is of paramount importance. What do you think?

Answers to these questions depend on our ethical perspectives and our attitudes toward nature and our place in it. In fact, ethical questions and dilemmas underlie many of the issues we will study throughout this book. Environmental leaders frequently claim that our environmental problems are evidence, not of technological failures, but of moral breakdowns in society. In this chapter, we present some tools that will help you understand and analyze these controversies as well as many others you will encounter throughout this text.

ENVIRONMENTAL ETHICS AND PHILOSOPHY

Ethics is a branch of philosophy concerned with **morals** (the distinction between right and wrong) and **values** (the ultimate worth of actions or things). Ethics evaluates the relationships, rules, principles, or codes that require or forbid certain conduct. Most Western ethicists consider the roots of their field to be the famous questions posed by Socrates and the Greek philosophers 2500 years ago: "What is the good life? How ought we, as moral beings, to behave?"

Environmental ethics asks about the moral relationships between humans and the world around us. Do we have special duties, obligations, or responsibilities to other species or to nature in general? Are there ethical principles that constrain how we use resources or modify our environment? If so, what are the foundations of those constraints and how do they differ from principles governing our relations to other humans? How are our obligations and responsibilities to nature weighed against human values and interests? Do some interests or values supercede others?

Are There Universal Ethical Principles?

The first question considered in ethics is whether *any* moral laws are objectively valid and independent of cultural context, history, or situation. How the question is answered depends largely on the philosophical disposition of the respondent. **Universalists,** such as Plato and Kant, assert that the fundamental principles of ethics are universal, unchanging, and eternal (fig. 2.2). These rules of right and wrong are valid regardless of our interests, attitudes, desires, or preferences. Some believe these rules are revealed by God, while others maintain they can be discovered through reason and knowledge.

Relativists, such as Plato's opponents, the Sophists, claim that moral principles always are relative to a particular person, society, or situation. Although there may be right and wrong—or at least better and worse—things to do, relativists assert that no transcendent, absolute principles apply regardless of circumstances. In this view, ethical values always are contextual. Friedrich Nietzsche's famous aphorism, "there are no facts, only interpretations" has been the banner for recent generations of relativists.

FIGURE 2.2 Plato and Aristotle debate moral philosophy in a painting by Raphael. Plato (*left*) motions upward, indicating a transcendent, universal moral truth, while Aristotle (*right*) motions downward to suggest grounded, situational ethics.

Nihilists, such as Schopenhauer, claim that the world makes no sense at all. Everything is completely arbitrary, and there is no meaning or purpose in life other than the dark, instinctive, unceasing struggle for existence. According to this view, there is no reason to behave morally. Only power, strength, and sheer survival matter. "Might is right; eat or be eaten." There is no such thing as a "good" life: we live in a world of uncertainty, pain, and despair. Nevertheless, Schopenhauer enjoyed living in a comfortable, civilized society where rules of normative behavior and good conduct prevailed.

Utilitarians hold that an action is right that produces the greatest good for the greatest number of people. This philosophy is usually associated with the English philosopher Jeremy Bentham (1748–1832); but something very similar was suggested by Plato, Socrates, Aristotle, and others. Bentham was an eccentric genius and a hedonist who equated goodness with happiness, and happiness with pleasure. He regarded pleasure as the only thing worth having in its own right. Thus, the good life is one of maximum pleasure. Insofar as people are moral animals, in his view, we should act to produce the greatest pleasure for the greatest number. To do so is good; not to do so is wrong.

Utilitarianism was modified and made less hedonistic by Bentham's brilliant protégé, John Stuart Mill (1806–1873). Mill believed that pleasures of the intellect are superior to pleasures of the body. He held that the greatest pleasure is to be educated and to act according to enlightened, humanitarian principles. This enlightened form of utilitarianism inspired Gifford Pinchot and the early conservationists (see chapter 1), who argued that the purpose of conservation is to protect resources for the "greatest good for the greatest number *for the longest time.*"

Although utilitarianism remains widely popular today, it has drawbacks. It can, for instance, be used to justify reprehensible acts. If ten thousand Romans greatly enjoyed watching a few Christians being eaten by lions, did that make it the right thing to do? Does the pleasure of the tormentors outweigh the suffering of victims? Most of us would conclude that it does not. Justice, freedom, morals, and loyalty take precedence over pleasure, or even happiness, although it could be argued that furthering moral ends and right action ought to bring the greatest happiness in the long run.

Modernism and Postmodernism

Much of the modern, Western worldview is based on theories about reason and progress of seventeenth-century Enlightenment philosophers and scientists such as René Descartes, Francis Bacon, and Isaac Newton. These positivists hoped to develop universal laws of morality through objective science that allows control of nature as well as an understanding of the world and of ourselves. The inevitable outcome, they believed, would be moral progress, universal justice, and, ultimately, the happiness of all humans.

The experiences of the twentieth century, especially the horrors of world wars, have led to a kind of cultural despair among many philosophers who see the quest for individual liberty, self-realization, and experience leading to hedonism, narcissism, and social disintegration. Poststructuralists and postmodernists, such as Jacques Derrida, Jean-Francois Lyotard, and Michel Foucault, claim there is no grand narrative of history and no universal philosophy. Focusing on marginalized, disempowered groups and local, contingent knowledge, they cast doubt on classical notions of truth, reality, and meaning. According to this perspective, reality is constructed by our discourses rather than revealed by them; we never can know anything unequivocally.

How, you may be wondering, does this apply to environmental ethics? From a postmodernist perspective, nature—or at least our perception of it—is an arbitrary, ever-shifting social construction. If you ask a diverse group of people what nature is, you will most likely get many different answers. To a city dweller, nature might be a park with a few trees and squirrels; to a farmer, it may mean a productive field; and to a backpacker, it probably means a pristine wilderness. On the other hand, a logger would see that same wilderness as merely a supply of lumber. From a postmodernist view, no single viewpoint is inherently better than any other.

While relativism recognizes unique differences in the human condition, it can lead to paralysis of will. If every opinion or interpretation is equally valid, how can we choose whom to believe or what to do? What do you think? Is nature whatever we believe it is? Can we replace original nature with a human-constructed "second nature" (fig 2.3) or virtual reality whenever we like? Are there ethical absolutes in our relations with nature or is everything relative and subjective?

Values, Rights, and Obligations

For many philosophers, only humans are **moral agents,** beings capable of acting morally or immorally and who can—and should—accept responsibility for their acts. Capacities that enable humans to form moral judgments include moral deliberation, the resolve to carry out decisions, and the responsibility to hold oneself answerable for failing to do what is right.

Of course, not all humans have all these capacities all the time. Children and those who are mentally retarded, mentally ill, or for some other reason, lacking a full use of reason, are not regarded as moral agents. If a child murders someone, we don't hold the child responsible. Nonetheless, she/he still has rights. Children are considered **moral subjects,** beings who are not moral agents themselves but who have moral interests of their own and can be treated rightly or wrongly by others.

Historically, the idea that weaker members of society deserve equal treatment with those who are stronger was not a universally held opinion. In many societies, women, children, outsiders, serfs, and others were treated as property by those who were more powerful. The Greek philosophers, for example, accepted slavery as natural and necessary. Gradually, we have come to believe that all humans have certain inalienable rights: life, liberty, and the pursuit of happiness, for example. No one can ethically treat another human as a mere object for their own pleasure, gratification, or profit. This gradually widening definition of whom we consider ethically significant is called **moral extensionism.**

FIGURE 2.3 Is nature only a social creation? Can the synthetic world of Disneyland be equated with the natural world?

Do Other Animals Have Rights?

Perhaps the most important question in environmental ethics is whether moral extensionism encompasses nonhumans. Do other species have rights as well? Are they moral agents or at least moral subjects? For many philosophers, the answer is no. Reason and consciousness—or at least a potential thereof—are essential for moral considerability in this view. René Descartes (1596–1650), for instance, claimed that animals are mere automata (machines) and can neither reason nor feel pain (fig. 2.4).

Most pet owners disagree, claiming that while animals may not have the same self-consciousness as humans, they are intelligent and clearly have feelings. As sentient (perceptive) beings, they deserve ethical treatment. But what about nonsentient beings? Should we extend moral consideration to bugs, rocks, landscapes? Some people think so. Let's see why.

Intrinsic and Instrumental Values

Rather than couch ethics strictly in terms of rights, some philosophers prefer to consider values. Value is a measure of the worth of something. But value can be either inherent or conferred. All

FIGURE 2.4 Do other organisms have inherent values and rights? These chickens are treated as if they are merely egg-laying machines. Many people argue that we should treat them more humanely.

FIGURE 2.5 Mineral King Valley at the southern border of Sequoia National Park was the focus of an important environmental law case in 1969. The Disney Corporation wanted to build a ski resort here, but the Sierra Club sued to protect the valley on behalf of the trees, rocks, and native wildlife.

humans, we believe, have **inherent value**—an intrinsic or innate worth—simply because they are human. They deserve moral consideration no matter who they are or what they do. Tools, on the other hand, have conferred, or **instrumental value.** They are worth something only because they are valued by someone who matters. If I hurt you without good reason, I owe you an apology. If I borrow your car and smash it into a tree, however, I don't owe the car an apology. I owe *you* the apology—or reimbursement—for ruining your car. The car is valuable only because you want to use it. It doesn't have inherent values or rights of itself.

Do Nonsentient Things Have Inherent Value?

How does this apply to nonhumans? Domestic animals clearly have an instrumental value because they are useful to their owners. But some philosophers would say they also have inherent values and interests. By living, breathing, struggling to stay alive, the animal carries on its own life independent of its usefulness to someone else. Some would draw a line of moral considerability at the limit of sentience (the ability to perceive sensations such as pain). Others argue that it is in the best interest of bugs, worms, and plants to be treated well, even if they aren't aware of it. In this perspective, just being alive gives things inherent value.

Some people believe that even nonliving things also have inherent worth. Rocks, rivers, mountains, landscapes, and certainly the earth itself, have value. These things were in existence before we came along and we couldn't recreate them if they are altered or destroyed. In a landmark 1969 court case, the Sierra Club sued the Disney Corporation on behalf of the trees, rocks, and wildlife of Mineral King Valley in the Sierra Nevada Mountains (fig. 2.5) where Disney wanted to build a ski resort. The Sierra Club argued that it represented the interests of beings that could not speak for themselves in court.

A legal brief entitled *Should Trees Have Standing?* written for this case by Christopher D. Stone proposed that organisms as well as ecological systems and processes should have standing (or rights) in court. After all, corporations—such as Disney—are treated as persons and given legal rights even though they are really only figments of our imagination. Why shouldn't nature have similar standing? The case went all the way to the Supreme Court but was overturned on a technicality. In the meantime, Disney lost interest in the project and the ski resort was never built. What do you think? Where would you draw the line of what deserves moral considerability? Are there ethical limits on what we can do to nature?

FIGURE 2.6 Do humans have a right to use or destroy other species or natural objects in any way we choose? Do we have duties, obligations, or responsibilities toward nature?

FIGURE 2.7 For many people a productive, domesticated landscape such as this mosaic of farmland and sugarbush in Vermont presents the ideal perspective. With careful stewardship—including a balance of population density and environmental resources—the land can be stable, harmonious, and fruitful.

WORLDVIEWS AND ETHICAL PERSPECTIVES

Our beliefs about our proper roles in the world are deeply conditioned by our ethical perspectives (What Do You Think?, p. 42). As historian Lynn White, Jr., said, "What people do about their ecology depends on what they think about themselves in relation to the things around them." In this section, we will look at how some different moral philosophies reflect our attitudes toward nature.

Domination

Throughout history, many cultures have claimed that humans hold a special place in creation. Pride in our power to reshape the world to our liking and a belief that we are superior to other creatures often have been used to justify **domination** of nature (fig. 2.6). In an influential 1967 paper entitled *The Historic Roots of Our Ecological Crisis,* Lynn White, Jr., traced this tradition to the biblical injunction to "Be fruitful, and multiply, and replenish the earth, and subdue it: and have dominion over the fish of the sea, and over the fowl of the air, and over every living thing that moveth upon the earth" (Genesis 1:28). In antiquity, White claimed, "every tree, every spring had its guardian spirit. Before one cut a tree, dammed a brook, or killed an animal, it was important to placate the spirit in charge of that particular situation. By destroying pagan animism, Christianity made it possible to exploit nature in a mood of indifference to feelings of natural objects." This view of humans as the focus of creation is termed **anthropocentric,** or human-centered.

Although many people agree with this analysis, others argue that this passage is translated and interpreted inaccurately. Genesis is really intended, they claim, to teach us to love and nurture creation rather than dominate and exploit it. White himself pointed out that Judeo-Christian culture has a long tradition of caring for nature. He recommended St. Francis of Assisi as an inspiration and patron saint for all environmentalists.

In recent years, religious groups have played important roles in nature protection. The National Religious Partnership for the Environment is made up of representatives of most major denominations in North America. "Although we may disagree on how the earth's creatures were created," they say, "we really need to work together to save what's left." Acting on this belief, a coalition of evangelical Christians has been instrumental in opposing Congressional attacks on the Endangered Species Act. The Central Conference of American Rabbis declared that desecration of the Headwaters Redwood Forest in California breaks our covenant with the Creator. And Orthodox Patriarch Bartholomew has said that "to commit a crime against nature is a sin."

Stewardship

Many tribal or indigenous people, both hunters and gatherers, as well as traditional agricultural societies have a strong sense of **stewardship** or responsibility to manage and care for a particular place. As custodians of resources, they see their proper role as working together with human and nonhuman forces to sustain life. Humility and reverence are essential in this worldview, where humans are seen as partners in natural processes rather than masters—not outside of nature but part of it.

This attitude also is held by many modern farmers (fig. 2.7) or others in close contact with nature. Authors René Dubos and Wendell Berry have written eloquently about the need to nurture and sustain both the rural landscape and culture. In their view,

what do you think?

THE WORLD AROUND YOU

Worldviews and Values

Worldviews are sets of basic beliefs, images, and values that make up a way of looking at and making sense of the world around us. They shape how we interpret and interact with our environment and other people. Most worldviews are learned early in life and are not easily changed. In cases where we encounter evidence that doesn't fit our worldview, we often reject the evidence and cling to prior beliefs. You have probably heard someone say, "I couldn't believe my eyes." What does that imply?

Our basic beliefs impact not only the way we think about ourselves and our place in the world, but they also determine what questions are valid to ask and who has a right to ask them. Worldviews are not usually arrived at rationally, nor are they necessarily accurate, but they strongly influence our behavior. People often cannot explain why they believe what they do. It is difficult to verbalize the underlying reasons for a particular set of ideas. When you find yourself saying, "I can't explain this, it's just the way things are," you are probably describing your worldview.

Values are a measure of the ultimate worth of a thing or an action. They can represent principles, standards, or rules of conduct derived from one's worldview. As you read this book, you will find that many of the issues discussed are highly contentious and have no single clear answer. What each of us thinks about these issues depends on our particular set of values.

Each of us has deeply held, core values that we generally learn as very young children or develop—often without being conscious of it—from life experiences. Think about the following list of attributes: friendship, safety, popularity, security, wisdom, frugality, beauty, creativity, fame, fulfilling duties, accomplishment, happiness, individuality, knowledge, kindness. If everything else were stripped away, what two or three basic characteristics would you want to be remembered by? Perhaps this list doesn't really capture who you think you are or who you'd like to be. Try completing the following sentence: "I'd give up everything if I could be ———." That's probably your real core value.

A principle of cooperative education is that you never really understand your own position on a particular issue until you understand the beliefs of someone who has a different point of view from your own. Environmental science provides an excellent opportunity to examine a variety of ways of looking at the world. This doesn't mean that you have to agree with everyone around you. But it's helpful to examine others' fundamental values as a way of gaining insight into your own. In discussions about environmental issues, try to see problems from the perspective of someone from an entirely different socioeconomic, religious, or political background from yours. You may find that your own values and worldviews become clearer.

In evaluating what you believe and why you believe what you do, think about your position on the following statements. Note that these are generally in opposing pairs but there is no intended association between any particular combination of statements. You may find that you agree only partially with some or all of the positions. The point of this exercise is to invite you to think about why you agree or don't agree with some of all of these propositions and what that says about your worldview. What social experiences, cultural beliefs, family background, or other personal influences may have shaped your views on these issues? If you were rewording some of all of these statements, how would you state them?

1. Humans are distinct and separate from nature. Our role is to subdue, command, and use other organisms and natural resources for our own benefit.
2. Every organism has rights and values of its own, regardless of its usefulness to us. We are merely part of nature but we may have special responsibilities to care for our neighbors because of our unique abilities.

humans can improve the world and make it a better place, both for themselves and for other organisms. As Voltaire said in *Candide,* "This may be the best of all possible worlds, but we must tend our garden."

Biocentrism, Animal Rights, and Ecocentrism

Many environmentalists criticize both stewardship and dominion as being too anthropocentric. They favor, instead, the **biocentric** (life-centered) egalitarianism of John Muir or Aldo Leopold, who claimed that all living organisms have intrinsic values and rights regardless of whether they are useful to us. Leopold wrote, "Of the 22,000 higher plants and animals native to Wisconsin, it is doubtful whether more than 5 percent can be sold, fed, eaten, or otherwise put to economic use. Yet these creatures are members of the land community, and if (as I believe) its stability depends on its integrity, they are entitled to continuance. . . A thing is right when it tends to preserve the integrity, stability, and beauty of the biotic community. It is wrong when it tends otherwise." For many biocentrists, biodiversity is the highest ethical value in nature. Species and populations, as the basic units of biodiversity, are the locus of inherent value.

Some animal rights advocates question the importance of species or populations, claiming that each individual organism is of value. They point out that the individual lives, reproduces, and experiences pleasure or pain, not the group. Many ecologists, in contrast, view larger-scale ecological processes such as evolution, adaptation, and the grand biogeochemical cycles as the most important aspects of nature. In this view, which is described as **ecocentric** (ecologically centered) because it claims moral values and rights for ecological processes and systems, the whole is considered more important than its individual parts. If you kill an individual organism, you deny it a few months or years of life, but if you eliminate an entire species or a whole landscape, you have destroyed something that took millions of years to create.

In the broadest ecocentric view, individuals could be seen to have only instrumental value while abiotic resources, ecological

3. There are some things that are always right or wrong regardless of the circumstances or situation.
4. Whether an act is right or wrong depends on how, when, why, and by whom it was done. There are shades of gray when it comes to ethics and morals.
5. The world is characterized by scarcity and competition for limited resources. Further growth of either human populations or standard of living will be disastrous.
6. The world is one of opportunity and potential. Continued economic and population growth will have more good than bad effects.
7. Technology has caused most of our environmental problems. Returning to a simpler, nontechnological way of life is the only way to solve these problems.
8. Further technological advances are not only beneficial but essential for solving the problems we now face.
9. Problems that occur in other parts of the globe are not our concern. We should pay attention to the issues in our own backyard.
10. Whether we like it or not, we are all part of a global system. What happens anywhere in the world affects us in some way.

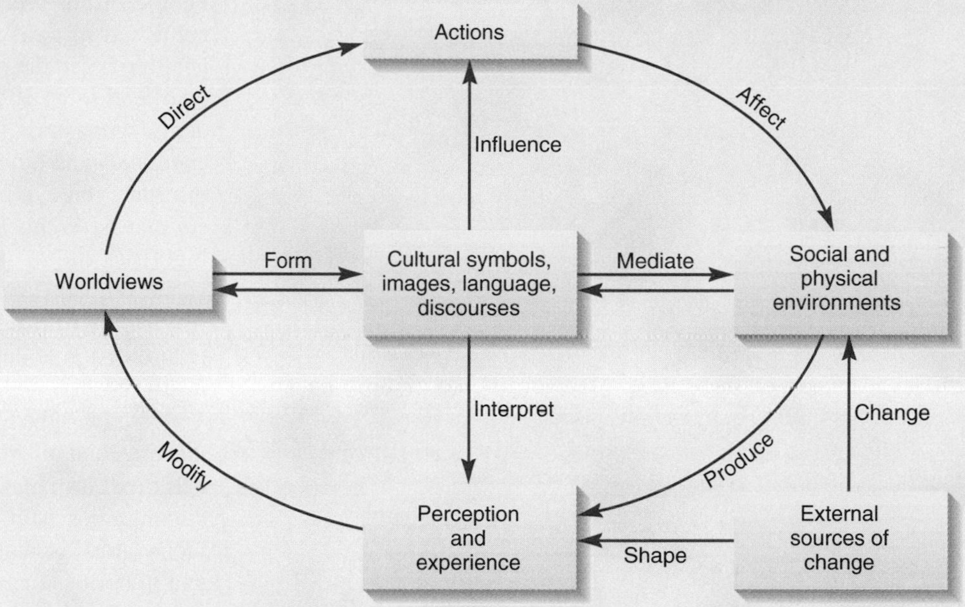

Each of us has a worldview that is shaped by, and in turn informs and interprets, our actions and experiences. We often aren't aware of our own worldview, only that others seem peculiar and obviously wrong. Note the similarity between this diagram and the concept maps in the Introduction.

11. I regard friendship, family, a sense of belonging, and fulfilling my duties as the highest or core values in my life.
12. I regard individuality, creativity, excitement, adventure, and accomplishment as the highest or core values in my life.

cycles, and the whole earth possess inherent value (table 2.1). Nature doesn't seem to care about individuals. Vastly more offspring are born than can ever survive. Even species come and go. What seems to have longevity in nature are processes like photosynthesis and evolution.

Ecofeminism

Many feminists believe that none of these philosophies is sufficient to solve environmental problems or to tell us how we ought to behave as moral agents. They argue that all these philosophies come out of a patriarchal system based on domination and duality that assigns prestige and importance to some things but not others. In a patriarchal worldview, men are superior to women, minds are better than bodies, and culture is higher than nature. Feminists contend that domination, exploitation, and mistreatment of women, children, minorities, and nature are intimately connected and mutually reinforcing. They reject all "isms" of domination: sexism, racism, classism, heterosexism, and speciesism.

TABLE 2.1 Worldviews and Ethical Perspectives—A Comparison

PHILOSOPHY	INTRINSIC VALUE	INSTRUMENTAL VALUE	ROLE OF HUMANS
Anthropocentric	Humans	Nature	Masters
Stewardship	Humans and Nature	Tools	Caretakers
Biocentric	Species	Abiotic nature	One of many
Animal rights	Individuals	Processes	Equals
Ecocentric	Processes	Individuals	Destroyers
Ecofeminist	Relationships	Roles	Caregivers

FIGURE 2.8 Shiva Nataraja, Lord of the Dance, dances the world into existence. In another manifestation, Shiva becomes the destroyer of worlds. The modern scientific view bears some resemblance to this image of a world in a constant cycle of creation and destruction.

Ecofeminism, a pluralistic, nonhierarchical, relationship-oriented philosophy that suggests how humans could reconceive themselves and their relationships to nature in nondominating ways, is proposed as an alternative to patriarchal systems of domination. It is concerned not so much with rights, obligations, ownership, and responsibilities as with care, appropriate reciprocity, and kinship. This worldview promotes a richly textured understanding or sense of what human life is and how this understanding can shape people's encounters with the natural world. Among ecofeminist leaders are Karen Warren, Vandana Shiva, Carolyn Merchant, Rosemary Ruether, and Ynestra King.

According to ecofeminist philosophy, when people see themselves as related to others and to nature, they will see life as bounty rather than scarcity, as cooperation rather than competition, and as a network of personal relationships rather than isolated egos. Like the postmodernists, ecofeminists reject the view of a single, ahistoric, context-free, neutral observation stance. Instead, they favor multiple understandings, complex relationships, and "embodied objectivity."

In *Healing the Wounds* Ynestra King wrote, "We can use [ecofeminism] as a vantage point for creating a different kind of culture and politics that would integrate intuitive, spiritual, and rational forms of knowledge, embracing both science and magic insofar as they enable us to transform the nature-culture distinction and to envision and create a free, ecological society." Like the Hindu god, Shiva, humans can be both destroyers and creators of nature (fig. 2.8).

ENVIRONMENTAL JUSTICE

People of color in the United States and around the world are subjected to a disproportionately high level of environmental health risks in their neighborhoods and on their jobs. Minorities, who tend to be poorer and more disadvantaged than other residents, work in the dirtiest jobs where they are exposed to toxic chemicals and other hazards. More often than not they also live in urban ghettos, barrios, reservations, and rural poverty pockets that have shockingly high pollution levels and are increasingly the site of unpopular industrial facilities, such as toxic waste dumps, landfills, smelters, refineries, and incinerators. **Environmental justice** combines civil rights with environmental protection to demand a safe, healthy, life-giving environment for everyone.

Among the evidence of environmental injustice is the fact that three out of five African-Americans and Hispanics, and nearly half of all Native Americans, Asians, and Pacific Islanders live in communities with one or more uncontrolled toxic waste sites, incinerators, or major landfills. A recent Greenpeace study found that minorities make up twice as large a population share in communities with these locally unwanted land uses **(LULUs)** as in communities without them. And the inequities are growing. Whereas in 1980 the average minority population near a landfill or hazardous waste facility was about 22 percent, in 1994 it was 36 percent.

Although it is difficult to distinguish between race, class, historical patterns of ethnic groups, economic disparities, and other social factors in these disputes, racial origins often seem to play a role in exposure to environmental hazards. While poor people in general are more likely to live in polluted neighborhoods than rich people, the discrepancy between the pollution exposure of middle class blacks and middle class whites is even greater than the difference between poorer whites and blacks. Where upper class whites can "vote with their feet" and move out of polluted and dangerous neighborhoods, blacks and other minorities are restricted by color barriers and prejudice (overt or covert) to the less desirable locations (fig. 2.9).

Environmental Racism

Racial prejudice is a belief that someone is inferior merely because of their race. Racism is prejudice with power. **Environmental racism** is inequitable distribution of environmental hazards based on race. Evidence of environmental racism can be seen in lead poisoning in children. The Federal Agency for Toxic Substances and Disease Registry considers lead poisoning to be the number one environmental health problem for children in the United States. Some 4 million children—many of whom are African-American, Latino, Native American, or Asian, and most of whom live in inner city areas—have dangerously high lead levels in their bodies. This lead is absorbed from old lead-based house paint, contaminated drinking water from lead pipes or lead solder, and soil polluted by industrial effluents and automobile exhaust. The evidence of racism

FIGURE 2.9 Poor people and people of color often live in the most dangerous and least desirable places. Here children play next to a chemical refinery in Texas City, Texas.

FIGURE 2.10 Living conditions in the *colonias* or unplanned settlements along the U.S./Mexican border often are appalling. Many U.S. businesses have been attracted to this area by low wages, compliant union bosses, and lax environmental protection.

is that at every income level, whether rich or poor, black children are two to three times more likely than whites to suffer from lead poisoning.

Dumping Across Borders

Because of their quasi-independent status, most Native-American reservations are considered sovereign nations that are not covered by state environmental regulations. Court decisions holding that reservations are specifically exempt from hazardous waste storage and disposal regulations have resulted in a land rush of seductive offers from waste disposal companies to Native-American reservations for onsite waste dumps, incinerators, and landfills. The short-term economic incentives can be overwhelming for communities in which adult unemployment runs between 60 and 80 percent. Uneducated, powerless people often can be tricked or intimidated into signing environmentally and socially disastrous contracts. Nearly every tribe in America has been approached with proposals for some dangerous industry or waste facility.

The practice of targeting poor communities of color in the Third or Fourth World for waste disposal and/or experimentation with risky technologies has been described as **toxic colonialism.** Internationally, the trade in toxic waste has mushroomed in recent years as wealthy countries have become aware of the risks of industrial refuse. Poor, minority communities at home and abroad are being increasingly targeted as places to dump unwanted wastes. Although a treaty regulating international shipping of toxics was signed by 105 nations in 1989, millions of tons of toxic and hazardous materials continue to move—legally or illegally—from the richer countries to the poorer ones every year. This issue is discussed further in chapter 23.

Another form of toxic colonialism is the flight of polluting industries from developed nations and states where control requirements are stringent to less developed areas where regulations are lax and local politicians are easily co-opted. For example, more than 2000 *maquiladoras,* or assembly plants operated by American, Japanese, or other foreign companies, are now located along the United States/Mexico border to take advantage of favorable import quotas, low wages, and weak pollution control laws.

Mexican laborers work under appalling conditions in some of these factories, assembling imported components into consumer goods to be exported to the United States. Although the jobs are demeaning, dangerous, and pay far less than the minimum U.S. wage, large populations have been attracted to squalid shantytowns along the border where industrial effluents poison the air and water (fig. 2.10).

In 1992, an Environmental Justice Act was introduced in the United States Congress to identify areas threatened by the highest levels of toxic chemicals, assess health effects caused by emissions of those chemicals, and ensure that groups or individuals residing within those areas have opportunities and resources to participate in public discussions concerning siting and cleanup of industrial facilities. Perhaps we need something similar worldwide.

Are "Green" Organizations Too White?

Although concern for the environment cuts across race and class lines, environmental activists traditionally have been individuals with above-average education, greater access to economic resources, and a greater sense of personal power than others in society. The agendas of national environmental organizations tend to focus on wilderness and wildlife preservation, wise resource management, outdoor recreation, and other issues that are not high among the concerns of those at the bottom of society who are

FIGURE 2.11 The great Louisiana Toxics March mobilized local residents from the notorious stretch of the Mississippi River between New Orleans and Baton Rouge known as "cancer alley" because of its high concentration of petrochemical industries and environmental illnesses. Environmental justice combines elements of the civil rights movement with environmental protection.

struggling for immediate survival. Relatively little attention has been paid by the mainline environmental groups to inner city problems. Considerable criticism has been leveled against these groups for their lack of minority staff members and inattention to environmental justice.

Sadly, an unanticipated result of successful lobbying by environmental organizations against polluting industries and LULUs in recent years has been increased vulnerability of minority communities to the siting of unpopular facilities such as municipal landfills, toxic waste dumps, metal smelters, power plants, and incinerators. Too often **NIMBY** (not in my backyard) **protests** result in dumping of pollution in someone else's backyard. And the minorities who live in polluted communities are often subjected to double jeopardy, in that they are exposed to increased environmental risks, while at the same time, they frequently have problems gaining access to health and medical care facilities.

A growing awareness of the inequitable share of environmental hazards borne by communities of color and the long-term risks to their health and survival has stimulated greater activism and organization around environmental protection for these groups (fig. 2.11). The first National People of Color Environmental Leadership Summit was held in 1991 in Washington, DC. More than 650 delegates, participants, and observers, representing all 50 states, Puerto Rico, Mexico, and the Marshall Islands attended this meeting. The stated purpose of the gathering was "to build a unified national voice and an effective agenda for environmental justice." In combining the civil rights movement with environmental protection, the environmental justice movement represents an important new direction in conservation history.

Is Nature Fragile or Resilient?

For many of our earlier ancestors, nature seemed vast, chaotic, and implacable. Wild lands were full of fierce, wild animals, unpredictable storms, and other life-threatening hazards. By comparison, humans seemed frail and endangered. Early writers described the wilderness as a place of danger and evil spirits. The word *panic* (fear of the nature god Pan) described the hysterical dread that early Greeks felt when forced to travel through wild nature. They believed that if they heard the music of Pan's pipes they would be lured away from the life of reason and civility in the city and doomed to live forever as wild brutes in the forest. This attitude persisted into fairly recent times. As the *Mayflower* approached New England in 1620, Governor William Bradford described the landscape as a "hideous and desolate wilderness . . . a land of spiritual darkness in which demons and wild beasts dwell."

Humans have struggled to control the forces of nature and to create havens of peace, comfort, and safety for themselves and their domestic livestock. *Paradise* is a Persian word describing a walled garden or oasis that offers shelter and respite from the harsh desert (fig. 2.12*a*). Many cultures had similar ideals of a place of perfect ease, abundance, and solace. The garden metaphor, in which nature is domesticated and cultivated by the human caretaker to produce useful products and aesthetic pleasure, still has a powerful appeal for many of us.

In recent years, however, as technological power to disrupt natural systems has increased, our view of ourselves with respect to nature has changed. We now can change the course of rivers, create vast artificial lakes, turn woodlands and grasslands into deserts, and literally move mountains. "Manmade" earthquakes have been triggered by pumping liquid wastes into deep disposal wells or by the weight of water and sediments accumulated in some artificial lakes. We already have changed regional climates and we now may be modifying the global climate as well.

Increasingly, we see nature as fragmented, threatened, and vulnerable (fig. 2.12*b*). Critics warn that our actions threaten to irreparably upset a delicate balance that maintains natural functions. But how fragile is that balance? How much can we perturb nature without threatening not only our own existence but that of all the rest of life as well? These may be questions that we don't want to answer through global experiments.

On the other hand, claims of fragility and delicate balances in nature sometimes are overstated. They may be statements of faith rather than fact. In many circumstances, nature is amazingly fecund and resilient. Great Mayan cities in Central America that once housed thousands of people and were surrounded by farms have been almost completely swallowed by the surrounding rainforest (fig. 2.13). Explorers are still finding evidence of this civilization under jungle-covered mounds. Perhaps 1000 years from now, explorers will poke around the weed-covered rubble of our great cities and wonder what happened to us.

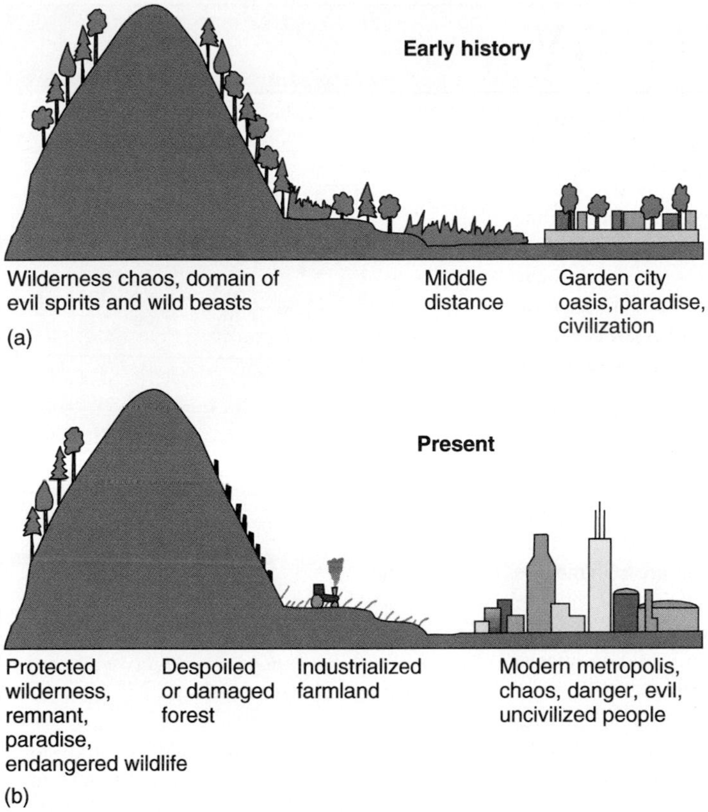

FIGURE 2.12 Our attitudes toward nature have changed dramatically as cities and domesticated landscapes have expanded and wilderness has diminished. In earlier times (a), the city was seen as a sacred place of comfort, peace, and order, while the wilderness was seen as the domain of chaos, evil, and danger. Now we see the remaining remnants of wild nature as endangered enclaves of paradise while the city represents chaos, evil, and danger (b).

FIGURE 2.13 Tikal, in what is now Guatemala, was one of the largest Mayan cities 1200 years ago, with a population that may have reached 100,000 people. Today, almost nothing of this great city remains except a few temples rising above the jungle.

FIGURE 2.14 In modern society, information, education, research, communication, and technology have replaced material goods or natural resources as the main sources of wealth and power.

SCIENCE AS A WAY OF KNOWING

Science and technology have become dominant forces in our lives. Many of us have become completely dependent on science and the machines it creates (fig. 2.14) and yet know very little about how they work. As Henry Thoreau said, "Lo! Men have become the tools of their tools." Understanding something about the history and philosophy of technology and the scientific process may help you not only in your current course of study but also in everyday life.

A Faustian Bargain?

What comes to mind when you think about science and technology? Do they seem mysterious, incomprehensible, and yet powerful? Do you expect science courses to be difficult, disagreeable, and full of obscure details? For many people today, science and the machines it creates seem to be a Faustian bargain (in Goethe's story Dr. Faustus sells his soul to the devil in exchange for power and wealth). One example is the atom bomb, which gives us terrible power to destroy. And yet science, through modern medicine using nuclear equipment, also has saved many lives and eliminated much suffering. Although there have been many misapplications of science and scientific knowledge, we hope to persuade you in this section that science and technology also have many benefits and that scientific methods—especially those of environmental science—can be useful in your everyday life.

CASE STUDY

SQUIRRELS OR SCIENCE?

Rising from the hot, dry Sonoran Desert, several small mountain ranges in southeastern Arizona create cool, moist forest refuges for a variety of rare species. Because the desert prevents many species from migrating between ranges, the mountains become islands on which isolated populations evolve along different paths. There are twenty-five known subspecies of montane red squirrels, for instance. Nearly every range has its own subspecies. One of them, the Mt. Graham Red Squirrel (*Tamiasciurus hudsonicus grahamensis*), has become a pawn in a battle between astronomers and preservationists that illustrates some important differences in our attitudes toward science and nature.

Mountain tops attract astronomers as well as wildlife and these are boom times for astronomy. Technological advances have made it possible to explore questions previously beyond our abilities, but the number of places suitable for high-resolution stargazing is declining rapidly. Urban smog, dust, light, and electromagnetic radiation interfere with observations in many areas. Some of the best locations in the world are in the southern Andes, but travel and logistics at these sites are difficult and expensive.

Mt. Graham, in Arizona's Pinaleno Range, has many attractive features for astronomy. Only about 100 km (60 mi) as the crow flies from Tucson, Mt. Graham is, at 3267 m (10,717 ft), the second highest mountain in the state. It has spectacularly clear air and dark nights. With an international airport nearby and all the modern telecommunications advantages of a modern American city at hand, astronomers can fly in for the weekend and enjoy the comforts of first-class hotels after a night of work.

Other competing uses for the mountain have made this a battleground, however. An unlikely coalition of animal rights activists, radical environmentalists, hunters, off-road motorcyclists, and traditional Native Americans have fought to keep telescopes off the mountain. For some, the issue is preserving rare and endangered species and their unique habitat; for others, it is outdoor recreation and personal freedom; for still others, it is a question of despoiling sacred religious sites; and for a few, it is simply a matter of opposing big science and technology on principle.

The institutions proposing new telescopes also make up a surprising alliance. The Smithsonian Institution, the Vatican, the Max Planck Institute from Germany, and the University of Arizona, have all built, or are planning to build, telescopes on Mt. Graham. Several other American universities—including Ohio State, Minnesota, and Texas—expressed an interest but pulled out when students protested.

The Mt. Graham Red Squirrel, which lives only in a small area of spruce-fir forest at the mountain peak—exactly where astronomers want to build telescopes—was declared an endangered subspecies by the Fish and Wildlife Service in 1987. A year later, the United States Congress exempted the project from laws protecting cultural sites or endangered species. Since then, thirty-five acres of mountain top have been designated for astronomy. Environmentalists still are outraged. Lobbying, legal proceedings, civil disobedience, and ecosabotage are ongoing.

While beautiful, the mountain is hardly untouched. Logging has gone on there for a century. More than 200,000 people visit every year to hunt, camp, hike, cut Christmas trees, ride dirt bikes, and ski. Some elders of the San Carlos Apache tribe claim that the mountain is a traditional religious site; others dispute that it has any special value. The rarity of finding a cool, moist forest in the midst of the dry rocky desert is probably what endears the mountain to most people.

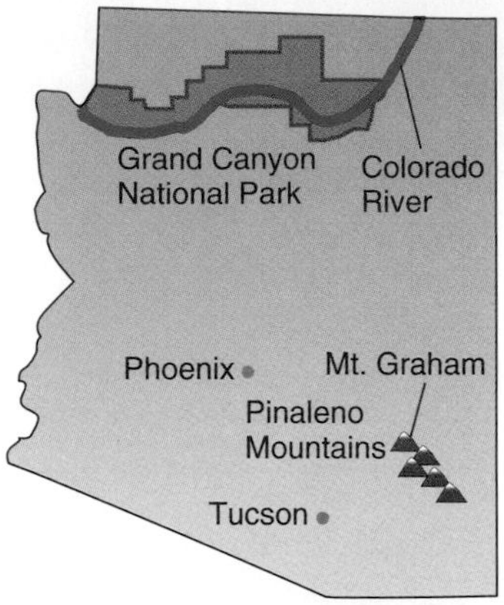

Mt. Graham in the Pinaleno Range of southeastern Arizona is the site of highly contested astronomical observatories.

Some wildlife managers believe that the two hundred existing squirrels will do better within the fenced grounds of the astronomical site than in a forest open to hunters and bikers. After all, other squirrels coexist well with humans if they have food and a place to live. Protesters feel that a high-tech facility of this sort is an affront to the wilderness and to free-living wild squirrels.

Ethical Considerations

What's your opinion about this situation? Does a telescope complex on the mountain top ruin the landscape? What rights or values would you ascribe to the squirrels? What value does pure science have? Is technology categorically inimical to nature or wilderness? Is nature synonymous with wilderness? Is this a matter of choosing between a unique ecosystem or a world-class scientific facility, or might there be a way to have both?

What Is Science?

In the best sense, **science** is a systematic, precise, objective way to study the natural world. It is often an exciting and satisfying enterprise that requires creativity, skill, and insight. Science takes many different forms and is done in assorted ways by widely diverse people. Scientists strive to be neutral and unbiased. Sometimes, however, it is impossible to remain aloof. The process of making observations changes the subject as well as the observer.

Even though you may not be a scientist, you use scientific technique without being aware of it. Suppose your flashlight doesn't work. The problem could be in the batteries, the bulb, or the switch—or all of them could be faulty. How can you distinguish between the possibilities? A series of methodical steps to test each component can be helpful. First, you might try new batteries. If that doesn't help, you might replace the bulb with one that you know works. If neither of these steps solves your problem, perhaps the flashlight has a faulty switch. You could try the original battery and bulb in a different flashlight. By testing the variables one at a time, you should be able to identify the problem.

What you have been following in this example is the **scientific method** for methodical inquiry. Science is a way of knowing about the physical world based on an ordered cycle of observations, methodical investigation, and interpretation of results. The general flow of an experimental scientific study is shown in figure 2.15. We start with observations: in this case, my flashlight doesn't work. From this we formulate a hypothesis, or a provisional explanation: the batteries must be dead. To be useful, our hypothesis must enable us to make predictions that we can test.

Testing Hypotheses

In experimental science, tests of our hypotheses must be controlled in the sense that only one variable is changed at a time. If we change both the bulb and the batteries simultaneously we won't know which is good or bad. We might be discarding perfectly good components. Testing one part with two others that we know work should show which is at fault. Finally, we carry out the experiment, collect data—in this case, look for light—and draw conclusions.

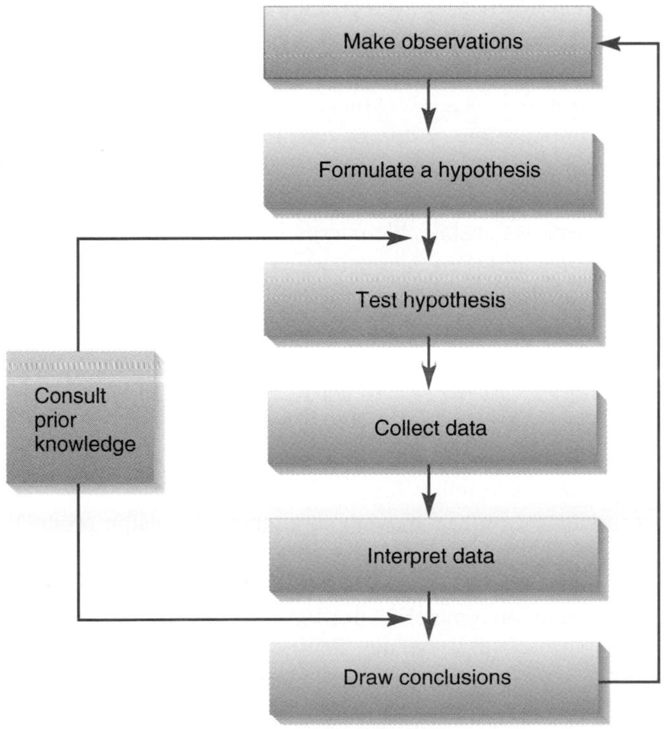

FIGURE 2.15 Ideally, scientific investigation follows a series of logical, orderly steps to formulate and test hypotheses.

Often the results of one experiment gives us information that leads to further hypotheses and additional experiments. If the batteries are OK, then we suspect the bulb is burned out and design a way to test that hypothesis. In every case, prior knowledge and experience help us design experiments and interpret results. Eventually, with evidence from a group of related investigations, we create a theory to explain a set of general principles.

Descriptive and Interpretive Science

Not every field of science is accessible to our basic senses—seeing, hearing, touching, smelling, and tasting—nor are all amenable to direct experimentation. We don't regard it as acceptable, for instance, to study the effects of toxins on human health by deliberately poisoning people, no matter how useful that information might be. Although geologists and evolutionary biologists can do limited kinds of experiments, they can hardly build mountain ranges in the laboratory or recreate the millions of years that gave rise to certain species. Scientists in these fields test their ideas and explanations indirectly by looking for historic evidence of support or contradiction. Or we use model systems such as laboratory animals as surrogates. While there are difficulties in interpreting these indirect results, they are both scientific and useful.

Academics in some areas go further and reject the positivist, reductionist approaches of modern hypothetico-deductive science (which holds that the world is knowable and that by studying individual parts of a phenomenon, we can deduce important truths about the whole). Drawing on the uncertainty principles of quantum physics, these critics argue that there is no such thing as an objective, uninvolved observer. Anytime you make an observation, you effect the observed, and by trying to subdivide objects or processes into constituent parts, you lose important characteristics of the whole. The best thing to do, in this perspective, is to recognize inherent biases and simply try to describe and interpret our subjects as accurately and faithfully as possible. This is sometimes called **holistic** or **interpretive science.** Most natural history is descriptive, inductive science that argues from generalities to specifics rather than the other way around.

Proof Is Elusive

A common misconception is that science proves theories. You probably have heard claims that something has been proven "scientifically." In fact, scientific interpretations are always conditional. The evidence from experiments may support a particular hypothesis or provide evidence contrary to it, but we can never be absolutely sure that we have a final proof. A possibility always exists that some new evidence will require a modification of our conclusions. We may even have to discard them and formulate an entirely new set of theories.

Because of these uncertainties, you can often find distinguished scientists who draw opposite conclusions from the same data. This may seem confusing and discouraging. What should we believe when the experts disagree? Learning to evaluate information systematically, keeping an open mind, and remaining alert for new data and alternative interpretations are important lessons from science that can help us understand many aspects of life. This is an excellent opportunity to practice the critical thinking skills described in the introduction.

Technology and Progress

For the past two centuries, a central tenet of Western culture has been an almost religious faith in progress: an inevitable march of human betterment. Originally formulated during eighteenth-century enthusiasm over the American and French Revolutions, this theory seemed to be proven by the increase in material wealth and standard of living provided by the Industrial Revolution. But while technology and development brought many benefits, they had a darker side as well (fig. 2.16). Pollution, rapid urbanization, inhumane working conditions for many, and vast disparities in wealth and power between classes still cause social and environmental crises. Whether the root causes of these problems are in technology or human nature, technology clearly allows us to make mistakes faster and on a larger scale than ever before.

FIGURE 2.16 Science and technology give us power to improve our lives, but they also make it possible to make bigger mistakes faster than ever before, as is the case with these industrial smokestacks shown in the 1890s.

A nineteenth-century English backlash against the excesses of industrialization led by Ned Ludd gave rise to the term *Luddites* for opponents of rampant technology. The Luddites smashed power looms and other machines that were threatening the craft guilds, cottage industries, and village networks that sustained traditional rural communities.

In this century, even more dangerous technologies, such as nuclear weapons, biological warfare, and the petrochemical industry, along with the problems caused by earlier technologies, such as biodiversity losses, global climate changes, and destruction of stratospheric ozone, have led many intellectuals and young people to question whether progress is either possible or desirable. **Neo-Luddites** now assert that all large-scale human endeavors eventually fail, that science and technology cause more problems than they solve, and that our only hope is to abandon modern life and go back to a low-tech pastoral or hunting-and-gathering society.

Some neo-Luddites resort to terrorist bombings and sabotage to try to bring down mainstream culture. Others flee to end-of-the-road refuges where they attempt to recreate a simpler, agrarian life. Rural life can be more resource intensive than urban living, however, especially if you demand all the modern conveniences. We probably couldn't all live off the land in remote places without very destructive environmental effects.

Appropriate Technology

As historian Lewis Mumford pointed out, technology consists of more than machines. It includes all the techniques, knowledge, and organization that we use to accomplish tasks. When you build a fire, or use a rock as a tool, you are employing some of the oldest and most revolutionary human technologies. Whether our technology is destructive or constructive depends, in part, on the tools themselves, but even more on our worldview about how and why we use them. As sustainable-energy expert Amory Lovins points out, some technologies such as nuclear power, genetic engineering, and nanotechnology might be benign in the hands of a "wise, far-seeing, and incorruptible people." Unfortunately that seems not to describe most of us.

In 1973, British economist E. F. Schumacher published a widely popular book entitled *Small Is Beautiful.* It introduced the concept of **appropriate technology,** which promotes machines and approaches suitable for local conditions and cultures. The appropriate-technology movement attempts to design productive facilities in places where people now live, not in urban areas. It looks for products that are affordably made by simple production methods from local materials for local use. It advocates safe, creative, environmentally sound, emotionally satisfying work conducted in conditions of human dignity and freedom that creates social bonds rather than breaking them down.

Rather than to try to convert local economies and tastes into copies of Western culture, appropriate technologists try to work with indigenous people to create sustainable livelihoods suitable for prevailing conditions (fig. 2.17). They hope that appropriate technology can help us avoid future environmental damage and to repair mistakes made in the past. Some advances have been made toward these goals, but the promise and power of the dominant Western paradigm are very seductive.

FIGURE 2.17 Appropriate technology, such as this multipurpose cultivator/tractor, is designed to suit local needs and conditions in developing countries. Small, simple, and affordable machines that are locally manufactured and repaired can reduce drudgery and improve productivity without reducing the number of jobs available.

SUMMARY

Are there universal, eternally valid ethical principles or moral laws? Universalists think so. Nihilists disagree. Relativists believe that everything is contextual. Utilitarians hold that something is good that brings the greatest good to the greatest number. Postmodernists argue that everything is socially constructed. Ecofeminists contend that patriarchal systems of domination and duality cause both environmental degradation and social dysfunction. They call for a more pluralistic, nonhierarchical, caring treatment of both nature and other people. Whichever of these worldviews you hold shapes your views about nature and our place in it.

Anthropocentrists claim that the world was made for our domination, and that only humans have inherent or intrinsic rights and values. Nature, from this perspective is only a source of materials for humans. Many people who live close to the land feel a sense of stewardship or responsibility to care for creation. They see themselves as caretakers rather than dominators. Biocentrists consider all living things to have inherent value. We are merely one of many species. Animal rights advocates place their emphasis on individual animals. Ecocentrics maintain that ecological processes such as evolution, adaptation, and biogeochemical cycles are the most important parts of nature. In their view, individuals don't count for much and humans are mostly a negative influence.

Science is a methodical, meticulous, dispassionate study of the natural world. It takes many different forms and is done in various ways by diverse people, but observation, hypothesis formation, testing, analysis, and re-evaluation of hypotheses in light of new data form the core of positivist scientific methodology. Descriptive or interpretive sciences take a more holistic, inductive approach to knowledge.

Technology brings us many benefits, but it also creates pollution, consumes resources, despoils nature, and allows us to separate individuals, classes, and nations into those who have and those who do not. It gives us the power to make mistakes faster and on a larger scale than ever before. Then again, appropriate technology may also provide options to avoid environmental damage in the future or to repair mistakes made in the past.

QUESTIONS FOR REVIEW

1. Define universalist, nihilist, relativist, utilitarian, postmodern, and ecofeminist ethics.
2. Explain how resource exploiters and social justice advocates use these ethical positions to support their causes.
3. Describe the differences between moral agents, moral subjects, and those who fail to qualify for moral considerability.
4. What are inherent and instrumental values? Who has them? Why?
5. Where is Mineral King Valley and why is it important in environmental history? What happened there?
6. Compare and contrast stewardship, anthropocentric, biocentric, ecocentric, and environmental justice worldviews. Which is closest to your own views?
7. What did Lynn White, Jr. say about the role of religion in environmental problems?
8. What recent events have countered the claim that religions oppose environmental protection?
9. Draw a diagram showing the scientific method and describe, in your own words, what it means.
10. Not every science is experimental. Explain how geologists or evolutionary biologists might test their hypotheses.

QUESTIONS FOR CRITICAL THINKING

1. Review a topic in this chapter—ecofeminism or the question of nature's fragility or resiliency, for example—and arrange the arguments for or against this idea as a series of short statements. Determine whether each is a statement of fact or opinion.
2. How would you verify or disprove fact statements given in question 1? Do the opinions or conclusions reached in this argument *necessarily* follow from the facts given? What alternatives could be proposed?
3. Reflect on the preconceptions, values, beliefs, contextual perspective that you bring to the discussion above. Does it coincide with those in the textbook? What different interpretation would your perspective impose on the argument?
4. Try to put yourself in the place of a person from a minority community, an underdeveloped nation, or a Third World country in discussing questions of environmental justice and environmental quality. What preconceptions, values, beliefs, and contextual perspective would you bring to the issue? What would you ask for from the majority society?
5. What is your environmental ethic? What experiences, cultural background, education, religious beliefs help shape your worldview?
6. Try some role-playing with a classmate, friend, or family member. Take the position of a universalist, biocentrist, ecocentrist, postmodernist, or ecofeminist and debate the merits of cutting trees in the National Forest. On what points would you agree and where would you disagree in this issue?
7. This chapter addresses many social issues and theories. Are these areas appropriate for an environmental science textbook? Why or why not? Try answering this question from the perspective of Gifford Pinchot, John Muir, the ecofeminist, or the person of color whom you identified in earlier questions. Would they agree?

KEY TERMS

anthropocentric 41
appropriate technology 50
biocentric 42
domination 41
ecocentric 42
ecofeminism 44
environmental ethics 37
environmental justice 44
environmental racism 44
holistic science 49
inherent value 40
instrumental value 40
interpretive science 49
LULUs 44
moral agents 39
moral extensionism 39
morals 37
moral subjects 39
neo-Luddites 50
nihilists 38
NIMBY protests 46
relativists 38
science 48
scientific method 49
stewardship 41
toxic colonialism 45
universalists 38
utilitarians 38
values 37

ADDITIONAL INFORMATION ON THE INTERNET

Visit our website at http://www.mhhe.com/environmentalscience/ for specific information about each of the topics below:

Mount Graham International Observatory
Forest Service Employees for Environmental Ethics
Environmental Justice
Environmental Justice: Environmental Health Information Service
Environmental Ethics
Environmental News Network
Environmental Philosophers
Handbook to a Better Future

PROFILE

Environmental Activist

Micaela Martinez

Micaela Martinez' concerns about the environment began when she was in high school. Back then, she found herself joining citizen campaigns on a variety of local issues, even helping stage a bike-a-thon to protest a developer's plans to build in habitat crucial to the survival of an endangered butterfly. She went off to college committed to pursuing a career that would enable her to make a contribution to environmental quality. She came to the conclusion that since habitat was so important to organisms' well-being, a major in wildlife biology was the way to make the biggest impact.

She began her college education at Skyline Junior College in a San Francisco suburb. She planned to complete two years there and then transfer to Humboldt State University to complete a major in wildlife management. Her ultimate goal was to work for the California Fish and Game Department. While in college, she continued to volunteer on behalf of assorted environmental causes.

It was during those years that her career goals underwent a sudden and fundamental change. As the result of a national election, a number of pro-environment incumbents lost their offices to candidates with decidedly less environmentally friendly viewpoints. As she watched policy proposals come forth from the new officeholders, she began to feel that effecting environmental change had less to do with biology and habitat and more to do with politics.

Since Micaela was already well on her way to completion of the requirements, she finished her wildlife degree despite the change of heart on her career choice. Soon after graduation she took a job as a canvasser with the California League of Conservation Voters. She participated in door-to-door canvassing, which is used by a number of environmental organizations to inform people about local or regional environmental issues and to identify voters who are concerned about environmental issues, as well as to seek donations.

After gaining canvassing experience with the League of Conservation Voters, Micaela joined the staff of Clean Water Action, a national organization that hires many college students for door-to-door canvassing. She currently serves as supervisor of eight canvass programs around the country, including efforts in California, Pennsylvania, Maryland, and Washington, DC.

Part of Micaela's work is directed at helping the organization set goals and priorities, and the strategies to achieve them. She handles job performance reviews, develops recruiting plans, and makes career day visits to college campuses. She also oversees her organization's efforts in voter identification, phone banks, and get-out-the-vote efforts on election day. Finally, she writes press releases, organizes press conferences, and helps build coalitions with allied organizations and community leaders. Her work generally intensifies during the summer months, and particularly in election years.

Despite not working directly in her area of technical training, Micaela finds her knowledge of the natural sciences helpful in her job. She feels that her scientific training gives her a good background understanding of the issues under debate. She also feels that her training in wildlife biology has helped her understand the mindset of people in both natural resource agencies and industry. She has found this understanding of diverse perspectives very helpful in building coalitions.

Micaela has found her technical writing course of particular value in helping her write succinctly and to present technical information in an understandable form. She wishes she had left college with a better knowledge of human social dynamics, however. Much of her time is spent working with people and she feels that courses in business management or the psychology of team building would have helped.

Her advice to students interested in a career in environmental activism is to learn as much as possible about environmental matters from a scientific or technological perspective. She also feels it is important to volunteer for environmental causes in their college community. Not only does it provide valuable experience, but it is also an excellent way to meet lots of people, which may lead to job opportunities as well. She suggests that during summer breaks students gain activism experience by either voluntary or paid positions as canvassers, campaign workers, or legislative aids.

Micaela enjoys working with people drawn to environmental activism. She also gets great satisfaction seeing the results of her work, knowing that she has an impact. She says it's a great feeling going home every day realizing that one person really can make a difference.

CHAPTER 3

Matter, Energy, and Life

One hill cannot shelter two tigers.

—Charles Elton quoting a Chinese proverb—

OBJECTIVES

After studying this chapter, you should be able to:

- describe matter, atoms, and molecules and give simple examples of the role of four major kinds of organic compounds in living cells.
- define *energy* and explain the difference between kinetic and potential energy.
- understand the principles of conservation of matter and energy and appreciate how the laws of thermodynamics affect living systems.
- know how photosynthesis captures energy for life and how cellular respiration releases that energy to do useful work.
- define *species, populations, communities,* and *ecosystems* and understand the ecological significance of these levels of organization.
- discuss food chains, food webs, and trophic levels in biological communities and explain why there are pyramids of energy, biomass, and numbers of individuals in the trophic levels of an ecosystem.
- recognize the unique properties of water and explain why the hydrologic cycle is important to us.
- compare the ways that carbon, nitrogen, sulfur, and phosphorus cycle within ecosystems.

ESSENTIAL STUDY PARTNER

The Essential Study Partner (ESP) CD-ROM that accompanies this textbook contains additional information on the following topics:

- atoms, bonding, and organic compounds: principles: matter and life: atoms to cells
- photosynthesis and respiration: principles: matter and life: energy
- species, population, community, and ecosystem: principles: ecosystems: introduction
- ecosystem trophic levels, food chains, and food webs: principles: ecosystems: introduction
- productivity, biomass, and energy flow: principles: ecosystems: energy flow
- nutrient cycles, sources, and human alterations: principles: ecosystems: nutrient cycles

Above: A puffin returns to its nest with a bill full of fish for its chick.

The Mystery of Lake Lebarge

Toxins Abound in a Boreal Lake

There are strange things done in the midnight sun
By the men who moil for gold;
The Arctic trails have their secret tales
That would make your blood run cold;
The Northern Lights have seen queer sights,
But the queerest they ever did see
Was that night on the marge of Lake Lebarge
I cremated Sam MaGee.

—Collected Poems of Robert Service, New York: Dood, Mead & Co., 1940.—

Chances are, if you've ever heard of Lake Lebarge, it was in this poem by Robert Service. Nevertheless, this remote, northern lake is the site of an ecological puzzle that could never have been imagined by the miners who swarmed through this region in the Klondike gold rush a century ago, but that reveals some important ecological interconnections in our modern environment.

Lake Lebarge is a narrow finger lake (5 km wide by 50 km long) just north of the town of Whitehorse in the headwaters of the fabled Yukon River (fig. 3.1). Beginning about a decade ago, scientists collected water and biological samples from the lake to use as controls in a study of uptake of toxic industrial chemicals. The intent was to compare toxin levels in organisms from this remote and supposedly pristine watershed with samples collected closer to industrial and agricultural areas. Surprisingly, the top game fish from Lake Lebarge turned out to have some of the highest levels of PCBs, toxaphene, DDT, and other industrial and agricultural chemicals found anywhere in Canada.

Located about 1600 km (1000 mi) north of Vancouver, BC, Lake Lebarge would seem to be well removed from major pollution sources. Far from being pure and pristine, however, the lake water was found to have up to 100 times the concentration of toxic organic chemicals of lakes much further south. And trout and pike, the top predators in the lake, accumulate these compounds at levels thousands of times that of their counterparts in other ecological systems. How could this be? The answer is that volatile organic chemicals are being transported thousands of kilometers by air currents and are falling into the lake in the form of toxic rain and snow. The transport apparently is not in a single leap but rather a series of shorter hops. These compounds evaporate from water and soil in warmer climates. They are carried aloft by rising air currents where they cool, condense, and precipitate out. On reaching the surface, the most volatile chemicals re-evaporate, and with each cycle, they move further north until they reach very cold places, such as Lake Lebarge, where they accumulate to extraordinary levels. It's a sort of distillation process on a continental scale.

Contaminants in the water are taken up first by tiny floating plants and animals known as plankton. Insects and crustaceans prey on plankton and store toxins they contain in tissues and organs. Minnows and small fish concentrate the toxins further. Finally, long-living top predators (game fish) in the lake accumulate the highest levels of all. This process of concentration through the food chain is especially intense in Lake Lebarge where clear, cold waters are relatively infertile. This means that the top predators eat relatively fewer invertebrates than they might in more nutritious, warmer lakes. Trout and pike in Lake Lebarge mostly consume other fish and, thus, have unusually high toxic burden.

If you ever have a chance to visit Lake Lebarge, you wouldn't be advised to make a meal of the fish you might catch there. We'll discuss the health effects and some of the possible political solutions for this contamination in subsequent chapters of this book. In this chapter, we'll examine the biogeochemical cycles and ecological processes that are responsible for the movement of materials through our environment and into and through living food chains.

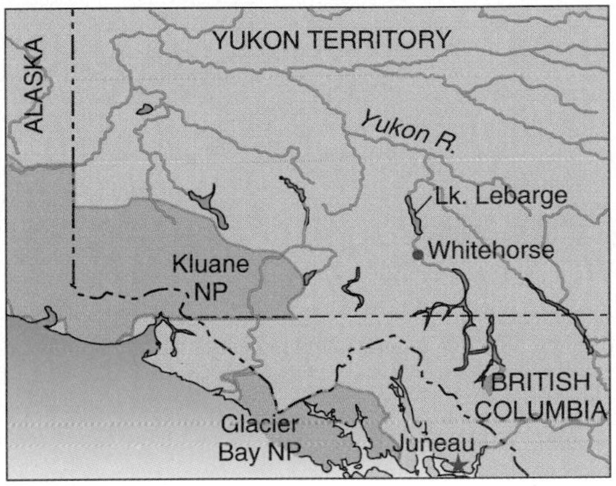

FIGURE 3.1 The Yukon River flows through Lake Lebarge just north of the historic town of Whitehorse. This was the route to the Klondike during the gold rush of 1897.

FROM ATOMS TO CELLS

How and why materials are cycled between the living and nonliving parts of our environment are the domain of **ecology**, the scientific study of relationships between organisms and their environment. Modern biology covers a wide range of inquiry, ranging from molecules to ecosystems to the entire planet. Ecology examines the life histories, distribution, and behavior of individual species, as well as the structure and function of natural systems at the level of populations, communities, ecosystems, and landscapes (fig. 3.2). The systems approach of ecology encourages us to think holistically and to see interconnections that make whole systems more than just the sum of their individual parts.

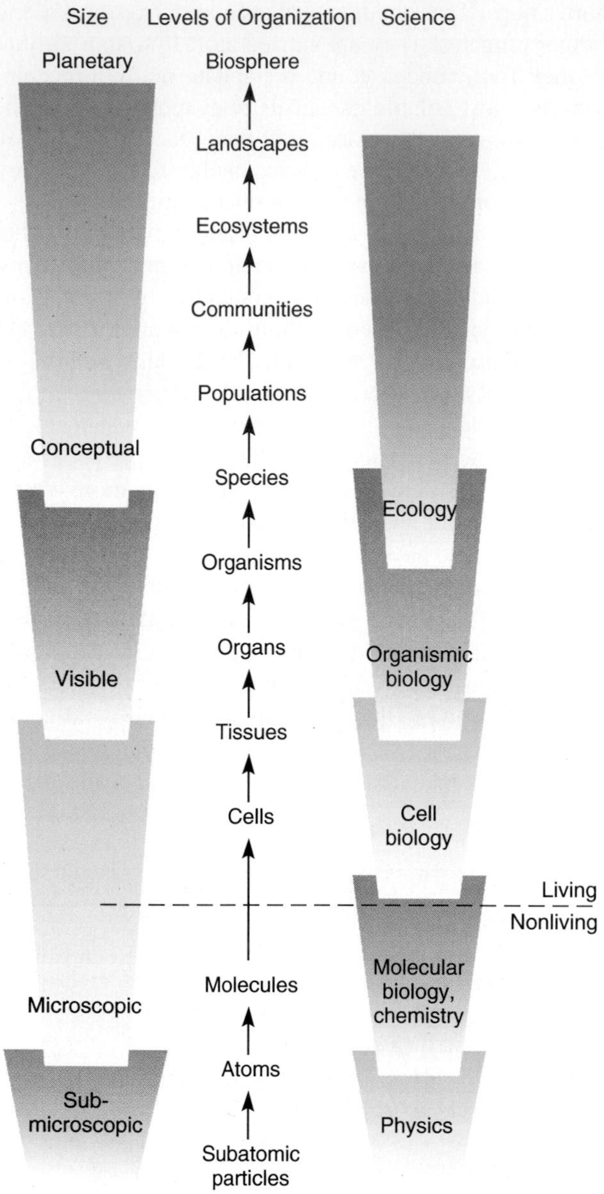

FIGURE 3.2 Levels of organization and the sciences that study them.

In this chapter, we will survey some fundamental aspects of energy flow and material recycling within ecosystems. We will look at why living organisms need a constant supply of these essential ingredients, how energy and materials are transferred from one organism to another, and how humans benefit from, and also disrupt, those ecological relationships.

In a sense, every organism is a chemical factory that captures matter and energy from its environment and transforms them into structures and processes that make life possible. To understand how these processes work, we need to understand something about the fundamental properties of matter and energy. This section presents a survey of some of these principles.

Atoms, Molecules, and Compounds

Everything that takes up space and has mass is **matter.** All matter has three interchangeable physical forms, or phases: gas, liquid, and solid. Water, for example, can exist as vapor (gas), fluid (liquid), or ice (solid). Matter also consists of unique chemical forms we call elements, molecules, and compounds. Each of the 115 known elements (65 of which are cycled biogenically) has distinct chemical characteristics. Among the more common elements in biology are carbon (represented by the symbol C), hydrogen (H), oxygen (O), nitrogen (N), and phosphorus (P).

All elements are composed of discrete units called **atoms,** which are the smallest particles that exhibit the characteristics of the element. Atoms are tiny units of matter composed of positively charged protons, negatively charged electrons, and electrically neutral neutrons. Protons and neutrons, which have approximately the same mass, are clustered in the nucleus in the center of the atom (fig. 3.3). Electrons, which are tiny in comparison to the other units, orbit the nucleus at high speed. Atoms that have an equal number of electrons and protons are electrically neutral. Those that have gained or lost electrons, and therefore are positively or negatively charged, are called **ions.**

Each element has a characteristic number of protons per atom, called its **atomic number.** The number of neutrons in different atoms of the same element can vary within certain limits. Thus, the atomic mass, which is the sum of the protons and neutrons in each nucleus, also can vary. We call forms of a single element that differ in atomic mass **isotopes.** For example, hydrogen, the lightest element, normally has only one proton (and no neutrons) in its nucleus. A small percentage of hydrogen atoms in nature have one proton and one neutron. We call this isotope deuterium (^2H). An even smaller percentage of natural hydrogen called tritium (^3H) has one proton plus two neutrons.

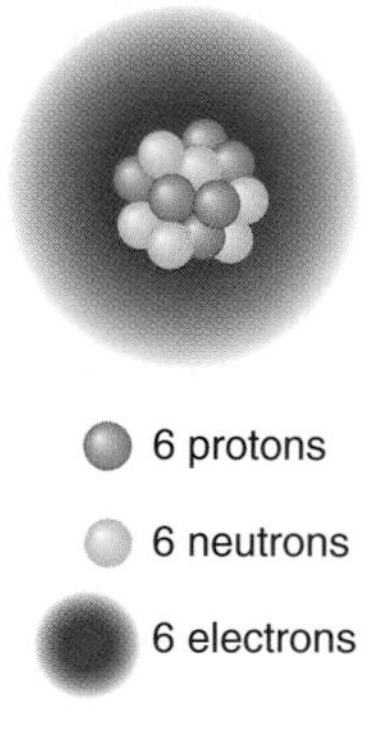

FIGURE 3.3 As difficult as it may be to imagine when you look at a solid object, all matter is composed of tiny, moving particles, separated by space and held together by energy. It is hard to capture these dynamic relationships in a drawing. This model represents carbon-12, with a nucleus of six protons and six neutrons; the six electrons are represented as a fuzzy cloud of potential locations rather than as individual particles.

Tritium is an example of a **radioactive isotope,** an unstable form that spontaneously emits either high-energy electromagnetic radiation or subatomic particles (or both), while changing into another isotope or even a different element. We call this process radioactive decay. Some isotopes decay very rapidly, while others decay very slowly. The **half-life**—the time required for half the atoms in a sample to decay—for tritium is about thirteen years. By contrast, some isotopes of iodine have half-lives measured in seconds, while plutonium, a waste product of nuclear power reactions, has a half-life of 24,000 years. The radioactive emissions from these atoms may be neutrons, beta particles (high-energy electrons), alpha particles (helium nuclei consisting of two protons and two neutrons), or gamma rays (very short-wavelength radiation).

In a nuclear fission reaction, an unstable isotope is bombarded with free neutron particles, causing it to split into two smaller atoms. This fission releases vast quantities of energy, which provides the power for atomic bombs and nuclear power plants. In a fusion reaction, light atoms of hydrogen or helium are slammed together at such high speeds that they fuse to become a single larger atom. This reaction also releases an enormous amount of energy that is the basis for a hydrogen bomb. Fusion and fission reactions are discussed further in chapter 21.

Atoms can join together to form **molecules.** A **compound** is a molecule containing different kinds of atoms. Water, for example, is a compound composed of two atoms of hydrogen attached to a single oxygen atom, shown by the formula H_2O. In a few cases, two atoms of the same element combine to form a molecule. Hydrogen gas (H_2), molecular oxygen (O_2), and molecular nitrogen (N_2) consist of such diatomic molecules. Most molecules are incredibly small, but some can be relatively large. The genetic information in your cells, for instance, is contained in deoxyribonucleic acid (DNA) molecules, each of which contains billions of atoms and can be seen as long linear chains in very high-power microscopes.

The forces that hold atoms together in molecules are called chemical bonds. Some bonds are very stable; that is, it takes a great deal of energy to create them or break them apart. Other bonds are weak and ephemeral. It's fortunate for us that this range exists. If all bonds were weak, we wouldn't be able to form solid structures such as bones, muscles, or nerves. If, on the other hand, all bonds were of the strongest variety, we wouldn't have any moving parts, and life would be impossible.

Organic Compounds

Organisms use some elements in abundance, others in trace amounts, and others not at all. Certain vital substances are concentrated within cells, while others are actively excluded. Carbon is a particularly important element because chains and rings of carbon atoms form the skeletons of **organic compounds,** the material of which biomolecules, and therefore living organisms, are made. A century ago, scientists thought that organic compounds had some mystical quality and were so complex that they could never be created outside of living cells. We now know this to be false. Many of the millions of known organic compounds can be synthesized fairly easily in the laboratory.

The four major categories of bioorganic compounds are lipids, carbohydrates, proteins, and nucleic acids. Most lipids belong to a family of molecules called hydrocarbons because they are composed of chains of carbon atoms, most of which have two hydrogens attached. Some common examples of lipids are fats and oils. They make up an important part of the membranes that surround cells as well as their internal organelles. Carbohydrates, as their name suggests, are composed of carbon, hydrogen, and oxygen $(CH_2O)_n$, where n represents many repeated units arranged in rings or linear chains. Some common examples of carbohydrates are sugars, starch, and cellulose.

Proteins are composed of long chains of subunits called amino acids, which are made primarily of carbon, hydrogen, and nitrogen (along with some other minor elements). Proteins make up much of both the structural and functional components of cells. Enzymes, for instance, are proteins. Nucleic acids are composed of combinations of a sugar molecule, a nitrogen-containing ring structure, and a phosphate bridge that can link many subunits (known as nucleotides) into long molecules such as DNA and RNA. Nucleic acids store the genetic information that directs the processes of life.

Cells: The Fundamental Units of Life

All living organisms are composed of **cells,** minute compartments within which the processes of life are carried out (fig. 3.4). Microscopic organisms such as bacteria, some algae, and protozoa are composed of single cells. Most higher organisms are multicellular, usually with many different cell varieties. Your body, for instance, is composed of several trillion cells of about two hundred distinct types. Every cell is surrounded by a thin but dynamic membrane of lipid and protein that receives information about the exterior world and regulates the flow of materials between the cell and its environment. Inside, cells are subdivided into tiny organelles and subcellular particles that provide the machinery for life. Some of these organelles store and release energy. Others manage and distribute information. Still others create the internal structure that gives the cell its shape and allows it to fulfill its role.

All of the chemical reactions required to create these various structures, provide them with energy and materials to carry out their functions, dispose of wastes, and perform other functions of life at the cellular level are carried out by a special class of proteins called **enzymes.** Enzymes are molecular catalysts, that is, they regulate chemical reactions without being used up or inactivated in the process. Think of them as tools: like hammers or wrenches, they do their jobs without being consumed or damaged as they work. There are generally thousands of different kinds of enzymes in every cell, all necessary to carry out the many processes on which life depends. Altogether, the multitude of enzymatic reactions performed by an organism is called its **metabolism.**

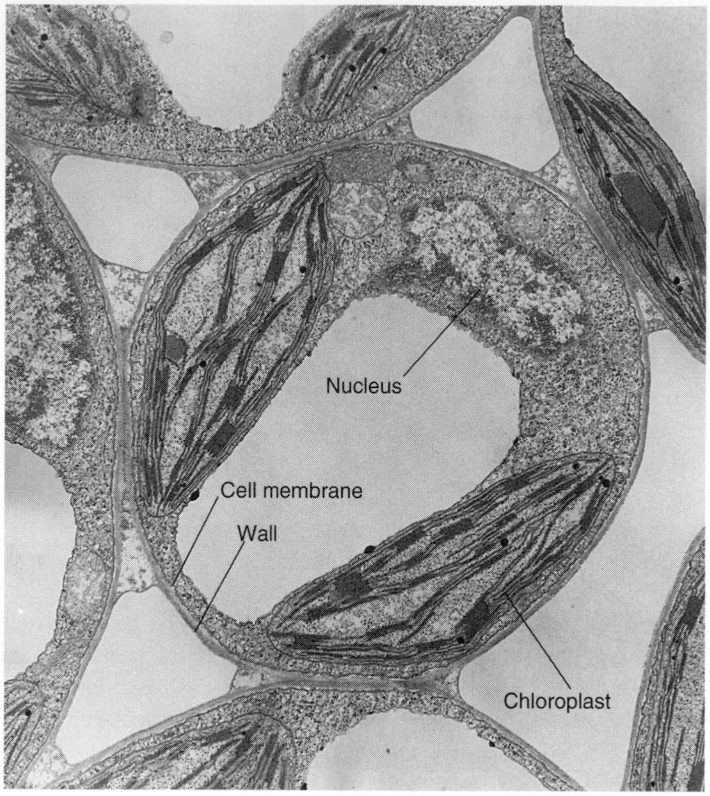

FIGURE 3.4 A single, spherical plant cell is enclosed within a heavy cellulose cell wall in the center of this high-magnification photograph. Parts of five other cells can be seen around it. The bean-shaped structures with multiple layers of internal membranes are chloroplasts. Chlorophyll inside the chloroplasts captures light energy and uses it to carry out photosynthesis, discussed later in this chapter.

ENERGY AND MATTER

Energy and matter are essential constituents of both the universe and living organisms. Matter, of course, is the material of which things are made. Energy provides the force to hold structures together, tear them apart, and move them from one place to another. In this section we will look at some fundamental characteristics of these components of our world.

Energy Types and Qualities

Energy is the ability to do work such as moving matter over a distance or causing a heat transfer between two objects at different temperatures. Energy can take many different forms. Heat, light, electricity, and chemical energy are examples that we all experience. The energy contained in moving objects is called **kinetic energy.** A rock rolling down a hill, the wind blowing through the trees, water flowing over a dam (fig. 3.5), or electrons speeding around the nucleus of an atom are all examples of kinetic energy. **Potential energy** is stored energy that is latent but available for use. A rock poised at the top of a hill and water stored

FIGURE 3.5 Water stored behind this dam represents potential energy. Water flowing over the dam has kinetic energy, some of which is converted to heat.

behind a dam are examples of potential energy. **Chemical energy** stored in the food that you eat and the gasoline that you put into your car are also examples of potential energy that can be released to do useful work. Energy is measured in calories, BTUs, or **joules** (J). One joule is the work done when one kg is accelerated 1 m per second per second ($1J = 1\ Kg \bullet m^2/s^2$). One calorie = 4.184 J.

Power is the rate of doing work. **Heat** describes the total kinetic energy of atoms or molecules in a substance not associated with bulk motion of the substance. **Temperature** is a measure of the speed of motion of a typical atom or molecule in a substance. Note that heat and temperature are not the same. A substance can have a low temperature (low average molecular speed) but a high heat content (much mass and many moving molecules or atoms). For example, a match is so hot that you can't touch it without getting burned, but it doesn't add much heat to your room. A radiator, on the other hand, is only warm to the touch but keeps your room a comfortable temperature.

Energy that is diffused, dispersed, and low in temperature is considered low-quality energy because it is difficult to gather and use for productive purposes. The heat stored in the oceans, for instance, is low-quality. Conversely, energy that is intense, concentrated, and high in temperature is high-quality energy because of its usefulness in carrying out work. The intense flames of a very hot fire or high-voltage electrical energy are examples of high-quality forms that are valuable to humans. These distinctions are important, because many of our most common energy sources are low-quality and must be concentrated or transformed into high-quality before they are useful to us.

Conservation of Matter

Under ordinary circumstances, matter is neither created nor destroyed but rather is recycled over and over again. Some of the molecules that make up your body probably contain atoms that once made up the body of a dinosaur and most certainly were part of many smaller prehistoric organisms, as chemical elements are used and reused by living organisms. Matter is transformed and combined in different ways, but it doesn't disappear; everything goes somewhere. These statements paraphrase the physical principle of **conservation of matter.**

How does this principle apply to human relationships with the biosphere? Particularly in affluent societies, we use natural resources to produce an incredible amount of "disposable" consumer goods. If everything goes somewhere, where do the things we dispose of go after the garbage truck leaves? As the sheer amount of "disposed-of stuff" increases, we are having greater problems finding places to put it. Ultimately, there is no "away" where we can throw things we don't want any more.

Thermodynamics and Energy Transfers

Organisms use gases, water, and nutrients and then return them to the environment in altered forms as by-products of their metabolic processes. Year after year, century after century, the same atoms find endless reincarnation in new molecules synthesized by succeeding organisms as they feed, grow, and die. This exchange and continuity are made possible, however, by something that cannot be recycled: energy. *Energy must be supplied from an external source to keep biological processes running.* Energy flows in a one-way path through biological systems and eventually into some low-temperature sink such as outer space. It can be used to accomplish work as it flows through the system, and it can be stored temporarily in the chemical bonds of organic molecules, but eventually it is released and dissipated.

The study of thermodynamics deals with how energy is transferred in natural processes. More specifically, it deals with the rates of flow and the transformation of energy from one form or quality to another. Thermodynamics is a complex, quantitative discipline, but you don't need a great deal of math to understand some of the broad principles that shape our world and our lives.

The **first law of thermodynamics** states that energy is *conserved;* that is, it is neither created nor destroyed under normal conditions. It may be transferred from one place or object to another, but the total amount of energy remains the same. Similarly, energy may be transformed, or changed from one form to another (for example, from the energy in a chemical bond to heat energy), but the total amount is neither diminished nor increased.

The **second law of thermodynamics** states that, with each successive energy transfer or transformation in a system, less energy is available to do work. This is not a contradiction of the first law; the energy is not lost or destroyed, merely degraded or dissipated from a higher-quality form to a lower-quality form. We can think of this process as an energy "expenditure," or the "cost" in terms of useful energy of doing work. The second law recognizes a tendency of all natural systems to go from a state of order (for example, high-quality energy) toward a state of increasing disorder (for example, low-quality energy, such as heat energy). As these systems move from order to disorder, their *entropy,* or amount of disorder, increases, reflecting the loss of energy. Stated simply, the second law of thermodynamics says there is always less useful energy available when you finish a process than there was before you started. Because of this loss, everything in the universe tends to fall apart, slow down, and get more disorganized.

How does the second law of thermodynamics apply to organisms and biological systems? Organisms are highly organized, both structurally and metabolically. Constant care and maintenance is required to keep up this organization, and a constant supply of energy is required to maintain these processes. Every time some energy is used by a cell to do work, some of that energy is dissipated or lost as heat. If cellular energy supplies are interrupted or depleted, the result—sooner or later—is death.

ENERGY FOR LIFE

Ultimately, most organisms depend on the sun for the energy needed to create structures and carry out life processes. A few rare biological communities live in or near hot springs or thermal vents in the ocean where hot, mineral-laden water provides energy-rich chemicals that form the basis for a limited and unique way of life. For most of us, however, the sun is the ultimate energy source. In this section, we will look at how green plants capture solar energy and use it to create organic molecules that are essential for life.

Solar Energy: Warmth and Light

Our sun is a star, a fiery ball of exploding hydrogen gas. Its thermonuclear reactions emit powerful forms of radiation, including potentially deadly ultraviolet and nuclear radiation (fig. 3.6), yet life here is nurtured by, and dependent upon, this searing, energetic source. Solar energy is essential to life for two main reasons.

First, the sun provides warmth. Most organisms survive within a relatively narrow temperature range. In fact, each species has its own range of temperatures within which it can function normally. At very high temperatures, biomolecules break down or become distorted and nonfunctional. At very low temperatures, the chemical reactions of metabolism occur too slowly to enable organisms to grow and reproduce. Other planets in our solar system are either too hot or too cold to support life as we know it. The earth's water and atmosphere help to moderate, maintain, and distribute the sun's heat.

Second, organisms depend on solar radiation for life-sustaining energy, which is captured by green plants, algae, and some bacteria in a process called **photosynthesis.** Photosynthesis converts radiant energy into useful, high-quality chemical energy in the bonds that hold together organic molecules.

How much potential solar energy is actually used by organisms? The amount of incoming, extraterrestrial solar radiation is enormous, about 1372 watts/m^2 at the top of the atmosphere (1 watt = 1 J per second). However, not all of this

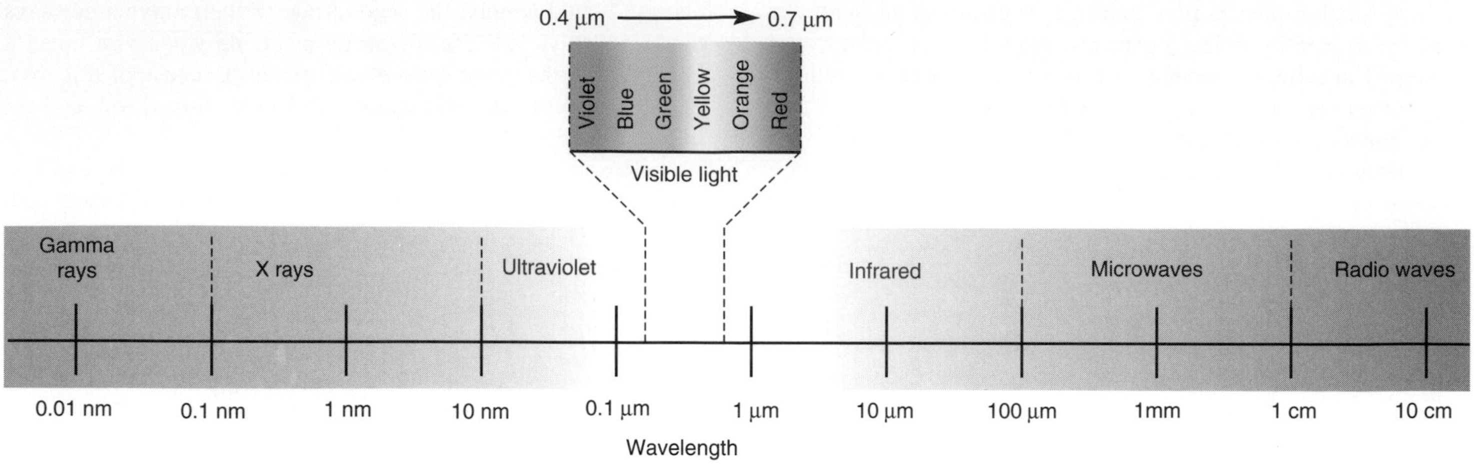

FIGURE 3.6 The electromagnetic spectrum.

radiation reaches the earth's surface. More than half of the incoming sunlight may be reflected or absorbed by atmospheric clouds, dust, and gases. In particular, harmful, short wavelengths are filtered out by gases (such as ozone) in the upper atmosphere; thus, the atmosphere is a valuable shield, protecting life-forms from harmful doses of ultraviolet and other forms of radiation. Even with these energy reductions, however, the sun provides much more energy than biological systems can harness, and more than enough for all our energy needs if technology can enable us to tap it efficiently.

Of the solar radiation that does reach the earth's surface, about 10 percent is ultraviolet, 45 percent is visible, and 45 percent is infrared. Most of that energy is absorbed by land or water or is reflected into space by water, snow, and land surfaces. (Seen from outer space, the earth shines about as brightly as Venus.)

Fortunately for us, some radiation is captured by organisms through photosynthesis. Even then, however, the amount of energy that can be used to build organic molecules is further reduced. Photosynthesis can use only certain wavelengths of solar energy that are within the visible light range of the electromagnetic spectrum. These wavelengths are in the ranges we perceive as red and blue light. Furthermore, half, or more, of the light energy absorbed by leaves is consumed as water evaporates. Consequently, only about 1 to 2 percent of the sunlight falling on plants is captured for photosynthesis. This small percentage represents the energy base for virtually all life in the biosphere!

How Does Photosynthesis Capture Energy?

Photosynthesis occurs in tiny membranous organelles called chloroplasts that reside within plant cells (see fig. 3.4). The most important key to this process is chlorophyll, a unique green molecule that can absorb light energy and use it to create high-energy chemical bonds in compounds that serve as the fuel for all subsequent cellular metabolism. Chlorophyll doesn't do this important job all alone, however. It is assisted by a large group of other lipid, sugar, protein, and nucleotide molecules. Together these components carry out two interconnected cyclic sets of reactions (fig. 3.7).

Photosynthesis begins with a series of steps called light-dependent reactions, because they occur only while light is being received by the chloroplast. During these reactions, water molecules are split, releasing molecular oxygen. This is the source of all the oxygen in the atmosphere on which all animals, including humans, depend for life. The other products of the light reactions are small, mobile, high-energy molecules that serve as the fuel or energy source for the next set of processes, the light-independent reactions. These reactions, as their name implies, can occur in the chloroplast after the light has been turned off. The enzymes in this complex use energy captured from light to add a carbon atom (from carbon dioxide) to a small sugar molecule.

In most temperate-zone plants, photosynthesis can be summarized in the following equation:

$$6H_2O + 6CO_2 + \text{solar energy} \xrightarrow[\text{chlorophyll}]{} C_6H_{12}O_6 \text{ (sugar)} + 6O_2$$

We read this equation as "water plus carbon dioxide plus energy produces sugar plus oxygen." The reason the equation uses six water and six carbon dioxide molecules is that it takes six carbon atoms to make the sugar product. If you look closely, you will see that all the atoms in the reactants balance with those in the products. This is an example of conservation of matter.

You might wonder how making a simple sugar benefits the plant. The answer is that glucose is an energy-rich compound that serves as the central, primary fuel for all metabolic processes of cells. The energy in its chemical bonds—the ones created by photosynthesis—can be released by other enzymes and used to make other molecules (lipids, proteins, nucleic acids, or other carbohydrates), or it can drive kinetic processes such as movement

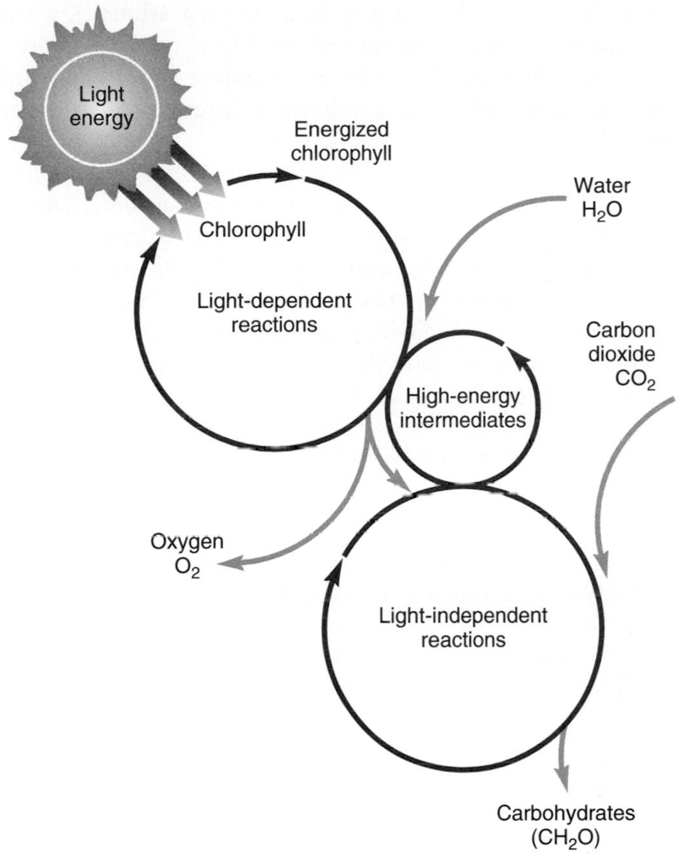

FIGURE 3.7 Photosynthesis represents the capture of light energy by chlorophyll molecules and the use of that energy to synthesize energy-rich chemical compounds that can be used by cells to drive the processes of life. Carbon dioxide and water are consumed in these reactions while oxygen and carbohydrates are produced.

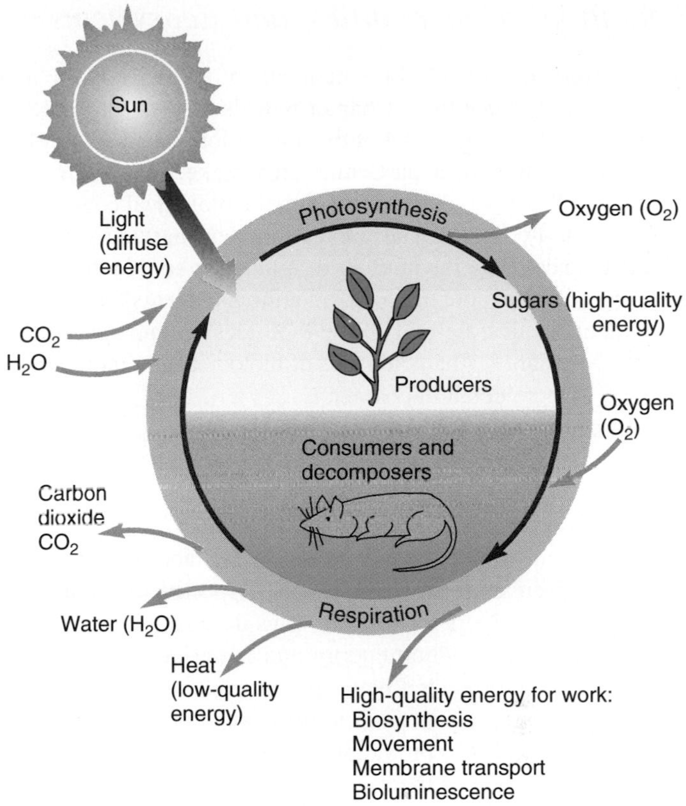

FIGURE 3.8 Energy exchange in an ecosystem. Plants take water and carbon dioxide from the environment and use the energy from sunlight to convert them into energy-rich sugars and other organic chemicals, releasing oxygen in the process. Consumers and decomposers take up oxygen and break down sugars during cellular respiration to release useful energy for living. In this process, water, carbon dioxide, and low-quality heat are released to the environment.

of ions across membranes, transmission of messages, changes in cellular shape or structure, or movement of the cell itself in some cases. This process of releasing chemical energy, called **cellular respiration,** involves splitting carbon and hydrogen atoms from the sugar molecule and recombining them with oxygen to re-create carbon dioxide and water. The net chemical reaction, then, is the reverse of photosynthesis:

$$C_6H_{12}O_6 + 6O_2 \rightarrow 6H_2O + 6CO_2 + \text{released energy}$$

Note that in photosynthesis, energy is *captured,* while in respiration, energy is *released.* Similarly, photosynthesis *consumes* water and carbon dioxide to *produce* sugar and oxygen, while respiration does just the opposite. In both sets of reactions, energy is stored temporarily in chemical bonds, which constitute a kind of energy currency for the cell.

We animals don't have chlorophyll and can't carry out photosynthetic food production. We do have the components for cellular respiration, however. In fact, this is how we get all our energy for life. We eat plants—or other animals that have eaten plants—and break down the organic molecules in our food through cellular respiration to obtain energy (fig. 3.8). In the process, we also consume oxygen and release carbon dioxide, thus completing the cycle of photosynthesis and respiration. Later in this chapter we will see how these feeding relationships work.

FROM SPECIES TO ECOSYSTEMS

While cellular and molecular biologists study life processes at the microscopic level, ecologists study interactions at the species, population, biotic community, or ecosystem level. In Latin, *species* literally means *kind.* In biology, **species** refers to all organisms of the same kind that are genetically similar enough to breed in nature and produce live, fertile offspring. There are several qualifications and some important exceptions to this definition of species (especially among bacteria and plants), but for our purposes this is a useful working definition.

Populations, Communities, and Ecosystems

A **population** consists of all the members of a species living in a given area at the same time. Chapter 6 deals further with population growth and dynamics. All of the populations of organisms living and interacting in a particular area make up a **biological community.** What populations make up the biological community of which you are a part? The population sign marking your city limits announces only the number of humans who live there, disregarding the other populations of animals, plants, fungi, and microorganisms that are part of the biological community within the city's boundaries. Characteristics of biological communities are discussed in more detail in chapter 4.

An ecological system, or **ecosystem,** is composed of a biological community and its physical environment. The environment includes abiotic factors (nonliving components), such as climate, water, minerals, and sunlight, as well as biotic factors, such as organisms, their products (secretions, wastes, and remains), and effects in a given area. Broadening our perspective from the biological community to the community plus its surroundings fosters study of the ways in which energy and materials are obtained, processed, stored, or cycled between components of the ecosystem. This systems approach tends to focus more on roles played by various members of the community rather than on the unique life stories of the individuals themselves. Looking at the organization and functions of a system can give us valuable insights into how it works.

For simplicity's sake, we think of ecosystems as fixed ecological units with distinct boundaries. If you look at a patch of woods surrounded by farm fields, for instance, a relatively sharp line separates the two areas, and conditions such as light levels, wind, moisture, and shelter are quite different in the woods than in the fields around them. Because of these variations, distinct populations of plants and animals live in each place. By studying each of these areas, we can make important and interesting discoveries about who lives where and why and about how conditions are established and maintained there.

The division between the fields and woods may not be as definite and impenetrable as we might imagine at first glance, however. Air, of course, moves freely from one to another, and the runoff after a rainfall may carry soil, leaf litter, and even live organisms between the areas. Birds may feed in the field during the day but roost in the woods at night, giving them roles in both places. Are they members of the woodland community or the field community? Is the edge of the woodland ecosystem where the last tree grows, or does it extend to every place that has an influence on the woods? Depending on how tenuous the connections you are willing to consider, the whole world might be part of an ecosystem.

As you can see, it may be difficult to draw clear boundaries around communities and ecosystems. To some extent we define these units by what we want to study and how much information we can handle. Thus, an ecosystem might be as large as a whole watershed or as small as a pond or even the surface of your skin. Even though our choices may be somewhat arbitrary, we still can make important discoveries about how organisms interact with each other and with their environment within these units. The woods are, after all, significantly different from the fields around them.

Like the woodland we just considered, most ecosystems are open, in the sense that they exchange materials and organisms with other ecosystems. A stream ecosystem is an extreme example. Water, nutrients, and organisms enter from upstream and are lost downstream. The species and numbers of organisms present may be relatively constant, but they are made up of continually changing individuals. Some ecosystems are relatively closed, in the sense that very little enters or leaves them. A balanced aquarium is a good example of a closed ecosystem. Aquatic plants, animals, and decomposers can balance material cycles in the aquarium if care is taken to balance their populations. Because of the second law of thermodynamics, however, every ecosystem must have a constant inflow of energy and a way to dispose of heat. Thus, at least with regard to energy flow, every ecosystem is open.

Many ecosystems have mechanisms that maintain composition and functions within certain limits. A forest tends to remain a forest for the most part and to have forestlike conditions if it isn't disturbed by outside forces. Some ecologists suggest that ecosystems—or perhaps all life on the earth—may function as superorganisms (What Do You Think?, p. 63).

Food Chains, Food Webs, and Trophic Levels

Photosynthesis is the base of the energy economy of all but a few special ecosystems, and ecosystem dynamics are based on how organisms share food resources. In fact, one of the major properties of an ecosystem is its **productivity,** the amount of **biomass** (biological material) produced in a given area during a given period of time. Photosynthesis is described as *primary productivity* because it is the basis for almost all other growth in an ecosystem. Manufacture of biomass by organisms that eat plants is termed *secondary productivity.* A given ecosystem may have very high total productivity but if decomposers decompose organic material as rapidly as it is formed, the *net productivity* will be low.

Think about what you have eaten today and trace it back to its photosynthetic source. If you have eaten an egg, you can trace it back to a chicken, which ate corn. This is an example of a **food chain,** a linked feeding series. Now think about a more complex food chain involving you, a chicken, a corn plant, and a grasshopper. The chicken could eat grasshoppers that had eaten leaves of the corn plant. You also could eat the grasshopper directly—some humans do. Or you could eat corn yourself, making the shortest possible food chain. Humans have several options of where we fit into food chains.

In ecosystems, some consumers feed on a single species, but most consumers have multiple food sources. Similarly, some species are prey to a single kind of predator, but many species in an ecosystem are beset by several types of predators and parasites.

what do you think? THE WORLD AROUND YOU

Chaos or Stability in Ecosystems?

The questions we ask are shaped by our conceptual frameworks or paradigms, models that serve as general explanations of how things work. You may be surprised to learn that not all ecologists agree about something as central to their science as the basic properties of biological communities and ecosystems.

One paradigm for nature is that of pioneer biogeographer F. E. Clements (1874–1945), who argued that biological communities behave as if they are superorganisms. Species and populations, in his view, work together something like the organs of the body to maintain a stable, well-balanced set of functions and composition. Clements borrowed the physiological term **homeostasis** (from the Greek *homeo*, same, and *stasis*, stationary) to describe the equilibrium achieved by nature when left undisturbed.

Clements also saw a similarity between the growth and development of an individual and the regular developmental stages of an ecosystem from bare ground or fallow field to mature forest (see chapter 4). He claimed that every landscape has a characteristic "climax" community toward which it will develop if free from external interference. The climax community, he believed, represents the maximum state of complexity and stability possible for a given set of environmental conditions.

Many people are attracted by this purposeful, deterministic view of nature. Others, however, believe that assuming intentional design or plan in nature is unscientific. H. A. Gleason, for instance, who was a contemporary of Clements, regarded biological communities as merely chance associations of species able to migrate into and live in a specific place at a particular time. The presence or absence of any one species, Gleason argued, is independent of all others. He suggested that we see ecosystems as stable and uniform only because our lifetimes are so short and our view so limited.

Clements' views were very influential in the early days of ecology, and for many years homeostasis generated by diversity was regarded as the most important principle in nature. More recently, however, the individualistic, probabilistic views of Gleason have attracted more supporters. Nevertheless, both of these viewpoints continue to have proponents.

Recent research on cybernetics and planetary systems has awakened a new interest in the self maintaining, equilibrating features of biological systems. The Gaia hypothesis, named by biophysicist James Lovelock after the ancient Greek goddess of the earth, originated in speculations about the unique environmental conditions of the earth. In contrast to our neighboring planets, which are either much too hot or much too cold for life as we know it, most of the earth is just right for our existence (some people call Earth the Goldilocks planet).

This fortunate set of circumstances is more than a happy accident, according to the Gaia hypothesis. Rather it is the result of active intervention by the living biota to create and sustain a livable environment. In this view, the whole earth operates as if it were a single superorganism. This homeostasis depends on active feedback processes operated automatically and unconsciously by the biota. Although Lovelock dissociates himself from any mystical or religious implications of his theory, many people believe it suggests design and meaning in the world.

The Gleasonian view of nature as individualistic and unpredictable also has modern supporters. Paleobiogeography (the study of distribution of plants and animals through history based on analysis of fossils and sediments) shows a much greater variability in landscapes than Clements thought possible. What appears to be a stable forest or prairie community may have had dramatically different species composition and ecosystem characteristics over the millennia. From this perspective, patchiness, variability, and randomness of species distribution in communities are more important and characteristic than homogeneity. Constant change seems to be the rule rather than the exception for much of nature.

What triggers these changes? Random, often catastrophic events appear to be major forces in shaping our world. Erratic climate change, irregular volcanic eruptions, giant asteroids crashing into the earth, and the slow but inexorable processes of continental drift, mountain building, and erosion modify landscapes in complex ways. The science of chaos and catastrophe allows us to find patterns—or at least limits—in what had hitherto seemed merely noise. An important revelation from this view is that complex systems can exhibit emergent properties, characteristics that could not have been predicted from a study of individual components.

Both of these contrasting views of nature have interesting implications. From the Gaia hypothesis we might infer that even though nature has self-correcting mechanisms, there may be thresholds of disruption beyond which it may not be able to recover. If nature has a plan for survival, it may not include us. From chaos and catastrophe theory, we realize that there are great uncertainties in nature. If we can't predict what may happen next, we might want to leave a margin of safety in how much we disturb nature.

Perhaps even more useful is the illustration of the importance of conceptual frameworks in shaping our views. If you look for examples of stability and order, you will find them. At the same time, if you look for change, you will find many examples of that as well. Probably neither of these paradigms is exclusively right, but both give useful insights. Which is closest to your worldview? Do you see the world in terms of harmony and stability? Or do you see more evidence of randomness and constant change? What scientific questions and policy implications would you tend to propose if you believed in one paradigm rather than the other?

In this way, individual food chains become interconnected to form a **food web.** Figure 3.9 shows feeding relationships among some of the larger organisms in a woodland and lake community. If we were to add all the insects, worms, and microscopic organisms that belong in this picture, however, we would have overwhelming complexity. Perhaps you can imagine the challenge ecologists face in trying to quantify and interpret the precise matter and energy transfers that occur in a natural ecosystem!

FIGURE 3.9 Each time an organism feeds, it becomes a link in a food chain. In an ecosystem, food chains become cross-connected when predators feed on more than one kind of prey, thus forming a food web. How many food chains make up the food web shown here? The arrows in this diagram and in figure 3.10 indicate the direction in which matter and energy are transferred through feeding relationships.

An organism's feeding status in an ecosystem can be expressed as its **trophic level** (from the Greek *trophe,* food). In our first example, the corn plant is at the **producer** level; it transforms solar energy into chemical energy, producing food molecules. Other organisms in the ecosystem are **consumers** of the chemical energy harnessed by the producers. An organism that eats producers is a primary consumer. An organism that eats primary consumers is a secondary consumer, which may, in turn, be eaten by a tertiary consumer, and so on. Most terrestrial food chains are relatively short (seeds → mouse → owl), but aquatic food chains may be quite long (microscopic algae → copepod → minnow → crayfish → bass → osprey). The length of a food chain also may reflect the physical characteristics of a particular ecosystem. A harsh arctic landscape has a much shorter food chain than a temperate or tropical one (fig. 3.10).

Organisms can be identified both by the trophic level at which they feed and by the *kinds* of food they eat (fig. 3.11). **Herbivores** are plant eaters, **carnivores** are flesh eaters, and **omnivores** eat both plant and animal matter. What are humans? We are natural omnivores, by history and by habit. Tooth structure is an important clue to understanding animal food preferences, and humans are no exception. Our teeth are suited for an omnivorous diet, with a combination of cutting and crushing surfaces that are not highly adapted for one specific kind of food, as are the teeth of a wolf (carnivore) or a horse (herbivore).

One of the most important trophic levels is occupied by the many kinds of organisms that remove and recycle the dead bodies and waste products of others. **Scavengers** such as crows, jackals, and vultures clean up dead carcasses of larger animals. **Detritivores** such as ants and beetles consume litter, debris, and dung, while **decomposer** organisms such as fungi and bacteria complete the final breakdown and recycling of organic materials. It could be argued that these microorganisms are second in importance only to producers because without their activity nutrients would remain locked up in the organic compounds of dead organisms and discarded body wastes, rather than being made available to successive generations of organisms.

Ecological Pyramids

If we arrange the organisms in a food chain according to trophic levels, they often form a pyramid with a broad base representing primary producers and only a few individuals in the highest trophic levels. This pyramid arrangement is especially true if we look at the energy content of an ecosystem (fig. 3.12). True to the

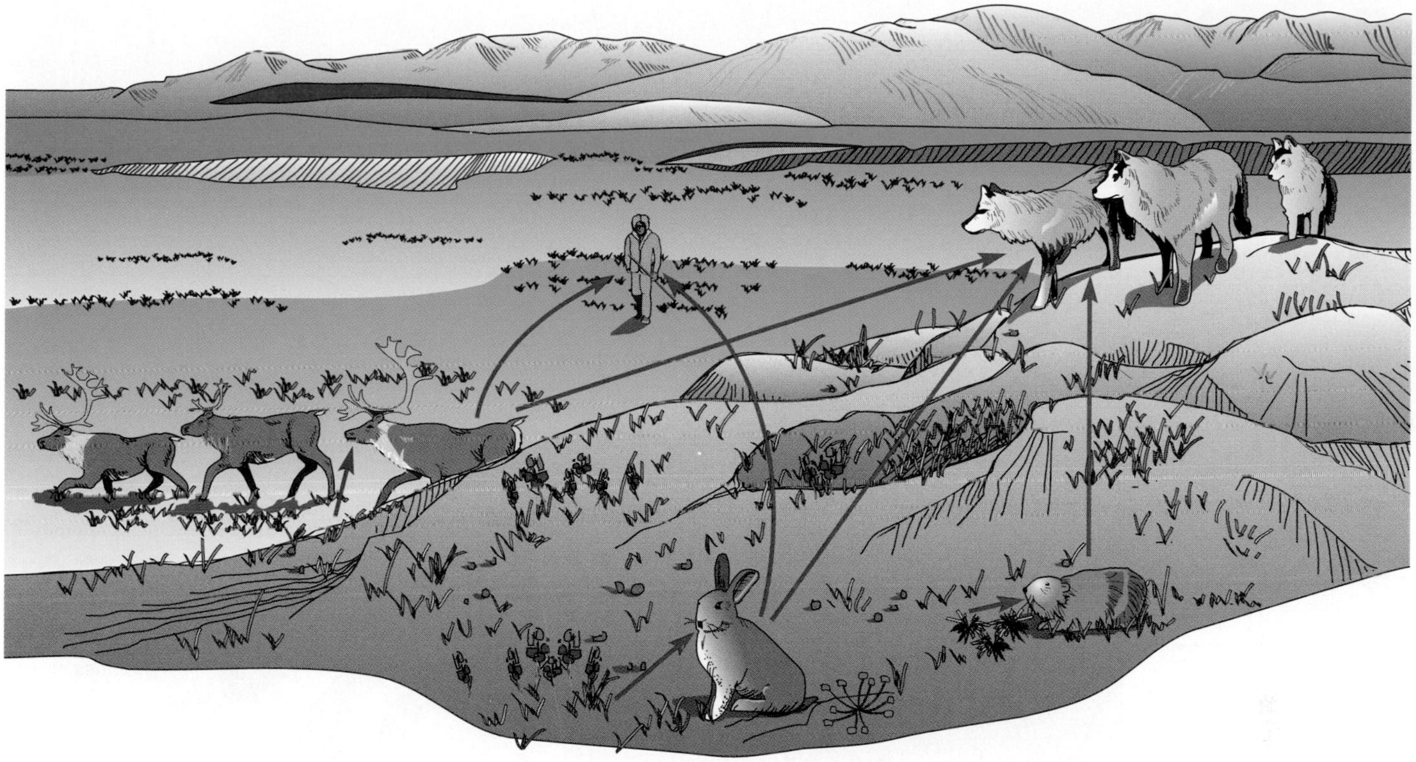

FIGURE 3.10 Harsh environments tend to have shorter food chains than environments with more favorable physical conditions. Compare the arctic food chains depicted here with the longer food chains in the food web in figure 3.9.

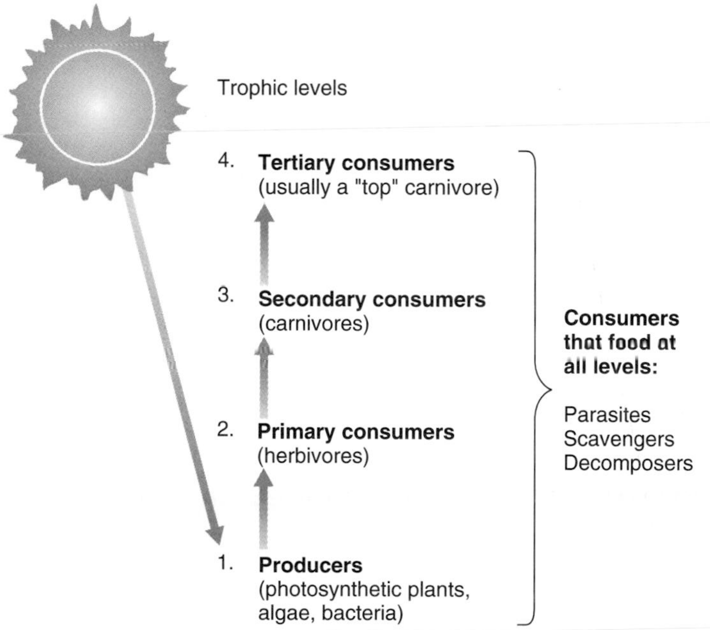

FIGURE 3.11 Organisms in an ecosystem may be identified by how they obtain food for their life processes (producer, herbivore, carnivore, omnivore, scavenger, decomposer, reducer) or by consumer level (producer; primary, secondary, or tertiary consumer) or by trophic level (1st, 2nd, 3rd, 4th).

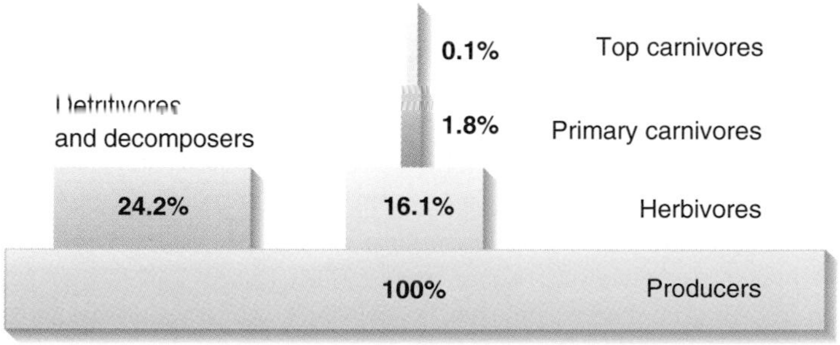

FIGURE 3.12 A classic example of an energy pyramid from Silver Springs, Florida. The numbers in each bar show the percentage of the energy captured in the primary producer level that is incorporated in the biomass of each succeeding level. Detritivores and decomposers feed at every level but are shown attached to the producer bar because this level provides most of their energy.
Source: Data from Howard T. Odum, "Trophic Structure and Productivity of Silver Springs, Florida" in *Ecological Monographs,* 27:55–112, 1957, Ecological Society of America.

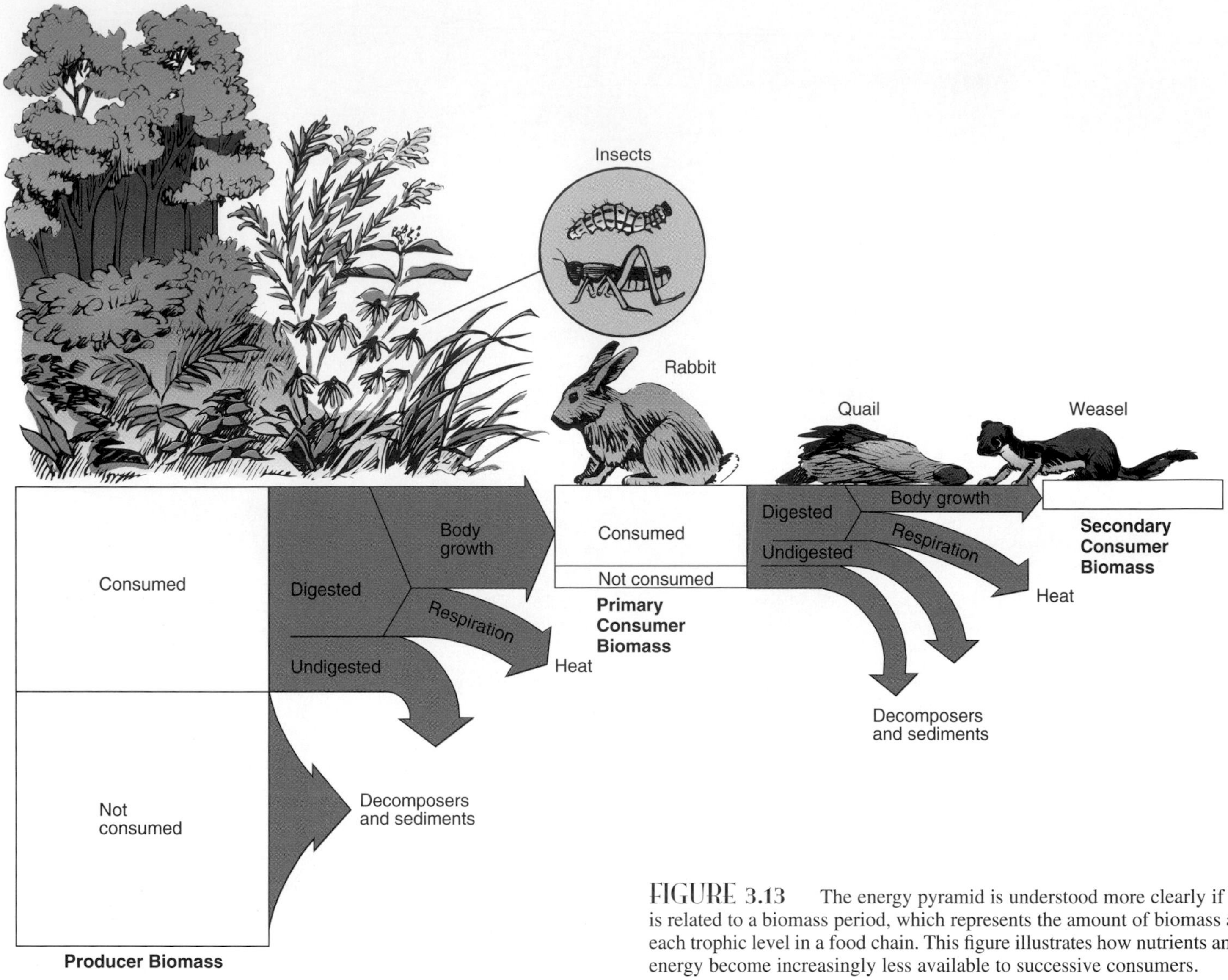

FIGURE 3.13 The energy pyramid is understood more clearly if it is related to a biomass period, which represents the amount of biomass at each trophic level in a food chain. This figure illustrates how nutrients and energy become increasingly less available to successive consumers.

second principle of thermodynamics, less food energy is available to the top trophic level than is available to preceding levels. For example, it takes a huge number of plants to support a modest colony of grazers such as prairie dogs. Several colonies of prairie dogs, in turn, might be required to feed a single coyote.

Why is there so much less energy in each successive level in this figure? In the first place, some of the food that organisms eat is undigested and doesn't provide usable energy. Much of the energy that is absorbed is used in the daily processes of living or lost as heat when it is transformed from one form to another and thus isn't stored as biomass that can be eaten.

Furthermore, predators don't operate at 100 percent efficiency. If there were enough foxes to catch all the rabbits available in the summer when the supply is abundant, there would be too many foxes in the middle of the winter when rabbits are scarce. A general rule of thumb is that only about 10 percent of the energy in one consumer level is represented in the next higher level (fig. 3.13). The amount of energy available is often expressed in biomass. For example, it generally takes about 100 kg of clover to make 10 kg of rabbit and 10 kg of rabbit to make 1 kg of fox.

The total number of organisms and the total amount of biomass in each successive trophic level of an ecosystem also may form pyramids (fig. 3.14) similar to those describing energy content. The relationship between biomass and numbers is not as dependable as energy, however. The biomass pyramid, for instance, can be inverted by periodic fluctuations in producer populations (for example, low plant and algal biomass present during winter in temperate aquatic ecosystems). The numbers pyramid also can be inverted. One coyote can support numerous tapeworms for example. Numbers inversion also occurs at the lower trophic levels (for example, one large tree can support thousands of caterpillars).

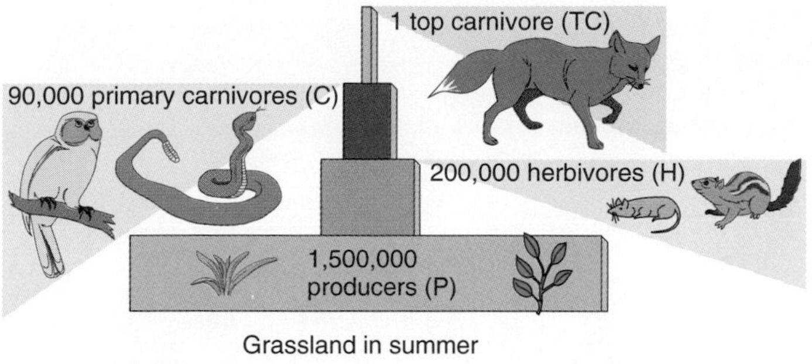

FIGURE 3.14 Usually, smaller organisms are eaten by larger organisms and it takes numerous small organisms to feed one large organism. The classic study represented in this pyramid shows numbers of individuals at each trophic level per 1000 m^2 of grassland, and reads like this: to support one individual at the top carnivore level, there were 90,000 primary carnivores feeding upon 200,000 herbivores that in turn fed upon 1,500,000 producers.

MATERIAL CYCLES AND LIFE PROCESSES

To our knowledge, the earth is the only planet in our solar system that provides a suitable environment for life as we know it. Even our nearest planetary neighbors, Mars and Venus, do not meet these requirements. Maintenance of these conditions requires a constant recycling of materials between the biotic (living) and abiotic (nonliving) components of ecosystems. The unique properties of water make it indispensable for life (In Depth, p. 70) and give water a central place in the cycling of many elements through ecosystems. Water itself also cycles in a planetwide process that shapes the land, regulates climate, and provides a constant supply of pure, fresh water to terrestrial life (see chapter 19).

The Carbon Cycle

Carbon serves a dual purpose for organisms: (1) it is a structural component of organic molecules, and (2) the energy-holding chemical bonds it forms represent energy "storage." The **carbon cycle** begins with the intake of carbon dioxide (CO_2) by photosynthetic organisms (fig. 3.15). Carbon (and hydrogen and oxygen) atoms are incorporated into sugar molecules during photosynthesis. Carbon dioxide is eventually released during respiration, closing the cycle.

The path followed by an individual carbon atom in this cycle may be quite direct and rapid, depending on how it is used in an organism's body. Imagine for a moment what happens to a simple sugar molecule you swallow in a glass of fruit juice. The sugar molecule is absorbed into your bloodstream where it is made available to your cells for cellular respiration or for making more complex biomolecules. If it is used in respiration, you may exhale the same carbon atom as CO_2 the same day.

Alternatively, that sugar molecule can be used to make larger organic molecules that become part of your cellular structure. The carbon atoms in it could remain a part of your body until it decays after death. Similarly, carbon in the wood of a thousand-year-old tree will be released only when the wood is digested by fungi and bacteria that release carbon dioxide as a by-product of their respiration.

Can you think of examples where carbon may not be recycled for even longer periods of time, if ever? Coal and oil are the compressed, chemically altered remains of plants or microorganisms that lived millions of years ago. Their carbon atoms (and hydrogen, oxygen, nitrogen, sulfur, etc.) are not released until the coal and oil are burned. Enormous amounts of carbon also are locked up as calcium carbonate ($CaCO_3$), used to build shells and skeletons of marine organisms from tiny protozoans to corals. Most of these deposits are at the bottom of the oceans. The world's extensive surface limestone deposits are biologically formed calcium carbonate from ancient oceans, exposed by geological events. The carbon in limestone has been locked away for millennia, which is probably the fate of carbon currently being deposited in ocean sediments. Eventually, even the deep ocean deposits are recycled as they are drawn into deep molten layers and released via volcanic activity. Geologists estimate that every carbon atom on the earth has made about thirty such round trips over the last 4 billion years.

How does tying up so much carbon in the bodies and byproducts of organisms affect the biosphere? Favorably. It helps balance CO_2 generation and utilization. Carbon dioxide is one of the so-called greenhouse gases because it blocks radiation of heat from the earth's surface, retaining it instead in the atmosphere. This phenomenon is discussed in more detail in chapter 17. Photosynthesis and deposition of $CaCO_3$ remove atmospheric carbon dioxide; therefore, vegetation (especially large forested areas such as the tropical rainforests) and the oceans are very important **carbon sinks** (storage deposits). Cellular respiration and combustion both release CO_2, so they are referred to as carbon sources of the cycle.

Presently, natural fires and human-created combustion of organic fuels (mainly wood, coal, and petroleum products) release huge quantities of CO_2 at rates that seem to be surpassing the pace of CO_2 removal. Scientific concerns over the linked problems of increased atmospheric CO_2 concentrations, massive deforestation, and reduced productivity of the oceans due to pollution are discussed in later chapters.

The Nitrogen Cycle

Organisms cannot exist without amino acids, peptides, and proteins, all of which are organic molecules containing nitrogen. The nitrogen atoms that form these important molecules are provided by producer organisms. Plants assimilate (take up) inorganic nitrogen from the environment and use it to build their own protein molecules, which are eaten by consumer organisms, digested, and used to build their bodies. Even though nitrogen is

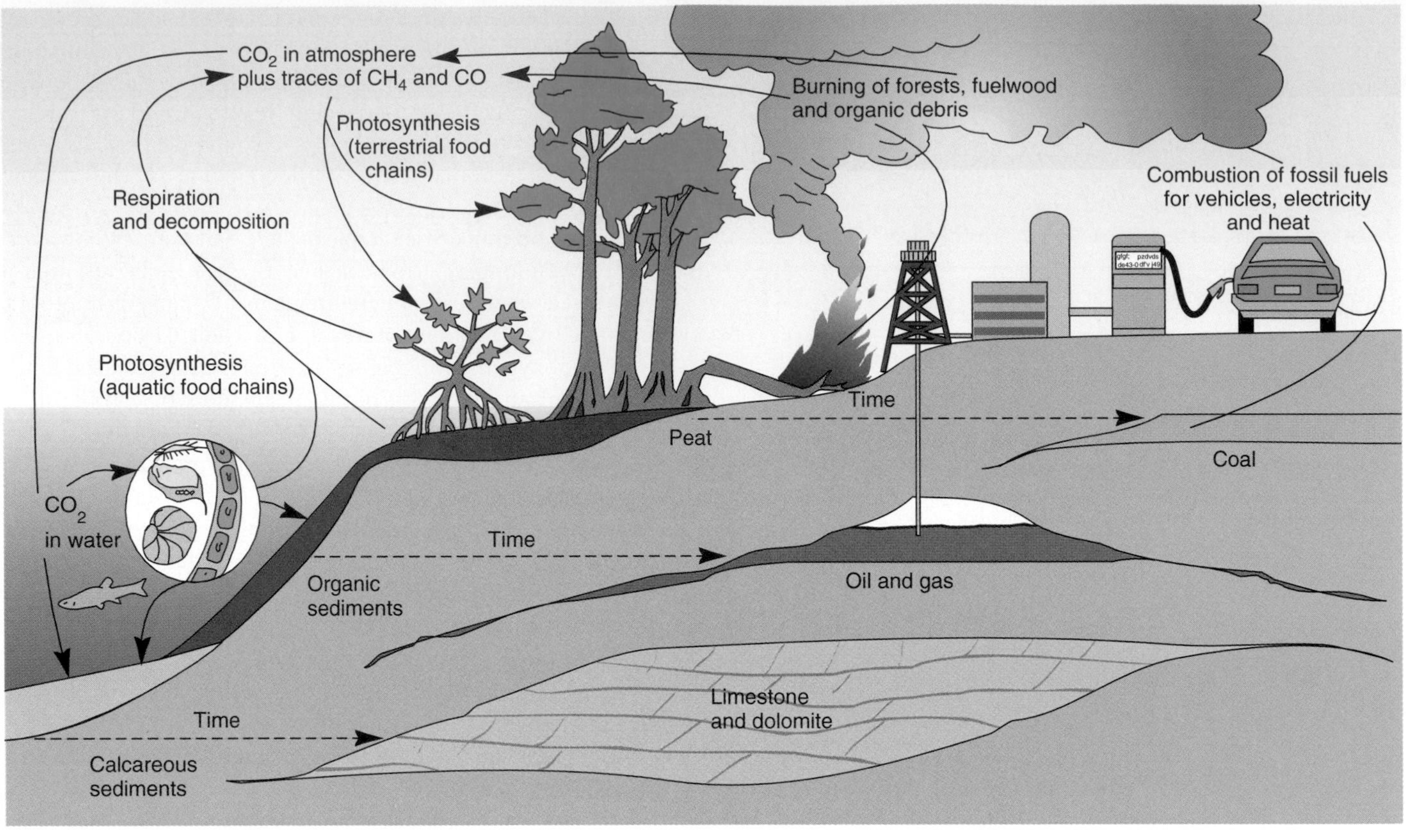

FIGURE 3.15 Atmospheric carbon dioxide is the "source" of carbon in the carbon cycle. It passes into ecosystems through photosynthesis and is captured in the bodies and products of living organisms. It is released to the atmosphere by weathering, respiration, and combustion. Carbon may be locked up for long periods in both organic (coal, oil, gas) and inorganic (limestone, dolomite) geological formations, which are, therefore, referred to as carbon "sinks."

the most abundant gas (about 78 percent of the atmosphere), however, plants cannot use N_2, the stable diatomic (2-atom) molecule in the air.

Where and how, then, do green plants get *their* nitrogen? The answer lies in the most complex of the gaseous cycles, the **nitrogen cycle.** Figure 3.16 summarizes the nitrogen cycle. The key natural processes that make nitrogen available are carried out by nitrogen-fixing bacteria (including some blue-green algae or cyanobacteria). These organisms have a highly specialized ability to "fix" nitrogen, meaning they change it to less mobile, more useful forms by combining it with hydrogen to make ammonia (NH_3).

Nitrite-forming bacteria combine the ammonia with oxygen, forming nitrites, which have the ionic form NO_2^-. Another group of bacteria then convert nitrites to nitrates, which have the ionic form NO_3^-, that can be absorbed and used by green plants. After nitrates have been absorbed into plant cells, they are reduced to ammonium (NH_4^+) which is used to build amino acids that become the building blocks for peptides and proteins.

Members of the bean family (legumes) and a few other kinds of plants are especially useful in agriculture because they have nitrogen-fixing bacteria actually living *in* their root tissues (fig. 3.17). Legumes and their associated bacteria enrich the soil, so interplanting and rotating legumes with crops such as corn that use but cannot replace soil nitrates are beneficial farming practices that take practical advantage of this relationship.

Nitrogen reenters the environment in several ways. The most obvious path is through the death of organisms. Their bodies are decomposed by fungi and bacteria, releasing ammonia and ammonium ions, which then are available for nitrate formation. Organisms don't have to die to donate proteins to the environment, however. Plants shed their leaves, needles, flowers, fruits, and cones; animals shed hair, feathers, skin, exoskeletons, pupal cases, and silk. Animals also produce excrement and urinary wastes that contain nitrogenous compounds. Urinary wastes are especially high in nitrogen because they contain the detoxified wastes of protein metabolism. All of these by-products of living organisms decompose, replenishing soil fertility.

How does nitrogen reenter the atmosphere, completing the cycle? Denitrifying bacteria break down nitrates into N_2 and nitrous oxide (N_2O), gases that return to the atmosphere; thus, denitrifying bacteria compete with plant roots for available nitrates. However, denitrification occurs mainly in waterlogged soils that have low oxygen availability and a high amount of decomposable organic matter. These are suitable growing

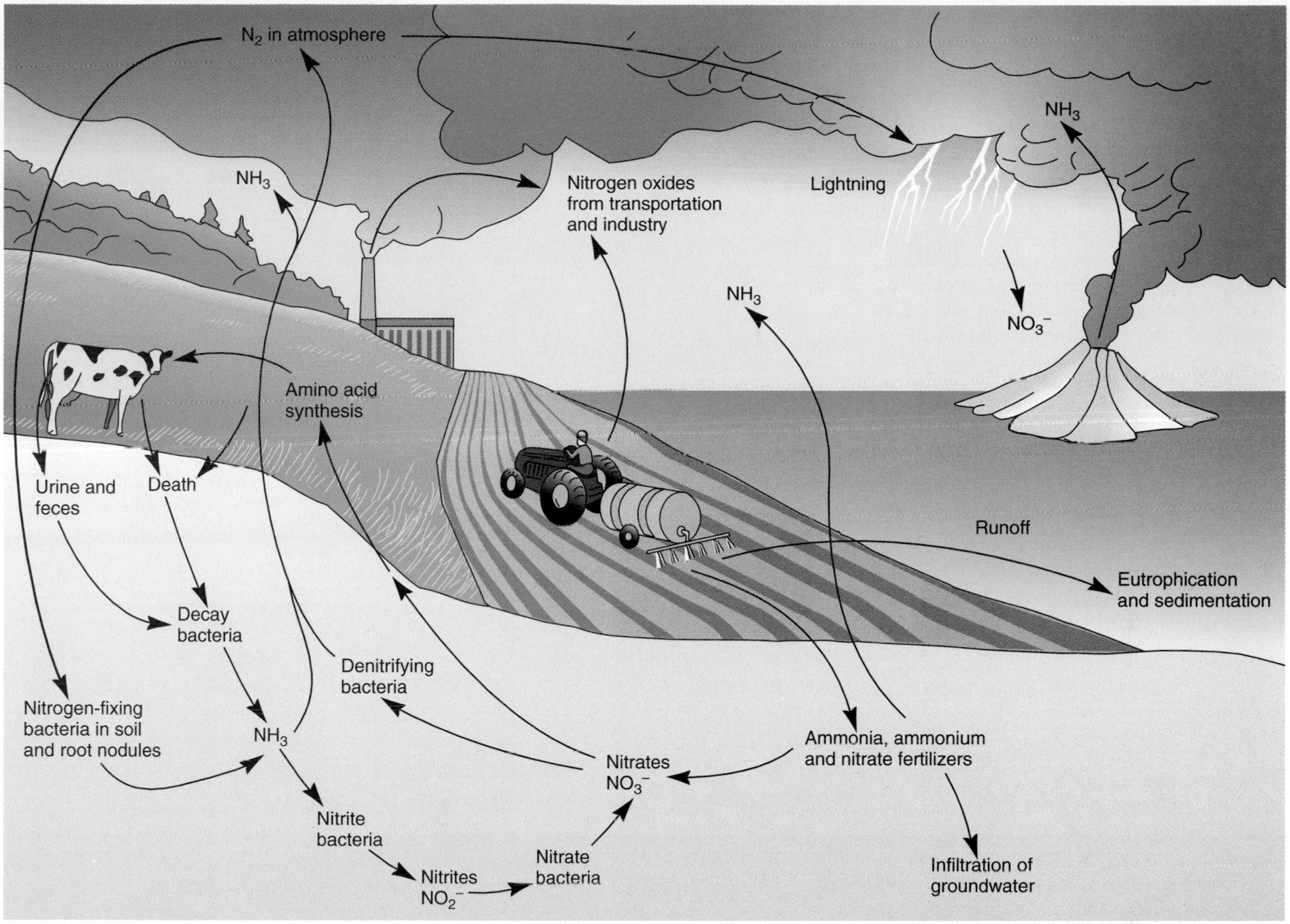

FIGURE 3.16 Nitrogen is incorporated into ecosystems when plants and bacteria use it to build their own amino acids and is released from ecosystems by bacterial decomposition. Both natural and human interactions with the nitrogen cycle are depicted here.

FIGURE 3.17 The roots of this adzuki bean plant are covered with bumps called nodules. Each nodule is a mass of root tissue containing many bacteria that help to convert nitrogen in the soil to a form the bean plants can assimilate and use to manufacture amino acids.

IN DEPTH. . . . .

A "Water Planet"

If travelers from other solar systems were to visit our lovely, cool, blue planet, they might call it Aqua rather than Terra because of its outstanding feature: the abundance of streams, rivers, lakes, and oceans of liquid water. The earth is the only place that we know of where water exists in liquid form in any appreciable quantity. Liquid water covers nearly three-fourths of the earth's surface, and during the winter, snow and ice cover a good deal of the rest. Not only is water essential for cell structure and metabolism, but water's unique physical and chemical properties directly impact the earth's surface temperatures, its atmosphere, and the interactions of life-forms with their environments. Water has many unique, almost magical qualities. Without the wonderful properties of water, life on earth would not be possible. Let's look in more detail at some of these properties and how they affect life.

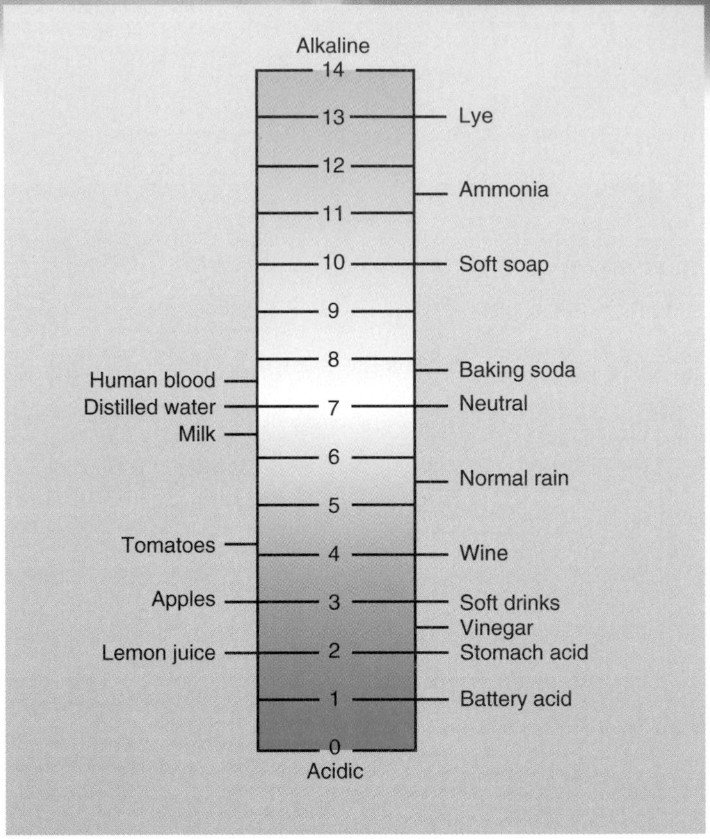

The pH scale. The numbers represent the negative logarithm of the hydrogen ion concentration in water.

1. Water is the primary component of cells and makes up 60 to 70 percent (on average) of the weight of living organisms. It fills cells, thereby giving form and support to many tissues. More than just being "filler," however, it has vital biological roles. Water is the medium in which all of life's chemical reactions occur, and it is an active participant in many of these reactions.
2. Water is the only inorganic liquid that exists under normal conditions at temperatures suitable for life, and it is the solvent in which most substances must be dissolved before cells can absorb, use, or eliminate them. These substances include food molecules, mineral nutrients, gases, hormones and other chemical communicators, and waste by-products of metabolism. When compounds dissolve in water, they often have a tendency to break into positively and negatively charged ions. The resulting electrical attractions and repulsions between ions are an important part of cellular chemistry.
3. Water molecules themselves can ionize, breaking into H^+ (hydrogen ions) and OH^- (hydroxyl ions). These ions help maintain the balance between acids and bases in cells, helping to offset (buffer) fluctuations caused by the release of other ions during

conditions for many wild plant species in swamps and marshes, but not for most cultivated crop species, except for rice, a domesticated wetlands grass.

In recent years, humans have profoundly altered the nitrogen cycle. By using synthetic fertilizers, cultivating nitrogen-fixing crops, and burning fossil fuels, we now convert more nitrogen to ammonia and nitrates than all natural land processes combined. However, this excess nitrogen input is causing serious loss of soil nutrients such as calcium and potassium, acidification of rivers and lakes, and rising atmospheric concentrations of the greenhouse gas, nitrous oxide. It also encourages the spread of weeds into areas such as prairies occupied by native plants adapted to nitrogen-poor environments. Blooms of toxic algae and dinoflagellates in coastal areas may also be linked to excess nitrogen carried by rivers.

IN DEPTH

(continued)

metabolism. The term **pH** refers to the relative abundance of H^+ ions in a solution. On a pH scale from 0 to 14, 7 is neutral; values lower than 7 are acidic, and those higher than 7 are basic. A proper pH balance is critical to healthy cellular functioning.

4. Water molecules are cohesive, tending to stick together tenaciously. You have experienced this property if you have ever done a belly flop off a diving board. Because of its cohesiveness, water has the highest surface tension of any common, natural liquid. These forces hold the water molecules at the surface together so strongly that they form a layer strong enough to support small insects. The same cohesion causes capillary action, which is the tendency of water to be drawn into small channels. Without capillary action, movement of water and nutrients into groundwater reservoirs and through living organisms might not be possible.
5. Because water has a high heat capacity, it helps protect us from temperature fluctuations. It can absorb or release large amounts of heat energy before its own temperature changes. For this reason, large bodies of water such as the oceans or the Great Lakes have a moderating effect on their local climates. Without the presence of liquid oceans, the surface temperature of the earth would undergo wide fluctuations between day and night as do the moon and Mars.
6. Water exists as a liquid over a wide temperature range that, for most of the world (at least during summer months), corresponds to the ambient temperature range. For most substances, the freezing point is only a few degrees lower than the boiling point. This means that they exist as either a solid or a gas, but only briefly as a liquid. Organisms synthesize organic compounds such as oils and alcohols that remain liquid at ambient temperatures because they are so valuable to life, but the original and predominant liquid in nature is water.
7. Water is unique in that it expands when it crystallizes. Most substances shrink as they change from liquid to solid. Ice floats because it is less dense than liquid water. When temperatures fall below freezing, the surface layers of lakes, rivers, and oceans cool faster and freeze before deeper water. Floating ice then insulates underlying layers, keeping most water bodies liquid throughout the winter in most places. Without this feature, lakes, rivers, and even oceans in high latitudes would freeze solid and never melt.
8. Water has a high heat of vaporization. Because of the amount of heat it absorbs in changing from a liquid to a vapor state, evaporating water is an effective way for organisms to shed excess heat. Many animals pant or sweat to moisten evaporative cooling surfaces. Why do you feel less comfortable on a hot, humid day than on a hot, dry day? Because the water vapor-laden air inhibits the rate of evaporation from your skin, thereby impairing your ability to shed heat. The heat absorbed when water vaporizes is released when condensation occurs. This accounts for a large transfer of heat from the oceans over the continents as water vapor turns into rain.

Altogether, these unique properties of water not only shape life at the molecular and cellular level, they also determine many of the features of both the biotic and abiotic components of our world.

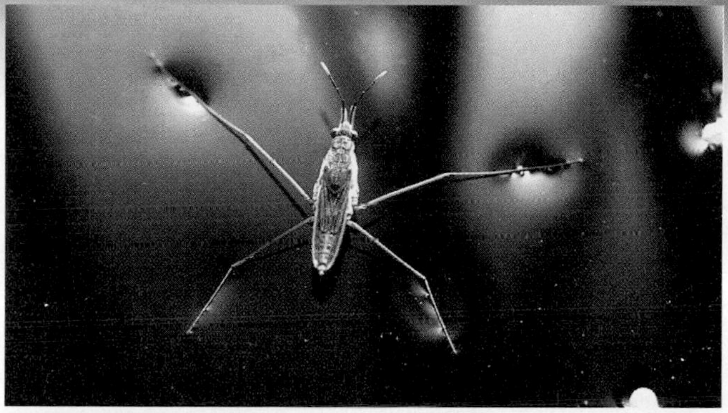

Surface tension is demonstrated by the resistance of a water surface to penetration, as when it is walked upon by a water strider.

The Phosphorus Cycle

Minerals become available to organisms after they are released from rocks. Two mineral cycles of particular significance to organisms are phosphorus and sulfur. Why do you suppose phosphorus is a primary ingredient in fertilizers? At the cellular level, energy-rich, phosphorus-containing compounds are primary participants in energy-transfer reactions, as we have discussed. The amount of available phosphorus in an environment can, therefore, have a dramatic effect on productivity. Abundant phosphorus stimulates lush plant and algal growth, making it a major contributor to water pollution.

The **phosphorus cycle** (fig. 3.18) begins when phosphorus compounds are leached from rocks and minerals over long periods

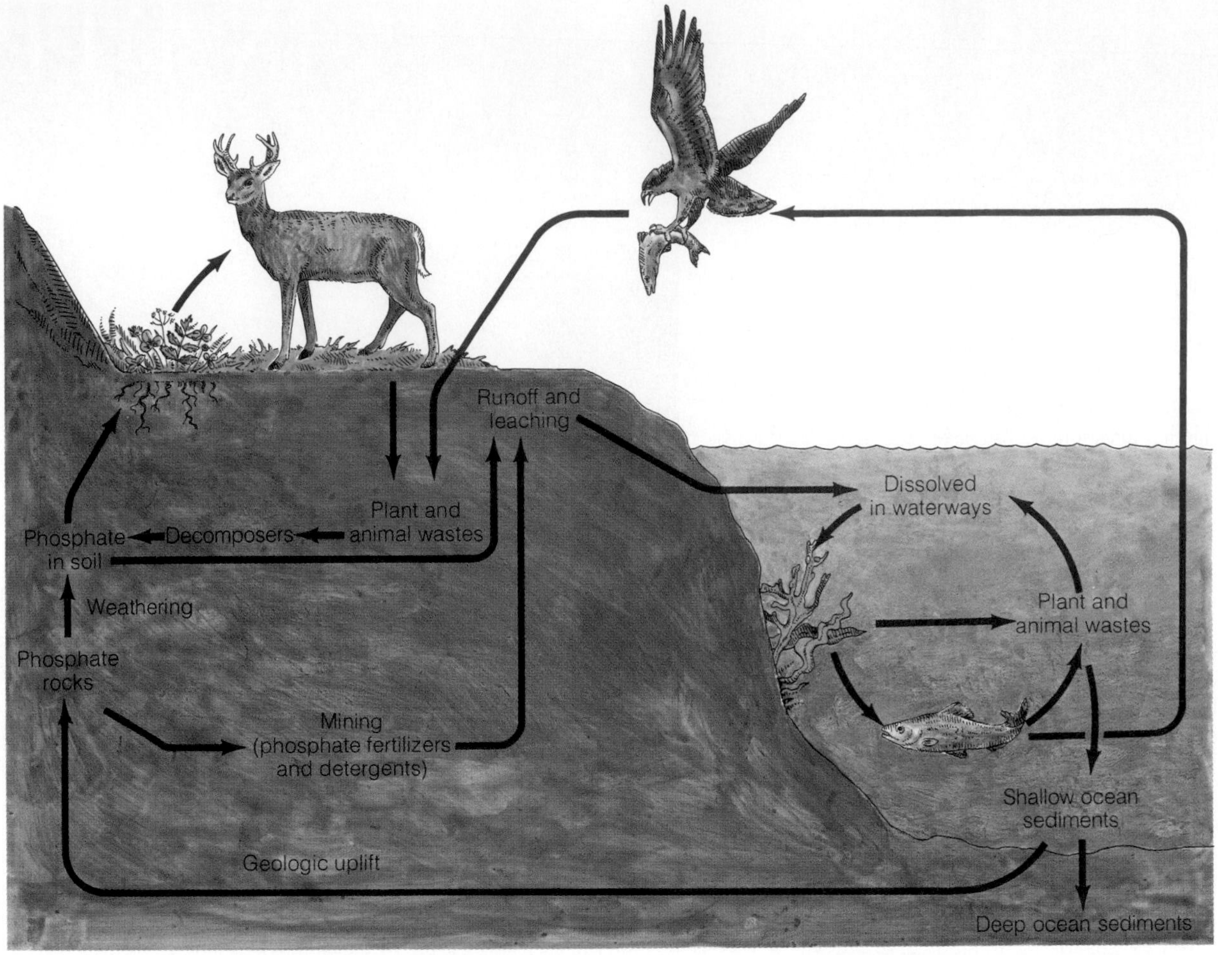

FIGURE 3.18 The phosphorus cycle carries this essential mineral from its source in rocks and soil, through plants and animals, and back to sediments as wastes and carrion.

of time. Because phosphorous has no atmospheric form, it is usually transported in aqueous form. Inorganic phosphorus is taken in by producer organisms, incorporated into organic molecules, and then passed on to consumers. It is returned to the environment by decomposition. An important aspect of the phosphorus cycle is the very long time it takes for phosphorus atoms to pass through it. Deep sediments of the oceans are significant phosphorus sinks of extreme longevity. Phosphate ores that now are mined to make detergents and inorganic fertilizers represent exposed ocean sediments that are millennia old. You could think of our present use of phosphates, which are washed out into the river systems and eventually the oceans, as an accelerated mobilization of phosphorus from source to sink. Aquatic ecosystems often are dramatically affected in the process because excess phosphates can stimulate explosive growth of algae and photosynthetic bacteria populations, upsetting ecosystem stability. Notice also that in this cycle, as in the others, the role of organisms is only one part of a larger picture.

The Sulfur Cycle

Sulfur plays a vital role in organisms, especially as a minor but essential component of proteins. Sulfur compounds are important determinants of the acidity of rainfall, surface water, and soil. In addition, sulfur in particles and tiny air-borne droplets may act as critical regulators of global climate. Most of the earth's sulfur is tied up underground in rocks and minerals such as iron disulfide (pyrite) or calcium sulfate (gypsum). This inorganic sulfur is released into air and water by weathering, emissions from deep seafloor vents, and by volcanic eruptions (fig. 3.19).

The **sulfur cycle** is complicated by the large number of oxidation states the element can assume, including hydrogen sulfide (H_2S), sulfur dioxide (SO_2), sulfate ion (SO_4^{-2}), and sulfur, among others. Inorganic processes are responsible for many of these transformations, but living organisms, especially bacteria, also sequester sulfur in biogenic deposits or release it into the environment. Which of the several kinds of sulfur bacteria prevail in

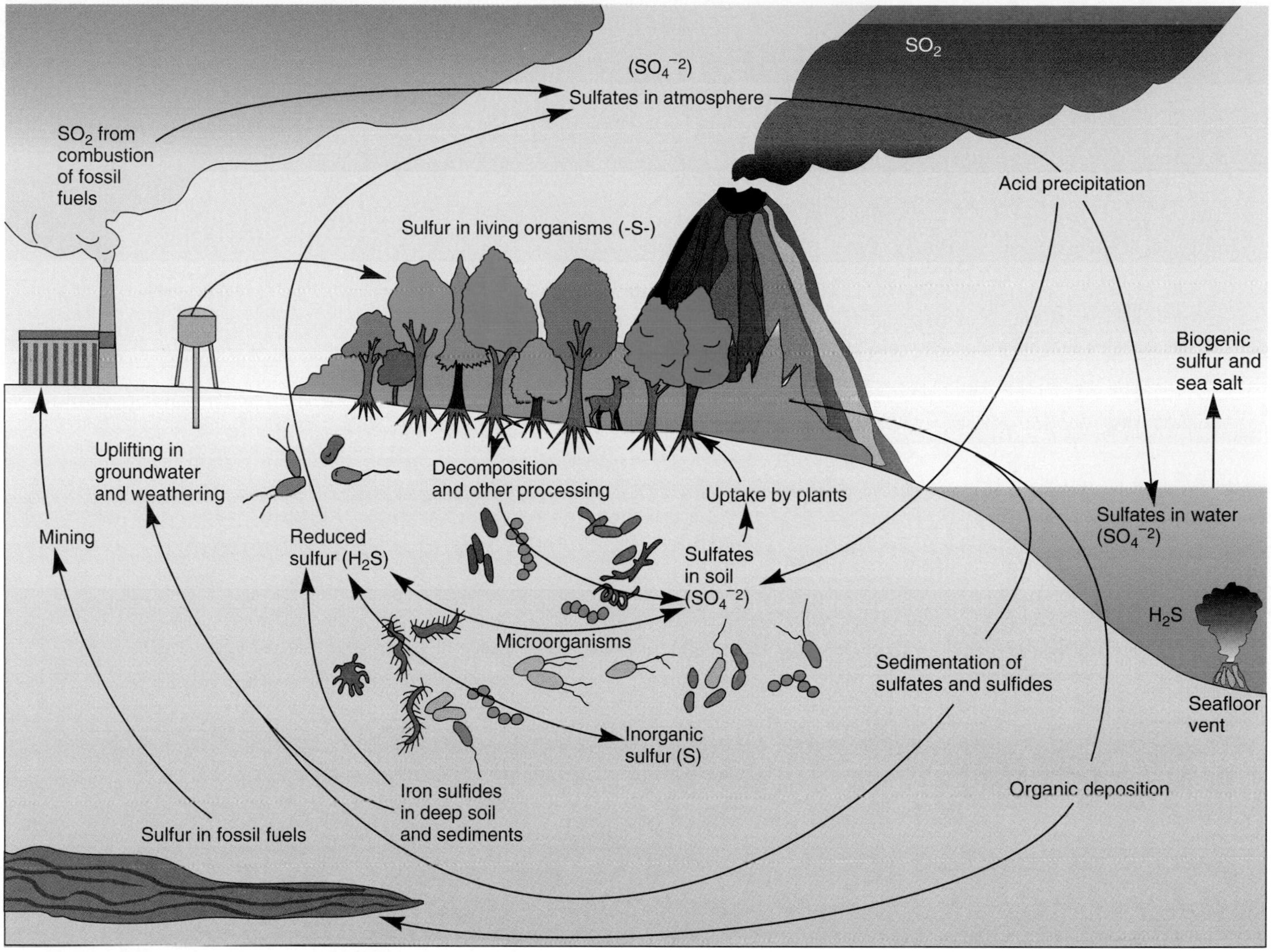

FIGURE 3.19 Sulfur is present mainly in rocks, soil, and water. It cycles through ecosystems when it is taken in by organisms. Combustion of fossil fuels causes increased levels of atmospheric sulfur compounds, which create problems related to acid precipitation.

any given situation depends on oxygen concentrations, pH, and light levels.

Human activities also release large quantities of sulfur, primarily through burning fossil fuels. Total yearly anthropogenic sulfur emissions rival those of natural processes, and acid rain caused by sulfuric acid produced as a result of fossil fuel use is a serious problem in many areas (see chapter 18). Sulfur dioxide and sulfate aerosols cause human health problems, damage buildings and vegetation, and reduce visibility. They also absorb UV radiation and create cloud cover that cools cities and may be offsetting greenhouse effects of rising CO_2 concentrations.

Interestingly, biogenic sulfur emissions by oceanic phytoplankton may play a role in global climate regulation. When ocean water is warm, tiny, single-celled organisms release dimethlysulfide (DMS) that is oxidized to SO_2 and then to SO_4 in the atmosphere. Acting as cloud droplet condensation nuclei, these sulfate aerosols increase the earth's albedo (reflectivity) and cool the earth (fig. 3.20). As ocean temperatures drop because less sunlight gets through, phytoplankton activity decreases, DMS production falls, and clouds disappear. Thus DMS, which may account for half of all biogenic sulfur emissions, could be a feedback mechanism that keeps temperature within a suitable range for all life.

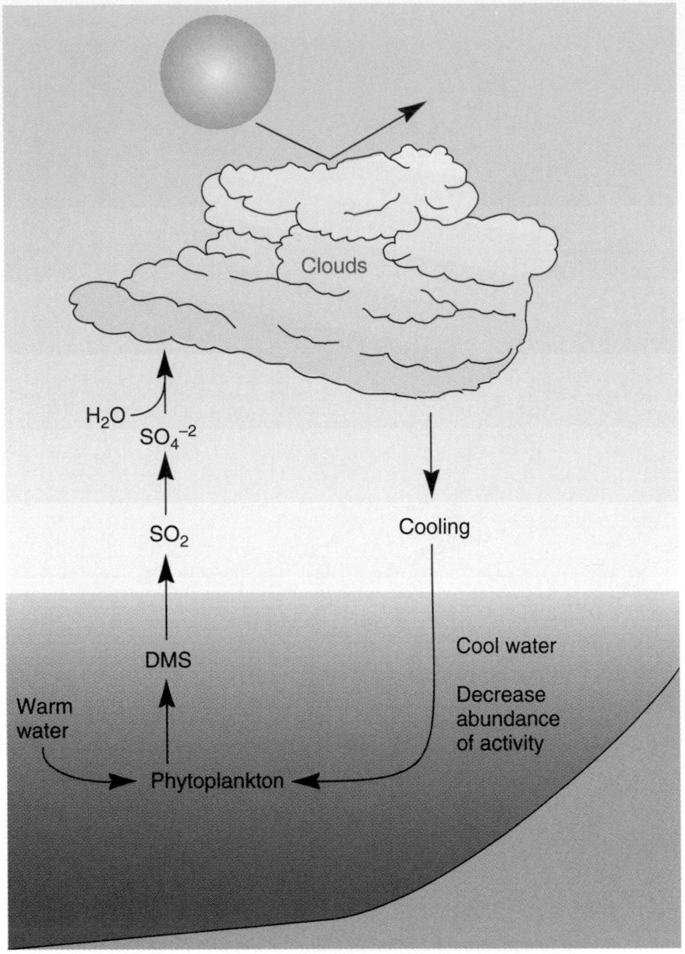

FIGURE 3.20 Dimethylsulfide (DMS), produced by oceanic phytoplankton (microscopic single-celled plants) may act as a governor, or feedback control, for global climate. As oceans warm, more DMS is produced, causing clouds to block sunlight. Less solar energy means a cooler earth, less active phytoplankton, and, thus, less DMS.

SUMMARY

Certain conditions, including the availability of required chemical elements, a steady influx of solar energy, mild surface temperatures, the presence of liquid water, and a suitable atmosphere, are essential for life on earth. Only the earth in our solar system has all these conditions. Life requires and makes possible highly organized exchanges of matter and energy between organisms and their environments.

Matter is the observable material of which the universe is composed. It exists in three interchangeable phases: gas, liquid, and solid. Matter is made up of atoms, which are composed of particles called protons, neutrons, and electrons. The energy that holds atoms together forms the basis for energy transfers in the bodies of living organisms and, therefore, in the biosphere.

A steady influx of solar radiation provides the heat and light energy needed to support life in the biosphere. Water, which covers approximately three-fourths of the earth's surface, is a remarkable substance. Because of its unique characteristics, it stabilizes the biosphere's temperature and provides the medium in which life processes occur. The earth's atmosphere provides gases necessary for life, helps maintain surface temperatures, and filters out dangerous radiation.

Ecosystem dynamics are governed by physical laws, including the law of conservation of matter and the first and second laws of thermodynamics. The recycling of matter is the basis of the cycles of elements that occur in ecosystems. Unlike matter, energy is not cycled. Energy always flows through systems in a one-way process in which some energy is converted from a high-quality, concentrated form to a lower-quality, less useful, dispersed form. We describe this increase in disorder as entropy, a fundamental limit to life. In ecosystems, solar energy enters the system and is converted to chemical energy by the process of photosynthesis. The chemical energy stored in the bonds that hold food molecules together is available for metabolism of organisms.

A species is all the organisms of the same kind that are genetically similar enough to breed in nature and produce live, fertile offspring. The populations of different species that live and interact within a particular area at a given time make up a biological community. An ecosystem is composed of a biological community together with all the biotic and abiotic factors that make up the environment in a defined area. Although ecosystem boundaries may be rather arbitrary, the holistic or systems approach to biology has provided rich insights into who lives where, when, how, and why.

Matter and energy are processed through the trophic levels of an ecosystem via food chains and food webs. At each energy transfer point less energy is available to work because of the laws of thermodynamics, so energy must be supplied to an ecosystem continuously. The relationships between producers and consumers in an ecosystem, often depicted as pyramids, demonstrate this principle. Most of the energy that enters an ecosystem comes, ultimately, from the sun.

The biosphere is a source of large quantities of essential elements. In a given ecosystem, these elements are constantly used and reused by living organisms. Water, carbon, nitrogen, sulfur, and phosphorus, for instance, are recycled in ecosystems through complex biogeochemical cycles.

QUESTIONS FOR REVIEW

1. Define *atom* and *element.* Are these terms interchangeable?
2. Your body contains vast numbers of carbon atoms. How is it possible that some of these carbon atoms may have been part of the body of a prehistoric creature?
3. In the biosphere, matter follows a circular pathway while energy follows a linear pathway. Explain.
4. The oceans store a vast amount of heat, but (except for climate moderation) this huge reservoir of energy is of little use to humans. Explain the difference between high-quality and low-quality energy.
5. Ecosystems require energy to function. Where does this energy come from? Where does it go? How does the flow of energy conform to the laws of thermodynamics?
6. Heat is released during metabolism. How is this heat useful to a cell and to a multicellular organism? How might it be detrimental, especially in a large, complex organism?
7. Photosynthesis and cellular respiration are complementary processes. Explain how they exemplify the laws of conservation of matter and thermodynamics.
8. What do we mean by carbon-fixation or nitrogen-fixation? Why is it important to humans that carbon and nitrogen be "fixed"?
9. The population density of large carnivores is always very small compared to the population density of herbivores occupying the same ecosystem. Explain this in relation to the concept of an ecological pyramid.
10. A species is a specific kind of organism. What general characteristics do individuals of a particular species share? Why is it important for ecologists to differentiate among the various species in a biological community?

QUESTIONS FOR CRITICAL THINKING

1. When we say that there is no "away" where we can throw things we don't want anymore, are we stating a premise or a conclusion? If you believe this is a premise, supply the appropriate conclusion. If you believe it is a conclusion, supply the appropriate premises. Does the argument change if this statement is a premise or a conclusion?
2. Suppose one of your classmates disagrees with the statement above, saying, "Of course there is an 'away.' It's anywhere out of *my* ecosystem." How would you answer?
3. A few years ago most laundry detergents contained phosphates for added cleaning power. Can you imagine any disadvantages to adding soluble phosphate to household products?
4. The first law of thermodynamics is sometimes summarized as "you can't get something for nothing." The second law is summarized as "you can't even break even." Explain what these phrases mean. Is it dangerous to oversimplify these important concepts?
5. The ecosystem concept revolutionized ecology by introducing holistic systems thinking as opposed to individualistic life history studies. Why was this a conceptual breakthrough?
6. Why is it important to recognize that ecosystems often are open and that boundaries may be fuzzy? Do these qualifications diminish the importance of the ecosystem study?
7. The holistic or systems approach to biology has sometimes been criticized as "black box" engineering. It allows us to make broad generalizations about what goes into or comes out of a system without knowing the precise details of how the system works. What do you think are the benefits and limitations of this approach?

8. Compare and contrast the views of F. E. Clements and H. A. Gleason concerning the concept of biological communities as superorganisms. How could these eminent biogeographers study the same communities and reach opposite interpretations? What evidence would be necessary to settle this question? Is lack of evidence the problem?
9. The properties of water are so unique and so essential for life as we know it that some people believe it proves that our planet was intentionally designed for our existence. What would an environmental scientist say about this belief?
10. The DMS feedback control of global climate is offered by some people as evidence for the Gaia hypothesis. Why might they take this position?

KEY TERMS

atomic number 56
atoms 56
biological community 62
biomass 62
carbon cycle 67
carbon sinks 67
carnivores 64
cells 57
cellular respiration 61
chemical energy 58
compound 57
conservation of matter 59
consumers 64
decomposers 64
detritivores 64
ecology 55
ecosystem 62
energy 58
enzymes 57
first law of thermodynamics 59
food chain 62
food web 63
half-life 57
heat 58
herbivores 64
homeostasis 63
ions 56
isotopes 56
joules 58
kinetic energy 58
matter 56
metabolism 57
molecules 57
nitrogen cycle 68
omnivores 64
organic compounds 57
pH 71
phosphorus cycle 71
photosynthesis 59
population 62
potential energy 58
producers 64
productivity 62
radioactive isotope 57
scavengers 64
second law of thermodynamics 59
species 61
sulfur cycle 72
temperature 58
trophic level 64

ADDITIONAL INFORMATION ON THE INTERNET

Visit our website at http://www.mhhe.com/environmentalscience/ for specific information about each of the topics below:

The Biology Project (activities and tutorials in cell biology)
Cells: functions and organelles
Biogeochemical cycles
What is photosynthesis (Arizona State University)
Bioweb: resources for cell biologists
Thermodynamics: general information
(Photosynthesis--) ASU Photosynthesis Center
The Hydrological Cycle

CHAPTER 4

Biological Communities and Species Interaction

Any species of bug is an irreplaceable marvel, equal to the works of art which we religiously preserve in our museums.

—Claude Levi-Strauss—

OBJECTIVES

After studying this chapter, you should be able to:

- describe how environmental factors determine which species live in a given ecosystem and where or how they live.
- understand how random genetic variation and natural selection lead to evolution, adaptation, niche specialization, and partitioning of resources in biological communities.
- compare and contrast interspecific predation, competition, symbiosis, commensalism, mutualism, and coevolution.
- discuss productivity, diversity, complexity, and structure of biological communities and how these characteristics might be connected to resilience and stability.
- explain how ecological succession results in ecosystem development and allows one species to replace another.
- give some examples of exotic species introduced into biological communities and describe the effects such introductions can have on indigenous species.

ESSENTIAL STUDY PARTNER

The Essential Study Partner (ESP) CD-ROM that accompanies this textbook contains additional information on the following topics:

- environmental gradients, survival, and tolerance: principles: communities: introduction
- habitat, niche, and resource partitioning: principles: communities: introduction/principles: communities: species interactions
- evolution, divergent and allopatric speciation: populations: biodiversity: species
- species interactions and the competitive exclusion principle: principles: communities: species interactions
- keystone predators: principles: communities: species interactions
- primary and secondary community succession: principles: communities: succession

Above: A field of shooting stars and Johnny-jump-ups add spring color to this open woodland community.

Orcas, Otters, Urchins, and Kelp:
Disrupting a Marine Food Web

What's happening to Alaskan sea otters? The sudden disappearance of thousands of sea otters along the Aleutian Island chain in recent years has surprised marine biologists and alerted them to a complex and disturbing ecological mystery story.

Once nearly driven into extinction in the northern Pacific by commercial fur hunters, sea otters (*Enhydra lutris*) made a remarkable recovery after Russia, Japan, Canada, and the United States finally agreed to stop hunting them in 1911. From only a few scattered individuals throughout their former range between Siberia and southern California, the otter population had rebounded to more than 100,000 in the Aleutians alone by the end of the twentieth century. The wreck of the *Exxon Valdez* in Prince William Sound in 1989 killed at least 1000 otters, but the population quickly recovered once the spilled crude oil dissipated. Why, wildlife specialists wondered, are these charismatic animals suddenly vanishing now over much of their former range? The answers to this puzzle seem to involve two highly popular marine mammals, overfishing by humans, and a complex ecological food web that affects a whole biological community.

Few animals are as cute and cuddly as sea otters. Their round fuzzy faces, dark button eyes, and huge walrus mustaches seem to express perpetual curiosity and good nature (fig. 4.1). They are beautiful swimmers, supple and swift in the water. Highly skilled hunters, they live mainly on shellfish and sea urchins from the ocean floor. Their thick, luxurious fur protects them from the cold ocean and was the prize sought by hunters who once nearly extirpated them entirely. With few natural enemies other than man, the otters generally seemed safe once hunting was outlawed. But changes in the arctic food web seem now to have changed the otter's future dramatically.

The other charismatic marine mammal in this story is the orca. Once known as killer whales, these bold, black and white animals have become widely popular, especially since the release of the movie, *Free Willy.* Whale watchers flock to the British Columbia and Alaskan coasts where they marvel at the graceful power of swimming orcas and listen to underwater recordings of their complex calls. Orcas inhabiting bays and estuaries eat mostly salmon, while those living around outer islands and the open ocean feed primarily on seals and sea lions. In both cases, overfishing by humans has seriously depleted food supplies on which these top predators depend.

Beginning in the late 1980s, seal and sea lion populations over much of the northern Pacific crashed, presumably because there weren't enough fish to sustain them. Although orcas prefer the high fat content of seals and sea lions, apparently some hungry whales have now turned to hunting otters. One piece of evidence for this is that otter populations in bays and inlets that are inaccessible to orcas have remained high, while those in exposed areas have declined by as much as 90 percent. Because otters are much smaller than seals and sea lions, it takes far more otters to sustain an orca. Researchers estimate that a single orca would need to eat 1825 otters a year to satisfy all its nutritional needs. If all the orcas in the northern Pacific were to turn exclusively to otters, the population would be decimated in just a few years.

Loss of sea otters seriously impacts many other organisms. The giant kelp "forests" in which the otters live (see fig. 4.11) are highly dependent on otters for protection. When otters are removed, sea urchin populations explode. Urchins eat kelp and too many of them can completely destroy a kelp bed and the whole biological community to which it offers shelter, food, and substrate. Without otters to control the urchins, up to 90 percent of the kelp has been eliminated in some areas of Alaska. Denuded urchin "barrens" support few fish, sea birds, or other normal inhabitants. Perhaps if we stop taking so many fish the whole system will return to normal, but it may take many years to restore original relationships. And what if the whales that have learned to hunt otters don't go back to their original prey? Have we permanently disrupted this prolific marine community?

How do we account for the amazing and complex interrelationships in an ecosystem such as this? How do organisms such as sea otters, orcas, kelp, and sea urchins become adapted to their particular environment and to each other? How can we understand the intricate interactions between various members of biological communities so as to avoid introducing other disastrous disruptions in the future? In this chapter we will examine competition, predation, symbiosis, tolerance limits, evolution, and other features of species, populations, and communities that affect both the organisms themselves and the ecosystems of which they are a part.

FIGURE 4.1 Almost driven into extinction a century ago by fur hunters, sea otters now face a threat from hungry killer whales.

WHO LIVES WHERE, AND WHY?

"Why" questions often are the stimulus for scientific research, but the research itself centers on "how" questions. Why, we wonder, does a particular species live where it does? More to the point, how is it *able* to live there? How does it deal with the physical resources of its environment and are some of its techniques unique? How does it interact with the other species present? And what gives one species an edge over another species in a particular habitat?

In this section we will examine some specific ways organisms are limited by the physical aspects of their environment. We then will discuss how members of a biological community interact, pointing out a few of the difficulties ecologists encounter when they attempt to discern patterns and make generalizations about community interactions and organization.

Critical Factors and Tolerance Limits

Every living organism has limits to the environmental conditions it can endure. Temperatures, moisture levels, nutrient supply, soil and water chemistry, living space, and other environmental factors must be within appropriate levels for life to persist. In 1840, Justus von Liebig proposed that the single factor in shortest supply relative to demand is the critical determinant in the distribution of that species. Ecologist Victor Shelford later expanded this principle of limiting factors by stating that each environmental factor has both minimum and maximum levels, called **tolerance limits,** beyond which a particular species cannot survive (fig. 4.2) or is unable to reproduce. The single factor closest to these survival limits, he postulated, is the critical limiting factor that determines where a particular organism can live.

At one time, ecologists accepted this concept so completely that they called it Liebig's or Shelford's Law and tried to identify unique factors limiting the growth of every population of plants and animals. For many species, however, we find that the interaction of several factors working together, rather than a single limiting factor, determines biogeographical distribution. If you have ever explored the rocky coasts of New England or the Pacific Northwest, for instance, you probably have noticed that mussels and barnacles endure extremely harsh conditions but generally are sharply limited to an intertidal zone where they grow so thickly that they often completely cover the substrate. No single factor determines this distribution. Instead, a combination of temperature extremes, drying time between tides, salt concentrations, competitors, and food availability limits the number and location of these animals.

For other organisms, there may be a specific *critical factor* that, more than any other, determines the abundance and distribution of that species in a given area. A striking example of cold intolerance as a critical factor is found in the giant saguaro cactus (*Carnegiea gigantea*), which grows in the dry, hot Sonoran desert of southern Arizona and northern Mexico (fig. 4.3). Saguaros are extremely sensitive to low temperatures. A single exceptionally cold winter night with temperatures below freezing for twelve hours or more will kill growing tips on the branches. Young saguaros are more susceptible to frost damage than adults, but seedlings typically become established under the canopy of small desert trees such as mesquite that shield the young cacti from the

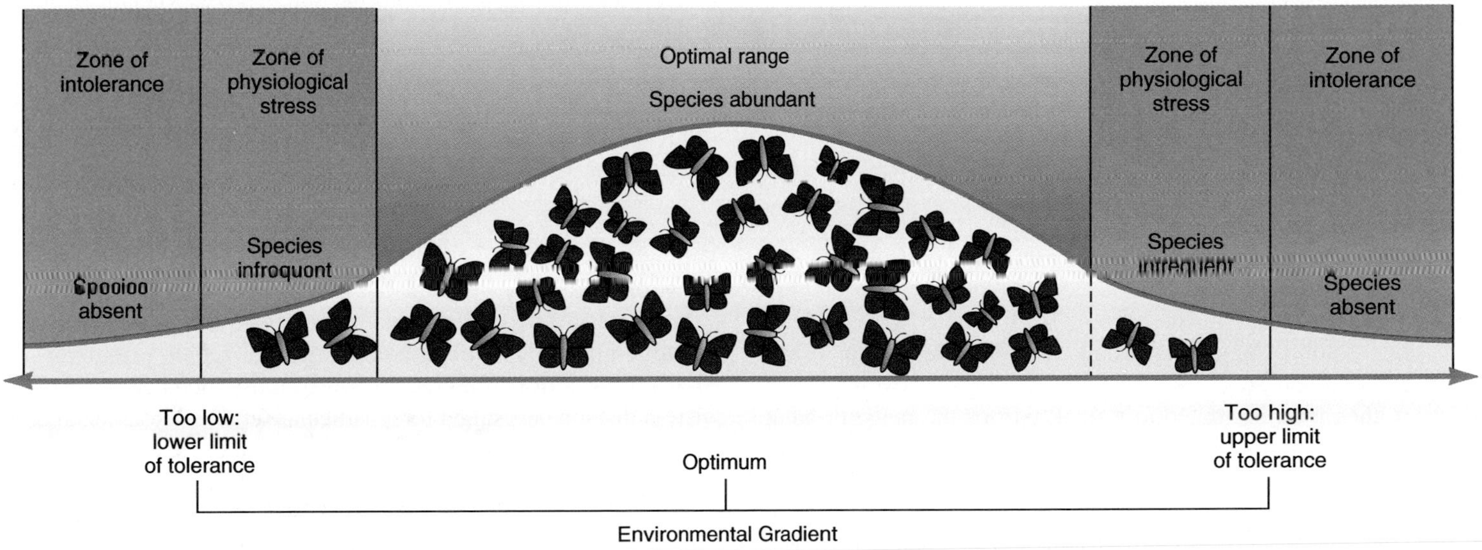

FIGURE 4.2 The principle of tolerance limits states that for every environmental factor, an organism has both maximum and minimum levels beyond which it cannot survive. The greatest abundance of any species along an environmental gradient is around the optimum level of the critical factor most important for that species. Near the tolerance limits, abundance decreases because fewer individuals are able to survive the stresses imposed by limiting factors.

FIGURE 4.3 Saguaro cacti, symbolic of the Sonoran desert, are an excellent example of distribution controlled by a critical environmental factor. Extremely sensitive to low temperatures, saguaros are found only where minimum temperatures never dip below freezing for more than a few hours at a time.

cold night sky. Unfortunately, the popularity of grilling with mesquite wood has caused extensive harvesting of the nurse trees that once sheltered small saguaros, adversely affecting reproduction of this charismatic species.

Animal species, too, exhibit tolerance limits that often are more critical for the young than for the adults. The desert pupfish (*Cyprinodon*), for instance, occurs in small isolated populations in warm springs in the northern Sonoran desert. Adult pupfish can survive temperatures between 0°C and 42°C (a remarkably high temperature for a fish) and are tolerant to an equally wide range of salt concentrations. Eggs and juvenile fish, however, can only live between 20°C and 36°C and are killed by high salt levels. Reproduction, therefore, is limited to a small part of the range of adult fish, which is often restricted anyway by the size of the small springs and desert seeps in which the species lives.

Sometimes the requirements and tolerances of species are useful indicators of specific environmental characteristics. The presence or absence of such species can tell us something about the community and the ecosystem as a whole. Locoweeds, for example, are small legumes that grow where soil concentrations of selenium are high. Because selenium often is found with uranium deposits, locoweeds have an applied economic value as **environmental indicators.** Such indicator species also may demonstrate the effects of human activities. Lichens and eastern white pine are less restricted in habitat than locoweeds, but are indicators of air pollution because they are extremely sensitive to sulfur dioxide and acid precipitation. Bull thistle is a weed that grows on disturbed soil but is not eaten by cattle; therefore, an abundant population of bull thistle in a pasture is a good indicator of overgrazing. Similarly, anglers know that trout species require clean, well-oxygenated water, so the presence or absence of trout can be an indicator of water quality.

Natural Selection, Adaptation, and Evolution

How is it that mussels have developed the ability to endure pounding waves, daily exposure to drying sun and wind, and seasonal threats of extreme cold or hot temperatures? What enables desert pupfish to tolerate hot, mineral-laden springs? How does the saguaro survive in the harsh temperatures and extreme dryness of the desert? We commonly say that each of these species is "adapted" to its special set of conditions, but what does that mean? In this section, we will examine one of the most important concepts in biology: how species acquire traits that allow them to live in unique ways in particular environments.

We use the term *adapt* in two ways. One is a limited range of *physiological modifications* (called acclimation) available to individual organisms. If you keep house plants inside all winter, for example, and then put them out in full sunlight in the spring, they get sunburned. If the damage isn't too severe, your plants will probably grow new leaves with a thicker cuticle and denser pigments that protect them from the sun. They can adapt to some degree, but the change isn't permanent. Another winter inside will make them just as sensitive to the sun as before. Furthermore, the changes they acquire are not passed on to their offspring. Although the potential to acclimate is inherited, each generation must develop its own protective epidermis.

Another type of adaptation operates at the population level and is brought about by inheritance of specific genetic traits that allow a species to live in a particular environment. This process is explained by the theory of **evolution,** developed by Charles Darwin and Alfred Wallace. According to this theory, species change gradually through competition for scarce resources and **natural selection,** a process in which those members of a population that are best suited for a particular set of environmental conditions will survive and produce offspring more successfully than their ill-suited competitors.

Natural selection acts on preexisting genetic diversity created by a series of small, random mutations (changes in genetic material) that occur spontaneously in every population. These mutations produce a variety of traits, some of which are more advantageous than others in a given situation. Where resources are limited or environmental conditions place some selective pressure

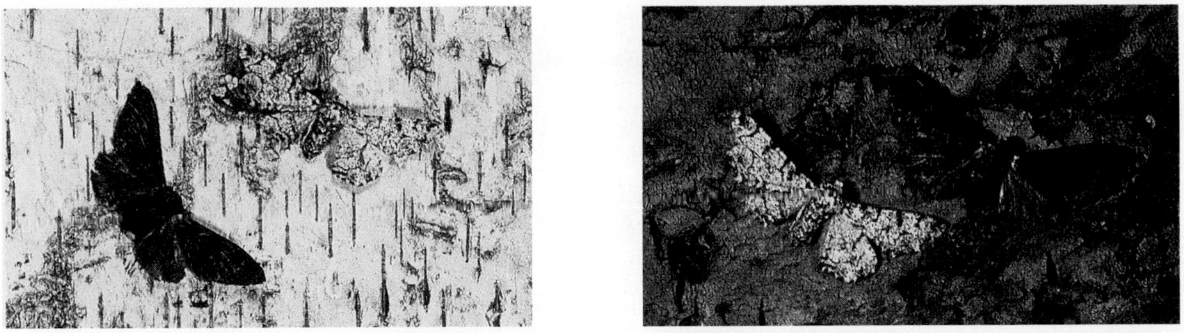

FIGURE 4.4 European peppered moths can have either light- or dark-colored wings. Under normal conditions, the lighter moths closely matched the coloration of the birch bark on which they rest during the day. When forests were blackened by industrial pollution in the nineteenth century, the light-colored moths were more visible to predators and the darker moths dominated the population. Since pollution has been reduced, however, and birch bark is once again white, the lighter moths have now become more plentiful.

on a population, individuals with those advantageous traits become more abundant in the population, and the species gradually evolves or becomes better suited to that particular environment. Although each change may be very slight, many mutations over a very long time have produced the incredible variety of different life-forms that we observe in nature.

A classic example of natural selection is the changing coloration of European peppered moths, *Biston betularia* (fig. 4.4). Until about 1850, nearly all of these moths had light-colored wings that closely matched the light bark of the trees on which they rest during the day. As industrial pollution darkened the trunks of trees around major urban centers, biologists noticed a dramatic increase in moths with dark coloration. Probably the dark-colored individuals had always been present in the population but were spotted and eaten more frequently by birds when the trees were white. As tree bark darkened, however, the light-colored moths were at a disadvantage. Recently, as factory smoke has been reduced, tree bark is whitening and white moths are becoming more abundant.

The variety of finches observed by Charles Darwin on the Galápagos Islands is another classic example of speciation driven by availability of different environmental opportunities (fig. 4.5). Originally derived from a single seed-eating species that somehow crossed the thousands of kilometers from the mainland, the finches have evolved into a dozen or more distinct species that differ markedly in appearance, food preferences, and habitats they occupy. Fruit eaters have thick parrotlike bills; seed eaters have heavy, crushing bills; insect eaters have thin probing beaks to catch their prey. One of the most unusual species is the woodpecker finch, which pecks at tree bark for hidden insects. Lacking the woodpecker's long tongue, however, the finch uses a cactus spine as a tool to extract bugs.

The amazing variety of colors, shapes, and sizes of dogs, cats, rabbits, fish, flowers, vegetables, and other domestic species are evidence of deliberate selective breeding. The various characteristics of these organisms arose through mutations. We simply kept the ones we liked.

What environmental factors cause selective pressure and influence fertility or survivorship in nature? They include (1) physiological stress due to inappropriate levels of some critical environmental factor, such as moisture, light, temperature, pH, or specific nutrients; (2) predation, including parasitism and disease; (3) competition; and (4) luck. In some cases the organisms that survive environmental catastrophes or find their way to a new habitat where they start a new population may simply be lucky rather than more fit or better suited to subsequent environmental conditions than their less fortunate contemporaries.

Be sure you understand that while selection affects individuals, evolution and adaptation work at the population level. Individuals don't evolve; species do. Each individual is locked in by genetics to a particular way of life. Most plants, animals, or microbes have relatively limited ability to modify their physical makeup or behavior to better suit a particular environment. Over time, however, random genetic changes and natural selection can change an entire population.

Given enough geographical isolation or selective pressure, the members of a population become so different from their ancestors that they may be considered an entirely new species that has replaced the original one. Alternatively, isolation of population subsets by geographical or behavioral factors that prevent exchange of genetic material can result in branching off of new species that coexist with their parental line. Suppose that two populations of the same species become separated by a body of water, a desert, or a mountain range that they cannot cross. Over a very long time—often millions of years—random mutations and different environmental pressures may cause the populations to evolve along such dissimilar paths that they can no longer interbreed successfully even if the opportunity to do so arises. They have now become separate species as in the case of the Galápagos finches. The barriers that divide subpopulations are not always physical. In some cases, behaviors such as when and where members of a population feed, sleep, or mate—or how they communicate—may separate them sufficiently for divergent evolution and speciation to occur even though they occupy the same territory.

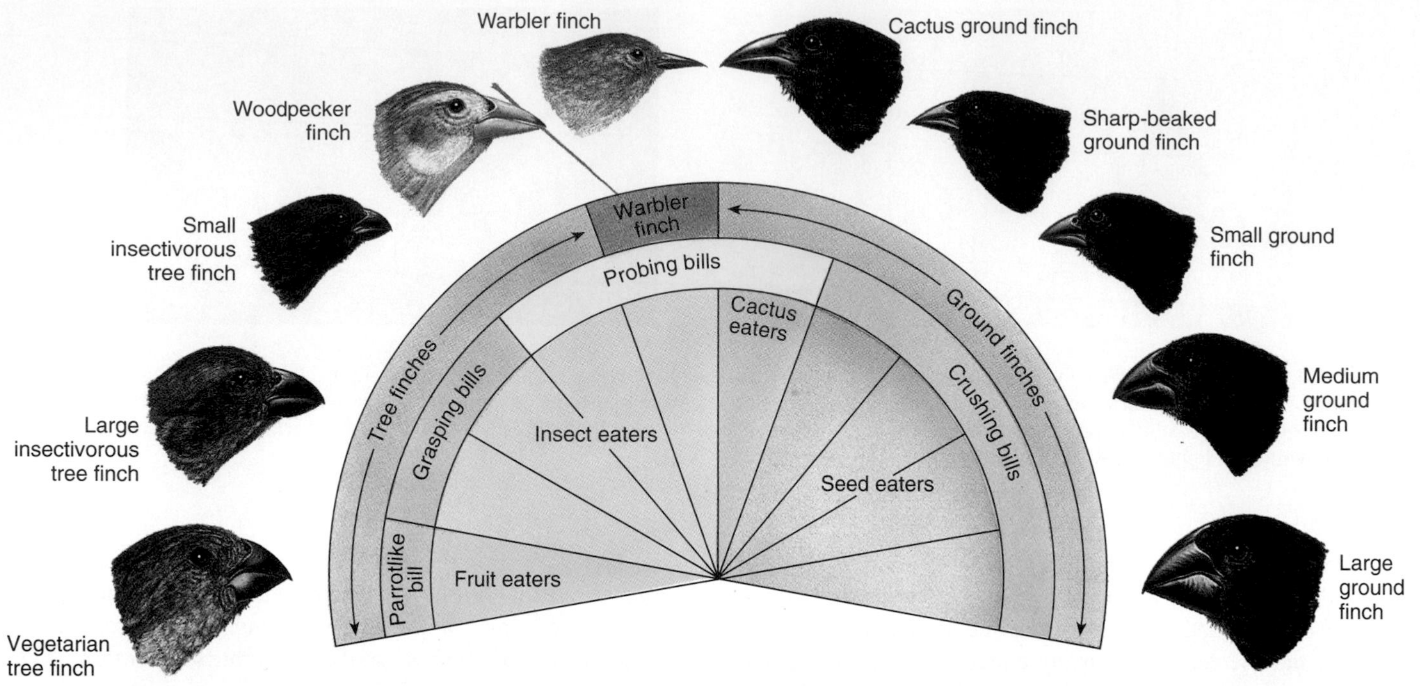

FIGURE 4.5 Some species of Galápagos Island finches. Although all are descendents of a common ancestor, they now differ markedly in appearance, habitat, and feeding behavior. Ground finches (*lower right*) eat cactus leaves; warbler finches (*upper left*) eat insects; others eat seeds or have mixed diets. The woodpecker finch (*upper left*) pecks tree bark as do woodpeckers, but lacks a long tongue. Instead, it uses cactus spines as tools to extract insects.

Natural selection and adaptation can cause organisms with a similar origin to become very different in appearance and develop different habits over time, but they can also result in unrelated organisms coming to look and act very much alike. We call this latter process *convergent* evolution. The cactus-eating Galápagos finches (fig. 4.5), for example, look and act very much like parrots even though they are genetically very dissimilar. The features that enable parrots to eat fruit successfully work well for these finches also.

A common mistake is to believe that organisms develop certain characteristics because they want or need them. This is incorrect. A duck doesn't have webbed feet because it wants to swim or needs to swim in order to eat; it has webbed feet because some ancestor happened to have a gene for webbed feet that gave it some advantage over other ducks in its particular pond and because those genes were passed on successfully to its offspring. A variety of different genetic types are always present in any population, and natural selection simply favors those best suited for particular conditions. Whether there is a purpose or direction to this process is a theological question rather than a scientific one and is beyond the scope of this book.

The Ecological Niche

Habitat describes the place or set of environmental conditions in which a particular organism lives. A more functional term, the **ecological niche,** is a description of either the role played by a species in a biological community or the total set of environmental factors that determine species distribution. Niches as community roles—describing how a species obtains food, what relationships it has with other species, and the services it provides its community, for example—were first described by the British ecologist, Charles Elton in 1927. Thirty years later, the American limnologist G. E. Hutchinson proposed a more biophysical definition of this concept. Every species, he pointed out, has a range of physical and chemical conditions (temperature, light levels, acidity, humidity, salinity, etc.) as well as biological interactions (predators and prey present, defenses, nutritional resources available, etc.) within which it can exist. Figure 4.2, for example, shows the abundance of a hypothetical species along a single factor gradient. If it were possible to graph simultaneously all of the factors that affect a particular species, a multidimensional space would result that describes the ecological niche available to that species.

Some species, like raccoons or coyotes, are generalists that eat a wide variety of food and live in a broad range of habitats (including urban areas). Others, such as the panda (fig. 4.6), are specialists that occupy a very narrow niche. Specialists often tend to be rarer than generalists and less resilient to disturbance or change.

A few species such as elephants, chimpanzees, and baboons learn how to behave from their social group and can invent new ways of doing things when presented with new opportunities or challenges. Most organisms, however, are limited by genetically determined physical structure and instinctive behavior to established niches.

FIGURE 4.6 The giant panda feeds exclusively on bamboo. Although its teeth and digestive system are those of a carnivore, it is not a good hunter, and has adapted to a vegetarian diet. To support its size, it must eat up to 40 lbs of bamboo leaves and stalks daily. This narrow specialization makes it extremely vulnerable to fluctuations in the bamboo population. In the 1970s, huge acreages of bamboo flowered and died, and many pandas starved.

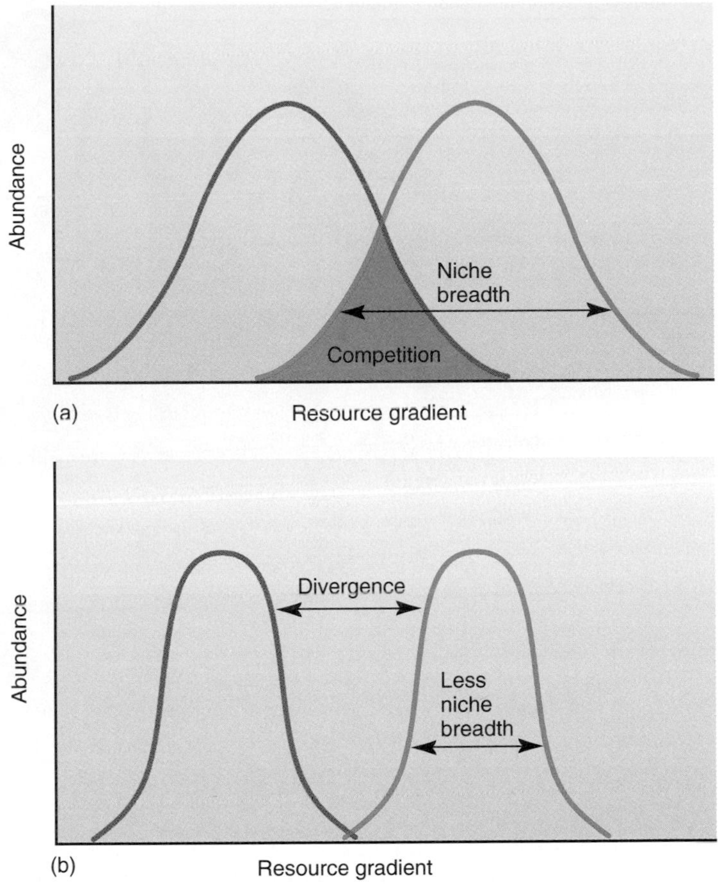

FIGURE 4.7 Resource partitioning and niche specialization caused by competition. Where niches of two species overlap along a resource gradient, competition occurs (shaded area in (*a*)). Individuals occupying this part of the niche are less successful in reproduction so that characteristics of the population diverge to produce more specialization, narrower niche breadth, and less competition between species (*b*).

Over time, though, niches can evolve, just as physical characteristics do. The law of competitive exclusion states that no two species will occupy the same niche and compete for exactly the same resources in the same habitat for very long. Eventually, one group will gain a larger share of resources while the other will either migrate to a new area, become extinct, or change its behavior or physiology in ways that minimize competition. We call this latter process of niche evolution **resource partitioning** (fig. 4.7). It can produce high levels of specialization that allow several species to utilize different parts of the same resource and coexist within a single habitat (fig. 4.8).

Niche specialization also can create behavioral separation that allows subpopulations of a single species to diverge into separate species. Why doesn't this process continue until there is an infinite number of species? The answer is that a given resource can be partitioned only so far. Populations must be maintained at a minimum size to avoid genetic problems and to survive bad times. This puts an upper limit on the number of different niches—and therefore the number of species—that a given community can support.

Perhaps you haven't thought of time as an ecological factor, but niche specialization in a community is a twenty-four hour phenomenon. Swallows and insectivorous bats both catch insects, but some insect species are active during the day and others at night, providing noncompetitive feeding opportunities for day-active swallows and night-active bats.

SPECIES INTERACTIONS AND COMMUNITY DYNAMICS

Predation and competition for scarce resources are major factors in evolution and adaptation. Not all biological interactions are competitive, however. Organisms also cooperate with, or at least

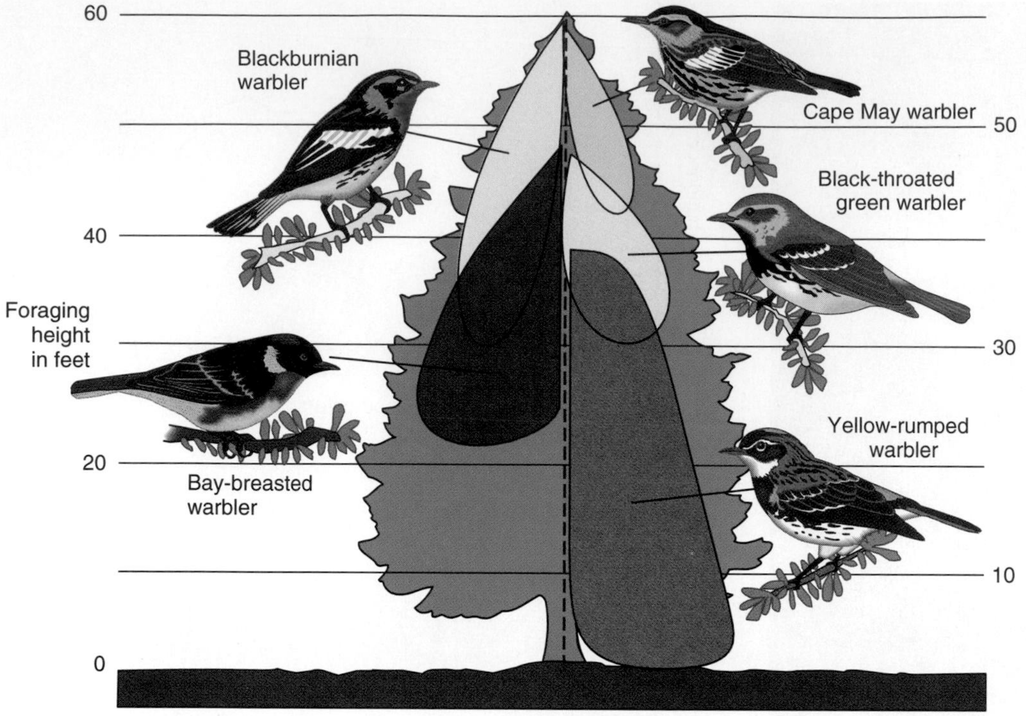

FIGURE 4.8 Resource partitioning and the concept of the ecological niche are demonstrated by several species of wood warblers that use different strata of the same forest. This is a classic example of the principle of competitive exclusion.
Source: Original observation by R. H. MacArthur.

tolerate, members of their own species as well as individuals of other species in order to survive and reproduce. In this section, we will look more closely at the different interactions within and between species that shape biological communities.

Predation

All organisms need food to live. Producers make their own food, and consumers eat organic matter created by other organisms. In most communities, as we saw in chapter 3, photosynthetic organisms are the producers. Consumers include herbivores, carnivores, omnivores, scavengers, and decomposers. With which of these categories do you associate the term *predator?* Ecologically, the term has a much broader meaning than you might expect. A **predator** in an ecological sense, is an organism that feeds directly upon another living organism, whether or not it kills the prey to do so (fig. 4.9). By this definition herbivores, carnivores, and omnivores that feed on live prey are predators, but scavengers, detritivores, and decomposers that feed on dead things are not. In this broad sense, **parasites** (organisms that feed on a host organism or steal resources from it without killing it) and even **pathogens** (disease-causing organisms) might also be considered predator organisms.

Predation is a potent and complex influence on the population balance of communities involving (1) all stages of the life cycles of predator and prey species, (2) many specialized food-obtaining mechanisms, and (3) specific prey-predator adaptations that either resist or encourage predation (Case Study, p. 86).

FIGURE 4.9 Insect herbivores are predators as much as are lions and tigers. In fact, insects consume the vast majority of biomass in the world. Complex patterns of predation and defense have often evolved between insect predators and their plant prey.

Predation throughout the life cycle is very pronounced in marine intertidal animals, for example. Many crustaceans, mollusks, and worms release eggs directly into the water, and the eggs and free-living larval and juvenile stages are part of the floating community, or **plankton** (fig. 4.10). Planktonic animals feed upon each other and are food for successively larger carnivores, including small fish. As prey species mature, their predators change. Barnacle larvae are planktonic and are eaten by fish. Adult barnacles, on the other hand, build hard shells that protect them from fish but can

FIGURE 4.10 Microscopic plants and animals form the basic levels of many aquatic food chains and account for a large percentage of total world biomass. Many oceanic plankton are larval forms that have very different habitats and feeding relationships than their adult forms.

FIGURE 4.11 Giant kelp is a massive alga that forms dense "forests" off the Pacific coast of California. It is a keystone species in that it provides food, shelter, and structure essential for a whole community. Removal of sea otters allows sea urchin populations to explode. When the urchins destroy the kelp, many other species suffer as well.

be crushed by limpets and other mollusks. Predators also change their feeding targets. Adult frogs, for instance, are carnivores, but the tadpoles of most species are grazing herbivores. Sorting out the trophic levels in these communities can be very difficult.

Predation is an important factor in evolution. Predators prey most successfully on the slowest, weakest, least fit members of their target population, thus reducing competition, preventing excess population growth, allowing successful traits to become dominant in the prey population, and making the prey population stronger and healthier. Aldo Leopold described this effect in his pioneering studies of predators and deer on the Kaibab Plateau in northern Arizona. When wolves and mountain lions were present, the deer were in better balance with their environment than after these predators were removed. Without predators, the deer overpopulated the area, exhausted their food supply, and starved to death.

Prey species have evolved many protective or defensive adaptations to avoid predation. In plants, for instance, this often takes the form of thick bark, spines, thorns, or chemical defenses. Animal prey may become very adept at hiding, fleeing, or fighting back against predators. Predators, in turn, evolve mechanisms to overcome the defenses of their prey. This process in which species exert selective pressure on each other is called **coevolution.**

Keystone Species

A **keystone species** is a species or set of species whose impact on its community or ecosystem is much larger and more influential than would be expected from mere abundance. Originally, keystone species were thought to be top predators, such as wolves, whose presence limits the abundance of herbivores and thereby reduces their grazing or browsing on plants. Recently, it has been recognized that less conspicuous species also play essential community roles. Certain tropical figs, for example, bear during seasons when no other fruit is available for frugivores (fruit-eating animals). If these figs were removed, many animals would starve to death during periods of fruit scarcity. With those animals gone, many other plant species that depend on them at other times of year for pollination and seed-dispersal would disappear as well.

Even microorganisms can play vital roles. In some forest ecosystems, mycorrhizae (fungi associated with tree roots) are essential for mineral mobilization and absorption. If the fungi die, so do the trees and many other species that depend on a healthy forest community. Rather than being a single species, mycorrhizae are actually a group of species that together fulfill a keystone function.

Often a number of species are intricately interconnected in biological communities so that it is difficult to tell which is the essential key. Consider the Pacific kelp forests and sea otters that opened this chapter (fig. 4.11). Giant kelp provide shelter for a number of fish and shellfish species and so could be regarded as the key to community structure. Sea urchins, however, feed on the kelp and determine their number and distribution while sea otters regulate urchins and kelp provides a resting place for dozing otters. Which of these species is the most important? Each depends on and affects the others. Perhaps we should think in terms of a "keystone set" of organisms in some ecosystems.

CASE STUDY

WHERE HAVE ALL THE SONGBIRDS GONE?

Every June, some 2200 amateur ornithologists and bird watchers across the United States and Canada join in an annual bird count called the Breeding Bird Survey. Organized in 1966 by the U.S. Fish and Wildlife Service to follow bird population changes, this survey has discovered some shocking trends. While birds such as robins, starlings, and blackbirds that prosper around humans have increased their number and distribution over the past thirty years, many of our most colorful and melodious forest birds have declined severely. The greatest decreases have been among the true songbirds such as thrushes, orioles, tanagers, catbirds, vireos, buntings, and warblers. These long-distance migrants nest in northern forests but spend the winters in South or Central America or in the Caribbean Islands. Scientists call them neotropical migrants.

In many areas of the eastern United States and Canada, three-quarters or more of the neotropical migrants have declined significantly since the survey was started. Some that once were common have become locally extinct. Grover Archbold Park and Rock Creek Park in Washington, DC, for instance, lost 75 percent of their songbird population and 90 percent of their long-distance migrant species in just twenty years. Nationwide, cerulean warblers, American redstarts, and ovenbirds declined about 50 percent in the single decade of the 1970s. Studies of radar images from National Weather Service stations in Texas and Louisiana suggest that only about half as many birds fly across the Gulf of Mexico each spring now compared to the 1960s. This could mean a loss of about half a billion birds in total.

Many of our most colorful and melodious woodland songbirds, such as this redstart, are vanishing due to habitat destruction, predation and nest parasitism.

What causes these devastating losses? Destruction of critical winter habitat is clearly a major issue. Birds often are much more densely crowded in the limited areas available to them during the winter than they are on their summer range. Unfortunately, forests throughout Latin America are being felled at an appalling rate. Central America, for instance, is losing about 1.4 million hectares (2 percent of its forests or an area about the size of Yellowstone National Park) each year. If this trend continues, there will be essentially no intact forest left in much of the region in fifty years.

But loss of tropical forests is not the only threat. Recent studies show that fragmentation of breeding habitat and nesting failures in the United States and Canada may be just as big a problem for woodland songbirds. Many of the most threatened species are adapted to deep woods and need an area of 10 hectares (24.7 acres) or more per pair to breed and raise their young. As our woodlands are broken up by roads, housing developments, and shopping centers, it becomes more and more difficult for these highly specialized birds to find enough contiguous woods to nest successfully.

Predation and nest parasitism also present a growing threat to many bird species. While birds have probably always lost eggs and nestlings to predators, there has been a startling increase in predation in the past thirty years. Raccoons, opossums, crows, bluejays, squirrels, and house cats thrive in human-dominated landscapes. They are protected from larger predators like wolves or owls and find abundant supplies of food and places to

Competition

Competition is another kind of antagonistic relationship within a community. For what do organisms compete? To answer this question, think again about what all organisms need to survive: energy and matter in usable forms, space, and specific sites for life activities (What Do You Think?, p. 88). Plants compete for growing space for root and shoot systems so they can absorb and process sunlight, water, and nutrients. Animals compete for living, nesting, and feeding sites, as well as for food, water, and mates. Competition among members of the same species is called **intraspecific competition,** whereas competition between members of different species is called **interspecific competition.**

If you look closely at a patch of weeds growing on good soil early in the summer, you likely will see several types of interspecific competition. First of all, many weedy species attempt to crowd out their rivals by producing prodigious numbers of seeds. After the seeds germinate, the plants race to grow the tallest, cover the most ground, and get the most sun. You may observe several strategies to do this. For example, vines don't build heavy stems of their own; they simply climb up over their neighbors to get to the light.

We often think of competition among animals as a bloody battle for resources. A famous Victorian description of the struggle for survival was "nature red in tooth and claw." In fact, a better metaphor is a race. Have you ever noticed that birds always eat fruits and berries just before they are ripe enough for us to pick? Having a tolerance for bitter, unripe fruit gives them an advantage in the race for these food resources. Many animals tend to avoid fighting if possible. It's not worth getting injured. Most confrontations are more noise and show than actual fighting.

Intraspecific competition can be especially intense because members of the same species have the same space and nutritional requirements; therefore, they compete directly for these environmental resources. How do plants cope with intraspecific

CASE STUDY

Continued

hide. Their numbers have increased dramatically, as have their raids on bird nests. A comparison of predation rates in the Great Smoky Mountain National Park and in small rural and suburban woodlands shows how devastating predators can be. In a 1000-hectare study area of mature, unbroken forest in the national park, only one songbird nest in fifty was raided by predators. By contrast, in plots of 10 hectares or less near cities, up to 90 percent of the nests were raided.

Nest parasitism by brown-headed cowbirds is one of the worst threats for woodland songbirds. Originally called buffalo birds, these small blackbirds were adapted to follow migratory bison herds picking up seeds and insects from the droppings. Because they didn't stay in one place long enough to raise a family, they developed the habit of depositing their eggs in the nests of other species, leaving their young to be raised by surrogate parents. The young cowbirds are generally larger and more aggressive than the resident chicks, which generally starve to death because they don't get enough food. Adult cowbirds also find a welcome source of food and shelter around humans. Once fairly uncommon in the United States, there are now about 150 million of these parasites.

A study in southern Wisconsin found that 80 percent of the nests of woodland species were raided by predators and that three-quarters of those that survived were invaded by cowbirds. Another study in the Shawnee National Forest in southern Illinois found that 80 percent of the scarlet tanager nests contained cowbird eggs and that 90 percent of the wood thrush nests were taken over by these parasites. The sobering conclusion of this latter study is that there probably is no longer any place in Illinois where scarlet tanagers and wood thrushes can breed successfully.

What can we do about this situation? First, we can support sustainable development in Third World countries so that people there can enjoy a better standard of living without destroying their forests and natural areas. A number of such projects are discussed elsewhere in this book. Next, we should identify and protect critical habitat at home and abroad on which especially endangered species depend. Buying up inholdings that fragment the forest and preserving corridors that tie together important areas will help. In areas where people already live, we could encourage clustering of houses to protect as much woods as possible. We also might discourage clearing underbrush and trees from yards and parks to leave shelter for the birds.

Could we reduce the number of predators or limit their access to critical breeding areas? Human residents might not like the idea of reintroducing wolves and bears, but they might accept fencing or trapping of small predators. A campaign to keep house cats inside during the breeding season would certainly help.

Ethical Considerations

Some wildlife managers already are trapping cowbirds. The Kirtland's warbler is one of the rarest songbirds in the United States. It nests only in young, fire-maintained jackpine forests in Michigan. Controlled burning to maintain habitat for this endangered species was started in the 1960s, but the population continued to decline. Studies showed that 90 percent of the nests were being parasitized by cowbirds. Since 1972, refuge managers have trapped and killed some 7000 cowbirds each year to protect the warblers. In the past two decades, the number of breeding pairs of warblers has risen from about 150 to nearly 400. Would it be possible to do something similar on a nationwide scale? Could we trap and kill 150 million cowbirds? Should we eliminate one species to save another? What do you think?

competition? The inability of seedlings to germinate in the shady conditions created by parent plants acts to limit intraspecific competition by favoring the mature, reproductive plants. Many plants have adaptations for dispersing their seeds to other sites by air, water, or animals. Undoubtedly you've seen dandelion seeds on their tiny parachutes (fig. 4.12) or had sticky or burred seeds become attached to your clothing. Some plants secrete substances that inhibit the growth of seedlings near them, including their own and those of other species. This strategy is particularly significant in deserts where water is a limiting factor.

FIGURE 4.12 Many plant species have mechanisms for seed dispersal over long distances, which enables them to reach new areas for growth. The ability to move rapidly into newly disturbed areas gives "weedy" species a competitive advantage in some circumstances.

what do you think?

THE WORLD AROUND YOU

Understanding Competition

Ecology is a relatively young science. Consequently, many ecological processes are incompletely understood. How a community comes to have its particular organization is one area of uncertainty. Some ecologists feel that physical factors are the most important determinants in community organization, while others feel that interspecific competition is most important.

How can we find out which view is correct? Ecologists employ the scientific method, as described in chapter 2, to better understand community dynamics. This process is mostly refined common sense and its basic elements can be useful in everyday life.

Once ecologists have decided on the concept to be investigated, they look for a specific situation that can either be observed or manipulated to provide relevant information. For example, ecologist Richard Karban was interested in how competition affected a community. He learned that larvae of two insect species, the meadow spittlebug and the calendula plume moth, both feed and develop on the seaside daisy, a common beach plant on the American west coast. The specific question to be investigated was: Does competition affect these two insect species, therefore impacting community organization?

Competition might reduce survival rate, larval growth, or both. Karban's procedure involved setting up four groups of plants at Bodega Bay, CA: one got both spittlebugs and moths, another got only spittlebugs, another only moths, and a fourth had neither. He compared survival rates of spittlebugs and moths when competitors were present and absent.

There are three important general considerations in designing scientific investigations:

1. Things need to be organized in such a way that the outcome can clearly be linked to a particular cause. In other words, differences in insect survival rates need to be clearly attributable to competition and not to other factors.

   Karban accomplished this by making his plant/insect groups as uniform as possible, except for the presence or absence of competitors. He eliminated genetic differences between plants by using plants from the same clone. He was careful to put the same numbers of insects on each plant to eliminate animal density as a factor, and so on.
2. The data collected must be a reliable representation of the larger situation and not simply the result of chance. This is usually accomplished by replicating the procedure many times. Instead of setting up just a few plants with one or both insects present, Karban set up 30 plants with each treatment. The procedure was repeated a second year. This gave him a cumulative total of 60 plants that had just spittlebugs, 60 plants having just moths, and 60 plants each having both or neither spittlebugs and moths. With such a large number of replications it was highly likely that differences in survival rates were, in fact, the result of competition and not simply chance occurrences.
3. Finally, conclusions must be justified by the data. Karban's statistical analysis revealed that spittlebug persistence was nearly 40 percent higher when the plume moths were absent. Plume moth persistence was not significantly affected by spittlebug presence, however.

His overall conclusion was:

> Evidence from this and other studies supports the contention that interspecific competition can play an important role in influencing densities of plant-feeding insects.

Notice the caution expressed in these words. He did not claim to have proven anything. Instead, his study "*supports* the contention." Second, he states competition "*can* play an important role," instead of using stronger language. And finally, he restricts these conclusions to plant-feeding insects. Karban carefully avoids drawing conclusions beyond the realm supported by his data.

Based on a healthy skepticism, clarity of language, critical evaluation of relationships and information, and caution in coming to judgment, critical thinking in science has been a very successful tool in enhancing understanding.

Spittlebugs produce mounds of foam under which they hide from predators while feeding on host plants.

Animals also have developed adaptive responses to intraspecific competition. Two major examples are varied life cycles and territoriality. The life cycles of many invertebrate species have juvenile stages that are very different from the adults in habitat and feeding. Compare a leaf-munching caterpillar to a nectar-sipping adult butterfly or a planktonic crab larva to its bottom-crawling adult form. In these examples, the adults and juveniles of each species do not compete because they occupy different ecological niches.

FIGURE 4.13 Plants compete for light and growing space in this Indonesian rainforest. Epiphytes, such as the ferns and bromeliads shown here, find a place to grow in the forest canopy by perching on the limbs of large trees. This may be a commensal relationship if the epiphytes don't hurt their hosts. Sometimes, however, the weight of epiphytes breaks off branches and even topples whole trees.

You may have observed robins chasing other robins during the mating and nesting season. Robins and many other vertebrate species demonstrate **territoriality,** an intense form of intraspecific competition in which organisms define an area surrounding their home site or nesting site and defend it, primarily against other members of their own species. Territoriality helps to allocate the resources of an area by spacing out the members of a population. It also promotes dispersal into adjacent areas by pushing grown offspring outward from the parental territory.

Territory size depends on the size of the species and the resources available. A pair of robins might make do with a suburban yard, but a large carnivore like a tiger may need thousands of square kilometers.

Symbiosis

In contrast to predation and competition, symbiotic interactions between organisms can be nonantagonistic. **Symbiosis** is the intimate living together of members of two or more species. **Commensalism** is a type of symbiosis in which one member clearly benefits and the other apparently is neither benefited nor harmed. Cattle often are accompanied by cattle egrets, small white shore birds who catch insects kicked up as the cattle graze

FIGURE 4.14 Lichens, such as the various species growing on this log, are a combination of algae and fungi in a classic example of mutualistic symbiosis.

through a field. The birds benefit while the cattle seem indifferent. Many of the mosses, bromeliads, and other plants growing on trees in the moist tropics are also considered to be commensals (fig. 4.13). These epiphytes get water from rain and nutrients from leaf litter and dust fall, and often neither help nor hurt the trees on which they grow. In a way, the robins and sparrows that inhabit suburban yards are commensals with humans.

Lichens are a combination of a fungus and a photosynthetic partner, either an alga or a cyanobacteria. Their association is a type of symbiosis called **mutualism,** in which both members of the partnership benefit (fig. 4.14). Some ecologists believe that cooperative, mutualistic relationships may be more important in evolution than we have commonly thought. Aggressive interactions often are dangerous and destructive, while cooperation and compromise may have advantages that we tend to overlook. Survival of the fittest often may mean survival of those organisms that can live best with one another.

Parasitism, described earlier as a form of predation, also could be considered a type of symbiosis, where one species benefits and the other is harmed. All of these relationships have a bearing on such ecological issues as resource utilization, niche specialization, diversity, predation, and competition. Symbiotic relationships often enhance the survival of one or both partners.

Symbiotic relationships often entail some degree of coadaptation or coevolution of the partners, shaping—at least in part—their structural and behavioral characteristics. An interesting case of mutualistic coadaptation is seen in Central and South American swollen thorn acacias and their symbiotic ants. Acacia ant colonies live within the swollen thorns on the acacia tree branches and feed on two kinds of food provided by the trees: nectar produced in glands at the leaf bases and special protein-rich structures produced on leaflet tips. The acacias thus provide shelter and food for the ants. Although they spend energy to provide these services, the trees are not physically harmed by ant feeding.

What do the acacias get in return and how does the relationship relate to community dynamics? Ants tend to be aggressive

FIGURE 4.15 Poison arrow frogs of the family Dendrobatidae use brilliant colors to warn potential predators of the extremely toxic secretions from their skin. Native people in Latin America use the toxin on blowgun darts.

defenders of their home areas, and acacia ants are no exception. They drive off herbivorous insects that attempt to feed on their home acacia, thus reducing predation. They also trim away vegetation that grows around their home tree, thereby reducing competition. This is a fascinating example of how a symbiotic relationship fits into community interactions. It is also an example of coevolution based on mutualism rather than competition or predation.

Defensive Mechanisms

Many species of plants and animals have toxic chemicals, body armor, and other ingenious defensive adaptations to protect themselves from competitors or predators. Arthropods, amphibians, snakes, and some mammals, for instance, produce noxious odors or poisonous secretions to induce other species to leave them alone. Plants also produce a variety of chemical compounds that make them unpalatable or dangerous to disturb. Perhaps you have brushed up against poison ivy or stinging nettles in the woods or you have encountered venomous insects or snakes and appreciate the wisdom of leaving them alone. Often, species possessing these chemical defenses will evolve distinctive colors or patterns to warn potential enemies (fig. 4.15).

Sometimes species that actually are harmless will evolve colors, patterns, or body shapes that mimic species that are unpalatable or poisonous. This is called **Batesian mimicry** after the English naturalist H. W. Bates, who described it in 1857. Wasps, for example, often have bold patterns of black and yellow stripes to warn off potential predators. The rarer longhorn beetle (fig. 4.16), although it has no stinger, looks and acts much like wasps and thus avoids predators as well. Another form of mimicry, called **Müllerian mimicry,** named for the German biologist Fritz Müller, who described it in 1878, involves two species, both of which are unpalatable or dangerous and have evolved to look alike. When predators learn to avoid either species, both benefit.

Species also evolve amazing abilities to avoid being discovered. You very likely have seen examples of insects that look exactly like dead leaves or twigs to hide from predators. Predators also use camouflage to hide as they lay in wait for their prey. The scorpion fish (fig. 4.17) blends in remarkably well with its surroundings as it waits for smaller fish to come within striking distance. Not all cases of mimicry are to avoid or carry out predation, however. Some tropical orchids have evolved flower structures that look exactly like female flies. Males attempting to mate unwittingly carry away pollen.

COMMUNITY PROPERTIES

The processes and principles that we have studied thus far in this chapter—tolerance limits, species interactions, resource partitioning, evolution, and adaptation—play important roles in determining the characteristics of populations and species. In this

FIGURE 4.16 An example of Batesian mimicry. The dangerous wasp (*left*) has bold yellow and black bands to warn predators away. The much rarer longhorn beetle (*right*) has no poisonous stinger, but looks and acts like a wasp and thus avoids predators as well.

FIGURE 4.17 This highly camouflaged scorpion fish lays in wait for its unsuspecting prey. Natural selection and evolution have created the elaborate disguise seen here.

section we will look at some fundamental properties of biological communities and ecosystems—productivity, diversity, complexity, resilience, stability, and structure—to learn how they are affected by these factors.

Productivity

A community's **primary productivity** is the rate of biomass production, an indication of the rate of solar energy conversion to chemical energy. The energy left after respiration is net primary production. Photosynthetic rates are regulated by light levels, temperature, moisture, and nutrient availability. Figure 4.18 shows approximate productivity levels for some major ecosystems. As you can see, tropical forests, coral reefs, and estuaries (bays or inundated river valleys where rivers meet the ocean) have high levels of productivity because they have abundant supplies of all these resources. In deserts, lack of water limits photosynthesis. On the arctic tundra or in high mountains, low temperatures inhibit plant growth. In the open ocean, a lack of nutrients reduces the ability of algae to make use of plentiful sunshine and water.

Some agricultural crops such as corn (maize) and sugar cane grown under ideal conditions in the tropics approach the productivity levels of tropical forests. Because shallow water ecosystems such as coral reefs, salt marshes, tidal mud flats, and other highly productive aquatic communities are relatively rare compared to the vast extent of open oceans—which are effectively biological deserts—marine ecosystems are much less productive *on average* than terrestrial ecosystems.

Even in the most photosynthetically active ecosystems, only a small percentage of the available sunlight is captured and used to make energy-rich compounds. Between one-quarter and three-quarters of the light reaching plants is reflected by leaf surfaces. Most of the light absorbed by leaves is converted to heat that is either radiated away or dissipated by evaporation of water. Only 0.1 to 0.2 percent of the absorbed energy is used by chloroplasts to synthesize carbohydrates.

In a temperate-climate oak forest, only about half the incident light available on a midsummer day is absorbed by the leaves. Ninety-nine percent of this energy is used to evaporate water. A large oak tree can transpire (evaporate) several thousand liters of water on a warm, dry, sunny day while it makes only a few kilograms of sugars and other energy-rich organic compounds.

Abundance and Diversity

Abundance is an expression of the total number of organisms in a biological community, while **diversity** is a measure of the number of different species, ecological niches, or genetic variation present. The abundance of a particular species often is inversely related to the total diversity of the community. That is, communities with a very large number of species often have only a few members of any given species in a particular area. As a general rule, diversity decreases but abundance within species increases as we go from the equator toward the poles. The arctic has vast numbers of insects such as mosquitoes, for example, but only a few species. The tropics, on the other hand, have vast numbers of species—some of which have incredibly bizarre forms and habits—but often only a few individuals of any particular species in a given area.

Consider bird populations. Greenland is home to 56 species of breeding birds, while Colombia, which is only one-fifth the size of Greenland, has 1395. Why so many species in Colombia and so few in Greenland?

Climate and history are important factors. Greenland has such a harsh climate that the need to survive through the winter or escape to milder climates becomes the single most important critical factor that overwhelms all other considerations and severely limits the ability of species to specialize or differentiate into new forms. Furthermore, because Greenland was covered by glaciers until about 10,000 years ago, there has been little time for new species to develop.

Many areas in the tropics, by contrast, have relatively abundant rainfall and warm temperatures year-round so that ecosystems there are highly productive. The year-round dependability of food, moisture, and warmth supports a great exuberance of life and allows a high degree of specialization in physical shape and behavior. Coral reefs are similarly stable, productive, and conducive to proliferation of diverse and exotic life-forms. The enormous abundance of brightly colored and fantastically shaped fish, corals, sponges, and arthropods in the reef community is one of the best examples we have of community diversity.

Productivity is related to abundance and diversity, both of which are dependent on the total resource availability in an ecosystem as well as the reliability of resources, the adaptations of the member species, and the interactions between species. You

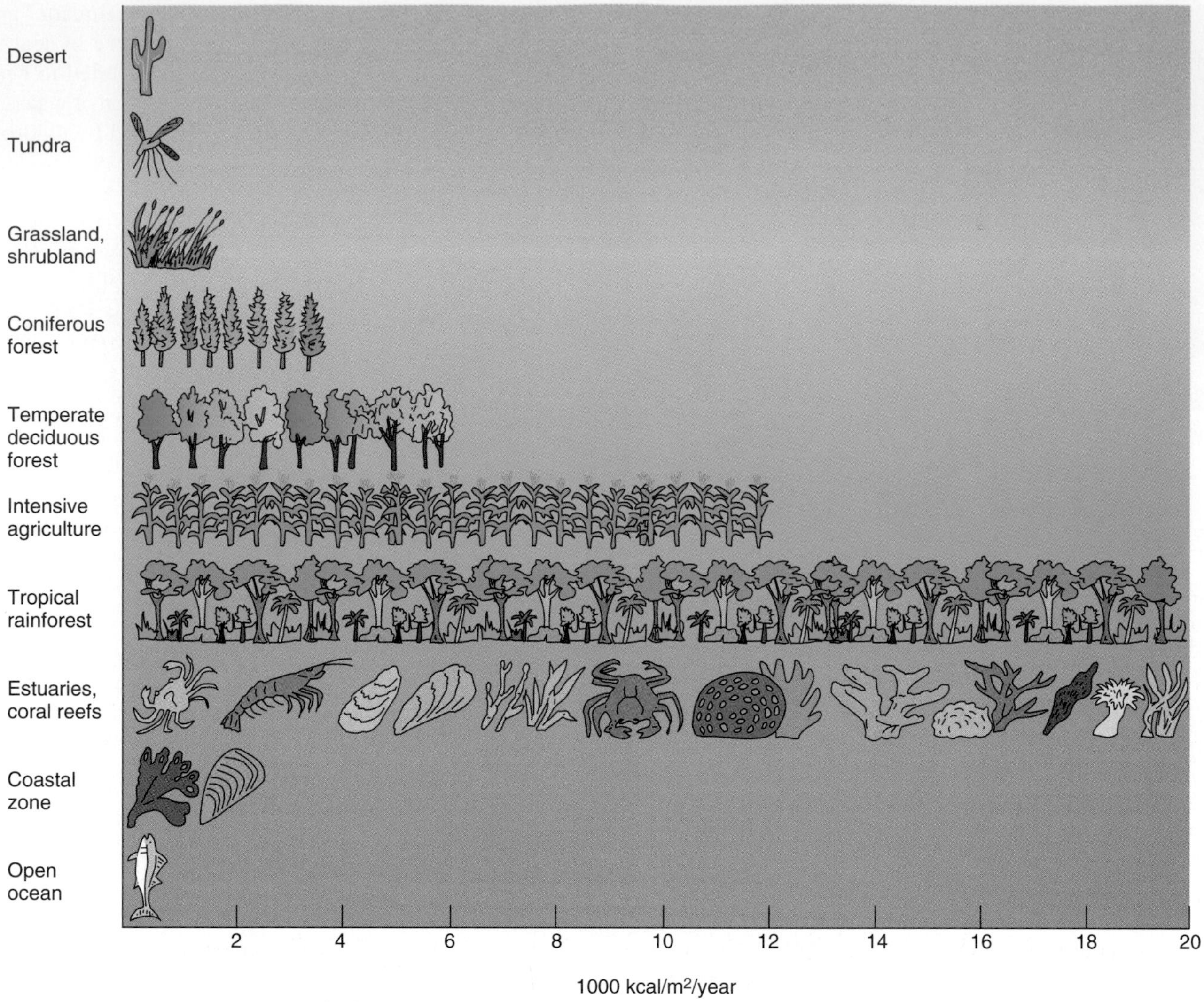

FIGURE 4.18 Relative biomass accumulation of major world ecosystems. Only plants and some bacteria capture solar energy. Animals consume biomass to build their own bodies.

shouldn't assume that all communities are perfectly adapted to their environment. A relatively new community that hasn't had time for niche specialization or a disturbed one where roles such as top predators are missing may not achieve maximum efficiency of resource use or reach its maximum level of either abundance or diversity.

Complexity and Connectedness

Community complexity and connectedness generally are related to diversity and are important because they help us visualize and understand community functions. **Complexity** in ecological terms refers to the number of species at each trophic level and the number of trophic levels in a community. A diverse community may not be very complex if all its species are clustered in only a few trophic levels and form a relatively simple food chain.

By contrast, a complex, highly interconnected community (fig. 4.19) might have many trophic levels, some of which can be compartmentalized into subdivisions. In tropical rainforests, for instance, the herbivores can be grouped into "guilds" based on the specialized ways they feed on plants. There may be fruit eaters, leaf nibblers, root borers, seed gnawers, and sap suckers, each composed of species of very different size, shape, and even biological kingdom, but that feed in related ways. A highly interconnected community such as this can form a very elaborate food web.

Resilience and Stability

Many biological communities tend to remain relatively stable and constant over time. An oak forest tends to remain an oak forest, for example, because the species that make it up have self-perpetuating

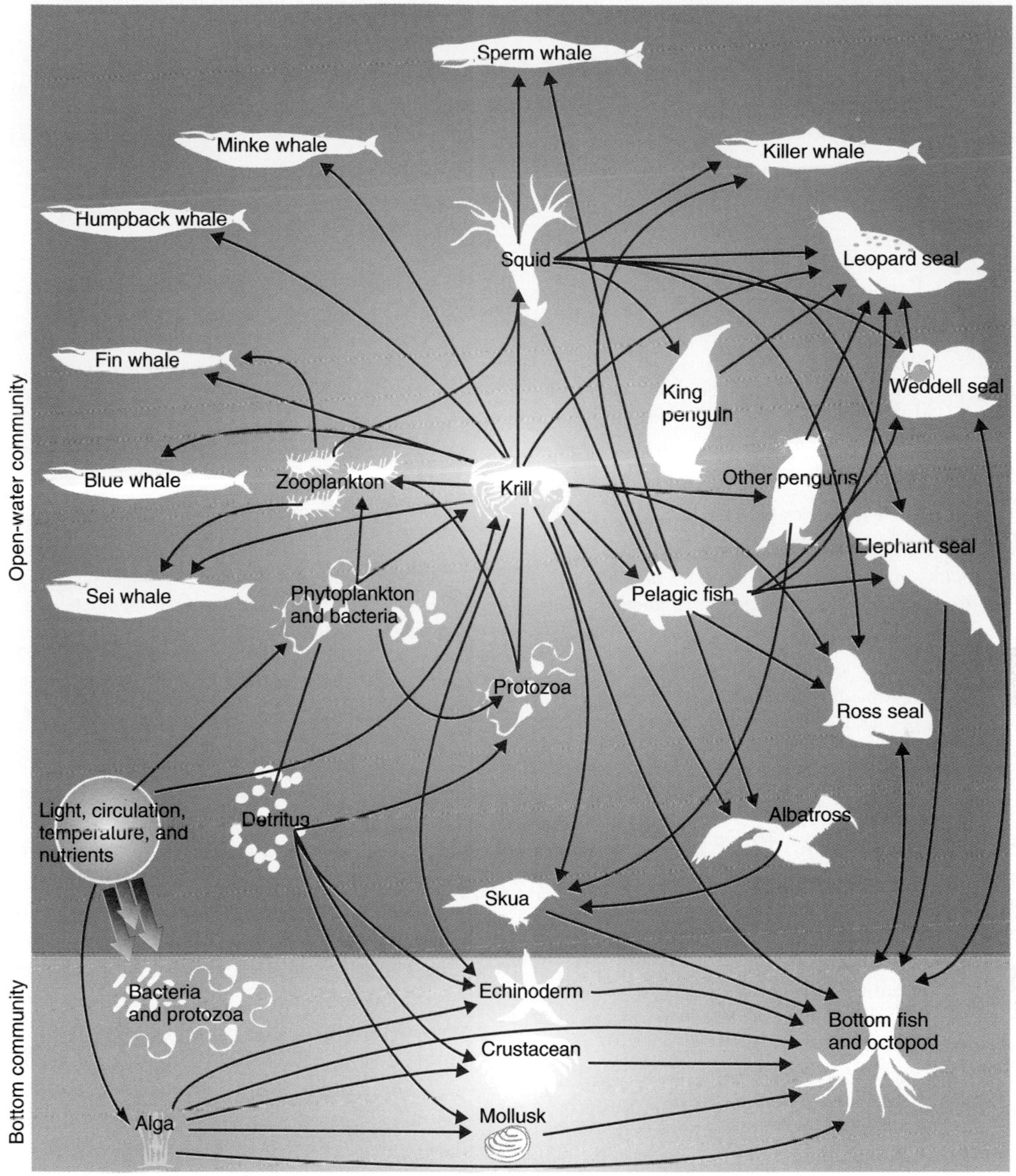

FIGURE 4.19 A complex and highly interconnected community can have many species at each trophic level and many relationships, as illustrated by this Antarctic marine food web

mechanisms. We can identify three kinds of stability or resiliency in ecosystems: *constancy* (lack of fluctuations in composition or functions), *inertia* (resistance to perturbations), and *renewal* (ability to repair damage after disturbance).

In 1955, Robert MacArthur, who was then a graduate student at Yale, proposed that the more complex and interconnected a community is, the more stable and resilient it will be in the face of disturbance. If many different species occupy each trophic level, some can fill in if others are stressed or eliminated by external forces, making the whole community resistant to perturbations and able to recover relatively easily from disruptions. This theory has been controversial, however. Some studies support it, while others do not. For example, Minnesota ecologist David Tilman, in studies of native prairie and recovering farm fields, found that plots with high diversity were better able to withstand and recover from drought than those with only a few species.

Yet, in a diverse and highly specialized ecosystem, removal of a few keystone members can eliminate many other associated species. Eliminating a major tree species from a tropical forest, for example, may destroy pollinators and fruit

what can you do?

Developing a Sense for Where You Live

Although we often feel insignificant and powerless in the face of the big problems facing society, there are things that individuals can do to help improve our environment. One of the first steps towards becoming a more effective environmental citizen is to learn something about the place where you live.

- What natural ecosystems and biological communities existed in your area before European settlement?
- What impact, if any, did indigenous people have on the flora, fauna, or topography of your locality?
- What are the dominant species (besides humans) in your neighborhood now and where did they originate?
- How much rain falls in your region in a given year? Is there generally more precipitation in one month or season than another? What role does moisture availability play in determining who lives where and how?
- What are the seasonal high, low, and average temperatures where you live? How do native plants and animals adapt to severe weather events and seasonal variations?
- Is there a keystone species or group of species especially important in determining the structure and functions of your local ecosystems? What factors might threaten those keystone components?
- Where do your drinking water, food, and energy come from? What local and regional environmental impacts are caused by production, use, and disposal of those resources? Could you lessen those impacts by changing your sources or use patterns of resources?
- Is there a park or wildlife refuge near where you live? Does it contain any rare, threatened, or endangered species? What makes them rare, threatened, or endangered?
- Are there opportunities for volunteer work to improve your local environment: planting trees, cleaning up a river or lake, restoring a wetland, recycling trash, helping to maintain a refuge or park?

distributors as well. We might replant the trees, but could we replace the whole web of relationships on which they depend? In this case, diversity has made the forest less resilient rather than more.

Community Structure

Ecological **structure** refers to patterns of spatial distribution of individuals and populations within the community as well as the relation of a particular community to its surroundings. At the local level, even in a relatively homogeneous environment, individuals in a single population can be distributed randomly, clumped together, or in highly regular patterns (fig. 4.20). In randomly arranged populations, individuals live wherever resources are available. Ordered patterns may be determined by the physical environment but are more often the result of biological competition. For example, competition for nesting space in a penguin colony is often fierce. Each nest tends to be just out of reach of the neighbors sitting on their own nests. Constant squabbling produces a highly regular pattern. Similarly, sagebrush releases toxins from roots and fallen leaves that inhibit the growth of competitors and create a circle of bare ground surrounding each bush. As neighbors fill in empty spaces up to the limit of this chemical barrier, a regular spacing results.

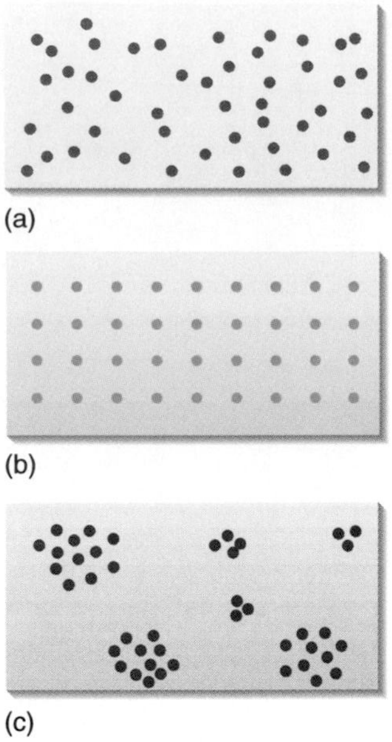

FIGURE 4.20 Distribution of members of a population in a given space can be random (*a*), ordered (*b*), or clustered (*c*). These patterns are determined both by the physical environment and by biological interactions. They may produce a graininess or patchiness in community structure.

Some other species cluster together (fig. 4.20*c*) for protection, mutual assistance, reproduction, or to gain access to a particular environmental resource. Dense schools of fish, for instance, cluster closely together in the ocean, increasing their chances of detecting and escaping predators (fig. 4.21). Similarly, predators, whether sharks, wolves, or humans, often hunt in packs to catch their prey. A flock of blackbirds descending on a cornfield or a troop of baboons traveling across the African savanna band together both to avoid predators and to find food more efficiently.

Plants, too, can be protected by clustering. A grove of wind-sheared evergreen trees is often found packed tightly together at the crest of a high mountain or along the seashore. They offer mutual protection from the wind not only to each other but also to other creatures that find shelter in or under their branches.

If we look closely at larger and more ecologically varied communities, we often find a mosaic of smaller units or subsets of the whole assemblage. These subunits develop because each species has a preference for specific, localized conditions. Ecologists describe these patterns as **patchiness** or graininess in spatial

FIGURE 4.21 Fish and birds often flock together in dense bands for protection and mutual feeding.

distribution. Some communities are coarse-grained with large patches; others are so fine-grained as to be almost indistinguishable from random distribution. Even though individual patches may be small and not very distinct from adjacent areas, the specific interactions in each patch may be more important in determining functions of the community than the overall composition.

Distribution in a community can be vertical as well as horizontal. The tropical forest, for instance, has many layers, each with different environmental conditions and combinations of species. While the larger trees are part of each layer, the communities of smaller plants, animals, and microbes that live in each part of this complex community generally are distinct and very different from each other. Similarly, aquatic communities are often stratified into layers based on light penetration in the water, temperature, salinity, pressure, or other factors.

FIGURE 4.22 Ecologists call the sharp edge, or boundary between the woods and sunny glade seen in this picture, an ecotone. Edge effects such as increased sunshine, wind, and temperature, as well as decreased humidity may extend a considerable distance into the forest. When fragmented into many small patches, the forest may become all edge, with no true interior core at all.

Edges and Boundaries

An important aspect of community structure is the boundary between one habitat and its neighbors. We call these relationships **edge effects.** In many cases, the edge of a patch of habitat is relatively sharp and distinct. In moving from a woodland patch into a grassland or cultivated field, you sense a dramatic change from the cool, dark, quiet forest interior to the windy, sunny, warmer, open space of the field or pasture. Species composition may change equally suddenly if organisms adapted to one set of conditions can't survive in the other (fig. 4.22).

Ecologists call the boundaries between adjacent communities **ecotones.** A community that is sharply divided from its neighbors is called a closed community. In contrast, communities with gradual or indistinct boundaries over which many species cross are called open communities. Often this distinction is a matter of degree or perception. As we saw earlier in this chapter, birds might feed in fields or grasslands but nest in the forest. As they fly back and forth, the birds interconnect the ecosystems by moving energy and material from one to the other, making both systems relatively open. Furthermore, the forest edge, while clearly different from the open field, may be sunnier and warmer than the forest interior, and may have a different combination of plant and animal species than either field or forest "core."

Depending on how far edge effects extend from the boundary, differently shaped habitat patches may have very dissimilar amounts of interior area (fig. 4.23). In Douglas fir forests of the Pacific Northwest, for example, increased rates of blowdown, decreased humidity, absence of shade-requiring ground cover, and other edge effects can extend as much as 200 m into a forest. A 40-acre block (200 meters square) surrounded by clear-cut would have essentially no true core habitat at all.

Many popular game animals, such as white-tailed deer and pheasants that are adapted to human disturbance, often are most plentiful in boundary zones between different types of habitat. Game managers once were urged to develop as much edge as possible to promote large game populations. Today, however, most wildlife conservationists recognize that the edge effects associated with habitat fragmentation are generally detrimental to biodiversity. Preserving large habitat blocks and linking smaller blocks with migration corridors are among the best ways to protect rare and endangered species (see chapter 15).

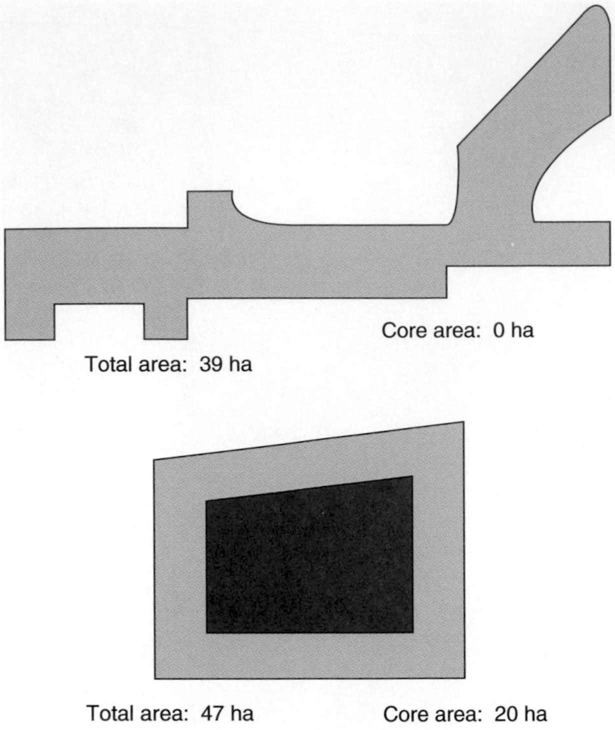

FIGURE 4.23 Shape can be as important as size in small preserves. While these areas are close to the same size, no place in the top figure is far enough from the edge to have characteristics of core habitat, while the bottom patch has a significant core.

COMMUNITIES IN TRANSITION

So far our view of communities has focused on the day-to-day interactions of organisms with their environments, set in a context of survival and selection. In this section, we'll step back to look at some transitional aspects of communities, including where communities meet and how communities change over time.

Ecological Succession

Biological communities have a history in a given landscape. The process by which organisms occupy a site and gradually change environmental conditions by creating soil, shade, shelter, or increasing humidity is called ecological succession or development. **Primary succession** occurs when a community begins to develop on a site previously unoccupied by living organisms, such as an island, a sand or silt bed, a body of water, or a new volcanic flow (fig. 4.24). **Secondary succession** occurs when an existing community is disrupted and a new one subsequently develops at the site. The disruption may be caused by some natural catastrophe, such as fire or flooding, or by a human activity, such as deforestation, plowing, or mining. Both forms of succession usually follow an orderly sequence of stages as organisms modify the environment in ways that allow one species to replace another.

In primary succession on a terrestrial site, the new site first is colonized by a few hardy **pioneer species,** often microbes, mosses, and lichens that can withstand harsh conditions and lack of resources. Their bodies create patches of organic matter in which protists and small animals can live. Organic debris accumulates in pockets and crevices, providing soil in which seeds can become lodged and grow. We call this process of environmental modification by organisms **ecological development** or facilitation. The community of organisms often becomes more diverse and increasingly competitive as development continues and new niche opportunities appear. The pioneer species gradually disappear as the environment changes and new species combinations replace the preceding community. In a global sense, the gradual changes brought about by living organisms have created many of the conditions that make life on earth possible. You could consider evolution to be a very slow, planetwide successional and developmental process.

Examples of secondary succession are easy to find. Observe an abandoned farm field or clear-cut forest (fig. 4.25) in a temperate climate. The bare soil first is colonized by rapidly growing annual plants (those that grow, flower, and die the same year) that have light, wind-blown seeds and can tolerate full sunlight and exposed soil. They are followed and replaced by perennial plants (those that live for several to many years), including grasses, various nonwoody flowering plants, shrubs, and trees. As in primary succession, plant species progressively change the environmental conditions. Biomass accumulates and the site becomes richer, better able to capture and store moisture, more sheltered from wind and climate change, and biologically more complex. Species that cannot survive in a bare, dry, sunny, open area find shelter and food as the field turns to prairie or forest.

Eventually, in either primary or secondary succession, a community develops that seemingly resists further change. Ecologists call this a **climax community** because it appears to be the culmination of the successional process. An analogy is often made between community succession and organism maturation. Beginning with a primitive or juvenile state and going through a complex developmental process, each progresses until a complex, stable, and mature form is reached. It's dangerous to carry this analogy too far, however, because no mechanism is known to regulate communities in the same way that genetics and physiology regulate development of the body.

As mentioned in chapter 3, the concept of succession to a climax community was first championed by the pioneer biogeographer F. E. Clements. He viewed this process as being like a parade or relay, in which species replace each other in predictable groups and in a fixed, regular order, and as being driven almost entirely by climate. This community-unit theory was opposed by Clements's contemporary, H. A. Gleason, who saw community history as a much more individualistic and random process driven by many environmental factors. He argued that temporary associations are formed according to the conditions prevailing at a particular time and the species available to colonize a given area. You might think of the Gleasonian model as a time-lapse movie of a busy railroad station. Passengers come and go; groups form and then dissipate. Patterns and assemblages that seem significant to us may not mean much in the long run.

FIGURE 4.24 Primary succession on a terrestrial site is shown in five stages (*left to right*), beginning with rocks that are initially colonized by a pioneer community of lichens and mosses and ending with a climax forest community.

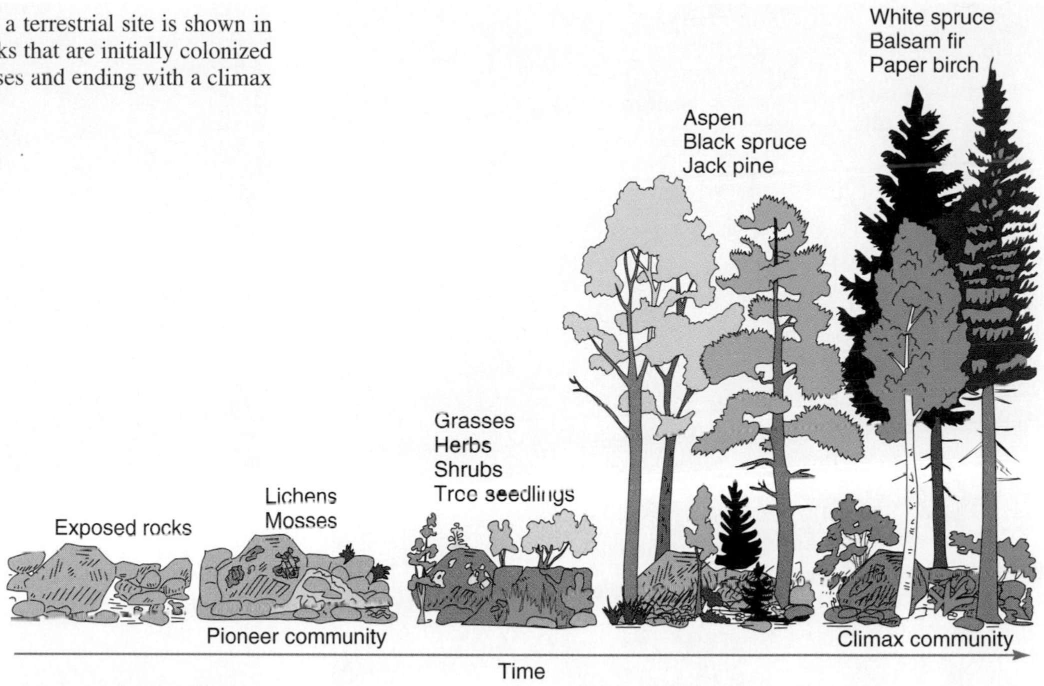

FIGURE 4.25 This area was once a cool, shady black spruce stand. The forest floor was covered by a deep, moist layer of sphagnum moss. Clear-cutting and burning have turned it into a dry, sunny, barren ground on which few of the former residents can survive. Secondary succession will eventually restore previous conditions if the climate doesn't change and further disturbance is prevented.

The process of succession may not be as deterministic as we once thought, yet mature or highly developed ecological communities may tend to be resilient and stable over long periods of time because they can resist or recover from external disturbances. Many are characterized by high species diversity, narrow niche specialization, well-organized community structure, good nutrient conservation and recycling, and a large amount of total organic matter. Community functions, such as productivity and nutrient cycling, tend to be self-stabilizing or self-perpetuating. What once were regarded as "final" climax communities, however, may still be changing. It's probably more accurate to say that the rate of succession is so slow in a climax community that, from the perspective of a single human lifetime, it appears to be stable.

Some landscapes never reach a stable climax in the traditional sense because they are characterized by, and adapted to, periodic disruption. They are called **equilibrium communities** or **disclimax communities.** Grasslands, the chaparral shrubland of California, and some kinds of coniferous forests, for instance, are shaped and maintained by periodic fires that have long been a part of their history. They are, therefore, often referred to as **fire-climax communities** (fig. 4.26). Plants in these communities are adapted to resist fires, reseed quickly after fires, or both. In fact, many of the plant species we recognize as dominants in these communities *require* fire to eliminate competition, to prepare seedbeds for germination of seedlings, or to open cones or thick seed coats. Without fire, community structure may be quite different.

Introduced Species and Community Change

Succession requires the continual introduction of new community members and the disappearance of previously existing species. New species move in as conditions become suitable; others die or move out as the community changes. New species also can be introduced after a stable community already has become established. Some cannot compete with existing species and fail to become established. Others are able to fit into and become part of the community, defining new ecological niches. If, however, an introduced species preys upon or competes more successfully with one or more populations that are native to the community, the entire nature of the community can be altered.

FIGURE 4.26 This lodgepole pine forest in Yellowstone National Park was once thought to be a climax forest, but we now know that this forest must be constantly renewed by periodic fire. It is an example of an equilibrium, or disclimax, community.

FIGURE 4.27 Mongooses were released in Hawaii in an effort to control rats. The mongooses are active during the day, however, while the rats are night creatures so they ignored each other. Instead, the mongooses attacked defenseless native birds and became as great a problem as the rats.

Human introductions of Eurasian plants and animals to non-Eurasian communities often have been disastrous to native species because of competition or overpredation. Oceanic islands offer classic examples of devastation caused by rats, goats, cats, and pigs liberated from sailing ships. All these animals are prolific, quickly developing large populations. Goats are efficient, nonspecific herbivores; they eat nearly everything vegetational, from grasses and herbs to seedlings and shrubs. In addition, their sharp hooves are hard on plants rooted in thin island soils. Rats and pigs are opportunistic omnivores, eating the eggs and nestlings of seabirds that tend to nest in large, densely packed colonies, and digging up sea turtle eggs. Cats prey upon nestlings of both ground- and tree-nesting birds. Native island species are particularly vulnerable because they have not evolved under circumstances that required them to have defensive adaptations to these predators.

Sometimes we introduce new species in an attempt to solve problems created by previous introductions but end up making the situation worse. In Hawaii and on several Caribbean Islands, for instance, mongooses were imported to help control rats that had escaped from ships and were destroying indigenous birds (fig. 4.27). Since the mongooses were diurnal (active in the day), however, and rats are nocturnal, they tended to ignore each other. Instead, the mongooses also killed native birds and further threatened endangered species. Our lessons from this and similar introductions have a new technological twist. Some of the ethical questions currently surrounding the release of genetically engineered organisms are based on concerns that they are novel organisms, and we might not be able to predict how they will interact with other species in natural ecosystems—let alone how they might respond to natural selective forces. It is argued that we can't predict either their behavior or their evolution.

SUMMARY

The principle of limiting factors states that for every physical factor in the environment, there are both maximum and minimum tolerable limits beyond which a given species cannot survive. Often the factor in shortest supply or closest to the tolerance limit for a particular species at a particular time is the critical factor that will determine the abundance and distribution of that species in that ecosystem. Random genetic variations create diversity in a population that gives some individuals advantages in a given set of circumstances. The best-suited organisms will survive and reproduce more successfully than the ill-suited ones. Eventually, the genes for these successful characteristics predominate in the population and the species becomes adapted to its environment and to a particular role. This process leads to evolution of a species either through gradual replacement of the original parental type or a splitting of a population into two species.

Habitat describes the place in which an organism lives; niche describes either the role an organism plays or the total set of conditions that control its distribution. Natural selection often leads to niche specialization and resource partitioning that reduce competition between species. Organisms interact within communities in many ways. Predation—feeding on another organism—involves pathogens, parasites, and herbivores as well as carnivorous predators. Competition is another kind of antagonistic relationship in which organisms vie for space, food, or other resources. Symbiosis is the intimate living together of two species. Mutualism means that both species benefit; commensalism means that one species benefits while the other is indifferent.

Some fundamental properties of biological communities are productivity, diversity, complexity, resilience, stability, and structure. Productivity is a measure of the rate at which photosynthesis produces biomass made of energy-rich compounds. Tropical rainforests are generally the most productive of all terrestrial communities; coral reefs and estuaries are generally the most productive aquatic communities. Diversity is a measure of the number of different species in a community, while abundance is the total number of individuals. Often the most productive and stable communities are highly diverse and profusely populated, but sometimes a high degree of specialization makes an ecosystem more, rather than less, susceptible to disturbance. Ecological complexity refers to the number of species at each trophic level as well as the total number of trophic levels in a community. Structure concerns the patterns of organization, both spatial and functional, in a community. Often a keystone species, or group of species, plays an unusually important role in determining community structure, composition, or function. All of these characteristics are affected both by physical and chemical factors as well as biological interactions between the organisms that make up the community. Edge effects at the boundaries between different habitat types can be important where landscapes are fragmented into isolated patches.

Ecological succession and development are processes by which organisms alter the environment in ways that allow some species to replace others. Primary succession starts with a previously unoccupied site. Secondary succession occurs on a site that has been disturbed by external forces. Often succession proceeds until a mature, diverse, climax community is established. These mature communities may have self-perpetuating processes that make them resistant to change and resilient to disturbance. Whether diversity always leads to stability, however, is controversial. Communities that are disrupted regularly by fires or other natural disasters sometimes establish dynamic equilibrium or disclimax communities dependent on constant renewal. Introduction of new species by natural processes, such as opening of a land bridge, or through human intervention can upset the natural relationships in a community and cause catastrophic changes for indigenous species.

QUESTIONS FOR REVIEW

1. Explain how tolerance limits to environmental factors determine distribution of a highly specialized species such as the desert pupfish. Compare this to the distribution of a generalist species such as cowbirds or starlings.
2. Productivity, diversity, complexity, resilience, and structure are exhibited to some extent by all communities and ecosystems. Describe how these characteristics apply to the ecosystem in which you live.
3. Describe the general niche occupied by a bird of prey, such as a hawk or an owl. How can hawks and owls exist in the same ecosystem and not adversely affect each other?
4. Define keystone species and explain their importance in community structure and function.
5. All organisms within a biological community interact with each other. The most intense interactions often occur between individuals of the same species. What concept discussed in this chapter can be used to explain this phenomenon?
6. Relationships between predators and prey play an important role in the energy transfers that occur in ecosystems. They also influence the process of natural selection. Explain how predators affect the adaptations of their prey. This relationship also works in reverse. How do prey species affect the adaptations of their predators?
7. Competition for a limited quantity of resources occurs in all ecosystems. This competition can be interspecific or intraspecific. Explain some of the ways an organism might deal with these different types of competition.
8. Each year fires burn large tracts of forestland. Describe the process of succession that occurs after a forest fire destroys an existing biological community. Is the composition of the final successional community likely to be the same as that which existed before the fire? What factors might alter the final outcome of the successional process? Why may periodic fire be beneficial to a community?
9. Explain the concept of climax community. Why does the climax community often exhibit a higher level of stability than that found in other successional stages?

10. Discuss the dangers posed to existing community members when new species are introduced into ecosystems. What type of organism would be most likely to survive and cause problems in a new habitat?

QUESTIONS FOR CRITICAL THINKING

1. Ecologists debate whether biological communities have self-sustaining, self-regulating characteristics or are highly variable, accidental assemblages of individually acting species. What outlook or worldview might lead scientists to favor one or the other of these theories?
2. The concepts of natural selection and evolution are central to how most biologists understand and interpret the world, and yet the theory of evolution is contrary to the beliefs of many religious groups. Why do you think this theory is so important to science and so strongly opposed by others? What evidence would be required to convince opponents of evolution?
3. What is the difference between saying that a duck has webbed feet because it needs them to swim and saying that a duck is able to swim because it has webbed feet?
4. The concept of keystone species is controversial among ecologists because most organisms are highly interdependent. If each of the trophic levels is dependent on all the others, how can we say one is most important? Choose an ecosystem with which you are familiar and decide whether it has a keystone species or keystone set.
5. Some scientists look at the boundary between two biological communities and see a sharp dividing line. Others looking at the same boundary see a gradual transition with much intermixing of species and many interactions between communities. Why such different interpretations of the same landscape?
6. The absence of certain lichens is used as an indicator of air pollution in remote areas such as national parks. How can we be sure that air pollution is really responsible? What evidence would be convincing?
7. We tend to regard generalists or "weedy" species as less interesting and less valuable than rare and highly specialized endemic species. What values or assumptions underlie this attitude?
8. What part of this chapter do you think is most likely to be challenged or modified in the future by new evidence or new interpretations?

KEY TERMS

abundance 91
Batesian mimicry 90
climax community 96
coevolution 85
commensalism 89
complexity 92
disclimax communities 97
diversity 91
ecological development 96
ecological niche 82
ecotone 95
edge effects 95
environmental indicators 80
equilibrium communities 97
evolution 80
fire-climax communities 97
habitat 82
interspecific competition 86
intraspecific competition 86
keystone species 85
Müllerian mimicry 90
mutualism 89
natural selection 80
parasites 84
patchiness 94
pathogens 84
pioneer species 96
plankton 84
predator 84
primary productivity 91
primary succession 96
resource partitioning 83
secondary succession 96
structure 94
symbiosis 89
territoriality 89
tolerance limits 79

ADDITIONAL INFORMATION ON THE INTERNET

Visit our website at http://www.mhhe.com/environmentalscience/ for specific information about each of the topics below:

Otters, Orcas, and other marine species in Alaska

Bird monitoring and research:
Pautuxent Observatory, Cornell Laboratory of Ornithology

Biological Resources Division, USGS

National Biological Information Infrastructure

Long-term Ecological Research Network

Atmosphere, Hydrosphere, and Biosphere

International Geosphere and Biosphere Program

Complexity in ecological systems

PROFILE

Environmental Advocate

Marion Taylor

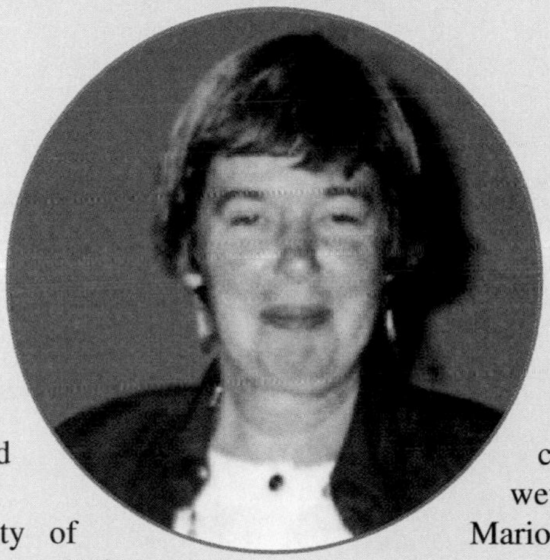

As is true for a growing number of people, Marion Taylor has discovered that life experiences can influence career plans, changing them unpredictably. Marion is employed as the Director of Environmental Affairs for the Federation of Ontario Naturalists, a membership-based organization headquartered in Toronto. She finds her work both stimulating and meaningful, but quite different from what she expected to be doing, given her formal college training.

Marion graduated from the University of Toronto with majors in English and French. She taught for a few years and then took time out to have a family. She developed an interest in nature and after a broken marriage, began to do volunteer work for environmental causes, including work with the Federation of Ontario Naturalists. When a position in the organization opened up, she applied for it. Although she lacked technical training in the natural sciences, her focus on language in college had provided her with equally important communication and critical thinking skills.

Besides her experience as a member and volunteer, which had familiarized Marion with the work of the organization, she feels her writing ability was a critical factor in her successful employment. Like many others, Marion discovered that job experience, whether gained as part of an internship program or as an unpaid volunteer, can often put you a step ahead of other applicants.

An important part of her work is to advocate for sound environmental management across a broad spectrum of natural resource issues throughout the province of Ontario. One project was to influence development of a new Crown Forest Sustainability Act to guide forest management policy for the extensive Ontario public lands. She, with other activists, successfully championed the incorporation of the use of sustainability indicators into the Act. This will help insure that wildlife and water quality, as well as biological diversity issues, are considered along with timber harvest in forest management plans.

Marion also worked with the Ministry of Natural Resources on boundary and policy issues for the million-hectare Wabakimi Provincial Park northwest of Lake Nipigon in western Ontario. She helped establish ecologic park boundaries, focusing attention on wilderness recreational values and on preserving a viable unit of boreal forest to serve as an undisturbed benchmark against which the effects of various forest management practices could be measured.

She has also worked with local conservation authorities in southern Ontario on flood control and other river issues, and is a citizen member on a committee of government, community, and industry representatives resolving conflicts between gravel extraction and water, wetland, and community values. In these roles Marion negotiates with government agents and industry representatives, who often arrive with conflicting points of view. While technical backgrounds help in these settings, to be truly effective requires strong interpersonal skills.

Another aspect of Marion's job includes writing a column for the organization's magazine, attending numerous meetings, occasionally testifying at public hearings, and spending lots of time on the telephone. Budget allowing, she travels to advise and encourage various grassroots citizen efforts.

For students interested in preparing themselves for advocacy, Marion advises getting as broad a base of expertise as possible. Although wishing she had formal training in computers and the biological sciences, Marion feels the most essential element of preparation is to develop critical thinking. She feels strongly that success in her job requires practicing skepticism, that is, seeking out the factual basis for claims and positions taken by others. Marion finds that writings and positions taken by either side of any environmental issue are often unreasonable, unfair, and sometimes, outright silly. She finds it very important to be able to identify, clarify, and, where necessary, question assumptions. Effective advocacy and resolution of many of the issues with which she works requires skills to clarify issues and language, to make the complex understandable without distortion. She is emphatic that students will be well-served by consciously working at being clear, concise, and fair in thought and language.

Marion also advises students to develop good interpersonal skills. The ability to listen and to understand the conceptual frameworks of other people is very important. Her experience illustrates that skills and attributes other than simple technical knowledge can be important in successfully establishing an environmental career.

CHAPTER 5

Biomes, Landscapes, Restoration, and Management

What is the use of a house if you haven't got a tolerable planet to put it on?

—Henry David Thoreau—

OBJECTIVES

After studying this chapter, you should be able to:

- recognize the characteristics of major aquatic and terrestrial biomes and understand the most important factors that determine the distribution of each type.
- describe ways in which humans disrupt or damage each of these ecosystem types.
- summarize the overall patterns of human disturbance of world biomes as well as some specific, important examples of losses obscured by broad aggregate categories.
- explain the principles and practices of landscape ecology and ecosystem management.
- evaluate the pros and cons of restoring, replacing, or substituting ecosystems and resources for those we have damaged.

ESSENTIAL STUDY PARTNER

The Essential Study Partner (ESP) CD-ROM that accompanies this textbook contains additional information on the following topics:

- temperature, precipitation, and vegetation: principles: biomes: climate
- terrestrial, aquatic, and high-elevation biome characteristics, productivity, and distribution: principles; biomes: land biomes/ principles: ecosystems: energy flow
- types of freshwater, marine, and wetland ecosystems and their structure: principles: biomes: aquatic systems

Above: Volunteers on a Texas beach stake down discarded Christmas trees to stabilize dunes and shelter beach grasses.

Integrity, Stability, and Beauty of the Land

In 1935, pioneering wildlife ecologist Aldo Leopold bought 32 ha (80 acres) of worn out, sandy farmland on the banks of the Wisconsin River not far from his home in Madison. Originally intended to be merely a hunting camp, the farm quickly became a year-round retreat from the city, as well as a laboratory in which Leopold could test his theories about conservation, environmental ethics, and ecologically based land management. A dilapidated chicken shack, the only remaining building from the original farm, was remodeled into a rustic cabin (fig. 5.1). The whole Leopold family participated in tree planting, bird watching, gardening, and exploring nature.

The old farm was not pristine wilderness nor were the Leopolds merely spectators. They regarded themselves as participating citizens of the land community, seeking to restore it to ecological health and beauty. Planting as many as 6000 trees and bushes each spring, they practiced "wild husbandry," using axes and shovels to reverse the abuses of previous owners and to revitalize the land through active management, care, and understanding. "Conservation," Leopold wrote, "is the positive exercise of skill and insight, not merely a negative exercise of abstinence or caution."

While building this relationship with the land—by which he meant all the plants and animals as well as the nonliving components of the landscape—Leopold mused on the ethics and meaning of conservation and the proper role of humans in nature. The first part of his much-beloved *Sand County Almanac* is a collection of essays about experiments and experiences at the farm. All of us, he claimed, should choose a piece of land on which we can practice stewardship and develop a sense of place. It doesn't have to be a beautiful place. In fact, it might be best to adopt a weedy, unwanted patch that needs our love and care. Both we and the land benefit from such connectedness, he maintained.

Leopold's essay on "The Land Ethic" is a cornerstone of the conservation movement and one of the most eloquent statements of environmental philosophy in American nature writing. In it, Leopold wrote, "We abuse the land because we regard it as a commodity belonging to us. When we see land as a community to which we belong, we may begin to use it with love and respect. . . . A land ethic, then, reflects the existence of an ecological conscience, and this in turn reflects a conviction of individual responsibility for the health of the land. Health is the capacity of the land for self-renewal. Conservation is our effort to understand and preserve this capacity. . . . A thing is right when it preserves the integrity, stability, and beauty of the biotic community. It is wrong when it does otherwise."

The story of Leopold's Sand County farm embodies many of the themes of this chapter. First, we survey some of the major biomes or biological communities around the world as a measure of what the components of a healthy landscape might be. Then we examine what the relatively new fields of landscape and restoration ecology say about our environment and the roles we might play in it. Finally, we look at the emerging goals of ecosystem management and some of the controversies and questions that have persisted from Aldo Leopold's day to our own about how we can and should care for nature.

FIGURE 5.1 Aldo Leopold's Sand County farm in central Wisconsin served as a refuge from the city and a laboratory to test theories about land conservation, environmental ethics, and ecologically based land management.

TERRESTRIAL BIOMES

Many places on the earth share similar climatic, topographic, and soil conditions, and roughly comparable communities have developed in response to analogous conditions in widely separated locations. These broad types of biological communities are called **biomes.** Although, as we pointed out in previous chapters, there can be considerable variation in the individual species that make up biological communities at two similar sites or even at a single site over time, some broad landscape categories exist. Recognizing these categories gives us insights about the general kinds of plants and animals that we might expect to find there and how they may be adapted to their particular environment.

Temperature and precipitation are among the most important determinants in biome distribution. If we know the general temperature range and precipitation level, we can predict what kind of biological community is likely to develop on a particular site if that site is free of disturbance for a sufficient time (fig. 5.2). Biome distribution also is influenced by the prevailing landforms of an area. Mountains, in particular, exert major influences on biological communities.

Figure 5.3 shows the distribution of major terrestrial biomes around the world. Because of its broad scope, this map ignores the

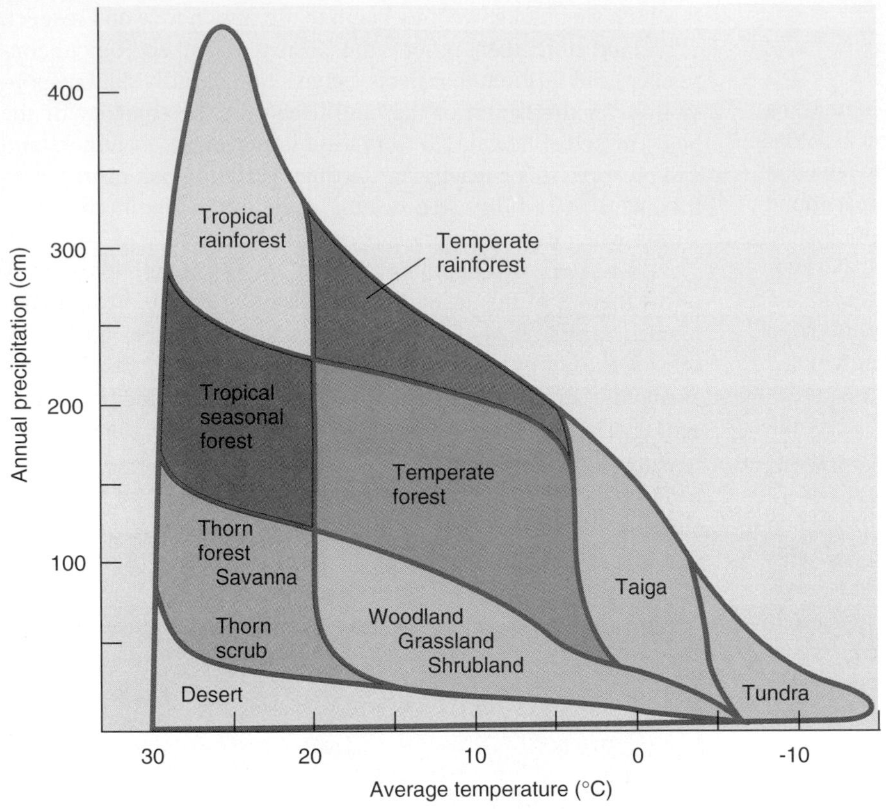

FIGURE 5.2 Biomes most likely to occur in the absence of human disturbance or other disruptions, according to average annual temperature and precipitation. *Note:* this diagram does not consider soil type, topography, wind speed, or other important environmental factors. Still, it is a useful general guideline for biome location.

From *Communities and Ecosystems,* 2/e by R. H. Whitaker, © 1975. Reprinted by permission of Prentice Hall, Upper Saddle River, New Jersey.

many variations present within each major category. Most terrestrial biomes are identified by the dominant plants of their communities (for example, grassland or deciduous forest). The characteristic diversity of animal life and smaller plant forms within each biome is, in turn, influenced both by the physical conditions and the dominant vegetation.

Deserts

Deserts are characterized by low moisture levels and precipitation that is both infrequent and unpredictable from year to year. With little moisture to absorb and store heat, daily and seasonal temperatures can fluctuate widely. Deserts that have less than 2.5 cm (1 in) of measurable precipitation support almost no vegetation. Deserts with 2.5 to 5 cm (1 to 2 in) annual precipitation have sparse vegetation (less than 10 percent of the ground is covered), and plants in this harsh climate need a variety of specializations to conserve water and protect tissues from predation. Seasonal leaf production, water-storage tissues, and thick epidermal layers help reduce water loss. Spines and thorns discourage predators while also providing shade (fig. 5.4).

Warm, dry, descending air (see chapter 17) creates broad desert bands in continental interiors at about 30° north in the American Southwest, North and South Africa, China, and Australia. These descending air currents also help create desert strips along the west coast of South America and Africa that are among the driest regions in the world. Although we think of deserts as hot, barren, and filled with sand dunes, those at high latitudes or high elevations are often cool or even cold and sand dunes are actually rather rare away from coastal areas. Most deserts around the world are gravelly or rocky scrubland, where 5 to 10 cm (2 to 4 in) of annual precipitation supports a sparse but often species-rich community dominated by shrubs or small trees. In rare years when winter rains are adequate, a breathtaking profusion of spring ephemerals can carpet the desert with flowers.

Animals of the desert have both structural and behavioral adaptations to meet their three most critical needs: food, water, and heat survival. Many desert animals escape the main onslaught of daytime heat by hiding in burrows or rocky shelters from which they emerge only at night. Pocket mice and kangaroo rats (and their Old World counterparts, gerbils) get most of the moisture they need from the seeds and grains they eat. They have many adaptations to conserve water such as producing highly concentrated urine and nearly dry feces that allow them to eliminate body wastes without losing precious moisture.

Deserts may seem formidable, but they are more vulnerable than you might imagine. Desert soils are easily disturbed by human activities and are slow to recover because the harsh desert climate severely reduces the ability of desert communities to recover from damage. Tracks left by army tanks practicing in the California desert during World War II, for instance, are still visible.

Many dry areas have been overgrazed, mainly by domestic livestock. Other areas are being converted to agricultural land in spite of uncertain future availability of water to sustain crops. Often after humans have degraded and abandoned desert areas they remain wastelands unsuitable for native vegetation or wildlife for a very long time.

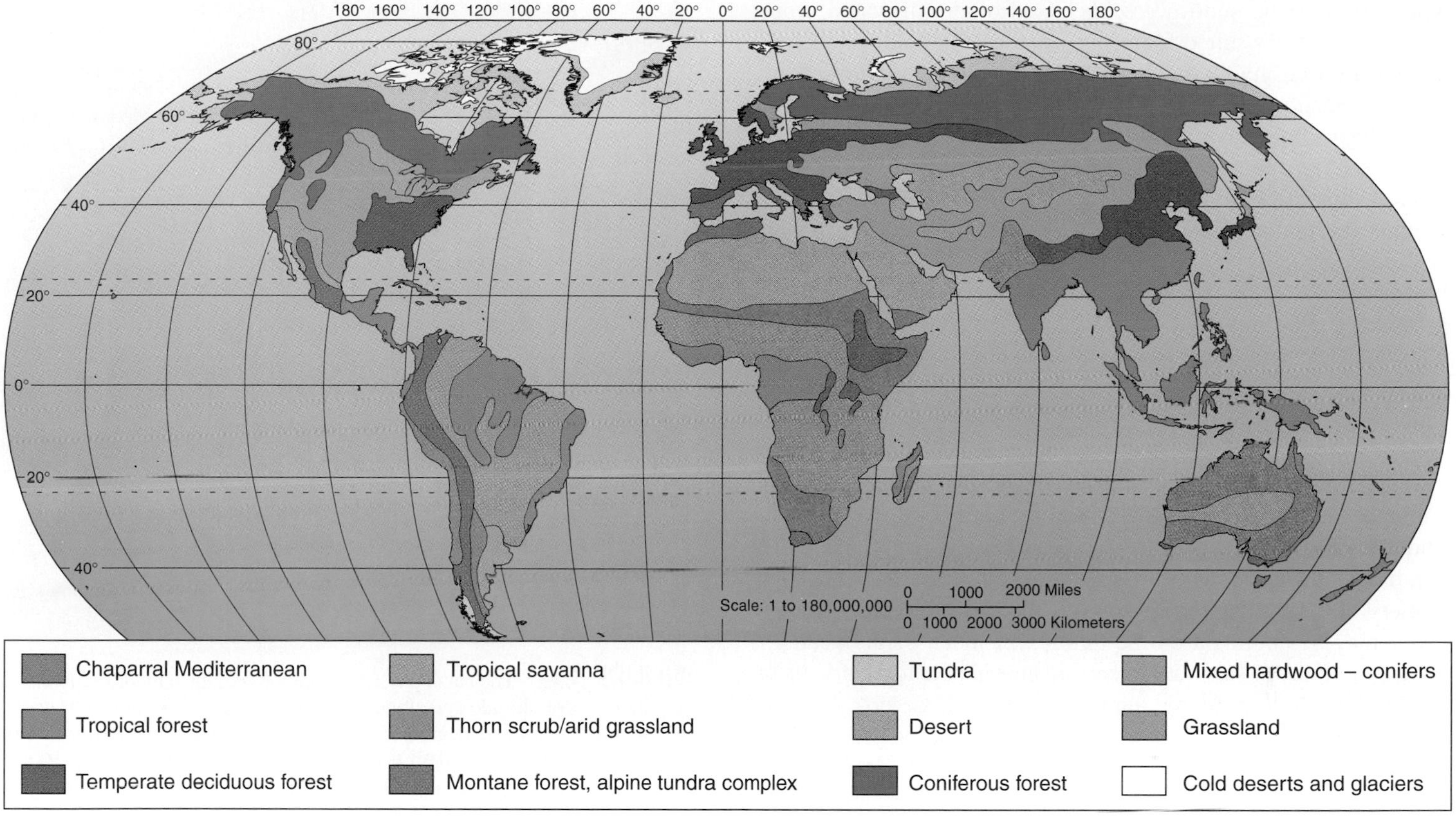

FIGURE 5.3 This generalized map of world biomes gives an overview of the vegetation types on different landmasses. At this scale, much detail cannot be shown, so such maps should be interpreted with that limitation in mind. Note that both moist evergreen and seasonally dry forests are included in tropical forests.

FIGURE 5.4 Joshua trees in the Joshua Tree National Park are really large yuccas—members of the lily family. Like other plants in this hot, dry, rocky landscape, they are adapted to conserve water and repel enemies.

Grasslands: Prairies and Savannas

The moderately dry continental climates of the Great Plains of central North America, the broad Russian steppes, the African veldt, and the South American pampas support **grasslands,** rich biological communities of grasses, seasonal herbaceous flowering plants, and open savannas. Seasonal cycles for temperature and precipitation contribute to abundant vegetative growth that both protects and enriches the soils of these prairies and plains, making them among the richest farmlands in the world.

Grasslands have few trees because inadequate rainfall, large daily and seasonal temperature ranges, and frequent grass fires kill woody seedlings. Exceptions are the narrow gallery forests that form corridors along rivers and streams through the grasslands.

In many parts of the world, grasslands are artificially created or maintained by native people using fire. They do this to improve hunting and ease of travel.

In contrast, fire suppression and conversion of the rich prairie soil to farmland have greatly reduced native grasslands elsewhere. The most productive North American prairie lands now grow wheat, corn, oats, sunflowers, flax, and other cultivated crops. Lands that are less suited for crops because of limited water availability are used mainly as rangeland, where overgrazing reduces natural diversity of the community and contributes to soil erosion.

Before humans converted North America's grasslands to croplands and rangelands, they contained herds of wildlife that rivaled those of Africa's Serengeti Plain. Vast herds of bison roamed the plains along with wolves, deer, elk, and pronghorn antelopes. Shorebirds and migratory waterfowl occupied the millions of ponds, potholes, and marshy spots that once dotted our grasslands. Hunting, fencing, wetland drainage, introduction of alien species, and other human modifications to the land have greatly diminished most wildlife populations since the last century.

Tundra

Climates in high mountain areas or at far northern or southern latitudes often are too harsh for trees. This treeless landscape, called **tundra,** is characterized by a very short growing season, cold, harsh winters, and the potential for frost any month of the year. Although water may be abundant on the tundra, for much of the year it is locked up in ice or snow and therefore unavailable to plants. As far as plants are concerned, the tundra is a very cold desert.

The *arctic* tundra is a biome of low productivity, low diversity, and low resilience. Winters are long and dark. Only the top several centimeters of the soil thaw out in the summer, and the lower soil is permafrost. This permanently frozen layer prevents snowmelt water from being absorbed into the soil, so the surface soil is waterlogged during the summer. Try to imagine the difficulties encountered by plants in this kind of soil. Most of the year it is completely frozen, and even during the brief growing season, the permafrost is an impenetrable barrier to deep root growth. In addition, the top layer buckles and heaves in response to cycles of freezing and thawing, toppling plants and disrupting root systems.

The *alpine* tundra (fig. 5.5) differs from the arctic tundra in several ways. Plants of the alpine tundra face different challenges than those of the arctic. The thin mountain air permits intense solar bombardment, especially by ultraviolet radiation; thus, many alpine plants have deep pigmentation that shields their inner cells. The glaring summer sun also causes very hot daytime ground temperatures, even though the night temperatures may return to freezing. Alpine soil is windswept and often gravelly or rocky. The sloping terrain causes moisture to drain quickly. Due to this combination of sun, soil, slope, and air currents, drought is a problem—as opposed to the wet conditions in the arctic tundra.

Although the tundra may swarm with life during the brief summer growing season, only a few species are able to survive the harsh winters or to migrate to warmer climates. Dominant tundra plants are dwarf shrubs, sedges, grasses, mosses, and lichens. Its larger lifeforms, such as arctic musk ox and caribou, or alpine mountain goats and mountain sheep, must be adapted to survive the harsh climate and sparse food supply. Many animals migrate or hibernate during winter. Flocks of migratory birds nest on the abundant summer arctic wetlands, which also nurture hordes of bloodsucking insects that feed upon the summer flocks and herds (and tourists!).

FIGURE 5.5 The harsh physical conditions of alpine tundra ecosystems are the result of altitude and slope. The growing season is short, ultraviolet radiation in the thin air is greater than at lower altitudes, and extreme temperature fluctuations are possible even during the summer. Air currents and the thin, rocky soil also contribute to the arid conditions of alpine tundra.

Damage to the tundra is slow to heal. At present, the greatest threat to this distinctive biome is oil and natural gas wells in the Arctic and mineral excavation in mountain regions. Because plants grow slowly during the brief summer at high altitudes or latitudes, truck ruts and bulldozer tracks on the tundra landscapes may take centuries to heal. Furthermore, some of the most promising sites for oil exploration or mining are summer feeding and breeding grounds for animals such as caribou, grizzly bears, or mountain sheep. Even seemingly minor incursions into this fragile ecosystem can cause great damage.

Conifer Forests

Several distinctive biomes are dominated by **conifer** (cone-bearing) trees. Where moisture is limited by sandy soil, low precipitation, or a short growing season, plants reduced water loss by evolving thin, needlelike evergreen leaves with a thick waxy coating. Although not as efficient in carrying out photosynthesis during summer months as the broad, soft leaves of deciduous trees, conifer needles and scales can survive harsh winters or extended droughts and do accomplish some photosynthesis even under poor conditions. In mountain areas, particularly, fire has been an important and fairly regular factor in maintaining the coniferous forest.

The **boreal forest,** or northern coniferous forest, stretches in a broad band of mixed coniferous and deciduous trees around the world between about 45° and 60° north latitude, depending on altitude and distance from coastlines or major rivers. Among the dominant conifers are pine, hemlock, spruce, cedar, and fir. Some common deciduous trees are birches, aspens, and maples, while mosses and lichens form much of the ground cover. In this moist, cool biome, streams and wetlands abound—especially on recently glaciated landscapes. As a result, there are many lakes, potholes, bogs, and fens (fig. 5.6). Insects that have aquatic stages in their life cycles, such as mosquitoes and biting flies, are particularly abundant, to the consternation of humans as well as wild birds and mammals.

The northernmost edge of the boreal forest is a species-poor black spruce and sphagnum moss (peat moss) woodland—often called by its Russian name, **taiga**—that forms a ragged border with the treeless arctic tundra (fig. 5.7). The harsh climate limits both productivity and resilience of the taiga community. Cold temperatures, very wet soil during the growing season, and acids produced by fallen conifer needles and sphagnum inhibit full decay of organic matter. As a result, thick layers of semidecayed organic material, called peat, form. Boreal peat deposits are being explored as energy sources; however, the environmental disturbance caused by peat mining in this boreal community could be severe and long-lasting, perhaps even permanent.

The southern pine forest, another kind of U.S. coniferous forest ecosystem, is characterized by a warm, moist climate and sandy soil and, in the past, was subjected to frequent fires. Now it is managed extensively for timber and such resinous products as turpentine and rosin. The undergrowth includes saw palmetto and various thorny bushes.

The coniferous forests of the Pacific coast represent yet another special set of environmental circumstances. Mild temperatures and abundant precipitation—up to 250 cm (100 in) per year—result in luxuriant plant growth and huge trees such as the California redwood, the largest tree in the world and the largest organism of any kind known to have ever existed. Redwoods formerly grew along the Pacific coast from California to Oregon, but their distribution has been greatly reduced by logging without regard to sustainable yield or restoration.

In its wettest parts, the coastal forest becomes a **temperate rainforest,** a cool, rainy forest that often is enshrouded in fog. Condensation from the canopy (leaf drip) becomes a major form of precipitation, and annual precipitation exceeds 250 cm (100 in) in some places. Mosses, lichens, and ferns cover tree branches, old stumps, and the forest floor itself. The largest, best-preserved area of temperate rainforest is Olympic National Park in Washington.

FIGURE 5.6 Mixed conifer and deciduous forests like this one in northern Minnesota are characteristic of the Great Lakes Region. Although other species assemblages are found elsewhere, temperate cool forests around the world share many common characteristics.

FIGURE 5.7 This boreal forest in Alaska's Chugach Mountains has small, widely spaced black spruce intermixed with willows and heather on a wet peatland.

Broad-Leaved Deciduous and Evergreen Forests

Forests of broad-leaved trees occur throughout the world where rainfall is plentiful. In temperate regions, the climate supports lush summer plant growth when water is plentiful but requires survival adaptations for the frozen season. A key adaptation of **deciduous** trees is the ability to produce summer leaves and then shed them at the end of the growing season. This rich and varied biome contains associations of many tree species, including oak, maple, birch, beech, elm, ash, and other hardwoods. These tall trees form a forest canopy over a diverse understory of smaller trees, shrubs, and herbaceous plants, including many annual spring flowers that grow, flower, set seed, and store carbohydrates before they are shaded by the canopy (fig. 5.8). Where the climate is warm year-round, forests are dominated by evergreen trees such as the live oaks and cypresses of the southern United States.

FIGURE 5.8 The deciduous forests of the eastern United States are stratified assemblages of tall trees, understory shrubs, and small, lovely, ground-covering species. A rich fauna is supported by the variety of niche opportunities that are present.

FIGURE 5.9 A giraffe feeds on acacias in the dry, thorn shrub land of Africa.

Most of the dense forest that once covered Central Europe was cleared a thousand years ago. When European settlers first came to North America, dense forests of broad-leaved deciduous trees covered most of the eastern half of what is now the United States. Much of that original deciduous forest was harvested for timber a century or more ago. Now, large areas of forest in the eastern United States have regrown and are approaching old-growth status, although with a different species composition than the original biome. Vast original deciduous forests remain in eastern Siberia, but they are being harvested at a rapid rate, perhaps the highest deforestation rate in the world. As the forests disappear, so too do Siberian tigers, bears, cranes, and a host of other unique and endangered species.

Mediterranean/Chaparral/Thorn Shrub

A Mediterranean climate is characterized by warm, dry summers and cool, moist winters. Evergreen shrubs with small, leathery, sclerophyllous (hard, waxy) leaves form dense thickets. Scrub oaks, drought-resistant pines, or other small trees often cluster in sheltered valleys. Fires burn fiercely in this fuel-rich plant assemblage and are a major factor in plant succession. Annual spring flowers often bloom profusely, especially after fires. In California, this landscape is called **chaparral,** Spanish for thicket. Some typical animals include jackrabbits, kangaroo rats, mule deer, chipmunks, lizards, and many bird species. Very similar landscapes are found along the Mediterranean coast as well as southwestern Australia, central Chile, and South Africa. Although this biome doesn't cover a very large total area, it contains a high number of unique species and often is considered a "hot-spot" for biodiversity. It also is highly desired for human habitation, often leading to conflicts with rare and endangered plant and animal species.

Areas that are drier year-round, such as the African Sahel (edge of the Sahara Desert), northern Mexico, or the American Intermountain West (or Great Basin) tend to have a more sparse, open shrubland, characterized by sagebrush (*Artemisia* sp.), chamiso (*Adenostoma* sp.), or saltbush (*Atriplex* sp.). In Africa, acacias and other spiny plants dominate this landscape, giving it the name **thorn shrub.** Some typical animals of this biome in America are a wide variety of snakes and lizards, rodents, birds, antelope, and mountain sheep. In Africa, this landscape is home to gazelle, rhinos, and giraffes (fig. 5.9).

Tropical Moist Forests

The humid tropical regions of South and Central America, Africa, Southeast Asia, and some of the Pacific Islands support one of the most complex and biologically rich biome types in the world (fig. 5.10). Although there are several kinds of moist tropical forests, they share common attributes of ample rainfall and uniform temperatures. Cool **cloud forests** are found high in the mountains where fog and mist keep vegetation wet all the time. **Tropical rainforests** occur where rainfall is abundant—more than 200 cm (80 in) per year—and temperatures are warm to hot year-round.

The soil of both these tropical moist forest types tends to be old, thin, acidic, and nutrient-poor, yet the number of species present can be mind-boggling. For example, the number of insect species found in the canopy of tropical rainforests has been estimated to be in the millions! It is estimated that one-half to two-thirds of all species of terrestrial plants and insects live in tropical forests.

The nutrient cycles of these forests also are unique. Almost all (90 percent) of the nutrients in the system are contained in the bodies of the living organisms. This is a striking contrast to temperate forests, where nutrients are held within the soil and made available for new plant growth. The luxuriant growth in tropical rainforests depends on rapid decomposition and recycling of dead organic material. Leaves and branches that fall to the forest floor decay and are incorporated almost immediately back into living biomass.

When the forest is removed for logging, agriculture, and mineral extraction, the thin soil cannot support continued cropping and cannot resist erosion from the abundant rains. And if the cleared area is too extensive, it cannot be repopulated by the rainforest community. Rapid deforestation is occurring in many tropical areas as people move into the forests to establish farms and ranches, but the land soon loses its fertility.

Tropical Seasonal Forests

Many areas in India, Southeast Asia, Australia, West Africa, the West Indies, and South America have tropical regions characterized by distinct wet and dry seasons instead of uniform heavy rainfall throughout the year, although temperatures are hot year-round. These areas have produced communities of **tropical seasonal forests:** semievergreen or partly deciduous forests tending toward open woodlands and grassy savannas dotted with scattered, drought-resistant tree species.

Tropical dry forests have typically been more attractive than wet forests for human habitation and have suffered greater degradation. Clearing a dry forest with fire is relatively easy during the dry season. Soils of dry forests often have higher nutrient levels and are more agriculturally productive than those of a rainforest. Finally, having fewer insects, parasites, and fungal diseases than a wet forest makes a dry or seasonal forest a healthier place for humans to live. Consequently, these forests are highly endangered in many places. Less than 1 percent of the dry tropical forests of the Pacific coast of Central America or the Atlantic coast of South America, for instance, remain in an undisturbed state.

FIGURE 5.10 Tropical rainforests, like this one in Java, support a luxuriant profusion of species and biomass. Notice the many vines and epiphytes clinging to the buttressed trunk of this giant forest tree.

AQUATIC ECOSYSTEMS

Water is essential to life in every biological community and is present to varying degrees in all of them. Although freshwater ecosystems and wetlands are often parts of other, larger biomes, they are so important biologically that they deserve individual attention here.

Freshwater and Saline Ecosystems

Freshwater ecosystems include the standing waters of ponds and lakes as well as the flowing waters of rivers and streams. There also are some unique freshwater ecosystems, including underground rivers and subterranean pools. Freshwater ecosystems are as varied as their individual sites because they are influenced not only by characteristics of local climate, soil, and resident communities but also by the surrounding terrestrial ecosystems and anything that happens uphill or upstream from them (fig. 5.11). As with terrestrial ecosystems, the biological communities of

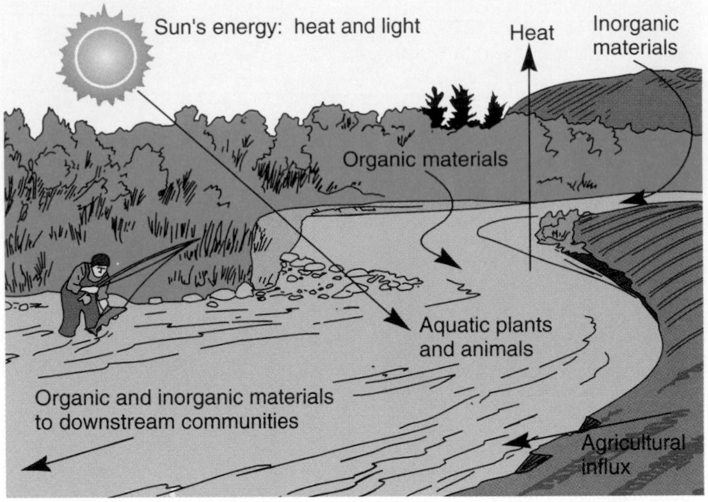

FIGURE 5.11 The character of freshwater ecosystems is greatly influenced by the immediately surrounding terrestrial ecosystem, and even by ecosystems far upstream or far uphill from a particular site.

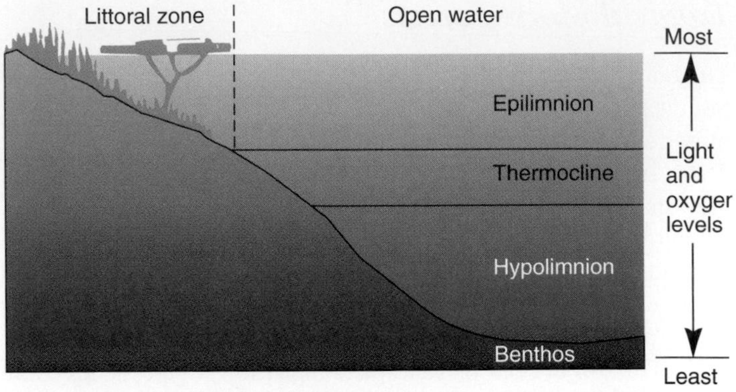

FIGURE 5.12 The layers of a deep lake are determined mainly by gradients of light, oxygen, and temperature. The epilimnion is affected by surface mixing from wind and thermal convections, while mixing between the hypolimnion and epilimnion is inhibited by a sharp temperature and density difference at the thermocline.

freshwater ecosystems are largely determined by the physical characteristics of the environment, except that the surrounding medium is water instead of air.

Aquatic organisms have the same basic needs as terrestrial organisms: carbon dioxide, water, and sunlight for photosynthesis; oxygen for respiration; and food and mineral nutrients for energy, growth, and maintenance. The availability of these necessities is influenced by such site characteristics as (1) substances that are dissolved in the water, such as oxygen, nitrates, phosphates, potassium compounds, and other by-products of agriculture and industry; (2) suspended matter, such as silt and microscopic algae, that affect water clarity and, therefore, light penetration; (3) depth; (4) temperature; (5) rate of flow; (6) bottom characteristics (muddy, sandy, rocky); (7) internal convective currents; and (8) connection to, or isolation from, other aquatic ecosystems.

Vertical stratification is an important aspect of standing water ecosystems, especially in regard to gradients of light, temperature, nutrients, and oxygen. Organisms tend to form distinctive vertical subcommunities in response to this stratification of physical factors. The plankton subcommunity consists mainly of microscopic plants, animals, and protists (single-celled organisms such as amoebae) that float freely within the water column. Some nonplanktonic organisms are specialized to live at the air-water interface (for example, insects such as water striders), and still others are able to swim freely in the open waters (such as fish).

Finally, many animal species are bottom dwellers (such as snails, burrowing worms and insect larvae, bacteria) that make up the **benthos,** or bottom subcommunity. Oxygen levels are lowest in the benthic environment. Anaerobic bacteria can live in the very low-oxygen bottom sediments. Emergent plants such as cattails and rushes are rooted in bottom sediments of the littoral (near shore) zone but spread their leaves above the water surface. They create important structural and functional links between strata of the ecosystem and often are the greatest producers of net primary productivity in the ecosystem.

Deeper lakes are characterized by the presence of a warmer, upper layer that is mixed by the wind (the epilimnion; *epi* = upon) and a colder, deep layer that is not mixed (the hypolimnion; *hypo* = below) (fig. 5.12). The two layers are separated by a distinctive temperature transition zone called the **thermocline,** or mesolimnion.

Lakes and ponds have a tendency to undergo succession, which includes changes in the biological community as a response to increases in nutrient levels. Human sources of nutrient input can radically increase the rate at which a lake "ages," as discussed in chapter 20 (eutrophication).

Humans utilize freshwater communities as sources of food and for recreation, waste disposal, cooling of nuclear power plants, and many other industrial uses. Waterways also are major transport lanes for barge and ship traffic, involving channel dredging and contamination from fuel and cargo leaks. Dams, canals, and other diversions have significant effects on freshwater resources. In these and other ways, humans have an enormous impact on the character of individual freshwater ecosystems and their biological communities.

Not all freshwater lakes contain what we would consider "fresh" water. The Caspian Sea, Dead Sea, Great Salt Lake, and numerous other saline (salty) lakes have been formed by evaporative shrinkage of large bodies of water. Deserts, particularly, may have alkaline or arsenic-containing lakes and potholes. Still other salty and mineral-rich ponds and small lakes are fed by mineral hot springs.

Estuaries and Wetlands: Transitional Communities

Estuaries are bays or semi-enclosed bodies of brackish (salty but less so than seawater) water that form where rivers enter the ocean. Estuaries usually contain rich sediments carried downriver, forming shoals and mudflats that nurture a multitude of aquatic life.

Estuaries are sheltered from the most drastic ocean action but do experience tidal ebbs and flows. Daily tides may even cause river levels to rise and fall far inland from the river mouth. The combination of physical factors in estuaries makes them very productive and of high species diversity. They are significant "nurseries" for economically important fish, crustaceans (such as crabs and shrimp), and mollusks (such as clams, cockles, and oysters).

Where the continental shelf is broad and shallow, an extensive fan-shaped sediment deposit called a **delta** may form at the river mouth. Deltas often are channeled by branches of the river, creating extensive coastal wetlands that are part of the larger estuarine zone.

Wetlands of several types form near bodies of water. **Wetlands** are ecosystems in which the land surface is saturated or covered with standing water at least part of the year and vegetation is adapted for growth under saturated conditions. There are special names for specific kinds of wetlands, but we can group them into three major categories: swamps, marshes, and bogs and fens. Defined pragmatically, **swamps** are wetlands with trees (fig. 5.13); **marshes** are wetlands without trees (fig. 5.14); and **bogs** and **fens** are areas that may or may not have trees, and have waterlogged soils that tend to accumulate peat. Swamps and marshes tend to be associated with flowing water. Fens are fed by groundwater and surface runoff, whereas bogs are fed solely by precipitation. Swamps and marshes generally have high productivity, while bogs and fens have low productivity.

The water in marshes and swamps usually is shallow enough to allow full penetration of sunlight and seasonal warming. These mild conditions favor great photosynthetic activity, resulting in high productivity at all trophic levels. In short, life is abundant and varied. Wetlands are major breeding, nesting, and migration staging areas for waterfowl and shorebirds.

Wetlands perform major ecosystem services, the importance of which cannot be overstated. As mentioned previously, they support a great diversity of life-forms. What may be less obvious is their role in planetary water relationships. Wetlands act as traps and filters for water that moves through them. Runoff water is slowed as it passes through the shallow, plant-filled areas, reducing flooding. As a result, sediments are deposited in the wetlands instead of traveling into rivers and, eventually, the oceans. In this way, wetlands both clarify surface waters and aid in the accumulation and formation of fertile land. Furthermore, chemical interactions in wetland ecosystems neutralize and detoxify substances in the water. Finally, water in wetlands seeps into the ground, helping to replenish underground water reservoirs called aquifers (see chapter 19).

Wetlands convert naturally to terrestrial communities largely through sedimentation, eutrophication, or stream cutting and draining. Human activities have greatly accelerated these processes in many places, and wetlands are being destroyed or degraded around the world at a disturbing rate (see chapter 15). This destruction is of great concern because it means loss of ecological services to the biosphere as well as loss of essential habitats for a myriad of species.

FIGURE 5.13 Biological communities, such as the Okefenokee Swamp in Georgia, can be amazingly diverse, complex, and productive. Myriad life-forms coexist in exuberant abundance, interlinked by manifold relationships. Who lives where, and why, are central ecological questions.

Shorelines and Barrier Islands

Ocean shorelines, including rocky coasts, sandy beaches, and offshore barrier islands, are particularly rich in life-forms. Rocky shorelines, in particular, support an incredible density and diversity of organisms that grow attached to any solid substrate, including each other. Sandy shorelines, on the other hand, provide homes for organisms that live among the sand grains and in burrows.

Sandy beaches are understandably popular to humans. Cities, resorts, and residences are built on beaches. Grasses and trees that hold the dunes are destroyed, destabilizing the soil system and thus increasing its susceptibility to wind and wave erosion. The constant battle to maintain such real estate is costly. Insurance is very expensive because of the hazards of natural erosion and storms. Building and maintaining sea walls and protecting and replenishing beaches are expensive, continuous processes. Protective structures often have their own effect on shore recontouring and may even increase the rate of natural shoreline loss.

FIGURE 5.14 Coastal wetlands such as this salt marsh are highly productive ecosystems and are vitally important for a wide variety of terrestrial and aquatic species.

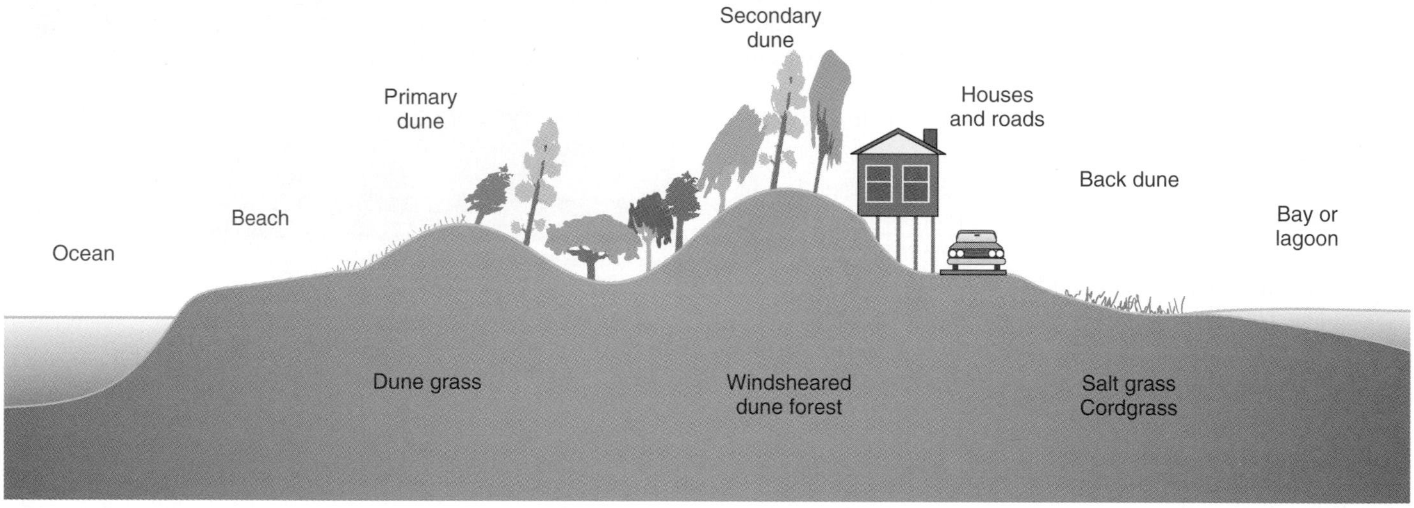

FIGURE 5.15 A cross section through a barrier island. Fragile dune grass communities on the primary dune must be protected to ensure stability of the dune complex. Roads and houses should be limited to the back dune area, both for safety and to relieve pressure on vulnerable dune vegetation. Only a few narrow footpaths should cross the dunes to the beach where recreational use is permitted.

Barrier islands are low, narrow, sandy islands that form offshore from a coastline. In North America, they are particularly characteristic of the Atlantic and Gulf coasts (fig. 5.15). Barrier islands protect inland shores from the onslaught of the surf, especially during severe storms. Because they are so lovely, barrier islands have been tempting targets for real estate development and about 20 percent of the barrier island surface in the United States has been developed. Unfortunately, human occupation of barrier islands often destroys the values that attract us there in the first place. Walking or driving vehicles over dune grass destroys stabilizing vegetation cover and triggers erosion. Cutting roads through dunes and building houses further destabilizes islands. A single, major storm can do massive damage, destroying houses and perhaps washing away most of an island.

Ignoring these environmental realities endangers many rare species and puts human lives and economic investment at risk. Although we now have coastal zoning and land use regulations, we still have many environmentally damaging developments in place.

Coral reefs form in clear, warm, tropical seas and are particularly well developed in the south Pacific (fig. 5.16). They are the accumulated calcareous skeletons of innumerable tiny colonial animals called corals. Each of the interconnected coral animals

FIGURE 5.16 Coral reefs harbor some of the most diverse biological communities on the earth, rivaling tropical rainforests in species diversity, and surpassing any biome in the number of phylla (broad taxonomic groups) represented. Surprisingly, almost all the organisms you see here are animals.

builds a calcium carbonate chamber on the surface of the accumulated secretions of previous generations of animals. Coral reefs usually form along the edges of shallow, submerged banks or shelves. The depth at which they form is limited by the depth of light penetration, due at least in part to the presence of symbiotic photosynthetic protists in their tissues. Because of the photosynthetic relationship, the presence of dead versus living coral growths can be an index of either previous ocean levels or decreased light penetration due to increased turbidity. Coral reef communities rival tropical forest communities in species diversity, numbers of individuals, brilliance of color, and interesting forms of both plants and animals.

Reefs also are among the most endangered biological communities on the earth. Destructive fishing practices (using dynamite or cyanide to stun fish, for instance) and harvesting of coral for building or the pet trade have damaged or destroyed about three-fourths of all reefs in the world. In the next section, we will discuss some other types of human disturbance of natural habitats.

HUMAN DISTURBANCE

Humans have become dominant organisms over most of the earth, damaging or disturbing more than half of the world's terrestrial ecosystems to some extent (fig. 5.17). By some estimates, humans preempt about 40 percent of the net terrestrial primary productivity of the biosphere either by consuming it directly, by interfering with its production or use, or by altering the species composition or physical processes of human-dominated ecosystems. Conversion of natural habitat to human uses is the largest single cause of biodiversity losses.

Researchers from the environmental group Conservation International have attempted to map the extent of human disturbance of the natural world (fig. 5.17). The greatest impacts have been in Europe, parts of Asia, North and Central America, and islands such as Madagascar, New Zealand, Java, Sumatra, and those in the Caribbean. Data from this study are shown in table 5.1.

Temperate broad-leaved forests are the most completely human-dominated of any major biome. The climate and soils that support such forests are especially congenial for human occupation. In eastern North America or most of Europe, for example, only remnants of the original forest still persist. Regions with a Mediterranean climate generally are highly desired for human habitation. Because these landscapes also have high levels of biodiversity, conflicts between human preferences and biological values frequently occur (fig. 5.18).

Temperate grasslands, temperate rainforests, tropical dry forests, and many islands also have been highly disturbed by human activities. If you have traveled through the American cornbelt states such as Iowa or Illinois, you have seen how thoroughly former prairies have been converted to farmlands. Intensive cultivation of this land exposes the soil to erosion and fertility losses (see chapter 11). Islands, because of their isolation, often have high numbers of endemic species. Many islands, such as Madagascar, Haiti, and Java have lost more than 99 percent of their original land cover.

Tundra and Arctic deserts are the least disturbed biomes in the world. Harsh climates and unproductive soils make these areas unattractive places to live for most people. Temperate conifer forests also generally are lightly populated and large areas remain in a relatively natural state. However, recent expansion of forest harvesting in Canada and Siberia may threaten the integrity of this biome. Large expanses of tropical moist forests still remain in the Amazon and Congo basins but in other areas of the tropics such as West Africa, Madagascar, Southeast Asia, and the Indo-Malaysian peninsula and archipelago, these forests are disappearing at a rapid rate (see chapter 14).

As mentioned earlier, wetlands have suffered severe losses in many parts of the world. About half of all original wetlands in the United States have been drained, filled, polluted, or otherwise degraded over the past 250 years. In the prairie states, small potholes and seasonally flooded marshes have been drained and converted to croplands on a wide scale. Iowa, for example, is estimated to have lost 99 percent of its presettlement wetlands. Similarly, California has lost 90 percent of the extensive marshes and deltas that once stretched across its central valley. Wooded swamps and floodplain forests in the southern United States have been widely disrupted by logging and conversion to farmland.

Similar wetland disturbances have occurred in other countries as well. In New Zealand, over 90 percent of natural wetlands have been destroyed since European settlement. In Portugal, some 70 percent of freshwater wetlands and 60 percent of estuarine habitats have been converted to agriculture and industrial areas. In

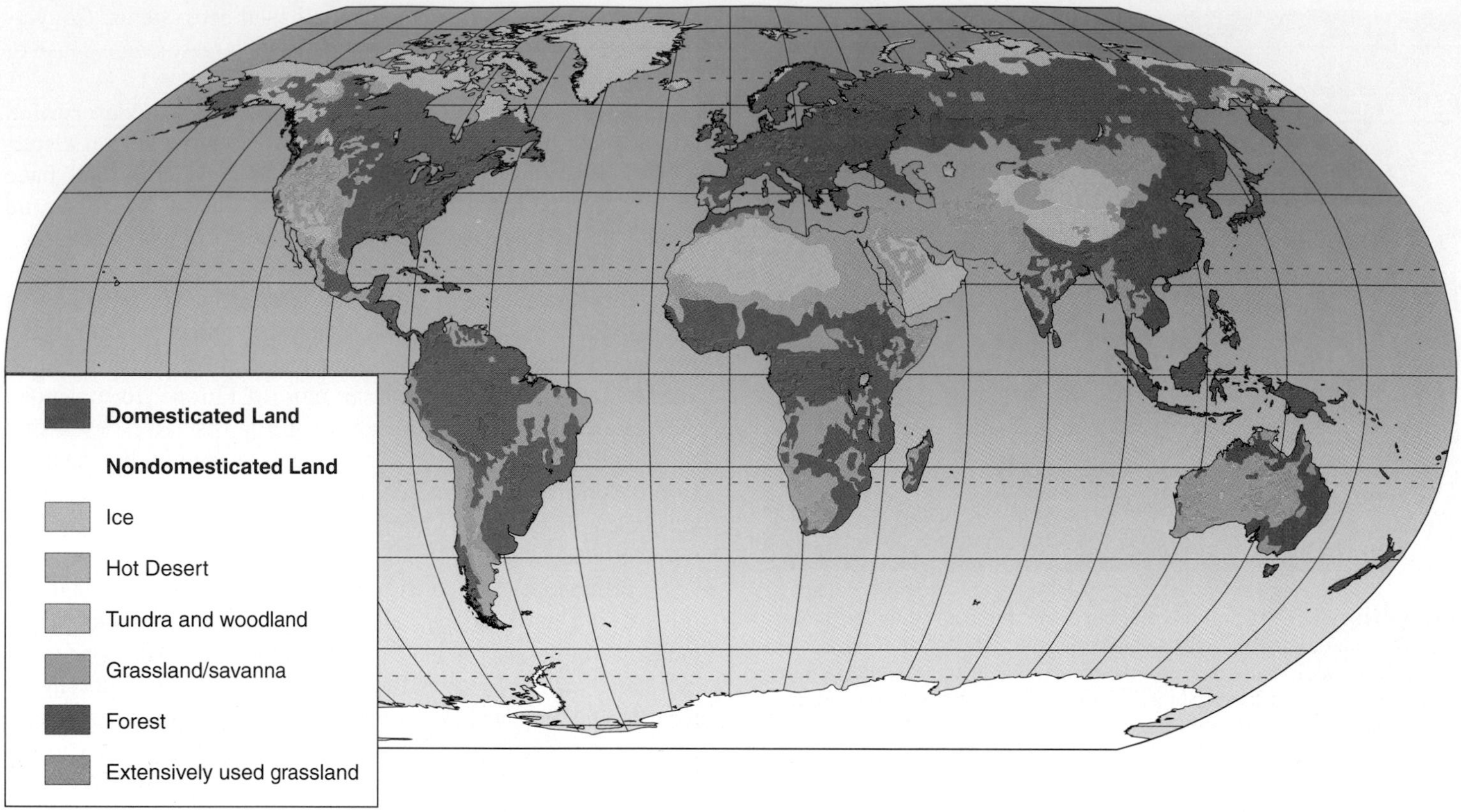

FIGURE 5.17 Domesticated land has replaced much of the earth's original land cover.

TABLE 5.1 Human Disturbance

BIOME	TOTAL AREA (10^6 KM^2)	% UNDISTURBED HABITAT	% HUMAN DOMINATED
Temperate broadleaved forests	9.5	6.1	81.9
Chaparral and thorn scrub	6.6	6.4	67.8
Temperate grasslands	12.1	27.6	40.4
Temperate rainforests	4.2	33.0	46.1
Tropical dry forests	19.5	30.5	45.9
Mixed mountain systems	12.1	29.3	25.6
Mixed island systems	3.2	46.6	41.8
Cold deserts/semideserts	10.9	45.4	8.5
Warm deserts/semideserts	29.2	55.8	12.2
Moist tropical forests	11.8	63.2	24.9
Tropical grasslands	4.8	74.0	4.7
Temperate conifer forests	18.8	81.7	11.8
Tundra and arctic desert	20.6	99.3	0.3

Note: Where undisturbed and human dominated areas do not add up to 100 percent, the difference represents partially disturbed lands.

Source: Data from Hannah, Lee, et al., "Human Disturbance and Natural Habitat: A Biome Level Analysis of a Global Data Set," in *Biodiversity and Conservation,* 1995, Vol. 4:128–155.

FIGURE 5.18 "There was an environment before human beings and there will be an environment after human beings. So what's the big deal?" © Peter Kohlsaat/Modern Times Syndicate.

Indonesia, almost all the mangrove swamps that once lined the coasts of Java have been destroyed, while in the Philippines and Thailand, more than two-thirds of coastal mangroves have been cut down for firewood or conversion to shrimp and fish ponds.

LANDSCAPE ECOLOGY

So far in this chapter, we have considered very broad ecological regions and humans solely as disturbing agents. This is a very Clemensian view (see chapter 3) of nature as static and deterministic. Most modern scientists see nature as much more diverse, dynamic, and complex than is suggested by this perspective. Modern technology, including remote sensing, powerful computers, and geographical information systems (GIS) allow us to study temporal and spatial complexity, heterogeneity, and dynamic fluxes in ecosystems that would previously have been ignored as random noise or unwelcome complications (fig. 5.19).

Landscape ecology is the study of reciprocal effects of spatial pattern on ecological processes. By reciprocal effects, we mean that complex spatial patterns shape, and are shaped by, the dynamic ecological processes that occur in them. In common usage, a landscape is as much as you can see at one time from a high vantage point. By this definition, it is smaller than a biome but generally larger than a single ecosystem. In landscape ecology, a landscape is a geographical unit with a history that shapes the features of the land and organisms in it as well as our reaction to, and interpretation of, the land.

Landscape ecology considers humans important elements of most landscapes, while other ecologists generally have tried to study areas devoid of human influence. Few places on the earth, however, are devoid of human impacts. Human disturbances include hunting, timber harvesting, farming, grazing, pollution, introduction of exotic species, and alteration of atmospheric chemistry and, perhaps, global climate. Restricting our definition of "nature" to a few pristine remnants will lead us either to ignore most of the earth's surface or demand that humans remove themselves entirely from large tracts, something not likely in poor countries that need access to resources.

Patchiness and Heterogeneity

The Amazonian rainforest is a good example of the new perspective provided by landscape ecology. Usually the rainforest is depicted as a vast, homogeneous block of forest that has remained stable and unchanging for hundreds of millions of years. A detailed analysis, however, shows that it actually is a complex mosaic of many distinct habitat types and species assemblages distributed according to soil type, topography, climate, and site history. These patches are in a constant state of change as big trees fall or fires open up new ground for succession.

Landscape ecologists claim that if we look closely, all landscapes consist of similar mosaics of discrete, bounded patches with different biotic or abiotic composition. Often a predominant or continuous cover type acts as a matrix in which other patch types appear to be embedded. Like individual habitats in the rainforest, these patches are dynamic and change components or functions over time. Among the causes of this patchiness are human and natural disturbances as well as underlying ecological factors such as successional processes.

Landscape heterogeneity can exist across a wide range of scale from burned patches measuring thousands of hectares in Yellowstone National Park to the effects of soil crumb size and insect burrows in a few square centimeters of soil. These spatial patterns exist in marine, freshwater, and wetland environments as well as terrestrial habitats, and thus the term "landscape" can apply to aquatic environments as well as dry land. A basic question in landscape ecology is whether a given phenomenon appears or applies across many different scales or is restricted to a particular scale. In other words, do patterns emerge, change, or disappear if we look at a landscape up close or from afar?

Landscape Dynamics

Time and space are of special concern in landscape ecology, for that matter, to the rest of us as well. As humorist Garrison Keillor says, "Time exists so everything doesn't happen at once; space exists so everything doesn't happen to us." How does this apply to ecology? A physiological ecologist might want to study the effects of temperature on the rate of photosynthesis in plants. To make the study

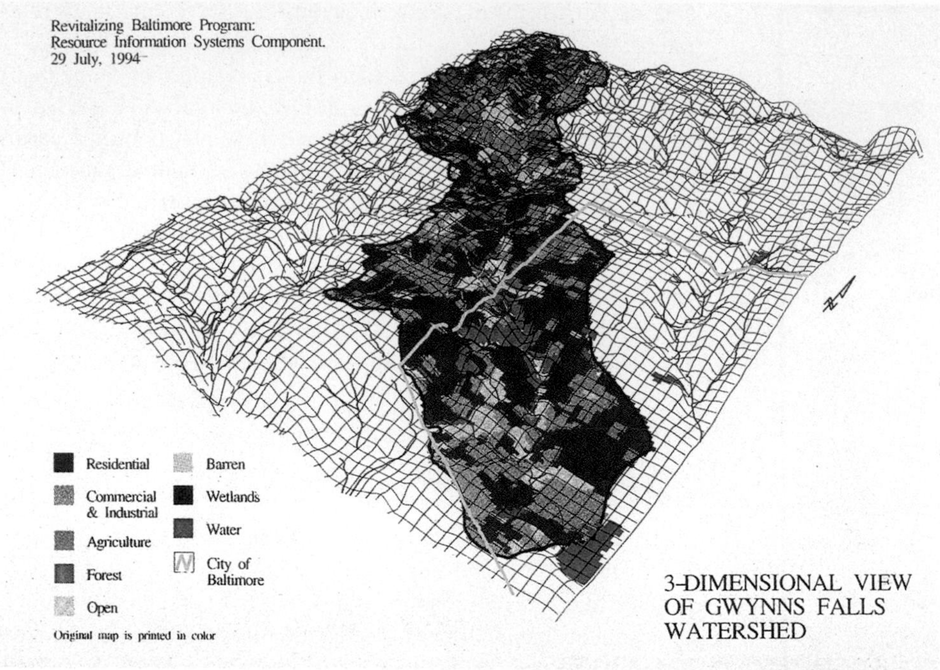

FIGURE 5.19 Landscape ecology uses geographic information systems (GIS) to map patch size, type, and configuration as in this 3-D map of the Gwynns Falls watershed near Baltimore, Maryland. This mapping assists planners in analyzing land use patterns.

manageable, she or he might ask, "What is the photosynthetic rate in this particular *plant,* at this *exact* time, at these *specific* temperatures?" A landscape ecologist, in contrast, might ask, "What is the effect of temperature on photosynthesis in this *place,* given its unique combination of history, composition, and characteristics?" Developing a deep relationship with, and understanding of, a particular landscape is sometimes referred to as having a "sense of place."

The boundaries between habitat patches are considered especially significant by landscape ecologists. Edges can induce, inhibit, or regulate movement of materials, energy, or organisms across a landscape. Thus the dynamics between patches may be of greater overall importance than what happens within each patch. As chapter 4 shows, wildlife managers are beginning to be very aware of edge effects as well as the size, shape, and distribution of habitat patches in fragmented landscapes.

This focus on interactions between neighboring communities is quite different from classical ecological focus on the structure and functions of discrete communities, populations, or ecosystems. In its interest in complexity and emergent properties of systems, landscape ecology draws on theories of chaos and complexity from mathematics and physics.

There also are many similarities between landscape ecology and the equally new discipline of conservation biology. Among their shared tenets are: (1) evolutionary change is a central feature of natural systems, (2) nonequilibrium dynamics and uncertainty are more characteristic of nature than are stability and determinism, (3) heterogeneity and diversity are good and ought to be maintained, (4) the context of the surrounding landscape is important, and (5) human needs, desires, abilities, and potential must be considered in efforts to understand and protect nature. All of these points are important issues in the design of nature preserves, parks, and wildlife refuges (see chapter 15).

RESTORATION ECOLOGY

Can the lessons of Aldo Leopold's Sand County farm be applied to disrupted and damaged landscapes elsewhere in the world? Advances in ecological understanding coupled with practical knowledge gained through hands-on field work have given birth to a new discipline of **restoration ecology** that seeks to repair or reconstruct ecosystems damaged by humans or natural forces.

Although land stewardship and husbandry have long been practiced by people in many land-based cultures, the field of ecological restoration suddenly captured public attention in 1988 when a conference called Restoring the Earth, held in Berkeley, California, drew an overflow crowd of more than 800 scientists, policymakers, and activists to share ideas and experiences from restoration projects. There is now a society for, and journal of, restoration ecology. This fledgling science has been given a boost in recent years as courts and regulatory agencies increasingly have required companies and developers to restore or replace illegally damaged wetlands or habitats of endangered species.

Defining Some Terms

Restoration generally means to bring something back to a former condition. Ecological restoration involves active manipulation of nature to re-create species composition and ecosystem processes as close as possible to the state that existed before human disturbance (fig. 5.20). Although there may be many similarities between restoration and conservation, stewardship, or management, the former often entails more direct intervention to achieve a predetermined end than do these other fields.

Rehabilitation refers to attempts to rebuild elements of structure or function in an ecological system without necessarily

FIGURE 5.20 Volunteers lay willow branches and brush matting to stabilize dunes and restore beach vegetation.

achieving complete restoration to its original condition. Often rehabilitation means to bring an area back to a useful state for human purposes rather than to a truly natural state. It aims to reverse deterioration of a resource even if it cannot be restored fully.

Remediation is a process of cleaning chemical contaminants from a polluted area by physical or biological methods as a first step toward protecting human and ecosystem health. Incineration, for instance, often is a cost-effective method of cleaning oil-contaminated soils. Volatile organic solvents in aquifers can be removed by pumping out groundwater and aerating it or passing it through absorbent materials.

Living organisms are highly effective cleaning agents for many contaminants. Water hyacinths, for example, have a great capacity for absorbing heavy metals and other toxins from polluted water. Microorganisms (bacteria and fungi) can be found in nature or engineered in the laboratory to destroy many dangerous chlorinated compounds. Chopped-up horseradish roots are very effective in removing phenols from industrial effluents. And common locoweeds can extract selenium from contaminated soils. Sometimes simply adding fertilizer to encourage plant and microbial growth is the best way to clean up surface pollutants.

Reclamation typically is used to describe chemical or physical manipulations carried out in severely degraded sites, such as open-pit mines or large-scale construction. The Surface Mining Control and Reclamation Act (SMCRA), for example, requires mine operators to use reclamation techniques to restore the shape of the land to the original contour and revegetate it to minimize impacts on local surface and ground waters.

Historically, reclamation meant irrigation projects that brought wetlands and deserts (considered useless wastelands) into agricultural production. Thus, the Bureau of Reclamation and the Army Corps of Engineers, in the early part of this century dredged, diked, drained, and provided irrigation water to convert millions of acres of wild lands into farmlands. Many of those projects were highly destructive to natural landscapes and we are now using reclamation techniques to restore formerly "reclaimed" land to a more natural state.

Re-creation attempts to construct a new biological community on a site so severely disturbed that there is virtually nothing left to restore. The new system may be modeled on what we think was there before human disturbance or it may be something that never existed on that site but that we think suits current conditions. Often developers and government are required to **mitigate** damage caused in one area by re-creating a comparable biological community in another place.

In what is probably the most ambitious ecological restoration project in history, the U.S. Army Corps of Engineers is now trying to reverse 50 years of draining, ditching, and damming in south Florida in an effort to repair damage to the Everglades. The scope of the project is staggering. Estimated to cost $8 billion and to take more than 20 years, the Corps proposes to rip out old levees, fill canals, return rivers to their original meandering channels (fig. 5.21), reconstruct drained wetlands, drill thousands of wells into underground aquifers, and elevate the Tamiami Trail freeway so that water and wildlife can move more freely into the Everglades National Park. Instead of flushing wet-season rainfall swiftly out to sea through artificial channels as is now done, the new plan will store millions of cubic meters of water in artificial lagoons and underground aquifers from which it can be recovered during dry seasons. This will not only benefit wild landscapes but will also provide a much-needed water source for rapidly growing urban centers. Thousands of hectares of former wetlands that had been drained and farmed are now being turned back into wetlands. Years of accumulated pesticides and fertilizers in the soil, however, are solubilized from flooded fields, and massive bird, fish, and reptile mortalities have been reported, especially around Lake Apopka in central Florida.

Many similar experiences make environmentalists suspicious of some mitigation and restoration projects. "We can't just move ecosystems around like furniture that we put wherever we want," says author Seth Zuckerman. And John Berger, organizer of the first Restoring the Earth Conference, said, "The purpose of restoration ecology is to repair previous damage, not to legitimize further destruction."

Conflicting Views of Restoration

Controversy has arisen between restorationists and preservationists over both the effectiveness and the ideology of different approaches to protecting nature. Preservationists claim that failures such as the Everglades drainage problems show that the best strategy is to avoid

FIGURE 5.21 The naturally meandering Kissimmee River (*right channel*) was straightened and drained by the Army Corps of Engineers (*left*) for flood control thirty years ago. Now the Corps is attempting to reverse its actions and restore the Kissimmee and its associated wetlands to their original state.

destructive projects in the first place. They argue that our faith in science and technology to provide cures for our mistakes is ill-advised. A belief that we can fix any messes we make just serves as an excuse for business as usual, in their opinion.

Restorationists counter that we are unlikely to preserve more than small areas in pristine parks and nature reserves. While nice, these areas can't solve our global ecological problems, especially in poor countries where immediate survival needs force people to use resources regardless of the impacts on nature. Furthermore, expecting a biological community to remain unchanging is contrary to our modern understanding of ecology. Evolution obviously entails change. If nature is always changing, and that change is one of the essential properties we want to sustain, nature can't simply be locked up in a museum like a piece of art. Restorationists ask, who is to say that the changes we make in restoring or re-creating ecosystems is unnatural?

This last point raises questions about the place of humans in nature. Are we members of the biological community or are we separate from it? Should we use our creative energies to try to improve on nature, or should we leave well enough alone? Many ecologists argue that we don't know enough about ecosystems to successfully tinker with them.

Some social scientists, on the other hand, claim that nature is merely a social creation and that we can and should invent ingenious and imaginative new forms of nature to achieve ecological and social goals. Humans have been, are now, and will continue to be, part of the land community for better or worse, they contend. In this view, we have an unavoidable influence on nature that gives us a large degree of responsibility for it. We ought to cultivate, manipulate, use, and even manufacture nature, according to some people. Needless to say, many preservationists regard this as unmitigated arrogance. What do you think?

Tools of Restoration

Some of the earliest examples of restoration have involved labor-intensive horticultural or animal control methods. The Curtis Prairie at the University of Wisconsin in Madison, for instance, was reclaimed by Civilian Conservation Corps workers (fig. 5.22) and student volunteers, starting in 1934. Native species from the area were collected from remnant prairies along railroad rights-of-way and in pioneer cemeteries, and then hand-planted and cultivated on old, abandoned farm fields. Prairie plants initially had difficulty getting established and competing against exotics until it was recognized that periodic fires were essential for maintaining this biological community. The prairie is now flourishing and serves as a seed source for other restoration projects (fig. 5.23).

An even more laborious restoration project in Bermuda involved replanting forests, digging ponds, and re-creating a nearly vanished ecosystem to provide habitat for the endangered Bermuda cahow (see Case Study, p. 120). Sometimes the key to rebuilding a community is to remove alien intruders. In the Coachella Valley of California, for example, thousands of exotic salt cedars that dry out riparian habitats and crowd out native vegetation are being removed to protect the endangered Coachella Valley fringe-toed lizard. In Hawaii, feral pigs that root out native plant species and eat native birds are being hunted down and removed. Similarly, on the Galápagos Islands, feral goats and rats are being shot, trapped, and poisoned to protect native species.

A less expensive approach has been adopted in restoring an oak savanna near Chicago. Seeds collected from remnant prairies are simply broadcast rather than being carefully planted. This is not only faster and cheaper, it is more like the natural process of wind-dispersal of seeds. Fire is used to discourage invasions of exotic species (fig. 5.24). This is called successional restoration because it depends on natural succession to determine the outcome.

In the Flint Hills of southern Kansas, the Nature Conservancy has reintroduced both fire and native bison (buffalo) to its 16,000 ha (40,000 acres) tall-grass prairie preserve. The bison disperse seeds and create disturbances necessary for survival of many other native species. And in Costa Rica, an almost vanished dry tropical forest is being rebuilt in the Guanacaste Conservation Area by preventing fires and allowing domestic livestock to act as seed dispersal agents since the native grazing animals have completely disappeared.

Letting Nature Heal Itself

Sometimes all we have to do to reestablish a healthy ecosystem is simply to walk away and let organisms recolonize an area. The demilitarized zone (DMZ) between North Korea and South Korea was totally devastated in 1953 when the Korean War

FIGURE 5.22 In 1935, workers from the Civilian Conservation Corps dig up old farm fields in preparation for planting the native grasses and annual plants that eventually became the University of Wisconsin's restored Curtis Prairie.

FIGURE 5.23 The Curtis Prairie as it looked in 1980. This restored prairie has taught us a great deal about restoration in general and about the dynamics and ecology of prairie ecosystems in particular.

ended. Shattered tree stumps and ruined villages lay strewn across a barren landscape pockmarked by bomb craters and littered with the debris of war. Because neither side was allowed to move back into this political buffer zone, the area has become a wildlife refuge and a luxuriant oasis for nature in an otherwise densely populated countryside. Scarred slopes have been reclothed with a dense hardwood forest where deer, lynx, and an occasional tiger find shelter. Former rice paddies have reverted

FIGURE 5.24 A burn-crew technician sets a back fire to control the prescribed fire that will restore a native prairie.

to marshland in which waterfowl flourish and the trumpeting cries of the endangered Manchurian cranes are heard once again.

Similarly, parts of the political no-man's-land between the former East and West Berlin became a haven for wildlife during the forty years that the Berlin Wall separated the two parts of the city. This is a heartening example of the regenerative power of nature. Now that the city is reunified, the question arises about what to do with this area. Should it be allowed to remain an unkempt urban wilderness or should it be turned into a park?

Authenticity

Authenticity is a contentious issue in restorations of all sorts. Is it necessary, or even desirable, to try to restore a particular place to an exact replica of its original ecosystem? If you could create a pleasing, healthy biological community that resembles what existed in

CASE STUDY

RESTORATION OF THE BERMUDA CAHOW

The cahow is a seabird endemic (restricted) to Bermuda and adjacent islands off the east coast of North America. A member of the petrel family, related to albatrosses, shearwaters, and other wide-ranging seabirds, cahows once formed dense, noisy colonies that fed on the rich fisheries around the island. When European sailors first landed on Bermuda 400 years ago, cahows were abundant. Like many endemic island species, the ground-nesting cahow had never experienced predation and had no defenses against the pigs, goats, and rats introduced by the first settlers. Overhunting and habitat destruction further decimated the species. By the late 1600s—about the same time that the last dodo was killed on Mauritius—cahows disappeared from Bermuda.

For three centuries, the cahow was assumed to be extinct. In 1951, though, scientists found a few living cahows on some tiny islands in the Bermuda harbor. A protection and recovery program was begun immediately, including establishment of a sanctuary on the 6-hectare (15-acre) Nonsuch Island, which has become an excellent example of environmental restoration.

The Bermuda cahow or hook-billed petrel was thought to be extinct for nearly three centuries.

Nonsuch was a near desert after centuries of abuse, neglect, and habitat destruction. All the native flora and fauna were gone, along with most of its soil. This was a case of re-creating nature rather than merely protecting what was left. Sanctuary superintendent David Wingate, who has devoted his entire professional life to this project, has brought about a remarkable transformation of this barren little island. Reestablishing a viable population of cahows has had the added benefit of rebuilding an entire biological community.

The first step in restoration was to reintroduce native vegetation and re-create habitat. Thousands of native tree and shrub seedlings were planted. Initial progress was slow as trees struggled to get a foothold; once the forest knit itself into a dense thicket that deflected the salt spray and ocean winds, however, the natural community began to reestablish itself.

It takes constant surveillance to remove volunteer exotic plant species and to exclude rats, cats, and toxic toads that swim from the main island.

The benefits of indigenous species became apparent in 1987 when Hurricane Emily roared across Bermuda. Up to 70 percent of nonnative trees were uprooted or snapped off by gale-force winds, littering streets and bringing down power lines. The dense, low-profile, native trees on Nonsuch were barely touched by the winds. Demands soared for hurricane-adapted species to replace those lost along streets and in gardens.

Just providing habitat for the cahows was not enough, however, to restore the population.

David Wingate examines a cahow nest on Bermuda's Castle Harbor Islands Nature Reserve.

Each pair lays only one egg per year and only about half survive under ideal conditions. It takes eight to ten years for fledglings to mature, giving the species a low reproductive potential. They also compete poorly against the more common long-tailed tropic birds that steal nesting sites and destroy cahow eggs and fledglings. Special underground burrows were built with baffled entrances designed to admit only cahows. Young birds were hand-raised by humans to ensure a proper diet and protection.

By 1997, the cahow population had rebounded to fifty-six nesting pairs. It is too early to know if this is enough to be stable over the long term, but the progress to date is encouraging. Perhaps more important than rebuilding this single species is that the island has become a living museum of precolonial Bermuda that benefits many species besides its most famous resident. It is a heartening example of what can be done with vision, patience, and some hard work.

the past, is that close enough, or should you strive for a precise duplicate? Suppose that the original ecosystem was made up of ugly, nasty, diseased plants and animals. Would it be acceptable to try to improve on nature? If you were restoring a northern wetland, would you deliberately reintroduce mosquitoes, black flies, leeches, and ticks? If your wetland was in the South, would you add poisonous water snakes and spiders?

Some ecologists contend that restorations should use only plants and animals that have evolved under the particular environmental conditions of the place to be rehabilitated. According to this viewpoint, you shouldn't introduce organisms—even of the same species—that come from some other place because they may lack unique adaptations that come from evolving in a specific location. Suppose you are restoring a cattail marsh. How far would you go to find cattails to transplant? What if there is a small marsh with cattails right next to the area on which you are working? These plants are probably closely related to the ones that once grew in your marsh. But if you take roots or seeds from the few plants remaining in the neighboring marsh, you may disrupt cattail regeneration there. Is having nearby plants that important, or would it be alright to get stock from farther away?

Back To What?

When humans take charge of ecosystems, questions sometimes are raised about what our goals should be. Suppose natural forces, such as hurricanes or fires, disrupt a wilderness area, should we use the principles of restoration ecology to tidy up or improve an area or leave it to natural processes? In some cases where humans have altered an area, there may be more than one historic state to which we could restore it. In The Nature Conservancy's Hassayampa River Preserve near Phoenix, for instance, pollen grains preserved in sediments reveal that 1000 years ago, the area was grassy marsh that was unique in the surrounding desert landscape. Some ecologists would try to rebuild a similar marsh. Corn pollen in the same sediments, however, show that 500 years ago Native Americans began farming the marsh. Which is more important, the natural or early agricultural landscape?

Unfortunately, it may not be possible to return to conditions of either 500 or 1000 years ago since climate changes and evolution may have made the communities existing at that time incompatible with current conditions. This creates a difficult question for restoration ecologists. If change is natural and inevitable, who is to say that present conditions—whatever they are or however they have come about—are bad? How should we distinguish between desirable and undesirable changes? If it is simply a matter of human preference, some people might prefer a golf course or a shopping mall on the site. What would you say to someone who claims that since humans are a part of nature, whatever changes we make to a landscape also are natural?

Creating Artificial Ecosystems

As we learn more about how ecosystems work and about the valuable ecological services they provide, we are coming to realize that we can use some natural principles in human-designed systems. The Arcata, California, marsh and wildlife sanctuary is a famous example of using nature to solve a human problem. In 1974, Arcata was faced with the prospect of spending $36 million to upgrade its municipal sewage treatment plant and reduce dumping of inadequately treated wastewater into Humboldt Bay. City residents and faculty from Humboldt State University devised a low-cost alternative that might serve as a model for many other communities.

An abandoned dump near the existing treatment plant was turned into an artificial wetland (fig. 5.25). Wastewater from conventional treatment processes flows through oxidation ponds where sunlight and air kill pathogenic microbes. The effluent is then filtered through the artificial marsh, where aquatic plants filter out nutrients and zooplankton eat microbes and fish eat the plankton. Finally, the water trickles into a saltwater slough where pelicans and cormorants gather to feed on the plentiful fish and where oysters perform a final filtering step before the "polished" water enters the bay. The marsh has become a favorite bird-watching and recreational site. It also saves millions of dollars and has turned what was formerly an eyesore into a source of civic pride.

FIGURE 5.25 Arcata, California, built an artificial marsh as a low-cost, ecologically based treatment system for sewage effluent.

ECOSYSTEM MANAGEMENT

As you probably already know, there is a growing sense among ecologists, and the public at large, that our environmental situation is rapidly reaching a crisis point. Many of our management policies, while well-meaning, have made matters worse rather than better. **Ecosystem management** is a relatively new discipline in environmental science that attempts to integrate ecological, economic, and social goals in a unified, systems approach. It recognizes that we cannot have sustained progress toward social goals in a deteriorating environment or economy, and vice versa. Each of these domains affects, and is affected by, the others. There are many definitions of ecosystem management. Guidelines suggested for U.S. Federal Agencies are shown in table 5.2.

A Brief History of Ecosystem Management

While the term "ecosystem management" is new, a few visionary ecologists such as Aldo Leopold had the foresight to advocate many specific elements of this science fifty years ago. Leopold's attempts to restore his Sand County farm to ecological health and beauty foreshadowed many of the ideas now widely espoused by ecosystem managers. Another pioneer in this field is environmental policy expert Lynton Caldwell, who wrote in 1970 that we should use ecosystems as the basis for public land policy. He

TABLE 5.2 Principles of Ecosystem Management

- Managing across whole landscapes, watersheds, or regions over an ecological time scale.
- Considering human needs and promoting sustainable economic development and communities.
- Maintaining biological diversity and essential ecosystem processes.
- Utilizing cooperative institutional arrangements.
- Integrating science and management.
- Generating meaningful stakeholder and public involvement and facilitating collective decision-making.
- Adapting management over time, based on conscious experimentation and routine monitoring.

understood that to do so "would require that the conventional [political] matrix be unraveled and rewoven in a new pattern." Unfortunately, although land use planning was a high priority of the burgeoning U.S. environmental movement in the 1970s, the forces of inertia and private interest thwarted attempts to manage ecologically.

Most federal and many state natural resource agencies in the United States are attempting to identify endangered landscapes and implement ecosystem management as their guiding policy. The U.S. Forest Service, the Bureau of Land Management, and the National Park Service, for instance, all have adopted versions of ecosystem management. This is a marked improvement in many cases over past practices. Traditionally, many of these agencies emphasized commodity production and the commercial or recreational use of natural resources as their first priority. Management objectives often were designed to expedite the development, extraction, and/or production of resources on public lands above all other uses. Wildlife and fish habitats, endangered species, and cultural, scenic, historic, or aesthetic values tended to be viewed mainly as inconveniences or hindrances to economic uses.

Principles and Goals of Ecosystem Management

Conservation biologist E. R. Grumbine, in a seminal article on ecosystem management, pointed out several important differences between this integrative approach and the traditional policies of the past:

1. *Hierarchical context:* A focus on any one level of the biodiversity hierarchy (genes, species, populations, ecosystems, landscapes) is insufficient. Ecosystem managers must see interconnections between all levels.

FIGURE 5.26 Ecosystem management requires a high level of routine monitoring and data collection to assess what works and what doesn't. Adaptive management means that every program is regarded as experimental and subject to change as new information becomes available.

2. *Ecological boundaries:* Rather than divide administrative units by political boundaries, the watersheds, ecosystems, or other natural units should be managed in an integrated fashion.
3. *Data collection and routine monitoring:* To function correctly, ecosystem management requires ongoing research and data collection so that successes or failures may be recognized and evaluated.
4. *Adaptive management:* Ecosystem management assumes that scientific knowledge is provisional and regards management plans as learning processes or continuous experiments where incorporating the results of previous actions allows managers to remain flexible and adapt to uncertainty (fig. 5.26).
5. *Organizational change:* Implementing ecosystem management requires changes in agency structure and ways of doing business. Accomplishing meaningful changes in entrenched bureaucracies may be the most challenging aspect of this new approach.
6. *Humans in nature:* People cannot be separated from nature. Humans inescapably affect ecological patterns and processes and are in turn affected by them.
7. *Values:* Regardless of the role of scientific knowledge, human values play a dominant role in ecosystem management goals.

Some commonly reported goals for ecosystem management are presented in table 5.3. In general, the criteria listed by Aldo

TABLE 5.3 Ecosystem Management Goals

- Maintain viable populations of native species *in situ.*
- Represent, within protected areas, all native ecosystem types across their natural range of variation.
- Protect essential ecological processes such as nutrient cycles, succession, hydrologic processes, etc.
- Manage over long enough time periods to sustain the evolutionary potential of species and ecosystems.
- Accommodate human use and occupancy within these constraints.

Leopold, as the basis of environmental ethics—integrity, stability, and beauty of biotic communities—are good benchmarks for ecosystem management.

Critiques of Ecosystem Management

Many of the criticisms of ecosystem management are similar to those mentioned earlier with respect to restoration ecology. Many ecologists argue that we don't understand ecosystems now and we probably never will, given the chaotic, catastrophic, and unpredictable qualities of nature. It is arrogant of us, therefore, to imagine that we can manage nature. Furthermore, in developing countries, political pressures and immediate needs of poor people make it difficult to enforce environmental protection laws now. This will become even more difficult as populations grow and resources diminish.

Often vision statements and organizational plans for ecosystem management are little more than empty slogans, while nothing really changes. Simply claiming that we will be organized, efficient, and effective doesn't necessarily make us so. Many resource managers worry that ecosystem management encourages us to believe we can have our cake and eat it too. We simply have to accept limits on what we can do and have, they argue. From this perspective we ought to set aside large blocks of untrammeled nature and free ourselves of the illusion that science or technology can provide solutions to our environmental or social problems.

The debate over ecosystem management reflects a deep division that runs through much of environmental science. The cleavage between preservation and management, between conservative and progressive policies, and between pessimism and optimism all are based on different worldviews about what is possible and what is likely to occur. Where do you stand in this spectrum of opinions? Do you think we can—or should even try to—manage ecosystems; restore, reclaim, or rehabilitate landscapes; or create new and imaginative forms of nature? Is it possible for us to do what is necessary to preserve the integrity, stability, and beauty of our environment?

SUMMARY

Major ecosystem types called biomes are characterized by similar climates, soil conditions, and biological communities. Among the major terrestrial biomes are deserts, tundra, grasslands, temperate deciduous forests, temperate coniferous forests, tropical moist forests, and tropical seasonal forests. Aquatic ecosystems include oceans and seas, rivers and lakes, estuaries, marshes, swamps, bogs, fens, and reefs. Moisture and temperature are generally the most critical determinants for terrestrial biomes. Periodic natural disturbances, such as fires, play a major role in maintaining some biomes.

Humans have disturbed, preempted, or damaged much—perhaps half or more—of all terrestrial biomes and now dominate about 40 percent of all net primary productivity on the land. Some of this disturbance and domination is permanent, but we have opportunities to apply ecological knowledge and practical experience to restoring and repairing ecosystems. In some cases, all we have to do is leave nature alone; in other cases, more active management is required to reestablish functioning ecosystems. Remediation usually means cleaning the soil of pollution, while reclamation is applied to physical repair of severely damaged land. Mitigation means to replace or re-create an ecosystem in restitution for damage caused by a construction project or some other human activity. Whether we can or should try to create new, synthetic forms of nature is controversial.

Landscape ecology is the study of reciprocal effects of spatial pattern on ecological processes. A landscape, in this sense, is a bounded geographical unit that includes both living and nonliving components. History is important in that it shapes the land and organisms as well as our reactions to, and interpretations of, the landscape. Humans are important elements of most landscapes and few places on the earth are devoid of human impacts. Restricting our definition of "nature" to a few pristine remnants will lead us either to ignore most of the earth's surface or to demand that humans remove themselves entirely from large tracts, something not very likely.

Patchiness and heterogeneity are characteristic of most landscapes. Movement of organisms and materials across boundaries between habitat patches play an important regulatory role in many ecosystems. Some landscape patterns apply across a wide range of scales while others are restricted to a specific scale.

Ecosystem management involves managing whole landscapes over long timescales, taking human needs into consideration, maintaining biodiversity and ecological processes, utilizing innovative institutional arrangements, integrating science and policy, encouraging public involvement, and adapting management based on conscious experimentation and routine monitoring. Ecosystem management, like sustainable development, recognizes that social goals, environmental quality, and economic health are all inextricably interlinked.

QUESTIONS FOR REVIEW

1. Who was Aldo Leopold and why is he considered important in the history of American conservation?
2. Throughout the central portion of North America is a large biome once dominated by grasses. Describe how physical conditions and other factors control this biome.
3. What is taiga and where is it found? Why might logging in taiga be more disruptive than in southern coniferous forests?
4. Why are tropical moist forests often less suited for agriculture and human occupation than tropical deciduous forests?
5. Describe four different kinds of wetlands and explain why they are important sites of biodiversity and biological productivity.
6. Which major biomes have been most heavily disturbed by human activities?
7. Define a landscape. Describe the major ecological and cultural features of the landscape in which you live.
8. Define *restoration, rehabilitation, remediation, reclamation, ecological re-creation,* and *mitigation.* Give an example of each.
9. List and explain either the seven principles or the five goals of ecosystem management.
10. Explain some of the major criticisms of ecosystem management.

QUESTIONS FOR CRITICAL THINKING

1. In which biome do you live? What physical and biological factors are most important in shaping your biological community? How do the present characteristics of your area differ from those 100 or 1000 years ago?
2. Could your biome be returned to something resembling its original condition? What tools (or principles) would you use to do so?
3. A beautiful stand of pine trees in a park in Connecticut was blown down by a windstorm and is now a fire hazard to nearby property as well as being an aesthetic mess. Some people argue that salvage loggers should clean out the area and replant pine trees. Others argue that it was a natural forest and should be left to natural forces. What do you think?
4. Historic records show that indigenous aboriginal people kept the forest described in question 3 as an open, park-like savanna by burning it regularly. Should prescribed fires be a part of the restoration ecology plan?
5. Some environmentalists worry that the science of restoration ecology may give us the arrogant attitude that we can do anything we want now because we can repair the damage later. How would you respond to that concern?
6. What do you think Aldo Leopold meant by integrity, stability, and beauty of the land?
7. Disney World in Florida wants to expand onto a wetland. It has offered to buy and preserve a large nature preserve in a different area to make up for the wetland it is destroying. Is that reasonable? What conditions would make it reasonable or unreasonable?
8. Suppose further that the wetland being destroyed in question 7 and its replacement area both contain several endangered species (but different ones). How would you compare different species against each other? How many plant or insect species would one animal species be worth?
9. Explain why there might be differences in philosophy or worldview between preservationists and restorationists. Which approach do you prefer?
10. Authenticity is a contentious issue in restoration. Is it necessary, or even desirable, to try to create an exact replica of an original ecosystem? Could the changes that humans make be considered part of natural change?

KEY TERMS

barrier islands 112
benthos 110
biomes 103
bogs 111
boreal forest 106
chaparral 108
cloud forest 109
conifer 106
coral reefs 112
deciduous 107
delta 111
deserts 104
ecosystem management 121
estuaries 110
fens 111
freshwater ecosystems 109
grasslands 105
landscape ecology 115
marshes 111
mitigate 117
reclamation 117
re-creation 117

ADDITIONAL INFORMATION ON THE INTERNET

Visit our website at http://www.mhhe.com/environmentalscience/ for specific information about each of the topics below:

Biomes
Restoration Ecology center, University of Wisconsin
Restoration and Reclamation Review
Aldo Leopold Nature Center
Landscape ecology and birds: slide show
Landscape ecology
GAP: Gap Analysis Program
Yellowstone Ecosystem Management Studies
Forest management and activism links
Modeling in Everglades management planning
Aquatic Ecosystem Restoration Foundation
Invasive Species
Environmental Issues Index
Sierra Club Critical Ecoregions Program
US Bureau of Reclamation
US Fish and Wildlife Service National Wetlands Inventory
Phytoremediation

PART TWO

POPULATION, ECONOMICS, POLICY, AND HEALTH

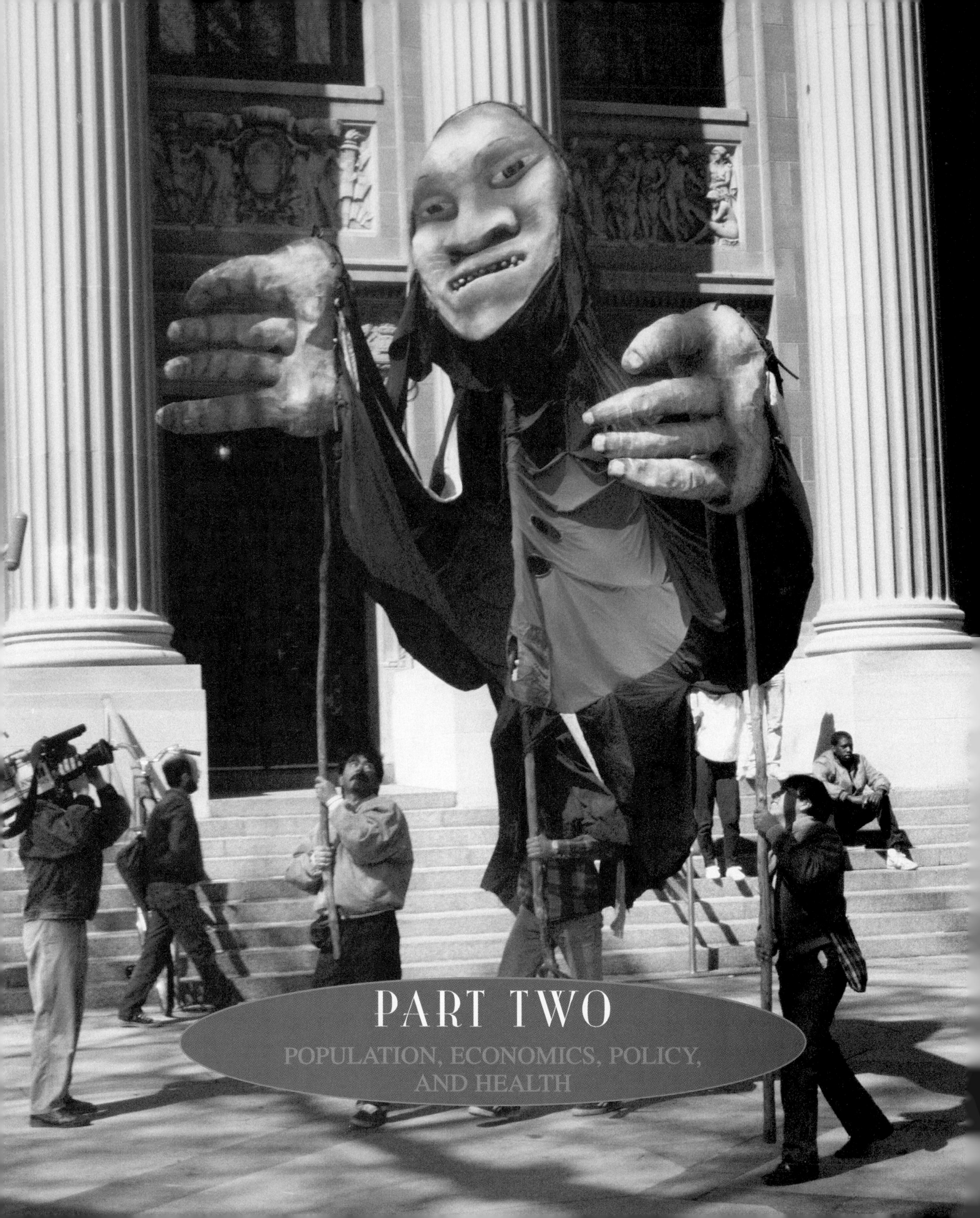

CHAPTER 6

Population Dynamics

Nature teaches more than she preaches.

—John Burroughs—

OBJECTIVES

After studying this chapter, you should be able to:

- appreciate the potential of exponential growth.
- draw a diagram of J and S curves and explain what they mean.
- explain who Thomas Malthus was and what he said about population growth.
- describe environmental resistance and discuss how it can lead to logistic or stable growth.
- define *fecundity, fertility, birthrates, life expectancy, death rates,* and *survivorship*.
- compare and contrast density-dependent and density-independent population processes.

ESSENTIAL STUDY PARTNER

The Essential Study Partner (ESP) CD-ROM that accompanies this textbook contains additional information on the following topics:

- exponential and logistic growth: populations: dynamics: growth
- environmental resistance and carrying capacity: populations: dynamics: growth
- reproductive strategies, life history characteristics, and survivorship curves: population: dynamics: strategies
- age structure and population momentum: populations: human population: demography
- density-independent and density-dependent population regulating factors: population: dynamics: regulation
- population dieback and predator regulation: population: dynamics: regulation

Opposite: Ordinary people can do extraordinary things when they work together.

Above: A herd of wildebeest (gnu) cross the Mara River in their annual migration across the Masri Mara National Park in Kenya.

Columbia River Salmon

Until about a century ago, as many as 16 million salmon and cutthroat trout migrated every year up the Columbia River system to their breeding grounds in small headwater streams and lakes over a watershed larger than Texas (fig. 6.1). This was probably the greatest anadromous fish (spend part of life cycle in freshwater and part in saltwater) migration in the world. Some fish swam up to 2000 km (1250 mi) from the mouth of the Columbia on the Pacific coast to its headwaters in British Columbia or up tributary streams such as the Snake and Salmon Rivers in what is now Idaho. The fish were marvelously adapted to this remarkable journey. At least 420 separate stocks (or ecotypes) differed in the timing of their run and the subtle chemical signals they followed to find the stream where they hatched at exactly the right time for water conditions and food supplies to support their offspring. Adult salmon die after spawning and their decaying bodies nourish the ecosystem on which fry and fingerlings (young fish) will depend. The adults, each weighing as much as 45 kg (100 lbs), represent an enormous influx of nutrients into the small streams where they breed. Studies have shown that up to half of the nitrogen in riparian (stream side) vegetation in some streams comes from migrating fish. Without dead fish, the whole ecosystem is impoverished.

Both native people and wildlife in the Pacific Northwest depended on this prodigious bounty. For a few months every year there were more fish than anyone could eat. The runs were probably never uniform, however. For reasons that we don't fully understand, the number of fish returning from the ocean often would vary as much as 50 percent from year to year. Even now, we don't know where salmon go during the 2 to 5 years adults spend in the ocean or exactly what they eat while growing to such enormous sizes. Undoubtedly, changes in ocean temperature, circulation patterns, and food supplies caused by phenomena such as the El Niño/Southern Oscillation (chapter 17) affect population sizes. There seems to be a 40-year cycle, for instance, when salmon runs in Alaska are high and Columbia runs are low or vice versa. Before European settlement, however, these variations didn't matter much because there were more than enough salmon for everyone.

Salmon runs in the Columbia have undergone a disastrous decline during the past 80 years (fig. 6.2). Total numbers are down nearly 90 percent from historic highs, and only about one-tenth are now wild stocks; the rest are hatchery-reared. There are many reasons for this decrease. Overfishing by commercial operations has taken a toll. Runoff from logging, agriculture, road building, and urban areas carries warm water, sediment, and pollution that kill eggs and fry. Irrigators pump water from the river, reducing its flow. Perhaps the greatest threat to the salmon are dams that block migration. Over the past century 124 huge dams have been built on the Columbia along with 55 more on its tributaries. The river has become a chain of reservoirs. Except at its mouth, only 70 km (44 mi) of the Columbia runs free today. Fish ladders (stair step pools) help some adults move upstream but smolt (young fish), which depend on river currents to help them downstream, get lost in the slack water behind dams. Whole populations of smolt are now being barged downstream to get them past dams and reservoirs but many fail to survive anyway.

Hatchery rearing of salmon began on the Columbia in the 1890s. Rather than stem the decline, unfortunately, releases of hatchery fish often have exacerbated problems. Only a few genetic strains are bred in hatcheries rather than the hundreds of wild stocks. Hatchery fish have been selected for fast growth, early maturity, and aggressive behavior. This allows them to interbreed with and outcompete native stocks. With much of their original genetic diversity gone, the fish now lack the subtle adaptations that are needed to migrate to remote streams at just the right time. At least half of the original wild salmon runs on the Columbia are now extinct and almost all that remain are in serious trouble.

In a population that fluctuates as widely as salmon, it is difficult to detect patterns until long after critical events occur. If fewer fish show up this year compared to last, is it just a natural variation or an omen that something is wrong? The last big run on the Columbia River occurred in 1924 when the commercial harvest was nearly 19 million kg (42 million lbs). By the time fishwheels were prohibited in 1935, or when seines, traps, and set nets were banned in 1950, the population was already in a catastrophic decline. By 1991, when the Snake River sockeye was added to the endangered species list, only four of these fish made it to their spawning grounds in Idaho's Sawtooth Mountains. Since then, five other runs (Snake River spring and fall chinook, Umpaqua River cutthroat, and lower Columbia chum and chinook) have been added to the endangered species list while others are under consideration. Fishers and recreationists are urging the government to destroy at least four of the dams on the Snake to allow the river to run free again. Industries and residents of the Columbia River basin, who enjoy some of the lowest electric rates in the nation oppose this plan.

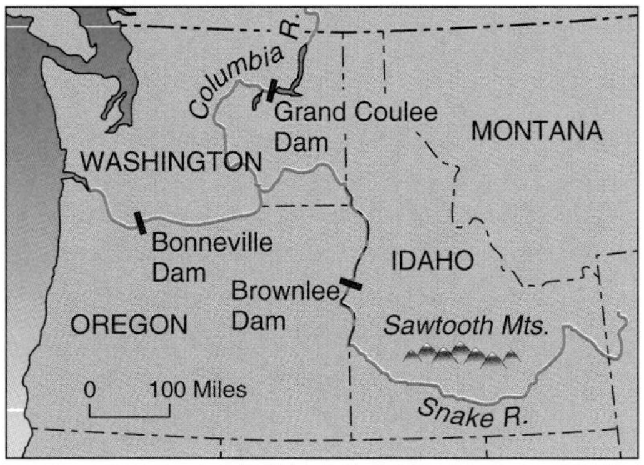

FIGURE 6.1 Up to 16 million fish per year once migrated up the Columbia River and its tributaries such as the Snake.

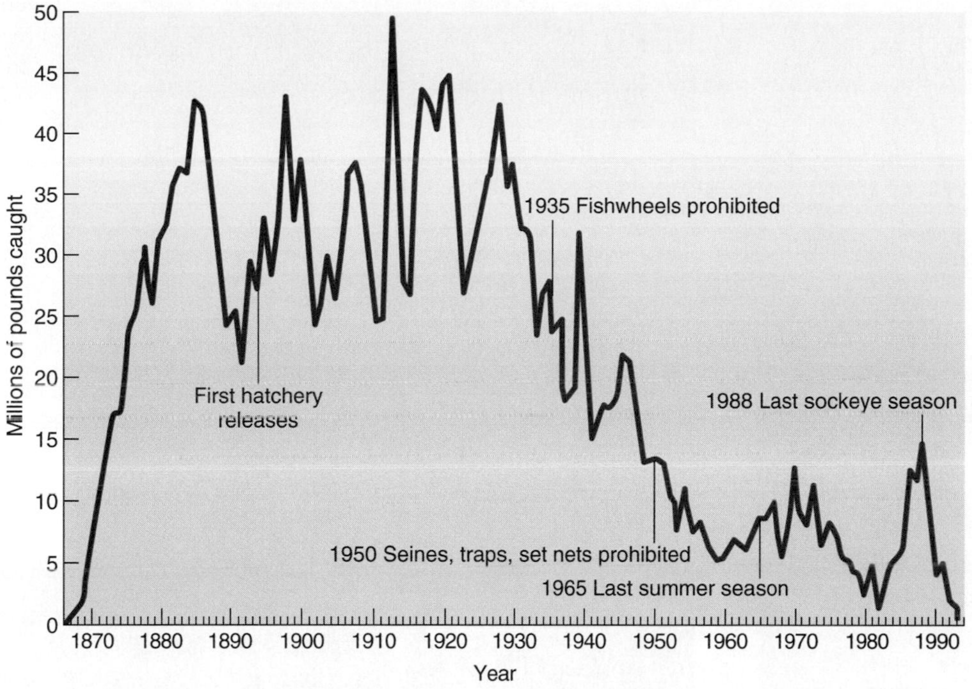

FIGURE 6.2 Commercial salmon harvest in the Columbia River 1855–1991. By the time fishwheels, seines, traps, and set nets were prohibited, the fish were already in a downward slide.

Source: Data "Columbia Fish Runs and Fisheries 1938–93" from Washington Department of Fish and Wildlife and Oregon Department of Fish and Wildlife 1994 Status Report, Olympia, WA.

DYNAMICS OF POPULATION GROWTH

Why do some populations grow to enormous size while others do not? What causes abrupt population increases or decreases? Where was there a turning point in the downward spiral of the salmon population that we might have recognized at the time? How many individuals does it take to make a viable population? What roles do competition, genetic diversity, and ecological adaptations play in maintaining or reducing wild species? These are some of the questions asked by population biologists. In this chapter, we will survey the factors that lead to population growth and decline. We also will consider the mechanisms that regulate total population size, distribution, and age structure. One of the most important concepts in population biology is that both density dependent and density independent factors play roles in determining how big a particular population in a given habitat can be.

Exponential Growth and Doubling Times

Given optimum environmental conditions, many biological species have the capacity for **exponential growth,** or growth at a constant *rate* of increase per unit of time. It is called exponential because the rate of increase can be expressed as a constant *fraction,* or exponent, by which the existing population is *multiplied.* This pattern also is called **geometric growth** because the sequence of growth follows a geometric pattern of increase, such as 2, 4, 8, 16, and so on. By contrast, a pattern of growth that increases at a constant *amount* per unit of time is called **arithmetic growth.** The sequence in this case might be 1, 2, 3, 4 or 1, 3, 5, 7. Notice in these examples that a constant amount is *added* to the population (fig. 6.3).

Organisms are able to grow exponentially, given an appropriate environment, because of the amplifying power of biological reproduction. Most parents have the ability to produce multiple offspring over their lifetime. If even a few of those offspring survive to reproductive age, the population will grow.

As you can see in figure 6.3, the number of individuals added to a population at the beginning of a geometric growth curve is rather small. But within a very short time, the numbers begin to increase quickly because a given percentage becomes a much larger amount as the population gets bigger. The growth curve produced by this constant rate of unfettered growth is called a **J curve** because of its shape. At 1 percent per year, a population or a bank account doubles in roughly seventy years (table 6.1). A useful rule of thumb is that if you divide seventy by the annual percentage growth, you will get the approximate doubling time in years. As a result, a population growing at 35 percent doubles every two years. You can also apply this rule to calculate doubling time of human populations. Countries growing at 4 percent per year will *double* their populations in 17.5 years. A country growing at a rate of 0.1 percent annually will double its population in 700 years. We will use this rule to discuss human population growth in chapter 7.

Biotic Potential

Neither salmon nor humans are the fastest reproducing of all organisms. Many species have amazingly high reproductive rates that give them the potential to produce enormous populations very

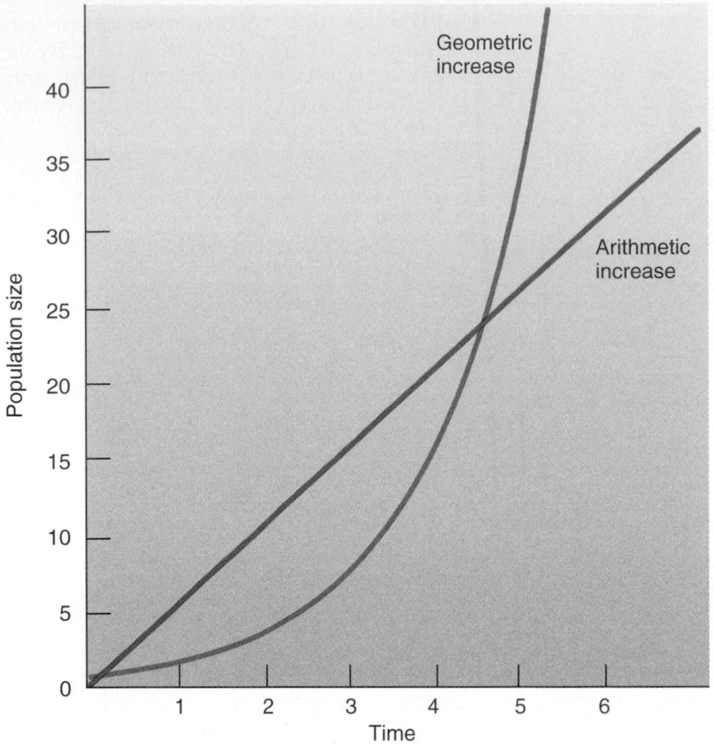

FIGURE 6.3 Arithmetic and geometric growth curves. Note that the geometric or exponential curve grows more slowly at first, but then accelerates past the arithmetic curve, which grows at a steady incremental pace throughout.

quickly, given unlimited resources and freedom from limiting factors (fig. 6.4). We call the maximum reproductive rate of an organism its **biotic potential.** Table 6.2 shows the potential number of offspring that a single female housefly (*Musca domestica*) and her offspring could produce in a year. The result is astounding. Each female fly lays an average of 120 eggs in each generation. The eggs hatch and mature into sexually active adults and lay their own eggs in fifty-six days. In one year (seven generations), if all its offspring survived long enough to reproduce, a single female could be the ancestor of 5.6 *trillion* flies. If this rate of reproduction continued for ten years, the whole earth would be covered several meters deep with houseflies! Fortunately, this has not happened because of factors that limit the reproductive success of houseflies. This example, however, illustrates the potential for biological populations to increase rapidly.

TABLE 6.1 Doubling Times at Various Compound Interest Rates

ANNUAL % INCREASE	DOUBLING TIME (YEARS)
0.1	700
0.5	140
1.0	70
2.0	35
5.0	14
7.5	9
10.0	7
100.0	0.7

FIGURE 6.4 Reproduction gives many organisms the potential to expand populations explosively. The cockroaches in this kitchen could have been produced in only a few generations. A single female cockroach can produce up to eighty eggs every six months. This exhibit is in the Smithsonian Institute's National Museum of Natural History.

Population Oscillations and Irruptive Growth

In the real world, there are limits to growth. When a population exceeds the carrying capacity of its environment or some other limiting factor comes into effect, death rates begin to surpass birth rates. The growth curve becomes negative rather than positive, and the population decreases as fast as, or faster, than it grew. We call this the population crash or **dieback.**

The extent to which a population exceeds the carrying capacity of its environment is called **overshoot,** and the severity of the dieback is generally related to the extent of the overshoot. This pattern of **population explosion** followed by a **population crash** is called **irruptive** or **Malthusian growth.** It is named after the eighteenth-century economist Thomas Malthus, who concluded that human populations tend to grow until they exhaust their resources and become subject to famine, disease, or war. Malthus and his theories are discussed further in chapter 7.

Populations may go through repeated oscillating cycles of exponential growth and catastrophic crashes, as shown in figure 6.5. These cycles may be very regular if they depend on a few sim-

TABLE 6.2 Biotic Potential of Houseflies (*Musca domestica*) in One Year

Assuming that:
—a female lays 120 eggs per generation
—half of these eggs develop into females
—there are seven generations per year

	TOTAL POPULATION	
GENERATION	IF ALL FEMALES IN EACH GENERATION LAY 120 EGGS AND THEN DIE	IF ALL GENERATIONS SURVIVE ONE YEAR AND ALL FEMALES REPRODUCE MAXIMALLY IN EACH GENERATION
1	120	120
2	7,200	7,320
3	432,000	446,520
4	25,920,000	27,237,720
5	1,555,200,000	1,661,500,920
6	93,312,000,000	101,351,520,120
7	5,598,720,000,000	6,182,442,727,320

Source: Data from E. J. Kormondy, *Concepts of Ecology,* 3d ed., © 1985 Harper & Row Publishers, Inc.

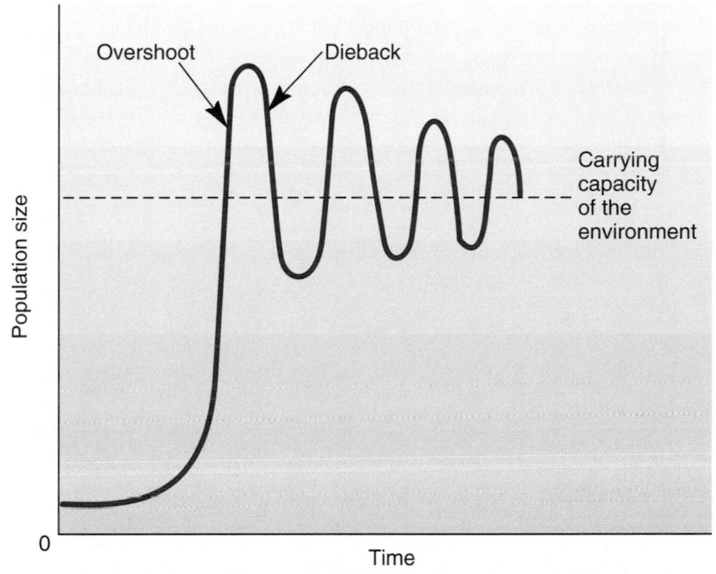

FIGURE 6.5 Population oscillations. Some species demonstrate a pattern of cyclic overshoot and dieback.

ple factors, such as the seasonal light- and temperature-dependent bloom of algae in a lake. They also may be very irregular if they depend on complex environmental and biotic relationships that control cycles, such as the population explosions of migratory locusts in the desert or tent caterpillars and spruce budworms in northern forests.

Growth to a Stable Population

Not all biological populations go through these cycles of irruptive population growth and catastrophic decline. The growth rates of many species are regulated by both internal and external factors so that they come into equilibrium with their environmental resources. These species may grow exponentially when resources are unlimited, but their growth slows as they approach the carrying capacity of the environment. This pattern is called **logistic growth,** a mathematical description of its constantly changing rate.

Together, factors that tend to reduce population growth rates are called **environmental resistance.** In later sections of this chapter, we will look in more detail at these factors and how they limit growth and regulate population size. First, we will see how logistic growth compares to the Malthusian growth and what kinds of organisms we are talking about in these different patterns.

How does the growth curve of a stable population differ from the J curve of an exploding population? Figure 6.6 shows an idealized comparison between biotic potential and sustainable growth. The J curve on the left in this figure represents the growth without restraint that we just discussed. It rises rapidly toward the maximum biotic potential of the species. The curve to the right represents logistic growth. We call this pattern an **S curve** because of its shape. It is also called a sigmoidal curve (for the Greek letter sigma). The area between these curves is the cumulative effect of environmental resistance. Note that the resistance becomes larger and the rate of logistic growth becomes smaller as the population approaches the carrying capacity of the environment.

Chaotic and Catastrophic Population Dynamics

Real populations under natural conditions rarely follow the linear dynamics and smooth growth curves shown in figures 6.5 and 6.6. Apparently haphazard fluctuations often characterize population dynamics. Look again at salmon catches between about 1885 and 1925 in figure 6.2. Does this represent an irregular oscillation, random variation, or some other pattern of population dynamics? Population ecology has begun to incorporate ideas from nonlinear mathematics to describe these patterns as being chaotic or catastrophic.

Chaotic systems exhibit variability that, while not necessarily random, can be of a complexity whose pattern is not observable over a normal human timescale. In these systems, minute differences in initial conditions can lead to dramatically different outcomes after many iterations.

A famous metaphor for chaotic dynamics is the "butterfly effect," first described by meteorologist Edward Lorenz in 1979. Lorenz observed that seemingly insignificant variations in the beginning parameters of computerized climate models resulted in

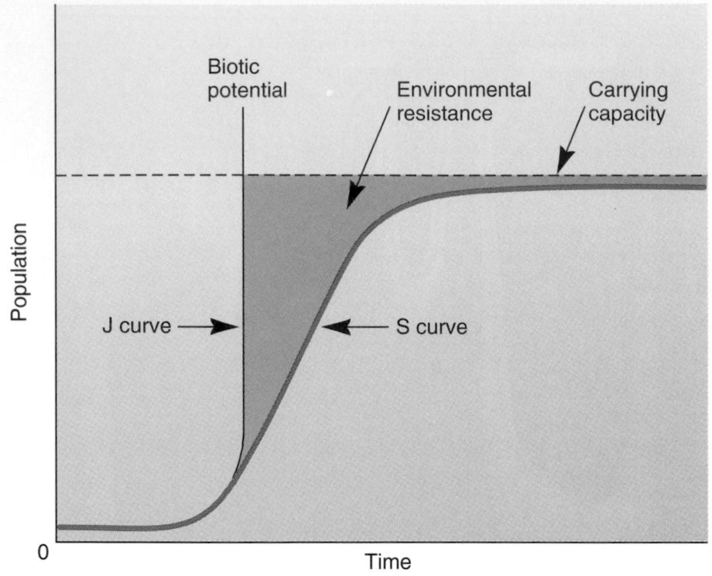

FIGURE 6.6 J and S population curves. The vertical J represents theoretical unlimited growth. The S represents population growth and stabilization in response to environmental resistance.

totally different final results. It is, he said, as if a butterfly flapping its wings in Rio de Janeiro could, through a long series of unpredictable multiplying effects in weather systems, eventually result in a tornado in Texas.

Catastrophic theory was first developed by mathematicians as a purely hypothetical exercise, but some biologists regard it as a good explanation of population dynamics displaying abrupt discontinuities. As conditions change, a **catastrophic system** may jump abruptly from one seemingly steady state to another without any intermediate stages. Balancing a chair on two legs is a good model of catastrophe. You may be stable for a long time with relatively minor corrective input, but, if you pass a certain threshold, suddenly find yourself flat on your back on the floor—and, once again, stable. Among populations, the sudden outbursts and diebacks of spruce budworm in the forests of eastern North America or of lemmings in the arctic tundra often are mentioned as examples of catastrophic dynamics.

Is there any practical difference between chaos or catastrophe and simple random behavior? Perhaps so. In a chaotic system, although you can't predict behavior on a short timescale with any certainty, you may be able to identify outer boundary conditions beyond which the system will not go. In a catastrophic system, you can avoid pushing a population close to the edge of an unstable flip point. Perhaps the collapse of Columbia salmon runs represents a catastrophic flip.

Strategies of Population Growth

There appear to be evolutionary advantages, as well as disadvantages, in both Malthusian and logistic growth patterns. Although we should avoid implying intention in natural systems where the controlling forces may be entirely mechanistic, it sometimes helps us see these advantages in terms of "strategies" of adaptation and "logic" in different modes of reproduction.

Malthusian "Strategies"

Organisms with Malthusian growth patterns often tend to occupy low trophic levels in their ecosystems (chapter 3), or to be pioneers in succession (chapter 4). As generalists or opportunists, they move quickly into disturbed environments, grow rapidly, mature early, and produce many offspring. They usually do little to care for their offspring or protect them from predation. They depend on sheer numbers and dispersal mechanisms to ensure that some offspring survive to adulthood (table 6.3). They have little investment in individual offspring, using their energy to produce vast numbers instead.

Many insects, rodents, marine invertebrates, parasites, and annual plants (especially the ones we consider weeds) follow this reproductive strategy. Their numbers generally are limited by predators or other controlling factors in the environment. They reproduce at the maximum rate possible to offset these losses. If the external factors that normally control their populations are inoperative, they tend to rapidly overshoot the carrying capacity of the environment and then die back catastrophically, as we have just seen. Among this group are the weeds, pests, or other species we consider nuisances that reproduce profusely, adapt quickly to environmental change, and survive under a broad range of conditions.

TABLE 6.3 Characteristics of Contrasting Reproductive Strategies

EXTERNALLY CONTROLLED GROWTH	INTRINSICALLY CONTROLLED GROWTH
1. Short life	1. Long life
2. Rapid growth	2. Slower growth
3. Early maturity	3. Late maturity
4. Many small offspring	4. Fewer large offspring
5. Little parental care or protection	5. High parental care and protection
6. Little investment in individual offspring	6. High investment in individual offspring
7. Adapted to unstable environment	7. Adapted to stable environment
8. Pioneers, colonizers	8. Later stages of succession
9. Niche generalists	9. Niche specialists
10. Prey	10. Predators
11. Regulated mainly by extrinsic factors	11. Regulated mainly by intrinsic factors
12. Low trophic level	12. High trophic level

what do you think?

THE WORLD AROUND YOU

What Is Earth's Carrying Capacity for Humans?

As you will learn in the next chapter, human numbers are doubling about every forty years. At that rate there would be an astounding 170 quadrillion people six hundred years from now. That would put one person on each square meter over the entire land surface of the earth. **Carrying capacity** reflects the limits imposed on population growth by finite space and finite resources. Earth's resources would never support such ludicrous numbers. So, what is the carrying capacity of the earth for humans? That question cannot be meaningfully addressed without clarifying a basic assumption: the type of lifestyle on which to develop the estimates.

Different lifestyles have different resource requirements. Are people to be vegetarians or will meat be a significant part of the diet? Will the earth's resources be counted upon to provide additional amenities beyond food? The answers to those questions have a profound effect on the numbers of people the earth can sustain. For illustration purposes imagine a miniature planet we could call Terrabase. Consider the following scenarios to see how lifestyle affects carrying capacity.

Scenario 1.

In this scenario, the humans living on Terrabase are vegetarians, and 100 percent of the planet's space is devoted to raising human food. Let's assume that under these conditions 1000 people can be sustained. Carrying capacity equals 1000.

Scenario 2.

Most people enjoy meat in the diet. Assume that the people of Terrabase do not live as vegetarians, but obtain half their calories by eating herbivores such as beef cattle. This would require a significant amount of plant mass to be fed to animals. But, as you learned in chapter 3, much energy is lost in transfer between trophic levels. It takes about 10 calories of plant food to produce one calorie of beef. After doing the calculations, it turns out that under these conditions, Terrabase will produce enough food to sustain only 180 people. Simply by eating meat, the carrying capacity has been reduced by over 80 percent.

Scenario 3.

In both scenarios 1 and 2, all of the land has been used exclusively for human food production. But people have consumer needs as well: cars, parking spaces, televisions, washers, dryers, clothing, shopping centers, and much more, all of which require space that would have to be subtracted from that used to produce food. Recreational space is also important to us. We want athletic fields, golf courses, bird sanctuaries, nature preserves, and hunting lands. All of which divert even more land away from food production.

Assume the residents of Terrabase had the high living standard of industrialized nations, requiring the immense and continuous input of chemicals, energy, paper, and other raw materials, as well as requiring land for waste disposal. How much of Terrabase would be devoted to these uses? If it is 20 percent, the carrying capacity is reduced to 140.

Wild organisms also play ecological roles important to our well-being. Leaving space for wild nature further reduces land available to produce human food.

"What is the earth's carrying capacity for humans?" is not a meaningful question until the cultural context within which people are to live has been clarified.

The magnitude of the cultural impact on carrying capacity is underlined by information provided by Robert Goodland and others. It is estimated that the continuous production of 10 to 15 acres are necessary to sustain just one person at our affluent lifestyle. To support all of the earth's 5.6 billion humans at such a lifestyle would require three times as much productive land as is available. Defined in those terms, the carrying capacity of the earth is around 1.8 billion people, one-third the world's current numbers. On the other hand, if carrying capacity is defined as the number of humans the planet could adequately feed without providing other amenities, that number is obviously much larger.

Logistic "Strategies"

While environmental resistance is a factor in controlling population growth in all species, those exhibiting logistic growth tend to grow more slowly and are more likely to be regulated by intrinsic characteristics than those with Malthusian patterns. These organisms are usually larger, live longer, mature more slowly, produce fewer offspring in each generation, and have fewer natural predators than the species below them in the ecological hierarchy. Some typical examples of this strategy are wolves, elephants, whales, and primates. Each of these species provides more care and protection for its offspring than do "lower" organisms.

Elephants, for instance, are not reproductively mature until they are 18 to 20 years old. During youth and adolescence, a young elephant is part of a complex extended family that cares for it, protects it, and teaches it how to behave. A female elephant normally conceives only once every four or five years after she matures. The gestation period is about eighteen months; thus, an elephant herd doesn't produce many babies in a given year. Since they have few enemies (except humans) and live a long life (often sixty or seventy years), however, this low reproductive rate produces enough elephants to keep the population stable, given appropriate environmental conditions.

An important underlying question to much of the discussion in this book is which of these strategies humans follow. Do we more closely resemble wolves and elephants in our population growth, or does our population growth pattern more closely resemble that of moose and rabbits? Will we overshoot the carrying capacity of our environment (or are we already doing so), or will our population growth come into balance with our resources (What Do You Think?, p. 133)?

FACTORS THAT INCREASE OR DECREASE POPULATIONS

Now that you have seen population dynamics in action, let's focus on what happens *within* populations, which are, after all, made up of individuals. In this section, we will discuss how new members are added to and old members removed from populations. We also will examine the composition of populations in terms of age classes and introduce terminology that will apply in subsequent chapters.

Natality, Fecundity, and Fertility

Natality is the production of new individuals by birth, hatching, germination, or cloning, and is the main source of addition to most biological populations. Natality is usually sensitive to environmental conditions so that successful reproduction is tied strongly to nutritional levels, climate, soil or water conditions, and—in some species—social interactions between members of the species. The maximum rate of reproduction under ideal conditions varies widely among organisms and is a species-specific characteristic. We already have mentioned, for instance, the differences in natality between several different species.

Fecundity is the physical ability to reproduce, while **fertility** is a measure of the actual number of offspring produced. Because of lack of opportunity to mate and successfully produce offspring, many fecund individuals may not contribute to population growth. Human fertility often is determined by personal choice of fecund individuals.

Immigration

Organisms are introduced into new ecosystems by a variety of methods. Seeds, spores, and small animals may be floated on winds or water currents over long distances. This is a major route of colonization for islands, mountain lakes, and other remote locations. Sometimes organisms are carried as hitchhikers in the fur, feathers, or intestines of animals traveling from one place to another. They also may ride on a raft of drifting vegetation. Some animals travel as adults—flying, swimming, or walking. In some ecosystems, a population is maintained only by a constant influx of immigrants. Salmon, for instance, must be important predators in some parts of the ocean, but their numbers are maintained only by recruitment from mountain streams thousands of kilometers away.

Mortality and Survivorship

An organism is born and eventually it dies; it is mortal. **Mortality,** or death rate, is determined by dividing the number of organisms that die in a certain time period by the number alive at the beginning of the period.

Since the number of survivors is more important to a population than is the number that died, mortality is often better expressed in terms of **survivorship** (the percentage of a cohort that survives to a certain age) or **life expectancy** (the probable number of years of survival for an individual of a given age). Table 6.4 shows U.S. life expectancies by age and sex. Notice that for each year you survive, your life expectancy increases slightly. For example, the average life expectancy for a female is 79. If she lives to 80, however, she can expect to live 8.9 years, on average. Obviously the balance between natality and mortality determine whether a population grows or declines. To understand how these factors interact, we need to define some more population terms.

Life span is the longest period of life reached by a given type of organism. The process of living entails wear and tear that eventually overwhelm every organism, but maximum age is dictated primarily by physiological aspects of the organism itself. There is an enormous difference in life span between different species. Some microorganisms live their whole life cycles in a matter of hours or minutes. Bristlecone pine trees in the mountains of California, on the other hand, have life spans up to 4600 years.

Most individuals in a population do not live anywhere near the maximum life span for their species. The major factors in early mortality are predation, parasitism, disease, accidents, fighting, and environmental influences, such as climate and nutrition. Important differences in relative longevity among various types of organisms are reflected in the survivorship curves shown in figure 6.7.

TABLE 6.4 Life Expectancies in the U.S. (all races)

AGE	REMAINING YEARS OF LIFE		
	BOTH SEXES	MALE	FEMALE
0	76.1	73.0	79.0
1	75.6	72.6	78.6
5	71.7	68.7	74.7
10	66.8	63.8	69.7
15	61.9	58.9	64.8
20	57.1	54.2	59.9
25	52.4	49.6	55.1
30	47.7	44.9	50.2
35	43.0	40.4	45.4
40	38.4	35.9	40.7
45	33.9	31.5	36.0
50	29.4	27.1	31.5
55	25.2	23.0	27.1
60	21.2	19.2	22.9
65	17.5	15.7	18.9
70	14.1	12.5	15.3
75	11.1	9.8	11.9
80	8.3	7.3	8.9
85	6.1	5.4	6.4

Source: Data from National Center for Health Statistics, 1997, U.S. Dept. of Health and Human Services.

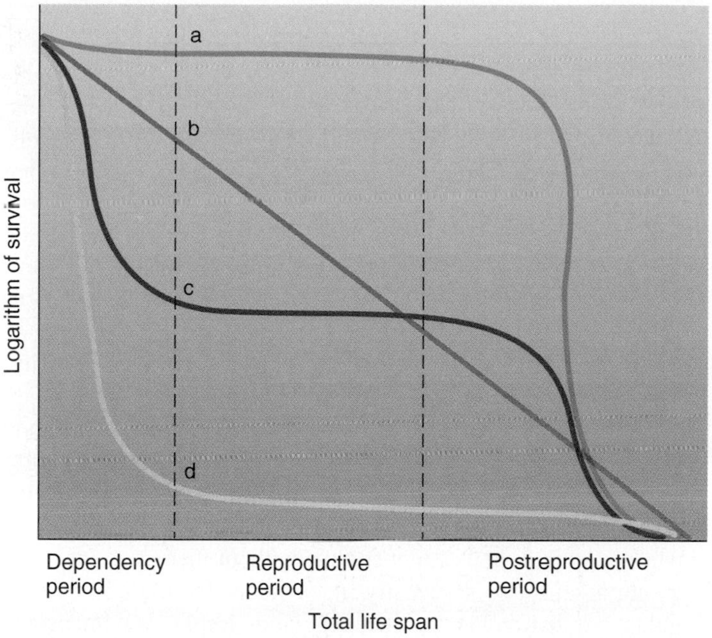

FIGURE 6.7 Four basic types of survivorship curves for organisms with different life histories. Curve (*a*) represents organisms, such as humans or whales, which tend to live out the full physiological life span if they survive early growth. Curve (*b*) represents organisms, such as sea gulls, in which the rate of mortality is fairly constant at all age levels. Curve (*c*) represents such organisms as white-tailed deer, moose, or robins, which have high mortality rates in early and late life. Curve (*d*) represents such organisms as clams and redwood trees, which have a high mortality rate early in life but live a full life if they reach adulthood.

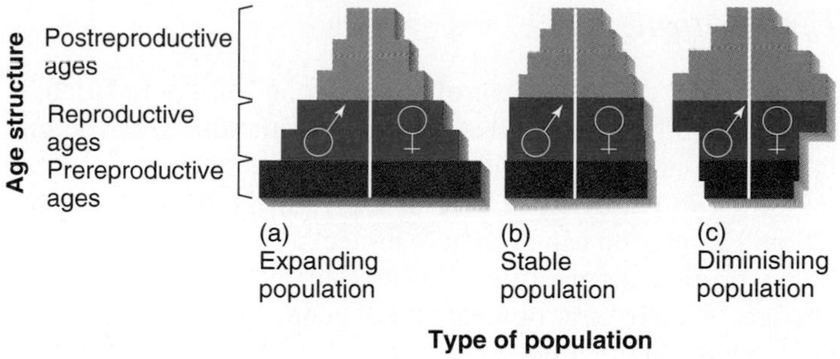

FIGURE 6.8 Typical age structure diagrams for hypothetical populations that are expanding, stable, or decreasing. Different numbers of individuals in each age class create these distinctive shapes. In each diagram, the left half represents males while the right half represents females.

Four general patterns of survivorship can be seen in this idealized figure. Curve (*a*) is the pattern of organisms that tend to live their full physiological life span if they reach maturity and then have a high mortality rate when they reach old age. This pattern is typical of many large mammals, such as whales, bears, and elephants (when not hunted by humans), as well as humans in developed countries. Interestingly, some very small organisms, including predatory protozoa and rotifers (small, multicellular, freshwater animals) have similar survivorship curves even though their maximum life spans may be hundreds or thousands of times shorter than those of large mammals. In general, curve (*a*) is the pattern for top consumers in an ecosystem, although many annual plants have a similar survivorship pattern.

Curve (*b*) represents the survivorship pattern for organisms for which the probability of death is generally unrelated to age. Sea gulls, for instance, die from accidents, poisoning, and other factors that act more or less randomly. Their mortality rate is generally constant with age, and their survivorship curve is a straight line.

Curve (*c*) is characteristic of many songbirds, rabbits, members of the deer family, and humans in less-developed countries (see chapter 7). They have a high mortality early in life when they are more susceptible to external factors, such as predation, disease, starvation, or accidents. Adults in the reproductive phase have a high level of survival. Once past reproductive age, they become susceptible again to external factors and the number of survivors falls quite rapidly.

Curve (*d*) is typical of organisms at the base of a food chain or those especially susceptible to mortality early in life. Many tree species, fish, clams, crabs, and other invertebrate species produce a very large number of highly vulnerable offspring, few of which survive to maturity. Those individuals that do survive to adulthood, however, have a very high chance of living most of the maximum life span for the species.

Age Structure

An outcome of the interaction between mortality and natality is that growing or declining populations usually will have very different proportions of individuals in various age classes. Figure 6.8 shows three hypothetical population profiles that are distinguished by differences in distribution among prereproductive, reproductive, and postreproductive age classes. Pattern (*a*) is characteristic of a rapidly expanding population. Young make up a large proportion of the population. This large number of prereproductive individuals represent **population momentum** because there is potential for rapid increase in natality once the youngsters reach reproductive age. The age structure of human populations in rapidly growing, developing countries generally resembles pattern (*a*).

When natality comes into balance with mortality, the population enters a stationary phase having an age structure pattern similar to (*b*). This pattern is characteristic of nations with a stable population size, such as many European countries.

The age structure shown by pattern (*c*) is characteristic of a diminishing population, where natality has fallen to a level lower than the replacement number. The bulge in the upper age classes represents adults still living but reproducing at a lower rate. When those individuals die, the population will be much smaller.

Emigration

Emigration, the movement of members out of a population, is the second major factor that reduces population size. The dispersal factors that allow organisms to migrate into new areas are important in removing surplus members from the source population. Emigration can even help protect a species. For instance, if the original population is destroyed by some catastrophic change in their environment, their genes still are carried by descendants in other places. Many organisms have very specific mechanisms to facilitate migration of one or more of each generation of offspring.

FACTORS THAT REGULATE POPULATION GROWTH

So far, we have seen that differing patterns of natality, mortality, life span, and longevity can produce quite different rates of population growth. The patterns of survivorship and age structure created by these interacting factors not only show us how a population is growing but also can indicate what general role that species plays in its ecosystem. They also reveal a good deal about how that species is likely to respond to disasters or resource bonanzas in its environment. But what factors *regulate* natality, mortality, and the other components of population growth? In this section, we will look at some of the mechanisms that determine how a population grows.

Various factors regulate population growth, primarily by affecting natality or mortality, and can be classified in different ways. They can be *intrinsic* (operating within individual organisms or between organisms in the same species) or *extrinsic* (imposed from outside the population). Factors can also be either **biotic** (caused by living organisms) or **abiotic** (caused by nonliving components of the environment). Finally, the regulatory factors can act in a *density-dependent* manner (effects are stronger or a higher proportion of the population is affected as population density increases) or *density-independent* manner (the effect is the same or a constant proportion of the population is affected regardless of population density).

In general, biotic regulatory factors tend to be density-dependent, while abiotic factors tend to be density-independent. There has been much discussion about which of these factors is most important in regulating population dynamics. In fact, it probably depends on the particular species involved, its tolerance levels, the stage of growth and development of the organisms involved, the specific ecosystem in which they live, and the way combinations of factors interact. In most cases, density-dependent and density-independent factors probably exert simultaneous influences. Depending on whether regulatory factors are regular and predictable or irregular and unpredictable, species will develop different strategies for coping with them.

Density-Independent Factors

In general, the factors that affect natality or mortality independently of population density tend to be abiotic components of the ecosystem. Often weather (conditions at a particular time) or climate (average weather conditions over a longer period) are among the most important of these factors. Extreme cold or even moderate cold at the wrong time of year, high heat, drought, excess rain, severe storms, and geologic hazards—such as volcanic eruptions, landslides, and floods—can have devastating impacts on particular populations.

Abiotic factors can have beneficial effects as well, as anyone who has seen the desert bloom after a rainfall can attest. Fire is a powerful shaper of many biomes. Grasslands, savannas, and some montane and boreal forests often are dominated—even created—by periodic fires. Some species, such as jack pine and Kirtland's warblers, are so adapted to periodic disturbances in the environment that they cannot survive without them.

In a sense, these density-independent factors don't really regulate population *per se,* since regulation implies a homeostatic feedback that increases or decreases as density fluctuates. By definition, these factors operate without regard to the number of organisms involved. They may have such a strong impact on a population, however, that they completely overwhelm the influence of any other factor and determine how many individuals make up a particular population at any given time.

Density-Dependent Factors

Density-dependent mechanisms tend to reduce population size by decreasing natality or increasing mortality as the population size increases. Most of them are the results of interactions *between* populations of a community (especially predation), but some of them are based on interactions *within* a population.

Interspecific Interactions

As we discussed in chapter 4, a predator feeds on—and usually kills—its prey species. While the relationship is one-sided with respect to a particular pair of organisms, the prey species as a whole may benefit from the predation. For instance, the moose that gets eaten by wolves doesn't benefit individually, but the moose *population* is strengthened because the wolves tend to kill old or sick members of the herd. Their predation helps prevent population overshoot, so the remaining moose are stronger and healthier.

Sometimes predator and prey populations oscillate in a sort of synchrony with each other as is shown in figure 6.9. This is a classic study of the number of furs brought into Hudson Bay Company trading posts in Canada between 1840 and 1930. As you can see, the numbers of Canada lynx fluctuate on about a ten-year cycle that is similar to, but slightly out of phase with, the population peaks of snowshoe hares. When the hare population is high and food is plentiful, lynx reproduction is very successful and lynx populations grow rapidly. Predation diminishes the number of hares

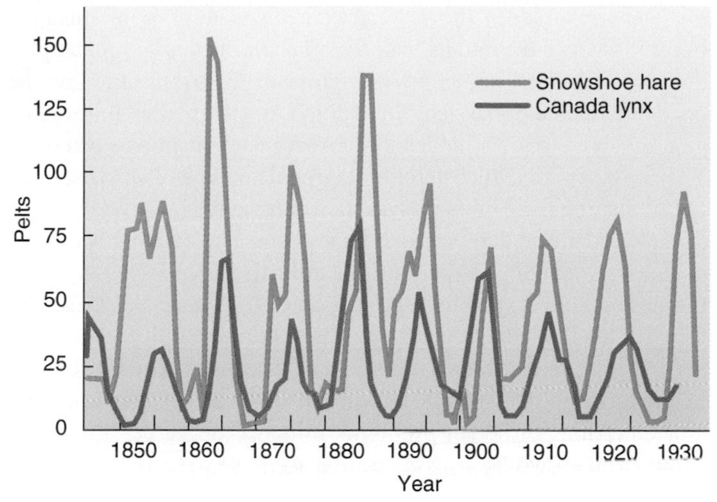

FIGURE 6.9 Oscillations in the populations of snowshoe hare and lynx in Canada suggest the close interdependency of this prey-predator relationship. This data is based on the number of pelts received by the Hudson Bay Company. Both predator and prey show a ten-year cycle in population growth and decline.

Source: Data from D. A. MacLulich, *Fluctuations in the Numbers of the Varying Hare (Lepus americus)*. Toronto: University of Toronto Press, 1937, reprinted 1974.

available until eventually lynx do not have enough to eat and their population sizes fall. As the predators die off, the prey begin to make a comeback and the cycle starts over again. This predator-prey oscillation is known as the Lotka-Volterra model after the scientists who first described it mathematically.

Parasitism is a form of predation in which the parasite takes nutrients or other benefits at the expense of its host but usually does not kill it. The mosquito or tick that helps itself to a blood meal from your epidermis is a good example. Often, a parasite and its host develop a sort of balance with each other, something like the predator/prey relationship we have just discussed. If the parasite is too virulent and kills its host, food for its offspring will be unavailable. Wherever a parasite and host combination have lived together for a long time, the host is likely to have developed some resistance to its parasite so that enough individuals survive to continue the population. If parasites are introduced to new ecosystems, however, or new host species are introduced into the parasite's ecosystem, the effects can be disastrous. On the plains of Africa, for instance, small biting flies called tsetse flies carry the trypanosome that causes sleeping sickness. The native species of hoofed mammals are mostly resistant to this protozoan parasite, but domestic cattle, which have not been part of the community long enough to adapt, may be killed or become seriously debilitated if they are exposed.

In interspecific competition, two species compete for the same environmental resources in an ecosystem, but this competition rarely involves outright struggles or confrontations. It more often involves a contest in which the species tend to eat faster or grow faster or grab a bigger share of the resources in some other way. As a result, competing populations may work out a balance of resource partitioning (chapter 4), or one population may eclipse the other.

Not all interspecific interactions are harmful to one of the species involved. Mutualism and commensalism, for instance, are interspecific interactions that are beneficial or neutral in terms of population growth (chapter 4).

Intraspecific Interactions

Individuals within a population also compete for resources. When population density is low, resources are likely to be plentiful and the population growth rate will approach the maximum possible for the species, assuming that individuals are not so dispersed that they cannot find mates. As population density approaches the carrying capacity of the environment, however, one or more of the vital resources becomes limiting. The stronger, quicker, more aggressive, more clever, or luckier members get a larger share, while others get less and then are unable to reproduce successfully or survive.

Territoriality is one principal way many animal species control access to environmental resources. The individual, pair, or group that holds the territory will drive off rivals if possible, either by threats, displays of superior features (colors, size, dancing ability), or fighting equipment (teeth, claws, horns, antlers). Members of the opposite sex are attracted to individuals that are able to seize and defend the largest share of the resources. From a selective point of view, these successful individuals presumably represent superior members of the population and the ones best able to produce offspring that will survive.

Stress and Crowding

Stress and crowding also are density-dependent population control factors. When population densities get very high, organisms often exhibit symptoms of what is called **stress shock** or **stress-related diseases.** These terms describe a loose set of physical, psychological, and/or behavioral changes that are thought to result from the stress of too much competition and too close proximity to other members of the same species. There is a considerable controversy about what causes such changes and how important they are in regulating natural populations. The strange behavior and high mortality of arctic lemmings or snowshoe hares during periods of high population density may be a manifestation of stress shock. On the other hand, they could simply be the result of malnutrition, infectious disease, or some other more mundane mechanism at work.

Some of the best evidence for the existence of stress-related disease comes from experiments in which laboratory animals, usually rats or mice, are grown in very high densities with plenty of food and water but very little living space (table 6.5). A variety of symptoms are reported, including reduced fertility, low resistance to infectious diseases, and pathological behavior, such as hypoactivity, hyperactivity, aggression, lack of parental instincts, sexual deviance, and cannibalism. Dominant animals seem to be affected least by crowding, while subordinate animals—the ones presumably subjected to the most stress in intraspecific interactions—seem to be the most severely affected.

TABLE 6.5 The Influence of Density on Fecundity in the House Mouse (*Mus musculus*)

	SPARSE	MEDIUM	DENSE	VERY DENSE
Average number/m^3	34	118	350	1600
Average percentage pregnant	58.3	49.4	51.0	43.4
Average number per litter	6.2	5.7	5.6	5.1

Source: Data from C. Southwick, "Population Characteristics of House Mice Living in English Corn Ricks: Density Relationships," *Proc. Zool. Soc. London,* 131:163–175, 1958, The Zoological Society of London.

The clinical symptoms reported in these studies include enlargement of the adrenal glands, reduction of the thymus and reproductive glands, and deterioration of the heart, blood vessels, kidney, and liver. The most probable explanation for these changes is that the adrenal gland is the source of epinephrine (adrenalin), the "fight-or-flight" hormone that raises energy levels and stimulates an animal to respond to danger or stress. Ordinarily, the burst of epinephrine released by fear, anger, or excitement is short-lived. When the stimulation is very frequent, however, the adrenal glands become enlarged in response to the continual call for epinephrine, and hormone production is elevated to a constant high level. This puts stress on the heart and circulatory system. In addition to epinephrine, the adrenal glands secrete sex hormones that affect metabolism and behavior. They also secrete corticosteroids that regulate mineral levels, energy metabolism, and the immune system. Abnormally high levels of these hormones damage the liver and kidneys and can lead to a variety of symptoms including hypertension, arthritis, arteriosclerosis, gastrointestinal ulcers, and reduced resistance to disease. All these factors will tend to increase mortality and decrease fertility.

Are there equivalent stress-related diseases that cause aberrant behavior in human societies? Some people suggest that the high concentrations of social problems in dense urban centers might result from crowding stresses similar to those observed in laboratory animals. They point out that inner cities have much higher rates of disease and social problems, such as mental illness, family dysfunction, alcohol and drug abuse, and criminal behavior, than do other areas. The inner city also has much higher levels of poverty, neglect, undereducation, unemployment, and other social conditions, however, that may be more important than the population density problems found there. While it is tempting to extrapolate conclusions from laboratory experiments to natural populations where a complex of factors are interacting, the conclusions can be misleading. In chapter 24, we discuss the socioeconomic disparities in our urban centers and their effects on human populations.

SUMMARY

Population dynamics play an important role in determining how ecosystems work. Biological organisms generally have the ability to produce enough offspring so populations can grow rapidly when resources are available. Given optimum conditions, populations of many organisms can grow exponentially; that is, they grow at a constant rate of increase so that the population size doubles in some regular interval of time. If conditions appropriate for this kind of growth persist and other factors don't intervene to reduce abundance, exponential growth can produce astronomical numbers of organisms. We describe the rapidly rising curve of an exponentially growing population as a J curve.

Some populations will grow exponentially until they overshoot the carrying capacity of the environment. Mortality rates rise as resources become limited and the population may crash, often dying back as rapidly as it rose. Other species have intrinsic mechanisms that regulate their population growth rate, resulting in an equilibrium at or near the carrying capacity of the environment.

The most important components of population dynamics are natality, fertility, fecundity, life span, longevity, mortality, immigration, and emigration. The sum of all additions to and subtractions from the population determines the net rate of growth. Mortality rates and longevity are often expressed as survivorship rates that reveal much about a species' place in its ecosystem and the kinds of hazards that eliminate members from the population.

The factors that regulate population dynamics can be either intrinsic or extrinsic to the population. They can be caused by biotic or abiotic forces, and they can act on the population in either a density-dependent or density-independent fashion. The most important abiotic regulatory factors are usually climate and weather. The most important biological factors are usually competition (both interspecific and intraspecific), predation, and disease.

Often, organisms develop specific behavioral patterns that reduce conflict and facilitate resource partitioning. In some cases of extreme crowding, it is thought that physiological responses to excessive intraspecific interaction can result in stress-related disease that can lead to aberrant behavior, reduced reproductive success, increased susceptibility to disease, and high rates of mortality. Whether these mechanisms operate in humans at high population densities is a matter of controversy.

QUESTIONS FOR REVIEW

1. What factors are associated with population declines of Columbia River salmon?
2. What is the difference between exponential and arithmetic growth?
3. Given a growth rate of 3 percent per year, how long will it take for a population of 100,000 individuals to double? How long will it take to double when the population reaches 10 million?
4. What is environmental resistance? How does it affect populations?
5. What is the difference between fertility and fecundity?
6. Describe the four major types of survivorship patterns and explain what they show about the role of the species in an ecosystem.
7. What are the main interspecific population regulatory interactions? How do they work?
8. What are the suspected causes and effects of stress shock or stress-related disease?

QUESTIONS FOR CRITICAL THINKING

1. Compare the advantages and disadvantages to a species that result from exponential or logistic growth. Why do you think hares have evolved to reproduce as rapidly as possible, while lynx appear to have intrinsic or social mechanisms to limit population growth?
2. If you were designing the Columbia River salmon recovery project, what threats would you tackle first? How would you monitor your progress to assess whether you're reaching your goals?
3. Are humans subject to environmental resistance in the same sense that other organisms are? How would you decide whether a particular factor that limits human population growth is ecological or social?
4. There obviously are vast differences in birth and death rates, survivorship, and life spans among species. There must be advantages and disadvantages in living longer or reproducing more quickly. Why hasn't evolution selected for the most advantageous combination of characteristics so that all organisms would be more or less alike?
5. Abiotic factors that influence population growth tend to be density-independent, while biotic factors that regulate population growth tend to be density-dependent. Explain.
6. Some people consider stress and crowding studies of laboratory animals highly applicable in understanding human behavior. Other people question the cross-species transferability of these results. What considerations would be important in interpreting these experiments?
7. What implications for human population control might we draw from our knowledge of basic biological population dynamics?

KEY TERMS

abiotic 136
arithmetic growth 129
biotic 136
biotic potential 130
carrying capacity 133
catastrophic system 132
chaotic systems 131
dieback 130
emigration 136
environmental resistance 131
exponential growth 129
fecundity 134
fertility 134
geometric growth 129
irruptive or Malthusian growth 130
J curve 129
life expectancy 134
life span 134
logistic growth 131
mortality 134
natality 134
overshoot 130
population crash 130
population explosion 130
population momentum 135
S curve 131
stress-related diseases 137
stress shock 137
survivorship 134

ADDITIONAL INFORMATION ON THE INTERNET

Visit our website at http://www.mhhe.com/environmental science/ for specific information about each of the topics below:

Washington State salmon listed as Endangered Species
Pacific Salmon recovery plan: NMFS
On-line course: Population Ecology
Biodiversity (World Resources Institute)
Chaos theory
Live model: Logistic population growth
Spatial and functional structure of populations

PROFILE

Environmental Engineer

Jim Bartlett

When Jim Bartlett went off to college at the University of Iowa he was not at all settled on a career direction. He had a considerable interest in the outdoors coupled with an inclination to tinker, so he considered engineering a possibility. After his first year, he declared a major in industrial engineering. Soon after, he found his heart wasn't really in his studies, so he switched to civil engineering, influenced, in part, by an emerging interest in the environment. He focused his efforts on building and design and got his bachelor degree in civil engineering with a balance of structural-related and environmental-related course work with the intention of seeking work with a consulting firm.

After graduation, Jim discovered that many consulting firms were looking for people with advanced degrees. A friendly professor encouraged him to enroll in graduate school. After completing a master's degree in environmental engineering, Jim moved to San Francisco to pursue a career in this field. He accepted a job with Brown and Caldwell, a large firm providing environmental engineering, consulting, and analytical services. His firm is involved in a wide variety of projects, including environmental remediation efforts as well as design work. They worked with the Walnut Valley Water District to design a system for Pomona, CA, to use treated wastewater for irrigation of golf courses, school yards, and other large commercial and public landscapes. Reuse of the treated wastewater not only saved businesses money but also avoided having to apply fertilizer to the sites being irrigated.

Jim's firm is also engaged in some exciting experimental work in the use of biofilter technology to remediate environmental contamination. In field tests, a common soil bacterium was able to degrade 98 percent of the commercial solvent, trichloroethylene, in some polluted groundwater. Efforts are now underway to commercialize that technology.

As a new employee, Jim's work consisted of performing lots of individual tasks on a variety of projects. One day might be spent doing calculations, while the next week found him doing library research on a piece of equipment. Tasks were typically limited in scope and piecemeal, but did give Jim a sense of whole projects and the importance of the smaller pieces.

After the first few years, Jim has shifted to energy-related projects, with much of this work involved in developing energy-saving techniques. A recent project was to replace an inefficient heat pump system at a community college with a highly efficient system that runs at night to produce and store chilled water for use in cooling the buildings during the day. He has also worked on the new cogeneration system that uses heat from turbines to recover steam, producing significant energy savings for a large research hospital.

Jim admits that he could have been better prepared for his job, finding communication skills, writing in particular, to be of tremendous importance. He says he began his job with poor writing skills because he didn't work at developing them while in school. He feels very indebted to several mentors at Brown and Caldwell who taught him the basics of technical writing. He now works as a project manager, where he finds person-to-person communication with clients, contractors, vendors, and members of the project team a significant part of the job.

What are the prospects for employment in a field like Jim's? He says environmental consulting is very competitive. He feels there will be continued growth, but not nearly as rapidly as in the past. His firm has done lots of work that was regulation-driven, which has leveled off as well. The growth area is now found in retrofitting and remediation, particularly on federal sites. An example is the closure of McClellan Air Force Base. In order to make McClellan and other sites suitable for other uses, numerous site contamination problems must be remedied, providing increased work opportunities for environmental consultants even as employment in other sectors falls due to the facility closures.

Looking back on his environmental interests during his early college days, he feels he simply wasn't aware of the opportunities that existed for anyone with his environmental interests. He feels that in making career decisions, students need to focus on what things are really important to them, since that is where their real job satisfaction is going to come from.

CHAPTER 7

Human Populations

Every child comes with the message that God is not yet tired of the man.

—Rabindranath Tagore—

OBJECTIVES

After studying this chapter, you should be able to:

- trace the history of human population growth
- summarize Malthusian and Marxian theories of limits to growth as well as why technological optimists and supporters of social justice oppose these theories.
- explain the process of demographic transition and why it produces a temporary population surge.
- understand how changes in life expectancy, infant mortality, women's literacy, standards of living, and democracy affect population changes.
- evaluate pressures for and against family planning in traditional and modern societies.
- compare modern birth control methods and prepare a personal family planning agenda.

Above: Thousands of giant stone heads on Easter Island in the South Pacific bear mute testimony to the dangers of excess population growth that exhausts resources and damages essential ecological processes.

ESSENTIAL STUDY PARTNER

The Essential Study Partner (ESP) CD-ROM that accompanies this textbook contains additional information on the following topics:

- past, present, and future human population trends: populations: human population: demography and control
- fertility, birth rate, death rate and standard of living: populations: human population: demography
- zero population growth and replacement rate: populations: human population: demography
- aging population in industrialized countries: populations: human population: demography
- demographic transitions of industrializing countries: populations: human population: demography
- family planning factors, population control, and world birth rates: populations: human population: control

The Saga of Easter Island

One of the most remote habitable places on the earth, Easter Island, lies about 3200 km west of South America, the nearest continent, and more than 2000 km from the closest occupied island (Pitcairn). With a mild climate and fertile volcanic soils, Easter Island should have been a tropical paradise, but when it was "discovered" by Dutch explorer Jacob Roggeveen in 1722, it resembled a barren wasteland more than a paradise. Covered by a dry grassland, the island had no trees and few bushes more than a meter tall. No animals inhabited the island except humans, chickens, rats, and a few insects.

The 2000 people living on the island at the time eked out a pitiful existence. Having no seaworthy canoes, they couldn't venture out on the ocean to fish. With no trees to provide building materials or firewood, the island's cool, wet, windy winters were miserable; meager gardens hardly produced enough food for subsistence.

And yet, scattered along the coastline were thousands of immense stone heads, some as large as 30 meters tall, weighing more than 200 metric tons (previous page). How could such a small population have carved, moved, and erected these enormous effigies? Was there once a larger and more advanced civilization on the island? If so, where did they go?

Historical studies have shown that conditions on the island were once very different than they are now. Until about 1500 years ago, the island was covered with a lush subtropical forest and the soil was deep and fertile. Polynesian people apparently reached Easter Island about A.D. 400. Anthropological and linguistic evidence suggests they sailed from the Marquesas Islands 3500 kilometers to the northwest. Excavations of archeological sites show that the early settlers' diet consisted mainly of porpoises, land-nesting seabirds, and garden vegetables. Populations soared, reaching as much as 20,000 on an island only about 15 km across.

By A.D. 1400 the forest appears to have disappeared completely—cut down for firewood and to make houses, canoes, and rollers for transporting the enormous statues. Without a protective forest cover, soil washed off steep hillsides. Springs and streams dried up, while summer droughts made gardens less productive. All wild land birds became extinct and seabirds no longer nested on the island. Lacking wood to build new canoes, the people could no longer go offshore to fish. Statues carved at this time show sunken cheeks and visible ribs suggesting starvation.

At this point, chaos and warfare seem to have racked the land. The main bones found in fireplaces were those of rats and humans. Cannibalism apparently was rampant as the population decreased by 90 percent. The few remaining people cowered in caves, a pitiful remnant of a once impressive civilization. When we try to imagine how people reached this condition, we wonder why they didn't control their population and conserve their resources. What were their thoughts as they cut down the last trees, stranding themselves on this island of diminishing possibilities?

Does this story have lessons for us? Is Easter Island an example of what could happen to the rest of us if our population grows and we use up our store of resources? The debate over the carrying capacity of the earth for humans remains one of the most contentious and important issues in environmental science. Some demographers warn that we are headed for a disaster similar to that of Easter Island. Others hope that we will be more clever and perceptive than the unhappy people who destroyed the resource base on which they depended. What do you think? How will we recognize and respond to excess population and consumption levels?

POPULATION GROWTH

Every second, on average, four or five children are born somewhere on the earth. In that same second, two other people die. This difference between births and deaths means a net gain of nearly 2.5 more humans per second in the world population. This means we are growing at a little less than 9000 per hour, 214,000 per day, or almost 78 million more people per year. In July, 1999, the world population passed 6 billion, making us the most numerous vertebrate species on the planet. For the families to whom these children are born, this may well be a joyous and long-awaited event (fig. 7.1). But is a continuing increase in humans good for the planet in the long run?

Many people worry that overpopulation will cause—or perhaps already is causing—resource depletion and environmental degradation that threaten the ecological life-support systems on which we all depend. These fears often lead to demands for immediate, worldwide birth control programs to reduce fertility rates and to eventually stabilize or even shrink the total number of humans.

Others believe that human ingenuity, technology, and enterprise can extend the world carrying capacity and allow us to overcome any problems we encounter. From this perspective, more people may be beneficial rather than disastrous. A larger population means a larger workforce, more geniuses, more ideas about what to do. Along with every new mouth comes a pair of hands. Proponents of this worldview—many of whom happen to be economists—argue that continued economic and technological growth can both feed the world's billions and enrich everyone enough to end the population explosion voluntarily. Not so, counter many ecologists. Growth is the problem; we must stop both population and economic growth.

Yet another perspective on this subject derives from social justice concerns. From this worldview, there are enough resources for everyone. Current shortages are only signs of greed, waste, and

FIGURE 7.1 A Mayan family in Guatemala with four of their six living children. Decisions on how many children to have are influenced by many factors, including culture, religion, need for old age security for parents, immediate family finances, household help, child survival rates, and power relationships within the family. Having many children may not be in the best interest of society at large, but may be the only rational choice for individual families.

oppression. The root cause of environmental degradation, in this view, is inequitable distribution of wealth and power rather than population size. Fostering democracy, empowering women and minorities, and improving the standard of living of the world's poorest people are what are really needed. A narrow focus on population growth only fosters racism and an attitude that blames the poor for their problems while ignoring the deeper social and economic forces at work.

Whether human populations will continue to grow at present rates and what that growth would imply for environmental quality and human life are among the most central and pressing questions in environmental science. In this chapter, we will look at some causes of population growth as well as how populations are measured and described. Family planning and birth control are essential for stabilizing populations. The number of children a couple decides to have and the methods they use to regulate fertility, however, are strongly influenced by culture, religion, politics, and economics, as well as basic biological and medical considerations. We will examine how some of these factors influence human demographics.

Human Population History

For most of our history, humans have not been very numerous compared to other species. Studies of hunting and gathering societies suggest that the total world population was probably only a few million people before the invention of agriculture and the domestication of animals around ten thousand years ago. The larger and more secure food supply made available by the agricultural revolution allowed the human population to grow, reaching perhaps 50 million people by 5000 B.C. For thousands of years, the number of humans increased very slowly. Archaeological evidence and historical descriptions suggest that only about 300 million people were living at the time of Christ (table 7.1).

Until the Middle Ages, human populations were held in check by diseases, famines, and wars that made life short and uncertain for most people (fig. 7.2). Furthermore, there is evidence that many early societies regulated their population size through cultural taboos and practices such as infanticide. Among the most destructive of natural population controls were bubonic plagues that periodically swept across Europe between 1348 and 1650. During the worst plague years (between 1348 and 1350), it is estimated that at least one-third of the European population perished. Notice, however, that this did not retard population growth for very long. In 1650, at the end of the last great plague, there were about 600 million people in the world.

As you can see in figure 7.2, human populations began to increase rapidly after A.D. 1600. Many factors contributed to this rapid growth. Increased sailing and navigating skills stimulated commerce and communication between nations. Agricultural developments, better sources of power, and better health care and hygiene also played a role. We are now in an exponential or J curve pattern of growth (What Do You Think?, p. 146).

It took all of human history to reach 1 billion people in 1804, but little more than 150 years to reach 3 billion in 1960. In 1999, only 39 years later, we passed 6 billion. Will this population explosion continue until we overshoot the carrying capacity of our environment—if we have not done so already—and experience a catastrophic dieback similar to those described in chapter 6? Or will our populations stabilize at some tolerable level? As you will see later in this chapter, there is some evidence that population growth already is slowing, but whether we will reach equilibrium soon enough and at a size that can be sustained over the long term remains a difficult but important question.

TABLE 7.1 World Population Growth and Doubling Times

DATE	POPULATION	DOUBLING TIME
5000 B.C.	50 million	?
800 B.C.	100 million	4200 years
200 B.C.	200 million	600 years
A.D. 1200	400 million	1400 years
A.D. 1700	800 million	500 years
A.D. 1900	1600 million	200 years
A.D. 1965	3200 million	65 years
A.D. 1990	5300 million	38 years
A.D. 2020 (estimate)	8230 million	55 years

Source: Data from Population Reference Bureau, Inc., Washington, DC.

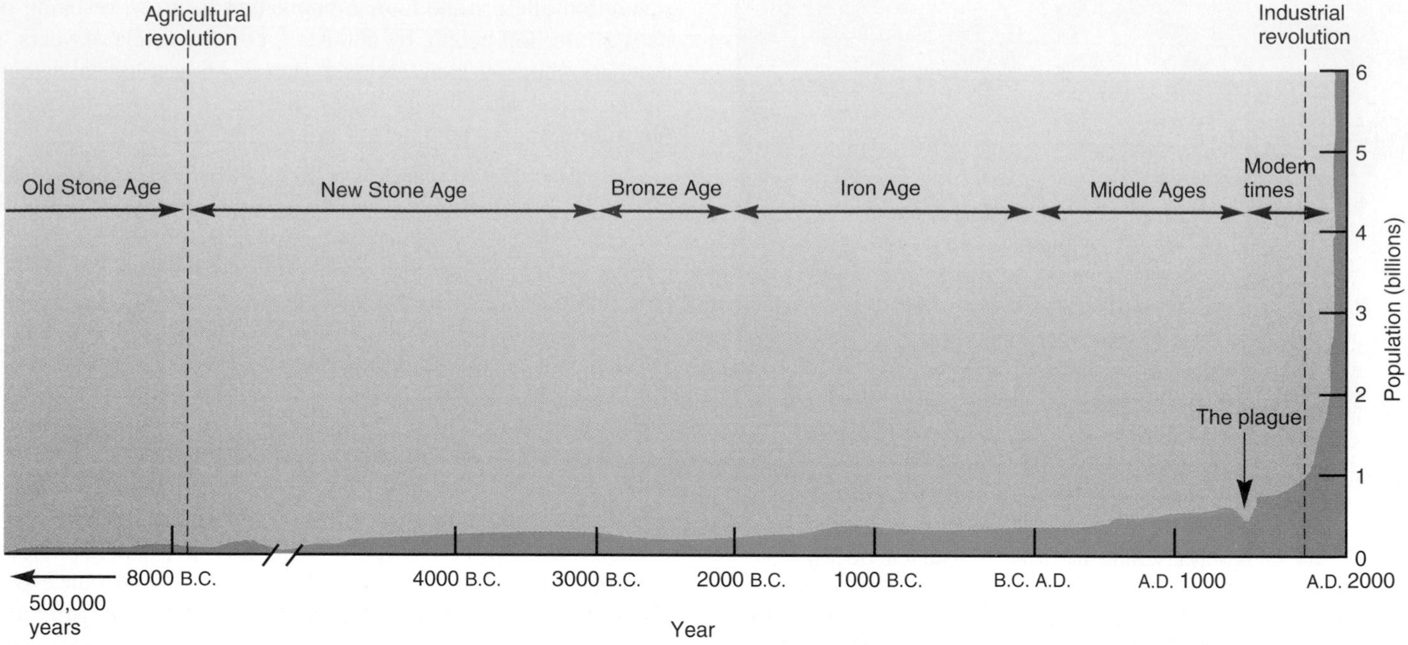

FIGURE 7.2 Human population levels through history. Since about A.D. 1000, our population curve has assumed a J shape. Are we on the upward slope of a population overshoot? Will we be able to adjust our population growth to an S curve? Or can we just continue the present trend indefinitely?

LIMITS TO GROWTH: SOME OPPOSING VIEWS

As with religion and politics, people have widely differing opinions about population and resources. Some believe that population growth is the ultimate cause of poverty and environmental degradation. Others argue that poverty, environmental degradation, and overpopulation are all merely symptoms of deeper social and political factors. In this section, we will examine some opposing worldviews and their implications.

Malthusian Checks on Population

In 1798, the Rev. Thomas Malthus wrote *An Essay on the Principle of Population as It Affects the Future Improvement of Society, with Remarks on the Speculations of Mr. Godwin, M. Condorcet, and Other Writers* to refute the views of progressives and optimists—including his father—who were inspired by the egalitarian principles of the French Revolution to predict a coming utopia. Malthus argued that human populations tend to increase at an exponential or compound rate while food production either remains stable or increases only slowly. The result, he predicted, is that human populations inevitably outstrip their food supply and eventually collapse into starvation, crime, and misery. Malthus's theory might be summarized by the graphic in figure 7.3*a*.

According to Malthus, the only ways to stabilize human populations are "positive checks," such as disease or famines that kill people, or "preventative checks," including all the factors that prevent human birth. Among the preventative checks he advocated were "moral restraint," including late marriage and celibacy until a couple can afford to support children. Many social scientists and biologists have been influenced by Malthus. Charles Darwin, for instance, derived his theories about the struggle for scarce resources and survival of the fittest after reading Malthus's essay.

If Malthus's views of the consequences of exponential population growth were dismal, the corollary he drew was even more bleak. He believed that most people are too lazy and immoral to regulate birth rates voluntarily. Consequently, he opposed efforts to feed and assist the poor in England because he feared that more food would simply increase their fertility and thereby perpetuate the problems of starvation and misery.

Not surprisingly, Malthus's ideas provoked a great social and economic debate. Karl Marx was one of his most vehement critics, claiming that Malthus was a "shameless sycophant of the ruling classes." According to Marx, population growth is a symptom rather than a root cause of poverty, resource depletion, pollution, and other social ills. The real causes of these problems, he believed, are exploitation and oppression (fig. 7.3*b*). Marx argued that workers always provide for their own sustenance given access to means of production and a fair share of the fruits of their labor. According to Marxians, the way to slow population growth *and* to alleviate crime, disease, starvation, misery, and environmental degradation is through social justice.

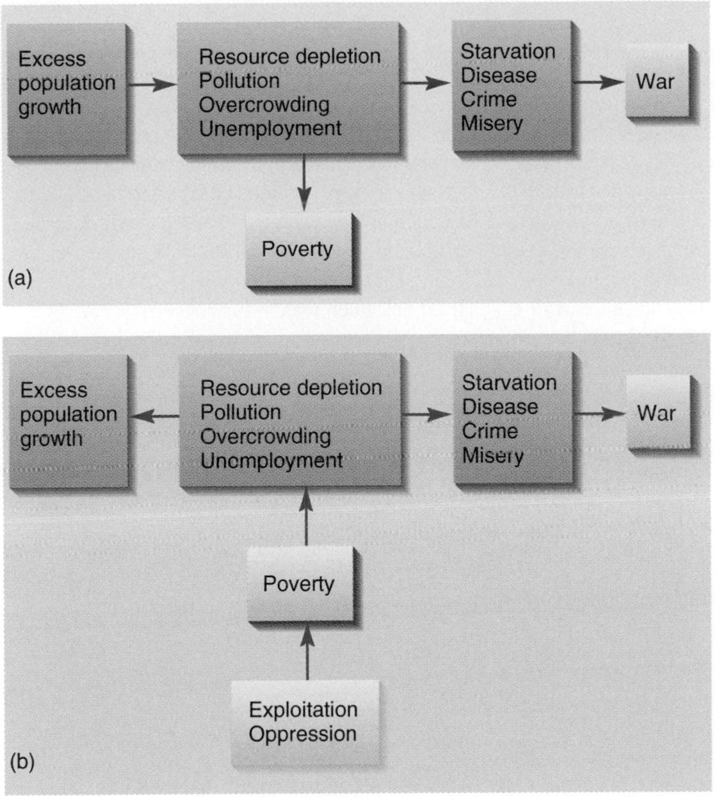

FIGURE 7.3 (*a*) Thomas Malthus argued that excess population growth is the ultimate cause of many other social and environmental problems. (*b*) Karl Marx argued that oppression and exploitation are the real causes of poverty and environmental degradation. Population growth in this view is a symptom or result of other problems, not the source.

FIGURE 7.4 Is this our future?

Malthus and Marx Today

Both Marx and Malthus developed their theories about human population growth in the nineteenth century when understanding of the world, technology, and society were much different than they are now. Still, the questions they raised are relevant today. While the evils of racism, classism, and other forms of exploitation that Marx denounced still beset us, it is also true that at some point available resources must limit the numbers of humans that the earth can sustain (fig. 7.4).

Those who agree with Malthus, that we are approaching—or may already have surpassed—the carrying capacity of the earth are called **neo-Malthusians.** In their view, we should address the issue of surplus population directly by making birth control our highest priority. An extreme version of this worldview is expressed by Cornell University entomologist David Pimentel, who claims that the "optimum human population" would be about 5 billion fewer than currently inhabit the planet. He fails to suggest how the 16 percent of us allowed to live would be chosen.

Neo-Marxians, on the other hand, believe that only eliminating oppression and poverty through technological development and social justice will solve population problems. Perhaps both these viewpoints have some validity. A compromise position is that population growth, poverty, and environmental degradation all are interrelated. No factor exclusively causes any other but each influences and, in turn, is influenced by the others.

Can Technology Make the World More Habitable?

Technological optimists argue that Malthus was wrong in his predictions of famine and disaster two hundred years ago because he failed to account for scientific progress. In fact, food supplies have increased faster than population growth since Malthus's time. There have been terrible famines in the past two centuries, but they were caused more by politics and economics than lack of resources or sheer population size. Whether this progress will continue remains to be seen, but technological advances have increased human carrying capacity more than once in our history.

The burst of growth of which we are a part, was stimulated by the scientific and industrial revolutions. Progress in agricultural productivity, engineering, information technology, commerce, medicine, sanitation, and other achievements of modern life have made it possible to support approximately one thousand times as many people per unit area as was possible ten thousand years ago.

Much of our growth in the past three hundred years has been based on availability of easily acquired natural resources, especially cheap, abundant fossil fuels. Whether we can develop alternative, renewable energy sources in time to avert disaster when current fossil fuels run out is a matter of great concern (chapter 21).

what do you think?
THE WORLD AROUND YOU

Looking for Bias in Graphs

Graphs, pictorial representations of information, can help others understand what we have to say. They are particularly helpful in conveying patterns of relationship. Why are graphs used so widely and why do they make such powerful impressions on us?

Not only are images expressive, but graphs come equipped with substantiating data. It is one thing to read that "the human population is rising explosively," but it is considerably more compelling to see the J curve pictured together with concrete numbers as evidence. The combination of picture and numbers often conveys powerful impressions beyond the numbers themselves.

An important question addressed in this chapter is: How serious is the worldwide population problem? Does it demand action? The graphs to follow all present information relevant to human population studies. Yet the impressions left with the reader vary considerably depending on the type of data used, and the design of the graph.

Reexamine figure 7.2, which presents an historical perspective on the growth in human numbers. What conclusions do you draw from it? Does it suggest we are experiencing an explosive rise in human numbers? That the current increase is unprecedented in our species history? Does it also imply, perhaps more indirectly, that the earth's population problem is serious, that there is an urgency to respond? In light of concepts of environmental resistance, carrying capacity, and overshoot and dieback, many people would conclude that the current growth, highlighted by the graph, is unsustainable.

Now examine figure 1.

It plots the same variables as figure 7.2, but suggests that population is rising at a modest rate. It does not produce the sense of explosiveness and urgency as figure 7.2. How can the impressions be so different?

Notice that the graph in figure 1 covers a much shorter time period. This greatly changes the time interval lengths on the horizontal scale. In figure 7.2, one millimeter represents about 50 years, but less than one year in figure 1. This changes the line's slope from nearly vertical in figure 7.2 to a modest incline in figure 1. Slope impacts the visual impression created by a graph, and therefore on its interpretation. How could you change the time axis in figure 1 to flatten the slope even more?

Next examine figure 2. This graph plots the stabilization ratio rather than population size over time. This graph reveals that the rate of growth for most regions except Africa has been in decline since about 1970. By itself, does the graph's downward-trending line suggest there really isn't much of a population problem, or at least not much reason for concern?

All three graphs present information relevant to a discussion of worldwide population growth, yet each gives a different impression of the seriousness of the problem. So, how can we analyze graphs and their messages in a thoughtful way? There is no single formula applicable to all situations, certainly, but a few questions are worth keeping in mind.

Is the time frame of reference appropriate or is it too restricted to allow a valid, comprehensive assessment of the issue? Are the unit intervals on the graph appropriately sized? Is the impression created by the graph real or simply an artifact of the graph's format? Does the data presented in a graph present only a partial, perhaps misleading, view of the whole?

Because graphs can create powerful impressions, they need to be interpreted with care.

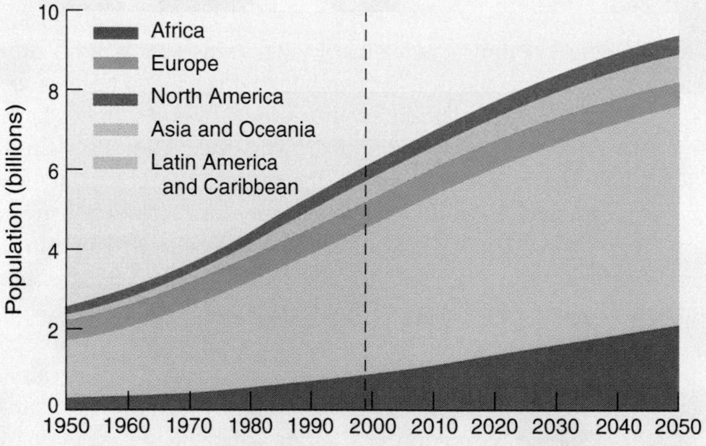

Figure 1. The U.N. Population Division projects continued population growth although at a gradually slowing rate over the next 50 years.
Source: World Resources Institute, *World Resources 1998–99.*

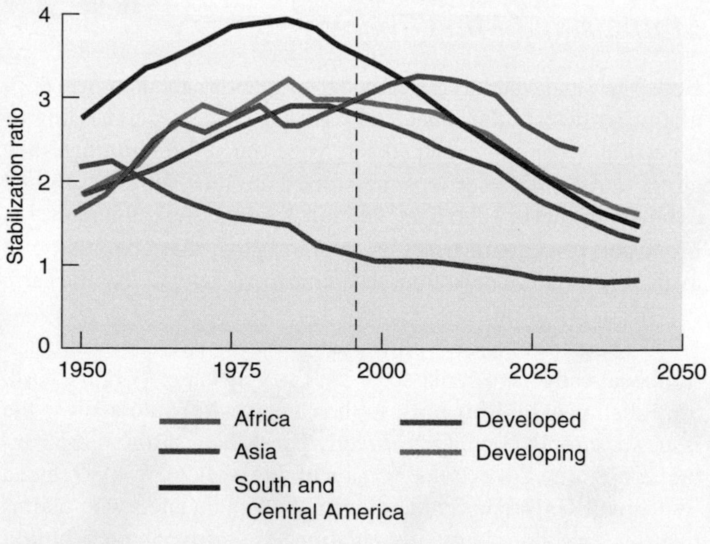

Figure 2. The stabilization ratio is measured by dividing crude birth rate by crude death rate. A ratio of 1 indicates zero population growth.
Source: United Nations (U.N.) Population Division, *World Population Prospects, 1950–2050 (The 1996 Revision),* on diskette (U.N., New York, 1996).

Can More People Be Beneficial?

There can be benefits as well as disadvantages in larger populations. More people mean larger markets, more workers, and efficiencies of scale in mass production of goods. Greater numbers also provide more intelligence and enterprise to overcome problems such as underdevelopment, pollution, and resource limitations. Human ingenuity and intelligence can create new resources through substitution of new materials and new ways of doing things for old materials and old ways. For instance, utility companies are finding it cheaper and more environmentally sound to finance insulation and energy-efficient appliances for their customers rather than build new power plants. The effect of saving energy that was formerly wasted is comparable to creating a new fuel supply.

Economist Julian Simon was one of the most outspoken champions of this rosy view of human history. People, he argued, are the "ultimate resource" and there is no evidence that pollution, crime, unemployment, crowding, the loss of species, or any other resource limitations will worsen with population growth. This outlook is shared by leaders of many developing countries who insist that instead of being obsessed with population growth, we should focus on the inordinate consumption of the world's resources by people in richer countries. What constitutes a resource and which resources might limit further human population growth are questions we will return to in subsequent chapters in this book. For now, we will move on to discuss how people are counted.

HUMAN DEMOGRAPHY

Demography is derived from the Greek words *demos* (people) and *graphos* (to write or to measure). It encompasses vital statistics about people, such as births, deaths, and where they live as well as total population size. In this section, we will survey ways human populations are measured and described, and discuss demographic factors that contribute to population growth.

How Many of Us Are There?

Even in this age of information technology and communication, we do not know exactly how many people there are in the world. Some countries have never even had a census taken, and even those that have been done may not be accurate. Governments may overstate or understate their populations to make their countries appear larger and more important or smaller and more stable than they really are. Individuals, especially if they are homeless, refugees, or illegal aliens, may not want to be counted or identified.

Although there is considerable uncertainty about the exact number of people in some places, most demographers agree that the human population passed 6 billion in 1999. As chapter 1 points out, only about 20 percent of that population lives in the more-developed or rich countries of the world. Four out of five humans live in the poorer countries of the less-developed world. Demographers estimate that 90 percent of the population growth expected to occur in the next century will take place in the Third World (fig. 7.5).

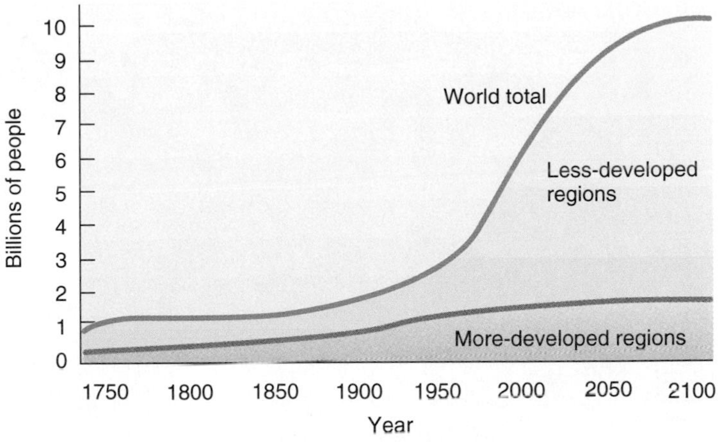

FIGURE 7.5 Estimated human population growth, 1750–2100, in less-developed and more-developed regions. More than 90 percent of all growth in this century and projected for the next is in the less-developed countries.

TABLE 7.2 Twelve Most Populous Countries in 2025 (estimated populations in millions)

COUNTRY	1950	1999	2050	RATIO*
India	358	976	1533	4.3
China	555	1255	1516	2.7
Pakistan	39	148	357	9.2
United States	152	274	347	2.2
Nigeria	33	122	338	10.2
Indonesia	80	207	318	4.0
Brazil	54	165	243	4.5
Bangladesh	42	124	218	5.2
Ethiopia	18	62	212	11.8
Iran	17	73	170	10.0
Mexico	28	96	154	5.5
Russia	102	147	114	1.1

*Ratio 2050 to 1950.
Source: Data from *World Resources 1998–99*, World Resources Institute.

Some countries, especially in the developing world, have experienced amazing growth rates and are expected to reach extraordinary population sizes by the middle of the next century. Table 7.2 shows the 12 largest countries in the world arranged by their projected size in 2050. Note that while China was the most populous country throughout the twentieth century, India is predicted to pass it in the next century because its population control

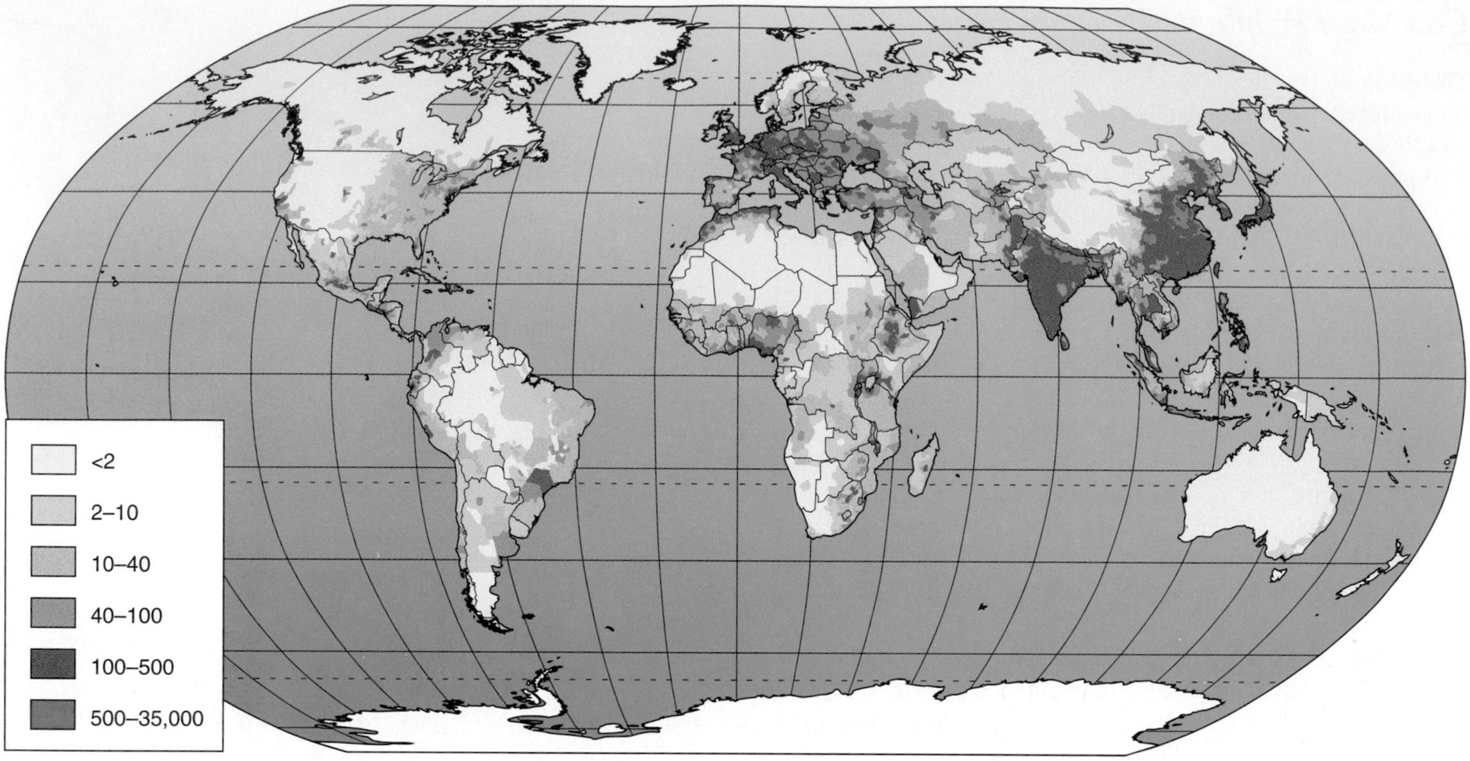

FIGURE 7.6 Population density in persons per square kilometer.

programs have been less successful than China's. Pakistan, which had only 39 million residents in 1950, is forecast to have the third largest population in 2050. Iran and Ethiopia, each with less than 20 million people 50 years ago, are likely to grow at least tenfold over a century. In many of these countries, rapid population growth is a serious problem. Bangladesh, about the size of Iowa, is already overcrowded at 124 million people. Adding another 94 million people will only add to current problems.

Not every country is growing. Russia, for instance, is now losing about 600,000 people per year. After having been the fourth largest country in the world in 1950, it entered a period of steep decline when the Soviet Union broke apart in 1992. Environmental pollution, hyperinflation, crime, corruption, disease, and despair have reduced the standard of living for most citizens so that birth rates have decreased by nearly half and death rates have risen dramatically. By 2050, Russia will probably have a smaller population than Vietnam, the Philippines, or the Democratic Republic of Congo. AIDS and other communicable diseases are taking a serious toll in many African countries. In Zimbabwe, Botswana, Zambia, and Namibia, for example, where between 20 and 25 percent of the population have AIDS or are HIV positive, declining populations are expected in the next few decades.

Figure 7.6 shows human population distribution around the world. Notice the high densities supported by fertile river valleys of the Nile, Ganges, Yellow, Yangtze, and Rhine Rivers and the well-watered coastal plains of India, China, and Europe. Historic factors such as technology diffusion and geopolitical power also play a role in geographic distribution.

Fertility and Birthrates

As we pointed out in chapter 6, fecundity is the physical ability to reproduce, while fertility describes the actual production of offspring. Those without children may be fecund but not fertile. The most accessible demographic statistic of fertility is usually the **crude birth rate,** the number of births in a year per thousand persons. It is statistically "crude" in the sense that it is not adjusted for population characteristics such as the number of women in reproductive age.

The **total fertility rate** is the number of children born to an average woman in a population during her entire reproductive life. Upper-class women in seventeenth- and eighteenth-century England, whose babies were given to wet nurses immediately after birth and who were expected to produce as many children as possible, often had twenty-five or thirty pregnancies. The highest recorded total fertility rates for working-class people is among some Anabaptist agricultural groups in North America who have averaged up to twelve children per woman. In most tribal or traditional societies, food shortages, health problems, and cultural practices limit total fertility to about six or seven children per woman even without modern methods of birth control.

Fertility is usually calculated as births per woman because, in many cases, it is difficult to establish paternity. Some demographers argue, nevertheless, that we should pay more attention to birth rates per male, because in some cultures men have far more children, on average, than do women. In Cameroon, for instance, due to multiple marriages, extramarital affairs, and a high rate of female mortality, men are estimated to have 8.1 children in their lifetime, while women average only 4.8.

The **zero population growth (ZPG)** rate (also called the replacement level of fertility) is the number of births at which people are just replacing themselves. In the more highly developed countries, where infant mortality rates are low, this rate is usually about 2.1 children per couple. It takes slightly more than two children per couple to stabilize the population because some people are infertile, have children who do not survive, or choose not to have children.

Fertility rates have declined dramatically in every region of the world except Africa over the past 50 years (fig. 7.7). Only a few decades ago, total fertility rates above 6 were common in many countries. The average family in Mexico in 1975, for instance, had 7 children. By 1998, however, the average Mexican woman had only 2.5 children. According to the World Health Organization, 61 out of the world's 190 countries are now at or below a replacement rate of 2.1 children per couple. The greatest fertility reduction has been in Southeast Asia, where rates have fallen by more than half. Most of this decrease has occurred in just the past few decades and, contrary to what many demographers expected, some of the poorest countries in the world have been remarkably successful in lowering growth rates. Bangladesh, for instance, reduced its fertility rate from 6.9 in 1980, to only 3.1 children per woman in 1998. China dropped from a fertility rate of 6 in 1970 to 1.8 in 1995. Although the world as a whole still has an average fertility rate of 2.8, growth rates are now lower than at any time since World War II, and if fertility declines like those in Bangladesh and China were to occur everywhere in the world, our total population might begin to decline by the end of the next century.

It's interesting to compare conditions in some of the fastest and slowest growing countries. Table 7.3 compares 11 countries with the highest total fertility rates in the world (Yemen, Oman, Niger, Uganda, Ethiopia, Afghanistan, Laos, Angola, Malawi, Burkina Faso, Liberia, and Congo, in descending order) with the 13 lowest fertility countries (Ukraine, Russia, Romania, Latvia, Hungary, Greece, Belarus, Austria, Estonia, Germany, Slovenia, and Italy, also in descending order). Notice that all of the highest fertility countries except Yemen, Oman, Afghanistan, and Laos are in sub-Saharan Africa. Oman is an unusual case; a conservative, relatively wealthy, Islamic state, its high fertility is linked to religious and cultural factors. All the other high fertility countries are among the poorest nations in the world with very low human development indices, low life expectancies, and high birth, death, and infant mortality rates. Collectively, these countries are growing at 2.9 percent per year, which means they are doubling their population every 24 years, on average. Note that most of these countries have had civil wars, coups, or other forms of political or economic crisis in recent years.

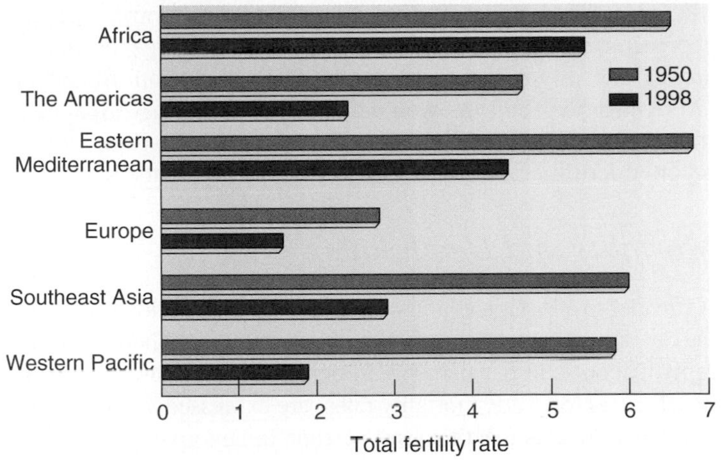

FIGURE 7.7 Declines in fertility rates by region, 1950 and 1998.
Source: Data from World Health Organization, *World Health Report 1999.*

TABLE 7.3 Countries with Highest and Lowest Fertility Rates

	HIGHEST	LOWEST (W. EUROPE)	LOWEST (E. EUROPE)
Total Fertility Rate	6.7	1.3	1.4
GNP/capita/yr	$225	$19,000	$3108
Human Development Index	0.282	0.741	0.928
Life expectancy	47.6 yr	77.6 yr	68.9 yr
Birth rate	47.6/1000	9.7/1000	9.9/1000
Death rate	17.6/1000	10.0/1000	13/1000
Infant mortality rate	120/1000 births	7/1000 births	22/1000 births
Annual growth rate	2.9%	0.1%	–0.4%

Source: Data from *World Resources 1998-1999,* World Resource Institute; *Demographic Indicators,* 1997 UN Population Division: and *World Development Report,* 1999 UN Development Program.

The list of low fertility countries has changed dramatically over the past decade. Where the most slowly growing countries once were in western Europe or Asian countries such as Japan, Hong Kong, and Singapore, a new group has joined this category since the breakup of the former Soviet Union. Some of the lowest fertility rates in the world are now found in former "Eastern Bloc" countries such as Belarus, Latvia, Romania, Russia, and Ukraine. These countries haven't reached the level of social modernization (high income, child survival, health care, education) considered necessary for birth rate reduction. In fact, many have slipped backward in their standard of living. All are now in the medium to low category of human development, with an HDI below that of

Malaysia, Thailand, or Belize. Notice that while birth rates are approximately the same in western and eastern Europe, death rates are significantly higher in the latter. This is true both for infants (22 deaths per 1000) as well as adults. Interestingly, Spain and Italy, although predominately Roman Catholic, have the lowest reported fertility rates (1.2 children per woman) in the world.

Mortality and Death Rates

A traveler to a foreign country once asked a local resident, "What's the death rate around here?" "Oh, the same as anywhere," was the reply, "about one per person." In demographics, however, **crude death rates** (or crude mortality rates) are expressed in terms of the number of deaths per thousand persons in any given year. Countries in Africa where health care and sanitation are limited may have mortality rates of twenty or more per thousand people. Wealthier countries generally have mortality rates around ten per thousand. The number of deaths in a population is sensitive to the age structure of the population. Rapidly growing, developing countries such as Belize or Costa Rica have lower crude death rates (4 per 1000) than do the more-developed, slowly growing countries, such as Denmark (12 per 1000). This is because there are proportionately more youths and fewer elderly people in a rapidly growing country than in a more slowly growing one.

Population Growth Rates

Crude death rate subtracted from crude birth rate gives the **natural increase** of a population. We distinguish natural increase from the **total growth rate,** which includes immigration and emigration, as well as births and deaths. Both of these growth rates are usually expressed as a percent (number per hundred people) rather than per thousand. A useful rule of thumb is that if you divide 70 by the annual percentage growth, you will get the approximate doubling time in years. Afghanistan, for example, which is growing 5.3 percent per year, is doubling its population every 13 years. The United States and Canada, which have natural increase rates of 0.8 percent per year, are doubling in 87.5 years. Actually, because of immigration, U.S. total growth is considerably faster than natural increase. Spain and the United Kingdom, with natural increase rates of 0.1 percent, are doubling in about 700 years. Most countries in eastern Europe have negative growth rates and declining populations. The fastest decline currently is Latvia, which at –1.1 percent per year will lose half its population in 64 years. The world growth rate is now 1.4 percent, which means that the population will double in about 50 years if this rate persists.

Life Span and Life Expectancy

Life span is the oldest age to which a species is known to survive. Although there are many claims in ancient literature of kings living for one thousand years or more, the oldest age that can be certified by written records was that of Jeanne Louise Calment of Arles, France, who was 122 years old at her death in 1997. The aging process is still a medical mystery, but it appears that cells in our bodies have a limited ability to repair damage and produce new components. At some point they simply wear out, and we fall victim to disease, degeneration, accidents, or senility.

Life expectancy is the average age that a newborn infant can expect to attain in any given society. It is another way of expressing the average age at death. For most of human history, we believe that life expectancy in most societies has been between thirty-five and forty years. This doesn't mean that no one lived past age forty, but rather that so many deaths at earlier ages (mostly early childhood) balanced out those who managed to live longer.

Declining mortality, not rising fertility, is the primary cause of most population growth in the past three hundred years. Crude death rates began falling in western Europe during the late 1700s. Most of this advance in survivorship came long before the advent of modern medicine and is due primarily to better food and better sanitation.

The twentieth century has seen a global transformation in human health unmatched in history. This revolution can be seen in the dramatic increases in life expectancy in most places. Worldwide, the average life expectancy has risen from about 40 to 65.5 years over the past century. Table 7.4 shows gains in some selected countries. The greatest progress has been in developing countries. Take the case of Chile, for example. In 1900, the average Chilean man could expect to live only 29 years, while the average woman would reach just 33 years. By 1998, although Chile had an annual per capita income less than $4000 (US), the average life expectancy for both men and women had more than doubled and was very close to that of countries with ten times its income level. Longer lives were due primarily to better nutrition, improved sanitation, clean water, and education rather than miracle drugs or high-tech medicine. While the gains were not as great for the already industrialized countries, residents of the United States, Italy, and Japan, for example, now live about half-again as long as they did at the beginning of the century.

TABLE 7.4 Life Expectancy at Birth, Selected Countries, around 1910 and in 1998

COUNTRY	AROUND 1910		1998	
	MALES	FEMALES	MALES	FEMALES
Australia	56	60	75	81
Chile	29	33	72	78
Italy	46	47	75	81
Japan	43	43	77	83
New Zealand	60	63	74	80
Norway	56	59	75	81
United States	49	53	73	80

Source: World Health Organization, *World Health Report 1999.*

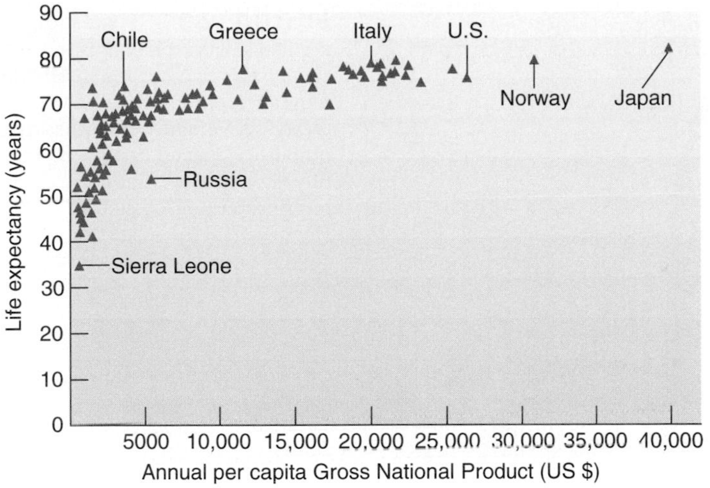

FIGURE 7.8 As incomes rise, so does life expectancy up to about $4000 (US). Russia is an exception, with a life expectancy nearly 20 years less than that of Chile, even though their GNP is about the same.
Source: The World Bank, *World Development Indicators 1997,* the World Bank, Washington, D.C., 1997.

As figure 7.8 shows, there is a good correlation between annual income and life expectancy up to about $4000 US per person. Beyond that level—which is generally enough for adequate food, shelter, and sanitation for most people—life expectancies level out at about 75 years for men and 80 for women. Russia is a striking exception to this pattern. With a GNP per person near $5000 (US), Russian life expectancy is only 58 years for men and 71 for women. Russian men now live about 14 years less, on average, than they did before the breakup of the USSR. As mentioned earlier, disastrous economy, alcoholism, poor nutrition, and substandard medical care all contribute to this decline.

Large discrepancies in how benefits of modernization and social investment are distributed within countries are revealed in differential longevity of various groups. The greatest life expectancy reported anywhere in the United States is for women in Stearns County, Minnesota, who live to an average age of 86. By contrast, Native American men on Pine Ridge Indian Reservation in neighboring South Dakota, live, on average, only to age 45. Only a few countries in Africa have a lower life expectancy. Sterns County is populated mainly by prosperous German-Catholic farmers. The Pine Ridge Reservation is the poorest area in America with an unemployment rate near 75 percent and high rates of poverty, alcoholism, drug use, and cultural alienation. Similarly, African-American men in Washington, DC live, on average, only 57.9 years, or less than men in Lesotho or Swaziland.

Some demographers believe that life expectancy is approaching a plateau, while others predict that advances in biology and medicine might extend longevity markedly. If our average age at death approaches one hundred years, as some expect, society will be profoundly affected. In 1970, the median age in the United States was thirty. By 2100, the median age could be over sixty. If workers continue to retire at sixty-five, half of the population could be unemployed, and retirees might be facing thirty-five or forty years of retirement. We may need to find new ways to structure and finance our lives.

Living Longer: Demographic Implications

A population that is growing rapidly by natural increase has more young people than does a stationary population. One way to show these differences is to graph age classes in a histogram as shown in figure 7.9. In Mexico, which is growing at a rate of 2.5 percent per year, 42 percent of the population is in the prereproductive category (below age fifteen). Even if total fertility rates were to fall abruptly, the total number of births, and population size, would continue to grow for some years as these young people enter reproductive age. This phenomenon is called population momentum.

A population that has recently entered a lower growth rate pattern, such as the United States, will have a bulge in the age classes for the last high-birth rate generation. A country that has had a stable population for many years, such as Sweden, will have approximately the same numbers in all age classes. Notice that there are more females than males in the older age group in the United States because of differences in longevity between the sexes. The United States has a high percentage of retired people because of long life expectancy.

Countries with a high percentage of children (such as Mexico) or a high percentage of old people (such as Sweden) have a problem with their **dependency ratio,** or the number of nonworking compared to working individuals in a population. Mexico has a high number of children to be supported by each working person. In Sweden, although there are fewer children, a large number of retired persons must be supported by a small working population. This changing age structure and shifting dependency ratio is occurring worldwide (fig. 7.10). There is considerable worry that not enough workers will be available to support retirement systems in countries where population growth has slowed sharply and life expectancies have increased. Retirement might have to be postponed or eliminated altogether.

Emigration and Immigration

Humans are highly mobile, so emigration and immigration play a larger role in human population dynamics than they do in those of many species. Currently, about 800,000 people immigrate legally to the United States each year, but many more enter illegally. Western Europe receives about 1 million applications each year for asylum from economic chaos and wars in former socialist states and the Middle East. The United Nations High Commission on Refugees estimated in 1997 that at least 38 million refugees left their countries for political or economic reasons, while another 30 million fled their homes but remained displaced persons in their own countries.

Immigration is a controversial issue in most wealthy countries. "Guest workers" often perform heavy, dangerous, or disagreeable work that citizens are unwilling to do. Many migrants

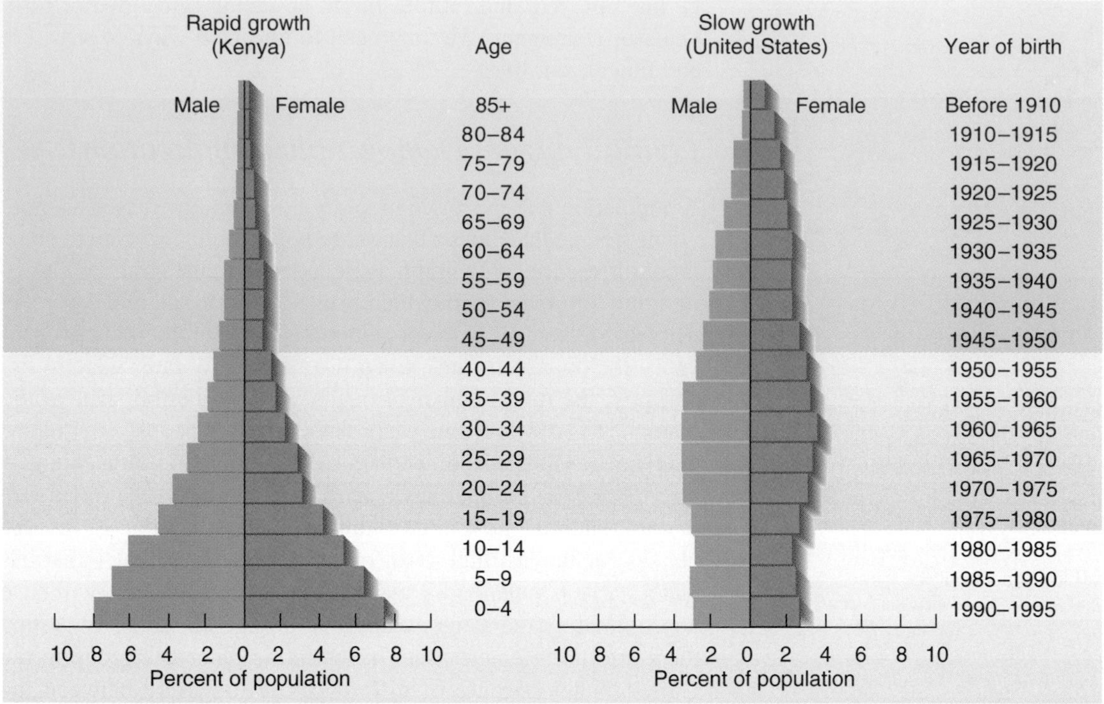

FIGURE 7.9 Population by age and sex in rapidly growing and slowly growing countries. Middle regions represent individuals of reproductive ages. Note the high proportion of children in the rapidly growing population and the high proportion of elderly in the slow-growing population.

Sources: Data from U.S. Bureau of the Census, the United Nations, and Population Reference Bureau.

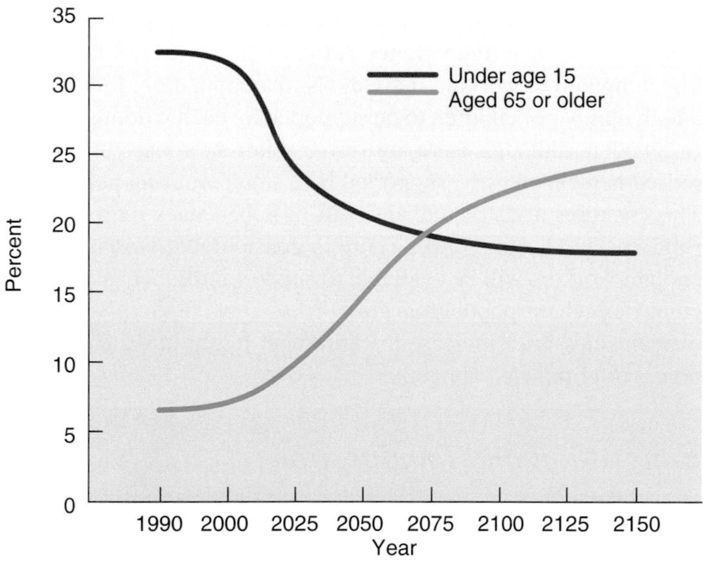

FIGURE 7.10 Changing age structure of world population. In the twenty-first century, children under 15 years of age will make up a smaller percentage of world population, while people over 65 years old will make up a rapidly rising share of the population.

and alien workers are of a different racial or ethnic background than the majority in their new home. They generally are paid low wages and given substandard housing, poor working conditions, and few rights. Local residents often complain that immigrants take away jobs, overload social services, and ignore established rules of behavior or social values.

Refugees seeking jobs and asylum in the richer countries have been subjected to increasingly violent racist and xenophobic attacks in many countries. Those who are out of work and alienated from society find these people an easy target for their anger. Ironically, those who are themselves immigrants or descendants of immigrants are sometimes the most adamant about closing doors to future immigration. Proponents of an open door policy claim that immigrants make a valuable contribution to society, bringing new energy, ingenuity, cultural diversity, and vigor.

Peaceful internal relocations also are demographically important. In the United States, for instance, a mass migration of rural southern people into northern industrialized cities in the 1930s and 1940s had profound effects on both the areas they left and the new places they settled. A similar shift of labor and business from the Rust Belt of the Great Lakes and mid-Atlantic States to the Sun Belt of the South and West in the 1960s and 1970s left many older industrialized cities empty, poverty-stricken wastelands.

Some countries encourage, or even force, internal mass migrations as part of a geopolitical demographic policy. In the 1970s, Indonesia embarked on an ambitious "transmigration" plan to move 65 million people from the overcrowded islands of Java and Bali to relatively unpopulated regions of Sumatra, Borneo, and New Guinea. Attempts to turn rainforest into farmland had disastrous environmental and social effects, however, and this plan was greatly scaled back. In 1990, China announced intentions to move 100 million people into a pristine region along the Amur River joining Siberia. This is a region of great biodiversity, home to endangered Siberian tigers, cranes, and other rare species. An influx of millions of people could be a major biological catastrophe.

CASE STUDY

FAMILY PLANNING IN THAILAND

One of the most successful birth control programs in the world is in Thailand, where the annual population growth rate fell from 3.3 to 1.2 percent between 1972 and 1995. The number of children an average woman would have in a lifetime dropped during this period from 5.8 to 2.2. Much of this progress was due to the leadership of Mechai Viravaidya, the founder and director of the Community-Based Family Planning Service (CBFPS) of Thailand.

Concerned about the environmental and economic effects of rapid population growth in his country, Viravaidya could see that the top-down approach to family planning used in Thailand was ineffective. A different approach—one focused on the wants and needs of poor people, especially in rural areas—was needed. Having been chosen to escort the Thai Miss Universe on a publicity tour in the early 1970s, Viravaidya was well-known in his country and had developed an expertise in using the media to influence public opinion. He decided to use his new-found fame and skills to do something positive for his country.

When Viravaidya started his campaign, birth control was a foreign concept to most Thai people. The whole subject was embarrassing and not something discussed in polite company. Few people knew anything about modern methods of contraception. The first thing CBFPS decided to do was to try to overcome the taboo of discussing sexual issues in public. Humorous billboards and signs on buses, public contests to build stacks of birth control pills and blow up condoms like balloons, songs and jingles on the radio, and free condom giveaways made the subject of family planning a familiar and even popular topic. Children learned about family planning in school and took material home to educate their parents.

Viravaidya became one of the most well-known figures in Thailand. His good-natured but indefatigable promotion of birth control earned him the nickname of Mr. Condom. Financial incentives were offered to community members willing to distribute contraceptives as well as to villagers who practice family planning. Having a poster in the front window proclaiming a household as one enrolled in the CBFPS program became a source of pride.

CBFPS worked on more than birth control, however. Viravaidya recognized a need to change the socioeconomic reasons that cause people to want large families. Rural development projects designed to increase family income and to provide educational opportunities were undertaken as an important part of family planning. In 1977, the program changed its name to Population and Community Development Association (PCDA) to reflect a belief that the solution to the population growth problem is located at the village level. People will tend to have fewer children if they are confident that the ones they have will survive. Among the greatest threats to children's health are malnutrition and lack of clean drinking water.

In 1970 when Viravaidya started his work, about 25 percent of all Thai children died before age five, mainly from infectious diseases. The PCDA now helps finance and build rainwater catchment systems in villages to ensure a year-round clean water supply. Ways to improve agricultural and livestock output are taught to ensure greater financial security and a better food supply. Today, infant mortality has been cut by 75 percent and parents have lowered their desired number of children from 7 or 8 to 2 or 3. This means both less pain over lost children and real progress toward a stable population.

Individual participation is a key factor in the success of PCDA programs. No handouts are given; each person is responsible for paying back loans. The PCDA approach has helped villages become self-sufficient. It encourages villagers to find a path to a better life and lets their own efforts and desires determine how far and how fast they will go.

Ethical Considerations

Trying to change people's beliefs about something as basic as sex and family planning can easily infringe on religious beliefs and social traditions. Does a society's goal of stabilizing population justify interfering with people's right to privacy and individual freedom?

Minority groups in Thailand and elsewhere believe their populations are already so low that they are in danger of disappearing in the larger majority culture. They claim a special right to have as many children as they like. Altogether these minority people make up a large population. If you could make world population policy, would you grant their request? Where would you draw the line?

POPULATION GROWTH: OPPOSING FACTORS

A number of social and economic pressures affect decisions about family size, which in turn affects the population at large. In this section we will examine both positive and negative pressures on reproduction.

Pronatalist Pressures

Factors that increase people's desires to have babies are called **pronatalist pressures.** Raising a family may be the most enjoyable and rewarding part of many people's lives. Children can be a source of pleasure, pride, and comfort. They may be the only source of support for elderly parents in countries without a social security system. Where infant mortality rates are high, couples may

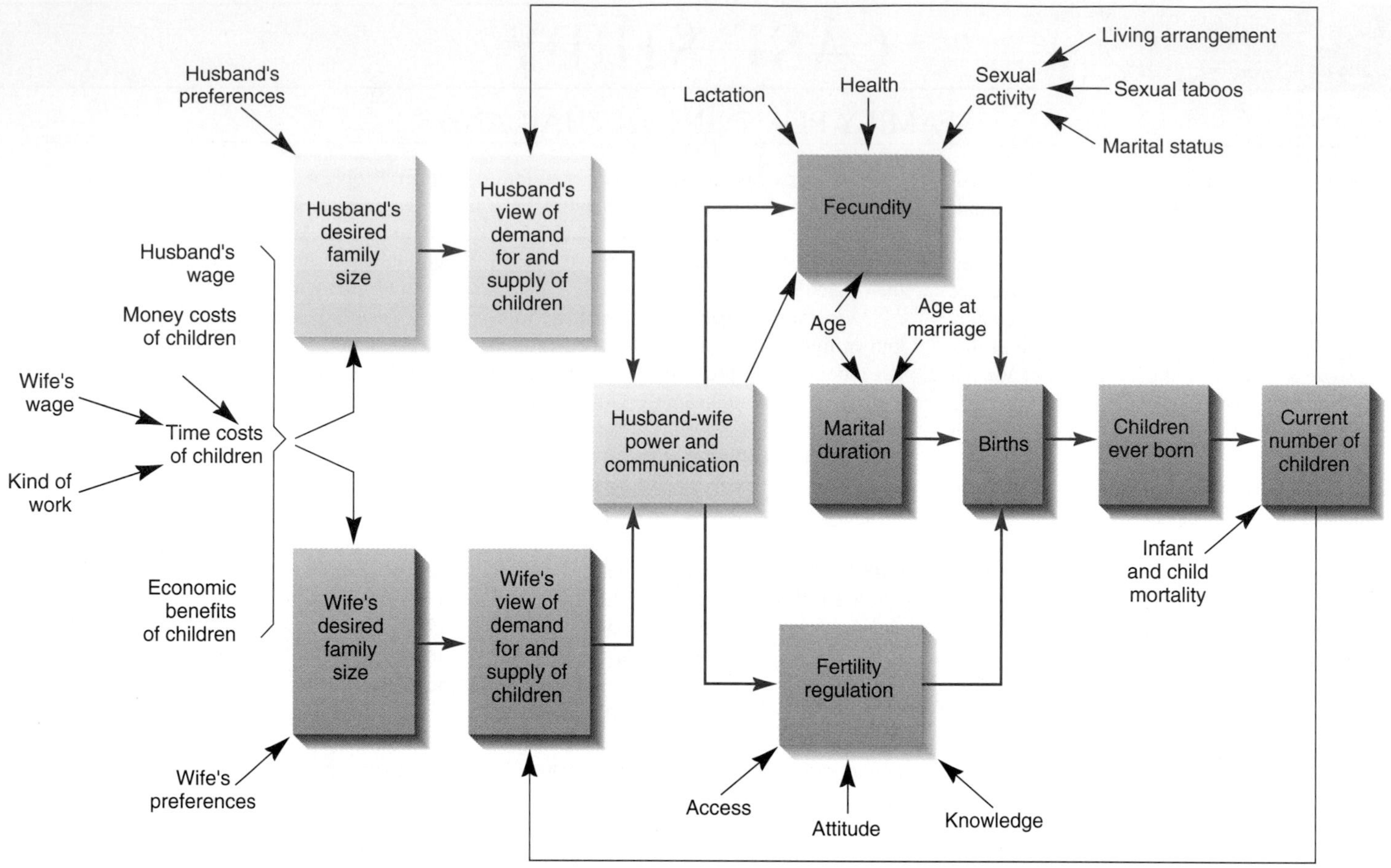

FIGURE 7.11 A model for the variables determining fertility. Three major factors interact in this model: the biological supply of children, the demand for children, and the regulation of fertility. Other intervening factors have either positive or negative effects on this interaction.

need to have many children to be sure that at least a few will survive to take care of them when they are old. Where there is little opportunity for upward mobility, children give status in society, express parental creativity, and provide a sense of continuity and accomplishment otherwise missing from life. Often children are valuable to the family not only for future income, but even more as a source of current income and help with household chores. In much of the developing world, children as young as six tend domestic animals and younger siblings, fetch water, gather firewood, and help grow crops or sell things in the marketplace. Parental desire for children rather than an unmet need for contraceptives may be the most important factor in population growth in many cases.

Society also has a need to replace members who die or become incapacitated. This need often is codified in cultural or religious values that encourage bearing and raising children. In some societies, families with few or no children are looked upon with pity or contempt. The idea of deliberately controlling fertility may be shocking, even taboo. Women who are pregnant or have small children are given special status and protection. Boys frequently are more valued than girls because they carry on the family name and are expected to support their parents in old age. Couples may have more children than they really want in an attempt to produce a son.

Male pride often is linked to having as many children as possible. In Niger and Cameroon, for example, men, on average, want 12.6 and 11.2 children, respectively. Women in these countries consider the ideal family size to be only about one-half that desired by their husbands. Even though a woman might desire fewer children, however, she may have few choices and little control over her own fertility. In many societies, a woman has no status outside of her role as wife and mother. Without children, she has no source of support.

Figure 7.11 shows a model for the variables determining fertility. Three primary factors interact in this model: the biological supply of children determined by fecundity and infant mortality, the demand for children determined by economics and social values, and the regulation of fertility determined by knowledge, attitudes, and access to birth control. The combined interaction of these factors and their intervening variables determine fertility.

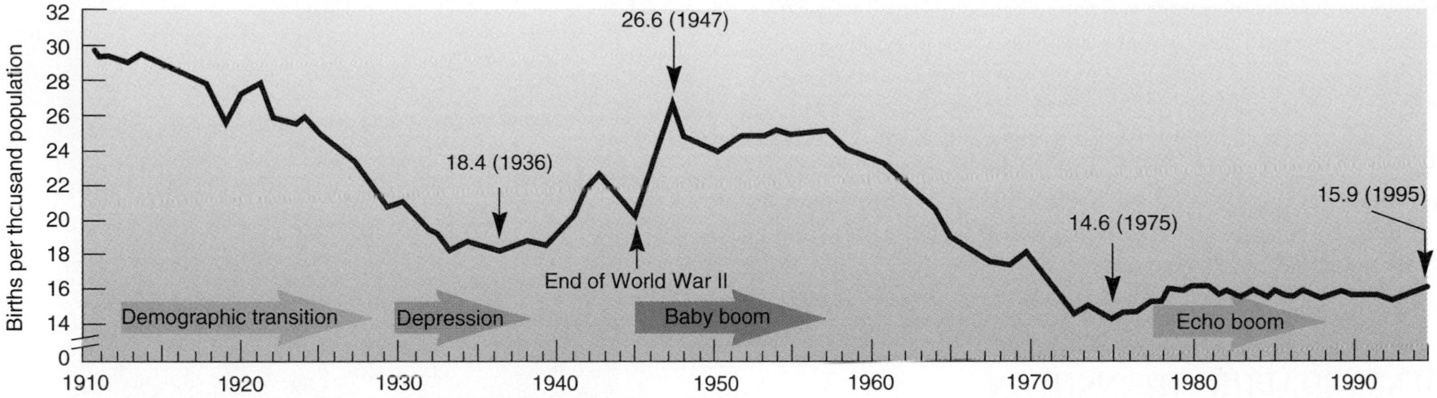

FIGURE 7.12 Birthrates in the United States, 1910–1995. The falling birthrate from 1910 to 1929 represents a demographic transition from an agricultural to an industrial society. Note that this decline occurred before the start of the Great Depression. The baby boom following the Second World War lasted from 1945 to 1957. A much smaller "echo boom" occurred around 1980 when the baby boomers started to reproduce, but it produced far fewer births than anticipated.

Sources: Data from Population Reference Bureau, Inc., and U.S. Bureau of the Census.

Birth Reduction Pressures

In more highly developed countries, many pressures tend to reduce fertility. Higher education and personal freedom for women often result in decisions to limit childbearing. The desire to have children is offset by a desire for other goods and activities that compete with childbearing and childrearing for time and money. When women have opportunities to earn a salary, they are less likely to stay home and have many children. Not only are the challenge and variety of a career attractive to many women, but the money that they can earn outside the home becomes an important part of the family budget. Thus, education and socioeconomic status are usually inversely related to fertility in richer countries. In developing countries, however, fertility is likely to increase as educational levels and socioeconomic status rise. With higher income, families are better able to afford the children they want; more money means that women are likely to be healthier, and therefore better able to conceive and carry a child to term.

In less-developed countries where feeding and clothing children can be a minimal expense, adding one more child to a family usually doesn't cost much. By contrast, raising a child in the United States can cost hundreds of thousands of dollars by the time the child is through school and is independent. Under these circumstances, parents are more likely to choose to have one or two children on whom they can concentrate their time, energy, and financial resources.

Figure 7.12 shows U.S. birth rates between 1910 and 1995. As you can see, birth rates have fallen and risen in a complex pattern. The period between 1910 and 1930 was a time of industrialization and urbanization. Women were getting more education than ever before and entering the workforce. The Great Depression in the 1930s made it economically difficult for families to have children, and birth rates were low. The birth rate increased at the beginning of World War II (as it often does in wartime). For reasons that are unclear, a higher percentage of boys are usually born during war years.

At the end of the war, there was a "baby boom" as couples were reunited and new families started. This high birth rate persisted through the times of prosperity and optimism of the 1950s, but began to fall in the 1960s. Part of this decline was caused by the small number of babies born in the 1930s. This meant fewer young adults to give birth in the 1960s. Part was due to changed perceptions of the ideal family size. Whereas in the 1950s women typically wanted four children or more, in the 1970s the norm dropped to one or two (or no) children. A small "echo boom" occurred in the 1980s as people born in the 1960s began to have babies, but changing economics and attitudes seem to have permanently altered our view of ideal family size in the United States.

Birth Dearth?

Most European countries now have birth rates below replacement rates, and Italy, Russia, Austria, Germany, Greece, and Spain are experiencing negative rates of natural population increase. In Asia, Japan, Singapore, and Taiwan are also facing a "child shock" as fertility rates have fallen well below the replacement level of 2.1 children per couple. There are concerns in all these countries about falling military strength (lack of soldiers), economic power (lack of workers), and declining social systems (not enough workers and taxpayers) if low birth rates persist or are not balanced by immigration.

Economist Ben Wattenberg warns that this "birth dearth" might seriously erode the powers of Western democracies in world affairs. He points out that Europe and North America accounted for 22 percent of the world's population in 1950. By the 1980s, this number had fallen to 15 percent, and by the year 2030, Europe and North America probably will make up only 9 percent of the world's population. Germany, Hungary, Denmark, and Russia now offer incentives to encourage women to bear children. Japan offers financial support to new parents, and Singapore provides a dating service to encourage marriages among the upper classes as a way of increasing population.

On the other hand, since Europeans and North Americans consume so many more resources per capita than most other people in the world, a reduction in the population of these countries will do more to spare the environment than would a reduction in population almost anywhere else.

DEMOGRAPHIC TRANSITION

In 1945, demographer Frank Notestein pointed out that there is a typical pattern of falling death rates and birth rates due to improved living conditions that usually accompanies economic development. He called this pattern the **demographic transition** between high birth and death rates to lower birth and death rates. Population histories of Sweden, Japan, and England and Wales are shown in figure 7.13.

Development and Population

The early stages of each graph in figure 7.13 represent the conditions in a premodern society. Food shortages, malnutrition, lack of sanitation and medicine, accidents, and other hazards keep death rates around 30 per 1000 people. Birth rates are correspondingly high to keep population densities relatively constant.

In the nineteenth century in Europe and in the twentieth century in Japan, better jobs and more income provided by industrialization improved the standard of living, causing death rates to fall—often very rapidly. Birth rates may actually rise at first (note the curve for Japan) but eventually fall as people see that societal improvements are permanent. It takes some years, however, to change attitudes and values. Note that populations grow rapidly during this phase as the rate of natural increase (birth rate minus death rate) rises. Depending on how long it takes to complete this phase, the population may go through one or more rounds of doubling before coming into balance again.

The right-hand side of each curve represents conditions in developed countries where the transition is complete and both birth rates and death rates are low, often one-third or less than those in the predevelopment era. The population comes into a new equilibrium in this phase, but at a much larger size than before. Most of the countries of northern and western Europe went through a demographic transition in the nineteenth or early twentieth century similar to the curves shown here.

Many of the most rapidly growing countries in the world such as Kenya, Yemen, Libya, and Jordan now are in the middle phase of this demographic transition. Their death rates have fallen close to the rates of the fully developed countries, but birth rates have not fallen correspondingly. In fact, both their birth rates and total population are higher than those in most European countries when industrialization began three hundred years ago. The large disparity between birth and death rates means that many developing countries now are growing at 3 to 4 percent per year. This high rate of growth in the Third World could boost total world population to 10 billion or more before the end of the next century. Perhaps the most important questions in this whole chapter are—Why are birth rates not yet falling in these countries? and what can be done about it?

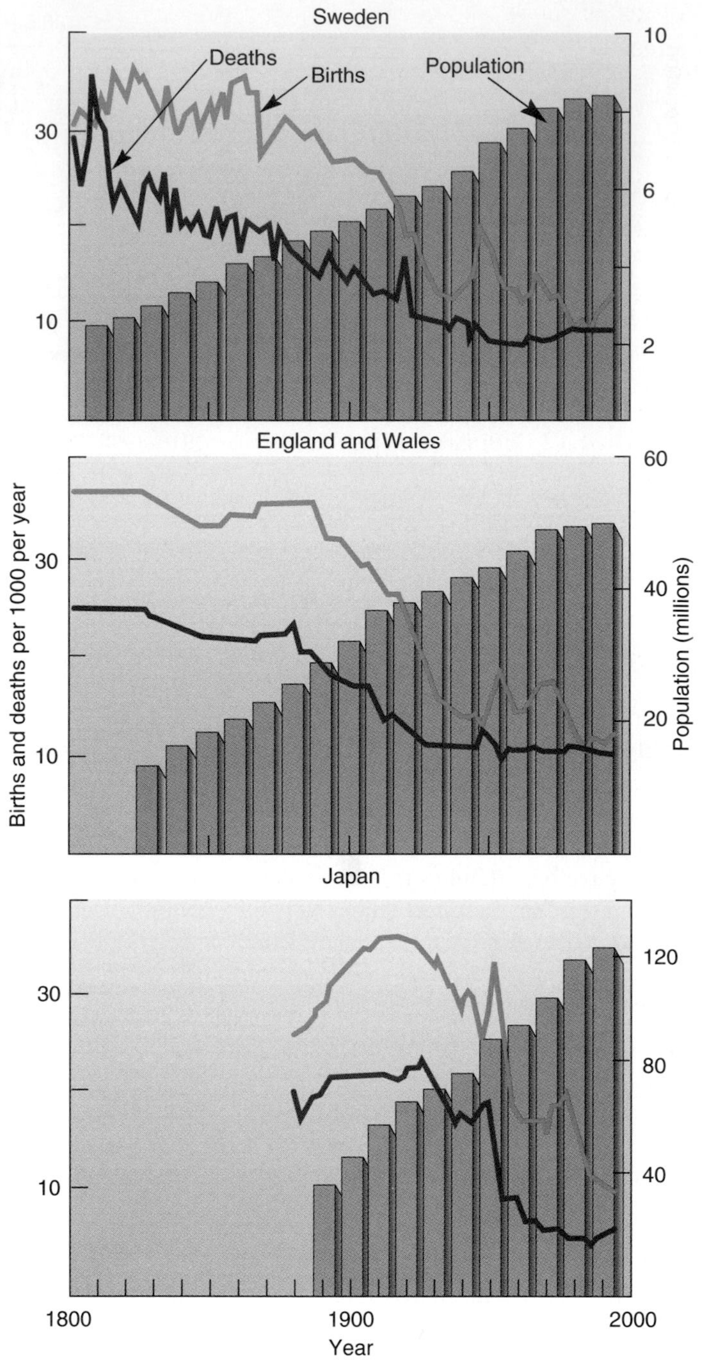

FIGURE 7.13 Demographic transitions in industrializing countries. Death rates (*bottom lines*) generally drop first as quality of life improves. Birth rates may actually increase as more women achieve wanted fertility and more children survive. Eventually, birth and death rates come into equilibrium. During the transition, the difference between births and deaths (*lines*) produces rapid population growth (*bars*).

An Optimistic View

Some demographers claim that a demographic transition already is in progress in most developing nations. Problems in taking censuses and a normal lag between falling death and birth rates may hide this for a time, but the world population should stabilize sometime in the next century. Some evidence supports this view. As we mentioned earlier in this chapter, fertility rates have fallen dramatically nearly everywhere in the world over the past half century.

Some countries have had remarkable success in population control. In Thailand, Indonesia, and Colombia, for instance, total fertility dropped by more than half in twenty years (see Case Study, p. 153). Morocco, Dominican Republic, Jamaica, Peru, and Mexico all have seen fertility rates fall between 30 percent and 40 percent in a single generation. The following factors contribute to stabilizing populations:

- Growing prosperity and social reforms that accompany development reduce the need and desire for large families in most countries.
- Technology is available to bring advances to the developing world much more rapidly than was the case a century ago, and the rate of technology transfer is much faster than it was when Europe and North America were developing.
- Less-developed countries have historic patterns to follow. They can benefit from our mistakes and chart a course to stability more quickly than they might otherwise do.
- Modern communications (especially television) have caused a revolution of rising expectations that act as a stimulus to spur change and development.

A Pessimistic View

Economist Lester Brown of the Worldwatch Institute takes a more pessimistic view. He warns that many of the poorer countries of the world appear to be caught in a "demographic trap" that prevents them from escaping from the middle phase of the demographic transition. Their populations are now growing so rapidly that human demands exceed the sustainable yield of local forests, grasslands, croplands, or water resources. The resulting resource shortages, environmental deterioration, economic decline, and political instability may prevent these countries from ever completing modernization. Their populations may continue to grow until catastrophe intervenes.

Many people argue that the only way to break out of the demographic trap is to immediately and drastically reduce population growth by whatever means are necessary. They argue strongly for birth control education and bold national policies to encourage lower birthrates. Some agree with Malthus that helping the poor will simply increase their reproductive success and further threaten the resources on which we all depend. Author Garret Hardin described this view as lifeboat ethics. "Each rich nation," he said, "amounts to a lifeboat full of comparatively rich people. The poor of the world are in other much more crowded lifeboats. Continuously, so to speak, the poor fall out of their lifeboats and swim for a while, hoping to be admitted to a rich lifeboat, or in some other way to benefit from the goodies on board . . . . We cannot risk the safety of all the passengers by helping others in need. What happens if you share space in a lifeboat? The boat is swamped and everyone drowns. Complete justice, complete catastrophe."

A Social Justice View

A third view is that **social justice** (a fair share of social benefits for everyone) is the real key to successful demographic transitions. The world has enough resources for everyone, but inequitable social and economic systems cause maldistributions of those resources. Hunger, poverty, violence, environmental degradation, and overpopulation are symptoms of a lack of social justice rather than a lack of resources. Although overpopulation exacerbates other problems, a narrow focus on this factor alone encourages racism and hatred of the poor. A solution for all these problems is to establish fair systems, not to blame the victims. Small nations and minorities often regard calls for population control as a form of genocide. Figure 7.14 expresses the opinion of many people in less-developed countries about the relationship between resources and population.

An important part of this view is that many of the rich countries are, or were, colonial powers, while the poor, rapidly growing countries were colonies. The wealth that paid for progress and security for developed countries was often extracted from colonies, which now suffer from exhausted resources, exploding populations, and chaotic political systems. Some of the world's poorest countries such as India, Ethiopia, Mozambique, and Haiti had rich resources and adequate food supplies before they were impoverished by colonialism. Those of us who now enjoy abundance may need to help the poorer countries not only as a matter of justice but because we all share the same environment.

An Ecojustice View

In addition to considering the rights of fellow humans, we should also consider those of other species. Rather than ask what is the maximum number of humans that the world can possibly support, perhaps we should think about the needs of other creatures. As we convert natural landscapes into agricultural or industrial areas, species are crowded out that may have just as much right to exist as we do. Perhaps we should seek the optimum number of people at which we can provide a fair and decent life for all humans while causing the minimum impact on nonhuman neighbors.

Infant Mortality and Women's Rights

Survival of children is one of the most critical factors in stabilizing population. When infant and child mortality rates are high, as they are in much of the developing world, parents tend to have high numbers of children to ensure that some will survive to adulthood. There has

FIGURE 7.14 Controlling our population and resources—there may be more than one side to the issue.
Used with permission of the Asian Cultural Forum on Development.

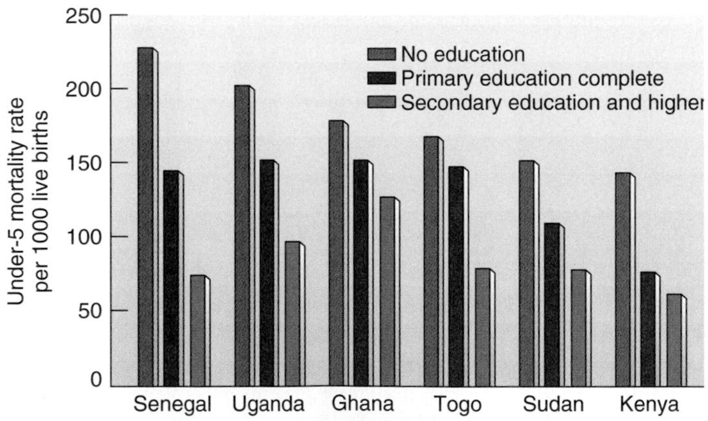

FIGURE 7.15 Childhood deaths drop as female education rises. Under-5 mortality rate is shown for selected African countries compared to level of female education.
Source: World Resources Institute, *World Resources* 1998–99.

never been a sustained drop in birth rates that was not first preceded by a sustained drop in infant and child mortality. One of the most important distinctions in our demographically divided world is the high infant mortality rates in the less developed countries. Better nutrition, improved health care, simple oral rehydration therapy, and immunization against infectious diseases (chapter 9) have brought about dramatic reductions in child mortality rates, which have been accompanied in most regions by falling birth rates. It has been estimated that saving 5 million children each year from easily preventable communicable diseases would avoid 20 or 30 million extra births.

Increasing family income does not always translate into better welfare for children since men in many cultures control most financial assets. Often the best way to improve child survival is to ensure the rights of mothers. Opportunities for women's education, for instance, as well as land reform, political rights, opportunities to earn an independent income, and improved health status of women often are better indicators of family welfare than rising GNP (fig. 7.15).

FAMILY PLANNING AND FERTILITY CONTROL

Family planning allows couples to determine the number and spacing of their children. It doesn't necessarily mean fewer children—people may use family planning to have the maximum number of children possible—but it does imply that the parents will control their reproductive lives and make rational, conscious decisions about how many children they will have and when those children will be born, rather than leaving it to chance. As the desire for smaller families becomes more common, birth control becomes an essential part of family planning in most cases. In this context, **birth control** usually means any method used to reduce births, including celibacy, delayed marriage, contraception, methods that prevent implantation of embryos, and induced abortions.

Traditional Fertility Control

Evidence suggests that people in every culture and every historic period have used a variety of techniques to control population size. Studies of hunting and gathering people, such as the Kung! or San of the Kalahari Desert in southwest Africa, indicate that our early ancestors had stable population densities, not because they killed each other or starved to death regularly, but because they controlled fertility.

For instance, San women breast-feed children for three or four years. When calories are limited, lactation depletes body fat stores and suppresses ovulation. Coupled with taboos against intercourse while breast-feeding, this is an effective way of spacing children. Other ancient techniques to control population size include celibacy, folk medicines, abortion, and infanticide. We may find some or all of these techniques unpleasant or morally unacceptable, but we shouldn't assume that other people are too ignorant or too primitive to make decisions about fertility.

Current Birth Control Methods

Modern medicine gives us many more options for controlling fertility than were available to our ancestors. Some of these techniques are safer, easier, or more pleasant to use (fig. 7.16). The major categories of birth control techniques include (1) avoidance of sex during fertile periods [celibacy; using changes in body temperature or cervical mucus color and viscosity to judge when ovulation will occur]; (2) mechanical barriers that prevent contact between sperm and egg [condoms, spermicides, diaphragm, cervical cap, and vaginal sponge]; (3) surgical methods that prevent release of sperm or egg [sterilization: tubal ligation or use of the Filshie clip in females, vasectomy in males] (fig. 7.17); (4) chemicals that prevent maturation or release of sperm or eggs or implantation of the embryo in the uterus [the pill: estrogen + progesterone, progesterone alone for females, gossypol for males]; (5) physical barriers to implantation [IUD]; and (6) abortion.

Norplant, the trade name for flexible, matchstick-sized, silicon-rubber implants containing a slow-release analog of progesterone, was approved for use in the United States. The implants are inserted under the skin where they will release hormones for up to five years (fig. 7.16*c*). Depo-Provera, an injectable progesterone analog, is given by injections four times a year. Both implants and injections have very low failure rates (0.3 percent compared to about 1 percent for oral contraceptives) and eliminate the need to keep track of and take daily pills. They also can be used without knowledge of one's partner, who may oppose birth control. Both injections and implants cause problems for some women who experience increased vaginal bleeding or absence of menstrual periods. Norplant may be linked to ovarian cysts and the inserts have been difficult to remove.

Condom use has more than doubled over the past 20 years, from about 3.5 million users in 1980 to 8 million in 1999. While condoms have about a 10 percent failure rate when used alone, their effectiveness can be increased by combining them with spermicidal creams. Condoms are important protection against sexually transmitted diseases. A condom for women called Reality consists of two flexible plastic rings connected by a strong, clear polyurethane sheath. One ring fits over the cervix much like a diaphragm, while the other remains outside the vagina. Each condom costs about two dollars and is designed to be discarded after a single use. The six-month failure rate is about 12 percent, which means that 12 percent of women using only this device will get pregnant in six months. A female condom gives women control over their own reproduction, but some women may find it difficult or unpleasant to use.

None of these methods is perfect and none suits every contraceptive need. Many require careful, conscientious use. Which choice is best for you depends on your life situation and your plans for the future (see What Can You Do?, p. 160).

For nearly 20 years, the French drugs, RU486 (mifepristone or mifegyne), has been used as a "morning after" drug in Europe. RU486 blocks the effects of progesterone in maintaining the lining of the uterine wall. It is usually administered together with misoprostol (a prostaglandin analog), which causes uterine contraction and expulsion of the fetus. Although for some women, RU486 is preferable to surgical abortion, neither procedure is to be taken lightly. Both require medical supervision and entail cramps, bleeding, and some risk. Interestingly, RU486 also appears to have promise in treating breast cancer, brain cancer, diabetes, and hypertension. Opposition by anti-abortion groups has kept this drug from being marketed in the United States.

Recently, a combination of two drugs already on the market for other uses have been shown to be as safe and effective as RU486. Methotrexate and misoprostol are administered a week apart to induce abortion. Equally opposed by abortion foes, these drugs could make the debate over whether to make RU486 available in the United States a moot point.

New Developments in Birth Control

More than 100 new contraceptive methods are now being tested, and some appear to have great promise Adhesive patches that dispense progesterone through the skin have shown promise. Each patch, about the size of a half dollar, lasts for about a week. They can be worn on the arm, abdomen, or buttocks. Soft silicone vaginal rings containing slow-release progesterone analogs also are being studied. These have all the advantages of Depo-Provera or Norplant but should be much cheaper and easier to use since they can be inserted and removed by the user. Biodegradable implants for both women and men are being tested that would not require removal. Development of simple, inexpensive, do-it-yourself tests for levels of estrogen and progesterone in urine may make the

what can you do?

Personal Fertility

What are the options for personal reproductive decisions? More alternatives are available now than were a generation ago, but more concerns about health, sexuality, and parenthood face us also. How can an individual know what to do? Each of us must examine our own values, situation, and choices to design a personal fertility plan.

First, examine your own life and your hopes for the future. How old are you? Are you sexually active now? Will you be in the future? How many children—if any—will you want and when would you like them to be born? Timing and number of births can critically affect education, jobs, travel, and other life aspirations. How many children will you be able to support and educate, given projections for future costs and income? As global citizens, we also should ask how many children it is responsible for us to bring into the world, given the impact we have.

Next, examine the alternatives available to meet your goals. Remember that you don't need to commit yourself to a single option for life. Among the choices are:

- *Abstinence*: One hundred percent effective in birth control and prevention of sexually transmitted diseases. Under some circumstances, this may be the best and safest course of action. As a long-term strategy, however, celibacy may inhibit intimate and fulfilling relationships for many people.
- *Oral contraceptives (pills)*: Ninety-nine percent effective. Easy to use and readily reversible. Not recommended for women over age 35 who smoke because of the risk of strokes and other cardiovascular problems. Pills with low estrogen levels or progesterone alone can reduce these risks. Some increase in breast cancer is associated with prolonged use of synthetic hormones, but they seem to give some protection from endometrial and ovarian cancer. Overall, cancer risks seem slightly lower with birth control pills than without.
- *Implantable or injectable progesterone analogs*: Ninety-nine percent effective and highly reversible. Doesn't require daily attention or approval of partner. May be suitable for women who cannot take oral contraceptives. Some women experience dizziness, fatigue, headaches, vaginal bleeding, or absent periods with these methods. May increase risks of osteoporosis. About the same breast cancer risk and other cancer benefits as oral pills.
- *Barrier methods*: The condom, diaphragm, cervical cap, and vaginal sponge are generally 85 to 95 percent effective. They help prevent sexually transmitted diseases, but don't bet your life on them. They require some planning ahead but not the daily attention of the pill. Any of these may be a good option for someone who cannot use hormone analogs or has sex only occasionally.
- *Spermicidal creams, jellies, douches*: Often not very effective by themselves in either birth control or disease prevention. May help, however, when used together with barrier methods.
- *Intrauterine device (IUD)*: Ninety-four percent effective. Doesn't require daily attention or alter natural hormone balance. May increase frequency or severity of sexually transmitted diseases as well as vaginal bleeding, cramps, and other side effects. Generally recommended only for mature women in monogamous relationships.
- *Natural methods*: Careful recordkeeping, temperature measurements, and examination of cervical mucus can be 70 to 90 percent effective in birth control. This method meets ethical or religious scruples but may be difficult and demanding to follow. Prohibits sex on some days.
- *Vasectomy, tubal ligation, Filshie clip:* These surgical methods are more than 99 percent effective in birth control but are not easily reversed. These may be good options for those who are sure they don't want any more children.

As you can see, no birth control method is perfect for everyone. All have some advantages and disadvantages. You should be aware that the failure rates under "perfect" conditions often are very different from the "typical" use. Spermicides, for example, are nearly 100 percent effective in clinical trials, but about one-third of women depending on this method alone get pregnant because of misuse. Almost all techniques require deliberate attention and planning. By thinking carefully and communicating clearly about values and desires, however, couples can make responsible decisions about their fertility and sexual behavior.

rhythm method easier to follow and more reliable for women who cannot or would rather not use other methods of birth control.

A group of drugs known as gonadotropin releasing-hormone agonists show promise in suppressing egg and sperm development. Some antipregnancy vaccines (immunization against chorionic gonadotropin—a hormone required to maintain the uterine lining) and antisperm vaccines are being tested that would use the immune system to prevent fertilization or embryonic implantation, but they are years away from the market. Clinical trials of two oral hormones for men (testosterone undeconate and cyproterone acetate) have shown some promise in suppressing sperm production but it may be years before they are ready to market or before men will be willing to take them.

THE FUTURE OF HUMAN POPULATIONS

How many people will be in the world a century from now? Most demographers believe that world population will stabilize sometime during the next century. The total number of humans, when we reach that equilibrium, is likely to be somewhere around 8 to 10 billion people, depending on the success of family planning programs and the multitude of other factors affecting human populations. Figure 7.18 shows three scenarios projected by the U.N. Population Division. The optimistic (low) projection shows that world population might stabilize at about

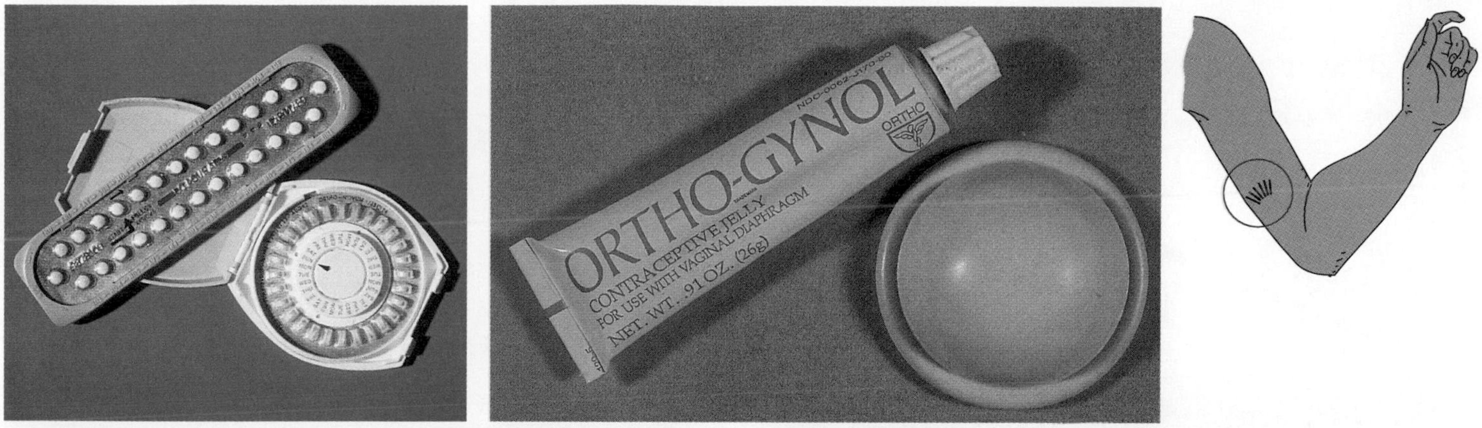

(a) (b) (c)

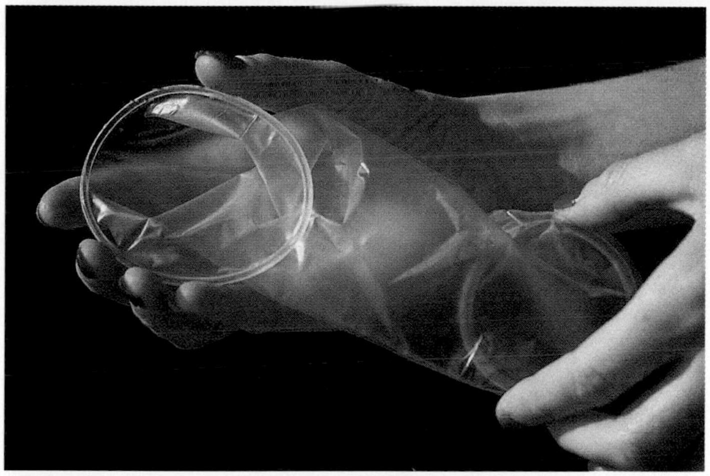

(d)

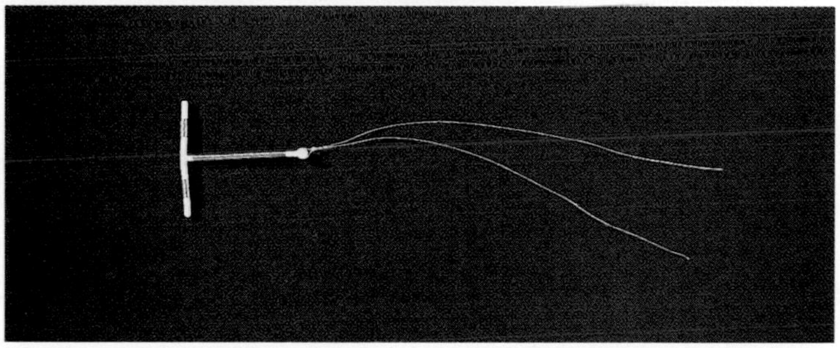

(e)

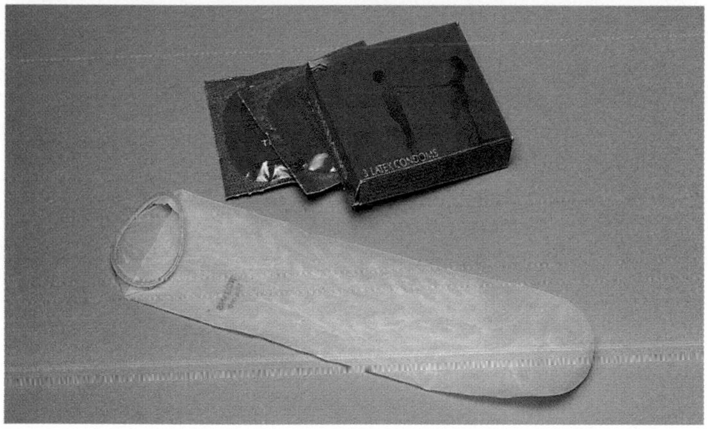

(f)

FIGURE 7.16 Some birth control methods include (*a*) daily birth control pills; (*b*) spermicidal jelly and diaphragm; (*c*) Norplant subdermal slow release hormones; (*d*) female condom; (*e*) intrauterine device (shown with applicator); (*f*) male condom.

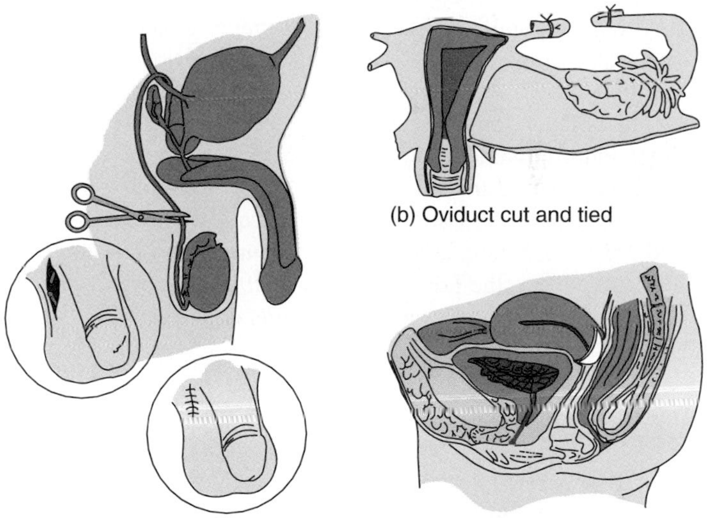

FIGURE 7.17 Surgical birth control procedures include vasectomy for men and tubal ligation for women. (*a*) Vasectomy requires only a small incision under local anesthesia to cut and tie the sperm-conducting vas deferens. (*b*) Tubal ligation in women is a more difficult operation and generally requires general anesthesia. Neither procedure affects sex life or sexual characteristics, but neither is easily reversible. (*c*) The cervical cap, or diaphragm, is an example of a mechanical barrier. It covers the entrance to the cervix and prevents sperm from entering the uterus. While not as dependable as surgical methods, mechanical barriers are easily reversible.

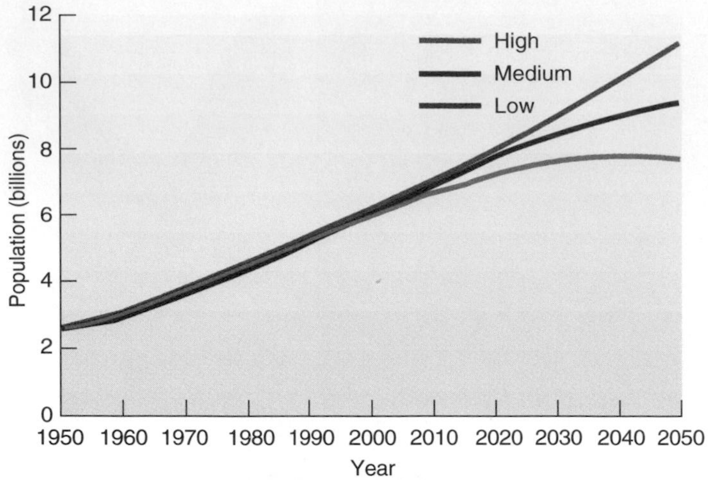

FIGURE 7.18 Population projections from the United Nations Population Division. The high projection assumes that total population will reach 11.2 billion by 2050. The medium projection assumes that total fertility rate will stabilize at 2.1 children per woman. The low projection assumes that lower-than-replacement rates of 1.6 will be reached by 2020 and that population will decline to 7.7 billion by 2050.
Source: Data from United Nations Population Division, *World Population Prospects,* 1996.

8 billion in the middle of the twenty-first century. The medium projection suggests that growth might continue to around 10 billion. The most pessimistic (high) projection shows no indication of slowing for at least 50 years.

Which of these scenarios will we follow? As you have seen in this chapter, population growth is a complex subject. To accomplish a stabilization or reduction of human populations will require substantial changes from business as usual. One impediment is that the United States Congress refuses to make payments to the United Nations Family Planning Fund because some of the 135 countries that receive UN aid include abortion as part of population control programs. Population issues were conspicuously absent from the agenda of the UN Conference on Environment and Development in Rio de Janeiro (the Earth Summit) in 1992 because of opposition from religious groups and developing countries. At a special UN Conference on Population held in Cairo, Egypt, in 1994, religion again was a complicating factor. This conference did, however, come out strongly in favor of the rights of women and children.

An encouraging sign is that worldwide contraceptive use has increased sharply in recent years. About half of the world's married couples used some family planning techniques in 1999, compared to only 10 percent 30 years earlier, but another 300 million couples say they want but do not have access to family planning. Contraceptive use varies widely by region (fig. 7.19), with high levels in Latin America and East Asia but relatively low use in much of Africa. Note the differences both in the percentage of couples using birth control in different regions and in the methods used. The IUD and male sterilization are common in Asia but rarely used in Africa or Latin America. What economic or cultural factors do you see in these patterns?

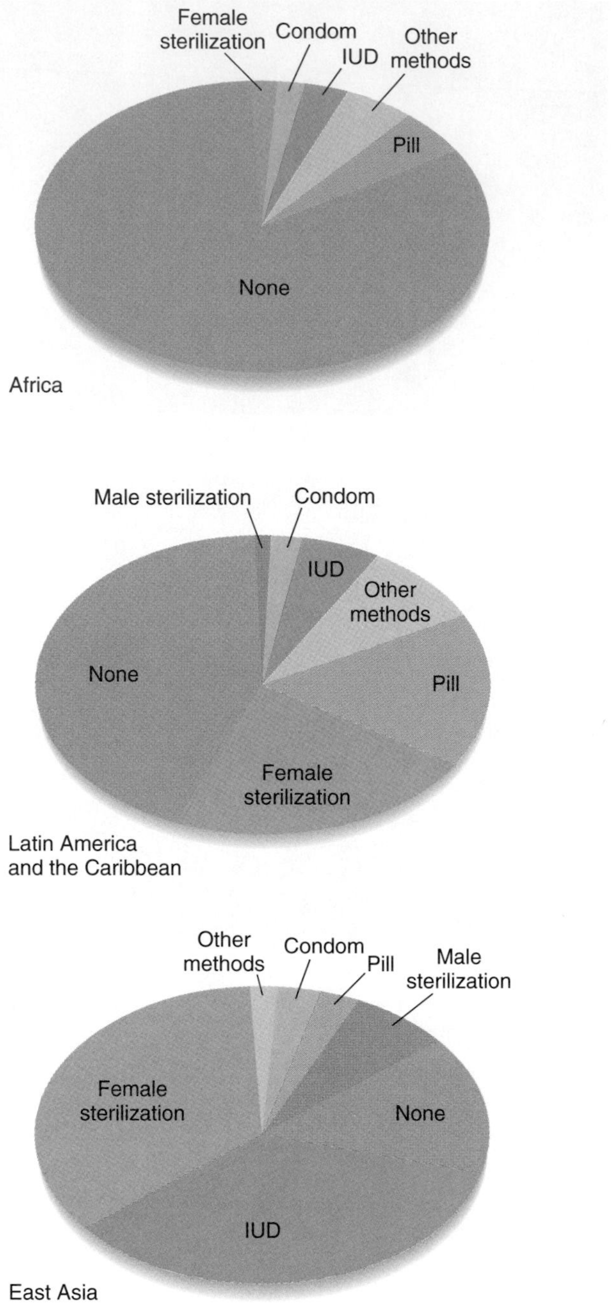

FIGURE 7.19 Variations in contraceptive use in different regions. Preferences for one form of contraception over another reflect both cultural attitudes and government programs.
Source: Data from U.N. Family Planning Agency, 1995.

The World Health Organization estimates that nearly 1 million conceptions occur daily around the world as a result of some 100 million sex acts. At least half of those conceptions are unplanned or unwanted. Still, birth rates already have begun to fall in East Asia and Latin America (fig. 7.20). Similar progress is expected in South Asia in a few years. Only Africa will probably continue to grow in the twenty-first century. Coordinated, effective family planning requires governmental action, ranging from policies to assistance.

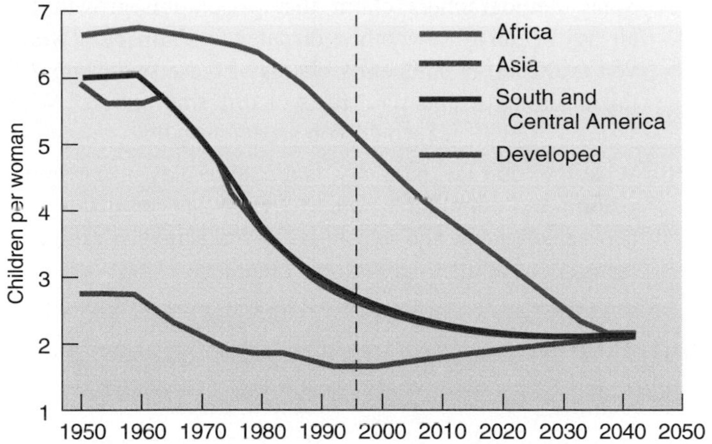

FIGURE 7.20 Fertility declines, real and projected, 1950–2050.
Source: Data from U.N. Population Division, *World Population Prospects, 1996.*

Deep societal changes are often required to make family planning programs successful. Among the most important of these are (1) improved social, educational, and economic status for women (birth control and women's rights are often interdependent); (2) improved status for children (fewer children are born as parents come to regard them as valued individuals rather than possessions); (3) acceptance of calculated choice as a valid element in life in general and in fertility in particular (belief that we have no control over our lives discourages a sense of responsibility); (4) social security and political stability that give people the means and the confidence to plan for the future; (5) knowledge, availability, and use of effective and acceptable means of birth control.

Concerted efforts to bring about these changes can be effective. Twenty years of economic development and work by voluntary family planning groups in Zimbabwe, for example, have lowered total fertility rates from 8.0 to 5.5 children per woman on average. Surveys showed that desired family sizes have fallen nearly by half (9.0 to 4.6) and that nearly all women and 80 percent of men in Zimbabwe use contraceptives. If similar progress can be sustained elsewhere, human populations may be restrained after all.

SUMMARY

Human populations have grown at an unprecedented rate over the past three centuries. By 1999, the world population stood at 6 billion people. If the current growth rate of 1.4 percent per year persists, the population will double in 50 years. Most of that growth will occur in the less-developed countries of Asia, Africa, and Latin America. There is a serious concern that the number of humans in the world and our impact on the environment will overload the life-support systems of the earth.

The crude birth rate is the number of births in a year divided by the average population. A more accurate measure of growth is the general fertility rate, which takes into account the age structure and fecundity of the population. The crude birthrate minus the crude death rate gives the rate of natural increase. When this rate reaches a level at which people are just replacing themselves, zero population growth is achieved.

In the more highly developed countries of the world, growth has slowed or even reversed in recent years so that without immigration from other areas, populations would be declining. The change from high birth and death rates that accompanies industrialization is called a demographic transition. Many developing countries have already begun this transition. Death rates have fallen, but birth rates remain high. Some demographers believe that as infant mortality drops and economic development progresses so that people in these countries can be sure of a secure future, they will complete the transition to a stable population. Others fear that excessive population growth and limited resources will catch many of the poorer countries in a demographic trap that could prevent them from ever achieving a stable population or a high standard of living.

While larger populations bring many problems, they also may be a valuable resource of energy, intelligence, and enterprise that will make it possible to overcome resource limitation problems. A social justice view argues that a more equitable distribution of wealth might reduce both excess population growth and environmental degradation.

We have many more options now for controlling fertility than were available to our ancestors. Some techniques are safer than those available earlier; many are easier and more pleasant to use. Sometimes it takes deep changes in a culture to make family planning programs successful. Among these changes are improved social, educational, and economic status for women; higher values on individual children; accepting responsibility for our own lives; social security and political stability that give people the means and confidence to plan for the future; and knowledge, availability, and use of effective and acceptable means of birth control.

QUESTIONS FOR REVIEW

1. At what point in history did the world population pass its *first* billion? What factors restricted population before that time, and what factors contributed to growth after that point?
2. How might growing populations be beneficial in solving development problems?
3. Why do some economists consider human resources more important than natural resources in determining the future of a country?
4. Where will most population growth occur in the next century? What conditions contribute to rapid population growth in some countries?
5. Define *crude birth rate, total fertility rate, crude death rate,* and *zero population growth.*
6. What is the difference between life expectancy and longevity?
7. What is dependency ratio, and how might it affect the United States in the future?
8. What pressures or interests make people want or not want to have babies?
9. Describe the conditions that lead to a demographic transition.
10. Describe the major choices in modern birth control.

QUESTIONS FOR CRITICAL THINKING

1. What do you think is the optimum human population? The maximum human population? Are the numbers different? If so, why?
2. Some people argue that technology can provide solutions for environmental problems; others believe that a "technological fix" will make our problems worse. What personal experiences or worldviews do you think might underlie these positions?
3. Karl Marx called Thomas Malthus a "shameless sycophant of the ruling classes." Why would the landed gentry of the eighteenth century be concerned about population growth of the lower classes? Are there comparable class struggles today?
4. Try to imagine yourself in the position of a person your age in a Third World country. What family planning choices and pressures would you face? How would you choose among your options?
5. Some demographers claim that population growth has already begun to slow; others dispute this claim. How would you evaluate the competing claims of these two camps? Is this an issue of uncertain facts or differing beliefs? What sources of evidence would you accept as valid?
6. What role do race, ethnicity, and culture play in our immigration and population policies? How can we distinguish between prejudice and selfishness on one hand and valid concerns about limits to growth on the other?

KEY TERMS

birth control 159
crude birth rate 148
crude death rates 150
demographic transition 156
demography 147
dependency ratio 151
family planning 158
life expectancy 150
natural increase 150
neo-Malthusians 145
pronatalist pressures 153
social justice 157
total fertility rate 148
total growth rate 150
zero population growth (ZPG) 149

ADDITIONAL INFORMATION ON THE INTERNET

Visit our website at http://www.mhhe.com/environmental science/ for specific information about each of the topics below:

World Demographic Data, Data by country: US Census Bureau
Population Clock, comparing current and historic numbers
Population Research Institute
Animated Demography demo
Family Planning
Population Council
International Planned Parenthood
Office of Population Research, Princeton University
Beyond Malthus: essay from Worldwatch Institute
Paul Ehrlich biography
Africa--Population Programs and Activities

CHAPTER 8

Ecological Economics

Gentlemen, there is no more money. We shall have to start using our brains.

—Ernest Rutherford—

OBJECTIVES

After studying this chapter, you should be able to:

- explain the difference between neoclassical and ecological economics and how each discipline views ecological processes and natural resources.
- distinguish between different types and categories of resources.
- understand how resource supply and demand affect price and technological progress.
- develop a position on limits to growth and economic carrying capacity of our environment.
- discuss internal and external costs, market approaches to pollution control, and cost-benefit analysis.
- define GNP and explain some alternative ways to measure values of natural resources and real social progress.
- analyze the role of business and some possible strategies for achieving future sustainability.

ESSENTIAL STUDY PARTNER

The Essential Study Partner (ESP) CD-ROM that accompanies this textbook contains additional information on the following topics:

- classical, neoclassical, natural resource, and ecological economics: populations: economics: worldviews
- nonrenewable, renewable, and proven resources: populations: economics: resources
- privatization and cooperative management of resources: populations: economics: resources
- resource sustainability, technology, and conservation: populations: economics: sustainability
- environmental costs and the Index of Sustainable Economic Welfare (ISW): populations: economics: sustainability
- cost-benefit analysis: populations: economics: sustainability

Above: Biosphere 2.

Creating Another Earth

In September 1993, eight tired, hungry people emerged from two years of isolation in a high-tech megaterrarium in the desert near Oracle, Arizona. Designed as an ecotechnological model for space exploration and colonization, this bioengineered facility was intended to grow food, cleanse the air, and recirculate and purify water for its eight inhabitants without exchange of materials (including atmospheric gases) with the outside world. Called Biosphere 2 (Biosphere 1 being the earth), the 13,000 square meter (3.15 acres) complex consisted of living quarters and greenhouses containing food crops and small, but photogenic, areas representing deserts, rainforests, savannas, and an ocean.

Almost from its beginning, Biosphere 2 experienced controversy and problems. By the end of the first year of their mission, the Biospherians reported deteriorating air and water quality. Oxygen concentrations in the air had fallen from 21 percent to 14 percent. This is equivalent to oxygen levels at an elevation of 5300 m (17,500 ft) and was barely sufficient to keep the occupants alive and functioning. At the same time, carbon dioxide concentrations were undergoing large daily and seasonal variations and nitrous oxide (N_2O) in the air had reached mind-numbing levels. In January 1993, fresh air was pumped in to replenish the dome's atmosphere and rescue the inhabitants.

Subsequent investigations showed that the missing oxygen was being consumed by microbes in the excessively rich soil in which food crops were being grown. At the same time, fresh concrete used in construction was absorbing carbon dioxide released by microbial metabolism. If this carbon dioxide sink hadn't been available, the air would have become unbreathable long before it did. Water systems also became polluted with excess nutrients, degrading aquatic habitats and contaminating drinking water supplies. Species losses were much higher than originally anticipated. Of the 25 introduced vertebrate species, for example, 18 became extinct. All insect pollinators also died, so that most plants were unable to produce seeds and food supplies dwindled to alarming levels. Weedy vines, particularly morning glories (*Ipomoea hederacea*), flourished in the carbon-rich atmosphere and threatened to choke out more desirable plants. Although the majority of the insects disappeared, ants, cockroaches, and katydids thrived and overran everything. Biosphere 2 cost nearly $200 million to build, with an additional cost of about $1 million per year for fossil fuel energy to keep all the systems running. That averages out to about $25 million per human occupant for the two-year experiment. How much more would it cost to produce a truly balanced ecological system that could maintain life indefinitely? Although the experiment failed to meet its original objectives, it illustrates the value of ecological services provided by the natural world that most of us take for granted. What will we have lost and what might it cost to replace resources and services on which we now depend if we deplete or destroy the natural world? In this chapter, we will examine how ecological economists assign prices and values to natural resources and ecological services.

ORIGINS OF OUR ECONOMIC WORLDVIEWS

Economy is the management of resources to meet household and community needs. It implies efficient and prudent use of time, effort, and material goods. Economists ask "what is this worth," and "how much does it cost?" Interestingly, ecology and economy are derived from the same root words and concerns. *Oikos* (ecos) is the Greek word for household. Economics is the *nomos,* or counting, of the household goods and services. Ecology is the *logos,* or logic, of how the household works. In both disciplines, the household is extended to include the whole world. Economic analysis helps us assess the relative values and choose between competing options. The values we assign, however, and the options we consider reasonable, depend very much on our worldview. In this section, we'll briefly survey some important economic worldviews.

CLASSICAL ECONOMICS

Classical economics originally was a branch of moral philosophy concerned with how individual interest and values intersect with larger social goals. The founder of modern western economics, Adam Smith (1723–1790), was an ethicist concerned with individual freedom of choice. Smith's landmark book *Inquiry into the Nature and Causes of the Wealth of Nations,* published in 1776, argued that

> Every individual endeavors to employ his capital so that its produce may be of the greatest value. He generally neither intends to promote the public interest, nor knows how much he is promoting it. He intends only his own security, only his own gain. And he is in this led by an invisible hand to promote an end which was no part of his intention. By pursuing his own interests he frequently promotes that of society more effectually than when he really intends to.

FIGURE 8.1 Under communism, the government set the price, quantity, and quality of teapots in China. Now market prices determine supply and demand.

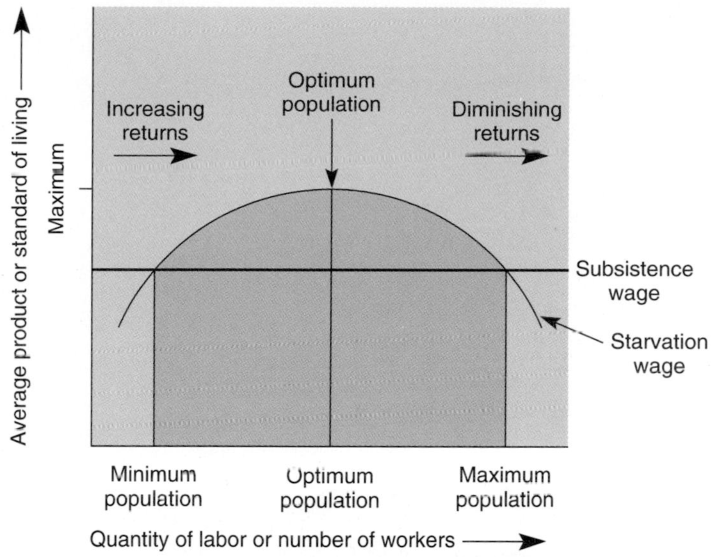

FIGURE 8.2 A hypothetical graph of diminishing returns. Increasing labor input raises productivity and wages until an optimum population is reached. Continued population growth creates a labor surplus that reduces wages to subsistence levels where workers starve or find something else to do.

This statement often is taken as justification for the capitalist system, in which market competition between willing sellers and buyers is believed to bring about the greatest efficiency of resource use, the optimum balance between price and quality, and to be critical for preserving individual liberty. As we will discuss later in this chapter, however, *laissez faire* market systems rarely incorporate factors such as social or environmental costs.

Much of classical economic theory dealt with the effects of resource scarcity. Thomas Malthus' views on population (see chapter 7), for example, gave economics a reputation as the "dismal science." In his theory of **diminishing returns,** Malthus argued that rising wages and improved well-being would lead to excess reproduction among the working class. A labor surplus would then cause wages to fall below subsistence levels, resulting in starvation, disease, and crime (fig. 8.2). Eventually, he thought, wages would equilibrate at the lowest possible survival level. In Malthus' view, land for food production was the limiting factor in both population growth and economic development.

David Ricardo (1772–1823), an important contemporary of Malthus, introduced a better understanding of the relation between supply and demand in economics. **Demand** is the amount of a product or service that consumers are willing and able to buy at various possible prices, assuming they are free to express their preferences. **Supply** is the quantity of that product being offered for sale at various prices, other things being equal. The inverse relationship between supply and demand is shown in figure 8.3. As prices rise, the supply increases and demand falls. The reverse holds true as price decreases. The difference between the cost of production and the price buyers are willing to pay, Ricardo called rent. Today we call it profit.

In a free market of independent and intelligent buyers and sellers, supply and demand should come into a **market equilibrium,** represented in figure 8.3 by the intersection of the two curves. In real life, prices are not determined strictly by total supply and demand as much as what economists called **marginal costs.** Sellers ask themselves, "What would it cost to produce one more unit of this product or service? Suppose I add one more worker or buy an extra supply of raw materials, how much profit could I make?" Buyers ask themselves similar questions, "How much would I benefit and what would it cost if I bought one more widget?" If both buyer and seller find the marginal costs and benefits attractive, a sale is made.

There are exceptions, however, to this theory of supply and demand. Consumers will buy some things regardless of cost. Raising the price of cigarettes, for instance, doesn't necessarily reduce demand. We call this **price elasticity.** Other items have price inelasticity; that is, they follow supply/demand curves exactly. When price goes up, demand falls and vice versa.

John Stuart Mill (1806–1873), another important classical philosopher/economist, believed that perpetual growth in material well-being is neither possible nor desirable. Economies naturally mature to a steady state, he believed, leaving people free to pursue nonmaterialistic goals. He didn't regard this equilibrium state to

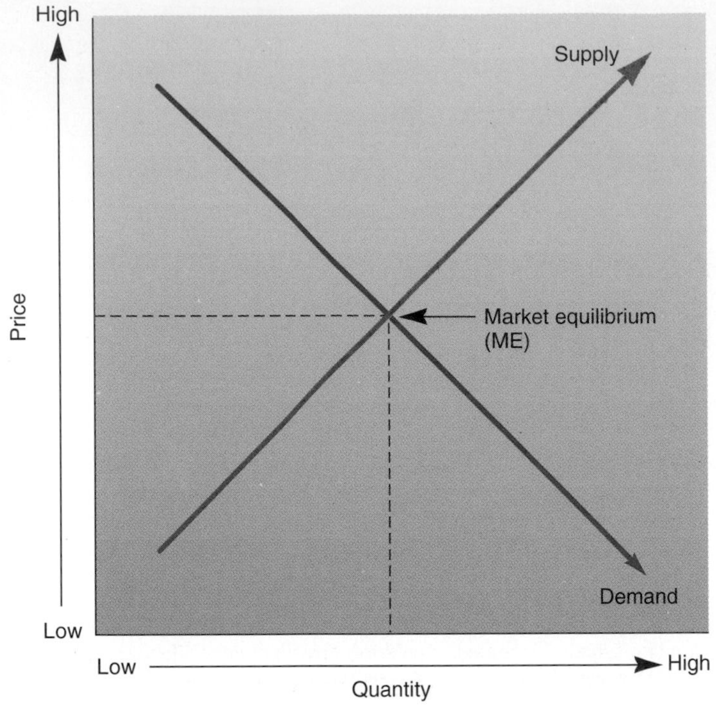

FIGURE 8.3 Classic supply/demand curves. When price is low, supply is low and demand is high. As prices rise, supply increases but demand falls. Market equilibrium is the price at which supply and demand are equal.

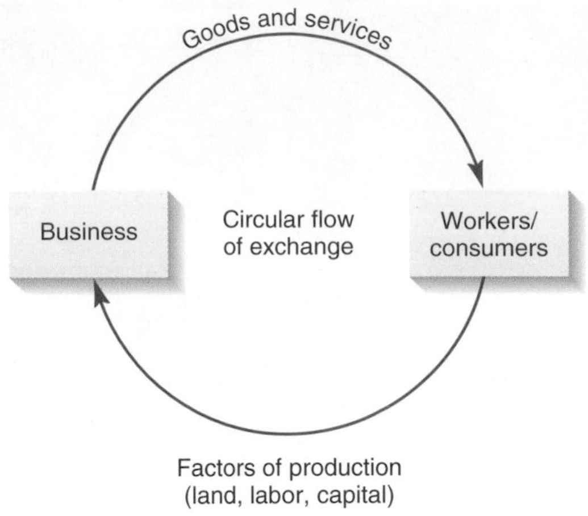

FIGURE 8.4 The neoclassical model of the economy focuses on the flow of goods, services, and factors of production (land, labor, capital) between business and individual workers/consumers. The social and environmental consequences of these relationships are irrelevant in this view.

be necessarily one of misery and want, however. In *Principles of Political Economy,* he wrote, "It is scarcely necessary to remark that a stationary condition of capital and population implies no stationary state of human improvement. There would be as much scope as ever for all kinds of mental culture, and moral and social progress; as much room for improving the art of living, and much more likelihood of its being improved when minds cease to be engrossed by the art of getting on." This view has much in common with the concepts of a steady-state economy and sustainable development that we will examine later in this chapter.

Neoclassical Economics

Toward the end of the nineteenth century, the field of economics divided into two broad camps. **Political economy** continued the tradition of moral philosophy and concerned itself with social structures, value systems, and relationships among the classes. This group included reformers such as Karl Marx and E. F. Schumacher, along with many socialists, anarchists, and utopians. The other camp, called **neoclassical economics,** adapted principles of modern science to economic analysis. This approach strives to be mathematically rigorous, noncontextual, abstract, and predictive. Neoclassical economists claim to be objective and value free, leaving social concerns to other disciplines. Like their classical predecessors, they retain an emphasis on scarcity and the interaction of supply and demand in determining prices and resource allocation (fig. 8.4).

Continued economic growth is considered to be both necessary and desirable in the neoclassical worldview. Growth is seen as the only way to maintain full employment and avoid class conflict arising from inequitable distribution of wealth. Natural resources are viewed as merely a factor of production rather than a critical supply of materials, services, and waste sinks by neoclassical economics. Because factors of production are thought to be interchangeable and substitutable, materials and services provided by the environment are not considered indispensable. As one resource becomes scarce, neoclassical economists believe, substitutes will be found.

Natural resource economics shares much in common with neoclassical attitudes but expands the value of nature as an important source of raw materials and a sink for wastes (fig. 8.5). Geological and ecological processes also recycle materials over long or short timescales, but these processes are considered to be external to the economic system. Because the environment and natural resources are seen as relatively abundant compared to "built" or human-made capital, the Riccardian rent for labor, technology, and money are expected to be higher than that for nature.

Ecological Economics

A relatively new discipline called **ecological economics** brings the insights of ecology to economic analysis. In contrast to previous approaches, this is a transdisciplinary field that intends to be holistic, contextual, value-sensitive, and ecocentric. It incorporates the principles of thermodynamics and coevolution into its worldview. A principle concern of ecological economics is the equitable distribution of resources and rights within current generations and between this generation and future ones. It also is mindful of rights of other species. An important difference between this new field and its predecessors is a recognition of the world as an open, dynamic system within which the human-economy is inextricably embedded (fig. 8.6).

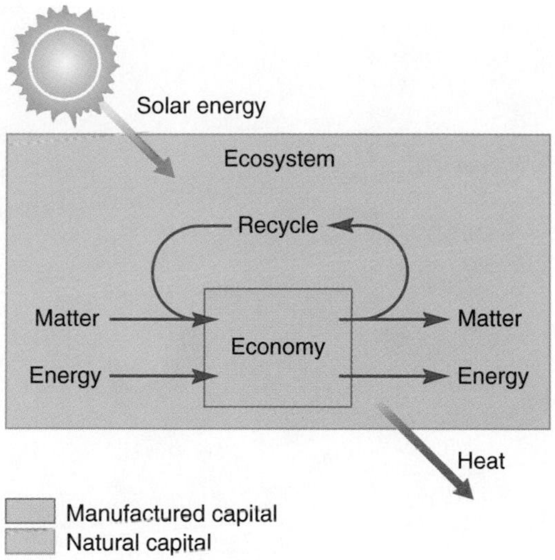

FIGURE 8.5 In natural resource economics, the economy is seen as an open subset of the ecosystem. Matter and energy are consumed in the form of raw resources and rejected as waste. Material is recycled by ecological processes while heat is ejected back to space. Manufactured or "built" capital is regarded as scarce—and therefore valuable—while natural capital is regarded as plentiful and, therefore, cheap.
Source: T. Prugh, *National Capital and Human Economic Survival* 1995, ISEE.

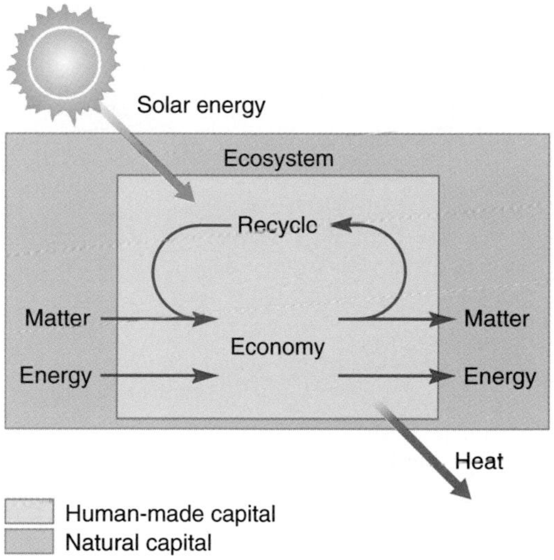

FIGURE 8.6 An ecological economic view reverses ideas of scarcity. Because material recycling is considered to be integral to the economy, human-made or built capital is seen to be large compared to limited supplies of natural capital.
Source: Herman Daly in A.M. Jansson, et al., *Investing in Natural Capital,* ISEE.

Ecological economics acknowledges our dependence of the essential life-support services provided by nature. Rather than regard natural and manufactured resources as completely interchangeable, ecological economists regard some aspects of nature as irreplaceable and essential. This places limits on human population growth and resource consumption and raises questions about the carrying capacity of our environment. Seeing nature as limited and relatively fragile compared to the human economy reverses our appraisal of which resources are scarce and, therefore, valuable. Ecological economists regard natural resources as important to our well-being and as worthy of protection and development as are human or technological resources.

The **steady-state economy** advocated by ecological economists is characterized by low human birth and death rates, use of renewable energy sources, material recycling, and emphasis on durability, efficiency, and stability rather than high throughput of materials. Building on the ideas of John Stuart Mill, some pioneers of this concept include the Rumanian economist Nicholas Georgescu-Rogen and American economists Kenneth Boulding and Herman E. Daly, all of whom have explored concepts of sustainability and limits to growth. Much of the rest of this chapter will be devoted to exploring the differences between neoclassical and ecological economics and a discussion of limits to growth.

RESOURCES, CAPITAL, AND RESERVES

A fundamental question in all economic analysis is whether one option or state of the world is better than another. Economics deals with questions of resource allocation or trade-offs either at the "micro" scale of individual decisions or at the "macro" scale of national policies and world trade. How can we determine the available amount and value of a specific resource, given a particular economic system and technology? In this section, we will look at some ways to categorize capital, resources, and reserves.

Resource Types

Capital is any form of wealth available for use in the production of more wealth. Economists distinguish between *natural capital* (goods and services provided by nature), *human or cultural capital* (knowledge, experience, and human enterprise), and *manufactured or built capital* (tools, infrastructure, and technology). Social scientists would add *social capital* (shared values, trust, cooperative spirit, and community organization) to this list. A **resource** is anything with potential use in creating wealth or giving satisfaction. Natural resources can be either renewable or nonrenewable, as well as tangible or intangible. Although we generally define these terms from a human perspective, remember that many resources we desire are important to other species as well.

In general, **nonrenewable resources** are the earth's geological endowment: the minerals, fossil fuels, and other materials present in fixed amounts in the environment (fig. 8.7). Although many of these resources are renewed or recycled by geological or ecological processes, the timescales to do so are so long by human standards that the resource will be gone once present supplies are exhausted. Predictions abound that we are in imminent danger of running out of one or another of these exhaustible resources. The actual available supplies of many commodities—such as metals—can be effectively extended, however, by more efficient use, recycling, substitution of one material for another, or better extraction from dilute or remote sources.

FIGURE 8.7 The geological resources of the earth's crust, such as oil from this well, are nonrenewable. Often the limit to our use of these resources is not so much the absolute amount available but the energy cost required to extract the resources and the environmental consequences of doing so.

FIGURE 8.8 Biological resources are renewable in that they replace themselves by reproduction, but if overused or misused, populations die. When a whole species is lost, it cannot be re-created. It is permanently lost as a component of its ecosystem and as a resource to humans. These northern fur seals were almost totally exterminated by overhunting in the nineteenth century.

Renewable resources are things that can be replenished or replaced. They include sunlight—our ultimate energy source—and the biological organisms and biogeochemical cycles that provide essential ecological services (fig. 8.8 and table 8.1). This is discussed later in this chapter under nonmarket values. Because biological organisms and ecological processes are self-renewing, we often can harvest surplus organisms or take advantage of ecological services without diminishing future availability if we do so carefully. Unfortunately, our stewardship of these resources often is less than ideal. Even once vast biological populations such as passenger pigeons, American bison (buffalo), and Atlantic cod, for instance, were exhausted by overharvesting in only a few years (see chapter 13). Similarly, we appear to be upsetting climatic systems with potentially disastrous results (see chapter 17). To our dismay, we have found that such mismanagement of many renewable resources may make them more ephemeral and limited than fixed geological resources.

Abstract or **intangible resources** include open space, beauty, serenity, wisdom, diversity, and satisfaction (fig. 8.9). Strangely, these resources can be both infinite *and* exhaustible. There is no upper limit to the amount of beauty, knowledge, or compassion that can exist in the world, yet they can be easily destroyed. A single piece of trash can ruin a beautiful vista or a single cruel remark can spoil an otherwise perfect day. On the other hand, unlike tangible resources that usually are reduced by use or sharing, intangible resources often are increased by use and multiplied by being shared. Nonmaterial assets can be important economically. Information management and tourism—both based on intangible resources—have become two of the largest and most powerful industries in the world.

TABLE 8.1 Important Ecological Services

1. *Regulate* global energy balance; chemical composition of the atmosphere and oceans; local and global climate; water catchment and groundwater recharge; production, storage, and recycling of organic and inorganic materials; maintenance of biological diversity.
2. *Provide* space and suitable substrates for human habitation, crop cultivation, energy conversion, recreation, and nature protection.
3. *Produce* oxygen, fresh water, food, medicine, fuel, fodder, fertilizer, building materials, and industrial inputs.
4. *Supply* aesthetic, spiritual, historic, cultural, artistic, scientific, and educational opportunities and information.

Source: R. S. de Groot, *Investing in Natural Capital,* 1994.

Economic Resource Categories

Although we have defined a resource as anything useful, we should distinguish between economic usefulness and total abundance or a material or service. Vast supplies of potentially important materials are present in the earth's crust, for instance, but they are useful to us only if they can be recovered in reasonable amounts with available technology and with acceptable environmental and social costs. Within the aggregate total of any natural

FIGURE 8.9 Many of the intangible resources of nature enrich the quality of our lives. Our enjoyment of these resources may be enhanced by sharing them with others. On the other hand, too many other people, or people engaged in conflicting activities, can destroy the very values we seek.

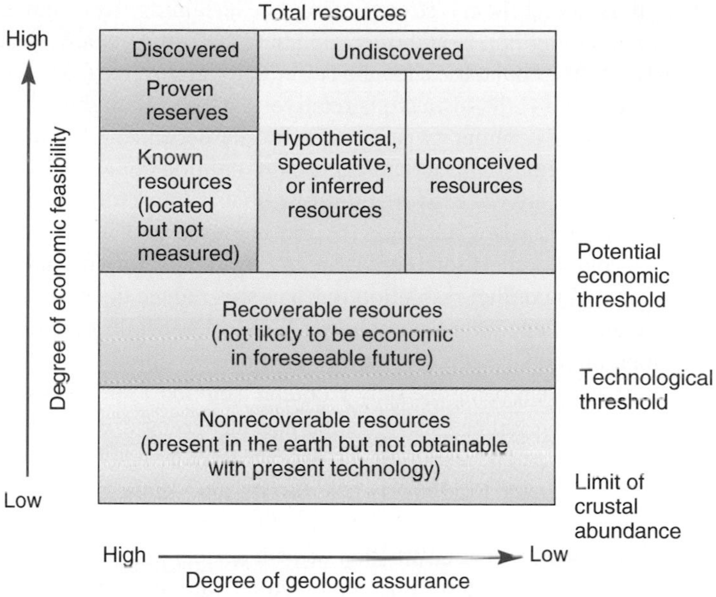

FIGURE 8.10 Categories of natural resources. These categories, based on degree of geological assurance and economic feasibility of recovery, were intended to describe mineral resources. With some modification, these categories can be applied to many other nonrenewable resources.

resource, we can distinguish categories on the basis of economic and technological feasibility, as well as the resource location and quantity (fig. 8.10).

You've probably read about fossil fuel proven reserves or recoverable resources. What do these terms mean? **Proven reserves** of a resource are those that have been thoroughly mapped and are economical to recover at current prices with available technology. **Known resources** are those that have been located but are not completely mapped. Their recovery or development may not be economical now, but they are likely to become economical in the foreseeable future. **Undiscovered resources** are only speculative or inferred from similarities with known deposits. There also may be unconceived resources—sometimes called "unknown-unknowns"—that we don't even know where to look for.

Recoverable resources are accessible with current technology but are not necessarily economical. Unrecoverable resources are so diffuse or remote that they are not likely ever to be technologically accessible. There can be vast differences between these categories. For instance, the U.S. Geological Survey estimates that only 0.01 percent of the minerals in the upper one kilometer of the earth's crust will ever be economically recoverable. That means that ten thousand times as much is not economically available and will probably never be technically feasible to recover. Chapters 16 and 21 give some practical examples of proven, known, and ultimately recoverable resources.

Communal Property Resources

In 1968, biologist Garret Hardin wrote a widely quoted article entitled "The **Tragedy of the Commons**" in which he argued that any commonly held resource inevitably is degraded or destroyed because the narrow self-interests of individuals tend to outweigh public interests. Hardin offered as a metaphor the common woodlands and pastures held by most colonial New England villages. In deciding how many cattle to put on the commons, Hardin theorized, each villager would attempt to maximize his or her own personal gain. Adding one more cow to the commons could mean a substantially increased income for an individual farmer. The damage done by overgrazing, however, would be shared among all the farmers. According to Hardin, the only solution would be either to give coercive power to the government or to privatize the resource.

Hardin intended this dilemma, known in economics as the "free-rider" problem, to warn about human overpopulation and resource availability. Other authors have used his metaphor to explain such diverse problems as African famines, air pollution, fisheries declines, and urban crime. What Hardin really was describing, however, was an **open access system** in which there are no rules to manage resource use. In fact, many communal resources have been successfully managed for centuries by cooperative arrangements among users. Some examples include Native American management of wild rice beds and hunting grounds; Swiss village-owned mountain forests and pastures; Maine lobster fisheries; communal irrigation systems in Spain, Bali, and Laos; and nearshore fisheries almost everywhere in the world. A large body of literature in economics and social sciences describes how these cooperative systems work. Among the features shared by **communal resource management systems** are: (1) Community members have lived on the land or used the resource for a long time and anticipate that their children and grandchildren will as well, thus giving them a strong interest in sustaining the resource and maintaining bonds with their neighbors; (2) the resource has clearly defined boundaries; (3) the community group size is known and enforced; (4) the resource is relatively scarce and highly variable so that the community is forced to be interdependent; (5) management strategies appropriate for local conditions have evolved over time and are collectively enforced; that is, those affected by the rules have a say in them; (6) the resource and its use are actively monitored, discouraging anyone from cheating or taking too much; (7) conflict resolution mechanisms reduce discord; and (8) incentives encourage compliance with rules, while sanctions for noncompliance keep community members in line.

Rather than being the only workable solution to problems in common pool resources, privatization and increasing external controls often prove to be disastrous. In places where small villages have owned and operated local jointly held forests and fishing grounds for generations, nationalization and commodification of resources generally have led to rapid destruction of both society and ecosystems. Where communal systems once enforced restraint over harvesting, privatization encouraged narrow self-interest and allowed outsiders to take advantage of the weakest members of the community.

A tragic example is the forced privatization of Indian reservations in the United States. Failing to recognize or value local knowledge and forcing local people to participate in a market economy allowed outsiders to disenfranchise native people and resulted in disastrous resource exploitation. Learning to distinguish between open access systems and communal property regimes is important in understanding how best to manage natural resources.

POPULATION, TECHNOLOGY, AND SCARCITY

Are we about to run out of essential natural resources? It stands to reason that if we consume a fixed supply of nonrenewable resources at a constant rate, eventually we'll use up all the economically recoverable reserves. There are many warnings in the environmental literature that our extravagant depletion of nonrenewable resources sooner or later will result in catastrophe. The dismal prospect of Malthusian diminishing returns and a life of misery, starvation, and social decay inspire many environmentalists to call for an immediate change to voluntary simplicity and lower consumption rates. Models for exploitation rates of nonrenewable resources—called Hubbert Curves, after Stanley Hubbert who developed them—often closely match historic experience for natural resource depletion (fig. 8.11).

Many economists, however, contend that neither supply/demand relationships nor economically recoverable reserves are rigidly fixed. Human ingenuity and enterprise often allow us to respond to scarcity in ways that postpone or alleviate the dire effects predicted by modern Jeremiahs. In the next section we will look at some of the arguments for and against limits to growth of the global economy.

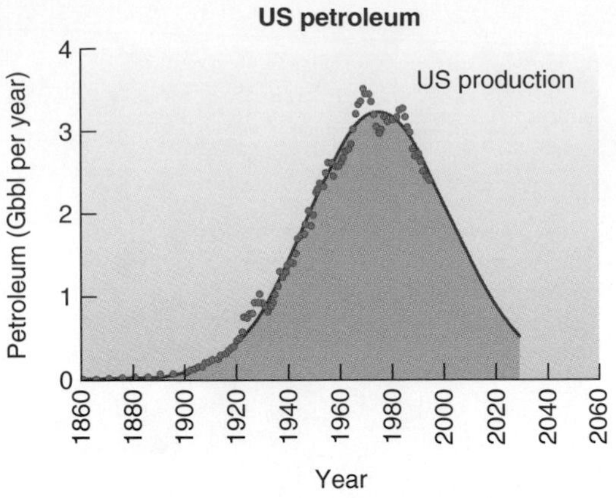

FIGURE 8.11 United States petroleum production 1860 to 1995. Dots indicate actual production. The bell-shaped curve is a theoretical Hubbert Curve for a nonrenewable resource. The shaded area under the curve, representing 220 Gbbl (Gbbl = Gigabarrels or billions of standard 42 gallon barrels), is an estimate of the total economically recoverable resource.

Market Efficiencies and Technological Development

In a pioneer or frontier economy, methods for harvesting resources and turning them into useful goods and services tend to be inefficient and wasteful. This may not matter, however, if the supply of resources exceeds the demand for them. The loggers, for example, who swarmed across the Great Lakes Forest at the beginning of the twentieth century, wasted a vast amount of wood. Their inefficiency didn't seem important, however, because the supply of trees appeared infinite, while labor, capital, and means for getting lumber to markets was scarce. As markets and societies develop, however, better technology and more efficient systems allow people to create the same amount of goods and services using far fewer resources. We now produce hundreds of times the crop yield with less labor from the same land that pioneers farmed.

This increasing technological efficiency can dramatically shift supply and demand relationships (fig. 8.12). As technology makes goods and services cheaper to produce, the quantity available at a given price can increase greatly. The market equilibrium or the point at which supply and demand equilibrate will shift to lower prices and higher quantities as a market matures.

Scarcity can actually serve as a catalyst for innovation and change (fig. 8.13). As materials become more expensive and difficult to obtain, it becomes cost-effective to discover new supplies or to use available ones more efficiently. The net effect is as if a new supply of resources has been created or discovered. Several important factors play a role in this cycle of technological development:

- Technical inventions can increase efficiency of extraction, processing, use, and recovery of materials.
- Substitution of new materials or commodities for scarce ones can extend existing supplies or create new ones. For instance,

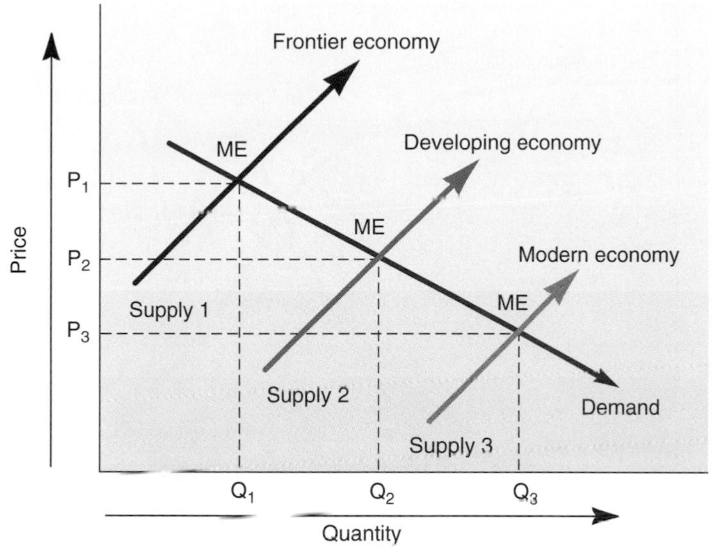

FIGURE 8.12 Supply and demand curves at three different stages of economic development. At each stage there is a market equilibrium point at which supply and demand are in balance. As the economy becomes more efficient, the equilibrium shifts so there is a larger quantity available at a lower price than before. (P = price, Q = quantity, ME = market equilibrium)

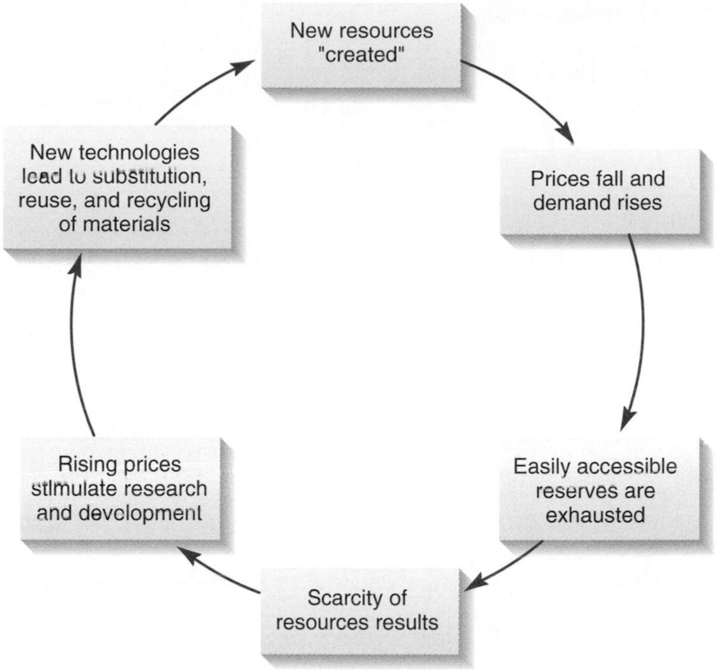

FIGURE 8.13 Scarcity/development cycle. Paradoxically, resource use and depletion of reserves can stimulate research and development, the substitution of new materials, and the effective creation of new resources.

substitution of aluminum for copper, concrete for structural steel, grain for meat, and synthetic fibers for natural ones all remove certain limits to growth.

- Trade makes remote supplies of resources available and may also bring unintended benefits in information exchange and cultural awakening.
- Discovery of new reserves through better exploration techniques, more investment, and looking in new areas becomes rewarding as supplies become limited and prices rise.
- Recycling becomes feasible and accepted as resources become more valuable. Recycling now provides about 37 percent of the iron and lead, 20 percent of the copper, 10 percent of the aluminum, and 60 percent of the antimony that is consumed each year in the United States.

Increasing Environmental Carrying Capacity

Despite repeated warnings that rapidly growing populations and increasing affluence are bound to exhaust natural resources and result in rapid price increases, technological developments of the sort described earlier have resulted in price decreases for most raw materials over the last hundred years. Consider copper for example. Twenty years ago worries about impending shortages led the United States to buy copper and store it in strategic stockpiles. Estimates of the amount of this important metal needed for electric motors, telephone lines, transoceanic cables, and other uses essential for industrialized society far exceeded known reserves. It looked as if severe shortages and astronomical price increases were inevitable. But then aluminum power lines, satellites, fiber optics, integrated circuits, microwave transmission, and other inventions greatly diminished the need for copper. Although prices are highly variable because of world politics and trade policies, the general trend for most materials has been downward in this century (fig. 8.14). It is as if the carrying capacity of our natural resources—at least in terms of copper—has been increased.

Economists generally believe that this pattern of substitutability and technological development is likely to continue into the future. Ecologists generally disagree. There are bound to be limits, they argue, to how many people our environment can support. An interesting example of this debate occurred in 1980. Ecologist Paul Ehrlich bet economist Julian Simon that increasing human populations and growing levels of material consumption would inevitably lead to price increases for natural resources. They chose a package of five metals—chrome, copper, nickel, tin, and tungsten—priced at the time at $1000. If, in ten years, the combined price (corrected for inflation) was higher than $1000, Simon would pay the difference. If the combined price had fallen, Ehrlich would pay. In 1990 Ehrlich sent Simon a check for $576.07; the price for these five metals had fallen 47.6 percent.

Does this prove that resource abundance will continue indefinitely? Hardly. Ehrlich claims that the timing and set of commodities chosen simply were the wrong ones. The fact that we haven't yet run out of raw materials doesn't mean that it will never happen. Many ecological economists now believe that some nonmarket resources such as ecological processes may be

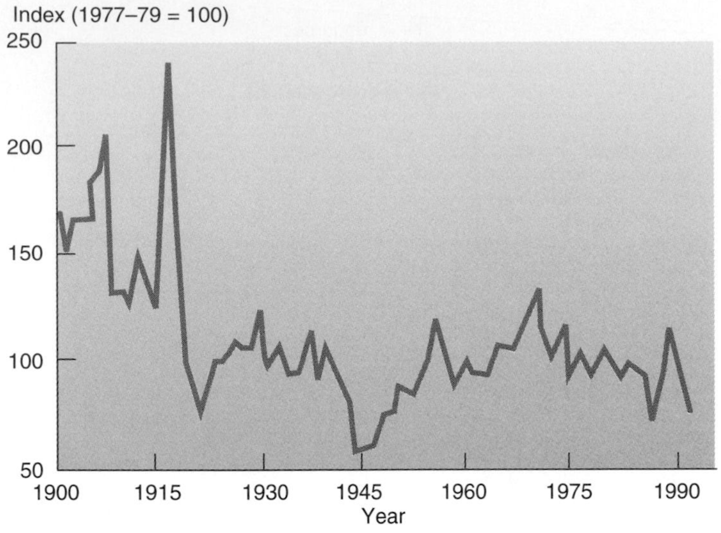

FIGURE 8.14 Long-run world prices for nonferrous metals (aluminum, copper, tin, and zinc) in the twentieth century, adjusted for inflation. Real prices fell rapidly in the beginning of the century due to technological advances and then remained relatively constant except for wars and temporary shortages or surpluses. There is no evidence that exhaustion of limited supplies is driving up prices.
Source: World Bank data.

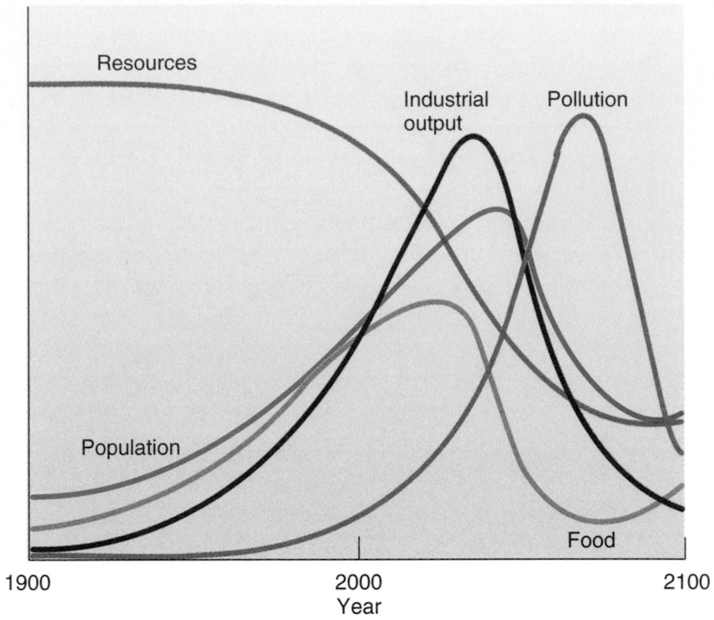

FIGURE 8.15 A run of one of the world models in *Limits to Growth.* This model assumes business-as-usual for as long as possible until Malthusian limits cause industrial society to crash. Notice that pollution continues to increase well after industrial output, food supplies, and population have all plummeted.

more irreplaceable than tangible commodities like metals. What do you think? Are we approaching the limits of nature to support more humans and more consumption? Which resources, if any, do you think are most likely to be limiting in the future?

Economic Models

In the early 1970s, an influential study of resource limitations was funded by the Club of Rome, an organization of wealthy business owners and influential politicians. The study was carried out by a team of scientists from the Massachusetts Institute of Technology headed by Donnela Meadows. The results of this study were published in the 1972 book *Limits to Growth.* A complex computer model of the world economy was used to examine various scenarios of different resource depletion rates, growing population, pollution, and industrial output. Given the Malthusian assumptions built into this model, catastrophic social and environmental collapse seemed inescapable.

Figure 8.15 shows one example of the world model. Food supplies and industrial output rise as population grows and resources are consumed. Once past the carrying capacity of the environment, however, a crash occurs as population, food production, and industrial output all decline precipitously. Pollution continues to grow as society decays and people die, but, eventually, it also falls. Notice the similarity between this set of curves and the "boom and bust" population cycles described in chapter 6.

Many economists criticized these results because they discount technological development and factors that might mitigate the effects of scarcity. In 1992, the Meadows group published updated computer models in *Beyond the Limits* that include technological progress, pollution abatement, population stabilization, and new public policies that work for a sustainable future. If we adopt these changes sooner rather than later, the computer shows an outcome like that in figure 8.16, in which all factors stabilize sometime in this century at an improved standard of living for everyone. Of course neither of these computer models shows what *will* happen, only what some possible outcomes *might* be, depending on the choices we make.

Why Not Conserve Resources?

Even if large supplies of resources are available or the technological advances to mitigate scarcity exist, wouldn't it be better to reduce our use of natural resources so they will last as long as possible? Will anything be lost if we're frugal now and leave more to be used by future generations? Many economists would argue that resources are merely means to an end rather than an end in themselves. They have to be used to have value.

If you bury your money in a jar in the backyard, it will last a long time but may not be worth much when you dig it up. If you invest it productively, you may have much more in the future than you do now. Furthermore, a window of opportunity for investment may be open now but not later. Suppose that 300 years ago our ancestors had decided that the Industrial Revolution should not be allowed to continue because it required too much resource

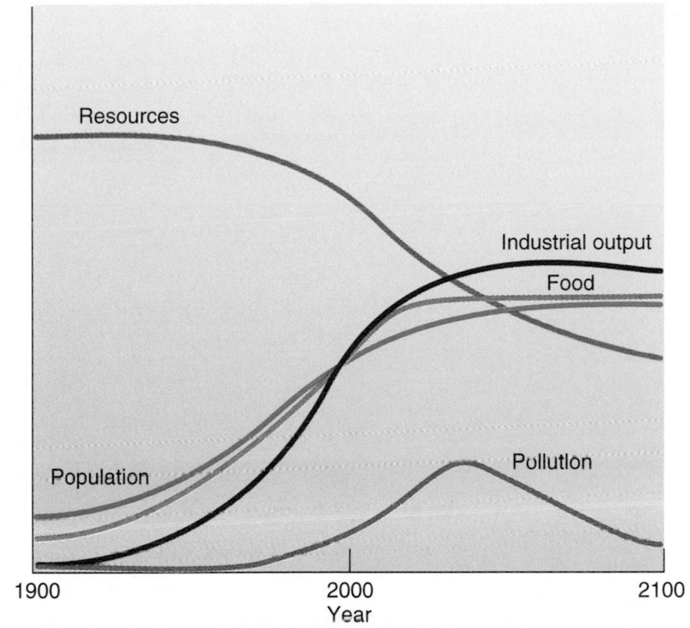

FIGURE 8.16 A run of the world model from *Beyond the Limits*. This model assumes that population and consumption are curbed, new technologies are introduced, and sustainable environmental policies are introduced immediately rather than after resources are exhausted.

consumption. There certainly would be more easily accessible resources available now, but would we be better off than we are? What do you think? Where is the proper balance between using our resources now or saving them for the future?

NATURAL RESOURCE ACCOUNTING

How can we determine the value of environmental goods and services? Some of the most crucial natural resources are not represented by monetary prices in the marketplace. Certain resource allocation decisions are political or social. Others are simply ignored. Groundwater, sunlight, clean air, biological diversity, and other assets that we all share in common often are treated as public goods that anyone can use freely. Until they are transformed by human activities, natural resources are regarded as having little value. Ecological economics calls for recognition of the real value of those resources in calculating economic progress. In this section we'll look at some suggestions for how we could do this.

Gross National Product

The most common way to measure a nations's output is **gross national product (GNP).** GNP can be calculated in two ways. One is the money flow from households to businesses in the form of goods and services purchased (see fig. 8.4). The other is to add up all the costs of production in the form of wages, rent, interest, taxes, and profit. In either case, a subtraction is made for capital depreciation, the wear and tear on machines, vehicles, and buildings used in production. Some economists prefer **gross domestic product (GDP),** which includes only the economic activity within national boundaries. Thus the vehicles made and sold by Ford in Europe don't count in GDP.

Both GNP and GDP are criticized as a measure of real progress or well-being because it doesn't attempt to distinguish between economic activities that are beneficial or harmful. People who develop cancer from smoking, for instance, contribute to the GNP by running up large hospital bills. The pain and suffering they experience doesn't appear on the balance sheets.

Ecological economists also criticize GNP because it doesn't account for natural resources used up or ecosystems damaged by economic activities. Robert Repeto of the World Resources Institute estimates that soil erosion in Indonesia, for instance, reduces the value of crop production about 40 percent per year. If natural capital were taken into account, Indonesian GNP would be reduced by at least 20 percent annually.

Similarly, Costa Rica experienced impressive increases in timber, beef, and banana production between 1970 and 1990. But decreased natural capital during this period represented by soil erosion, forest destruction, biodiversity losses, and accelerated water runoff add up to at least $4 billion or about 25 percent of annual GNP.

Alternatives to GNP or GDP

A number of systems have been proposed as alternatives to GNP that reflect genuine progress and social welfare. In their 1989 book, Herman Daly and John Cobb proposed an Index of Sustainable Economic Welfare (ISW) that takes into account real per capita income, distributional equity, natural resource depletion, environmental damage, and the value of unpaid labor. They point out that while per capita GNP in the United States increased significantly between 1950 and 1986, per capita ISW was relatively flat (fig. 8.17). Some social service organizations would add to this index the costs of social breakdown and crime, which would decrease ISW even further over this time span.

The United Nations Development Program (UNDP) uses a benchmark called the Human Development Index (HDI) to track social progress. HDI incorporates life expectancy, educational attainment, and standard of living as critical measures of development. Gender issues are accounted for in the Gender Development Index (GDI), which is simply HDI adjusted or discounted for inequality or achievement between men and women.

In its annual Human Development Report, the UNDP compares country-by-country progress. As you might expect, the highest development levels are generally found in North America, Europe, and Japan. In 1999, Canada ranked first in the world in both HDI and GDI. The United States ranked fourth after France and Norway. The fifteen countries with the lowest HDI in 1999 were all in Africa. Haiti ranks the lowest in the western hemisphere; Bhutan, Nepal, and Yemen rank just above Haiti.

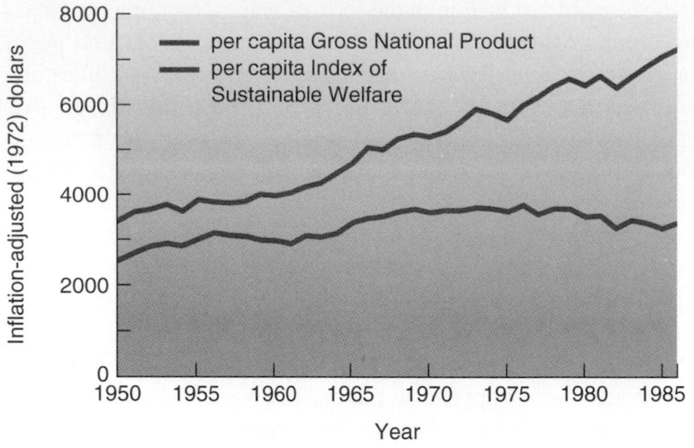

FIGURE 8.17 Although per capita GNP in the United States more than doubled between 1950 and 1986 in inflation-adjusted dollars, an index of sustainable welfare that takes into account natural resource depletion, environmental damage, and options for future generations has hardly increased at all.
Source: H. Daly and J. B. Cobb, *For the Common Good,* 1989, Beacon Press.

Although poverty remains widespread in many places, encouraging news also can be found in development statistics. Poverty has fallen more in the past fifty years, the UNDP reports, than in the previous five hundred years. Child death rates in developing countries as a whole have been more than halved. Average life expectancy has increased by 30 percent while malnutrition rates have declined by almost a third. The proportion of children who lack primary school has fallen from more than half to less than a quarter. And the share of rural families without access to safe water has fallen from nine-tenths to about one-quarter.

Some of the greatest progress has been made in Asia. China and a dozen other countries with populations that add up to more than 1.6 billion, have decreased the proportion of their people living below the poverty line by half. Still, between 1987 and 1997 the number of people with incomes less than $1 per day increased by almost 100 million to 1.3 billion—and the number appears to be growing in every region except Southeast Asia and the Pacific. Even in industrial countries, more than 100 million people live below the poverty line and 37 million are chronically unemployed.

Economic growth can be a powerful means of reducing poverty, but its benefits are not automatic. The GNP of Honduras, for instance, grew 2 percent per year in the 1980s and yet poverty doubled. To combat poverty, the UNDP calls for "pro-poor growth" designed to spread benefits to everyone. Specifically, some key elements of this policy would be to: (1) create jobs that pay a living wage, (2) lessen inequality, (3) encourage small-scale agriculture, microenterprises, and the informal sector, (4) foster technological progress, (5) reverse environmental decline in marginal regions, (6) speed demographic transitions, and (7) provide education for all. Since about three-quarters of the world's poorest people live in rural areas, raising agricultural productivity and incomes is a high priority for these actions.

Measuring Nonmarket Values

New tools and new approaches are needed to represent nature in national accounting and human development. Among the natural resource characteristics that ecological economists suggest be taken into account include:

- Use values: the price we pay to use or consume a resource
- Option value: preserving options for the future
- Existence value: those things we like to know still exist even though we may never use or even see them
- Aesthetic values: aspects we appreciate for their beauty
- Cultural values: factors important for cultural identity
- Scientific and educational values: information or experience-rich aspects of nature.

How can we measure these values of natural resources and ecological services when they are not represented in market systems? Ecological economists often have to resort to "shadow pricing" or other indirect valuation methods for natural resources. For instance, what is the worth of a day of canoeing on a wild river? We might measure opportunity costs such as how much we pay to get to the river or to rent a canoe. The direct out-of-pocket costs might represent only a small portion, however, of what it is really worth to participants. Another approach is contingent valuation in which potential resource users are asked, "how much would you be willing to pay for this experience?" or "what price would you be willing to accept to sell your access or forego this opportunity?" These approaches are controversial because people may report what they think they ought to pay rather than what they would really pay for these activities.

In 1997, a group of ecological economists attempted to put a price on the goods and services provided by natural ecosystems. Although many ecological processes have no direct market value, we can estimate replacement costs, contingent values, shadow prices, and other methods of indirect assessment to determine a rough value. For instance, we now dispose of much of our wastes by letting nature detoxify them. How much would it cost if we had to do this ourselves? In their study, the economists looked at everything from recreational beaches to forest lumber to hidden services such as the ocean's regulation of atmospheric carbon dioxide to pollination of crops by insects.

The estimated annual value of all ecological goods and services provided by nature ranged from $16 trillion to $54 trillion, with a median worth of $33 trillion, or half again the annual GNPs of all national economies in the world (table 8.2). These estimates are probably understated because they omit ecosystem services from several biomes, such as deserts and tundra that are poorly understood in terms of their economic contributions. The most valuable ecosystems in terms of biological processes are wetlands and coastal estuaries because of their high level of biodiversity and their central role in many biogeochemical cycles.

Three comprehensive goals for managing natural capital are presented in table 8.3.

TABLE 8.2 Estimated Annual Value of Ecological Services

ECOSYSTEM SERVICES	VALUE (TRILLION $US)
Soil formation	17.1
Recreation	3.0
Nutrient cycling	2.3
Water regulation and supply	2.3
Climate regulation (temperature and precipitation)	1.8
Habitat	1.4
Flood and storm protection	1.1
Food and raw materials production	0.8
Genetic resources	0.8
Atmospheric gas balance	0.7
Pollination	0.4
All other services	1.6
Total value of ecosystem services	**33.3**

Source: Adapted from R. Costanza et al., "The Value of the World's Ecosystem Services and Natural Capital," *Nature,* Vol. 387 (1997).

TABLE 8.3 Goals for Sustainable Natural Resource Use

- Harvest rates for renewable resources (those like organisms that regrow or those like fresh water that are replenished by natural processes) should not exceed regeneration rates.
- Waste emissions should not exceed the ability of nature to assimilate or recycle those wastes.
- Nonrenewable resources (such as minerals) may be exploited by humans, but only at rates equal to the creation of renewable substitutes.

Cost-Benefit Analysis

One way to evaluate public projects is to analyze the costs and benefits they generate in a **cost-benefit analysis (CBA).** This process attempts to assign values to resources as well as to social and environmental effects of carrying out or not carrying out a given undertaking. CBA is one of the main conceptual frameworks of resource economics and is used by decision makers around the world as a way of justifying the building of dams, roads, and airports, as well as in considering what to do about biodiversity loss, air pollution, and global climate change. Deeply entrenched in bureaucratic practice and administrative culture, this technique has become much more widespread in American public affairs since the Reagan administration's executive orders in the 1980s calling for the application of CBA to all regulatory decisions and legislative proposals. Many conservatives see CBA as a way of eliminating what they consider to be unnecessary and burdensome requirements to protect clean air, clear water, human health, or biodiversity.

The first step in CBA is to identify who or what might be affected by a particular plan. What are the potential outcomes and results? What alternative actions might be considered? After identifying and quantifying all the contingent factors, an attempt is made to assign monetary costs and benefits to each component. Usually the direct expenses of a project are easy to ascertain. How much will you have to pay for land, materials, and labor? The monetary worth of lost opportunities—to swim or fish in a river, or to see birds in a forest—on the other hand, are much harder to appraise. How would you put a price on good health or a long life? It's also important to ask who will bear the costs and who will reap the benefits of any proposal. Are there consequences that cannot be given a monetary price? What is a bug or a bird worth, for instance, or the opportunity for solitude or inspiration? Eventually, the decision maker compares all the costs and benefits to see whether the project is justified or whether some alternative action might bring more benefits at less cost.

At the same time that some environmentalists are using CBA as a way of obtaining recognition for environmental issues within the terms of mainstream discourse, others are challenging this method of resource accounting as amoral and deeply flawed. Grassroots opponents of roads and hydroelectric dams around the world have repeatedly contested the ways that CBA values land, forests, streams, fisheries and livelihoods, as well as its reliance on unaccountable experts and neglect of equity issues. Ordinary people often refuse to answer questions about how much money they would pay to save a wilderness or how much they would accept to allow it to be destroyed. Third World delegates to the Intergovernmental Panel on Climate Change angrily rejected a cost-benefit analysis of policy options regarding global warming that assigned a higher value to lives of people in industrialized countries compared to those in less developed nations.

An example of the CBA approach that enrages many environmental and human rights activists was contained in a memo circulated in 1991 by World Bank chief economist Lawrence Summers (currently U.S. Treasury Secretary). Summers pointed out that "the costs of health-impairing pollution depends on the foregone earnings from increased morbidity and mortality. From this point of view, a given amount of health-impairing pollution should be done in the country with the lowest cost, which will be the country with the lowest wages. I think the economic logic behind dumping a load of toxic waste in the lowest-wage country is impeccable. . . I've always thought that the under-populated countries in Africa are vastly *under*-polluted. Only the lamentable fact that so much pollution is generated by non-tradable industries (transportation, electrical generation) and that the unit transport costs of solid waste are so high prevent world-welfare-enhancing trade in air pollution and waste." While this memo may have been tongue-in-cheek and intended mainly to stimulate discussion, it represents the worst fears of many people about the indiscriminate application of CBA.

Some other criticisms of CBA include its absence of standards, inadequate attention to alternatives, and the placing of monetary values on intangible and diffuse or distant costs and benefits. Who judges how costs and benefits will be estimated? How can we compare things as different as the economic gain from cheap power with loss of biodiversity or the beauty of a free-flowing river? Critics claim that placing monetary values on everything could lead to a belief that only money and profits count and that any behavior is acceptable as long as you can pay for it. Sometimes speculative or even hypothetical results are given specific numerical values in CBA and then treated as if they are hard facts. Risk-assessment techniques (see chapter 9) may be more appropriate for comparing uncertainties.

Figure 8.18 shows an example of a cost-benefit analysis for reducing particulate air pollution (soot) in Poland. As you can see, removing the highest 40 percent of particulates is highly cost effective. Approaching 70 percent particulate removal, however, the costs may exceed benefits. Data such as this can be useful to decision makers. On the other hand, this same study showed negligible benefits compared to high costs for sulfur removal, perhaps for some of the reasons detailed above.

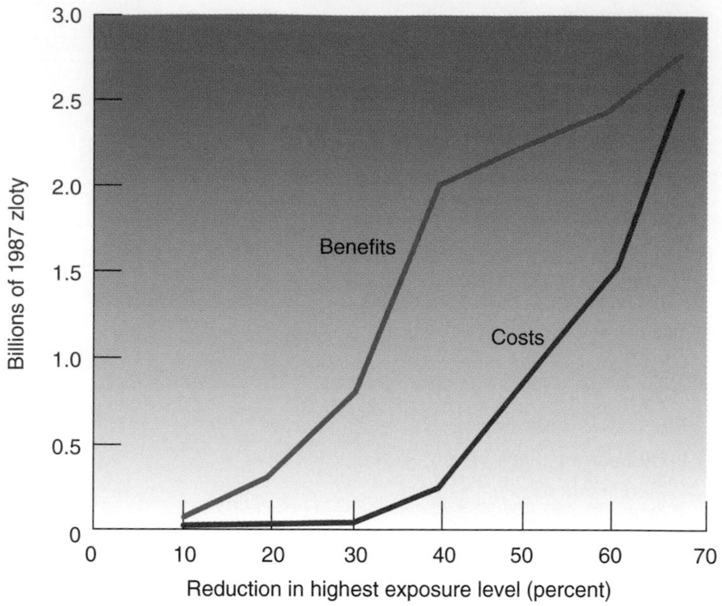

FIGURE 8.18 Total estimated benefits and costs of reducing particulate air pollution in Tarnobrzeg, Poland. Removing the highest 40 to 50 percent of particulates costs relatively little compared to the large benefits that would be gained. Removing 70 percent of particulates may cost more than the benefits are worth.

Source: J. Cofala, T. Lis, and H. Balandynowicz, *Cost-Benefit Analysis of Regional Air Pollution Control: Case Study for Tarnobrzeg,* 1991, Polish Academy of Sciences, Warsaw.

Intergenerational Justice and Discount Rates

"A bird in the hand is worth two in the bush." All of us are familiar with sayings that suggest it is better to have something now than in the distant future. **Discount rates** are the economists' way to introduce a time factor in accounting. It is a recognition that having something today is worth more to most of us than it will be in the future. In general, discount rates are equivalent to interest rates. When you buy something on credit, for instance, you are deciding that having the use of that item now is worth 10 or 20 percent more to you than the money you paid would be worth in a year.

The choice of discount rates to apply to future benefits becomes increasingly problematic with intangible resources or long time frames. How much will a barrel of oil or a four-thousand-year-old redwood tree be worth a couple of centuries from now? Maybe there will be substitutes for oil by then so the barrel of oil won't be worth much. On the other hand, if we don't find substitutes, that oil or tree may be priceless someday. This forecasting is complicated by the fact that we are making decisions not only for ourselves, but also for future generations.

Although having access to clean groundwater or biological diversity one hundred years from now isn't worth much to us—assuming that we will be long gone by that time—those resources might be quite valuable to our descendants. Future citizens will be affected by the choices we make today, but they don't have a vote. Our decisions about how to use resources raise difficult questions about justice between generations. How should we weigh their interests in the future against ours right now?

Which discount rate should be used for future benefits and costs is often a crucial question in large public projects such as dams and airports. Proponents may choose to use low discount rates that make a venture seem attractive while opponents prefer higher rates that show the investment to be questionable. These questions are especially difficult in comparing future environmental costs to immediate financial returns. Environmental activists often need to understand economic nuances when they are fighting environmentally destructive schemes.

Internal and External Costs

Internal costs are the expenses (monetary or otherwise) that are borne by those who use a resource. Often, internal costs are limited to the direct out-of-pocket expenses involved with gaining access to the resource and turning it into a useful product or service.

External costs are the expenses (monetary or otherwise) that are borne by someone other than the individuals or groups who use a resource. External costs often are related to public goods and services derived from nature. Some examples of external costs are the environmental or human health effects of using air or water to dispose of wastes. Since these effects usually are diffuse and difficult to quantify, they do not show on the ledgers of the responsible parties. They are likely to be ignored in private decisions about the costs and benefits of a purchase or a project. One way to use the market system to optimize resource use is to make sure that those who reap the benefits of resource use also bear all the external costs. This is referred to as **internalizing costs.**

A controversial aspect of U.S. proposals for curbing greenhouse gas emissions in the most efficient and least costly way is to use market incentives similar to those authorized by the 1990 Clean Air Act. This would have the effect of internalizing external costs but is regarded by its opponents as merely a license to pollute.

TRADE, DEVELOPMENT, AND JOBS

Trade can be a powerful tool in making resources available and raising standards of living. Think of the things you now enjoy that might not be available if you had to live exclusively on the resources available in your immediate neighborhood. Too often, the poorest, least powerful people suffer in this global marketplace. To balance out these inequities, nations can deliberately invest in economic development projects. In this section, we'll look at some aspects of trade, development, business, and jobs that have impacts on our environment and welfare.

International Trade

International issues further complicate questions of resource management. Much of the vast discrepancy between richer and poorer nations is related to past economic and political history, as well as to current international trade relations. According to the economic theory of comparative advantage, each place has goods or services it can supply in better quality or at better prices than its neighbors. International trade allows us to take advantage of all the best products from around the world. The banking and trading systems that regulate credit, currency exchange, shipping rates, and commodity prices were set up by the richer and more powerful nations in their own self-interest. The General Agreement on Tariffs and Trade (GATT) and World Trade Organization (WTO) agreements, for example, negotiated primarily between the largest industrial nations, regulate 90 percent of all international trade (see chapter 25).

These systems tend to keep the less-developed countries in a perpetual role of resource suppliers to the more-developed countries. The producers of raw materials, such as mineral ores or agricultural products, get very little of the income generated from international trade (fig. 8.19). Furthermore, they suffer both from low commodity prices relative to manufactured goods and from wild "yo-yo" swings in prices that destabilize their economies and make it impossible to either plan for the future or to accumulate the capital for further development.

In 1974, the United Nations General Assembly adopted Resolution Number 3281 calling for a "new international economic order" in which "every State has and shall freely exercise full permanent sovereignty, including possession, use and disposal over all its wealth, natural resources and economic activities." This idea of a new relationship between the richer nations of the North and the poorer South was developed further by the United Nations Commission chaired in 1980 by Willy Brandt, former mayor of West Berlin, and at the 1992 "Earth Summit." As you might suppose, most rich nations have been less enthusiastic than the poor ones about changing present economic relationships.

International Development

No single institution has more influence on the financing and policies of developing countries than the World Bank. Of some $25 billion loaned each year for Third World projects by multinational development banks, about two-thirds comes from the World Bank. For every dollar invested by the World Bank, two dollars are attracted from other sources. If you want to have an impact on what is happening in the developing countries, it is imperative to understand how this huge enterprise works.

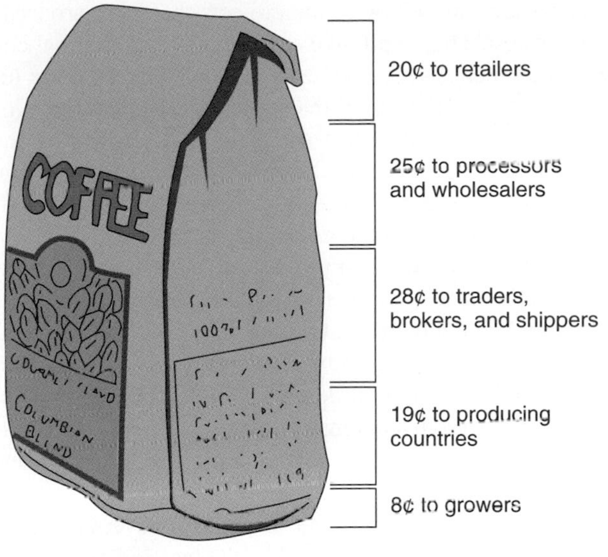

FIGURE 8.19 What do we really pay for when we purchase a dollar's worth of coffee?

The World Bank was founded in 1945 to provide aid to war-torn Europe and Japan. In the 1950s, its emphasis shifted to development aid for Third World countries. This aid was justified on humanitarian grounds, but providing markets and political support for Western capitalism was an important by-product.

The World Bank is jointly owned by 150 countries, but one-third of its support comes from the United States. Its president has always been an American. The bulk of its $66.8 billion capital comes from private investors who buy instruments (bonds and debentures) on the open market. Loan applications are first screened by the professional staff and then voted on by a council of all member countries. No loan has ever been turned down once it makes it to the full council. So far, one hundred countries have borrowed some $140 billion from the World Bank for a wide variety of development projects.

Many World Bank projects have been environmentally destructive and highly controversial. In Botswana, for example, $18 million was provided to increase beef production for export by 20 percent, despite already severe overgrazing on fragile grasslands. The project failed, as did two previous beef production projects in the same area. In Ethiopia, rich floodplains in the Awash River Valley were flooded to provide electric power and irrigation water for cash export crops. More than 150,000 subsistence farmers were displaced and food production was seriously reduced. In India, the sacred Narmada River is being transformed by 30 large dams and 135 medium-sized ones financed by the World Bank. About 1.5 million hill people and farmers will be displaced as a result.

These are only a few of the projects that have aroused concern and protest. They typify the environmental and social costs of many development programs. Loans from the World Bank tend to favor large-scale agricultural, transportation, and energy projects, which make up more than half of all funds it distributes. In part, this is because both the World Bank and debtor nations find it easier to manage one big project than many small ones. Furthermore, both the World Bank and the governments borrowing money prefer large, impressive, modern projects to show how the money is being spent. Small cooperatives or cottage industries might do more for the people than a big dam, but they don't look as impressive.

In a response to criticisms of World Bank policies, the U.S. Congress now insists that all loans for international development be reviewed for environmental and social effects before being approved. It asks for assurance that each project (a) uses renewable resources and does not exceed the regenerative capacity of the environment, (b) does not cause severe or irreversible environmental deterioration, and (c) does not displace indigenous people.

In 1999, leaders from the Group of Seven industrial nations (Germany, Japan, France, Canada, Italy, Britain, and the United States) promised to cut the debt burden of the most heavily indebted poor countries in what they described as a decisive step toward alleviating poverty. Altogether, the amount of debt relief could approach $90 billion, and 41 countries (32 of which are African) could benefit from this decision. The share of debt for every man, woman, and child in many of these countries far exceeds the annual per capita income. In Mozambique, for example, where the average income is only $80 per year, the accumulated foreign debt is $10,424 per person. The amount owed to the United States by all the heavily indebted poor countries totals $6.8 billion, or slightly more than the cost of just three of our fleet of 21 B-2 Stealth Bombers.

A number of religious and social justice groups urge the rich countries to go further by declaring a "jubilee for the millennium," in which all debt for foreign assistance or agricultural development programs would be forgiven. Priority should be given to countries that have demonstrated a sustained commitment to poverty alleviation or that have recently suffered a natural disaster, according to these groups. They also would exclude countries that support international terrorism, or which have excessive military spending, gross violations of human rights, or drug trafficking. Under this more generous plan, up to 90 percent of the $127 billion owed by the most heavily indebted countries might be canceled, and the internal spending for education, health, and nutrition by those countries could be quadrupled.

Microlending

An encouraging alternative to the huge development projects financed by the World Bank is provided by small-scale, community-based banking pioneered by economist Muhammad Yunus of Bangladesh. A network of Grammeen (village) Banks makes small loans (averaging $67) directly to the poorest people in the community. Ninety percent of the customers are women who have no collateral and no steady source of income. Loans are enough to buy a used sewing machine, a bicycle, a loom, a cow, some garden tools, or other means of providing needed family income (fig. 8.20).

FIGURE 8.20 A small loan to buy a used sewing machine can provide a sustainable livelihood for a family in the developing world.

Compared to a loan repayment rate of only 30 percent in conventional banks, the recovery rate on Grammeen accounts is a remarkable 98 percent. Peer lending is the key to this success. Borrowers are organized into five-member groups that serve both as mutual aid societies and collection agencies. Payments must be made in regular weekly installments. If one member of the group defaults, the others repay the loan. Creating a sense of dignity, respect, and mutual support within the group encourages individual responsibility and enterprise.

Microlending programs similar to the Grammeen Banks are now in place around the world. More than one hundred organizations in the United States assist microenterprises by providing loans, grants, or training. The Women's Self-Employment Project in Chicago, for instance, teaches job skills to single mothers in housing projects. Similarly, "tribal circle" banks on Native American reservations successfully finance microscale economic development projects. Perhaps the biggest banks could learn from the smallest ones how to really help people.

GREEN BUSINESS

During the first Industrial Revolution 200 years ago, raw materials such as lumber, minerals, and clean water seemed inexhaustible, while nature was regarded as something to be tamed and civilized. The "solution to pollution is dilution," was a common approach for disposal of unwanted wastes. Recently, however, some leading industrialists and business executives have realized that business as usual may not be sustainable over the long term. Global pollution, resource depletion, loss of biological diversity, and the threat of global climate change all suggest that change is essential. As the leading sources of much pollution and resource depletion, these leaders realized that they had a role to play in redesigning systems and processes to ensure that their enterprises have a long-term future.

Known by a variety of names, including eco-efficiency, clean production, pollution prevention, industrial ecology, natural capitalism, restorative technology, the natural step, environmentally preferable products, design for the environment, and the next industrial revolution, this movement has had some remarkable successes and presents an encouraging pathway for how we might achieve both environmental protection and social welfare. Some of the leaders in this new approach to business include Paul Hawken, William McDonough, Ray Anderson, Amory Lovins, David Crockett, and John and Nancy Todd.

Operating in a socially responsible manner consistent with the principles of sustainable development and environmental protection, they have shown, can be good for employee morale, public relations, and the bottom line simultaneously. Environmentally conscious or "green" companies such as the Body Shop, Patagonia, Aveda, Malden Mills, Johnson and Johnson, and Interface, Inc. (Case Study, p. 182) consistently earn high marks from community and environmental groups. Conserving resources, reducing pollution, and treating employees and customers fairly may cost a little more initially, but can save money and build a loyal following in the long run.

Paul Hawken's 1993 book, *The Ecology of Commerce,* was a seminal influence in convincing many people to reexamine the role of business and economics in environmental and social welfare. Basing his model for a new industrial revolution on the principles of ecology, Hawken points out that almost nothing is discarded or unused in nature. The wastes from one organism become the food of another. Industrial processes, he argues, should be designed on a similar principle (table 8.4). Rather than a linear pattern in which we try to maximize the throughput of material and minimize labor, products and processes should be designed to

- be energy efficient;
- use renewable materials;
- be durable and reusable or easily dismantled for repair and remanufacture, nonpolluting throughout their entire life cycle;
- provide meaningful and sustainable livelihoods for as many people as possible;
- protect biological and social diversity;
- use minimum and appropriate packaging made of reusable or recyclable materials.

We can do all this and at the same time increase profits, reduce taxes, shrink government, increase social spending, and restore our environment, Hawken claims. Recently, Hawken has served as chairperson for The Natural Step in America, a movement started in Sweden by Dr. Robert, a physician concerned about the increase in environmentally related cancers. Through a consensus process, a group of 50 leading scientists endorsed a description of the living systems on which our economy and lives depend. More than 60 major European corporations and 55 municipalities have incorporated sustainability principles (table 8.5) into their operations.

TABLE 8.4 Goals for an Eco-Efficient Economy

- Introduce no hazardous materials into the air, water, or soil.
- Measure prosperity by how much natural capital we can accrue in productive ways.
- Measure productivity by how many people are gainfully and meaningfully employed.
- Measure progress by how many buildings have no smokestacks or dangerous effluents.
- Make the thousands of complex governmental rules unnecessary that now regulate toxic or hazardous materials.
- Produce nothing that will require constant vigilance from future generations.
- Celebrate the abundance of biological and cultural diversity.
- Live on renewable solar income rather than fossil fuels.

Corporations committed to eco-efficiency and clean production include such big names as Monsanto, 3M, Dupont, Duracell, and Johnson and Johnson. Following the famous three Rs—reduce, reuse, recycle—these firms have saved money and gotten welcome publicity. Savings can be substantial. Pollution prevention programs at 3M, for example, have saved more than $750 million. In a major public relations achievement, Du Pont has cut its emissions of airborne cancer-causing chemicals almost 75 percent since 1987. Small operations can benefit as well. Stanley Selegut, owner of three eco-tourist resorts in the U.S. Virgin Islands, attributes $5 million worth of business to free press coverage about the resorts' green building features and sustainable operating practices.

Design for the Environment

Our current manufacturing system is incredibly wasteful. On average, for every truckload of products delivered in the United States, 32 truckloads of waste are produced along the way. The automobile is a typical example. Industrial ecologist, Amory Lovins, calculates that for every 100 gallons (380 l) of gasoline burned in your car engine, only one percent (1 gal or 3.8 l) actually moves passengers. All the rest is used to move the vehicle itself. The wastes produced—carbon dioxide, nitrogen oxides, unburned hydrocarbons, rubber dust, heat—are spread through the environment where they pollute air, water, and soil.

CASE STUDY

ECO-EFFICIENT CARPETING FROM INTERFACE, INC.

In 1994, in response to customers' concerns about health problems caused by chemical fumes from new carpeting, wall coverings, and other building materials, Ray Anderson, founder and CEO of Interface, Inc., a billion-dollar-a-year interior furnishing company, decided to review company environmental policy. What he found was that the company really didn't have an environmental vision other than to obey all relevant laws and comply with regulations. He also learned that carpeting—of which Interface was the world's third-largest manufacturer—is one of the highest volume and longest lasting components in landfills. A typical carpet is made of nylon embedded in fiberglass and polyvinyl chloride. After a useful life of about 12 years, most carpeting is ripped up and discarded. Every year, more than 770 million m^2 (920 million yd^2) of carpet weighing 1.6 billion kg (3.5 billion lbs) ends up in U.S. landfills. The only recycling that most manufacturers do is to shave off some of the nylon for remanufacture. Everything else is buried in the ground where it will last at least 20,000 years.

Ray Anderson

At about the same time that Interface was undergoing its environmental audit, Anderson was given a copy of Paul Hawken's book *The Ecology of Commerce.* Reading it, he said, was like "a spear through the chest." He vowed to turn his company around, to make its goal sustainability instead of simply maximizing profits. Rather than sell materials, Interface would focus on selling service. The key is what Anderson calls an "evergreen lease." First of all, the carpet is designed to be completely recyclable. Where most flooring companies merely sell carpet, Interface offers to lease carpets to customers. As carpet tiles wear out, old ones are removed and replaced as part of the lease. The customer pays no installation or removal charges, only a monthly fee for constantly fresh-looking and functional carpeting. Everything in old carpet is used to make new product. Only after many reincarnations as carpet, are materials finally sent to the landfill.

Dramatic changes have been made at Interface's 26 factories. Toxic air emissions have been reduced significantly by changing manufacturing processes and substituting nontoxic materials for more dangerous ones. Solar power is replacing fossil fuel use. Less waste is produced as more material is recycled and products are designed for ecoefficiency. The total savings from pollution prevention and recycling in 1999 was $87 million.

Not only has Interface continued to be an industry leader, it was named one of the "100 Best Companies to Work For in America" by *Fortune* magazine. Ray Anderson has become a popular speaker on the topic of eco-efficiency and clean production. He cochairs the President's Council on Sustainable Development, was named Entrepreneur of the Year by Ernest & Young, and was the Georgia Conservancy's Conservationist of the Year in 1998. Anderson's book, *Mid-Course Correction: Toward a Sustainable Enterprise,* published in 1999 by Chelsea Green, has won critical acclaim.

Transforming an industry as large as interior furnishing has not been an overnight success. "Like aircraft carriers," Anderson says, "big businesses don't turn on a dime." Still, he has shown that the principles of sustainability and financial success can coexist and can lead to a new prosperity that includes both environmental and human dividends. His motto, that we should "put back more than we take and do good to the Earth, not just no harm," has become a vision for a new industrial revolution that now is reaching many companies beyond his own.

Ethical Considerations

What responsibilities do businesses have to protect the environment or save resources beyond the legal liabilities spelled out in the law? None whatever, according to conservative economist Milton Friedman. In fact, Friedman argues, it would be unethical for corporate leaders to consider anything other than maximizing profits. To spend time or resources doing anything other than making profits and increasing the value of the company is a betrayal of their duty. What do you think? Should social justice, sustainability, or environmental protection be issues of concern to corporations?

Architect William McDonough urges us to rethink design approaches (table 8.6). In the first place, he says, we should question whether the product is really needed. Could we provide the same service in a more eco-efficient manner? According to McDonough, products should be divided into three categories:

1. *Consumables* are products like food, natural fabrics, or paper that can harmlessly go back to the soil as compost.
2. *Service products* are durables such as cars, TVs, and refrigerators. These products should be leased to the customer to provide their intended service, but would always belong to the manufacturer. Eventually they would be returned to the maker who would be responsible for recycling or remanufacturing the product. Knowing that they will have to dismantle the product at the end of its life will encourage manufacturers to design for easy disassembly and repair.
3. *Unmarketables* are compounds like radioactive isotopes, persistent toxins, and bioaccumulative chemicals. Ideally, no one would make or use these products. But because

TABLE 8.5 The Natural Step: System Conditions for Sustainability

1. Minerals and metals from the earth's crust must not systematically increase in nature.
2. Materials produced by human society must not systematically increase in nature.
3. The physical basis for biological productivity must not be systematically diminished.
4. The use of resources must be efficient and just with respect to meeting human needs.

eliminating their use will take time, McDonough suggests that in the meantime these materials should belong to the manufacturer and be molecularly tagged with the maker's mark. If they are discovered to be discarded illegally, the manufacturer would be held liable.

Following these principles, McDonough Bungart Design Chemistry has created nontoxic, easily recyclable materials to use in buildings and for consumer goods. Among some important and innovative "green office" projects designed by the McDonough + Partners architectural firm are the Environmental Defense Fund headquarters in New York City, the Environmental Studies Center at Oberlin College in Ohio, the European Headquarters for Nike in Hilversum, the Netherlands, and the Gap Coporate Offices in San Bruno, California (fig. 8.21). Intended to promote employee well-being and productivity as well as eco-efficiency, the Gap building has high ceilings, abundant skylights, windows that open, a full-service fitness center (including pool), and a landscaped atrium for each office bay that brings the outside in. The roof is covered with native grasses. Warm interior tones and natural wood surfaces (all wood used in the building was harvested by certified sustainable methods) give a friendly feeling. Paints, adhesives, and floor coverings are low toxicity and the building is one-third more energy efficient than strict California laws require. A pleasant place to work, the offices are help in recruiting top employees and improve both effectiveness and retention. As for the bottom line, Gap, Inc. estimates that the increased energy and operational efficiency will have a four- to eight-year payback.

Green Consumerism

Consumer choice can play an important role in persuading businesses to produce eco-friendly goods and services (What Can You Do? p. 184). Increasing interest in environmental and social sustainability has caused an explosive growth of green products. The National Green Pages published by Co-Op America currently lists more than two thousand green companies. You can find ecotravel agencies, telephone companies that donate profits to environmental groups, entrepreneurs selling organic foods, shade-grown coffee, straw-bale houses, geodesic-dome kits, paint thinner made from orange peels, sandals made from recycled auto tires, earthworms for composting, and a plethora of hemp products including burgers, ale, clothing, shoes, rugs, balm, shampoo, and insect repellent. Although these eco-entrepreneurs represent a tiny sliver of the $7 trillion per year U.S. economy, they often serve as pioneers in developing new technologies and offering innovative services.

In some industries eco-entrepreneurs have found profitable niches (organic, naturally colored clothing, for example) within a larger market. In other cases, once a consumer demand has built up, major companies add green products or services to their inventory. Natural foods, for instance, have grown from the domain of a few funky, local co-ops to a $7 billion market segment. Most supermarket chains now carry some organic food choices. Similarly, natural-care health and beauty products reached $2.8 billion in sales in 1999 out of a $33 billion industry. By supporting these products, you can ensure that they will continue to be available and, perhaps, even help expand their penetration into the market.

FIGURE 8.21 The award-winning Gap, Inc. corporate offices in San Bruno, CA, demonstrate some of the best features of environmental design. A roof covered with native grasses provides insulation and reduces runoff. Natural lighting, an open design, and careful relation to its surroundings make this a good place to work.

TABLE 8.6 McDonough Design Principles

Inspired by the way living systems actually work, Bill McDonough offers three simple principles for redesigning processes and products:

1. *Waste equals food.* This principle encourages elimination of the concept of waste in industrial design. Every process should be designed so that the products themselves, as well as leftover chemicals, materials, and effluents, can become "food" for other processes.
2. *Rely on current solar income.* This principle has two benefits: First, it diminishes, and may eventually eliminate, our reliance on hydrocarbon fuels. Second, it means designing systems that sip energy rather than gulping it down.
3. *Respect diversity.* Evaluate every design for its impact on plant, animal, and human life. What effects do products and processes have on identity, independence, and integrity of humans and natural systems? Every project should respect the regional, cultural, and material uniqueness of its particular place.

what can you do?

Personally Responsible Consumerism

There are many things that each of us can do to lower our ecological impacts and support green businesses through responsible consumerism and ecological economics.

- Practice living simply. Ask yourself if you really need more material goods to make your life happy and fulfilled.
- Rent, borrow, or barter when you can. Do you really need to have your own personal supply of tools, machines, and other equipment that you use, at most, once per year?
- Recycle or reuse building materials: doors, windows, cabinets, appliances. Shop at salvage yards, thrift stores, yard sales, or other sources of used clothes, dishes, appliances, etc.
- Consult the National Green Pages from Co-Op America for a list of eco-friendly businesses. Write one letter each month to a company from which you buy goods or services and ask them what they are doing about environmental protection and human rights.
- Buy green products. Look for efficient, high quality materials that will last and that are produced in the most environmentally friendly manner possible. Subscribe to clean energy programs if they are available in your area. Contact your local utility and ask that they provide this option if they don't now.
- Buy products in bulk if you can or look for the least amount of packaging. Choose locally-grown or locally-made products made under humane conditions by workers who receive a fair wage.
- Think about the total life-cycle costs of the things you buy, including environmental impacts, service charges, energy use, and disposal costs as well as initial purchase price.
- Stop junk mail. Demand that your name be removed from mass-mailing lists.
- Invest in socially and environmentally responsible mutual funds or green businesses when you have money for investment.

Jobs and the Environment

For years business leaders and politicians have portrayed environmental protection and jobs as mutually exclusive. Pollution control, protection of natural areas and endangered species, limits on use of nonrenewable resources, they claim, will strangle the economy and throw people out of work. Ecological economists dispute this claim, however. Their studies show that only 0.1 percent of all large-scale layoffs in the United States in recent years were due to government regulations (fig. 8.22). Environmental protection, they argue, is not only necessary for a healthy economic system, it actually creates jobs and stimulates business.

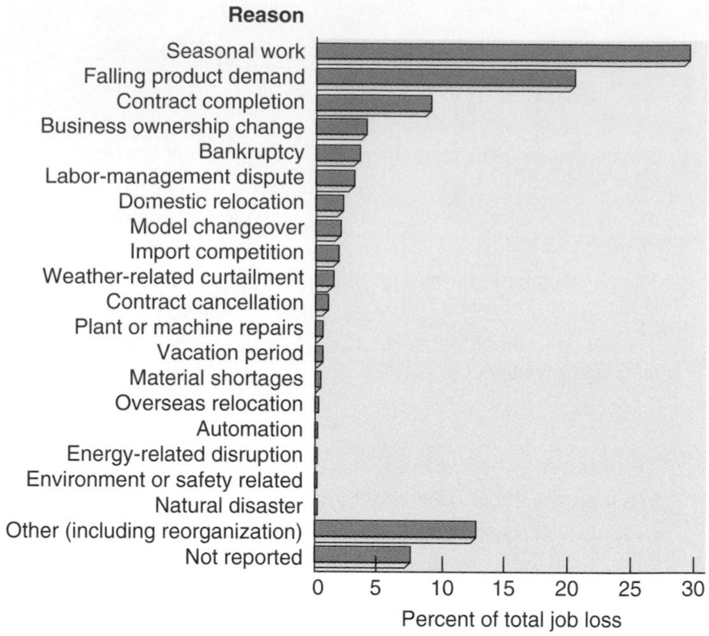

FIGURE 8.22 Although opponents of environmental regulation often claim that protecting the environment costs jobs, studies by economist E. S. Goodstein show that only 0.1 percent of all large-scale layoffs in the United States were the result of environmental laws.
Source: Data from E. S. Goodstein, Economic Policy Institute, Washington, D.C.

Recycling, for instance, takes much more labor than extracting virgin raw materials. This doesn't necessarily mean that recycled goods are more expensive than those from virgin resources. We're simply substituting labor in the recycling center for energy and huge machines used to extract new materials in remote places.

Japan, already a leader in efficiency and environmental technology, has recognized the multibillion dollar economic potential of "green" business. The Japanese government is investing $4 billion (U.S.) per year on research and development that targets seven areas ranging from utilitarian projects such as biodegradable plastics and heat-pump refrigerants to exotic schemes such as carbon-dioxide-fixing algae and hydrogen-producing microbes.

By the year 2000, Japan expects to be selling $12 billion worth of equipment and services per year worldwide. They are already marketing advanced waste incinerators, pollution control equipment, alternative energy sources, and water-treatment systems. Unfortunately, the United States has been resisting international pollution control conventions rather than recognizing the potential for economic growth *and* environmental protection in the field of "green" business.

SUMMARY

In this chapter, we have reviewed the origins of some western economic worldviews. Neoclassical and market approaches now dominate the world economy but they often do a poor job of incorporating nonmarket resources or protecting the environment. Ecological economics brings the insights of ecology into economic analysis. It calls for consideration of the role of natural capital and ecological services in planning and accounting. A resource is anything with potential use for creating wealth or giving satisfaction. Minerals and fossil fuels are examples of nonrenewable resources. Biological organisms and ecological processes are renewable and self-reproducing but can be exhausted by overuse. Resources can be defined by their economic and technological feasibility, as well as their location and physical size. We distinguished between known, proven, inferred, and unconceived resources, and between economically important and technically accessible reserves.

Questions about the scarcity of resources and their effects on economic development are important in determining what kind of society we have. Open-access systems encourage narrow self-interests and resource exploitation. In communal property resource systems, self-governing, locally based community management has successfully sustained natural resources for centuries in many cases.

Cost-benefit analysis can be a useful tool in environmental management. Ultimately, our aim should be to internalize costs that are now treated as externalities. GNP is used as a measure of economic progress but new measures for human well-being and environmental health are needed. The index of sustainable welfare and the human development index are proposed as alternatives.

Some business leaders believe their only responsibility is to maximize profits. Others regard operating in a socially responsible manner consistent with the principles of sustainable development and environmental protection to be good for employee morale, public relations, and the bottom line. The principles of eco-efficiency and design for the environment could help us attain sustainability.

REVIEW QUESTIONS

1. Define economics and distinguish between classical, neoclassical, and ecological economics.
2. Define resources and give some examples of renewable, nonrenewable, and intangible resources.
3. List four economic categories of resources and describe the differences among them.
4. Describe the relationship between supply and demand.
5. What causes diminishing returns in natural resource use? How does population growth affect this phenomenon?
6. Describe how cost-benefit ratios are determined and how they are used in natural resource management.
7. Explain how scarcity and technological progress can extend resource availability and extend the carrying capacity of the environment.
8. Describe how GNP is calculated and explain why this may fail to adequately measure human welfare and environmental quality. Discuss some alternative measures of national progress.
9. Why does the marketplace sometimes fail to optimally allocate natural resource values?
10. List some of the characteristics of an eco-efficient economic system.

QUESTIONS FOR CRITICAL THINKING

1. If you could retroactively stabilize economic growth or population growth at some point in the past, when would you choose? What assumptions or values shape your choice?
2. When the ecologist warns that we are using up irreplaceable natural resources and the economist rejoins that ingenuity and enterprise will find substitutes for most resources, what underlying premises and definitions shape their arguments?
3. How can intangible resources be infinite and exhaustible at the same time? Isn't this a contradiction in terms? Can you find other similar paradoxes in this chapter?
4. What is the difference between hypothetical and unconceived (or unknown-unknown) resources? How can we plan for resources that we haven't even thought of yet? Are there costs in assuming that there are no unknown-unknowns?
5. What would be the effect on the developing countries of the world if we were to change to a steady-state economic system? How could we achieve a just distribution of resource benefits while still protecting environmental quality and future resource use?
6. Resource use policies bring up questions of intergenerational justice. Suppose you were asked: "What has posterity ever done for me?" How would you answer?

7. If you were doing a cost-benefit study, how would you assign a value to the opportunity for good health or the existence of rare and endangered species in faraway places? Is there a danger or cost in simply saying some things are immeasurable and priceless and therefore off limits to discussion?
8. Why not conserve resources to the maximum extent possible? Discuss the costs and benefits of resource investment versus conservation.
9. Is it right for business to consider ethics and the welfare of future generations? Why or why not?
10. If natural capitalism or eco-efficiency has been so good for some entrepreneurs, why haven't all businesses moved in this direction?

KEY TERMS

capital 169
classical economics 166
communal resource management systems 171
cost-benefit analysis 177
demand 167
diminishing returns 167
discount rates 178
ecological economics 168
external costs 178
gross domestic product (GDP) 175
gross national product (GNP) 175
intangible resources 170
internal costs 178
internalizing costs 178
known resources 171
marginal costs 167
market equilibrium 167
neoclassical economics 168
nonrenewable resources 169
open access system 171
political economy 168
price elasticity 167
proven reserves 171
recoverable resources 171
renewable resources 170
resource 169
steady-state economy 169
supply 167
Tragedy of the Commons 171
undiscovered resources 171

ADDITIONAL INFORMATION ON THE INTERNET

Visit our website at http://www.mhhe.com/environmental science/ for specific information about each of the topics below:

International Society for Ecological Economics
Trade and the Environment
African workshop on Trade and the Environment
Further reading on Business, the Economy, and the Environment
Paul Hawken
The Natural Step
Ecological Design
Ecological Economics
Agricultural Economics: virtual library
Biosphere II
Microlending
Sustainable Development
Center for Development and the Environment
World Bank
Network for Environmental and Economic Responsibility

PROFILE

Environmental Consultant

Jeremy Sterk

Despite growing up in the Midwest, Jeremy Sterk has been fascinated with oceans since childhood. His lifelong dream has been to become a marine biologist. But a hobby and a bit of serendipity intervened to have an important, and unplanned, influence on his career direction.

Jeremy attended college in the Midwest, where he pursued a major in aquatic biology. Following graduation, he planned to apply for graduate school at a coastal university to work on an advanced degree in marine biology. In addition to his interest in aquatics, however, Jeremy's earlier fascination with computers had become a serious hobby. Although he did not realize it at the time, the hours spent on the computer hobby would turn out to be an investment in his future employment.

One day, one of Jeremy's college biology professors described a project he had undertaken to create a computerized identification key to the freshwater algae. The project was laborious, however, because of the computer system and language with which the professor was working. Because of his hobby, Jeremy quickly grasped what the professor was attempting to do and realized that a simple hypertext-linked database provided a simpler approach. Jeremy produced a demonstration model of the alternative that impressed the professor. Working in collaboration, the two of them completed the computerized key that has now been produced commercially by a major publisher.

After college, Jeremy decided to move to Florida. Although he had no immediate job prospects or contacts, Jeremy searched newspaper ads and followed other leads and found an opening with a small environmental consulting firm. The position involved field work, including plant identification, as well as other responsibilities. While he didn't have a strong botany background, his computer skills were an important asset and he got the job.

Much of the company's work involves helping individuals and land developers comply with various permitting requirements. As such, Jeremy typically serves as the go-between for company clients and the permitting agencies, such as the South Florida Water Management District and the Army Corps of Engineers.

Jeremy spends about half his time in the field doing a variety of tasks. He commonly does surveys of clients' property to determine if they contain wetlands and frequently has to identify wetlands boundaries. His responsibilities also include helping clients with mitigation plans to offset negative environmental impacts the project would otherwise have.

He performs transect surveys of plant and animal species, paying particular attention to those species listed as threatened or endangered. He has done quite a bit of work with gopher tortoises, eagles, and red-cockaded woodpeckers. He also does vegetation mapping for environmental assessment reports. He occasionally works with wildlife monitoring, such as installing infrared sensors in the wildlife crossings under Florida highways to obtain information on use of these crossings by wildlife, including the endangered Florida panther.

The other half of the time Jeremy is involved with paperwork: writing reports and corresponding with regulatory agencies and clients over permit conditions. The need for writing and communication skills was one of the abilities highlighted by the interviewer when Jeremy applied for his job. He has also updated his firm's computer operation.

Jeremy enjoys his job very much. He likes spending so much time in the woods and swamps, experiencing a side of Florida that few people ever see. For students interested in this line of work, Jeremy advises them to get a solid background in biology and as broad a base of other preparation as they can. He wishes he had taken more botany and formal training in soil science. He feels coursework in statistics is helpful in his and other related fields. A course in technical writing would have been useful as well. Not surprisingly, he also feels a person cannot have too much background in computers.

Jeremy enjoys his work so much that he has put his marine biology plans on hold. He is a bit uneasy about not pursuing a lifelong dream, and he will eventually go on to graduate school. But for now, he feels his present work could easily become a lifetime career.

CHAPTER 9

Environmental Health and Toxicology

Do I dare to eat a peach? Do I dare?

—T. S. Eliot—

OBJECTIVES

After studying this chapter, you should be able to:

- define *health* and *disease* in terms of some major environmental factors that affect humans.
- identify some major infectious organisms and hazardous agents that cause environmental diseases.
- distinguish between toxic and hazardous chemicals and between chronic and acute exposures and responses.
- compare the relative toxicity of some natural and synthetic compounds as well as report on how such ratings are determined and what they mean.
- evaluate the major environmental risks we face and how risk assessment and risk acceptability are determined.

ESSENTIAL STUDY PARTNER

The Essential Study Partner (ESP) CD-ROM that accompanies this textbook contains additional information on the following topics:

- infectious and parasitic disease: populations: toxicology: types of hazards
- ignitable and corrosive hazardous chemicals: populations: toxicology: types of hazards
- mutagenic, teratogenic, and carcinogenic toxic chemicals: populations: toxicology: types of hazards
- biomagnification of toxins: populations: toxicology: toxicity
- sensitivity, lethal and effective doses: populations: toxicology: toxicity
- actual versus perceived risk: populations: toxicology: risk
- risk assessment, risk management, and public policy: populations: toxicology: risk

Above: A Polish mother gives an oxygen treatment to her child who suffers from air pollution-related respiratory disease. In some parts of Eastern Europe and the former USSR, up to 90 percent of all children suffer from environmentally linked diseases.

Outbreak

Ebola Hemorrhagic Fever

In mid-April, 1995, a 36-year-old laboratory technician named Kimfumu checked into the medical clinic in Kikwit, Zaire (now the Democratic Republic of Congo) complaining of a severe headache, stomach pains, fever, dizziness, weakness, and exhaustion. Surgeons did an exploratory operation to try to find the cause of his illness. To their horror, they found his entire gastrointestinal tract was necrotic and putrefying. He bled uncontrollably, and within hours was dead. By the next day, the five medical workers who cared for him, including an Italian nun who assisted in the operation, began to show similar symptoms, including high fevers, fatigue, bloody diarrhea, rashes, red and itchy eyes, vomiting, and bleeding from every body orifice. Less than 48 hours later, they, too, were dead, and the disease was spreading like wildfire throughout the city of 600,000.

As panicked residents fled into the bush, government officials responded to calls for help by closing off all travel—including humanitarian aid—into or out of Kikwit, about 400 km (250 mi) from Kinshasa, the national capital (fig. 9.1). Fearful neighboring villages felled trees across the roads to seal off the pestilent city. No one dared enter houses where dead corpses rotted in the intense tropical heat. Boats plying the adjacent Kwilu River refused to stop to take on or discharge passengers or cargo. Food and clean water became scarce. Hospitals could hardly function as medicines and medical personnel became scarce.

Deadly tropical fevers are an unfortunate fact of life in Central Africa but rarely are they this contagious or lethal. The plague that afflicted Kikwit was Ebola hemorrhagic fever, a viral disease for which there is no known treatment. Within a few weeks, about 400 people in Kikwit had contracted the virus and 350 were dead. Where a 10 percent mortality rate is considered high for most infectious diseases, Ebola kills up to 90 percent of its victims, usually within only a few days after exposure.

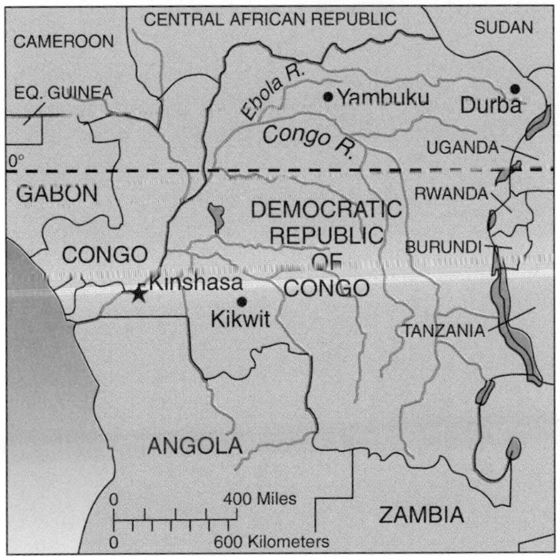

FIGURE 9.1 Ebola fever is named for the Ebola River in the Democratic Republic of Congo (formerly Zaire) where the disease was first reported.

The Kikwit Ebola outbreak was neither the first nor last appearance of this dread disease. The first recognized Ebola epidemic occurred in 1976 in Yambuku, Zaire (near the Ebola River, after which the virus was named), where at least 280 people died. Three years later, 22 patients died in Sudan from a slightly different and less virulent form of Ebola. In 1996, about 100 people were killed by Ebola in two separate episodes in Gabon, and in 1999 an outbreak in the gold mining town of Durba in the Democratic Republic of Congo killed at least 63 people.

Ebola is one of two members of a family of RNA viruses called the Filoviridae (fig. 9.2). The other filovirus causes Marburg fever, an equally lethal and contagious hemorrhagic disease, named after a German town where it was first contracted by laboratory workers who handled imported monkeys infected with the virus. Together with members of three other families (arenaviruses, bunyanviruses, and flaviviruses), these viruses cause a group of terrible, episodic diseases including Lassa fever, Rift Valley fever, Bolivian fever, and Hanta or Four-Corners fever (named after the region of the southwestern United States where it was first reported).

The viruses associated with most of these emergent, hemorrhagic fevers are zoonotic. That means a reservoir of viruses naturally resides in an animal host or arthropod vector. No host or vector is known for Ebola but we know that monkeys and other primates can contract related diseases. Why viruses remain peacefully in their hosts for many years without causing much more trouble than a common cold, but then erupt sporadically and unpredictably into terrible human epidemics is a new and growing question in environmental health and risk assessment. In this chapter we will look at some of the changes humans are bringing about in our environment as well as how a variety of environmental factors affect our health. While you're not likely to encounter Ebola unless you venture into the African jungle, studying this material should give you some ideas about what we might do to ensure a safer environment and how you can adopt a healthier lifestyle.

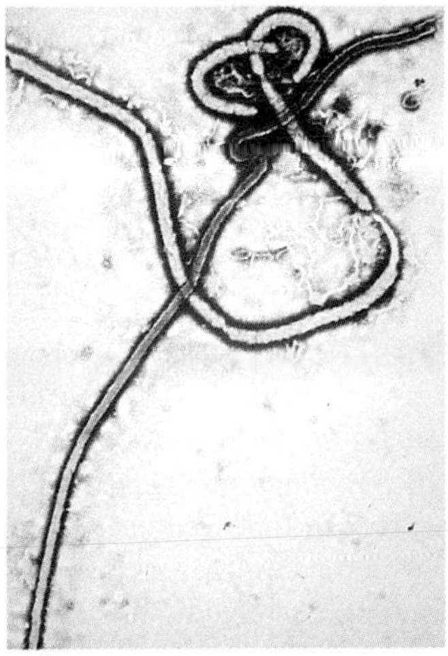

FIGURE 9.2 The Ebola virus as seen in a transmission electron microscope.

TYPES OF ENVIRONMENTAL HEALTH HAZARDS

What is health? The World Health Organization defines **health** as a state of complete physical, mental, and social well-being, not merely the absence of disease or infirmity. By that definition, we all are ill to some extent. Likewise, we all can improve our health to live happier, longer, more productive, and more satisfying lives if we pay attention to what we do.

What is a disease? A **disease** is a deleterious change in the body's condition in response to an environmental factor that could be nutritional, chemical, biological, or psychological. Diet and nutrition, infectious agents, toxic chemicals, physical factors, and psychological stress all play roles in **morbidity** (illness) and mortality (death). To understand how these factors affect us, let's look at some of the major categories of environmental health hazards.

Infectious Organisms

For most of human history, the greatest health threats have always been pathogenic (disease-causing) organisms. Although cardiovascular diseases (heart attacks and strokes), cancer, injuries (intentional and unintentional), and other ills of modern life now have become the leading killers almost everywhere in the world, infectious diseases still kill at least 22 million people every year, or about 43 percent of all disease-related deaths. A majority of these deaths are in the poorer countries of the world where better nutrition, improved sanitation, and inexpensive vaccinations could save millions of lives each year. How does that affect us? As you will learn in this chapter our world is increasingly interconnected. New, extremely virulent diseases like Ebola as well as drug-resistant forms of more familiar diseases such as tuberculosis and cholera can arise in remote areas and then spread rapidly throughout the world.

Before you learn about all the terrible illnesses that afflict people, it's worth pointing out that in many ways most people are healthier than ever before. One evidence of this is the fact that people are living longer almost everywhere. As chapter 7 points out, life expectancies have increased around the world from about 40 years to 65 years on average over the past century. Although developing countries lag behind the wealthier, industrialized nations, progress in education and health care is being made among some of the world's poorest people.

There are many different ways to count sicknesses and deaths. Often even within a single report you can find different numbers attributed to the same diseases. Nevertheless, it is clear that contagious illnesses account for seven of the eleven top categories in global mortalities (table 9.1). Of the broad categories of communicable diseases, chronic obstructive lung infections such as tuberculosis cause the most deaths each year, followed closely by acute respiratory infections such as pneumonia, influenza, and pertussis (whooping cough). Perinatal conditions including infections and other problems associated with birth take some 3.2 million lives every year. Some four *billion* new cases of diarrhea each year are responsible for about 3 million deaths, mostly in children under five years old. Both bacteria and protozoans can cause diarrhea (fig. 9.3). Many children who suffer severe diarrhea early in life subsequently show signs of permanent mental and developmental retardation.

TABLE 9.1 Leading Causes of Death Worldwide

DISEASE	MILLIONS OF DEATHS/YEAR
Cardiovascular disease	9.7
Cancers	6.0
Chronic lung disease*	5.5
Acute lung diseases	4.1
Injuries	4.0
Perinatal conditions	3.2
Diarrhea	3.0
HIV/AIDS	2.3
Malaria	2.0
Vaccine-preventable infections	1.7
Other known disease	3.6
Unknown causes	5.9
Total	51.0

*See text for description.
Source: Data from World Health Organization, 1997.

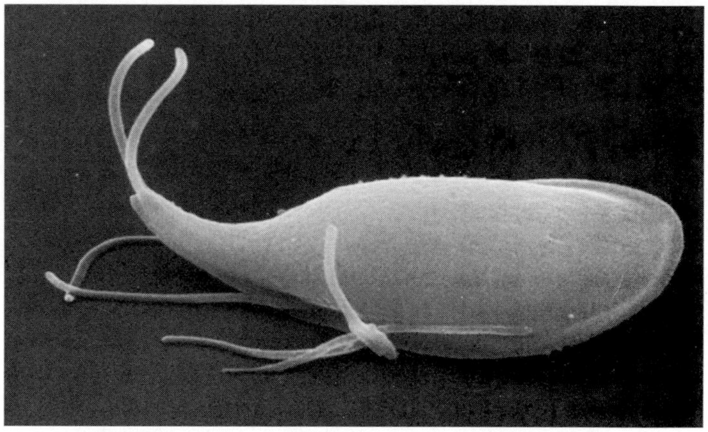

FIGURE 9.3 *Giardia,* a parasitic intestinal protozoan, is reported to be the largest single cause of diarrhea in the United States. It is spread from human feces through food and water. Even pristine wilderness areas have *Giardia* outbreaks due to careless campers.

Although virtually unknown 15 years ago, acquired immune deficiency syndrome (AIDS) has become the fifth greatest cause of infectious deaths. The World Health Organization (WHO) estimates that 30 million people are now infected with this virus and 2.3 million die each year. Two-thirds of all AIDS patients are in sub-Saharan Africa. In countries such as Zimbabwe and Zambia, 25 percent of all adults are infected. In spite of high birth rates, some African countries may have declining populations because of this disease. Combinations of powerful new drugs have reduced the AIDS death rate in the United States by half in the past few years. Treating a single patient, however, costs about $10,000 per year, an impossible amount for poor countries such as Ethiopia, where total health expenditures are less than $1 per person per year. AIDS is now spreading rapidly in South and Southeast Asia, where as many as half of all prostitutes are infected.

Malaria may well be the most deadly of any individual communicable diseases, causing around 500 million new infections and at least 2 million deaths per year. As we will discuss shortly, malaria, which once seemed about to be eradicated, is now resurging. A cluster of vaccine-preventable diseases including measles, polio, hepatitis B, and tetanus claim some 1.7 million lives each year, mostly among children in developing countries. Considerable progress is being made in eliminating many of these diseases. Smallpox, for instance, was eradicated in 1977, and polio might be eliminated in 2000, according to the WHO. A massive inoculation campaign in 1998 and 1999 immunized more than 450 million children against polio. An encouraging step in the crusade to eliminate many infectious diseases is the creation of edible vaccines by transferring antigenic genes into crop plants by Dr. Charles Arntzen and his colleagues at Cornell University. These transgenic crops could be grown very cheaply and would avoid the need for refrigeration, needles, and inoculations. Unfortunately, total U.S. expenditures for research on all infectious diseases amount to only $75 million per year. To put that in perspective, we spend $225 million per year to support military bands.

Morbidity and Quality of Life

Death rates don't tell us everything about the burden of disease. Obviously many people suffer from illness who don't die but whose quality of life is severely diminished. When people are sick, work doesn't get done, crops don't get planted or harvested, meals aren't cooked, children can't study and learn. The billion people who suffer from various worms, flukes, and other internal parasites at any given time rarely die from their affliction but the suffering and debilitation is real, nonetheless. Statistics on the total economic and social consequences of diseases are difficult to obtain, but the annual toll is clearly very high (table 9.2).

Over the years health agencies have attempted to develop new ways to measure disability or loss of quality life along with mortality. One of the newest—and still controversial—measures is the **Disability-Adjusted Life Year** or **DALY**. DALYs combine premature deaths and loss of a healthy life resulting from illness or disability. This is an attempt to evaluate the total burden of disease, not just how many people die. Clearly many more years of expected life are lost when a child dies of neonatal tetanus than when a 70-year-old dies of pneumonia. Similarly, a child permanently paralyzed by polio will have many more years of suffering and lost potential than will a senior citizen who has a stroke. According to the World Health Organization, communicable diseases are responsible for nearly half of all the 2.5 billion DALYs lost each year, and children under age 15 account for at least half of that total.

The heaviest burden of illness is borne by the poorest people who can afford neither a healthy environment nor adequate health care. About 90 percent of all DALY losses occur in the developing world where less than one-tenth of all health care dollars is spent (fig. 9.4). The 1.3 billion people who live on less than $1 per day generally lack access to adequate housing, sanitation, and safe drinking water, all of which increase their exposure to pathogens such as those responsible for the roughly 4 billion cases of diarrhea each year. Malnutrition exacerbates many diseases. Well-fed children rarely die from diarrhea or tuberculosis. Better garbage and human waste disposal, clean water, inexpensive childhood inoculations, and a number of other interventions could feasibly prevent somewhere between 10 and 40 percent of this disease burden (table 9.2).

Somewhere around 2 billion people at any given time suffer from worms, flukes, protozoa, and other internal parasites. While victims rarely die from these infections, their quality of life clearly suffers. People whose nutrition is marginal at best can ill afford to donate blood to uninvited guests. The combination of malnutrition

TABLE 9.2 Annual Disease Burden in Poor Households

DISEASE GROUP	MAIN SOURCE	DALY BURDEN* (MILLIONS/YEAR)	FEASIBLE REDUCTION (PERCENT)
Respiratory infections	Indoor air pollution, crowding	119	15
Diarrhea	Water supply, sanitation	99	40
Tuberculosis	Crowding, malnutrition	46	10
Chronic respiratory diseases and cancer	Indoor air pollution, crowding	45	10
Parasites, worms	Garbage disposal, sanitation	26	30
Trachoma	Water supply, hygiene	8	10

*Disability-Adjusted Life Years = years lost from premature death and disability.
Source: Data from World Bank, *World Development Report 1998.*

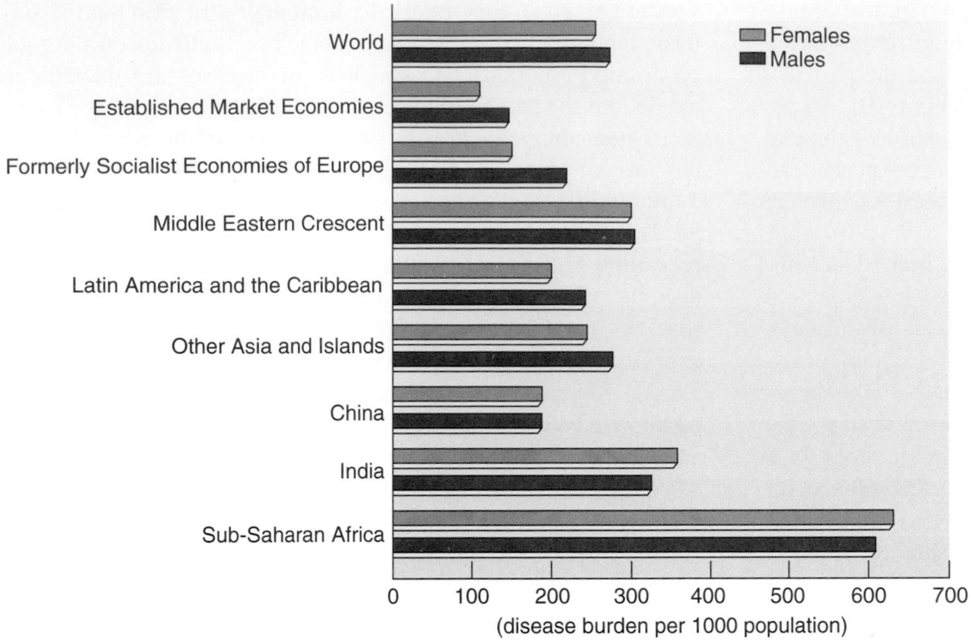

FIGURE 9.4 Disease burden by gender and demographic region.
Source: Data from Christopher J.L. Murray and Alan D. Lopez, eds., *The Global Burden of Disease, Volume 1* (World Health Organization, Harvard School of Public Health, and The World Bank, Geneva, 1996) pp. 541–612.

and anemia from blood loss can retard normal childhood growth and development. They also can impair the immune system, leaving afflicted children and adults less able to battle common diseases such as measles, diarrhea, respiratory infections, tuberculosis, and malaria.

Progress is being made, however, in eradicating some dreadful parasitic diseases. Onchocerciasis (river blindness) is caused by tiny nematodes (round worms) transmitted by the bite of black flies. Masses of dead worms accumulate in the eyeball, destroying vision. This dreadful disease affects some 18 million people and permanently blinds about 500,000 every year. In some African villages, nearly every adult over thirty is blind from onchocerciasis (fig. 9.5) In multinational effort, governments of affected countries working together with the United Nations and the World Bank have begun to control the blackfly by insecticide sprays. In addition, the U.S. pharmaceutical manufacturer Merck & Co. has committed to providing ivermectin—a drug that safely and effectively kills the larvae in the body—free of charge as long as river blindness exists. Similarly, another pharmaceutical company, SmithKline Beecham, has pledged to freely distribute the drug albendazole to about one billion people in an effort to eliminate the grossly disfiguring disease known as elephantiasis (see fig. 12.8) that now afflicts about 120 million people. This effort will cost the company about $500 million over the next two decades. Drancunculiasis or guinea worm, another dreadful tropical diseases, is being eliminated through community health education and provision of safe drinking water (Case Study, p. 193).

FIGURE 9.5 A child leads a blind adult in West Africa. River blindness, caused by tiny nematodes that are transmitted by biting flies, affects 18 million people. In some African villages, nearly every adult over age thirty is blind due to this disease.

Emergent Diseases and Environmental Change

Although many diseases such as measles, pneumonia, and pertussis (whooping cough) have probably inflicted humans for millenia, at least 30 new infectious diseases have appeared in the past two decades while many well-known diseases have reappeared in more virulent, drug-resistant forms. Frightening maladies such as Ebola and Marburg fever have made sudden and highly lethal outbreaks in many places around the world (fig. 9.6) An **emergent disease** is one never known before or one that has been absent for at least 20 years.

CASE STUDY

FIGHTING THE FIERY SERPENT

In 1986, the world health community began a campaign to eliminate the guinea worm (*Dracunculus medinensis*) from the entire world by the year 2000. If successful, this will be only the second global disease ever completely eradicated (smallpox, which was abolished in 1977 was first), and the only time that a human parasite will have been totally exterminated worldwide. Known as the fiery serpent, the guinea worm is a terrible scourge of many tropical countries. Infection starts by drinking stagnant water contaminated with tiny water fleas (called cyclops) containing guinea worm larvae. Inside the human body, the worms grow to as long as one meter (3 ft). After a year of migrating through the body, a threadlike adult worm emerges slowly through a painful skin blister. Most worms come out of the legs or feet but they can appear anywhere on the body. The 8 to 12 weeks of continuous emergence are accompanied by burning pain, fever, nausea, and vomiting. Many victims bathe in a local pond or stream to soothe their fever and pain. When the female worm senses water, she releases tens of thousands of larvae, starting the cycle once again.

As the worm emerges from the wound, it can be rolled around small stick and pulled out a few centimeters each day. Sometimes the entire worm can be extracted in a few days, but the process usually takes weeks. If you pull too fast and the worm breaks off, the part left in your body can die and fester, leading to serious secondary infections. If the worm comes out through a joint, permanent crippling can occur. There is no cure for guinea worm disease once the larvae are ingested. There is no vaccine, and having been infected once doesn't give you immunity. The only way to break the cycle is through behavioral changes. Community health education, providing clean water from wells or by filtering or boiling drinking water, eliminating water fleas by chemical treatment, and teaching infected victims to stay out of drinking supplies are the only solutions to this dreadful problem.

When the eradication campaign was started in 1986, guinea worms were endemic to 16 countries in sub-Saharan Africa as well as Yemen, India, and Pakistan. Every year about 3.5 million people were stricken and at least 100 million people were at risk. With the leadership of former U.S. President Jimmy Carter, a consortium of agencies, institutions, and organizations—including the WHO, UNICEF, the United Nations Development Program (UNDP), the World Bank, bilateral aid agencies, and the governments of many developed countries—banded together to fight this disease. Although complete success has not yet occurred, encouraging progress has been made. Already the guinea worm infections are down more than 96 percent. Pakistan was the first country to be declared completely free of these parasites. Infection rates in Kenya, Senegal, Cameroon, Chad, India, and Yemen are down below 100 cases per year. More than 80 percent of all remaining cases occur in Sudan, where civil war, poverty, drought, and governmental resistance to outside aid have made treatment difficult.

An encouraging outcome of this crusade is the demonstration that public health education and community organization can be effective, even in some of the poorest and most remote areas. Village-based health workers and volunteers conduct disease surveillance and education programs allowing funds and supplies to be distributed in an efficient manner. Once people understand how the disease spreads and what they need to do to protect themselves and their families, they do change their behavior. And when the campaign is completed and guinea worms are completely vanquished, the health workers and volunteers will be available for further community development projects.

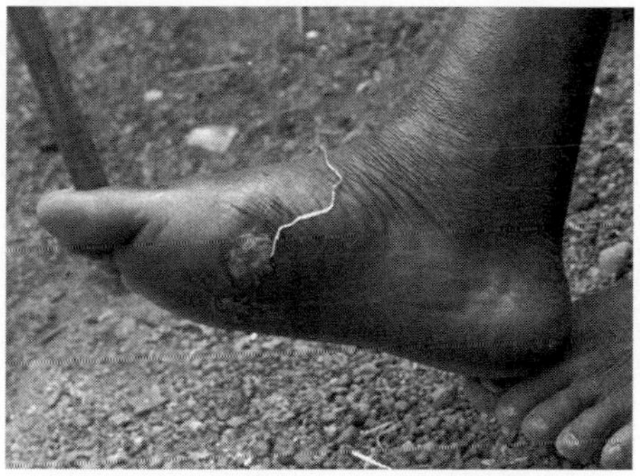

Guinea worms burrow under the skin of human hosts, where they live for a year while growing up to 1 m (3 ft) in length. When mature, the worms emerge at night to lay their eggs then retreat back into a festering wound. There is hope that better sanitation and medical care can completely eradicate this terrible parasite by the year 2000.

Probably the largest loss of life from an individual disease in a single year was the great influenza epidemic of 1918. Somewhere between 30 and 40 million people succumbed to this virus in less than 12 months. This was more than twice the total number killed in all the battles of World War I, which was occurring at the time. Crowded, unsanitary troop ships carrying American soldiers to Europe started the epidemic. War refugees, soldiers from other nations returning home, and a variety of other travelers quickly spread the virus around the globe. Flu is especially contagious, spreading either by direct contact with an infected person or by breathing airborne particles released by coughing or sneezing. Most flu strains are zoonotic (transmitted from an animal host to humans). Pigs, birds, monkeys, and rodents often serve as reservoirs from which viruses can jump to humans. Although new flu strains seem to appear nearly every year, we're fortunate that no epidemic has been as deadly as that of 1918, but we never know what might be waiting for the right moment to strike.

A number of factors contribute currently to the appearance and spread of contagious diseases. With a population of 6 billion, human densities are much higher, enabling germs to spread further and faster than ever before. Expanding populations push into remote areas, hunting animals that once would have never seen humans, converting lands to agriculture, and being exposed to exotic pathogens and parasites. We have caused environmental change on a massive scale: cutting forests, polluting air and water, creating unhealthy urban surroundings, causing local—and perhaps global—climate change.

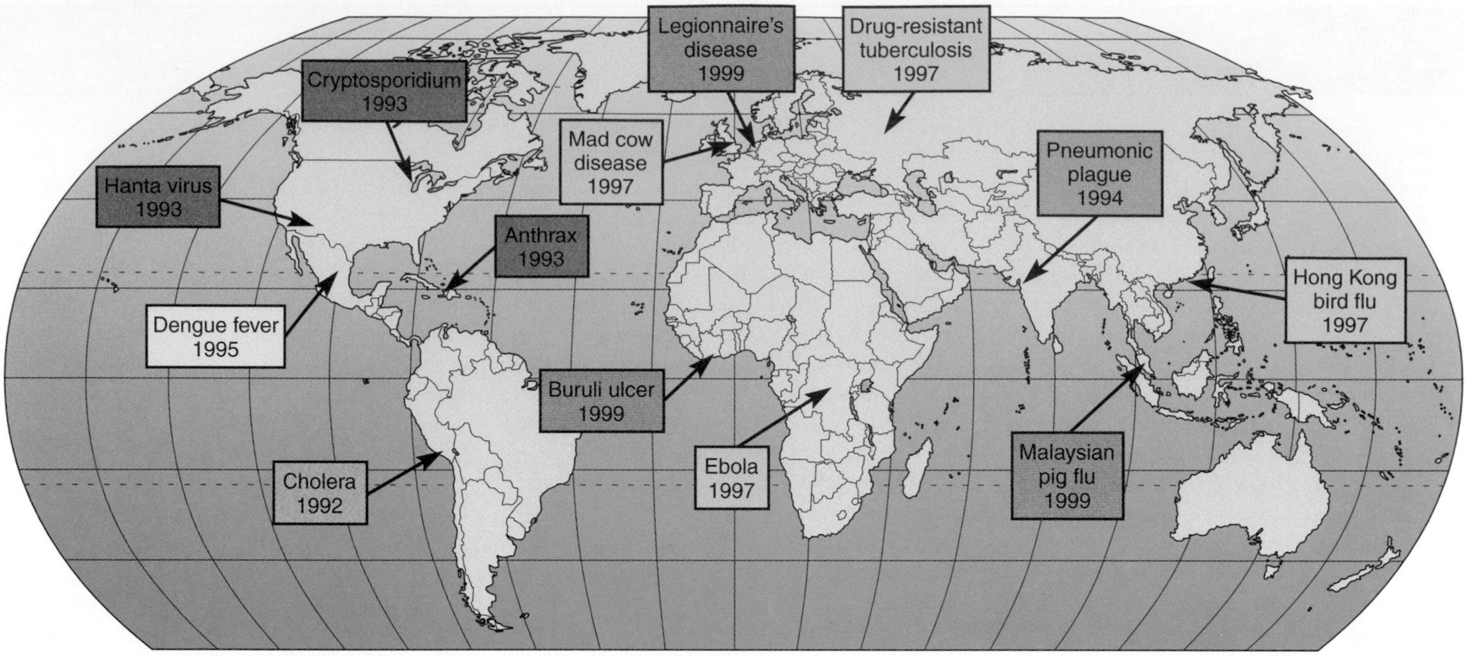

FIGURE 9.6 Some outbreaks of highly lethal, infectious diseases in the 1990s. Why are supercontagious organisms emerging in so many different places?
Source: Data from United States Center for Disease Control and Prevention.

Nutrients and pesticides running off farm fields cause blooms of toxic algae in lakes and near-shore ocean bodies. Elimination of predators and habitat changes favor disease-carrying organisms such as mice, rats, cockroaches, and mosquitoes.

The outbreak of Hanta virus in the American Southwest in 1993 is an example of how environmental change helps spread disease. Sometimes called the Four-Corners disease, after the area where the disease was first recognized, Hanta is a hemorrhagic fever that causes massive fluid accumulation in the lungs. Apparently healthy, young people can be stricken and die within a matter of hours, essentially drowning from pulmonary edema. The primary host appears to be deer mice. Heavier-than-normal rainfall in the early 1990s (probably related to El Niño events), together with agricultural expansion in the desert Southwest, provided food supplies that supported a mouse population explosion. The virus is spread in mouse droppings and urine. Rapidly growing cities and more people camping, hunting, hiking, and working outdoors, brought increasing numbers of people and mice in contact. Although the number of deaths from Hanta pulmonary syndrome is relatively small, it's wise to avoid deer mouse nesting areas if you're in their habitat.

Dengue or breakbone fever is spread by mosquitoes. In the 1960s about 30,000 cases were reported each year. The WHO believes that some 20 million new infections now occur annually and that at least 2.5 billion people live in areas where mosquitoes carry this disease. An important factor in the spread of many diseases is the speed and frequency of modern travel. Millions of people go every day from one place to another by airplane, boat, train, or automobile. Very few places on earth are more than 24 hours by jet plane from any other place. Think of the consequences if someone with Ebola or some other highly virulent disease is put on a plane to London's Gatwick or Chicago's O'Hare Airport through which hundreds of thousands of people pass each day on their way to every continent.

In 1992, after being absent for nearly a century, a cholera epidemic swept down the west coast of South America. More than 1 million people were infected and at least 12,000 died from this severe diarrheal disease. A few years later, the cholera epidemic hit Russia and the number of cases jumped from just 23 in the prior year to more than 1000. Why was it reemerging after a long absence and how could it move so quickly? Cholera is generally spread by contact with water or food contaminated with human waste containing *Vibro cholera* bacteria. Sewage-contaminated water from the bilges of ocean-going vessels dumped into harbors of cities with inadequate water purification facilities is thought to be the means of international transmission. A dormant form of the choleral bacteria has recently been shown to inhabit tiny crustaceans called copepods. Warm water—as occurs during an El Niño episode in the Pacific Ocean off South America—stimulates copepod populations and the bacterium emerges in its infectious form.

Antibiotic and Pesticide Resistance

Malaria, the most deadly of all insect-borne diseases, is an example of the return of a disease that once was thought to be nearly vanquished. Malaria now claims about 3 million lives every year—90 percent in Africa and most of them children. With the advent of modern medicines and pesticides, malaria had nearly been wiped

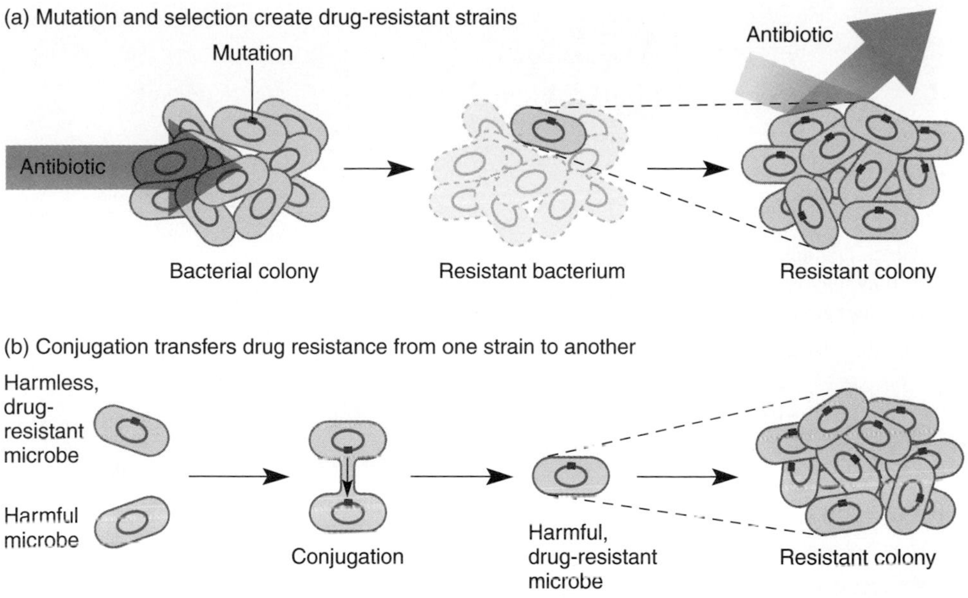

FIGURE 9.7 How microbes acquire antibiotic resistance. Random mutations make a few cells resistant. When challenged by antibiotics, only those cells survive to give rise to a resistant colony. Sexual reproduction (conjugation) or plasmid transfer moves genes from one strain or species to another.

out in many places but recently has come roaring back. The protozoan parasite that causes the disease is now resistant to most antibiotics, while the mosquitoes that transmit it have developed resistance to many insecticides. Spraying of DDT in India and Sri Lanka reduced malaria from millions of infections per year to only a few thousand in the 1950s and 1960s. Now South Asia is back to its pre-DDT level of some 2.5 million new cases of malaria every year. Other places that never had malaria or dengue now have them as a result of climate change and habitat alteration. Gulf-coast states in America, for example are now home to the *Aedes aegypti* mosquito that carries these diseases.

Why have vectors such as mosquitoes and pathogens such as the malaria parasite become resistant to pesticides and antibiotics? Part of the answer is natural selection and the ability of many organisms to evolve rapidly. Another factor is the human tendency to use control measures carelessly. When we discovered that DDT and other insecticides could control mosquito populations, we spread them indiscriminately without much thought to ecological considerations. In the same way, antimalarial medicines such as chloroquine were given to millions of people, whether they showed symptoms or not. This was a perfect recipe for natural selection. Many organisms were exposed only minimally to control measures. This allowed those with natural resistance to outcompete others and spread their genes through the population (fig. 9.7). After repeated cycles of exposure and selection, many microorganisms and their vectors are insensitive to almost all our weapons against them.

There are many examples of drug resistance in pathogens. Tuberculosis, once the foremost cause of death in the world, had nearly been eliminated—at least from the developed world—by the end of the twentieth century. Drug-resistant varieties of TB are now spreading rapidly, however. One of the places these strains arise is in Russia, where crowded prisons with poor sanitation, little medical care, gross overcrowding, and inadequate nutrition serve as a breeding ground for this deadly disease. Inmates who are treated with antibiotics rarely get a complete dose. Those with TB aren't segregated from healthy inmates. Patients with active TB are released from prison and sent home to spread the disease further. And migrants fleeing from Russian economic and social chaos carry the disease to other countries.

Another frightening development is the appearance of drug-resistant strains of *Staphylococcus aureus,* the most common form of hospital-acquired infections. Staph A has many forms, some of which are extremely toxic. You may have heard of toxic-shock syndrome where staphylococcus toxins spread through the body and sometimes bring death in a matter of just hours. Another strain of staphylococcus, sometimes called flesh-eating bacteria, causes massive necrosis that destroys skin, connective tissue, and muscle. For 40 years vancomycin has been the last recourse against staph infections. Strains resistant to everything else could be controlled by this antibiotic. Now vancomycin-resistant staph strains are being reported in many places.

At least half of the 100 million antibiotic doses prescribed every year in the United States are unnecessary or are the wrong ones. And many people who start a course of antibiotic treatment fail to carry it out for the time prescribed. In addition, antibiotics are fed routinely to farm animals in low doses to stimulate weight gain. All of these practices select for resistant organisms. Furthermore, plasmids (small, circular, naked nucleic acid molecules) move from one bacterial cell to another carrying genes. Often a plasmid will transfer a group of antibiotic-resistance genes in a package so a cell that has never encountered a particular drug will already be insensitive to it.

Toxic Chemicals

Dangerous chemical agents are divided into two broad categories: hazardous and toxic. **Hazardous** means dangerous. This category includes flammables, explosives, irritants, sensitizers, acids, and caustics. Many chemicals that are hazardous in high concentrations are relatively harmless when dilute. **Toxins** are poisonous. This means they

what can you do?

Tips for Staying Healthy

- Eat a balanced diet with plenty of fresh fruits, vegetables, legumes, and whole grains. Wash fruits and vegetables carefully, they may well have come from a country where pesticide and sanitation laws are lax.
- Use unsaturated oils such as olive or cannola rather than hydrogenated or semisolid fats such as margarine.
- Cook meats and other foods at temperatures high enough to kill pathogens; clean utensils and cutting surfaces; store food properly.
- Wash your hands frequently. You transfer more germs from hand to mouth than any other means of transmission.
- When you have a cold or flu, don't demand antibiotics from your doctor—they aren't effective against viruses.
- If you're taking antibiotics, continue for the entire time prescribed—quitting as soon as you feel well is an ideal way to select for antibiotic-resistant germs.
- Practice safe sex.
- Don't smoke and avoid smoky places.
- If you drink, do so in moderation. Never drive when your reflexes or judgment are impaired.
- Exercise regularly: walk, swim, jog, dance, garden. Do something you enjoy that burns calories and maintains flexibility.
- Get enough sleep. Practice meditation, prayer, or some other form of stress reduction.
- Make a list of friends and family who make you feel more alive and happy. Spend time with one of them at least once a week.

react with specific cellular components to kill cells. Because of this specificity, they often are harmful even in dilute concentrations. Toxins can be either general poisons that kill many kinds of cells, or they can be extremely specific in their target and mode of action. Ricin, for instance, is a protein found in castor beans and one of the most toxic organic compounds known. Three hundred picograms (trillionths of a gram) injected intravenously is enough to kill an average mouse. A single molecule can kill a cell. This is about two hundred times more lethal than any dioxin, which often are claimed to be the most toxic substances known. Table 9.3 shows some of the toxic chemicals and elements of greatest concern to the EPA.

Allergens are substances that activate the immune system. Some allergens act directly as **antigens;** that is, they are recognized as foreign by white blood cells and stimulate the production of specific antibodies. Other allergens act indirectly by binding to other materials and changing their structure or chemistry so they become antigenic and cause an immune response.

Formaldehyde is a good example of a widely used synthetic chemical that is a powerful sensitizer. It is both directly and indirectly allergenic. Some people who are exposed to formaldehyde in plastics, wood products, insulation, glue, fabrics, and a variety of other products become hypersensitive not only to formaldehyde itself but also to many other materials in their environment, sometimes called the "sick building" syndrome. These individuals may have to go to great lengths to protect themselves from these allergenic substances.

TABLE 9.3 Toxic Chemicals in the United States Causing the Greatest Risk to Human Health

Benzene	Methyl ethyl ketone
Cadmium	Methyl isobutyl ketone
Carbon tetrachloride	Nickel
Chloroform	Tetrachloroethylene
Chromium	Toluene
Cyanides	Trichloroethane
Dichloromethane	Trichloroethylene
Lead	Xylene(s)
Mercury	

Source: Data from Environmental Protection Agency.

Immune system depressants are pollutants that seem to suppress the immune system rather than activate it. Little is known about how this occurs or which chemicals are responsible. Immune system failure is thought to have played a role, however, in widespread deaths of seals in the North Atlantic and of dolphins in the Mediterranean. These dead animals generally contain high levels of pesticide residues, polychlorinated biphenyls (PCBs), and other contaminants, that may disrupt normal endocrine hormone functions (see What Do You Think, chapter 12) and make them susceptible to a variety of opportunistic infections. Similarly, some humans with "sick house" syndrome or other environmental illnesses seem to have defective immune responses. Demonstrating a clear cause-and-effect relationship in these cases usually is difficult, however. Exactly which pollutants affect which people often remains unclear.

Neurotoxins are a special class of metabolic poisons that specifically attack nerve cells (neurons). The nervous system is so important in regulating body activities that disruption of its activities is especially fast-acting and devastating. Different types of neurotoxins act in different ways. Heavy metals such as lead and mercury kill nerve cells and cause permanent neurological damage. Anesthetics (ether, chloroform, halothane, etc.) and chlorinated hydrocarbons (DDT, Dieldrin, Aldrin) disrupt nerve cell membranes necessary for nerve action. Organophosphates (Malathion, Parathion) and carbamates (carbaryl, zeneb, maneb) inhibit acetylcholinesterase, an enzyme that regulates signal transmission between nerve cells and the tissues or organs they innervate (for example, muscle). Most neurotoxins are both extremely toxic and fast-acting.

Mutagens are agents, such as chemicals and radiation, that damage or alter genetic material (DNA) in cells. This can lead to birth defects if the damage occurs during embryonic or fetal growth. Later in life, genetic damage may trigger neoplastic (tumor) growth.

When damage occurs in reproductive cells, the results can be passed on to future generations. Cells have repair mechanisms to detect and restore damaged genetic material, but some changes may be hidden, and the repair process itself can be flawed. It is generally accepted that there is no "safe" threshold for exposure to mutagens. Any exposure has some possibility of causing damage.

Teratogens are chemicals or other factors that specifically cause abnormalities during embryonic growth and development. Some compounds that are not otherwise harmful can cause tragic problems in these sensitive stages of life. One of the most well-known examples of teratogenesis is that of the widely used sedative thalidomide. In the 1960s, thalidomide (marketed under the trade name Cantergan) was the most widely used sleeping pill in Europe. It seemed to have no unwanted effects and was sold without prescription. When used by pregnant women, however, it caused abnormal fetal development resulting in phocomelia (meaning seal like limbs), in which there is a hand or foot, but no arm or leg (fig. 9.8). There is evidence that taking a single thalidomide pill in the first weeks of pregnancy is sufficient to cause these tragic birth defects. Altogether, at least 12,000 children were affected before this drug was withdrawn from the market. Fortunately, thalidomide was not approved for sale in the United States because the Food and Drug Administration was not satisfied with the laboratory tests of its safety.

Ironically, thalidomide has positive as well as negative features. The drug has been found to be effective in treating leprosy and is being tested against AIDS, cancer, retinal degeneration, and tissue rejection in organ transplants. Tragically, these beneficial applications continue to have a dark side. In Brazil, where thalidomide has been used widely to treat leprosy, some doctors failed to warn patients about the dangers of becoming pregnant while on the drug. Other people, hearing about miraculous cures with thalidomide, obtained it from unlicensed laboratories without knowing about its side effects. In 1994, more than 50 cases of thalidomide-related birth defects were reported in Brazil.

Perhaps the most prevalent teratogen in the world is alcohol. Drinking during pregnancy can lead to **fetal alcohol syndrome**—a cluster of symptoms including craniofacial abnormalities, developmental delays, behavioral problems, and mental defects that last throughout a child's life. Even one alcoholic drink a day during pregnancy has been associated with decreased birth weight.

Carcinogens are substances that cause **cancer**, invasive, out-of-control cell growth that results in malignant tumors. Cancer rates rose in most industrialized countries during the twentieth century, and cancer is now the second leading cause of death in the United States, killing 539,533 people in 1999. According to the American Cancer Society, 1 in 2 males and 1 in 3 females in the United States will have some form of cancer in their lifetime. Some authors blame this cancer increase on toxic synthetic chemicals in our environment and diet. Others argue that it is attributable mainly to lifestyle (smoking, sunbathing, alcohol) or simply living longer.

If the number of deaths from cancer is adjusted for age, the only major types that have become more prevalent in recent years are prostate cancer in men and lung cancer in women (fig. 9.9). If we look at total incidence rather than death rates, the picture changes considerably. Recently, the number of new cases of breast, testis, and skin cancer, for example, have increased significantly, but so has the success in treating these diseases. Better diagnosis and treatment, rather than lower incidence, is probably the reason that mortality for cancers such as stomach, uterus, and colon have decreased over the past thirty years.

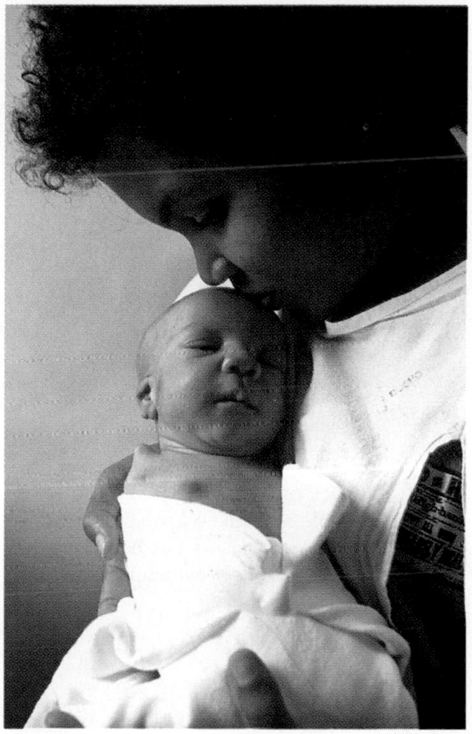

FIGURE 9.8 Development of this baby's arms and legs was blocked when its mother took the sedative thalidomide early in her pregnancy. Although the drug has been banned in Europe and North America for twenty years, it is still used to treat leprosy in some countries. Unfortunately, some of this potent teratogen is used by pregnant women who are unaware of its tragic side effects.

Natural and Synthetic Toxins

There has been so much bad news lately about the dangers of industrial chemicals that some people assume that all human-made compounds are poisonous while all natural materials must be benign and innocuous. In fact, many natural chemicals are very dangerous while many synthetic ones are relatively harmless. Since most plants and many animal species can't escape from predators or defend themselves by fighting back, many of them have evolved a kind of chemical warfare, secreting or storing in their tissues a vast arsenal of irritants, toxins, metabolic disrupters, and other chemicals that discourage competitors and predators. Some of the chemical defenses employed by organisms are very sophisticated and specific. Both plants and animals make chemicals similar to—or even identical with—neurotransmitters, hormones, or regulatory molecules of predators or potential enemies. Our cells don't distinguish whether these chemicals are natural or synthetic.

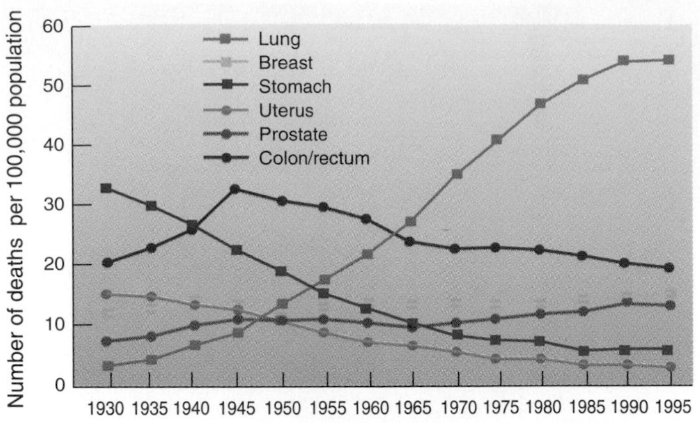

FIGURE 9.9 Age-adjusted cancer death rates in the United States. Although the total incidence of cancer is up in recent years, better treatment means that more people survive. When adjusted for an aging population, mortality for most major cancers has been stable or falling. One exception is lung cancer, 90 percent of which is attributable to increased smoking, especially among women.

You may have heard the argument that we evolved together with natural toxins whereas synthetic chemicals are so new that we haven't had time to adapt to them. But, in fact, natural chemicals such as arsenic and cyanide have been around longer than we have but are still toxic to us. Organic compounds like solanine and oxalic acid are common in many of the foods we eat but we haven't developed an immunity to them.

Toxicologist Bruce Ames claims that there are 10,000 times as many natural pesticides in our diets as synthetic ones. He argues that our fear of synthetic chemicals may divert our attention from more important issues. He finds natural compounds in crops such as potatoes, tomatoes, coffee, celery, and mushrooms, for instance, that are more carcinogenic than some commercial products. Ironically, plants attacked by insects may synthesize natural toxins that are more dangerous than the residues left from protective treatment with synthetic pesticides. Similarly, treatment of crops with fungicides (many of which are carcinogenic) may prevent growth of molds that are even more carcinogenic. Simply because food is raised "organically" may not necessarily make it safer than food raised by current commercial practices.

Other environmental health specialists, however, argue that we should not underrate the dangers of toxic synthetic chemicals. Natural chemicals in our diet are significantly different when mixed with fiber and a multitude of other substances than they may appear as pure chemicals in laboratory tests. Broccoli, for example, clearly contains carcinogens, but reduces the risk of cancer when added to the diet of laboratory animals because it also contains anticancer factors.

Physical Agents, Trauma, and Stress

Physical agents, such as radiation, also are serious environmental health hazards. Radiation associated with nuclear power is discussed in chapter 21. Although the data are often contradictory, some researchers believe that ubiquitous low-frequency electromagnetic fields associated with power lines and ordinary household appliances may pose a health risk. Noise, another important physical health threat, is discussed in chapter 24.

Trauma, injury caused by accidents or violence, has surely always been a life-threatening environmental factor for humans. The incidence of trauma-related deaths per 1000 persons is very similar in developed and less-developed countries. Accidents and violence cause far fewer deaths in the United States than infectious diseases do, even though media coverage might make it seem otherwise. Still, trauma is the principal cause of death for people between the ages of 1 and 38 in the United States, and it is the leading cause of years of life lost before age 65 in all industrialized countries. Every year there are about 100,000 premature deaths from accidents and violence in the United States and twice as many cases of permanent disability. More than half of these deaths and injuries are caused by motor vehicle collisions. About 90 percent of the accidents involve private automobiles, trucks, or motorcycles, and at least half are caused by drivers under the influence of alcohol or other drugs.

Stress and lifestyle once were considered factors that might cause unhappiness but were not life-threatening. We now know, however, that stress is clearly related to physical diseases, such as heart attack, stroke, and atherosclerosis, which are the leading causes of nontrauma-related death in the United States and most other industrialized countries. Since stress is a component of our cultural environment, it deserves a place in a discussion of environmental health.

In medical terms, **stress** refers to physical, chemical, or emotional factors that place a strain on an organism for which there is inadequate adaptation. This can result in physical responses that contribute to disease. Adverse stress responses are not unique to humans. Plants show signs of environmental stress, as do most animals. When laboratory or zoo animals are kept in cages next to especially aggressive members of the same species, they often show many signs of anxiety and stress. Even though they are separated by glass walls so that no physical contact is possible, the stressed animals often die prematurely of cardiovascular diseases. Gastrointestinal disturbances, such as ulcers, are common human responses to stress. Stress also contributes to susceptibility to infectious diseases, as many students realize at exam time.

Diet

Diet also has an important effect on health. For instance, there is a strong correlation between cardiovascular disease and the amount of salt and animal fat in one's diet. Fat intake was once thought to be associated with breast cancer, but long-term studies in several countries suggest that the linkage may not be correct. Still, fat intake may be linked to other cancers (fig. 9.10). Highly processed foods, fat, and smoke-cured, high-nitrate meats also seem to be associated with cancer.

Fruits, vegetables, whole grains, complex carbohydrates, and dietary fiber (plant cell walls), on the other hand, often have

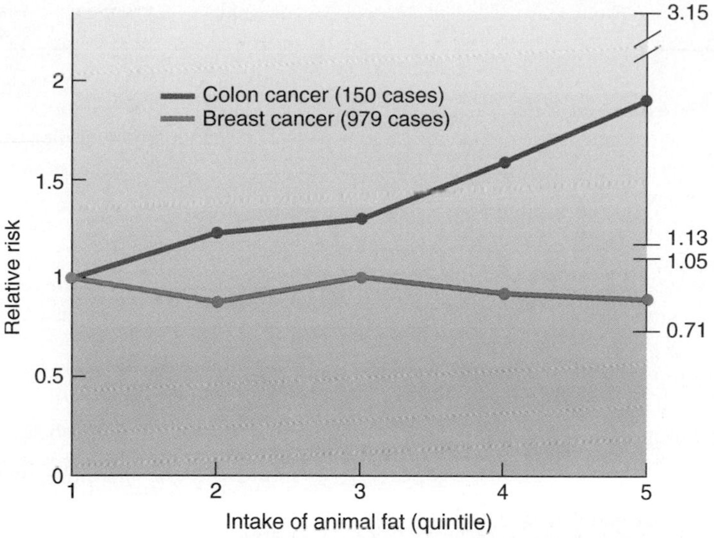

FIGURE 9.10 Relative risk of colon and breast cancer for different levels of animal fat intake in Japan. Bars represent 95 percent confidence intervals for women in the highest quintile (fifth of the population).
Source: Data from *Food and Nutrition Paper #57,* Fig. 11.2, UNFAO.

beneficial health effects. Certain dietary components, such as pectins; vitamins A, C, and E; substances produced in cruciferous vegetables (cabbage, broccoli, cauliflower, brussels sprouts); and selenium, which we get from plants, seem to have anticancer effects.

Eating too much food is a significant dietary health factor in developed countries and among the well-to do everywhere. At least one-fourth of all Americans are considered overweight. Cutting back on the number of calories consumed reduces the strain on bones, muscles, and other organs and has additional beneficial effects, including reduction of cardiovascular disease, diabetes, and—perhaps—cancer.

In some areas of the world, people seem to live exceptionally long lives. The Abkhasian people in the Caucasus Mountains of Soviet Georgia, the Hunzans in the mountains of Pakistan, and the Vilcabama villagers in Ecuador, for instance, appear to be among the longest-lived people in the world. Many claim to be 120 to 140 years old, although it is difficult to substantiate when they were born. Still, they do appear to live longer and to be more physically active later in life than do most of us.

These people have several factors in common that probably contribute to longevity. They live at moderately high elevations where the climate is cool, dry, and sunny. They lead active, vigorous lives in low-pressure, nonindustrialized settings. Life in their small villages is uniform and predictable. Stress levels are low. All ages work together in the fields and in the home, doing practical, physical work that directly benefits their community. The elderly are respected and live a useful, active life. The whole family shares in decision making, recreation, and religion. Their diet is usually simple, with low fat and salt, high fiber content, little meat, and lots of fruits and vegetables. They enjoy clean air and pure water. Their culture and uncomplicated lifestyles reduce conflict and anxiety. We might benefit from incorporating some aspects of their lives into our own (table 9.4).

TABLE 9.4 National Health Recommendations and Diet Goals

1. Balance the food you eat with physical activity to maintain or improve your weight.
2. Choose a diet with plenty of grain products, vegetables, and fruits.
3. Choose a diet low in fat, saturated fat, and cholesterol.
4. Eat a variety of foods.
5. Choose a diet moderate in salt and sodium.
6. Choose a diet moderate in sugars.
7. If you drink alcoholic beverages, do so in moderation.

Source: Data from U.S. Department of Health and Human Services, 1995.

MOVEMENT, DISTRIBUTION, AND FATE OF TOXINS

There are many sources of toxic and hazardous chemicals in the environment and many factors related to each chemical itself, its route or method of exposure, and its persistence in the environment, as well as characteristics of the target organism (table 9.5), that determine the danger of the chemical. We can think of an ecosystem as a set of interacting compartments between which a chemical moves, based on its molecular size, solubility, stability, and reactivity (fig. 9.11). The routes used by chemicals to enter our bodies also play important roles in determining toxicity (fig. 9.12). In this section, we will consider some of these characteristics and how they affect environmental health.

Solubility

Solubility is one of the most important characteristics in determining how, where, and when a toxic material will move through the environment or through the body to its site of action (What Do You Think? p. 200). Chemicals can be divided into two major groups: those that dissolve more readily in water and those that dissolve more readily in oil. Water-soluble compounds move rapidly and widely through the environment because water is ubiquitous. They also tend to have ready access to most cells in the body because aqueous solutions bathe all our cells. Molecules that are oil- or fat-soluble (usually organic molecules) generally need a carrier to move through the environment, into, and within, the body. Once inside the body, however, oil-soluble toxins penetrate readily into tissues

what do you think?

THE WORLD AROUND YOU

Soft Vinyl Toys and Medical Supplies

Are soft vinyl baby toys and medical products safe or should they be banned? Rival groups on opposite sides of this question have recently issued numerous reports denouncing each other and attacking the motives, methods, integrity, and conclusions of their opponents. How can we know whom to believe or what to do?

At stake in this controversy is a wide range of useful products ranging from medical IV bags, catheters, disposable gloves, and surgical equipment, to squeezable children's toys, baby bottles, rattles, and teething rings. What these things have in common is that they are made of polyvinyl chloride (PVC) softened with phthalate (pronounced "thalate") plasticizers. There are many phthalates, but the two most common are di (2-ethylhexyl) phthalate (DEHP)—found generally in medical devices—and diisononyl phthalate (DINP), which is used primarily in baby toys. These compounds can make up as much as half the weight of some soft vinyls. Because plasticizers don't bind chemically to the polymer, they can diffuse out of the final product under the right conditions.

Both DEHP and DINP are known to be toxic at high doses to laboratory animals, having been linked to a variety of illnesses including reproductive abnormalities as well as kidney and liver damage and possibly some cancers. The question is whether high enough levels of plasticizers are likely to leech out of PVC products to be dangerous. According to the U.S. Consumer Product Safety Commission, there is little to fear. "Based on scientific studies currently available," they reported, "the staff concludes that few, if any, children are at risk from liver or other organ toxicity from the release of DINP from these products. This is because the amount they might ingest does not reach a level that would be harmful." They also suggested, however, that we should "continue to work on better estimates of the amount of phthalate released when products are mouthed by children."

Not everyone agrees that phthalates are safe. A group of American health workers and environmentalists calling themselves Health Care Without Harm (HCWH) claim soft PVC in medical products and children's toys represents an unacceptable and unnecessary risk. They argue that the "precautionary principle" says that we shouldn't be exposed to toxic chemicals in consumer products and that the burden of proof of harmlessness lies with manufacturers, not the general public. In 1999, the European Union banned all sales of soft PVC toys.

"Nonsense," says former Surgeon General C. Everett Koop. "This is just the latest phony chemical scare. . . The ceaseless obsession with ousting the frequently nonexistent bogeymen from our chemical cornucopia does quite a lot to strengthen the ranks of consumer groups but very little to actually improve the health and quality of our lives. And while it provides television newsmagazines with a well-worn story line, it ultimately diverts our attention from real opportunities to enhance life and longevity. . . Highly inflammatory scare campaigns and dubious news stories have unnecessarily frightened families into believing their baby's teething rings and vinyl toys were conduits of cancer-causing chemicals. A distinguished panel of 17 scientists and physicians, under my chairmanship, concluded that the chemicals DINP and DEHP used to make toys and medical devices soft and flexible, are safe and pose no harm to adults or children."

HCWH, on the other hand, led by the environmental group Greenpeace, charges that Koop's "blue ribbon" panel of experts, which was funded primarily by the chemical industry, rejected important scientific studies and ignored safe, cost-competitive substitutes for every current soft PVC product. Claiming that the chlorine industry is trying to bog down the debate in a "dueling risk assessment strategy perfected long ago by the tobacco industry," HCWH argues that it is simply immoral to knowingly expose babies and sick people to toxic materials. Although they acknowledge being considered "on the fringe" by some mainstream environmental groups, HCWH points out that a number of products such as asbestos, tobacco, DDT, and tetraethyl lead were promoted by industry as perfectly safe but subsequently turned out to be quite dangerous.

Consumer pressure and fears of lawsuits have lead many manufacturers to quickly pull soft PVC products—especially baby toys and teethers—off store shelves. Most soft plastics in stores today are polyethylene or polypropylene, which do not contain plasticizers. What do you think of this debate? Are manufacturers caving in to mass hysteria, or should they have known better than to expose us to soft PVC in the first place? How would you evaluate the claims of the various sides of this argument? Can you think of reasons that one side or the other might be trying to mislead you? Is this a moral issue or simply a question of scientific facts? What additional information would you need to make an informed judgment about the safety of soft PVC?

---

and cells because the membranes that enclose cells are themselves made of similar oil-soluble chemicals. Once they get inside cells, oil-soluble materials are likely to be accumulated and stored in lipid deposits where they may be protected from metabolic breakdown and persist for many years.

Bioaccumulation and Biomagnification

Cells have mechanisms for **bioaccumulation,** the selective absorption and storage of a great variety of molecules. This allows them to accumulate nutrients and essential minerals, but at the same time, they also may absorb and store harmful substances through these same mechanisms. Toxins that are rather dilute in the environment can reach dangerous levels inside cells and tissues through this process of bioaccumulation.

The effects of toxins also are magnified in the environment through food webs. **Biomagnification** occurs when the toxic burden of a large number of organisms at a lower trophic level is accumulated and concentrated by a predator in a higher trophic level. Phytoplankton and bacteria in aquatic ecosystems, for instance, take up heavy metals or toxic organic molecules from water or sediments (fig. 9.13). Their predators—zooplankton and

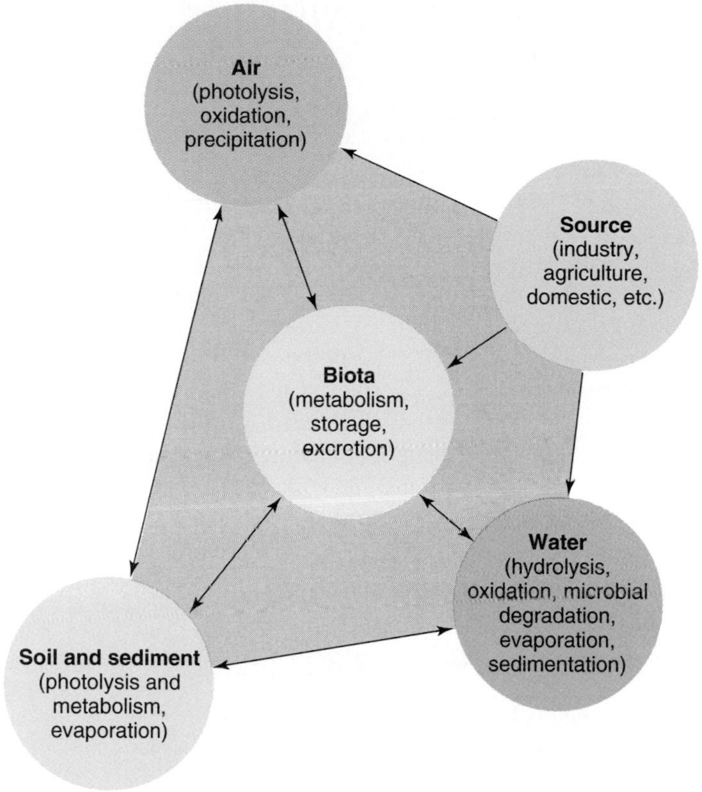

FIGURE 9.11 Movement and fate of chemicals in the environment. Processes that modify, remove, or sequester compounds are shown in parentheses. Toxins also move directly from a source to soil and sediment.

TABLE 9.5 Factors in Environmental Toxicity

FACTORS RELATED TO THE TOXIC AGENT

1. Chemical composition and reactivity
2. Physical characteristics (such as solubility, state)
3. Presence of impurities or contaminants
4. Stability and storage characteristics of toxic agent
5. Availability of vehicle (such as solvent) to carry agent
6. Movement of agent through environment and into cells

FACTORS RELATED TO EXPOSURE

1. Dose (concentration and volume of exposure)
2. Route, rate, and site of exposure
3. Duration and frequency of exposure
4. Time of exposure (time of day, season, year)

FACTORS RELATED TO ORGANISM

1. Resistance to uptake, storage, or cell permeability of agent
2. Ability to metabolize, inactivate, sequester, or eliminate agent
3. Tendency to activate or alter nontoxic substances so they become toxic
4. Concurrent infections or physical or chemical stress
5. Species and genetic characteristics of organism
6. Nutritional status of subject
7. Age, sex, body weight, immunological status, and maturity

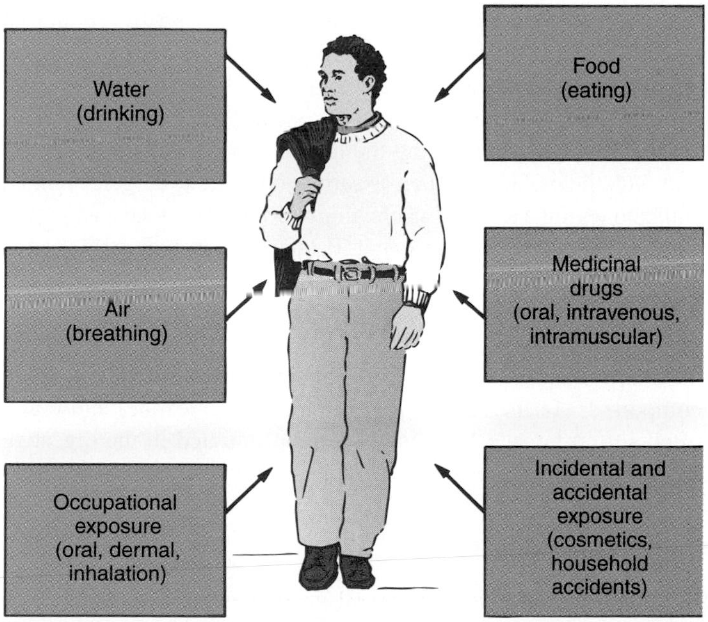

FIGURE 9.12 Routes of exposure to toxic and hazardous environmental factors.

small fish—collect and retain the toxins from many prey organisms, building up higher concentrations of toxins. The top carnivores in the food chain—game fish, fish-eating birds, and humans—can accumulate such high toxin levels that they suffer adverse health effects (chapter 20). One of the first known examples of bioaccumulation and biomagnification was DDT, which accumulated through food chains so that by the 1960s it was shown to be interfering with reproduction of peregrine falcons, brown pelicans, and other predatory birds at the top of their food chains (chapter 12).

Persistence

Some chemical compounds are very unstable and degrade rapidly under most environmental conditions so that their concentrations decline quickly after release. Some of the modern herbicides, for instance, quickly lose their toxicity. Other substances are more persistent and last for long times. Some of the most useful chemicals such as chlorofluorocarbons, PVC plastics, chlorinated hydrocarbon pesticides, and asbestos are valuable because they are resistant to degradation. This stability also causes problems because

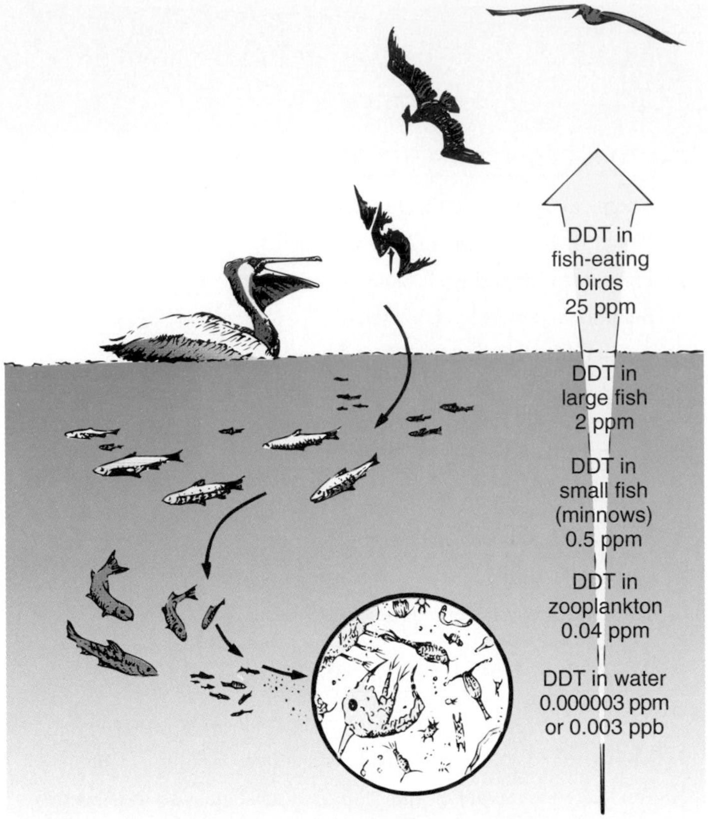

FIGURE 9.13 Bioaccumulation and biomagnification. Organisms lower on the food chain take up and store toxins from the environment. They are eaten by larger predators, who are eaten, in turn, by even larger predators. The highest members of the food chain can accumulate very high levels of the toxin.

these materials persist in the environment and have unexpected effects far from the sites of their original use. DDT, for instance, is a useful pesticide because it breaks down very slowly and doesn't have to be reapplied very often. Its toxic effects may spread to unintended victims, however, and it may be stored for long periods of time in organisms that lack mechanisms to destroy it.

Chemical Interactions

Some materials produce *antagonistic* reactions. That is, they interfere with the effects or stimulate the breakdown of other chemicals. For instance, vitamins E and A can reduce the response to some carcinogens. Other materials are *additive* when they occur together in exposures. Rats exposed to both lead and arsenic show twice the toxicity of only one of these elements. Perhaps the greatest concern is *synergistic* effects. Synergism is an interaction in which one substance exacerbates the effects of another. For example, occupational asbestos exposure increases lung cancer rates 20-fold. Smoking increases lung cancer rates by the same amount. Asbestos workers who also smoke, however, have a 400-fold increase in cancer rates. How many other toxic chemicals are we exposed to that are below threshold limits individually but combine to give toxic results?

MECHANISMS FOR MINIMIZING TOXIC EFFECTS

A fundamental concept in toxicology is that every material can be poisonous under some conditions, but most chemicals have some safe level or threshold below which their effects are undetectable or insignificant. Each of us consumes lethal doses of many chemicals over the course of a lifetime. One hundred cups of strong coffee, for instance, contain a lethal dose of caffeine. Similarly, one hundred aspirin tablets, or ten kilograms (22 lbs) of spinach or rhubarb, or a liter of alcohol would be deadly if consumed all at once. Taken in small doses, however, most toxins can be broken down or excreted before they do much harm. Furthermore, damage they cause can be repaired. Sometimes, however, mechanisms that protect us from one type of toxin or at one stage in the life cycle become deleterious with another substance or in another stage of development. Let's look at how these processes help protect us from harmful substances as well as how they can go awry.

Metabolic Degradation and Excretion

Most organisms have enzymes that process waste products and environmental poisons to reduce their toxicity. In mammals, most of these enzymes are located in the liver, the primary site of detoxification of both natural wastes and introduced poisons. Sometimes, however, these reactions work to our disadvantage. Compounds, such as benzepyrene, for example, that are not toxic in their original form are processed by these same enzymes into cancer-causing carcinogens. Why would we have a system that makes a chemical more dangerous? Evolution and natural selection are expressed through reproductive success or failure. Defense mechanisms that protect us from toxins and hazards early in life are "selected for" by evolution. Factors or conditions that affect postreproductive ages (like cancer or premature senility) usually don't affect reproductive success or exert "selective pressure."

We also reduce the effects of waste products and environmental toxins by eliminating them from our body through excretion. Volatile molecules, such as carbon dioxide, hydrogen cyanide, and ketones are excreted via breathing. Some excess salts and other substances are excreted in sweat. Primarily, however, excretion is a function of the kidneys, which can eliminate significant amounts of soluble materials through urine formation. Accumulation of toxins in the urine can damage this vital system, however, and the kidneys and bladder often are subjected to harmful levels of toxic compounds. In the same way, the stomach, intestine, and colon often suffer damage from materials concentrated in the digestive system and may be afflicted by diseases and tumors.

Repair Mechanisms

In the same way that individual cells have enzymes to repair damage to DNA and protein at the molecular level, tissues and organs that are exposed regularly to physical wear-and-tear or to toxic or hazardous materials often have mechanisms for damage repair. Our skin and the epithelial linings of the gastrointestinal tract, blood vessels, lungs, and urogenital system have high cellular reproduction rates to replace injured cells. With each reproduction cycle, however, there is a chance that some cells will lose normal growth controls and run amok, creating a tumor. Thus any agent, such as smoking or drinking, that irritates tissues is likely to be carcinogenic. And tissues with high cell-replacement rates are among the most likely to develop cancers.

MEASURING TOXICITY

In 1540, the German scientist Paracelsus said "the dose makes the poison," by which he meant that almost everything is toxic at some level. This remains the most basic principle of toxicology. Sodium chloride (table salt), for instance, is essential for human life in small doses. If you were forced to eat a kilogram of salt all at once, however, it would make you very sick. A similar amount injected into your bloodstream would be lethal. How a material is delivered—at what rate, through which route of entry, and in what medium—plays a vitally important role in determining toxicity.

This does not mean that all toxins are identical, however. Some are so poisonous that a single drop on your skin can kill you. Others require massive amounts injected directly into the blood to be lethal. Measuring and comparing the toxicity of various materials is difficult because not only do species differ in sensitivity, but individuals within a species respond differently to a given exposure. In this section, we will look at methods of toxicity testing and at how results are analyzed and reported.

Animal Testing

The most commonly used and widely accepted toxicity test is to expose a population of laboratory animals to measured doses of a specific substance under controlled conditions. This procedure is expensive, time-consuming, and often painful and debilitating to the animals being tested. It commonly takes hundreds—or even thousands—of animals, several years of hard work, and hundreds of thousands of dollars to thoroughly test the effects of a toxin at very low doses. More humane toxicity tests using computer simulation of model reactions, cell cultures, and other substitutes for whole living animals are being developed. However, conventional large-scale animal testing is the method in which we have the most confidence and on which most public policies about pollution and environmental or occupational health hazards are based.

In addition to humanitarian concerns, there are several problems in laboratory animal testing that trouble both toxicologists and policymakers. One problem is differences in sensitivity to a toxin of the members of a specific population. Figure 9.14 shows a typical dose/response curve for exposure to a hypothetical toxin. Some individuals are very sensitive to the toxin, while others are insensitive. Most, however, fall in a middle category forming a bell-shaped curve. The question for regulators and politicians is whether we should set pollution levels that will protect everyone, including the most sensitive people, or only aim to protect the average person. It might cost billions of extra dollars to protect a very small number of individuals at the extreme end of the curve. Is that a good use of resources?

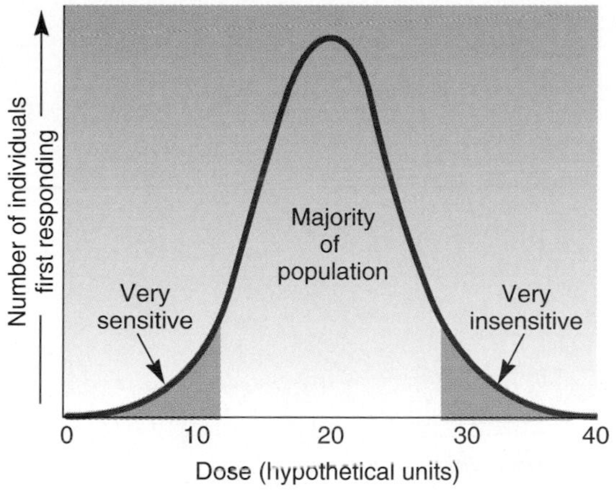

FIGURE 9.14 Probable variations in sensitivity to a toxin within a population. Some members of a population may be very sensitive to a given toxin, while others are much less sensitive. The majority of the population falls somewhere between the two extremes.

Dose/response curves are not always symmetrical, making it difficult to compare toxicity of unlike chemicals or different species of organisms. A convenient way to describe toxicity of a chemical is to determine the dose to which 50 percent of the test population is sensitive. In the case of a lethal dose (LD), this is called the **LD50** (fig. 9.15).

Unrelated species can react very differently to the same toxin, not only because body sizes vary but also because of differences in physiology and metabolism. Even closely related species can have very dissimilar reactions to a particular toxin. Hamsters, for instance, are nearly 5000 times less sensitive to some dioxins than are guinea pigs. Of 226 chemicals found to be carcinogenic in either rats or mice, 95 caused cancer in one species but not the other. These variations make it difficult to estimate the risks for humans since we don't consider it ethical to perform controlled experiments in which we deliberately expose people to toxins.

Toxicity Ratings

It is useful to group materials according to their relative toxicity. A moderate toxin takes about one gram per kilogram of body weight (about two ounces for an average human) to make a lethal dose. Very toxic materials take about one-tenth that amount, while extremely toxic substances take one-hundredth as much (only a few drops) to kill most people. Supertoxic chemicals are extremely potent; for some, a few micrograms (millionths of a gram—an amount invisible to the naked eye) make a lethal dose. These materials are not all synthetic. One of the most toxic chemicals known, for instance, is ricin, a protein found in castor bean seeds. It is so toxic that 0.3 billionths of a gram given intravenously will generally kill a mouse. If aspirin were this toxic, a single tablet, divided evenly, could kill 1 million people. Table 9.6 shows some representative LD50s for a

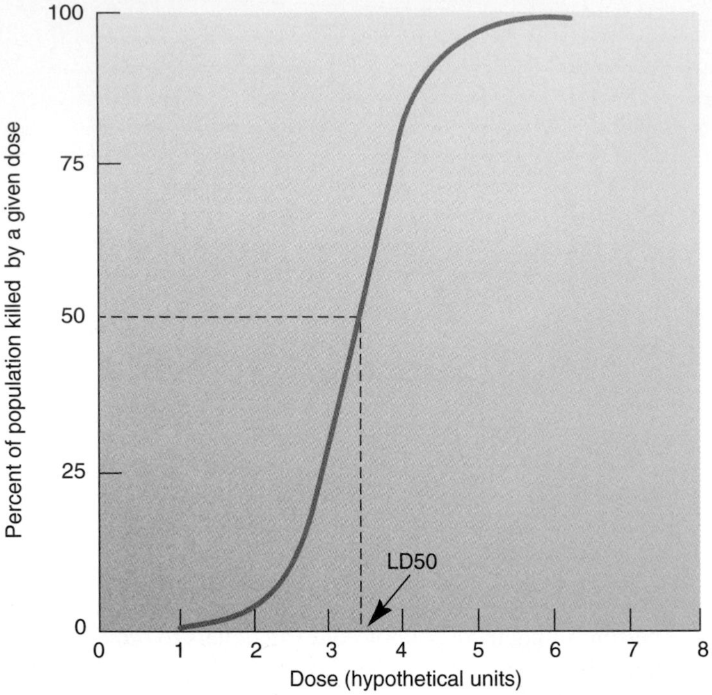

FIGURE 9.15 Cumulative population response to increasing doses of a toxin. The LD50 is the dose that is lethal to half the population.

TABLE 9.6 Acute Lethal Dose for Some Toxic Organic Chemicals

CHEMICAL	EXPOSURE	LD50
Ricin (castor bean)	Ivn-mus	3 ng/kg
	Orl-rat	100 mg/kg
Botulism toxin	Ipr-mus	160 ng/kg
Dioxin (tetrachlorodioxin)	Orl-gpg	600 ng/kg
	Orl-hmstr	3 mg/kg
Muscarine (mushroom poison)	Ivn-mus	250 μg/kg
Parathion (insecticide)	Ipr-rat	1.5 mg/kg
Aflatoxin (fungal toxin)	Orl-mky	1.75 mg/kg
Nicotine	Ivn-cat	2 mg/kg
	Orl-rat	53 mg/kg
DDT (dichlorodiphenyl-trichloroethane)	Orl-hum	50 mg/kg
Toxaphene	Orl-rat	60 mg/kg
2,4-D (dichlorophenoxyacetic acid)	Orl-hum	80 mg/kg

orl = oral, ivn = intravenous, ipr = intraperitoneal, mus = mouse, mky = monkey, hmstr = hamster, hum = human, gpg = guinea pig
nanogram (ng) = 1×10^{-9} g
microgram (μg) = 1×10^{-6} g
milligram (mg) = 1×10^{-3} g
Source: Data from *Registry of Toxic Effects of Chemical Substances,* National Institute for Occupational Safety and Health, 1985.

variety of exposure routes and test species. This is not meant to be an exhaustive table but rather a sampling to show relative toxicities and a few of the organisms and routes of administration used in these tests.

Many carcinogens, mutagens, and teratogens are dangerous at levels far below their direct toxic effect because abnormal cell growth exerts a kind of biological amplification. A single cell, perhaps altered by a single molecular event, can multiply into millions of tumor cells or an entire organism. Just as there are different levels of direct toxicity, however, there are different degrees of carcinogenicity, mutagenicity, and teratogenicity. Methanesulfonic acid, for instance, is highly carcinogenic, while the sweetener saccharin is a suspected carcinogen whose effects may be vanishingly small.

Acute versus Chronic Doses and Effects

Most of the toxic effects that we have discussed so far have been **acute effects.** That is, they are caused by a single exposure to the toxin and result in an immediate health crisis of some sort. Often, if the individual experiencing an acute reaction survives this immediate crisis, the effects are reversible. **Chronic effects,** on the other hand, are long lasting, perhaps even permanent. A chronic effect can result from a single dose of a very toxic substance, or it can be the result of a continuous or repeated sublethal exposure.

We also describe long-lasting *exposures* as chronic, although their effects may or may not persist after the toxin is removed. It usually is difficult to assess the specific health risks of chronic exposures because other factors, such as aging or normal diseases, act simultaneously with the factor under study. It often requires very large populations of experimental animals to obtain statistically significant results for low-level chronic exposures. Toxicologists talk about "megarat" experiments in which it might take a million rats to determine the health risks of some supertoxic chemicals at very low doses. Such an experiment would be terribly expensive for even a single chemical, let alone for the thousands of chemicals and factors suspected of being dangerous.

An alternative to enormous studies involving millions of animals is to give massive amounts—usually the maximum tolerable dose—of a toxin being studied to a smaller number of individuals and then to extrapolate what the effects of lower doses might have been. This is a controversial approach because it is not clear that responses to toxins are linear or uniform across a wide range of doses.

Figure 9.16 shows three possible results from low doses of a toxin. Curve (*a*) shows a baseline level of response in the population, even at zero dose of the toxin. This suggests that some other factor in the environment also causes this response. Curve (*b*) shows a straight-line relationship from the highest doses to zero exposure. Many carcinogens and mutagens show this kind of response. Any exposure to such agents, no matter how small, carries some risks. Curve (*c*) shows a threshold for the response where some minimal dose is necessary before any effect can be observed. This generally suggests the presence of some defense mechanism that prevents the toxin from reaching its target in an active form or

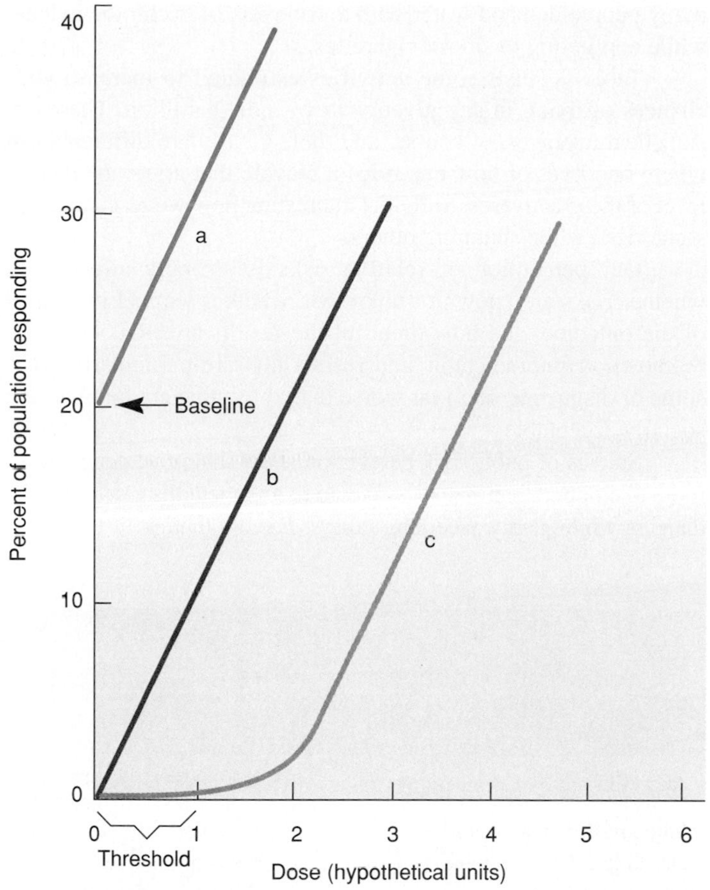

FIGURE 9.16 Three possible dose-response curves at low doses. (*a*) Some individuals respond, even at zero dose, indicating that some other factor must be involved. (*b*) Response is linear down to the lowest possible dose. (*c*) Threshold must be passed before any response is seen.

repairs the damage that it causes. Low levels of exposure to the toxin in question may have no deleterious effects, and it might not be necessary to try to keep exposures to zero.

Which, if any, environmental health hazards have thresholds is an important but difficult question. The 1958 Delaney Clause to the U.S. Food and Drug Act forbids the addition of *any* amount of known carcinogens to food and drugs, based on the assumption that any exposure to these substances represents unacceptable risks. This standard was replaced in 1996 by a *no reasonable harm* requirement, defined as less than one cancer for every million people exposed over a lifetime. This change was supported by a report from the National Academy of Sciences concluding that synthetic chemicals in our diet are unlikely to represent an appreciable cancer risk. We will discuss risk analysis in the next section.

Detection Limits

You may have seen or heard dire warnings about toxic materials detected in samples of air, water, or food. A typical headline announced recently that 23 pesticides were found in 16 food samples. What does that mean? The implication seems to be that any amount of dangerous materials is unacceptable and that counting the numbers of compounds detected is a reliable way to establish danger. We have seen, however, that the dose makes the poison. It matters not only what is there, but how much, where it is located, how accessible it is, and who is exposed. At some level, the mere presence of a substance is insignificant.

Toxins and pollutants may seem to be more widespread now than in the past, and this is surely a valid perception for many substances. The daily reports we hear of new materials found in new places, however, are also due, in part, to our more sensitive measuring techniques. Twenty years ago, parts per million were generally the limits of detection for most chemicals. Anything below that amount was often reported as zero or absent rather than more accurately as undetected. A decade ago, new machines and techniques were developed to measure parts per billion. Suddenly, chemicals were found where none had been suspected. Now we can detect parts per trillion or even parts per quadrillion in some cases. Increasingly sophisticated measuring capabilities may lead us to believe that toxic materials have become more prevalent. In fact, our environment may be no more dangerous; we are just better at finding trace amounts.

RISK ASSESSMENT AND ACCEPTANCE

Even if we know with some certainty how toxic a specific chemical is in laboratory tests, it still is difficult to determine **risk** (the probability of harm times the probability of exposure) if that chemical is released into the environment. As you already have seen, many factors complicate the movement and fate of chemicals both around us and within our bodies. Furthermore, public perception of relative dangers from environmental hazards can be skewed so that some risks seem much more important than others.

Assessing Risks

A number of factors influence how we perceive relative risks associated with different situations.

- People with social, political, or economic interests—including environmentalists—tend to downplay certain risks and emphasize others that suit their own agendas. We do this individually as well, building up the dangers of things that don't benefit us while diminishing or ignoring the negative aspects of activities we enjoy or profit from.
- Most people have difficulty understanding and believing probabilities. We feel that there must be patterns and connections in events, even though statistical theory says otherwise. If the coin turned up heads last time, we feel certain that it will turn up tails next time. In the same way, it is difficult to understand the meaning of a 1-in-10,000 risk of being poisoned by a chemical.
- Our personal experiences often are misleading. When we have not personally experienced a bad outcome, we feel it is more rare and unlikely to occur than it actually may be. Furthermore, the anxieties generated by life's gambles make us want to deny uncertainty and to misjudge many risks.

- We have an exaggerated view of our own abilities to control our fate. We generally consider ourselves above-average drivers, safer than most when using appliances or power tools, and less likely than others to suffer medical problems, such as heart attacks. People often feel they can avoid hazards because they are wiser or luckier than others.
- News media give us a biased perspective on the frequency of certain kinds of health hazards, overreporting some accidents or diseases while downplaying or underreporting others. Sensational, gory, or especially frightful causes of death like murders, plane crashes, fires, or terrible accidents occupy a disproportionate amount of attention in the public media. Heart diseases, cancer, and stroke kill nearly 15 times as many people in the United States as do accidents and 75 times as many people as do homicides, but the emphasis placed by the media on accidents and homicides is nearly inversely proportional to their relative frequency compared to either cardiovascular disease or cancer. This gives us an inaccurate picture of the real risks to which we are exposed.
- We tend to have an irrational fear or distrust of certain technologies or activities that leads us to overestimate their dangers. Nuclear power, for instance, is viewed as very risky, while coal-burning power plants seem to be familiar and relatively benign; in fact, coal mining, shipping, and combustion cause an estimated 10,000 deaths each year in the United States, compared to none known so far for nuclear power generation. An old, familiar technology seems safer and more acceptable than does a new, unknown one.

Accepting Risks

How much risk is acceptable? How much is it worth to minimize and avoid exposure to certain risks? Most people will tolerate a higher probability of occurrence of an event if the harm caused by that event is low. Conversely, harm of greater severity is acceptable only at low levels of frequency. A 1-in-10,000 chance of being killed might be of more concern to you than a 1-in-100 chance of being injured. For most people, a 1-in-100,000 chance of dying from some event or some factor is a threshold for changing what we do. That is, if the chance of death is less than 1 in 100,000, we are not likely to be worried enough to change our ways. If the risk is greater, we will probably do something about it. The Environmental Protection Agency generally assumes that a risk of 1 in 1 million is acceptable for most environmental hazards. Critics of this policy ask, acceptable to whom?

For activities that we enjoy or find profitable, we are often willing to accept far greater risks than this general threshold. Conversely, for risks that benefit someone else we demand far higher protection. For instance, your chance of dying in a motor vehicle accident in any given year is about 1 in 5000, but that doesn't deter many people from riding in automobiles. Your chances of dying from lung cancer if you smoke one pack of cigarettes per day is about 1 in 1000. By comparison, the risk from drinking water with the EPA limit of trichloroethylene is about 2 in 1 billion. Strangely, many people demand water with zero levels of trichloroethylene, while continuing to smoke cigarettes.

Table 9.7 lists some activities estimated to increase your chances of dying in any given year by 1 in 1 million. These are statistical averages, of course, and there clearly are differences in where one lives or how one rides a bicycle that affect the danger level of these activities. Still, it is interesting how we readily accept some risks while shunning others.

Our perception of relative risks is strongly affected by whether risks are known or unknown, whether we feel in control of the outcome, and how dreadful the results are. Risks that are unknown or unpredictable and results that are particularly gruesome or disgusting seem far worse than those that are familiar and socially acceptable.

Studies of public risk perception show that most people react more to emotion than statistics. We go to great lengths to avoid some dangers while gladly accepting others. Factors that are involuntary,

TABLE 9.7 Activities Estimated to Increase Your Chances of Dying in Any Given Year by 1 in 1 Million

ACTIVITY	RESULTING DEATH RISK
Smoking 1.4 cigarettes	Cancer, heart disease
Drinking 0.5 liter of wine	Cirrhosis of the liver
Spending 1 hour in a coal mine	Black lung disease
Living 2 days in New York or Boston	Air pollution
Traveling 6 minutes by canoe	Accident
Traveling 10 miles by bicycle	Accident
Traveling 150 miles by car	Accident
Flying 1000 miles by jet	Accident
Flying 6000 miles by jet	Cancer caused by cosmic radiation
Living 2 months in Denver	Cancer caused by cosmic radiation
Living 2 months in a stone or brick building	Cancer caused by natural radioactivity
One chest X ray	Cancer caused by radiation
Living 2 months with a cigarette smoker	Cancer, heart disease
Eating 40 tablespoons of peanut butter	Cancer from aflatoxin
Living 5 years at the site boundary of a typical nuclear power plant	Cancer caused by radiation from routine leaks
Living 50 years 5 miles from a nuclear power plant	Cancer caused by accidental radiation release
Eating 100 charcoal-broiled steaks	Cancer from benzopyrene

Source: From William Allman, "Staying Alive in the Twentieth Century," *Science 85,* 5(6): 31, October 1985. Used by permission of the author.

unfamiliar, undetectable to those exposed, catastrophic, or that have delayed effects or are a threat to future generations are especially feared while those that are voluntary, familiar, detectable, or immediate cause less anxiety. Even though the actual number of deaths from automobile accidents, smoking, or alcohol, for instance, are thousands of times greater than those from pesticides, nuclear energy, or genetic engineering, the latter preoccupy us far more than the former.

ESTABLISHING PUBLIC POLICY

Risk management combines principles of environmental health and toxicology together with regulatory decisions based on socioeconomic, technical, and political considerations (fig. 9.17). The biggest problem in making regulatory decisions is that we are usually dealing with many sources of harm to which we are exposed, often without being aware of them. It is difficult to separate the effects of all these different hazards and to evaluate their risks accurately, especially when the exposures are near the threshold of measurement and response. In spite of often vague and contradictory data, public policymakers must make decisions.

The case of the sweetener saccharin is a good example of the complexities and uncertainties of risk assessment in public health. Studies in the 1970s at the University of Wisconsin and the Canadian Health Protection Branch suggested a link between saccharin and bladder cancer in male rats. Critics of these studies pointed out that humans would have to drink 800 cans of diet soda *per day* to get a saccharin dose equivalent to that given to the rats. Furthermore, they argued that people are not just very large rats. Scientists countered that while the doses given to rats was high, the results can be extrapolated to doses actually encountered by humans. Although the Food and Drug Act forbids the addition of any substance that causes cancer in any amount in any animal, Congress has repeatedly exempted saccharin from being banned from food because of the uncertainty about its risks.

Experiments testing the toxicity of saccharin in rats merely give a range of probable toxicities in humans. The lower end of this range indicates that only 1 person in the United States would die from using saccharin every 1000 years. That is clearly inconsequential. The higher estimate, however, indicates that 3640 people would die each year from the same exposure. Is that too high a cost for the benefits of having saccharin available to people who must restrict sugar intake? How does the cancer risk compare to the dangers of obesity, cardiovascular disease, and other problems caused by eating too much sugar? What other alternatives might there be to saccharin? A popular but more expensive alternative (aspartame, derived from the amino acid aspartic acid), which bears the trade name Nutrasweet, also is controversial because of uncertainties about its safety.

In 1989, a near panic swept the United States after the Natural Resources Defense Council issued a report claiming that children were being exposed to dangerous chemicals in fruits and vegetables. The worst of these was believed to be Alar (daminozoide), used to treat apples to promote even ripening and reduce surface blemishes. The apple industry claimed that you would have to eat 28,000 pounds of treated apples every day for 70 years to approach the exposures that caused cancer in laboratory animals. Pesticide opponents, however, claimed that the risks were much greater, especially for children.

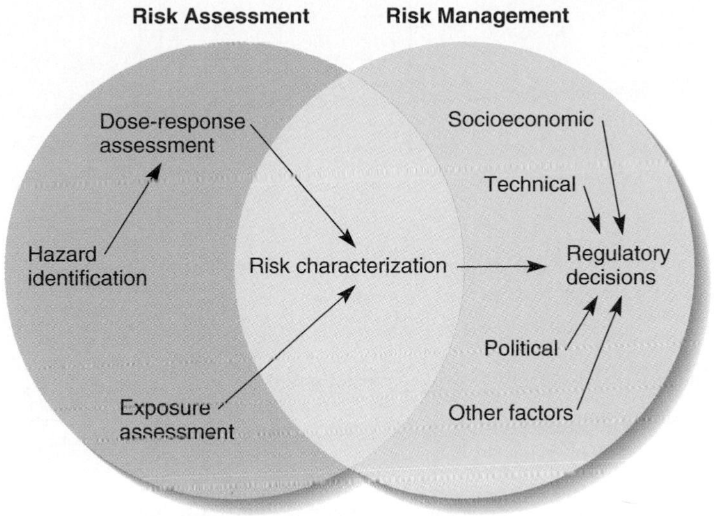

FIGURE 9.17 Risk assessment organizes and analyzes data to determine relative risk. Risk management sets priorities and evaluates relevant factors to make regulatory decisions.
Source: Data from D.E. Patton, "USEPA's Framework for Ecological Risk Assessment" in *Human Ecological Risk Assessment,* Vol. 1 No. 4.

In setting standards for environmental toxins, we need to consider (1) combined effects of exposure to many different sources of damage, (2) different sensitivities of members of the population, and (3) effects of chronic as well as acute exposures. Some people argue that pollution levels should be set at the highest amount that does *not* cause measurable effects. Others demand that pollution be reduced to zero if possible, or as low as is technologically feasible. It may not be reasonable to demand that we be protected from every potentially harmful contaminant in our environment, no matter how small the risk. As we have seen, our bodies have mechanisms that enable us to avoid or repair many kinds of damage so that most of us can withstand some minimal level of exposure without harm (fig. 9.18).

On the other hand, each challenge to our cells by toxic substances represents stress on our bodies. Although each individual stress may not be life-threatening, the cumulative effects of all the environmental stresses, both natural and human-caused, to which we are exposed may seriously shorten or restrict our lives. Furthermore, some individuals in any population are more susceptible to those stresses than others. Should we set pollution standards so that no one is adversely affected, even the most sensitive individuals, or should the acceptable level of risk be based on the average member of the population?

Finally, policy decisions about hazardous and toxic materials also need to be based on information about how such materials affect the plants, animals, and other organisms that define and maintain our environment. In some cases, pollution can harm or destroy whole ecosystems with devastating effects on the life-supporting cycles on which we depend. In other cases, only the

FIGURE 9.18 "Do you want to stop reading those ingredients while we're trying to eat?"
Reprinted with permission of the *Star-Tribune,* Minneapolis-St. Paul.

most sensitive species are threatened. Table 9.8 shows the Environmental Protection Agency's assessment of relative risks to human welfare. This ranking reflects a concern that our exclusive focus on reducing pollution to protect human health has neglected risks to natural ecological systems. While there have been many benefits from a case-by-case approach in which we evaluate the health risks of individual chemicals, we have often missed broader ecological problems that may be of greater ultimate importance.

TABLE 9.8 Relative Risks to Human Welfare

RELATIVELY HIGH-RISK PROBLEMS
Habitat alteration and destruction
Species extinction and loss of biological diversity
Stratospheric ozone depletion
Global climate change
RELATIVELY MEDIUM-RISK PROBLEMS
Herbicides/pesticides
Toxics and pollutants in surface waters
Acid deposition
Airborne toxics
RELATIVELY LOW-RISK PROBLEMS
Oil spills
Groundwater pollution
Radionuclides
Thermal pollution

Source: Data from Environmental Protection Agency.

SUMMARY

Health is a state of physical, mental, and social well-being, not merely the absence of disease or infirmity. The cause or development of nearly every human disease is at least partly related to environmental factors. For most people in the world, the greatest health threat in the environment is still, as it always has been, from pathogenic organisms. Bacteria, viruses, protozoans, parasitic worms, and other infectious agents probably kill more people each year than any other cause of death. Highly lethal emergent diseases such as Ebola and AIDS along with newly drug-resistant forms of old diseases are an increasing worry everywhere in the world.

Stress, diet, and lifestyle also are important health factors. Our social or cultural environment may be as important as our physical environment in determining the state of our health. People in some areas in the world live exceptionally long and healthful lives. We might be able to learn from them how to do so as well.

Estimating the potential health risk from exposure to specific environmental factors is difficult because information on the precise dose, length and method of exposure, and possible interactions between the chemical in question and other potential toxins to which the population may have been exposed is often lacking. In addition, individuals have different levels of sensitivity and response to a particular toxin and are further affected by general health condition, age, and sex.

The distribution and fate of materials in the environment depend on their physical characteristics and the processes that transport, alter, destroy, or immobilize them. Uptake of toxins into organisms can result in accumulation in tissues and transfer from one organism to another.

Estimates of health risks for large, diverse populations exposed to very low doses of extremely toxic materials are inexact because of biological variation, experimental error, and the necessity of extrapolating from results with small numbers of laboratory animals. In the end, we are left with unanswered questions. Which are the most dangerous environmental factors that we face? How can we evaluate the hazards of all the natural and synthetic chemicals that now exist? What risks are acceptable? We have not yet solved these problems or answered all the questions raised in this chapter, but it is important that these issues be discussed and considered seriously.

QUESTIONS FOR REVIEW

1. What is the difference between toxic and hazardous? Give some examples of materials in each category.
2. What are some of the most serious infectious diseases in the world? How are they transmitted?
3. How do stress, diet, and lifestyle affect environmental health? What diseases are most clearly related to these factors?
4. How do the physical and chemical characteristics of materials affect their movement, persistence, distribution, and fate in the environment?
5. Define LD50. Why is it more accurate than simply reporting toxic dose?
6. What is the difference between acute and chronic toxicity?
7. Define *carcinogenic, mutagenic, teratogenic,* and *neurotoxic.*
8. Why are soft PVC toys considered dangerous to children?
9. How do organisms reduce or avoid the damaging effects of environmental hazards?
10. What are the relative risks of smoking, driving a car, and drinking water with the maximum permissible levels of trichloroethylene? Are these relatively equal risks?

QUESTIONS FOR CRITICAL THINKING

1. What consequences (positive or negative) do you think might result from defining health as a state of complete physical, mental, and social well-being? Who might favor or oppose such a definition?
2. How would you feel or act if your child were dying of diarrhea? Why do we spend more money on heart or cancer research than childhood illnesses?
3. Some people seem to have a poison paranoia about synthetic chemicals. Why do we tend to assume that natural chemicals are benign while industrial chemicals are evil?
4. Analyze the claim that we are exposed to thousands of times more natural carcinogens in our diet than industrial ones. Is this a good reason to ignore pollution?
5. Are good health and a clean environment a basic human right or merely something for which we should strive?
6. What are the premises in the discussion of assessing risk? Could conflicting conclusions be drawn from the facts presented in this section? What is your perception of risk from your environment?
7. Table 9.7 equates activities such as smoking 1.4 cigarettes, having one chest X ray, and riding 10 miles on a bicycle. How was this equation derived? Do you agree with it? Do some items on this list require further clarification?
8. Who were the stakeholders in the saccharin controversy, and what were their interests or biases? Was Congress justified in refusing to ban saccharin? Should soft drink cans have warning labels similar to those on cigarettes?
9. Should pollution levels be set to protect the average person in the population or the most sensitive? Why not have zero exposure to all hazards?
10. What level of risk is acceptable to you? Are there some things for which you would accept more risk than others?

KEY TERMS

acute effects 204
allergens 196
antigens 196
bioaccumulation 200
biomagnification 200
cancer 197
carcinogens 197
chronic effects 204
Disability-Adjusted Life Year (DALY) 191
disease 190
emergent disease 192
fetal alcohol syndrome 197
hazardous 195
health 190
LD50 203
morbidity 190
mutagens 196
neurotoxins 196
risk 205
stress 198
teratogens 197
toxins 195
trauma 198

ADDITIONAL INFORMATION ON THE INTERNET

Visit our website at http://www.mhhe.com/environmentalscience for specific information about each of the topics below:

Pan-American Health Organization
World Health Organization
Center for Disease Control
National Institute of Health
National Library of Medicine
Environmental Chemistry
Endocrine Disruptors
Occupational Safety and Health Administration
Emerging Diseases
Toxic Substances and Disease Registry
Environmental Hazards in your Area
Sierra Club Toxic Substances Info
Tips for Laundry and Cleaning: avoiding use and exposure to toxic household chemicals

PART THREE

FOOD, LAND, AND BIOLOGICAL RESOURCES

CHAPTER 10

Environmental Policy, Law, and Planning

The significant problems we face cannot be solved at the same level of thinking we were at when we created them.

—Albert Einstein—

OBJECTIVES

After studying this chapter, you should be able to:

- understand the cycle by which policy is established.
- follow the path of a bill through the legislature.
- explore the differences between civil, criminal, and administrative law.
- judge the effectiveness of litigation in environmental issues.
- consider the reasons that international treaties have or have not been successful.
- appreciate the importance of wicked problems, resilience, and adaptive management in environmental planning.
- scrutinize collaborative, community-based planning methods.

ESSENTIAL STUDY PARTNER

The Essential Study Partner (ESP) CD-ROM that accompanies this textbook contains additional information on the following topics:

- environmental policy pathways: populations: societal issues: public action
- legislation for environmental education and careers: populations: societal issues: public action
- international treaties and conventions: populations: societal issues: public action
- environmental organizations: populations: societal issues: public action

Opposite: Food displayed in this floating market in Thailand shows some of the diversity and abundance that we derive from agriculture.

Above: The Capitol in Washington, DC, houses both the Senate and House of Representatives and is a major source of both environmental policy and law.

A Civil Action

Woburn, Massachusetts is a small, industrial city on the outskirts of Boston (fig. 10.1). Throughout the eighteenth and nineteenth centuries Woburn was a major leather manufacturing center with as many as 20 tanneries in operation at one time. Between 1853 and 1929, the Woburn Chemical Works was one of the largest industrial complexes in America. Centuries of careless toxic waste disposal badly contaminated the soil and groundwater under much of the city, and for many years residents complained that their well water tasted and looked terrible. In 1958, when the city drilled two new wells (designated G and H) to serve the growing population, the city engineer warned that the water was contaminated but the wells were used anyway for domestic consumption.

In 1971, young Jimmy Anderson, who lived in the part of Woburn served by these wells was diagnosed with leukemia. In talking with neighbors, Jimmy's mother, Anne, discovered that 11 other children within a few blocks of her house also had cancer. Depending on how you calculate the sample size, this was between 2.5 and 12 times the expected rate of childhood cancers. Was this merely a statistical anomaly or an ominous pattern? What might be causing these tragic illnesses? Could it be something in the water?

Although their suspicions were initially dismissed as emotional and unscientific, Woburn residents finally learned in 1979 that the water from wells G and H was indeed contaminated with a variety of metals and organic solvents including several suspected carcinogens. Coming just a year after a similar revelation about chemical contamination and links to childhood diseases at Love Canal in Niagara, NY, this discovery encouraged Anne Anderson and others to begin to ask who was responsible for polluting their neighborhood. In 1982, just a year after Jimmy died, Anne Anderson and seven other families whose children also had cancer sued the W. R. Grace Company and Beatrice Foods for damages caused by negligent disposal of toxic wastes on their properties near the wells. The families were represented by attorney Jan Schlichtmann of the public-interest law firm, Trial Lawyers for Public Justice of Boston.

Both Grace and Beatrice immediately filed motions for dismissal, claiming that even if they had dumped toxic wastes, the plaintiffs couldn't prove which of the many pollution sources in Woburn was cause of a specific disease in a particular person. These two corporations were chosen as targets out of all the possible industries in Woburn, their lawyers argued, simply because of their deep pockets. The judge ruled, however, that the case had sufficient merit to proceed to a jury trial. After four years of interrogatories, deposition of hundreds of witnesses, thousands of pages of documents, extensive examination of medical histories and physical conditions of each of the plaintiff families, and investigation of the industrial sites, the case finally went to trial in 1986. In the discovery process, the plaintiffs conducted their own on-site investigation that revealed drums of toxic chemicals buried on the Grace property. Conviction on charges that the company had lied to the EPA about when and where wastes had been disposed didn't help the defense in the civil trial.

After a five-month trial and seven days of jury deliberation, the case against Beatrice was dismissed, but the jury found that Grace had negligently contaminated the Woburn wells. The jury could not decide, however, when contaminants from the Grace property might have reached the wells. Was it before or after the children developed cancer? This uncertainty led the judge to dismiss the verdict and order a new trial. Rather than go through the process all over again, both sides agreed to settle for $8 million. Grace also agreed to participate in a $68 million cleanup of the wells, the most expensive Superfund project in Massachusetts at the time.

Because the case was settled out of court, it doesn't create a legal precedent, but it was one of the first times that plaintiffs succeeded in gaining compensation in an environmental injury lawsuit. As a story of a few local families challenging corporate giants, it gained national attention, and served as a warning to corporations that they can be held liable for personal injuries from negligent disposal of toxic wastes. A 1995 novel by Jonathan Harr about this case, titled *A Civil Action,* was turned into a movie by the same name starring John Travolta and Robert Duvall. Check out the novel or movie if you'd like to see more about how the drama unfolded.

This case illustrates both changing attitudes in the United States towards waste disposal and environmental liability as well as use of the courts to redress personal environmental injuries. In this chapter, we'll examine both how environmental policy is formed as well as how the legislative, legal, and administrative systems work to accomplish policy goals.

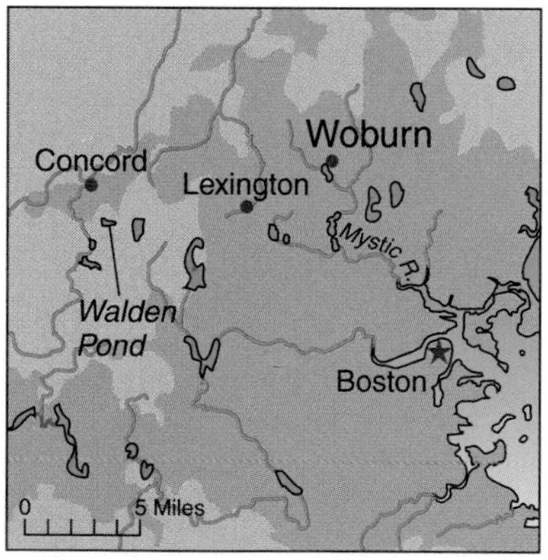

FIGURE 10.1 Woburn, a small industrial city within the Boston metropolitan area, was the site of an important environmental law case.

ENVIRONMENTAL POLICY

The term "policy" is used in many different ways to indicate both formal and informal decisions or intentions at a personal, community, national, or international level. You might have an informal policy never to accept telemarketing calls; your church may have an open-door policy for visitors; and the United States has a policy not to negotiate with terrorists. At the same time, the Clean Air Act is a formal statement of our national policy on acceptable air quality, while the Convention on Global Climate Change represents the official intentions of many nations to curb greenhouse gases. Interestingly, policy can describe both an actual document as well as a contractual agreement, such as when you buy insurance. At its core, then, **policy** is a plan or statement of intentions—either written or stated—about a course of action or inaction intended to accomplish some end. Some political scientists limit the term public policy to the principles, laws, executive orders, codes, or goals established by some government body or institution. For the purposes of this chapter, **environmental policy** will be taken as those official rules and regulations concerning the environment that are adopted, implemented, and enforced by some governmental agency as well as general public opinion about environmental issues (fig. 10.2).

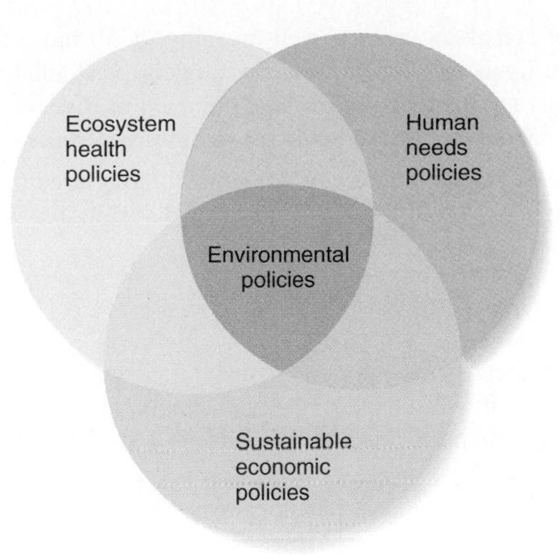

FIGURE 10.2 The best environmental policies incorporate economic, ecological, and social/cultural considerations.
Source: Modified from Ray Grizzle, *Bioscience*, vol. 44(4), April 1994.

Political Decision Making

The policies we establish depend to a great extent on the system within which they operate. For many of us, the ideal political system is one that is open, honest, transparent, reaches the best possible decisions to maximize benefits to everyone. In a pluralistic, democratic society, we aim to give everyone an equal voice in policy making. Ideally, many separate interests put forward their solutions to public problems that are discussed, debated, and evaluated fairly and equally. Facts and access are open to everyone. Policy choices are made democratically but compassionately; implementation is reasonable, fair, and productive. Unfortunately, this isn't always the way our system works. What are some alternate explanations of how we create policy?

Politics as Power

According to some observers, politics is really the struggle among often unequal, competing interest groups as they strive to shape public policy to suit their own agendas. The political system, in this view, manages group conflict by (1) establishing rules to ensure civil competition, (2) encouraging compromises and balancing interests to the extent possible, (3) codifying compromises as public policy, and (4) enforcing laws and rules based on that policy. Where this form of power politics is operational, it often results in a tyranny of a powerful elite over the impotent masses. Those elites manipulate public opinion and give up only as much power and wealth as necessary to maintain overall control.

However, while self-interest and power politics clearly are important forces in American public life, they can't account for all of the civil rights and environmental movements, the war on poverty, and public interest that characterized much of the 1960s and 1970s. The force of ideals, values, and altruism sometimes carry the day even in our winner-take-all system. We seem to go through periods of public spirit, optimism, and openness to change every few decades that give our public life a sense of generosity and good will.

Rational Choice

Another model for public decision making is **rational choice** and science-based management. In this utilitarian approach, no policy should have greater total costs than benefits. In choosing between policy alternatives, we should always prefer those with the greatest cumulative welfare and the least negative impacts. Professional administrators would weigh various options and make an objective, methodical decision that would bring maximum social gain. As Chapters 2 and 8 illustrate, there are many arguments against utilitarian, cost-benefit approaches that are generally applicable to public decision making.

- Many conflicting values and needs cannot be compared because they aren't comparable or we don't have perfect information.
- There are few generally agreed-upon broad societal goals but rather benefits to specific groups and individuals, many of which are in conflict.
- Policymakers generally are not motivated to make decisions on the basis of societal goals, but rather to maximize their own rewards: power, status, money, or reelection.
- Large investments in existing programs and policies create "path dependence" and "sunken costs" that prevent policymakers from considering good alternatives foreclosed by previous decisions.
- Uncertainty about consequences of various policy options compels decision makers to stick as closely as possible to previous policies to reduce the likelihood of adverse, calamitous, unanticipated consequences.

- Policymakers, even if well-meaning, don't have sufficient intelligence or adequate data or models to calculate accurate costs and benefits when large numbers of diverse political, social, economic, and cultural values are at stake.
- The segmented nature of policy making in large bureaucracies makes it difficult to coordinate decision making.

The Policy Cycle

How do policy issues and options make their way onto the stage of public debate? In this section, we will look at the **policy cycle** by which problems are identified and acted upon in the public arena (fig. 10.3). The first stage in this process is problem identification. Sometimes the government identifies issues for groups that have no voice or don't recognize problems themselves. In other cases, the public identifies a problem such as loss of biodiversity or health effects exposure to toxic waste and demands redress by the government. In either case, proponents describe the issue—either privately or publicly—and characterize the risks and benefits of their preferred course of action. Seizing the initiative in issue identification often allows leaders to define terms, set the agenda, organize stakeholders, choose tactics, aggregate related issues, and legitimate (or de-legitimate) issues and actors. It can be a great advantage to define the terms or choose the location of a debate. Next, stakeholders develop proposals for preferred policy options, often in the form of legislative proposals or administrative rules. Proponents build support for their position through media campaigns, public education, and personal lobbying of decision makers. By following the legislative or administrative process through its many steps, interest groups ensure that their proposals finally get enacted into law or established as a rule or regulation.

The next step is implementation. Ideally, government agencies will faithfully carry out policy directives as they organize bureaucracies, provide services, and enforce rules and regulations, but often it takes continued monitoring to make sure the system works as it should. Evaluating the results of policy decisions is as important as establishing them in the first place. Measuring impacts on target and nontarget populations shows us whether the intended goals, principles, and course of action are being attained. Finally, suggested changes or adjustments are considered that will make the policy fairer or more effective.

There generally are two different routes by which this cycle is carried out. Special economic interest groups such as industry associations, labor unions, or wealthy and powerful individuals don't need (or often want) much public attention or support for their policy initiatives. They generally carry out the steps of issue identification, agenda setting, and proposal development privately because they can influence legislative or administrative processes directly through their contacts with decision makers. Public interest groups, on the other hand, often lack direct access to corridors of power and need to rally broad general support to legitimate their proposals. An important method for getting their interests on the table is to attract media attention. Organizing a dramatic protest or media event can generate a lot of free publicity. Announcing some

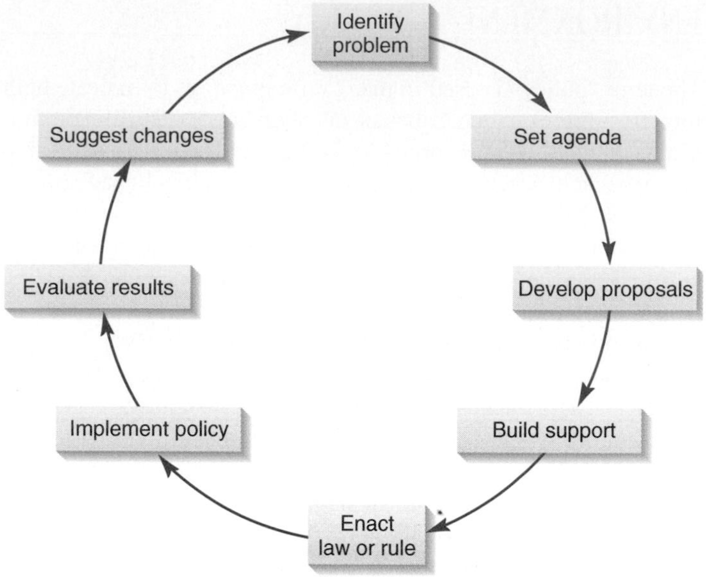

FIGURE 10.3 The policy cycle.

dire threat or sensational claim is a good way to gain attention. The problem is that it takes ever increasing levels of hysteria and hyperbole to get yourself heard in the flood of shocking news with which we are bombarded every day. Ironically, many groups that bemoan the overload of rhetorical overstatement that engulfs us, contribute to it in order to be heard above the din.

NEPA and EIS

The National Environmental Policy Act (NEPA) forms the cornerstone of both U.S. environmental policy and environmental law. Passed in 1969 but signed by President Nixon in 1970, NEPA does three important things: (1) It authorizes the Council on Environmental Quality (CEQ), the oversight board for general environmental conditions; (2) It directs federal agencies to take environmental consequences into account in decision making; and (3) It requires an Environmental Impact Statement (EIS) be prepared for every major federal project having a significant impact on the quality of the human environment. When NEPA was being debated, there were suggestions that it should be a constitutional amendment guaranteeing the right of a clean environment to everyone. Difficulties in defining what "clean" means, along with worries about how we could achieve this ambitious goal, limited NEPA to a statute with more limited, but still important, powers than a constitutional amendment. What do you think? Does everyone have an inherent right to a clean environment? How would you define clean?

The EIS process has proven to be one of the most powerful tools in the environmental arsenal. It requires more open and environmentally sensitive planning in both the agencies themselves and in private corporations seeking to do business with the government. An EIS can bring to light adverse aspects of a project that might otherwise remain hidden. It can provide valuable information about a proposal to opponents who can't afford to do their own research.

FIGURE 10.4 Every major federal project in the United States must be preceded by an Environmental Impact Statement. This is the Glen Canyon Dam on the Colorado River in Arizona.

An EIS doesn't forbid environmentally destructive activities if they comply otherwise with relevant laws, but it demands that we admit what we're really doing. Once embarrassing information is revealed, however, few agencies will bulldoze ahead oblivious to public opinion.

There are several threshold considerations that determine whether a project requires an EIS. The activity must be federal and it must be major, with a significant impact on the human environment (fig. 10.4). Evaluations are always somewhat subjective as to whether specific activities meet these characteristics. Each case is unique and depends on context, geography, the balance of beneficial versus harmful effects, and whether any areas of cultural, scientific, or historical importance might be affected. To do a complete EIS for a project is usually time-consuming and costly. The final document is often hundreds of pages long and generally takes six to nine months. Sometimes just requesting an EIS is enough to sideline a questionable project. In other cases, it gives adversaries time to rally public opposition and information with which to criticize what's being proposed. If agencies don't agree to prepare an EIS voluntarily, environmentalists can petition the courts to force them to do so.

Every EIS must contain the following elements: (1) purpose and need for the project, (2) alternatives to the proposed action (including taking no action), (3) a statement of positive and negative environmental impacts of the proposed activities. In addition, an EIS should make clear the relationship between short-term resources and long-term productivity, as well as any irreversible commitment of resources resulting from project implementation.

ENVIRONMENTAL LAW

Laws are rules set by authority, society or custom. Church laws, social morés, administrative regulations, and a variety of other codes of behavior can be considered laws if they are backed by some enforcement power. Government laws are established by federal, state, or local legislative bodies or administrative agencies. **Environmental law** constitutes a special body of official rules, decisions, and actions concerning environmental quality, natural resources, and ecological sustainability. Each of the three branches of government play a role in establishing the rules of law. **Statute law** consists of formal documents or decrees enacted by the legislative branch of government declaring, commanding, or prohibiting something. It represents the formal will of the legislature. **Case law** is derived from court decisions in both civil and criminal cases. **Administrative law** rises from executive orders, administrative rules and regulations, and enforcement decisions in which statutes passed by the legislature are interpreted in specific applications and individual cases. Because every country has different legislative and legal processes, this chapter will focus primarily on the U.S. system in the interest of simplicity and space.

A Brief Environmental History

Before looking at how each of these branches of government function it might be useful to review how the underlying attitudes towards our environment and the policies that underlie our environmental laws have changed over the past century. As the Woburn story shows, for most of its history, U.S. environmental policy has had a laissez-faire or hands-off attitude toward business and private property. Pollution and environmental degradation were regarded as the unfortunate but necessary cost of doing business. If you didn't like the smell of a tannery or the sight of a waste dump, you were free to go somewhere else. While there were some early laws forbidding gross interference with another person's property or rights—the Rivers and Harbors Act of 1899, for example, made it illegal to dump so much refuse in waterways that navigation was blocked—in general, everyone was free to do whatever they wanted on their own property. People were either unaware of, or didn't pay much attention to, the fact that pollutants can move through the air, soil, and water to endanger people distant from the source.

The emergence of the modern environmental movement in the 1960s and 1970s marked a dramatic turning point in our understanding of the dangerous consequences of pollution and our demands to be protected from it. Rachel Carson's *Silent Spring* (1962) and Barry Commoner's *Closing Circle* (1971) alerted the public to the ecological and health risks of pesticides, hazardous wastes, and toxic industrial effluents. Public activism in the civil rights and antiwar movements was carried over to environmental protests and demands for environmental protection. Emergence of new media—especially television—provided access to environmental news and made events in faraway places seem immediate and important.

FIGURE 10.5 Beach cleanup efforts after the Santa Barbara oil spill in 1969 made excellent media material and had an important role in U.S. environmental policy.

The 1969 blowout of an oil well in the Santa Barbara Channel just off the coast of southern California is a good example of how the convergence of actors, events, timing, and media attention can shape public opinion and influence the policy cycle. For many weeks, black, gooey crude oil washed up onto beautiful southern California beaches. The oil spill made a perfect story for TV. The continuing saga was ideal for nightly updates. The setting guaranteed good photos and was readily recognized by the viewing public as an important place (fig. 10.5). National news networks had just developed the capacity for live satellite feeds and were hungry to use their new technology. Los Angeles was one of few locations with reliable uplinks and Santa Barbara was close enough for a film crew to go out every day to get some good footage and be back in the studio in time for the five o'clock news.

Because the story was ongoing, it fit well in the 30-second spots characteristic of TV news. The audience was familiar with both the issue and the images that described it. Reporters didn't have to spend precious seconds explaining what was happening, but could just give a sporting-event-like update on which side was winning today. A policy debate in Congress might be more important, but is too complex to explain in a few seconds and doesn't provide exciting visuals. The wealthy residents of Santa Barbara were media-savvy, and had the influence and contacts to publicize the oil spill. While some of the cleanup efforts were not very effective, they made great visual footage. Attractive young people, smudged with oil, trying vainly to sweep gooey crud off a beautiful beach made ideal TV footage. Although the Santa Barbara oil spill wasn't nearly as big as others around the world, it played an important role in mobilizing public opinion and was an major factor in passage of the 1972 Clean Water Act.

As a result of awakened public concern about environmental issues, more than 27 major federal laws for environmental protection and hundreds of administrative regulations were established in the United States in the environmental decade of the 1970s. The statutes, case law, rules, precedents, and agencies resulting from that period created the foundation on which much of our current environmental protection rests. In the initial phase of this environmental revolution, the main focus was on direct regulation and litigation to force malefactors to obey the law. In recent years, attention has shifted from end-of-the-pipe command and control to pollution prevention and collaborative methods that can provide win/win solutions for all stakeholders. We'll look at some of those alternative dispute resolution approaches later in this chapter. But next, let's look more closely at how environmental law is established and administered.

Statutory Law: The Legislative Branch

Establishing laws at either the state or federal level is one of the most important ways of protecting our environment. Many environmental groups spend a good deal of their time and resources trying to influence the legislative process. In this section we'll look at how that system works.

Federal laws (statutes) are enacted by Congress and must be signed by the president. They originate as legislative proposals called bills, which are usually drafted by the congressional staff, often in consultation with representatives of various interest groups. Thousands of bills are introduced every year in Congress. Some are very narrow, providing funds to build a specific section of road or to help a particular person, for instance. Others are extremely broad, perhaps overhauling the social security system or changing the entire tax code. Similarly, environmental legislation might deal with a very specific local problem or a national or international issue. Often a number of competing bills on a single issue may be introduced as proponents from different sides attempt to incorporate their views into law. A bill may have a single sponsor if it is the pet project of a particular legislator, or it may have 100 or more coauthors if it is an issue of national importance.

A Convoluted Path

After introduction, each bill is referred to a committee or subcommittee with jurisdiction over the issue for hearings and debate (fig. 10.6). Most hearings take place in Washington, but if the bill is controversial or legislators want to attract publicity for themselves or the issue, they may conduct field hearings closer to the site of the controversy. The public often has an opportunity to give testimony at field hearings (fig. 10.7). Although it's not likely that you will change the opinions of many legislators no matter how fervent or cogent your testimony, these events can be a good place to gain attention and educate the public about a topic. Hearings and debates also build a record of legislative intent that can be valuable

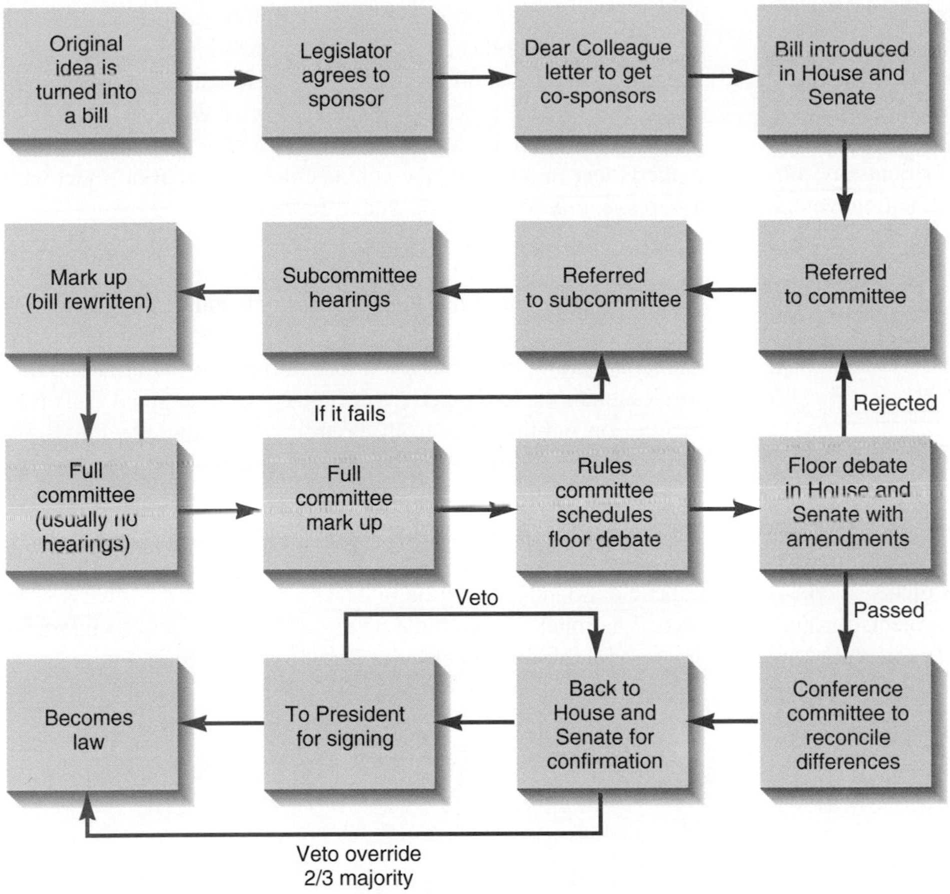

FIGURE 10.6 The path of a bill from inception to actual law.

FIGURE 10.7 Citizens line up to testify at a legislative hearing. By getting involved in the legislative process, you can be informed and have an impact on governmental policy.

in later interpretation and implementation of laws by courts and administrative agencies.

If a bill has sufficient support within the subcommittee, its language will be "marked up" or revised and modified to be more widely acceptable and to improve its chances of passage. At this stage several competing bills might be combined into a single compromise version. If the compromise bill garners sufficient support, it is forwarded to the full committee for more hearings, debate, and a vote. If it fails in the full committee, the bill is sent back to the subcommittee for more work and further compromise. A bill that succeeds in the full committee is reported to the full House or Senate for a floor debate. Often opponents of a bill will attempt to amend it during each of these stages to lessen its impact or to make it so unpalatable that even the original authors can no longer support it. Some amendments add completely unrelated material or even reverse the intent of a bill. As bills move through this convoluted pathway, interested parties can follow their progress in the *Congressional Quarterly Weekly,* a publication both in print and online that keeps track of proposed legislation. Many environmental groups also maintain websites with up-to-date information on events in Congress.

By the time an issue has passed through both the House and Senate, the versions approved by the two bodies are likely to be different. They go then to conference committee to iron out any differences between them. After going back to the House and Senate for confirmation, the final bill goes to the president, who may either sign it into law or veto it. If the president vetoes the bill, it may still become law if two-thirds of the House and Senate vote to override the veto. If the president takes no action within 10 days of receiving a bill from Congress, the bill becomes law without his signature.

One exception to this procedure is that if Congress adjourns before the 10-day period elapses, the bill does not become law. The president, by doing nothing, is said to have exercised a "pocket veto."

The Thomas website, maintained by Congress, also has current information about the progress of legislation. You can find out how your Senator and Representative voted on critical environmental issues by consulting The League of Conservation Voters, which ranks each member of Congress on their voting record. The Defenders of Wildlife maintains a daily e-mail environmental news service called Greenwire that has up-to-date information about what's happening in Washington. Table 10.1 lists some of the most important recent federal environmental legislation.

Legislative Riders

There are two types of legislation: authorizing bills become law, while appropriation bills provide funds for federal agencies and programs. Appropriation bills can have language attached expressing the intent of Congress, but, in theory, at least, are not supposed to make policy, merely fund existing plans and projects. Legislators who can't muster enough votes to pass pet projects through regular channels often will try to add authorizing amendments called **riders** into completely unrelated funding bills. Even if they oppose the riders, other members of Congress have a difficult time voting against an appropriation package for disaster relief or to fund programs that benefit their districts. Often this happens in conference committee because when the conference report goes back to the House and Senate, the vote is either to accept or reject with no opportunity to debate or amend further.

Starting with the 104th Congress, antienvironmental forces began using this tactic to roll back environmental protections and gain access to natural resources. Environmental groups were outraged, for instance, when riders were attached to 1996 supplemental spending bills that put a moratorium on listing additional species under the Endangered Species Act and exempted "salvage" logging on public lands from environmental laws. In the 105th Congress, numerous antienvironmental riders were attached to appropriation bills. The Omnibus spending bill in fiscal year 1999, for example, was festooned with 41 special-interest amendments to allow grazing on public land without environmental review, impede parkland acquisition in Alaska, increase timber harvest on several national forests, build new roads through some wilderness areas and wildlife refuges, eliminate protection for certain coastal

TABLE 10.1 Major U.S. Environmental Laws

LEGISLATION	PROVISIONS
National Environmental Policy Act of 1969	Declared national environmental policy, required environmental impact statements, created Council on Environmental Quality.
Clean Air Act of 1970	Established national primary and secondary air quality standards. Required states to develop implementation plans. Major amendments in 1977 and 1990.
Clean Water Act of 1972	Set national water quality goals and created pollutant discharge permits. Major amendments in 1977 and 1996.
Federal Pesticides Control Act of 1972	Required registration of all pesticides in U.S. commerce. Major modifications in 1996.
Marine Protection Act of 1972	Regulated dumping of waste into oceans and coastal waters.
Coastal Zone Management Act of 1972	Provided funds for state planning and management of coastal areas.
Endangered Species Act of 1973	Protects threatened and endangered species, directs FWS to prepare recovery plans.
Safe Drinking Water Act of 1974	Set standards for safety of public drinking water supplies and to safeguard groundwater. Major changes made in 1986 and 1996.
Toxic Substances Control Act of 1976	Authorized EPA to ban or regulate chemicals deemed a risk to health or the environment.
Federal Land Policy and Management Act of 1976	Charged the BLM with long-term management of public lands. Ended homesteading and most sales of public lands.
Resource Conservation and Recovery Act of 1976	Regulated hazardous waste storage, treatment, transportation, and disposal. Major amendments in 1984.
National Forest Management Act of 1976	Gave statutory permanence to national forests. Directed USFS to manage forests for "multiple use."
Surface Mining Control and Reclamation Act of 1977	Limited strip mining on farmland and steep slopes. Required restoration of land to original contours.
Alaska National Interest Lands Act of 1980	Protected 40 million ha (100 million acres) of parks, wilderness, and wildlife refuges.
Comprehensive Environmental Response, Compensation and Liability Act of 1980	Created $1.6 billion "Superfund" for emergency response, spill prevention, and site remediation for toxic wastes. Established liability for cleanup costs.
Superfund Amendments and Reauthorization Act of 1994	Increased Superfund to $8.5 billion. Shares responsibility for cleanup among potentially responsible parties. Emphasizes remediation and public "right to know."

Source: N. Vig and M. Kraft, *Environmental Policy in the 1990s*, 3rd Congressional Quarterly Press.

barrier islands, and prevent grizzly bear reintroduction in Idaho and Montana, among other things. In another case, riders to delay mining reforms and forbid designation of critical habitat under the Endangered Species Act were attached to relief bills to aid war victims in Kosovo and Hurricane Mitch victims in Honduras. Congressional leaders pledged to end this practice, but little has been done so far to stop it.

Lobbying

Groups or individuals with an interest in pending legislation can often cause a great deal of influence by **lobbying,** or using personal contacts, public pressure, and political action to persuade legislators to vote or act in their favor. The term derives from the habit of partisans and professional lobbyists to lurk in the hallways and lobbies of Congress hoping to snare a passing legislator to urge them to vote in a specific way. We also use the same term to describe efforts to influence administrative agencies.

Most major environmental organizations maintain offices in Washington from which they monitor legislative and administrative programs and policies. Hundreds of professional environmental lobbyists and volunteer activists attend hearings, meet with legislators and agency personnel, draft proposed legislation and administrative rules, and attempt in a variety of ways to shape the national environmental agenda. They join thousands of other amateur and professional lobbyists representing industry and business organizations, workers, property owners, religious groups, ethnic associations, and just about every other kind of interest group that you can imagine. Walking the halls of Congress or those of the House or Senate Office Buildings, you see an amazing mixture of people attempting to be heard. It's fascinating to be part of this process.

In a survey of professional lobbyists, a majority agreed that personal contacts were the most effective way to influence decision makers. Undoubtedly the best way to make contact is through personal friends of a legislator or someone to whom they owe political allegiance. Having a famous movie star, a person of great power or wealth, or some other celebrity represent your group also can help open doors for your ideas. Your own senator or representative is more likely to be interested in your views than someone with whom you have no connection. They place a high priority on responding to their own constituents who can vote them in (or out) of office. But even if you aren't rich or illustrious or politically connected, you can often get a fair hearing from legislative staff—if not their bosses—if you have a persuasive case concerning an important topic.

What can those of us do to have an impact on legislation who can't afford to go to Washington to be directly involved? Getting involved in local election campaigns can greatly increase your access to legislators. Writing letters or making telephone calls also are highly effective ways to get your message across. You'd be surprised at how few letters or calls legislators receive even on important national issues. Your voice can have an important impact. How to write an effective letter is described further in chapter 25. All legislators now have e-mail addresses, although it isn't clear how much weight this form of communication carries.

FIGURE 10.8 Protests, marches, and public demonstrations can be an effective way to get your message out and to influence legislators.

Getting media attention can sway the opinions of decision makers. Organizing protests, marches, demonstrations, or other kinds of public events can call attention to your issue (fig. 10.8). Public education campaigns, press conferences, TV ads, and a host of other activities can be helpful. Tax-exempt (503c) organizations can't lobby directly or engage in politics, but there is a murky line between educational ads and outright campaigning and lobbying. Joining together with other like-minded groups can greatly increase your clout and ability to get things done. It's hard for a single individual or even a small group to have much impact, but if you can organize a mass movement, you may be very effective.

Case Law: The Judicial Branch

Over the past 30 years, appeals to the judicial system have often been the most effective ways for seeking redress for environmental damage and forcing changes in how things are done. Activist judges and sympathetic juries in both federal and state court systems have been willing to take a stand where legislatures have been too timid or conservative to do so. Many groups spend a great deal of their time and energy bringing lawsuits that will shape environmental policy. The Environmental Defense Fund, for example, operates primarily in this arena. In the early days of the organization, their motto was, "Sue the bastards, that will get their attention." Even if you're not interested in environmental issues, it's worthwhile knowing something about how this system works. You may find yourself in court someday.

The judicial branch of government establishes environmental law by ruling on the constitutionality of statutes and interpreting their meaning. We describe the body of legal opinions built up by many court cases as case law. Often legislation is written in vague and general terms so as to make it widely enough accepted to gain passage. Congress, especially in the environmental area, often leaves it to the courts to "fill in the gaps." As one senator said when Congress was about to pass the Superfund legislation, "All we know is that the

American people want these hazardous waste sites cleaned up . . . Let the courts worry about the details." When trying to interpret a law, the courts depend on the legislative record from hearings and debates to determine congressional intent. What was a particular statute meant to do by those who wrote and passed it?

The Court System

The United States is divided into 96 federal court districts, each of which has at least one trial court. Over these district courts are the circuit court of appeals, which hears disputes arising from questions about procedural issues and interpretations of the law in district courts. There are 12 geographic regions for the appeals courts. The federal courts have jurisdiction over federal criminal prosecutions, claims against the federal government, claims arising under federal statutes or treaties, and cases in which defendants or plaintiffs come from two or more states. The residence of the defendant of location of the property in dispute usually determines the venue, or the court in which each case is heard. Each state has its own courts that generally parallel the federal system. These courts have jurisdiction over cases arising from state laws. The U.S. Supreme Court is the court of last resort for appeals for both federal and state court systems.

A trial court judge presides over trials, rules on motions made by attorneys, and decides questions of law, such as what evidence is admissible, and what law applies to the case (fig. 10.9). The judge controls the pace of the proceedings and maintains decorum in the court room. Although Thomas Jefferson said, "Ours is a government of laws, not men," in reality, the judge has tremendous power over the outcome of a trial. Certain courts earn a reputation of being pro- or antienvironmental. Litigants always try to know something about the judge's ideology before bringing a case to trial.

The first judge to hear a case arising from a particular statute or situation has the greatest latitude to interpret the law and set a **precedent** to be used as an example in subsequent trials. These decisions are binding, however, only on those courts on a lower level and in the same system. Precedents from the California courts, for instance, are not binding in Arizona courts. Furthermore, if a judge *distinguishes* a case—determines it is different from other cases—she or he is not obliged to adhere to prior precedents. Or a judge can always simply overturn a clearly applicable precedent as a matter of "correcting the law" where technology or changing community values has made prior decisions outdated.

Legal Thresholds

Before a trial can start, the litigants must satisfy certain threshold requirements. The first of these is **standing,** or whether the participants have a right to stand before the bar and be heard. The main criteria for standing is a valid interest in the case. Plaintiffs must show that they are materially affected by the situation they petition the court to redress. This is an important point in environmental cases. Groups or individuals often want to sue a person or corporation for degrading the environment. But unless they can show that they personally suffer from the degradation, courts are likely to deny standing.

FIGURE 10.9 A trial court judge presides over trials, rules on motions, and interprets the law.

In a landmark 1969 case, for example, the Sierra Club challenged a decision of the Forest Service and the Department of the Interior to lease public land in California to Walt Disney Enterprises for a ski resort. The land in question was a beautiful valley that cut into the southern boundary of Sequoia National Park (see fig. 2.5). Building a road into the valley would have necessitated cutting down a grove of giant redwood trees within the park. The Sierra Club argued that it should be granted standing in the case to represent the trees, animals, rocks, and mountains that couldn't defend their own interests in court. After all, the club, pointed out, corporations—such as Disney Enterprises—are treated as persons and represented by attorneys in the courts. Why not grant trees the same rights? The case went all the way to the Supreme Court, which ruled that the Sierra Club failed to show that it or any of its members would be materially affected by the development.

In addition to standing, plaintiffs must show that their case represents a "live" legal dispute that is likely to result in a final and meaningful judgment, and that there is a present controversy for which a decision is needed. In other words, you can't bring a suit over a hypothetical situation or one that is no longer cogent. For example, suppose you want to stop a development project that would destroy endangered species habitat. If the defendant can show that the species couldn't inhabit the habitat (if it were already extinct, for instance), the case is moot (of no practical importance) and the courts will refuse to hear it.

Criminal Law

Criminal law derives from those federal and state statutes that prohibit wrongs against the state or society, such as arson, rape, murder, and robbery. Serious crimes, like murder or rape, that are punishable by long-term incarceration or heavy fines are called felonies. Lesser crimes, such as shoplifting or vandalism, that result in smaller fines or shorter sentences in a county or city jail are labeled misdemeanors. Definitions vary from state to state. What may be a felony crime in one state may be only a misdemeanor in another.

A criminal case is always initiated by a government prosecutor. Guilt or innocence of the defendant is determined by a jury of peers, but the sentence is imposed, often in consultation with the jury, by the judge. The judge is responsible for keeping order in the hearings and for determining points of law. The jury acts as a fact-finding body that weighs the truth and reliability of the witnesses and evidence. Figure 10.10 shows the course of events in a lawsuit.

Violation of many environmental statutes constitutes criminal offenses. In 1975, the U.S. Supreme Court ruled that corporate officers can be held criminally liable for violations of environmental laws if they were grossly negligent, or the illegal actions can be considered willful and knowing violations. In 1982, the EPA created an Office for Criminal Investigation. Recently as many as 100 criminal prosecutions per year have been brought to court for environmental crimes. In one of the toughest criminal sentences imposed so far, the president of a Colorado company was sentenced, in 1999, to 14 years in prison for knowingly dumping chlorinated solvents that contaminated the water table. The company itself was fined $950,000 and put on probation for ten years. Also in 1999, two Connecticut men were convicted of three counts of abandoning 40 drums of hazardous waste in a parking lot in New Haven. The fact that they stripped off or painted over the identifying labels was taken as willful intent to commit a crime. They face up to seven years in prison and a $250,000 fine for each count.

Civil Law

Civil law is defined as a body of laws regulating relations between individuals or between individuals and corporations. Issues such as property rights, and personal dignity and freedom are protected by civil law. In some cases, legislative statutes, such as the Civil Rights Act, establish specific aspects of civil law. In other cases, where no particular statute exists, custom and a body of previous court decisions, collectively called **common law,** establish precedents that constitute a working definition of individual rights and responsibilities. The Woburn case that opened this chapter, is an example of **tort law** (tort derives from the Middle English word for injury) that seeks compensation for damages. This kind of civil action is usually initiated by the attorney representing the injured party (the plaintiff). The defendant in a civil case has a right to be tried by a jury, but in highly technical issues, this right often is waived and the case is heard only by the judge. Being found guilty of a civil offense can result in financial penalties but not jail time.

In contrast to a criminal case where the burden of proof lies with the prosecution, and defendants are considered innocent until proven guilty, civil cases can be decided on a "preponderance of evidence." This makes civil cases considerably easier to win than criminal cases where the evidence is ambiguous. A number of mitigating factors also are taken into account in determining guilt and assigning penalties in civil cases. Culpability is based on whether the defendant could reasonably have anticipated and avoided the offense. A "good faith effort" to comply or solve the problem can be a factor. The compliance history is important. Is this a first offense or a habitual repeater? Finally, is there evidence of economic benefit to the perpetrator? That is, did the violator gain personally from the action? If so, it is more likely that willful intent was involved.

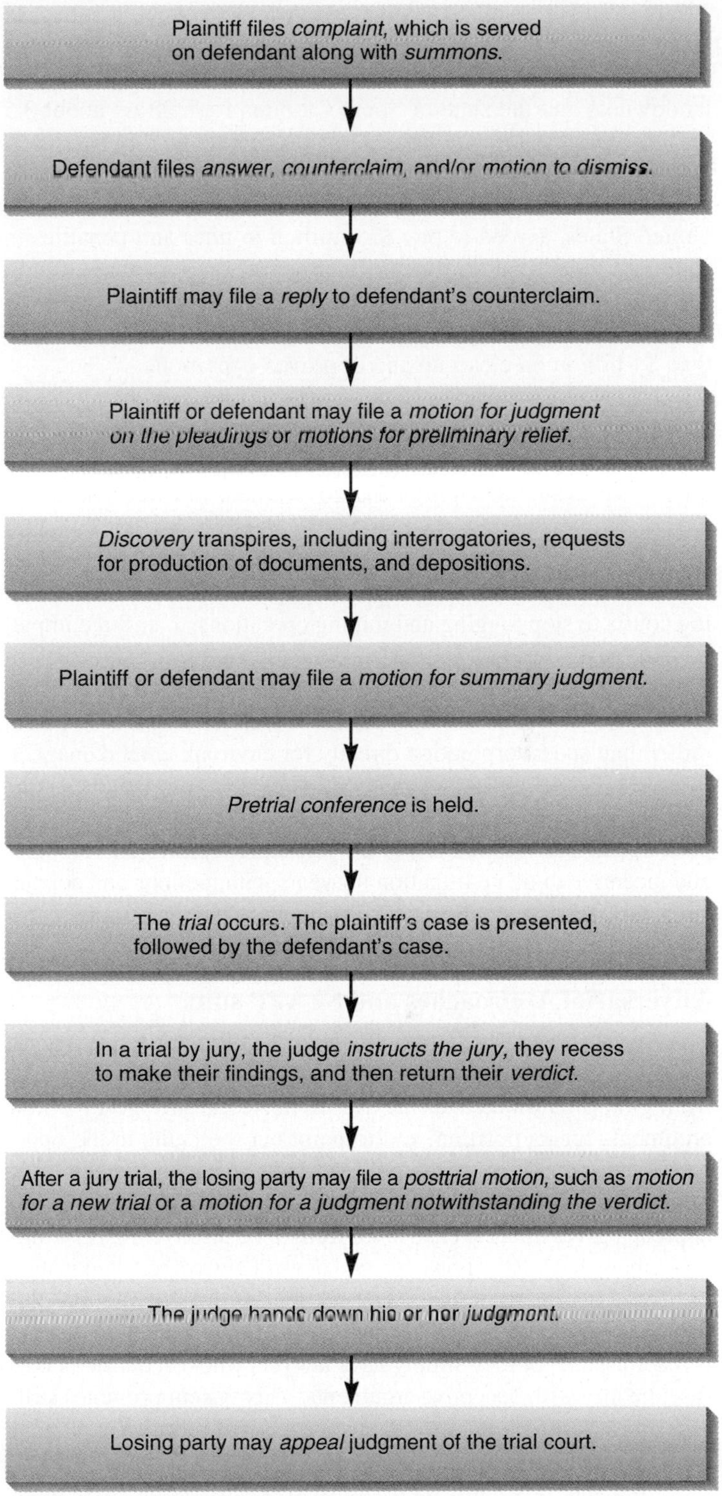

FIGURE 10.10 Events in a civil lawsuit.

Most people consider being convicted of a criminal offense much more serious than losing a civil case, because the former can lead to incarceration while the latter only costs money. Civil judgments can be costly, however. A group of Alaskan fishermen won $5 billion from the Exxon oil company for damages caused by the

1989 *Exxon Valdez* oil spill. Civil cases can be brought in both state and federal court. In 1999, the Atlanta-based Colonial Pipeline Company was ordered to pay over $20 million in damages to local landowners and the state of South Carolina for spilling about 3.5 million liters (nearly a million gallons) of diesel fuel into the Reedy River near Simpsonville, SC. And in January 2000, the Koch Oil Company, one of the largest pipeline and refinery operators in the United States, agreed to pay $35 million in fines and penalties to state and federal authorities for negligence in more than 300 oil spills in Texas, Oklahoma, Kansas, Alabama, Louisiana, and Missouri between 1990 and 1997. Koch also agreed to spend more than $1 billion on clean-up and improved operations.

Sometimes the purpose of a civil suit is to seek an injunction or some other form of equitable relief from the actions of an individual, a corporation, or a governmental agency. You might ask the courts, for example, to order the government to cease and desist from activities that are in violation of either the spirit or the letter of the law. This sort of civil action is heard only by a judge; no jury is present. Environmental groups have been very successful in asking courts to stop logging and mining operations, to enforce implementation of the endangered species act, to require agencies to enforce air and water pollution laws, and a host of other efforts to protect the environment and conserve natural resources. Often, rather than sue a corporation directly for environmental damage, it is more effective to sue the government for not enforcing laws that would have prevented the damage. A big corporation with deep pockets can afford squadrons of lawyers and may have the resources and incentive to tie up litigation for years with motions and counter suits. Federal or state agencies may be more inclined to agree that you are right and to be willing to settle the matter quickly.

Adversarial Approaches and SLAPP suits

The American legal system is adversarial, pitting one side against the other in an effort to distinguish right from wrong, or innocence from guilt. In a trial, each side tries to make the strongest possible arguments for its position, and to point out the faults in the opponent's case. The jury, as neutral fact-finder, hears arguments from both sides and makes an objective decision. We believe that this approach gives the best chance for truth to be discovered. It is time-consuming and costly, however, and doesn't always result in justice. Everyone has heard of cases where it seems perfectly obvious that the defendant is guilty, and yet they get off because certain evidence is inadmissible, or they simply have lawyers who can dazzle or confuse the jury with deceptive arguments. This system promotes strife and confrontation. It doesn't encourage compromise and often leaves lasting hatreds that make future cooperation impossible.

Because defending a lawsuit is so expensive, the mere threat of litigation can be a chilling deterrent. Increasingly, environmental activists are being harassed with Strategic Lawsuits Against Political Participation **(SLAPP)**. Citizens who criticize businesses that pollute or government agencies that are derelict in their duty to protect the environment are often sued in retaliation. While most of these preemptive strikes are groundless and ultimately dismissed, defending yourself against them can be exorbitantly expensive and take up time that might have been spent working on the original issue. Public interest groups and individual activists—many of whom have little money to defend themselves—often are intimidated from taking on polluters. For example, a West Virginia farmer wrote an article about a coal company's pollution of the Buckhannon River. The company sued him for $200,000 for defamation. Similarly, citizen groups fighting a proposed incinerator in upstate New York were sued for $1.5 million by their own county governments. A Texas woman called a nearby landfill a dump—and her husband was named in a $5 million suit for failing to "control his wife." Of course these suits also are expensive for the company or agency that initiates them, but they may be far cheaper than paying a fine or scrapping a big project.

Administrative Law: The Executive Branch

More than 100 federal agencies and thousands of state and local boards and commissions have environmental oversight. They usually have power to set rules, adjudicate disputes, and investigate misconduct. Federal agencies often delegate power to a matching state agency in order to decentralize authority. The enabling legislation to create each agency is called an "organic" act because it establishes a basic unit of governmental organization. In the federal government, most executive agencies come under the jurisdiction of cabinet-level departments such as Agriculture, Interior, or Justice (fig. 10.11).

Agency rule-making and standard-setting can be either formal or informal. In an informal case, notice and background for proposed rules are published in the Federal Register. Opportunities for all interested parties to submit comments are provided. This is often an important avenue for environmentally concerned citizens and public interest groups to have an impact on environmental policy. In formal rule making, a public hearing is held with witnesses and testimony much like a civil trial. Witnesses can be cross-examined. A complete transcript is made and final findings are published in the public record. It is generally more difficult for individuals to intervene in a formal hearing, although sometimes there is an opportunity to submit written comments. Rule making is often a complex, highly technical process that is difficult for citizen groups to understand and monitor. The proceedings are usually less dramatic and colorful than criminal trials, and yet can be very important for environmental protection.

Regulatory Agencies

The EPA is the primary agency with responsibility for protecting environmental quality. Created in 1970, at the same time as NEPA, the EPA is a cabinet-level department, with more than 18,000 employees and 10 regional offices. Often in conflict with Congress, other agencies of the Executive Branch, and environmental groups, the EPA has to balance many competing interests and conflicting opinions. Greatly influenced by politics, the agency changes dramatically depending on which party is in power and what attitudes toward the environment prevail at any given time. Under the Nixon and Carter administrations, the EPA grew rapidly and enforced air and water quality standards vigorously. When Ronald Reagan was elected to office in 1980, however, the EPA took a sharp downward turn. Reagan cut the

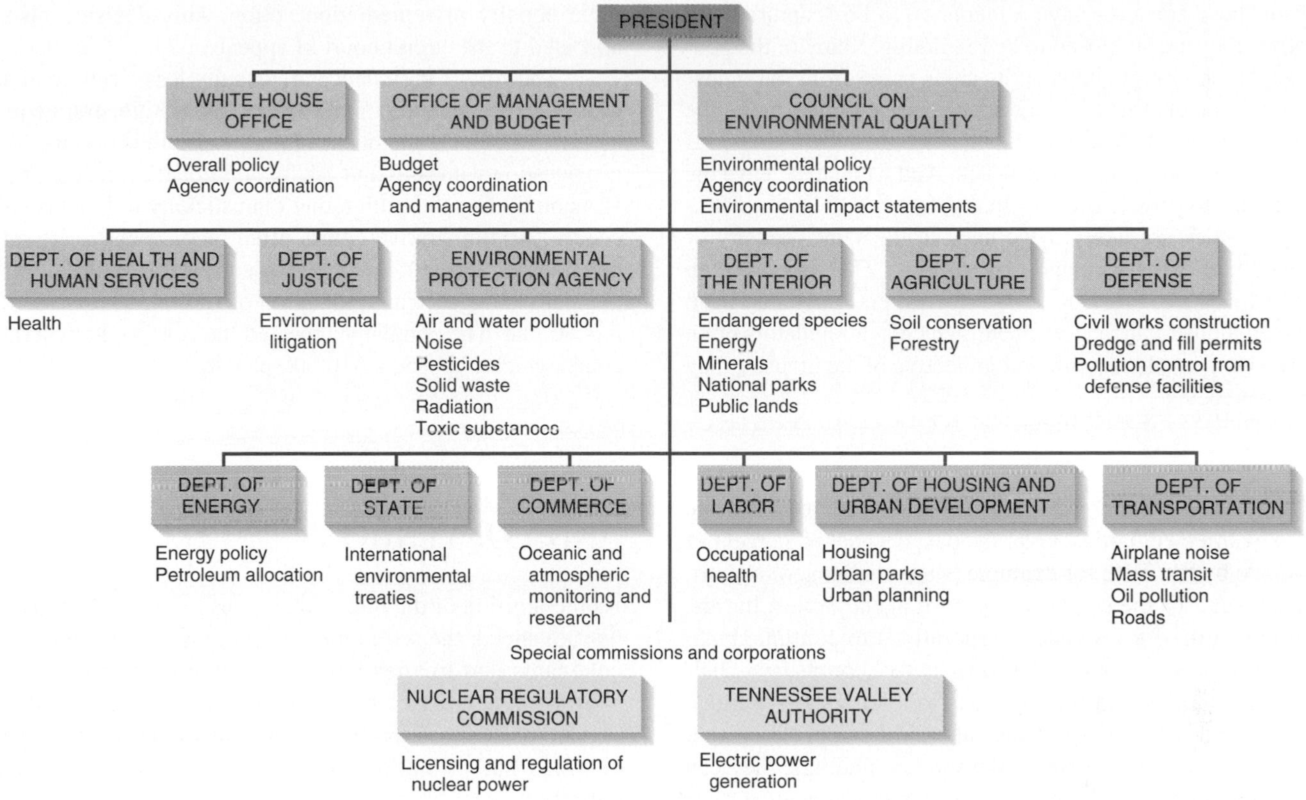

FIGURE 10.11 Major agencies of the Executive Branch of the U.S. federal government with responsibility for resource management and environmental protection.
Source: Data from U.S. General Accounting Office.

research and enforcement budgets by 50 percent. He appointed Anne Gorsuch (Burford)—who openly opposed environmental protection—as head of the agency. During a time when the EPA was being asked by Congress to do more, enforcement actions dropped by more than 70 percent. The agency staff was decimated, with many of the most experienced and effective employees either fired or driven out by frustration. With the election of Bill Clinton in 1992 and the appointment of Carol Browner as administrator, the EPA once again regained its concern for environmental quality.

The Departments of the Interior and Agriculture are to natural resources what the EPA is to pollution. Interior is home to the National Park Service, which is responsible for more than 376 national parks, monuments, historic sites, and recreational areas. It also houses the Bureau of Land Management (BLM), which administers some 140 million ha (350 million acres) of land, mostly in the western United States. In addition, Interior is home to the U.S. Fish and Wildlife Service, which operates more than 500 national wildlife refuges and administers endangered species protection.

The Department of Agriculture is home to the U.S. Forest Service, which manages about 175 national forests and grasslands, totaling some 78 million ha (193 million acres). With 39,000 employees, the Forest Service is nearly twice as large as the EPA (fig. 10.12). The Department of Labor houses the Occupational Health and Safety Agency (OSHA), which oversees workplace safety. Research that forms the basis for OSHA standards is carried out by the National Institute for Occupational Safety and Health (NIOSH). In addition, several independent agencies that are not tied to any specific department also play a role in environmental protection and public health. The Consumer Products Safety Commission passes and enforces regulations to protect consumers, and the Food and Drug Administration is responsible for the purity and wholesomeness of food and drugs.

FIGURE 10.12 A forest ranger explains wilderness rules to a hiker.

All of these agencies have a tendency to be "captured" by the industries they are supposed to be regulating. Many of the people with expertise to regulate specific areas came from the industry or sector of society that their agency oversees. Furthermore, the people they work most closely with and often develop friendships with are those they are supposed to watch. And when they leave the agency to return to private life—as many do when the administration changes—they are likely to go back to the same industry or sector where their experience and expertise lies. The effect is often what's called a "revolving door," where workers move back and forth between industry and government. As a result regulators often become overly sympathetic with and protective of the industry they should be overseeing.

Administrative Courts

Administrative regulations and rule-making procedures often are objected to by affected parties. Over the past decade, 80 percent of the rules made by the EPA, for example, were challenged in court. It is very unusual for the courts to overturn agency rules, but the challenge buys time for regulated corporations to continue business as usual for a while longer. **Administrative courts** hear challenges to agency rules and regulations. An administrative judge can consider both the validity of the rule and its application to a specific case. If the parties dispute the judge's findings, they can appeal to a district court. The courts rarely overturn agency rules unless (1) the enabling act is too vague or unconstitutional, (2) the agency has gone beyond the scope of power granted by the legislature, or (3) the agency didn't follow proper procedures.

Administrative courts also hear enforcement cases where an individual or corporation has violated an agency rule or standard. Suppose, for instance, that a factory is found to be exceeding allowable air pollution emissions. After an investigation, a complaint is filed with an administrative judge. A hearing is scheduled and the judge listens to both sides of the case and issues an opinion, which is usually a recommendation to the head of the responsible agency for a penalty or remediation action. This decision also can be appealed to the circuit court of appeals.

The rules for evidence are usually less strict in an administrative case than in a criminal action. The administrative judge acts as both fact finder and decision maker. There is no jury. The judge can question witnesses and ask for additional evidence. The emphasis is on finding the truth rather than sticking to strict rules of procedure. Administrative courts often recommend relatively small penalties as a way of encouraging early settlement. In one case, a company charged with a violation of EPA standards was assessed a $500 fine. The company appealed the case to the federal district court, which imposed a $10,000 penalty.

INTERNATIONAL TREATIES AND CONVENTIONS

As recognition of the interconnections in our global environment has advanced, the willingness of nations to enter into protective covenants and treaties has grown concomitantly. Figure 10.13 shows the number of parties participating in some of the most important of these treaties. Note that the earliest of these conventions has no nations as participants; they were negotiated entirely by panels of experts. Not only the number of parties taking part in these negotiations has grown, but the rate at which parties are signing on and the speed at which agreements take force also have increased rapidly. The Convention on International Trade in Endangered Species (CITES), for example, was not enforced until 14 years after ratification, but the Convention on Biological Diversity was enforceable after just one year, and had 160 signatories only four years after introduction. Over the past 25 years, more than 170 treaties and conventions have been negotiated to protect our global environment. Designed to regulate activities ranging from intercontinental shipping of hazardous

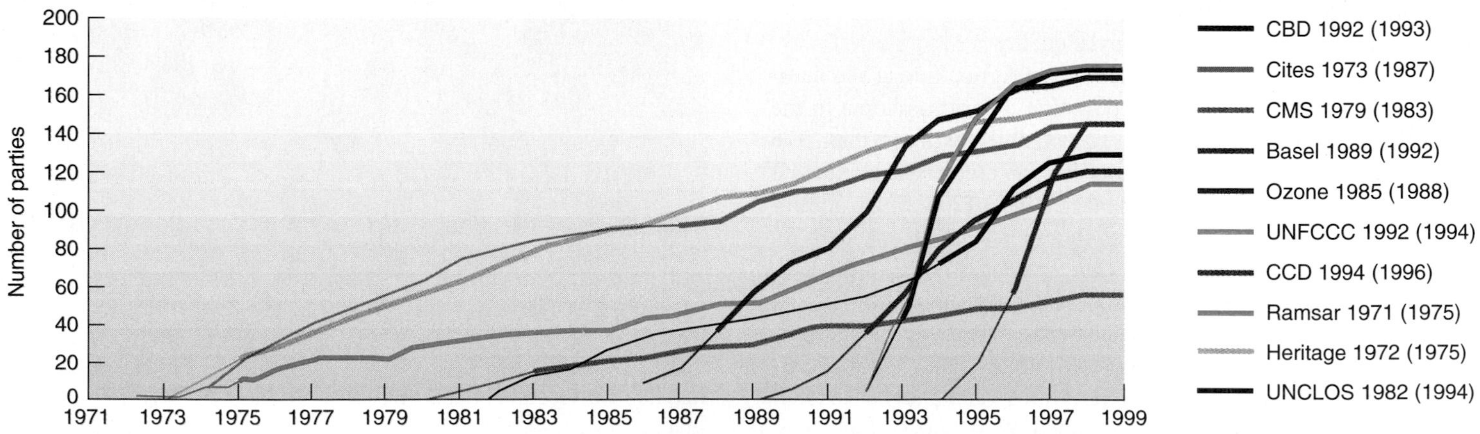

FIGURE 10.13 Additions of participating parties to some major international environmental treaties. The thick portion of each line shows when the agreement went into effect (date in parentheses). See p. 233 for complete treaty names.
Source: United Nations Environment Programme from Global Environment Outlook-2000.

waste, to deforestation, overfishing, trade in endangered species, global warming, and wetland protection, these agreements theoretically cover almost every aspect of human impacts on the environment.

Unfortunately, many of these environmental treaties constitute little more than vague, good intentions. In spite of the fact that we often call them laws, there is no body that can legislate or enforce international environmental protection. The United Nations and a variety of regional organizations bring stakeholders together to negotiate solutions to a variety of problems but the agreed-upon solutions generally rely on moral persuasion and public embarrassment for compliance. Most nations are unwilling to give up sovereignty. There is an international court, but it has no enforcement power. Nevertheless, there are creative ways to strengthen international environmental protection.

One of the principal problems with most international agreements is the tradition that they must be by unanimous consent. A single recalcitrant nation effectively has veto power over the wishes of the vast majority. For instance, more than 100 countries at the U.N. Conference on Environment and Development (UNCED), held in Rio de Janeiro in 1992, agreed to restrictions on the release of greenhouse gases. At the insistence of U.S. negotiators, however, the climate convention was reworded so that it only *urged*—but did not require—nations to stabilize their emissions.

As a way of avoiding this problem, some treaties incorporate innovative voting mechanisms. When a consensus cannot be reached, they allow a qualified majority to add stronger measures in the form of amendments that do not need ratification. All members are legally bound to the whole document unless they expressly object. This approach was used in the Montreal Protocol, passed in 1987 to halt the destruction of stratospheric ozone by chlorofluorocarbons (CFCs). The agreement allowed a vote of two-thirds of the 140 participating nations to amend the protocol. Although initially the protocol called for only a 50 percent reduction in CFC production, subsequent research showed that ozone was being depleted faster than previously thought. The protocol was strengthened by amendment to an outright ban on CFC production in spite of the objection of a few countries.

Where strong accords with meaningful sanctions cannot be passed, sometimes the pressure of world opinion generated by revealing the sources of pollution can be effective. NGOs and others can use this information to expose violators. For example, the environmental group Greenpeace discovered monitoring data in 1990 showing that Britain was disposing of coal ash in the North Sea. Although not explicitly forbidden by the Oslo Convention on ocean dumping, this evidence proved to be an embarrassment, and the practice was halted.

Trade sanctions can be an effective tool to compel compliance with international treaties. The Montreal Protocol, for example, bound signatory nations not to purchase CFCs or products made using them from countries that refused to ratify the treaty. Because many products employed CFCs in their manufacture, this stipulation proved to be very effective. On the other hand, trade agreements also can work against environmental protection. The World Trade Organization was established to make international trade more fair and to encourage development. It has been used, however, to subvert national environmental laws. In a ruling in 1998, the WTO forbid the United States from restricting imports of shrimp from Thailand, Malaysia, India, and Pakistan that were caught with nets that trap endangered sea turtles. Similarly, under provisions of the North American Free Trade Agreement, the Ethyl Corporation of the United States sued Canada for $250 million in compensation for banning the manganese-based gasoline additive MMT, which is suspected to be a neurotoxin. Environmentalists are concerned about a powerful international treaty currently being negotiated by the Organization of Economic Cooperation and Development, a group of 29 of the wealthiest countries in the world. The Multilateral Agreement on Investments, or MAI, may be more threatening to social justice and environmental protection than either the WTO or NAFTA.

DISPUTE RESOLUTION AND PLANNING

The adversarial approach of our current legal system often fails to find good solutions for many complex environmental problems. Identifying an enemy and punishing him for transgressions seems more important to us than finding win/win compromises. Gridlocks occur in which conflicts between adversaries breed mutual suspicion and decision paralysis. The result is continuing ecosystem deterioration, economic stagnation, and growing incivility and confrontation. The complexity of many environmental problems arises from the fact that they are not purely ecological, economic, or social, but a combination of all three. They require an understanding of the interrelations between nature and people. Are there ways to break these logjams and find creative solutions? In this section we will look at some new developments in mediation, dispute resolution, and alternative procedures for environmental decision making.

Wicked Problems and Adaptive Management

Rational choice theories of planning and decision making assume that if we just collect more data, buy faster computers to crunch numbers, build more complex models, and spend more money, any problem should be resolvable. More information, it's assumed, will lead automatically to better management. This "bigger hammer" approach may be effective in problems that are difficult but relatively straightforward. Increasingly, however, we have come to recognize that many of the most important problems we face don't fit this pattern. Questions like what ecosystem health means, or how clean is clean, don't have simple right or wrong answers. They depend on your worldview and how you define these terms. Different people come to different conclusions even if they share the same information.

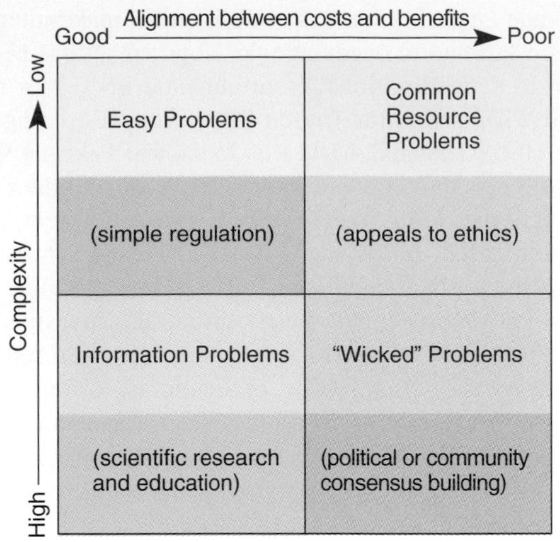

FIGURE 10.14 The difficulty of environmental decision making increases as problems become more complex and the congruence between those who bear costs and enjoy benefits decreases.

Environmental scientists describe problems with no simple right or wrong answers as being **wicked problems,** not in the sense of having malicious intent, but rather as obstinate or intractable. These problems often are nested within other sets of interlocking issues. The definition of both the problem and its solutions differ for various stakeholders. There are no value-free, objective answers for these dilemmas, only choices that are better or worse depending on your viewpoint. Wicked problems are important and have serious consequences, but also are complex and have a poor match between who bears the costs and who bears the benefits on any proposed solution (fig. 10.14). They usually can't be solved by simple rules and regulations, more scientific research, or appeals to ethics. Often the best solution comes from community-based planning and consensus building. Inherent uncertainty gives these questions no clear end point. You cannot know when all possible solutions have been explored.

Recent advancement in understanding how ecological systems work gives us some insight into many wicked problems. Like biological organisms, social problems often change and evolve over time. Their history unfolds in complex ways, depending on chance interactions and unpredictable events. Like ecological systems, there may never be a stable equilibrium in many environmental issues. Each involves an assemblage of issues and actors that are unique in time and place. They can't be standardized. There are no good precedents from previous experience. Their solutions are unique, and what may work today, may not be applicable tomorrow. How can we learn to cope with such uncertainty?

One promising approach to solving wicked environmental problems comes from the work of ecologists C. S. Holling and Lance Gunderson, and planners Steven Light and Kai Lee, among others. Starting with the observation that human understanding of nature is imperfect, this group believes that all human interactions with nature should be experimental. They suggest that environmental policies should incorporate **adaptive management,** or "learning by doing," designed from the outset to test clearly formulated hypotheses about the ecological, social, and economic impacts of the actions being undertaken. Rather than assume that what seemed the best initial policy option will always remain so, we need to carefully monitor how conditions are changing and what effects we are having on both target and nontarget elements of the system. If our policy succeeds, the hypothesis is affirmed. But if the policy fails, an adaptive design still permits learning, so that future decisions can proceed from a better base of understanding. The goal of adaptive management and experimental design is to enable us to live with the unexpected. They aim to yield understanding as much as to produce answers or solutions (table 10.2). This approach to natural resources is similar to—but more explicitly experimental than—ecosystem management (see chapter 2).

Resilience in Ecosystem and Institutions

The great economist Joseph Schumpeter described "waves of creative destruction" that transform economic systems. Another insight from Holling and his collaborators is that similar cycles of destructive creation operate in both ecological systems and in policy institutions (fig. 10.15). This is a familiar process that occurs in secondary succession (chapter 4). The release phase of the cycle occurs when factors such as fires, storms, or pests disturb a biological community, mobilizing nutrients and making space available for new growth. During the reorganization phase, pioneer and opportunist species colonize the new habitat. These species grow rapidly on the accessible carbon, nutrient, and energy sources during the exploitation stage. As the community matures, both the

TABLE 10.2 Institutional Conditions for Adaptive Management

1. There is a mandate to take action in the face of uncertainty.
2. Decision makers are aware they are experimenting anyway.
3. Decision makers care about improving outcomes over biological timescales.
4. Preservation of pristine environments is no longer an option, and human intervention cannot produce desired outcomes predictably.
5. Resources are sufficient to measure ecosystem-scale behavior.
6. Theory, models, and field methods are available to estimate and infer ecosystem-scale behavior.
7. Hypotheses can be formulated.
8. Organizational culture encourages learning from experience.
9. There is sufficient stability to measure long-term outcomes; institutional patience is essential.

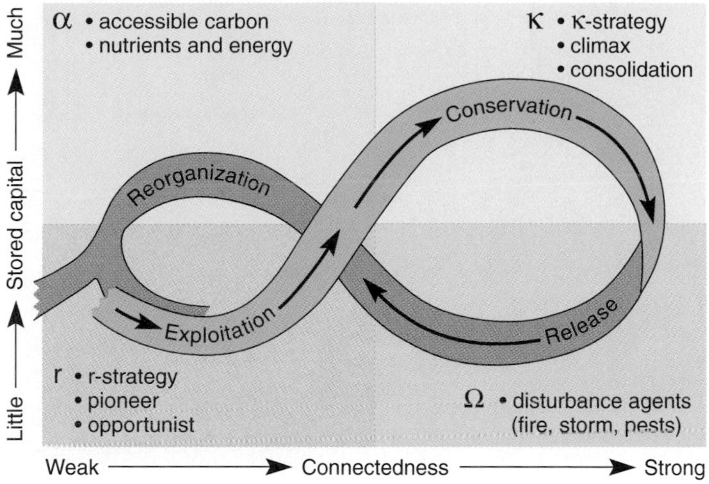

FIGURE 10.15 The creative-destruction cycle. Resilience, or the ability to reorganize and recover from disturbance, is the most important characteristic of both natural and human systems.

stored capital and connectedness increase until the ecosystem reaches a stage at which the system is poised for some new disturbance that starts the cycle again.

The most important characteristic of natural systems is their **resilience,** or ability to recover from disturbance. This doesn't imply that the ecosystem always returns to the exact condition it was in before the disturbance. It may have a new assemblage of species, or different set of physical conditions, but if it is resilient, the system has the ability to reorganize itself in creative and constructive ways. "Environmental quality is not achieved by attempting to eliminate change or surprises," Holling observed. The goal, instead, is resilience in the face of surprise. Surprise can be counted on. Resilience comes from adaptation to stress, from survival of the fittest in a turbulent environment.

In studying a variety of natural resource management regimes, Holling and others observed that human institutions also follow a similar pattern. In studying a variety of natural resource management issues ranging from restoration of the Florida everglades, to control of spruce budworm in New England forests, to cattle grazing in South Africa, to protection of salmon in the Pacific Northwest, they observed that every attempt to manage ecological variables one factor at a time inexorably leads to less resilient ecosystems, more rigid management systems, and more dependent societies. Initial success sets the conditions for eventual collapse. Take the example of forest fire suppression. For 70 years, the U.S. Forest Service has had a very effective policy of putting out all forest fires. The result has been that flammable debris has built up in the forests so that major conflagrations are now inevitable. During this time, however, people have felt safe moving to the borders of the forests and now there is a large population with a huge investment in property that needs to be protected from fire. Furthermore, a big bureaucracy has built up whose raison d'etre is to fight fires. It takes more and more money to forestall a calamity that becomes increasingly likely because of our efforts to prevent it.

What happens in each of these cases is that our goal to control variability in ecological systems leads us to a narrow purpose and to focus exclusively on solving a single problem. But elements of the system change gradually as a consequence of our management success in ways that we did not anticipate. As more homogenous ecosystems develop over a landscape scale, resilience decreases, and it becomes more likely that the system will flip suddenly into a new regime. What can we do to avoid this trap? Table 10.3 suggests some important lessons for ecological managers.

The Precautionary Principle

One response to the uncertainty of wicked problems and chaotic, nonlinear, discontinuous systems is to plan a margin of safety for error or surprises. Drawing on studies of ecological systems, many conservation biologists advocate a **precautionary principle** that says that when an activity raises threats of harm to human health or the environment, precautionary measures should be taken even if some cause and effect relationships are not fully established scientifically. At a meeting at the Wingspread Center in 1998, an international group of scientists, government officials, lawyers, and grassroots environmental activists agreed on four basic tenets of precautionary action:

- People have a duty to take anticipatory steps to prevent harm. If you have a reasonable suspicion that something bad might be going to happen, you have an obligation to try to stop it.

TABLE 10.3 Planning for Resilience

1. Interdisciplinary, integrated modes of inquiry are needed for adaptive management of wicked problems.
2. We must recognize that these problems are fundamentally nonlinear and that we need nonlinear approaches to them.
3. Interactions between slow ecological processes such as global climate change or soil erosion in the American cornbelt are difficult to study together with the fast processes that bring creative destruction such as potential collapse of Antarctic ice sheets or appearance of a dead zone in the Gulf of Mexico, but we need to look for connections.
4. The spatial and temporal scales of our concerns are widening. We now need to consider global connections and problems in our planning.
5. Both ecological and social systems are evolutionary and are not amenable to simple solutions based on knowledge of small parts of the whole or on assumptions of constancy or stability of fundamental relationships.
6. We need adaptive management policies that focus on building resilience and the capacity of renewal in both ecosystems and human institutions.

Source: L. Gunderson, C. Holling, and S. Light, *Barriers and Bridges to the Renewal of Ecosystems and Institutions,* 1995, Columbia University Press.

- The burden of proof of carelessness of a new technology, process, activity, or chemical lies with the proponents, not with the general public.
- Before using a new technology, process, or chemical, or starting a new activity, people have an obligation to examine a full range of alternatives, including the alternative of not using it.
- Decisions applying the precautionary principle must be open, informed, and democratic, and must include the affected parties.

Opponents of this approach claim that it could prevent us from doing anything productive or innovative. What do you think? Is this just common sense or an invitation to decision paralysis?

Arbitration and Mediation

Another set of alternatives to the adversarial nature of litigation and administrative challenges is the growing field of dispute resolution. Increasingly used to avoid the time, expense, and winner-take-all confrontation inherent in tort law, these techniques encourage compromise and workable solutions with which everyone can live.

Arbitration is a formal process of dispute resolution somewhat like a trial. There are stringent rules of evidence, cross-examination of witnesses, and the process results in a legally binding decision. The arbitrator takes a more active role than a judge, however, and is not as constrained by precedent. The arbitrator is more interested in resolving the dispute rather than strict application or interpretation of the law. Arbitrators must have formal training and be certified by the Federal Mediation and Conciliation Service or the American Arbitration Association. Arbitration is usually an attractive prospect if you don't think you could win a formal lawsuit, but why would anyone agree to arbitration if they think they can win the whole enchilada in court? They might take this route just to avoid disagreeable surprises. Juries can be fickle. Furthermore, they might want to get an unpleasant process over with sooner rather than later. In addition, arbitration often is written into contracts so that the disputants have no choice over the matter.

There are disadvantages to arbitration. It doesn't create a legally binding precedent, something that often is the main motivation for a lawsuit. There is less opportunity to appeal if you don't like the decision you get. There also is less protection from self-incrimination, false witnesses, or evidence you didn't expect. You don't generate nearly as much publicity because the proceedings and record are not public. For some litigants, the publicity generated by a trial is more valuable than the settlement itself. Finally, you are less likely to win the whole thing. Some sort of compromise is the most likely outcome.

Mediation is a process in which disputants are encouraged to sit down and talk to see if they can come up with a solution by themselves (fig. 10.16). The mediator makes no final decision but is simply a facilitator of communication. This process is especially useful in complex issues where there are multiple stakeholders with different interests, as is often the case in environmental controversies. For example, a mediation was attempted to work out policy and management disputes about the Boundary Waters Canoe Area Wilderness in northern Minnesota. Local property owners, resort operators, anglers who wanted to use motorboats on wilderness lakes, wildlife protection groups, wilderness advocates, and the U.S. Forest Service all were represented around the table. Each group had its own agenda, although they all quickly coalesced into pro-motor and anti-motor factions. Interestingly, although we usually think of environmentalists as being relatively disadvantaged in these issues, in this case, the local folks felt they were unfairly outgunned by the high-powered lawyers who represented some of the pro-wilderness and wildlife groups. Although the participants agreed on many points and it seemed as if the mediation was about to come up with a workable compromise, at the last minute everything fell apart and the groups refused to agree about anything. Each side accused the other of intransigence and bad will, and the issues they disagreed upon will likely end up in court.

FIGURE 10.16 Mediation encourages stakeholders to discuss issues and try to find a workable compromise.

This example illustrates both the promises and perils of mediation. It can be quicker and cheaper than court battles. It can lead to compromise and understanding that will lead to further cooperation, and that will solve problems faster than endless appeals. And it may find creative solutions that satisfy multiple parties and interests. On the other hand no one can be forced to mediate or to do so in good faith. Rancorous participants can tie up the process in long, pointless arguments that only make others more angry. Ultimately, there can be a tyranny of the minority. A single person can veto an agreement that everyone else wants. Furthermore, mediation represses or denies certain irreconcilable structural conflicts, giving the impression of equality between disputants when none really exists. Unequal negotiating skills of the participants can lead to unfair outcomes and even more rancor and paranoia than before the mediation was attempted. As is the case in arbitration, mediation doesn't generate the publicity and complete victory that some groups may desire.

Collaborative Approaches to Community-Based Planning

Over the past several decades, natural resource managers have come to recognize the value of holistic, adaptive, multiuse, multivalue approach to planning. Involving all stakeholders and interest groups early in the planning process can help avoid the "train wrecks" in which adversaries become entrenched in non-negotiable positions. Working with local communities can tap into traditional knowledge and gain acceptance for management plans that finally emerge from policy planning. This approach is especially important in nonlinear, nonequilibrium systems and wicked problems. Among the more important reasons to use collaborative approaches are:

- The way wicked problems are formulated depends on your worldview. Incorporating a variety of perspectives early in the process is more likely to lead to the development of acceptable solutions in the end.
- People have more commitment to plans they have helped develop. The first stage is therefore to identify those involved and to engage them in the process.
- There is truth in the old adage that "two heads are better than one." Involving multiple stakeholders and multiple sources of information enriches the process.
- Community-based planning provides access to situation-specific information and experience that can often only be obtained by active involvement of local residents.
- Participation is an important management tool. Project-threatening resistance on the part of certain stakeholders can be minimized by inviting active cooperation of all stakeholders throughout the planning process.
- The knowledge and understanding needed by those who will carry out subsequent phases of a project can only be gained through active participation.

A good example of community-based planning can be seen in the Atlantic Coastal Action Programme (ACAP) in eastern Canada. The purpose of this project is to develop blueprints for the restoration and maintenance of environmentally degraded harbors and estuaries in ways that are both biologically and socially sustainable. Officially established under Canada's Green Plan and supported by Environment Canada, this program created 13 community groups, some rural and some urban, with membership in each dominated by local residents. Federal and provincial government agencies are represented primarily as nonvoting observers and resource people. Each community group is provided with core funding for full-time staff who operate an office in the community and facilitate meetings.

FIGURE 10.17 The Bay of Fundy has the greatest tidal range in the world. It is the site of innovative community-based environmental planning process.

Four of the 13 ACAP sites are in the Bay of Fundy, an important and unique estuary lying between New Brunswick and Nova Scotia. Approximately 270 km (168 mi) long, and with an area of more than 12,000 sq km (4600 sq mi), the bay, together with the nearby Georges Bank and the Gulf of Maine form one of the richest fisheries in the world. With the world's highest recorded tidal range (up to 16 m or 52 ft at maximum spring tide), the bay sustains a great variety of fishery and wildlife resources, and provides habitat for a number of rare or endangered species. Now home to more than one million people, the coastal region is an important agricultural, lumbering, and paper-producing region (fig. 10.17).

Since European settlement began in 1604, the Bay of Fundy region has experienced great changes in population growth, resource use, and human-induced ecosystem change. More than 80 percent of the saltmarshes present in 1604 have been eliminated or degraded. Pollution and sediment damage harbors and biological communities. Overfishing and introduction of exotic species have resulted in endemic species declines. The collapse of cod, halibut, and haddock fishing has had devastating economic effects on the regional economy and the livelihoods of local residents. Aquaculture is now a more valuable activity than all wild fisheries.

To cope with these complex, intertwined social and biological problems, ACAP is bringing together different stakeholders from around the bay to create comprehensive plans for ecological, economic, and social sustainability. Through citizen monitoring and adaptive management, the community builds social capital (knowledge, cooperative spirit, trust, optimism, working relations), develops a sense of ownership in the planning process, and eliminates some of the fears and sectorial rivalry that often divides local groups, outsiders, and government agents.

On the other hand, giving a greater voice and increased power to local communities isn't always seen as a positive step by those from the outside (What Do You Think? p. 230).

what do you think?

THE WORLD AROUND YOU

The Quincy Library Group

Northern California has had intense debates about forest management policy for many years. On one hand, many small rural communities are almost totally dependent on the forest industry for economic survival. On the other hand, as native, old-growth forests become increasingly scarce, many environmental groups call for a sharp decrease in logging on public lands, and campaign to preserve as much as possible of what wilderness is left in its original state. The Forest Service is caught in the middle of this debate, with a mandate for both environmental protection and economic production. How can it reconcile these competing demands?

An experiment in community-based environmental planning in the small Sierra Nevada town of Quincy, California, has recently generated a great deal of interest and controversy. Some people praise it as an exciting model for cooperative management, while others deplore it as a fraud and a hoax designed to circumvent existing environmental controls and give away precious natural resources to local industry.

Several wicked problems had led the citizens of Quincy to feel that they were facing a crisis that demanded some drastically new approaches. Decades of fire suppression had left the forest surrounding the town choked with dead trees and woody debris. In the dry climate of the northern Sierras, a catastrophic fire seemed inevitable. Calls for protection of roadless areas and old-growth-associated species such as the marbled murlet and the northern spotted owl, worried the forest products industry that it might not have a continuing supply of wood. With a single-industry economy, and timber harvest down as much as 80 percent from historic highs, Quincy, like many of its neighboring communities, felt like an endangered species itself. Attempts at dialog among the various factions in town usually ended up in shouting matches.

One day in the early 1990s, three Quincy residents with very different backgrounds—a local timber industry employee, an ardent environmentalist, and a county supervisor—got together to talk about their common concerns. Agreeing to meet in the only building in town where they weren't allowed to shout at each other, the Quincy Library Group was born. As other residents joined the conversation, a consensus began to emerge about how the forest could be managed, environmental quality could be protected, and the local economy could survive.

The Quincy Library Plan calls for new management plans on 1 million ha (2.47 million acres) of land on the Plumas, Lassen, and Tahoe National Forests. "Fuel breaks" created by thinning out dense sections of forest would allow harvesting of up to 24,000 ha (60,000 acres) of public land each year. In addition, another 24,000 ha of forest would be harvested to support the local forest industry. Roadless areas, riparian zones, and habitat for endangered species such as the spotted owl would be protected. Hailed as a breakthrough in innovative, cooperative planning, the Library Group's proposal was introduced in Congress by California Senator Dianne Feinstein and Representative Vic Fazio, as an experimental, adaptive management strategy. Each year of the "pilot project" the Forest Service must report on the economic, social, and environmental effects of its actions. After five years of environmental monitoring and science-based assessment, the whole plan will be reexamined.

Failing to muster enough support in Congress to pass on its own, the Quincy Library Plan was attached as a rider on the 1999 Omnibus Appropriation Bill, which was signed into law by President Clinton in 1998. Environmental groups denounced both the method of passage as well as the content of the plan, claiming that it will double the timber harvest in the affected forest region, and subvert existing environmental laws. They claim it is a corporate welfare handout to the Sierra Pacific Industries, and will open the door to privatizing national forests. Interestingly, a split occurred between some national environmental groups, which opposed excess local control, and their northern California chapters, which defended local knowledge and autonomy.

What do you think? Is community-based planning a recognition of the wisdom and practical experience of local residents, or simply a way to give special favors to local industry? Is this courageous innovation, or simply sleeping with the enemy? Is the fact that participants come to understand and like each other a healthy development or the beginning of a sellout? Would it be better to maintain an adversarial stance in this case, or seek compromise?

Located in the northern Sierra Nevada mountains, the small town of Quincy, California, is the site of a controversial experiment in community-based resource planning.

Green Plans

Several national governments have undertaken integrated environmental planning that incorporates community round-tables for vision development. Canada, New Zealand, Sweden, and Denmark all have so-called **green plans** or comprehensive, long-range national environmental strategies. The best of these plans weave together complex systems, such as water, air, soil, and energy, and mesh them with human factors such as economics, health, and carrying capacity. Perhaps the most thorough and well-thought-out green plan in the world is that of the Netherlands.

Developed in the 1980s through a complex process involving the public, industry, and government, the 400-page Dutch plan contains 223 policy changes aimed at reducing pollution and establishing economic stability. Three important mechanisms have been adopted for achieving these goals: integrated life-cycle management, energy conservation, and improved product quality. These measures should make consumer goods last longer and be more easily recycled or safely disposed of when no longer needed. For example, auto manufacturers are now required to design cars so they can be repaired or recycled rather than being discarded.

Among the guiding principles of the Dutch green plan are: (1) the "stand-still" principle that says environmental quality will not deteriorate, (2) abatement at the source rather than cleaning up afterward, (3) the "polluter pays" principle that says users of a resource pay for negative effects of that use, (4) prevention of unnecessary pollution, (5) application of the best practicable means for pollution control, (6) carefully controlled waste disposal, and (7) motivating people to behave responsibly.

The Netherlands have invested billions of guilders in implementing this comprehensive plan. Some striking successes already have been accomplished. Between 1980 and 1990, emissions of sulfur dioxide, nitrogen oxides, ammonia, and volatile organic compounds were reduced 30 percent. By 1995, pesticide use had been reduced 25 percent from 1988 levels, and chlorofluorocarbon use had been virtually eliminated. By 1998, industrial wastewater discharge into the Rhine River was 70 percent less than a decade earlier. Some 250,000 ha (more than 600,000 acres) of former wetlands that had been drained for agriculture are being restored as nature preserves and 40,000 ha (99,000 acres) of forest are being replanted. This is remarkably generous and foresighted in such a small, densely populated country, but the Dutch have come to realize they cannot live without nature (fig. 10.18).

Not all goals have been met so far. Planned reductions in CO_2 emissions failed to materialize when cheap fuel prices encouraged fuel-inefficient cars. Currently a carbon tax is being considered. A sudden population increase caused by immigration from developing countries and Eastern Europe also complicates plan implementation, but the basic framework of the Dutch plan has much to recommend it, nevertheless. Other countries would be more sustainable and less environmentally destructive if they were to adopt a similar plan.

FIGURE 10.18 Under the Dutch Green Plan, 250,000 ha (600,000 acres) of drained agricultural land is being restored to wetland and 40,000 ha (99,000 acres) are being replanted as woodland.

SUMMARY

Although "policy" can have multiple meanings, environmental policy, in this chapter, is taken to mean both public opinion as well as official rules and regulations concerning our environment. Although we would like to think that fairness or rational choice and science characterize our political system, power often is the most important factor. The policy cycle describes the steps by which problems are identified and defined, and solutions are proposed, debated, enacted into law, and monitored.

The National Environmental Policy Act (NEPA) forms the cornerstone of both environmental policy and law in the United States. One of its most important provisions is the requirement of Environmental Impact Statements (EIS) for all major federal projects and programs. Laws are rules established through legislation (statutes), judicial decisions (case law), custom (common law), or administrative decisions (administrative law). Over the past century, U.S. policy and law has shifted from a hands-off attitude toward industry and private property, to end-of-the-pipe command and control, to more collaborative, pragmatic approaches.

Passage of bills in Congress is a convoluted process with many opportunities for amendments and riders to be added. Competing interests "lobby" throughout this process to try to change the outcome. Over the past 30 years, litigation has often been the most effective route for environmental protection. Understanding the differences between civil, criminal, and administrative courts is important in our quest for environmental quality. Establishing standing or the right to be heard in court may be the crucial stumbling block in this effort. SLAPP suits, while often baseless, are intended to intimidate or inactivate environmental groups.

Although many international treaties and conventions have been passed to protect our global environment, most are vague or toothless. Some innovative measures have been devised to compel compliance. Some alternatives to adversarial litigation include arbitration, mediation, and community-based planning. These techniques are useful in complex, unpredictable, multistakeholder, multivalue issues. We call these wicked problems. Learning from ecological systems, we see that some of the main goals for environmental policy and planning should be ecological and institutional resilience and adaptive management. The Netherlands has done one of the best jobs of any nation in using these approaches to formulate a comprehensive "green plan."

QUESTIONS FOR REVIEW

1. What is the policy cycle, and how does it work?
2. Describe the path of a bill through Congress. When are riders and amendments attached?
3. What are the differences and similarities between civil, criminal, and administrative law?
4. List some of the major acts of environmental legislation over the past 30 years.
5. Why have some international environmental treaties and conventions been effective while most have not? Describe two such treaties.
6. What are wicked problems? Why are they difficult?
7. What is resilience? Why is it important?
8. Describe adaptive management.
9. What is collaborative, community-based planning?
10. What is unique about the Dutch green plan?

QUESTIONS FOR CRITICAL THINKING

1. Suppose you had been on the jury in the Woburn case. What evidence would persuade you that the W. R. Grace company was responsible for Jimmy Anderson's leukemia? If other companies also contaminated the Woburn wells, does that exonerate W. R. Grace?
2. What role do you believe fairness, power, and rational choice play in our political process? Would our environmental policies be different if we had a different political system?
3. Which is the most important step in the policy cycle? If you were leader of a major environmental group, where would you put your efforts in establishing policy?
4. Do you believe that trees, wild animals, rocks, or mountains should have legal rights and standing in the courts? Why or why not? Are there partial rights or some other form of protection you would favor for nature?
5. It's sometimes difficult to determine whether a lawsuit is retaliatory or based on valid reason. How would you define a SLAPP suit, and differentiate it from a legitimate case?
6. Try creating a list of arguments for and against an international body with power to enforce global environmental laws. Can you see a way to create a body that could satisfy both reasons for and against this power?
7. Think of a familiar example of a wicked environmental problem. What are the most important elements that make it wicked? What institutional changes could we implement to make this issue less wicked?
8. The Holling diagram for the creative-destruction cycle (fig. 10.15) is described in terms of ecological change. Try applying this model to cycles of change in human institutions. Describe the actors, conditions, and forcing factors in each of the four quadrants of the model.
9. Take a current wicked environmental problem. If you were an environmental leader trying to resolve this problem, would you choose litigation, arbitration, or mediation? What are your reasons for favoring or rejecting each one?
10. Based on the information you have now, do you favor or oppose the Quincy Library Plan? What additional information would you need to make a good decision about this case?

KEY TERMS

adaptive management 226
administrative courts 224
administrative law 215
arbitration 228
case law 215
civil law 221
common law 221
criminal law 220
environmental law 215
environmental policy 213

ADDITIONAL INFORMATION ON THE INTERNET

Visit our website at /http://www.mhhe.com/environmentalscience for specific information about each of the topics below:

League of Conservation Voters

Center for International Ecological Law

Environmental Defense Fund

Natural Resource Defense Fund: legislative watch

West Coast Environmental Law (British Columbia)

Earth Justice

Natural Resources Defense Council

Environmental Law sites

Union of Concerned Scientists: activist scientists on public policy

Congress News, from C-SPAN

SOME IMPORTANT INTERNATIONAL TREATIES

CBD: Convention on Biological Diversity 1992 (1993)

CITES: Convention on International Trade on Endangered Species of Wild Fauna and Flora 1973 (1987)

CMS: Convention on the Conservation of Migratory Species of Wild Animals 1979 (1983)

Basel: Basel Convention on the Transboundary Movements of Hazardous Wastes and their Disposal 1989 (1992)

Ozone: Vienna Convention for the Protection of the Ozone Layer and Montreal Protocol on Substances that Deplete the Ozone Layer 1985 (1988)

UNFCCC: United Nations Framework Convention on Climate Change 1992 (1994)

CCD: United Nations Convention to Combat Desertification in those Countries Experiencing Serious Drought and/or Desertification, Particularly in Africa 1994 (1996)

Ramsar: Convention on Wetlands of International Importance especially as Waterfowl Habitat 1971 (1975)

Heritage: Convention Concerning the Protection of the World Cultural and Natural Heritage 1972 (1975)

UNCLOS: United Nations Convention on the Law of the Sea 1982 (1994)

PROFILE

Environmental Affairs Coordinator

Kristin Groce

Growing up in southeast Chicago, Kristin Groce experienced some of the urgent social and environmental problems of an aging urban industrial center (see p. 547 Chapter 24). A high school environmental science course piqued her interest in improving environmental quality in her neighborhood. It also showed her that not all businesses are bad; they can be valuable partners in community development. Tutoring and participating in the African American Association in high school increased her social awareness. Being a member of the dance line introduced her to racial diversity and cooperation.

After high school, Kristin looked at many different colleges. She chose Jackson State University in Alabama because of the friendly people she met there, and the beautiful forests surrounding the campus. She majored in biology, with a concentration in environmental science and a minor in business. This unusual combination has proven to be great for her present job at the Southeast Chicago Development Commission (SCDC) where, as Environmental Affairs Coordinator, she works with businesses, community, and environmental groups to revitalize the local economy and clean up the environment.

As a senior in college, Kristin was beginning to think about jobs. She visited her college career office and searched the Internet and newspapers for job opportunities, but her best lead was from her grandmother, who saw an announcement advertising openings at SCDC. Kristin applied and was offered a job. One of the projects she has worked on is native landscaping of an industrial corridor that both beautifies the community and also gives a better image to local businesses. She also coordinates an environmental education program for local school children, and oversees a service learning program for high school students. She also is involved in wetland cleanup projects, preserving local natural areas, pollution prevention plans for industry, and purchasing of new energy-efficient equipment. Community development for SCDC means revitalizing the local economy, as well as improving social and natural capital.

An understanding of how business works has turned out to be invaluable for Kristin. Among the other college courses she finds very useful are communications, writing, computer skills, and biology. She wishes now that she had taken more courses to learn about native species and ecosystems, but she is learning a great deal on the job. Being able to work with a wide variety of people—from school kids, to business people, to grandmothers—has proven to be a key to her position.

When asked what she likes best about her present job, Kristin says that she loves being able to work with so many different people in the community in which she grew up. “I didn’t realize,” she says, “that I could come back home to improve the situation here and create jobs.” When asked what she likes least about her job, Kristin replies that she often feels caught in the middle between community factions and businesses that just don’t get along. She wishes that cooperation and understanding between different groups were easier to attain.

At some point, Kristin says she probably will go back to school for an MBA or engineering degree so she can work for a consulting firm or become an environmental compliance officer in a corporation. She might also be interested in an environmental career in government, but for now, she feels good about being able to help her community and do something positive for both the people and environment in her own neighborhood.

CHAPTER 11

Food and Agriculture

We abuse the land because we regard it as a commodity belonging to us.
When we see land as a community to which we belong,
we may begin to use it with love and respect.

—Aldo Leopold—

OBJECTIVES

After studying this chapter, you should be able to:

- describe world food supplies and some causes of chronic hunger in the midst of growing food surpluses.
- explain some major human nutritional requirements as well as the consequences of deficiencies in those nutrients.
- differentiate between famine and chronic undernutrition and understand the relation between natural disasters and social or economic forces in triggering food shortages.
- analyze some of the promises and perils of genetic engineering.
- sketch the role of living organisms, physical forces, and other factors in creating and maintaining fertile soil.
- differentiate between the sources and effects of land degradation including erosion, nutrient depletion, waterlogging, and salinization.
- explain the need for water, energy, and nutrients for sustained crop production as well as some limits on our use of these resources.
- recognize the potential for low-input, sustainable, regenerative agriculture.

ESSENTIAL STUDY PARTNER

The Essential Study Partner (ESP) CD-ROM that accompanies this textbook contains additional information on the following topics:

- human nutritional needs and diet deficiencies: living resources: food: food needs
- world food consumption, crop production, and famines: living resources: food: food production
- green and blue revolution: living resources: food: food production/living resources: food: future prospects
- soil composition, types, and organisms: living resources: agriculture: soils
- land degradation and loss of topsoil: living resources: agriculture: land degradation
- energy use on farms: living resources: agriculture: other problems
- sustainable agriculture: living resources: agriculture: other problems

Above: A sunny counterpane landscape in Devon, England, illustrates the bounty of nature. With careful stewardship, the land represents a sustainable resource for humans and other species.

Are Shrimp Safe to Eat?

If you've bought shrimp recently at a restaurant or grocery store, chances are very good that they came from a commercial shrimp farm in a developing country such as Thailand, Ecuador, or Mexico. Once considered a luxury food, shrimp has become much more affordable in recent years, and now competes with tuna as the most popular seafood in the United States. As the world's leading shrimp-consuming country, the U.S. imports around 500,000 metric tons of farm-raised shrimp every year, or about half the total world production. Although this plentiful supply of a reasonably priced, highly desirable food is a boon to diners, there are social and environmental costs associated with its production that aren't widely known.

Shrimp aquaculture or farming first became profitable about 20 years ago and has since mushroomed into a major industry in the developing world. While total catches of wild shrimp have remained relatively stable at about 2 million metric tons per year over the past two decades, farm-raised production has exploded from less than 80,000 metric tons in 1980 to nearly one million metric tons in 1998. These shrimp are raised in shallow ponds ranging in size from a few hundred square meters to many hectares, generally constructed on or near the coastline of a tropical country. Asia has by far the largest area devoted to shrimp farming with more than 1.2 million hectares (3 million acres) in Thailand, Indonesia, China, India, Vietnam, and Bangladesh. Ecuador, with 130,000 hectares of ponds, raises about 60 percent of all shrimp in the Western Hemisphere, and is second in the world (after Thailand) in total production.

Hailed as a "blue revolution" 25 years ago, shrimp farming and other types of aquaculture were promoted as a way to provide a nutritious, inexpensive source of protein for the growing world population as well as to reduce the pressures on already dwindling supplies of wild seafood. While much commercial fish farming has been devoted to high-price, export species such as salmon, shrimp, and oysters, cultivation of some 10 million metric tons of less expensive freshwater fish such as carp and tilapia for local consumption has, indeed, increased the protein supply available in many developing countries. Culture of saltwater species such as shrimp, however, has caused considerable damage both to wild stocks and also to ecosystems that support them.

One of the biggest problems is that flooded mangrove forests and coastal wetlands often are destroyed to build shrimp ponds. Mangrove forests are extremely important as nurseries for a wide variety of ocean species. They absorb excess nutrients and sediment that would otherwise pollute nearshore waters and threaten coral reefs. About half of all mangrove forests in the world already have been destroyed. Shrimp farms are thought to be responsible for about one-fourth of that destruction. Furthermore, because shrimp farms often are stocked in very high densities, fresh seawater is flushed regularly through the ponds to wash out uneaten food, dead animals, feces, ammonia, phosphorus, and carbon dioxide. To prevent diseases among the teeming shrimp populations, most farmers also treat the ponds with antibiotics and chemicals such as formalin and calcium hypochlorite to kill pathogens and pests. All these wastes can leak into freshwater aquifers or overload coastal waters into which they are dumped, causing eutrophication and poisoning large numbers of resident organisms. Similar problems often occur in salmon-raising operations.

Most Asian shrimp farms are stocked with hatchery-produced young shrimp that can be certified free of diseases. In Latin America, however, many shrimp farmers prefer to stock their ponds with juvenile shrimp caught in the wild because they are cheaper and are thought to be stronger and have a higher survival rate. Shrimp harvesters scour estuaries and tidal wetlands to collect young shrimp to sell to farmers. Their fine-mesh nets catch large numbers of unwanted "by-catch" species. Although evidence is sparse, there are concerns that this harvest depletes populations of both wild shrimp and many other species. Furthermore, carnivorous species like salmon and shrimp often are fed high-protein fish meal made from wild ocean fish (sardines, anchovies, pilchard, and other low-value species). Because it takes roughly 2 kg of fish meal to produce a kilogram of farmed fish or shrimp, the result is a net loss of protein.

Not all aquaculture operations are environmentally harmful. With conscientious, scientific management, excess feeding can be minimized, diseases can be controlled without harmful chemical or antibiotic releases, water use can be minimized, and polluted effluent can be treated before being discharged into the environment. Rather than abandon contaminated ponds and start over building expensive new facilities on virgin land, farmers are learning to be careful in how they manage their operations. There isn't yet a certification process, however, so consumers can't tell whether the seafood products they buy have been obtained in an ecologically sound and sustainable manner.

So, while eating shrimp is probably safe for you, it may not be good for the environment. This example is only one of many dilemmas we face with respect to food and agriculture. In this chapter we will look at global food supplies and some of the problems associated with production and distribution of food.

NUTRITION AND FOOD SUPPLIES

Despite repeated dire predictions that runaway population growth would soon lead to terrible famines (see chapter 7), world food supplies have more than kept up with increasing human numbers over the past two centuries. In 1950, when there were 2.5 billion people on earth, the average daily diet provided less than 2000 calories per person. More than a billion people were considered chronically **undernourished,** meaning that they had less than the caloric intake needed for an active healthy life. In 1999, by contrast, even though the population had more than doubled to 6 billion people, the world food supply was sufficient to provide more than 2500 calories per person per day, or enough for a healthy and productive life for everyone if it were equitably distributed.

On the surface, at least, there is plenty of food available; in fact, there are surpluses in many areas. For those of us in the richer countries, the most common dietary problem is **overnutrition,** or too many calories. The average daily caloric intake in North America and Europe is about 3500 calories, nearly one-third more than is needed for healthy living. At least 20 percent of Americans are seriously overweight, a condition that contributes to high blood pressure, heart attacks, strokes, and other cardiovascular diseases that have become the leading causes of death in most developed countries.

Asia has experienced the most rapid increase in crop production, increasing food supplies per person by more than half over the past 40 years (fig. 11.1). China and Indonesia have had some of the most impressive food gains in history, tripling their output in little more than a decade. In only four years, Indonesia moved from the world's largest importer of rice to a country not only self-sufficient but boasting the world's biggest rice reserves. This success was due to a variety of factors including increased use of irrigation, fertilizers, and pesticide; expanded croplands; and new high-yielding crop varieties. Among the biggest questions in environmental science is whether those successes can be extended throughout the world, and whether we can feed another two or three billion people likely to be born in this century.

In some regions, notably in sub-Saharan Africa, food production has not kept pace with rapid population growth. Thirty-five of the 40 countries in this region have had decreasing per capita food production over the past two decades. The worst declines have been in countries such as Angola, Ethiopia, Sudan, Somalia, and Mozambique, where drought, war, grinding poverty, and governmental mismanagement have combined to produce heart-rending scenes of starvation and misery. The collapse of the former Soviet Union in 1992 also led to a precipitous collapse in food production. While privatization of collective farms in China led to increased output, it had the opposite effect in the former USSR. Crops rot in the fields where there is no money to pay harvesters, no transportation to get commodities to market, and too few customers with funds to buy goods if they do make it to the city.

Chronic Hunger and Food Security

As you might imagine, since many of us get up to one-third more calories than the average world diet, many people—especially in the poorer countries of the world—get far less. Although the

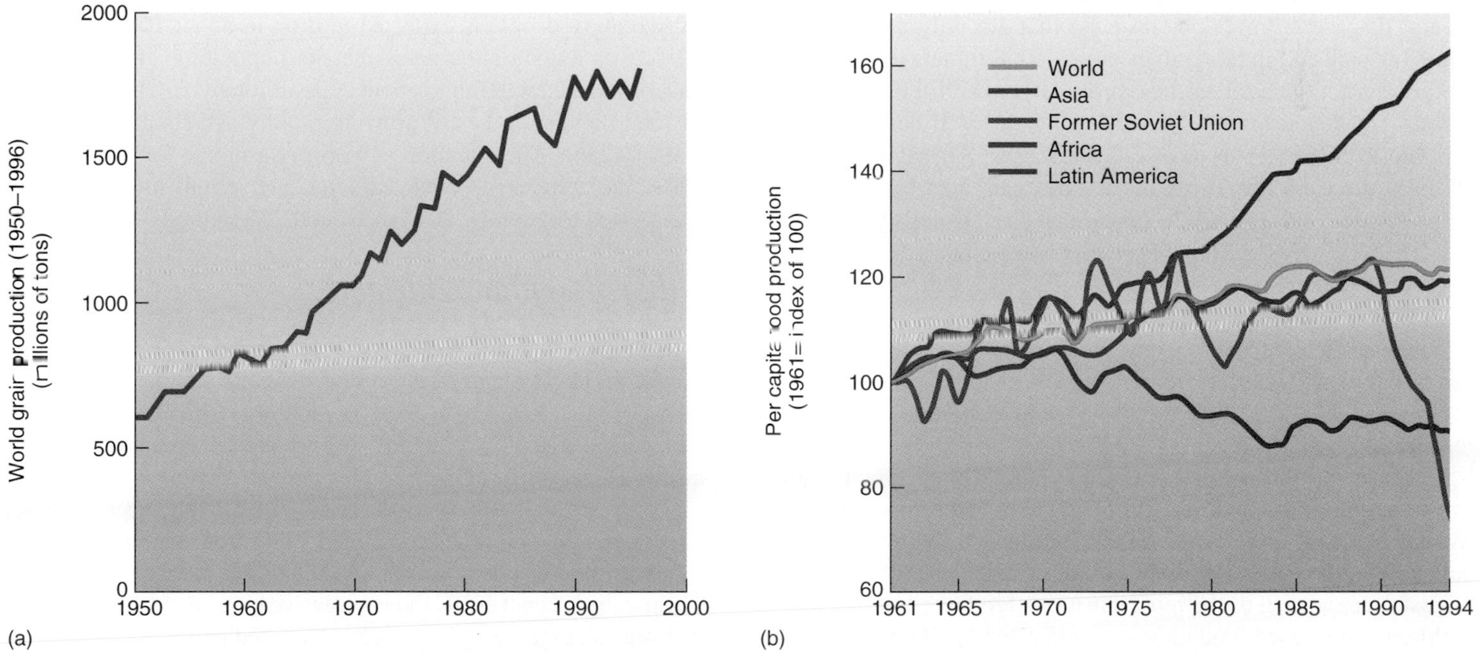

FIGURE 11.1 (*a*) Total world grain production has increased dramatically since 1950. (*b*) Relative per capita food production by region. Index is based on 1961. Notice general decline in Africa and sudden collapse of agricultural production in the former Soviet Union since 1990.
Source: (a) Data from USDA. (b) Data from Food and Agriculture Organization of the United Nations (FAO), FAOSTAT-PC on diskette (FAO, Rome, 1995).

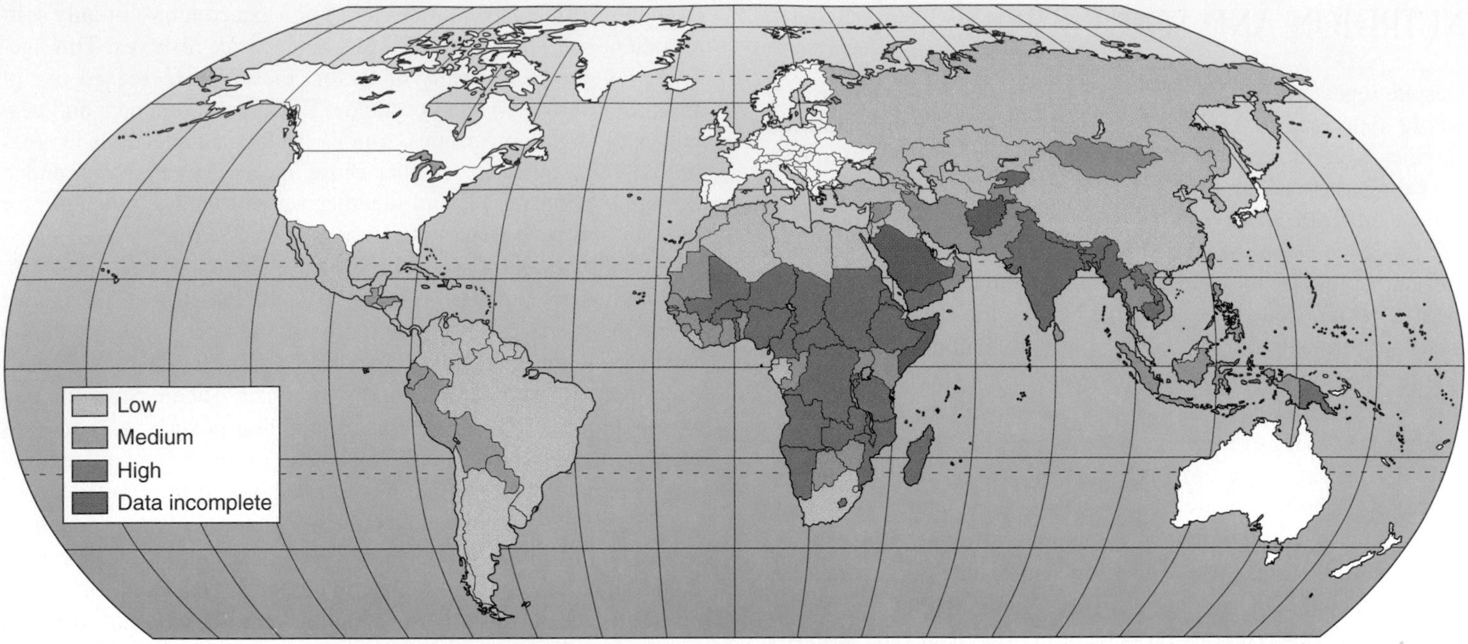

FIGURE 11.2 Countries with populations at risk of inadequate nutrition. The United States, Canada, Europe, Japan, and Australia have little risk.

global proportion of chronically hungry people has declined over the past half century, the Food and Agricultural Organization (FAO) says the total number remains at about 800 million people, or about one person in five in the developing world. At least 200 million of those who don't have enough food are children. Chronic undernourishment during childhood years leads to permanently stunted growth, mental retardation, and other social and developmental disorders. Infectious diseases that are only an inconvenience for well-fed individuals become lethal threats to those who are poorly nourished. Diarrhea rarely kills a well-fed person, but a child weakened by nutritional deficiencies will be highly susceptible to this and a host of other diseases. At least 11 million children die each year from diseases exacerbated by nutritional deficiencies. And around 20 percent of the Disability-Adjusted Life Years (DALYs) suffered every year (see chapter 9) are related to chronic nutritional deficits.

Poverty is the greatest threat to **food security,** or the ability to obtain sufficient food on a day-to-day basis. The 1.4 billion people in the world who have to live on less than $1 per day all too often can't buy the food they need and don't have access to resources to grow it for themselves. Food security occurs at multiple scales. In the poorest countries, hunger may affect nearly everyone. In other countries, although the average food availability may be good, some individual communities or families may not have enough to eat. And within families, males often get both the largest share as well as the most nutritious food, while women and children—who need food most—all too often get the poorest diet.

Where does this persistent hunger occur? Figure 11.2 shows countries with the greatest risk of food shortages. As you can see, most of sub-Saharan Africa, South and Southeast Asia, and parts of Latin America fall in this category. The largest number of hungry people in 1999, according to the FAO, was in East and Southeast Asia, where some 275 million people lacked an adequate diet. The second highest incidence of chronic undernutrition was in South Asia, where at least 250 million went hungry on a regular basis. Africa was third with about 225 million hungry people. Remarkable progress is being made in increasing food production in much of East and Southeast Asia, where the number of chronically hungry people is expected to drop to about 100 million by 2010. South Asia anticipates that its population of chronically undernourished will fall to about 200 million by 2010 even though its total population is still growing rapidly and soon will surpass that of China. The situation is worsening in sub-Saharan Africa, where the number of people experiencing serious food insecurity is expected to grow by 2010 to around 275 million.

Other Essential Nutrients

In addition to energy (calories), we need specific nutrients in our diet, such as proteins, vitamins, and certain minerals. It is possible to have plenty of calories and still suffer from **malnourishment,** a nutritional imbalance caused by a lack of specific dietary components or an inability to absorb or utilize essential nutrients. People in richer countries often eat too much meat, salt, and fat and too little fiber, vitamins, trace minerals, and other components lost from highly processed foods. In poorer countries, people often lack necessary nutrients because they cannot afford more expensive foods such as meat, fruits, and vegetables that would provide a balanced diet.

The two most widespread human protein deficiency diseases are kwashiorkor and marasmus. **Kwashiorkor** is a West African word meaning "displaced child." It occurs when people—especially children—eat a diet that is low in protein or has poor-quality protein,

FIGURE 11.3 The swollen abdomen of this child is characteristic of kwashiorkor, a protein-deficiency disease. A diet containing a balanced mixture of amino acids can reverse these symptoms.

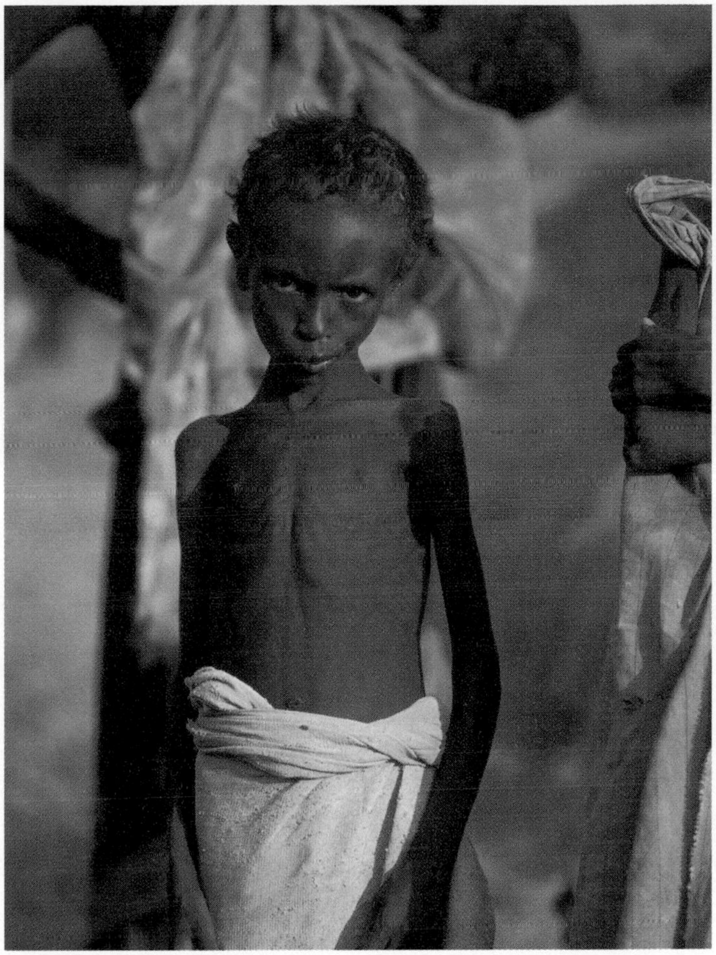

FIGURE 11.4 Marasmus is caused by combined energy (calorie) and protein deficiencies. Children with marasmus have the wizened look and dry, flaky skin of an old person.

even though it may have sufficient calories. Children with kwashiorkor often have reddish-orange hair and puffy, discolored skin and a bloated belly (fig. 11.3). **Marasmus** (from the Greek "to waste away") is caused by a diet low in both calories and protein. A child suffering from severe marasmus is generally thin and shriveled like a tiny, very old starving person (fig. 11.4). Children with both these deficiencies become anemic and listless and have low resistance to infections. Even if they survive childhood infectious diseases, they are likely to suffer from stunted growth, mental retardation, and other developmental problems.

Starchy foods like maize (corn), polished rice, and manioc (tapioca), which form the dietary basis for many poor people, tend to be low in several essential vitamins. Folic acid (found in dark green leafy vegetables) is essential for early fetal development. Folic acid deficiencies have been linked to terrible neurological problems in babies such as microencephaly (abnormally small heads) or even anencephaly (lacking a brain). Vitamin A (found in green and yellow vegetables and dairy products) is essential for good vision. Globally, some 42 million children under age 6 have mild to moderate vitamin A deficiency. In its severe form, this deficiency can cause blindness. About 250,000 children in developing countries go blind annually and up to 80 percent of those children die within a year.

Iron deficiency is the most common dietary mineral problem in the world. According to the FAO, at least 3.6 billion people (mostly women and children) suffer from iron deficiency, which leads to **anemia** (low levels of hemoglobin in the blood). The problem is most severe in India, where more than 80 percent of all pregnant women are anemic. Anemia increases the risk of death from hemorrhage in childbirth and also affects the baby's development. Red meat, eggs, legumes, and green vegetables all are good sources of iron. Meat and tempe (fermented soybeans) are generally the only dietary sources of vitamin B_{12}, which is necessary for proper absorption of iron from foods.

Iodine is essential for synthesis of thyroxine, an endocrine hormone that regulates metabolism and brain development and function, among other things. Chronic iodine deficiency causes goiter (a swollen thyroid gland, fig. 11.5) and hyperthyroidism. UNESCO estimates that 26 million children suffer brain damage each year as a result of iodine deficiency. Around 600 million people—mainly in South and Southeast Asia—suffer from goiter, and an estimated 1.5 billion people are at risk of this disease.

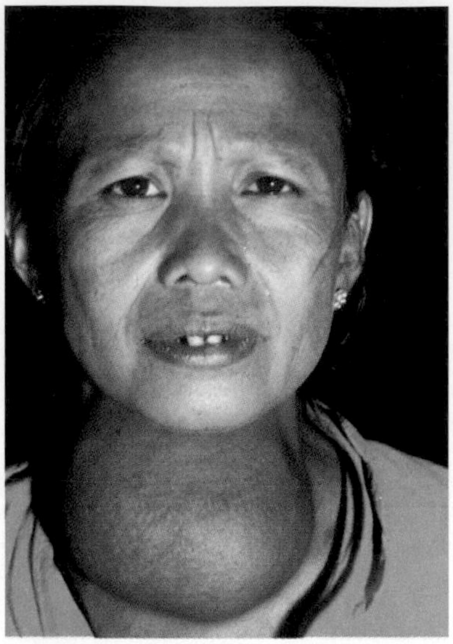

FIGURE 11.5 Goiter, a swelling of the thyroid gland at the base of the neck, is often caused by an iodine deficiency. It is a common problem in many parts of the world, particularly where soil iodine is low and seafood is unavailable.

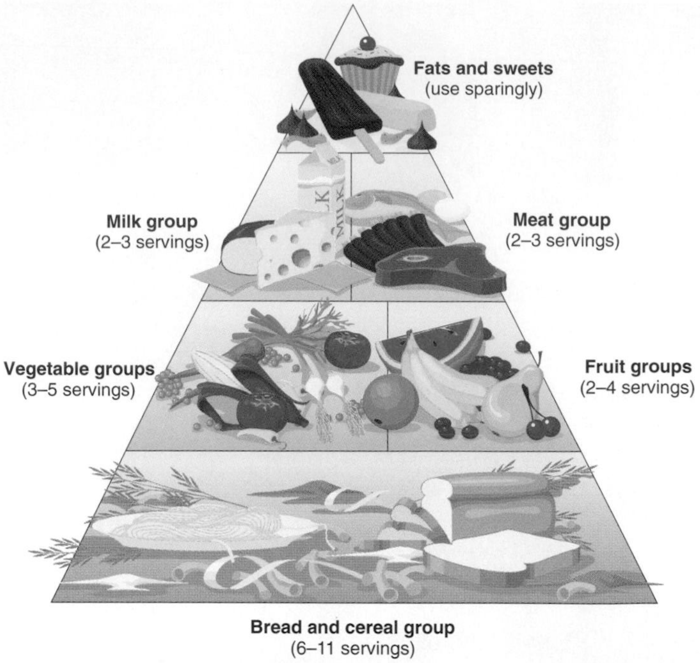

FIGURE 11.6 The food pyramid recommended by the U.S. Department of Agriculture. You should eat two to four times as much bread, cereal, rice, and pasta as milk, meat, eggs, or nuts. Fats, oils, and sweets should be eaten sparingly, if at all.

Eating a Balanced Diet

What's the best way to be sure we're getting a healthy diet? Generally it isn't necessary to take synthetic dietary supplements. Eating a balanced diet with plenty of whole grains, fruits, and vegetables should give you all the nutrients you need. For years Americans were advised to eat daily servings of four major food groups: meat, dairy products, grains, and fruits and vegetables. Placing meat and dairy products first was the result of lobbying by those industries. More recently, the U.S. Department of Agriculture has revised recommendations for a balanced diet (fig. 11.6) emphasizing grains, fruits, and vegetables, with only sparing servings of meat, dairy, fats, and sweets.

Famines

The chronic hunger and malnutrition described previously is silent and generally invisible, affecting individuals, families, and communities, often in the midst of general plenty, but **famines** are characterized by large-scale food shortages, massive starvation, social disruption, and economic chaos. Starving people eat their seed grain and slaughter their breeding stock in a desperate attempt to keep themselves and their families alive (fig. 11.7). Even if better conditions return, they have sacrificed their productive capacity and will take a long time to recover. Famines are characterized by mass migrations as starving people travel to refugee camps in search of food and medical care. Many die on the way or fall prey to robbers.

What causes these terrible tragedies? Environmental conditions are usually the immediate trigger, but politics and economics are often equally important in preventing people from getting the food they need. Adverse weather, insect infestations, and other natural disasters cause crop failures and create food shortages. But as economist Amartya K. Sen—whose work won the Nobel Prize in 1998—points out, these factors have generally been around for a

FIGURE 11.7 Children wait for their daily ration of porridge at a feeding station in Somalia. When people are driven from their homes by hunger or war, social systems collapse, diseases spread rapidly, and the situation quickly becomes desperate.

long time, and local people usually have adaptations that get them through hard times if they are allowed to follow traditional patterns of migration and farming. Arbitrary political boundaries, however, along with wars and land seizures by the rich and powerful, block access to areas that once served as refuge during droughts, floods, and other natural disasters. Poor people can neither grow their own food nor find jobs to earn money to buy the food they need.

In 1974, for instance, a terrible famine struck Bangladesh and thousands died. The immediate cause was floods in June through August that interfered with rice planting, eliminating jobs on which many farm workers depended. Fears about impending rice shortages triggered panic buying, hoarding by speculators, and rapidly rising prices. The government began relief efforts, but it was too little, too late. By October it became apparent that rice harvests were actually higher than in previous years; prices dropped and so did mortality rates. The irony was that more food was available in 1974 than in any other year in the decade. Poor people simply couldn't afford to buy it. The United States played a role in this tragedy. We cut off aid to Bangladesh as punishment for selling jute fibers to Cuba. Withdrawal of food aid on which Bangladesh had become dependent only fueled the panic and price gouging that occurred later.

By contrast, the state of Maharashtra in central India suffered a similar drought in 1972–73 that reduced crop yields by 50 percent. As in Bangladesh, farm laborers were thrown out of work. The Indian government moved quickly, however, to employ workers building roads, wells, and other public projects. Although wages were low, food prices didn't rise and people could afford an adequate if meager diet. Although the amount of food available per person was less than half that in Bangladesh, few Indians starved. Since they remained in their villages and rural infrastructure was improved during the drought by public works projects, farmers recovered quickly once the rains returned.

Similarly, the world was horrified by terrible photographs of starving children in Ethiopia and the Sudan in the 1980s. More than 1 million people died in the Horn of Africa. Droughts triggered this famine, but neighboring countries that experienced comparable weather did not have problems as severe as these unfortunate countries did. In Sudan and Ethiopia, where total food supplies dropped about 10 percent, the famine was severe. Zimbabwe and Cape Verde, by contrast, which suffered worse droughts and lost about 40 percent of their normal harvest, had no famine. In fact, social welfare programs brought about mortality *decreases* in Zimbabwe and Cape Verde during this time. Author Amartya Sen points out that armed conflict and political oppression almost always are at the root of famine. No democratic country with a relatively free press, he says, has ever had a major famine.

The aid policies of rich countries often serve more to get rid of surplus commodities and make us feel good about our generosity than to get at the root causes of starvation (fig. 11.8). Herding people into feeding camps generally is the worst thing to do for them. The stress of getting there kills many of them, and the crowding and lack of sanitation in the camps exposes them to epidemic diseases. There are no jobs in the refugee camps, so people can't support themselves if they try. Social chaos and family breakdown expose those who are weakest to robbery and violence. Having left their land and tools behind, people can't replant crops when the weather returns to normal.

FIGURE 11.8 We rarely learn much from the media about the underlying causes of famine. Often our attention is focused on images of people from wealthy nations generously helping those who are suffering. This makes us feel good, but doesn't get to the root of the problems.

MAJOR FOOD SOURCES

Of the thousands of edible plants and animals in the world, only about a dozen types of seeds and grains, three root crops, twenty or so common fruits and vegetables, six mammals, two domestic fowl, and a few fish and other forms of marine life make up almost all of the food humans eat. Table 11.1 shows annual production of some important foods in human diets. In this section, we will highlight sources and characteristics of those foods.

Major Crops

The three crops on which humanity depends for the majority of its nutrients and calories are wheat, rice, and maize. Together, nearly 1600 million metric tons of these three grains are grown each year. Wheat and rice are especially important since they are the staple foods for most of the 5 billion people in the developing countries of the world. These two grass species supply around 60 percent of the calories consumed directly by humans.

Potatoes, barley, oats, and rye are staples in mountainous regions and high latitudes (northern Europe, north Asia) because they grow well in cool, moist climates. Cassava, sweet potatoes, and other roots and tubers grow well in warm, wet areas and are staples in Amazonia, Africa, Melanesia, and the South Pacific. Sorghum and millet are drought resistant and are staples in the dry regions of Africa.

TABLE 11.1 Some Important Food Resources

CROP	1998 YIELD (MILLION METRIC TONS)
Wheat	586
Rice	384
Maize (corn)	597
Potatoes	281
Barley and oats	170
Soybeans	155
Cassava and sweet potato	286
Sugar (cane and beet)	119
Sorghum and millet	84
Vegetable oils	91
Vegetables and fruits	884
Meat and milk	432
Fish and seafood	94

Source: Data from Worldwatch Institute 1999.

Fruits and vegetables—including vegetable oils—make a surprisingly large contribution to human diets. They are especially welcome because they typically contain high levels of vitamins, minerals, dietary fiber, and complex carbohydrates.

Meat, Milk, and Seafood

Meat and milk are prized by people nearly everywhere, but their distribution is highly inequitable. Although the industrialized, more-developed countries of North America, Europe, and Japan make up only 20 percent of the world population, they consume 80 percent of all meat and milk in the world. The 80 percent of the world's people in less-developed countries raise 60 percent of the 3 billion domestic ruminants and 6 billion poultry in the world but consume only 20 percent of all animal products.

About 90 percent of the grain grown in North America is used to feed dairy and beef cattle, hogs, poultry, and other animals. It is used especially to fatten beef cattle during the last three months before they are slaughtered. As chapter 3 shows, there is a great loss of energy with each step up the food chain. This means we could feed far more people if we ate more grain directly rather than feeding it to livestock. Every 16 kg of grain and soybeans fed to beef cattle in feedlots produce about 1 kg of edible meat. The other 15 kg are used by the animal for energy or body parts we do not eat or they are eliminated. If we were to eat the grain directly, we would get twenty-one times more calories and eight times more protein than we get by eating the meat it produces. Hogs and poultry are about two and four times as efficient, respectively, as cattle in converting feed to edible meat.

Fish and other seafood contribute about 94 million metric tons of high-quality food to the world's diet. This is an important protein source in many countries, making up about one-half of the animal protein and one-fourth of total dietary protein in Japan, for instance.

Unfortunately, overharvesting and habitat destruction threaten most of the world's wild fisheries. Annual catches of ocean fish rose by about 4 percent annually between 1950 and 1988. Since 1989, however, 13 of 17 major marine fisheries have declined dramatically or become commercially inviable. According to the United Nations, 70 percent of the world's edible ocean fish, crustaceans, and mollusks are declining and in urgent need of managed conservation.

An explosion of fishing technology in the 1950s and 1960s made fishers both more efficient and more lethal. Sonar, radar, remote sensing, and global positioning systems have turned fishing from an art into a science. Longlines extending 100 km and carrying 60,000 hooks, bag-shaped trawl nets large enough to engulf a dozen jumbo jets, and drift nets covering two million square meters, tended by floating factories that can process and freeze 500 metric tons of fish per day, make it possible to exhaust entire populations in just a few years. Between 1970 and 1990, the number and average vessel size of the world fishing fleet doubled, now representing twice the capacity needed to extract the total annual sustainable harvest. To catch $70 billion worth of fish each year, the fishing industry incurs costs totaling $124 billion. Direct and indirect subsidies make up most of the $54 billion deficit.

This intense fishing pressure destroys nontarget species as well as those sought commercially. Worldwide, one animal in four taken from the ocean is unwanted "by-catch." In 1998, for instance, high seas drift nets killed at least 30 million nontarget animals including diving sea birds and marine mammals. Trawl nets dragged across the ocean floor also destroy plants and animals that provide spawning habitat and food sources for rebuilding commercial populations. Fish farming has replaced hunting for wild animals in many cases, but serious environmental problems in this industry cause concerns.

SOIL: A RENEWABLE RESOURCE

Growing the food and fiber needed to support human life is a complex enterprise that requires knowledge from many different fields and cooperation from many different groups of people. In this section, we'll survey some of the principles of soil science and look at some of the inputs necessary for continued agricultural production.

Of all the earth's crustal resources, the one we take most for granted is soil. We are terrestrial animals and depend on soil for life, yet most of us think of it only in negative terms. English is unique in using "soil" as an interchangeable word for earth and excrement. "Dirty" has a moral connotation of corruption and impurity. Perhaps these uses of the word enhance our tendency to abuse soil without scruples; after all, it's only dirt.

The truth is that **soil** is a marvelous substance, a living resource of astonishing beauty, complexity, and frailty. It is a complex mixture of weathered mineral materials from rocks, partially decomposed organic molecules, and a host of living organisms. It can be considered an ecosystem by itself. Soil is an essential component of the biosphere, but it must be nurtured and cultivated to bring it to its highest potential.

There are at least 20,000 different soil types in the United States and many thousands more worldwide. They vary because of the influences of parent material, time, topography, climate, and organisms on soil formation. There are young soils that, because they have not weathered much, are rich in soluble nutrients. There are old soils, like the red soils of the tropics, from which rainwater has washed away most of the soluble minerals and organic matter, leaving behind clay and rust-colored oxides.

To understand the potential for feeding the world on a sustainable basis we need to know how soil is formed, how it is being lost, and what can be done to protect and rebuild good agricultural soil. With careful husbandry, soil can be replenished and renewed indefinitely. Many farming techniques deplete soil nutrients, however, and expose the soil to the erosive forces of wind and moving water. As a result, in many places we are essentially mining this resource and using it much faster than it is being replaced.

Building good soil is a slow process. Under the best circumstances, good topsoil accumulates at a rate of about 10 tons per hectare (2.5 acres) per year—enough soil to make a layer about 1 mm deep when spread over a hectare. Under poor conditions, it can take thousands of years to build that much soil. Perhaps one-third to one-half of the world's current croplands are losing topsoil faster than it is being replaced. In some of the worst spots, erosion carries away about 2.5 cm (1 in) of topsoil per year. With losses like that, agricultural production has already begun to fall in many areas.

Soil Composition

Most soil is about half mineral. The rest is air and water together with a little organic matter from plant and animal residue. The mineral particles are derived either from the underlying bedrock or from materials transported and deposited by glaciers, rivers, ocean currents, windstorms, or landslides. The weathering processes that break rocks down into soil particles are described in chapter 16.

Particle sizes affect the characteristics of the soil (table 11.2). The spaces between sand particles give sandy soil good drainage and usually allow it to be well aerated, but also cause it to dry out quickly when rains are infrequent. Tight packing of small particles in silty or clay soils makes them less permeable to air and water than sandy soils. Tiny capillary spaces between the particles, on the other hand, store water and mineral ions better than more porous soils. Because

TABLE 11.2 Soil Particle Sizes

CLASSIFICATION	SIZE
Gravel	2 to 64 mm
Sand	0.05 to 2 mm
Silt	0.002 to 0.05 mm
Clay	Less than 0.002 mm

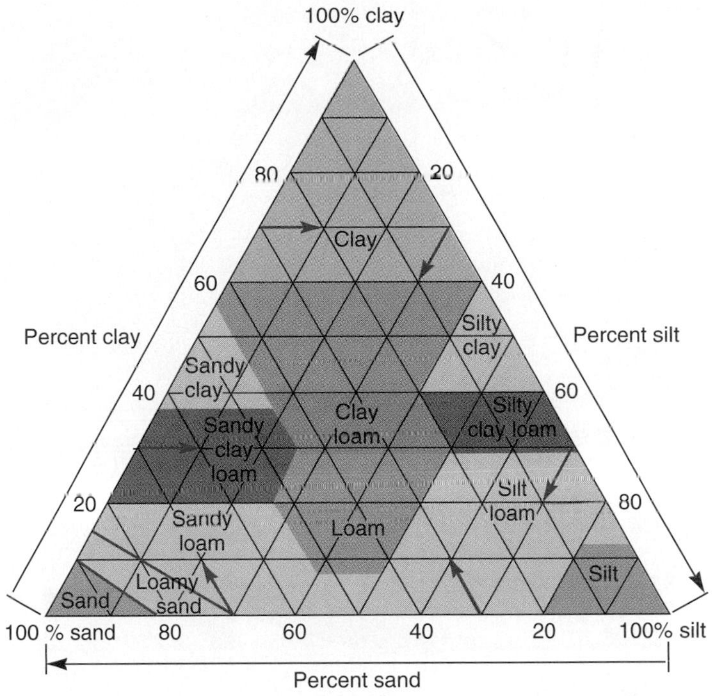

FIGURE 11.9 Soil texture is determined by the percentages of clay, silt, and sand particles in the soil. Soils with the best texture for most crops are loams, which have enough larger particles (sand) to be loose, yet enough smaller particles (silt and clay) to retain water and dissolved mineral nutrients.
Source: Data from Soil Conservation Service.

clay particles have a large surface area and a high ionic charge, they stick together tenaciously, giving clay its slippery plasticity, cohesiveness, and impermeability. Soils with a high clay content are called "heavy soils," in contrast to easily worked "light soils" that are composed mostly of sand or silt. Varying proportions of these mineral particles occur in each soil type (fig. 11.9). Farmers usually consider sandy loam the best soil type for cultivating crops.

The organic content of soil can range from nearly zero for pure sand, silt, or clay, to nearly 100 percent for peat or muck, which is composed mainly of partly decomposed plant material. Much of the organic material in soil is **humus,** a sticky, brown, insoluble residue from the partially decomposed bodies of dead plants and animals. Humus is the most significant factor in the development of "structure," a description of how the soil particles clump together. Humus coats mineral particles and holds them together in loose crumbs, giving the soil a spongy texture that holds water and nutrients needed by plant roots, and maintains the spaces through which delicate root hairs grow.

Soil Organisms

Without soil organisms, the earth would be covered with sterile mineral particles far different from the rich, living soil ecosystems on which we depend for most of our food. The activity of the myriad organisms living in the soil help create structure, fertility, and tilth (condition suitable for tilling or cultivation) (fig. 11.10).

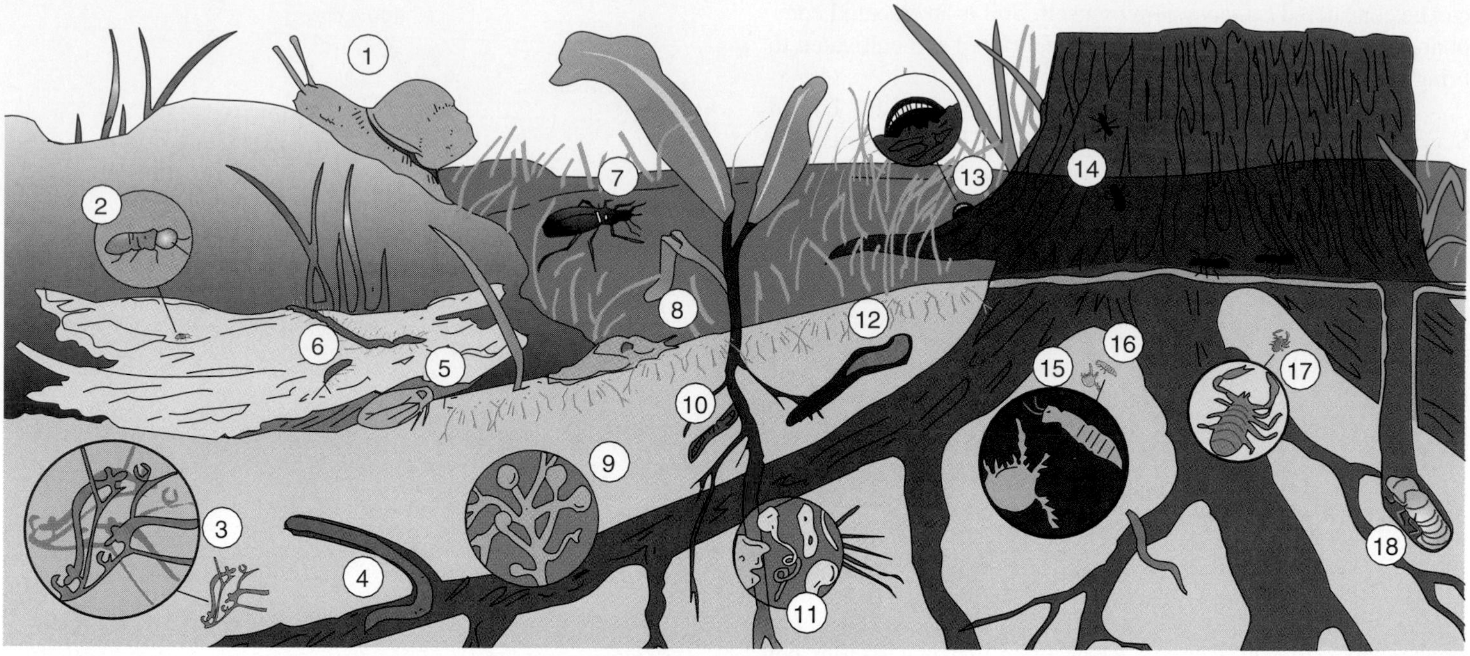

FIGURE 11.10 Soil ecosystems include numerous consumer organisms, as depicted here: (1) snail, (2) termite, (3) nematodes and nematode-killing constricting fungus, (4) earthworm, (5) wood roach, (6) centipede, (7) carabid (ground) beetle, (8) slug, (9) soil fungus, (10) wireworm (click beetle larva), (11) soil protozoans, (12) earthworm, (13) sow bug, (14) ants, (15) mite, (16) springtail, (17) pseudoscorpion, and (18) cicada nymph.

Soil organisms usually stay close to the surface, but that thin living layer can contain thousands of species and billions of individual organisms per hectare. Algae live on the surface, while bacteria and fungi flourish in the top few centimeters of soil. A single gram of soil (about one-half teaspoon) can contain hundreds of millions of these microscopic cells. Algae and blue-green bacteria capture sunlight and make new organic compounds. Bacteria and fungi decompose organic detritus and recycle nutrients that plants can use for additional growth. The sweet aroma of freshly turned soil is caused by actinomycetes, bacteria that grow in funguslike strands and give us the antibiotics streptomycin and tetracyclines.

Roundworms, segmented worms, mites, and tiny insects swarm by the thousands in that same gram of soil from the surface. Some of them are herbivorous, but many of them prey upon one another. Soil roundworms (nematodes) attack plant rootlets and can cause serious crop damage. A carnivorous fungus snares nematodes with tiny loops of living cells that constrict like a noose when a worm blunders into it. Burrowing animals, such as gophers, moles, insect larvae, and worms, tunnel deeper in the soil, mixing and aerating it. Plant roots also penetrate lower soil levels, drawing up soluble minerals and secreting acids that decompose mineral particles. Fallen plant litter adds new organic material to the soil, returning nutrients to be recycled.

Soil Profiles

Most soils are stratified into horizontal layers called **soil horizons** that reveal much about the history and usefulness of the soil. The thickness, color, texture, and composition of each horizon are used to classify the soil. A cross-sectional view of the horizons in a soil is called a soil profile. Figure 11.11 shows the series of horizons generally seen in a soil profile. Soil scientists give each horizon a letter or descriptive name. Soils usually have one to seven or more horizons with different properties depending on the soil type and history of a specific area.

The soil surface is often covered with a layer of leaf litter, crop residues, or other fresh or partially decomposed organic material (O horizon). Under this organic layer is the surface horizon, usually an A horizon or **topsoil,** composed of mineral particles mixed with organic material. The A horizon ranges from a thickness of several meters under virgin prairie to zero in some deserts. The surface horizon contains most of the living organisms and organic material in the soil, and it is in this layer that most plants spread their roots to absorb water and nutrients. The surface horizon often blends into another horizon (E horizon) that is subject to leaching (removal of soluble nutrients) by water percolating through it. This zone of leaching may have a very different appearance and composition from horizons above or below it.

Beneath the surface horizons the subsurface horizons, or **subsoil,** usually have a lower organic content and higher concentrations of mineral particles. Soluble compounds and clay particles carried by water percolating down from the layers above normally accumulate in the B horizon. Subsoil particles can become cemented together to form hardpan, a dense, impermeable layer that blocks plant root growth and prevents water from draining properly.

Beneath the subsoil is the parent material, or C horizon, made of weathered rock fragments with very little organic material.

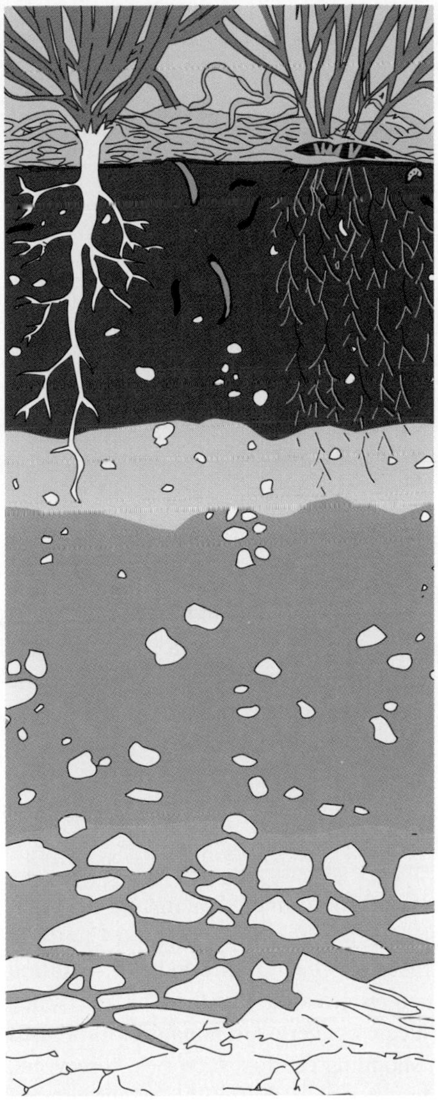

FIGURE 11.11 Soil profile showing possible soil horizons. The actual number, composition, and thickness of these layers varies in different soil types.

Weathering of this layer produces new soil particles and allows downward expansion of the horizons above. About 70 percent of all the parent horizon material in the United States was transported to its present site by geologic forces (glaciers, wind, and water) and is not directly related to the bedrock below it.

Soil Types

Soils are classified according to their structure and composition into orders, suborders, great groups, subgroups, families, and series. The richest farming soils are the mollisols (formed under grasslands) and alfisols (formed under moist, deciduous forests). North America is fortunate to have extensive areas of these fertile soils.

FIGURE 11.12 In many areas, soil or climate constraints limit agricultural production. These farm workers in Chad weed a field that is too dry for good crop yields.

WAYS WE USE AND ABUSE SOIL

Only about 11 percent of the earth's land area (14.66 million sq km out of a total of 132.4 million sq km) is currently in agricultural production. Perhaps four times as much land could potentially be converted to cropland, but much of this land serves as a refuge for cultural or biological diversity or suffers from constraints, such as steep slopes, shallow soils, poor drainage, tillage problems, low nutrient levels, metal toxicity, or excess soluble salts or acidity, that limit the types of crops that can be grown there (fig. 11.12).

Land Resources

Table 11.3 shows the distribution of cropland by region. In parts of Canada and the United States, temperate climates, abundant water, and high soil fertility produce high crop yields that contribute to high standards of living. Other countries, although rich in land area, lack suitable soil, topography, water, or climate to sustain our levels of productivity.

If current population projections are correct, the current world average of 0.27 ha (0.7 acre) of cropland per person will decline to 0.17 ha (0.42 acre) by the year 2025. In Asia, cropland will be even more scarce—0.09 ha (0.22 acre) per person—25 years from now. If you live on a typical quarter-acre suburban lot, look at your yard and imagine feeding yourself for a year on what you could produce there.

In the developed countries, 95 percent of recent agricultural growth in the twentieth century has come from improved crop varieties or increased fertilization, irrigation, and pesticide use, rather than from bringing new land into production. In fact, less land is

TABLE 11.3 Cropland by Region and per Capita

REGION	POPULATION (MILLIONS)	CROPLAND (MILLIONS OF HECTARES)	CROPLAND/CAPITA
Africa	778	190	0.24
Europe	729	316	0.43
North America	304	233	0.77
Central America	131	41	0.31
South America	332	115	0.35
Asia	3589	622	0.17
Oceania*	29	52	1.79
World	5892	1569	0.27

*Oceania includes Australia, New Zealand, and many small-island nations.
Source: Data from *World Resources 1998–99*, World Resource Institute.

being cultivated now than one hundred years ago in North America, or six hundred years ago in Europe. As more effective use of labor, fertilizer, and water and improved seed varieties have increased in the more developed countries, productivity per unit of land has increased, and marginal land has been retired, mostly to forests and grazing lands. In many developing countries, land continues to be cheaper than other resources, and new land is still being brought under cultivation, mostly at the expense of forests and grazing lands. Still, at least two-thirds of recent production gains have come from new crop varieties and more intense cropping rather than expansion into new lands.

The largest increases in cropland over the last 30 years occurred in South America and Oceania where forests and grazing lands are rapidly being converted to farms. Many developing countries are reaching the limit of lands that can be exploited for agriculture without unacceptable social and environmental costs, but others still have considerable potential for opening new agricultural lands. East Asia, for instance, already uses about three-quarters of its potentially arable land. Most of its remaining land has severe restrictions for agricultural use. Further increases in crop production will probably have to come from higher yields per hectare. Latin America, by contrast, uses only about one-fifth of its potential land, and Africa uses only about one-fourth of the land that theoretically could grow crops. However, there would be serious ecological trade-offs in putting much of this land into agricultural production.

While land surveys tell us that much more land in the world *could* be cultivated, not all of that land necessarily *should* be farmed. Much of it is more valuable in its natural state. The soils over much of tropical Asia, Africa, and South America are old, weathered, and generally infertile. Most of the nutrients are in the standing plants, not in the soil. In many cases, clearing land for agriculture in the tropics has resulted in tragic losses of biodiversity and the valuable ecological services that it provides. Ultimately, much of this land is turned into useless scrub or semidesert.

On the other hand, there are large areas of rich, subtropical grassland and forest that are well watered, have good soil, and could become productive farmland without unduly reducing the world's biological diversity. Argentina, for instance, has pampas grasslands about twice the size of Texas that closely resemble the American Midwest a century ago in climate and potential for agricultural growth. Some of this land could probably be farmed with relatively little ecological damage if it were done carefully.

Land Degradation

Agriculture both causes and suffers from environmental degradation. The International Soil Reference and Information Centre in the Netherlands estimates that every year 3 million ha (7.4 million acres) of cropland are ruined by erosion, 4 million ha are turned into deserts, and 8 million ha are converted to nonagricultural uses such as homes, highways, shopping centers, factories, reservoirs, etc. Over the past fifty years, some 1.9 *billion* ha of agricultural land (an area greater than that now in production) have been degraded to some extent. About 300 million ha of this land is strongly degraded (deep gullies, severe nutrient depletion, crops grow poorly, restoration is difficult and expensive), while 910 million ha—about the size of China—are moderately degraded. Nearly 9 million ha of former croplands are so degraded that they no longer support any crop growth at all. The causes of this extreme degradation vary: In Ethiopia it is water erosion, in Somalia it is wind, and in Uzbekistan salt and toxic chemicals are responsible.

Definitions of degradation are based on both biological productivity and our expectations about what the land should be like. Often this is a subjective judgment and it is difficult to distinguish between human-caused deterioration and natural processes like drought. We generally consider the land degraded when the soil is impoverished or eroded, water runs off or is contaminated more than is normal, vegetation is diminished, biomass production is decreased, or wildlife diversity diminishes. On farmlands this results in lower crop yields. On ranchlands it means fewer livestock can be supported per unit area. On nature reserves it means fewer species.

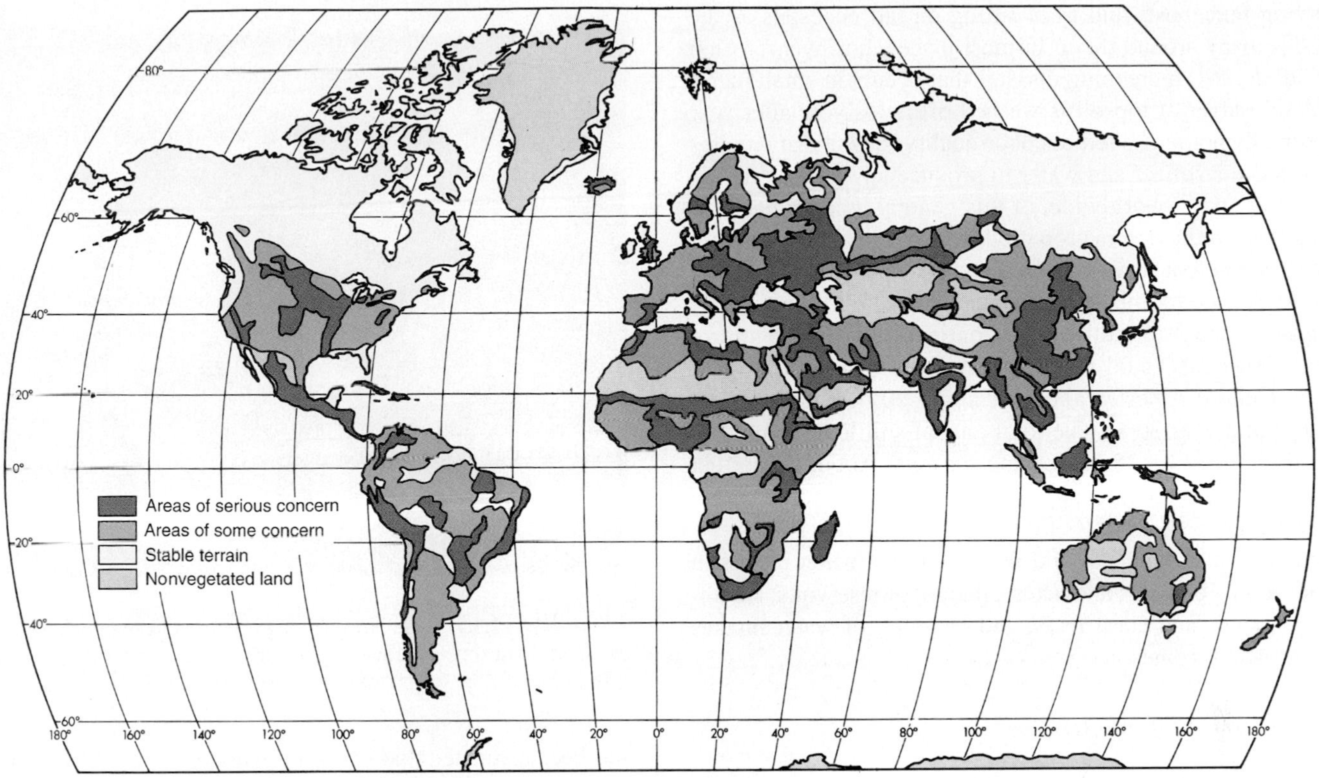

FIGURE 11.13 Areas of concern for soil degradation. In areas of serious concern, widespread moderate or localized severe degradation has already occurred, and resources are lacking or environmental conditions make rehabilitation difficult or impossible. In areas of some concern, current degradation is lighter and potential for rehabilitation is greater. How is your home area classified?

From Jerome Fellmann, et al., *Human Geography,* 4th ed., from data in *World Resources 1992–1993,* International Source Reference and Information Center.

The amount and degree of land degradation varies by region and country. About 20 percent of land in Africa and Asia is degraded, but most is in either the light or moderate category. In Central America and Mexico, by contrast, 25 percent of all vegetated land suffers moderate to extreme degradation. Figure 11.13 shows some areas of greatest concern for soil degradation, and figure 11.14 shows the mechanisms for this problem. Water and wind erosion provide the motive force for the vast majority of all soil degradation, worldwide. Chemical deterioration includes nutrient depletion, salinization (salt accumulation), acidification, and pollution. Physical deterioration includes compaction by heavy machinery or trampling by cattle, waterlogging—water accumulation—from excess irrigation and poor drainage, and laterization—solidification of iron and aluminum-rich tropical soil when exposed to sun and rain.

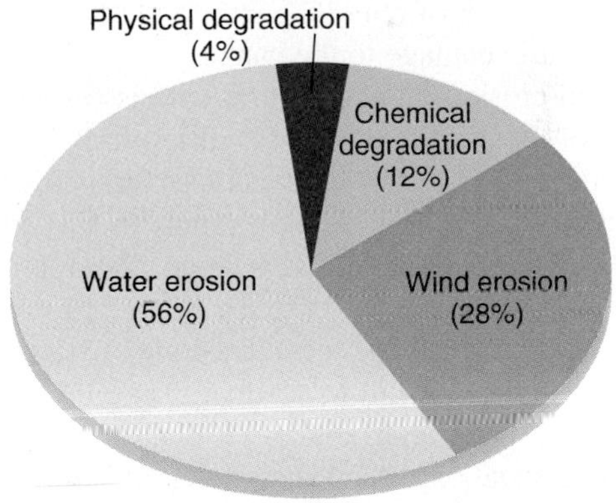

FIGURE 11.14 Mechanisms of soil degradation worldwide for all land uses.
Source: Data from UNEP 1993.

Erosion: The Nature of the Problem

Erosion is an important natural process, resulting in the redistribution of the products of geologic weathering, and is part of both soil formation and soil loss. The world's landscapes have been sculpted by erosion. When the results are spectacular enough, we enshrine them in national parks as we did with the Grand Canyon. Where erosion has worn down mountains and spread soil over the plains, or deposited rich alluvial silt in river bottoms, we gladly farm it. Erosion is a disaster only when it occurs in the wrong place at the wrong time.

In some places, erosion occurs so rapidly that anyone can see it happen. Deep gullies are created where water scours away the soil, leaving fenceposts and trees sitting on tall pedestals as the land erodes away around them. In most places, however, erosion is more subtle. It is a creeping disaster that occurs in small increments. A thin layer of topsoil is washed off fields year after year until eventually nothing is left but poor-quality subsoil that requires more and more fertilizer and water to produce any crop at all.

The net effect, worldwide, of this general, widespread topsoil erosion is a reduction in crop production equivalent to removing about 1 percent of world cropland each year. Many farmers are able to compensate for this loss by applying more fertilizer and by bringing new land into cultivation. Continuation of current erosion rates, however, could reduce agricultural production by 25 percent in Central America and Africa and 20 percent in South America by the year 2000. The total annual soil loss from croplands is thought to be 25 billion metric tons. About twice that much soil is lost from rangelands, forests, and urban construction sites each year.

In addition to reduced land fertility, this erosion results in sediment-loading of rivers and lakes, siltation of reservoirs, smothering of wetlands and coral reefs, and clogging of water intakes and waterpower turbines.

FIGURE 11.15 Severe gullying is cutting deep trenches in this pastureland. Lost topsoil reduces productivity here while it causes siltation downstream.

Mechanisms of Erosion

Wind and water are the main agents that move soil around (fig. 11.14). When little rivulets of running water gather together and cut small channels in the soil, the process is called **rill erosion.** When rills enlarge to form bigger channels or ravines that are too large to be removed by normal tillage operations, we call the process **gully erosion** (fig. 11.15). Streambank erosion refers to the washing away of soil from the banks of established streams, creeks, or rivers, often as a result of removing trees and brush along streambanks and by cattle damage to the banks.

Most soil erosion on agricultural land is rill erosion. Large amounts of soil can be transported a little bit at a time without being very noticeable. A farm field can lose 20 metric tons of soil per hectare during winter and spring runoff in rills so small that they are erased by the first spring cultivation. That represents a loss of only a few millimeters of soil over the whole surface of the field, hardly apparent to any but the most discerning eye. But it doesn't take much mathematical skill to see that if you lose soil twice as fast as it is being replaced, eventually it will run out.

Wind can equal or exceed water in erosive force, especially in a dry climate and on relatively flat land. When plant cover and surface litter are removed from the land by agriculture or grazing, wind lifts loose soil particles and sweeps them away. Windborne dust is sometimes transported from one continent to another. Scientists in Hawaii can tell when spring plowing begins in China because dust from Chinese farmland is carried by winds all the way across the Pacific Ocean. Similarly, summer dust storms in the Sahara Desert of North Africa carry about one billion tons of soil in massive airborne dust plumes over the Atlantic and Mediterranean every year. This dust creates a hazy atmosphere over islands in the Caribbean Sea, 5000 km (3000 mi) away and has measurable regional climatic effects. Similarly, it has been estimated that winds blowing over the Mississippi River basin have one thousand times the soil-carrying capacity of the river itself.

Some of the highest erosion rates in the world occur in the United States and Canada. The U.S. Department of Agriculture reports that 69 million hectares (170 million acres) of U.S. farmland and range are eroding at rates that reduce long-term productivity. Five tons per acre (11 metric tons per hectare) is generally considered the maximum tolerable rate of soil loss because that is generally the highest rate at which soil forms under optimum conditions. Some farms lose soil at twice that rate or more.

Intensive farming practices are largely responsible for this situation. Row crops, such as corn and soybeans, leave soil exposed for much of the growing season (fig. 11.16). Deep plowing and heavy herbicide applications create weed-free fields that look neat but are subject to erosion. Because big machines cannot easily follow contours, they often go straight up and down the hills, creating ready-made gullies for water to follow. Farmers sometimes plow through grass-lined watercourses and have pulled out windbreaks and fencerows to accommodate the large machines and to get every last square meter into production. Consequently, wind and water carry away the topsoil.

Pressed by economic conditions, many farmers have abandoned traditional crop rotation patterns and the custom of resting land as pasture or fallow every few years. Continuous monoculture cropping can increase soil loss tenfold over other farming patterns. A soil study in Iowa showed that a three-year rotation of corn, wheat, and clover lost an average of only 2.7 short tons per acre (6 metric tons per hectare). By comparison, continuous wheat production on the same land caused nearly four times as much erosion, and continuous corn cropping resulted in seven times as much soil loss as the rotation with wheat and clover.

FIGURE 11.16 Annual row crops leave soil bare and exposed to erosion for most of the year, especially when fields are plowed immediately after harvest, as this one always is.

Erosion Hotspots

Data on soil condition and soil erosion often is incomplete, but it is evident that many places have problems as severe as, or perhaps worse than ours. China, for example, has a large area of loess (windblown silt) deposits on the North China Plain that once was covered by forest and grassland. The forests were cut down and the grasslands were converted to cropland. This plateau is now scarred by gullies 30 to 40 m (100 to 130 ft) deep, and the soil loss is thought to be at least 480 metric tons per hectare per year. This would be equivalent to 3 cm (1.2 in) of topsoil per year.

One way to estimate soil loss is to measure the sediment load carried by rivers draining an area. The highest concentration of sediment in any river is in the Huang (Yellow) River that originates in the loess plateau of China. Although its drainage basin is only one-fifth as big as that of the Mississippi River, the Huang carries more than four times as much soil each year. This suggests that the average soil loss *per hectare* in China may be twenty times that in the United States.

Haiti is another country with severely degraded soil. Once covered with lush tropical forest, the land has been denuded for firewood and cropland. Erosion has been so bad that some experts now say the country has absolutely *no* topsoil left, and poor peasant farmers have difficulty raising any crops at all. Economist Lester Brown of Worldwatch Institute warns that the country may never recover from this ecodisaster.

OTHER AGRICULTURAL RESOURCES

Soil is only part of the agricultural resource picture. Agriculture is also dependent upon water, nutrients, favorable climates to grow crops, productive crop varieties, and upon the mechanical energy to tend and harvest them.

Water

All plants need water to grow. Agriculture accounts for the largest single share of global water use. Some 73 percent of all fresh water withdrawn from rivers, lakes, and groundwater supplies is used for irrigation (chapter 19). Although estimates vary widely (as do definitions of irrigated land), about 15 percent of all cropland, worldwide, is irrigated.

Some countries are water rich and can readily afford to irrigate farmland, while other countries are water poor and must use water very carefully. The efficiency of irrigation water use is rather low in most countries. High evaporative and seepage losses from unlined and uncovered canals often mean that as much as 80 percent of water withdrawn for irrigation never reaches its intended destination. Farmers often tend to over-irrigate because water prices are relatively low and because they lack the technology to meter water and distribute just the amount needed.

Excessive use not only wastes water; it often results in **waterlogging.** Waterlogged soil is saturated with water, and plant roots die from lack of oxygen. **Salinization,** in which mineral salts accumulate in the soil, occurs particularly when soils in dry climates are irrigated with saline water. As the water evaporates, it leaves behind a salty crust on the soil surface that is lethal to most plants. Flushing with excess water can wash away this salt accumulation but the result is even more saline water for downstream users.

Worldwide, irrigation problems are a major source of land degradation and crop losses. The Worldwatch Institute reports that 60 million ha (150 million acres) of cropland have been damaged by salinization and waterlogging. Water conservation techniques can greatly reduce problems arising from excess water use. Conservation also makes more water available for other uses or for expanded crop production where water is in short supply (see chapter 19).

Fertilizer

In addition to water, sunshine, and carbon dioxide, plants need small amounts of inorganic nutrients for growth. The major elements required by most plants are nitrogen, potassium, phosphorus, calcium, magnesium, and sulfur. Calcium and magnesium often are limited in areas of high rainfall and must be supplied in the form of lime. Lack of nitrogen, potassium, and phosphorus even more often limits plant growth. Adding these elements in fertilizer usually stimulates growth and greatly increases crop yields. A good deal of the doubling in worldwide crop production since 1950 has come from increased inorganic fertilizer use. In 1950, the average amount of fertilizer used was 20 kg per ha. In 1990, this had increased to an average of 91 kg per ha worldwide.

Farmers may overfertilize because they are unaware of the specific nutrient content of their soils or the needs of their crops. While European farmers use more than twice as much fertilizer per hectare as do North American farmers, their yields are not proportionally higher. Phosphates and nitrates from farm fields and cattle feedlots are a major cause of aquatic ecosystem pollution. Nitrate levels in groundwater have risen to dangerous levels in many areas where intensive farming is practiced. Young children are especially sensitive to the presence of nitrates. Using nitrate-contaminated water to mix infant formula can be fatal for newborns.

What are some alternative ways to fertilize crops? Manure and green manure (crops grown specifically to add nutrients to the soil) are important natural sources of soil nutrients. Nitrogen-fixing bacteria living symbiotically in root nodules of legumes are valuable for making nitrogen available as a plant nutrient (chapter 3). Interplanting or rotating beans or some other leguminous crop with such crops as corn and wheat are traditional ways of increasing nitrogen availability.

There is considerable potential for increasing world food supply by increasing fertilizer use in low-production countries if ways can be found to apply fertilizer more effectively and reduce pollution. Africa, for instance, uses an average of only 19 kg of fertilizer per ha (17 lb per acre), or about one-fourth of the world average. It has been estimated that the developing world could at least triple its crop production by raising fertilizer use to the world average.

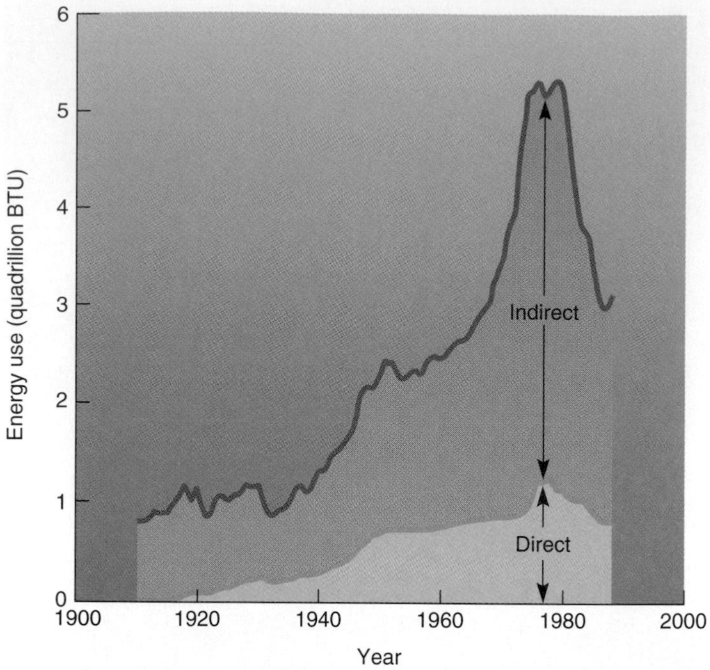

FIGURE 11.17 Total energy use in U.S. agriculture. Direct use includes liquid fuels for vehicles as well as electricity for lights, motors, and cooling or heating. Indirect energy represents manufactured inputs such as fertilizers and pesticides.
Source: Data from R. Constanza (ed.), *Ecological Economics,* 1995, Society for Ecological Economics.

Energy

Farming as it is generally practiced in the industrialized countries is highly energy-intensive. Fossil fuels supply almost all of this energy. Between 1920 and 1980, direct energy use on farms rose as gasoline and diesel fuels were consumed by increasing mechanization of agriculture. An even greater increase in indirect energy use, in the form of synthetic fertilizers, pesticides, and other agricultural chemicals, also occurred, especially after World War II (fig. 11.17).

After crops leave the farm, additional energy is used in food processing, distribution, storage, and cooking. It has been estimated that the average food item in the American diet travels 2000 km (1250 mi) between the farm that grew it and the person who consumes it. The energy required for this complex processing and distribution system may be five times as much as is used directly in farming. Altogether the food system in the United States consumes about 16 percent of the total energy we use. Most of our foods require more energy to produce, process, and get to market than they yield when we eat them.

Clearly, unless we find some new sources of energy, our present system is unsustainable. As fossil fuels become more scarce, we may need to adopt farming methods that are self-supporting. Is it possible that we may need to go back to using draft animals that can eat crops grown on the farm? Could we reintroduce natural methods of pest control, fertilization, crop drying, and irrigation? Or can we develop alternative energy sources to run our farming enterprise?

NEW CROPS AND GENETIC ENGINEERING

Although at least 3000 species of plants have been used for food at one time or another, most of the world's food now comes from only 16 widely grown crops. Many new or unconventional varieties might be valuable human food supplies, however, especially in areas where conventional crops are limited by climate, soil, pests, or other problems. Among the plants now being investigated as potential additions to our crop roster are winged beans (fig. 11.18), a perennial plant that grows well in hot climates where other beans will not grow. It is totally edible (pods, mature seeds, shoots, flowers, leaves, and tuberous roots), resistant to diseases, and enriches the soil. Another promising crop is tricale, a hybrid between wheat (*Triticum*) and rye (*Secale*) that grows in light, sandy, infertile soil. It is drought resistant, has nutritious seeds, and is being tested for salt tolerance for growth in saline soils or irrigation with seawater. Some traditional crop varieties grown by Native Americans, such as tepary beans, amaranth, and Sonoran panicgrass are being collected by seed conservator Gary Nabhan both as a form of cultural revival for native people and as a possible food crop for harsh environments.

FIGURE 11.18 Winged beans bear fruit year round in tropical climates and are resistant to many diseases that prohibit growing other bean species. Whole pods can be eaten when they are green, or dried beans can be stored for later use. It is a good protein source in a vegetarian diet.

Green Revolution

So far, however, the major improvements in farm production have come from technological advances and modification of a few well-known species. Yield increases often have been spectacular. A century ago, when all maize (corn) in the United States was open pollinated, average yields were about 25 bushels per acre. In 1999, average yields from hybrid maize were around 130 bushels per acre, and under optimum conditions, 250 bushels per acre are possible. Most of this gain was accomplished by conventional plant breeding: geneticists laboriously hand-pollinating plants, moving selected genes from one variety to another.

Starting about 50 years ago, agricultural research stations began to breed tropical wheat and rice varieties that would provide food for growing populations in developing countries. The first of the "miracle" varieties was a dwarf, high-yielding wheat (fig. 11.19) developed by Norman Borlaug (who received a Nobel Peace Prize for his work) at a research center in Mexico. At about the same time, the International Rice Institute in the Philippines developed dwarf rice strains with three or four times the production of varieties in use at the time. The dramatic increases obtained as these new varieties spread around the world has been called the **green revolution**. It is one of the main reasons that world food supplies have more than kept pace with the growing human population over the past few decades.

Most green revolution breeds really are "high responders," meaning that they yield more than other varieties if given optimum levels of fertilizer, water, and protection from pests and diseases. Under suboptimum conditions, on the other hand, high responders may not produce as well as traditional varieties (fig. 11.20). Poor farmers who can't afford the expensive seed, fertilizer, water, etc., which are required to become part of this movement, usually are left out of the green revolution. In fact they may be driven out of farming altogether as rising land values and falling commodity prices squeeze them from both sides.

Genetic Engineering

Discovery of the structure of DNA by James Watson and Francis Crick in 1953, and the subsequent unraveling of the genetic code, set the stage for a further revolution that may bring even more spectacular advances in agriculture and medicine. Once the genetic machinery was understood, scientists began to dream of **genetic engineering** or using molecular techniques to move genes from one organism to another. In 1973, scientists managed to clip out a gene from the chromosome of one microbe and splice it into another microbe. In the early 1980s, several laboratories succeeded in transferring genes from bacteria to plants and growing adult "transgenic" plants from single, artificially transformed cells. It is now possible to build entirely new genes, taking a bit from here, a bit from there, even synthesizing artificial sequences to create desired characteristics.

There could be some dramatic benefits from this new technology. It's conceivable that crops could be genetically engineered to tolerate salty, waterlogged, or low-nutrient soils, thus allowing degraded farmland to once again become productive. Plants that produce their own pesticides might reduce the need for toxic chemicals, while engineering for improved protein or vitamin content could make our food more nutritious. Among the first genetically engineered crops was a bacterium that made strawberries frost-resistant (fig. 11.21), and a tomato with genes that gave it a longer shelf-life in the grocery store. Unfortunately—or maybe fortunately, depending on your perspective—the frost-free bacteria didn't work that well, and the "flavr savr" tomato didn't taste very good, wasn't a good producer, and disappeared quickly from the market.

Still, the rate at which transgenic crop varieties are being produced is increasing rapidly. Between 1987 and 1997, some 3300 field releases were conducted for more than 60 different crops ranging from maize to apples to canola oil seed (fig. 11.22). Some crops have reached mainstream status. About one-third of the U.S. maize (corn) and soybean crops are now transgenic, as is one-fifth of all cotton. It's estimated that 70 percent of all processed food in the United States contains some transgenic material. The chances are very good that everyone in the U.S. has eaten something made from genetically engineered crops, whether they know it or not. The only other countries with sizable amounts of genetically engineered crops are Argentina and Canada. Currently about half of all Canadian canola is transgenic as are an equal amount of all Argentinean soybeans. The total area planted with genetically engineered crops in the United States is about 20 million hectares, or three-quarters of all such cropland in the world. Argentina and Canada account for almost all other transgenic cropland.

Pest Resistance

Biotechnologists recently have created plants with genes for endogenous insecticides. *Bacillus thuringiensis* (Bt), a bacterium, makes toxins lethal to Lepidoptera (butterfly family) and Coleoptera (beetle family). The genes for some of these toxins have

FIGURE 11.19 Short-stemmed "semidwarf" wheat (*center*) developed by Nobel prize winner, Norman Borlaug, is a high responder; that is, it can utilize high levels of fertilizer and water to produce high yields without growing so tall that the stalks fall over and are impossible to harvest. Notice how much denser the seed heads are on the semidwarf than on its normal size relative (*left*).

been transferred into crops such as maize (to protect against European cut worms), potatoes (to fight potato beetles), and cotton (for protection against boll weevils). This allows farmers to reduce insecticide spraying. In 1998, some 8 million hectares—or about 29 percent of all transgenic crops in the world—were planted to Bt-containing varieties, mostly in North America. Most European nations have not approved Bt-containing varieties. Fear of transgenic products has been a sticking point in agricultural trade between the United States and the European Union.

Entomologists worry that Bt plants churn out toxin throughout the growing season, regardless of the level of infestation, creating perfect conditions for selection of Bt resistance in pests. The effectiveness of this natural pesticide—one of the few available to organic growers—is likely to be destroyed within a few years. One solution is to plant at least a part of every field in non-Bt crops that will act as a refuge for nonresistant pests. The hope is that interbreeding between these "wild-type" bugs and those exposed to Bt will dilute out recessive resistance genes. Deliberately harboring pests and letting them munch freely on crops is something that many farmers find hard to do. In addition, devoting a significant part of their land to nonproductive crops lowers the total yield and counteracts the profitability of engineered seed.

There also is a concern about the effects on nontarget species. In laboratory tests, about half of a group of monarch butterfly caterpillars died after being fed on plants dusted with pollen from Bt corn. Under field conditions, critics point out, corn pollen can drift long distances and might contaminate the milkweeds on which monarchs normally feed. Geneticists consider this a trivial problem. They'll just create varieties in which the Bt gene is inactive in pollen, they claim. Still, others wonder if this whole enterprise is a good idea.

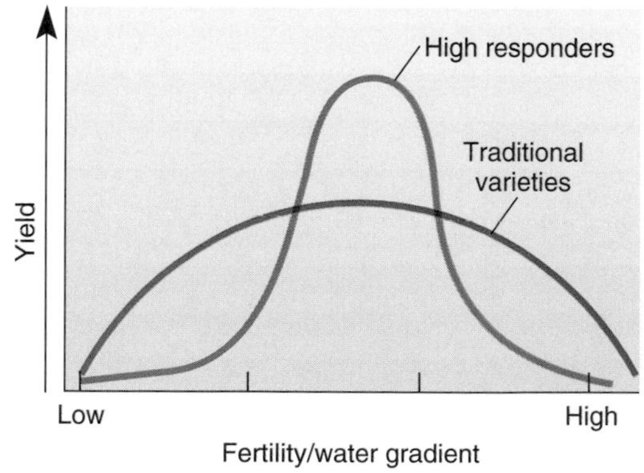

FIGURE 11.20 Green revolution miracle crops are really high responders, meaning that they have excellent yields under optimum conditions. For poor farmers who can't afford the fertilizer and water needed by high responders, traditional varieties may produce better yields.

FIGURE 11.21 A researcher measures growth of genetically-engineered rice plants. Superior growth and yield, pest resistance, and other useful traits can be moved from one species to another by genetic engineering.

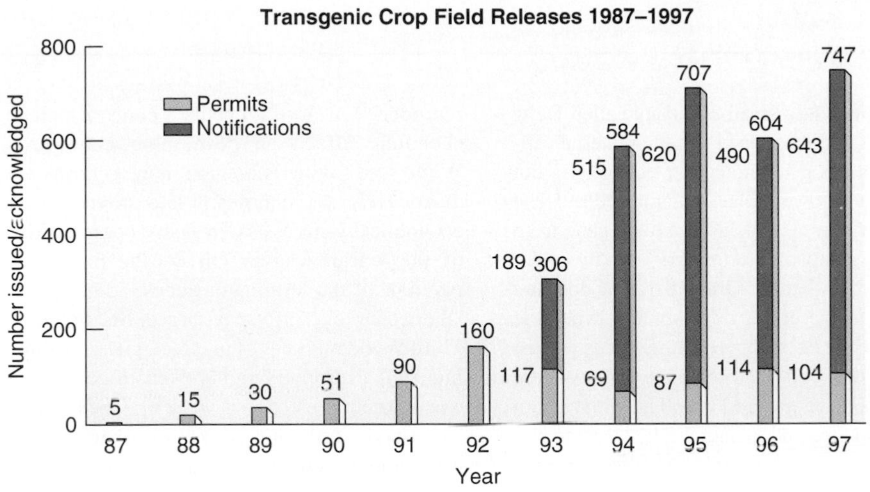

FIGURE 11.22 Field releases of genetically modified field crops increased rapidly in the 1990s. Most releases now only require notification of intent rather than an official permit.
Source: Data from *USDA*'s Animal and Plant Health Inspection Service (*APHIS*).

Weed Control

The most popular transgenic crops are not insect-resistant but those engineered to tolerate high doses of herbicides. Currently these crops occupy 20 million hectares worldwide, or about three-quarters of all genetically engineered acreage. The two main products in this category are Monsanto's "Roundup Ready" crops—so-called because they can withstand treatment with Monsanto's best-selling herbicide, Roundup (glyphosate)—and AgrEvo's "Liberty Link" crops, which resist that company's Liberty (glufosinate) herbicide. Because crops with these genes can grow in spite of high herbicide doses, farmers can spray fields heavily to exterminate weeds. This allows for conservation tillage and leaving more crop residue on fields to protect topsoil from erosion—both good ideas—but it also means using much more herbicide in higher doses than might otherwise be done.

Widespread use of these crops could create herbicide-resistant "super weeds" if genes jump from domestic species to wild relatives. This isn't much of a risk in North America where wild relatives for common crops are rare. It could be a serious problem, however, in the high biodiversity regions where our domestic crops originated, and where closely related, wild species remain common. The chances for "escaped" genes is, in fact, pretty good. In 1997, just one year after Roundup Ready canola was first planted in Canada, wild mustard plants (a canola relative) were found to have interbred and obtained the herbicide-resistant gene.

Public Opposition

Although many people in agribusiness are very enthusiastic about the promise of biotechnology, public opposition to bioengineered food components has been growing. Consumers have refused to buy transgenic products such as growth-hormone-treated (BST) milk. Some of this reaction is more emotional than scientific. Proponents of BST (bovine somatatrophin) point out that some growth hormone is always present in milk. We have modified cows through ordinary breeding and nutrition so that a typical holstein is now twice the size, grows twice as fast, and produces hundreds of times as much milk in a year as its wild progenitors. All this undoubtedly requires extra growth hormone. The real argument, some claim, is economic and social, even if those aspects aren't often mentioned. Only large industrial dairy farms can afford BST treatment along with the extra feed and other inputs it requires. This is just another step in the demise of small family farms and the consolidation of agribusiness into a few giant conglomerates.

Protests against bioengineered field crops also have been widespread. In 1999 the seven largest grocery chains in Europe made a commitment to stock only genetically unmodified maize, potatoes, soybeans, and wheat. At about the same time, the Supreme Court of India upheld a ban on testing of transgenic varieties, and activists burned fields suspected of harboring genetically engineered plants. A special target of this anger are varieties endowed with "terminator" genes (What Do You Think? p. 254).

SUSTAINABLE AGRICULTURE

How, then, shall we feed the world? Can we make agriculture compatible with sustainable ecological and social systems? Having discussed some of the problems that beset modern agriculture, we will now consider some suggested ways to overcome problems and make farming and food production just and lasting enterprises. This goal is usually termed **sustainable agriculture,** or **regenerative farming,** both of which aim to produce food and fiber on a sustainable basis and repair the damage caused by destructive practices. Some alternative methods are developed through scientific research; others are discovered in traditional cultures and practices nearly forgotten in our mechanization and industrialization of agriculture.

Soil Conservation

With careful husbandry, soil is a renewable resource that can be replenished and renewed indefinitely. Since agriculture is the area in which soil is most essential and also most often lost through erosion, agriculture offers the greatest potential for soil conservation and

what do you think?

THE WORLD AROUND YOU

Terminator Genes

Research in biotechnology and genetic engineering is very expensive. Monsanto is reported to have spent $500 million developing Roundup Ready genes, or about as much as the entire annual USDA research budget. Naturally, they want to protect potential profits from this valuable property. Farmers who buy Monsanto seeds are required to sign a contract that stipulates what kinds of pesticides can be used on fields as well as an agreement not to save seed or allow patented crops to cross with other varieties. Seed sleuths investigate to ensure that contracts are fulfilled. By inserting unique hidden sequences in their synthetic genes, forensic molecular biologists can detect the presence of patented genetic material in fields for which royalties weren't paid. Already Monsanto has taken legal action against more than 300 farmers for replanting proprietary seeds. Farmers claim they can't prevent transgenic pollen from blowing onto their fields and introducing genes against their will. A whole new set of legal precedents is likely to be established by these suits.

A new weapon has recently been introduced in this struggle that many people regard as quite sinister. Using genetic research of a USDA scientist, a small company called Delta and Pine Land developed genetic material officially entitled "gene protection technology" but commonly known as "terminator" genes. The terminator complex includes a toxic gene from a noncrop plant stitched together with two other bits of coding that keep the killer gene dormant until late in the crop's development, when the toxin affects only the forming seeds. Thus, although the crop yield is about normal, there is no subsequent generation and no worry about farmers saving and replanting. They have to buy new seed every year. Delta was quickly purchased by Monsanto for $1 billion, or hundreds of times the small company's book value. This may have been the only time a whole company was purchased just to get a gene complex.

Engineered sterility is not uncommon; it is widely used in producing hybrid crops such as maize. What is unusual about this gene-set is that it can be moved easily from one species to another, and it can be packaged in every seed sold by the parent company. It's also unique to deliberately introduce a toxin into the part that people eat. So what's wrong with a company trying to protect its research investment? For one thing, there's a worry that the toxins might be harmful to consumers, even though toxicity tests so far show no danger. Furthermore these genes may escape. What if some of our major crops become self-sterile and can no longer reproduce? A more immediate concern is the economic effects in developing countries. While seed saving is not common on farms in most developed countries, it is customary and economically necessary in many poorer parts of the world. Melvin Oliver, the principal inventor of the terminator genes, admits that "the technology primarily targets Second and Third World markets"—in effect, guaranteeing intellectual property rights even in countries where patent protection is weak or nonexistent.

Large corporations like Monsanto argue that without patent protection, they can't afford to do the research needed to provide further advances in biotechnology. Critics charge that these companies make enough profit in developed countries to pay back their costs. Targeting less-developed countries and introducing something as potentially dangerous as the terminator gene, they claim is immoral. International protests caused Monsanto to announce in 1999 that it was suspending plans to release crops with terminator genes "for the time being." Still, biotechnology research continues at a furious pace and other genetically-modified organisms are sure to be available soon. What do you think? Are those who protest this technology simply afraid of things that are new and unfamiliar, or are there legitimate reasons for concern? How can we assess risks in a novel and unknown technologies such as these?

---

rebuilding. Some rice paddies in Southeast Asia, for instance, have been farmed continuously for a thousand years without any apparent loss of fertility. The rice-growing cultures that depend on these fields have developed management practices that return organic material to the paddy and carefully nurture the soil's ability to sustain life.

While American agriculture hasn't reached that level of sustainability, there is evidence that soil conservation programs are having a positive effect. In one Wisconsin study, erosion rates in one small watershed were 90 percent less in 1975–1993 than they were in the 1930s. Among the most important elements in soil conservation are land management, ground cover, climate, soil type, and tillage system.

Managing Topography

Water runs downhill. The faster it runs, the more soil it carries off the fields. Comparisons of erosion rates in Africa have shown that a 5 percent slope in a plowed field has three times the water runoff volume and eight times the soil erosion rate of a comparable field with a 1 percent slope. Water runoff can be reduced by leaving grass strips in waterways and by **contour plowing,** that is, plowing across the hill rather than up and down. Contour plowing is often combined with **strip-farming,** the planting of different kinds of crops in alternating strips along the land contours (fig. 11.23). When one crop is harvested, the other is still present to protect the soil and keep water from running straight downhill. The ridges created by cultivation make little dams that trap water and allow it to seep into the soil rather than running off. In areas where rainfall is very heavy, tied ridges are often useful. This method involves a series of ridges running at right angles to each other, so that water runoff is blocked in all directions and is encouraged to soak into the soil.

Terracing involves shaping the land to create level shelves of earth to hold water and soil (fig. 11.24). The edges of the terrace are planted with soil-anchoring plant species. This is an expensive procedure, requiring either much hand labor or expensive machinery, but makes it possible to farm very steep hillsides. The rice terraces in the Chico River Valley in the Philippines rise as much as 300 m (1000 ft) above the valley floor. They are considered one of the wonders of the world.

FIGURE 11.23 Contour plowing and strip-cropping on these Wisconsin farms protect the soil from erosion and help maintain fertility, as well as providing a beautiful landscape. With care and good stewardship, we can increase the carrying capacity of the land and create a sustainable environment.

FIGURE 11.24 Rice terraces on Java, Indonesia. Some rice paddies have been cultivated for hundreds or even thousands of years without any apparent loss of productivity.

Planting **perennial species** (plants that grow for more than two years) is the only suitable use for some lands and some soil types. Establishing forest, grassland, or crops such as tea, coffee, or other crops that do not have to be cultivated every year may be necessary to protect certain unstable soils on sloping sites or watercourses (low areas where water runs off after a rain).

Providing Ground Cover

Annual row crops such as corn or beans generally cause the highest erosion rates because they leave soil bare for much of the year (table 11.4). Often, the easiest way to provide cover that protects soil from erosion is to leave crop residues on the land after harvest. They not only cover the surface to break the erosive effects of wind and water, but they also reduce evaporation and soil temperature in hot climates and protect ground organisms that help aerate and rebuild soil. In some experiments, 1 ton of crop residue per acre (0.4 ha) increased water infiltration 99 percent, reduced runoff 99 percent, and reduced erosion 98 percent. Leaving crop residues on the field also can increase disease and pest problems, however, and may require increased use of pesticides and herbicides.

Where crop residues are not adequate to protect the soil or are inappropriate for subsequent crops or farming methods, such **cover crops** as rye, alfalfa, or clover can be planted immediately after harvest to hold and protect the soil. These cover crops can be plowed under at planting time to provide green manure. Another method is to flatten cover crops with a roller and drill seeds through the residue to provide a continuous protective cover during early stages of crop growth.

In some cases, interplanting of two different crops in the same field not only protects the soil but also is more efficient use of the land, providing double harvests. Native Americans and pioneer farmers, for instance, planted beans or pumpkins between the corn rows. The beans provided nitrogen needed by the corn, pumpkins crowded out weeds, and both crops provided foods that nutritionally balance corn. Traditional swidden (slash-and-burn) cultivators in Africa and South America often plant as many as twenty different crops together in small plots. The crops mature at different times so that there is always something to eat, and the soil is never exposed to erosion for very long.

Mulch is a general term for a protective ground cover that can include manure, wood chips, straw, seaweed, leaves, and other natural products. For some high-value crops, such as tomatoes, pineapples, and cucumbers, it is cost-effective to cover the ground with heavy paper or plastic sheets to protect the soil, save water, and prevent weed growth. Israel uses millions of square meters of plastic mulch to grow crops in the Negev Desert.

TABLE 11.4 Soil Cover and Soil Erosion

CROPPING SYSTEM	AVERAGE ANNUAL SOIL LOSS (TONS/HECTARE)	PERCENT RAINFALL RUNOFF
Bare soil (no crop)	41.0	30
Continuous corn	19.7	29
Continuous wheat	10.1	23
Rotation: corn, wheat, clover	2.7	14
Continuous bluegrass	0.3	12

Source: Based on 14 years' data from Missouri Experiment Station, Columbia, MO.

CASE STUDY

ORGANIC FARMING IN CUBA

The biggest experiment in low-input, sustainable agriculture in world history is occurring now in Cuba. The sudden collapse of the socialist bloc, upon which Cuba had been highly dependent for trade and aid, has forced an abrupt and difficult conversion from conventional agriculture to organic farming on a nationwide scale. Methods developed in Cuba could help other countries find ways to break their dependence on synthetic pesticides and fossil fuels.

Between the Cuban revolution in 1959 and the breakdown of trading relations with the Soviet Union in 1989, Cuba experienced rapid modernization, a high degree of social equity and welfare, and a strong dependence on external aid. Cuba's economy was supported during this period by the most modern agricultural system in Latin America. Farming techniques, levels of mechanization, and output often rivaled those in the United States. The main crop was sugarcane, almost all of which was grown on huge state farms and sold to the former Soviet Union at premium prices. More than half of all food eaten by Cubans came from abroad, as did most fertilizers, pesticides, fuel, and other farm inputs on which agricultural production depended.

Under the theory of comparative advantage, it seemed reasonable for Cuba to rely on international trade. With the collapse of the socialist bloc, however, Cuba's economy also fell apart. In 1990, wheat and grain imports decreased by half and other foodstuffs declined even more. At the same time, fertilizer, pesticide, and petroleum imports were down 60 to 80 percent. Farmers faced a dual challenge: how to produce twice as much food using half the normal inputs.

The crisis prompted a sudden turn to a new model of agriculture. Cuba was forced to adopt sustainable, organic farming practices based on indigenous, renewable resources. Typically, it takes three to five years for a farmer in the United States to make the change from conventional to organic farming profitable. Cuba, however, didn't have that long; it needed food immediately.

Cuba's agricultural system is based on a combination of old and new ideas. Broad community participation and use of local knowledge is essential. Scientific, adaptive management is another key. Diverse crops suitable to local microclimates, soil types, and human nutritional needs have been adopted. Natural, renewable energy sources such as wind, solar, and biomass fuels are being substituted for fossil fuels. Oxen and mules have replaced some 500,000 tractors idled by lack of fuel.

Soil management is vital for sustainable agriculture. Organic fertilizers substitute for synthetic chemicals. Livestock manure, green manure crops, composted municipal garbage, and industrial-scale cultivation of high-quality humus in earthworm farms all replenish soil fertility. In 1995 more than 100,000 metric tons of worm compost were produced and spread on fields.

Pests are suppressed by crop rotation and biological controls rather than chemical pesticides. For example, the parasitic fly (*Lixophaga diatraeae*) controls sugarcane borers; wasps in the genus *Trichogramma* feed on the eggs of grain weevils; while the predatory ant (*Pheidole megacephala*) attacks sweet potato weevils. Pest control also involves innovative use of biopesticides, such as *Bacillis thuringiensis,* that are poisonous or repellent to crop pests. Finally, integrated pest management includes careful monitoring of crops and measures to build populations of native beneficial organisms and to enhance the vigor and defenses of crop species.

Worker brigades from schools and factories help provide farm labor during harvest season. In addition to state farms and rural communes, urban gardening provides a much-needed supplement to city diets. Individual gardens are encouraged, but community or institutional gardens—schools, factories, and mass organizations—also produce large amounts of food.

Although food supplies in Cuba still are limited and diets are austere, the crisis wasn't as bad as many feared. In some ways, this draconian transition is fortunate. Cuba is now on a sustainable path and is a world leader in sustainable agriculture. It could serve as a model for others who surely will face a similar transition when our supplies of fossil fuels run out.

Cuban farmers use oxen as replacement for tractors idled by fuel shortages. Cuba is now carrying out the world's largest experiment ever in low-input, sustainable agriculture.

Reduced Tillage

Farmers have traditionally used a moldboard plow to till the soil, digging a deep trench and turning the topsoil upside down. In the 1800s, it was shown that tilling a field fully—until it was "clean"—increased crop production. It helped control weeds and pests, reducing competition; it brought fresh nutrients to the surface, providing a good seedbed; and it improved surface drainage and aerated the soil. This is still true for many crops and many soil types, but it is not always the best way to grow crops. We are finding that less plowing and cultivation often makes for better water management, preserves soil, saves energy, and increases crop yields.

There are three major **reduced tillage systems.** *Minimum till* involves reducing the number of times a farmer disturbs the soil by plowing, cultivating, etc. This often involves a disc or chisel plow rather than a traditional moldboard plow. A chisel plow is a curved chisel-like blade that doesn't turn the soil over but creates ridges on which seeds can be planted. It leaves up to 75 percent of plant debris on the surface between the rows, preventing erosion (fig. 11.25). *Conserv-till* farming uses a coulter, a sharp disc like a pizza cutter, which slices through the soil, opening up a furrow or slot just wide enough to insert seeds. This disturbs the soil very little and leaves almost all plant debris on the surface. *No-till* planting is accomplished by drilling seeds into the ground directly through mulch and ground cover. This allows a cover crop to be interseeded with a subsequent crop.

Farmers who use these conservation tillage techniques often must depend on pesticides (insecticides, fungicides, and herbicides) to control insects and weeds. Increased use of toxic agricultural chemicals is a matter of great concern. Massive use of pesticides is not, however, a necessary corollary of soil conservation. It is possible to combat pests and diseases with integrated pest management that combines crop rotation, trap crops, natural repellents, and biological controls (chapter 12).

FIGURE 11.25 In ridge-tilling, a chisel plow is used to create ridges on which crops are planted and shallow troughs filled with crop residue. Less energy is used in plowing and cultivation, weeds are suppressed and moisture is retained by the ground cover, or crop residue, left on the field.

Low-Input Sustainable Agriculture

Throughout the United States, Canada, and other crop-exporting countries such as Brazil and New Zealand, huge, highly mechanized farms are becoming increasingly common. Giant grain fields stretch over thousands of acres, while large numbers of cows, hogs, or chickens live their entire life in enormous, warehouse-size confinement barns. Often owned by multinational agricultural conglomerates, these corporate farms are operated by scientific standards that might make your local hospital seem backward. Milk cows, for instance, are tracked by computer, each one automatically given an individual diet, antibiotics, and hormone injections according to their pedigree, growth rate, or milk production. Farm managers travel by helicopter to survey vast spreads. One "farm" in California covers 2000 ha (almost 50,000 acres) and has an annual cash-flow of around $50 million.

In contrast to this trend towards industrialized agriculture, other farmers are going back to a size and style that their grandparents might have used a century ago. Finding that they can't—or don't want to—compete in capital-intensive production, these folks are making money and staying in farming by returning to small-scale, low-input farming. Rather than milking 1000 cows in a highly mechanized operation, some farmers now raise only a couple of dozen cows and let them graze freely in pastures. Animals live outdoors in the winter (fig. 11.26). Milk production follows natural yearly cycles rather than being controlled by synthetic hormones. The yield is lower, but so are costs.

Where 20,000 chickens living in a single immense shed—or 1000 hogs raised in an industrial warehouse—must constantly be dosed with antibiotics to keep them healthy, animals allowed free range tend to have fewer diseases and require less care. Vast amounts of manure generated by industrial-scale factory farming are stored in football-field-sized lagoons and spread on fields. Odors from ammonia, hydrogen sulfide (rotten egg odor), and other air pollutants make life miserable for neighbors. In 1999, heavy rains from Hurricane Floyd caused hundreds of North Carolina hog manure lagoons to overflow. Around 10 billion liters (2.5 billion gal) of manure washed down rivers to the sea, creating a large dead zone in Palmico Sound (see chapter 20). In 1999, the EPA announced that it would require air pollution permits for factory farms.

In contrast to these gigantic operations, one hog farmer in Iowa whose forty sows graze in the pasture along with his eighty dairy cows, spends 30 percent less per animal for feed, 70 percent less for veterinary bills, and half as much for buildings and equipment as neighboring confinement operations.

Although there isn't complete agreement on what sustainable actually means, low-input farming generally involves more crop rotation and less pesticides (or none), as well as less inorganic fertilizer, water, heavy machinery, and fossil fuel energy than the high-intensity, monoculture operation practiced by conventional farms.

FIGURE 11.26 On the Minar family's 230-acre dairy farm near New Prague, MN, cows and calves spend the winter outdoors in the snow, bedding down on hay. Dave Minar is part of a growing counterculture that is seeking to keep farmers on the land and bring prosperity to rural areas.

Total revenue from this type of agriculture often is less than that of conventional operations, but profits may actually be higher. A study in North Dakota, for instance, found that conventional farms averaged $20 more per acre in total yield but low-input farming had a net income that was $2 per acre *higher,* because total costs were less.

These alternative farming methods are good for our environment. By using fewer chemicals and less manipulation of the soil, erosion and pollution are reduced and the long-range health of the land is improved. Although this movement is too new to yet have good human health statistics, it seems likely that low-input farming will be good for the farmers too. And it is good for taxpayers, as well. A study in Minnesota found that conventional farmers received an average payment of $9214 in government subsidies in 1990, while sustainable farmers received an average of $3597 that same year.

Preserving small-scale, family farms also helps preserve rural culture. As Marty Strange of the Center for Rural Affairs in Nebraska asks, "Which is better for the enrollment in rural schools, the membership of rural churches, and the fellowship of rural communities—two farms milking 1000 cows each or twenty farms milking 100 cows each?" Family farms help keep rural towns alive by purchasing machinery at the local implement dealer, gasoline at the neighborhood filling station, and groceries at the mom-and-pop grocery store.

One way to support local agriculture is to shop at a farmer's market. The produce is fresh, and profits go directly to the person who grows the crop. A local food co-op or owner-operated grocery store also is likely to buy from local farmers and to feature pesticide-free foods. Many co-ops and buyers' associations sign contracts directly with producers to grow the types of food they want to eat. This benefits both parties. Producers are guaranteed a local market for organic food or specialty items. Consumers can be assured of quality and can even be involved in production.

If you belong to a co-op or grow your own food you may have to accept fruits and vegetables that are not absolutely perfect. You can trim away a few bad spots or wash off an insect or two. If you overlook a worm or a bug, it may be less harmful to you than the toxic chemicals that are used to make supermarket foods sterile and cosmetically perfect!

SUMMARY

Over the past 30 years, the total amount of food in the world has increased faster than the average rate of population growth, so there is now more food per person than there was in the 1970s, even though the total number of people has doubled.

There now is enough food to supply everyone in the world with more than the minimum daily food requirements, but food is inequitably distributed. The FAO estimates that 800 million people are chronically undernourished or malnourished, and at least 11 million children die each year from diseases related to malnutrition. Additional millions survive on a deficient diet, suffering from resulting stunted growth, mental retardation, and developmental disorders.

Among the essential dietary ingredients for good health are adequate calories, proteins, vitamins, and minerals. Marasmus and kwashiorkor are protein-deficiency diseases; anemia and goiter are caused by mineral deficiencies that affect millions of people worldwide.

The three major crops that are the main source of calories and nutrients for most of the world's people are rice, wheat, and maize. Some new crops or unrecognized traditional crops hold promise for increasing the nutritional status of the poorer people of the world. Scientific improvement of existing crops and modernization of agriculture (irrigation, fertilizer, and better management) are potential sources of greater agricultural production, but also raise some troubling questions about safety and equity.

Fertile, tillable soil for growing crops is an indispensable resource for our continued existence on earth. Soil is a complex system of inorganic minerals, air, water, dead organic matter, and a myriad of different kinds of living organisms. There are hundreds of thousands of different kinds of soils, each produced by a unique history, climate, topography, bedrock, transported material, and community of living organisms.

It is estimated that 25 billion tons of soil are lost from croplands each year because of wind and water erosion. Perhaps twice as much is lost from rangelands and permanent pastures. This erosion causes pollution and siltation of rivers, reservoirs, estuaries, wetlands, and offshore reefs and banks. The net effect of this loss is worldwide crop reduction equivalent to losing 15 million ha (37 million acres), or 1 percent of the world's cropland each year.

The United States has some of the highest rates of soil erosion in the world. Soil erosion exceeds soil formation on at least 40 percent of U.S. cropland. About one-half of the topsoil that existed in North America before European settlement has been lost.

It is possible that food production could be expanded considerably, even on existing farmland, given the proper inputs of fertilizer, water, high-yield crops, and technology. This will be essential if human populations continue to grow as they have during this century. Whether it will be possible to supply agricultural inputs and expand crop production remains to be seen. Global climate change could have devastating effects on world food supplies and might necessitate conversion of forests and grasslands to feed the world's population.

Many new and alternative methods could be used in farming to reduce soil erosion, avoid dangerous chemicals, improve yields, and make agriculture just and sustainable. Returning to low-input, regenerative, "organic" farming may be more sustainable and more healthful than our current practices. Growing your own food or buying locally grown food at co-ops, farmers' markets, or through a producers' or buyers' association can provide healthy, wholesome food and also support sustainable agriculture.

QUESTIONS FOR REVIEW

1. How many people in the world are chronically undernourished and how many die each year from starvation and nutritionally related diseases?
2. How many calories does the average person need per day to maintain a healthy, active life?
3. What are proteins, vitamins, and minerals, and why do we need them in our diet?
4. What do we mean by green revolutions?
5. How are interfamily relationships important in food security?
6. What is the composition of soil? What is humus? Why are soil organisms so important?
7. What are four kinds of erosion? Why is it a problem?
8. What are some possible effects of over-irrigation?
9. What can farmers do to increase agricultural production without increasing land use?
10. What is sustainable agriculture?
11. What is genetic engineering, or biotechnology, and how might they help or hurt agriculture?

QUESTIONS FOR CRITICAL THINKING

1. What worldviews might make people believe that there are already too many people to be fed or that technological progress may allow us to feed double or triple current populations? Which side of this argument do you support?
2. Suppose that a seafood company wants to start a fish farming operation in a lake near your home. What regulations or safeguards would you want to see imposed on its operation? How would you weigh the possible costs and benefits of this operation?
3. Debate the claim that famines are caused more by human actions (or inactions) than by environmental forces. What is the critical element or evidence in this debate?
4. Should farmers be forced to use ecologically sound techniques that serve farmers' best interests in the long run, regardless of short-term consequences? How could we mitigate hardships brought about by such policies?
5. In a crisis, when small farms are in danger of being lost, should preserving the soil be the farmer's first priority?
6. Should we encourage (and subsidize) the family farm? What are the advantages and disadvantages (economic and ecological) of the small farm and the corporate farm?
7. Should we try to increase food production on existing farmland, or should we sacrifice other lands to increase farming areas?
8. Some rice paddies in Southeast Asia have been cultivated continuously for a thousand years or more without losing fertility. Could we, and should we, adapt these techniques to our own country?
9. Do you think that agribusinesses should be allowed to insert lethal "terminator" genes in their crop varieties? Why or why not?

KEY TERMS

anemia 239
contour plowing 254
cover crops 255
famines 240
food security 238
genetic engineering 251
green revolution 251
gully erosion 248

humus 243
kwashiorkor 238
malnourishment 238
marasmus 239
mulch 255
overnutrition 237
perennial species 255
reduced tillage systems 257
regenerative farming 253
rill erosion 248
salinization 249
soil 242
soil horizons 244
strip-farming 254
subsoil 244
sustainable agriculture 253
terracing 254
topsoil 244
undernourished 237
waterlogging 249

ADDITIONAL INFORMATION ON THE INTERNET

Visit our website at http://www.mhhe.com/environmentalscience for specific information about each of the topics below:

Natural Resources Conservation Service
Institute for Agriculture and Trade Policy
The Earth Network for Sustainable Development. Information
National Society of Consulting Soil Scientists, Inc.
Genetically Modified Foods—Greenpeace
International Agricultural Research (CGIAR)
US Department of Agriculture
USDA Sustainable Development and Small Farms Home Page
Aquaculture Network Information Center
United Nations Food and Agriculture Organization
Sustainable Agriculture Research and Education Program, University of California at Davis
Endangered Oceans: Seafood Watch
The Hunger Web, Brown University
Biotechnology and Food Security
Institute for Food and Development Policy
The Hunger Project, global organization working for the end of world hunger
International Institute for Ecological Agriculture
Endocrine Disruptors
"Terminator" Genes

CHAPTER 12

Pest Control

It ain't the things we know that cause all the trouble; it's the things we think we know that ain't so.

—Will Rogers—

OBJECTIVES

After studying this chapter, you should be able to:

- define the major types of pesticides and describe the pests they are meant to control.
- outline the history of pest control, including the changes in pesticides that have occurred in the last half of the twentieth century.
- appreciate the benefits of pest control.
- relate some of the problems of pesticide use.
- explain some alternative methods of pest control.
- discuss pesticide regulation and the acceptability of small amounts of weak carcinogens in our diet.

ESSENTIAL STUDY PARTNER

The Essential Study Partner (ESP) CD-ROM that accompanies this textbook contains additional information on the following topics:

- pesticide types: living resources: pest control: pesticides
- DDT: living resources: pest control: pesticides
- diminishing effectiveness and resistance to pesticides: living resources: pest control: pesticides
- pesticide effects on humans: living resources: pest control: pesticides
- crop rotation, burning crop residue, and biological control with predatory species: living resources: pest control: alternatives
- Integrated Pest Management (IPM): living resources: pest control: alternatives

Above: A helicopter sprays pesticides over a sugar beet field in California.

DDT and Fragile Eggshells

During the 1960s, peregrine falcons (fig. 12.1), bald eagles, osprey, brown pelicans, shrikes, and several other predatory bird species suddenly disappeared from former territories in eastern North America. What caused this sudden decline? Studies revealed that eggs laid by these birds had thin, fragile shells that broke before hatching. Eventually, these reproductive failures were traced to residues of DDT and its degradation product, DDE, which had concentrated through food webs until reaching toxic concentrations in top trophic levels such as these bird species.

DDT (dichloro-diphenyl-trichloroethane), an inexpensive and highly effective insecticide, had been used widely to control mosquitoes, biting flies, codling moths, potato beetles, corn earworms, cotton bollworms, and a host of other costly and irritating pest species around the world. First produced commercially in 1943, more than 50 million pounds of DDT were sprayed on fields, forests, and cities in the United States by 1950.

But as we have since discovered, there are disadvantages to widespread release of these toxic compounds in the environment. In the case of falcons, eagles, and other top predators, DDT and DDE inhibit enzymes essential for deposition of calcium carbonate in eggshells, resulting in soft, easily broken eggs. As we will discuss later in this chapter, other compounds chemically related to DDT are now thought to be disrupting endocrine hormone functions and causing reproductive losses in many species other than birds (What Do You Think?, p. 268).

These effects, coupled with the discovery of a pervasive presence of various chlorinated hydrocarbons in human tissues worldwide, led to the banning of DDT in most industrialized countries in the early 1970s. Peregrine falcons, which had declined to only about 120 birds in the United States (outside of Alaska) in the mid-1970s, now number about 1400, most of them bred in captivity and then released into the wild. Bald eagle reproduction has increased from an average of 0.46 young per nest in 1974 to 1.21 young per nest in 1994 in eastern Canada. Peregrine falcons and bald eagles had recovered enough to be removed from the endangered species list in the eastern United States in 1994.

What have we learned from this experience? Chemical pesticides offer a quick, convenient, and relatively inexpensive way to eliminate annoying or destructive organisms. At the same time, however, excessive pesticide use can kill beneficial organisms and upset the natural balance between predator and prey species. Modern pesticides undoubtedly have saved millions of human lives by killing disease-causing insects and by increasing food supplies. But we must understand what these powerful chemicals are doing and use them judiciously. In this chapter, we will study the major types of pests and pesticides along with some of the benefits and problems involved in our battle against pests.

FIGURE 12.1 Peregrine falcons disappeared from the eastern United States in the 1960s as a result of excess pesticide use.

WHAT ARE PESTS AND PESTICIDES?

A pest is something or someone that annoys us, detracts from some resource that we value, or interferes with a pursuit that we enjoy. In this chapter, we will concentrate on **biological pests,** organisms that reduce the availability, quality, or value of resources useful to humans. What's annoying or undesirable depends, of course, on your perspective. The mosquitoes that swarm in clouds over a marsh in the summer may be irritating to us, but they are an essential food source for birds and bats that feed on them. You may regard dandelions in your yard as tenacious and obnoxious weeds, but in some countries dandelions are cultivated as beautiful, exotic flowers and as a food source. Of the millions of species of organisms only about one hundred plants, animals, fungi, and microbes cause 90 percent of all crop damage worldwide.

Insects tend to be the most frequent pests, in part, because they make up at least three-quarters of all species on the earth. Most pest organisms tend to be generalists, the opportunistic species that reproduce rapidly, migrate quickly into disturbed areas, and that are pioneers in ecological succession. They compete aggressively against more specialized endemic species and can often take over a biotic community, especially where humans have disrupted natural conditions and created an opening into which they can slip. Most Americans are familiar with dandelions, ragweed, English sparrows, starlings, European pigeons, and other "weedy" species that survive well in urban habitats. Chapter 13 describes how exotic aliens brought in by humans are crowding out native species in many places.

A **pesticide** is a chemical that kills pests. We generally think of toxic substances in this category, but chemicals that drive away pests or prevent their development are sometimes included as well.

FIGURE 12.2 Synthetic chemicals can eliminate pests quickly and efficiently, but what are the long-term costs to us and to our environment?

FIGURE 12.3 Geese make good biological control agents. They eat weeds, grass, and insects but leave many crops alone. Their droppings enrich the soil, their down can be used to make garments and pillows, and a goose dinner makes a welcome protein source for many people.

Pest control can also include activities such as killing pests by burning crop residues or draining wetlands to eliminate breeding sites. A broad-spectrum pesticide that kills all living organisms is called a **biocide** (fig. 12.2). Fumigants, such as ethylene dibromide or dibromochloropropane, used to protect stored grain or sterilize soil fall into this category. Generally, we prefer narrower spectrum agents that attack a specific type of pest: **herbicides** kill plants; **insecticides** kill insects; **fungicides** kill fungi; acaricides kill mites, ticks, and spiders; nematicides kill nematodes (microscopic roundworms); rodenticides kill rodents; and avicides kill birds. Pesticides can also be defined by their method of dispersal (fumigation, for example) or their mode of action, such as an ovicide, which kills the eggs of pests. In a sense, the antibiotics used in medicine to fight infections are pesticides as well.

A BRIEF HISTORY OF PEST CONTROL

Humans have probably always used chemicals to protect themselves from pests, but in the past 50 years we have entered a new era of pesticide use. How did we develop these chemical agents and how do current uses differ from previous practices?

Early Pest Controls

Using chemicals to control pests may well have been among our earliest forms of technology. People in every culture have known that salt, smoke, and insect-repelling plants can keep away bothersome organisms and preserve food. The Sumerians controlled insects and mites with sulfur 5000 years ago. Chinese texts 2500 years old describe mercury and arsenic compounds used to control body lice and other pests. Greeks and Romans used oil sprays, ash and sulfur ointments, lime, and other natural materials to protect themselves, their livestock, and their crops from a variety of pests.

In addition to these metals and inorganic chemicals, people have used organic compounds, biological controls, and cultural practices for a long time. Alcohol from fermentation and acids in pickling solutions prevent growth of organisms that would otherwise ruin food. Spices were valued both for their flavors and because they deterred spoilage and pest infestations. Romans burned fields and rotated crops to reduce crop diseases. The Chinese developed plant-derived insecticides and introduced predatory ants in orchards to control caterpillars 1200 years ago. Many farmers still use ducks and geese to catch insects and control weeds (fig. 12.3).

Synthetic Chemical Pesticides

The modern era of chemical pest control began in 1934 with the discovery of the insecticidal properties of DDT (*D*ichloro-*D*iphenyl-*T*richloroethane) by Swiss chemist Paul Müller. DDT was used to control potato beetles in Switzerland in 1939 and commercial production began in 1943. It became extremely important during World War II in areas where tropical diseases and parasites posed greater threats to soldiers than enemy bullets.

DDT seemed like a wonderful discovery. It is cheap, stable, soluble in oil, and easily spread over a wide area. It is highly toxic to insects but relatively nontoxic to mammals. The oral LD50 for DDT for humans is 113 mg/kg, compared to 50 mg/kg for nicotine and 366 mg/kg for caffeine. Where other control processes act slowly and must be started before a crop is planted, DDT can save a crop even when pests already are well established. Its high toxicity for target organisms makes DDT very effective, often producing 90 percent control with a single application. DDT seemed like the magic bullet for which science had been searching. It was sprayed on crops and houses, dusted on people and livestock, and used to combat insects all over the world (fig. 12.4). Paul Müller received a Nobel prize in 1948 for his discovery.

FIGURE 12.4 Before we realized the toxicity of DDT, it was sprayed freely on people to control insects as shown here at Jones Beach, NY, in 1948.

As you will learn in this chapter, however, indiscriminate and excessive use of synthetic chemical pesticides has caused serious ecological damage and long-term harm to human health, and it has turned relatively innocuous species into serious pests.

PESTICIDE USES AND TYPES

Information on total world pesticide use is spotty and uncertain. The United Nations Environment Program reports sales totals rather than quantity because of difficulty in obtaining data from developing countries. In any case, it is clear that pesticide use has increased dramatically since World War II. Sales have grown from almost nothing in 1950 to approximately $33 billion for 2.6 million metric tons in 1999. About 90 percent of all pesticides worldwide are used in agriculture or food storage and shipping. The wealthier countries of the developed world consume some three-fourths of all pesticides, but rates of use in developing countries are rising 7 to 8 percent per year compared to 2 to 4 percent annual increases in the more-developed countries.

Pesticide Use in the United States and Canada

The U.S. Environmental Protection Agency reports that 650,000 metric tons of pesticides were used in the United States in 1999. Germany and Italy—which have the world's next highest consumption after the United States—each use about one-fifth as much. Canada uses one-tenth as much total pesticide as does the United States but has doubled its consumption over the past decade, whereas U.S. consumption has remained relatively stable. About 20 percent of all agricultural lands in Canada was treated with herbicides in 1970 and only 2 percent was sprayed with insecticides. Fifteen years later herbicide use had more than doubled to over half of all cropland, and insecticide use had grown tenfold, especially in the prairie provinces of Alberta and parts of Saskatchewan.

Nearly three-quarters of all pesticides applied in the United States are used in agriculture. Herbicides account for about 59 percent of the total quantity, insecticides 22 percent, fungicides 11 percent, and all other types together about 8 percent. Row crops such as corn, soybeans, and cotton account for about three-quarters of all herbicides used in the United States. When these crops are grown year after year in vast monoculture fields, they require heavy doses of insecticides as well. Fungicides are used mainly on fruits and vegetables. Household applications represent about 12 percent of all pesticide use in the United States but are almost 23 percent of insecticide use—higher than any other place in the world.

Pesticide Types

One way to classify pesticides is by their chemical structure. This is useful because environmental properties—such as stability, solubility, and mobility—and toxicological characteristics of members of a particular chemical group are often similar.

Inorganic pesticides include compounds of arsenic, copper, lead, and mercury. These broad-spectrum poisons are generally highly toxic and essentially indestructible, remaining in soil forever. Seeds are sometimes coated with a mercury or arsenic powder to deter insects and rodents during storage or after planting. Handling such seeds with bare hands can be very dangerous for farmers or gardeners. They are generally neurotoxins and even a single dose can cause permanent damage.

Natural organic pesticides, or "botanicals," generally are extracted from plants. Some important examples are nicotine and nicotinoid alkaloids from tobacco; rotenone from the roots of derris and cubé plants; pyrethrum, a complex of chemicals extracted from the daisylike *Chrysanthemum cinerariaefolium* (fig. 12.5); and turpentine, phenols, and other aromatic oils from conifers. All are toxic to insects, but nicotine is also toxic to a broad spectrum of organisms including humans. Rotenone is commonly used to kill fish. Turpentine, phenols, and other natural hydrocarbons are effective pesticides, but synthetic forms such as pentachlorophenol are more stable and more toxic than natural forms. They penetrate surfaces well and are used to prevent wood decay.

Fumigants are generally small molecules such as carbon tetrachloride, carbon disulfide, ethylene dichloride, ethylene dibromide, methylene bromide, and dibromochloropropane that gasify easily and penetrate rapidly into a variety of materials. They are used to sterilize soil and prevent decay or rodent and insect infestation of stored grain. Because these compounds are extremely dangerous for workers who apply them, use has been curtailed or banned altogether.

Chlorinated hydrocarbons, or organochlorines such as DDT, chlordane, aldrin, dieldrin, toxaphene, paradichlorobenzene (mothballs), and lindane, are synthetic organic insecticides that inhibit

FIGURE 12.5 Harvesting pyrethrum-containing flowers in Kenya to extract natural pesticides.

FIGURE 12.6 The United Farm Workers of America estimates that 300,000 farmworkers in the United States suffer from pesticide-related illnesses each year.

nerve membrane ion transport and block nerve signal transmission. They are fast acting and highly toxic in sensitive organisms. Toxaphene, for instance, kills goldfish at 5 parts per billion (5 µg/liter), one of the highest toxicities for any compound in any organism. Chlorinated hydrocarbons persist in soil up to 15 years, become concentrated through food chains, and are stored in fatty tissues of a variety of organisms. The chloriphenoxy herbicides 2,4 D and 2,4,5 T have hormone-like growth-regulating properties and are selective for broad-leaved flowering plants.

Organophosphates, such as parathion, malathion, dichlorvos, dimethyldichlorovinylphosphate (DDVP), and tetraethylpyrophosphate (TEPP), are an outgrowth of nerve gas research during World War II. They inhibit cholinesterase, an enzyme essential for removing excess neurotransmitter from synapses in the peripheral nervous system. They are extremely toxic to mammals, birds, and fish (generally 10 to 100 times more poisonous than most chlorinated hydrocarbons). A single drop of TEPP on your skin can be lethal. Because they are quickly degraded, they are much less persistent in the environment than organochlorines, generally lasting only a few hours or a few days. These compounds are very dangerous for workers such as grape pickers who often are sent into fields too soon after they have been sprayed (fig. 12.6).

Carbamates, or urethanes, such as carbaryl (Sevin), aldicarb (Temik), aminocarb (Zineb), carbofuran (Baygon), and Mirex share many organophosphate properties, including mode of action, toxicity, and lack of environmental persistence and low bioaccumulation. Carbamates generally are extremely toxic to bees and must be used carefully to prevent damage to these beneficial organisms.

Microbial agents and *biological controls* are living organisms or toxins derived from them used in place of pesticides. Bacteria such as *Bacillus thuringiensis* or *Bacillus popilliae* kill caterpillars or beetles by producing a toxin that ruptures the digestive tract lining when eaten. Parasitic wasps such as the tiny *Trichogramma* genus attack moth caterpillars and eggs, while lacewings and ladybugs control aphids. Viral diseases also have been used against specific pests.

PESTICIDE BENEFITS

Like all organisms, humans compete with other species for food and shelter and struggle to protect ourselves from diseases and predators. Synthetic chemical pesticides are important weapons in this fight for survival.

Disease Control

Insects and ticks serve as vectors in the transmission of a number of disease-causing pathogens and parasites. Consider malaria, for example. About 500 million people suffer from this disease at any given time, and about 2 million die each year from *Plasmodium* protozoans spread to humans by *Anopheles* mosquitoes (fig. 12.7). It is estimated that insecticidal mosquito control has prevented at least 50 million deaths from malaria over the past 50 years. Sri Lanka is a classic example of pesticide benefits. In the early 1950s, more than 2 million cases of malaria were reported in Sri Lanka each year. After DDT spraying began in 1954, new malaria cases almost completely disappeared. When DDT spraying was discontinued in 1964, however, malaria reappeared almost immediately.

FIGURE 12.7 Malaria, spread by the *Anopheles* mosquito, is one of the largest causes of human disease and premature death in the world. By controlling mosquitoes, pesticides save 1 million lives per year in tropical countries.

FIGURE 12.8 Elephantiasis is caused by parasitic worms (filaria) that block lymph vessels and cause fluid accumulation in various parts of the body.

Within three years the annual incidence was more than 1 million cases per year. The Sri Lankan government resumed DDT spraying in 1968 and continues limited use of this insecticide despite environmental concerns. They concluded that the reductions in medical expenses, social disruption, pain and suffering, and lost work resulting from malaria outweighed the direct costs of insecticide spraying by a thousand to one.

Some other diseases spread by biting insects include yellow fever and related viral diseases such as encephalitis, also carried by mosquitoes, and trypanosomiasis or sleeping sickness caused by protozoans transmitted by the tsetse fly. Onchocerciasis (river blindness) and filariasis (one form of which is elephantiasis, fig. 12.8) are caused by tiny worms spread by biting flies that afflict hundreds of millions of people in tropical countries. Bubonic plague (or Black Death) is caused by bacteria transmitted by fleas shared by rats and humans. Typhus, caused by a rickettsia, is spread by body lice. All of these terrible diseases can be reduced by judicious use of pesticides. If you were faced with a choice of going blind before age 30 because of masses of worms accumulating in your eyeballs or a small chance of cancer due to pesticide exposures if you live to age 50 or 60, which would you choose?

Crop Protection

Although reliable data on crop losses are difficult to obtain, it is thought that plant diseases, insect and bird predation, and competition by weeds reduce crop yields worldwide by at least one-third. Postharvest losses to rodents, insects, and fungi may be as high as another 20 to 30 percent. Without modern chemical pesticides, these losses might be much higher. Although we said in chapter 11 that there is more than enough food in the world to adequately feed everyone now living—if food were equitably distributed—this most certainly would not be true, given current methods of production, if modern chemical pesticides were unavailable.

A commonly quoted estimate is that farmers save \$3 to \$5 for every \$1 spent on pesticides. This means lower costs and generally better quality for consumers. In some cases, insects and fungal diseases cause only small losses in terms of the total crop quantity, but the cosmetic damage they cause greatly reduces the economic value of crops. For example, although codling moth larvae consume very little of the apples they infest, the brown trails they leave as they crawl through the fruit and the possibility of biting into a living worm often make an unsprayed crop unsalable. Some consumers have come to believe, however, that it may be better to trim a few visible worms or wormholes from their apples than to eat invisible but perhaps more dangerous pesticide residues.

PESTICIDE PROBLEMS

While synthetic chemical pesticides have brought us great economic and social benefits, they also cause a number of serious problems. In this section we will examine some of the worst of those problems.

Effects on Nontarget Species

It is estimated that up to 90 percent of the pesticides we use never reach their intended targets (fig. 12.9). Many beneficial organisms are poisoned unintentionally as a result. For instance, about 20 percent of all honeybee colonies in the United States are destroyed

FIGURE 12.9 This machine sprays insecticide on orchard trees—and everything else in its path. Up to 90 percent of all pesticides never reach target organisms.

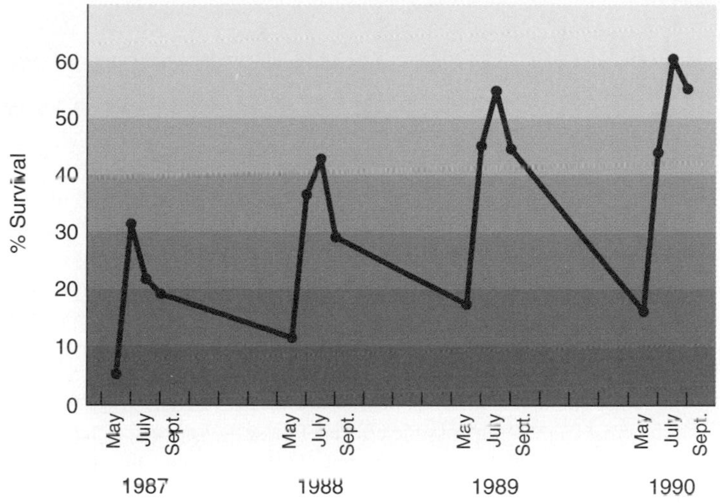

FIGURE 12.10 Survival of tobacco budworms to a constant dose of pyrethroid at different times of the year. Highest survival percentages are in July of each year, but maximum survival grows as worms become resistant.
Source: Data from S. Micinski, et al., *Resistant Pest Management Newsletter,* Vol. 3, No. 2, July 1991.

each year and another 15 percent are damaged by pesticide spray drift or residues on the flowers they visit. Direct losses to beekeepers amount to several million dollars per year. Losses to crops the bees would have pollinated may be ten times higher.

In some cases, the effects of poisoning nontarget species are immediate and unmistakable. In one episode in 1972, a single application of the insecticide Azodrin to combat potato aphids on a farm in Dade County, Florida, killed 10,000 migrating robins in three days. Similarly, a 1991 derailment of a Southern Pacific tanker car on a tricky canyon bridge just north of Dunsmuir, California, dumped 75,000 liters (20,000 gallons) of highly toxic metam sodium herbicide into the Sacramento River. The entire river ecosystem—including aquatic plants, insects, amphibians, and at least 100,000 trout—was completely wiped out for 45 kilometers (27 miles) downstream.

In other cases, the effects are more difficult to pin down, although not less serious. A study published in 1999 linked insecticide spraying on Canadian forests with dramatic declines (up to 77 percent) in Atlantic salmon. The chemical suspected in this case is 4-nonylphenol, a powerful endocrine-hormone disrupter (What Do You Think? p. 268).

Pesticide Resistance and Pest Resurgence

Pesticides almost never kill 100 percent of a target species even under the most ideal conditions. As we discussed in chapter 4, every population contains some diversity in tolerance to adverse environmental factors. The most resistant members of a population survive pesticide treatment and produce more offspring like themselves with genes that enable them to withstand further chemical treatment (fig. 12.10). Because most pests propagate rapidly and produce many offspring, the population quickly rebuilds with pesticide-resistant individuals. We call this phenomenon **pest resurgence** or rebound. The United Nations Environment Program

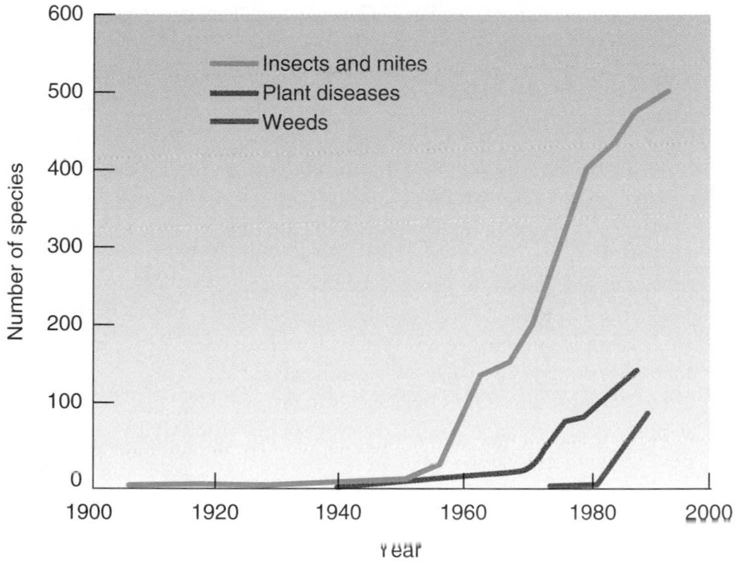

FIGURE 12.11 Many pests have developed resistance to pesticides. Because insecticides were the first class of pesticide to be used widely, selection pressures led insects to show resistance early. More recently, plant pathogens and weeds are also becoming insensitive to pesticides.
Source: Data from *State of the World,* 1996.

reports that at least 500 insect pest species and another 250 or so weeds and plant pathogens worldwide have developed chemical resistance (fig. 12.11). Of the 25 most serious insect pests in California, three-quarters are now resistant to one or more insecticides. This resistance means that it takes constantly increasing doses to get the same effect or that farmers who are caught on a **pesticide treadmill** must constantly try newer and more toxic chemicals in

what do you think?

THE WORLD AROUND YOU

Environmental Estrogens

What might alligators in Florida, seals in the North Sea, salmon in the Great Lakes, and you have in common? All are at the top of their respective food chains and all appear to be accumulating threatening levels of toxic environmental chemicals in their body tissues. One of the most frightening possible effects of those chemicals is that they seem to be able to disrupt endocrine hormones that regulate many important bodily functions. Evidence for this seems quite convincing in some wildlife populations, but whether it also is true for humans is one of the most contentious and important questions in environmental toxicology today.

One of the first examples of hormone-disrupting chemicals in the environment was a dramatic decline in alligators a decade ago in Florida's Lake Apopka. Surveys showed that 90 percent of the alligator eggs laid each year were infertile and that of the few that hatched, only about half survived more than two weeks. Male hatchlings had shrunken penises and unusually low levels of the male hormone testosterone. Female alligators, meanwhile, had highly elevated estrogen levels and abnormal ovaries. The explanation seems to be that a DDT spill in the lake in the 1980s, along with pesticide-laden runoff from adjacent farm fields, has led to high levels of DDE (a persistent breakdown product of DDT) in the reptiles' tissues and eggs. Because of a similarity in chemical structure, DDE appears to interfere with the action of androgens and estrogens, the normal sex hormones.

Researchers have begun to suspect that mysterious outbreaks of health and reproductive problems in other wildlife populations may have similar origins. Immune-system failures that killed thousands of seals along the coast of Europe and Scandinavia in 1992, for instance, are thought to have been caused by high levels of pesticides, PCBs, dioxins, and other toxins in their diet. Similarly, reproductive failures in fish and bird populations in the Great Lakes, fewer turtle hatchlings in farm ponds, abnormal thyroids and dramatic increases in tumors in fish, all are now thought to be related to hormone disturbances by exogenous chemicals.

But are humans affected as well? It is quite clear that people everywhere in the world have accumulated many of these same toxic chemicals in their bodies. Women who eat lots of fish from contaminated waters have been shown to have babies with elevated rates of mental, developmental, and behavioral disorders. Studies of women with estrogen-sensitive breast and vaginal cancers were found to have higher than normal levels of pesticides such as DDE in their tissues. Sperm counts in men appear to have decreased by about 50 percent over the past fifty years, while testicular and prostate cancers have increased dramatically during that same time.

Good evidence exists from controlled laboratory experiments that rats and mice exposed *in utero* or through mother's milk to very low levels of estrogen-like compounds develop physical, reproductive, and behavioral problems. We know that some of these chemicals act as synthetic hormones, others are antagonists that block normal hormone function. Furthermore, there can be striking synergy between some compounds. When endosulfan and DDT or chlordane are applied together, for example, the combination is 1600 times more estrogenic than either chemical alone.

The question is whether these chemicals are linked to human health problems. Many of these compounds are hundreds or thousands of times less active than normal hormones, leading skeptics to doubt that they have any noticeable effects except in animals exposed to extremely high levels from a chemical spill. Since some effects are positive while others are negative, they could cancel each other out. Furthermore, we may have protective mechanisms that are lacking in highly inbred laboratory rodents, and we can eat a highly varied diet that includes protective factors as well as toxins.

The bottom line is that we don't know (and we may never know for sure) whether falling sperm counts, increasing cancers, birth defects, immune diseases, and behavioral disorders in humans are caused by endocrine-disrupting environmental chemicals. Of course, we should do more research and testing of the physiological actions of these chemicals. In 1996, the EPA ordered pesticide manufacturers to begin testing for disrupting effects. Given the continuing uncertainty about the dangers we face, what more do you think we should do? Is this threat serious enough to warrant drastic steps to reduce our risk? If you were head of the Environmental Protection Agency or the Food and Drug Administration, how much certainty would you demand before acting to protect our environment and ourselves from this frightening potential threat?

A shocking decrease in fertility and hatchling survival of alligators in Lake Apopka has been linked to pesticide pollution. Are humans being affected by similar toxins?

an attempt to stay ahead of the pests. Despite a 33-fold increase in pesticide use in the United States since 1944, crop losses caused by pests have actually increased (table 12.1); some people argue that we are losing the battle to control pests.

One of the most ominous developments in this race to find effective pesticides is that we increasingly find pests that are resistant to chemicals to which they have never been exposed. Apparently, genes for pesticide resistance are being transferred from one species to another by means of vectors such as viruses and plasmids (naked pieces of viral-like DNA). Often a whole cluster of genes jump between species so that multiple chemical-tolerance is inherited before a particular pest is ever exposed to any of the chemicals. Widespread use of crops genetically engineered to produce pesticides endogenously (chapter 11) is likely to cause even more pesticide resistance.

In a way, every pesticide has a very limited useful life span before target species become resistant to it or the pesticide builds up intolerable environmental concentrations. We have made a big mistake in broadcasting pesticides recklessly and extravagantly. DDT, for instance, is such a helpful insecticide that we should have used it sparingly and carefully, so that it would still be effective against the worst insect pests. It has been spread so widely that 50 of the 60 malaria-carrying mosquitoes are now resistant to it and the environmental side effects outweigh its benefits. You can think of the useful life of DDT as a nonrenewable resource that we squandered.

Creation of New Pests

Often the worst side effect of broadcast spraying a pesticide is that we kill beneficial predators that previously kept a number of pests under control. Many of the agricultural pests that we want to eliminate, for instance, are herbivores such as aphids, grasshoppers, or moth larvae that eat crop plants. Under natural conditions, their populations are kept under control by predators such as wasps, lady bugs, and praying mantises. As we discussed in chapter 3, there are generally far fewer predators in a food web than the species they prey upon. When we use a broad-spectrum pesticide, higher trophic levels are more likely to be knocked out than lower ones. This means that species that previously were insignificant can be released from natural controls and suddenly become major pests.

TABLE 12.1 Percentage of Crops Lost Annually to Pests in the United States

YEAR	INSECTS	DISEASES	WEEDS
1944	7.1	10.5	13.8
1974	13.0	12.0	8.0
1989	13.0	12.0	12.0

Source: Data from D. Pimentel, et al., "Environmental and Economic Effects of Reducing Pesticides," *Bioscience,* 41:6-12, June 1991.

FIGURE 12.12 The cotton boll weevil caused terrible devastation in southern U.S. cotton fields in the 1930s and 1940s. DDT and other insecticides controlled this pest briefly, but the beetle quickly became resistant.

Consider the case of the Canete Valley in Peru. Before DDT was introduced in 1949, cotton yields were about 500 kg per ha. By 1952 yields had risen to nearly 750 kg per ha, but DDT-resistant boll weevils also had appeared (fig. 12.12). Toxaphene replaced DDT, but within two years it also became ineffective against boll weevils, which rebounded to higher levels than ever. Even worse, *Heliothis* worms—which had not previously been a problem—began increasing rapidly. The wasps that earlier had kept both organisms in check were poisoned by the increasing pesticide doses. By 1955, cotton yields were down to 330 kg per ha, one-third less than before any pesticides were used.

Similarly, in California in 1995, cotton yields plunged by 20 percent despite a doubling of the number of insecticide applications and a 43 percent increase in pesticide costs. With the advent of chemical pest controls, farmers tend to abandon mixed crops, rotation regimes, and other traditional methods of management. This creates a greater potential for pest damage and pest resistance and thus makes it economical to use even higher levels of pesticide treatment. The application of large amounts of pesticides over wide expanses of land, together with changes in the ways the land is managed, have undermined the biological and ecological forces and interactions that previously governed population dynamics among species. A lack of understanding of the ways pesticides adversely affect beneficial organisms is one reason pest managers stick with chemical-based systems despite declining pesticide efficiency.

Persistence and Mobility in the Environment

The qualities that make DDT and other chlorinated hydrocarbons so effective—stability, high solubility, and high toxicity—also make them environmental nightmares. Because they persist for years, even decades in some cases, and move freely through air, water, and soil, they often show up far from the point of original application. Some of these compounds have been discovered far from any possible source and long after they most likely were used. Because they have an affinity for fat, many chlorinated hydrocarbons are bio-concentrated and stored in the bodies of predators—such as porpoises, whales, polar bears, trout, eagles, ospreys, and humans—that feed at the top of food webs. In a study of human pesticide uptake and storage, Canadian researchers found the levels of chlorinated hydrocarbons in the breast milk of Inuit mothers living in remote arctic villages was five times that of women from Canada's industrial region some 2500 km (1600 mi) to the south. Inuit people have the highest levels of these persistent pollutants of any human population except those contaminated by industrial accidents.

These compounds accumulate in polar regions by what has been called the "grasshopper effect," in which they evaporate from water and soil in warm areas and then condense and precipitate in colder regions. In a series of long-distance hops, they eventually collect in polar regions, where they accumulate in top predators. Polar bears, for instance, have been shown to have concentrations of certain chlorinated compounds 3 billion times greater than the seawater around them. In Canada's St. Lawrence estuary, beluga (white whales), which suffer from a wide range of infectious diseases and tumors thought to be related to environmental toxins, have such high levels of chlorinated hydrocarbons that their carcasses must be treated as toxic waste.

Although DDT hasn't been used legally in the United States or Canada for more than 20 years, in 1999 researchers reported discovery of p,p'-DDE, a DDT breakdown by-product, in the amniotic fluid of 30 percent of a sample of pregnant women in Los Angeles, California. A growing baby is surrounded by amniotic fluid during its nine months in the mother's womb. The levels found in some women—up to 0.6 ppm—are sufficient to cause concern. It was once thought that these compounds wouldn't cross the placental barrier, but this apparently is not the case.

Atrazine and alochlor are the most widely used herbicides in North America. More than 96 percent of the maize (corn) and soybean crops in the United States are treated with these compounds. The 31 million kg (70 million lbs) of atrazine used every year in the U.S. makes it the biggest commercial pesticide. Surveys of wells have found that up to 30 percent of all community wells and as much as 60 percent of all private wells in the midwestern U.S. corn belt are contaminated with atrazine and alochlor. In a related study, researchers measured levels of these two compounds in rain falling over much of Europe that would be illegal if it were supplied as drinking water.

The ubiquity of these contaminants has led to widespread movement to ban them altogether. At an international meeting in Montreal in 1998, nearly 100 nations agreed on a process that should eventually lead to a binding, worldwide treaty to outlaw the worst of these **persistent organic pollutants (POPs).** Table 12.2 lists the "dirty dozen" targeted by the first draft of this treaty. All of these compounds are chlorinated and all are suspected of disrupting normal endocrine hormone responses. Although almost all these chemicals are banned or severely restricted in North America and Europe, between 1994 and 1996 more than 100,000 metric tons annually of chemicals in these categories was shipped from U.S. ports, mostly to developing countries where regulations are more lax. Ironically, because they are so persistent, many come back to us as residues on foods and other commodities we buy from those countries.

TABLE 12.2 The "Dirty Dozen" Persistent Organic Pollutants

COMPOUND(S)	USES
Aldrin	Insecticide used on corn, potatoes, cotton, and for termite control
Clordane	Insecticide used on vegetables, small grains, maize, sugarcane, fruits, nuts, and cotton
Dieldrin	Insecticide used on cotton, corn, potatoes, and for termite control
DDT	Insecticide, now used primarily for disease vector control
Endrin	Insecticide used on field crops such as cotton and grains and as a rodenticide
Hexachlorobenzene (HCB)	Fungicide used for seed treatment and as an industrial chemical
Heptachlor	Insecticide used against soil insects, termites, and grasshoppers
Mirex	Insecticide used to combat fire ants, termites, mealybugs, and as a fire retardant
Toxaphene	A mixture of chemicals used as an insecticide on cotton as well as tick and mite control in livestock and fish eradication
Polychlorinated biphyenyls (PCBs)	Industrial chemicals used as insulators in electrical transformers, solvents, paper coatings, and plasticizers
Dioxins	A large family of by-products of chlorinated chemical production and incineration
Furans	A large group of by-products of chlorinated chemical production and incineration

Human Health Problems

Pesticide effects on human health can be divided into two categories: (1) short-term effects, including acute poisoning and illnesses caused by relatively high doses and accidental exposures, and (2) long-term effects suspected to include cancer, birth defects, immunological problems, Parkinson's disease, and other chronic degenerative diseases.

The World Health Organization (WHO) estimates that between 3.5 and 5 million people suffer acute pesticide poisoning and at least 20,000 die each year. At least two-thirds of this illness and death results from occupational exposures in developing countries where people use pesticides without proper warnings or protective clothing (fig. 12.13). A tragic example of occupational pesticide exposure is found among workers in the Latin American flower industry. Fueled by the year-round demand in North America for fresh vegetables, fruits, and flowers, a booming export trade

FIGURE 12.13 Farmworkers in developing countries often apply dangerous pesticides without any protective clothing or other safeguards. Warnings and instructions printed in English are often of little help.

has developed in countries such as Guatemala, Colombia, and Ecuador. To meet demands in North American markets for perfect flowers and produce, growers use high levels of pesticides, often spraying daily with fungicides, insecticides, nematicides, and herbicides. Working in warm, poorly ventilated greenhouses with little protective clothing, the workers—70 to 80 percent of whom are women—find it hard to avoid pesticide contact. Nearly two-thirds of nearly 9000 workers surveyed in Colombia experienced blurred vision, nausea, headaches, conjunctivitis, rashes, and asthma. Although harder to document, they also reported serious chronic effects such as stillbirths, miscarriages, and neurological problems.

Long-term health effects caused by chronic pesticide exposures are difficult to document conclusively. As we discussed in chapter 9, isolating a specific pesticide-related disease from the myriad other risks we encounter is complex. One well-known outcome of chronic pesticide exposure is that farmers who use 2,4-D on their crops are up to eight times as likely to contract a cancer known as non-Hodgkin's lymphoma as those who don't use herbicides.

In an ongoing study of prenatal exposure to PCBs and other persistent contaminants, researchers from Wayne State University in Detroit, Michigan, have documented significant learning and attention problems in children whose mothers ate Lake Michigan fish regularly in the years prior to pregnancy. At age 11, the most highly exposed children had difficulty paying attention, suffered from poorer short- and long-term memory, were twice as likely to be below average in reading comprehension, and were three times as likely to have low IQ scores as their age-matched peers.

Researchers in a similar study in Mexico showed striking differences in behavior and motor skills in children exposed to pesticides compared to peers with minimal pesticide risk. Two groups of Yaqui children from northwestern Mexico were tested in this study. The children were similar in every respect except for their pesticide exposure. Children who lived in the foothills come from ranches where pesticides are used sparingly, if at all. Children from the Yaqui valley, live in a farming area where pesticide use has been heavy for the past 50 years. Samples of mother's milk and umbilical cord blood taken from valley families contained high amounts of several persistent pesticides including aldrin, endrin, dieldrin, heptachlor, and DDE. In tests designed to measure growth and development, the valley children fell far behind their foothill-dwelling peers (fig. 12.14). Farm children also exhibited diminished memory, decreased physical stamina, higher irritability, a greater tendency to fight, and poorer eye-hand coordination.

FIGURE 12.14 Representative examples of drawings of people by 4- and 5-year-old Yaqui children relatively unexposed to pesticides (foothills) and those heavily exposed (valley).
Source: From E.A. Guillette, et al., "An Anthropological Approach to the Evaluation of Preschool Children Exposed to Pesticides in Mexico" in *Environmental Health Perspective,* 106(6):347-353, 1998. U.S. Dept. of Health and Human Services.

ALTERNATIVES TO CURRENT PESTICIDE USES

Can we avoid using toxic pesticides or lessen their environmental and human-health impacts (fig. 12.15)? In many cases, improved management programs can cut pesticide use between 50 and 90 percent without reducing crop production or creating new diseases. Some of these techniques are relatively simple and save money

FIGURE 12.15 Our approach to nature is to beat it into submission, but at what cost?

while maintaining disease control and yielding crops with just as high quality and quantity as we get with current methods. In this section, we will examine behavioral changes, biological controls, and integrated pest-management systems that could substitute for current pest-control methods.

Behavioral Changes

Crop rotation (growing a different crop in a field each year in a two- to six-year cycle) keeps pest populations from building up. For instance, a soybean/corn/hay rotation is effective and economical against white-fringed weevils. Mechanical cultivation can substitute for herbicides but increases erosion. Flooding fields before planting or burning crop residues and replanting with a cover crop can suppress both weeds and insect pests. Habitat diversification, such as restoring windbreaks, hedgerows, and ground cover on watercourses, not only prevents soil erosion but also provides perch areas and nesting space for birds and other predators that eat insect pests. Growing crops in areas where pests are absent makes good sense. Adjusting planting times can avoid pest outbreaks, while switching from huge monoculture fields to mixed polyculture (many crops grown together) makes it more difficult for pests to find the crops they like (see Case Study, p. 273). Tillage at the right time can greatly reduce pest populations. For instance, spring or fall plowing can help control overwintering corn earworms.

An important behavioral adjustment can occur in our attitudes and preferences. Farmers can be persuaded that the bare earth look in row crops made possible by powerful herbicides is not good for the soil or for themselves. Allowing a few weeds to creep in or planting a cover crop between rows may make more sense in the long run. Consumers may have to learn to accept fruits and vegetables that are less than perfect.

FIGURE 12.16 The praying mantis looks ferocious and is an effective predator against garden pests, but it is harmless to humans. They can even make interesting and useful pets.

FIGURE 12.17 Ladybird beetles (ladybugs) prey on a variety of pests both as larvae and adults. For a few dollars you can buy several thousand of these hardy and colorful little garden protectors from organic gardening supply stores.

Biological Controls

Biological controls such as predators (wasps, ladybugs, praying mantises; figs. 12.16 and 12.17) or pathogens (viruses, bacteria, fungi) can control many pests more cheaply and safely than broad-spectrum, synthetic chemicals. *Bacillus thuringiensis* or Bt, for example, is a naturally occurring bacterium that kills the larvae of lepidopteran (butterfly and moth) species but is harmless to mammals. A number of important insect pests such as tomato hornworm, corn rootworm, cabbage loopers, and others can be controlled by spraying bacteria on crops. Larger species are effective as well. Ducks, chickens, and geese, among other species, are used to rid fields of both insect pests

CASE STUDY

REGENERATIVE AGRICULTURE IN IOWA

Dick and Sharon Thompson operate a diversified crop and livestock farm near Boone, Iowa. Originally, the Thompsons practiced high-intensity, monocrop farming using synthetic pesticides and fertilizers just as all their neighbors did. But they felt that something was wrong. Their hogs and cattle were sick. Fertilizer, pesticide, and petroleum prices were rising faster than crop prices. They began looking for a better way to farm. Through 30 years of careful experimentation and meticulous recordkeeping, they have developed a set of alternative farming techniques they call "regenerative agriculture" because it relies on natural processes to rebuild and protect soil.

Rather than depend on synthetic chemical herbicides and pesticides to keep their fields clean of weeds and pests, the Thompsons use a variety of old and new techniques including crop rotation, cover crops, and mechanical cultivation. Instead of growing corn and beans over and over again in the same fields as most of their neighbors do, the Thompsons change crops every year so that no one weed species can become dominant and all species remain relatively easy to control. In the fall, nitrogen-fixing cover crops are planted to hold soil against wind erosion and to keep down weeds.

Before planting, animal manure is spread on fields to rebuild fertility. During the summer, cattle are pastured on fallow land, using intensive grazing techniques that discourage weed growth and spread of manure over the whole field. The soil organic content—the sentinel indicator of soil health—registers at 6 percent, which is more than twice that of their neighbors. Untouched Midwestern prairie usually has about 7 percent organic content. The capacity to store extra carbon in soil might allow farmers to bid on carbon set-aside contracts.

The high levels of organic matter and available nutrients in the Thompsons' fields, coupled with the absence of pesticides that might harm beneficial microbes and pathogens, help crops compete against weeds and insects. Weed control specialists predict that in the future more farmers will follow the Thompsons' lead and concentrate on microbial biocontrol rather than depend on conventional herbicide-dependent systems, some of which can impair soil quality and lead to carryover injury to crops.

Among the cultivation techniques used by the Thompsons are chisel plowing, ridge-tilling, and rotary hoe cultivation. These techniques leave more crop residue on the surface to protect the soil than does conventional moldboard plowing. Chisel plowing merely scratches the surface rather than turning the soil upside down. The rotary hoe is a tool used just after crops germinate to skim the soil surface and remove recently germinated weeds. In ridge tilling, a small plow scrapes weeds out of shallow valleys and mounds up soil into small ridges where crops grow.

More is known about the Thompson operation—production methods, yields, costs and returns, weed counts, soil quality, and environmental impacts—than any other similar farm in the United States. Through 30 years of on-farm experiments, the Thompsons have collaborated with scientists from a variety of institutions. Dozens of research reports and articles have been written about how the Thompsons' diversified farming system affects land fertility, erosion, and livestock health. Every year a field day is held on the farm to give neighbors and others a chance to see how the diversified system works.

While yields on the Thompsons' land is comparable to those of their neighbors, lower reliance on off-farm inputs—including pesticides, fertilizers, and animal drugs—keeps the Thompsons' production costs significantly lower than those in conventional cropping systems. Growing corn costs the Thompsons $1.50 per bushel compared to $2.11 per bushel on neighboring farms. Similarly, soybeans cost the Thompsons $3.90 per bushel compared to $4.80 per bushel for their neighbors. In addition to favorable financial returns, the Thompsons benefit in other ways from their innovative system. The quality of their soil is significantly better than that under conventional agriculture and is steadily improving in fertility, tilth, and health.

Through their innovative work, Dick and Sharon Thompson are helping find ways to profitably produce high yields without degrading the land or the environment. In 1996, the Thompsons were selected by the *Des Moines Register* as Iowa's "Farm Leaders of the Year" in recognition of their contributions to the science of sustainable agriculture.

Dick and Sharon Thompson of Boone, Iowa, practice sustainable farming to produce healthy crops that are economically successful.

and weeds. These biological organisms are self-reproducing and often have wide prey tolerance. A few mantises or ladybugs released in your garden in the spring will keep producing offspring and protect your fruits and vegetables against a multitude of pests for the whole growing season. Plants, too, can be effective weapons against insect pests. Home gardeners often plant a border of insect-repelling plants, such as garlic or marigolds, around their crops. On a larger scale, farmers might also benefit from this approach.

Herbivorous insects also have been used to control weeds (fig. 12.18). For example, the prickly pear cactus was introduced to Australia about 150 years ago as an ornamental plant. This hardy cactus escaped from gardens and found an ideal home in the dry soils

FIGURE 12.18 Biological pest control is illustrated by the Klamath weed (*foreground*), which had invaded millions of hectares of California rangeland in the 1940s. In the background, the weeds have been eliminated by the introduction of a natural predator, the *Chrysolina* beetle. Within a few years, the weeds were entirely eradicated.
Source: Data from Natural Resources Development Council.

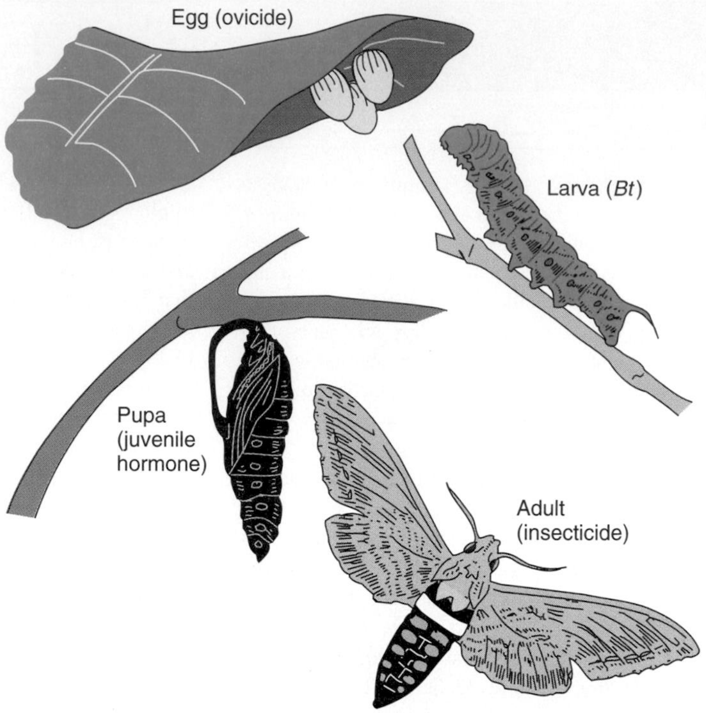

FIGURE 12.19 Stages in the life cycle of an insect and the control agents that kill or inhibit each stage. *Bacillus thuringiensis* (Bt) kills caterpillars when they eat leaves with these bacteria on the surface. The presence of excess juvenile hormone in the environment prevents maturation of pupae. Predators attack at all stages.

of the outback. It quickly established huge, dense stands that dominated 25 million ha (more than 60 million acres) of grazing land. A natural predator from South America, the cactoblastis moth, was introduced into Australia in 1935 to combat the prickly pear. Within a few years, cactoblastis larvae had eaten so much prickly pear that the cactus has become rare and is no longer economically important.

Genetics and bioengineering can help in our war against pests. Traditional farmers have long known to save seeds of disease-resistant crop plants or to breed livestock that tolerate pests well. Modern science can speed up this process through selection regimes or by using biotechnology to transfer genes between closely related or even totally unrelated species. There are concerns about the ecological dangers of moving too fast in this new area. Releasing sterile males to interfere with insect pest reproduction has been used successfully in some cases. Screwworms, for example, are the flesh-eating larvae of flies that lay their eggs in scratches or skin wounds of livestock. They were a terrible problem for ranchers in Texas and Florida in the 1950s, but release of massive numbers of radiation-sterilized males disrupted reproduction and eliminated this pest in Florida. In Texas, where flies continue to cross the border from Mexico, control has been more difficult, but continual vigilance keeps the problem manageable.

Other promising approaches are to use hormones that upset development or sex attractants to bait traps containing toxic pesticides. Many municipalities control mosquitoes with these techniques rather than aerial spraying of insecticides because of worries about effects on human health. Briquettes saturated with insect juvenile hormone are scattered in wetlands where mosquitoes breed. The presence of even minute amounts of this hormone prevent larvae from ever turning into biting adults (fig. 12.19). Unfortunately, many beneficial insects as well as noxious ones are affected by the hormone, and birds or amphibians that eat insects may be adversely affected when their food supply is reduced. Some communities that formerly controlled mosquitoes have abandoned these programs, believing that having naturally healthy wetlands is worth getting a few bites in the summer.

Integrated Pest Management

Integrated pest management (IPM) is a flexible, ecologically based pest-control strategy that uses a combination of techniques applied at specific times, aimed at specific crops and pests. It often uses mechanical cultivation and techniques such as vacuuming bugs off crops as an alternative to chemical application (fig. 12.20). IPM doesn't give up chemical pest controls entirely but rather tries to use the minimum amount necessary only as a last resort and avoids broad-spectrum, ecologically disruptive products. IPM relies on preventive practices that encourage growth and diversity of beneficial organisms and enhance plant defenses and vigor. Careful, scientific monitoring of pest populations to determine **economic thresholds,** the point at which potential economic damage justifies pest control expenditures, and the precise time, type, and method of pesticide application is critical in IPM.

FIGURE 12.20 This machine, nicknamed the "salad vac," vacuums bugs off crops as an alternative to treating them with toxic chemicals.

Trap crops, small areas planted a week or two earlier than the main crop, are also useful. This plot matures before the rest of the field and attracts pests away from other plants. The trap crop then is sprayed heavily with enough pesticides so that no pests are likely to escape. The trap crop is destroyed so that workers will not be exposed to the pesticide and consumers will not be at risk. The rest of the field should be mostly free of both pests and pesticides.

IPM programs are already in use all over the United States on a variety of crops. Massachusetts apple growers who use IPM have cut pesticide use by 43 percent in the past ten years while maintaining per-acre yields of marketable fruit equal to that of farmers who use conventional techniques. Some of the most dramatic IPM success stories come from the developing world. Cuba, for example (Case Study, p. 273) has turned almost entirely to organic farming. In Brazil, pesticide use on soybeans has been reduced up to 90 percent with IPM. In Costa Rica, use of IPM on banana plantations has eliminated pesticides altogether in one region. In Africa, mealybugs were destroying up to 60 percent of the cassava crop (the staple food for 200 million people) before IPM was introduced in 1982. A tiny wasp that destroys mealybug eggs was discovered and now controls this pest in over 65 million ha (160 million acres) in 13 countries.

A successful IPM program that could serve as a model for other countries is found in Indonesia, where brown planthoppers developed resistance to virtually every insecticide and threatened the country's hard-won self-sufficiency in rice. In 1986, President Suharto banned 56 of 57 pesticides previously used in Indonesia and declared a crash program to educate farmers about IPM and the dangers of pesticide use. Researchers found that farmers were spraying their fields habitually—sometimes up to three times a week—regardless of whether fields were infested. By allowing natural predators to combat pests and spraying only when absolutely necessary with chemicals specific for planthoppers, Indonesian farmers using IPM have had higher yields than their neighbors using normal practices and they cut pesticide costs by 75 percent. In 1988, only two years after its initiation, the program was declared a success. It is being extended throughout the whole country. Since nearly half the people in the world depend on rice as their staple crop, this experiment could have important implications elsewhere (fig. 12.21).

REDUCING PESTICIDE EXPOSURE

The total numbers and amounts of different chemicals to which we are potentially exposed is overwhelming. The 650,000 metric tons of pesticides used in the United States every year contain about 600 active ingredients combined with more than 1200 presumably inactive carriers, solvents, preservatives, and other ingredients in about 25,000 commercial products. Less than 10 percent of the active pesticide ingredients have been subjected to a full battery of ten chronic health-effect tests, and studies of the inactive ingredients have started only recently (fig. 12.22). Of the 321 pesticides screened so far, the EPA reports that 146 are probable human carcinogens. Since 1972, only 40 pesticides have been banned. Critics charge that this slow pace puts all of us at risk and that assessment and regulation must proceed more quickly.

Regulating Pesticides

Three federal agencies share responsibility for regulating pesticides used in food production in the United States: the Environmental Protection Agency (EPA), the Food and Drug Administration (FDA), and the Department of Agriculture (USDA). The EPA regulates the sale and use of pesticides under the Federal Insecticide, Fungicide, and Rodenticide Act (FIFRA), which mandates the "registration" (licensing) of all pesticide products. Based on scientific studies, the EPA determines which pesticides will not pose significant risks to human health or the environment when used according to label directions. Under the Federal Food, Drug, and Cosmetic Act (FFDCA), the EPA also sets "tolerance levels" or limits for the amount of pesticide residues that lawfully may remain in or on foods marketed in the United States, whether grown domestically or imported from abroad. The FDA and USDA enforce pesticide use and tolerance levels set by the EPA. These agencies can seize and destroy food shipments found to contain pesticide residues in violation of limits set by the EPA.

In Canada, the Pest Control Products Act and Regulations, administered by Agriculture Canada, performs similar functions. Individual Canadian provinces also set local standards. The Ontario Ministry of Agriculture and Food, for instance, introduced a comprehensive program in 1988 that sets a goal of 50 percent reduction in pesticide use by the year 2002. Nonchemical pest control methods, on-farm education, and biotechnology and the development of pest-resistant crop varieties are all planned to play a role in this program. Savings on chemical costs are expected to be about $100 million per year.

A controversial amendment to the U.S. FFDCA called the **Delaney Clause** was added in 1958, prohibiting the addition of any known cancer-causing agent to processed foods, drugs, or cosmetics.

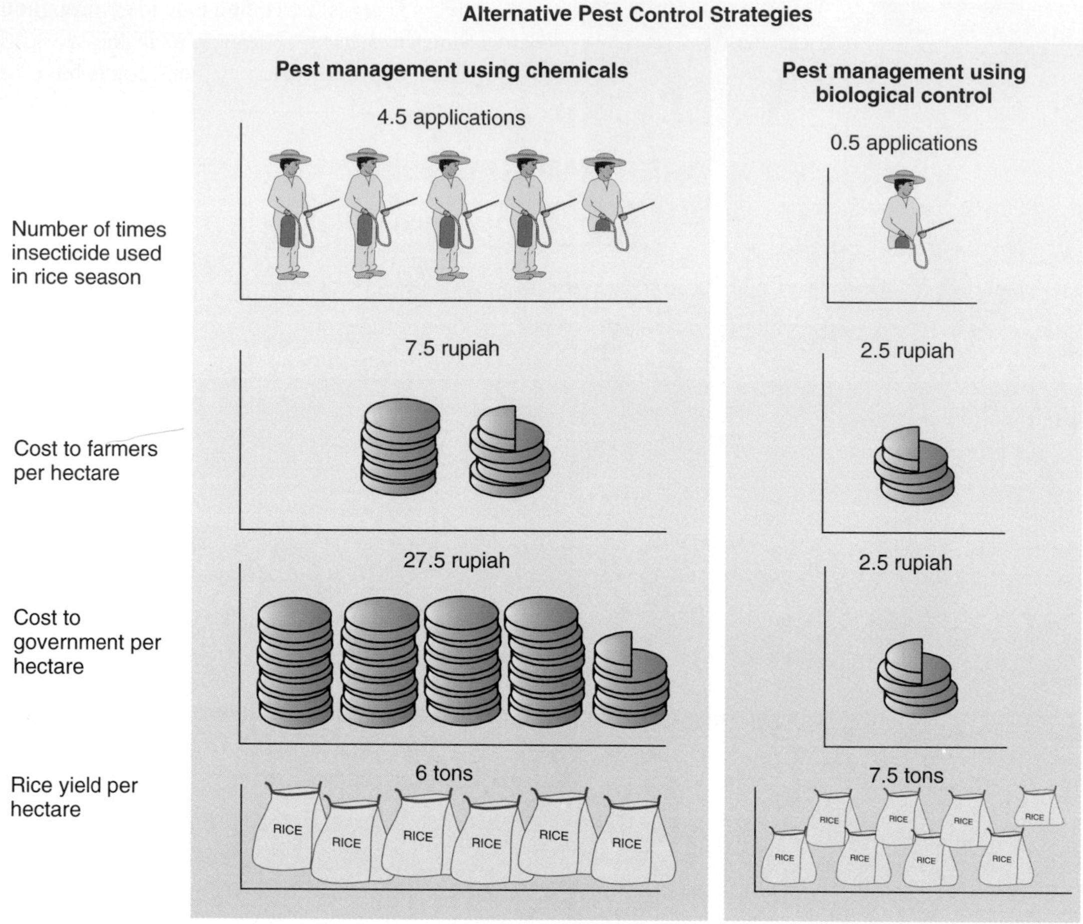

FIGURE 12.21 Indonesia has one of the world's most successful integrated pest management (IPM) programs. Switching from toxic chemicals to natural pest predators has saved money while also increasing rice production.
Source: Data from Tolba, et al., *World Environment, 1972–1992,* p. 307, Chapman & Hall, 1992 United Nations Environment Programme.

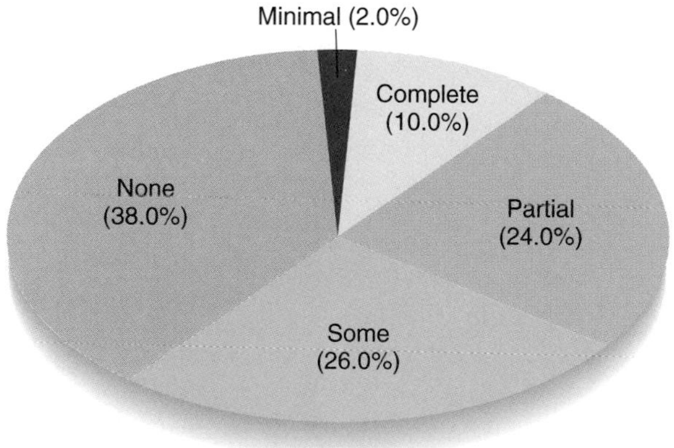

FIGURE 12.22 Level of testing for health effects of 3350 active and inert pesticide ingredients. Complete testing includes a battery of ten tests for both chronic and acute toxicity and genetic effects.
Source: Data from National Academy of Sciences.

Processed foods include those that are canned, preserved, dehydrated, or frozen. Although this amendment was based on noble instincts, its absolute zero-risk approach produced some anomalous results. Because it is impossible to test every one of the millions of natural and synthetic chemicals in our food, the EPA took the position that additives in use before 1954 and "generally recognized as safe" at that time were permissible. Furthermore, pesticides registered after 1978 were subject to the Delaney Clause while those registered earlier had to meet less stringent standards. One bizarre effect was that you could consume all the coffee—which has about 20 known carcinogens—you want, but, in theory, a food processor couldn't add a single drop of coffee to processed food.

Since the Delaney Clause went into effect nearly 40 years ago, however, scientists have developed exquisitely sensitive techniques for measuring chemicals. Today some tests can detect one part per quintillion—roughly equal to a tablespoon of liquid in all the Great Lakes combined. At this level of sensitivity, probably everything we eat contains some carcinogens. The question becomes not one of zero risk but rather which risks are worth worrying about.

what can you do?

Food Safety Tips

To reduce the amount of pesticide residues and other toxic chemicals in your diet, follow these simple rules.

- Wash and scrub all fresh fruits and vegetables thoroughly under running water.
- Peel fruits and vegetables when possible. Throw away the outer leaves of leafy vegetables such as lettuce and cabbage.
- Store food carefully so it doesn't get moldy or pick up contaminants from other foods. Use food as soon as possible to ensure freshness.
- Cook or bake foods that you suspect have been treated with pesticides to break down chemical residues.
- Trim the fat from meat, chicken, and fish. Eat lower on the food chain where possible to reduce bioaccumulated chemicals.
- Don't pick and eat berries or other wild foods that grow on the edges of roadsides where pesticides may have been sprayed.
- Grow your own fruits and vegetables without using pesticides or with minimal use of dangerous chemicals.
- Ask for organically grown food at your local grocery store or shop at a farmer's market or co-op where you can get such food.

Concern about the possible effects of pesticide residues in our diet has led many consumers to request organically grown food that is produced without pesticides. Farmers' markets, such as the one shown here, can be a good place to find locally produced organic food.

In 1996, Congress passed, and the president signed, legislation amending both FFDCA and FIFRA to replace Delaney's zero risk requirement with a less stringent standard of reasonable certainty of no harm, defined as no more than one case of cancer for every million people exposed over a lifetime. Pesticide risks are to be weighed against benefits. Even the relaxed standard of one-in-a-million cancer risk can be exceeded if cancellation of a particular chemical would adversely affect consumers or disrupt the food supply. This new standard of acceptable risk applies to both raw and processed foods.

Industry and farm groups were pleased by this cost-benefit approach to risk assessment as well as the establishment of uniform state regulations on pesticides unless an individual state can show "compelling local conditions" and unreasonable dietary risks to its residents from national standards. Environmental and consumer groups were pleased by rules that pesticide registrations must be reviewed every 15 years. New pesticides that would replace more dangerous existing ones will receive expedited registration so they can be available sooner than they otherwise would be.

In 1999, following these new rules, the EPA banned the use of methyl parathion on all fruits and many vegetables and limited the quantity of azinphos methyl that can be used on foods common in children's diets, such as apples, peaches, and pears. Citing studies showing that infants and children are more susceptible than adults to toxic pesticides because their internal organs are still developing and their enzymatic, metabolic, and immune systems provide less protection, the EPA concluded that children need an extra measure of protection. The EPA also is studying hormone-disrupting chemicals and will probably regulate products on this basis in the near future.

It is wise to be aware of potential hazards from pesticides in our diet and environment, but it is important to keep these risks in perspective. While you may have heard the statement that 80 percent of all cancer is environmental, most of that risk comes from personal lifestyle choices such as smoking, unhealthy diets, and excess sun exposure. British epidemiologists Richard Doll and Richard Peto estimate that the cancer risk from commercial pesticides in food is "unimportant." Biochemist Bruce Ames agrees, claiming that 99.9 percent of all carcinogens to which we are exposed are natural.

A Personal Plan

Even though pesticide exposure may not be the greatest threat that most of us face, why not take reasonable steps to reduce your use of and exposure to toxic materials? There are many things that each of us can do to minimize pesticides in our food and environment.

- Plant ground cover in your yard that can compete successfully with weeds. Learn to appreciate a mixture of different plants rather than striving for a perfect, lush carpet of grass. What's wrong with a few dandelions, anyway?
- Install or repair screens on doors and windows to keep out insects. Caulk around windows, doors, and baseboards to seal routes of entry for ants, roaches, mice, and other pests. Put a few bay leaves in containers of flour or cereal to discourage flour weevils.

- Clean up spilled food and empty garbage regularly to eliminate food for ants or roaches. Keep food in tightly sealed containers. Sprinkle borax around drains and ducts where pests may be lurking. Use sticky flypaper to catch flies rather than spray dangerous chemicals inside your house or apartment.
- Get rid of aphids, scales, mites, and other houseplant pests by washing leaves and stems individually with rubbing alcohol, or spray plants with a dilute solution of dish soap in water. Simply washing vegetation with pure water will remove many pests and improve the health and appearance of your plants.
- A saucer of stale beer in your garden will attract and drown slugs. If you have only a few worms or bugs, pick them off by hand. Purchase some predators such as ladybugs or praying mantises to protect garden plants. If you have a lot of bugs in your house or apartment, a praying mantis can make an interesting house pet.
- Drain stagnant water in or near your yard that might serve as breeding sites for mosquitoes. Pick up dead branches, fallen leaves and other litter that provide a haven for termites and other pests. Put up houses for bug-eating purple martins or bats.
- Accept slightly blemished fruits and vegetables. If you garden, trim away bad parts or give up a portion of your crop rather than saturate the environment with dangerous chemicals. Interplant insect-repellent plants such as marigolds, basil, peppermint, or garlic together with your sensitive garden plants.
- If you must use toxic chemical pesticides, use them in the smallest possible amounts and apply them only where and when it is necessary. Rather than spray chemicals, use traps or applicators that limit dispersal into the environment.
- Read *Organic Gardening and Farming* or some other magazine that reports regularly on environmental issues and alternative approaches to pest control.

Remember that being "natural" or organic does not necessarily make a chemical good, nor does being synthetic necessarily make one bad. Some natural products such as nicotine are highly toxic. Others such as mercury and lead are just as likely to bioconcentrate and persist in the environment as are synthetic compounds.

SUMMARY

Biological pests are organisms that reduce the availability, quality, or value of resources useful to humans. Pesticides are chemicals intended to kill or drive away pests. Many nonchemical pest-control approaches perform these same functions more safely and cheaply than do toxic chemicals. Of the millions of species in the world, only about one hundred kinds of animals, plants, fungi, and microbes cause most crop damage. Many beneficial organisms are injured by indiscriminate pesticide use, including natural predators that serve a valuable function in keeping potential pests under control.

Humans have probably always known of ways to protect themselves from annoying creatures, but our war against pests entered a new phase with the invention of synthetic organic chemicals such as DDT. These chemicals have brought several important benefits, including increased crop production and control of disease-causing organisms. Indiscriminate and profligate pesticide use also has caused many problems, such as killing nontarget species, creating new pests of organisms that were previously not a problem, and causing widespread pesticide resistance among pest species. Often highly persistent and mobile in the environment, many pesticides move through air, water, and soil and bioaccumulate or bioconcentrate in food chains causing serious ecological and human health problems.

A number of good alternatives offer ways to reduce our dependence on dangerous chemical pesticides. Among these are behavioral changes such as crop rotation, cover crops, mechanical cultivation, and planting mixed poly-cultures rather than vast monoculture fields. Consumers may have to learn to accept less than perfect fruits and vegetables. Biological controls such as insect predators, pathogens, or natural poisons specific for a particular pest can help reduce chemical use. Genetic breeding and biotechnology can produce pest-resistant crop and livestock strains, as well. Integrated pest management (IPM) combines all of these alternative methods together with judicious use of synthetic pesticides under precisely controlled conditions.

Regulating pesticide use is a controversial subject. Many people fear that we are exposed to far too many dangerous chemicals. Industry claims that it could not do business without these materials. The Delaney Clause, an amendment to the Federal Food, Drug, and Cosmetic Act, prohibits willful addition of any known carcinogen to foods, drugs, or cosmetics. Although noble in intention, this amendment has become difficult to enforce as we find that many commonly used materials are carcinogens. Should we weaken the law and allow some carcinogens as long as the risk is "negligible"?

Many of the procedures and approaches suggested for agriculture and industry also work at home to protect us from pests and toxic chemicals alike. By using a little common sense, we can have a healthier diet, lifestyle, and environment.

QUESTIONS FOR REVIEW

1. What is a pest and what are pesticides? What is the difference between a biocide, a herbicide, an insecticide, and a fungicide?
2. How much pesticide is used worldwide and in your country? In your country, which of the general categories of use and which specific type accounts for the greatest use? Has use been increasing or decreasing in recent years?
3. What is DDT and why was it considered a "magic bullet"? What are its benefits and disadvantages?
4. Describe fumigants, botanicals, chlorinated hydrocarbons, organophosphates, carbamates, and microbial pesticides.
5. Explain why pests often resurge or rebound after treatment with pesticides and how they become pesticide resistant. What is a pesticide treadmill?
6. Identify three major categories of alternatives to synthetic pesticides and describe, briefly, how each one works.
7. How did Australia fight prickly pear cactus? How did Florida eradicate screwworms?
8. What is the Delaney Clause and why is it controversial?
9. List nine things you could do to reduce pesticide use in your home.
10. List eight things you could do to reduce your dietary exposure to pesticides.

QUESTIONS FOR CRITICAL THINKING

1. In retrospect, do you think Paul Müller should have received a Nobel prize for discovering the insecticidal properties of DDT?
2. If you were a public health official in a country in which malaria, filariasis, or onchocerciasis were rampant, would you spray DDT to eradicate vector organisms? Would you spray it in your own house?
3. Pesticide treadmill, "dirty dozen" pesticides, and environmental estrogens are all highly emotional terms. Why would some people choose to use or not use these terms? Can you suggest alternative terms for the same phenomena that convey different values?
4. Many farmworkers who suffer from pesticide poisoning are migrants or minorities. Is this evidence of environmental racism? What evidence would you look for to determine whether environmental justice is being served?
5. Suppose that a developing country believes that it needs a pesticide banned in the United States or Canada to feed or protect the health of its people. Are we right to refuse to sell that pesticide?
6. How much extra would you pay for organically grown food? How would you define organic in this context?
7. If alternative pest control methods are so effective and so much safer, why aren't farmers and consumers adopting them more rapidly?
8. If you were a member of Congress, would you vote to repeal the Delaney Clause to the FFDCA? Why or why not?
9. What would you personally consider a "negligible" risk? Would you eat grapes if you knew they had a measurable amount of some pesticide? How small would the amount have to be?
10. Why do you suppose that the Thompsons' neighbors haven't adopted regenerative agriculture?

KEY TERMS

biocide 263
biological controls 272
biological pests 262
Delaney Clause 275
economic thresholds 274
fungicides 263
herbicides 263
insecticides 263
integrated pest management (IPM) 274
pesticide 262
pesticide treadmill 267
pest resurgence 267
persistent organic pollutants (POPs) 270

ADDITIONAL INFORMATION ON THE INTERNET

Visit our website at http://www.mhhe.com/environmentalscience for specific information about each of the topics below:

Crop Production and Pest Control: US Department of Agriculture
Biological Pest Control
Gypsy Moth
National Agriculture Pest Information Network
National Parks Service IPM Manual. Information on everything from thistles to ticks, ants to aphids
Pesticide Education Resources, University of Nebraska
UC Statewide Integrated Pest Management Program
Biocontrol Network (buy your own biocontrol products—a commercial site)
Biological Control: A Guide to Natural Enemies in Nature
Urban Integrated Pest Management, North Carolina Cooperative Extension Service
National Integrated Pest Management Network, news, stories, and more
Endocrine Disrupters

PROFILE

Conservation Fund Raiser

Arnold Phifer

Arnold (Arnie) Phifer, is Director of Development for Bat Conservation International (BCI), a non-profit organization that works to preserve bats and bat habitats around the world. With 14,000 members in 68 countries, BCI works with landowners, governmental agencies, private donors, foundations, researchers, and a host of others to protect the most misunderstood mammal on earth. Trying to get people to understand that bats aren't scary or threatening, but play vital ecological roles, and need to be protected rather than persecuted, occupies the 30 staff members—about half biologists and half support and administrative personnel—at BCI. Arnie's main job is to help raise the funds to keep this operation going.

Like many people in environmental careers, Arnie has no direct training in wildlife or ecology. As an undergraduate at Notre Dame, he majored in history. For his Masters degree in geography at the University of Massachusetts, Arnie focused on natural hazards and disasters such as earthquakes. Classes in biogeography gave him some science background that has proven useful in his current job. He also feels that a strong liberal arts background is valuable in providing skills in writing, communication, and understanding people, all of which are important in any job. A part-time position fund-raising for a library introduced him to the development world that has become his current career. He says that experience proved to be much like an internship, teaching him practical skills that opened other doors in ways he never anticipated.

Arnie's work schedule is highly varied. One day he might be working with biologists to translate scientific knowledge into a story to help people understand bats. Another day, he might be visiting field sites and catching bats for population and ecological studies. Later, he might be in the boardroom of a big corporation or foundation, making a pitch for funding. A big part of his job is writing grant proposals, supervising fund-raising campaigns, meeting with donors, talking to landowners, and other activities that support BCI projects. He also spends time teaching the financial or business side of the organization—paperwork, auditing, documentation—to the field scientists.

Located in Austin, Texas, BCI is in a perfect place to educate the public about bats. The Congress Avenue Bridge across Town Lake in the heart of downtown Austin has tiny spaces between structural elements that happen to be just right for bats to roost during the day. More than 1.5 million bats live under the bridge during the summer. When they emerge at twilight like a swirling cloud of smoke, it's an awesome sight. Bat watching has become a major tourist attraction for Austin, and standing on the bridge talking with visitors is an excellent way to further bat conservation.

When asked what he likes best about his job, Arnie replies that he likes working with people and having a foot in both science and public relations. He enjoys the chance to travel to see bat habitats and learn about the ecology and politics of these fascinating creatures. What does he dislike? The mounds of paperwork necessary to keep track of funds and to report progress can get tiring. What skills are essential for this kind of work? Arnie says that being able to work effectively with scientists, donors, and the public takes good communication and an ability to understand different perspectives. His job requires multitasking, and aptitude for detail, and a wide knowledge base.

What advice would he give to students who might like to follow a similar path? Arnie suggests students develop skills in writing and communication, get a good background in biology, work in an internship or volunteer job to get practical experience, and above all, pick a field that you care about so you can convince people that your project is important. Doing something you enjoy brings great benefits both in how you do your job and how you feel about it. Arnie says, "I love coming to work and know I'm doing something good for the world."

CHAPTER 13

Biodiversity

The first rule of intelligent tinkering is to save all the pieces.

—Aldo Leopold—

OBJECTIVES

After studying this chapter, you should be able to:

- define *biodiversity* and *species*.
- report on the total number and relative distribution of living species on the earth.
- summarize some of the benefits we derive from biodiversity.
- describe the ways humans cause biodiversity losses.
- evaluate the effectiveness of the Endangered Species Act and CITES in protecting endangered species.
- understand how gap analysis, ecosystem management, and captive breeding can contribute to preserving biological resources.
- propose ways we could protect endangered habitats and communities through large-scale, long-range, comprehensive planning.

ESSENTIAL STUDY PARTNER

The Essential Study Partner (ESP) CD-ROM that accompanies this textbook contains additional information on the following topics:

- biological species concept and biodiversity: living resources: biodiversity: species
- megadiversity and conservation areas: living resources: biodiversity: species
- extinction, habitat destruction, wildlife trade and overexploitation: living resources: biodiversity: extinction
- endangered, threatened, and vulnerable species: living resources: biodiversity: endangered species
- exotic species: living resources: biodiversity: extinction
- conservation legislation, recovery plans, zoos, and captive breeding: living resources: biodiversity: endangered species

Above: Big cats, like this South American jaguar, are endangered everywhere in the world. Will there be room for both humans and big, fierce animals?

Too Many Geese

Rising and then settling again in swirling clouds of glittering white, millions of snow geese move from field to field every winter, gleaning rice and other grain along the Gulf Coast of Texas and Louisiana (fig. 13.1). While these vast flocks delight hunters and bird watchers, they spell trouble for their summer nesting grounds along Canada's arctic coast (fig. 13.2). Beautiful snow-white birds with black wing-tips and pink bills and feet (a dark variant called the blue goose, has a blue-gray body and white head), snow goose populations have exploded over the past 30 years, especially those that follow the mid-continent flyway along the Mississippi River in their annual north and south migration. In the 1960s, the mid-continent population of "light" geese—a collective term for lesser snow geese and a smaller relative known as Ross' goose—was estimated to be around 800,000. Three decades later, that number had grown to between 5 and 10 million birds.

Biologists believe the rapid spread of soybean and rice farming in gulf coast wintering grounds is a large part of the explosive growth of snow goose populations. Millions of hectares that once were coastal marshes are now farm fields where mechanical harvesting leaves a bounty of grain on the ground. After a few months of feasting on this banquet, the geese are fat and healthy as they begin their annual spring migration north. Wildlife refuges and restored prairie potholes along migration routes provide safe places for the geese to rest and feed. And conversion of pastures to corn, barley, oats, rye, and wheat in northern states and provinces offers a valuable source of high-energy food during the migration. As a result, natural mortality has decreased and the population has climbed 5 to 7 percent per year for several decades.

This huge number of birds is having a devastating effect on the fragile tundra ecosystem in the far north where they breed and raise their young. Snow geese feed by grubbing, or pulling up and devouring plant roots. When the mid-continent population was less than a million birds spread out along the vast coastline from Hudson Bay to the Northwest Territories, plant-life on the salt marshes where they feed could keep up with summer goose grazing. But as their numbers grew, the birds began to eat more than could be replenished during the brief arctic growing season. Now, where the cord grasses and sedges preferred by geese once flourished, the ground is bare, salt-encrusted, and pitted with holes. Biologists question whether the feeding grounds will ever recover, because without plants to remove salt, the soil has become too saline for seeds to germinate.

A 1996 survey of 2000 km (1250 mi) of snow goose nesting area along the west coast of Hudson Bay estimated that 35 percent of the habitat was destroyed, another 30 percent was severely damaged, and the remaining 35 percent was seriously overgrazed. At least 30 other bird species dependent on the same habitat already have been adversely affected, including semipalmated sandpipers, red-necked phalaropes, dowitchers, Hudsonian godwits, wimbrels, stilt sandpipers, yellow rails, American wigeons, northern shovelers, oldsquaws, red-breasted mergansers, parasitic jaegers, and Lapland longspurs.

What can be done about this situation? One option would be to just wait for nature to take its course. Eventually, the geese will starve to death and their numbers will fall. But in the meantime, habitat will continue to degrade and many other species, some already rare, could suffer irreparably. Both Canadian and American wildlife services have taken steps to reduce the burgeoning goose population, including harassment to prevent nesting and "egging," or destruction of eggs. The nesting grounds are so spread out and difficult to access, however, that these activities aren't very effective. Hunting seasons have been expanded—including a spring season for northern states—and other rule changes such as reduced bag limits, use of electronic calls, more shells in each shotgun, and perhaps even hunting over bait have been proposed. It is hoped that the annual harvest of mid-continent light geese might be doubled or tripled with a goal of reducing the population to a sustainable level of about 1.5 million birds in a few years. Taking 2.5 million birds per year would represent a 25 percent annual harvest, up from the current level of about 8 percent. Increasing the "take" will be difficult, however, because snow geese are smart birds and notoriously hard to hunt.

This case illustrates several important issues of wildlife and biodiversity. Humans have changed the environment in ways that make some species overabundant while others have decreased or been driven into extinction. Whole ecosystems, even entire biomes, can be affected. In some cases the results are plain to see; in others they occur over such a wide area or in such remote places that we are unaware of what we have done until it is too late. In this chapter, we will look at some important types of biodiversity, how human actions disturb it, and ways we can protect and restore wild species and ecosystems on which we all rely.

FIGURE 13.1 Great flocks of lesser snow geese benefit from human-caused changes in their wintering areas and migration routes but are overwhelming the fragile arctic tundra where they nest during the summer.

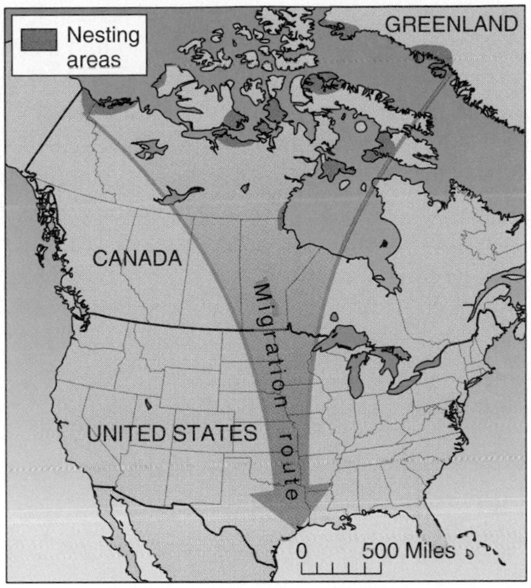

FIGURE 13.2 Lesser snow geese nest in fragile arctic tundra in Canada and Greenland. Rapidly growing goose populations are destroying vegetation in nesting areas and threatening other species that depend on this habitat.

BIODIVERSITY AND THE SPECIES CONCEPT

As far as we know, our planet is the only place in the universe that supports life, yet there are very few places on the earth that are not home to some kind of organism. From the driest desert to the dripping rainforests, from the highest mountain peaks to the deepest ocean trenches, life occurs in a marvelous spectrum of sizes, colors, shapes, life cycles, and interrelationships. Think for a moment how remarkable, varied, abundant, and important the other living creatures are with whom we share this planet. And consider how our lives would be impoverished if this biological diversity diminishes.

What Is Biodiversity?

Previous chapters of this book have described some of the fascinating varieties of organisms and complex ecological relationships that give the biosphere its unique, productive characteristics. Three kinds of **biodiversity** are essential to preserve these ecological systems: (1) *genetic diversity* is a measure of the variety of different versions of the same genes within individual species; (2) *species diversity* describes the number of different kinds of organisms within individual communities or ecosystems; and (3) *ecological diversity* assesses the richness and complexity of a biological community, including the number of niches, trophic levels, and ecological processes that capture energy, sustain food webs, and recycle materials within this system.

What Are Species?

As you can see from the preceding discussion, species occupy a crucial position in the concept of biodiversity, but what, exactly, do we mean by the term? In chapter 3, we defined species as all the organisms of the same kind able to breed in nature and produce live, fertile offspring. Underlying this commonly used definition is the idea that reproductive isolation caused by geography, physiology, or behavior prevents groups of otherwise similar organisms from exchanging genes, and therefore, gives them separate identities and evolutionary histories.

There are problems with species definitions based on reproductive isolation. Mating between species (hybridization) occurs in nature and may produce fertile offspring. Furthermore, investigators often cannot determine whether two groups that live in different places are capable of interbreeding. Sometimes, in fact, the only specimens available for study are dead. Species identification, therefore, is often based on morphological characteristics such as size, shape, color, skeletal structure, or on either chemical or genetic traits.

Determining how much difference is necessary before two similar groups of organisms can be considered separate species is highly subjective and taxonomists (scientists who study classification) debate which organisms are most closely related and which should be classified as separate species. Often these debates pit "lumpers" against "splitters." Given the same group of a 100 closely related organisms, one scientist might split them into 50 or a 100 different species, while another might lump them together into only 10 or 20 species.

DNA sequencing technology is revolutionizing biological taxonomy. Some groups that were thought to be far apart now appear to be closely related, while others may be moved into entirely different families or even orders. Understanding molecular evolution not only helps us see how species originate, but also helps us evaluate genetic diversity.

How Many Species Are There?

At the end of the great exploration era of the nineteenth century, some scientists confidently declared that every important kind of living thing on earth would soon be found and named. Most of those explorations focused on charismatic species such as birds and mammals. Recent studies of less conspicuous organisms such as insects and fungi suggest that millions of new species and varieties remain to be studied scientifically.

We now believe that the 1.4 million species presently known (table 13.1) represent only a small fraction of the total number that exist. Based on the rate of new discoveries by research expeditions—especially in the tropics—taxonomists estimate that there may be somewhere between 3 million and 50 million different species alive today. In fact, there may be 30 million species of tropical insects alone (fig. 13.3). About 70 percent of all known species are invertebrates (animals without backbones such as insects, sponges, clams, worms, etc.). This group probably makes up the vast majority of organisms yet to be discovered and may constitute

TABLE 13.1 Approximate Numbers of Known Living Species by Taxonomic Group

Bacteria and cyanobacteria	5000
Protozoa (single-celled animals)	31,000
Algae (single-celled plants)	27,000
Fungi (molds, mushrooms)	45,000
Multicellular plants	250,000
Sponges	5000
Jellyfish, corals, anemones	10,000
Flatworms (tapeworms, flukes)	12,000
Roundworms (nematodes, hookworms)	12,000
Earthworms and leeches	12,000
Clams, snails, slugs, squids, octopuses	70,000
Insects	750,000
Mites, ticks, spiders, crabs, shrimp, centipedes, other noninsect arthropods	120,000
Starfish, sea urchins	6000
Fish and sharks	22,000
Amphibians	4000
Reptiles	6000
Birds	9000
Mammals	4000
Total	1,400,000

Source: World Resources Institute, 1999.

FIGURE 13.3 Insects make up more than half of all known species and may be more than 90 percent of all species on earth. There may be more than a million species of beetles alone.

95 percent of all species. What constitutes a species in bacteria and viruses is even less certain than for other organisms, but there are large numbers of physiologically or genetically distinct varieties of these organisms, and it seems likely that even more exist in nature that we have not yet discovered.

Of all the world's species, only 10 to 15 percent live in North America and Europe. It is a rare occurrence to find a new species of higher plant or animal in the developed countries of the world. By contrast, the centers of greatest biodiversity tend to be in the tropics, especially tropical rainforests and coral reefs (fig. 13.4). Many of the organisms in these megadiversity countries have never been studied by scientists. The Malaysian Peninsula, for instance, has at least 8000 species of flowering plants, while Britain, with an area twice as large, has only 1400 species. There may be more botanists in Britain than there are species of higher plants. South America, on the other hand, has fewer than 100 botanists to study perhaps 200,000 species of plants.

HOW DO WE BENEFIT FROM BIODIVERSITY?

We benefit from other organisms in many ways, some of which we don't appreciate until a particular species or community disappears. Even seemingly obscure and insignificant organisms can play irreplaceable roles in ecological systems or be the source of genes or drugs that someday may be indispensable.

Food

All of our food comes from other organisms. Many wild plant species could make important contributions to human food supplies either as they are or as a source of genetic material to improve domestic crops. Noted tropical ecologist Norman Meyers estimates that as many as 80,000 edible wild plant species could be utilized by humans. Villagers in Indonesia, for instance, are thought to use some 4000 native plant and animal species for food, medicine, and other valuable products. Few of these species have been explored for possible domestication or more widespread cultivation. A 1975 study by the National Academy of Science (U.S.) found that Indonesia has 250 edible fruits, only 43 of which have been cultivated widely (fig. 13.5).

Unfortunately, overgrazing, forest clearing, conversion of natural landscapes to agriculture, and other forms of human disturbance are destroying potentially valuable food species and the wild ancestors of our domestic crops before they can be identified and their genes can be preserved. Later in this chapter, we will look at some of the programs underway to find useful wild species and preserve them in gene banks, botanical gardens, zoos, and nature preserves.

Drugs and Medicines

Living organisms provide us with many useful drugs and medicines (table 13.2). More than half of all prescriptions contain some natural products. The United Nations Development Programme estimates the value of pharmaceutical products derived from developing world plants, animals, and microbes to be more than $30 billion per year.

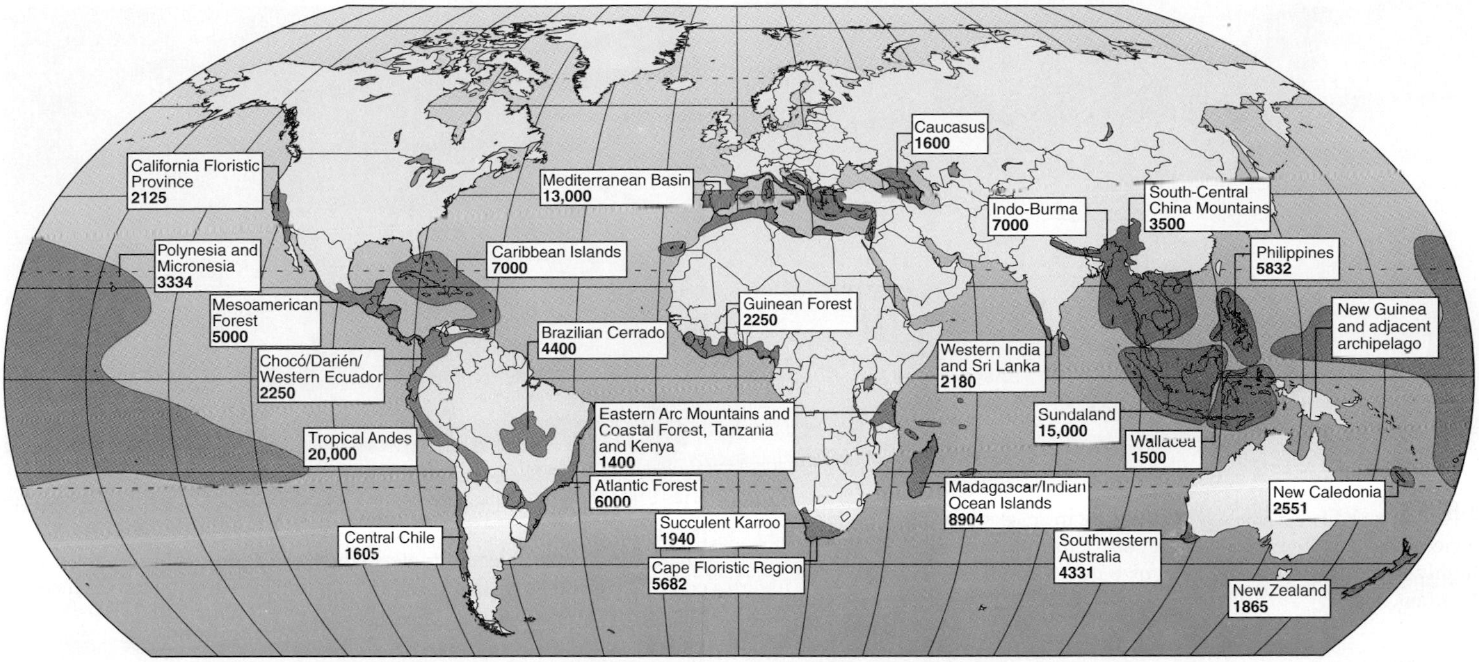

FIGURE 13.4 Biodiversity "hot spots" identified by Conservation International tend to be in tropical or Mediterranean climates and on islands, coastlines, or mountains where many habitats exist and physical barriers encourage speciation.

FIGURE 13.5 Mangosteens from Indonesia have been called the world's best-tasting fruit, but they are practically unknown beyond the tropical countries where they grow naturally. There may be thousands of other traditional crops and wild food resources that could be equally valuable but are threatened by extinction.

Indigenous communities that have protected and nurtured the biodiversity on which these products are based are rarely acknowledged—much less compensated—for the resources extracted from them. Many consider this expropriation "biopiracy" and call for royalties to be paid for folk knowledge and natural assets.

TABLE 13.2 Some Natural Medicinal Products

PRODUCT	SOURCE	USE
Penicillin	Fungus	Antibiotic
Bacitracin	Bacterium	Antibiotic
Tetracycline	Bacterium	Antibiotic
Erythromycin	Bacterium	Antibiotic
Digitalis	Foxglove	Heart stimulant
Quinine	Chincona bark	Malaria treatment
Diosgenin	Mexican yam	Birth-control drug
Cortisone	Mexican yam	Anti-inflammation treatment
Cytarabine	Sponge	Leukemia cure
Vinblastine, vincristine	Periwinkle plant	Anticancer drugs
Reserpine	Rauwolfia	Hypertension drug
Bee venom	Bee	Arthritis relief
Allantoin	Blowfly larva	Wound healer
Morphine	Poppy	Analgesic

Consider the success story of vinblastine and vincristine. These anticancer alkaloids are derived from the Madagascar periwinkle (*Catharanthus roseus*) (fig. 13.6). They inhibit the growth of cancer cells and are very effective in treating certain kinds of cancer. Twenty years ago, before these drugs were introduced, childhood leukemias were invariably fatal. Now the remission rate

FIGURE 13.6 The rosy periwinkle from Madagascar provides anticancer drugs that now make childhood leukemias and Hodgkin's disease highly remissible. You may recognize it as a popular flowering ornamental plant.

FIGURE 13.7 Costa Rican taxonomists study insect collections as part of an ambitious project to identify and catalog all the species in this small, but highly diverse, tropical country. The knowledge gained may contribute toward valuable commercial products that will provide funds to help preserve biodiversity.

for some childhood leukemias is 99 percent. Hodgkin's disease was 98 percent fatal a few years ago, but is now only 40 percent fatal, thanks to these compounds. The total value of the periwinkle crop is roughly $15 million per year, although Madagascar gets little of those profits.

Pharmaceutical companies are actively prospecting for useful products in many tropical countries. Merck, the world's largest biomedical company, is paying $1 million to the Instituto Nacional de Biodiversidad (INBIO) of Costa Rica for plant, insect, and microbe samples to be screened for medicinal applications. INBIO, a public/private collaboration, trains native people as practical "parataxonimists" to locate and catalog all the native flora and fauna—between 500,000 and 1 million species—in Costa Rica (fig. 13.7). Selling data and specimens will finance scientific work and nature protection. This may be a good model both for scientific information gathering and as a way for developing countries to share in the profits from their native resources.

Ecological Benefits

Although technological optimists may dream of someday living in completely artificial environments, so far, human life is inextricably linked to ecological services provided by other organisms. Soil formation, waste disposal, air and water purification, nutrient cycling, solar energy absorption, and management of biogeochemical and hydrological cycles all depend on the biodiversity of life (chapters 3 and 8). Total value of these ecological services is at least $33 trillion per year, or more than double total world GNP. The earth's ecosystems represent the culmination of historic evolutionary processes of immense antiquity and majesty. They have resulted from billions of years of evolution under conditions that may never have occurred anywhere else in the universe. Nature maintains ecological processes at no cost to us, and represents a genetic library of information we could never reproduce.

There has been a great deal of controversy about the role of biodiversity in ecosystem stability. Mathematical models suggest that simple ecosystems can be just as stable and resilient as more complex ones. Field studies by Minnesota ecologist David Tillman have shown, however, that some diverse biological plant communities withstand environmental stress better and recover more quickly than those with fewer species.

Because we don't fully understand the complex interrelationships between organisms, we often are surprised and dismayed at the effects of removing seemingly insignificant members of biological communities. For instance, wild species provide a valuable but often unrecognized service in suppressing pests and disease-carrying organisms. It is estimated that 95 percent of the potential pests and disease-carrying organisms in the world are controlled by other species that prey upon them or compete with them in some way. We find out how valuable natural predators are when we try to control systems with synthetic chemicals, because broad-spectrum biocides kill both pests and natural predators. As a result, pest populations often surge to higher levels than before (chapter 12). By preserving natural areas and conserving wild species, we utilize the stabilizing diversity of nature that keeps pest organisms in balance.

Aesthetic and Cultural Benefits

The diversity of life on this planet brings us many aesthetic and cultural benefits, and cultural diversity is inextricably linked to biodiversity (chapter 1). Millions of people enjoy hunting, fishing, camping, hiking, wildlife watching, and other outdoor activities based on nature. These activities provide invigorating physical exercise and allow us to practice pioneer living skills. Contact with nature also can be psychologically and emotionally restorative. In

FIGURE 13.8 Birdwatching and other wildlife observation contribute more than $29 million each year to the U.S. economy.

some cultures, nature carries spiritual connotations, and a particular species or landscape may be inextricably linked to a sense of identity and meaning. Observing and protecting nature has religious or moral significance for many people.

Nature appreciation is economically important. The U.S. Fish and Wildlife Service estimates that Americans spend $104 billion every year on wildlife-related recreation (fig. 13.8). This compares to $81 billion spent each year on new automobiles. Forty percent of all adults enjoy wildlife, including 39 million who hunt or fish and 76 million who watch, feed, or photograph wildlife. Ecotourism can be a good form of sustainable economic development, but we have to be careful that we don't abuse the places and cultures we visit (see chapter 15).

For many people, the value of wildlife goes beyond the opportunity to shoot or photograph, or even see, a particular species. They argue that **existence value,** based on simply knowing that a species exists, is reason enough to protect and preserve it. We contribute to programs to save bald eagles, redwood trees, whooping cranes, whales, and a host of other rare and endangered organisms because we like to know they still exist somewhere, even if we may never have an opportunity to see them.

From a biocentric perspective (see chapter 2) species deserve to exist for their own right, regardless of their value to us. Many religious groups such as the North American Coalition for Christianity and Ecology, the Southern Baptist Convention, the Orthodox Church, and the Coalition on the Environment and Jewish Life, have called for protecting nature because it is God's creation. Calling for a "Biotic Bill of Rights," some religious congregations argue that humans have a duty to keep nature intact and functioning as we received it from the creator.

WHAT THREATENS BIODIVERSITY?

Extinction, the elimination of a species, is a normal process of the natural world. Species die out and are replaced by others, often their own descendants, as part of evolutionary change. In undisturbed ecosystems, the rate of extinction appears to be about one species lost every decade. In this century, however, human impacts on populations and ecosystems have accelerated that rate, causing hundreds or perhaps even thousands of species, subspecies, and varieties to become extinct every year. If present trends continue, we may destroy *millions* of kinds of plants, animals, and microbes in the next few decades. In this section, we will look at some ways we threaten biodiversity.

Natural Causes of Extinction

Studies of the fossil record suggest that more than 99 percent of all species that ever existed are now extinct. Most of those species were gone long before humans came on the scene. Species arise through processes of mutation and natural selection and disappear the same way (chapter 4). Evolution can proceed gradually over millions of years or may occur in large jumps when new organisms migrate into an area or environmental conditions change rapidly. New forms replace their own parents or drive out less well-adapted competitors. In a sense, species that are replaced by their descendants are not completely lost. The tiny *Hypohippus,* for instance, has been replaced by the much larger modern horse, but most of its genes probably still survive in its distant offspring.

Periodically, mass extinctions have wiped out vast numbers of species and even whole families (table 13.3). The best studied of these events occurred at the end of the Cretaceous period when dinosaurs disappeared, along with at least 50 percent of existing genera and 15 percent of marine animal families. An even greater disaster occurred at the end of the Permian period about 250 million years ago when two-thirds of all marine species and nearly half of all plant and animal families died out over a period of about 10,000 years—a short time by geological standards. Current theories suggest that these catastrophes were caused by climate changes, perhaps triggered when large asteroids struck the earth. Many ecologists worry that global climate change caused by our release of "greenhouse" gases in the atmosphere could have similarly catastrophic effects (see chapter 17).

TABLE 13.3 Mass Extinctions

HISTORIC PERIOD	TIME (BEFORE PRESENT)	EFFECTS
Ordovician	444 million	25% of all families extinct
Devonian	370 million	19% of all families extinct
Permian	250 million	54% of families, 90% of species extinct
Triassic	210 million	23% of families, 1/2 of species extinct
Cretaceous	65 million	17% of families, 1/2 of species extinct (including dinosaurs but not mammals)
Quaternary	Present	1/3 to 2/3 of all species extinct if present trends continue?

Human-Caused Reductions in Biodiversity

The rate at which species are disappearing appears to have increased dramatically over the last 150 years. Between A.D. 1600 and 1850, human activities appear to have been responsible for the extermination of two or three species per decade. By some estimates, we are now losing species at thousands of times natural rates. If present trends continue, biologist Paul Ehrlich warns, somewhere between one-third to two-thirds of all current species could be extinct by the middle of this century. This would be a biodiversity loss of geologic proportions. Conservation biologists call this the sixth mass extinction, but note that this time it's not asteroids or volcanoes, but human impacts that are responsible.

Habitat Destruction

The biggest reason for the current increase in extinctions is habitat loss (fig. 13.9). Habitat fragmentation divides populations into isolated groups that are vulnerable to catastrophic events. Very small populations may not have enough breeding adults to be viable even under normal circumstances (fig. 13.10). Destruction of forests, wetlands, and other biologically rich ecosystems around the world threatens to eliminate thousands or even millions of species in a human-caused mass extinction that could rival those of geologic history. By destroying habitat, we eliminate not only prominent species, but also many obscure ones of which we may not even be aware. For further discussion, see chapters 14 and 15.

Hunting and Fishing

Overharvesting is responsible for depletion or extinction of many species. A classic example is the extermination of the American passenger pigeon (*Ectopistes migratorius*). Even though it inhabited only eastern North America, 200 years ago this was the world's most abundant bird with a population of between 3 and 5 billion animals. It once accounted for about one-quarter of all birds in North America. In 1830, John James Audubon saw a single flock

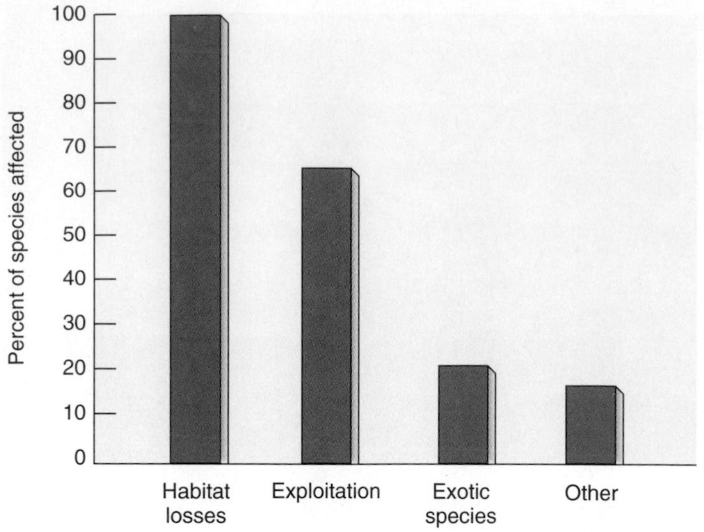

FIGURE 13.9 Threats to endangered mammals in Australasia and the Americas. For fish, exotic species represent a greater threat than shown here. For plants, trampling, grazing, collecting, flooding, and other human actions constitute serious dangers. Many species are endangered by multiple threats so that all these mammals are threatened by habitat loss but only 20 percent are affected by exotic species.
Source: Data from UNEP, 1994.

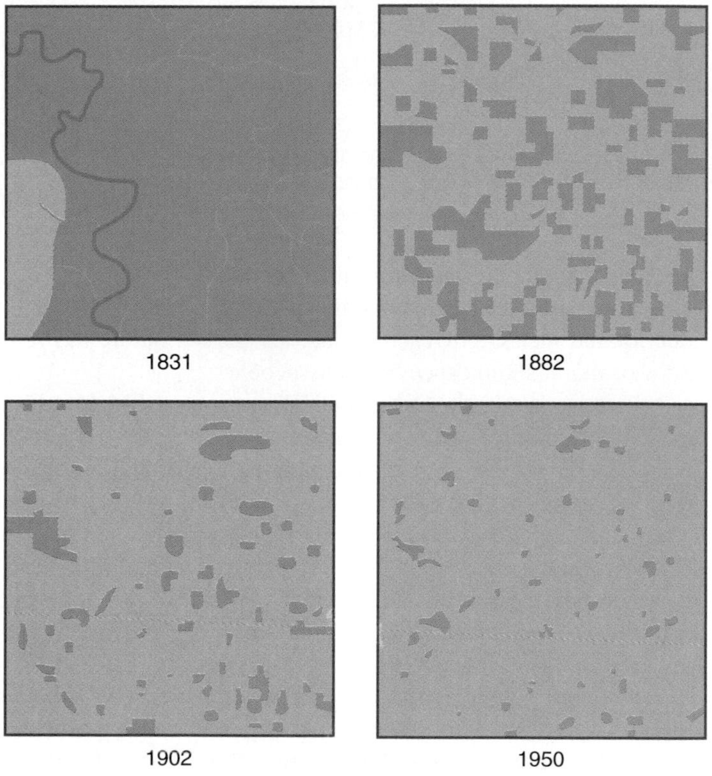

FIGURE 13.10 Decrease in wooded area of Cadiz Township in southern Wisconsin during European settlement. Shaded areas represent the amount of land in forest each year.

FIGURE 13.11 A stuffed passenger pigeon (*Ectopistes migratorius*) in a museum. The last member of this species died in the Cincinnati Zoo in 1914. Overhunting and habitat destruction caused their extinction.

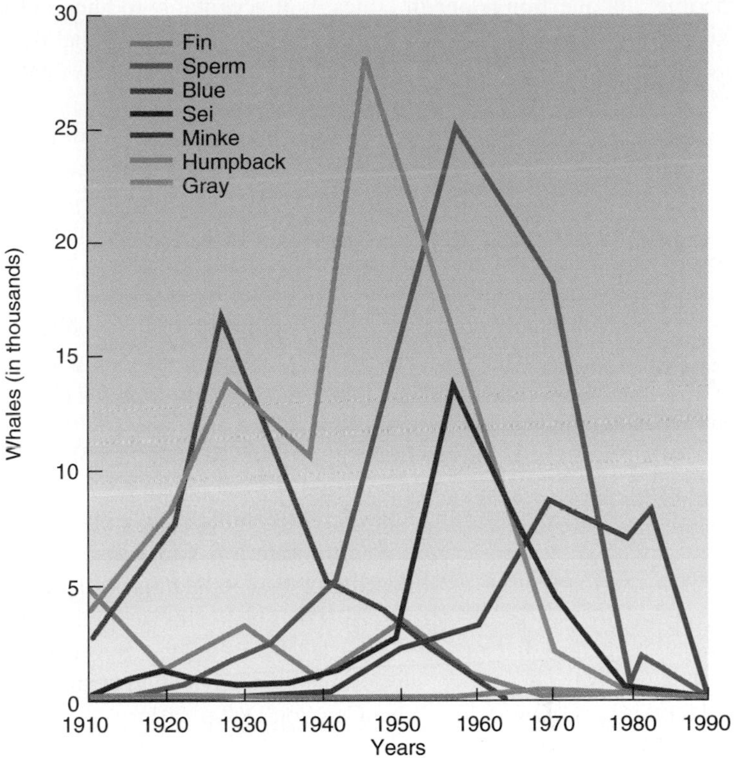

FIGURE 13.12 This world catch of whales shows a series of sharp peaks as each species in turn is hunted until commercially unprofitable. Some species such as minke, gray, and bowhead, have rebounded since commercial whaling stopped. Other species, such as blue and fin, remain rare and endangered.
Source: Data from *UNEP Environmental Data Report 1993–1994.*

of birds estimated to be ten miles wide, hundreds of miles long, and thought to contain perhaps a billion birds. In spite of this vast abundance, market hunting and habitat destruction caused the entire population to crash in only about 20 years between 1870 and 1890. The last known wild bird was shot in 1900 and the last existing passenger pigeon, a female named Martha, died in 1914 in the Cincinnati Zoo (fig. 13.11).

Some other well-known overhunting cases include the near extermination of the great whales and the American bison (buffalo). In 1850, some 60 million bison roamed the western plains. Many were killed only for their hides or tongues, leaving millions of carcasses to rot. Much of the bison's destruction was carried out by the U.S. Army to deprive native peoples who depended on bison for food, clothing, and shelter of these resources, thereby forcing them onto reservations. After 40 years, there were only about 150 wild bison left and another 250 in captivity.

We think that about 2.5 million great whales once inhabited the world's oceans. Their blubber was highly prized as a source of oil. At first, only the slowest whales such as the right or bowhead could be caught by human-powered dories and hand-thrown harpoons. Introduction of steamships and explosive harpoons in the nineteenth century, however, made it possible to catch and kill even the fastest whales and one species after another was driven into near extinction (fig. 13.12).

For the past decade, the International Whaling Commission has prohibited taking of great whales. This hunting ban seems to be having positive effects. With the exception of the North Atlantic right whales and southern blue whales, most populations appear to be recovering. There are now about 26,000 California gray whales, more than twice as many as 20 years ago and close to the prehunting number (see chapter 1). Around 3400 humpbacks now visit Hawaii each year compared to only one-third that many in the mid-1980s. Nearly one million minke whales—perhaps twice the pre-whaling number—now inhabit Arctic and Antarctic waters.

Iceland, Japan, and Norway argue that whale populations have rebounded enough to support limited hunting. They kill hundreds of whales each year mostly under the guise of "scientific research," although the meat and blubber of animals taken in these programs are still sold at a handsome profit.

It remains very difficult to count whales accurately. Whalers tend to come up with the highest estimates while environmental groups insist on more conservative numbers. For many people, the question is one of ethics. Is it acceptable to hunt and kill large, rare, intelligent, wild creatures like whales? What do you think? Where should we draw the line between what we eat and what we don't?

Fish stocks have been seriously depleted by overharvesting in many parts of the world. A huge increase in fishing fleet size and efficiency in recent years has led to a crash of many oceanic populations. Worldwide, 13 of 17 principal fishing zones are now reported to be commercially exhausted or in steep decline. At least three-quarters of all commercial oceanic species are overharvested. The collapse of the Atlantic cod fishery is symptomatic of the problem. Once present in such vast schools off the coast of New England, Nova Scotia, and Newfoundland that they could simply be dipped up in baskets, cod seemed inexhaustible. More than 300,000 metric tons per year were harvested in the 1980s, but then the population crashed. In 1992, Canada banned Atlantic cod fishing, putting 20,000 people out of work and closing down a $700-million-per-year industry. Trawlers have scoured the bottom, killing fish and destroying habitat, so that some biologists doubt the population will ever recover. Other species devastated by overharvesting include top predators such as sharks and billfish as well as slow-growing, deep-water species like orange roughy. Think about what you're eating if seafood is part of your diet (What Can You Do? p. 291).

Commercial Products and Live Specimens

In addition to harvesting wild species for food, we also obtain a variety of valuable commercial products from nature. Much of this represents sustainable harvest that may be a better option for some areas than agriculture or industry. Some forms of commercial exploitation are highly destructive, however, and represent a serious threat to certain rare species (fig. 13.13). Despite international bans on trade in products from endangered species, smuggling of furs, hides, horns, live specimens, and folk medicines amounts to millions of dollars each year.

Developing countries in Asia, Africa, and Latin America with the richest biodiversity in the world are the main sources of wild animals and animal products, while Europe, North America, and some of the wealthy Asian countries are the principal importers. Japan, Taiwan, and Hong Kong buy three-quarters of all cat and snake skins, for instance, while European countries buy a similar percentage of live birds. The United States imports 99 percent of all live cacti and 75 percent of all orchids sold each year.

The profits to be made in wildlife smuggling are enormous. Tiger or leopard fur coats can bring $100,000 in Japan or Europe. The population of African black rhinos dropped from approximately 100,000 in the 1960s to about 3000 in the 1980s because of a demand for their horns. In Asia, where it is prized for its supposed medicinal properties, powdered rhino horn fetches $28,000 per kg. In Yemen, a rhino horn dagger handle can sell for up to $1000.

Similarly, bird collectors will pay $10,000 for a rare hyacinth macaw from Brazil or up to $12,000 for a pair of golden-shouldered parakeets from Australia. An endangered albino python might bring $20,000 in Germany. The mortality rate in this live animal trade is enormous. It is generally estimated that 50 animals are caught or killed for every live animal that gets to market. Buyers, many of whom say they love animals, keep this trade going. If you collect rare plants, fish, or birds, make sure that what you buy is commercially grown, not wild-caught. If you travel abroad, don't buy products made from rare or endangered species. You not only contribute to their decline by doing so; you also expose yourself to confiscation and serious fines when you get home.

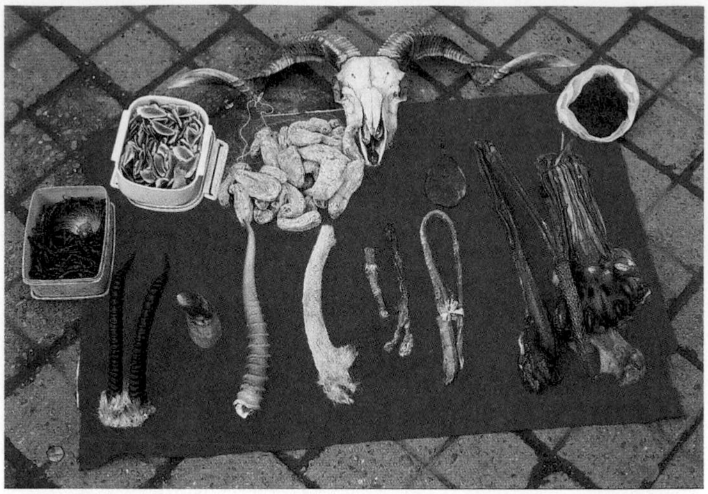

FIGURE 13.13 Parts from rare and endangered species for sale on the street in China. Use of animal products in traditional medicine and prestige diets is a major threat to many species.

Elephants are an important example of the problems and complexities of wildlife trade (fig. 13.14). In 1980, there were about 1.3 million African elephants; a decade later only half were left. Poachers killed more than 600,000 of these giant animals in the 1980s, mainly for their ivory tusks worth up to $100 per kilogram. An international ban on ivory trade in 1989 dried up the ivory trade and greatly reduced poaching.

In contrast to most of the continent, southern Africa has been very successful in conserving elephants. So successful, in fact, that herds have had to be thinned regularly to keep them from destroying their habitat. Namibia, Botswana, and Zimbabwe were allowed to sell some of their stockpiled ivory for conservation projects and to compensate citizens whose property and livelihoods are disrupted by marauding animals. Other countries opposed lifting the ban because of the fear that any ivory sales would rekindle poaching. In fact, in 1999, more than 350 elephants were killed in Zimbabwe alone by poachers who hope to slip ivory into legal sales routes. What do you think? Should we reward those nations that have protected their elephants or continue the ban?

Plants also are threatened by overharvesting. Wild ginseng has been nearly eliminated in many areas because of the Asian demand for roots that are used as an aphrodisiac and folk medicine. Cactus "rustlers" steal cacti by the ton from the American southwest and Mexico. With prices ranging as high as $1000 for rare specimens, it's not surprising that many are now endangered.

The trade in wild species for pets is an enormous business. Worldwide, some 5 million live birds are sold each year for pets, mostly in Europe and North America. Currently, pet traders import

what can you do?

Don't Eat Endangered Seafood

A good source of protein and rich with unsaturated oils, seafood is a healthy addition to your diet. But while some species of seafood are in plentiful supply, many ocean fish populations are seriously threatened by overharvesting. Which species should we avoid eating?

Shark	These top predators are slow growing and have low reproductive rates. Most commercial populations are overfished and declining rapidly.
Billfish	Swordfish, marlins, and sailfish are among the fastest and most highly adapted animals in the ocean. Like sharks, they mature late and reproduce slowly. Becoming rare in the Atlantic, their status elsewhere is unknown.
Shrimp	Plentiful in some regions but depleted in the Gulf of Mexico and the Gulf of California, shrimp harvest can result in 10 kg of unwanted species (including rare sea turtles) killed for every kilogram of shrimp caught. Shrimp farming causes serious land degradation and water pollution problems (see chapter 11).
Orange roughy	These deep-water fish live on seamounts off New Zealand and Australia, where they take 20 years to reach maturity. They can live more than 100 years, but excess harvesting has decimated populations in only a few decades.
Groupers	This large tribe of predominantly tropical fish has been over exploited in many places. Groupers become females as they age, so heavy fishing—which takes most of the old fish—can eliminate the breeding stock and prevent reproduction.
Groundfishes	Atlantic cod, haddock, pollack, yellowtail flounder, and monkfish once were abundant but now have been driven to the brink of extinction by overfishing. The collapse of these species probably ranks as the world's greatest fishery-management disaster. Pollack, which once were considered trash fish, are now the main harvest from U.S. waters. They make up most of the artificial crab, fish sticks, fish balls, and other generic fish products we eat.
Sea scallops	While farm-raised scallops and bay scallops from shallow estuaries generally remain plentiful, deep-sea Atlantic scallops are generally overfished. Dredging for scallops takes many other species and severely disrupts habitat.
Bluefin tuna	These magnificent predators—most of which are sold as sushi or high-price steaks—are severely depleted. Most canned "white" tuna is albacore, which is becoming rare. "Chunk light" tuna is yellowfin (caught by some fishers with nets that also trap dolphins) or skipjack, which remains plentiful. Ask your grocer what "dolphin-safe" means.
Red snapper	The popularity of this Gulf Coast fish in Cajun cooking caused its depletion in only a few years. Yellowtail snapper, however, appears to be plentiful.

What, then, can you eat, if you like seafood? Farm-raised catfish, tilapia, trout, and salmon have relatively little environmental impact. Most mackerel, Pacific pollack, dolphinfish (mahimahi), squids, crabs, and crayfish (crawdads) generally are in good shape. Herring and sardines have been overfished in some areas but remain abundant elsewhere. Wild freshwater species like bass, sunfish, pike, catfish, and carp are better managed than most ocean fish, but you have to be aware of local pollution problems such as mercury accumulation. In general, avoid the top predators both because they reproduce slowly and because they are more likely to store toxins. You don't have to give up eating fish entirely, but think about the species you consume.

(often illegally) into the United States some 2 million reptiles, 1 million amphibians and mammals, 500,000 birds, and 128 million tropical fish each year. About 75 percent of all saltwater tropical aquarium fish sold come from coral reefs of the Philippines and Indonesia.

Many of these fish are caught by divers using plastic squeeze bottles of cyanide to stun their prey (fig. 13.15). Far more fish die with this technique than are caught. Worst of all, it kills the coral animals that create the reef. A single diver can destroy all of the life on 200 square meters of reef in a day. Altogether, thousands of divers currently destroy about 50 km^2 of reefs each year. Net fishing would prevent this destruction, and it could be enforced if pet owners would insist on net-caught fish. More than half the world's coral reefs are potentially threatened by human activities, with up to 80 percent at risk in the most populated areas.

Predator and Pest Control

Some animal populations have been greatly reduced, or even deliberately exterminated, because they are regarded as dangerous to humans or livestock or because they compete with our use of resources. Every year, U.S. government animal control agents trap, poison, or shoot thousands of coyotes, bobcats, prairie dogs, and other species considered threats to people, domestic livestock, or crops. Many others are killed unintentionally by poisoned bait or misplaced traps or intentionally by private individuals for bounty or sport.

These animal control programs are controversial. Every year the Animal Damage Control Program spends about $20 million in federal and state funds to kill some 700,000 birds and mammals, about 100,000 of which are coyotes. Defenders of

FIGURE 13.14 Elephant populations have been decimated throughout the world. The biggest threat is from poachers who kill for their valuable ivory.

FIGURE 13.15 A diver uses cyanide to stun tropical fish being caught for the aquarium trade. Many fish are killed by the method itself, while others die later during shipment. Even worse is the fact that cyanide kills the coral reef itself.

wildlife regard this program as cruel, callous, and mostly ineffective in reducing livestock losses. Ranchers and trappers, on the other hand, argue that without predator control, western livestock operations would be uneconomical. Protecting flocks and herds with guard dogs or herders or keeping livestock out of areas that are home range of wild species would be a better solution (fig. 13.16), they believe.

Exotic Species Introductions

Exotic organisms—aliens introduced into habitats where they are not native—are one of the greatest threats to native biodiversity. Exotics can be thought of as biological pollution. Freed from the predators, parasites, pathogens, and competition that kept them in check in their native home, formerly mild-mannered species can turn into superaggressive "weedy" invaders in a new habitat. There are now more than 4500 alien species in the United States. A few of those causing most trouble include:

FIGURE 13.16 Sarah, a Hungarian Komondor guarding dog, protects a flock of goats from predators. Guarding dogs are not trained to herd sheep or goats; they merely stay with them to keep predators away. This not only saves livestock but avoids trapping, shooting, or poisoning the predators.

- Kudzu vine (*Pueraria lobata*) has blanketed large areas of the southeastern United States (fig. 13.17). Long cultivated in Japan for edible roots, medicines, and fibrous leaves and stems used for paper production, kudzu was introduced by the U.S. Soil Conservation Service in the 1930s to control erosion. Unfortunately, it succeeded too well. In the ideal conditions of its new home, kudzu can grow 18 to 30 m (60 to 100 ft) in a single season. Smothering everything in its path, it kills trees, pulls down utility lines, and causes millions of dollars in damage every year.
- Leafy spurge (*Euphorbia esula*), a relative of the familiar Christmas poinsettia, has spread across millions of hectares of the northern Great Plains since it was first introduced in the 1870s in the seeds of Mennonite farmers from Russia. A short, bushy, green weed with small yellow flowers, the entire plant—leaves, stems, roots, flowers, and seeds—is poisonous. It reduces the carrying capacity of western range by three-quarters when it invades because cattle can't eat it and it crowds out native grasses and forbs. North Dakota alone loses $7 million per year in forage and beef production due to spurge, and spends another $6 million annually in control programs.
- Purple loosestrife (*Lythrum salicaria*) grows in any wet soil. Originally cultivated by gardeners for its beautiful purple flower spikes, this tall wetland plant escaped into New England

FIGURE 13.17 Bioinvaders.

marshes about a century ago. Spreading rapidly across the Great Lakes, it now fills wetlands across much of the northern United States and southern Canada. Because it crowds out indigenous vegetation and has few native predators or symbionts, it tends to reduce biodiversity wherever it takes hold.

- Zebra mussels (*Dreissena polymorpha*) probably made their way from their home in the Caspian Sea to the Great Lakes in ballast water of transatlantic cargo ships, arriving sometime around 1985. Gluing themselves to any solid surface, zebra mussels reach enormous densities—up to 70,000 animals per square meter—covering fish spawning beds, smothering native mollusks, and clogging utility intake pipes. Found in all the Great Lakes, zebra mussels have recently been spotted in the Mississippi River and its tributaries. Public and private costs for zebra mussel removal now amount to some $400 million per year. On the good side, mussels have improved water clarity in Lake Erie at least fourfold by filtering out algae and particulates.

- The round goby (*Neogobius melanostomus*), a small, pugnacious freshwater fish from the Black Sea, probably also hitched a ride in ships coming into the Great Lakes. Devouring the eggs and fry of any species that shares their territory, these tough little fish drive off native gobies and baffle native game fish with their erratic swimming and ability to hide. The round goby's single, round pelvic fin distinguishes it from native species. One possible benefit is that round gobies voraciously devour zebra mussels.
- Asian long-horned beetles (*Anoplophora galabripennis*) are one of the most recent exotic threats in the United States. Probably transported in imported logs or wooden packing crates from Asia, these wood-eating insects were first spotted in Amityville, N.Y., in 1996. The beetle larvae burrow into living tree trunks where they cut off sap flow between leaves and roots. Partial to oaks, maples, poplars, and other hardwood species, these shiny black bugs with distinctive white spots and enormous feelers could cause billions of dollars in damage if they spread widely. Chicago suburbs cut and burned 2000 shade trees after discovering a long-horned beetle infestation in 1999.

The flow of organisms isn't just into America; we also send exotic species to other places. The Leidy's comb jelly, for example, a jellyfish native to the western Atlantic coast, has devastated the Black Sea, now making up more than 90 percent of all biomass at certain times of the year. Similarly, the bristle worm from North America has invaded the coast of Poland and now is almost the only thing living on the bottom of some bays and lagoons. A tropical seaweed named *Caulerpa taxifolia,* originally grown for the aquarium trade, has escaped into the northern Mediterranean, where it covers the shallow seafloor with a dense, meter-deep shag carpet from Spain to Croatia. Producing more than 5000 leafy fronds per square meter, this aggressive weed crowds out everything in its path. Rarely growing in more than scattered clumps less than 25 cm (10 in) high in its native habitat, this alga was transformed by aquarium breeding into a supercompetitor that grows over everything and can withstand a wide temperature range. Getting rid of these alien species once they dominate an ecosystem is difficult if not impossible.

Diseases

Disease organisms, or pathogens, may also be considered predators. To be successful over the long term, a pathogen must establish a balance in which it is vigorous enough to reproduce, but not so lethal that it completely destroys its host. When a disease is introduced into a new environment, however, this balance may be lacking and an epidemic may sweep through the area.

The American chestnut was once the heart of many Eastern hardwood forests. In the Appalachian Mountains, at least one of every four trees was a chestnut. Often over 45 m (150 ft) tall, 3 m (10 ft) in diameter, fast growing, and able to sprout quickly from a cut stump, it was a forester's dream. Its nutritious nuts were important for birds (like the passenger pigeon), forest mammals, and humans. The wood was straight grained, light, rot-resistant and used for everything from fence posts to fine furniture and its bark was used to tan leather. In 1904, a shipment of nursery stock from China brought a fungal blight to the United States, and within forty years, the American chestnut had all but disappeared from its native range. Efforts are now underway to transfer blight-resistant genes into the few remaining American chestnuts that weren't reached by the fungus or to find biological controls for the fungus that causes the disease.

An infection called whirling disease is decimating trout populations in many western states. It is caused by an exotic microorganism, *Myxobolus cerebralis,* which destroys cartilage in young fish causing them to swim erratically. Myxobolus spores, released when infected fish die, are ingested by Tubifex worms living in stream-bottom mud. Once inside the worm, the parasite transforms into a stage capable of infecting young salmonids. The parasite is thought to have come into the United States in 1956 in a shipment of frozen fish.

Pollution

We have known for a long time that toxic pollutants can have disastrous effects on local populations of organisms. Pesticide-linked declines of fish-eating birds and falcons was well documented in the 1970s. Marine mammals, alligators, fish, and other declining populations suggest complex interrelations between pollution and health (chapter 12). Mysterious, widespread deaths of thousands of seals on both sides of the Atlantic in recent years are thought to be linked to an accumulation of persistent chlorinated hydrocarbons, such as DDT, PCBs, and dioxins, in fat, causing weakened immune systems that make animals vulnerable to infections. Similarly, mortality of Pacific sea lions, beluga whales in the St. Lawrence estuary, and striped dolphins in the Mediterranean are thought to be caused by accumulation of toxic pollutants.

Lead poisoning is another major cause of mortality for many species of wildlife. Bottom-feeding waterfowl, such as ducks, swans, and cranes, ingest spent shotgun pellets that fall into lakes and marshes. They store the pellets, instead of stones, in their gizzards and the lead slowly accumulates in their blood and other tissues. The U.S. Fish and Wildlife Service (USFWS) estimates that 3000 metric tons of lead shot are deposited annually in wetlands and that between 2 and 3 million waterfowl die each year from lead poisoning. Scavengers, such as condors and bald eagles, eat birds and mammals that have lead shot or bullet fragments in their bodies. In 1987, the last five wild California condors were trapped and put into zoos (fig. 13.18). Captive breeding programs have rebuilt the condor population to about 160 birds. Around 110 birds are in captive breeding facilities in California and Idaho, while some 50 birds have been reintroduced into two wild populations in southern California and northern Arizona.

Genetic Assimilation

Some rare and endangered species are threatened by **genetic assimilation** because they crossbreed with closely related species that are more numerous or more vigorous. Opportunistic plants or animals that are introduced into a habitat or displaced from their

FIGURE 13.18 The California condor has the largest wingspan of any North American bird. Loss of habitat and food sources, shooting by hunters, and poisoning by lead shot and predator-control bait eliminated the entire wild population by 1987. Successful captive breeding in zoos has allowed successful reintroduction into their original habitat.

normal ranges by human actions may genetically overwhelm local populations. For example, hatchery-raised trout often are introduced into streams and lakes where they genetically dilute indigenous stocks. Similarly, black ducks have declined severely in the eastern United States and Canada in recent years. Hunting pressures and habitat loss are factors in this decline, but so is interbreeding with mallards forced into black duck habitat by destruction of prairie potholes in the west. As discussed earlier in this chapter, hybridization of endangered timber wolves with much more numerous coyotes or domestic dogs introduces foreign genes into an already dangerously small gene pool and reduces the likelihood that purebred animals will mate.

ENDANGERED SPECIES MANAGEMENT AND BIODIVERSITY PROTECTION

Over the years, we have gradually become aware of the harm we have done—and continue to do—to wildlife and biological resources. Slowly, we are adopting national legislation and international treaties to protect these irreplaceable assets. Parks, wildlife refuges, nature preserves, zoos, and restoration programs have been established to protect nature and rebuild depleted populations. There has been encouraging progress in this area, but much remains to be done. While most people favor pollution control or protection of favored species such as whales or gorillas, surveys show that few understand what biological diversity is or why it is important.

Hunting and Fishing Laws

In 1874, a bill was introduced in the United States Congress to protect the American bison, whose numbers were already falling to dangerous levels. This initiative failed, however, because most legislators believed that all wildlife—and nature in general—was so abundant and prolific that it could never be depleted by human activity. As we discussed earlier in this chapter, however, by the end of the nineteenth century, bison had plunged from some 60 million to only a few hundred animals.

By the 1890s most states had enacted some hunting and fishing restrictions. The general idea behind these laws was to conserve the resource for future human use rather than to preserve wildlife for its own sake. The wildlife regulations and refuges established since that time have been remarkably successful for many species. At the turn of the century, there were an estimated half a million white-tailed deer in the United States; now there are some 14 million—more in some places than the environment can support. Wild turkeys and wood ducks were nearly all gone fifty years ago. By restoring habitat, planting food crops, transplanting breeding stock, building shelters or houses, protecting these birds during breeding season, and other conservation measures, populations of these beautiful and interesting birds have been restored to several million each. Snowy egrets, which were almost wiped out by plume hunters 80 years ago, are now common again.

The Endangered Species Act

Establishment of the U.S. Endangered Species Act (ESA) of 1973 and the Committee on the Status of Endangered Wildlife in Canada (COSEWIC) in 1976 represented powerful new approaches to wildlife protection. Where earlier regulations had been focused almost exclusively on "game" animals, these programs seek to identify all endangered species and populations and to save as much biodiversity as possible, regardless of its usefulness to humans. **Endangered species** are those considered in imminent danger of extinction, while **threatened species** are those that are likely to become endangered—at least locally—within the foreseeable future. Bald eagles, gray wolves, brown (or grizzly) bears, and sea otters, for instance, together with a number of native orchids and other rare plants are considered to be locally threatened even though they remain abundant in other parts of their former range. **Vulnerable species** are naturally rare or have been locally depleted by human activities to a level that puts them at risk. They often are candidates for future listing. For vertebrates, protected categories include species, subspecies, and local races or ecotypes.

The ESA regulates a wide range of activities involving endangered species including "taking" (harassing, harming, pursuing, hunting, shooting, trapping, killing, capturing, or collecting) either accidentally or on purpose; importing into or exporting out of the United States; possessing, selling, transporting, or shipping; and selling or offering for sale any endangered species. Prohibitions apply to live organisms, body parts, and products made from endangered species. Violators of the ESA are subject to fines up to

$100,000 and one year imprisonment. Vehicles and equipment used in violations may be subject to forfeiture. In 1995 the Supreme Court ruled that critical habitat—habitat essential for a species survival—must be protected whether on public or private land.

Currently, the United States has 1500 species on its endangered and threatened species lists and about 500 candidate species waiting to be considered. Canada, which generally has less diversity because of its boreal location, has designated a total of 46 endangered and 50 threatened species. The number of listed species in different taxonomic groups reflects much more about the kinds of organisms that humans consider interesting and desirable than the actual number in each group. Compare tables 13.1 and 13.4 for instance. Although invertebrates make up about three-quarters of all known species, they account for only 9 percent of those deemed worthy of protection. The total number of endangered species worldwide is at least five times the number officially listed in the United States.

Listing of new species in the United States has been very slow, generally taking several years from the first petition to final determination. Limited funding, political pressures, listing moratoria, and changing administrative policies have created long delays. Hundreds of species are classified as "warranted (deserving of protection) but precluded" for lack of funds or local support. At least 18 species have gone extinct since being nominated for protection. Both environmentalists and property-rights advocates tend to believe that the "single-species" approach is flawed and that efforts should be focused on multiple species and habitat protection.

When Congress passed the original ESA, it probably intended to protect only a few charismatic species like birds and big game animals. Sheltering obscure species such as the Delhi Sands flower-loving fly, the Coachella Valley fringe-toed lizard, Mrs. Furbisher's lousewort, or the orange-footed pimple-back mussel most likely never occurred to those who voted for the bill. This raises some interesting ethical questions about the rights and values of seemingly minor species. Although uncelebrated, these species may be indicators of environmental health. Protecting them usually preserves habitat and a host of unlisted species.

Canada doesn't have an Endangered Species Act but federal legislation was introduced in 1999 that would use criminal law powers to protect both species at risk as well as critical habitat on both public and private land.

TABLE 13.4 Endangered and Threatened Species, U.S. and Foreign, Listed by USFWS

Mammals	331
Birds	267
Reptiles	112
Amphibians	22
Fish	113
Invertebrates	135
Plants	520

Source: Data from United States Fish and Wildlife Service, 1999.

FIGURE 13.19 Only 40 Florida panthers remain in the Everglades. We spend millions of dollars to protect them but the population may be too small to be viable.

Recovery Plans

Once a species is officially listed as endangered, the Fish and Wildlife Service is required to prepare a recovery plan detailing how populations will be rebuilt to sustainable levels. It usually takes years to reach agreement on specific recovery plans. Among the difficulties are costs, politics, interference on local economic interests, and the fact that once a species is endangered, much of its habitat and ability to survive is likely compromised. The total cost of recovery plans for all currently listed species is estimated to be nearly $5 billion.

The United States currently spends about $150 million per year on endangered species protection and recovery. About half that amount is spent on a dozen charismatic species like the Florida panther and grizzly bear, which receive around $13 million per year (fig. 13.19). By contrast, the 137 endangered invertebrates and 532 endangered plants get less than $5 million per year altogether. Our funding priorities are based more on emotion and politics than biology.

Some recovery plans have been gratifyingly successful. The American alligator was listed as endangered in 1967 because hunting (for meat, skins, and sport) and habitat destruction had reduced populations to precarious levels. Protection has been so effective that the species is now plentiful throughout its entire southern range. Florida alone estimates that it has at least 1 million alligators.

Twenty years ago, due mainly to DDT poisoning, only 800 bald eagles remained in the contiguous United States. In 1994, the population had rebounded to more than 8000 birds and the eagle's status was reduced from endangered to threatened. In 1999, bald eagles were declared recovered and removed from the list throughout the

United States. Similarly, peregrine falcons, which had been down to only 39 breeding pairs in the 1970s, had rebounded to 1650 pairs in 1999 and also were taken off the endangered species list. Declaring the ESA a success, Interior Secretary Bruce Babbit announced that 29 species, including mammals, fish, reptiles, birds, plants, and even one insect (the Tinian monarch) in addition to eagles and falcons have been removed or downgraded from the endangered species list.

Opponents of the ESA have repeatedly tried to require that economic costs and benefits be incorporated into endangered species planning. An important test of the ESA occurred in 1978 in Tennessee where construction of the Tellico Dam threatened a tiny fish called the snail darter. As a result of this case, a Federal Committee (the so-called "God Squad") was given power to override the ESA for economic reasons. Another important example is the case of the northern spotted owl (chapter 14), whose protection depends on preserving old-growth forest in the Pacific Northwest. Economists estimate that saving a population of 1600 to 2400 owls would cost $33 billion, with most of the losses borne by local companies and residents of Washington and Oregon. Environmentalists dispute these numbers and claim the owl is an umbrella species whose protection would aid many other organisms and resources (fig. 13.20).

An even more costly recovery program may be required for Columbia River salmon and steelhead endangered by hydropower dams and water storage reservoirs that block their migration to the sea. Opening floodgates to allow young fish to run downriver and adults to return to spawning grounds would have high economic costs to barge traffic, farmers, and electric rate payers who have come to depend on abundant water and cheap electricity. On the other hand, commercial and sport fishing for salmon is worth $1 billion per year and employs about 60,000 people directly or indirectly.

FIGURE 13.20 Endangered species often serve as a barometer for the health of an entire ecosystem and as surrogate protector for a myriad of less well-known creatures.
Copyright 1990 by Herblock in the *Washington Post.*

Private Land and Critical Habitat

Private land is essential in endangered species protection. Eighty percent of the habitat for more than half of all listed species is on nonpublic property. The Supreme Court has ruled that destroying habitat is as harmful to endangered species as directly taking (killing) them. Many people are very resistant, however, to restrictions on how they use their own property to protect what they perceive to be insignificant or worthless organisms. This is especially true when the land has potential for economic development. If property is worth millions of dollars as the site of a housing development or shopping center, most owners don't want to be told they have to leave it undisturbed to protect some rare organism. Landowners may be tempted to "shoot, shovel, and shut up," if they discover endangered species on their property. Many feel they should be compensated for lost value caused by ESA regulations.

Recently, to avoid what Interior Secretary Bruce Babbitt calls "environmental train wrecks" like the controversy over northern spotted owls, the Fish and Wildlife Service has been negotiating agreements called **habitat conservation plans** (HCP) with private landowners. Under these plans, landowners are allowed to harvest resources or build on part of their land as long as the species benefits overall. By improving habitat in some areas, funding conservation research, removing predators and competitors, or other steps that benefit the endangered species, developers are allowed to destroy habitat or even "take" endangered organisms.

So far, about 250 HCPs had been approved and 200 more in 19 states were in progress. A total of 7 million ha (18.5 million acres) of land are covered by these plans. Scientists and environmentalists often are critical of HCPs, claiming these plans often are based more on politics than biology, and that the potential benefits are frequently overstated. Defenders argue that by making the ESA more landowner-friendly, HCPs benefit wildlife in the long run.

Among the more controversial proposals for HCPs are the so-called Safe Harbor and No-Surprises Policies. Under the Safe Harbor clause, any increase in an animal's population resulting from a property owner's voluntary good stewardship would not increase their responsibility or affect future land use decisions. As long as the property owner complies with the terms of the agreement, he or she can make any use of the property. The No-Surprises provision says that the property owner won't be faced with new requirements or regulations after entering into an HCP. Scientists warn that change, uncertainty, dynamics, and flux are characteristic of all ecosystems. We can't say that natural catastrophes or environmental events won't make it necessary to modify conservation plans in the future.

Reauthorizing the Endangered Species Act

The ESA officially expired in 1992. Since then, Congress has debated many alternative proposals ranging from outright elimination to substantial strengthening of the act. Perhaps no other environmental issue divides Americans more strongly than the ESA. In the western United States, where traditions of individual liberty and freedom are strong and the federal government is viewed with considerable suspicion and hostility, the ESA seems to many to be a diabolical plot to take away private property and trample on individual rights. Many people believe that the law puts the welfare of plants and animals above that of humans. Farmers, loggers, miners, ranchers, developers, and other ESA opponents repeatedly have tried to scuttle the law or greatly reduce its power. Environmentalists, on the other hand, see the ESA as essential to protecting nature and maintaining the viability of the planet. They regard it as the single most effective law in their arsenal and want it enhanced and improved.

Proposals for a new ESA generally fall into one of two general categories. Environmentalists have repeatedly introduced versions that encourage an ecosystem and habitat protection approach rather than focusing on individual species. Reinforcing critical habitat protection and placing deadlines for listing and recovery plan completion, these bills would require that proposed HCPs be independently peer-reviewed by scientists. They also would expand citizen ability to enforce the law and increase civil and criminal penalties.

On the other side, ESA opponents want to reinforce "safe harbor" and "no surprises" policies and to allow exceptions to critical habitat designation. All recovery plan teams in their versions of the ESA would be required to include individuals with vested economic interests. Under these proposals, only the least costly, most cost effective or least burdensome measures would be taken to protect endangered organisms. Federal agencies would be allowed to do "self-consultation"—that is to avoid consulting wildlife specialists from the Fish and Wildlife Service before undertaking projects. ESA opponents also would require notification and hearings in all states with resident candidate species and would give small-parcel landowners expedited permits for HCPs.

Minimum Viable Populations

A critical question in all recovery programs is the minimum population size required for long-term viability of rare and endangered species. As chapter 6 shows, a species composed of a small number of individuals can undergo catastrophic declines due to environmental change, genetic problems, or simple random events when isolated in a limited geographic range. This phenomenon was elegantly described as **island biogeography** in the work of R. H. MacArthur and E. O. Wilson in 1967. Noticing that small islands far from a mainland have fewer terrestrial species than larger, nearer islands, MacArthur and Wilson proposed that species diversity is a balance between colonization and extinction rates (fig. 13.21). An island far from a population source naturally has a lower rate of colonization than a nearer island because it is harder for terrestrial organisms to reach. At the same time, a large island can support more individuals of a given species and is, therefore, less likely to suffer extinction due to natural catastrophes, genetic problems, or demographic uncertainty—the chance that all the members of a single generation will be of the same sex.

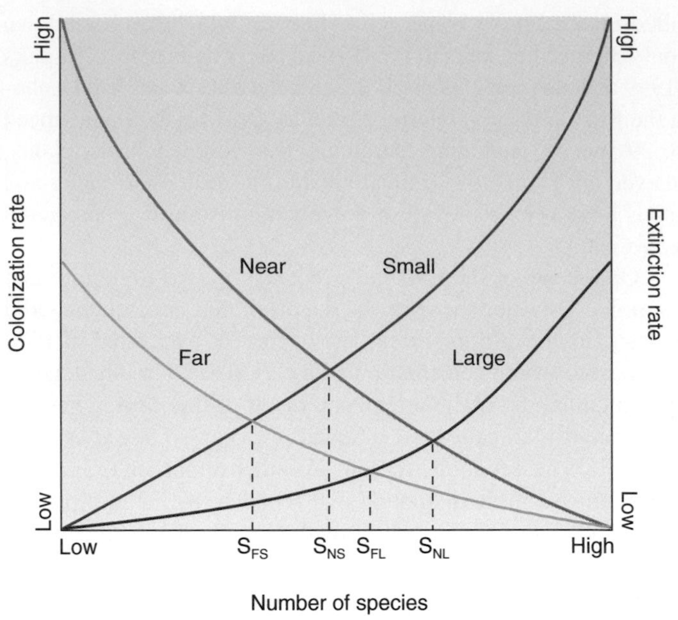

FIGURE 13.21 Predicted species richness on an island resulting from a balance between colonization (immigration) and extinction by natural causes. This island biogeography theory of MacArthur and Wilson (1967) is used to explain why large islands near a mainland (S_{NL}) tend to have more species than small, far islands (S_{FS}).
Source: Based on MacArthur and Wilson, *The Theory of Island Biogeography*, 1967, Princeton University Press.

Island biogeographical effects have been observed in many places. Cuba, for instance, is one hundred times as large and has about ten times as many amphibian species as its Caribbean neighbor, Monserrat. Similarly, in a study of bird species on the California Channel Islands, Jared Diamond observed that on islands with fewer than ten breeding pairs, 39 percent of the populations went extinct over an 80-year period, while only 10 percent of populations numbering between 10 and 100 pairs went extinct in the same time (fig. 13.22). Only one species numbering between 100 and 1000 pairs went extinct and no species with over 1000 pairs disappeared over this time.

Small population sizes affect species in isolated landscapes other than islands. Grizzly bears (*Ursus arctos horribilis*) once roamed across most of western North America. Hunting and habitat destruction reduced the number of grizzlies in the lower 48 states from an estimated 100,000 in 1800 to less than 1000 animals in six separate subpopulations that now occupy less than 1 percent of the historic range. Recovery target sizes—based on estimated environmental carrying capacities—are less than 100 animals for some subpopulations. Conservation biologists predict

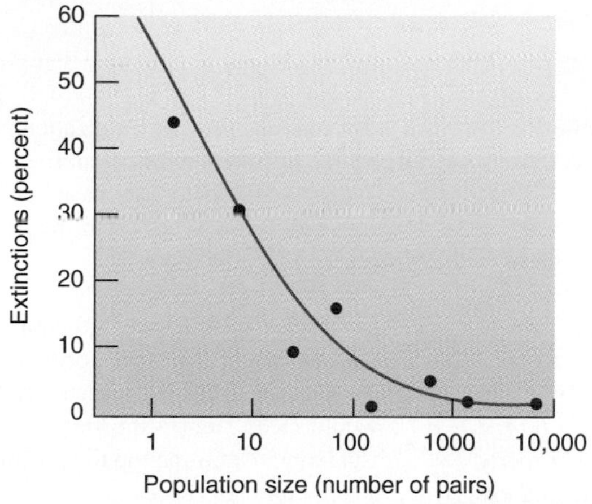

FIGURE 13.22 Extinction rates of bird species on the California Channel Islands as a function of population size over eighty years.
Source: Data from H.L. Jones and J. Diamond, "Short-term-base Studies of Turnover in Breeding Bird Populations on the California Coast Island," in *Condor,* vol. 78: 526–549, 1976.

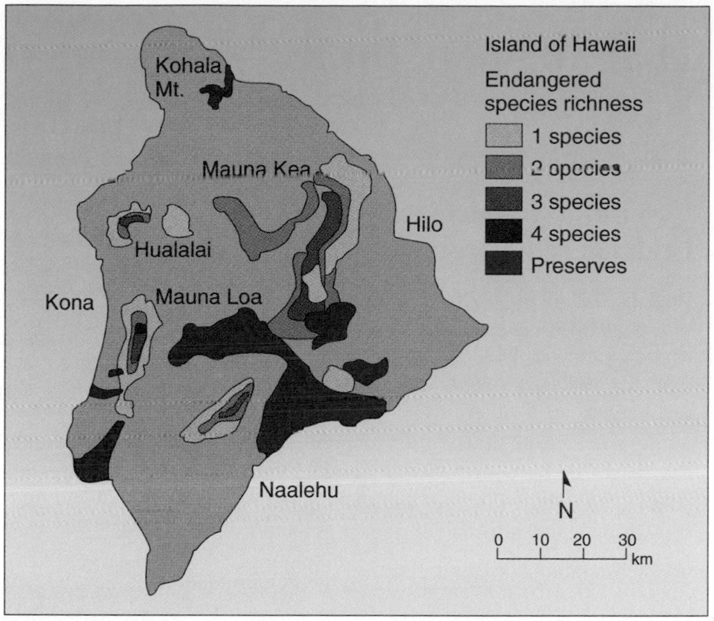

FIGURE 13.23 An example of the biodiversity maps produced by J. Michael Scott and the U.S. Fish and Wildlife Service. Notice that few of the areas of endangered species richness are protected in preserves, which were selected more for scenery or recreation than for biology.

that a completely isolated population of 100 bears cannot be maintained for more than a few generations. Even the 200 bears in Yellowstone National Park will be susceptible to genetic problems if completely isolated. Interestingly, computer models suggest that translocating only two unrelated bears into small populations every decade or so could greatly increase population viability.

For many species, loss of genetic diversity causes a variety of harmful effects that limit adaptability, reproduction, and species survival. How is diversity lost in small populations? (1) A *founder effect* occurs when a few individuals establish a new population. The limited genetic diversity from those original founders may not be enough to sustain the population. (2) A *demographic bottleneck* arises when only a few individuals survive some catastrophe. As the population replenishes itself, a limited genetic diversity similar to that in founder effect results. (3) *Genetic drift* is a reduction in gene frequency in a population due to unequal reproductive success, for example, some individuals breed more than others and their genes gradually come to dominate the population. (4) *Inbreeding* is mating of closely related individuals. Random, recessive, deleterious mutations—what we consider genetic diseases such as hemophilia in humans—that are usually hidden in a widely outcrossed population can be expressed when inbreeding occurs. This is why we have laws and cultural taboos prohibiting mating between siblings or first cousins in humans.

Some species seem not to be harmed by lack of genetic diversity. The northern elephant seal, for example, was reduced by overharvesting, a century ago, to fewer than 100 individuals. Today there are more than 150,000 of these enormous animals along the Pacific coast of Mexico and California. No marine mammal has come closer to extinction and then made such a remarkable recovery. All northern elephant seals today appear to be essentially genetically identical and yet they seem to be getting along just fine. Although interpretations are controversial, in highly selected populations where only the most fit individuals reproduce, or in which there are few deleterious genes, inbreeding may not be such a negative factor.

Habitat Protection

Over the past decade, growing numbers of scientists, land managers, policymakers, and developers have been making the case that it is time to focus on a rational, continent-wide preservation of ecosystems that support maximum biological diversity rather than a species-by-species battle for the rarest or most popular organisms. By focusing on populations already reduced to only a few individuals, we spend most of our conservation funds on species that may be genetically doomed no matter what we do. Furthermore, by concentrating on individual species we spend millions of dollars to breed plants or animals in captivity that have no natural habitat where they can be released. While flagship species such as mountain gorillas or Indian tigers are reproducing well in zoos and wild animal parks, the ecosystems that they formerly inhabited have largely disappeared.

A leader of this new form of conservation is J. Michael Scott, who was project leader of the California condor recovery program in the mid-1980s and had previously spent ten years working on endangered species in Hawaii. In making maps of endangered species, Scott discovered that even Hawaii, where more than 50 percent of the land is federally owned, has many vegetation types completely outside of natural preserves (fig. 13.23). The gaps between protected areas may contain more endangered species than are preserved within them.

This observation has led to an approach called **gap analysis** in which conservationists and wildlife managers look for unprotected landscapes that are rich in species. Computers and geographical information systems (GIS) make it possible to store, manage, retrieve, and analyze vast amounts of data and create

what do you think?

THE WORLD AROUND YOU

Economic Impacts of the Endangered Species Act

Reauthorization of the Endangered Species Act, originally scheduled for 1992, has been delayed because of the heated debate that has developed over the Act's provisions. Some people want the Act repealed, others want it strengthened, while some think it should be modified to accommodate the concerns of both sides. Most people support the idea that decisions ought to flow from factual information. Not surprisingly, advocates on both sides of the Endangered Species Act debate attempt to present factual information to convince others of the soundness of their position.

Sound arguments and sound judgments can only be built on representative information, however. Unfortunately, anecdotal information is often used to project generalized conclusions. Because of the inherent variability of events, arguments based on anecdotes or small samples may not reflect generalized reality at all. That is why all sound scientific investigations include multiple replications—to avoid basing conclusions on a few outcomes that are not truly representative of the whole.

Use of the "generalization from anecdote" method of constructing an argument is common on both sides of many issues, including the Endangered Species Act debate.

Charles Oliver feels the Endangered Species Act ought to be repealed. His position is based, in part, on the argument that the Act significantly damages our economic well-being. He opens his argument with the following statements:

> In Washington state, logger Dean Hurn is worried. He could be among the estimated 40,000 to 100,000 timber workers who will lose their jobs to protect the Northern spotted owl. . . . In Florida's Polk and Highland counties, 600 landowners can't build the homes they planned because their property is home to the Florida scrub jay. In these and many other cases, the Endangered Species Act has halted the plans and dreams of thousands of ordinary Americans. The act is up for renewal this year, and one might expect a piece of legislation that has harmed so many people to be itself endangered. . . .

The generalization implied is that the Endangered Species Act is responsible for massive economic dislocations. But other information calls this idea into question. An evaluation of 18,211 cases reviewed under ESA provisions between 1987 and 1991 revealed that 16,161 projects did not even require a biological opinion and proceeded without interference. In 1869 other cases, the biological opinion was that the proposal did not threaten an endangered species and the project could proceed as planned. Of the 181 remaining cases, most were completed, although with modifications in design. In all, over 99 percent of the projects examined under Endangered Species Act provisions proceeded unhindered or with marginal added cost or time delay.

By contrast, what generalization might be implied about the economic consequences of the Endangered Species Act based on the following anecdote as reported in the newsletter *Common Ground*?

> In Oregon, contrary to prediction, the sky isn't falling. Protecting northern spotted owls and old-growth forests, said many, would mean ruin to the state's economy. But Oregon, still the top timber-producing state despite less access to national forests, enjoys its lowest unemployment rate in years, well below the national average. Formerly timber-dependent counties report rising property values and a net increase in jobs. The 15,000 timber industry jobs lost by the state over 5 years have been offset by nearly 20,000 high tech jobs. Sony Corp., for instance, is coming to Springfield, Ore., lured in part by money from President Clinton's recovery program. "Owls versus jobs was just plain false," Mayor Bill Morrisette told *The New York Times*. "What we've got here is quality of life. As long as we don't screw that up, we'll always be able to attract people and business."

Can we generalize from this that the Endangered Species Act stimulates local economies? Hardly. Critical thinkers distrust anecdotes, regardless of the source, because valid generalizations cannot be built on small samples.

---

detailed, high-resolution maps relatively easily. This broad-scale, holistic approach seems likely to save more species than a piecemeal approach.

Conservation biologist, R. E. Grumbine suggests four remanagement principles for protecting biodiversity in a large-scale, long-range approach:

1. Protect enough habitat for viable populations of all native species in a given region.
2. Manage at regional scales large enough to accommodate natural disturbances (fire, wind, climate change, etc.).
3. Plan over a period of centuries so that species and ecosystems may continue to evolve.
4. Allow for human use and occupancy at levels that do not result in significant ecological degradation.

International Wildlife Treaties

The 1975 Convention on International Trade in Endangered Species (CITES) was a significant step toward worldwide protection of endangered flora and fauna. It regulated trade in living specimens and products derived from listed species, but has not been foolproof. Species are smuggled out of countries where they are threatened or endangered, and documents are falsified to make it appear they have come from areas where the species are still common. Investigations and enforcement are especially difficult in developing countries where wildlife is disappearing most rapidly. Still, eliminating markets for endangered wildlife is an effective way of stopping poaching. Appendix I of CITES lists 700 species threatened with extinction by international trade.

ZOOS, BOTANICAL GARDENS, AND CAPTIVE BREEDING PROGRAMS

Records of wildlife and exotic plant collections date back to the beginning of history. The ancient Chinese, Egyptians, Romans, and others kept menageries of exotic and bizarre wildlife. Many of these collections included humans as well as animals. War captives and strangers from distant lands were displayed in cages along with alien animals. Fights between animals or between animals and humans were common entertainment in earlier times.

In some places, low-life circuses and seedy roadside tourist attractions still exhibit sad collections of animals in cramped, unhealthy cages (fig. 13.24). In developed countries, however, these cruel and callous conditions have generally been eliminated by animal welfare laws. Most modern zoos and wildlife parks now keep animals in clean, humane conditions and have a strong commitment to species preservation and public education. Although they are held captive, many animals live longer in zoos and raise more young than they would in the wild. Often, animals that escape from their cages will voluntarily return because conditions are better there than on the outside.

Botanical gardens, such as the Kew Gardens in England, and research stations, such as the International Rice Institute in the Philippines, are repositories for rare and endangered plant species that sometimes have ceased to exist in the wild. Valuable genetic traits are preserved in these collections, and in some cases, plants with unique cultural or ecological significance may be reintroduced into native habitats after being cultivated for decades or even centuries in these gardens and seed banks.

New Zoos and Game Parks

Around 1900, Karl Haugenbeck, a German wildlife dealer, introduced a new style of displaying animals. He created naturalistic enclosures that attempted to show creatures in something like their native habitat. By using moats rather than fences to contain animals, Haugenbeck gave patrons a better view and a more intimate experience of the wildlife. Gradually, these principles have spread to other zoos. Clever landscaping can create uncannily authentic terrain, and inconspicuous barriers confine humans rather than animals. Large enclosures and multi-species groupings allow more normal social behavior and encourage reproduction (fig. 13.25). Under these conditions, zoos can be important places for scientific research in wildlife biology.

Today in the United States some 150 municipal zoos provide the only experience that most people ever have with large wild animals. More than 110 million visitors per year learn about nature through visits to zoos. Many new zoos de-emphasize the crowd-pleasing exotic animals such as lions, tigers, and elephants, concentrating instead on species that normally inhabit their climatic regime. Zoo visitors, however, often object that they want to see lions and tigers rather than indigenous species. They also complain that spacious, vegetated, modern enclosures hide animals from sight. If you were a zookeeper how would you balance entertainment and conservation?

FIGURE 13.24 Many zoos, such as this one in Central America, still house animals in cramped, unsanitary cages, where they are tormented by insensitive onlookers and deprived of any semblance of a normal, healthy life.

FIGURE 13.25 Newer zoos and game parks are making special efforts to provide comfortable, naturalistic habitats. It can be difficult, however, to find a balance between accessibility to visitors and privacy for animals.

Captive Breeding and Species Survival Plans

Until fairly recently, zoos depended on primarily wild-caught animals for most of their collections. This was a serious drain on wild populations, because up to 80 percent of the animals caught died from the trauma of capture and shipping. With better understanding of reproductive biology and better breeding facilities, most mammals in North American zoos now are produced by captive breeding programs. These programs have limitations, however. Bats, whales, and many reptiles rarely reproduce in captivity and still come mainly from the wild. Furthermore, we will never be able to protect the complete spectrum of biological variety in zoos. Worldwide, zoos house about 500,000 individual animals, but only about 900 species in total. Few of these animals are maintained in populations large enough to preserve the species indefinitely if wild populations were lost. According to one estimate, if all the space in U.S. zoos were used for captive breeding, only about one hundred species of large mammals could be maintained on a long-term basis.

These limitations lead to what is sometimes called the "Noah question:" how many species can or should we save? How much are we willing to invest to protect the slimy, smelly, crawly things? Would you favor preserving disease organisms, parasites, and vermin or should we use our limited resources to protect only beautiful, interesting, or seemingly useful organisms?

Even given adequate area and habitat conditions to perpetuate a given species, continued inbreeding of a small population in captivity can lead to the same kinds of fertility and infant survival problems described earlier for wild populations. To reduce genetic problems, zoos often exchange animals or ship individuals long distances to be bred. It sometimes turns out, however, that zoos far distant from each other unknowingly obtained their animals from the same source. Computer data bases operated by the International Species Information System located at the Minnesota Zoo, now keep track of the genealogy of many species. This system can tell the complete reproductive history of every animal in every zoo in the world for some species. Comprehensive species survival plans based on this genealogy help match breeding pairs and project resource needs.

Saving Rare Species in the Wild

Renowned zoologist George Schaller says that ultimately "zoos need to get out of their own walls and put more effort into saving the animals in the wild." An interesting application of this principle is a partnership between the Minnesota Zoo and the Ujung Kulon National Park in Indonesia, home to the world's few remaining Javanese rhinos. Rather than try to capture rhinos and move them to Minnesota, the zoo is helping to protect them in their native habitat by providing patrol boats, radios, housing, training, and salaries for Indonesian guards (fig. 13.26). There are no plans to bring any rhinos to Minnesota and chances are very slight that any of us will ever see one, but we can gain satisfaction that, at least for now, a few Javanese rhinos still exist in the wild.

FIGURE 13.26 The *KM Minnesota* anchored in Tamanjaya Bay in west Java. Funds raised by the Minnesota Zoo paid for local construction of this boat, which allows wardens to patrol Ujung Kulon National Park and protect rare Javanese rhinos from poachers.

SUMMARY

In this chapter, we have briefly surveyed world biodiversity and the ways humans both benefit from and threaten it. Natural causes of wildlife destruction include evolutionary replacement and mass extinction. Among the threats from humans are overharvesting of animals and plants for food and commercial products. Millions of live wild plants and animals are collected for pets, houseplants, and medical research. Among the greatest damage we do to biodiversity are habitat destruction, the introduction of exotic species and diseases, pollution of the environment, and genetic assimilation.

The potential value of the species that may be lost if environmental destruction continues could be enormous. It is also possible that the changes we are causing could disrupt vital ecological services on which we all depend for life.

The first hunting and fishing laws in the United States were introduced more than a century ago to restrict overexploitation and to preserve species for future uses. The Endangered Species Act and CITES represent a new attitude toward wildlife in which we protect organisms just because they are rare and endangered. Now we are expanding our concern from individual species to protecting habitat, threatened landscapes, and entire biogeographical regions. Social, cultural, and economic factors must also be considered if we want to protect biological resources on a long-term, sustainable basis.

Zoos can be educational and entertaining while still serving important wildlife conservation and scientific functions. Modern zoos have greatly improved the living conditions for captive animals, resulting in improved breeding success; still, there are limits to the number and types of species that we could maintain under captive conditions. Zoos need to get out of their own walls and help save animals and plants in the wild.

QUESTIONS FOR REVIEW

1. What is the range of estimates of the total number of species on the earth? Why is the range so great?
2. What group of organisms has the largest number of species?
3. Define *extinction.* What is the natural rate of extinction in an undisturbed ecosystem?
4. What are rosy periwinkles and what products do we derive from them?
5. Describe some foods we obtain from wild plant species.
6. List nine seafood groups that are reduced or endangered by human activities.
7. What is the current rate of extinction and how does this compare to historic rates?
8. Compare the scope and effects of the Endangered Species Act and CITES.
9. Describe eight ways that humans directly or indirectly cause biological losses.
10. What is gap analysis and how is it related to ecosystem management and design of nature preserves?

QUESTIONS FOR CRITICAL THINKING

1. One reviewer said that this chapter is the most biased in this book. Do you agree? How much moral outrage is appropriate in an issue such as this? Does emotion interfere with rational analysis or effective communication? What is the proper balance between emotion and objectivity in a subject such as this?
2. Many ecologists would like to move away from protecting individual endangered species to concentrate on protecting whole communities or ecosystems. Others fear that the public will only respond to and support glamorous "flagship" species such as gorillas, tigers, or otters. If you were designing conservation strategy, where would you put your emphasis?
3. Put yourself in the place of a fishing industry worker. If you continue to catch many species they will quickly become economically extinct if not completely exterminated. On the other hand, there are few jobs in your village and welfare will barely keep you alive. What would you do?
4. Only a few hundred grizzly bears remain in the contiguous United States, but populations are healthy in Canada and Alaska. Should we spend millions of dollars for grizzly recovery and management programs in Yellowstone National Park and adjacent wilderness areas?
5. How could people have believed a century ago that nature is so vast and fertile that human actions could never have a lasting impact on wildlife populations? Are there similar examples of denial or misjudgment occurring now?
6. Suppose you're having dinner with a friend who orders swordfish. What would you say? What are the ethical and biological arguments for or against eating endangered species?
7. In the past, mass extinction has allowed for new growth, including the evolution of our own species. Should we assume that another mass extinction would be a bad thing? Could it possibly be beneficial to us? To the world?
8. Some captive breeding programs in zoos are so successful that they often produce surplus animals that cannot be released into the wild because no native habitat remains. Plans to euthanize surplus animals raise storms of protests from animal lovers. What would you do if you were in charge of the zoo?
9. Debate with a friend or classmate the ethics of keeping animals captive in a zoo. After exploring the subject from one side, debate the issue from the opposite perspective. What do you learn from this exercise?

KEY TERMS

biodiversity 283
endangered species 295
existence value 287
exotic organisms 292
extinction 287
gap analysis 299
genetic assimilation 294
habitat conservation plans 297
island biogeography 298
threatened species 295
vulnerable species 295

ADDITIONAL INFORMATION ON THE INTERNET

Visit our website at http://www.mhhe.com/environmentalscience/ for specific information about each of the topics below:

World Biodiversity Map
Images of Snow Goose Damage, American Museum of Natural History
Snow Geese: US Fish and Wildlife Service
Virtual Library of Ecology and Biodiversity
Biodiversity in India
Biogeography
Marine Biodiversity
Endangered Species Information
Convention on International Trade in Endangered Species (CITES)
Invasive Species Web Page, from the National Biological Information Infrastructure
Biodiversity and Biological Collections: worldwide biodiversity studies
Biodiversity. A fact sheet on biodiversity from the Ecological Society of America
Endangered Species, US Fish and Wildlife Service
Request a species recovery plan from the USFWS

CHAPTER 14

Land Use: Forests and Rangelands

Forests precede people; deserts succeed them.

—François-René de Chateubriand—

OBJECTIVES

After studying this chapter, you should be able to:

- discuss the major world land uses and how human activities impact these areas.
- summarize some forest types and the products we derive from them.
- report on how and why tropical forests are being disrupted as well as how they might be better used.
- understand the major issues concerning forests in more highly developed countries such as the United States and Canada.
- outline the extent, location, and state of grazing lands around the world.
- describe how overgrazing causes desertification of rangelands.
- evaluate landownership patterns and explain why land reform and recognition of indigenous rights are essential for social justice as well as environmental protection.

ESSENTIAL STUDY PARTNER

The Essential Study Partner (ESP) CD-ROM that accompanies this textbook contains additional information on the following topics:

- forest distribution, diversity, services and products: living resources: use of resource: introduction
- clear-cutting, slash and burn, swidden agriculture, settlement, and forest fragmentation: living resources: use of resource: tropical forest
- forest preservation and debts-for-nature swap: living resources: use of resource: tropical forest
- natural fire and prescribed burns: living resources: use of resource: temperate forests
- reforestation, old- and second-growth forests: living resources: use of resource: temperate forests
- ranching, overgrazing, and desertification of rangeland: living resources: use of resource: rangelands

Above: Golden monarch butterflies fill the air in their overwintering sanctuaries in the mountains west of Mexico City.

Disappearing Butterfly Forests

Every fall, in one of the most remarkable spectacles in nature, somewhere around a quarter of a billion monarch butterflies (*Danaus plexippus*) flutter southward from Canada and the United States into the mountains of central Mexico, where they spend the winter in cool, high-altitude, oyamel fir forests (fig. 14.1). Incredibly, some of these delicate insects travel up to 4000 km (2500 mi) to reach their winter refuge. Discovered by entomologists only 25 years ago, about a dozen small patches of cloud-forest about 3000 m (9750 ft) in altitude in the mountains west of Mexico City offer exactly the right conditions for monarch hibernation. With night temperatures just above freezing, and cool, foggy days, these forests allow insects to conserve energy and avoid desiccation as they await spring. The average colony contains about 20 million monarchs; some larger ones are thought to have three times this many.

From November to March, tall trees in the mariposa (butterfly) forests are completely covered with bright orange and black blankets as millions of butterflies cling to tree bark and to each other, often breaking off branches with their collective weight (fig. 14.2). On sunny days when the butterflies awaken and flutter down to streams to drink, the air is filled with floating specks of color (see opposite page). As temperatures warm in the spring, more and more butterflies become active until sometime toward the end of March golden streams of insects pour down out of the Sierra Transvolcanica to begin their journey north once again. You can follow the migration and learn more about monarchs on the web page of Journey North, an interactive environmental science project for school children.

As they move northward through Mexico and then spread out from the Rocky Mountains to the Atlantic Coast, the monarchs follow spring flower emergence, sipping nectar and laying their eggs on milkweed plants providing toxins that protect caterpillars and adults from predators. Summer offspring of spring migrants live only about a month, but those hatched in the fall somehow sense the need to fly south, over a route they have never seen, to the cloud forests where they will survive until the next spring.

Unfortunately, the oyamel forests on which this whole cycle depends are rapidly disappearing. Representing less than two percent of all Mexican forests, this unique ecosystem is one of the rarest and the most endangered in the whole country. Wood harvesting and fires—both accidental and deliberately set to clear land for agriculture—are the greatest threats. In 1986 a presidential decree created the "Reserva de la Biosfera Mariposa Monarca" protecting 161,100 ha (about 60 sq mi) including 5 of the 12 known monarch overwintering areas. The decree provided two levels of protection: a "nuclear" zone, in which no timber harvest is allowed, and a buffer zone, in which limited cutting is permitted.

Most of the land from which the reserve was created was "ejido" or communal property, and local ejido members were never compensated properly for lost income where the logging ban was enforced. Consequently, logging—both legal and illegal—has continued in the sanctuary. Increasing numbers of ecotourists, who come to see the amazing monarch concentrations, provide some income for local communities, but tourism lasts only for about five winter months. Few peasant families can make enough money in this short season to last a whole year. Some conservationists are undertaking economic development projects to compensate residents for the costs of protecting this officially designated "threatened phenomenon" and the forests on which it depends. The Monarch Butterfly Sanctuary Foundation, for instance, is paying an ejido with land rights in the Sierra Chincua sanctuary not to log in the reserve buffer zone.

This case study exemplifies issues affecting forests around the world. Growing human populations with increasing standards of living put pressures on forests that are home to threatened species. In some cases, forest destruction is the work of large, transnational corporations; in others, poor peasants cut trees to clear farmland or to make a little money from forest products. In either case, events happening in places we've never heard of affect species and essential ecological processes we consider important. Finding alternative sources of forest commodities and ways to bring about economic development that provides sustainable livelihoods not dependent on forest depletion are worldwide challenges.

In this chapter we'll look at the state of world forests and grazing lands, examining how we use these lands, some problems our uses create, and ways in which threatened ecosystems can be managed more sustainably.

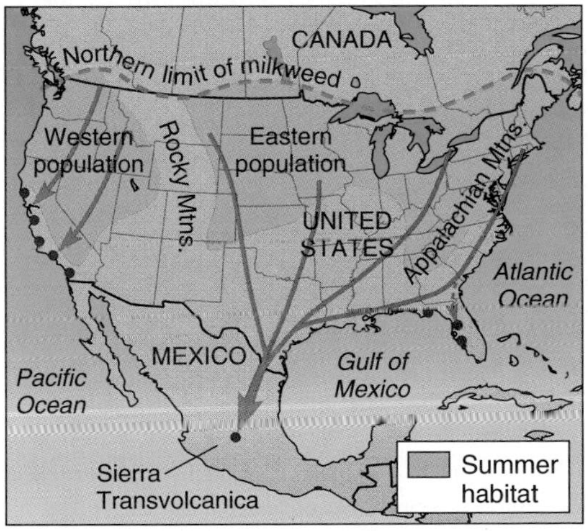

FIGURE 14.1 Fall migration routes and wintering sites for monarch butterflies. Some monarchs winter in Florida and along the Gulf Coast, but most of the population east of the Rockies migrates to Mexico. Western monarchs winter in California.
Source: Data from Brower and Malcolm, *Animal Migration: Endangered Phenomenon.*

FIGURE 14.2 Overwintering monarch butterflies cling to the branches of oyamel firs in the cool cloud forest.

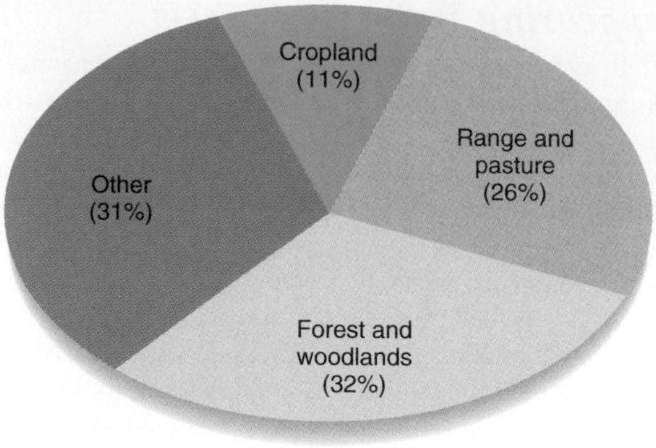

FIGURE 14.3 World land use. The "other" category includes tundra, desert, wetlands, and urban areas.
Source: Data from FAO, 1999.

WORLD LAND USES

The earth's total land area is about 132.4 million sq km (55.9 million sq mi), or about 29 percent of the surface of the globe. Figure 14.3 shows the area devoted to four major land-use categories. Much of the land that falls into the residual "other" category is tundra, marsh, desert, scrub forest, urban areas, bare rock, and ice or snow. About one-third of this land is so barren that it lacks plant cover altogether. While deserts and other unproductive lands are generally unsuitable for intensive human use, they play an important role in biogeochemical cycles and as a refuge for biological diversity. Presently, only about 4 percent of the world's land surface is formally protected in parks, wildlife refuges, and nature preserves (chapter 15).

Notice that approximately 11 percent of the earth's landmass is now used for crops. Some agricultural experts claim that as much as half of the 7.2 billion ha (18 billion acres) of present forests and grazing lands—especially in Africa and South America—could be converted to crop production, given the proper inputs of water, fertilizer, erosion control, and mechanical preparation. Although this land could feed a vastly larger human population (perhaps ten times the present number), sustained intensive agriculture could result in serious environmental and social problems (chapter 11).

Rapidly increasing human populations and expanding forestry and agriculture have brought about extensive land-use changes throughout the world. Humans have affected every part of the globe, and we now dominate most areas with temperate climates and good soils. Over the past 10,000 years billions of hectares of forests, woodlands, and grasslands have been converted to cropland or permanent pasture, but overharvesting, erosion, pollution, and other forms of degradation also havc turned large areas to desert or useless scrub. Biodiversity losses resulting from disruption of natural ecosystems is of great concern (chapter 13). Air pollution-related forest declines are discussed in chapter 18.

Cutting down forests or plowing grasslands have immediate and obvious destructive impacts on landscapes and wildlife. Given enough time, however, nature can be surprisingly resilient. New England, for example, lost most of its native forests to agriculture by the mid-nineteenth century, but today the region is largely reforested. Vermont, which had only 35 percent of its woods standing in 1850, is now 80 percent forested. Upstate New York has three times the population today than it had 150 years ago, yet it also has three times as much forest. Some rare species have been lost, but most of us would have difficulty distinguishing between second growth and primeval woods. Large mammals such as bears, moose, coyotes, and even wolves and mountain lions that had been absent for centuries now are reappearing. While this doesn't give us license to do anything we want to natural landscapes, there is hope that damage can be repaired, as we will discuss later in this chapter.

WORLD FORESTS

Forests play a vital role in regulating climate, controlling water runoff, providing shelter and food for wildlife, and purifying the air. They produce valuable materials, such as wood and paper pulp, on which we all depend. Furthermore, forests have scenic, cultural, and historic values that deserve to be protected. In this section, we will look at forest distribution, use, and management.

Forest Distribution

Before large-scale human disturbances of the world began many thousands of years ago, forests and woodlands probably covered about 6 billion ha (15 billion acres). Since then, about one-third of that area has been converted to cropland, pasture, settlements, or unproductive wastelands. The 4.2 billion ha of forests and woodlands covers around 32 percent of the earth's land surface, nearly three times as much as all croplands. About four-fifths of the forest is classified as **closed canopy** (where tree crowns spread over 20 percent or more of the ground) and has potential for commercial timber harvests. The rest is **open canopy** forest or **woodland,** in which tree crowns cover less than 20 percent of the ground.

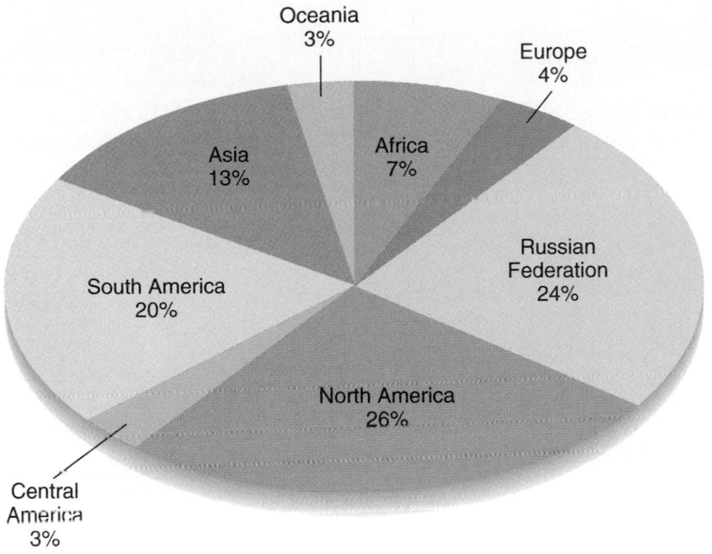

FIGURE 14.4 Current closed-canopy forests as percent of world total.
Source: Data from *World Resources Institute 1998–99,* World Resource Institute.

Figure 14.4 shows the distribution of forest by world region, while figure 14.5 presents the world's main vegetation zones (see chapter 5 for further description of these biomes). Table 14.1 describes closed-canopy forests as a percentage of original forest area by world regions. Among the forests of greatest concern are the remnants of original, primeval forests that are home to much of the world's biodiversity, endangered species, and indigenous human cultures. Sometimes called frontier forests, **old-growth forests** are those that cover a large enough area and have been undisturbed by human activities for a long enough time so that trees can live out a natural life cycle and ecological processes can occur in relatively normal fashion. That doesn't mean that all trees need be enormous or thousands of years old. In some old-growth forests most trees may live less than a century before being killed by disease or some natural disturbance like a fire. Nor does it mean that humans have never been present. Where human occupation entails relatively little impact, an old-growth forest may have been inhabited by people for a very long time. Even forests that have been logged or converted to cropland often can revert to old-growth characteristics if left alone long enough.

While forests still cover roughly two-thirds the area they once did worldwide, only 22 percent of those forests still retain old-growth features. The largest remaining areas of old-growth are in Russia, North America (mainly Canada), South America (mainly Brazil),

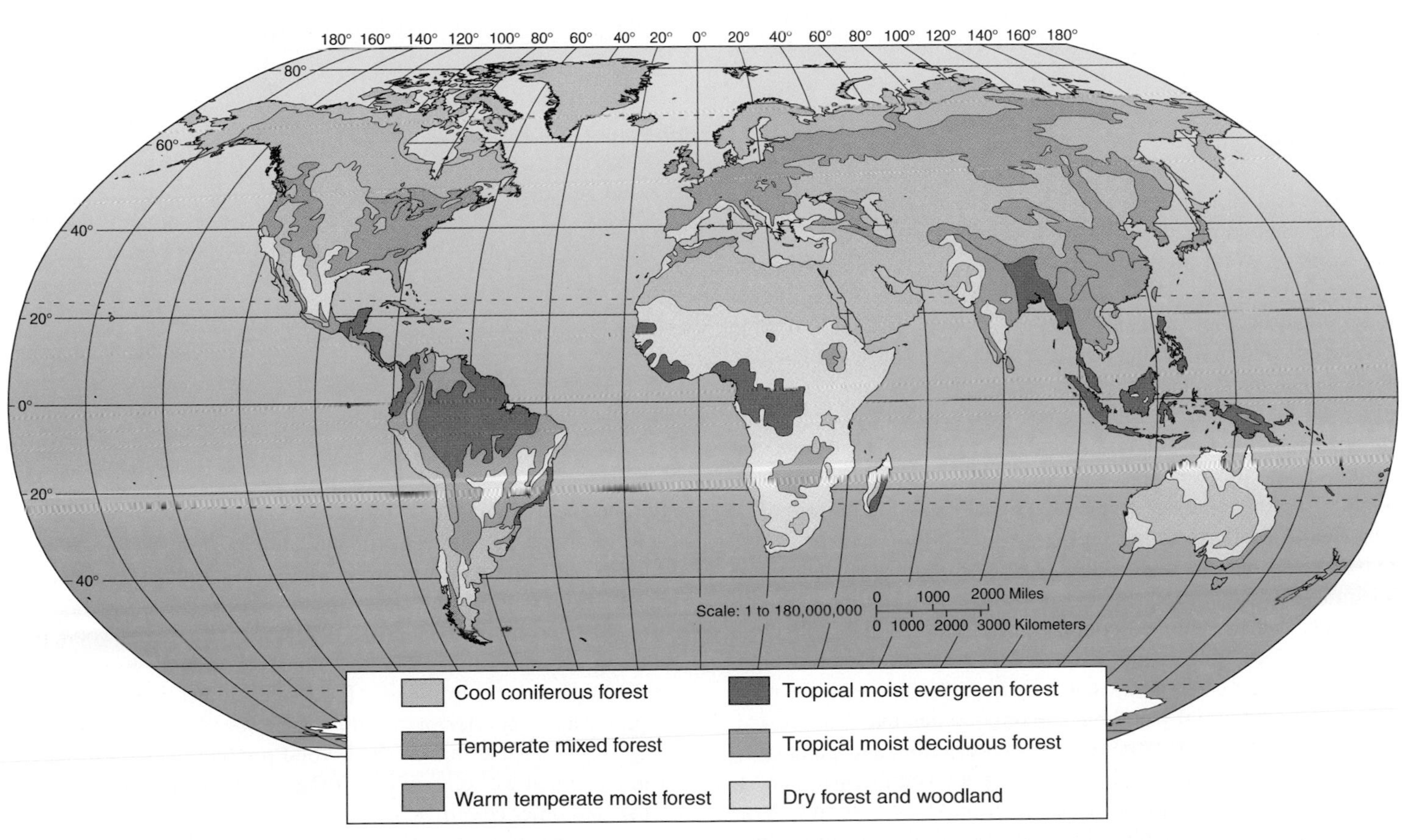

FIGURE 14.5 Main vegetation zones of the world's forests under natural conditions. Tan regions are grassland, tundra, desert, or ice.

TABLE 14.1 Current Closed-Canopy Forest Cover and Threats to Old-Growth Forests

REGION	CURRENT FOREST AS PERCENT OF ORIGINAL	CURRENT OLD-GROWTH AS PERCENT OF ORIGINAL	PERCENT OF OLD-GROWTH THREATENED
Africa	33.9	7.8	77
Asia	28.2	5.3	60
North America	77.3	34.1	26
Central America	54.5	9.7	87
South America	69.1	45.6	54
Europe	30.5	0.1	100
Russia	68.7	29.3	19
Oceania*	64.9	22.3	76
World	53.6	21.7	39

*Oceania consists of Papua New Guinea, Australia, New Zealand, and many small-island nations.
Source: Data from *World Resources Institute 1998–99,* World Resource Institute.

and Oceania (mainly Papua New Guinea). Together these four countries account for more than three-quarters of all relatively undisturbed forests in the world. In general, remoteness rather than laws protect those forests. Although table 14.1 describes only about one-fifth of Russian old-growth as threatened, rapid deforestation—both legal and illegal—especially in the Russian Far East, probably put a much greater area at risk. Note that all the remnants of ancient forest in Europe (mainly in Finland and Sweden) are considered threatened.

Forest Products

Wood plays a part in more activities of the modern economy than does any other commodity. There is hardly any industry that does not use wood or wood products somewhere in its manufacturing and marketing processes. Think about the amount of junk mail, newspapers, photocopies, and other paper products that each of us in developed countries handles, stores, and disposes of in a single day. Total world wood consumption is about 3.7 billion metric tons or somewhat more than 3.7 billion cubic meters annually (table 14.2).

Industrial timber and roundwood (unprocessed logs) are used to make lumber, plywood, veneer, particleboard, and chipboard. Together, they account for slightly less than one-half of worldwide wood consumption (about 1.57 billion tons per year). This exceeds the use of steel and plastics combined. International trade in wood and wood products amounts to more than $100 billion each year. Developed countries produce less than half of all industrial wood but account for about 80 percent of its consumption. Less-developed countries, mainly in the tropics, produce more than half of industrial wood but use only 20 percent.

The United States, the former Soviet Union, and Canada are the largest producers of both industrial wood (lumber and panels) and paper pulp. Although old-growth, virgin forest with trees large enough to make plywood or clear furniture lumber is diminishing everywhere, much of the industrial logging in North America and Europe occurs in managed forests where cut trees are replaced by new seedlings. In contrast, tropical hardwoods in Southeast Asia, Africa, and Latin America are being cut at an unsustainable rate, mostly from virgin forests.

TABLE 14.2 Annual Production and Trade of Wood Products (millions of cubic meters)

REGION	FUEL AND CHARCOAL	INDUSTRIAL TIMBER AND ROUNDWOOD	PAPER	NET TRADE
Africa	481	59	3	–4.2
Asia	850	273	64	48.0
North and Central America	156	575	95	–25.7
South America	244	119	8	–7.6
Europe	51	268	68	13.7
Russia	57	188	5	–11.1
Oceania	9	35	3	–13.5
World	1856	1556	247	—

Notes: A negative number indicates net exports, while a positive number indicates net imports. Roundwood includes lumber, panels, timbers, and other industrial products.
Source: Data from *World Resources Institute 1998–99,* World Resource Institute.

Japan is by far the world's largest net importer of wood, purchasing about 46 million cubic meters per year or 96 percent of net Asian imports. Ironically, the United States is both a major exporter and importer of wood because we buy wood and paper pulp from Canada and finished wood products from Asia at the same time that we sell raw logs, rough lumber, and waste paper to Japan and other countries.

More than half of the people in the world depend on firewood or charcoal as their principal source of heating and cooking

FIGURE 14.6 Firewood accounts for almost half of all wood harvested worldwide and is the main energy source for one-quarter of all humans.

fuel (fig. 14.6). Consequently, **fuelwood** accounts for slightly more than half of all wood harvested worldwide. Unfortunately, burgeoning populations and dwindling forests are causing wood shortages in many less-developed countries. About 1.5 billion people who depend on fuelwood as their primary energy source have less than they need. At present rates of population growth and wood consumption, the annual deficit is expected to increase from 400 million cubic meters (m^3) in 1995, to 2600 m^3 in 2025. At that point, the demand will be twice the available fuelwood supply. The average amount of wood used for cooking and heating in 63 less-developed countries is about 1 m^3 per person per year, roughly equal to the amount of wood that each American consumes as paper products alone.

Many people in poorer countries cook over open fires that deliver only about one-tenth of the available heat to cooking pots. Inexpensive metal stoves can double this efficiency, while locally made ceramic stoves can be four times as efficient as open fires (fig. 14.7). These stoves can save up to 20 percent of household income for urban families.

FIGURE 14.7 Locally made of inexpensive materials, this ceramic-lined metal Jiko stove and clay pot from Kenya are outstanding examples of appropriate technology. Using only one-fourth as much wood as an open fire, these stoves save forests, reduce air pollution, and save up to 20 percent of household income.

Forest Management

Approximately 25 percent of the world's forests are managed scientifically for wood production. **Forest management** involves planning for sustainable harvests, with particular attention paid to forest regeneration (see Case Study, p. 310). Fires, insects, and diseases damage up to one-quarter of the annual growth in temperate forests. Recently, reduced forest growth and sudden die-off of certain tree species in industrialized countries have caused great concern. It is thought that long-range transport of air pollutants (chapter 18) is contributing to this sudden forest death, but not all the causes and solutions are yet understood.

Most countries replant far less forest than is harvested or converted to other uses, but there are some outstanding examples of successful reforestation. China, for instance, cut down most of its forests one thousand years ago and has suffered centuries of erosion and terrible floods as a consequence. Recently, however, a massive reforestation campaign has been started. An average of 4.5 million ha per year were replanted during the last decade. South Korea also has had very successful forest restoration programs. After losing nearly all its trees during the civil war thirty years ago, the country is now about 70 percent forested again.

In spite of being the world's largest net importer of wood, Japan has increased forests to approximately 68 percent of its land area. Strict environmental laws and constraints on the harvesting of local forests encourage imports so that Japan's forests are being preserved while it uses those of its trading partners. It is estimated that two-thirds of all tropical hardwoods cut in Asia are shipped to Japan.

CASE STUDY

FORESTRY FOR THE SEVENTH GENERATION

Until the nineteenth century, the Menominee Nation occupied nearly half of what is now Wisconsin and northern Michigan. A woodland people, the Menominee hunted, fished, and gathered wild rice. Their name for themselves, "Mano'min ini'niwuk," means wild rice people. In 1854, besieged by smallpox, alcohol, and pressure from land-hungry European settlers, the tribe was forced onto a reservation representing less than 3 percent of their ancestral lands. Although the reservation—which lies along the Wolf River about 50 miles northwest of Green Bay, WI—is in the poorest county of the state, it represents a unique treasure. The forests covering 98 percent of the land make up the densest, most diverse woodlands in the Great Lakes region and the longest-running operation for sustained-yield forestry in the country.

Timber harvesting started in 1854, when 20 million board feet were cut for lumber, planks, firewood, and fence rails. Greedy lumber barons tried to gain control of valuable white pine holdings but the tribe resisted. While the rest of the state was clear-cut, burned over, and turned into farmland, the Menominee insisted on careful, selective cutting of individual trees. By 1890, the tribe had built its own sawmill and carried out the first sustainable harvest management plan in the country. When first inventoried in 1890, the 89,000 ha (234,000 acres) reservation contained 1.3 billion board feet of lumber. Today, after 107 years in which a total of 2.25 billion board feet were harvested, the forest stock has increased to 1.7 billion board feet.

Ironically, the successful forestry operations almost brought about an end to the tribe. By 1959, the Menominee had accumulated a $10 million surplus. The Bureau of Indian affairs declared them too wealthy for continued protected status. Congress officially terminated the tribe in 1960, distributing trust funds and land allotments to individual tribal members. Forestry and mill operations were turned over to a private corporation, which immediately began liquidating reserves and racking up debts. Tribal leaders fought termination and successfully restored reservation status in 1973. Although the forest remains largely undivided, tribal enterprises still are plagued by debts incurred under privatization.

Wise elders set up a simple forest management plan when they began operations a century ago. Rather than manage for short-term yields, as is the case for lands around them, the tribe aims for maximum quantity and quality of native species. They say, "instead of cutting the best, we cut the worst first. We're managing our resources to last forever." They have one of the few lumber operations in the country that preserves old-growth characteristics.

The forest is northern hardwood type with a mixture of sugar maple, beech, hemlock, basswood, yellow birch, white pine, jack pine, and aspen. The heart of the management plan is a continuous forest inventory to determine optimum growth, species balance, ecological health, and cutting cycles. Land is divided into 109 compartments based on 11 species combinations, topography, stand history, and management goals. Two-thirds of the forest is managed for mixed species and ages, with selective cutting on a fifteen-year cycle. About 20 percent is devoted to aspen and jack pine in even-age (clear-cut) stands of no more than 12 ha (30 acres) each. Nearly 400 ha (1000 acres) of white pine utilize a two-step shelterwood program that mimics the natural fire-succession sequence by artificially manipulating the balance of sunlight, competition, and soil disturbance. Judicial use of herbicides, prescribed burns, selective cutting, and rock raking maintain optimum growth and regeneration of this valuable species, which has largely disappeared elsewhere in the Great Lakes forest.

Some 300 people are now employed in the tribal forestry and sawmill operations. Logging is carried out by both Indian and non-Indian private contractors. Current harvest levels are about 30 million board feet per year. Lumber from the tribal mill is certified by the Green Cross organization as "good wood" harvested in socially and environmentally responsible manner.

As Aldo Leopold said in *Sand County Almanac,* the best definition of conservation "is written not with a pen, but with an axe. It is a matter of what a man thinks about while chopping, or while deciding what to chop." He was describing the kind of stewardship practiced on the Menominee reservation. In addition to welcome economic returns, the sustainable harvesting has brought them an aesthetically pleasing forest, spiritual rejuvenation, clean water, and a sense of pride in being Menominee.

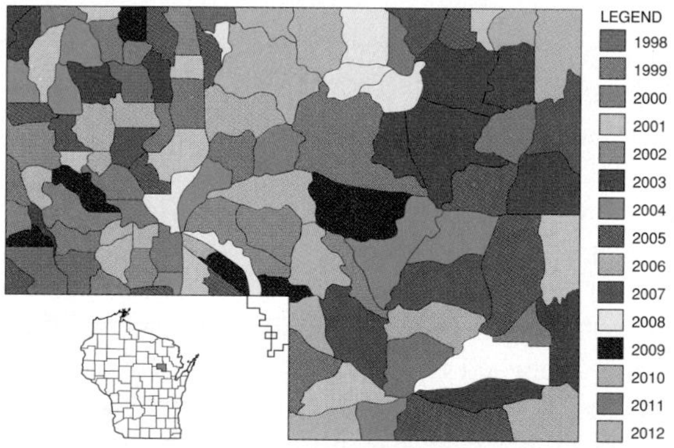

Menominee forestlands are divided into 109 compartments based on species composition, topography, stand history, and management goals. Legend shows planned year for harvest in 15-year cycle.
Courtesy of Menominee Tribal Enterprises. Reprinted by permission.

Many reforestation projects involve large plantations of single-species, single-use, intensive cropping called **monoculture forestry.** Although this produces high profits, a dense, single-species stand encourages pest and disease infestations. This type of management lends itself to mechanized clear-cut harvesting, which saves money and labor but tends to leave soil exposed to erosion. Monocultures eliminate habitat for many woodland species and often disrupt ecological processes that keep forests healthy and productive. When profits from these forest plantations go to absentee landlords or government agencies, local people have little incentive to prevent fires or keep grazing animals out of newly planted areas. In some countries, such as the Philippines, Israel, and El Salvador, government reforestation projects have been targets for destruction by antigovernment forces, with devastating environmental impacts.

Promising alternative agroforestry plans are being promoted by conservation and public service organizations such as The New Forest Fund and Oxfam. These groups encourage planting of mixed species, community woodlots including fruit and nut trees as well as fast-growing, multipurpose trees such as *Leucaena.* Millions of seedlings have been planted in hundreds of self-help projects in Asia, Africa, and Latin America. *Leucaena* is a legume, so it fixes nitrogen and improves the soil. Its nutritious leaves are good livestock fodder. It can grow up to 3 meters per year and quickly provides shade, forage for livestock, firewood, and good lumber for building. A well-managed *Leucaena* woodlot can yield up to 50 tons per hectare on a sustained basis but pest problems have developed in some areas and *Leucaena* can be an aggressive, weedy exotic if it escapes from cultivation. Community woodlots can be planted on wasteland or along roads or slopes too steep to plow so they do not interfere with agriculture. They protect watersheds, create windbreaks, and, if composed of mixed species, also provide useful food and forest products such as fruits, nuts, mushrooms, or materials for handicrafts on a sustained-yield basis.

TROPICAL FORESTS

The richest and most diverse terrestrial ecosystems on the earth are the tropical forests. Although they now occupy less than 10 percent of the earth's land surface, these forests are thought to contain more than two-thirds of all higher plant biomass and at least one-half of all plant, animal, and microbial species in the world.

Diminishing Forests

While many temperate forests are expanding slightly due to reforestation and abandonment of marginal farmlands, tropical forests are shrinking rapidly. In 1900, an estimated 12.5 million square kilometers of tropical lands were covered with closed-canopy forest. This was an area larger than the entire United States. The Food and Agriculture Organization of the United Nations estimates that about 0.8 percent of the remaining tropical forest is cleared each year (fig. 14.8). The species extinction due to this loss of habitat is discussed in chapter 13.

There is considerable debate about current rates of deforestation in the tropics. Scientists at Brazil's National Space Research Institute (INPE) use satellite data to estimate the number of fires and the amount of forest clearing in the Amazon. In 1999, more than 31,000 fires were spotted in a single month (fig. 14.9). Assuming that 40 percent of these fires occurred on recently cleared forest, remote sensing experts calculated that 8 million ha (about 20 million acres) per year were being cut. This number has led to widespread criticism of Brazil for allowing a

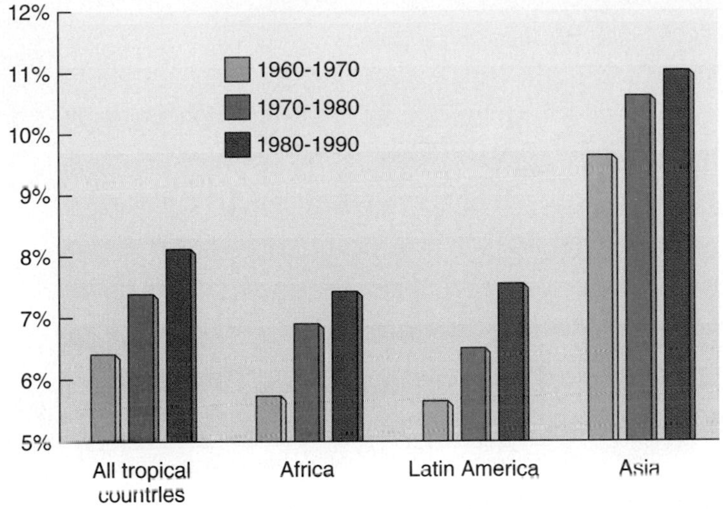

FIGURE 14.8 Estimated rate of tropical forest losses per decade 1960–1990. During this period about one-fifth of all natural tropical forest was cleared.
Source: Data from *World Resources* 1996–97, World Resources Institute.

FIGURE 14.9 Cutting and burning of tropical rainforest results in wildlife destruction, habitat loss, soil erosion, rapid water runoff, and waste of forest resources. It also contributes to global climate change.

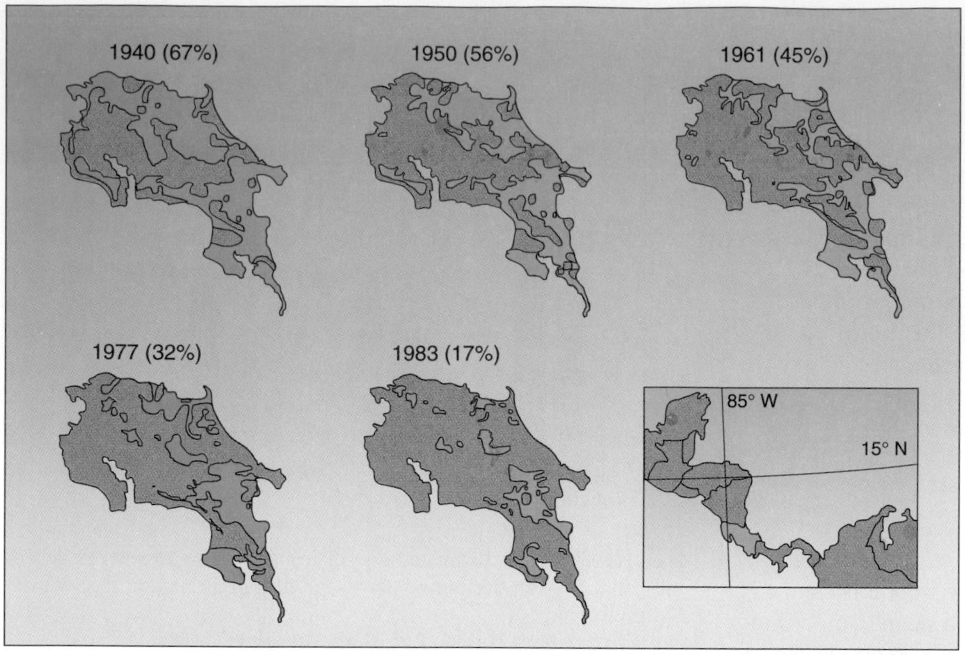

FIGURE 14.10 Loss of primary forest in Costa Rica 1940–1983. Percentages show forestland as a proportion of total land area. Since 1983 almost all remaining primary forest is in National Parks or other protected areas.
Sources: Data from Sader and Joyce, "Deforestation Rates and Trends in Costa Rica, 1940–1983," *Biotropica* 20:11–19; and T.C. Whitmore and G.T. Prance, *Biogeography and Quaternary History in Tropical America,* 1987, Clarendon Press, Oxford.

priceless world heritage to be destroyed. Brazilian scientists reevaluated their data and concluded that losses are far less than originally thought. Critics claim, however, that even the original estimate may be too low.

Similar uncertainty about rates of forest destruction exist in many places, partly because of divergent definitions of deforestation. Some scientists and politicians insist that it means a complete change from forest to agriculture, urban areas, or desert. Others include any area that has been logged, even if the cut is selective and regrowth will be rapid. There also are difficulties in interpreting satellite images. Savannas, open woodlands, and early successional stages following natural disturbance are hard to distinguish from logged areas. Countries also have economic and political reasons to hide or exaggerate the extent of their activities. Consequently, estimates for total tropical forest losses range from about 5 million to more than 20 million ha per year. The FAO estimates of 14.5 million ha are generally the most widely accepted.

By most accounts, Brazil has the highest rate of deforestation in the world, but it also has by far the largest tropical forests. Indonesia and Malaysia together may be losing as much primary forest each year as Brazil even though their original amount was far lower. In 1997, forest fires on Borneo and Sumatra produced clouds of smoke that made air unbreathable in Singapore, the Philippines, and even parts of Thailand and China. Other areas have lost a greater proportion of their original forests.

In Africa, the coastal forests of Ecuador, Sierra Leone, Ghana, Madagascar, Cameroon, and Liberia, already have been mostly destroyed. Haiti was once 80 percent forested; today, essentially all that forest has been destroyed and the land lies barren and eroded. India, Burma, Kampuchea, Thailand, and Vietnam all have little old-growth lowland forest left. In Central America, nearly two-thirds of the original moist tropical forest has been destroyed, mostly within the last 30 years and primarily due to conversion of forest to cattle range (fig. 14.10).

A variety of causes lead to this deforestation. In Costa Rica, in addition to cattle ranching (fig. 14.11), banana plantations consume large areas of forest. In Brazil, a combination of land clearing for cattle ranching coupled with an invasion of immigrants from the south are responsible for forest losses. Although land-hungry farmers often are blamed for setting destructive forest fires, the culprits often are big corporations. In Indonesia and Malaysia, for example, fires are set to cover illegal logging operations or to clear land for large-scale oil palm plantations.

Swidden Agriculture

Indigenous forest people often are blamed for tropical forest destruction because they carry out shifting agriculture that requires forest clearing. Actually, this ancient farming technique can be an ecologically sound way of obtaining a sustained yield from fragile tropical soils if it is done carefully and in moderation. This practice is sometimes called "slash and burn" by people who don't realize how complex and carefully balanced this method of farming actually can be when practiced sustainably. The preferred terms of **milpa** or **swidden agriculture** are taken from local names for *field.*

In this system, farmers clear a new plot of about a hectare (2.5 acres) each year. Small trees are felled and large trees are killed by girdling (cutting away a ring of bark) so that sunlight can penetrate through to the ground. After a few weeks of drying, the branches, leaf litter, and fallen trunks are burned to prepare a rich seedbed of ashes. Fast-growing crops, such as bananas and papayas, are planted immediately to control erosion and to shade root crops, such as cassava and sweet potato, which

FIGURE 14.11 Cattle graze on recently cleared tropical rainforest land in Costa Rica. About two-thirds of the forest in Central America has been destroyed, mostly in the past few decades as land is converted to pasture or cropland. Unfortunately, the soil is poorly suited to grazing or farming, and these ventures usually fail in a few years.

anchor the soil. Maize, rice, and up to 80 other crops are planted in a riotous profusion. Although they would not recognize the terms, these indigenous people are practicing **mixed perennial polyculture.**

The diversity of the milpa plot mimics that of the jungle itself, even though the species representation is more restricted. When managed well, the soil is covered with vegetation. This variety means that crops mature in a staggered sequence and there is almost always something to eat from the plot. It also helps prevent eruptive insect infestations that would plague a monoculture crop. Annual yields from a single hectare can be as high as 6 tons of grain (maize or rice) and another 5 tons of roots, vegetable crops, nuts, and berries. This yield is comparable to the best results with intensive row cropping, and about 1000 times as much food is produced from the same land as when it is converted to cattle pasture.

After a year or two, the forest begins to take over the garden plot again. The farmer will continue to harvest perennial crops for a while and will hunt for small animals that are attracted to the lush vegetation. Ideally, the land then will be allowed to remain covered with jungle vegetation for 10 to 15 years while nutrients accumulate before it is cleared and replanted again. In many places, population growth and displacement of farmers from other areas have forced shifting agriculturalists to reuse their traditional plots on shorter and shorter rotation. When plots are farmed every year or two, nutrients are lost faster than they can be replaced, the forest doesn't regrow as vigorously as it once did, and additional species are lost. Eventually, erosion and overuse reduce productivity so much that the land is practically useless.

Logging and Land Invasions

The other major source of forest destruction is usually the result of logging and subsequent invasion by land-hungry people from other areas. The loggers often are interested only in "creaming" the most valuable hardwoods, such as teak, mahogany, sandalwood, or ebony. Although only one or two trees per hectare might be taken, widespread devastation usually results. Because the canopy of tropical forests is usually strongly linked by vines and interlocking branches, felling one tree can easily bring down a dozen others. Tractors dragging out logs damage more trees, and construction of roads takes large land areas. Insects and infections invade wounded trees. Tropical trees, which usually have shallow root systems, are easily toppled by wind and erosion when they are no longer supported by their neighbors. Up to three-fourths of the canopy may be destroyed for the sake of a few logs. Obviously, the complex biological community of the layered canopy (chapter 4) is severely disrupted by this practice.

What happens next? Bulldozed roads make it possible for large numbers of immigrants to move into the forest in search of land to farm. People with little experience or understanding of the complex rainforest ecosystem try to turn it into farms and ranches (fig. 14.12). Too often, the result is ecological disaster. Rains wash away the topsoil and the tropical sun bakes the exposed subsoil into an impervious hardpan that is nearly useless for farming.

Degradation of rivers is another disastrous result of forest clearing. Tropical rivers carry two-thirds of all freshwater runoff in the world. In an undisturbed forest, rivers are usually clear, clean, and flow year-round because of the "sponge effect" of the thick root mat created by the trees. When the forest is disrupted and the thin forest soil is exposed, erosion quickly carries away the soil, silting river bottoms, filling reservoirs, ruining hydroelectric and irrigation projects, filling estuaries, and smothering coral reefs offshore. In Malaysia, sediment yield from an undisturbed primary forest was about 100 cubic meters per square kilometer per year. After forest clearing, the same river carried 2500 cubic meters per square kilometer per year.

Forest Protection

What can be done to stop this destruction and encourage careful management? While much of the news is discouraging, there are some hopeful signs for tropical forest conservation. Many tropical countries have realized that forests represent a valuable resource and they are taking steps to protect them. Indonesia has announced plans to preserve 100,000 square kilometers, one-tenth of its original forest. Zaire and Brazil each plan to protect 350,000 square kilometers (about the size of Norway) in parks and forest preserves. Costa Rica has one of the best plans for forest protection in the world. Attempts are being made there to not only rehabilitate the land (make an area useful to humans), but also restore the ecosystems to naturally occurring associations. One of the best known of these projects is Dan Janzen's work restoring the dry tropical forest of Guanacaste National Park.

FIGURE 14.12 Logging roads open up access to landless settlers, miners, and hunters who drive away native species and indigenous people. There often is a marked difference between new migrants and indigenous swidden agriculturists.

People on the grassroots level also are working to protect and restore forests. Reforestation projects build community pride while also protecting the land (fig. 14.13). India, for instance, has a long history of nonviolent, passive resistance to protest unfair government policies. The *satyagrahas* go back to the beginning of Indian culture and often have been associated with forest preservation. Gandhi drew on this tradition in his protests of British colonial rule in the 1930s and 1940s. During the 1970s, commercial loggers began large-scale tree felling in the Garhwal region in the state of Uttar Pradesh in northern India. Landslides and floods resulted from stripping the forest cover from the hills. The firewood on which local people depended was destroyed, and the way of life of the traditional forest culture was threatened. In a remarkable display of courage and determination, the village women wrapped their arms around the trees to protect them, sparking the *Chipko Andolan* movement (literally, movement to hug trees). They prevented logging on 12,000 square kilometers of sensitive watersheds in the Alakanada basin. Today, the *Chipko Andolan* movement has grown to more than four thousand groups working to save India's forests.

FIGURE 14.13 School children plant trees in a community reforestation project in Haiti. Grassroots efforts, such as this, benefit both local people and their environment. Many top-down development projects benefit only the ruling elite and multinational corporations.

Debt-for-Nature Swaps

Those of us in developed countries can make a contribution toward saving tropical forests as well. Financing nature protection is often a problem in developing countries where the need is greatest. One promising approach is called **debt-for-nature swaps.** Banks, governments, and lending institutions now hold nearly $1 trillion in loans to developing countries. There is little prospect of ever collecting much of this debt, and banks are often willing to sell bonds at a steep discount—perhaps as little as ten cents on the dollar. Conservation organizations buy debt obligations on the secondary market at a discount and then offer to cancel the debt if the debtor country will agree to protect or restore an area of biological importance.

There have been many such swaps. Conservation International, for instance, bought $650,000 of Bolivia's debt for $100,000—an 85 percent discount. In exchange for canceling this debt, Bolivia agreed to protect nearly 1 million ha (2.47 million acres) around the Beni Biosphere Reserve in the Andean foothills. Ecuador and Costa Rica have had a different kind of debt-for-nature swap. They exchanged debt for local currency bonds that are used to fund activities of local private conservation organizations in the country. This has the dual advantage of building and supporting indigenous environmental groups while protecting the land.

Agreements have been reached with Madagascar and Zambia to swap debts for nature, and negotiations are underway with Peru, Mexico, and Tanzania. Critics charge that these swaps compromise national sovereignty and that they will do little to reduce Third World debt or change the situations that led to environmental destruction in the first place.

TEMPERATE FORESTS

Tropical countries are not unique in harvesting forests at unsustainable rates and in an ecologically damaging fashion. Northern countries have a long history of liquidating forest resources that continues today in many places. Perhaps the largest and most destructive harvest in the world today is taking place in Eastern Russia. Siberia is larger than Amazonia and contains one-fourth of the world's timber reserves. Four million ha (10 million acres) of Siberian taiga and deciduous forests are being felled annually—primarily by Korean loggers—to shore up Russia's faltering economy. China has announced plans for a giant hydroelectric dam on the Amur River, which forms its border with Siberia. This power source will allow China to move 100 million settlers into the region, which is home to the endangered Siberian tiger and several species of cranes. The results are likely to be similar to those we have just discussed in reference to tropical forests.

In the United States and Canada, the two main issues in timber management are (1) cutting of the last remnants of old-growth forest and (2) methods used in timber harvest.

Ancient Forests of the Pacific Northwest

Only a century ago, most of the coastal ranges of Washington, Oregon, northern California, British Columbia, and southeastern Alaska were clothed in a lush forest of huge, ancient trees (fig. 14.14). The moist, mild climate and rich soil of the lowland valleys nurtured magnificent stands of redwood in California and of western red cedar, Douglas fir, hemlock, and Sitka spruce along the rest of the coast. Everyone knows that redwoods can be huge and very old, but did you know that these other species can reach 3 to 4 m (9 to 12 ft) in diameter, 90 m in height (as high as a twenty-story building), and 1000 or more years in age? These temperate rainforests are probably second only to tropical rainforests in terms of terrestrial biodiversity, and they accumulate more total biomass in standing vegetation per unit area than any other ecosystem on earth.

These old-growth temperate rainforests are extremely complex ecologically. Only in recent years have we begun to realize how many different species live there and how interrelated their life cycles are. Many endemic species such as the northern spotted owl (fig. 14.15), Vaux's swift, and the marbled murrelet are so highly adapted to the unique conditions of these ancient forests that they live nowhere else.

Before loggers and settlers arrived, there were probably 12.5 million ha (31 million acres) of virgin temperate rainforest in the Pacific Northwest. Less than 10 percent of that forest in the United States still remains, and 80 percent of what is left is scheduled to be cut down in the near future. British Columbia has felled at least 60 percent of its richest and most productive ancient forests and is now cutting some 240 million ha (600 million acres) annually, about ten times the rate of old-growth harvest in the United States. At these rates, the only remaining ancient forests in North America in 50 years will be a fringe around the base of the mountains in a few national parks.

FIGURE 14.14 This forest in Oregon's Opal Creek Valley is an excellent example of the old-growth temperate rainforests of the Pacific Northwest.

FIGURE 14.15 Only about 2000 pairs of northern spotted owls remain in the old-growth forests of the Pacific Northwest. Cutting old-growth forests threatens the endangered species, but reduced logging threatens the jobs of many timber workers.

Wilderness and Wildlife Protection

Many environmentalists would like to save all remaining virgin forest in the United States as a refuge for endangered wildlife, a laboratory for scientific study, and a place for recreation and spiritual renewal. Economic pressures to harvest the valuable giant trees are considerable, however. The forest products industry

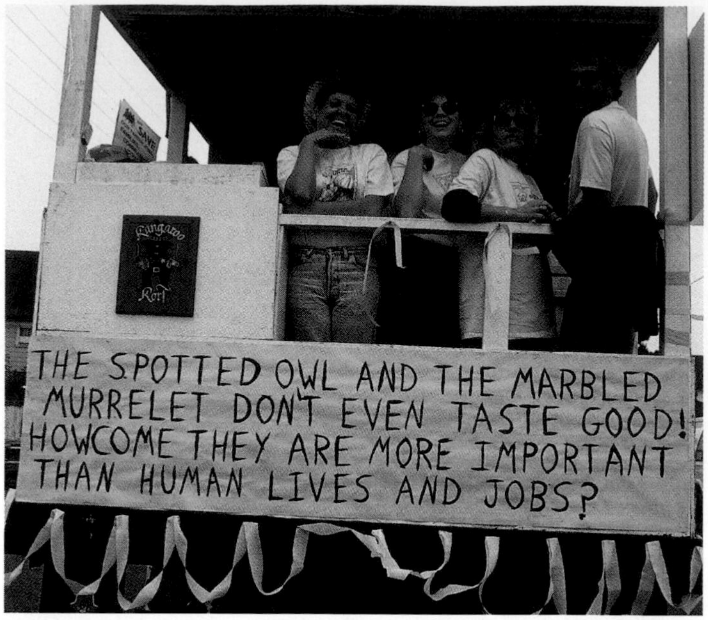

FIGURE 14.16 What worldviews, values, and attitudes are expressed here? When issues become polarized, rational dialogue becomes difficult. Suppose you were an environmentalist trying to save the northwest forest, where might you start to try to find common ground with these folks?

FIGURE 14.17 This huge clear-cut in Washington's Gilford Pinchot National Forest threatens species dependent on old-growth forest and exposes steep slopes to soil erosion. Restoring something like the original forest will take hundreds of years.

employs about 150,000 people in the Pacific Northwest and adds nearly $7 billion annually to the economy. In Oregon, forest products account for one-fifth of the gross state product and many small towns depend almost entirely on logging for their economic life. On the other hand, an economic development study of Washington and Oregon suggests that recreation could provide 16 jobs for every one lost by logging. The biggest problem is how people who are currently employed in the timber industry can make a living during a transition period.

In 1989, environmentalists sued the U.S. Forest Service over plans to clear-cut most of the remaining old-growth forest, arguing that spotted owls are endangered and must be protected under the Endangered Species Act. Federal courts agreed and ordered some 1 million ha (2.5 million acres) of ancient forest set aside to preserve the last 1000 pairs of owls. This is about half the remaining virgin forest in Washington and Oregon. The timber industry claims that 40,000 jobs have been lost, although environmentalists dispute this number. Outrage in the logging communities was loud and clear. Convoys of logging trucks converged on protest sites while angry crowds burned environmentalists in effigy. Bumper stickers urged "Save a logger; eat an owl" (fig. 14.16).

Environmentalists agree that logging jobs are disappearing but claim it is due mostly to mechanization, a naturally dwindling resource base, and the shipping of raw logs to Japan. They argue that the big trees are disappearing anyway. The question is when to stop cutting—now while a few remain or in a few years when they are all gone? The workers will have to be retrained anyway; why not sooner rather than later? In spite of complaints that logging restrictions on public land hurt business, companies with private land holdings have benefited from rising timber prices. Weyerhauser, for instance, reported a $90 million (81 percent) increase in the value of its timber holdings in 1992 as a result of the reduced cut in federal forests.

A compromise forest management plan will allow some continued cutting in the Pacific Northwest but also will protect a high percentage of prime ancient forests. In this plan, 0.9 million ha (2.2 million acres) of riparian (streamside) preserves are set aside to protect salmon and other aquatic wildlife, while 2 million ha (4.9 million acres) of "old-growth" reserves are established on national forest lands. The maximum annual allowable cut on public lands in the region is 1.2 billion board feet, more than environmentalists want, but far below the peak of 5 billion per year in the 1980s. Some of this logging is selective salvage and thinning in old-growth preserves.

Although current plans protect only a fraction of the original forest, local residents in many areas resent *any* restrictions on their access to resources on public or private lands. "Wise Use" groups have brought lawsuits and introduced federal, state, and local ordinances to prohibit programs that limit land use or to require repayment of losses that result from land-use regulations (What Do You Think?, p. 317).

Harvest Methods

Most lumber and pulpwood in the United States and Canada currently is harvested by **clear-cutting,** in which every tree in a given area is cut regardless of size (fig. 14.17). This method enables large machines to fell, trim, and skid logs rapidly, but wastes many small trees, often increases soil erosion, and eliminates habitat for many forest species. Early successional species, such as Douglas fir, jack pine, lodgepole pine, or loblolly pine, on the other hand, flourish after clear-cutting. In a forest managed for these trees, or for raspberries, blueberries, deer, or grouse, this is a good method.

what do you think?

THE WORLD AROUND YOU

Regulations and Property Rights

Should private property owners be entitled to compensation when environmental regulations restrict how their land may be used? That is the subject of a growing debate in Congress and across the country. The issue is referred to as "takings." The matter is rooted in the Fifth Amendment to the United States Constitution, which states that ". . . nor shall private property be taken for public use, without just compensation."

This requirement imposed on the government was designed to make sure that when private land was needed for a public purpose, such as a highway, the government could not simply seize the land, but had to pay for it. A growing number of landowners are claiming that governmental restrictions, environmental or otherwise, that reduce the potential economic gain from their land have the effect of taking away their rights. They argue that this is comparable to "taking" their property and that the Fifth Amendment prohibits this "without just compensation."

Over the years, landowners have brought a variety of lawsuits seeking compensation for lost profits. Pennsylvania passed a mine safety act requiring coal companies to leave part of the coal seams in place underground beneath homes and other surface users to prevent highways, buildings, and gas pipelines from collapsing. A coal company claimed in court that this constituted a "taking" because of the profits lost on unmined coal, and that they should be compensated for these losses.

A rancher in Nevada sued the government to recover the value of grass and water eaten on public land by wild elk herds, claiming this use deprived his cattle of these resources. In California, agricultural interests are concerned about plans to use water to reestablish salmon runs. They are preparing to sue if this reduces subsidized irrigation water available to them. An Arkansas tavern owner even sued the state for income lost because its drunk driver checkpoints caused people to drink less!

To date, courts have rejected most of these suits. Property owners have never been viewed as having unrestricted rights of use. Court decisions have held that property owners do not have a right to use property in ways that harm the community interest as a whole, that the Fifth Amendment does not guarantee owners the right to use land in the most profitable way conceivable, and that use restrictions may be imposed for a wide range of environmental and land-use purposes.

In a precedent-setting decision, the U.S. Supreme Court ruled in 1994 that a regulation denying *all* economic use of a property constitutes a taking. The case involved a South Carolina real-estate developer, named Lucas, who was building homes on beach-front property. When the state passed regulations designed to protect barrier islands from erosion, two of Lucas' lots became unbuildable. He sued and won. The state paid $1.2 million for his two lots. Although a century of legal decisions had held that the government has authority to eliminate harmful land uses even if regulations have an adverse effect on property owners, a newly conservative court set a new direction in this case.

More recently mining, logging, and land development interests, unhappy with court actions on takings, have sought legislation to restrict the scope of government regulations. Such bills have passed in a number of state legislatures, and federal legislation has been introduced in several recent congressional sessions.

Some of these bills would require government agencies to do a thorough analysis of any action that would affect either the use or value of privately owned property. Other bills would require government to pay owners when regulations reduce property value in excess of some determined percentage.

These efforts are opposed by many environmental, consumer, civil rights, and labor groups. They contend that many of the proposals go far beyond constitutional provisions and would paralyze the government's ability to protect the public interest in a clean environment and healthful surroundings. They fear that the bills being proposed would prevent enforcement of many antipollution laws, zoning ordinances, and even obscenity laws (one dial-a-porn company sued the federal government for damages due to rules that restrict the access of children to its services).

In many respects, the controversy over takings represents a classic conflict over two kinds of freedoms: the freedom to act as you like and the right to be protected from the damaging actions of others.

Where should the balance point be between these two types of conflicting rights? To what extent should property rights include the right to negatively impact the rights of others? How would you decide?

Size and shape of clear-cuts vary depending on topography and management policies. Clear-cuts can be in individual strips, alternating rows, small scattered patches, or areas as large as thousands of acres. Natural regeneration is better in small patches or strips that resemble natural forest openings than in huge clear-cuts. Small patches also are less disruptive for wildlife than are huge denuded spaces. It was once thought that good forest management required immediate removal of all dead trees and logging residue. Research has shown, however, that standing snags and coarse woody debris play important ecological roles, including soil protection, habitat for a variety of organisms, and nutrient recycling.

Other harvest practices offer variations on, or substitutes to, clear-cutting. *Coppicing* is used to encourage stump sprouts from species such as aspen, red oak, beech, or short-leaf pine and is usually accomplished by clear-cutting. In *seed tree harvesting,* some mature trees (generally two to five trees per hectare) are left standing to serve as a seed source in an otherwise clear-cut patch. *Shelterwood harvesting* involves removing mature trees in a series of two or more cuts. This encourages regeneration of wind- and sun-sensitive species such as spruce and fir. **Strip cutting** entails harvesting all the trees in a narrow corridor. For many forest types, the least disruptive harvest method is **selective cutting,** in which only a small percentage of the mature trees are taken in each 10- or 20-year rotation. Ponderosa pine, for example, are usually selectively cut to thin stands and improve growth of the remaining trees. A forest managed by selective cutting can retain many of the characteristics of age distribution and ground cover of a mature, old-growth forest.

Logs are skidded out of the forest by a variety of means, depending on slope, economics, and value of remaining trees in the forest. In clear-cuts with grades up to 30 percent, tractors or articulated skidders drag out logs. On steeper slopes, high-lead cables are used. In extremely rough or sensitive terrain, helicopters or balloons may lift out high-grade logs. Some loggers are finding it profitable and much less damaging to the forest to go back to old-fashioned horse or mule skidding (fig. 14.18). Animals make far narrower trails through the forest than big machines and cause less soil compaction, erosion, and scarring of trees that are left.

FIGURE 14.18 Selective logging with draft animals is much less damaging to forest soils and vegetation than using mechanical skidders.

The lush Douglas fir and redwood forests of the rainy Pacific Coast Range provide a vivid illustration of why clear-cutting in old-growth forests is controversial. The rich soil and mild, moist climate of the region promote rapid regeneration after clear-cutting, but many of these forests are on steep slopes where erosion is a serious problem. Hillsides are stripped of soil, which fills streams and smothers aquatic life. British Columbia has felled at least 60 percent of its richest and most productive ancient forests. About 5 percent of the province—but only 2 percent of the temperate rainforest—is preserved in parks, ecological reserves, or recreational areas.

Canadian First Nation people have blocked roads and brought lawsuits to protest destruction of traditional lands and subsistence ways of life. People concerned about commercial and sport fishing have joined the battle, both in British Columbia and in the United States. They argue that salmon spawning depends on the clear cold streams of the native forests. Harvesting timber often destroys this valuable resource. The income from a single year's salmon run can outweigh all the profits from timber harvesting, and the salmon return year after year, while 1000-year-old trees will never be seen again.

Below-Cost and Salvage Sales

Many people are surprised to learn that logging is permitted in national forests, but through most of its history, the U.S. Forest Service has regarded its primary job as providing a steady supply of cheap logs to the nation's timber industry. In many cases, prices charged for timber sales have not been high enough to repay management costs, cleanup of logging debris, or replanting. The result is a hidden subsidy that has cost taxpayers billions of dollars. The nonpartisan U.S. Congressional Research Service reports that in 1997 the net loss on U.S. Forest Service timber sales was at least $1.2 billion, not counting flood damage and wildlife habitat losses. Of the 104 national forests, 83 lost money on timber sales. We could end all timber sales on federal land and provide $30,000 per year for retraining or ecological work for every public land timber worker and still be ahead financially.

The Forest Service admits that costs were higher than income, but claims that the difference was far less than what critics calculate. A part of the discrepancy is how roads are entered on the balance sheet. In the past 40 years, the Forest Service has expanded its system of logging roads more than tenfold to a current total of 547,000 km (340,000 mi) or more than ten times the length of the interstate highway system. Government economists regard building roads into *de-facto* wilderness as a benefit because it opens up the country to motorized recreation and industrial exploitation. Wilderness enthusiasts see this road-building program as destructive and costly.

In 1995, Congress passed, and President Clinton signed, a budget bill containing a special rider allowing timber salvage sales that bypass ordinary environmental constraints or management quotas in areas where fires or diseases have killed large numbers of trees. Proponents claim this emergency action is necessary to protect forest health, to prevent waste of valuable timber, and to replace jobs lost to restrictions such as the northern spotted owl protection plan. Environmental groups are outraged, calling this "logging without laws." The Forest Service, bowing to local pressure for increased economic activity, quickly approved hundreds of sales under this special rider. Many are in old-growth forests that would otherwise be off-limits, and many perfectly healthy trees have been cut in the name of salvage.

A major change in national forest policy was announced in 1999 by new Forest Service chief Mike Dombeck—the first head to come from a forest recreation background. Dombeck declared a moritorium on new road-building in de facto wilderness. He said management decisions will be based on preserving ecosystem integrity, health, and sustainability. An increased emphasis will be placed on recreation, which now accounts for 75 percent of all USFS jobs, compared to 3 percent for logging. Planning will be based on scientific information, and citizens will have an opportunity to participate in decision-making processes. How this new approach plays out remains to be seen.

Fire Management

For more than 70 years, firefighting has been a high priority for forest managers. Smokey the Bear has appeared on posters, brochures, and even postage stamps to tell us of the horrors of wildfires and to warn us that "only you can prevent forest fires." To most people, a blackened, smoking, burned forest appears ruined forever. We envision raging flames devouring helpless wildlife, threatening homes, and wasting valuable timber resources (fig. 14.19). Given such frightening images, it's no wonder that the public demands fire protection and government agencies make every effort to provide it. In 1998 alone, the United States spent over $1 billion and lost 33 lives in efforts to stop forest fires.

Recent studies of the ecological role of fire in forests, however, suggest that much of our horror of fire and our attempts to suppress it may be misguided. As discussed in chapter 4, many biological communities are fire-adapted and require periodic fires for regeneration. In the western United States, for instance, dry montane forests originally were dominated by big trees such as ponderosa pine, Douglas fir, and giant sequoias, whose thick, fire-resistant bark and lack of branches close to the ground protected them from frequent creeping ground fires. Historic accounts describe these forests as open and parklike, with little underbrush, luxuriant grass, and abundant wildlife.

Eliminating fire from these forests has allowed shrubs and small trees to fill the forest floor, crowding out grasses and forbs (herbs that are not grasses). As woody debris accumulates, the chances of a really big fire increase. Small trees act as "fire ladders" to carry flames up into the crowns of forest giants. By preventing low-intensity fires that once kept the forest open and free of fuel, we actually threaten the trees we intend to protect.

Our attempts to put fires out often cause more ecological damage than the fires themselves. Firefighters bulldoze firebreaks through sensitive landscapes such as tundra or wetlands, leaving scars that last far longer than the effects of the fire. Often the only thing that extinguishes a major fire is a change in the weather. Millions of dollars spent to dig fire lines and bomb outbreaks with chemicals and water have little effect as the fire goes where it will.

For 30 years, the U.S. National Parks has had a policy of allowing some natural fires to burn and of setting prescribed fires in ecosystems where scientists consider periodic burns beneficial. The U.S. Forest Service has considered a similar policy for wilderness areas but has been criticized by a public raised on Smokey's fearsome images of fire.

FIGURE 14.19 Is this fire in Yellowstone National Park destroying the forest or renewing it? By suppressing fires and allowing fuel to accumulate, we make major fires such as this more likely. The safest and most ecologically sound management policy for some forests may be to allow natural or prescribed fires that don't threaten property or human life to burn periodically.

We are faced with a dilemma in many forests. After 70 years of fire suppression, fuel has now built up to a point where the next fire could be truly disastrous. The problem is how to remove excess fuel through controlled burning and thinning in a way that will return the forest to more natural conditions and yet protect property, human life, and important biological communities. We may need zoning regulations that restrict building or require fireproof construction methods and land management in highly flammable areas. Public reeducation programs to teach people the role of fire in natural systems is another important factor. Rather than seeing a burned forest as ruined, we can appreciate it as a natural stage in regeneration. This may require a very different environmental ethic than the one Smokey has been preaching and a modified view of our proper role in nature.

RANGELANDS

Pasture (generally enclosed domestic meadows or managed grasslands) and **open range** (unfenced, natural prairie and open woodlands) occupy about 26 percent of the world's land surface. Much of the western Great Plains and the Prairie Provinces of Canada fall in this category (fig. 14.20). The 3.8 billion ha (12 million sq mi) of permanent grazing lands in the world make up about twice the area of all agricultural crops. When you add to this about 4 billion ha of other lands (forest, desert, tundra, marsh, and thorn scrub) used seasonally for raising livestock, more than half of all land is used at least occasionally as grazing lands. More than 3 billion

what can you do?

Lowering Our Forest Impacts

For most urban residents, forests—especially tropical forests—seem far away and disconnected from our everyday life. There are things that each of us can do, however, to protect forests.

- Reuse and recycle paper. Make double-sided copies. Save office paper and use the back for scratch paper.
- Use e-mail. Store information in digital form rather than making hard copies of everything.
- If you build, conserve wood. Use wafer board, particle board, laminated beams or other composites rather than plywood and timbers made from old-growth trees.
- Buy products made from "good wood" or other certified sustainably-harvested wood.
- Don't buy products made from tropical hardwoods such as ebony, mahogany, rosewood, or teak unless the manufacturer can guarantee they were harvested from agroforestry plantations or sustainable-harvest programs.
- Don't patronize fast-food restaurants that purchase beef from cattle grazing on deforested rainforest land. Don't buy coffee, bananas, pineapples or other cash crops if their production contributes to forest destruction.
- Do buy Brazil nuts, cashews, mushrooms, rattan furniture, and other non-timber forest products harvested sustainably by local people from intact forests. Remember that tropical rainforest is not the only biome under attack. Contact the Taiga Rescue Network (http://www.sll.fi/TRN/Taiga News) for information about boreal forests.
- If you hike or camp in forested areas, practice minimum-impact camping. Stay on existing trails, don't build more or bigger fires than you absolutely need to. Use only downed wood for fires. Don't carve on trees or drive nails into them.
- Write to your Congressional representatives and ask them to support forest protection and environmentally-responsible government policies. Contact the United States Forest Service and voice your support for recreation and non-timber forest values.

cattle, sheep, goats, camels, buffalo, and other domestic animals turn plants into protein-rich meat and milk that make a valuable contribution to human nutrition. Sustainable pastoralism can increase productivity while maintaining biodiversity in a grassland ecosystem.

Because grasslands and open woodlands are attractive for human occupation, they frequently are converted to cropland, urban areas, or other human-dominated landscapes. You probably have heard a great deal about tropical rainforest destruction, but worldwide the rate of grassland losses each year is three times that of tropical forest, while the area lost is nearly six times as much. Although they may appear to be uniform and monotonous to the

FIGURE 14.20 This abandoned homestead on the American Great Plains symbolizes the difficulties of living in a harsh, semiarid climate. Some people have suggested that we ought to abandon agriculture in this area and create a large commons for wild animals, native people, and tourism.

untrained eye, native prairies can be highly productive and species-rich. More threatened American plant species occur in rangelands than any other major biome (fig. 14.21).

Range Management

By carefully monitoring the numbers of animals and the condition of the range, ranchers and pastoralists (people who live by herding animals) can adjust to variations in rainfall, seasonal plant conditions, and nutritional quality of forage to keep livestock healthy and avoid overusing any particular area. Conscientious management can actually improve the quality of the range.

Some nomadic pastoralists who follow traditional migration routes and animal management practices produce admirable yields from harsh and inhospitable regions. They can be ten times more productive than dryland farmers in the same area and come very close to maintaining the ecological balance, diversity, and productivity of wild ecosystems on their native range. Nomadic herding requires large open areas, however, and wars, political problems, travel restrictions, incursions by agriculturalists, growing populations, and changing climatic conditions on many traditional ranges have combined to disrupt an ancient and effective way of life. The social and environmental consequences often are tragic.

Overgrazing and Land Degradation

About one-third of the world's range is severely degraded by overgrazing, making this the largest cause of soil degradation (fig. 14.22). Among the countries with the most damage and the greatest area at risk are Pakistan, Sudan, Zambia, Somalia, Iraq, and Bolivia. Usually, the first symptom of improper range management is elimination of the most palatable herbs and grasses. Grazing animals tend to select species they prefer and leave the tougher, less

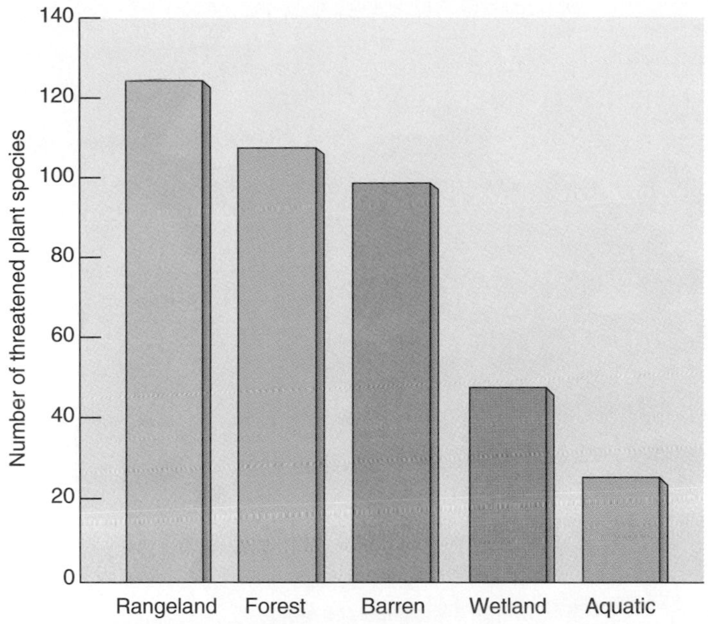

FIGURE 14.21 Threatened plant species in the United States, by ecosystem type, in 1990. Rangelands have more biodiversity and more threatened and endangered species than any other major biome.
Source: Data from Curtis A. Flather, et al., "Species Endangerment Patterns in the United States," *USDA, Forest Service Technical Report* 241, 1994, Table 3, Fort Collins, CO.

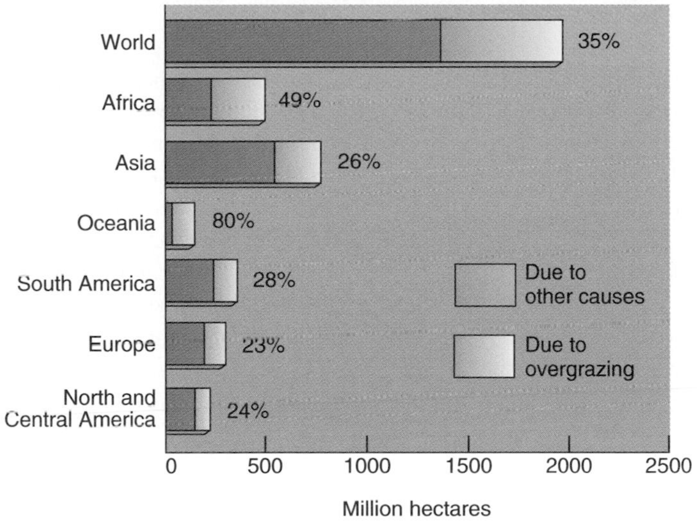

FIGURE 14.22 Rangeland soil degradation due to overgrazing and other causes. Notice that in Europe, Asia, and the Americas, farming, logging, mining, urbanization, etc., are responsible for about three-quarters of all soil degradation. In Africa and Oceania, where more grazing occurs and desert or semiarid scrub make up much of the range, grazing damage is higher.
Source: Data from *World Resources* 1994–95, World Resource Institute.

tasty plants. When native plant species are removed from the range, weedy invaders move in. Gradually, the nutritional value of the available forage declines. As overgrazing progresses, hungry animals strip the ground bare and their hooves pulverize the soil, hastening erosion (fig. 14.23).

FIGURE 14.23 Sheep graze on land in Guatemala that was once tropical forest. Soils in the tropics are often thin and nutrient-poor. When forests are felled, heavy rains carry away the soil and fields quickly degrade to barren scrubland. In some areas, land clearing has resulted in climate changes that have turned once lush forests into desert.

The process of denuding and degrading a once-fertile land initiates a desert-producing cycle that feeds on itself and is called **desertification.** With nothing to hold back surface runoff, rain drains off quickly before it can soak into the soil to nourish plants or replenish groundwater. Springs and wells dry up. Trees and bushes not killed by browsing animals or humans scavenging for firewood or fodder for their animals die from drought. When the earth is denuded, the microclimate near the ground becomes inhospitable to seed germination. The dry barren surface reflects more of the sun's heat, changing wind patterns, driving away moisture-laden clouds, and leading to further desiccation. Because of this process, deserts have been called the footprints of civilization.

This process is ancient, but in recent years it has been accelerated by expanding populations and political conditions that force people to overuse fragile lands. Those places that are most severely affected by drought are the desert margins, where rainfall is the single most important determinant in success or failure of both natural and human systems (fig. 14.24). In good years, herds and farms prosper and the human population grows. When drought comes, there is no reserve of food or water and starvation and suffering are widespread. Can we reverse this process? In some places, people are reclaiming deserts and repairing the effects of neglect and misuse.

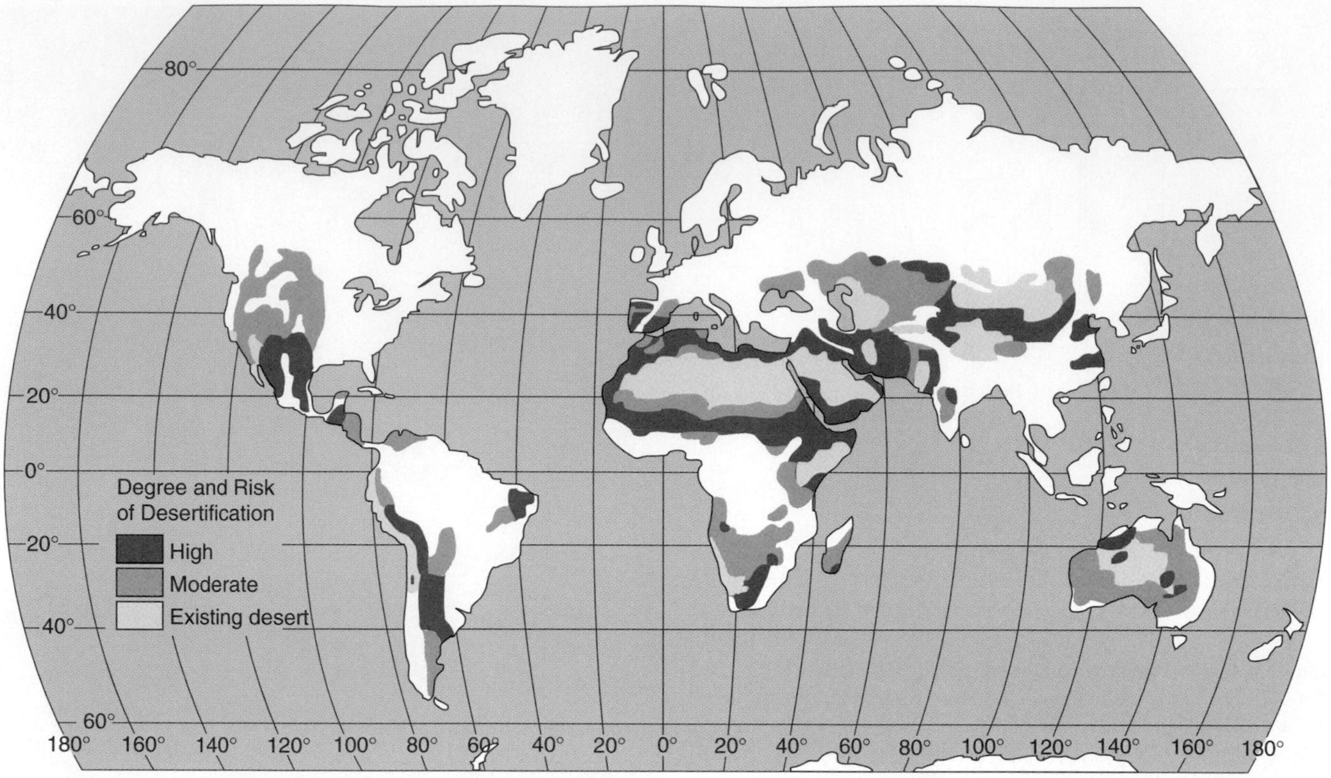

FIGURE 14.24 Degree and risk of desertification of arid lands due to human activities. Note that some existing deserts such as the Sahara in North Africa are so dry and lifeless that only a small amount of additional damage is possible.

As is the case in tropical forests, estimates of the extent of desertification around the world vary widely. Many arid lands are prone to highly variable rainfall with long periods of drought in which natural vegetation disappears and the land lies barren for months or years, making it difficult to distinguish between human-caused changes and normal climatic variations. A recent survey by the United Nations Environment Program found that about half of Africa, for instance, has become significantly drier over the past half century and that arid and hyperarid areas—which are most susceptible to desertification—have increased by about 50 million hectares, while humid and semiarid lands have decreased by the same amount. According to the International Soil Reference and Information Centre in the Netherlands, nearly three-quarters of all rangelands in the world show signs of either degraded vegetation or soil erosion. The highest percentage of moderate, severe, and extreme land degradation is in Mexico and Central America.

Forage Conversion by Domestic Animals

Ruminant animals, such as cows, sheep, goats, buffaloes, camels, and llamas, are especially efficient at turning plant material into protein because bacterial digestion in their multiple stomachs allows them to utilize cellulose and other complex carbohydrates that many mammals (including humans) cannot digest. As a result, they can forage on plant material from which we could otherwise extract little food value. Many grazers have very different feeding preferences and habits. Often the most effective use of rangelands is to maintain small mixed-species herds so that all vegetation types are utilized equally and none is overgrazed. Cattle and sheep, for instance, prefer grass and herbaceous plants, goats will browse on low woody shrubs, and camels can thrive on tree leaves and larger woody plants.

Worldwide, 85 percent of the forage for ruminants comes from native rangelands and pasture. In the United States, however, only 15 percent of livestock feed comes from native grasslands. The rest is made up of crops grown specifically for feed, particularly alfalfa, corn, and oats. Grain surpluses and our taste for well-marbled (fatty) meat have shifted livestock growing in the developed countries to feedlot confinement and high-quality diets. In the United States, roughly 90 percent of our total grain crop is used for livestock feed.

Harvesting Wild Animals

A few people in the world still depend on wild animals for a substantial part of their food. About one-half of the meat eaten in Botswana, for instance, is harvested from the wild. There are good reasons to turn even more to native species for a meat source. On the African savanna, researchers are finding that springbok, eland, impala, kudu, gnu, oryx, and other native animals forage more efficiently, resist harsh climates, are more pest- and disease-resistant, and fend off predators better than domestic livestock. Native species also are members of natural biological communities and demonstrate niche diversification, spreading feeding pressure over numerous plant populations in an area. A study

FIGURE 14.25 Native species, such as these American bison, are a valuable part of grassland ecosystems; they also may be the best way to harvest biomass for human consumption. Native species often forage more efficiently, resist harsh climates, and are more pest- and disease-resistant than domestic livestock.

FIGURE 14.26 On the right side of this fence is a lightly grazed pasture consisting of a dense growth of grasses and small, broad-leaved plants. On the left side is overgrazed rangeland, where the vegetation has become sparse.

by the U.S. National Academy of Sciences concluded that the semi-arid lands of the African Sahel can support only about 20 to 28 kg (44 to 62 lb) of cattle per hectare but can produce nearly three times as much meat from wild ungulates (hooved mammals) in the same area. In the United States, native bison and elk are raised as "novelty" meat sources; might there be a greater future role for ranching of wild animals on our own rangelands (fig. 14.25)?

Rangelands in the United States

The United States has approximately 319 million ha (788 million acres) of rangeland. Most of this rangeland is in the West, and about 60 percent is privately owned. Of the 120 million cattle and 20 million sheep in the United States, only about 2 percent of the cattle and 10 percent of the sheep graze on public rangelands. Federal lands are not very important, overall, in livestock production, but they do have important local economic and environmental ramifications. The Bureau of Land Management (BLM) controls 84 million ha (200 million acres) of grazing lands and the U.S. Forest Service manages about one-fourth as much.

The BLM manages more land than any other agency in the United States, but it is little known outside of the western states where most of its lands are located. It was created in 1946 by a merger of the Grazing Service and the General Land Office. Its formation signaled an end to public land disposal and a commitment to permanent management by the government. The BLM has such a strong inclination toward resource utilization that critics claim the initials really stand for the "bureau of livestock and mining." While only 25 percent of BLM land is considered suitable for grazing, its policies have an important effect on the economy and environment of western states.

State of the Range

The health of most public grazing lands in the United States is not good. Political and economic pressures encourage managers to increase grazing allotments beyond the carrying capacity of the range. Lack of enforcement of existing regulations and limited funds for range improvement have resulted in overgrazing, loss of native forage species, and erosion. The Natural Resources Defense Council claims that only 30 percent of public rangelands are in fair condition, and 55 percent are poor or very poor (fig. 14.26).

Overgrazing has allowed populations of unpalatable or inedible species, such as sage, mesquite, cheatgrass, and cactus, to build up on both public and private rangelands. Furthermore, competing herbivores, such as grasshoppers, jackrabbits, prairie dogs, and feral burros and horses, further damage the range and reduce the available forage. (A **feral** animal is a domestic animal that has taken up a wild existence.) Control programs using poison baits, traps, and hunting to reduce the numbers of these competing native and feral herbivores have been highly controversial. They are seen as both inhumane and dangerous. The case of wild burros and mustangs is especially difficult. They reduce critical winter forage needed by both native wildlife and domestic stock. Most ecologists and land managers believe that the herds of feral animals must be controlled, but attempts to reduce their numbers or remove them from the range meet with vigorous opposition from those who view them as a symbol of the freedom and romance of the West.

Several wildlife conservation groups regard cattle grazing as the most ubiquitous form of ecosystem degradation and the greatest threat to endangered species in the southwestern United States. They call for a ban on cattle and sheep grazing on all public lands, noting that it provides only 2 percent of the total forage consumed by beef cattle and supports only 2 percent of all livestock producers.

Grazing Fees

Another controversial aspect of federal range management is that fees charged for grazing on public lands are far below market value and represent an enormous hidden subsidy to western ranchers.

The 1999 minimum charge for grazing permits on BLM or USFS land was $1.35 per animal unit month (AUM). (One AUM is enough to feed an average cow or five sheep or goats for one month.) Comparable private land rents for $4.86 to $10.26 a head. According to the Congressional Research Service, it costs $3.21/AUM just to administer grazing on USFS or BLM land. The 31,000 permits on federal range bring in only $11 million in grazing fees but cost $47 million per year for administration and maintenance. The $32 million difference amounts to a massive "cow welfare" system of which few people are aware.

In 1993, the Clinton administration proposed raising grazing fees gradually to a minimum of $3.96/AUM. This proposal was defeated by a group of western legislators and rates were rolled back to 1986 levels. Just 10 percent of permit holders on BLM land account for almost two-thirds of the animals allowed to graze those lands; 6 percent of the permit holders on Forest Service lands are responsible for about half of the livestock. The Office of Management and Budget concluded that increasing grazing fees to a fair market value would reduce the livestock herd less than 10 percent and would result in removing animals only from the most marginal areas, which probably should not be grazed anyway.

Rotational Grazing

Where a small number of livestock are free to roam a large area, they generally eat the tenderest, best-tasting grasses and forbs first (wouldn't you?) leaving the tough, unpalatable species to flourish and gradually dominate the vegetation. In some places, farmers and ranchers find that short-term, intensive grazing helps to maintain forage quality. As South African range specialist, Allan Savory, observed, wild ungulates (hoofed animals) like gnus or zebras in Africa or bison (buffalo) in America often tend to form dense herds that graze briefly but intensively in a particular location before moving on to the next area. Rest alone doesn't necessarily improve pastures and rangelands. Short duration, **rotational grazing**—confining animals to a small area for a short time (often only a day or two) before shifting them to a new location—simulates the effects of wild herds. Forcing livestock to eat everything equally, to trample the ground thoroughly, and to fertilize heavily with manure before moving on helps to keep weeds in check and encourages growth of more desirable forage species. This approach doesn't work everywhere, however. Many plant communities in the U.S. desert Southwest, for example, apparently evolved in the absence of large, hoofed animals and can't withstand intensive grazing.

LANDOWNERSHIP AND LAND REFORM

Many of the problems discussed in this chapter have their roots in landownership and public policies concerning resource use. Some land tenure patterns have supported or even created inequality, ignorance, and environmental degradation, while other systems of land use have promoted equality, liberty, and progress. In this final section, we will look at how fair access to the land and its benefits can help bring about better conditions for humans and their environment.

FIGURE 14.27 Landless rural peasants still work most of the land in Asia, Latin America, and Africa while they live in poverty.

Who Owns How Much?

Whatever political and economic system prevails in a given area, the largest landowners usually wield the most power and reap the most benefits, while the landless and near landless suffer at the bottom of the scale. The World Bank estimates that nearly 800 million people live in "absolute poverty . . . at the very margin of existence." Three-fourths of these people are rural poor who have too little land to support themselves. Most of the other 200 million absolute poor are urban slum dwellers, many of whom migrated to the city after being forced out of rural areas (chapter 24).

In many countries, inequitable landownership is a legacy of colonial estate systems in which landless peasants still work in virtual serfdom (fig. 14.27). For instance, the United Nations reports that only 7 percent of the landowners in Latin America own or control 93 percent of the productive agricultural land. By far the largest share of landless poor in the world are in South Asia. In Haiti, the poorest and most environmentally degraded country in the Western Hemisphere, 1 percent of the population owns 90 percent of the land.

The 86 million landless households in India comprise about 500 million people; they are two-thirds of both the total population of India and of all the landless rural people in the world. Not surprisingly, the countries with the widest disparity in wealth are often the most troubled by social unrest and political instability. As Adam Smith said in *The Wealth of Nations* (1776), "No society can surely be flourishing and happy, of which the far greater part of the members are poor and miserable."

Land tenure is not just a question of social justice and human dignity. Political, economic, and ecological side effects of inequitable land distribution affect those of us far from the immediate location of the problem. For example, people in developing countries forced from their homes by increasing farm mechanization and cash crop production often try to make a living on marginal land that should not be cultivated. The local ecological damage caused by their struggle for land can, collectively, have a global impact. The contribution of concentrated landownership to these environmental problems, however, receives less attention than do the threatening ecological trends themselves.

Land Reform

Throughout history, some **land reform** movements seeking redistribution of landownership have been successful and others have not. Significant land reforms have occurred in many countries only after violent revolutionary movements or civil wars. An important factor in revolutions in Cuba, Mexico, China, Peru, the former Soviet Union, and Nicaragua was the breaking of feudal or colonial landownership patterns. This struggle continues in many countries. Hundreds of large and small peasant movements demand land redistribution, economic security, and political autonomy in Latin America, Africa, and Asia. Entrenched powers respond with economic reprisals, intimidation, "disappearances," and even open warfare. Millions of people have been killed in such struggles in recent centuries alone.

Consider Brazil: 340 rich landowners possess 46.8 million ha (117 million acres) of cropland, while 9.7 million families (70 percent of all rural people) are landless or nearly landless. Studies have shown that 13 percent of the land on the biggest estates is completely idle, while another 76 percent is unimproved pasture. These inequities and the tensions they create are important factors in rainforest invasions described earlier in this chapter.

In some countries, land reform has been relatively peaceful and also successful. In Taiwan, only 33 percent of the farm families owned the land they worked in 1949, whereas nearly 60 percent of rural families farm their own land today. South Korea also has carried out sweeping land redistribution. In the 1950s, more than half of all farmers were landless, but now 90 percent of all South Korean farmers own at least part of the land they till. In general, countries that have had successful land reform also have stabilized population growth and have begun industrialization and development.

What are the ecological implications of landownership? Absentee landlords have little personal contact with the land, may not know or care about what happens to it, and often won't let sharecroppers cultivate the same land from year to year for fear they may lay claim to it. This gives tenant farmers little incentive to protect or improve the land because they can't count on reaping any long-term benefits. In fact, if tenants do improve the land or increase their yields, their rents may be raised. Also, where land is aggregated into collective farms or large estates worked by landless peasants, productivity and health of the soil tend to suffer.

Far from being a costly concession to the idea of equality, land reform often can provide a key to agricultural modernization. In many countries, the economic case for land reform rivals the social case for redistributive policies. Many studies have shown that the productivity of owner-operated farms is significantly higher than that of corporate or absentee landowner farms. Independent farmers, especially those who have only a few hectares to grow crops, tend to lavish a great deal of effort and attention on their small plots, and their yields per hectare often are twice those of larger farms.

FIGURE 14.28 George Gillette, chairman of the Fort Berthold Indian Tribe Business Council, weeps as he watches the signing of a 1948 contract to sell the tribe's best land along the Missouri River for the Garrison Dam and Reservoir Project. After signing the contract, Gillette said, "Right now, the future does not look good to us."

Indigenous Lands

No place on the earth is completely unoccupied. Every place, even Antarctica, is claimed by someone, although many places are very sparsely populated. Indigenous (native or tribal) people make up about 10 percent of the world's population but occupy about 25 percent of the land. Many of the approximately 5000 indigenous cultures that remain today possess ecological knowledge of their ancestral lands that is of vital importance to all of us. According to author Alan Durning, "encoded in indigenous languages, customs, and practices may be as much understanding of nature as is stored in the libraries of modern science." And indigenous people continue to play an important role as guardians of wildlife and forests. The Kuna Indians of Panama say, "Where there are forests there are indigenous people, and where there are indigenous people there are forests."

Nearly every country in the world has a sad history of decimation of indigenous cultures and exploitation or annexation of ancestral lands. Time after time, native people have been stripped of their rights through economic pressure, cunning and deceit, or outright theft (fig. 14.28). In the past, colonial governments appropriated tribal lands under the excuse that sparsely populated or uncultivated land was free for the taking. Many nations still fail to recognize indigenous rights. In the Philippines, Cameroon, and Tanzania, for example, the government claims ownership of all forest lands, and indigenous people are considered squatters, even on their ancestral territories.

Our appetite for natural resources and land puts both native people and the ecosystems in which they live at risk. Decimated by plagues, violence, and corruption—and overwhelmed by the onrush of materialistic Western culture—at least half of the world's indigenous cultures are at risk of disappearing within the next century.

FIGURE 14.29 Native tribal people, such as these Yanomami from Brazil, are threatened by tropical forest destruction. These people have lived in harmony with nature for thousands of years. If their culture is lost, valuable knowledge about the forest will be lost as well.

Consider the case of the 9000 Yanomami people who live along the border between Brazil and Venezuela (fig. 14.29). In 1991, Brazil recognized Yanomami claims to 17.6 million ha of land, but invasions by *garimpeiros* (miners) into this territory have introduced malaria, tuberculosis, flu, and respiratory diseases to which native people have no immunity. Placer mining and mercury effluents kill fish and poison the rivers. Wildlife is killed or driven off by miners and hundreds of the Yanomami themselves have been hunted down and shot.

From the Philippines to Labrador, indigenous people are fighting for their ancestral territories. Some countries with large native populations, such as Papua New Guinea, Ecuador, Canada, and Australia, have acknowledged indigenous titles or rights to extensive areas. Often, years of history and modern development complicate these indigenous claims. One aboriginal band in Australia, for instance, claims ownership of Circular Quay, some of the most valuable land in the heart of Sidney. What would be a fair compensation after 150 years of occupation and building?

In April 1999, a Canadian territory called Nunavut (meaning "our home" in the native Inukatat language) was created (fig. 14.30). Spanning nearly 2 million square kilometers or almost 20 percent of Canada's land mass, this vast area of arctic tundra and glaciated mountains will be managed by its 25,000 residents, 90 percent of whom are Inuit (see book cover). Nunavut has some of the harshest weather anywhere in the world. Elesmere National Park at the northern extreme of the territory typically has one week of spring, one week of summer, and one week of fall. The rest of the year is winter, and five months are completely dark. With a population spread among 25 widely scattered villages and only one short paved road in the entire territory, Nunavut faces some difficult challenges. Foremost among these is the difficulty of preserving indigenous culture, skills, and values while also becoming part of the modern world. But the right to self-rule and control over their ancestral territory has brought a new sense of energy and purpose to native people. This settlement might be a good model for resolution of native land claims elsewhere in the world.

SUMMARY

Land has traditionally been a source of wealth and power. The ways we use this limited resource shape our lives and futures. About one-third of the earth's surface is too inhospitable for agriculture, livestock, or forestry but is vital for such purposes as wilderness preservation and recreation. We grow crops on about 11 percent of the total land area. With proper preparation, we could expand cropland in some areas. Most of the earth's land, however, is inappropriate for agriculture and is ruined by attempts to cultivate it.

Forests and woodlands cover about 32 percent of the earth's land area, providing a variety of useful products such as lumber, pulpwood, and firewood. Northern forests are growing faster in most areas of the world than they are being cut and seem in little danger of being exhausted. Tropical forests, on the other hand, are in critical danger. Irreplaceable ecosystems that are home to as many as half of all biological species are being destroyed. Quick profits encourage this exploitation, but hidden costs such as lost wildlife habitat, erosion and devaluation of exposed land, and other disastrous environmental damage will follow from small short-term gains.

A little more than one-quarter of the earth's land is used as range and pasture. Three billion grazing animals convert roughage to protein on poor land that could not otherwise be used to produce food for humans. When herds are managed properly, they actually can improve the quality of their pasture. Unfortunately, about one-third of the world's rangelands are degraded by overgrazing, with disastrous environmental consequences similar to forest destruction.

Land reform is an essential part of sound land-use management. Fair distribution of land and its benefits encourages good stewardship, increased food production, sustainable agriculture, and social justice. Inequitable land ownership, so common today in much of the world, forces the poor to use land unsuited to agriculture, while good land is monopolized by the rich. Dividing land more fairly could increase agricultural productivity and sustainability of the land because farmers who own their own land generally use it more efficiently and more carefully than do absentee landlords. Recognizing indigenous land rights is important both for preserving endangered cultures and for protecting ecological values.

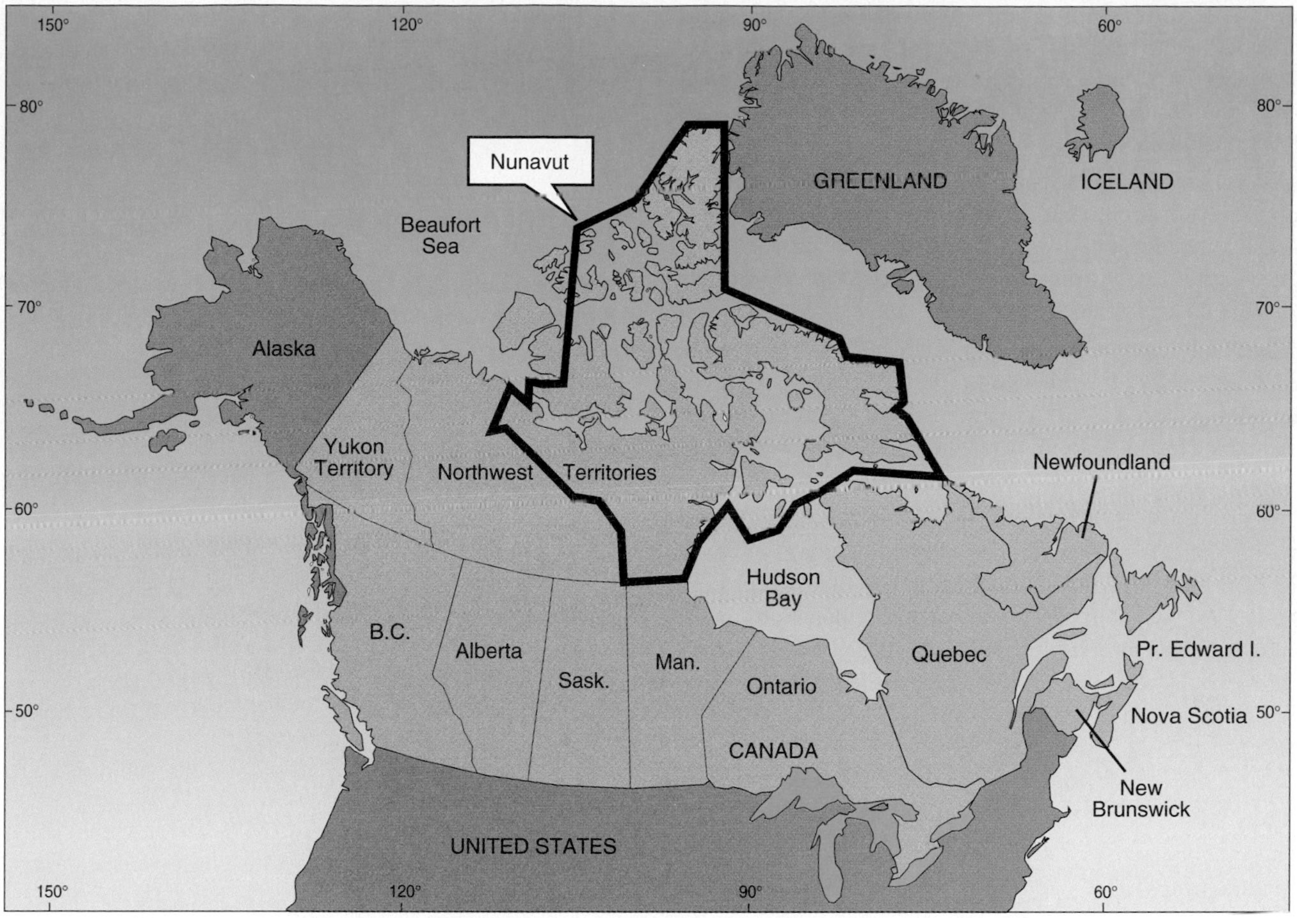

FIGURE 14.30 Nunavut, home to Canada's eastern Inuit people, gained its independence in April 1999. A harsh but beautiful land, most of which lies above the Arctic Circle, Nunavut stretches across nearly 2 million square kilometers of tundra, glaciers, and rocky islands.

QUESTIONS FOR REVIEW

1. Which type of land use occupies the greatest land area?
2. List some products that we derive from forests.
3. What are the advantages and disadvantages of monoculture forestry?
4. Describe milpa (or swidden) agriculture. Why are these techniques better for some fragile rainforest ecosystems than other types of agriculture?
5. What are some results of deforestation?
6. What are clear-cutting and below-cost timber sales, and why are they controversial?
7. How does overgrazing encourage undesirable forage species to flourish?
8. Why can grazing animals generate food on land that would otherwise be unusable for human occupation?
9. What is the relationship between fair land distribution and appropriate land use?
10. Give some examples of recognition of indigenous land titles.

QUESTIONS FOR CRITICAL THINKING

1. Some forestland and rangeland would be suitable as cropland. Do you think it should be converted? Why or why not?
2. Conservationists argue that we could reduce or redirect the demand for wood products; timber companies claim that continued production is essential for jobs and the economy. What evidence would you need to make a decision in this case and where might you get it?
3. Brazil needs cash to pay increasing foreign debts and to fund needed economic growth. Why shouldn't it harvest its forests and mineral resources to gain the foreign currency it wants and needs? If we want Brazil to save its forests, what can or should we do to encourage conservation?
4. Thousands of landless peasants are mining gold or cutting trees to create farms in officially protected Brazilian rainforest. What might the government do to stop these practices? Should it try to stop them?
5. What lessons do you think milpa (or swidden) agriculture has to offer large-scale commercial farming?

6. The U.S. government has kept timber sale prices and grazing leases low to maintain low prices of lumber and meat for consumers and to help support rural communities and traditional ways of life. How would you weigh those human interests against the ecological values of forests and rangelands?
7. Suppose that you live in an area of the Great Plains that is proposed for return to native prairie or that you are a logger in the Pacific Northwest whose job is threatened by a diminishing supply of old-growth timber. How would you feel about the changes discussed in this chapter?
8. Native Americans often were cheated out of their land or paid ridiculously low prices for it. Present owners, however, may not have been a part of earlier land deals and may have invested a great deal to improve the land. What would be a fair and reasonable settlement of these competing land claims?
9. There is considerable uncertainty about the extent of desertification of grazing lands or destruction of tropical rainforests. Put yourself in the place of a decision maker evaluating this data. What evidence would you want to see, or how would you appraise conflicting evidence?

KEY TERMS

clear-cutting 316
closed canopy 306
debt-for-nature swaps 314
desertification 321
feral 323
forest management 309
fuelwood 309
industrial timber 308
land reform 325
milpa agriculture 312
mixed perennial polyculture 313
monoculture forestry 310
old-growth forests 307
open canopy 306
open range 319
pasture 319
rotational grazing 324
selective cutting 318
strip cutting 318
swidden agriculture 312
woodland 306

ADDITIONAL INFORMATION ON THE INTERNET

Visit our website at http://www.mhhe.com/environmentalscience/ for specific information about each of the topics below.

Monarch Butterflies
Commercial Logging on Public Lands
Below-cost Timber sales
Congressional Research Service Report on Below-Cost Timber Sales
Trees for the Future—Forest and Forestry Links
The World's Biomes: Forests
The World's Biomes: Grasslands
Rangeland Management: Bureau of Land Management
U.S. Bureau of Land Management
Rainforest Action Network
American Forests—People Caring for Trees and Forests
Forests for Life Campaign: World Wildlife Federation
ICRAF: International Centre for Agroforestry
Lunatree: One person's crusade to save the redwoods
Quincy Library

PROFILE

Wetland Delineator

Andi Moffatt

Andrea (Andi) Moffatt is a wetland delineator for a Minnesota engineering company. In many construction and development projects, a determination must be made about possible impacts on wetlands. You might think it would be simple to determine what is a wetland and what isn't. A werland should be wet, right? A duck should splash when it lands in one, shouldn't it? Well, not always. Areas that are wet only part of the year may still qualify as wetlands if they contain plants characteristic of water-saturated soils. Andi surveys the plant community and soil conditions on the construction site to determine where the upland ends and the wetland starts. She also looks at hydrology of the area to see how surface runoff might affect adjacent wetlands.

Education is an important part of Andi's job. She works with city councils and municipal officials to help them understand wetland types and values. She also fills out permit applications and environmental assessment worksheets. So far, none of the projects her company has worked on has required a full-blown environmental impact assessment, but if one comes up, Andi will probably be involved in its preparation.

Another aspect of Andi's job is called wetland function and value assessment. She looks at groundwater quality, plant species present, and other indicators of ecological health. In helping cities plan for smart development, Andi sometimes has to make judgements about which wetlands can and should be saved and which ones are too degraded to be repaired. When a project damages a wetland, a mitigation project usually is required to restore or recreate an equal or greater area. Andi helps plan and supervise mitigation and restoration projects.

How does one become a wetland delineator? Andi majored in biology as an undergraduate at the College of St. Catherine, a private women's college. Although her college gave her excellent critical thinking and communication skills, most of her biology classes were in areas she didn't want to pursue as a career. A three-month internship with the Minnesota Department of Natural Resources doing vegetation mapping taught her some plant taxonomy, and also helped her decide to pursue jobs in field biology. Feeling that she needed more field biology courses, Andi applied to a master's progam in plant pathology where she studied wild rice pathogens and their hosts. The experience she gained in identifying wetland plants proved the key to her present job.

What are the best aspects of environmental engineering? Andi likes the combination of field and office work. She enjoys problem solving and working with people. Most of all, she says she feels good about contributing to a balance between environmental protection and development. People need homes and jobs, but those needs can be met while still maximizing environmental quality. What does she dislike? Sometimes it's hard to be the only biologist in a company full of engineers. Coming from different educational backgrounds, they often see projects in different ways than Andi does. The pay scale is lower for biologists as well. Sometimes she finds herself caught in the middle of battles between city officials and agency personnel. Although she'd rather save every wetland, Andi does feel good about being able to work with cities and agencies to make the best possible choices for both the environment and city residents.

How could you prepare for a job like this? Andi suggests a strong background in plants, chemistry, and soils. Good writing and communication skills are essential to help you explain ideas accurately and effectively. Ultimately, the product for most of her work is a written or oral report. Getting as much practical experience as possible is helpful. When she interviewed for her job, Andi was expected to present a portfolio of previous work. Fortunately she had examples of projects to show. Be outgoing. Every person you meet can help you in some way. "The more people you talk to," Andi says, "the better your chances will be."

Although her job is fast paced and sometimes stressful, Andi likes the fact that she can influence people who actually make real-world decisions. Looking at a wetland she has saved or helped restore is very rewarding. You can't stop all development, but there are ways to make city growth environmentally beneficial. Andi feels she is contributing to that process.

CHAPTER 15

Preserving Nature

Never doubt that a small group of thoughtful, committed people can change the world; indeed it is the only thing that can.

—Margaret Meade—

OBJECTIVES

After studying this chapter, you should be able to:

- understand the origins and current problems of national parks in America and other countries.
- recount some of the current problems in national parks in the United States and elsewhere in the world.
- evaluate the tension between conservation and economic development and how the Man and Biosphere program and ecotourism projects address this tension.
- explain the need for, and problems with, wildlife refuges and wilderness areas in the United States.
- demonstrate why wetlands are valuable ecologically and culturally and why they are currently threatened worldwide.
- report on current management policies and problems concerning floodplains, coastlines, and barrier islands.

ESSENTIAL STUDY PARTNER

The Essential Study Partner (ESP) CD-ROM that accompanies this textbook contains additional information on the following topics:

- parks, preserves, and Yellowstone National Park: living resources: natural areas: parks and preserves
- wilderness, wildlife refuges, and ecotourism: living resources: natural areas: parks and preserves
- preserve size, extinction, and minimum viable population size: living resources: natural areas: parks and preserves
- preserving biome types and loss of wetlands: living resources: natural areas: wetlands/living resources: natural areas: parks and preserves

Above: Reflection Lake, Mt. Rainier National Park, WA. Most national parks were chosen for scenic beauty rather than biological diversity.

Ecotourism on the Roof of the World

Rising dramatically from the steamy southern jungles of the Ganges River Valley to the icy peaks of the Himalayan mountains on the Tibetan border, Nepal is one of the most scenic countries in the world. Tourists savor the exotic culture of Katmandu or Namche Bazar or hike through lush mountain forests of rhododendron and pine. Offering spectacular scenery, friendly people, and low prices, this charismatic country has become a premiere destination for adventure travelers.

With an annual per capita income of only $170, Nepal is among the poorest countries in the world. The phenomenal increase in visitors over the past 20 years has brought much-needed income but also has caused severe environmental degradation. Forests along popular trekking trails have been decimated to provide firewood for cooking and heating of water for the numerous wealthy outsiders, while tons of garbage and discarded gear litter popular campsites.

One of the most popular Nepalese trekking routes is a three- to four-week circuit of the Annapurna Range in the center of the Himalayan Range (fig. 15.1). Crossing rushing rivers on swaying suspension bridges, passing between the 8167-m Dhaulagiri and the 8091-m Annapurna I (the seventh and tenth highest mountains in the world, respectively), this ancient pilgrim trail follows the Kali Gandaki Valley to holy shrines at Muktinath. Surmounting the 5416-m (17,769-ft) Thorung La pass north of Annapurna, hikers follow the Marsyangdi valley back to the regional center at Pokhara. First opened to foreigners in 1977, this trail now attracts over 45,000 visitors each year.

Most Nepalese benefit very little from tourists who congest their villages, consume resources, and snap photographs incessantly, but the Annapurna region is different. An innovative project was launched in 1985 to alleviate the destructive impact of masses of trekkers and to maximize the income-generating potential of ecotourism. The Annapurna Conservation Area Project (ACAP) is a 2590-km^2 (1000-mi^2) biosphere reserve that serves as an encouraging model for conservation and development in the Third World.

Far different from Western ideals of parks composed of empty, virgin land, the ACAP is home to more than 100,000 people who continue to use resources in traditional ways. The area is divided into five different zones; intensive farming lands around the periphery, protected forest and seasonal grazing areas in the foothills, special management zones along tourist routes, protected regions with high biological or cultural richness, and wilderness areas in the high peaks.

Recognizing that there can be no meaningful conservation without the active involvement of local people, fees paid by visitors to ACAP go directly to residents to manage the preserve. About $500,000 per year finances a variety of conservation, education, and development projects. More than 700 local entrepreneurs have been trained in lodge management, hygiene, and marketing. Forest guards have been hired, latrines built, trails repaired, and schools and clinics built for local people. Trekkers now are required to use kerosene rather than wood. Local tree nurseries provide stock for reforestation projects. Solar panels and water turbines provide renewable energy for both tourists and residents. The area is cleaner, healthier, and more enjoyable for everyone.

This unique and successful experiment gives us a different view of the meaning and purpose of parks and nature preserves than the ideal of pristine, unchanging nature conveyed by most American national parks. It raises some interesting questions about competing needs of human and nonhuman residents and how they might be balanced sustainably. It also provides a model of how protected areas might be designed and managed in other developing countries.

In this chapter, we will study the history of parks, preserves, and wildlife refuges around the world. We will examine other success stories as well as problems facing efforts to protect nature. Because of their great ecological importance, we will pay special attention to wetlands, floodplains, and coastal regions and how they need protection in many places.

FIGURE 15.1 The Annapurna Conservation Area Project directs money from visitors into development and environmental programs directed by, and of benefit to, local residents. This may be a model for preserving nature in other developing countries.

PARKS AND NATURE PRESERVES

Throughout the world, much of the most biologically productive land is in private hands. If we hope to preserve a meaningful sample of natural biodiversity and ecological functions, private lands will have to play a major role. Furthermore, the world is rapidly urbanizing. Already more than half of all humans live or work in urban areas, so if we are to have contact with nature on a regular basis, urban open space will be vitally important. In many cultures, people have set aside special places as parks, preserves, or sacred groves. We can learn much about environmental values and attitudes by studying those places. Furthermore, they serve as benchmarks to gauge what we have lost and what we might aspire to recover in nature.

Park Origins and History

Since ancient times, sacred groves have been set aside for religious purposes and hunting preserves or pleasuring grounds for royalty. As such, they have been reserved primarily for elite members of society. The imperial retreat of the Han emperors of China, for example, built in the second century B.C. near their capital Ch'ang-an, is the earliest landscaped park of which we have detailed description. Large enough to encompass mountains, forests, and marshes as well as palaces and formal gardens, the park reflected Taoist beliefs about the ideal landscape and our place in the cosmos. Great towers and mountaintop pavilions served as retreats from which the emperor could contemplate nature in tranquility. Not all was peace and serenity, however; the emperor and his entourage also enjoyed hunting herds of wildlife maintained for their enjoyment. Although much of the park appeared natural, it required just as much engineering and earth moving to achieve and maintain this wild appearance as in the formal geometric gardens of Renaissance Europe.

Natural landscaping became popular in England during the eighteenth century under the leadership of architects such as Lancelot Brown (known as Capability Brown because he saw a capability for improving nature everywhere). This new organic design rejected the straight lines and rigid symmetry of earlier gardens, opting for sweeping vistas over rolling hills, meadows, forests, and natural-looking ponds and marshes. The illusion of wild nature was carefully contrived, however. Clumps of trees were sculpted to create vistas and miniature buildings were strategically placed to emphasize the receding perspective. Brown built moats and fences hidden in ditches to control access to private property without interrupting the vista, a concept rediscovered by modern zoos (chapter 13).

Perhaps the first public parks open to ordinary citizens were the grand esplanades and the tree-sheltered agora that served as a gathering place in the planned Greek city. Central Park in New York City is an important successor to both this democratic ideal and the naturalistic principles of romantic landscaping. Promoted in 1844 by newspaper editor William Cullen Bryant as a "pleasuring ground in the open air for the benefit of all," the park was to provide healthful open space and contact with nature for the crowded masses of the city (fig. 15.2). A worldwide competition for design of the park was won by Frederick Law Olmstead, who became the father of landscape architecture in the United States.

FIGURE 15.2 Central Park in New York City was one of the first large urban parks designed to provide nature experience and healthful recreation for common people.

Olmstead left New York in 1864 to become the original commissioner of Yosemite Park in California, the first area set aside to protect wild nature in the United States. Yosemite was authorized by President Abraham Lincoln in the midst of the Civil War to protect its resources from the unbridled exploitation common in frontier areas. Because there was no mechanism for running a park at the national level at that time, Yosemite was deeded to the State of California. It was transferred back to the federal government as a national park in 1890.

In 1872, President Ulysses S. Grant signed an act designating about 800,000 ha (almost 2 million acres) of land in the Wyoming, Montana, and Idaho territories as Yellowstone National Park, the first *national* park in the world. Although the initial interest of both the founders and visitors to Yellowstone was the spectacular "curiosities" and natural "wonders" of the geysers, hot springs, and canyons (fig. 15.3), the park was large enough to encompass and preserve real wilderness. Because the territories had no means to manage the area, Yellowstone was made a national park and guarded by the army until the National Park Service was founded in 1916.

As it became apparent that wild nature and places of scenic beauty and cultural importance were rapidly disappearing with the closing of the North American frontier, the drive to set aside more national parks accelerated. Canada's Banff National Park was established in 1885. In the United States, Mount Rainier was authorized in 1899, Crater Lake in 1902, Mesa Verde in 1906, Grand Canyon in 1908, Glacier Park in 1910, and Rocky Mountain National Park in 1915.

FIGURE 15.3 Yellowstone National Park, established in 1872, is regarded as the first national park in the world. Although focused initially on the spectacle of natural curiosities and wonders, it has come to be appreciated for its beauty and wilderness values.

North American Parks

Parks serve a variety of purposes. They can teach us about our past and provide sanctuaries where nature is allowed to evolve in its own way. They are havens not only for wildlife but also for the human spirit. Canada and the United States have greater total amounts of land dedicated to protected areas than any country except Denmark (which protects vast areas of Greenland's ice and snow) and Australia (which has designated great expanses of outback as aboriginal lands and parks). Although Mexico's parks are newer and less extensive than its wealthy neighbors', they contain far more biological and cultural diversity than do parks in either the United States or Canada.

Existing Systems

The U.S. national park system has grown to more than 280,000 sq km (108,000 sq mi) in 376 parks, monuments, historic sites, and recreation areas. Each year about 300 million visitors enjoy this system. The most heavily visited units are the urban recreation areas, parkways, and historic sites. The jewels of the park system, however, and what most people imagine when they think of a national park, are the great wilderness parks of the West. Passage of the Alaska Lands Act of 1980 nearly doubled the national park system. State and local parks occupy only about one-sixteenth as much area as national parks yet have about twice as many visitors.

Canada has a total of 1471 parks and protected areas occupying about 150,000 sq km. Among this group are national parks, provincial parks, outdoor recreation parks, and historic parks. They range in size from vast wilderness expanses such as Wood Buffalo National Park in northern Manitoba or Ellesmere Island National Park Reserve in the Northwest Territories, to tiny pockets of cultural or natural history occupying only a few hectares. Kluane National Park in the Yukon, the new Tatshenshini-Alsek Wilderness in British Columbia, the adjoining Wrangell-St. Elias National Park, and Glacier Bay National Park in Alaska together encompass an area of about 10 million ha, roughly the size of Belgium or ten times as big as Yellowstone National Park. While many ecological reserves in Canada enforce strictly controlled access, other protected areas encourage intensive recreation, allow hunting, logging, or mining, and permit environmental manipulation for management purposes.

Park Problems

Originally, the great wilderness parks of Canada and the United States were seen as fortresses protected from development or exploitation by legal boundaries and diligent park rangers. Most were buffered from human impacts by their remote location and the wild lands surrounding them. Today, the situation has changed. Many parks have become islands of nature surrounded and threatened by destructive land uses and burgeoning human populations that crowd park boundaries. Forests are clear-cut right up to the edges of some parks, while mine drainage contaminates streams and groundwater. Garish tourist traps clustered at park entrances detract from the beauty and serenity that most visitors seek.

Threats to park values come from within as well. Roads, trails, and buildings in U.S. parks have suffered from years of underfunding and neglect. While the number of park visitors increased by one-third over the past decade, park budgets have decreased by about 25 percent. Hobbled by an unresponsive bureaucracy, inadequate budgets, and deteriorating infrastructure, many of the 12,000 employees are overworked, underappreciated, and thoroughly frustrated. The General Accounting Office estimates that the parks need $6 to 8 billion merely for overdue repair and restoration. Closed campgrounds, dirty toilets, uncollected garbage, storm debris on trails and beaches, and fewer nature lectures by rangers are some of the effects visitors notice from park underfunding. Proposals to close some remote parks or to sell corporate sponsorships for parks offend many people. How would you like to have Old Faithful brought to you by a brewing company or the giant redwoods presented by the tobacco industry?

Yosemite National Park exemplifies these problems. On popular three-day weekends, as many as 25,000 visitors crowd into the 18-square-kilometer (7 square-mile) valley floor, which is less than 1 percent of the total park area (fig. 15.4). They fill the valley with noise and smoke, trample fragile meadows and riverbanks, and spend hours in traffic jams. You can buy a pizza, play video games, do your laundry, play golf or tennis, and shop for curios and souvenirs in Yosemite Valley today, but you are less and less likely to experience the solitude and peace of nature extolled by John Muir. Park rangers have become traffic cops and crowd-control specialists rather than naturalists. A general management plan for Yosemite directed that both Park Service and Curry Company—the concessionaire for tourist services—headquarters be removed from the valley along with 370 buildings, employee housing, 17 percent of existing guest rooms, and all automobiles. Little progress has been made toward meeting these goals, however.

FIGURE 15.4 Visitors crowd popular trails in Yosemite National Park on summer weekends, making quiet and solitude impossible to find.

FIGURE 15.5 Off-road machines, such as motorcycles, dune buggies, and four-wheel drive vehicles can cause extensive and long-lasting damage to sensitive ecosystems. Tracks can persist for decades in deserts and wetlands where recovery is slow.

Other parks experience similar difficulties. The four contiguous mountain parks of the Canadian Rockies (Banff, Jasper, Yoho, and Kootenay) receive about 10 million visitors each year. Four towns are located within their boundaries and pressures are mounting to expand hotels, shops, downhill ski facilities, convention centers, and condominiums. In 1999 a proposal was rejected for an open-pit coal mine adjacent to Jasper National Park. On Cape Cod National Seashore and in the new California Desert Park, dune buggies, dirt bikes, and off-road vehicles (ORV) run over fragile sand dunes, disturbing vegetation and wildlife and destroying the aesthetic experience of those who come to enjoy nature (fig. 15.5). In Florida's Everglades National Park, water flow through the "river of grass" has been disrupted and polluted by encroaching farms and urban areas. Wading-bird populations have declined by 90 percent, down from 2.5 million in the 1930s to 250,000 now.

Air pollution is a serious threat to many parks. The haze over the Blue Ridge Parkway is no longer blue but gray-brown because of air pollution carried in by long-range transport. Sulfate concentrations in Shenandoah and Great Smoky Mountains National Parks are five times human health standards, and ozone levels in Acadia National Park in Maine exceed primary air quality standards by as much as 50 percent on some summer days. Visitors to the Grand Canyon once could see mountains 160 km (100 mi) away; now the air is so smoggy you can't see from one rim to the other during one-third of the year. The main culprits are power plants in Utah and Arizona that supply electricity to Los Angeles, Phoenix, and other urban areas. Acid rain threatens sensitive lakes in the high mountains of the West, as well as in New England and eastern Canada. Photochemical smog is damaging the giant redwoods in California's Sequoia National Park and contributing to forest declines in the Adirondacks.

Mining and oil interests continue to push for permission to dig and drill in the parks, especially on the 3 million acres of private inholdings (private lands) in the parks. These forces were successful in excluding mineral lands in the Misty Fjords and Cape Kruzenstern National Monuments in Alaska. Uranium mines at the edge of the Grand Canyon threaten to contaminate the Colorado River and the park's water supply with radioactive contaminants.

Some 80,000 ha (200,000 acres) of **inholdings**—land already in private ownership when parks were established—are in danger of conversion to incompatible uses. Conservation groups urge Congress to appropriate the hundreds of millions of dollars it would take to buy this property to prevent it from being turned into luxury housing and condominiums.

Wildlife

Wildlife is at the center of many arguments regarding whether the purpose of the parks is to preserve nature or to provide entertainment for visitors (fig. 15.6). In the early days of the parks, "bad" animals (such as wolves and mountain lions) were killed so that populations of "good" animals (such as deer and elk) would be high. Rangers cut trees to improve views, put out salt blocks to lure animals to good viewing points close to roads, and otherwise manipulated nature to provide a more enjoyable experience for the guests.

Critics of this policy claim that favoring some species over others has unbalanced ecosystems and created a sad illusion of a natural system. They claim that excessively large elk populations in Yellowstone and Grand Teton National Parks, for instance, have degraded the range so badly that other species such as mice and ground squirrels are being crowded out. Park rangers tried hiring professional hunters to reduce the elk herd, but a storm of

FIGURE 15.6 Wild animals have always been one of the main attractions in national parks. Many people lose all common sense when interacting with big, dangerous animals. This is not a petting zoo.

protest was raised. Sportsmen want to be able to hunt the elk themselves, animal lovers don't want them to be killed at all, and wilderness advocates don't like the precedent of hunting in national parks. The Park Service has retreated to a policy of "natural regulation," intended to let nature take its course. When elk starve to death, as thousands do in a hard winter, however, many people are appalled.

In Yellowstone National Park, unusually cold temperatures and deep snow during the winter of 1996–97 drove bison out of the high country and into surrounding national forest lands in search of food. More than 2000 animals—two-thirds of the entire park herd either starved to death or were gunned down by hunters and game wardens as they crossed park boundaries. Ranchers claimed that bison carry brucellosis and could infect domestic cattle, although there is no record of this ever happening. Native American tribes and animal rights activists protested this slaughter as cruel and unnecessary. Part of the problem is that a herd of 3000 bison is probably more than the carrying capacity of the land within the park, which has become an island of habitat encircled by lands claimed for other purposes by humans. Another reason for bison migration is that heavy snowmobile traffic packs down trails that invite animals to move through deep snow. In 1998, the Fish and Wildlife Service spent nearly $500,000 to build fences and holding corrals to prevent bison from leaving the park. Ironically, the fees paid by grazing rights in the national forests from which bison were excluded amount to about $15,000 per year. Proposals to reduce elk and bison populations by reintroducing predators such as wolves and mountain lions have been highly controversial (see What Do You Think? p. 336).

TABLE 15.1 IUCN Categories of Protected Areas

CATEGORY	DEGREE OF HUMAN IMPACT OR INTERVENTION
1. Ecological reserves and wilderness areas	Little or none
2. National parks	Low
3. Natural monuments and archeological sites	Low to medium
4. Habitat and wildlife management areas	Medium
5. Cultural or scenic landscapes, recreation areas	Medium to high

Source: Data from World Conservation Union, 1990.

New Directions

What else is being done to enhance the visitor's experience of a national park while maintaining the natural ecosystem as much as possible? Several parks have removed facilities that conflict with natural values. In Yellowstone, a big laundry and a cabin ghetto next to Old Faithful have been torn down. Shabby hotels and filling stations that once stood at Norris Junction and Yancy's Hole are gone. The golf course, slaughterhouse, and tent camps at Mammoth Hot Springs also have been removed. In Yosemite, Grand Canyon, and Denali, tourists park their cars and take shuttle buses into the park to reduce congestion and pollution. A poorly performing concessionaire was denied renewal of a contract for Yosemite, something that had never before happened.

There are proposals that a number of parks be closed to cars, and some areas might be closed to tourists altogether to protect wildlife and fragile ecosystems. The International Union for the Conservation of Nature and Natural Resources (IUCN) divides protected areas into five categories with increasing levels of protection and decreasing human impacts (table 15.1). Many of our parks have tried to meet all these goals simultaneously but may have to select those of highest value. Most parks limit the number of overnight visitors. The time may come when park permits will need to be reserved years in advance and visits to certain parks will be limited to once in a lifetime! How would you feel about such a policy? Would you rather visit a pristine, uncrowded park only once or a less perfect place whenever you wanted?

One of the biggest problems with managing parks and nature preserves is that boundaries usually are based on political rather than ecological considerations. Airsheds, watersheds, and animal territories or migration routes often extend far beyond official boundaries and yet profoundly affect communities that we are attempting to preserve. Yellowstone and Grand Teton Parks in northwestern Wyoming are examples of this concept. Although about 1 million ha (2.5 million acres) in total size, these parks probably cannot preserve viable populations of large predators such as

what do you think?

THE WORLD AROUND YOU

Yellowstone Wolves

On a bright moonlit night, a chilling chorus of howls drifts across the Lamar River Valley at the northern edge of Yellowstone National Park in Wyoming. Elk and deer shift nervously at this signal that a new predator—the gray or timber wolf—now roams their territory. Part of a group of about 120 gray wolves in the park, this pack is among the first of their species to inhabit Yellowstone in nearly 70 years. Hunted to extinction throughout the western United States, gray wolves have been making a comeback in the northern Rocky Mountains since the U.S. Fish and Wildlife Service transplanted 66 Canadian wolves into Yellowstone and central Idaho in 1995 and 1996. This reintroduction program is part of a long-discussed effort to restore ecosystems deprived of natural predators. The project has become mired in controversy, however, and the fate of the wolves—and their prey—now depends on decisions of judges far from Yellowstone wilderness.

Few animals arouse as much animosity and admiration as does the wolf. Children's bedtime tales from many cultures teach us to beware of the big, bad wolf. Ancestor to all our pet dogs, the wolf embodies some of our deepest fears and strongest impulses. It is a potent symbol of much that we both love and fear in wild nature. Less than a century ago, an estimated 100,000 wolves roamed the western United States. As settlers moved west, however, wolves were poisoned, shot, trapped, clubbed, or killed wherever and whenever they could be found.

Ecosystems in many predator-less places show clear signs of overpopulation by prey species such as elk and deer. The Yellowstone Park elk herd had grown to some 25,000 animals, probably four or five times the habitat's carrying capacity. Vegetation was overgrazed and populations of smaller animals such as ground squirrels were declining. Several decades ago, ecologists recommended that keystone predators such as wolves be reintroduced to control prey populations. These proposals bring howls of angry protest from local ranchers, who regard wolves as sinister killers that threaten children, pets, livestock and their whole way of life. "Shoot, shovel, and shut-up" is the preferred wolf management of many Westerners. It took more than 20 years to get approval for wolf reintroduction. One of the key compromises was that the transplanted animals were classified as a "nonessential, experimental population," which meant that any animals that threatened livestock could be shot.

Surprisingly, once back in the park, wolves became established very quickly. Taking advantage of the abundant food supply, they tripled their population in just three years. The effects on the ecosystem were immediate and striking. Biodiversity increased noticeably. Fewer elk, deer, and moose meant more food for squirrels, gophers, voles, and mice. Abundant small prey, in turn, led to increased numbers of eagles, hawks, fox, pine martens, and weasels. Large animal carcasses left by the wolves provided a feast for scavengers such as bears, ravens, and magpies. Coyotes, which had become common in the wolf's absence, suffered a 50 percent decline when wolves hunted them down and killed them. This helped small mammals that once were coyote prey. Plants such as grasses, forbs, willows, and aspen flourished in the absence of grazing and browsing pressure. Rangers and naturalists were delighted that the ecosystem was balanced once again. And tourists were thrilled to catch a glimpse of a wolf or to hear them howl.

Are wolves beautiful, thrilling symbols of wild nature or ruthless killers? Reintroduction of these top predators into Yellowstone National Park has enthusiastic support from environmental groups but passionate opposition from local ranchers and hunters.

Most wolf supporters believe that better livestock management and payments to ranchers for any authenticated losses can take care of any problems wolves might cause. Not everyone was happy, however, with the reintroduction program. The American Farm Bureau sued to force removal of the transplanted wolves, claiming they were a different subspecies from the former (extinct) residents. A group of environmental organizations, including the Audubon Society and the Sierra Club, also filed a suit asking that any "native" wolves that migrate into the park on their own not be part of the experimental population. A federal judge joined the two lawsuits. Most environmental groups then withdrew, but a few remained litigants because they opposed killing wolves under any circumstances. In 1998, the U.S. district court ruled that the nonessential experimental designation violates the Endangered Species Act. The judge ordered the transplanted wolves removed from Yellowstone, which really means they will be shot, because there is no other place to which they could be moved and not be an experimental population. This decision was overturned, however, by the Tenth Circuit Court of Appeals.

Wolf reintroduction raises some important questions about the purposes of parks. It also reveals different attitudes toward nature. Which of the ethical perspectives described in chapter 2 do you see represented here? Why do you think some animal rights groups would regard the suffering of individual wolves, who may be shot because they threaten livestock, as more important than the interests of the species as a whole? What role should science play in this dispute? Suppose you were a mediator charged with bringing all the stakeholders together to try to design an acceptable wolf management program in Yellowstone. What strategies would you use and where would you start?

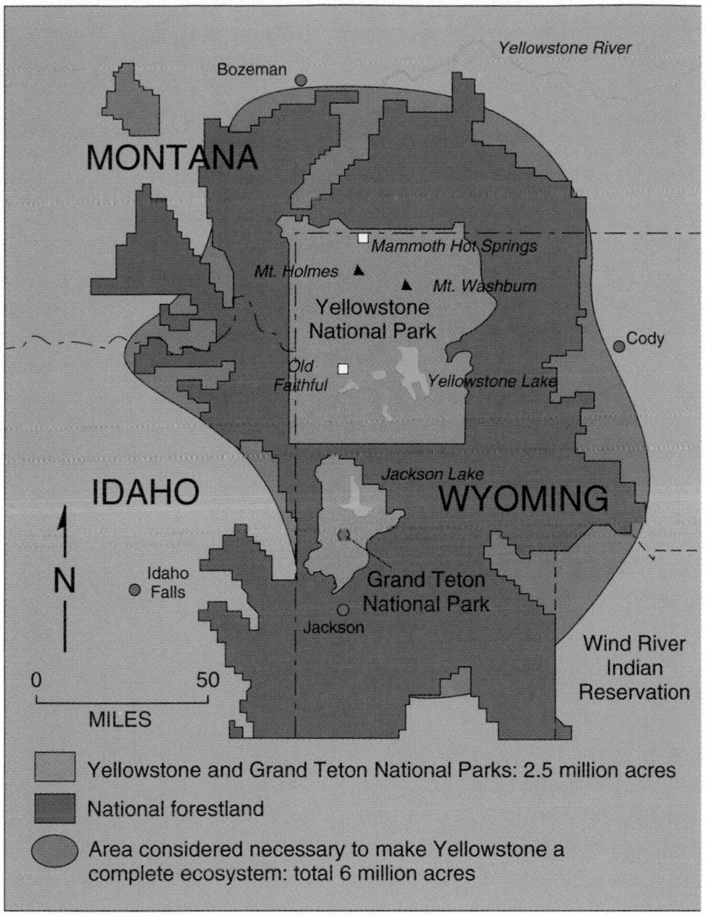

FIGURE 15.7 This map shows the Yellowstone ecosystem complex or biogeographical region, which extends far beyond the park boundaries. Park managers and ecologists believe that it is necessary to manage the entire region if the park itself is to remain biologically viable.

grizzly bears. Management policies in the surrounding national forests and private lands seriously affect conditions in the park (fig. 15.7). The natural **biogeographical area** (an entire self-contained ecosystem and its associated land, water, air, and wildlife resources) must be managed as a unit if we are to preserve all its values.

New Parks

One solution to congestion and overuse of our parks is to create new ones to distribute the load. This would also allow us to protect and enjoy other magnificent areas that deserve preservation in their own right.

The most recent addition to the U.S. park and preserve system is the Grand Staircase-Escalante National Monument, a 660,000 ha (1.7 million acres) tract of desert canyonlands in southern Utah that many people consider the most beautiful place in America (fig. 15.8). Established in 1997 by presidential decree, the monument is a remote wilderness of deep twisting canyons, rock spires, and lofty arches sculpted from the brilliantly colored sedimentary rock by the Escalante River and other small streams as they carve their way

FIGURE 15.8 The Escalante River joins the Colorado in the new Grand Staircase-Escalante National Monument in southern Utah.

through the Kaiparowits Plateau to join the Colorado River. The Grand Staircase is a series of gigantic "steps" carved by the Colorado as the plateau gradually rose in elevation. Each step is about 300 m (900 ft) high and is named for the characteristic color of its rock strata: Pink, Gray, White, or Vermilion Cliffs.

Environmentalists were delighted by protection of an area that had long been a high priority for wilderness status. Local residents, however, were outraged by what they considered to be outright larceny in denying them access to the area's rich resources. Beneath the multihued sandstone is thought to lie 62 billion tons of coal, 2 trillion cubic feet of natural gas, and 2 billion barrels of oil. Total value of the energy resources could be trillions of dollars, although extracting and getting these fuels to market would be difficult and expensive.

As soon as National Monument status was announced, local residents began a rearguard action, cutting roads to nowhere across the monument, hoping to disqualify it as wilderness. Congress also tried to declassify the area or to deny funds for management and protection. Conoco Oil has threatened to start exploratory oil drilling in the monument under terms of preexisting leases. The lesson is that although we love our existing parks, it's not easy to establish new ones.

Some other areas that have been proposed for park status are the Columbia Gorge in Washington and Oregon, Big Sur and Lake Tahoe in California, some of the Florida Keys, and the Hudson Valley and Thousand Islands in New York.

Canada's Green Plan, released in 1990, called for a doubling of protected areas to a total of 12 percent of total land surface. Representative samples of every ecoregion should be included in this network, including cultural landscapes and marine ecosystems. The proposed Matamek Ecological Reserve in Quebec will be over 700 sq km (270 sq mi) in size and will protect an entire watershed.

World Parks and Preserves

The idea of setting aside nature preserves has spread rapidly over the past 50 years as people around the world have become aware of the growing scarcity of wildlife and wild places. Still, in many cultures wilderness and wild lands are regarded as useless wastelands that should be put to some productive use if possible. So far, more than 530 million ha (nearly 4 percent of the earth's land) is designated as parks, wildlife refuges, and nature preserves worldwide (table 15.2).

The largest number of protected areas is in tropical dry forests and savannas—principally in Africa—and temperate deciduous forests—mainly in North America and Europe—but many of these preserves are too small to maintain significant biological populations over the long term. Vast stretches of arctic tundra in a few parks and preserves in Alaska, Canada, Greenland, and Scandinavia make up about 26 percent of all protected areas (fig. 15.9). Deserts and tropical humid forests also are well represented, but grasslands, aquatic ecosystems, and islands are badly underrepresented.

According to the United Nations Environment Programme, North and Central America have the largest fraction—33 percent of all protected land or nearly 10 percent of their land area—designated for protection of any continent. The former Soviet Union, with 17 percent of the world's land, has only 3 percent of the officially protected area. The rapid destruction of Siberian forests (see chapter 14) and terrible pollution problems in Russia and its former allies raises serious concerns about these natural areas. The IUCN has identified an additional 3000 areas, totaling about 3 billion hectares worthy for national park or wildlife refuge status. The most significant of these areas are designated world **biosphere reserves** or world heritage sites. Currently, about 300 of these special refuges have been designated in 75 countries.

TABLE 15.2 World Nature Preserves by Area and Type

BIOME	NUMBER OF AREAS	PERCENT OF PROTECTED AREA
Tropical dry forests	907	21
Tropical humid forests	355	12
Temperate deciduous forests	682	6
Temperate coniferous forests	114	7
Deserts	215	17
Tundra	31	26
Grasslands	116	3
Mountain regions	318	7
Islands	74	<1
Lakes and wetlands	10	<1

Source: Norman Meyers, 1998.

FIGURE 15.9 Hikers explore Ellesmere National Park in Canada's new Nunuvat Territory. Although rich in scenery and solitude, this park has very little biodiversity.

Among the individual countries with the most admirable plans to protect natural resources are Costa Rica, Tanzania, Rwanda, Botswana, Benin, Senegal, Central Africa Republic, Zimbabwe, Butan, and Switzerland, each of which has designated 10 percent or more of its land as ecological protectorates. Brazil has even more ambitious plans, calling for some 231,600 sq km, or 18 percent of the country to be protected in nature preserves. So far, however, many of these areas are parks in name only. Lacking guards, visitor centers, administrative personnel, or even boundary fences, they are open to vandals and thieves to loot as they will.

Protecting Natural Heritage

Even parks with systems in place for protection and management are not always safe from exploitation or changes in political priorities. Many problems threaten natural resources and environmental quality in the parks. In Greece, the Pindos National Park is threatened by plans to build a hydroelectric dam in the center of the park. Furthermore, excessive stock grazing and forestry exploitation in the peripheral zone are causing erosion and loss of wildlife habitat. In Colombia, the Paramillo National Park also is threatened by dam building. Oil exploration along the border of the Yasuni National Park in Ecuador pollutes water supplies, while miners and loggers in Peru have invaded portions of Huascaran National Park. In Palau, coral reefs identified as a potential biosphere reserve are damaged by dynamiting, while on some beaches in Indonesia every egg laid by endangered sea turtles is taken by egg hunters. These are just a few of the many problems in parks around the world. Often countries with the most important biomes lack funds, trained personnel, and experience to manage some of the areas under their control.

The IUCN has developed a **world conservation strategy** for natural resources that includes the following three objectives: (1) to maintain essential ecological processes and life-support systems (such as soil regeneration and protection, recycling of nutrients, and cleansing of waters) on which human survival and development depend; (2) to preserve genetic diversity, which is the foundation of breeding programs necessary for protection and improvement of cultivated plants and domesticated animals; (3) to ensure that any utilization of species and ecosystems is sustainable.

These goals are further elaborated in the ecological plan of action adopted by the IUCN and shown in table 15.3. A promising approach for financing these objectives is debt-for-nature swaps (chapter 14).

Size and Design of Nature Preserves

What is the optimum size and shape of a wildlife preserve? For many years, conservation biologists have disputed whether it is better to have a *s*ingle *l*arge *o*r *s*everal *s*mall reserves (the SLOSS debate). Ideally, a reserve should be large enough to support viable populations of endangered species, keep ecosystems intact, and isolate critical core areas from damaging external forces (see chapter 13). But as gaps are opened in habitat by human disturbance (fig. 15.10), and eventually areas are fragmented into isolated islands, edge effects may eliminate core characteristics everywhere.

To satisfy the conflicting needs and desires of humans and nature, we may need a spectrum of preserves with decreasing levels of interference and management ranging from: (1) *recreation areas,* designed primarily for human entertainment, aesthetics, and enjoyment; (2) *historic areas,* intended to preserve a landscape as we imagine it looked in some previous time—such as pre-settlement or pioneer days; (3) *conservation reserves,* set aside to maintain essential ecological functions, preserve biodiversity, or protect a particular species or group of organisms; (4) *pristine research areas,* to serve as a baseline of undisturbed nature; and (5) *inviolable preserves,* for sensitive species from which all human entrance is strictly prohibited.

TABLE 15.3 IUCN Ecological Plan of Action

1. Launch a consciousness-raising exercise to bring the issue of biological resources to the attention of policymakers and the public at large.
2. Design national conservation strategies that take explicit account of the values at stake.
3. Expand our network of parks and preserves to establish a comprehensive system of protected areas.
4. Undertake a program of training in the fields relevant to biological diversity to improve the scientific skills and technological grasp of those charged with its management.
5. Work through conventions and treaties to express the interest of the community of nations in the collective heritage of biological diversity.
6. Establish a set of economic incentives to make species conservation a competitive form of land use.

Source: Data from International Union for Conservation and National Resources.

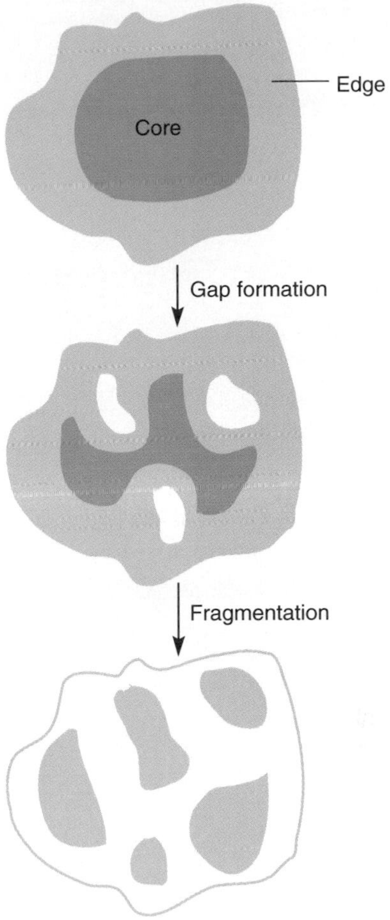

FIGURE 15.10 A patch or "island" of habitat becomes reduced and fragmented as vegetation gaps, or other human disturbances, expand until only small fragments remain. Note that core regions disappear early in this process. Finally, only edges are left in a matrix of disturbed habitat.

For some species with small territories, several small isolated refuges can support viable populations and provide insurance against a disease or other calamity that might wipe out a single population. But small preserves can't support species such as elephants or tigers that need large amounts of space. Given human needs and pressures, however, big preserves aren't always possible. Establishing **corridors** of natural habitat to allow movement of species from one area to another (fig. 15.11) can help maintain genetic exchange and prevent the high extinction rates often characteristic of isolated and fragmented areas.

An interesting experiment funded by the World Wildlife Fund and the Smithsonian Institution is being carried out in the Brazilian rainforest to determine the effects of shape and size on biological reserves. Some 23 test sites, ranging in size from one hectare (2.47 acres) to 10,000 hectares have been established. Some areas are surrounded by clear-cuts and newly created pastures (fig. 15.12), while others remain connected to the surrounding forest. Selected species are regularly inventoried to monitor their dynamics after disturbance. As was expected, some species disappear very quickly, especially from small areas. Sun-loving

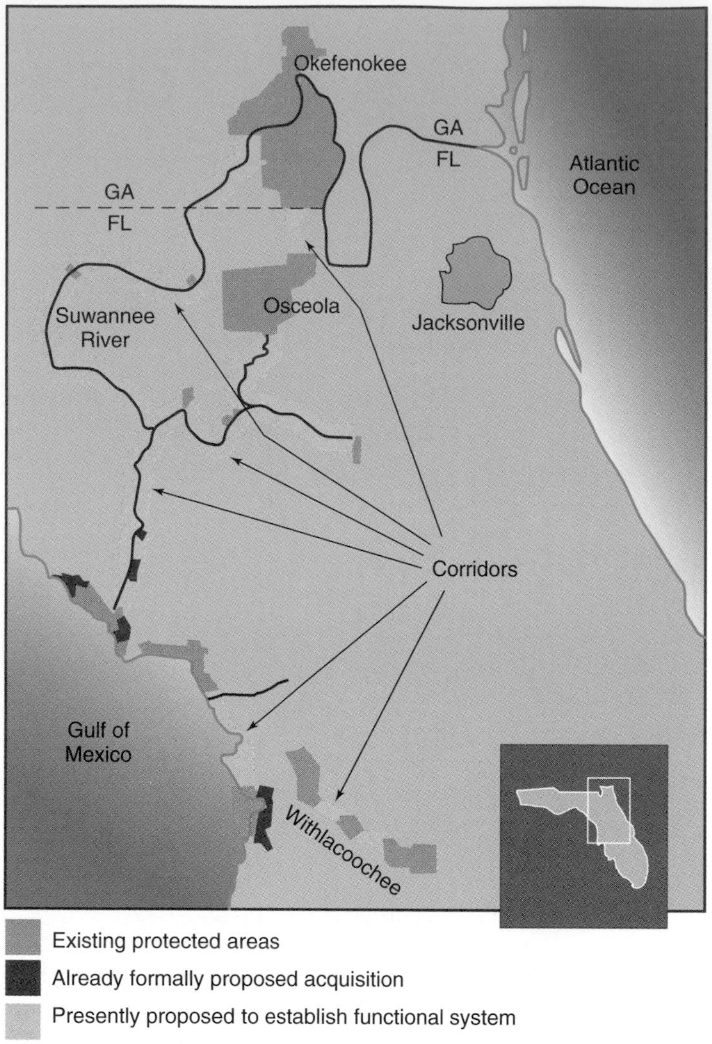

FIGURE 15.11 Corridors serve as routes of migration, linking isolated populations of plants and animals in scattered nature preserves. Although individual preserves may be too small to sustain viable populations, connecting them through river valleys and coastal corridors can facilitate interbreeding and provide an escape route if local conditions become unfavorable.

Source: R.F. Noss and L.D. Harris, "Nodes, Networks and MUMs: Preserving Diversity at All Levels," in *Environmental Management,* vol. 10:299–309, 1986.

FIGURE 15.12 How small can a nature preserve be? In an ambitious research project, scientists in the Brazilian rainforest are carefully tracking wildlife in plots of various sizes, either connected to existing forests or surrounded by clear-cuts. As you might expect, the largest and most highly specialized species are the first to disappear.

species flourish in the newly created forest edges, but deep-forest, shade-loving species move out, particularly when size or shape reduces the distance from the edge to the center below a certain minimum. This demonstrates the importance of surrounding some reserves with buffer zones that maintain the balance of edge and shade species.

Conservation and Economic Development

Many of the most seriously threatened species and ecosystems of the world are in the developing countries, especially in the tropics. This situation concerns us all because these countries are the guardians of biological resources that may be vital to all of us. Unfortunately, where political and economic systems fail to provide people with land, jobs, and food, disenfranchised citizens turn to legally protected lands, plants, and animals for their needs. Immediate human survival always takes precedence over long-term environmental goals. Clearly the struggle to save species and unique ecosystems cannot be divorced from the broader struggle to achieve a new world order in which the basic needs of all are met.

The tropics are suffering the greatest destruction and species loss in the world, especially in humid forests and coastal ecosystems. People in some of the affected countries are beginning to realize that the biological richness of their environment may be their most valuable resource and that its preservation is vital for sustainable development. Ecotourism can be more beneficial to many of these countries over the long term than extractive industries such as logging and mining (fig. 15.13 and Case Study, p. 344). Table 15.4 suggests some ways to ensure that tourism is ecologically and socially beneficial.

At the 1982 World Congress on National Parks in Bali, 500 scientists, managers, and politicians discussed the design and location of biological reserves and the ecological, economic, and social factors that impinge on wildlife preservation. They concluded that conservation and rural development are not necessarily incompatible. In many cases, sustainable production of food, fiber, medicines, and water in rural areas depends on ecosystem services derived from adjacent conservation reserves. Tourism associated with wildlife watching and outdoor recreation can be a welcome source of income for underdeveloped countries. If local people

FIGURE 15.13 Ecotourism can be a sustainable resource use. If local communities share in the revenue, it gives them an incentive to value and protect biodiversity and natural beauty.

TABLE 15.4 Ecotourism Suggestions

1. *Pretrip preparation.* Learn about the history, geography, ecology, and culture of the area you will visit. Understand the do's and don't's that will keep you from violating local customs and sensibilities.
2. *Environmental impact.* Stay on designated trails and camp in established sites, if available. Take only photographs and memories and leave only goodwill wherever you go.
3. *Resource impact.* Minimize your use of scarce fuels, food, and water resources. Do you know where your wastes and garbage go?
4. *Cultural impact.* Respect the privacy and dignity of those you meet and try to understand how you would feel in their place. Don't take photos without asking first. Be considerate of religious and cultural sites and practices. Be as aware of cultural pollution as you are of *environmental pollution.*
5. *Wildlife impact.* Don't harass wildlife or disturb plant life. Modern cameras make it possible to get good photos from a respectful, safe distance. Don't buy products such as ivory, tortoise shell, animal skins, or feathers from endangered species.
6. *Environmental benefit.* Is your trip strictly for pleasure or will it contribute to protecting the local environment? Can you combine ecotourism with work on cleanup campaigns or delivery of educational materials or equipment to local schools or nature clubs?
7. *Advocacy and education.* Get involved in letter writing, lobbying, or educational campaigns to help protect the lands and cultures you have visited. Give talks at schools or to local clubs after you get home to inform your friends and neighbors about what you have learned.

share in the benefits of saving wildlife, they probably will cooperate and the programs will be successful. To reformulate Thoreau's famous dictum, "In broadly shared economic progress is preservation of the wild."

Indigenous Communities and Biosphere Reserves

Areas chosen for nature preservation are often traditional lands of indigenous people who cannot simply be ordered out. Finding ways to integrate human needs with those of wildlife is essential for local acceptance of conservation goals in many countries. In 1986, UNESCO initiated its **Man and Biosphere (MAB) program** that encourages division of protected areas into zones with different purposes. Critical ecosystem functions and endangered wildlife are protected in a central core region where limited scientific study is the only human access allowed. Ecotourism and research facilities are located in a relatively pristine buffer zone around the core, while sustainable resource harvesting and permanent habitation are allowed in multiple-use peripheral regions (fig. 15.14).

Mexico's 545,000-hectare (2100-square-mile) Sian Ka'an Reserve on the Caribbean coast is a good example of a MAB reserve. The core area includes 528,000 ha (1.3 million acres) of coral reef and adjacent bays, marshes, and lowland tropical forest. More than 335 bird species have been observed within the reserve, along with endangered manatees, five types of jungle cats, spider and howler monkeys, and four species of increasingly rare sea turtles. Approximately 25,000 people live in communities in peripheral regions around the reserve, and the resort developments of Cancun are located just to the north. In addition to tourism, the economic base of the area includes lobster fishing, small-scale farming, and coconut cultivation.

The Amigos de Sian Ka'an, a local community organization, played a central role in establishing the reserve and is working to protect the resource base while it improves living standards for local people. New intensive farming techniques and sustainable harvesting of forest products enable people to make a living without destroying the resource base. Better lobster harvesting techniques developed at the reserve have improved the catch without depleting native stocks. Local people now see the preserve as a benefit rather than an imposition from outside. Unfortunately, the government has very limited funds to develop or patrol the reserve.

An even grander plan called *Passeo Pantera* (path of the panther) envisions a thousand-mile-long series of reserves and protected areas interconnected by natural corridors and managed buffer zones that would preserve both wildlife and native cultures along the entire Caribbean coast of Central America from the Yucatan to Panama. Whether the politically unstable and financially troubled

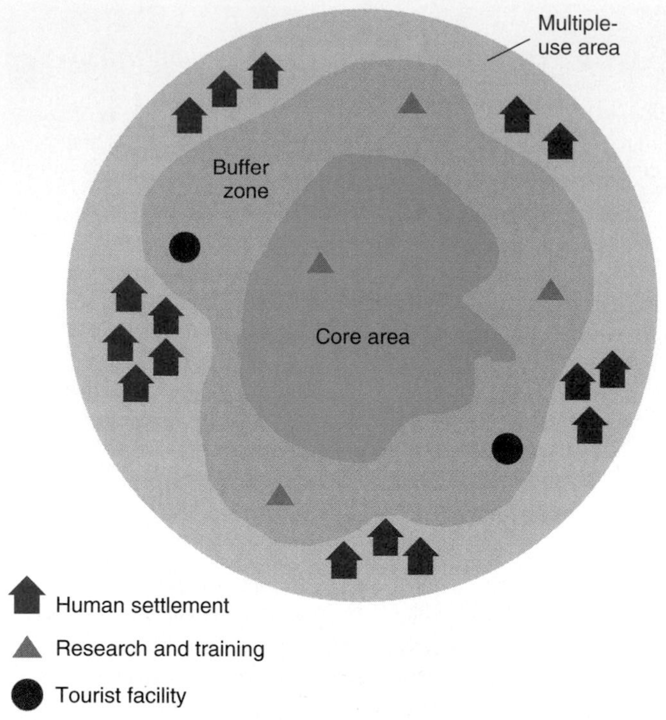

FIGURE 15.14 A model biosphere reserve. Traditional parks and wildlife refuges have well-defined boundaries to keep wildlife in and people out. Biosphere reserves, by contrast, recognize the need for people to have access to resources. Critical ecosystem is preserved in the core. Research and tourism are allowed in the buffer zone, while sustainable resource harvesting and permanent habitations are situated in the multiple-use area around the perimeter.

countries involved could get together to establish such a plan is questionable. Competing goals and objectives of conservation groups, international development banks, and powerful neighbors such as the United States also complicate the picture, but the idea is a noble one whose time perhaps has come.

In some cases, human occupancy can help maintain natural values. The Keolado National Park in India is such a case. After years of struggle, native villagers were able to show that grazing by domestic water buffalo is essential to maintaining wetlands and protecting biodiversity.

WILDERNESS AREAS

Although indigenous people had lived in the Americas for thousands of years before the first Europeans arrived, introduced diseases killed up to 90 percent of the existing population so that the continent appeared a vast, empty wilderness to early explorers. As historian Frederick Jackson Turner pointed out in a series of articles and speeches around the turn of the century, a belief that wilderness was not only a source of wealth but also the origin of strength, self-reliance, wisdom, and character is deeply embedded in our culture. The frontier was seen as a place for continuous generation of democracy, social progress, economic growth, and national energy. A number of authors, including Henry Thoreau, Aldo Leopold, Sigurd Olson, Edward Abbey, and Wallace Stegner, have written about the physical, mental, and social benefits for modern people of rediscovering solitude and challenge in wilderness (fig. 15.15).

Beginning with designated primitive and roadless areas in a few national forests in the 1920s, the United States has established a system of 264 wilderness areas encompassing nearly 40 million ha (100 million acres). The 1964 Wilderness Act, which is the basis for much of this system, defined **wilderness** as "an area of undeveloped land affected primarily by the forces of nature, where man is a visitor who does not remain; it contains ecological, geological, or other features of scientific or historic value; it possesses outstanding opportunities for solitude or a primitive and unconfined type of recreation; and it is an area large enough so that continued use will not change its unspoiled, natural conditions." Most of the areas meeting these standards are in the western states and Alaska.

Additional wilderness areas continue to be evaluated for protected status. Conservation groups argue that another 80 million ha (200 million acres), half of it in Alaska, should be added to the wilderness system. For more than 30 years, the Forest Service has been working on a roadless area review and evaluation (RARE) on all *de facto* wilderness areas under its jurisdiction. Using a deliberately "pure" interpretation that excluded all lands with any history of roads or development (even if all traces of human impact were long gone), it concluded that only about one-fourth of its 23 million ha (56 million acres) of roadless areas qualified for protection.

A prolonged battle has been waged over these *de facto* wilderness areas, pitting environmental groups who want more wilderness against loggers, miners, ranchers, and others who want less wilderness. The arguments for saving wilderness are that it provides (1) a refuge for endangered wildlife, (2) an opportunity for solitude and primitive recreation, (3) a baseline for ecological research, and (4) an area where we have chosen simply to leave things in their natural state. The arguments against more wilderness are that timber, energy resources, and critical minerals contained on these lands are essential for economic development.

To people who live in remote areas, jobs, personal freedom, and local control of resources seem more important than abstract values of wilderness. They often see themselves as an embattled minority trying to protect an endangered, traditional way of life against a wealthy elite who want to lock up huge areas for recreation or aesthetic purposes. Wilderness proponents point out that 96 percent of the country already is open for resource exploitation; the remaining 4 percent is mostly land that developers didn't want anyway.

The last large areas still being studied for wilderness preservation in the United States are on Bureau of Land Management (BLM) land. Until 1997, only 1.7 million acres out of a total of 276 million acres managed by the BLM had been designated as wilderness. Establishment of the Grand Staircase-Escalante National Monument doubled that total. Still, environmental groups argue that ten times as much should be given wilderness protection.

FIGURE 15.15 The Absaroka-Beartooth Wilderness, north of Yellowstone National Park, preserves a piece of relatively untouched natural ecosystem. Areas such as this serve as a refuge for endangered wildlife, a place for outdoor recreation, a laboratory for scientific research, and a source of wonder, awe, and inspiration.

For many people, especially those in developing countries, the idea of pristine wilderness untouched by humans is regarded as neither very important nor very interesting. In most places, all land is occupied fully—if sparsely—by indigenous people. To them the area is home no matter how empty it may look to outsiders. From this perspective, preserving biological diversity, scenic beauty, and other natural resources may be a good idea, but excluding humans and human features from the land does not necessarily make it more valuable. In fact, saving cultural heritage, working landscapes, and historical evidence of early human occupation can often be among the most important reasons to protect an area.

FIGURE 15.16 Wildlife refuges have contributed to the recovery of both game and non-game wildlife populations. In some cases, such as these elk on the National Elk Refuge near Jackson Hole, WY, protected species become so numerous that surplus population must be harvested or translocated.

WILDLIFE REFUGES

In 1901 President Teddy Roosevelt established 51 national **wildlife refuges,** the first in an important but troubled system for wildlife preservation in the United States (fig. 15.16). There are now 511 wildlife refuges in this system, encompassing nearly 40 million hectares of land and water and representing every major biome in North America. They range in size from less than 1 ha (2.47 acres)

CASE STUDY

ZIMBABWE'S "CAMPFIRE" PROGRAM

Stretching across Zimbabwe from Victoria Falls to the Mozambique border is a broad swath of acacia thornbrush known as the Zambezi River Valley. Home to some of the world's greatest numbers and diversity of big game animals, this area is a magnet for tourists who come to hunt, photograph, or just watch elephants, cape buffalo, antelope, lions, leopards, and other charismatic megafauna. Tourism is a lucrative business in Africa. Every year, tourists spend more than $250 million on wildlife-related activities, but little of that wealth makes its way to local residents. Foreign tour operators pocket the vast proportion of the profits, while native people get little or nothing.

Unfortunately, both the survival of the vast herds of wildlife and the tourist business based upon them are threatened as habitat is converted to farmland to support growing human populations. In response to declining wildlife numbers and threats to the lucrative tourist trade, the government has set aside about half of the valley as reserves and national parks. The other half remains communal land divided into tribal areas. Many of these communal lands border protected reserves. Because of this close proximity, poaching and illegal settlement has been a persistent problem on protected lands.

Many rural residents view the wildlife of Zambezi as, at best, a food source, and, at worst, an enemy. Local people receive little benefit from their proximity to parks and reserves. No compensation is provided when crops are trampled by elephants or when livestock is killed by lions. Buffalo, crocodiles, and hippopotamuses wander off reserve lands and endanger human lives in villages. Endemic poverty exacerbated by wildlife depredations and restricted hunting rights create serious problems for parks. As one tribal leader said, "African elephants have no chance until Africans have a chance."

In an attempt to alleviate both social and conservation issues around the parks and preserves, a new program was established entitled the Communal Areas Management Program for Indigenous Resources (CAMPFIRE). This program gives local people authority over wildlife on communal lands. Each village negotiates with safari companies to establish hunting and tourist concessions. A single hunter may spend more than $200,000 on a hunt. Half of this goes to the local community. The World Wildlife Fund estimates that CAMPFIRE has increased local household incomes up to 25 percent. At the end of the year, each village decides independently how it will use its income. Some villages divide it among households, giving each a cash dividend; others put it into community projects such as new schools or clinics or to pay children's school fees.

The communities involved in the CAMPFIRE program have gained a real sense of ownership and control over their local resources. They willingly take responsibility for the management and protection of wildlife. Poaching has declined and residents are actually starting to complain there are not enough animals. Wildlife is now viewed as a valuable commodity—a financial asset to the communities rather than a threat to their well-being.

Everyone wins in the CAMPFIRE program. Local communities receive funds and food from the wildlife in their region; the governmental costs of reserve management have declined; tourists get to view the unique animals and landscapes of the Zambezi River Valley; and illegal poaching has declined, allowing wildlife to thrive again in Zimbabwe.

Members of the Hikwaka, Zimbabwe Sewing and Bakery Co-Op display clothes made with a sewing machine financed by profits from ecotourism.

for the tiny Mille Lacs Refuge in Minnesota to 7.3 million ha (18 million acres) in the Arctic Wildlife Refuge of Alaska. Altogether, about 1 percent of U.S. surface area is designated as wildlife refuge. Every state has at least one refuge.

Major additions to the refuge system were made by President Franklin D. Roosevelt and Harold Ickes, his Secretary of the Interior, who took advantage of low prices during the Depression to add additional areas. The largest additions of land protected for wildlife came in 1980 when President Jimmy Carter signed the Alaska National Interest Land Act and added 22 million ha (54 million acres) of new refuges to the 12.5 million ha (31 million acres) already existing, about two-thirds of which was in Alaska.

Refuge Management

Although refuges were originally intended to be sanctuaries in which wildlife would be protected from hunting and other disturbances, a 1948 compromise allowed hunting in refuges in exchange

FIGURE 15.17 Hunters on their way to duck blinds in a wildlife refuge. Although it seems a contradiction, half of all refuges allow hunting.

FIGURE 15.18 A guard protects black rhinoceroses from poachers. His ancient rifle is no match, however, for the powerful automatic weapons with which the poachers are now armed.

for an agreement by hunters to purchase special duck stamps to raise money for wetland protection. Although a refuge that allows hunting seems like an oxymoron, and in spite of the fact that only 10 percent of refuge lands have been acquired with duck stamp funds, hunting has become firmly established in most units (fig. 15.17). Critics charge that the Fish and Wildlife Service, which administers the refuge system, is so strongly oriented toward hunting that many units have become little more than duck and goose farms.

Over the years, a number of improbable and incompatible uses have become accepted in wildlife refuges including oil drilling, cattle grazing, snowmobiling, motorboating, off-road vehicle use, timber harvesting, hay cutting, trapping, and camping. One Nevada refuge has a bombing and gunnery range, while another is the site of a brothel. A General Accounting Office report found that 60 percent of all refuges allow activities that are harmful to wildlife.

Refuges also face threats from external activities. More than three-quarters of all refuges in the United States have water pollution problems, two-thirds of which are serious enough to affect wildlife. A notorious example is the former Kesterson Wildlife Refuge in California, where selenium-contaminated irrigation water drained from farm fields turned the marsh into a death trap for wildlife rather than a sanctuary. Eventually, the marsh had to be drained and capped with clean soil to protect wildlife. Subsequent research has shown that at least twenty wildlife refuges in western states have toxic metal pollution caused by agriculture and industrial activities outside the refuge.

The biggest current battle over wildlife refuges concerns proposals for oil and gas drilling in the Arctic National Wildlife Refuge on the north slope of Alaska's Brooks Range (see Case Study, chapter 21).

International Wildlife Preserves

As we saw earlier in this chapter, most developing countries rarely have separate systems of parks and wildlife refuges. Many nature preserves are set up primarily to protect wildlife, however. An outstanding example of both the promise and the problems in managing parks in the less-developed countries is seen in the Serengeti ecosystem in Kenya and Tanzania. This area of savanna, thorn woodland, and volcanic highland lying between Lake Victoria and the Great Rift Valley in East Africa is home to the highest density of ungulates (hoofed grazing animals) in the world. Over 1.5 million wildebeests (or gnus) graze on the savanna in the wet season, when grass is available, and then migrate through the woodlands into the northern highlands during the dry season. The ecosystem also supports hundreds of thousands of zebras, gazelles, impalas, giraffes, and other beautiful and intriguing animals. The herbivores, in turn, support lions and a variety of predators and scavengers, such as leopards, hyenas, cheetahs, wild dogs, and vultures. This astounding diversity and abundance is surely one of the greatest wonders of the world.

Tanzania's Serengeti National Park was established in 1940 to protect 15,000 sq km (5700 sq mi), an area about the size of Connecticut or twice as big as Yellowstone Park. It is bordered on the east by the much smaller Ngorongoro Conservation Area and Lake Manyara National Park. Kenya's Masai Mara National Reserve borders the Serengeti on the north. Rapidly growing human populations push against the boundaries of the park on all sides. Herds of domestic cattle compete with wild animals for grass and water. Agriculturalists clamor for farmland, especially in the temperate highlands along the Kenya-Tanzania border. So many tourists flock to these parks that the vegetation is ground to dust by hundreds of sight-seeing vans, and wildlife find it impossible to carry out normal lives.

Perhaps the worst problem in Africa is **poachers,** illegal hunters who massacre wildlife for valuable meat, horns, and tusks. Where there once were about one million rhinos in Africa, the population had dropped by the mid-1980s to less than 10,000 animals. Antipoaching efforts have allowed the population to recover to about 15,000 currently (fig. 15.18).

Elephants are under a similar assault. Thirty years ago there were no elephants in the Serengeti, but perhaps 3 million in all of Africa. Since then, about 80 percent of the African elephants have been killed—mainly for their ivory—at a rate of 100,000 each year. The 2000 elephants now in Serengeti National Park have been driven there by hunting pressures elsewhere. Fortunately, the elephants find refuge in the park and add to the pleasure of tourists who come to see the wildlife; however, they are changing the ecosystem. Crowded into inadequate space, the elephants smash down the acacia trees, turning the woodland and mixed savanna into continuous grassland. This is beneficial for some animals, but not for others.

The poachers continue to pursue the elephants and rhinos, even in the park. Armed with high-powered rifles and even machine guns and bazookas from the many African wars in the last decade, the poachers take a terrible toll on the wildlife. Park rangers try to stop the carnage, but they often are outgunned by the poachers. The parks themselves are beginning to resemble war zones, with fierce, lethal firefights rather than peace and tranquility.

FIGURE 15.19 Wetlands provide irreplaceable ecological services. They are among the most diverse and productive of all ecosystems. Many species spend at least part of their life cycles in freshwater or saltwater wetlands.

WETLANDS, FLOODPLAINS, AND COASTAL REGIONS

Although many wetlands are protected to some extent in wildlife refuges and nature preserves, this land category is so ecologically important and so underrepresented among world protected areas that it deserves special discussion here. What is a wetland? Although you might think it easy to do, defining wetlands is a subject of intense debate because not all lands that have important wetland characteristics are wet all the time. Roughly defined, however, a **wetland** is an area often covered by shallow water or one in which the ground is wet enough long enough to support plants specialized to grow under saturated soil conditions (fig. 15.19). Chapter 5 describes some important types of wetlands. Worldwide, Canada and Russia have the largest wetland areas.

Wetland Values

Wetlands provide a number of irreplaceable ecological services. They often are highly productive and provide food and habitat for a wide variety of species. Although wetlands currently occupy less than 5 percent of the 0.9 billion ha (2.2 billion acres) of land in the United States, the Fish and Wildlife Service estimates that one-third of all endangered species spend at least part of their lives in wetlands. Because open oceans are often biological deserts, coastal wetlands are even more important. Nearly two-thirds of all marine shellfish and finfish rely on salt marshes and estuaries for spawning and juvenile development. Wetland storage of flood waters is worth an estimated $3 to 4 billion per year. Wetlands also improve water quality by acting as natural water purification systems, removing silt and absorbing nutrients and toxins. The flow of groundwater through coastal marshes prevents saltwater intrusion that would otherwise contaminate wells. Coastal wetlands help stabilize shorelines and reduce storm damage. Finally, wetlands provide important recreational opportunities for hunters, fishers, birdwatchers, and many others who enjoy nature.

Wetland Destruction

For many people, wetlands are worthless, disagreeable, and dangerous places full of insects, spiders, snakes, and mud. This attitude was reflected in public policies such as the U.S. Swamp Lands Act of 1850 that allowed individuals to buy swamps and marshes for as little as 10 cents per acre. During the 1930s and 1940s, the U.S. government subsidized wetland drainage for conversion to farmland. Until 1977, dumping "landfill" into wetlands was considered a convenient way to dispose of waste as well as a good way to create building space for highways and housing developments. The first protective measures for wetlands in the United States were passed in 1899, to prevent dumping rubbish into navigable waterways. The 1972 Clean Water Act required discharge permits (called Section 404 permits) intended to protect surface water quality. The act was interpreted by the courts in 1977 to prohibit both pollution and filling (but not drainage) of wetlands unless an exception is granted. The 1985 Farm Bill went even further in its "swampbuster" provisions that withdraw agricultural subsidies from farmers who drain, fill, or damage wetlands.

As a result of our earlier land-use policies, only about 40 million ha (100 million acres) of wetlands remain out of the original 90 million ha in the lower 48 states (fig. 15.20). Some states have lost more wetlands than others. By some estimates, at least two-thirds of the millions of prairie potholes (small temporary ponds that furnish breeding habitat for waterfowl and shorebirds across the northern

FIGURE 15.20 Over the past two centuries, more than half of the original wetlands in the lower 48 states have been drained, filled, polluted, or otherwise degraded. Some of the greatest losses have been in midwestern farming states where up to 99 percent of all wetlands have been lost.

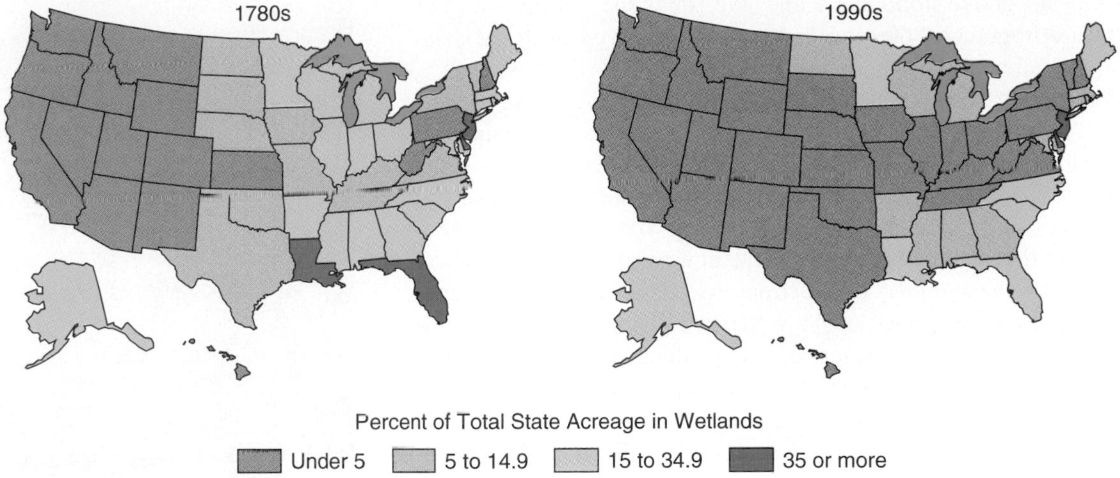

FIGURE 15.21 Louisiana has nearly 40 percent of the remaining coastal wetlands in the United States, but diversion of river sediments that once replenished these marshes and swamps, together with channels and boat wakes, are causing the loss of about 50 sq mi annually.

FIGURE 15.22 The Mississippi River inundates downtown Davenport, IA, during the summer floods of 1993. Although sympathetic with the heartbreak and economic losses caused by this flooding, many people argue that floodplains such as this should never have been settled in the first place.

plains) have been drained or filled. In Louisiana, which has about 40 percent of the remaining coastal wetlands in the United States, canal dredging for oil and gas production and navigation—coupled with diversion of river sediment that once replenished marshes—are causing losses of about 13,000 ha (50 sq mi) per year (fig. 15.21). Between 1950 and 1980, wetland losses in the United States were around 200,000 ha (500,000 acres) per year. A common estimate is that 87 percent of wetland losses are due to conversion to farmland, while 8 percent are due to urban development.

Although Section 404 permit applications are almost always approved and exemptions to the swampbuster provisions of the Farm Bill are almost always granted, these regulations are detested by farmers, real estate developers, miners, energy companies, and others who regard rules about what they can do with private lands as an infringement on owner's rights and an impediment to business. Although the 404 program has had little deterrent effect on wetland losses, other government programs have been surprisingly successful in restoring degraded wetlands or building new ones. Partners for Wildlife and North American Waterfowl programs administered by the Fish and Wildlife Service, and the Department of Agriculture's Wetland Reserve program have restored a total of 210,000 acres of wetland since 1995. The net gain is almost 70,000 acres.

Floods and Flood Control

The summer of 1993 was the wettest on record over much of the Mississippi Valley (fig. 15.22). Torrential rains falling on already saturated ground caused enormous flooding of the Missouri and upper Mississippi Rivers. In mid-July, the Mississippi and its tributaries had inundated several million hectares of land in the upper basin; it was as if a new great lake had suddenly appeared. The Mississippi reached a historic high of 47 ft at St. Louis, 17 ft

(5.23 m) above flood stage and five times the normal summer flow. Altogether, 8000 buildings were destroyed in the Mississippi Valley, 30,000 people were displaced, 26 lives were lost, and total losses exceeded $10 billion as a result of these floods.

The immediate cause of this disastrous flood was an unusual weather pattern—probably resulting from an El Niño event in the South Pacific (chapter 17)—that kept the jet stream low over the Midwest and pumped warm humid air up from the Gulf of Mexico and the Pacific to produce rains that were at least twice normal levels for the summer. Destruction of wetlands and natural vegetation cover over much of the watershed played a role in magnifying floods, as did a system of levees, flood walls, and channels that protect low-lying areas in normal years but failed catastrophically under these extreme conditions.

Before the natural vegetation of the Mississippi drainage basin was cut down, plowed up, and paved over by advancing civilization, the prairie grasses, forests, and natural wetlands acted like sponges to hold back rainfall and allow it to seep into the ground. Rivers tended to flow clear and uninterrupted, fed by a continuous supply released from this temporary storage. City streets, bare cropland, and drained wetlands lose their water-holding capacity, however, and rain that falls on them runs off immediately to create floods downstream.

Floodplains—low lands along riverbanks, lakes, and coastlines subjected to periodic inundation—are among the most valuable lands in many areas because of their rich soil, level topography, convenient water supply, access to river shipping, and recreational potential. For all these reasons, many cities are located at least in part in flood-prone areas. Public flood control efforts on the Mississippi began in 1879 when construction was started on levees (earthen berms along the river banks) to protect towns and low-lying farm fields. After a particularly disastrous flood in 1927, the Army Corps of Engineers began work on an extensive system of locks, dams, and channel dredging to improve navigation and provide further flood control (fig. 15.23).

The $25 billion river-control system built on the Mississippi and its tributaries has protected many communities over the past century. For others, however, this elaborate system has helped turn a natural phenomenon into a natural disaster. Deprived of the ability to spill out over floodplains, the river is pushed downstream to create faster currents and deeper floods until eventually a levee gives way somewhere. Hydrologists calculate that the floods of 1993 were about 3 m (10 ft) higher than they would have been given the same rainfall in 1900 before the flood-control structures were in place.

Under current rules, the government is obligated to restore any of the nearly 500 levees that meet federal standards and were damaged by flooding. Many people think that it would be much better to spend this money to restore wetlands, replace ground cover on water courses, build check dams on small streams, move buildings off the floodplain, and undertake other nonstructural ways of reducing flood danger. According to this view, floodplains should be used for wildlife habitat, parks, recreation areas, and other uses not susceptible to flood damage.

The National Flood Insurance Program administered by the Federal Emergency Management Agency (FEMA) was intended to aid people who cannot buy insurance at reasonable rates, but its effects have been to encourage building on the floodplains by making people feel that whatever happens, the government will take care of them. Many people would like to relocate homes and businesses out of harm's way after the recent floods or to improve them so they will be less susceptible to flooding, but owners of damaged property can collect only if they rebuild in the same place and in the same way as before. This serves to perpetuate problems rather than solve them.

FIGURE 15.23 Locks, dams, levees, and other flood control and navigational structures such as this complex on the Mississippi River at Dubuque, IA, have been a godsend for commercial shippers but have also been disastrous for riparian ecosystems, backwater sloughs, marshes, and the wildlife that depends on them.

Beaches, Barrier Islands, and Estuaries

Another group of lands of great biological importance threatened by our preference for living, working, and recreating near the water are beaches, barrier islands, coastal wetlands, and estuaries. An estuary is a bay or drowned valley where a river empties into the sea. Fresh water mingling with saltwater brings in sediment and nutrients and creates a gradient of salinity that makes estuaries among the most diverse and biologically productive ecosystems on the earth. As we pointed out earlier, the shoals and shallow reefs, salt marshes, tidal mudflats, mangrove swamps, and other coastal wetlands built up by these sediments are used during part of their life cycle by about two-thirds of all marine shellfish and finfish.

Where the continental shelf is relatively shallow, river sediments form beaches, sandbars, and barrier islands parallel to the shore. One of the world's longest and most spectacular sand beaches runs down the Atlantic coast of North America from New England to Florida and around the Gulf of Mexico. Much of this

FIGURE 15.24 Winter storms have eroded the beach and undermined the foundations of homes on this barrier island. Breaking through protective dunes to build such houses damages sensitive plant communities and exposes the whole island to storm sand erosion. Coastal zone management attempts to limit development on fragile sites.

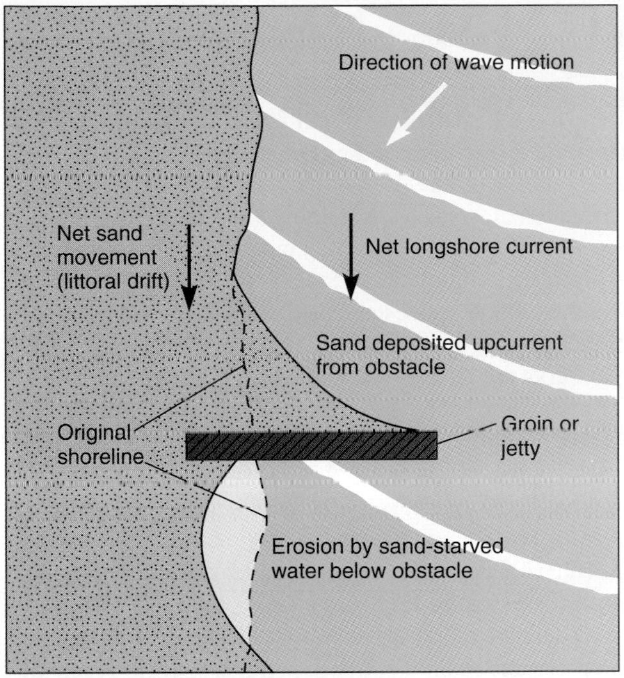

FIGURE 15.25 Groins are structures built perpendicular to a beach to slow erosion and trap sand. They build-up beaches toward prevailing wind and wave direction but starve beaches and increase erosion downstream. Dashed line indicates original shoreline.

beach lies on some 350 long, thin barrier islands that stand between the mainland and the open sea. Behind these barrier islands lie shallow bays or brackish lagoons fringed by marshes or swamps. The sediments that make up these landforms are constantly moved and reshaped by winds, waves, and currents. A 50-meter-wide beach can be created or removed by a single violent storm. Similarly, islands and sandbars appear and disappear over time.

Early inhabitants recognized that the shore was a hazardous place to live and settled on the bay side of barrier islands or as far upstream on coastal rivers as was practical. Modern residents, however, place a high value on living where they have an ocean view and ready access to the beach. And they assume that modern technology makes them immune from natural forces. The most valuable and prestigious property is closest to the shore. Over the past 50 years more than 1 million acres of estuaries and coastal marshes have been filled to make way for housing or recreational developments.

Construction directly on beaches and barrier islands can cause irreparable damage to the whole ecosystem. Damaging the vegetation that holds the shifting sand in place often not only puts houses and human life at risk but also eliminates habitat for rare and endangered species. Breaching dune systems to create roads or construction sites provides an avenue through which storms surge to wash away beaches or even whole islands. A single severe storm in 1962 caused $300 million in property damage along the East Coast and left hundreds of beach homes tottering into the sea (fig. 15.24).

Cities and individual property owners often spend millions of dollars to protect beaches from erosion. Sand is dredged from the ocean floor or hauled in by the truckload, only to wash away again in the next storm. Building artificial barriers such as groins or jetties can trap migrating sand and build beaches in one area, but they often starve downstream beaches and make erosion there even worse (fig. 15.25).

As is the case for inland floodplains, government policies often encourage people to build where they probably shouldn't. Subsidies for road building and bridges, support for water and sewer projects, tax exemptions for second homes, flood insurance, and disaster relief are all good for the real estate and construction business but invite people to build in risky places. Flood insurance typically costs $300 per year for $80,000 of coverage. In 1998, FEMA paid out $40 billion in claims, 80 percent of which were flood related. Settlement usually requires that structures be rebuilt exactly where and as they were before. There is no restriction on how many claims can be made, and policies are never canceled, no matter what the risk. Some beach houses have been rebuilt—at public expense—three times in a decade. The General Accounting Office found that 2 percent of federal flood policies were responsible for 30 percent of the claims.

The Coastal Barrier Resources Act of 1982 prohibited federal support, including flood insurance, for development on sensitive islands and beaches. In 1992, however, the Supreme Court ruled that ordinances forbidding floodplain development amount to an unconstitutional "taking" or confiscation of private property (see What Do You Think?, p. 317).

As part of the general rejection of governmental control over resources by the "wise use" movement, dozens of "taking" bills have been introduced in state legislatures prohibiting regulations or restrictions on how private property can be used. This may be a tactic to repeal not only coastal and floodplain zoning but also park and greenbelt establishment, wildlife refuges, wilderness designation, and almost every other land use we have discussed in this chapter.

SUMMARY

Parks, wildlife refuges, wilderness areas, and nature preserves occupy a small percentage of our total land area but protect valuable cultural resources and representative samples of the earth's species and ecosystems. We can discern much about people's environmental ideals by examining the gardens, parks, and recreation areas that they create. Yosemite in California was probably the first park in the world set aside to protect wild nature. Yellowstone, which was established in 1872, was the first official national park in the world.

Parks are havens for wildlife and places for healthful outdoor recreation. Many are overcrowded, misused, and neglected, however. Pollution and incompatible uses outside parks threaten the values that we seek to protect. Wildlife is at the center of many park controversies. Is the park's purpose to preserve wild nature or provide entertainment for visitors? How much management is acceptable? When should we intervene and when should we let nature take its course?

There are proposals to remove distracting or damaging uses from parks and to limit entrance permits. We may reach the point where each of us might be allowed to visit some of the more popular parks only once in a lifetime. One solution to congestion and overuse is to create additional parks in some of the many deserving but unprotected areas.

Worldwide, only about 4 percent of total land area has been protected in parks, wildlife refuges, and nature preserves. Some biomes such as dry tropical forests and tundra are well represented in this network, but others such as grassland and wetlands are underrepresented. The optimum size for nature preserves depends on the terrain and the values they are intended to protect, but—in general—the larger a reserve is, the more species it can protect. Establishing corridors to link separate areas can be a good way to increase effective space and to allow migration from one area to another. Economic development and nature protection can go hand in hand. Ecotourism may be the most lucrative and long-lasting way to use resources in many developing countries.

Areas chosen for preservation often are lands of indigenous people. Careful planning and zoning can protect nature and also allow sustainable use of resources. Man and Biosphere (MAB) reserves provide for multiple use in some areas but strict conservation in others. Wildlife refuges were intended to be sanctuaries for wildlife, but over the years many improbable and damaging uses have become established in them.

Wetlands are areas periodically covered with water that support unique assemblages of plants that can survive in water-saturated soil. Much of our wildlife, including many endangered species, depend on wetlands for a part of their life cycles. Historically, we have regarded wetlands as worthless, disagreeable places and we have encouraged people to drain or fill them. About half our original wetlands in the United States have been destroyed or degraded. Flood control and flood insurance encourage people to build homes and businesses on floodplains, barrier islands, beaches, and other fragile, flood-prone areas. Human and ecological costs of these misguided policies have been high.

QUESTIONS FOR REVIEW

1. Why is Yosemite called the first park to preserve wilderness, while Yellowstone is the first national park?
2. List some problems and threats from inside and outside our national parks.
3. Why is the reintroduction of wolves into Yellowstone a controversial issue?
4. Describe the IUCN categories of protected areas and the amount of human impact or intervention allowed in each. Can you name some parks or preserves in each category?
5. Which biomes or landscape types are best represented among the world's protected areas and which are least represented?
6. List the three main points in the IUCN World Conservation Strategy and six steps in the action plan to meet these goals.
7. Draw a diagram of an ideal MAB reserve. What activities would be allowed in each zone?
8. What is the legal definition of wilderness in the United States?
9. What role did Teddy Roosevelt and his cousin Franklin play in establishing wildlife refuges?
10. Describe some ecological values of wetlands and how they are threatened by human activities.

QUESTIONS FOR CRITICAL THINKING

1. Is "contrived" naturalness a desirable feature in parks and nature preserves? How much human intervention do you think is acceptable in trying to make nature more beautiful, safe, comfortable, or attractive to human visitors? Think of some specific examples that you would or would not accept.
2. Suppose you were superintendent of Yellowstone National Park. How would you determine the carrying capacity of the park for elk? How would you weigh having more elk or more ground squirrels? If there are too many elk, how would you thin the herd?
3. Suppose that as park manager you know that building tourist facilities brings in needed funds to protect nature but at the

same time more tourists destroy the natural values you want to protect. How do you balance these competing interests?

4. Why do you suppose that dry tropical forests and tundra are well represented in protected areas, whereas grasslands and wetlands are rarely protected? Consider social, cultural, and economic as well as biogeographical reasons in your answer.
5. Suppose that preserving healthy populations of grizzly bears and wolves requires that we set aside some large fraction of our national parks as a zone into which no humans will ever again be allowed to enter for any reason. Would you support protecting bears even if no one ever sees them or could even be sure that they still existed?
6. Suppose that you had trespassed into the bear sanctuary and were attacked by a bear. Should the rangers shoot the bear or let it eat you?
7. Are there any conditions under which you would permit oil drilling in the Arctic National Wildlife Refuge?
8. People have built homes and businesses on floodplains, beaches, or barrier islands under the assumption that the government will protect them from floods and storms. Can we now tell them that we can't or won't protect them?
9. Is a zoning ordinance that says you can't use your land in ecologically damaging ways an unfair intrusion into private property rights?

KEY TERMS

biogeographical area 337
biosphere reserves 338
corridors 339
floodplains 348
inholdings 334
Man and Biosphere (MAB) program 341
natural landscaping 332
poachers 345
wetland 346
wilderness 342
wildlife refuges 343
world conservation strategy 338

ADDITIONAL INFORMATION ON THE INTERNET

Visit our website at http://www.mhhe.com/environmentalscience/ for specific information about each of the topics below:

National Parks Conservation Association
National Wildlife Refuge system—US Fish and Wildlife Service
Man and the Biosphere: international biosphere reserve program
IUCN Red List of Threatened Animals
The Wilderness Society
The Nature Conservancy
The National Park Service
Sierra Club Critical Ecoregions Program
Captive breeding of endangered species
Great Outdoor National Parks and Preserves: National Park Guides
National Wildlife Federation
Ecotourism.org
Ecotourism: Easy things you can do to be an ecologically prudent tourist
Refuge planning on the Pacific Coast
Missouri Botanical Gardens research programs
World Conservation Monitoring Center: Protected Areas Virtual Library

PART FOUR

PHYSICAL RESOURCES

CHAPTER 16

Environmental Geology

Learn geology or die!

—Louis Agassiz—

OBJECTIVES

After studying this chapter, you should be able to:

- understand some basic geologic principles, including how tectonic plate movements affect conditions for life on the earth.
- explain how the three major rock types are formed and how the rock cycle works.
- summarize economic mineralogy and strategic minerals.
- discuss the environmental effects of mining and mineral processing.
- recognize the geologic hazards of earthquakes, volcanoes, and tsunamis.

ESSENTIAL STUDY PARTNER

The Essential Study Partner (ESP) CD-ROM that accompanies this textbook contains additional information on the following topics:

- earth's crust, mantle, and core: physical resources: the earth: geologic processes
- tectonic plates and volcanoes, earthquakes, mountains and trenches: physical resources: the earth: geologic processes
- continental drift: physical resources: the earth: geologic processes
- igneous, sedimentary, and metamorphic rock: physical resources: the earth: geologic processes
- environmental effects of mining: physical resources: the earth: geologic processes
- geologic hazards—effects of geologic processes on humans: physical resources: the earth: geologic processes

Opposite: Tower Falls in Yellowstone National Park cuts through consolidated volcanic ash.

Above: These glacier-carved peaks in Chile's Paine del Torres National Park show the effects of geologic forces.

Earthquake in Turkey

At 3:02 A.M. local time on August 17, 1999, a major earthquake with a magnitude of 7.4 on the Richter scale jarred western Turkey. At the epicenter, near the port city of Izmit and about 100 km (60 mi) east of Istanbul, the ground moved sideways as much as 5 m (16 ft). Shocks were felt as far away as the Turkish capital of Ankara, 480 km (300 mi) to the west. Poorly built apartment buildings and private homes collapsed, burying their sleeping residents under tons of cement and rubble (fig. 16.1). At least 15,000 people were confirmed dead. 45,000 were injured, and another 35,000 people were missing and believed to be buried by debris. Around 500,000 people were homeless after the quake, many of whom refused to reenter damaged buildings as the ground continued to shake with more than 250 aftershocks over the next few weeks. Although Izmit and its neighboring cities near the epicenter sustained the greatest damage, most deaths and injuries occurred in the heavily populated metropolitan area of Istanbul, home to some 12 million people.

In addition to building damage, the quake collapsed bridges on major highways, snapped electric lines and water pipes, and started fires when gas lines and oil tanks ruptured. Relief efforts were hindered by poor information, government red-tape, travel problems, and lack of emergency preparedness. Although Turkey lies in an highly active seismic zone and suffers earthquakes regularly, only one emergency rescue crew was available when this earthquake struck, and even they lacked the sophisticated videocameras and sensitive listening devices available in other countries. Hot weather, a general lack of sanitation and clean drinking water, and the large number of dead bodies believed still to be buried raised worries about the spread of disease in devastated communities.

Among the reasons that the death toll from this earthquake was so high is the poor quality of construction in many Turkish buildings. Less prosperous than other quake-prone regions like California or Japan, Turkey can ill afford the expensive reinforcing and building stabilization called for by wealthier city's building codes. Still, even Turkey's low official building standards are rarely enforced. Corrupt politicians who allowed unethical contractors to bypass even minimal safety codes are blamed for many deaths. In many places a well-constructed building with no damage at all could be seen standing next to others that had collapsed like houses of cards.

Although this was not by any means the largest or most deadly earthquake in history, it reminds us of the powerful forces that shape the earth's crust, our need to understand something about geology and earth science, and the wisdom of preparing for natural disasters. In this chapter, we will study some of the ways in which geological processes provide materials and create landscapes. We'll also look at some geological hazards and what we can do to avoid their dangers.

FIGURE 16.1 Residents of Izmit dig through a collapsed building looking for earthquake survivors.

A DYNAMIC PLANET

Although we think of the ground under our feet as solid and stable, the earth is a dynamic and constantly changing structure. Titanic forces stir inside the earth, causing continents to split, drift apart, and then crash into each other in slow but inexorable collisions. In this section, we will look at the structure of our planet and some of the forces that shape it.

A Layered Sphere

The **core,** or interior of the earth (fig. 16.2), is composed of a dense, intensely hot mass of metal—mostly iron—thousands of kilometers in diameter. Solid in the center but more fluid in the outer core, this immense mass generates the magnetic field that envelops the earth.

Surrounding the molten outer core is a hot, pliable layer of rock called the **mantle.** The mantle is much less dense than the core because it contains a high concentration of lighter elements, such as oxygen, silicon, and magnesium.

The outermost layer of the earth is the cool, lightweight, brittle **crust** of rock that floats on the mantle something like the "skin" on a bowl of warm chocolate pudding. The oceanic crust, which forms the seafloor, has a composition somewhat like that of the mantle, but it is richer in silicon. Continents are thicker, lighter regions of crust rich in calcium, sodium, potassium, and aluminum. The continents rise above both the seafloor and the ocean surface. Table 16.1 compares the composition of the whole earth (dominated by the dense core) and the crust.

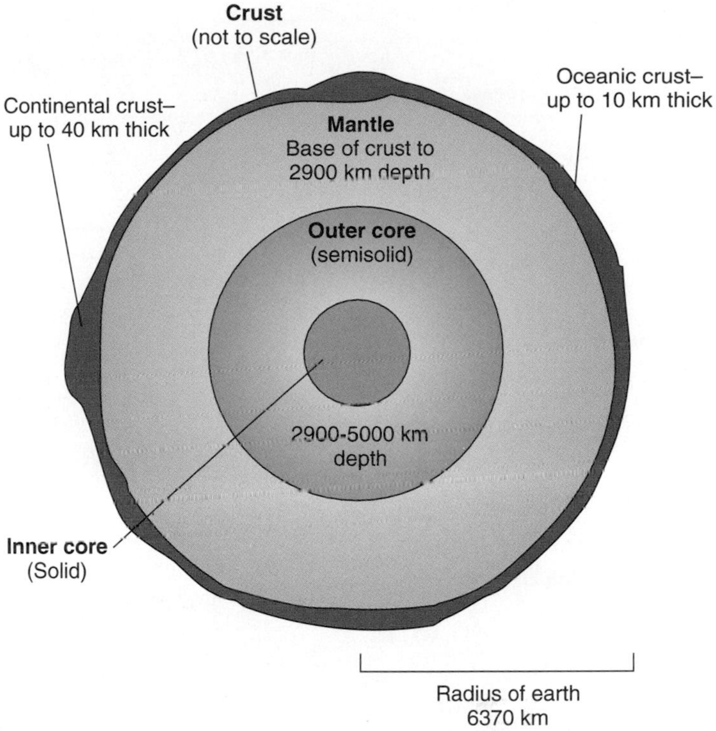

FIGURE 16.2 The layered Earth. The intensely hot, liquid or semisolid core is made mostly of molten metal. Around it is a solid but pliable mantle hot enough to bend and flow like warm taffy. Floating on top of the mantle is a thin crust of rock that breaks up into large, slowly moving tectonic plates. The crust appears ten times thicker in this drawing than it is in reality.

Tectonic Processes and Shifting Continents

Hot enough to flow, the upper layer of the mantle has huge convection currents that break the overlying crust into a mosaic of huge blocks called **tectonic plates** (fig. 16.3). These plates slide slowly across the earth's surface like immense icebergs, in some places breaking up into smaller pieces, in other places crashing ponderously into each other to create new, larger landmasses. Ocean basins form where continents crack and pull apart. **Magma** (molten rock) forced up through the cracks forms new oceanic crust that piles up underwater in mid-oceanic ridges. Creating the largest mountain range in the world, these ridges wind around the world for 74,000 km (46,000 mi). Although concealed from our view, this jagged range boasts higher peaks, deeper canyons, and sheerer cliffs than any continental mountains. Slowly spreading from these fracture zones, ocean plates push against continental plates.

Earthquakes are caused by the grinding and jerking as plates slide past each other. Mountain ranges are pushed up at the margins of colliding continental plates. The Atlantic Ocean is growing slowly as Europe and Africa drift away from the Americas. The Himalayas are still rising as the Indian subcontinent smashes into Asia. Southern California is slowly sailing north toward Alaska. In about 30 million years, Los Angeles will pass San Francisco, if either still exists by then.

When an oceanic plate collides with a continental landmass, the continental plate will ride up over the seafloor, and the oceanic plate is subducted, or pushed down into the mantle, where it melts and rises back to the surface as magma (fig. 16.4). Deep ocean

TABLE 16.1 Eight Most Common Chemical Elements (percent)

WHOLE EARTH		CRUST	
Iron	33.3	Oxygen	45.2
Oxygen	29.8	Silicon	27.2
Silicon	15.6	Aluminum	8.2
Magnesium	13.9	Iron	5.8
Nickel	2.0	Calcium	5.1
Calcium	1.8	Magnesium	2.8
Aluminum	1.5	Sodium	2.3
Sodium	0.2	Potassium	1.7

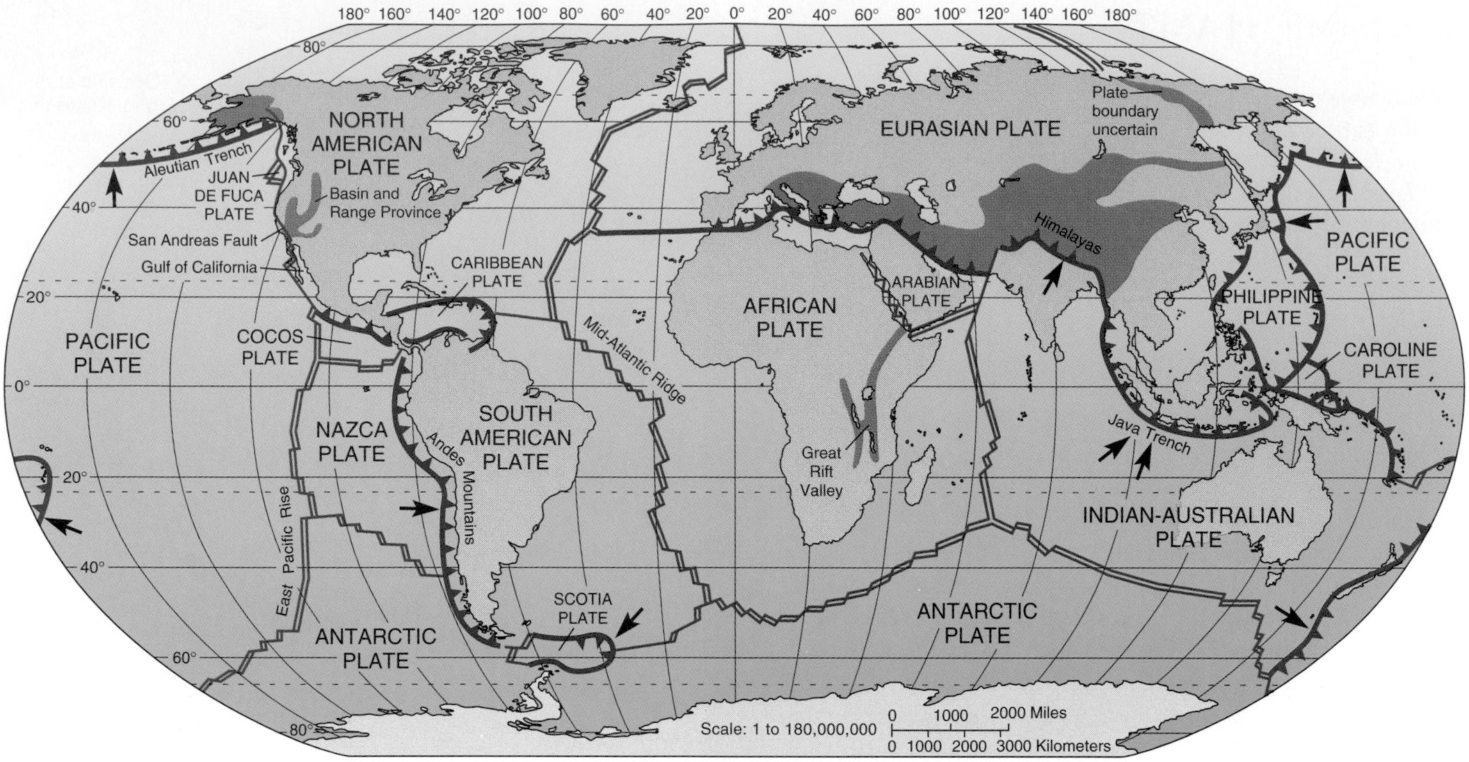

FIGURE 16.3 Map of tectonic plates. Plate boundaries are dynamic zones, characterized by earthquakes and volcanism and the formation of great rifts and mountain ranges. Arrows indicate direction of subduction where one plate is diving beneath another. These zones are sites of deep trenches in the ocean floor and high levels of seismic and volcanic activity.
Sources: U.S. Department of the Interior, U.S. Geological Survey.

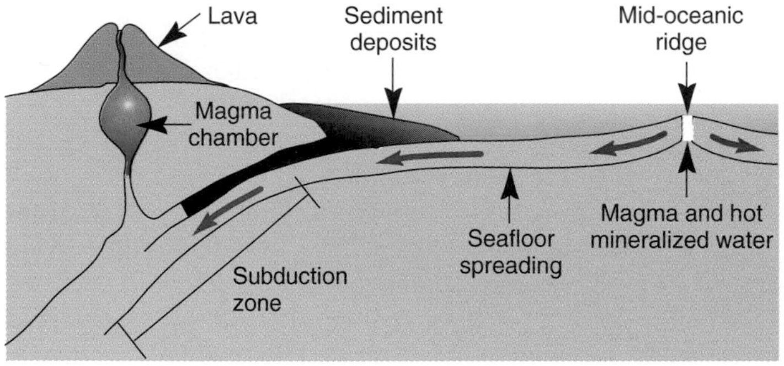

FIGURE 16.4 The rock cycle is driven by plate-tectonic movements. Old seafloor and sediment deposits are melted in subduction zones. The magma rises to erupt through volcanoes or to recrystallize at depth into igneous rocks. Weathering breaks down surface rocks and erosion deposits residue in sedimentary formations. The pressure and heat caused by tectonic movements cause metamorphism of both sedimentary and igneous rocks.

trenches mark these subduction zones, and volcanoes form where the magma erupts through vents and fissures in the overlying crust. All around the Pacific Ocean rim from Indonesia to Japan to Alaska and down the West Coast of the Americas is the so-called "ring of fire" where oceanic plates are being subducted under the continental plates. This ring is the source of more earthquakes and volcanic activity than any other place on the earth.

Over millions of years, the drifting continents can move long distances. Antarctica and Australia once were connected to Africa, for instance, somewhere near the equator, and supported luxuriant forests. Geologists suggest that several times in the earth's history most or all of the continents have gathered to form a single supercontinent, Pangaea, surrounded by a single global ocean (fig. 16.5). The rupture of this supercontinent and redistribution of its pieces probably has profound effects on the earth's climate and may help explain the periodic mass extinctions of organisms marking the divisions between many major geologic periods.

ROCKS AND MINERALS

A **mineral** is a naturally occurring, inorganic, solid element or compound with a definite chemical composition and a regular internal crystal structure. Naturally occurring, as opposed to synthetic, means not created by humans. Organic materials, such as coal, produced by living organisms or biological processes are not minerals. A mineral must be solid, therefore ice is a mineral but liquid water is not. The two fundamental characteristics of a mineral that distinguish it from all other minerals are its chemical composition and its crystal structure. No two minerals are identical in both

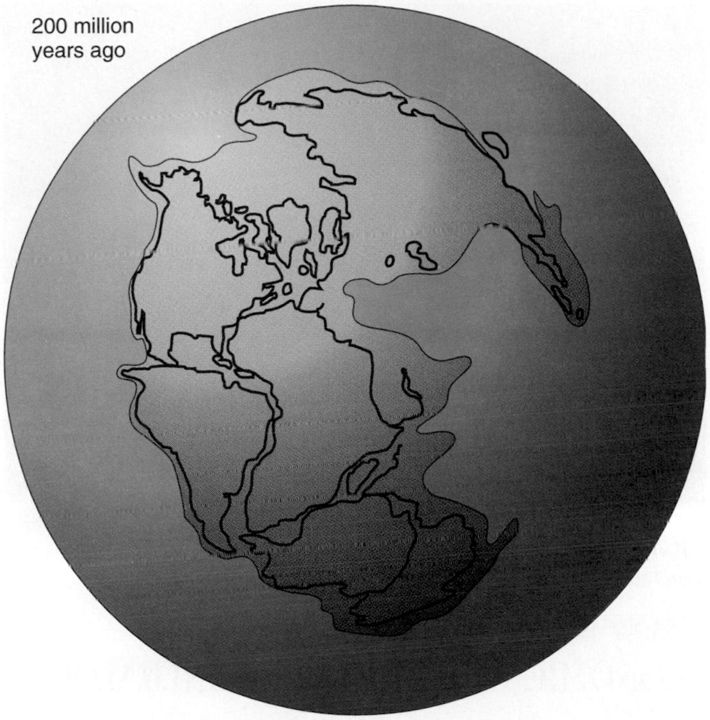

FIGURE 16.5 Pangaea, the ancient supercontinent of 200 million years ago, combined all the world's continents in a single landmass.

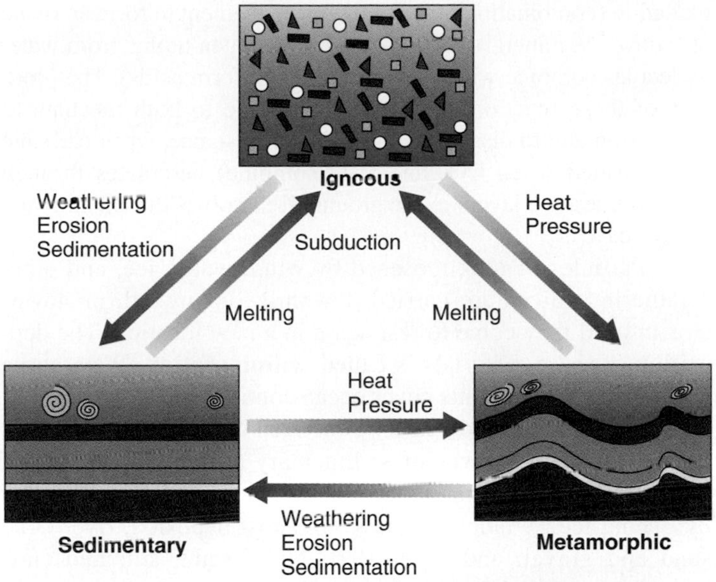

FIGURE 16.6 The rock cycle consists of processes of creation, destruction, and metamorphosis. Each of the three major rock types can be converted to either of the other types.

respects. Once purified, metals are solid, noncrystalline elements, and thus are not minerals. The ores from which they are extracted, however, are minerals and make up an important part of economic mineralogy.

A **rock** is a solid, cohesive, aggregate of one or more minerals. Within the rock, individual mineral crystals (or grains) are mixed together and held firmly in a solid mass. The grains may be large or small, depending on how the rock was formed, but each grain retains its own unique mineral qualities. Each rock type has a characteristic mixture of minerals (and therefore of different chemical elements), grain sizes, and ways in which the grains are mixed and held together. Granite, for example, is a mixture of quartz, feldspar, and mica crystals. Different kinds of granite have distinct percentages of these minerals and particular grain sizes depending on how quickly the rock solidified. These minerals, in turn, are made up of a few elements such as silicon, oxygen, potassium, and aluminum.

Rock Types and How They Are Formed

What could be harder and more permanent than rocks? Like the continents they create, rocks are also part of a relentless cycle of formation and destruction. They are made and then torn apart, cemented together by chemical and physical forces, crushed, folded, melted, and recrystallized by dynamic processes related to those that shape the large-scale features of the crust. We call this cycle of creation, destruction, and metamorphosis the **rock cycle** (fig. 16.6). Understanding something of how this cycle works helps explain the origin and characteristics of different types of rocks, as well as how they are shaped, worn away, transported, deposited, and altered by geologic forces.

There are three major rock classifications: igneous, sedimentary, and metamorphic. In this section, we will look at how they are made and some of their properties.

Igneous Rocks

The most common rock-type in the earth's crust is solidified from magma, welling up from the earth's interior. These rocks are classed as **igneous rocks** (from *igni,* the Latin word for fire). Magma extruded to the surface from volcanic vents cools quickly to make basalt, rhyolite, andesite, and other fine-grained rocks. Magma that cools slowly in subsurface chambers or is intruded between overlying layers makes granite, gabbro, or other coarse-grained crystalline rocks, depending on its specific chemical composition.

Weathering and Sedimentation

Most of these crystalline rocks are extremely hard and durable, but exposure to air, water, changing temperatures, and reactive chemical agents slowly breaks them down in a process called **weathering.** *Mechanical weathering* is the physical breakup of rocks into smaller particles without a change in chemical composition of the constituent minerals. You have probably seen mountain valleys scraped by glaciers or river and shoreline pebbles that are rounded from being rubbed against one another as they are tumbled by waves and currents. *Chemical weathering* is the selective removal or alteration of specific components that leads to weakening and disintegration of rock. Among the more important chemical weathering processes are oxidation (combination of oxygen with an element to form an oxide or hydroxide mineral) and hydrolysis (hydrogen atoms from water molecules combine with other chemicals to form acids). The products of these reactions are more susceptible to both mechanical weathering and to dissolving in water. For instance, when carbonic acid (formed when CO_2 and H_2O combine) percolates through porous limestone layers in the ground, it dissolves the calcium and creates caves.

Particles of rock loosened by wind, water, ice, and other weathering forces are carried downhill, downwind, or downstream until they come to rest again in a new location. The deposition of these materials is called **sedimentation.** Waterborne particles from sediments cover ocean continental shelves and fill valleys and plains. Most of the American Midwest, for instance, is covered with a layer of sedimentary material hundreds of meters thick in the form of glacierborne till (rock debris deposited by glacial ice), windborne loess (fine dust deposits), riverborne sand and gravel, and ocean deposits of sand, silt, and clay. Deposited material that remains in place long enough, or is covered with enough material to compact it, may once again become rock. Some examples of **sedimentary rock** are shale (compacted mud), sandstone (cemented sand), tuff (volcanic ash), and conglomerates (aggregates of sand and gravel).

Sedimentary rocks are formed from crystals that precipitate out of, or grow from a solution. An example is rock salt, made of the mineral halite, which is the name for ordinary table salt (sodium chloride). Salt deposits often form when a body of salt water dries up and salt crystals are left behind.

Sedimentary formations often have distinctive layers that show different conditions when they were laid down. Relatively soft sedimentary rocks such as sandstone can be shaped by erosion into striking features (fig. 16.7). Geomorphology is the study of the processes that shape the earth's surface and the structures they create.

Humans have become a major force in shaping landscapes. Geomorphologist Rodger Hooke, of the University of Maine, looking only at housing excavations, road building, and mineral production, estimates that we move somewhere around 30 to 35 gigatons (billion tons or Gt) per year worldwide. When combined with the 10 Gt each year that we add to river sediments through erosion, our earth-moving prowess is comparable to, or greater than, any other single geomorphic agent except plate tectonics.

Metamorphic Rocks

Preexisting rocks can be modified by heat, pressure, and chemical agents to create new forms called **metamorphic rock.** Deeply buried strata of igneous, sedimentary, and metamorphic rocks are subjected to great heat and pressure by deposition of overlying sediments or while they are being squeezed and folded by tectonic processes. Chemical reactions can alter both the composition and structure of the rocks as they are metamorphosed. Some common metamorphic rocks are marble (from limestone), quartzite (from sandstone), and slate (from mudstone and shale). Metamorphic rocks are often the host rock for economically important minerals such as talc, graphite, and gemstones.

FIGURE 16.7 Soft sedimentary rock layers can be sculpted by erosion into remarkable shapes such as Rainbow Bridge in Utah.

ECONOMIC GEOLOGY AND MINERALOGY

Economic mineralogy is the study of minerals that are heavily used in manufacturing (mainly metal ores) and are, therefore, an important part of domestic and international commerce. Nonmetallic economic minerals are mostly graphite, some feldspars, quartz crystals, diamonds, and many other crystals that are valued for their usefulness, beauty, and/or rarity. Metals have been so important in human affairs that major epochs of human history are commonly known by the dominant materials and the technology to use them (Stone Age, Bronze Age, Iron Age, etc.). The mining, processing, and distribution of these materials have broad and varied implications both for culture and our environment. Most economically valuable crustal resources exist everywhere in small amounts; the important thing is to find them concentrated in economically recoverable levels.

Public policy in the United States has encouraged mining on public lands as a way of boosting the economy and utilizing natural resources. Today many people think these laws seem outmoded and in need of reform (What Do You Think?, p. 361).

Metals

How has the quest for minerals and metals affected global development? We will focus first on world use of metals, earth resources that always have received a great deal of human attention. The availability of metals and the methods to extract and use them have determined technological developments, as well as economic and political power for individuals and nations. We still are strongly dependent on the unique lightness, strength, and malleability of metals.

The metals consumed in greatest quantity by world industry include iron (740 million metric tons annually), aluminum (40 million metric tons), manganese (22.4 million metric tons), copper and chromium (8 million metric tons each), and nickel (0.7 million metric tons). Most of these metals are consumed in the United States,

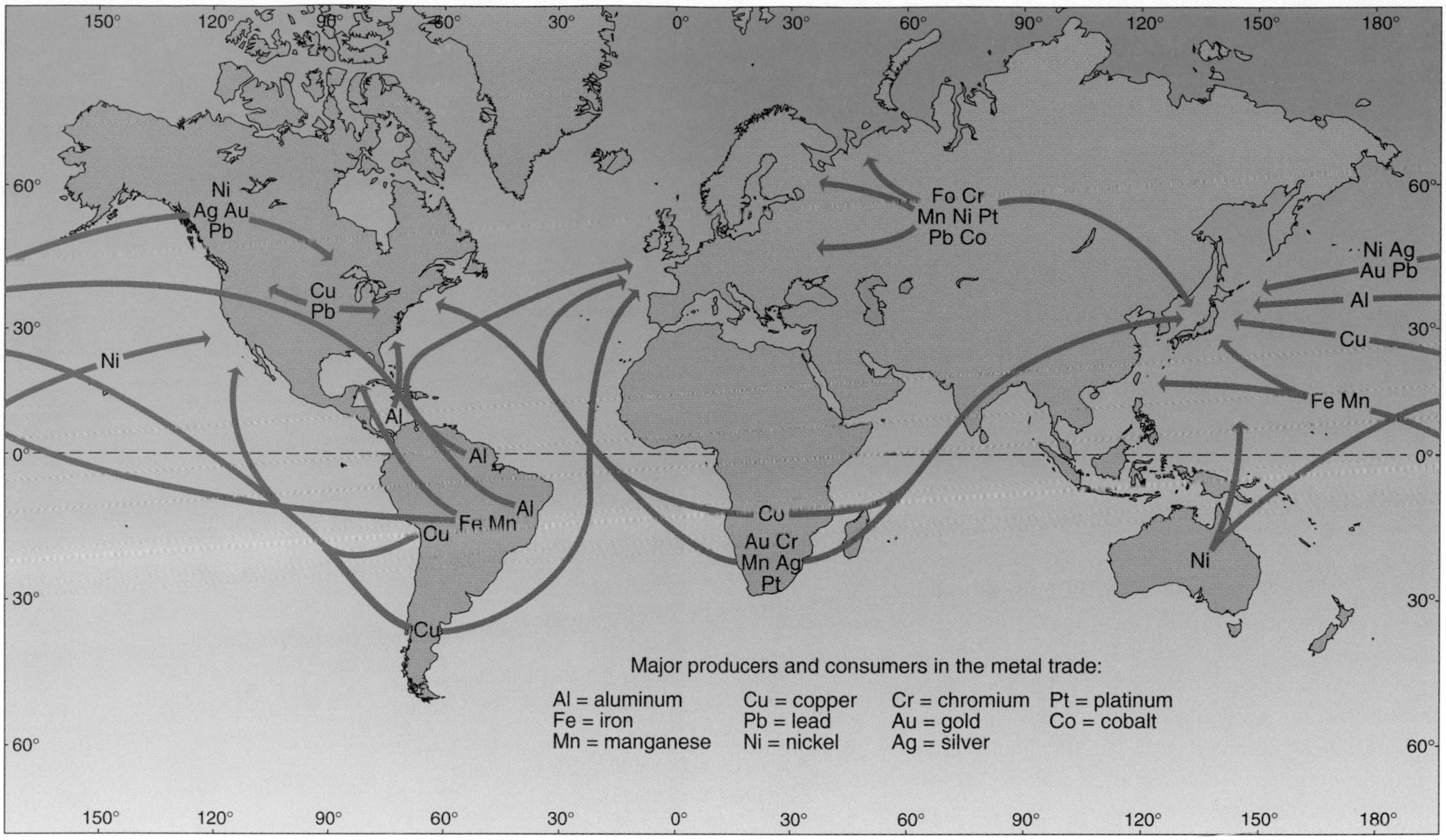

FIGURE 16.8 Global metal trade. Metals produced in South Africa, South America, and the former Soviet Union are shipped to markets in the United States, Europe, and Japan, creating a global economic network on which both consumers and producers depend.

Japan, and Europe, in that order. They are produced primarily in South America, South Africa, and the former Soviet Union. It is easy to see how these facts contribute to a worldwide mineral trade network that has become crucially important to the economic and social stability of all nations involved (fig. 16.8). Table 16.2 shows the primary uses of these metals.

Nonmetal Mineral Resources

Nonmetal minerals are a broad class that covers resources from silicate minerals (gemstones, mica, talc, and asbestos) to sand, gravel, salts, limestone, and soils. Sand and gravel production comprise by far the greatest volume and dollar value of all nonmetal mineral resources and a far greater volume than all metal ores. Sand and gravel are used mainly in brick and concrete construction, paving, as loose road filler, and for sandblasting. High-purity silica sand is our source of glass. These materials usually are retrieved from surface pit mines and quarries, where they have been deposited by glaciers, winds, or ancient oceans.

Limestone, like sand and gravel, is mined and quarried for concrete and crushed for road rock. It also is cut for building stone, pulverized for use as an agricultural soil additive that neutralizes acidic soil, and roasted in lime kilns and cement plants to make plaster (hydrated lime) and cement.

TABLE 16.2 Primary Uses of Some Major Metals Consumed in the United States

METAL	USE
Aluminum	Packaging foods and beverages (38%), transportation, electronics
Chromium	High-strength steel alloys
Copper	Building construction, electric and electronic industries
Iron	Heavy machinery, steel production
Lead	Leaded gasoline, car batteries, paints, ammunition
Manganese	High-strength, heat-resistant steel alloys
Nickel	Chemical industry, steel alloys
Platinum-group	Automobile catalytic converters, electronics, medical uses
Gold	Medical, aerospace, electronic uses; accumulation as monetary standard
Silver	Photography, electronics, jewelry

Evaporites (materials deposited by evaporation of chemical solutions) are mined for halite, gypsum, and potash. These are often found at or above 97 percent purity. Halite, or rock salt, is used for water softening and melting ice on winter roads in some northern areas. Refined, it is a source of table salt. Gypsum (calcium sulfate) now makes our plaster wallboard, but it has been used for plaster ever since the Egyptians plastered the walls of their frescoed tombs along the Nile River some 5000 years ago. Potash is an evaporite composed of a variety of potassium chlorides and potassium sulfates. These highly soluble potassium salts have long been used as a soil fertilizer.

Sulfur deposits are mined mainly for sulfuric acid production. In the United States, sulfuric acid use amounts to more than 200 lbs per person per year, mostly because of its use in industry, car batteries, and some medicinal products.

Strategic Metals and Minerals

World industry depends on about 80 minerals and metals, some of which exist in plentiful supplies. Three-fourths of these resources are abundant enough to meet all of our anticipated needs or have readily available substitutes. At least 18 metals, however, including tin, platinum, gold, silver, and lead, are in short supply.

Of these 80 metals and minerals, between one-half and one-third are considered "strategic" resources. **Strategic metals and minerals** are those that a country uses but cannot produce itself. As the term strategic suggests, these are materials that a government considers capable of crippling its economy or military strength if unstable global economics or politics were to cut off supplies. For this reason, wealthy industrial nations stockpile strategic resources, especially metals, in times when prices are low and a supply is available. Figure 16.9 shows some of the major stockpiles of strategic metals and ores in the United States.

For less wealthy mineral- and metal-producing nations, there is another side to strategic minerals. Many less-developed countries depend on steady mineral exports for most of their foreign exchange. Zambia, for instance, relies on cobalt production for 50 percent of its national income. If a steady international market is not maintained, such producer nations could be devastated. From the small-nation producer's point of view, metal or mineral exports, like concentration on any single product or industry, are an unstable economic foundation. Often no option exists, however, if a producer is to participate in the world economy. Environmental consequences of mining may not be a high priority under these circumstances.

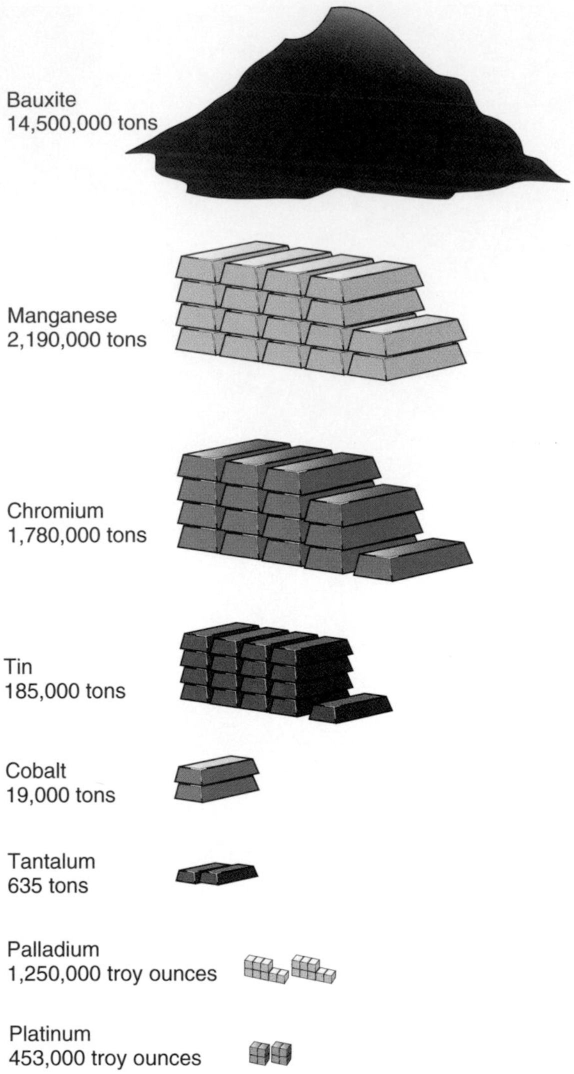

FIGURE 16.9 United States' stockpiles of strategic metals and metal ores.

ENVIRONMENTAL EFFECTS OF RESOURCE EXTRACTION

Geologic resource extraction involves physical processes of mining and physical or chemical processes of separating minerals, metals, and other geologic resources from ores or other materials. An ore is a rock in which a valuable or useful metal occurs at a concentration high enough to make mining it economically attractive. For metals like copper, a concentration close to 1 percent is necessary to make an ore worth mining. For precious metals, like gold or platinum, 0.0001 percent may be enough to make mining and smelting worthwhile.

Mining

Extracting geologic materials is done by several different techniques depending on the accessibility of the resource and the content or concentration of the material sought. All of these methods have environmental hazards. Native metals deposited in the gravel of streambeds can be washed out hydraulically in a process called placer mining. This not only destroys streambeds but fills the water with suspended solids that smother aquatic life. Larger or deeper

what do you think?

THE WORLD AROUND YOU

Should We Revise Mining Laws?

In 1872, the U.S. Congress passed the General Mining Law intended to encourage prospectors to open up the public domain and promote commerce. This law, which has been in effect more than a century, allows miners to stake an exclusive claim anywhere on public lands and to take—for free—any minerals they find. Claim holders can "patent" (buy) the land for \$2.50 to \$5 per acre (0.4 hectares) depending on the type of claim. Once the patent fee is paid, the owners can do anything they want with the land, just like any other private property. Although \$2.50 per acre may have been a fair market value in 1872, many people regard it as ridiculously low today, amounting to a scandalous give-away of public property.

In Nevada, for example, a mining company is buying federal land for \$9000 that contains an estimated \$20 billion worth of precious metals. Similarly, Colorado investors bought about 7000 ha (17,000 acres) of rich oil-shale land in 1986 for \$42,000 and sold it a month later for \$37 million. You don't actually have to find any minerals to patent a claim. A Colorado company paid a total of \$400 for 65 ha (160 acres) it claimed would be a gold mine. Ten years later, no mining has been done, but the property—which just happens to border the Keystone Ski Area—is being subdivided for condos and vacation homes.

According to the Bureau of Land Management (BLM), some \$4 billion in minerals are mined each year on U.S. public lands. Under the 1872 law, mining companies don't pay a penny for the ores they take. Furthermore, they can deduct a depletion allowance from taxes on mineral profits. Senator Dale Bumpers of Arkansas, who calls the 1872 mining law "a license to steal," estimates that the government could derive \$320 million per year by charging an 8 percent royalty on all minerals and probably could save an equal amount by requiring a bond to be posted to clean up after mining is finished.

On the other hand, mining companies argue they would be forced to close down if they had to pay royalties or post bonds. Many people would lose jobs and the economies of western mining towns would collapse if mining becomes uneconomic. We provide subsidies and economic incentives to many industries to stimulate economic growth. Why not mining for metals essential for our industrial economy? Mining is a risky and expensive business. Without subsidies, mines would close down and we would be completely dependent on unstable foreign supplies.

Mining critics respond that other resource-based industries have been forced to pay royalties on materials they extract from public lands. Coal, oil, and gas companies pay 12.5 percent royalties on fossil fuels obtained from public lands. Timber companies—although they don't pay the full costs of the trees they take—have to bid on logging sales and clean up when they are finished. Even gravel companies pay for digging up the public domain. Ironically, we charge for digging up gravel, but give gold away free.

Over the past decade, numerous mining bills have been introduced in Congress. Those supported by environmental groups, generally would require companies mining on federal lands to pay a higher royalty on their production. They also would eliminate the patenting process, impose stricter reclamation requirements, and give federal managers authority to deny inappropriate permits. In contrast, bills offered by Western legislators, and enthusiastically backed by mining supporters, tend to leave most provisions of the 1872 bill in place. They would charge a 2-percent royalty, but only after exploration, production, and other costs were deducted. Permit processes would consider local economic needs before environmental issues in this version. What do you think we should do about this mining law? How could we separate legitimate public-interest land use from private speculation and profiteering? Are current subsidies necessary and justifiable or are they just a form of corporate welfare?

Thousands of abandoned mines on public lands poison streams and groundwater with acid, metal-laced drainage. This old mine in Montana drains into the Blackfoot River, the setting of Norman Maclean's book A River Runs Through It.

FIGURE 16.10 Some giant mining machines stand as tall as a 20-story building and can scoop up thousands of cubic meters of rock per hour.

ore beds are extracted by strip mining or open-pit mining where overlying material is removed by large earth-moving equipment (fig. 16.10). The resulting pits can be many kilometers across and hundreds of meters deep (fig. 16.11). In some cases whole mountaintops or the surface of entire islands have been removed by surface mining (Case Study p. 363). Even deeper deposits are reached by underground tunneling, an extremely dangerous process for mine workers.

Old tunnels occasionally collapse, or subside. In coal mines, natural gas poses dangers of explosion. Uncontrollable fires, producing noxious smoke and gases, sometimes ignite coal-bearing scrap heaps stored inside or outside the mine. Surface waste deposits called tailings can cause acidic or otherwise toxic runoff when rainwater percolates through piles of stored material. Tailings from uranium mines give rise to wind scattering of radioactive dust.

Water leaking into mine shafts also dissolves metals and other toxic material. When this water is pumped out or allowed to seep into groundwater aquifers (chapter 20), pollution occurs. The Mineral Policy Center in Washington, DC, estimates that 550,000 abandoned mines pockmark the U.S. landscape. Some 19,000 km (12,000 mi) of rivers and streams in the United States are contaminated by mine drainage. The Environmental Protection Agency estimates that cleaning up mines and mine drainage will cost \$32 billion to \$72 billion.

The process of strip-mining involves stripping off the vegetation, soil, and rock layers, removing the minerals, and replacing the fill. The fill is usually replaced in long ridges, called spoil banks, because this is the easiest way to dump it cheaply and quickly. Spoil banks are very susceptible to erosion and chemical weathering. Rainfall leaches numerous chemicals in toxic concentrations from

FIGURE 16.11 The world's largest open-pit mine is Bingham Canyon, near Salt Lake City, UT. More than 5 billion tons of copper ore and waste material have been removed since 1906 to create a hole 800 m (2640 ft) deep and nearly 4 km (2.5 mi) wide at the top.

CASE STUDY

MINING A TROPICAL PARADISE

The tiny island-nation of Nauru (pronounced NAH-roo) in the western Pacific is the smallest and most remote republic in the world. It also is a case study in humanity's ability to plunder its environment. Located on the equator some 500 km west of its nearest neighbor in the Marshall Islands, Nauru has been inhabited by Polynesian people for thousands of years. When first visited by European explorers in the eighteenth century, the island was a lush tropical paradise of swaying coconut palms and white coral beaches. Sailors called it Pleasant Island, but today the name is a bitter joke.

Compared to its former condition, Nauru is probably the most environmentally devastated nation on earth. So much land has been devoured by strip-mining that residents now face the prospect of having to abandon the whole island and move elsewhere. What the miners sought was guano, a thick phosphate-rich layer of bird droppings prized by industrialized countries as fertilizer. Billions of dollars worth of this treasure have been exported, first by colonial powers and then, since independence in 1968, by the Nauruans themselves.

After a century of mining, Nauru's 7500 residents are among the richest people in the world, but their environment has been almost totally wrecked. Eighty percent of the 21 sq km (8 sq mi) island has been stripped, leaving a bleak, barren moonscape of jagged coral pinnacles, some as tall as 25 meters. With all soil washed away, almost nothing lives in this wasteland. Traveling across it is impossible. To make things even worse, removing the vegetation has changed the climate. Heat waves rising from the sun-baked rock drive away rain clouds and the island now is plagued by constant drought.

Not only the island is ravaged. Nauruans may be among the world's most affluent people, but they are also among the most unhealthy, plagued by cardiovascular disease, diabetes, and obesity brought about by a lifestyle of idleness and imported junk food.

After a century of surface mining of rich phosphate deposits, four-fifths of the tiny Pacific island of Nauru is a barren moonscape of jagged coral pinnacles. Although mining has made Nauruians rich, they may soon have no place to live.

Few islanders live past the age of 60. Since most mining is done by imported workers, Nauruans generally lack job skills to apply elsewhere. This case study is a sad example of how easy it is for modern technology and lack of foresight to degrade both a society and its environment.

The guano deposits are expected to last only a few more years. After that, residents will be left with only a thin sliver of habitable coastline. Some efforts have been made to import new soil and to restore vegetation to the desolate interior, but attempts have been unsuccessful. It may be too late to reverse the damage. The people may be able to use some of their accumulated trust fund to buy another island, but will they find one as comfortable and beautiful as what they once had? One village leader says wistfully, "I wish Nauru could be like it was before. I remember it was so beautiful and green everywhere. We could eat coconuts and breadfruit. It makes me cry when I see what has been done. I wish we'd never discovered the phosphate."

Could Nauru's example be a warning for all of us? Humans have a long history of depleting resources and then moving on. Could we find ourselves in a similar situation someday, having exhausted our natural resources but then having an uninhabitable world?

Ethical Considerations

If the Nauruans appeal for help in finding another place to live, would we be morally obligated to assist them? Do countries that bought fertilizer from Nauru bear a special responsibility for what has happened? If we see other examples of environmental destruction occurring elsewhere, would we be ethically justified to try to intervene and stop it?

the freshly exposed earth, and the water quickly picks up a heavy sediment load. Chemical- and sediment-runoff pollution becomes a major problem in local watersheds. Acid runoff had contaminated 10,000 km (6250 mi) of streams in the United States by 1980. Problems are made worse by the fact that the steep spoil banks are very slow to revegetate. Since the spoil banks do not have natural topsoil, succession, soil development, and establishment of a natural biological community occur very slowly.

The 1977 federal Surface Mining Control and Reclamation Act (SMCRA) requires better restoration of strip-mined lands, especially of land classed as prime farmlands, but restoration is difficult and expensive. Even if soil is carefully replaced, it will take centuries, if not millenia, to regain its former fertility by natural processes. Topsoil is dispersed and often buried by the activity of heavy machinery working to resculpture the land. Compaction disrupts air and water flow through soil, restricts root growth, and causes poor drainage, resulting in wet, stagnant soil conditions. The difficulties of reestablishing vegetation in dry climates means that reclamation is essentially impossible where rainfall is less than 25 cm per year.

The monetary expense of reclamation is also high. Minimum reclamation costs about $1000 per acre (0.4 ha), while "complete" restoration (where it is possible) costs five times that much. Nevertheless, 50 percent of U.S. coal (225-270 million metric tons per year) is strip-mined. Nearly a million acres (400,000 ha) of land in the United States have been devastated by strip-mining, often on public lands for which mining companies pay little or nothing.

Processing

Metals are extracted from ores by heating or treatment with chemical solvents. These processes often release large quantities of toxic materials that can be even more environmentally hazardous than mining. Smelting—roasting ore to release metals—is a major source of air pollution. One of the most notorious examples of ecological devastation from smelting is a wasteland near Ducktown, Tennessee (fig. 16.12). In the mid-1800s, mining companies began excavating the rich copper deposits in the area. To extract copper from the ore, they built huge open-air wood fires using timber from the surrounding forest. Dense clouds of sulfur dioxide released from sulfide ores poisoned the vegetation and acidified the soil over a 50-square-mile (13,000-hectare) area. Rains washed the soil off the denuded land, creating a barren moonscape where nothing could grow. Siltation of reservoirs on the Ocoee River impaired electric generation by the Tennessee Valley Authority (TVA).

Sulfur emissions from Ducktown smelters were reduced in 1907 after the U.S. Supreme Court ruled in Georgia's favor in a suit to stop interstate transport of air pollution. In the 1930s, the TVA began treating the soil and replanting trees to cut down on erosion. Recently, upwards of $250,000 per year has been spent on this effort. While the trees and other plants are still spindly and feeble, more than two-thirds of the area is considered "adequately" covered with vegetation; only 4 percent is still totally denuded. Similarly, smelting of copper-nickel ore in Sudbury, Ontario, a century ago caused widespread ecological destruction that is slowly being repaired following pollution control measures (see fig. 18.18).

FIGURE 16.12 A luxuriant forest once grew on this now barren hillside near Ducktown, TN. Smelter fumes killed all the vegetation nearly a century ago, and erosion has washed away all the topsoil. Restoration projects are slowly bringing back ground cover and rebuilding soil.

A recent technique called **heap-leach extraction** (fig. 16.13) makes it possible to separate gold from extremely low-grade ores but has a high potential for water pollution. Typically, the process involves piling crushed ore in huge heaps and spraying it with a dilute alkaline-cyanide solution, which percolates through the pile to dissolve the gold. The gold-containing solution is pumped to a processing plant where gold is removed by electrolysis. A thick clay pad and plastic liner beneath the ore heap is supposed to keep the poisonous cyanide solution from contaminating surface or groundwater, but leaks are common.

Once all the gold is recovered, mine operators may simply walk away from the operation, leaving vast amounts of toxic effluent in open ponds behind earthen dams. A case in point is the Summitville mine near Alamosa, Colorado. After extracting $98 million in gold, the absentee owners declared bankruptcy in 1992, abandoning millions of tons of mine waste and huge leaking ponds of cyanide. The Environmental Protection Agency may spend more than $100 million trying to clean up the mess and keep the cyanide pool from spilling into the Alamosa River.

In 2000, an enormous cyanide spill from a gold mining operation near Baia Mare in Romania poisoned millions of fish and threatened drinking water supplies along about 640 km (400 mi) of the Szamos, Tisza, and Danube Rivers in Hungary and Yugoslavia.

CONSERVING GEOLOGIC RESOURCES

There is great potential for extending our supplies of economic minerals and reducing the effects of mining and processing through recycling. The advantages of recycling are significant: less waste to dispose of, less land lost to mining, and less consumption of money, energy, and water resources.

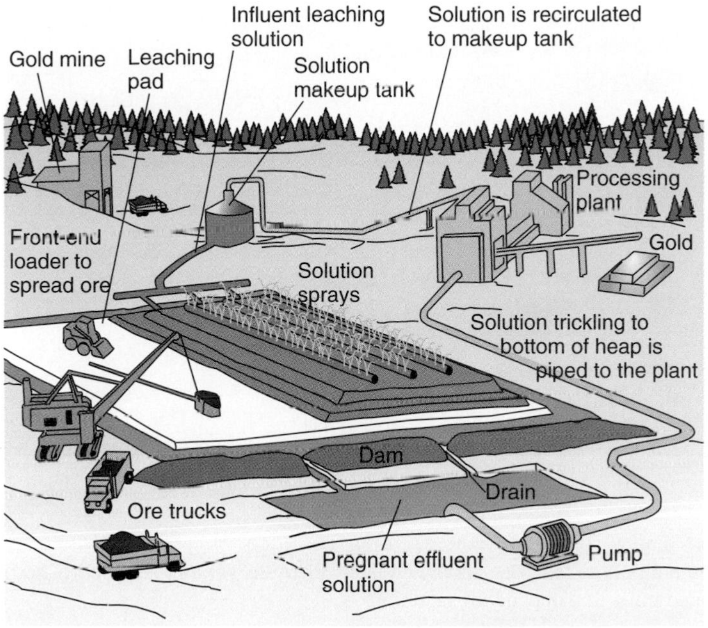

FIGURE 16.13 In a heap-leach operation, huge piles of low-grade ore are heaped on an impervious pad and sprayed continuously with a cyanide solution. As the leaching solution trickles through the crushed ore, it extracts gold and other precious metals. The "pregnant" effluent solution is then pumped to a processing plant where metals are extracted and purified. This technique is highly profitable but carries large environmental risks. *Source:* George Laycock, *Audubon Magazine,* vol. 91(7), July 1989, p. 77.

FIGURE 16.14 The richest metal source we have—our mountains of scrapped cars—offers a rich, inexpensive, and ecologically beneficial resource that can be "mined" for a number of metals.

TABLE 16.3 Energy Requirements in Producing Various Materials from Ore and Raw Source Materials

	ENERGY REQUIREMENT (MJ/kg)	
PRODUCT	NEW	FROM SCRAP
Glass	25	25
Steel	50	26
Plastics	162	n.a.
Aluminum	250	8
Titanium	400	n.a.
Copper	60	7
Paper	24	15

Source: Data from E. T. Hayes, *Implications of Materials Processing,* 1997.

Recycling

Some waste products already are being exploited, especially for scarce or valuable metals. Aluminum, for instance, must be extracted from bauxite by electrolysis, an expensive, energy-intensive process. Recycling waste aluminum, such as beverage cans, on the other hand, consumes one twentieth of the energy of extracting new aluminum. Today, nearly two-thirds of all aluminum beverage cans in the United States are recycled, up from only 15 percent 20 years ago. The high value of aluminum scrap ($650 a ton versus $60 for steel, $200 for plastic, $50 for glass, and $30 for paperboard) gives consumers plenty of incentive to deliver their cans for collection. Recycling is so rapid and effective that half of all the aluminum cans now on a grocer's shelf will be made into another can within two months. The energy cost of extracting other materials is shown in table 16.3.

Platinum, the catalyst in automobile catalytic exhaust converters, is valuable enough to be regularly retrieved and recycled from used cars (fig. 16.14). Other metals commonly recycled are gold, silver, copper, lead, iron, and steel. The latter

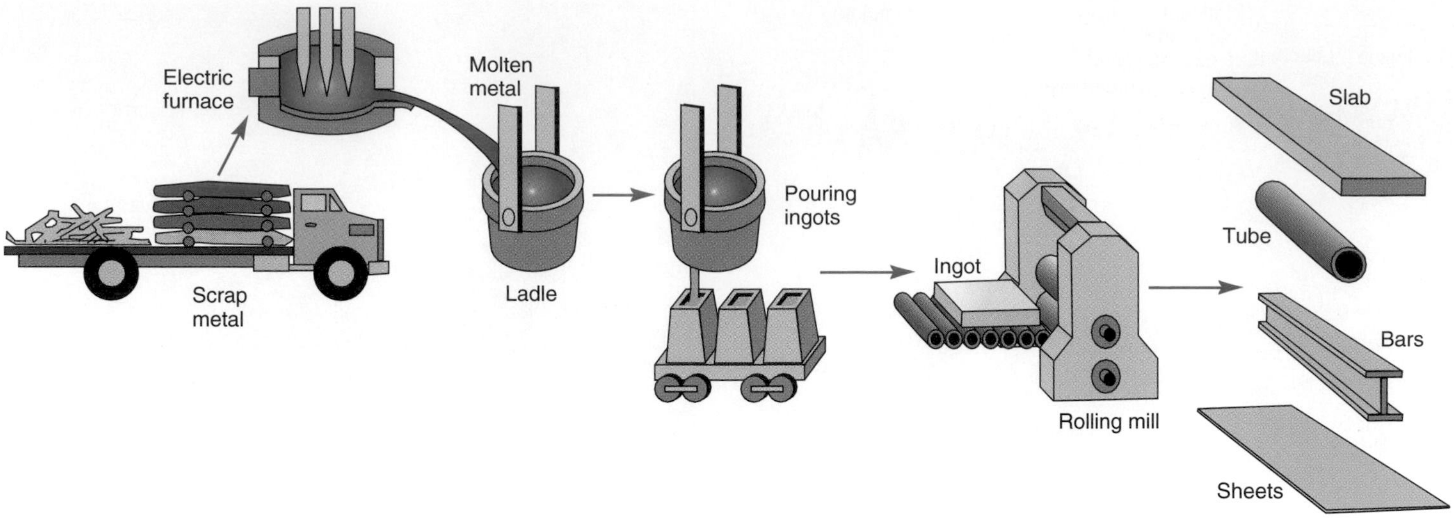

FIGURE 16.15 "Minimills" remelt and reshape scrap iron and steel. They not only extend our geologic resources by recycling discarded materials, they also conserve energy and are cheaper to operate than traditional integrated mills that depend on virgin ore.

four are readily available in a pure and massive form, including copper pipes, lead batteries, and steel and iron auto parts. Gold and silver are valuable enough to warrant recovery, even through more difficult means. See chapter 23 for further discussion of this topic.

Steel and Iron Recycling: Minimills

While total U.S. steel production has fallen in recent decades—largely because of inexpensive supplies from new and efficient Japanese steel mills—a new type of mill subsisting entirely on a readily available supply of scrap/waste steel and iron is a growing industry. Minimills, which remelt and reshape scrap iron and steel, are smaller and cheaper to operate than traditional integrated mills that perform every process from preparing raw ore to finishing iron and steel products. Minimills produce steel at between \$225 and \$480 per metric ton, while steel from integrated mills costs \$1425 to \$2250 per metric ton on average (fig. 16.15). The energy cost is likewise lower in minimills: 5.3 million BTU/ton of steel compared to 16.08 million BTU/ton in integrated mill furnaces. Minimills now produce about half of all of U.S. steel production. Recycling is slowly increasing as raw materials become more scarce and wastes become more plentiful.

Substituting New Materials for Old

Mineral and metal consumption can be reduced by new materials or new technologies developed to replace traditional uses. This is a long-standing tradition, for example, bronze replaced stone technology and iron replaced bronze. More recently, the introduction of plastic pipe has decreased our consumption of copper, lead, and steel pipes. In the same way, the development of fiber-optic technology and satellite communication reduces the need for copper telephone wires.

Iron and steel have been the backbone of heavy industry, but we are now moving toward other materials. One of our primary uses for iron and steel has been machinery and vehicle parts. In automobile production, steel is being replaced by polymers (long-chain organic molecules similar to plastics), aluminum, ceramics, and new, high-technology alloys. All of these reduce vehicle weight and cost, while increasing fuel efficiency. Some of the newer alloys that combine steel with titanium, vanadium, or other metals wear much better than traditional steel. Ceramic engine parts provide heat insulation around pistons, bearings, and cylinders, keeping the rest of the engine cool and operating efficiently. Plastics and glass fiber-reinforced polymers are used in body parts and some engine components.

Electronics and communications (telephone) technology, once major consumers of copper and aluminum, now use ultrahigh-purity glass cables to transmit pulses of light, instead of metal wires carrying electron pulses. Once again, this technology has been developed for its greater efficiency and lower cost, but it also affects consumption of our most basic metals.

GEOLOGIC HAZARDS

Earthquakes, volcanoes, floods, and landslides are normal earth processes, events that have made our earth what it is today. However, when they occur in proximity to human populations, their consequences can be among the worst and most feared disasters that befall us. For thousands of years people have been watching and recording these hazards, trying to understand them and learn how to avoid them.

Earthquakes

Earthquakes are sudden movements in the earth's crust that occur along faults (planes of weakness) where one rock mass slides past another one. When movement along faults occurs gradually and relatively smoothly, it is called creep or seismic slip and may be undetectable to the casual observer. When friction prevents rocks from slipping easily, stress builds up until it is finally released with a sudden jerk. The point on a fault at which the first movement occurs during an earthquake is called the epicenter.

Earthquakes, such as the one in Turkey that opened this chapter, have always seemed mysterious, sudden, and violent, coming without warning and leaving in their wake ruined cities and dislocated landscapes (table 16.4). Cities such as Kobe, Japan, or Mexico City, parts of which are built on soft landfill or poorly consolidated soil, usually suffer the greatest damage from earthquakes (fig 16.16). Water-saturated soil can liquify when shaken. Buildings sometimes sink out of sight or fall down like a row of dominoes under these conditions.

Earthquakes frequently occur along the edges of tectonic plates, especially where one plate is subducting, or diving down, beneath another. Earthquakes also occur in the centers of continents, however. In fact, one of the largest earthquakes ever recorded in North America was one of an estimated magnitude 8.8 that struck the area around New Madrid, Missouri, in 1812. Fortunately, few people lived there at the time and the damage was minimal. If a similar quake occurred along that fault today the damage in Memphis, Nashville, and St. Louis would be horrendous. A calamitous earthquake was one thought to have killed 242,000 people in China in 1976.

Modern contractors in earthquake zones are attempting to prevent damage and casualties by constructing buildings that can withstand tremors. The primary methods used are heavily reinforced structures, strategically placed weak spots in the building that can absorb vibration from the rest of the building, and pads or floats beneath the building on which it can shift harmlessly with ground motion.

One of the most notorious effects of earthquakes is the **tsunami.** These giant seismic sea swells (sometimes improperly called tidal waves) can move at 1000 km/hr (600 mph), or faster, away from the center of an earthquake. When these swells approach the shore, they can easily reach 15 m or more and some can be as high as 65 m (nearly 200 ft). A 1960 tsunami coming from a Chilean earthquake still caused 7-meter breakers when it reached Hawaii fifteen hours later. Tsunamis also can be caused by underwater volcanic explosions or massive seafloor slumping. The eruption of the Indonesian volcano Krakatoa in 1883 created a tsunami 40 m (130 ft) high that killed 30,000 people on nearby islands.

While most earthquakes occur in known earthquake-prone areas, sometimes they strike in unexpected places. In the United States, major quakes have occurred in South Carolina, Missouri, Massachusetts, Nevada, Texas, Utah, Arizona, and Washington, as well as in California and Alaska.

Volcanoes

Volcanoes and undersea magma vents are the sources of most of the earth's crust. Over hundreds of millions of years, gaseous emissions from these sources formed the earth's earliest oceans and atmosphere. Many of the world's fertile soils are weathered volcanic materials. Volcanoes have also been an ever-present threat to human populations. One of the most famous historic volcanic eruptions was that of Mount Vesuvius in southern Italy, which buried the cities of Herculaneum and Pompeii in A.D. 79. The mountain had been giving signs of activity before it erupted, but many citizens chose to stay and take a chance on survival. On August 24, the mountain buried the two towns in ash. Thousands

TABLE 16.4 Worldwide Frequency and Effects of Earthquakes of Various Magnitudes

RICHTER SCALE MAGNITUDE*	DESCRIPTION	AVERAGE NUMBER PER YEAR	OBSERVABLE EFFECTS
2–2.9	Unnoticeable	300,000	Detected by instruments, but not usually felt by people.
3–3.9	Smallest felt	49,000	Hanging objects swing, vibrations like passing of light truck felt.
4–4.9	Minor earthquake	6200	Dishes rattle, doors swing, pictures move, walls creak.
5–5.9	Damaging earthquake	800	Difficult to stand up. Windows, dishes break. Plaster cracks, loose bricks and tile fall, small slides along sand or gravel banks.
6–6.9	Destructive earthquake	120	Chimneys and towers fall, masonry walls damaged, frame houses move on foundations, small cracks in ground. Broken gas pipes start fires.
7–7.9	Major earthquake	18	General panic. Frame houses split and fall off foundations, some masonry buildings collapse, underground pipes break, large cracks in the ground.
8–8.9	Great earthquake	1 or 2	Catastrophic damage. Most masonry and frame structures destroyed. Roadways, dams, dikes collapse. Large landslides. Rails twist and bend. Underground pipes rupture.

*For every unit increase in the Richter scale, ground displacement increases by a factor of 10, while energy release increases by a factor of 30. There is no upper limit to the scale, but the largest earthquakes recorded have been 8.9.

Source: Data from B. Gutenberg in *Earth* by F. Press and R. Seiver, 1978, W. H. Freeman & Company.

FIGURE 16.16 An elevated freeway collapsed and buckled as a result of the 1995 earthquake in Kobe, Japan.

were killed by the dense, hot, toxic gases that accompanied the ash flowing down from the volcano's mouth. It continues to erupt from time to time.

Nuees ardentes (French for "glowing clouds") are deadly, denser-than-air mixtures of hot gases and ash like those that inundated Pompeii and Herculaneum. Temperatures in these clouds may exceed 1000°C, and they move at more than 100 km/hour (60 mph). Nuees ardentes destroyed the town of St. Pierre on the Caribbean island of Martinique on May 8, 1902. Mount Pelee released a cloud of nuees ardentes that rolled down through the town, killing somewhere between 25,000 and 40,000 people within a few minutes. All the town's residents died except a single prisoner being held in the town dungeon.

Mudslides are also disasters sometimes associated with volcanoes. The 1985 eruption of Nevado del Ruiz, 130 km (85 mi) northwest of Bogotá, Colombia, caused mudslides that buried most of the town of Armero and devastated the town of Chinchina. An estimated 25,000 people were killed. Heavy mudslides also accompanied the eruption of Mount St. Helens in Washington in 1980. Sediments mixed with melted snow and the waters of Spirit Lake at the mountain's base and flowed many kilometers from their source. Extensive damage was done to roads, bridges, and property, but because of sufficient advance warning, there were few casualties.

Volcanic eruptions often release large volumes of ash and dust into the air. Mount St. Helens expelled 3 km^3 of dust and ash, causing ash fall across much of North America. This was only a minor eruption. An eruption in a bigger class of volcanoes was that of Tambora, Indonesia, in 1815, which expelled 175 sq km of dust and ash, more than fifty-eight times that of Mount St. Helens. These dust clouds circled the globe and reduced sunlight and air temperatures enough so that 1815 was known as the year without a summer.

It is not just a volcano's dust that blocks sunlight. Sulfur emissions from volcanic eruptions combine with rain and atmospheric moisture to produce sulfuric acid (H_2SO_4). The resulting droplets of H_2SO_4 interfere with solar radiation and can significantly cool the world climate. In 1991, Mt. Pinatubo in the Philippines emitted 20 million tons of sulfur dioxide that combined with water to form tiny droplets of sulfuric acid (fig. 16.17). This acid aerosol reached the stratosphere where it circled the globe for two years. This thin haze cooled the entire earth by 1°C and postponed global warming for several years. It also caused a 10 to 15 percent reduction in stratospheric ozone, allowing increased ultraviolet light to reach the earth's surface. One of several theories about the extinction of the dinosaurs 65 million years ago is that they were killed by acid rain and climate changes caused by massive volcanic venting in the Deccan Plateau of India.

Floods

In most moderately humid climates, stream channels adjust to accommodate average maximum stream flows. Much of the year, the water level may be well below the stream bank height, but heavy rains or sudden snow melt can deliver more water than the stream can carry. Excess water that overflows stream banks and covers adjacent land is considered a **flood.** The severity of floods can be described by the depth of water above the normal stream banks or by how frequently a similar event normally occurs—on average—for a given area. Note that these are *statistical averages* over long time periods. A "10-year flood" would be expected to occur once in every ten years; a "100-year flood" would be expected to occur once every century. But two 100-year floods can occur in successive years or even in the same year.

The biggest economic loss from floods is usually not the buildings and property they carry away, but rather the contamination they cause. Virtually everything flood waters touch in a house—carpets, furniture, drapes, electronics, even drywall and insulation—must be removed and discarded. It's not just a matter of drying out. Although floods usually don't take as many lives as some other natural disasters, they are among the most costly.

Many human activities increase both the severity of floods and their adverse consequences. Paving roads and parking lots reduces water infiltration into the soil and speeds the rate of runoff into streams and lakes. Clearing land for agriculture and filling cities with buildings similarly increase both the volume and rate of water discharge after a storm. Cities need water and usually are

FIGURE 16.17 Mt. Pinatubo in the Philippines erupted in 1991, spewing hundreds of millions of tons of ash, rocks, and molten lava. Some 20 million tons of sulfur from the volcano created an acidic aerosol that circled the stratosphere for two years and cooled the global climate at least 1°C.

FIGURE 16.18 Parts of an expensive house slide down the hillside in Pacific Palisades, CA. Building at the edge of steep slopes made of unconsolidated sediment in an earthquake-prone region is a risky venture.

located near rivers or lakes for both practical and aesthetic reasons. Floodplains—the land near a stream inundated during normal floods—are generally the most fertile and easily farmed land in an area. These human preferences for riverfront or shoreline locations tend to increase the economic losses and human suffering when floods inevitably occur.

We often try to control floods and reduce their destructive impacts by building levees and flood walls to contain water within riverbanks or by channelization, that is, deepening and straightening stream channels to increase the velocity or volume of water it carries. These measures may protect one area but they often simply accentuate downstream problems (see chapter 15 for further discussion).

Landslides

Gravity constantly pulls downward on every material everywhere on earth, causing a variety of phenomena collectively termed mass wasting or mass movement, in which geological materials are moved downslope from one place to another. The resulting movement is often slow and subtle, but some slope processes such as rockslides, avalanches, and land slumping can be swift, dangerous, and very obvious. **Landslide** is a general term for rapid downslope movement of soil or rock. In the United States alone, over one billion dollars in property damage is done every year by landslides and related mass wasting.

In some areas, active steps are taken to control landslides or to limit the damage they cause. On the other hand, many human activities such as road construction, forest clearing, agricultural cultivation, and building houses on steep, unstable slopes increase both the frequency and the damage done by landslides. In some cases, people are unaware of the risks they face by locating on or under unstable hillsides. In other cases, they simply deny clear and obvious danger. Southern California, where people build expensive houses on steep hills of relatively unconsolidated soil, is often the site of large economic losses from landslides. Chapparal fires expose the soil to heavy winter rains. Resulting mudslides carry away whole neighborhoods and bury downslope areas in debris flows (fig. 16.18). Generally these processes are slow enough that few lives are lost, but the property damage can be high.

SUMMARY

The earth is a complex, dynamic system. Although it seems stable and permanent to us, the crust is in constant motion. Huge blocks called tectonic plates slide over the surface of the ductile mantle. They crash into each other in ponderous slow motion, crumpling their margins into mountain ranges and causing earthquakes. Sometimes one plate will slide under another, carrying rock layers down into the mantle where they melt and flow back toward the surface to be formed into new rocks.

Rocks are classified according to composition, structure, and origin. The three basic types of rock are igneous, metamorphic, and sedimentary. These rock types can be transformed from one to another by way of the rock cycle, a continuous process of weathering, transport, burying in sediments, metamorphism, melting, and recrystallization.

During the cooling and crystallization process that forms rock from magma, minerals and metals can become concentrated enough to become economically important reserves if they are close enough to the surface to be reached by mining. Having a reliable supply of strategically important minerals and metals is vital in industrialized societies.

A few places in the world are especially rich in mineral deposits. South Africa and the former Soviet Union contain most of the world's supply of several strategic minerals. Less-developed countries, most of which are in the tropics or the Southern Hemisphere, are often the largest producers of ores and raw mineral resources for the strategic materials on which the industrialized world depends. The major consumers of these resources are the industrialized countries.

Worldwide, only a small percentage of metals are recycled, although it is not a difficult process technically. Recycling saves energy and reduces environmental damage caused by mining and smelting. It reduces waste production and makes our metal supplies last much longer. Substitution of materials usually occurs when mineral supplies become so scarce that prices are driven up. Many of the strategic metals that we now stockpile may become obsolete when newer, more useful substitutes are found.

Both mining and processing of metals and mineral resources can have negative environmental effects. Mine drainage has polluted thousands of kilometers of streams and rivers. Fumes from smelters kill forests and spread pollution over large areas. Surface mining results in removal of natural ecosystems, soil disruption, creation of trenches and open pits, and accumulation of tailings. It is now required that strip-mined areas be recontoured, but revegetation is often difficult and limited in species composition. Smelting and chemical extraction processes also create pollution problems.

Earthquakes and volcanic events are natural geological hazards that are a result of movements of the earth's restless crust, core, and mantle. Big earthquakes are among the most calamitous natural disasters that befall people, sometimes killing hundreds of thousands in a single cataclysm.

QUESTIONS FOR REVIEW

1. Describe the layered structure of the earth.
2. Define mineral and rock.
3. What are tectonic plates and why are they important to us?
4. Why are there so many volcanoes and earthquakes along the "ring of fire" that rims the Pacific Ocean?
5. Describe the rock cycle and name the three main rock types that it produces.
6. Give some examples of strategic metals. Where are the largest supplies of these resources located?
7. Give some examples of nonmetal mineral resources and describe how they are used.
8. Describe some ways we recycle metals and other mineral resources.
9. What are some environmental hazards associated with mineral extraction?
10. Describe some of the leading geologic hazards and their effects.

QUESTIONS FOR CRITICAL THINKING

1. Look at the walls, floors, appliances, interior, and exterior of the building around you. How many earth materials were used in their construction?
2. What is the geologic history of your town or county?
3. Is your local bedrock igneous, metamorphic, or sedimentary? If you don't know, who might be able to tell you?
4. What would life be like without the global mineral trade network? Can you think of advantages as well as disadvantages?
5. Suppose a large mining company is developing ore reserves in the small, underdeveloped country where you live. How will revenues be divided fairly between the foreign company and local residents?
6. How could we minimize the destruction caused by geologic hazards? Should people be discouraged from building on floodplains or in volcanic or earthquake-prone areas?

7. How would your life be affected if your country were to run out of strategic minerals?
8. What effect do you think our need for strategic minerals has had on our foreign policy toward South Africa and the former Soviet Union?
9. How could our government encourage more recycling and more efficient use of minerals?
10. What can you do to protect yourself from geologic hazards?

KEY TERMS

core 355
crust 355
earthquakes 366
flood 368
heap-leach extraction 364
igneous rocks 357
landslide 369
magma 355
mantle 355
metamorphic rock 358
mineral 356
rock 357
rock cycle 357
sedimentary rock 358
sedimentation 358
strategic metals and minerals 360
tectonic plates 355
tsunami 367
volcanoes 367
weathering 357

ADDITIONAL INFORMATION ON THE INTERNET

Visit our website at http://www.mhhe.com/environmentalscience for specific information about each of the topics below:

National Geophysical Data Center
Geology: US Geological Survey
Worldwide Volcanic Activity Map
Earthquake Information from Around the World
Cascades Volcano Observatory
Plate Tectonics (USGS)
Geology of Hawaii Volcanoes
Savage Earth: PBS site on geological forces
Mining Claims on Public Lands
US Bureau of Mines: a posthumous commentary
Leasing: oil and gas
Office of Surface Mining (US Department of the Interior)
Information for Mining Activists

PROFILE

Fisheries Biologist

Robyn Angliss

Would you like a job that includes field surveys of whales and porpoises in Alaska and developing marine policy? Robyn Angliss has done both in her job at the United States National Marine Fisheries Service (NMFS). Although her title is fisheries biologist, Robyn specializes in mammals, especially marine mammals.

How can you get work that takes you to beautiful places to study charismatic animals? Robyn was in her junior year of college and was thinking about a career in marine intertidal ecology when a friend invited her to be part of a summer field study in Prince William Sound, Alaska, to photograph whales and examine the effects of the *Exxon Valdez* oil spill on killer whales and humpback whales. Although the offer came just days before she was scheduled to go home for summer vacation, she accepted and headed north for a wilderness adventure. While working in Alaska, Robyn met the researcher from NMFS who was funding her project. The next fall, back in Seattle for her final year at the University of Washington, Robyn offered to volunteer for the NMFS researcher she met in Alaska to get some experience and to continue to learn more about marine wildlife. After a few months of volunteering, a part-time paid position became available, and by the time she graduated, Robyn was working full time as a biological technician (wildlife) at NMFS.

For three years, Robyn spent about half of each year along the Alaska coast from Ketchikan to Point Barrow, photographing bowhead, gray, and beluga whales from the air, and surveying harbor porpoise and killer whales from boats. During the winters back in Seattle, she cataloged and analyzed photos, did population studies, and wrote reports. While employed nearly full time at NMFS, Robyn also worked on a masters of science in fisheries at the University of Washington, which she finished in 1994.

Although she loved her job on the West Coast, especially the summer field work in Alaska, and the variety of projects in which she was involved, Robyn decided that gaining experience in policy and administration would be a good idea. She transferred to a job at NMFS headquarters in Washington, DC, where she worked on regulations to manage interactions between marine mammals and U.S. commercial fisheries, briefed staff members of the U.S. House and Senate on marine mammal issues, and held workshops on policy issues for scientists, environmental advocates, and representatives of the commercial fishing industry. Robyn enjoyed being involved in policy formulation and she feels that her improved understanding of the policy process and important legislation will be invaluable during her future career.

Currently, Robyn is back in Seattle as a fisheries biologist for NMFS. In addition to her continuing work on harbor seal populations in Alaska, she has recently been involved in a project on wildlife management funded by Yellowstone National Park. Ranchers who raise cattle on lands surrounding Yellowstone worry that bison migrating out of the park during the winter may carry brucellosis, a disease that causes cattle to abort their calves. Part of Robyn's duty is to analyze management alternatives designed to reduce the probability that bison will transmit the disease to local cattle herds.

How could you get a job like Robyn's? A solid biology eduction is key, especially including ecology, population dynamics, and marine zoology. Robyn says that a field course that she attended at the Friday Harbor Laboratory in Washington was the best thing she did as an undergraduate. It was intense, but fun and rewarding. There are many other field stations around the country and Robyn strongly suggests that you attend one if you're interested in studying ecology or wildlife. Volunteering or working part-time to gain practical experience can be very valuable. Much of what she does in her job requires writing and good communication skills. "Keep an open mind," Robyn suggests, "and be willing to pursue new opportunities and adventures, because sometimes you may be pleasantly surprised at the direction it takes you."

CHAPTER 17

Air, Climate, and Weather

The trouble with our times is that the future is not what it used to be.

—Paul Valéry—

OBJECTIVES

After studying this chapter, you should be able to:

- describe how tornadoes form and why they are dangerous.
- summarize the structure and composition of the atmosphere.
- explain how the jet streams, prevailing winds, and frontal systems determine local weather.
- comprehend how El Niño cycles change ocean surface temperatures and affect continental climate.
- evaluate previous climatic catastrophes and the driving forces thought to bring about climatic change.
- debate the hypothesis that human actions may bring about global climate change.

ESSENTIAL STUDY PARTNER

The Essential Study Partner (ESP) CD-ROM that accompanies this textbook contains additional information on the following topics:

- atmospheric layers: principles: biomes: climate/living resources: the earth: weather
- ozone and greenhouse gases: physical resources: the earth: weather
- air movements, convection currents, jet streams, monsoons, and weather fronts: physical resources: the earth: weather
- Milankovitch cycles, ice ages, and global warming: physical resources: the earth: climate
- sources and effects of climatic change on humans: physical resources: the earth: climate

Above: Lightning flashes over the city as a summer thunderstorm approaches.

A Swarm of Deadly Tornadoes Sweeps across the Great Plains

Although they live in the midst of "Tornado Alley," where more violent storms occur than any other place in America, most people in Oklahoma City assumed they'd never actually see a tornado. Believing that deadly twisters occur only in remote, rural areas, or that modern technology protects us from threats of extreme weather, most urban residents give little thought to storm safety. Storm-warning sirens are routinely ignored as many people consider themselves immune from the forces of nature. This complacency proved to be mistaken, however, on May 3, 1999, when an extremely powerful frontal system sweeping across the prairies spawned more than 40 killer tornadoes that scoured a broad path of destruction from Texas, across Oklahoma, and into Kansas.

The average funnel cloud remains on the ground for only a few minutes, and usually is less than a hundred meters wide, but the twister that roared across Oklahoma City that afternoon gouged a path more than 1.6 km (1 mi) wide and 130 km (80 mi) long. Even more remarkable, wind speeds in the funnel were measured by Doppler radar at more than 510 km/hr (318 mph), the highest ever recorded at ground level anywhere on earth. This made the storm a level F5 on the Fujita scale. An F1 tornado breaks tree limbs and damages billboards. An F2 tornado tears roofs off houses and destroys mobile homes. An F3 tornado uproots trees and demolishes ordinary frame houses. An F4 tornado turns building materials into wall-piercing missiles. An F5 tornado hurls large animals, cars, and even houses long distances.

Whole neighborhoods were simply mowed down by this monster storm, with nothing left but jumbled scrap piles (fig. 17.1). Altogether 49 people were killed and nearly a thousand were injured. Total property losses were more than $1 billion, making this the nation's most destructive family of tornadoes in more than a decade.

What can you do to protect yourself from severe weather? In tornado or hurricane country, houses should be storm-proofed by having a safe room with reinforced walls and stocked with emergency supplies. Perhaps even more important, everyone should learn something about how weather systems work and how to recognize the signs of approaching storms. Paying attention to warning systems is valuable, but being aware of what's happening outside, and being able to make your own judgements about the risks you face, may be even more important.

A larger question is what these super storms, as well as unusual droughts, heat waves, ice storms, and hurricanes, tell us about overall weather patterns and global climate. Is this an omen of a new era of fierce storms and abnormal conditions? Have we changed our environment in ways that spell trouble for all of us in the future? Some atmospheric scientists think so, while others argue that there is no evidence yet of long-term human impacts on global climate. In this chapter, we will study the air around us. And we will look at some of the ways it affects us and is, in turn, affected by living organisms—including humans.

FIGURE 17.1 Thousands of homes and businesses were destroyed as a swarm of tornadoes rampaged across Oklahoma on May 3, 1999.

Weather is a description of the physical conditions of the atmosphere (moisture, temperature, pressure, and wind), all of which play a vital role in shaping ecosystems. The dynamism of the atmosphere is maintained by a ceaseless flow of solar energy. Winds generated by pressure gradients push large air masses of differing temperature and moisture content around the globe. Our daily weather is created by the movement of these air masses (fig. 17.2).

Climate is a description of the long-term pattern of weather in a particular area. Climates often undergo cyclic changes over decades, centuries, and millenia. Determining where we are in these cycles and predicting what may happen in the future is an important, but difficult, process. As human activities change the properties of the atmosphere, it becomes more difficult and more important to understand how the atmosphere works and what future weather and climate conditions may be.

FIGURE 17.2 The atmospheric processes that purify and redistribute water, moderate temperatures, and balance the chemical composition of the air are essential in making life possible. The earth is the only planet we know with a habitable atmosphere. To a large extent, living organisms have created, and help to maintain, the atmosphere on which we all depend.

Weather and climate are important, not only because they affect human activities, but because they are primary determinants of biomes and ecosystem distribution. Generalized, large-scale climate often is not as important in the life of an individual organism as is the microclimate in its specific habitat. Geographic boundaries that separate communities and ecosystems are established primarily by climatic boundaries created by temperature, moisture, and wind-distribution patterns (chapter 4). The movement and effects of pollutants also are strongly linked to weather and climatic conditions (chapter 18).

COMPOSITION AND STRUCTURE OF THE ATMOSPHERE

We live at the bottom of a virtual ocean of air. Extending upward about 1600 km (1000 mi), this vast, restless envelope of gases is far more turbulent and mobile than the oceans of water. Its currents and eddies are the winds.

Past and Present Composition

The composition of the atmosphere has changed drastically since it first formed as the earth cooled and condensed from interstellar gases and dust. Most geochemists believe that the earth's earliest atmosphere was made up mainly of hydrogen and helium and was hundreds of times more massive than it is now. Over billions of years, most of that hydrogen and helium diffused into space. At the same time, volcanic emissions have added carbon, nitrogen, oxygen, sulfur, and other elements to the atmosphere.

The current composition of the earth's atmosphere is unique in our solar system. This is the only place we know of with free oxygen and water vapor. We believe that virtually all the molecular oxygen in the air was produced by photosynthesis in blue-green bacteria, algae, and green plants. If that oxygen were not present, heterotrophs (like us) who oxidize organic compounds as an energy source could not exist. As chapter 3 discusses, producers and consumers create a balance between carbon dioxide and oxygen levels in the atmosphere. The Gaia hypothesis, first proposed by British chemist James Lovelock, suggests that the balance between atmospheric carbon dioxide and oxygen maintained by living organisms is responsible not only for creating a unique atmospheric chemical composition but also for other environmental characteristics that make life possible. We will discuss how carbon dioxide levels regulate temperatures later in this chapter. Organisms—especially humans—continue to modify the atmosphere, but at a dramatically accelerated rate, and in ways that may be disastrous for other forms of life.

Table 17.1 presents the main components of clean, dry air. Water vapor concentrations vary from near zero to 4 percent, depending on air temperature and available moisture. Small but important concentrations of minute particles and droplets—collectively called

TABLE 17.1 Present Composition of the Lower Atmosphere*

GAS	SYMBOL OR FORMULA	PERCENT BY VOLUME
Nitrogen	N_2	78.08
Oxygen	O_2	20.94
Argon	Ar	0.934
Carbon dioxide	CO_2	0.035
Neon	Ne	0.00182
Helium	He	0.00052
Methane	CH_4	0.00015
Krypton	Kr	0.00011
Hydrogen	H_2	0.00005
Nitrous oxide	N_2O	0.00005
Xenon	Xe	0.000009

*Average composition of dry, clean air

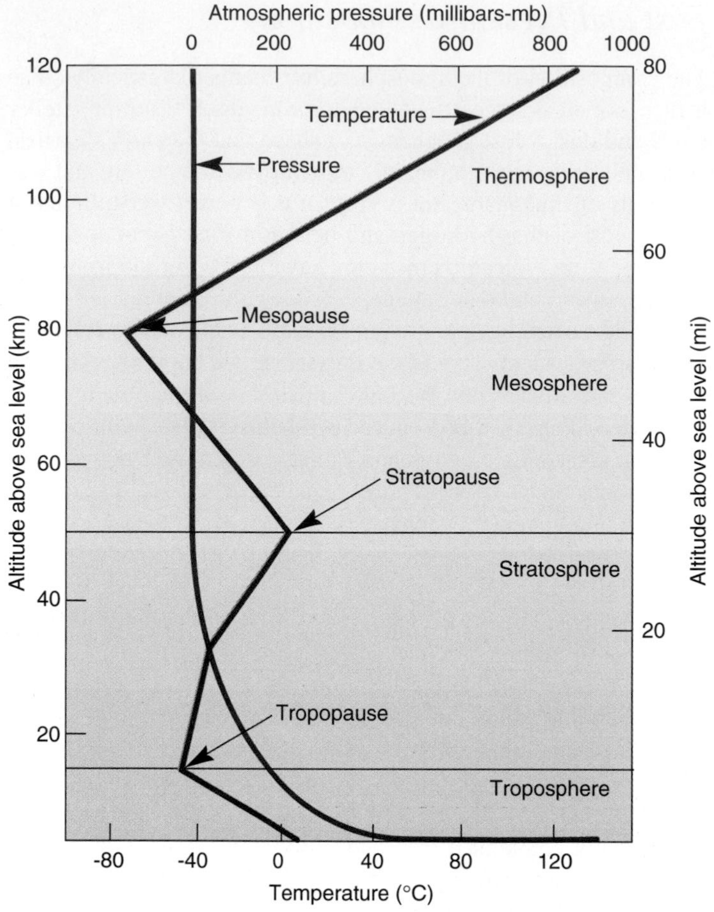

FIGURE 17.3 The atmosphere is layered into four distinct zones of contrasting temperature due to differential absorption of solar energy. Mixing between layers is inhibited by this temperature gradient. Although temperatures rise in the thermosphere, there are too few molecules present to measure much pressure rise.

aerosols—also are suspended in the air. They can have serious effects on human health and natural ecosystems as discussed later in this chapter and in chapter 18.

A Layered Envelope

The atmosphere is layered into four distinct zones of contrasting temperature due to differential absorption of solar energy (fig. 17.3). Understanding how these layers differ and what creates them helps us understand atmospheric functions.

The layer of air immediately adjacent to earth's surface is called the **troposphere.** Ranging in depth from about 18 km (10 mi) over the equator to about 8 km (5 mi) over the poles, this zone is where almost all weather events occur. Due to the force of gravity and the compressibility of gases, the troposphere contains about 75 percent of the total mass of the atmosphere. The troposphere's composition is relatively uniform over the entire planet because this zone is strongly stirred by winds. Air temperature drops rapidly with increasing altitude in this layer, reaching about –60°C (–76°F) at the top of the troposphere. A sudden reversal of this temperature gradient creates a sharp boundary, the tropopause, that limits mixing between the troposphere and upper zones.

The **stratosphere** extends from the tropopause up to about 50 km (31 mi). Air temperature in this zone is stable or even increases with higher altitude. Although more dilute than the troposphere, the stratosphere has a very similar composition except for two important components: water and ozone (O_3). The fractional volume of water vapor is about 1000 times lower, and ozone is nearly 1000 times higher than in the troposphere. Ozone is produced by lightning and solar irradiation of oxygen molecules and would not be present if photosynthetic organisms were not releasing oxygen. Ozone protects life on the earth's surface by absorbing most incoming solar ultraviolet radiation.

Recently discovered depletion of stratospheric ozone over Antarctica (and to a lesser extent over the whole planet) are of serious concern (see chapter 18). If these trends continue, we could be exposed to increasing amounts of dangerous ultraviolet rays, resulting in higher rates of skin cancer, genetic mutations, crop failures, and disruption of important biological communities. Unlike the troposphere, the stratosphere is relatively calm. There is so little mixing in the stratosphere that volcanic ash or human-caused contaminants can remain in suspension there for many years (chapter 16).

Above the stratosphere, the temperature diminishes again, creating the **mesosphere** or middle layer. The minimum temperature reached in this region is about –80°C (–120°F). At an altitude of 80 km, another abrupt temperature change occurs. This is the beginning of the **thermosphere,** a region of highly ionized gases, extending out to about 1600 km (1000 mi). Temperatures are very high in the thermosphere because molecules there are constantly bombarded by high-energy solar and cosmic radiation. There are so few molecules per unit volume, however, that if you were cruising through in a spaceship, you wouldn't notice the temperature increase.

The lower part of the thermosphere is called the **ionosphere.** This is where the aurora borealis (northern lights) appears when showers of solar or cosmic energy cause ionized gases to emit visible light. There is no sharp boundary that marks the end of the atmosphere. Pressure and density decrease gradually as one travels away from the earth until they become indistinguishable from the near vacuum of intrastellar space. The composition of the thermosphere also gradually merges with that of intrastellar space, being made up mostly of extremely dilute helium and hydrogen.

THE GREAT WEATHER ENGINE

The atmosphere is a great weather engine in which a ceaseless flow of energy from the sun causes global cycling of air and water that creates our climate and distributes energy and material through the environment.

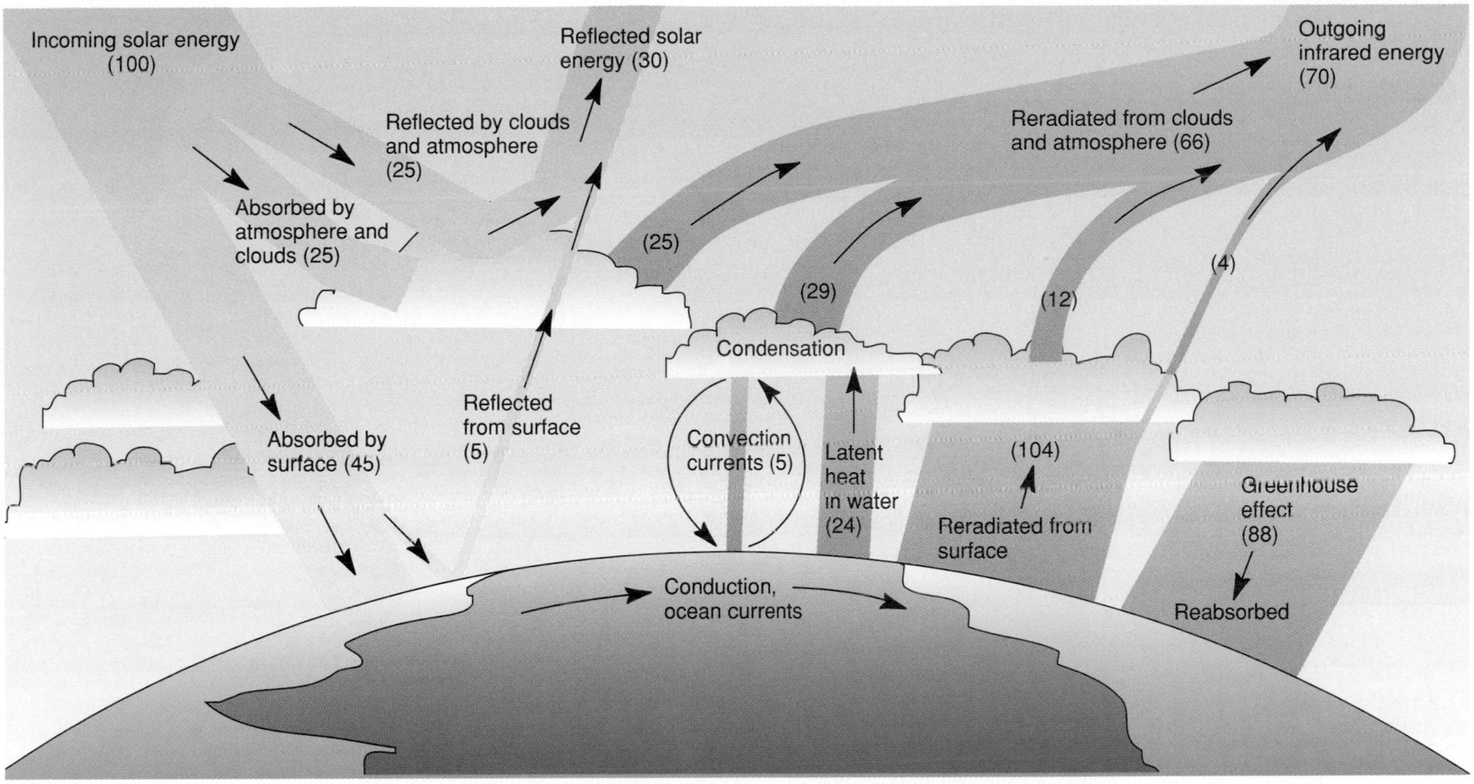

FIGURE 17.4 Energy balance between incoming and outgoing radiation. The atmosphere absorbs or reflects about half of the solar energy reaching the earth. Most of the energy reemitted from the earth's surface is long-wave, infrared energy. Most of this infrared energy is absorbed by aerosols and gases in the atmosphere and is reradiated toward the planet, keeping the surface much warmer than it would otherwise be. This is known as the greenhouse effect. The numbers shown are arbitrary units. Note that for 100 units of incoming solar energy, 100 units are reradiated to space, but more than 100 units are radiated from the earth's surface because of the greenhouse effect.

Solar Radiation Heats the Atmosphere

The sun supplies the earth with an enormous amount of energy. Although it fluctuates from time to time, incoming solar energy at the top of the atmosphere averages about 1330 watts per sq meter. About half of this energy is reflected or absorbed by the atmosphere, and half the earth faces away from the sun at any given time. Still, the amount reaching the earth's surface is at least 10,000 times greater than all installed electric capacity in the world.

The absorption of solar energy by the atmosphere is selective. Visible light (see fig. 3.6) passes through almost undiminished, whereas ultraviolet light is absorbed mostly by ozone in the stratosphere. Infrared radiation is absorbed mostly by carbon dioxide (CO_2) and water (H_2O) in the troposphere. Scattering of light by water droplets, ice crystals, and dust in the air also is selective. Short wavelengths (blue) are scattered more strongly than long wavelengths (red). The blue of a clear sky or clean, deep water at midday, and the spectacular reds of sunrise and sunset are the result of this differential scattering. On a cloudy day, as much as 90 percent of insolation is absorbed or reflected by clouds. Figure 17.4 shows the relative energy fluxes in the atmosphere.

Some solar energy is reflected from the earth's surfaces. **Albedo** is the term used to describe reflectivity. Fresh, clean snow can have an albedo of 90 percent, meaning that 90 percent of incident radiation falling on its surface is reflected. Dark surfaces, such as black topsoil or a dark forest canopy, absorb energy efficiently and might have an albedo of only 2 or 3 percent. The net average global albedo of the earth is about 30 percent. Clouds are responsible for most of that reflection. The earth's surface has a low average albedo (5 percent) due to the high energy absorbency of the oceans covering most of the globe.

Eventually, all the energy absorbed at the earth's surface is reradiated back into space. There is an important change in properties between incoming and outgoing radiation, however. Most of the solar energy reaching the earth is visible light, to which the atmosphere is relatively transparent; the energy reemitted by the earth is mainly infrared radiation (heat energy). These longer wavelengths are absorbed rather effectively in the lower levels of the atmosphere, trapping much of the heat close to the earth's surface. If the atmosphere were as transparent to infrared radiation as it is to visible light, the earth's surface temperature would be about 35°C (63°F) colder than it is now.

This phenomenon is called the "greenhouse effect" because the atmosphere, like the glass of a greenhouse, transmits sunlight while trapping heat inside. (The analogy is not totally correct, however, because glass is much more transparent to infrared radiation than is air; greenhouses stay warm mainly because the glass blocks air movement.) Increasing atmospheric carbon dioxide due to human activities appears to be causing a global warming that could cause major climatic changes, which we will discuss later in this chapter.

Because of cycling of infrared energy between the atmosphere and the planet, the amount of energy emitted from the earth's surface is about 30 percent greater than the total incoming solar radiation. The amount of energy reflected or reradiated from the top of the atmosphere must balance with the total insolation if the earth is to remain at a constant temperature.

Convection Currents and Latent Heat

Air currents, especially those carrying large amounts of water vapor, also play an important role in shaping our weather and climate. As the sun heats the earth's surface, some of that heat is transferred to adjacent air layers, causing them to expand and become less dense. This lighter air rises and is replaced by cooler, heavier air, resulting in vertical **convection currents** that stir the atmosphere and transport heat from one area to another.

Much of the solar energy absorbed by the earth is used to evaporate water. Because of the unique properties of water, it takes a significant amount of energy to change water from liquid to vapor state, and this energy is stored in the water vapor as latent or potential energy. That latent energy subsequently is released as heat when the water condenses.

Water vapor carried into the atmosphere by rising convection currents transports large amounts of energy and plays an important role in the redistribution of heat from low to high altitudes, and from the oceans to the continental landmasses (fig. 17.5). As warm, moist air rises, it expands (due to lower air pressures at higher altitudes) and cools. If condensation nuclei are present or if temperatures are low enough, the water will condense to form water droplets or ice crystals and precipitation will occur. Releasing latent heat causes the air to rise higher, cool more, and lose more water vapor. Rising, expanding air creates an area of relatively high pressure at the top of the convection column.

Air flows out of this high pressure zone toward areas of low pressure where cool, dry air is sinking (subsiding). This subsiding air is compressed (and therefore warmed) as it approaches the earth's surface, where it piles up and creates a region of relatively high pressure at the surface. Air flows out of this region back toward the area of low surface pressure caused by rising air, thus closing the cycle. These are the driving forces of the hydrologic cycle (chapter 19).

The convection currents just described can be as small and as localized as a narrow column of hot air rising over a sun-heated rock, or as large as the desert low-pressure cell that covers the U.S. Southwest most of the summer. The circulation patterns they create can be as mild as a gentle onshore breeze moving from the warm ocean toward the cooling shoreline as the sun goes down in the evening, or they can create monster cyclonic storms that drive hurricanes hundreds of kilometers wide across oceans and over continents. The force generated by such a storm (powered by latent heat released from condensing water vapor) can be equivalent to hundreds of megaton-sized nuclear bombs. These circulation systems—both large and small—are the driving force behind our weather, the moment-by-moment changes in the atmosphere.

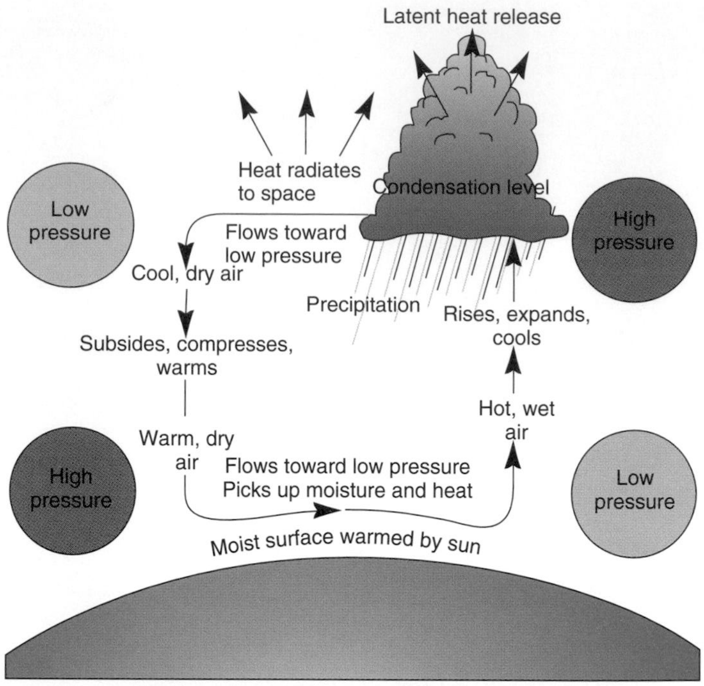

FIGURE 17.5 Convection currents and latent energy cause atmospheric circulation and redistribute heat and water around the globe.

WEATHER

Even though we have been watching the skies for thousands of years, trying to forecast what the weather will be, the science of meteorology (weather studies) is still rather imprecise and uncertain. Weather forces still are so important in our lives that it makes sense to learn as much as we can about how the atmosphere makes weather.

Energy Balance in the Atmosphere

Solar energy doesn't strike the whole globe equally. At the equator, the sun is almost directly overhead all year long. Its rays are very intense because it shines through a relatively short column of air (straight down), and energy flux (flow) is high. At the poles, however, sunlight comes in at an oblique angle. The long column of air through which light must pass before it reaches the surface causes much greater energy losses from absorption and scattering. Moreover, when the light does reach the ground, it is spread over a larger area because of its angle of incidence, reducing surface heating even more.

Furthermore, the tilt of the earth's axis means that there is no sunlight at the poles during much of the winter. The equator, by contrast, has days about the same length all year long; thus, compared to the poles, the equatorial regions have an energy surplus.

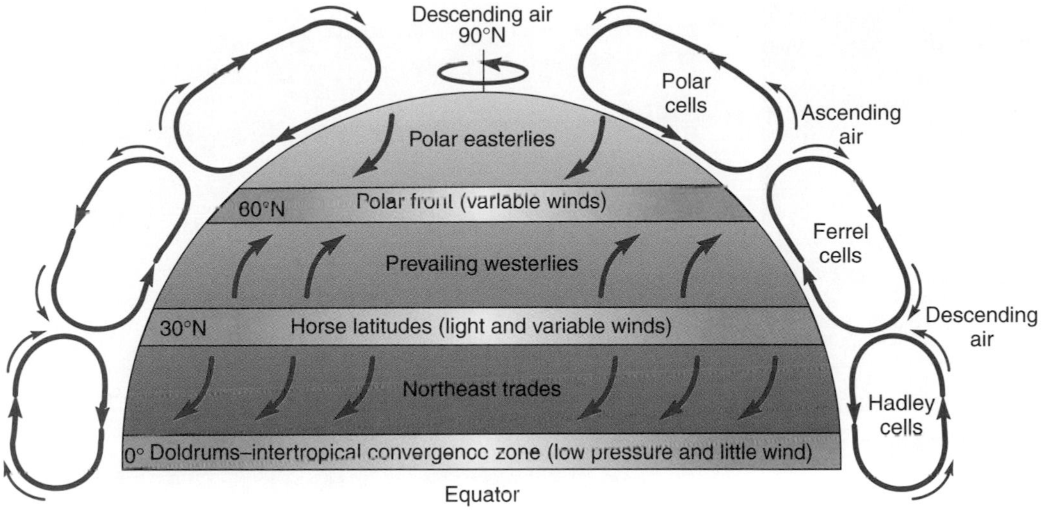

FIGURE 17.6 General circulation patterns over the Northern Hemisphere. The actual boundaries of the circulating convection cells vary from day to day and season to season, as do the local directions of surface winds. Surface topography also complicates circulation patterns, but within these broad regions, winds usually have a predominant and predictable direction.

This energy imbalance is evened out by movement of air and water vapor in the atmosphere, and by liquid water in rivers and ocean currents. Warm, tropical air moving toward the poles and cold, polar air moving toward the equator account for about half of this energy transfer. Latent heat in water vapor (mainly from the oceans) makes up about 30 percent of the global energy redistribution. The remaining 20 percent is carried mainly by ocean currents.

Convection Cells and Prevailing Winds

As air warms at the equator, rises, and moves northward, it doesn't go straight to the pole in a single convection current. Instead, this air sinks and rises in several intermediate bands, forming circulation patterns called cells (fig. 17.6). Nor do the returning surface flows within these cells run straight north and south. Friction, drag, and momentum cause air layers close to the earth's surface to be pulled in the direction of rotation. This deflection is called the **Coriolis effect.** In the Northern Hemisphere, the Coriolis effect deflects winds about 30° to the right of their expected path, creating clockwise or anticyclonic spiraling patterns in winds flowing out of a high-pressure center, and cyclonic or counterclockwise winds spiraling into a low-pressure area. In the Southern Hemisphere, winds and water movements shift in the opposite direction. The earth's rotation affects only large-scale movements, however. Contrary to popular beliefs, water draining out of a bathtub in the Southern Hemisphere is controlled by the shape of the drain, and will not necessarily swirl in the opposite direction from what it would in the north.

A major zone of subsidence occurs at about 30° north latitude. Air flows into this region of upper atmosphere low pressure both from the north and south. Air flowing back toward the equator is turned toward the west by the Coriolis effect, creating the steady northeast "trade winds" of subtropical oceans. Their name comes from the dependable routes they provided for merchant sailing ships in earlier days. Where this dry, subsiding air falls on continents, it creates broad, subtropical desert regions (chapter 19). Air flowing north from this region of subsidence turns eastward, giving rise to the prevailing westerlies of middle latitudes. (Notice that an eastward flowing wind is called a west wind or a westerly, due to the direction from which it originates.)

Winds directly under regions of subsiding air often are light and variable. They create the so-called horse latitudes because sailing ships bringing livestock to the New World were often becalmed here and had to throw the bodies of dead horses overboard. Rising air at the equator creates doldrums where the winds may fail for weeks at a time. Another band of variable winds at about 60° north, called the polar front, tends to block the southward flow of cold polar air. As we will see in the next section, however, all these boundaries between major air flows wander back and forth, causing great instability in our weather patterns, especially in midcontinent areas. The Southern Hemisphere has more stable wind patterns because it has more ocean and less landmass than the Northern Hemisphere.

Jet Streams

Superimposed on the major circulation patterns and prevailing surface winds are variations caused by large-scale upper air flows and shifting movements of the large air masses that they push and pull. The most massive of these rivers of air are the **jet streams,** powerful winds that circulate in shifting flows rivaling the oceanic currents in extent and effect. Generally following meandering paths from west to east, jet streams can be as much as 50 km wide and 5 km deep. The number, flowing speed, location, and size of jet streams all vary from day to day and place to place.

Wind speeds at the center of a jet stream are often 200 km/hr (124 mph) and may reach twice that speed at times. Located 6 to 12 km (3.7–7.5 mi) above the earth's surface, jet streams follow discontinuities in the tropopause (the boundary between the troposphere and the stratosphere), where they are broken into large, overlapping plates that fit together like shingles on a roof. The jet streams are probably generated by strong temperature contrasts where adjacent plates overlap.

There are usually two main jet streams over the Northern Hemisphere. The subtropical jet stream generally follows a sinuous path about 30° north latitude (the southern edge of the United

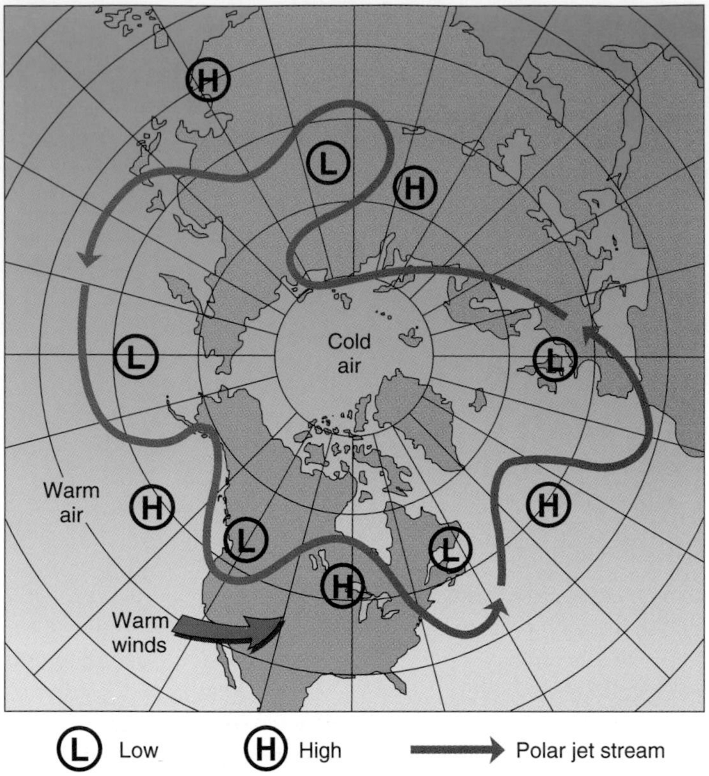

FIGURE 17.7 A typical pattern of the arctic circumpolar vortex. This large, circulating mass of cold air sends "fingers," or lobes, across North America and Eurasia, spreading storms in their path. If the vortex becomes stalled, weather patterns stabilize, causing droughts in some areas and excess rain elsewhere.

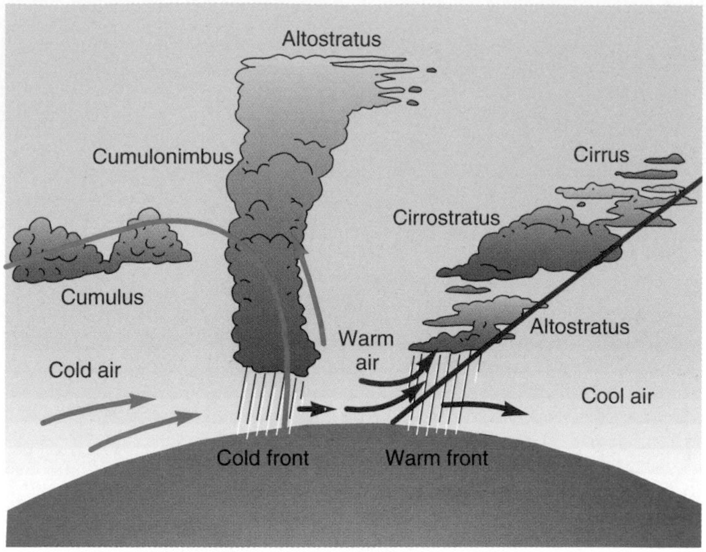

FIGURE 17.8 A cold front assumes a bulbous, "bull-nose" appearance because ground drag retards forward movement of surface air. As warm air is lifted up over the advancing cold front, it cools, producing precipitation. When warm air advances, it slides up over cooler air in front and produces a long, wedge-shaped zone of clouds and precipitation. The high cirrus clouds that mark the advancing edge of the warm air mass may be 1000 km and 48 hours ahead of the front at ground level.

States), while the northern jet stream follows a more irregular path along the edge of a huge cold air mass called the circumpolar vortex (fig. 17.7) that covers the earth's top like a cap with scalloped edges. This whole polar vortex rotates from west to east slightly faster than the planet's rotation. As it moves, the lobes, or fingers, of cold air that protrude south from the vortex sweep across Canada and the United States. The clash between cold, dry arctic air masses pushing south against warm, wet air masses moving north from the Gulf of Mexico or the Pacific Ocean causes winds, rains, and storms across the continent.

During the winter, as the Northern Hemisphere tilts away from the sun and the atmosphere cools, the polar air masses become stronger and push farther south, bringing snow and low temperatures across much of the United States. During the summer, as we tilt back toward the sun, warm air from the South pushes the polar jet stream back toward the pole.

Occasionally, the circumpolar vortex slows so that it rotates at nearly the same speed as the earth, stalling the motion of the lobes or air masses, and locking a huge ridge of hot, dry air over mid-America for months at a time. What causes air flow to be stalled like this—or to resume normal circulation patterns—is unknown, but the amount of heat in the atmosphere surely plays a role.

Frontal Weather

The boundary between two air masses of different temperature and density is called a front. When cooler air displaces warmer air, we call the moving boundary a **cold front.** Since cold air tends to be more dense than warm air, a cold front will hug the ground and push under warmer air as it advances. As warm air is forced upward, it cools adiabatically (without loss or gain of energy), and its cargo of water vapor condenses and precipitates. Upper layers of a moving cold air mass move faster than those in contact with the ground because of surface friction or drag, so the boundary profile assumes a curving, "bull-nose" appearance (fig. 17.8, left). Notice that the region of cloud formation and precipitation is relatively narrow. Cold fronts generate strong convective currents and often are accompanied by violent surface winds and destructive storms. An approaching cold front generates towering clouds called thunderheads that reach into the stratosphere where the jet stream pushes the cloud tops into a characteristic anvil shape. The weather after the cold front passes is usually clear, dry, and invigorating.

If the advancing air mass is warmer than local air, a **warm front** results. Since warm air is less dense than cool air, an advancing warm front will slide up over cool, neighboring air parcels, creating a long, wedge-shaped profile with a broad band of clouds and precipitation (fig. 17.8, right). Gradual uplifting and cooling of air in the warm front avoids the violent updrafts and strong convection currents that accompany a cold front. A warm front will have many layers of clouds at different levels. The highest layers are often wispy cirrus (mare's tail) clouds that are composed mainly of ice crystals. They may extend 1000 km (621 mi) ahead

FIGURE 17.9 As hurricane Hugo approached the Florida coast in 1989, the hollow, cloud-free eye could clearly be seen surrounded by the spiral wall cloud generated by intense warm, moist updrafts.

FIGURE 17.10 Tornadoes are local cyclonic storms caused by rapid mixing of cold, dry air and warm, wet air. Wind speeds in the swirling funnel can be more than 160 km/hr (100 mph) and can be very destructive where the storm touches down on the ground.

of the contact zone with the ground and appear as much as 48 hours before any precipitation. A moist warm front can bring days of drizzle and cloudy skies.

Cyclonic Storms

Few people experience a more powerful and dangerous natural force than cyclonic storms spawned by low-pressure cells over warm tropical oceans. As we discussed earlier in this chapter, low pressure is generated by rising warm air. Winds swirl into this low-pressure area, turning counterclockwise in the Northern Hemisphere due to the Coriolis effect. When rising air is laden with water vapor, the latent energy released by condensation intensifies convection currents and draws up more warm air and water vapor. As long as a temperature differential exists between air and ground and a supply of water vapor is available, the storm cell will continue to pump energy into the atmosphere.

Called **hurricanes** in the Atlantic and eastern Pacific, typhoons in the western Pacific, or cyclones in the Indian Ocean, winds near the center of these swirling air masses can reach hundreds of kilometers per hour and cause tremendous suffering and destruction. Often hundreds of kilometers across, these giant storms can generate winds as high as 320 kph (200 mph) and push walls of water called storm surges far inland (fig. 17.9). In 1970, a killer typhoon brought torrential rains, raging winds, and a storm surge up to 5 m high that flooded thousands of square kilometers of the flat coastal area of Bangladesh, drowning more than a half million people.

Tornadoes, swirling funnel clouds that form over land, also are considered to be cyclonic storms although their rotation isn't generated by Coriolis forces. While never as large or powerful as hurricanes, tornadoes can be just as destructive in the limited areas where they touch down (fig. 17.10). Some meteorologists consider smaller, less destructive phenomena such as waterspouts, dust devils, and smoke funnels over forest fires to be miniature tornadoes, but others think they have very different origins.

Tornadoes are generated on the American Great Plains by giant "supercell" frontal systems where strong, dry air cold fronts from Canada collide with warm humid air moving north from the Gulf of Mexico. Greater air temperature differences cause more powerful storms, which is why most tornadoes occur in the spring, when Arctic cold fronts penetrate far south over the warming plains. As warm air rises rapidly over dense, cold air, intense vertical convection currents generate towering thunderheads with anvil-shaped leading edges and domed tops up to 20,000 m (65,000 ft) high (fig. 17.11). Water vapor cools and condenses as it rises, releasing latent heat and accelerating updrafts within the supercell. Penetrating into the stratosphere, the tops of these clouds can encounter jet streams, which help create even stronger convection currents.

But what sets the air mass spinning? Scientists debate this point, but one theory is that differential air speeds called shear forces—fast-moving above but slower near the ground—roll the air ahead of the advancing front in invisible horizontal tubes (or vortices) much as you might roll modeling clay between your hands. As these rolling vortex tubes are sucked into the vertical convective currents, they tilt upright and set the air column in motion, creating a spiral structure called a mesocyclone. Initially, the swirling air mass may be many kilometers wide, but cold downdrafts from the rear flank of the storm can spiral around the mesocyclone, increasing its velocity and narrowing its diameter much as a skater spins faster as she pulls in her arms. As the opening story of this chapter shows,

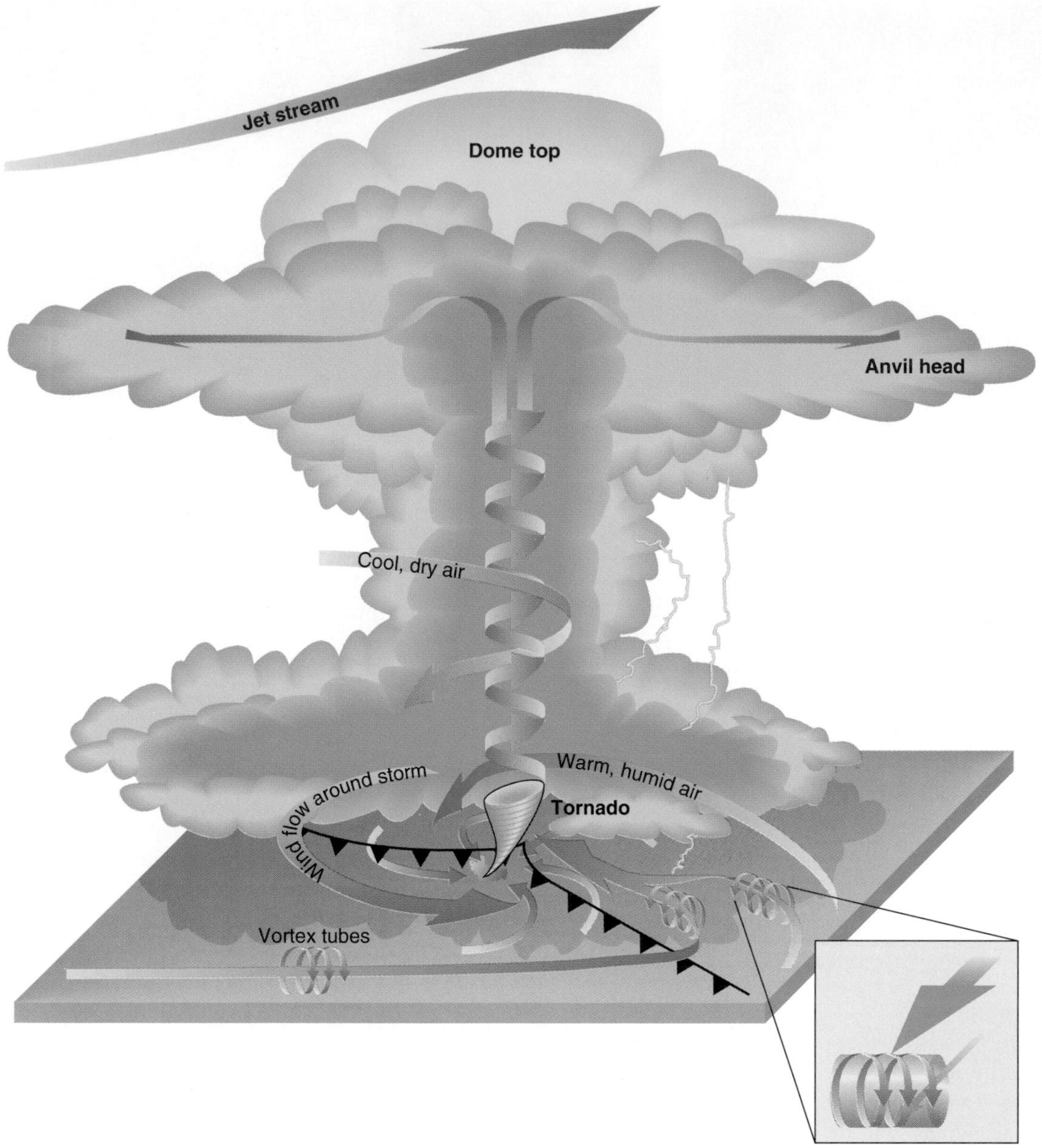

FIGURE 17.11 Tornadoes are generated by rapidly rising thermal updrafts where warm humid air is drawn up into a cold supercell thunderstorm. Rolling vortex tubes generated by differential air currents are drawn into the updraft and start it rotating. Cool, dry air spiraling down from the rear tightens the spiral and intensifies its speed.
Source: Data from National Center for Atmospheric Research, May 20, 1996.

winds at the edge of the rapidly spinning air column can reach speeds up to 510 kph (318 mph) that rip trees out of the ground and drive straws through telephone poles. In addition, low pressure in the center of the funnel cloud implodes houses, sucks people out of windows, and helps hurl large objects across the countryside.

Sometimes a supercell isn't organized enough to spawn tornadoes, but strong downdrafts, called **downbursts,** can generate straight-line winds well over 160 k/hr (100 mph). On July 4, 1999, a powerful straight-line wind flattened about 25 million trees in a swath 16 km by 48 km (10 by 30 mi) across the heart of the Boundary Waters Canoe Area in northern Minnesota. Witnesses said that as the winds hit, the forest began rippling like a field of wheat, and then in seconds, was knocked down as if by a steamroller.

The same strong updrafts that create tornadoes can also generate hail when falling water droplets are blown back up to higher altitudes where they freeze. Each time the droplets bounce up and down, they pick up more water and become larger until finally the ice balls are heavy enough to resist being lofted by updrafts. If they remain frozen all the way to the ground, we call them hail. If you cut open large hail stones, you often can see concretion rings where successive layers of ice were added.

Seasonal Winds

A **monsoon** is a seasonal reversal of wind patterns caused by different heating and cooling rates of the ocean and continents. The most dramatic examples of monsoon weather are found in tropical or subtropical countries where a large land area is cut off from continental air masses by mountain ranges and surrounded by a large volume of water. The Indian subcontinent is a good example

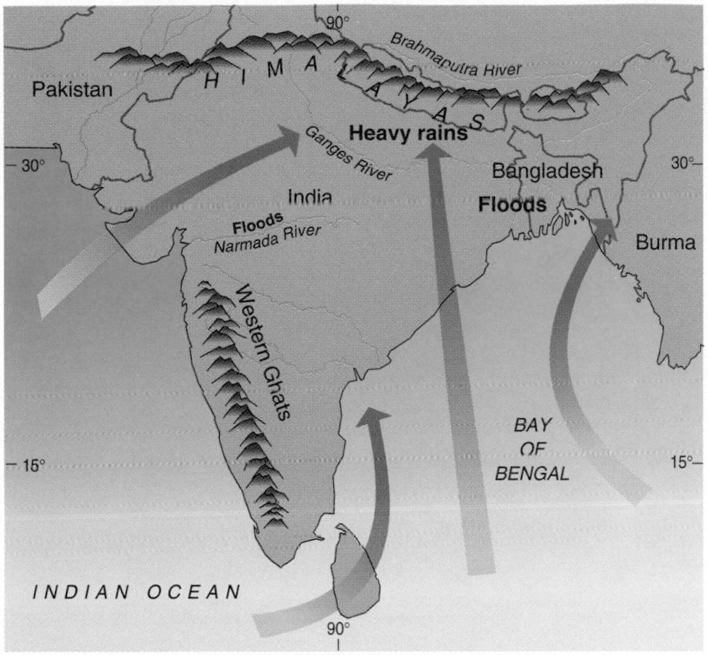

FIGURE 17.12 Summer monsoon air flows over the Indian subcontinent. Warming air rises over the plains of central India in the summer, creating a low-pressure cell that draws in warm, wet oceanic air. As this moist air rises over the Western Ghats or the Himalayas, it cools and heavy rains result. These monsoon rains flood the great rivers, bringing water for agriculture, but also causing much suffering.

FIGURE 17.13 Failure of monsoon rains brings drought, starvation, and death to both livestock and people in the Sahel desert margin of Africa. Although drought is a fact of life in Africa, many governments fail to plan for it, and human suffering is much worse than it needs to be.

(fig. 17.12). As the summer sun heats the Indian plains, a strong low-pressure system develops. Flow of cool air from the north is blocked by the Himalayan mountains. A continuous flow of moisture-laden air from the subtropical high-pressure area over the Arabian Sea sweeps around the tip of India and up into the Ganges Plain and the Bay of Bengal. As the onshore winds are driven against the mountains and rise, the air cools and drops its load of water.

During the four months of the monsoon, which usually lasts from May until August, 10 m (390 in) of rain may fall on Nepal, Bangladesh, and western India. The highest rainfall ever recorded in a season was 25 m (82 ft), which fell in about five months on the southern foothills of the Himalayas in 1970. These heavy rains result in high rates of erosion and enormous floods in the flat delta of the Ganges River, but they also irrigate the farmlands that feed the second most populous country in the world. In winter, the Indian landmass is cooler than the surrounding ocean and the wind flow reverses. The northeast winds pick up moisture as they blow over the Bay of Bengal and bring winter monsoon rains to Indonesia, Australia, and the Philippines.

Africa also experiences strong seasonal monsoon winds. Hot air rising over North Africa in the summer pulls in moist air from the Atlantic Ocean along the Gulf of Guinea. The torrential rains released by this summer circulation nourish the tropical forests of Central Africa and sometimes extend as far east as the Arabian Ocean. How far north into the Sahara Desert these rain clouds penetrate determines whether crops, livestock, and people live or die in the desert margin (fig. 17.13). In the winter, the desert cools and air flows back out over the now warmer Atlantic. These shifting winds allowed Arabian sailors, like Sinbad, to sail from Africa to India in the summer and return in the winter.

There seems to be a connection between strong monsoon winds over West Africa and the frequency of killer hurricanes in the Caribbean and along the Atlantic coast of North and Central America. In years when the Atlantic surface temperatures are high, a stronger than average monsoon trough forms over Africa. This pulls in moist maritime air that brings rain (and life) to the Sahel (the margin of the Sahara). This trough gives rise to tropical depressions (low pressure storms) that follow one another in regular waves across the Atlantic. The weak trade winds associated with "wet" years allow these storms to organize and gain strength. Evaporation from the warm surface water provides energy.

During the 1970s and 1980s when the Sahel had devastating droughts, the weather was relatively quiet in North America. Only one killer hurricane (winds over 70 km/hr or 112 mph) reached the United States in two decades. By contrast, in 1995 when ocean surface temperatures were high and rain returned to the Sahel, seven hurricanes and six named tropical storms were spawned in the Atlantic. During the summer of 1996, satellite images of the Atlantic showed five active hurricanes roaring across the Caribbean at the same time. Are we witnessing a long-term climatic change or merely a temporary abberation?

Weather Modification

As author Samuel Clemens (Mark Twain) said, "Everybody talks about the weather, but nobody does anything about it." People probably always have tried to influence local weather through religious ceremonies, dancing, or sacrifices. During the drought of the 1930s in the United States, "rainmakers" fleeced desperate farmers of thousands of dollars with claims that they could bring rain.

Some recent developments appear to be effective in local weather modification, at least in some circumstances, but they are not without drawbacks and controversy. Seeding clouds with dry ice or ionized particles, such as silver iodide crystals, can initiate precipitation if water vapor is present and air temperatures are near the condensation point (chapter 19). Dry ice also is effective at dispersing cold fog (where supercooled water droplets are present). Warm fog (air temperatures above freezing) and ice fog (ice crystals in the air) are not usually amenable to weather modification. Hail suppression by cloud seeding also can be effective, but dissipation of the clouds that generate hail diverts rain from areas that need it, as well. There are concerns that materials used in cloud seeding could cause air, ground, and water pollution. North Dakota recently agreed to pay for an environmental impact assessment to convince Montana farmers that North Dakota is not stealing rain by seeding clouds. Similarly, Wyoming and Idaho have been arguing over cloud seeding to increase winter snowpack in the Tetons that would provide summer irrigation water downstream in Idaho.

CLIMATE

If weather is a description of physical conditions in the atmosphere (humidity, temperature, pressure, wind, and precipitation), then climate is the *pattern* of weather in a region over long time periods. The interactions of atmospheric systems are so complex that climatic conditions are never exactly the same at any given location from one time to the next. While it is possible to discern patterns of average conditions over a season, year, decade, or century, complex fluctuations and cycles within cycles make generalizations difficult and forecasting hazardous (fig. 17.14). We always wonder whether anomalies in local weather patterns represent normal variations, a unique abnormality, or the beginnings of a shift to a new regime.

Climatic Catastrophes

Major climatic changes, such as those of the Ice Ages, have drastic effects on living organisms. When climatic change is gradual, species may have time to adapt or migrate to more suitable locations. Where climatic change is relatively abrupt, many organisms are unable to respond before conditions exceed their tolerance limits. Whole communities may be destroyed, and if the climatic change is widespread, many species may become extinct.

Perhaps the most well-studied example of this phenomenon is the great die-off that occurred about 65 million years ago at the end

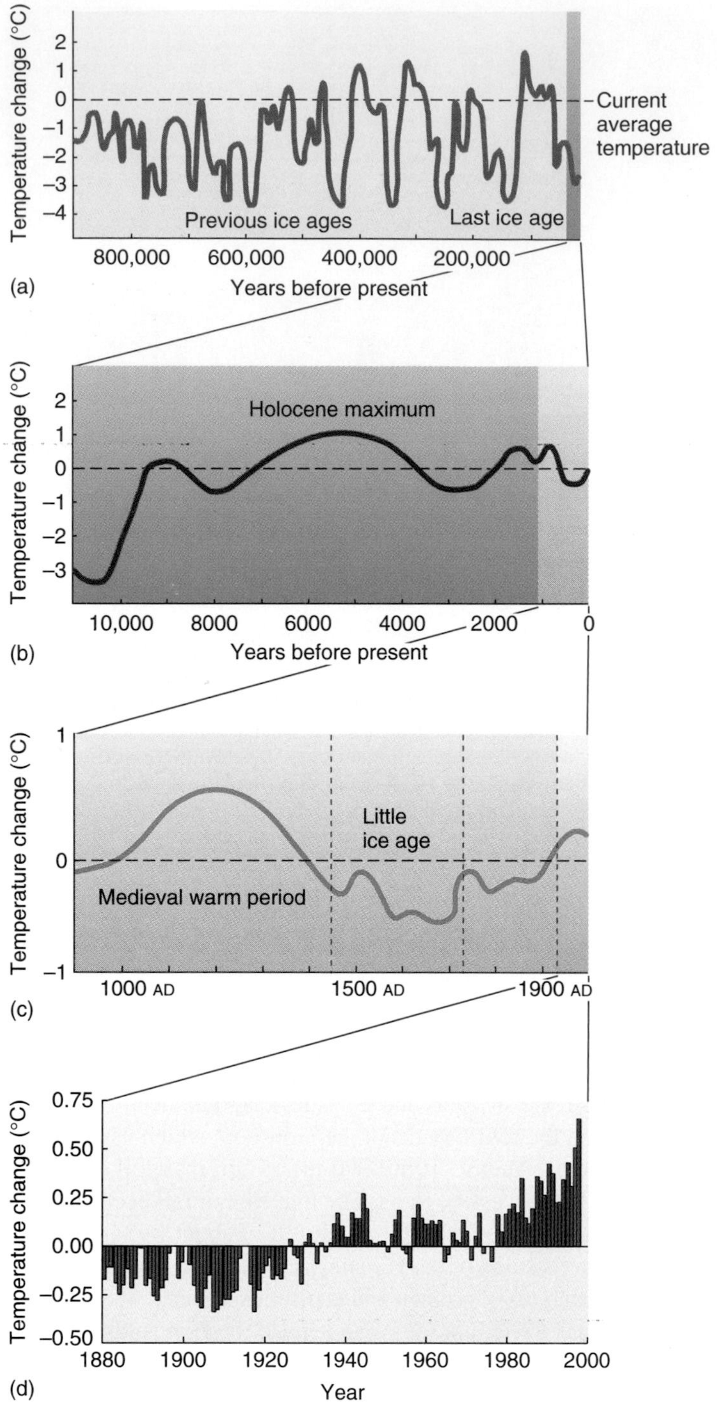

FIGURE 17.14 Global temperature variations over four time periods: (*a*) the past 900,000 years, (*b*) the past 10,000 years, (*c*) the past 1000 years, and (*d*) the most recent 120 years. The dotted lines in each represent the mean temperature for the twentieth century. Note that both horizontal and vertical scales change in each figure.

Source: (*d*) Data from National Climate Data Center/NESDIS/NOAA, 1998.

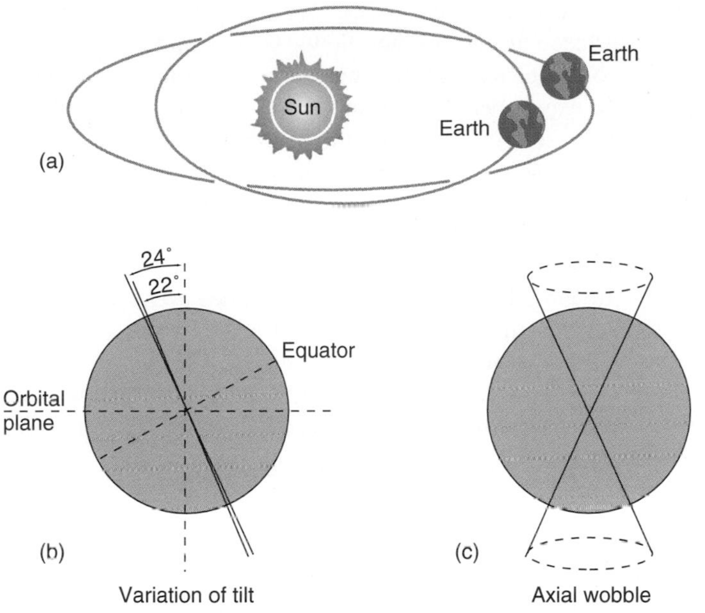

FIGURE 17.15 Milankovitch cycles, which may affect long-term climate conditions: (*a*) changes in the occupancy of the earth's orbit, (*b*) shifting tilt of the axis, and (*c*) wobble of the earth.

FIGURE 17.16 Milankovitch cycle periodicity? The light and dark bands in these seafloor sediments were laid down under different climatic regimes, each lasting 20,000 to 40,000 years. Now exposed along the French coast, these layers suggest the regularity of Milankovitch cycles.

of the Cretaceous period. Most dinosaurs—along with 75 percent of all previously existing plant and animal species—became extinct, apparently as a result of sudden cooling of the earth's climate. Geologic evidence suggests that this catastrophe was not an isolated event. There appear to have been several great climatic changes, perhaps as many as a dozen, in which large numbers of species were exterminated (see table 13.3, p. 288).

Driving Forces and Patterns in Climatic Changes

What causes catastrophic climatic changes? Some scientists believe that long-term climatic changes follow a purely random pattern brought about by chance interaction of unrelated events, such as asteroid impacts, cosmic radiation from exploding supernovas, massive volcanic eruptions, abrupt flooding of glacier meltwater into the ocean, and tectonic ocean spreading that changes patterns of ocean and wind circulation.

Other scientists find periodic patterns in weather cycles. One explanation is that changes in solar energy associated with 11-year sunspot cycles or 22-year solar magnetic cycles might play a role. Another theory is that a regular 18.6-year cycle of shifts in the angle at which our moon orbits the earth alters tides and atmospheric circulation in a way that affects climate. A theory that has received attention in recent years suggests that orbital variations as the earth rotates around the sun might be responsible for cyclic weather changes.

Milankovitch cycles, named after Serbian scientist Milutin Milankovitch, who first described them in the 1920s, are periodic shifts in the earth's orbit and tilt (fig. 17.15). The earth's elliptical orbit stretches and shortens in a 100,000-year cycle, while the axis of rotation changes its angle of tilt in a 40,000-year cycle. Furthermore, over a 26,000-year period, the axis wobbles like an out-of-balance spinning top. These variations change the distribution and intensity of sunlight reaching the earth's surface and, consequently, global climate. Bands of sedimentary rock laid in the oceans (fig. 17.16) seem to match both these Milankovitch cycles and the periodic cold spells associated with worldwide expansion of glaciers every 100,000 years or so.

A historical precedent for global climate change that had disastrous effects on humans was the "little ice age" that began in the 1400s. Temperatures dropped so that crops failed repeatedly in parts of northern Europe that once were good farmland. Scandinavian settlements in Greenland founded during the warmer period around A.D. 1000 lost contact with Iceland and Europe as ice blocked shipping lanes. It became too cold to grow crops, and fish that once migrated along the coast stayed farther south. The settlers slowly died out, perhaps in battles with Inuit people who were driven south from the high Arctic by colder weather.

Evidence from ice cores drilled in the Greenland ice cap suggests that world climate may change much more rapidly than previously thought. During the last major interglacial period 135,000 to 115,000 years ago, it appears that temperatures flipped suddenly from warm to cold or vice versa over a period of years or decades rather than centuries. One possible explanation is that ocean currents acting as a conveyor belt to transport heat from the equator to the North Atlantic may have suddenly stopped or even reversed course when surges of fresh water diluted salty ocean waters. Volcanic eruptions may have played a role as well. When Mt. Pinatubo in the Philippines erupted in 1991, it ejected enough dust, aerosols, and gas into the atmosphere to cool the climate by at least 1°C. Larger volcanoes may have had greater effects.

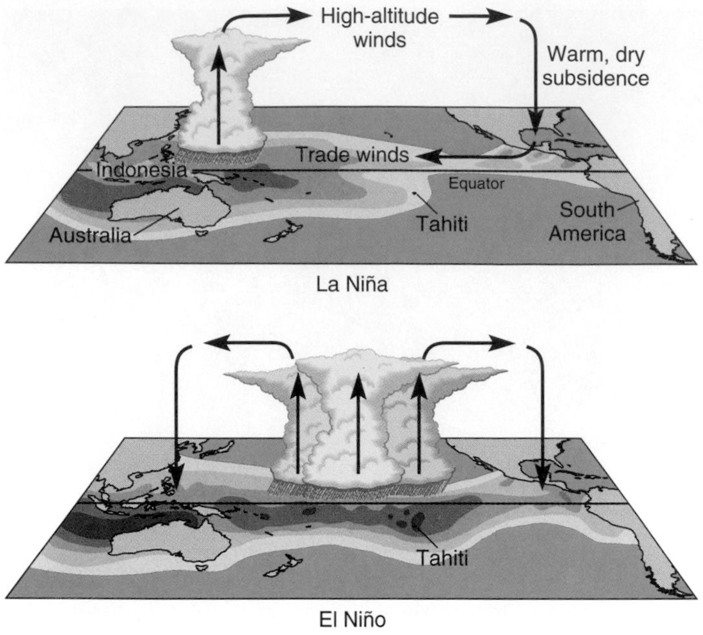

FIGURE 17.17 The El Niño/La Niña Southern Oscillation Cycle. Every three to seven years surface trade winds that normally push warm water westward toward Indonesia weaken and allow this pool of water to flow eastward toward South America.

El Niño/Southern Oscillations

What do sea surface temperatures near Indonesia, anchovy fishing off the coast of Peru, and the direction of tropical trade winds have to do with rainfall and temperatures over the United States? Strangely enough, they all may be interrelated. If the terms El Niño, La Niña, and Southern Oscillation are not yet a part of your vocabulary, perhaps they should be. They describe a connection between the ocean and atmosphere that appears to affect these factors as well as weather patterns throughout the world.

The core of this climatic system is a huge pool of warm surface water in the Pacific Ocean that sloshes slowly back and forth between Indonesia and South America like water in a giant bathtub. Most years, this pool is held in the western Pacific by steady equatorial trade winds pushing ocean surface currents westward (fig. 17.17). These surface winds are generated by a huge low-pressure cell formed by upwelling convection currents of moist air warmed by the ocean. Towering thunderheads created by rising air bring torrential summer rains to the tropical rainforests of Northern Australia and Southeast Asia. Winds high in the troposphere carry a return flow back to the eastern Pacific where dry subsiding currents create deserts from Chile to southern California. Surface waters driven westward by the trade winds are replaced by upwelling of cold, nutrient-rich, deep waters off the west coast of South America that support dense schools of anchovies and other finfish.

Every three to five years, for reasons that we don't fully understand, the Indonesian low collapses and the mass of warm surface water surges back east across the Pacific. One theory is that the high cirrus clouds atop the cloud columns absorb enough solar radiation to cool the ocean surface and reverse trade winds and ocean surface currents so they flow eastward rather than westward. Another theory is that eastward-flowing deep currents called baroclinic waves periodically interfere with coastal upwelling, warming the sea surface off South America and eliminating the temperature gradient across the Pacific. At any rate, the shift in position of the tropical depression sets off a chain of events lasting a year or more with repercussions in weather systems across North and South America and perhaps around the world.

Peruvian fishermen were the first to notice irregular cycles of rising ocean temperatures that resulted in disappearance of the anchovy schools on which they depended. They named these events **El Niño** (Spanish for the Christ child) because they often occur around Christmas time. We have come to call the intervening years La Niña (or little girl). Together, this cycle is called the El Niño Southern Oscillation (ENSO).

Actually, sea surface temperature measurements in the central Pacific near Tahiti are the best measurement of this giant oscillation. In May 1998, as the most recent El Niño was ending, surface temperatures in this region plunged 7°C (12.5°F) in a little over a week, a dramatic event considering the enormous heat-storing capacity of ocean waters.

How does the ENSO cycle affect us? During an El Niño year, the northern jet stream—which normally is over Canada—splits and is drawn south over the United States. This pulls moist air from the Pacific and Gulf of Mexico inland, bringing intense storms and heavy rains from California across the midwestern states. The intervening La Niña years bring hot, dry weather to these same areas. An unusually long El Niño event from 1991 to 1995 broke the seven-year drought over the western United States and resulted in floods of the century in the Mississippi Valley. Oregon, Washington, and British Columbia, on the other hand, tend to have warm, sunny weather in El Niño years rather than their usual rain. Droughts in Australia and Indonesia during El Niño episodes cause disastrous crop failures and forest fires, including one in Borneo in 1983 that burned 3.3 million ha (8 million acres). An even stronger El Niño in 1997 spread health-threatening forest fire smoke over much of Southeast Asia.

Are ENSO events becoming stronger or more irregular because of global climate change? It is difficult to say, but there are signs that warm ocean surface temperatures are spreading. In addition to the pool of warm water in the western Pacific associated with La Niña years, oceanographers recently discovered a similar warm region in the Indian Ocean. High sea surface temperatures spawn larger and more violent storms such as hurricanes and typhoons. On the other hand, increased cloud cover would raise the albedo while upwelling convection currents generated by these storms could pump heat into the stratosphere. This might have an overall cooling effect and act as a safety valve for global warming.

Human-Caused Global Climate Change

For the past decade, the Intergovernmental Panel on Climate Change (IPCC), a United Nations-sponsored working group of some 2400 scientists from around the world, studied anthropogenic (human-caused) impacts on global climate. Among the

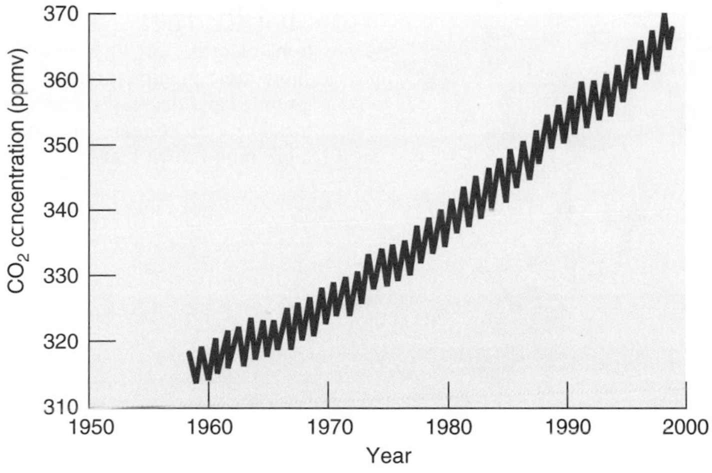

FIGURE 17.18 Carbon dioxide concentrations at Mauna Loa Observatory in Hawaii have increased dramatically since 1958. Annual fluctuations reflect differences in photosynthesis and respiration between winter and summer.
Source: Data from Dave Keeling and Tim Whorf, Scripps Institute of Oceanography.

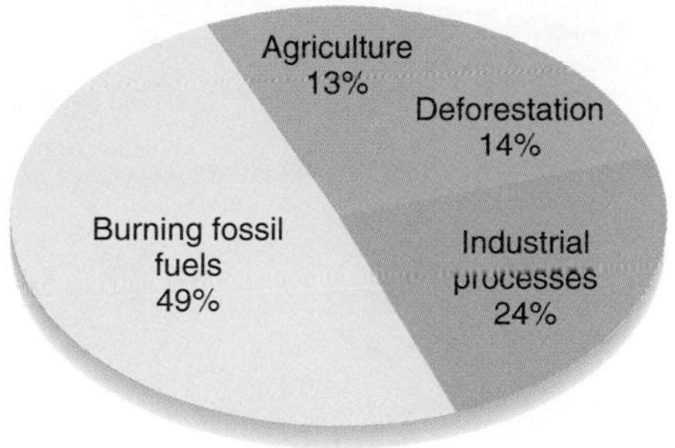

FIGURE 17.19 Contributions to global warming by different types of human activities.
Source: Data from World Resources Institute.

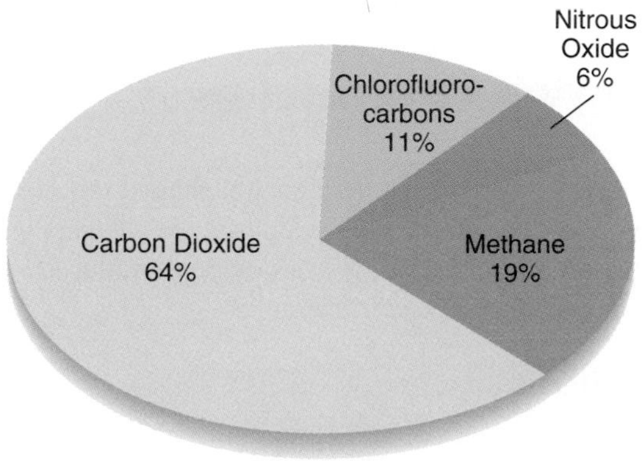

FIGURE 17.20 Relative contribution to global warming (percent of expected climate change) by human-caused releases of greenhouse gases.
Source: Data from J.T. Houghton, et al., eds. *Climate Change 1995: The Science of Climate Change,* published for the IPCC, in collaboration with the World Meteorological Organization and the United Nations Environment Programme, 1996.

more important conclusions reached by this panel are: (1) world climate has changed significantly over the past century; (2) the balance of evidence suggests a discernible human influence on global climate; and (3) climate models suggest that—if current trends continue—global mean surface air temperatures will increase between 1°C and 4.5°C (2° and 8°F) by A.D. 2100. This may not seem like much change, but the difference between current temperatures and the last ice age when glaciers covered much of North America was only about 5°C (10°F).

While a majority of scientists now support this analysis, many critics claim we don't yet know enough about global climate to make predictions. Opposition is especially strong among those with an interest in maintaining current patterns of energy consumption. But if even part of the scenarios portrayed by climate modelers is correct, this may well be *the* most crucial topic in environmental science. Global climate change could have a greater potential to change life on the planet than anything else we can do outside of nuclear war. In this section, we'll look at some of the evidence for, and possible consequences of, anthropogenic climate change.

Greenhouse Gases

The first evidence for anthropogenic atmospheric changes resulted from the International Geophysical Year in 1957. As part of that study, an observatory was established on top of Mauna Loa volcano in Hawaii to study air chemistry in a remote environment assumed to be relatively pristine and unaffected by human activities. To everyone's surprise, however, CO_2 measurements showed two striking trends. The first trend is a strong annual CO_2 cycle, even in the middle of the ocean, as plants take up carbon during the summer and release it in the winter (fig. 17.18). Because a majority of the world's land and vegetation are in the Northern Hemisphere, northern seasons dominate this signal. The other trend in the Mauna Loa data is a steady CO_2 increase from about 315 ppm by volume in 1958 to nearly 370 ppm in 1999.

The extra CO_2 added to the atmosphere each year is thought to come primarily from human actions. Burning fossil fuels, making cement, burning biomass, and other activities release about 6.3 billion metric tons of CO_2 per year into the air (fig. 17.19) (see chapter 18). The result is a 0.5 percent annual increase in atmospheric CO_2 concentrations. If current trends continue, CO_2 concentrations could reach about 500 ppm (approaching twice the preindustrial level of 280 ppm) by the end of the twenty-first century. Because CO_2 is a **greenhouse gas,** one that absorbs infrared radiation and warms surface air, this change could increase mean global temperatures significantly.

Carbon dioxide is not the only anthropogenic gas causing global warming (fig. 17.20). Methane (CH_4), chlorofluorocarbons (CFCs), and nitrous oxide (N_2O) also absorb infrared radiation and warm the atmosphere. Although rarer than CO_2, some of these gases trap heat more effectively. Methane, for instance, absorbs 20 to 30 times as much infrared and is accumulating in the atmosphere about twice as fast as CO_2. Methane is released by ruminant animals, wet-rice paddies, coal mines, landfills, and pipeline leaks. CFC releases

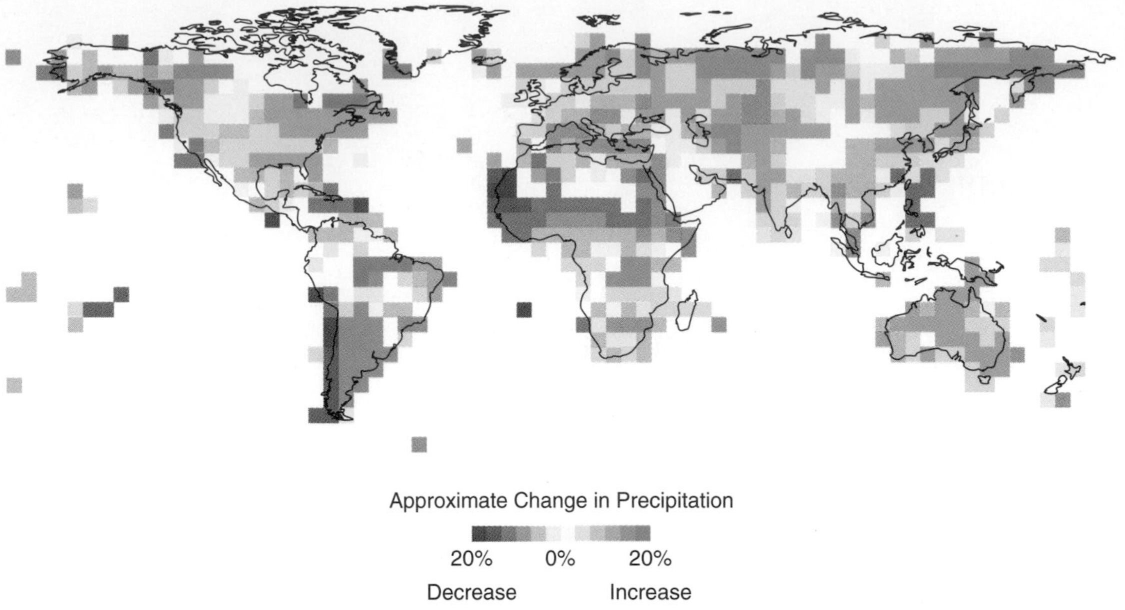

FIGURE 17.21 Precipitation records for this century show more precipitation at high latitudes and decreases in equatorial zones. Green indicates more rain; brown less.

have declined in recent years as substitutes have replaced them in many uses. The CFCs already in the atmosphere will persist for many decades or centuries, however. Nitrous oxide is produced by burning organic material and soil denitrification.

Aerosol Effects

Aerosols have a tendency to reflect sunlight and cool surface air temperatures. Locally, aerosols can more than offset warming caused by greenhouse gases, but aerosols tend to be short-lived in the air and their effects are temporary. Many industrial cities have experienced a distinct cooling trend in this century. Both anthropogenic and natural aerosols can be important in our climate. The 1991 eruption of Mt. Pinatubo in the Philippines ejected enough ash and sulfate particles to cool the global climate about 1°C for nearly a year.

Climate Changes in Twentieth Century

Evidence from several different types of measurement suggests that global mean surface air temperatures have increased about 0.5°C (0.9°F) over the past hundred years. Globally, the twentieth century has been the warmest since at least A.D. 1400. Twelve of the past 16 years were the warmest globally in the previous century, and 1998 was the warmest year since reliable instrument records began 120 years ago (see fig. 17.14*d*). Warming has been greatest over midlatitude continents during the winter and spring. Spring now occurs about a week earlier than normal across much of the United States and Canada, while the frost-free season in northern regions is now 11 days longer, on average, than it was in 1950. Night temperatures generally have increased more than day temperatures as greenhouse gases prevent heat from escaping at night but don't affect incoming solar radiation during the day. We can't be sure that all—or any—of this increase is due to greenhouse effects, but patterns of change fit those predicted by climate models.

Precipitation rates also have increased recently, especially at high latitudes during the winter. Parts of the United States, Europe, and Canada, especially at upper latitudes, were 20 percent wetter in the 1990s than in the previous 50 years (fig. 17.21). Extreme weather events—exceptionally heavy rains or prolonged drought, record high and low temperatures, and severe storms—also seem to be more common in many areas. Floods or storms that once occurred only every 100 to 500 years now may happen two or three times in a decade. It is difficult to distinguish normal weather variations from new patterns caused by human activities, but increased evaporation due to elevated air temperatures and warmer ocean surface waters are the immediate causes of the increased hurricanes and midwestern floods mentioned earlier in this chapter. Some oceanographers and climatologists suggest that the unusual frequency and length of ENSO events in recent years may signal a shift to a new climate regime in a warmer world.

Effects of Climate Change

Local climate changes could well have severe effects on human societies, agriculture, and natural ecosystems. Terrestrial and aquatic ecosystems provide many goods and services on which we depend. In many cases, both organisms and human infrastructure will not be able to move or adapt quickly enough to accommodate the rates at which we expect climate to change. Tropical areas will probably not get much warmer than they are now, but middle and high latitudes could experience unaccustomed heat. Chicago, for instance, might go from an average of 15 to 48 days each year above 32°C (90°F). Dallas could have temperatures that are hot 162 days per year rather than 100 days as it does now. Calcutta, however, where it is always steamy, will probably not get much worse. High latitudes would generally be wetter than now, but lands along the Tropic of Cancer could be drier than they are now. Some agricultural areas would benefit from these changes, but others like the Sahel region of Africa might suffer.

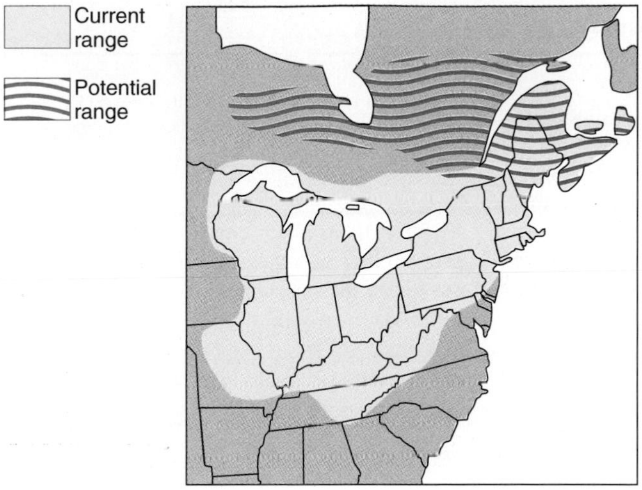

FIGURE 17.22 Change in suitable range for sugar maple, according to the climate-change model.
Source: Data from Margaret Davis and Catherine Zabriskie in *Global Warming and Biological Diversity,* ed. by Peters and Lovejoy, 1991, Yale University Press.

One of the more ominous outcomes of higher temperatures in the Arctic might be the release of vast stores of methane hydrate now locked under permafrost and in sediments in the ocean floor. Together with oxidation of high-latitude peat lands, release of these carbon stores could add as much CO_2 to the atmosphere as all the fossil fuels we have ever burned. We could trigger a disastrous escalating spiral in which the effects of warming cause even more warming.

Many wild plant and animal species are likely to be forced out of their current ranges as the climate warms (fig. 17.22). Given enough time and a route for migration, many species could adapt to new conditions, but we now are forcing them to move at least ten times the rate many achieved at the end of the last ice age. Biogeographical changes already are being observed. In Europe and North America, 57 butterfly species have either died out at the southern end of their range, or extended the northern limits, or both. On average theses populations have moved about 100 km (160 mi) north. Polar bears around Hudson Bay in Canada are up to 100 kg (220 lbs) lighter, on average, than they were 30 years ago. An earlier ice break-up gives them less time to hunt for seals. And bleaching of corals throughout tropical oceans also has been blamed on warming ocean temperatures.

Coastlines may be heavily impacted by rising sea levels and heavier storm damage. Sea levels appear to be increasing about 2 mm per year. Thermal expansion alone could multiply that rate five-fold, resulting in a one-meter rise over the next century. This could completely submerge some low-lying island nations and would displace about 70 million people each in China and Bangladesh. An even more dire possibility is that warming could accelerate the rate of glacial melting. It already is clear that mountain glaciers are retreating much faster than in the past. Professor Mauri S. Pelto of the North Cascades Glacier Project reports that all 47 glaciers in his study were retreating in 1995. By contrast, two-thirds of the glaciers studied in this region in 1967 were either stable or growing. This pattern seems to be widespread around the world.

Some scientists warn that Antarctic ice sheets, which hold by far the greatest volume of fresh water on the globe, also could melt, adding another five to six meters of fresh water to the oceans. The west Antarctic Ice Sheet is reported to have shrunk 25 percent over the past 25 years, including two giant icebergs, each about the size of Delaware, that calved off in 1987 and 1999. Arctic sea ice is reported to be 33 percent thinner and 500 km further north than usual. In 1999, for only the second time since monitoring began in 1912, there were no icebergs in the North Atlantic shipping lanes where the Titanic sank.

If this melting continues, about half of the world's megacities would need massive dikes to prevent flooding. Insurance companies worry that the $2 trillion in insured property along U.S. coastlines could be at risk from a combination of high seas and catastrophic storms. Weather-related losses already are climbing sharply, but some of that rise is due to more people living in risky places.

Finally, infectious diseases could become more common as the climate warms. Higher temperatures would allow insect vectors to expand their ranges as well as to be more active in areas where they now live. For example, malaria could expand from about 40 percent to 60 percent of the planet. Places like Nairobi, Kenya, and Harare, Zimbabwe, where malaria is now absent, could be invaded by mosquitoes. Furthermore, heat, increased UV radiation, and pollutants like chlorinated hydrocarbons all suppress disease-fighting immune systems in humans and other animals. This could exacerbate increased disease exposure.

Winners and Losers

Does anyone win in these scenarios? Residents of northern Canada, Siberia, and Scandinavia probably would enjoy warmer temperatures and longer growing seasons. Stronger monsoons bringing more moisture to parts of Central Africa and South Asia could result in higher crop yields. But soils in some areas will not support farming, no matter how agreeable the climate. Construction companies could be busy replacing buildings and infrastructure destroyed by storms and flooding, if we can afford to rebuild. About half of Florida would be flooded by a one-meter rise in sea level. The U.S. Gulf Coast could lose 5000 sq km, an area the size of Delaware.

Among those with the strongest concerns about global climate change is the Alliance of Small Island States founded by the president of the Maldive Islands (off the coast of India) about a decade ago. The Maldives, none of which rises more than a meter above the ocean, already experience more frequent flooding and erosion than in the past. They could become completely uninhabitable unless something is done soon to reverse current trends. Rather than agree to curb their own emissions, some industrialized nations have suggested that the 245,000 Maldivians abandon their country and move somewhere else.

Climate Skeptics

Not everyone agrees with the gloomy predictions of the IPCC. While much of the dissension comes from industry lobbying groups, a number of prominent scientists also argue that dire predictions about climate change are premature (see What Do You Think?, p. 390). Among the arguments raised by greenhouse skeptics are

what do you think?

THE WORLD AROUND YOU

Reasoned Judgment and Scientific Uncertainty

Critical thinking is characterized by its skepticism and its reliance on factual information before drawing conclusions. You read in this chapter about the potential consequences if significant global warming were to occur. You have also read that data from computer models and other sources are equivocal, creating a degree of uncertainty about the timing and precise nature of the warming event.

Some people use this lack of absolute certainty to advocate that we not undertake preventive changes until the uncertainty has been removed. They claim that to do so without absolute proof would be unscientific. They say that reducing the production of greenhouse gases would impose significant economic costs, their argument being that such actions are unwise because they could turn out to have been unnecessary. This position fails to acknowledge that some degree of uncertainty surrounds the vast majority of decisions and judgments society must make, including issues in science. By casting decision making as purely a process of separating fact from opinion, this position omits an important element of critical thinking—reasoned judgment.

The threat of significant, rapid, global climate change is but one of many issues confronting us in which factual evidence does not establish absolute certainty. Will your home be destroyed by fire? Will you develop lung cancer if you smoke cigarettes? Like global warming, available facts are insufficient to establish that any of these events will be visited upon someone with certainty. However, that they are possible, or even probable, is not in doubt.

The generalized dilemma could perhaps be framed like this: In issues such as these, should we do nothing until certainty is established? The downside risk, of course, is that disaster may be unavoidable if we wait. An insurance company is unlikely to underwrite fire coverage on your house when the fire trucks are on their way. Any serious effort to curb release of greenhouse gases would require time to implement. There would also be a lag time before such efforts could influence global climate. Measures undertaken only when scientific certainty was finally established may be too late to be of much help.

Or do we take preventive measures before an event is certain to occur, running the risk that our action may not have been needed? It certainly is true that your house may never burn down. And if it does not, the money spent for insurance premiums appears wasted.

In instances like these, critical thinking must come to a reasoned judgment. To do otherwise is to paralyze actions, preventing timely intervention. Society has long used reasoned judgment to address issues of uncertainty similar to global warming. Arriving at a reasoned judgment for most people involves a thought process that is very applicable to global warming. Two questions are evaluated:

1. What is the likelihood that the nasty event actually will occur?
2. How serious would it be if the nasty event occurs?

The more likely a nasty event is to happen, or the more serious the effect of the event, the greater the advisability of taking preventive action. This concept is modeled in the graph in this box with some examples. Where on the graph would you plot global warming?

Thomas Lovejoy expresses the reasoned judgment of many ecologists when he argues against doing "total planet experiments that bet the biosphere, if there is even a small chance we may regret the result. After all, there is not even an experimental control planet to colonize if we lose at biosphere roulette."

The stakes seem high? Does reason require us to act or wait?

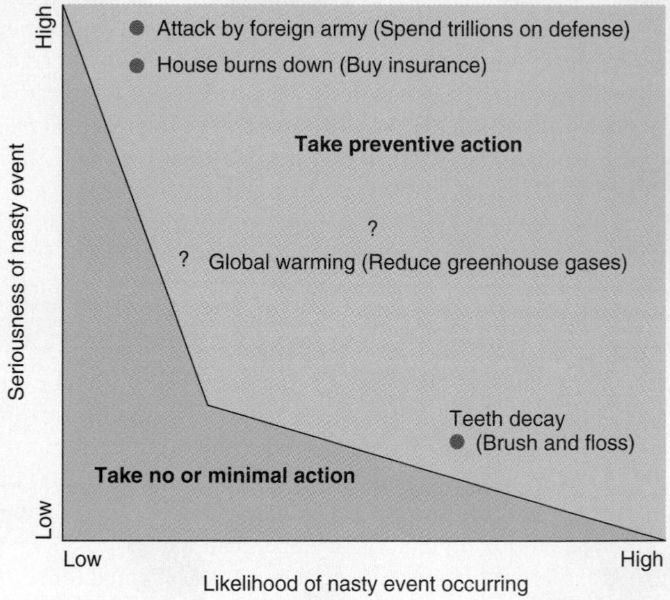

Society regularly takes preventive action to avoid or mitigate consequences of nasty events that may or may not happen.

(1) carbon dioxide may be less potent in warming than thought, (2) atmospheric methane may be declining, (3) models fail to account for water vapor, and (4) computer models generally predict that more warming should already have taken place than actually has. Water vapor may be the wild card in this game. It is a far more important greenhouse gas than CO_2. Doubling CO_2 is equivalent to only a 4 percent increase in relative humidity. Furthermore, if temperatures do increase, water evaporation should increase and more clouds should form. Clouds raise the earth's albedo and should cool the atmosphere and help keep temperatures in balance.

Furthermore, some skeptics point out that elevated CO_2 could stimulate plant growth and make water use more efficient. They claim we should celebrate fossil-fuel burning because it heralds "rebirth of the biosphere" and a return to a "paradise lost" of lush growth and higher productivity. If we weren't artificially adding CO_2 to the air, some believe, the climate would probably now be moving toward a new ice age. While these doubting scientists raise some valid points, the number of skeptics seems to be decreasing rapidly as evidence for climate change accumulates. The dominant scientific view is that global warming is a very serious threat and that humans must respond quickly and decisively if we are to avoid disaster.

Cutting Emissions

One of the centerpieces of the United Nations "Earth Summit" meeting in 1992 in Rio de Janeiro was the Framework Convention on Climate Change, which set an objective of stabilizing "greenhouse gas concentrations at a level that would prevent dangerous anthropogenic interference with the climate system." Industrial nations pledged to return to 1990 emissions by 2000. By 1999, only Fiji, a small Pacific island nation, which is worried about rising sea levels, had ratified the treaty. Unfortunately, few countries have actually reduced greenhouse gas emissions. The United States, which now produces about one-fifth of the extra 22 million tons of CO_2 added to the atmosphere each year, is by far the largest anthropogenic source of this gas, both in total and per capita. China is second in total emissions but ranks fourteenth in per capita releases because of its large population (fig. 17.23).

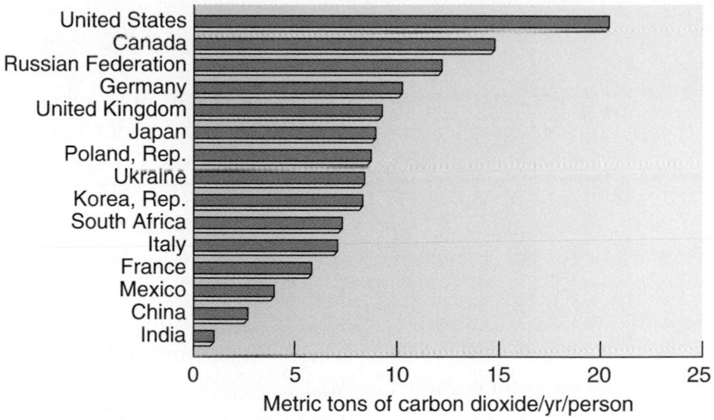

FIGURE 17.23 Annual per capita CO_2 emissions for 15 countries with highest industrial output. Although China is fourteenth in per capita emissions, it is second to the United States in total emissions due to its large population.
Source: Data from *World Resources 1998–99,* World Resources Institute.

At a follow-up conference on global climate change in Kyoto, Japan, in December 1997, 160 nations finally agreed on a new treaty called the Kyoto Protocol. Collectively, the industrialized countries pledged to roll back CO_2, methane, and nitrous oxide emissions about 5 percent below 1990 levels by 2012. Three other greenhouse gases, hydrofluorocarbons, perfluorocarbons, and sulfur hexafluoride, will also be reduced, although from what level was not decided. Individual nations have different limits depending on their output before 1990. Although conference participants acknowledged that penalties for violating quotas are essential, no specific mechanisms to enforce the protocol were established.

The United States lobbied hard in Kyoto for treating emissions as commodities and allowing trading, banking, and borrowing from target quotas as a part of this treaty. This approach is based on a successful program for controlling acid rain. Countries that beat their assigned reductions will be allowed to sell the excess to those countries that fall short. New forests can count as emission offsets. The United States may find it cheaper, for instance, to pay a developing country to plant a few million trees than to put emission controls on its power plants. Opponents of this approach consider it likely to cause backsliding. What is your opinion?

Many people oppose limits on greenhouse gas emissions, claiming that either global climate change hasn't been proven scientifically, or that even if it has, the costs of making changes are greater than the benefits. Industries and labor unions fear that strict greenhouse gas controls could eliminate 10,000 jobs in the United States and could cost $30 billion per year. Ecological economists, on the other hand, estimate that we could cut emissions by 30 percent without any additional expenditures or job losses, and that we could probably cut an additional 20 percent without excessive costs. From this perspective, positive benefits from reduced emissions—such as clean air, renewable energy, and stabilized weather—would more than offset costs of controls.

Developing nations were exempted from emission limits at Kyoto because of fears that restrictions would hold back standards of living. Wealthy countries created the problem, the poorer countries argued; now is it their responsibility to deal with it? What do you think would be fair in this situation? Can we ask poorer people to forego the benefits from energy consumption and other modern technology that we already enjoy? On the other hand, if the five billion people in the developing world continue to increase fossil fuel consumption at current rates, the result could be catastrophic.

Are there technological solutions for this problem? One suggestion would be to develop renewable energy sources and make them available to the developing world. Some researchers are studying the possibility of storing excess CO_2 in deep aquifers or in deep ocean trenches. The Alberta Research Council is pumping CO_2 into a deep coal seam where it binds to coal and releases methane which can be recovered and burned. There also have been suggestions that we fertilize the oceans to stimulate phytoplankton growth. Given that we don't know what ecological complications might arise from such actions, however, perhaps we should wait until we know more before adopting some of these risky solutions.

Another group of scientists argue that we should stop focusing on reducing emissions and think instead about adjusting to a warmer world. Even if the Kyoto Protocol is ratified, they maintain, it probably will only shave 0.1°F from the 2.5°F increase expected by 2050. Changing crop mixtures and cultivation methods, water conservation and efficient irrigation, replanting forests with new species, and helping wildlife relocate to new habitats may be necessary in the new climatic regime. In addition, cities may need new approaches to architecture and flood protection. And on an individual basis, we may want to rethink clothing styles and daily schedules. What changes might occur in your life if global warming becomes a fact of life?

SUMMARY

The atmosphere and living organisms have evolved together so that the present chemical composition of the air is both suitable for, and largely the result of, biological processes. Compression concentrates most gas molecules in a thin layer (the troposphere) near the earth's surface. Upper layers of the atmosphere, play an important role in protecting the earth's surface by intercepting dangerous ultraviolet radiation from the sun. The atmosphere is relatively transparent to visible light that warms the earth's surface and is captured by photosynthetic organisms and stored as potential energy in organic chemicals.

Heat is lost from the earth's surface as infrared radiation, but fortunately for us, carbon dioxide and water vapor naturally present in the air capture the radiation and keep the atmosphere warmer than it would otherwise be. When air is warmed by conduction or radiation of heat from the earth's surface, it expands and rises, creating convection currents. These vertical updrafts carry water vapor aloft and initiate circulation patterns that redistribute energy and water from areas of surplus to areas of deficit. Pressure gradients created by this circulation drive great air masses around the globe and generate winds that determine both immediate weather and long-term climate.

Earth's rotation causes wind deflection called the Coriolis effect, which makes air masses circulate in spiraling patterns. Strong cyclonic convection currents fueled by temperature and pressure gradients and latent energy in water vapor can create devastating storms. Tornadoes, while classified as cyclonic storms, are not set spinning by Coriolis forces. Instead, shear forces caused by differential wind speeds, together with rapidly rising warm convection currents and cold downdrafts are thought to create intensely focused spinning vortices. Although top wind speeds in tornadoes can be higher than those in hurricanes, total damage in the former is usually smaller because the area covered is smaller.

Another source of storms are the seasonal winds, or monsoons, generated by temperature differences between the ocean and a landmass. Monsoons often bring torrential rains and disastrous floods, but they also bring needed moisture to farmlands that feed a majority of the world's population. When the rains fail, as they do in drought cycles, ecosystem disruption and human suffering can be severe.

The El Niño/Southern Oscillation is a complex interaction between oceans and atmosphere that has far-reaching climatic, ecological, and social effects. ENSO cycles can affect things as widely different as forest fires in Indonesia, anchovy fishing in Peru, rainfall in the Sahara Desert, and how the corn grows in Iowa. Knowing something about how weather works can be helpful in our everyday life.

Many procedures claiming to control the weather are ineffectual, but some human actions—both deliberate and inadvertent—may change local weather and long-term climate. It is not yet clear what the future of our weather and climate will be. Many scientists warn that the gaseous pollutants we release into the atmosphere may trap radiant energy and cause a global warming trend that could drastically disrupt human activities and natural ecosystems. Understanding and protecting this complex, vital aspect of our world is clearly essential.

QUESTIONS FOR REVIEW

1. What are the main constituents of air? What are their sources?
2. Name and describe the four layers of air in the atmosphere.
3. What is the greenhouse effect?
4. What is the ENSO cycle?
5. What are the jet streams? How do they influence weather patterns?
6. How do tornadoes form and why are they so destructive?
7. Describe the Coriolis effect. What causes it?
8. What is the difference between weather and climate?
9. What are some theories explaining major climatic changes?
10. Would changes in the earth's orbit affect climate? How?

QUESTIONS FOR CRITICAL THINKING

1. What questions would you ask a climate modeler about the assumptions, compromises, and limitations of mathematical models?
2. Should humans try to control the weather? What would be the positive effects? What would be the dangers?
3. Can we avoid great climatic changes, such as another ice age or a greenhouse effect? How can we decide what to do when the science is uncertain and the risk is great?
4. Has there been a major change recently in the weather where you live? Can you propose any reasons for such changes?
5. What forces determine the climate in your locality? Are they the same for neighboring states?
6. What was the weather like when your parents were young? Do you believe the stories they tell you? Why or why not?

7. Have you ever experienced a tornado or hurricane? What was it like? What omens or warnings told you it was coming?
8. From which direction does bad weather come in your region? Why?
9. What would you do to adapt to a permanent drought? What effects would it have on your life?
10. What should we do about coastal cities threatened by rising oceans—rebuild, enclose in dikes, or just move?
11. Would you favor building nuclear power plants to reduce CO_2 emissions?

KEY TERMS

aerosols 376
albedo 377
climate 374
cold front 380
convection currents 378
Coriolis effect 379
downbursts 382
El Niño 386
greenhouse gas 387
hurricanes 381
ionosphere 376
jet streams 379
mesosphere 376
Milankovitch cycles 385
monsoon 382
stratosphere 376
thermosphere 376
tornadoes 381
troposphere 376
warm front 380
weather 374

ADDITIONAL INFORMATION ON THE INTERNET

Visit our website at http://mhhe.com/environmentalscience for specific information about each of the topics below:

Tornado Information
Hurricanes: Information from the US Geological Survey
CEARS Information on Climate
Weather and Climate Data, University of Wisconsin
National Weather Service Home Page—U.S. Government Warnings and Forecasts
National Snow and Ice Data Center
EPA Global Warming Site
Climate Change Facts
Canadian Joint Global Ocean Flux Study
Sierra Club Global Warming Campaign
Global Warming: Focus on the Future. Environmental Defense Fund
Global Warming: A list of links related to global warming.
The El Niño Theme Page, NOAA
El Niño Home Page from the Jet Propulsion Laboratory
El Niño Research, Forecasts, and Observation: NOAA El Niño page
Links to El Niño-Southern Oscillation Information
US Global Change Research Information Office Home Page
Global Change Master Directory (NASA)
Atlas of Antarctic Research

PROFILE

Interpretive Naturalist

Tom Moffatt

Chances are, if you've thought about environmental careers, one of the first jobs to come to mind might be working in a nature center. Tom Moffatt has such a job, and he loves it. As an interpretive naturalist at Wood Lake Nature Center in Richfield, Minnesota, Tom spends about half his time working on natural-history-based environmental education, mostly with elementary school groups. His classes involve lots of outdoor activities, hands-on exploration of the marshes, woods, and wildlife, and on-site environmental monitoring. The other half of Tom's work includes a great variety of duties such as planning and construction of exhibits, trails, and nature center facilities; curriculum writing; school scheduling; live animal care; forest, wetland, and wildlife management; and intern supervision. Tom also manages the center's sugarbush and apiary.

What does Tom like best about his job? The daily variety at the nature center is great. He loves being outdoors. Most of all, he enjoys working with a wide variety of people, and contributing to children's understanding and appreciation of nature. "Don't think," he warns others who might be interested in being a naturalist, "that this job just involves working outside and being good at plant and animal identification." An interpretive naturalist is first and foremost an educator. Your people skills and your enthusiasm for teaching are probably more important than your knowledge of biology. You have to like people and be interested in how they learn and how best to communicate your knowledge.

Tom hadn't thought about becoming a naturalist until his senior year in college. As an undergraduate at St. Olaf College, a small, private, liberal arts college, Tom majored in biology and took all the ecology classes he could. A career day that he attended just before graduation inspired him to think about nature centers. Without any previous experience in the field and lacking much background in natural history, Tom didn't find a permanent job right away. He got in touch with the State Association of Naturalists to see what opporturnities were available.

He found that although full-time jobs were scarce and highly competitive, there were many opportunities for internships and part-time, seasonal positions. A couple of years in these positions gave him a chance to learn the background and skills he needed for a full-time job. Now, at Wood Lake, he is the director of interns and can help others learn on the job as he did. He finds that learning practical skills and gaining experience is just one of the benefits of an internship. Tom recommends an internship before graduation for anyone considering a career in this field. About half of the interns he works with discover that they don't like being a naturalist as much as they thought they would. Finding out where your aptitudes and interests lie before you commit your life to a particular profession is a valuable thing.

When asked what he likes least about his job, Tom says that naturalists often aren't given credit for being professional educators and often aren't paid on the same scale as classroom teachers. He feels that his teaching isn't just fun and games for the kids. It's just as valuable a part of their education as any other subject; it just happens to take place outdoors. He also regrets that jobs in his field are so competitive and hard to find. Many people find themselves spending years in low-paying internships and seasonal work before they finally land a permanent position. Tom points out, however, that pay scales are now higher and job prospects are much better than when he entered the field.

How should you prepare if you'd like to be a naturalist? Tom advises getting a good solid grounding in natural history. He also suggests taking some educational theory courses, since this job is as much about teaching as it is about nature. Being willing to relocate to a new geographical area may also help. You may not find a job in your hometown or where you went to college, but there are many other opportunities that you could explore.

CHAPTER 18

Air Pollution

Politics is the art of making good decisions on insufficient evidence.

—Lord Kennet—

OBJECTIVES

After studying this chapter, you should be able to:

- describe the major categories and sources of air pollution.
- distinguish between conventional or "criteria" pollutants and unconventional types as well as explain why each is important.
- analyze the origins and dangers of some indoor air pollutants.
- relate why atmospheric temperature inversions occur and how they affect air quality.
- evaluate the dangers of stratospheric ozone depletion and radon in indoor air.
- understand how air pollution damages human health, vegetation, and building materials.
- compare different approaches to air pollution control and report on clean air legislation.
- judge how air quality around the world has improved or degraded in recent years and suggest what we might do about problem areas.

ESSENTIAL STUDY PARTNER

The Essential Study Partner (ESP) CD-ROM that accompanies this textbook contains additional information on the following topics:

- air pollution types and sources: physical resources: air pollution: sources
- temperature inversion and mountains: physical resources: air pollution: sources
- volatile organic compounds (VOCs): physical resources: air pollution: sources
- radon and indoor air pollution: physical resources: air pollution: sources
- ozone hole and UV: physical resources: air pollution: sources
- acid precipitation causes and aquatic effects: physical resources: air pollution: acid rain
- high smokestacks, scrubbers, and catalytic converters: physical resources: air pollution: control
- Clean Air Act: physical resources: air pollution: control

Above: Burning tropical forests to clear land for crops can create clouds of smoke thousands of kilometers across.

A Plague of Smoke

For several months during the unusually dry El Niño winter of 1997–98, a thick pall of smoke covered much of Southeast Asia. Generated primarily by thousands of forest fires on the Indonesian islands of Kalimantan (Borneo) and Sumatra, the smoke spread over eight countries and 75 million people, covering an area larger than Europe (fig. 18.1). The air quality in Singapore and the city of Kuala Lumpur, Malaysia, just across the Strait of Malacca from Indonesia, was worse than any industrial region in the world. In towns such as Palembang, Sumatra, and Banjarmasin, Kalimantan, in the heart of the massive conflagration, the air pollution index frequently passed 800, twice the level classified in the United States as an air quality emergency, hazardous to human health.

An estimated 20 million people were treated for illnesses such as asthma, bronchitis, emphysema, eye irritation, and cardiovascular diseases, while those who couldn't afford medical care went uncounted. The number of excess deaths from this months-long episode is unknown. Unable to see through the thick haze, several boats collided in the busy Straits of Malacca, and a plane crashed on Sumatra, killing 234 passengers. Cancelled airline flights, aborted tourist plans, lost workdays, medical bills, and ruined crops are estimated to have cost countries in the afflicted area several billion dollars. Wildlife suffered as well. In addition to the loss of habitat destroyed by fires, breathing the noxious smoke was as hard on wild species as it was on people. At the Pangkalanbuun Conservation Reserve, weak and disoriented orangutans were found suffering from respiratory diseases much like those of humans.

Indonesia has the second largest expanse of tropical forest and the highest number of endemic species in the world, so fires there are of special concern. The dry season in tropical Southeast Asia has probably always been a time of burning vegetation and smoky skies. Tinder-dry forests are ignited by lightning or fires set by farmers to clear cropland. Generally burning only a hectare or two at a time, traditional shifting cultivators often help preserve plant and animal species by opening up space for early successional forest stages. Globalization and the advent of large, commercial plantations, however, have changed everything.

Although Indonesian government officials blamed the fires on small-scale farmers and indigenous people, environmental groups gathered evidence that most of the burning was caused by large agribusiness conglomerates with close ties to the government and military. Clear-cutting hardwoods for sale abroad and burning what's left to make way for huge oil-palm plantations and fast-growing pulpwood trees, these companies ignore forest protection laws. Altogether, a couple of dozen businesses owned by wealthy entrepreneurs with friends in high places are thought to be responsible for burning some 20,000 km^2 (8000 mi^2) in 1997, or an area about the size of New Jersey.

Conditions in Indonesia in 1997 were extreme, but forest fires and other human activities in many places around the world also cause degraded air quality. Since we all breathe air, it is important to understand how air pollution arises, why it is dangerous, and what we might do about it. Studying this chapter will help you learn about this vital topic.

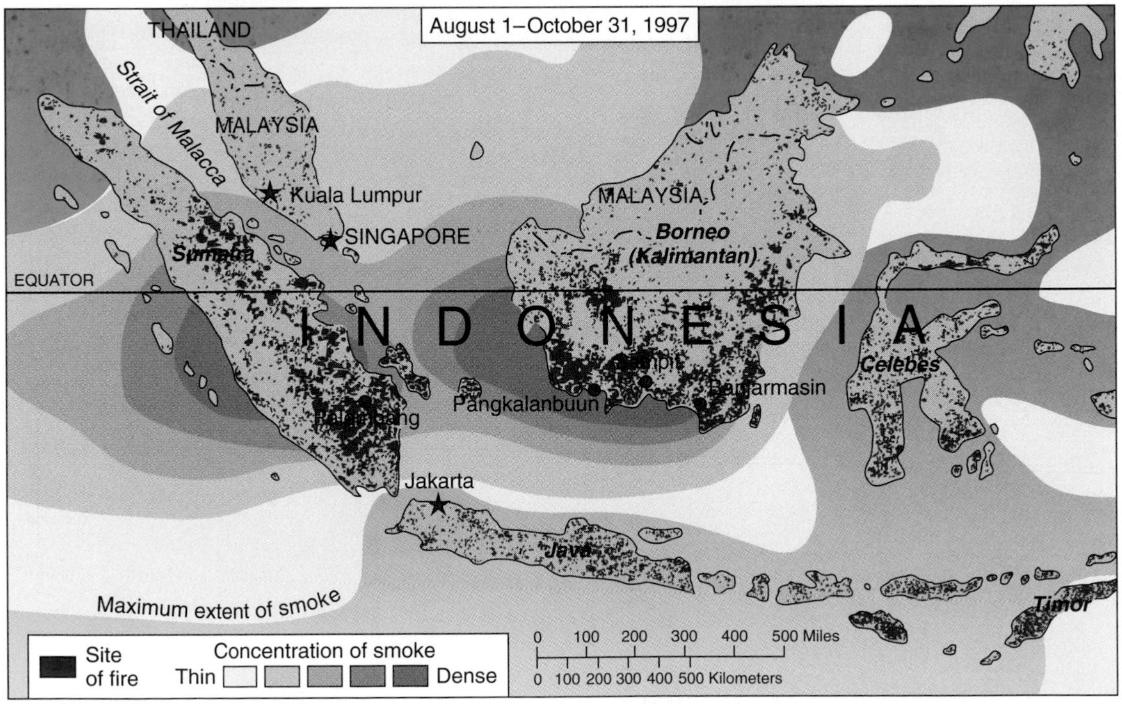

FIGURE 18.1 Thousands of fires (red dots) in Indonesia in 1997 spread a thick cloud of smoke over an area larger than Europe. *Source:* Data from NOAA and NASA.

FIGURE 18.2 Los Angeles on a smoggy day (*left*) is very different from days when the air is clear (*right*). Although much remains to be done, air quality in most American cities is significantly clearer now than when the Clean Air Act was passed in 1972.

THE AIR AROUND US

How does the air taste, feel, smell, and look in your neighborhood? Chances are that wherever you live, the air is contaminated to some degree. Smoke, haze, dust, odors, corrosive gases, noise, and toxic compounds are present nearly everywhere, even in the most remote, pristine wilderness. Air pollution is generally the most widespread and obvious kind of environmental damage. According to the Environmental Protection Agency (EPA), some 147 million metric tons of air pollution (not counting carbon dioxide or wind-blown soil) are released into the atmosphere each year in the United States by human activities (fig. 18.2). Total worldwide emissions of these pollutants are around 2 billion metric tons per year.

Over the past 20 years, however, air quality has improved appreciably in most cities in Western Europe, North America, and Japan. Many young people might be surprised to learn that a generation ago most American cities were much dirtier than they are today. This is an encouraging example of improvement in environmental conditions. Our success in controlling some of the most serious air pollutants gives us hope for similar progress in other environmental problems.

While developed countries have been making progress, however, air quality in the developing world has been getting much worse. Especially in the burgeoning megacities of rapidly industrializing countries (chapter 24), air pollution often exceeds World Health Organization standards all the time. In Lahore, Pakistan, and Xian, China, for instance, airborne dust, smoke, and dirt often are ten times higher than levels considered safe for human health.

In this chapter, we will examine the major types and sources of air pollution. We will study how they enter and move through the atmosphere and how they are changed into new forms, concentrated or dispersed, and removed from the air by physical and chemical processes. We also will look at some of the major effects of air pollution on human health, ecosystems, and materials. Finally, we will survey some of the control methods available to reduce air pollution or mitigate its effects, and the results of air pollution control efforts on ambient air quality in the United States and elsewhere.

NATURAL SOURCES OF AIR POLLUTION

It is difficult to give a simple, comprehensive definition of pollution. The word comes from the Latin *pollutus,* which means made foul, unclean, or dirty. Some authors limit the use of the term to damaging materials that are released into the environment by human activities. There are, however, many natural sources of air quality degradation. Some of the smoke that inflicted Southeast Asia in the opening story for this chapter came from natural fires. Volcanoes spew out ash, acid mists, hydrogen sulfide, and other toxic gases. Sea spray and decaying vegetation are major sources of reactive sulfur compounds in the air. Trees and bushes emit millions of tons of volatile organic compounds (terpenes and isoprenes), creating, for example, the blue haze that gave the Blue Ridge Mountains their name. Pollen, spores, viruses, bacteria, and other small bits of organic material in the air cause widespread suffering from allergies and airborne infections. Storms in arid regions raise dust clouds that transport millions of tons of soil and can be detected half a world away. Bacterial metabolism of decaying vegetation in swamps and of cellulose in the guts of termites and ruminant animals is responsible for as much as two-thirds of the methane (natural gas) in the air.

Does it make a difference whether smoke comes from a natural forest fire or one started by humans? In many cases, the chemical compositions of pollutants from natural and human-related sources are identical, and their effects are inseparable. Sometimes, however, materials in the atmosphere are considered innocuous at naturally occurring levels, but when humans add to these levels, overloading of natural cycles or disruption of essential processes can occur. While the natural sources of suspended particulate material in the air outweigh human sources at least tenfold worldwide, in many cities more than 90 percent of the airborne particulate matter is anthropogenic (human-caused).

FIGURE 18.3 Smokestacks of this Chinese cement factory are classed as primary or point pollution sources.

HUMAN-CAUSED AIR POLLUTION

What are the major types of air pollutants and where do they come from? In this section, we will define some general categories and sources of air pollution and survey the characteristics and emission levels of the seven conventional pollutants regulated by the Clean Air Act.

Primary and Secondary Pollutants

Primary pollutants are those released directly from the source into the air in a harmful form (fig. 18.3). **Secondary pollutants,** by contrast, are modified to a hazardous form after they enter the air or are formed by chemical reactions as components of the air mix and interact. Solar radiation often provides the energy for these reactions. Photochemical oxidants and atmospheric acids formed by these mechanisms are probably the most important secondary pollutants in terms of human health and ecosystem damage. We will discuss several important examples of such pollutants in this chapter.

Fugitive emissions are those that do not go through a smokestack. By far the most massive example of this category is dust from soil erosion, strip mining, rock crushing, and building construction (and destruction). In the United States, natural and anthropogenic sources of fugitive dust add up to some 100 million metric tons per year. The amount of CO_2 released by burning fossil fuels and biomass is nearly equal in mass to fugitive dust. Fugitive industrial emissions are also an important source of air pollution. Leaks around valves and pipe joints contribute as much as 90 percent of the hydrocarbons and volatile organic chemicals emitted from oil refineries and chemical plants.

Conventional or "Criteria" Pollutants

The U.S. Clean Air Act of 1970 designated seven major pollutants (sulfur dioxide, carbon monoxide, particulates, hydrocarbons, nitrogen oxides, photochemical oxidants, and lead) for which maximum **ambient air** (air all around us) levels are mandated. These seven **conventional** or **criteria pollutants** contribute the largest volume of air-quality degradation and also are considered the most serious threat of all air pollutants to human health and welfare. Figure 18.4 shows the major sources of the first six criteria pollutants. Table 18.1 shows an estimate of the total annual worldwide emissions of some important air pollutants. Now let's look more closely at the sources and characteristics of each of these major pollutants.

Sulfur Compounds

Natural sources of sulfur in the atmosphere include evaporation of sea spray, erosion of sulfate-containing dust from arid soils, fumes from volcanoes and fumaroles, and biogenic emissions of hydrogen sulfide (H_2S) and organic sulfur-containing compounds, such as dimethylsulfide, methyl mercaptan, carbon disulfide, and carbonyl sulfide. Total yearly emissions of sulfur from all sources amounts to some 114 million metric tons (fig. 18.5). Worldwide, anthropogenic sources represent about two-thirds of the total sulfur flux, but in most urban areas they contribute as much as 90 percent of the sulfur in the air. The predominant form of anthropogenic sulfur is sulfur dioxide (SO_2) from combustion of sulfur-containing fuel (coal and oil), purification of sour (sulfur-containing) natural gas or oil, and industrial processes, such as smelting of sulfide ores. China and the United States are the largest sources of anthropogenic sulfur, primarily from coal burning.

Sulfur dioxide is a colorless corrosive gas that is directly damaging to both plants and animals. Once in the atmosphere, it can be further oxidized to sulfur trioxide (SO_3), which reacts with water vapor or dissolves in water droplets to form sulfuric acid (H_2SO_4), a major component of acid rain. Very small solid particles or liquid droplets can transport the acidic sulfate ion (SO_4^{-2})

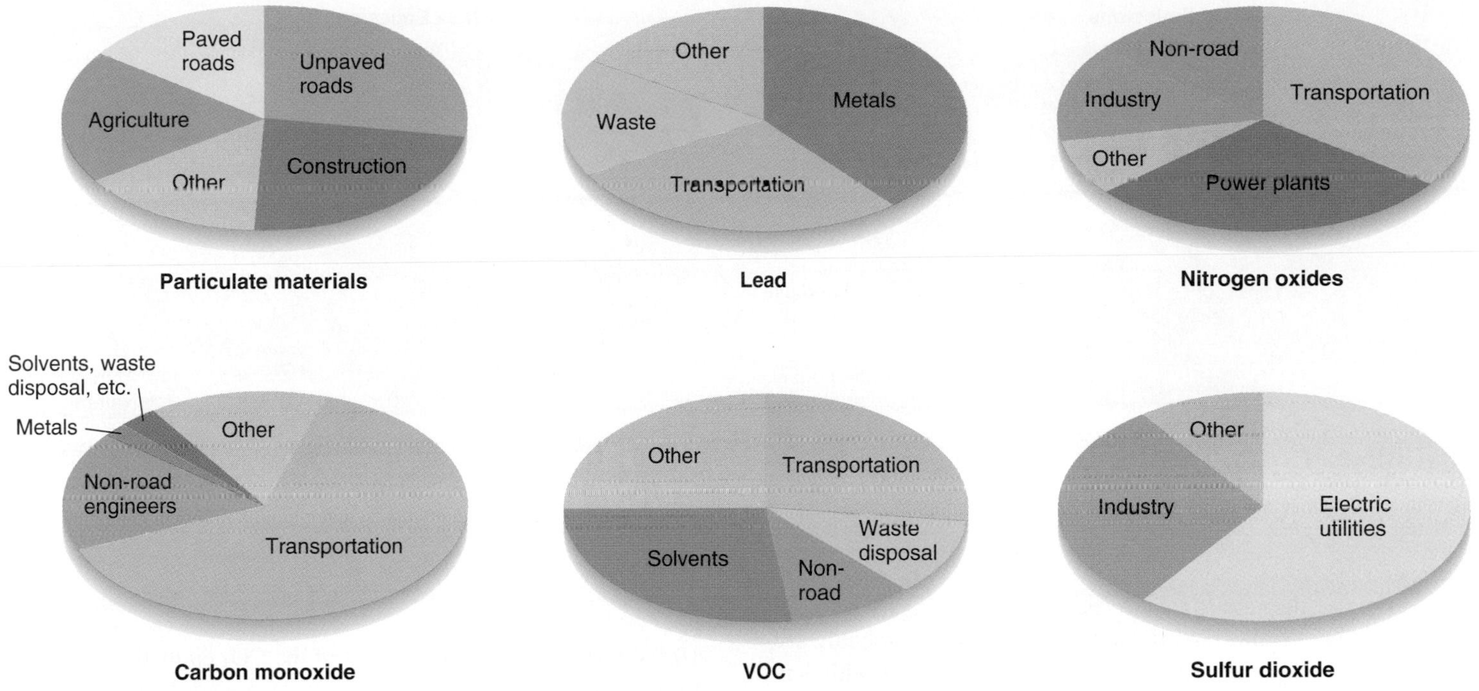

FIGURE 18.4 Anthropogenic sources of six of the primary "criteria" air pollutants in the United States.

Source: Data from Joyce E. Penner, "Atmospheric Chemistry and Air Quality" in W. B. Meyer and B. L. Turner (eds.), *Changes in Land Use and Land Cover: A Global Perspective,* 1994. Cambridge University Press and UNEP 1999.

TABLE 18.1 Estimated Fluxes of Pollutants and Trace Gases to the Atmosphere

SPECIES	SOURCES	APPROXIMATE ANNUAL FLUX	
		NATURAL (MILLIONS OF METRIC TONS/YR)	ANTHROPOGENIC
CO_2 (carbon dioxide)	Respiration, fossil fuel burning, land clearing, industrial processes	370,000	23,900*
CH_4 (methane)	Rice paddies and wetlands, gas drilling, landfills, animals, termites	155	350
CO (carbon monoxide)	Incomplete combustion, CH_4 oxidation, biomass burning, plant metabolism	1500	930
NMHC (nonmethane hydrocarbons)	Fossil fuels, industrial uses, plant isoprenes and other biogenics	860	92
NO_x (nitrogen oxides)	Fossil fuel burning, lightning, biomass burning, soil microbes	90	140
SO_x (sulfur oxides)	Fossil fuel burning, industry, biomass burning, volcanoes, oceans	35	79
SPM (suspended particulate materials)	Biomass burning, dust, sea salt, biogenic aerosols, gas to particle conversion	583	362

*About half of this is taken up again by plant photosynthesis while the rest accumulates in the atmosphere.

Source: Data from Joyce E. Penner, "Atmospheric Chemistry and Air Quality" in W. B. Meyer and B. L. Turner (eds.), *Changes in Land Use and Land Cover: A Global Perspective,* 1994. Cambridge University Press and UNEP 1999.

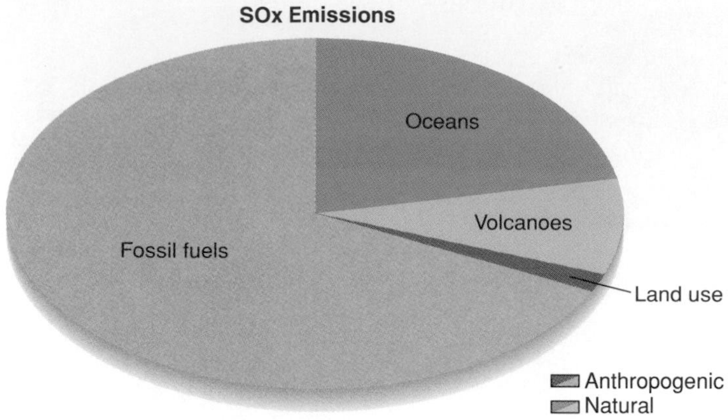

FIGURE 18.5 Sulfur fluxes into the atmosphere.
Source: Data from Joyce E. Penner, "Atmospheric Chemistry and Air Quality" in W. B. Meyer and B. L. Turner (eds.), *Changes in Land Use and Land Cover: A Global Perspective,* 1994. Cambridge University Press and UNEP 1999.

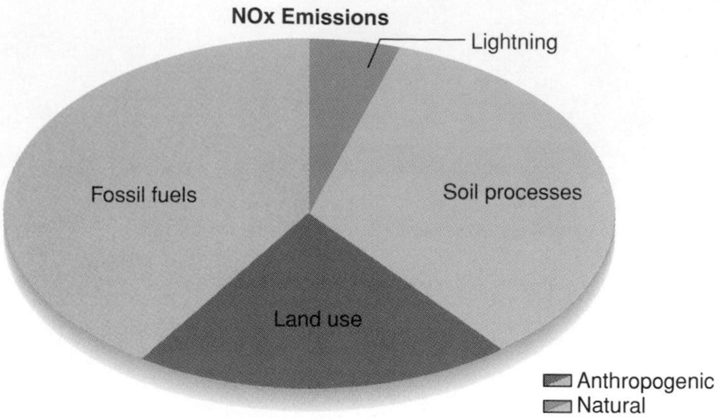

FIGURE 18.6 Worldwide sources of reactive nitrogen gases in the atmosphere.
Source: Data from Joyce E. Penner, "Atmospheric Chemistry and Air Quality" in W. B. Meyer and B. L. Turner (eds.), *Changes in Land Use and Land Cover: A Global Perspective,* 1994. Cambridge University Press and UNEP 1999.

long distances through the air or deep into the lungs where it is very damaging. Sulfur dioxide and sulfate ions are probably second only to smoking as causes of air pollution-related health damage. Sulfate particles and droplets reduce visibility in the United States as much as 80 percent.

Nitrogen Compounds

Nitrogen oxides are highly reactive gases formed when nitrogen in fuel or combustion air is heated to temperatures above 650°C (1200°F) in the presence of oxygen, or when bacteria in soil or water oxidize nitrogen-containing compounds. The initial product, nitric oxide (NO), oxidizes further in the atmosphere to nitrogen dioxide (NO_2), a reddish brown gas that gives photochemical smog its distinctive color. Because of their interconvertibility, the general term NO_x is used to describe these gases. Nitrogen oxides combine with water to make nitric acid (HNO_3), which is also a major component of atmospheric acidification.

The total annual emissions of reactive nitrogen compounds into the air are about 230 million metric tons worldwide (table 18.1). Anthropogenic sources account for 60 percent of these emissions (fig. 18.6). About 95 percent of all human-caused NO_x in the United States is produced by fuel combustion in transportation and electric power generation. Nitrous oxide (N_2O) is an intermediate in soil denitrification that absorbs ultraviolet light and plays an important role in climate modification (chapter 17). Excess nitrogen is causing fertilization and eutrophication of inland waters and coastal seas. It also may be adversely affecting terrestrial plants both by excess fertilization and by encouraging growth of weedy species that crowd out native varieties.

Carbon Oxides

The predominant form of carbon in the air is carbon dioxide (CO_2). It is usually considered nontoxic and innocuous, but increasing atmospheric levels (about 0.5 percent per year) due to human activities appear to be causing a global climate warming that may have disastrous effects on both human and natural communities. As table 18.1 shows, more than 90 percent of the CO_2 emitted each year is from respiration (oxidation of organic compounds by plant and animal cells). These releases are usually balanced by an equal uptake by photosynthesis in green plants.

For many years scientists have been puzzled about exactly how much CO_2 we produce and where it goes. Global processes can't be measured directly, so we estimate local contributions and then extrapolate to the whole world. This can lead to some big errors if our local examples are wrong. In 1914, Richard A. Houghton of the Woods Hole, MA Research Center calculated that burning vegetation and converting forests to pastures and crop fields in the tropics added 5.9 billion tons of CO_2 (or 1.6 billion tons of carbon) to the atmosphere each year. This number has been used by researchers for the past 85 years, but recently was challenged by a team from NASA's Goddard Institute, whose computer model shows only about half as much carbon emitted from tropical forests as Houghton calculated. Burning of fossil fuels is estimated to add between 5 and 5.5 billion tons of carbon to the atmosphere each year. Most authorities in this field currently accept an estimate of 6.3 billion tons of carbon per year from all anthropogenic sources, but that number could easily be off by 10 percent.

At the same time, there is equal uncertainty about where the extra carbon goes. Oceans are thought to take up about 2 billion tons both through photosynthesis by phytoplankton (algae) and shell formation by mollusks. The total amount of carbon in the air appears to be increasing at about 3.4 billion tons per year, leaving nearly 1 billion tons unaccounted for. A group of Princeton researchers recently estimated that northern forests and prairie soils may be accumulating as much as 2.2 billion tons of carbon per year (if this is true, either emissions are higher than currently thought, or the ocean is taking up less than previously calculated). A rival group from Woods Hole, however, claims that carbon storage in northern forests is far less than the Princeton researchers think. This debate has important political ramifications. In the climate change negotiations (chapter 17), the United States claimed credit for forest uptake of carbon that offset industrial emissions. Other countries were skeptical, however, and demanded further U.S. emission reductions.

Carbon monoxide (CO) is a colorless, odorless, nonirritating but highly toxic gas produced by incomplete combustion of fuel (coal, oil, charcoal, or gas), incineration of biomass or solid waste, or partially anaerobic decomposition of organic material. CO inhibits respiration in animals by binding irreversibly to hemoglobin. About 1 billion metric tons of CO are released to the atmosphere each year, half of that from human activities. In the United States, two-thirds of the CO emissions are created by internal combustion engines in transportation. Land-clearing fires and cooking fires also are major sources. About 90 percent of the CO in the air is consumed in photochemical reactions that produce ozone.

FIGURE 18.7 This coke oven produces both particulate material, carbon monoxide, carbon dioxide, heavy metals, and toxic hydrocarbons. Putting it in the middle of an urban area was poor land-use planning.

Metals and Halogens

Many toxic metals are mined and used in manufacturing processes or occur as trace elements in fuels, especially coal. These metals are released to the air in the form of metal fumes or suspended particulates by fuel combustion, ore smelting, and disposal of wastes. Worldwide lead emissions amount to about 2 million metric tons per year, or two-thirds of all metallic air pollution. Most of this lead is from leaded gasoline. Lead is a metabolic poison and a neurotoxin that binds to essential enzymes and cellular components and inactivates them. An estimated 20 percent of all inner-city children suffer some degree of mental retardation from high environmental lead levels.

Mercury is another dangerous neurotoxin that is widespread in the environment. The two largest sources of atmospheric mercury appear to be coal-burning power plants and waste incinerators. Mercuric fungicides in house paint were once a major source of this deadly pollutant but now are restricted. Long-range transport of lead and mercury through the air is causing bioaccumulation in aquatic ecosystems far from the emission sources. It is now dangerous to eat fish from some once-pristine lakes and rivers because of toxic metal contamination.

Other toxic metals of concern are nickel, beryllium, cadmium, thallium, uranium, cesium, and plutonium. Some 780,000

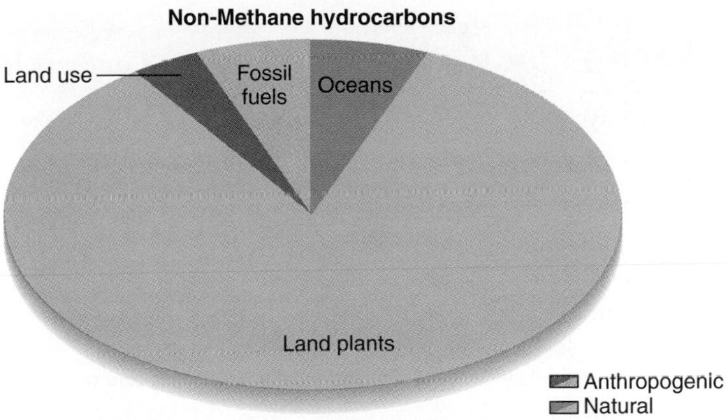

FIGURE 18.8 Sources of non-methane hydrocarbons in the atmosphere.
Source: Data from Joyce E. Penner, "Atmospheric Chemistry and Air Quality" in W. B. Meyer and B. L. Turner (eds.), *Changes in Land Use and Land Cover: A Global Perspective,* 1994. Cambridge University Press and UNEP 1999.

tons of arsenic, a highly toxic metalloid, are released from metal smelters, coal combustion, and pesticide use each year. Halogens (fluorine, chlorine, bromine, and iodine) are highly reactive and generally toxic in their elemental form. About 600 million tons of highly persistent chlorofluorocarbons (CFCs) are used annually worldwide in spray propellants, refrigeration compressors, and for foam blowing. They diffuse into the stratosphere where they release chlorine and fluorine atoms that destroy the ozone shield that protects the earth from ultraviolet radiation. We'll return to this topic later.

Particulate Material

An **aerosol** is any system of solid particles or liquid droplets suspended in a gaseous medium. For convenience, we generally describe all atmospheric aerosols, whether solid or liquid, as **particulate material.** This includes dust, ash, soot, lint, smoke, pollen, spores, algal cells, and many other suspended materials (fig. 18.7). Anthropogenic particulate emissions amount to about 362 million metric tons per year worldwide. Wind-blown dust, volcanic ash, and other natural materials may contribute considerably more suspended particulate material.

Particulates often are the most apparent form of air pollution since they reduce visibility and leave dirty deposits on windows, painted surfaces, and textiles. Respirable particles smaller than 2.5 micrometers are among the most dangerous of this group because they can be drawn into the lungs, where they damage respiratory tissues. Asbestos fibers and cigarette smoke are among the most dangerous respirable particles in urban and indoor air because they are carcinogenic.

Volatile Organic Compounds

Volatile organic compounds (VOCs) are organic chemicals that exist as gases in the air. Plants are the largest source of VOCs, releasing an estimated 350 million tons of isoprene (C_5H_8) and 450 million tons of terpenes ($C_{10}H_{15}$) each year (fig. 18.8). About 400

TABLE 18.2 Urban Air Toxics of Greatest Concern

Acetaldehyde	Coke oven emissions	Manganese compounds
Acrolein	Dioxins	Mercury compounds
Acrylonitrile	1,2-dibromoethane	Methylene chloride
Arsenic compounds	1,3-dichloropropane	Nickel compounds
Benzene	Propylene dichloride	Polychlorinated biphenyls
Beryllium compounds	Ethylene dichloride	Polycyclic organic matter
1,3-butadiene	Ethylene oxide	Quinoline
Cadmium compounds	Formaldehyde	1,1,2,2-tetrachloroethane
Carbon tetrachloride	Hexachlorobenzene	Tetrachloroethylene (perchloroethylene)
Chloroform	Hydrazine	Trichloroethylene
Chromium compounds	Lead compounds	Vinyl chloride

Source: EPA 1999.

million tons of methane (CH_4) are produced by natural wetlands and rice paddies and by bacteria in the guts of termites and ruminant animals. These volatile hydrocarbons are generally oxidized to CO and CO_2 in the atmosphere.

In addition to these natural VOCs, a large number of other synthetic organic chemicals, such as benzene, toluene, formaldehyde, vinyl chloride, phenols, chloroform, and trichloroethylene, are released into the air by human activities. About 28 million tons of these compounds are emitted each year in the United States, mainly unburned or partially burned hydrocarbons from transportation, power plants, chemical plants, and petroleum refineries. These chemicals play an important role in the formation of photochemical oxidants.

Of the 188 air toxics listed in the Clean Air Act, about two-thirds are VOCs and most of the rest are metal compounds. As part of the National Air Toxics Program, the EPA has identified 33 chemical compounds considered to be the greatest threat to public health in urban areas (table 18.2). Since 1990, when regulation of these materials began, the EPA estimates that air toxics emissions have been reduced more than one million tons per year, or almost 10 times greater reductions than were achieved in the previous 20 years. Air toxics standards now affect over 70 categories of major industries, such as chemical plants, oil refineries, and steel mills (fig. 18.9), as well as smaller sources such as dry cleaners, automobiles, and trucks.

Photochemical Oxidants

Photochemical oxidants are products of secondary atmospheric reactions driven by solar energy (fig. 18.10). One of the most important of these reactions involves formation of singlet (atomic) oxygen by splitting nitrogen dioxide (NO_2). This atomic oxygen then reacts with another molecule of O_2 to make **ozone** (O_3). Ozone formed in the stratosphere provides a valuable shield for the biosphere by absorbing incoming ultraviolet radiation. In ambient air, however, O_3 is a strong oxidizing reagent and damages vegetation, building materials (such as paint, rubber, and plastics), and sensitive tissues (such as eyes and lungs). Ozone has an acrid, biting odor that is a distinctive characteristic of photochemical smog. Hydrocarbons in the air contribute to accumulation of ozone by removing NO in the formation of compounds, such as peroxyacetyl nitrate (PAN), which is another damaging photochemical oxidant.

Unconventional Pollutants

The EPA has authority under the Clean Air Act to set **emission standards** (regulating the amount released) for certain **unconventional** or **noncriteria pollutants** that are considered especially toxic or hazardous. Among the materials regulated by emission standards are asbestos, benzene, beryllium, mercury, polychlorinated biphenyls (PCBs), and vinyl chloride. Most of these materials have no natural source in the environment (to any great extent) and are, therefore, only anthropogenic in origin.

In addition to these toxic air pollutants, some other unconventional forms of air pollution deserve mention. **Aesthetic degradation** includes any undesirable changes in the physical characteristics or chemistry of the atmosphere. Noise, odors, and light pollution are examples of atmospheric degradation that may not be life-threatening but reduce the quality of our lives. This is a very subjective category. Odors and noise (such as loud music) that are offensive to some may be attractive to others. Often the most sensitive device for odor detection is the human nose. We can smell styrene, for example, at 44 parts per billion (ppb). Trained panels of odor testers often are used to evaluate air samples. Factories that emit noxious chemicals sometimes spray "odor maskants" or perfumes into smokestacks to cover up objectionable odors.

In most urban areas, it is difficult or impossible to see stars in the sky at night because of dust in the air and stray light from buildings, outdoor advertising, and streetlights. This light pollution has become a serious problem for astronomers.

FIGURE 18.9 Annual toxic air pollution released in the United States. More than 300 toxic chemicals are released into the air. The 1990 Clean Air Amendments regulate these emissions for the first time.
Source: Data from Environmental Protection Agency.

Atmospheric oxidant production:

1. $NO + VOC \longrightarrow NO_2$ (nitrogen dioxide)
2. $NO_2 + UV \longrightarrow NO + O$ (nitric oxide + atomic oxygen)
3. $O + O_2 \longrightarrow O_3$ (ozone)
4. $NO_2 + VOC \longrightarrow$ PAN, etc. (peroxyacetyl nitrate)

Net results:

$NO + VOC + O_2 + UV \longrightarrow O_3$, PAN, and other oxidants

FIGURE 18.10 Secondary production of urban smog oxidants by photochemical reactions in the atmosphere.
Source: Data from B.J. Finlayson-Pitts and J.N. Pitts, *Atmospheric Chemistry,* 1986, John Wiley & Sons, Inc.

Indoor Air Pollution

We have spent a considerable amount of effort and money to control the major outdoor air pollutants, but we have only recently become aware of the dangers of indoor air pollutants. The EPA has found that indoor concentrations of toxic air pollutants are often higher than outdoors—up to twenty times higher for some toxins. Furthermore, people generally spend more time inside than out and therefore are exposed to higher doses of these pollutants.

Smoking is without doubt the most important air pollutant in the United States in terms of human health. The Surgeon General estimates that 430,000 people die each year in the United States from emphysema, heart attacks, strokes, lung cancer, or other diseases caused by smoking. These diseases are responsible for 20 percent of all mortality in the United States, or four times as much as infectious agents. Total costs for early deaths and smoking-related illnesses are estimated to be $100 billion per year. Eliminating smoking probably would save more lives than any other pollution-control measure.

In some cases, indoor air in homes has concentrations of chemicals that would be illegal outside or in the workplace. The EPA has found that concentrations of such compounds as chloroform, benzene, carbon tetrachloride, formaldehyde, and styrene can be seventy times higher in indoor air than in outdoor air. "Green design" principles can make indoor spaces both healthier and more pleasant (In Depth, p. 404).

In the less-developed countries of Africa, Asia, and Latin America where such organic fuels as firewood, charcoal, dried dung, and agricultural wastes make up the majority of household energy, smoky, poorly ventilated heating and cooking fires represent the greatest source of indoor air pollution (fig. 18.11). The World Health Organization (WHO) estimates that 2.5 billion people—nearly half the world's population—are adversely affected by pollution from this source. Women and small children spend long hours each day around open fires or unventilated stoves in enclosed spaces. The levels of carbon monoxide, particulates, aldehydes, and other toxic chemicals can be one hundred times higher than would be legal for outdoor ambient concentrations in the United States. Designing and building cheap, efficient, nonpolluting energy sources for the developing countries would not only save shrinking forests but would make a major impact on health as well.

IN DEPTH. . . . . .

Indoor Air

How safe is the air in your home, office, or school room? As we decrease air-infiltration into buildings to conserve energy, we often trap indoor air pollutants within spaces where most of us spend the vast majority of our time. In what has come to be known as "sick building syndrome," people complain of headaches, fatigue, nausea, upper-respiratory problems, and a wide variety of allergies from workplace or home exposure to airborne toxins. While these symptoms often are vague and difficult to verify scientifically, the U.S. Environmental Protection Agency estimates that sick building syndrome may cost $60 billion a year in medical expenses, absenteeism, and reduced productivity.

What might be making us sick? Mold spores are probably the greatest single cause of allergic reactions in indoor air. Moisture trapped in air-tight houses often accumulates in walls where molds flourish. Air ducts provide both a good environment for growth of pathogens such as Legionnaire's disease bacteria as well as a path for their dispersal. Legionnaire's pneumonia is much more prevalent than most people realize in places like California and Australia where air-conditioning is common. Uranium-bearing rocks and sediment are widespread across North America. When uranium decays, it produces carcinogenic radon gas that can seep into buildings. The EPA warns that one home in ten in the United States may exceed the recommended maximum radon concentration of 4 picocuries per liter.

In addition, we are exposed to a variety of synthetic chemicals emitted from carpets, wall coverings, building materials, and combustion gases. You might be surprised to learn how many toxic, synthetic compounds are used to construct buildings and make furniture. Formaldehyde, for instance is a component of more than 3000 products, including building materials such as particle board, waferboard, and urea-formaldehyde foam insulation. Vinyl chloride is used in plastic plumbing pipe, floor and wall coverings, and countertops. Volatile organic solvents make up as much as half the volume of some paint. New carpets and drapes typically contain up to two dozen chemical compounds designed to kill bacteria and molds, resist stains, bind fibers, and retain colors.

What can you do if you suspect that your living spaces are exposing you to materials that make you sick? Probably few students will be in a position anytime soon to build a new house with nontoxic materials, but there are some principles from the emerging field of "green design" that you might apply if you're house hunting, redecorating your apartment, or interviewing for a job (see chapter 8). Low-volatile paint is now available for indoor use. Nontoxic, formaldehyde-free plywood, particle board, and insulation can be used in new construction. Nonallergenic carpets, drapes, and wall coverings are available, but some architects recommend natural wood, stone, and plaster surfaces that are easier to clean and less allergenic than any fabric.

High rates of air exchange can help rid indoor air of moisture, odors, mold spores, radon, and toxins. Does that mean energy inefficiency? Not necessarily. Air-to-air heat exchangers keep heat in during the winter and out during the summer, while still providing a healthy rate of fresh-air flow. Bathrooms and kitchens should have outdoor vents. Gas or oil furnaces should be checked for carbon monoxide production. Although many cooks prefer gas stoves because they heat quickly, they can produce toxic carbon monoxide and nitrogen oxides. Contact your city housing authority or county extension service for further tips on how to make your home, work, or study environment healthier. No matter what your situation, there are things that each of us can do to make our indoor air cleaner and safer.

CLIMATE, TOPOGRAPHY, AND ATMOSPHERIC PROCESSES

Topography, climate, and physical processes in the atmosphere play an important role in transport, concentration, dispersal, and removal of many air pollutants. Wind speed, mixing between air layers, precipitation, and atmospheric chemistry all determine whether pollutants will remain in the locality where they are produced or go elsewhere. In this next section, we will survey some environmental factors that affect air pollution levels.

Inversions

Temperature inversions occur when a stable layer of warmer air overlays cooler air, reversing the normal temperature decline with increasing height and preventing convection currents from dispersing pollutants. Several mechanisms create inversions. When a cold front slides under an adjacent warmer air mass or when cool air subsides down a mountain slope to displace warmer air in the valley below, an inverted temperature gradient is established. These inversions are usually not stable, however, because winds accompanying these air exchanges tend to break up the temperature gradient fairly quickly and mix air layers.

The most stable inversion conditions are usually created by rapid nighttime cooling in a valley or basin where air movement is restricted. Los Angeles is a classic example of the conditions that create temperature inversions and photochemical smog (fig. 18.12). The city is surrounded by mountains on three sides and the climate is dry and sunny. Extensive automobile use creates high pollution levels. Skies are generally clear at night, allowing rapid radiant heat loss, and the ground cools quickly. Surface air layers are cooled by conduction, while upper layers remain relatively warm. Density differences retard vertical mixing. During the night, cool, humid, onshore breezes slide in under the contaminated air, squeezing it up against the cap of warmer air above and concentrating the pollutants accumulated during the day.

Morning sunlight is absorbed by the concentrated aerosols and gaseous chemicals of the inversion layer. This complex mixture quickly cooks up a toxic brew of hazardous compounds. As

FIGURE 18.11 Smoky cooking and heating fires may cause more ill health effects than any other source of indoor air pollution except tobacco smoking. Some 2.5 billion people, mainly women and children, spend hours each day in poorly ventilated kitchens and living spaces where carbon monoxide, particulates, and cancer-causing hydrocarbons often reach dangerous levels.

the ground warms later in the day, convection currents break up the temperature gradient and pollutants are carried back down to the surface where more contaminants are added. Nitric oxide (NO) from automobile exhaust is oxidized to nitrogen dioxide. As nitrogen oxides are used up in reactions with unburned hydrocarbons, the ozone levels begin to rise. By early afternoon, an acrid brown haze fills the air, making eyes water and throats burn. On summer days, ozone concentrations in the Los Angeles basin can reach 0.34 ppm or more by late afternoon and the pollution index can be 300, the stage considered a health hazard.

Dust Domes and Heat Islands

Even without mountains to block winds and stabilize air layers, many large cities create an atmospheric environment quite different from the surrounding conditions. Sparse vegetation and high levels of concrete and glass in urban areas allow rainfall to run off quickly and create high rates of heat absorption during the day and radiation at night. Tall buildings create convective updrafts that sweep pollutants into the air. Temperatures in the center of large cities are frequently 3° to 5°C (5° to 9°F) higher than the surrounding countryside. Stable air masses created by this "heat island" over the city concentrate pollutants in a "dust dome." Rural areas downwind from major industrial areas often have significantly decreased visibility and increased rainfall (due to increased condensation nuclei in the dust plume) compared to neighboring areas with cleaner air. In the late 1960s, for instance, areas downwind from Chicago and St. Louis reported up to 30 percent more rainfall than upwind regions.

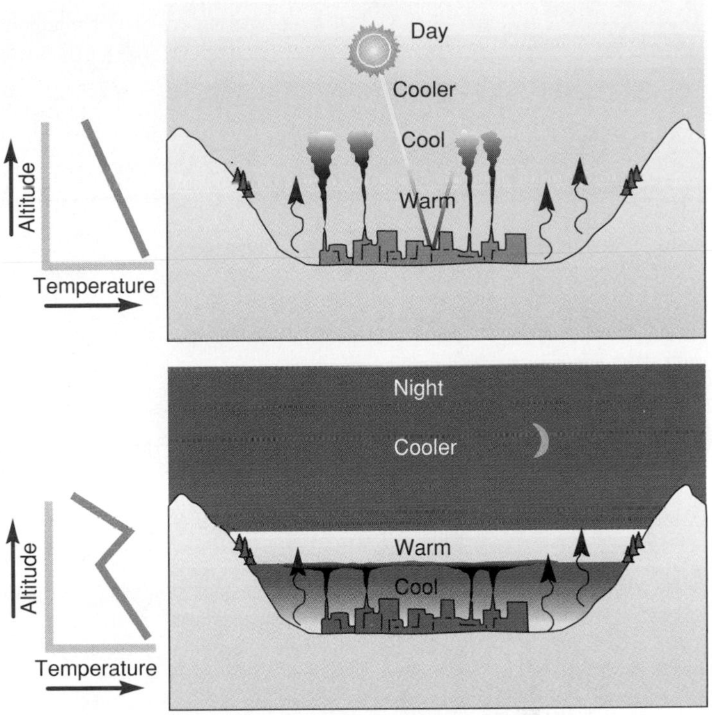

FIGURE 18.12 Atmospheric temperature inversions occur where ground level air cools more quickly than upper levels. This temperature differential prevents mixing and traps pollutants close to the ground.

Long-Range Transport

Fine aerosols can be carried great distances by the wind. Florida researchers recently showed that as much as half of the fine reddish dust visible in Miami's air during summer months is blown across the Atlantic from the Sahara Desert. Forests in Hawaii are nourished by dust from China's Takla Makan Desert nearly 6000 km (3750 mi) away.

Industrial pollutants are also transported great distances by wind currents. Some of the most toxic and corrosive materials delivered by long-range transport are secondary pollutants (such as sulfuric and nitric acids or ozone), produced by the mixing and interaction of atmospheric contaminants as they travel through the air. In 1999, atmospheric scientists were surprised to find a dense haze of soot and sulfate aerosols over the Indian Ocean. Covering an area larger than the continental United States and reaching from the Bay of Bengal to the Arabian Sea, this huge cloud of pollutants is produced in South Asia and blown south by the winter monsoon. In summer it is blown back north to fall as dirty acid rain.

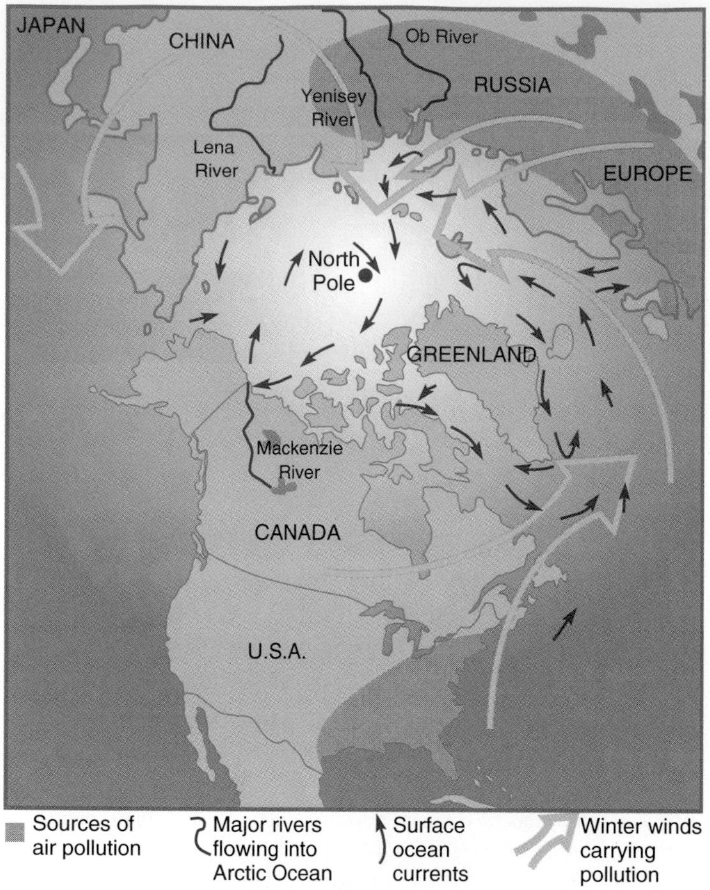

FIGURE 18.13 Air pollution from heavily industrialized regions in Europe and America are transported by circumpolar winds to the Arctic, where high levels of smog accumulate. The average transit time from Russia to Canada is only about three days.

Increasingly sensitive monitoring equipment has begun to reveal industrial contaminants in places usually considered among the cleanest in the world. Samoa, Greenland, and even Antarctica and the North Pole, all have heavy metals, pesticides, and radioactive elements in their air. Since the 1950s, pilots flying in the high Arctic have reported dense layers of reddish-brown haze clouding the arctic atmosphere. Aerosols of sulfates, soot, dust, and toxic heavy metals such as vanadium, manganese, and lead travel to the pole from the industrialized parts of Europe and Russia (fig. 18.13). These contaminants, trapped by winds that circle the pole, concentrate at high latitudes and eventually, falling out in snow and ice, enter the food chain. The Inuit people of Broughton Island, well above the Arctic Circle, have higher levels of polychlorinated biphenyls (PCBs) in their blood than any other known population, except victims of industrial accidents. Far from any source of this industrial by-product, these people accumulate PCBs from the flesh of fish, caribou, and other animals they eat.

Stratospheric Ozone

In 1985, the British Antarctic Atmospheric Survey announced a startling and disturbing discovery: ozone levels in the stratosphere over the South Pole were dropping precipitously during September

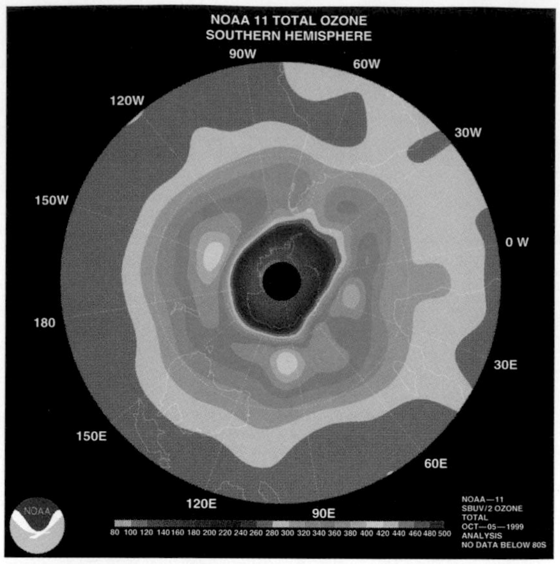

FIGURE 18.14 Ozone depletion over the South Pole is shown in this satellite image from October 5,1999. Outside the polar vortex, green and yellow areas show elevated ozone levels. Over Antarctica (outlined in white), however, stratospheric ozone levels are reduced by 80 percent or more from normal levels. The dark circle over the South Pole is an artifact of the measuring system.

and October every year as the sun reappears at the end of the long polar winter (fig. 18.14). This ozone depletion has been occurring at least since the 1960s but was not recognized because earlier researchers programmed their instruments to ignore changes in ozone levels that were presumed to be erroneous.

The 1998 ozone "hole" over Antarctia was the largest ever recorded, covering 27.3 million km^2 (10.5 million mi^2, or about twice the size of the entire United States) in which all the ozone between 14 and 20 kilometers altitude was destroyed. Ominously, this phenomenon is now spreading to other parts of the world as well. About 10 percent of all stratospheric ozone worldwide was destroyed during the spring of 1998 and levels over the Arctic averaged 40 percent below normal.

Why are we worried about stratospheric ozone? At ground level, ozone is a harmful pollutant, damaging plants, building materials, and human health; in the the upper atmosphere, however, where it screens out dangerous ultraviolet (UV) rays from the sun, ozone is an irreplaceable resource. Without this shield, organisms on the earth's surface would be subjected to life-threatening radiation burns and genetic damage. A 1 percent loss of ozone results in a 2 percent increase in UV reaching the earth's surface and could result in about a million extra human skin cancers per year worldwide if no protective measures are taken. Thus it is urgent that we learn what is attacking the ozone layer and find ways to reverse these trends if possible.

The exceptionally cold temperatures (–85 to –90°C) in Antarctica play a role in ozone losses. During the long, dark winter months, the strong circumpolar vortex (chapter 17) isolates Antarctic air and allows stratospheric temperatures to drop low enough to create ice crystals at high altitudes—something that rarely happens elsewhere over the world. Ozone and chlorine-containing molecules are

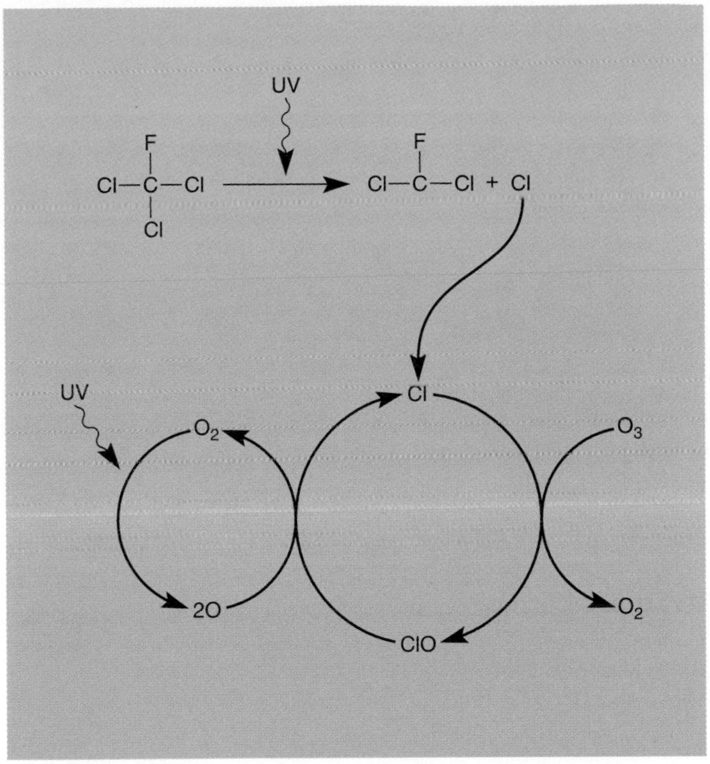

FIGURE 18.15 Destruction of stratospheric ozone by chlorine atoms. When exposed to UV radiation, CFCs release highly reactive chlorine atoms that react with ozone to produce chlorine monoxide. Singlet oxygen atoms, also produced by UV irradiation, react with ClO to start the cycle over again.

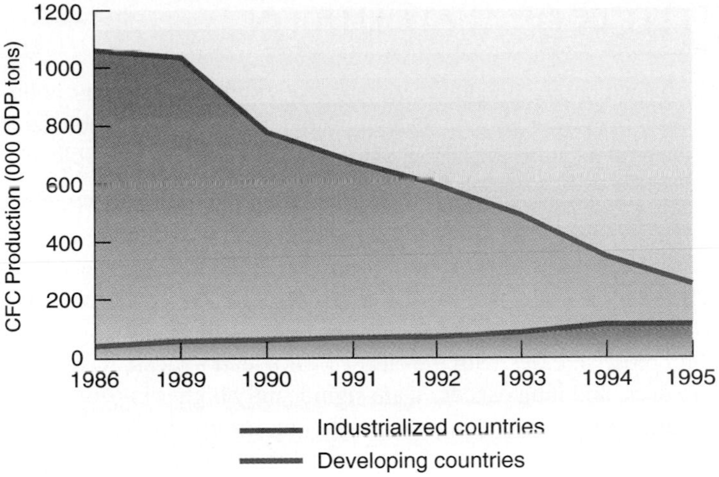

FIGURE 18.16 Production of chlorofluorocarbons in industrialized countries has fallen sharply since the Helsinki convention was passed in 1989. *Source:* Data from Sebastian Oberthur, Production and Consumption of Ozone-Depleting Substances, 1986–1995 (Deutsche Gesellschaft for Technische Zusammenarbeit, Bonn Germany, 1997, p. 30).

absorbed on the surfaces of these ice particles. When the sun returns in the spring and provides energy to liberate chlorine ions, destructive chemical reactions proceed quickly (fig. 18.15).

Humans release a variety of chlorine-containing molecules into the atmosphere. The ones suspected of being most important in ozone losses are **chlorofluorocarbons (CFCs)** and halon gases. CFCs were invented in 1928 by scientists at General Motors who were searching for a less toxic refrigerant than ammonia. Commonly known by the trade name Freon, CFCs were regarded as wonderful compounds. They are nontoxic, nonflammable, chemically inert, cheaply produced, and useful in a wide variety of applications.

Because these molecules are so stable, however, they persist for decades or even centuries once released. When they diffuse out into the stratosphere, the intense UV irradiation releases chlorine atoms that destroy ozone. Since the chlorine atoms are not themselves consumed in these reactions, they continue to destroy ozone for years until they finally precipitate or are washed out of the air.

Until 1978, aerosol spray cans used more CFCs than any other product. Although we didn't know about the special conditions in the Antarctic at that time, it was suspected that CFCs might threaten stratospheric ozone, so laws were passed in the United States, Canada, and some European countries to ban nonessential uses. Still, some 320,000 metric tons of CFCs were used worldwide every year until 1988 as refrigerants, solvents, spray propellants, and foam-blowing agents.

The discovery of stratospheric ozone losses has brought about a remarkably quick international response. At a 1989 conference in Helsinki, 81 nations agreed to phase out CFC production by the end of the century. As evidence accumulated showing that losses were larger and more widespread than previously thought, the deadline for the elimination of all CFCs (halons, carbon tetrachloride, and methyl chloroform) was moved up to 1996 and a $500 million fund was established to assist poorer countries to switch to non-CFC technologies. Fortunately, alternatives to CFCs for most uses already exist. The first substitutes will be hydrochlorofluorocarbons (HCFCs), which release much less chlorine per molecule. Eventually, we hope to develop halogen-free molecules that work just as well and are no more expensive than CFCs.

There is some evidence that the CFC ban is already having an effect. CFC production in industrialized countries has fallen nearly 80 percent since 1989 (fig. 18.16). The buildup of CFCs in the atmosphere is declining more rapidly than expected. In 70 years or so, stratospheric ozone levels are expected to be back to normal. Unfortunately, there may be a downside to stopping ozone destruction. Ozone is a potent greenhouse gas (chapter 17). Some models suggest that lower ozone levels have been offsetting the effects of increased CO_2. When ozone is restored, global warming may be accelerated. As so often is the case, when we disturb one environmental factor, we affect others as well.

In 1995, chemists Sherwood Rowland, Mario Molina, and Paul Crutzen shared the Nobel Prize for their work on atmospheric chemistry and stratospheric ozone. This was the first Nobel Prize for an environmental issue.

EFFECTS OF AIR POLLUTION

So far we have looked at the major types and sources of air pollutants. Now we will turn our attention to the effects of those pollutants on human health, physical materials, ecosystems, and global climate.

Human Health

The EPA estimates that people in the most polluted cities in the United States are 15 to 17 percent more likely to die prematurely than those in cities with the cleanest air. Heart attacks, respiratory diseases, and lung cancer all are significantly higher in people who breathe dirty air, compared to matching groups in cleaner environments. This can mean as much as a 5- to 10-year decrease in life expectancy if you live in the worst parts of Los Angeles or Baltimore, compared to a place with clean air. Of course your likelihood of suffering ill health from air pollutants depends on the intensity and duration of exposure as well as your age and prior health status. You are much more likely to be at risk if you are very young, very old, or already suffering from some respiratory or cardiovascular disease. Some people are super-sensitive because of genetics or prior exposure. And those doing vigorous physical work or exercise are more likely to succumb than more sedentary folks.

Conditions are often much worse in other countries than Canada or the United States. The United Nations estimates that at least 1.3 billion people around the world live in areas where the air is dangerously polluted. In the "black triangle" region of Poland, Hungary, the Czech Republic, and Slovakia, for example, respiratory ailments, cardiovascular diseases, lung cancer, infant mortality, and miscarriages are as much as 50 percent higher than in cleaner parts of those countries. And in China, city dwellers are 4 to 6 times more likely than country folk to die of lung cancer. As mentioned earlier, the greatest air quality problem is often in poorly ventilated homes in poorer countries where smoky fires are used for cooking and heating. Billions of women and children spend hours each day in these unhealthy conditions. The World Health Organization estimates that 4 million children under 5 die each year from acute respiratory diseases exacerbated by air pollution.

How does air pollution cause these health effects? The most common route of exposure to air pollutants is by inhalation, but direct absorption through the skin or contamination of food and water also are important pathways. Because they are strong oxidizing agents, sulfates, SO_2, NO_x, and O_3 act as irritants (see chapter 9) that damage delicate tissues in the eyes and respiratory passages. Fine suspended particulate materials (less than 10 mm) penetrate deep into the lungs and are both irritants and fibrotic agents. Inflammatory responses set in motion by these irritants impair lung function and trigger cardiovascular problems as the heart tries to compensate for lack of oxygen by pumping faster and harder. If the irritation is really severe—see the example of Bhopal in chapter 9—so much fluid seeps into lungs through damaged tissues that the victim actually drowns.

Carbon monoxide binds to hemoglobin and decreases the ability of red blood cells to carry oxygen. Asphyxiants such as this cause headaches, dizziness, heart stress, and can even be lethal if concentrations are high enough. Lead also binds to hemoglobin and reduces oxygen-carrying capacity at high levels. At lower levels, lead causes long-term damage to critical neurons in the brain that results in mental and physical impairment and developmental retardation.

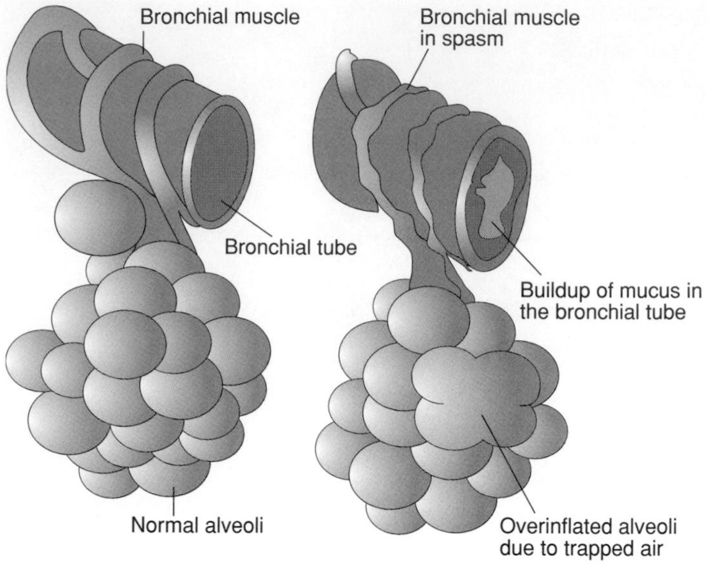

FIGURE 18.17 When an asthma victim encounters a "trigger," such as dust, cold air, or irritating chemicals, muscles around the bronchial tubes contract and secretory cells in the bronchioles produce a thick mucus that blocks airways. Wheezing and difficulty in breathing result.

Some important chronic health effects of air pollutants include bronchitis and emphysema.

Bronchitis is a persistent inflammation of bronchi and bronchioles (large and small airways in the lung) that causes a painful cough and involuntary muscle spasms that constrict airways. Severe bronchitis can lead to **emphysema,** an irreversible obstructive lung disease in which airways become permanently constricted and alveoli are damaged or even destroyed. Stagnant air trapped in blocked airways swells the tiny air sacs in the lung (alveoli), blocking blood circulation. As cells die from lack of oxygen and nutrients, the walls of the alveoli break down, creating large empty spaces incapable of gas exchange (fig. 18.17). Thickened walls of the bronchioles lose elasticity and breathing becomes more difficult. Victims of emphysema make a characteristic whistling sound when they breathe. Often they need supplementary oxygen to make up for reduced respiratory capacity.

Irritants in the air are so widespread that about half of all lungs examined at autopsy in the United States have some degree of alveolar deterioration. The Office of Technology Assessment (OTA) estimates that 250,000 people suffer from pollution-related bronchitis and emphysema in the United States, and some 50,000 excess deaths each year are attributable to complications of these diseases, which are probably second only to heart attack as a cause of death.

Smoking is undoubtedly the largest cause of obstructive lung disease and preventable death in the world. The World Health Organization says that tobacco kills some 3 million people each

FIGURE 18.18 Sulfur dioxide emissions and acid precipitation from the International Nickel Company copper smelter (*background*) killed all vegetation over a large area near Sudbury, Ontario. Even the pink granite bedrock has burned black. The installation of scrubbers has dramatically reduced sulfur emissions. The ecosystem farther away from the smelter is slowly beginning to recover.

FIGURE 18.19 Soybean leaves exposed to 0.8 parts per million sulfur dioxide for 24 hours show extensive chlorosis (chlorophyll destruction) in white areas between leaf veins.

year. This makes it rank with diarrhea and malaria as one of the world's leading killers. Because of cardiovascular stress caused by carbon monoxide in smoke and chronic bronchitis and emphysema, about twice as many people die of heart failure as die from lung cancer associated with smoking.

Plant Pathology

In the early days of industrialization, fumes from furnaces, smelters, refineries, and chemical plants often destroyed vegetation and created desolate, barren landscapes around mining and manufacturing centers. The copper-nickel smelter at Sudbury, Ontario, is a spectacular and notorious example of air pollution effects on vegetation and ecosystems. In 1886, the corporate ancestors of the International Nickel Company (INCO) began open-bed roasting of sulfide ores at Sudbury. Sulfur dioxide and sulfuric acid released by this process caused massive destruction of the plant community within about 30 km (18.6 mi) of the smelter. Rains washed away the exposed soil, leaving a barren moonscape of blackened bedrock. Super-tall, 400 m smokestacks were installed in the 1950s and sulfur scrubbers were added twenty years later. Emissions were reduced by 90 percent and the surrounding ecosystem is beginning to recover. The area near the factory is still a grim, empty wasteland, however (fig. 18.18). Similar destruction occurred at many other sites during the nineteenth century. Copperhill, Tennessee; Butte, Montana; and the Ruhr Valley in Germany are some well-known examples, but these areas also are showing signs of recovery since corrective measures were taken.

There are two probable ways that air pollutants damage plants. They can be directly toxic, damaging sensitive cell membranes much as irritants do in human lungs. Within a few days of exposure to toxic levels of oxidants, mottling (discoloration) occurs in leaves due to chlorosis (bleaching of chlorophyll), and then necrotic (dead) spots develop (fig. 18.19). If injury is severe, the whole plant may be killed. Sometimes these symptoms are so distinctive that positive identification of the source of damage is possible. Often, however, the symptoms are vague and difficult to separate from diseases or insect damage.

Another mechanism of action is exhibited by chemicals, such as ethylene, that act as metabolic regulators or plant hormones and disrupt normal patterns of growth and development. Ethylene is a component of automobile exhaust and is released from petroleum refineries and chemical plants. The concentration of ethylene around highways and industrial areas is often high enough to cause injury to sensitive plants. Some scientists believe that the devastating forest destruction in Europe and North America may be partly due to volatile organic compounds.

Certain combinations of environmental factors have **synergistic effects** in which the injury caused by exposure to two factors together is more than the sum of exposure to each factor individually. For instance, when white pine seedlings are exposed to subthreshold concentrations of ozone and sulfur dioxide individually, no visible injury occurs. If the same concentrations of pollutants are given together, however, visible damage occurs. In alfalfa, however, SO_2 and O_3 together cause less damage than either one alone. These complex interactions point out the unpredictability of future effects of pollutants. Outcomes might be either more or less severe than previous experience indicates.

Pollutant levels too low to produce visible symptoms of damage may still have important effects. Field studies using open-top chambers and charcoal-filtered air show that yields in some sensitive crops, such as soybeans, may be reduced as much as 50 percent by currently existing levels of oxidants in ambient air. Some plant pathologists suggest that ozone and photochemical oxidants are responsible for as much as 90 percent of agricultural, ornamental, and forest losses from air pollution. The total costs of this damage may be as much as $10 billion per year in North America alone.

Acid Deposition

Most people in the United States became aware of problems associated with **acid precipitation** (the deposition of wet acidic solutions or dry acidic particles from the air) within the last decade or so, but English scientist Robert Angus Smith coined the term "acid rain" in his studies of air chemistry in Manchester, England, in the 1850s. By the 1940s, it was known that pollutants, including atmospheric acids, could be transported long distances by wind currents. This was thought to be only an academic curiosity until it was shown that precipitation of these acids can have far-reaching ecological effects.

pH and Atmospheric Acidity

We describe acidity in terms of pH (the negative logarithm of the hydrogen ion concentration in a solution). The pH scale ranges from 0 to 14, with 7, the midpoint, being neutral (chapter 3). Values below 7 indicate progressively greater acidity, while those above 7 are progressively more alkaline. Since the scale is logarithmic, there is a tenfold difference in hydrogen ion concentration for each pH unit. For instance, pH 6 is 10 times more acidic than pH 7; likewise, pH 5 is 100 times more acidic, and pH 4 is 1000 times more acidic than pH 7.

Normal, unpolluted rain generally has a pH of about 5.6 due to carbonic acid created by CO_2 in air. Volcanic emissions, biological decomposition, and chlorine and sulfates from ocean spray can drop the pH of rain well below 5.6, while alkaline dust can raise it above 7. In industrialized areas, anthropogenic acids in the air usually far outweigh those from natural sources. Acid rain is only one form in which acid deposition occurs. Fog, snow, mist, and dew also trap and deposit atmospheric contaminants. Furthermore, fallout of dry sulfate, nitrate, and chloride particles can account for as much as half of the acidic deposition in some areas.

Aquatic Effects

It has been known for about 30 years that acids—principally H_2SO_4 and HNO_3—generated by industrial and automobile emissions in northwestern Europe are carried by prevailing winds to Scandinavia where they are deposited in rain, snow, and dry precipitation. The thin, acidic soils and oligotrophic lakes and streams in the mountains of southern Norway and Sweden have been severely affected by this acid deposition. Some 18,000 lakes in Sweden are now so acidic that they will no longer support game fish or other sensitive aquatic organisms.

Generally, reproduction is the most sensitive stage in fish life cycles. Eggs and fry of many species are killed when the pH drops to about 5.0. This level of acidification also can disrupt the food chain by killing aquatic plants, insects, and invertebrates on which fish depend for food. At pH levels below 5.0, adult fish die as well. Trout, salmon, and other game fish are usually the most sensitive. Carp, gar, suckers, and other less desirable fish are more resistant.

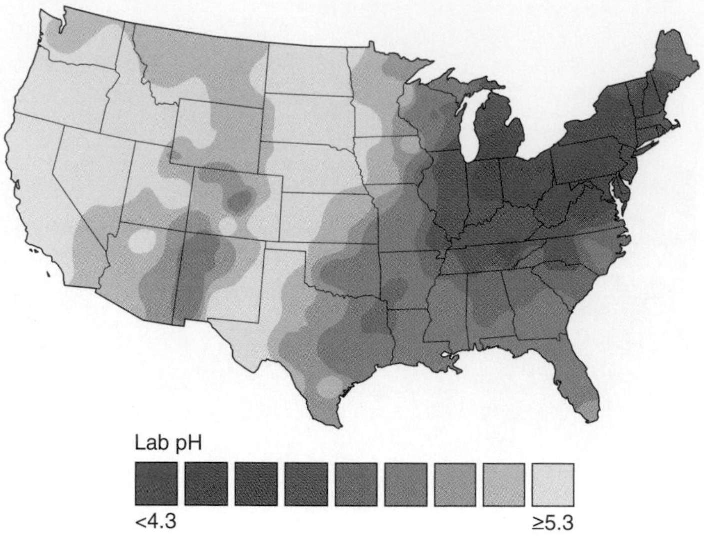

FIGURE 18.20 Acid precipitation over the United States.
Source: National Acid Deposition Program, 1997.

There are several ways acids kill fish. Acidity alters body chemistry, destroys gills and prevents oxygen uptake, causes bone decalcification, and disrupts muscle contraction. Another dangerous effect (for us as well as fish) is that acid water leaches toxic metals, such as mercury and aluminum, out of soil and rocks.

In the early 1970s, evidence began to accumulate suggesting that air pollutants are acidifying many lakes in North America. Studies in the Adirondack Mountains of New York revealed that about half of the high-altitude lakes (above 1000 m or 3300 ft) are acidified and have no fish. Areas showing lake damage correlate closely with average pH levels in precipitation (fig. 18.20). Some 48,000 lakes in Ontario are endangered and nearly all of Quebec's surface waters, including about 1 million lakes, are believed to be highly sensitive to acid deposition.

Much of the western United States has relatively alkaline bedrock and carbonate-rich soil, which counterbalance acids from the atmosphere. Recent surveys of the Rocky Mountains, the Sierra Nevadas in California, and the Cascades in Washington, however, have shown that many high mountain lakes and streams have very low buffering capacity (ability to resist pH change) and are susceptible to acidification.

Sulfates account for about two-thirds of the acid deposition in eastern North America and most of Europe, while nitrates contribute most of the remaining one-third. In urban areas, where transportation is the major source of pollution, nitric acid is equal to or slightly greater than sulfuric acids in the air. A vigorous program of pollution control has been undertaken by both Canada and the United States. Although SO_2 and NO_x emissions have decreased dramatically over the past three decades over much of Europe and eastern North America as a result of pollution control measures, rain falling in these areas remains acidic. Damage to natural ecosystems also continues to be greater than scientists expected. Apparently, alkaline dust that would once have neutralized acids in air has been depleted by years of acid rain and is

FIGURE 18.21 A Frasier fir forest on Mount Mitchell, NC killed by acid rain, insect pests, and other stressors.

now no longer effective. This may also lead to a loss of cations such as calcium, magnesium, sodium, and potassium essential to plant growth.

Forest Damage

In the early 1980s, disturbing reports appeared of rapid forest declines in both Europe and North America. One of the earliest was a detailed ecosystem inventory on Camel's Hump Mountain in Vermont. A 1980 survey showed that seedling production, tree density, and viability of spruce-fir forests at high elevations had declined about 50 percent in 15 years. By 1990, almost all the red spruce, once the dominant species on the upper part of the mountain, were dead or dying. A similar situation was found on Mount Mitchell in North Carolina where almost all red spruce and Fraser fir above 2000 m (6000 ft) are in a severe decline. Nearly all the trees are losing needles and about half of them are dead (fig. 18.21).

European forests also are dying at an alarming rate. West German foresters estimated in 1982 only 8 percent of their forests showed air pollution damage. By 1983, some 34 percent of the forest was affected, and in 1985, more than 4 million hectares (about half the total forest) were reported to be in a state of decline. The loss to the forest industry is estimated to be about one billion DM (Deutsche marks) per year.

Similar damage is reported in Czechoslovakia, Poland, Austria, and Switzerland. Again, high-elevation forests are most severely affected. This is a disaster for mountain villages in the Alps that depend on forests to prevent avalanches in the winter. Sweden, Norway, the Netherlands, Romania, China, and the former Soviet Union also have evidence of growth reduction, defoliation, root necrosis, lack of seedling growth, and premature tree death. The species afflicted vary from place to place, but the overall picture is of widespread forest destruction.

This complex phenomenon probably has many contributing factors, but air pollution and deposition of atmospheric acids are thought to be leading causes of forest destruction in many areas. In the longest-running forest-ecosystem monitoring record in North America, researchers at the Hubbard Brook Experimental Forest in New Hampshire have shown that forest soils have become depleted of natural buffering reserves of basic cations such as calcium and magnesium through years of exposure to acid rain. Replacement of these cations by hydrogen and aluminum ions seems to be one of the main causes of plant mortality.

Other mechanisms may also play a role in forest decline. Overfertilization by nitrogen compounds may make trees sensitive to early frost. Toxic metals, such as aluminum, may be solubilized by acidic groundwater. Plant pathogens and insect pests may damage trees or attack trees debilitated by air pollution. Fungi that form essential mutualistic associations (called mycorrhizae) with tree roots may be damaged by acid rain. Other air pollutants, such as sulfur dioxide, ozone, or toxic organic compounds may damage trees. Repeated harvesting cycles in commercial forests may remove nutrients and damage ecological relationships essential for healthy tree growth. Perhaps the most likely scenario is that all these environmental factors act cumulatively but in different combinations in the deteriorating health of individual trees and entire forests.

Buildings and Monuments

In cities throughout the world, some of the oldest and most glorious buildings and works of art are being destroyed by air pollution. Smoke and soot coat buildings, paintings, and textiles. Limestone and marble are destroyed by atmospheric acids at an alarming rate. The Parthenon in Athens, the Taj Mahal in Agra, the Colosseum in Rome, frescoes and statues in Florence, medieval cathedrals in Europe (fig. 18.22), and the Lincoln Memorial and Washington Monument in Washington, DC, are slowly dissolving and flaking away because of acidic fumes in the air. Medieval stained glass windows in Cologne's gothic cathedral are so porous from etching by atmospheric acids that pigments disappear and the glass literally crumbles away. Restoration costs for this one building alone are estimated at three to four billion German marks ($1.5 to $2 billion).

On a more mundane level, air pollution also damages ordinary buildings and structures. Corroding steel in reinforced concrete weakens buildings, roads, and bridges. Paint and rubber deteriorate due to oxidization. Limestone, marble, and some kinds of sandstone flake and crumble. The Council on Environmental Quality estimates that U.S. economic losses from architectural damage caused by air pollution amount to about $4.8 billion in direct costs and $5.2 billion in property value losses each year.

Visibility Reduction

Foul air obscuring the skies above industrialized cities has long been recognized as a problem, but we have realized only recently that pollution affects rural areas as well. Even supposedly pristine places like our national parks are suffering from air pollution. Grand Canyon National Park, where maximum visibility used to be 300 km (185 mi), is now so smoggy on some winter days that visitors can't see the opposite rim only 20 km (12.5 mi) across the canyon. Mining operations, smelters, and power plants (some of which were moved to the desert to improve air quality in cities like Los Angeles) are the main culprits. Similarly, the vistas from Shenandoah National Park just outside Washington, DC, are so hazy that summer visibility is often less than 1.6 km (1 mi) because of smog drifting in from nearby urban areas.

Historical records show that over the past four or five decades human-caused air pollution has spread over much of the United States. John Trijonis of the Santa Fe Corporation reports that a gigantic "haze blob" as much as 3000 km (about 2000 mi) across covers much of the eastern United States in the summer, cutting visibility as much as 80 percent. Smog and haze are so prevalent, Trijonis says, that it's hard for people to believe that the air once was clear. Studies indicate, however, that if all human-made sources of air pollution were shut down, the air would clear up in a few days and there would be about 150 km (90 mi) visibility nearly everywhere rather than the 15 km to which we have become accustomed.

FIGURE 18.22 Atmospheric acids, especially sulfuric and nitric acids, have almost completely eaten away the face of this medieval statue. Each year, the total losses from air pollution damage to buildings and materials amounts to billions of dollars.

AIR POLLUTION CONTROL

What can we do about air pollution? In this section we will look at some of the techniques that can be used to avoid creating pollutants or to clean up effluents before they are released. We also will look at some legislation that regulates pollutant emissions and ambient air quality.

Moving Pollution to Remote Areas

Among the earliest techniques for improving local air quality was moving pollution sources to remote locations and/or dispersing emissions with smokestacks. These approaches exemplify the attitude that "dilution is the solution to pollution." One electric utility, for example, ran newspaper and magazine ads in the early 1970s, claiming to be a "pioneer" in the use of tall smokestacks on its power plants to "disperse gaseous emissions widely in the atmosphere so that ground level concentrations would not be harmful to human health or property." The company claimed that their smoke would be "dissipated over a wide area and come down finally in harmless traces." Far from being harmless, however, those "traces" are the main source of many of our current problems. We are finding that there is no "away" to which we can throw our unwanted products. A far better solution to pollution is to prevent its release. We will now turn our attention to emission-control technology.

Particulate Removal

Filters remove particles physically by trapping them in a porous mesh of cotton cloth, spun glass fibers, or asbestos-cellulose, which allows air to pass through but holds back solids. Collection efficiency is relatively insensitive to fuel type, fly ash composition, particle size, or electrical properties. Filters are generally shaped into giant bags 10 to 15 meters long and 2 or 3 meters wide. Effluent gas is blown into the bottom of the bag and escapes through the sides much like the bag on a vacuum cleaner (fig. 18.23*a*). Every few days or weeks, the bags are opened to remove the dust cake. Thousands of these bags may be lined up in a "baghouse." These filters are usually much cheaper to install and operate than electrostatic filters.

Electrostatic precipitators (fig. 18.23*b*) are the most common particulate controls in power plants. Fly ash particles pick up an electrostatic surface charge as they pass between large electrodes in the effluent stream. This causes the particle to migrate to and accumulate on a collecting plate (the oppositely charged electrode). These precipitators consume a large amount of electricity, but maintenance is relatively simple and collection efficiency can be as high as 99 percent. Performance depends on particle size and chemistry, strength of the electric field, and flue gas velocity.

The ash collected by all of these techniques is a solid waste (often hazardous due to the heavy metals and other trace components of coal or other ash source) and must be buried in landfills or other solid waste disposal sites.

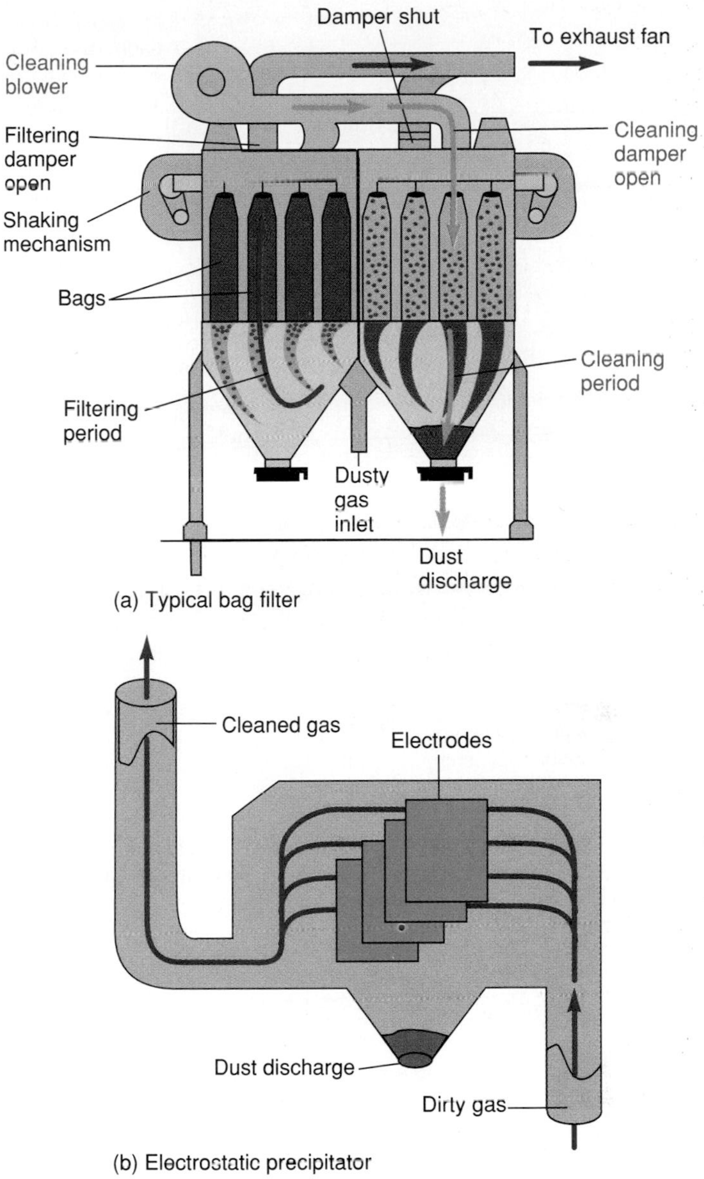

FIGURE 18.23 Typical emission-control devices: (*a*) bag filter and (*b*) electrostatic precipitator. Note that two stages in the operational cycle are shown in (*a*), the filtering period on the left and the period of cleaning the filter bag on the right.

Sulfur Removal

As we have seen earlier in this chapter, sulfur oxides are among the most damaging of all air pollutants in terms of human health and ecosystem damage. It is important to reduce sulfur loading. This can be done either by using low-sulfur fuel or by removing sulfur from effluents.

Fuel Switching and Fuel Cleaning

Switching from soft coal with a high sulfur content to low-sulfur coal can greatly reduce sulfur emissions. This may eliminate jobs, however, in such areas as Appalachia that are already economically depressed. Changing to another fuel, such as natural gas or nuclear energy, can eliminate all sulfur emissions as well as those of particulates and heavy metals. Natural gas is more expensive and more difficult to ship and store than coal, however, and many people prefer the sure dangers of coal pollution to the uncertain dangers of nuclear power (chapter 21). Alternative energy sources, such as wind and solar power, are preferable to either fossil fuel or nuclear power, and are becoming economically competitive (chapter 22) in many areas. In the interim, coal can be crushed, washed, and gassified to remove sulfur and metals before combustion. This improves heat content and firing properties but may replace air pollution with solid waste and water pollution problems.

Limestone Injection and Fluidized Bed Combustion

Sulfur emissions can be reduced as much as 90 percent by mixing crushed limestone with coal before it is fed into a boiler. Calcium in the limestone reacts with sulfur to make calcium sulfite ($CaSO_3$), calcium sulfate ($CaSO_4$), or gypsum ($CaSO_4.2H_2O$). In ordinary furnaces, this procedure creates slag, which fouls burner grates and reduces combustion efficiency.

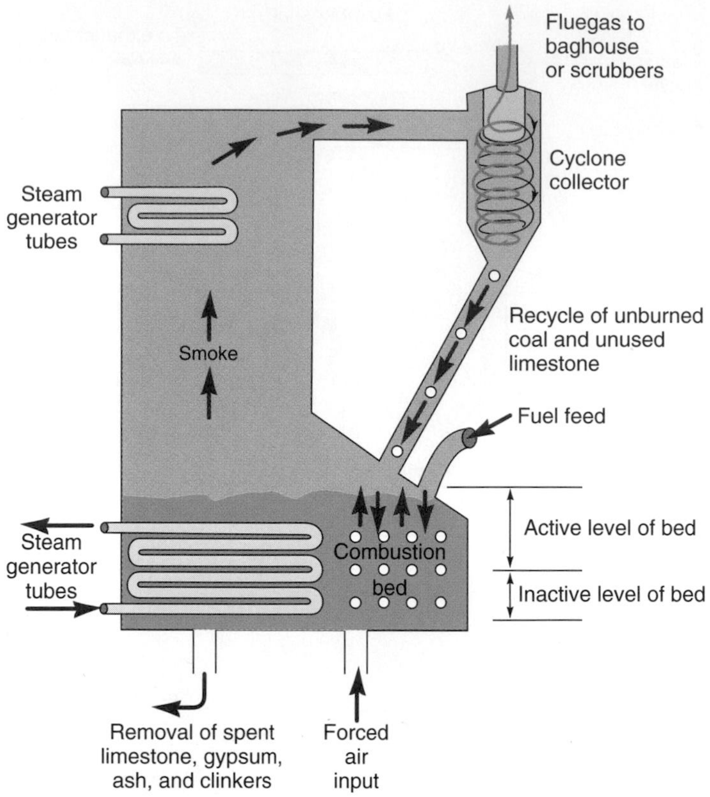

FIGURE 18.24 Fluidized bed combustion. Fuel is lifted by strong air jets from underneath the bed. Efficiency is good with a wide variety of fuels, and SO_2, NO_x, and CO emissions are much lower than with conventional burners.

A relatively new technique for burning, called fluidized bed combustion, offers several advantages in pollution control. In this procedure, a mixture of crushed coal and limestone particles about a meter (3 ft) deep is spread on a perforated distribution grid in the combustion chamber (fig. 18.24). When high-pressure air is forced through the bed, the surface of the fuel rises as much as one meter and resembles a boiling fluid as particles hop up and down. Oil is sprayed into the suspended mass to start the fire. During operation, fresh coal and limestone are fed continuously into the top of the bed, while ash and slag are drawn off from below. The rich air supply and constant motion in the bed make burning efficient and prevent buildup of large slag clinkers. Steam generator pipes are submerged directly into the fluidized bed, and heat exchange is more efficient than in the water walls of a conventional boiler. More than 90 percent of SO_2 is captured by the limestone particles, and NO_x formation is reduced by holding temperatures around 800°C (1500°F) instead of twice that figure in other boilers. These low temperatures also preclude slag formation, which aids in maintenance. The efficient burning of this process makes it possible to use cheaper fuel, such as lignite or unwashed subbituminous coal, rather than higher priced hard coal.

Flue Gas Desulfurization

Crushed limestone, lime slurry, or alkali (sodium carbonate or bicarbonate) can be injected into a stack gas stream to remove sulfur after combustion. These processes are often called flue gas scrubbing. Spraying wet alkali solutions or limestone slurry is relatively inexpensive and effective, but maintenance can be difficult. Rock-hard plaster and ash layers coat the spray chamber and have to be chipped off regularly. Corrosive solutions of sulfates, chlorides, and fluorides erode metal surfaces. Electrostatic precipitators don't work well because of fouling and shorting of electrodes after wet scrubbing.

Dry alkali injection (spraying dry sodium bicarbonate into the flue gas) avoids many of the problems of wet scrubbing, but the expense of appropriate reagents is prohibitive in most areas. A hybrid procedure called spray drying has been tested successfully in pilot plant experiments. In this process, a slurry of pulverized limestone or slaked lime is atomized in the stack gas stream. The spray rate and droplet size are carefully controlled so that the water flash evaporates and a dry granular precipitate is produced. Passage through a baghouse filter removes both ash and sulfur very effectively.

As with coal washing, scrubbing often results in a trade-off of an air pollution problem for a solid waste disposal problem. Sulfur slag, gypsum, and other products of these processes can amount to three or four times as much volume as fly ash. A large power plant can produce millions of tons of waste per year.

Sulfur Recovery Processes

Instead of making a throwaway product that becomes a waste disposal problem, sulfur can be removed from effluent gases by processes that yield a usable product, such as elemental sulfur, sulfuric acid, or ammonium sulfate. Catalytic converters are used in these recovery processes to oxidize or reduce sulfur and to create chemical compounds that can be collected and sold. Markets have to be reasonably close for economic feasibility, and fly ash contamination must be reduced as much as possible.

Nitrogen Oxide Control

Undoubtedly the best way to prevent nitrogen oxide pollution is to avoid creating it. A substantial portion of the emissions associated with mining, manufacturing, and energy production could be eliminated through conservation (chapter 21).

Staged burners, in which the flow of air and fuel are carefully controlled, can reduce nitrogen oxide formation by as much as 50 percent. This is true for both internal combustion engines and industrial boilers. Fuel is first burned at high temperatures in an oxygen-poor environment where NO_x cannot form. The residual gases then pass into an afterburner where more air is added and final combustion takes place in an air-rich, fuel-poor, low-temperature environment that also reduces NO_x formation. Stratified-charge engines and new orbital automobile engines use this principle to meet emission standards without catalytic converters.

what can you do?

Saving Energy and Reducing Pollution

- Be aware of weather patterns in your area. Prepare for storms and emergencies.
- Conserve energy: carpool, bike, walk, use public transport, buy compact fluorescent bulbs, and energy-efficient appliances (see chapter 22 for other suggestions).
- Buy refrigerators and air conditioners designed for CFC alternatives. If you have old appliances or other CFC sources, dispose of them responsibly.
- Plant a tree and care for it (every year).
- Write to your Congressional representatives and support a transition to an energy-efficient economy.
- If green-pricing options are available in your area, buy renewable energy.
- Have your car tuned every 10,000 miles (16,000 km) and make sure that its antismog equipment is working properly. Turn off your engine when waiting longer than one minute. Start trips a little earlier and drive slower—it not only saves fuel but it's safer, too.
- Use latex-based, low-volatile paint rather than oil-based (alkyd) paint.
- Don't top off your fuel tank when you buy gasoline; stop when the automatic mechanism turns off the pump. Don't dump gasoline or used oil on the ground or down the drain.
- Turn off lights, computers, TVs, and other appliances when you're not using them.

The approach adopted by U.S. automakers for NO_x reductions has been to use selective catalysts to change pollutants to harmless substances. Three-way catalytic converters use platinum-palladium and rhodium catalysts to remove up to 90 percent of NO_x, hydrocarbons, and carbon monoxide at the same time (fig. 18.25). Unfortunately, this approach doesn't work on diesel engines, power plants, smelters, and other pollution sources because of problems with back pressure, catalyst life, corrosion, and production of unwanted by-products, such as ammonium sulfate (NH_4SO_4), that foul the system.

Raprenox (*rap*id *re*moval of *n*itrogen *ox*ides) is a new technique for removing nitrogen oxides that was developed by the U.S. Department of Energy Sandia Laboratory in Livermore, California. Exhaust gases are passed through a container of common, nonpoisonous cyanuric acid. When heated to 350°C (662°F), cyanuric acid releases isocyanic acid gas, which reacts with NO_x to produce CO_2, CO, H_2O, and N_2. In small-scale diesel engine tests, this system eliminated 99 percent of the NO_x. Whether it will work in full-scale applications, especially in flue gases contaminated with fly ash, remains to be seen.

Hydrocarbon Controls

Hydrocarbons and volatile organic compounds are produced by incomplete combustion of fuels or solvent evaporation from chemical factories, painting, dry cleaning, plastic manufacturing, printing, and other industrial processes that use a variety of volatile organic chemicals. Closed systems that prevent escape of fugitive gases can reduce many of these emissions. In automobiles, for instance, positive crankcase ventilation (PCV) systems collect oil that escapes from around the pistons and unburned fuel and channels it back to the engine for combustion. Modification of carburetor and fuel systems prevents evaporation of gasoline. In the same way, controls on fugitive losses from valves, pipes, and storage

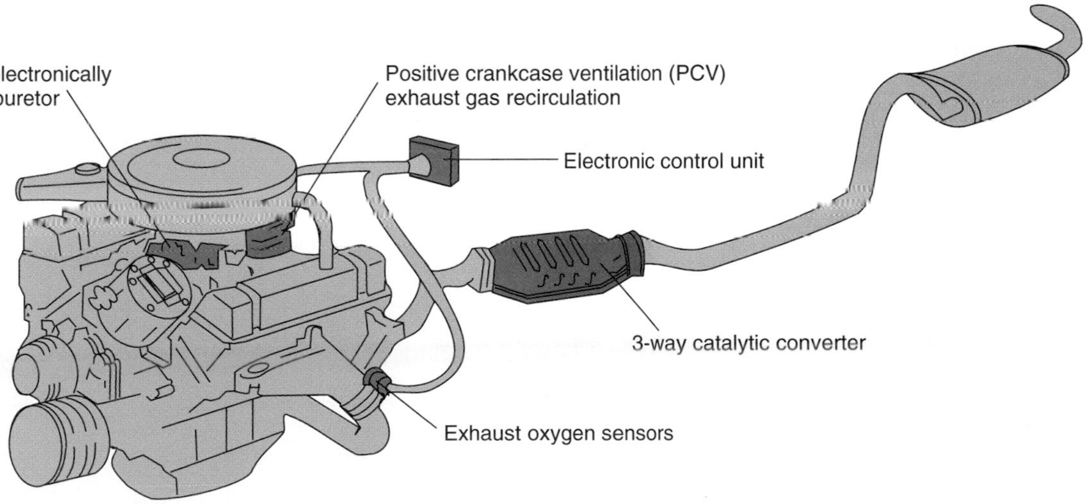

FIGURE 18.25 Elements of a modern automobile emission-control system. A closed-loop, electronically controlled carburetor or fuel-injector carefully meters fuel-air ratios to optimize combustion. Exhaust oxygen sensors measure completeness of fuel burning. Positive crankcase ventilation captures oil "blow-by" and unburned exhaust gases and recycles them to the cylinder.

tanks in industry can have a significant impact on air quality. Afterburners are often the best method for destroying volatile organic chemicals in industrial exhaust stacks. High air-fuel ratios in automobile engines and other burners minimize hydrocarbon and carbon monoxide emissions, but also cause excess nitrogen oxide production. Careful monitoring of air-fuel inputs and oxygen levels in exhaust gases can minimize all these pollutants.

CLEAN AIR LEGISLATION

Throughout history, there have been countless ordinances prohibiting emission of objectionable smoke, odors, and noise. Air pollution traditionally has been treated as a local problem, however, to be regulated by local authorities. The Clean Air Act of 1963 was the first national legislation in the United States aimed at air pollution control. Federal grants were provided to states to combat pollution, but the act was careful to preserve states' rights to set and enforce air quality regulations. It soon became obvious that some pollution problems cannot be solved on a local basis.

In 1970, an extensive set of amendments essentially rewrote the Clean Air Act. These amendments identified the "criteria pollutants" discussed earlier in this chapter, and established national ambient air quality standards. These standards are divided into two categories. **Primary standards** (table 18.3) are intended to protect human health, while **secondary standards** are set to protect materials, crops, climate, visibility, and personal comfort. Primary and secondary standards are generally the same for most pollutants. Ambient standards assume that pollutants have no adverse effects beyond certain thresholds. They also assume that pollutants arising from numerous diverse sources are more reasonably and effectively regulated by setting maximum total levels in the atmosphere than by regulating individual emissions. Some environmentalists disagree with both of these assumptions. These standards are the basis of a warning system called the Air Pollutant Standards Index (table 18.4).

In 1990, after many years of acrimonious debate and political maneuvering, the Clean Air Act was extensively rewritten and updated. Important provisions included the following:

- *Acid rain.* Sulfur dioxide releases will be cut from 24 million tons in 1990 to 10 million tons in 2000 by requiring the 111 largest sulfur emitters to meet strict standards. Nitrogen oxide emissions were reduced from 6 million tons in 1990 to 4 million tons per year.
- *Urban smog.* Motor vehicle tailpipe emissions of hydrocarbons and nitrogen oxides were reduced 35 percent and 60 percent, respectively, in all new cars. Oil companies are required to offer alternative fuels, such as methanol or

TABLE 18.3 National Ambient Air Quality Standards (NAAQS)

POLLUTANT	PRIMARY (HEALTH-BASED) AVERAGING TIME	STANDARD CONCENTRATION
TSP[a]	Annual geometric mean[b]	50 μg/m^3
	24 hours	150 μg/m^3
SO_2	Annual arithmetic mean[c]	80 μg/m^3 (0.03 ppm)
	24 hours	120 μg/m^3 (0.14 ppm)
CO	8 hours	10 mg/m^3 (9 ppm)
	1 hour	40 mg/m^3 (35 ppm)
NO_2	Annual arithmetic mean	80 μg/m^3 (0.05 ppm)
O_3	Daily max 1 hour avg	235 μg/m^3 (0.12 ppm)
Lead	Maximum quarterly avg	1.5 μg/m^3

[a]Total suspended particulate material.
[b]The geometric mean is obtained by taking the *n*th root of the product of *n* numbers. This tends to reduce the impact of a few very large numbers in a set.
[c]An arithmetic mean is the average determined by dividing the sum of a group of data points by the number of points.

TABLE 18.4 Air Pollutant Standards Index

RATING	DESCRIPTION	HEALTH EFFECTS	SUGGESTED ACTIONS
500	Disaster	Very hazardous to all; serious injury and excess deaths, especially in sensitive persons	Stay inside with doors and windows closed; avoid all physical activity
400	Emergency	Hazardous to general population; grave injury possible	Avoid outdoor exercise; young, elderly, and ill should reduce all activity
300	Warning	Very unhealthy for all; serious threat to young, elderly, or ill	Elderly or those with heart or lung disease stay indoors
200	Alert	Irritation of eyes and lungs; aggravation of existing disease	Sensitive persons stay indoors
100	Moderate	NAAQS maximum permissible levels	Avoid traffic and congestion
50	Good	No known short-term effects	No restrictions

ethanol (sometimes called oxygenated fuels), hydrogen, or compressed natural gas (methane) in cities with the worst pollution problems. Cities not meeting air quality standards for ozone and smog are divided into five categories (marginal, moderate, serious, severe, and extreme); deadlines for attaining standards are set for three, six, nine, fifteen, and twenty years, respectively.

- *Toxic air pollutants.* Since the 1990 Clean Air revisions, 188 air toxins have been listed in about 250 source categories (chemical factories, dry cleaners, coke furnaces, printing plants, etc.). The largest polluters will be required to install the best available technology to reduce emissions 90 percent by 2003. The EPA estimates that toxic emissions already have been reduced by about 1 million tons per year by these standards.
- *Ozone protection.* Chlorofluorocarbons and carbon tetrachloride have been phased out. Recovery and recycling programs for existing CFCs have been instituted. Methyl chloroform will be outlawed by 2002. Hydrochlorofluorocarbons in aerosol cans and insulation will be phased out by 2030. Although these regulations already have reduced chlorine emissions, CFC smuggling from developing countries such as Mexico—where they are still cheap and legal—has become a serious problem.
- *Marketing pollution rights.* Corporations are allowed to offset emissions by buying, selling, and "banking" pollution rights from other factories at an expected savings of $2 billion to $3 billion per year. This is a controversial free-market approach that may make economic sense for industry and environmental sense on average but may be disastrous for some localities.
- *Volatile organic compounds.* Rather than specify a single number for maximum admissible emissions or ambient air concentrations of volatile organic chemicals, these compounds now are regulated by hundreds of pages of detailed standards specifying designs of manufacturing or storage facilities, operating conditions, sampling methods, control or mitigation techniques, recordkeeping, and maintenance. These rules stipulate minute details such as specific types of valves and pumps to be used or the color of paint for storage tank tops.

In 1997, yet another set of changes to the U.S. Clean Air Act was approved by Congress and signed into law. Ambient ozone standards will be lowered from 0.12 ppm (1 hr avg.) to 0.08 ppm (8 hr avg.). Soot and dust emission standards will apply to particles 2.5 microns or more in diameter rather than the previous 10 micron limit. To meet these new requirements, factories and power plants will have to clean up their smokestacks; auto pollution will have to be reduced either by getting cars off the road or by switching to new technologies such as electric vehicles. Oil companies will have to cut the sulfur content of gasoline by 90 percent.

New rules have been proposed requiring that sport-utility vehicles, pickups, and minivans meet the same emission standards as cars. In addition, outboard motors, motorcycles, lawnmowers, and personal watercraft (jet-skis and wave runners), snowmobiles, and all-terrain vehicles will probably be regulated. These 2-cycle engines emit more unburned hydrocarbons than all cars and trucks put together.

The EPA estimates that costs of these measures could be as high as $8.5 billion per year, but that they should save 15,000 lives, cut hospital admissions for respiratory illnesses by 9000, and reduce chronic bronchitis cases by 60,000 each year. Industry groups complain that costs will be up to ten times as high and that evidence for health benefits is shaky. Many cities and counties that had finally reached accord with earlier, less stringent standards, will now find themselves once more out of compliance.

The EPA won't fully implement these latest standards for ozone and fine soot until 2008 to give states a chance to set up monitoring systems and to find ways to eliminate pollution in the most cost effective manner. During this time, controversy and political maneuvering over air regulations undoubtedly will continue. As you can see from this history of air pollution control, environmental protection tends to be a long, convoluted process. Regulations often are applied in an incremental and somewhat erratic manner.

California has gone further than the federal government in making specific plans for air pollution control. The South Coast Air Quality Management District has adopted 160 rules to clean the air in the Los Angeles Basin. If these measures are successful, smog-causing emissions could be reduced by 70 percent. The number of days when the air is considered hazardous to breathe would decrease from 150 per year to 0 per year.

Reaching these goals will require substantial lifestyle changes for most Californians. Aerosol hair sprays, deodorants, charcoal lighter fluid, gasoline-powered lawnmowers, and drive-through burger stands could be banned. More than 3000 consumer products including automotive polishes, spot removers, herbicides, lubricants, and floor wax strippers must meet new pollution limits. Paints and cleaning solutions would have to contain fewer volatile solvents. Radial tires and more stringent emission controls would be mandated for automobiles. Clean-burning oxygenated fuels or electric motors would be required for all vehicles. Car pooling would be encouraged, parking lots would be restricted, and limits would be placed on the number of cars a family could have.

California's land-use zones and housing codes might have to be changed to accommodate new commuting patterns. Substantial relocations could result. The cost is estimated to be about 60 cents per person per day. Opponents argue that the price tag could be as high as $15 billion a year and 30,000 lost jobs. Whether Californians, or any of us, care enough about health and the environment to make these changes and pay these costs remains to be seen.

CURRENT CONDITIONS AND FUTURE PROSPECTS

Although we have not yet achieved the Clean Air Act goals in many parts of the United States, air quality has improved dramatically in the last decade in terms of the major large-volume pollutants. For 23 of the largest U.S. cities, the number of days in which

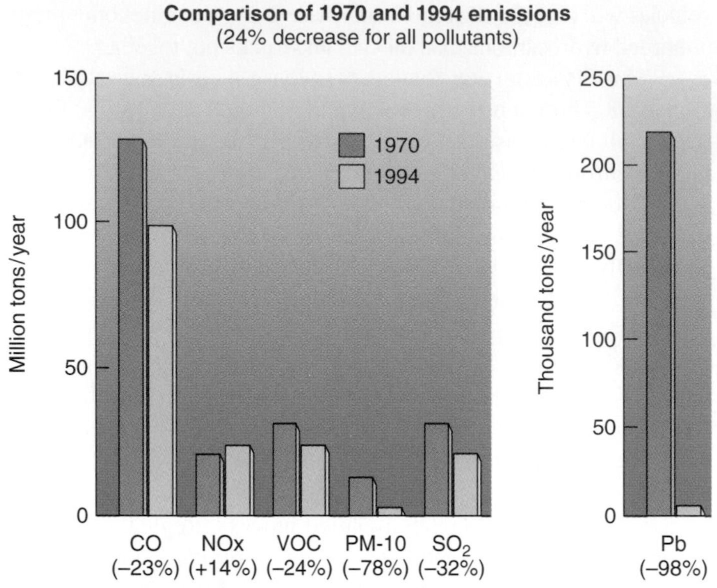

FIGURE 18.26 Air pollution trends in the United States, 1970–1994. Although population and economic activity increased during this period, emissions of all "criteria" air pollutants except for nitrogen oxides decreased significantly.
Source: Data from Environmental Protection Agency.

FIGURE 18.27 Soot from factories and refineries has turned buildings, streets, and even this shepherd's sheep black in Copsa Mica, Romania.

air quality reached the hazardous level (PSI greater than 300) is down 93 percent from an average of 1.8 days/year a decade ago to 0.13 days/year now. Of 97 metropolitan areas that failed to meet clean air standards in the 1980s, 41 were in compliance in 1991-92. For many cities, this was the first time they met air quality goals in 20 years. Still, the EPA estimates that some 86 million Americans breathe unhealthy air at least part of the time.

The EPA estimates that emissions of particulate materials decreased 78 percent, lead fell 98 percent, SO_2 declined 32 percent, and CO shrank 23 percent between 1970 and 1994 (fig. 18.26). Industrial cities, such as Chicago, Pittsburgh, and Philadelphia, that suffered "smokestack" pollution have had 90 percent reductions in number of days exceeding NAAQS maxima. Filters, scrubbers, and precipitators on power plants and other large stationary sources are responsible for most of the particulate and SO_2 reductions. Catalytic converters on automobiles are responsible for most of the CO and O_3 reductions. The only conventional "criteria" pollutant that has not dropped significantly is NO_x, which had risen 300 percent since 1940, and is up 14 percent since 1970.

Because automobiles are the main source of NO_x, cities where pollution is largely from traffic still have serious air quality problems. Los Angeles, Anaheim, and Riverside, California, are among the smoggiest cities in the United States. Industrial cities such as Baltimore, New York City, Chicago, Gary, Houston, Milwaukee, Muskegon, Philadelphia, and San Diego also have continuing problems. In 1999, for the first time in many years, Houston passed Los Angeles as the worst air quality in the country. Eighty-five other urban areas are still considered nonattainment regions. In spite of these local failures, however, 80 percent of the United States now meets the NAAQS goals. This improvement in air quality is perhaps the greatest environmental success story in our history.

The outlook is not so encouraging in other parts of the world, however. The major metropolitan areas of many developing countries are growing at explosive rates to incredible sizes (chapter 24), and environmental quality is abysmal in many of them. The composite average annual levels of SO_2 in Tehran, Iran, for instance, are more than 150 mg/m^3, and peak levels can be up to ten times higher. Mexico City remains notorious for bad air. Pollution levels exceed WHO health standards 350 days per year and more than half of all city children have lead levels in their blood sufficient to lower intelligence and retard development. Its 131,000 industries and 2.5 million vehicles spew out more than 5500 tons of air pollutants daily. Santiago, Chile, averages 299 days per year on which suspended particulates exceed WHO standards of 90 mg/m^3.

While there are few statistics on China's pollution situation, it is known that many of China's 400,000 factories have no air pollution controls. Experts estimate that home coal burners and factories emit 10 million tons of soot and 15 million tons of sulfur dioxide annually and that emissions have increased rapidly over the past 20 years. Sheyang, an industrial city in northern China, is thought to have the world's worst continuing particulate problem with peak winter concentrations over 700 mg/m^3 (nine times U.S. maximum standards). Airborne particulates in Sheyang exceed WHO standards on 347 days per year. Beijing, Xian, and Guangzhou are nearly as bad. The high incidence of cancer in Shanghai is thought to be linked to air pollution. (fig. 18.27).

As political walls came down across Eastern Europe and the Soviet Union at the end of the 1980s, horrifying environmental conditions in these centrally planned economies were revealed. Inept industrial managers, a rigid bureaucracy, and lack of democracy have created ecological disasters. Where governments own, operate, and regulate industry, there are few checks and balances or incentives to clean up pollution. Much of the Eastern bloc depends heavily on soft brown coal for its energy and pollution controls are absent or highly inadequate.

Southern Poland, northern Czech Republic, and Slovakia are covered most of the time by a permanent cloud of smog from factories and power plants. Acid rain is eating away historic buildings and damaging already inadequate infrastructures. The haze is so dark that drivers must turn on their headlights during the day. Residents complain that washed clothes turn dirty before they can dry. Zabrze, near Katowice in southern Poland, has particulate emissions of 3600 metric tons per square kilometer. This is more than seven times the emissions in Baltimore, Maryland, or Birmingham, Alabama, the dirtiest cities (for particulates) in the United States. Home gardening in Katowice has been banned because vegetables raised there have unsafe levels of lead and cadmium.

For miles around the infamous Romanian "black town" of Copsa Mica, the countryside is so stained by soot that it looks as if someone had poured black ink over everything. Birth defects afflict 10 percent of infants in northern Bohemia. Workers in factories there get extra hazard pay—burial money, they call it. Life expectancy in these industrial towns is as much as ten years less than the national average. Espenhain, in the industrial belt of the former East Germany, has one of the world's highest rates of sulfur dioxide pollution. One of every two children has lung problems, and one of every three has heart problems. Brass doorknobs and name plates have been eaten away by the acidic air in just a few months.

Not all is pessimistic, however. There have been some spectacular successes in air pollution control. Sweden and West Germany (countries affected by forest losses due to acid precipitation) cut their sulfur emissions by two-thirds between 1970 and 1985. Austria and Switzerland have gone even further. They even regulate motorcycle emissions. The Global Environmental Monitoring System (GEMS) reports declines in particulate levels in 26 of 37 cities worldwide. Sulfur dioxide and sulfate particles, which cause acid rain and respiratory disease, have declined in 20 of these cities.

Ten years ago, Cubatao, Brazil, was described as the "Valley of Death," one of the most dangerously polluted places in the world. A steel plant, a huge oil refinery, and fertilizer and chemical factories churned out thousands of tons of air pollutants every year. Trees died on the surrounding hills. Birth defects and respiratory diseases were alarmingly high. Since then, however, the citizens of Cubatao have made remarkable progress in cleaning up their environment. The end of military rule and restoration of democracy allowed residents to publicize their complaints. The environment became an important political issue. The state of São Paulo invested about $100 million, and the private sector spent twice as much to clean up most pollution sources in the valley. Particulate pollution was reduced 75 percent. Ammonia emissions were reduced 97 percent, hydrocarbons that cause ozone and smog were cut 86 percent, and sulfur dioxide production fell 84 percent. Fish are returning to the rivers, and forests are regrowing on the mountains. Progress is possible! We hope that similar success stories will be obtainable elsewhere.

SUMMARY

In this chapter, we have looked at major categories, types, and sources of air pollution. We have defined air pollution as chemical or physical changes brought about by either natural processes or human activities, resulting in air quality degradation. Air pollution has existed as long as there has been an atmosphere. Perhaps the first major human source of air pollution was fire. Burning forests, fossil fuels, and wastes continues to be the largest source of anthropogenic (human-caused) air pollution. The seven conventional large-volume pollutants are NO_x, SO_2, CO, lead, particulates, volatile organic compounds, and photochemical oxidants. The major sources of air pollution are transportation, industrial processes, stationary fuel combustion, and solid waste disposal.

We also looked at some unconventional pollutants. Indoor air pollutants, including formaldehyde, asbestos, toxic organic chemicals, radon, and tobacco smoke may pose a greater hazard to human health than all of the conventional pollutants combined. Odors, visibility losses, and noise generally are not life-threatening but serve as indicators of our treatment of the environment.

Some atmospheric processes play a role in distribution, concentration, chemical modification, and elimination of pollutants. Among the most important of these processes are long-range transport of pollutants and photochemical reactions in trapped inversion layers over urban areas.

Encouraging improvements have been made in ambient outdoor air quality over most of the United States in the last few decades. We have made considerable progress in designing and installing pollution-control equipment to reduce the major conventional pollutants. There are many types of scrubbers, filters, catalysts, fuel modification processes, and new burning techniques for controlling pollution. The Clean Air Act regulates air quality in the United States through both ambient standards and emission limits, and new amendments promise a dramatic improvement in our atmosphere. There is much yet to be done, especially in developing countries and in Eastern Europe, but air pollution control is, perhaps, our greatest success in environmental protection and an encouraging example of what can be accomplished in this field.

QUESTIONS FOR REVIEW

1. Define primary and secondary air pollutants.
2. What are the seven "criteria" pollutants in the original Clean Air Act? Why were they chosen?
3. What are the two main categories of materials in the EPA's list of 33 air toxics of greatest concern?
4. What is acid deposition? What causes it?
5. What is an atmospheric inversion and how does it trap air pollutants?
6. How do electrostatic precipitators, baghouse filters, flue gas scrubbers, and catalytic converters work?
7. What is the difference between ambient standards and emission limits?
8. Describe the health effects and suggested actions for each of the levels of the pollution standards index (PSI).
9. Which of the conventional pollutants has decreased most in the recent past and which has decreased least?

QUESTIONS FOR CRITICAL THINKING

1. Some authors limit pollution to human-caused materials. If a natural source releases the same toxic chemicals as a factory, is one pollution and the other not?
2. What might be done to improve indoor air quality? Should the government mandate such changes?
3. Why do you suppose that air pollution is so much worse in Eastern Europe than in the West?
4. Suppose air pollution causes a billion dollars in crop losses each year but controlling the pollution would also cost a billion dollars. Should we insist on controls?
5. Utility managers once claimed that it would cost $1000 per fish to control acid precipitation in the Adirondack lakes and that it would be cheaper to buy fish for anglers than to put scrubbers on power plants. Suppose that was true. Does it justify continuing pollution?
6. The ban on chlorofluorocarbons may mean that billions of people won't be able to have refrigerators and that millions will die each year from food poisoning. Is that justified by saving stratospheric ozone?
7. Is it possible to have zero emissions of pollutants? What does zero mean in this case?
8. If there are thresholds for pollution effects (at least as far as we know now), is it reasonable or wise to depend on environmental processes to disperse, assimilate, or inactivate waste products?
9. Should the EPA ban all 2-cycle engines (chain saws, outboard motors, lawnmowers, personal watercraft)? Why or why not?
10. Do you think that we should continue to use ambient air quality standards or change to absolute emission standards for all pollutants?

KEY TERMS

acid precipitation 410
aerosol 401
aesthetic degradation 402
ambient air 398
bronchitis 408
carbon monoxide 401
chlorofluorocarbons 407
conventional or criteria pollutants 398
dry alkali injection 414
electrostatic precipitators 413
emission standards 402
emphysema 408
filters 413
fugitive emissions 398
nitrogen oxides 400
ozone 402
particulate material 401
photochemical oxidants 402
primary pollutants 398
primary standards 416
secondary pollutants 398
secondary standards 416
sulfur dioxide 398
synergistic effects 409
temperature inversions 404
unconventional or noncriteria pollutants 402
volatile organic compounds 401

ADDITIONAL INFORMATION ON THE INTERNET

Visit our website at http://www.mhhe.com/environmentalscience for specific information about each of the topics below:

Air Pollution and Health
Environmental Protection Agency
National Air and Radiation Environmental Laboratory
National Vehicle and Fuel Emissions Laboratory
Ozone Depletion: Imagery
Ozone Depletion, from the Environmental Protection Agency (EPA)
Plain English Guide to the Clean Air Act. Information from the EPA
EPA's Indoor Air Quality (AIQ) Home Page
EPA Office of Mobile Sources (Motor Vehicles, Fuels, Engines)
10 Tips for Drivers. 10 things you can do to prevent pollution from your vehicle

CHAPTER 19

Water Use and Management

"If there is magic on this planet, it is in water."

—Loren Eisley—

OBJECTIVES

After studying this chapter, you should be able to:

- summarize how the hydrologic cycle delivers fresh water to terrestrial ecosystems and how the cycle balances over time.
- contrast the volume and residence time of water in the earth's major compartments.
- describe the important ways we use water and distinguish between withdrawal, consumption, and degradation.
- appreciate the causes and consequences of water shortages around the world and what they mean in people's lives in water-poor countries.
- debate the merits of proposals to increase water supplies and manage demand.
- apply some water conservation methods in your own life.

ESSENTIAL STUDY PARTNER

The Essential Study Partner (ESP) CD-ROM that accompanies this textbook contains additional information on the following topics:

- freshwater availability: physical resources: water: overview
- water cycle: physical resources: water: quantity
- groundwater layers, and depletion of aquifers: physical resources: water: quantity
- precipitation and human habitation: physical resources: water: quantity
- global and domestic water use: physical resources: water: quantity
- water waste, practices and habits: physical resources: water: quantity
- dams, reservoirs, desalinization and drip irrigation: physical resources: water: quantity

Above: Yellowstone Lake glistens on a calm summer morning.

Where Has the River Gone?

For more than 5000 years, the Huang He, or Yellow River, has been the center of Chinese civilization. The major water supply for the arid North China Plain, the river also brings fertile soil from the mountains to replenish crop fields (fig. 19.1). But the river is known, too, as "China's sorrow" because of its frequent, catastrophic floods. More than 1500 times in the last five millennia, the river has breached its dikes, turning the surrounding plain into a broad, shallow lake. These floods often resulted in dramatic realignments of the river channel. The last of these was in 1947, when the river shifted its mouth from the Yellow Sea, near Gansu, more than 320 km (200 mi) north to the Gulf of Zhili.

Water to irrigate paddy rice has long been the key to feeding China's people. With only 10 percent of its land arable, China could feed only a fraction of its population of 1.2 billion without irrigation. But with rapidly growing urban populations and increasing industrialization that compete for scarce water supplies, it isn't clear that that there will be enough of this precious resource to go around. This is especially true in the arid North, where the water table beneath much of the farm land that currently supplies about 40 percent of China's food and feeds nearly 500 million people has dropped an average of 1.5 m (roughly 5 ft) per year for the last decade.

FIGURE 19.1 Flowing more than 4800 km (3000 mi) from the high Tibetian Plateau, across the southern Gobi Desert, to the Yellow Sea, the Huang He, or Yellow River is the major water source for much of the arid North China Plain. Currently, agricultural, industrial, and domestic water withdrawals drain the river dry for several months each year.

So much water is being diverted from the Huang He, that the river now frequently dries up completely in its lower stretches. Since 1985, the river has failed to reach the sea every year for at least a few months. In 1996, the river failed for 133 days. In 1997, a year plagued by drought in north China, there was no water in the lower Huang He for 226 days. Unless ways are found to conserve water or augment river flows, drastic changes will have to be made in Chinese agriculture, industry, and domestic water use. Some cities may have to be relocated. Farms and factories will close. Grain shortages in China could drive up world food prices with disastrous consequences for poor people everywhere.

The Huang He isn't the only river in the world that sometimes dries up completely because of human diversions and excess consumption. The mighty Colorado that roars through Arizona's Grand Canyon, is so depleted by the time it reaches the Mexican border that it seeps across the desert in a salty trickle that disappears entirely long before it reaches the Sea of Cortez. Similarly, the Amu Dar'ya and Syr Dar'ya in Kazakhstan were diverted completely in the 1970s, causing the Aral Sea to lose about two-thirds of its volume. Even the mighty Nile, the longest river in the world, and the sacred Ganges, are so overexploited that at some times of the year, hardly any of their water reaches the ocean.

The disappearance of these once mighty rivers is only a symptom of a larger dilemma. In many countries around the world, water shortages are becoming increasingly common. The United Nations warns that water supplies are likely to become one of the most pressing resource issues of the twenty-first century. Already, some 2 billion people—one-third of the world's population—live in countries with moderate to high water stress. By 2025, two-thirds of all humans could be living in countries where water supplies are inadequate. Some sociologists warn that we could have water wars in the near future; others claim that we already have. In this chapter, we will look at the processes that supply fresh water and how humans access and use it. We will survey major water compartments of the environment and see how they are depleted by human activities. Finally, we will examine some ways that we can conserve and extend this vital resource.

WATER RESOURCES

Water is a marvelous substance—flowing, rippling, swirling around obstacles in its path, seeping, dripping, trickling—constantly moving from sea to land and back again. Water can be clear, crystalline, icy green in a mountain stream (fig. 19.2) or black and opaque in a cypress swamp. Water bugs skitter across the surface of a quiet lake; a stream cascades down a stairstep ledge of rock; waves roll endlessly up a sand beach, crash in a welter of foam, and recede. Rain falls in a gentle mist, refreshing plants and animals. A violent thunderstorm floods a meadow, washing away stream banks.

Water is essential for life (chapter 3). It is the medium in which all living processes occur. Water dissolves nutrients and distributes them to cells, regulates body temperature, supports structures, and removes waste products. About 60 percent of your body is water. You could survive for weeks without food, but only a few days without water.

The earth is the only place in the universe, as far as we know, where liquid water exists in substantial quantities. Oceans, lakes, rivers, glaciers, and other bodies of liquid or solid water cover more than 70 percent of our world's surface. The total amount of water on our planet is immense—more than 1404 million cu km (370

FIGURE 19.2 Why do we find water so pleasing and so fascinating? Why are its many forms so central to natural and human systems? And what would life be like if we didn't have abundant supplies of clean, pure water?

billion billion gal) (table 19.1). If the earth had a perfectly smooth surface, an ocean about 3 km (1.9 mi) deep would cover everything. Fortunately for us, continents rise above the general surface level, creating dry land over about 30 percent of the planet.

What's the source of all this water? Most geologists believe it comes from rock weathering and volcanic activity. Some meteorologists claim that thousands of small, icy comets enter the earth's atmosphere each day, adding significant amounts of water to our supplies. Evidence for this hypothesis has been controversial, however.

The Hydrologic Cycle

> All rivers run into the sea, yet the sea is not full: Unto the place from which rivers come, thither they return again.
>
> —Ecclesiastes 1:7—

The hydrologic cycle (water cycle) describes the circulation of water as it evaporates from land, water, and organisms; enters the atmosphere; condenses and is precipitated to the earth's surfaces; and moves underground by infiltration or overland by runoff into rivers, lakes, and seas (chapter 3). The total amount of water on earth remains about the same from year to year, and the hydrologic cycle simply moves it from one place to another (fig. 19.3). This process supplies fresh water to the landmasses while also playing a vital role in creating a habitable climate and moderating world temperatures. Movement of water back to the sea in rivers and glaciers is a major geological force that shapes the land and redistributes material. Plants play an important role in the hydrologic cycle, absorbing groundwater and pumping it into the atmosphere by transpiration (transport plus evaporation). In tropical forests, as much as 75 percent of annual precipitation is returned to the atmosphere by plants.

TABLE 19.1 Some Units of Water Measurement

One cubic kilometer (km^3) equals one billion cubic meters (m^3), one trillion liters, or 264 billion gallons.

One acre-foot is the amount of water required to cover an acre of ground one foot deep. This is equivalent to 325,851 gallons, or 1.2 million liters, or 1234 m^3, about the amount consumed annually by a family of four in the United States.

One cubic foot per second of river flow equals 28.3 liters per second or 449 gallons per minute.

Solar energy drives the hydrologic cycle by evaporating surface water. **Evaporation** is the process in which a liquid is changed to vapor (gas phase) at temperatures well below its boiling point. Water also can move between solid and gaseous states without ever becoming liquid in a process called **sublimation.** On bright, cold, windy winter days, when the air is very dry, snowbanks disappear by sublimation, even though the temperature never gets above freezing. This is the same process that causes "freezer burn" of frozen foods.

In both evaporation and sublimation, molecules of water vapor enter the atmosphere, leaving behind salts and other contaminants and thus creating purified fresh water. This is essentially distillation on a grand scale. We used to think of rainwater as a symbol of purity, a standard against which pollution could be measured. Unfortunately, increasing amounts of atmospheric pollutants are picked up by water vapor as it condenses into rain.

The amount of water vapor in the air is called humidity. Warm air can hold more water than cold air. When a volume of air contains as much water vapor as it can at a given temperature, we say that it has reached its **saturation point. Relative humidity** is the amount of water vapor in the air expressed as a percentage of the maximum amount (saturation point) that could be held at that particular temperature.

When the saturation concentration is exceeded, water molecules begin to aggregate in the process of **condensation.** If the temperature at which this occurs is above 0°C, tiny liquid droplets result. If the temperature is below freezing, ice forms. For a given amount of water vapor, the temperature at which condensation occurs is the **dew point.** Tiny particles, called **condensation nuclei,** float in the air and facilitate this process. Smoke, dust, sea salts, spores, and volcanic ash all provide such particles. Even apparently clear air can contain large numbers of these particles, which are generally too small to be seen by the naked eye. Sea salt is an excellent source of such nuclei, and heavy, low clouds frequently form in the humid air over the ocean. Some nucleating agents are so efficient at accumulating water that they can cause precipitation even when the air is far below its saturation point.

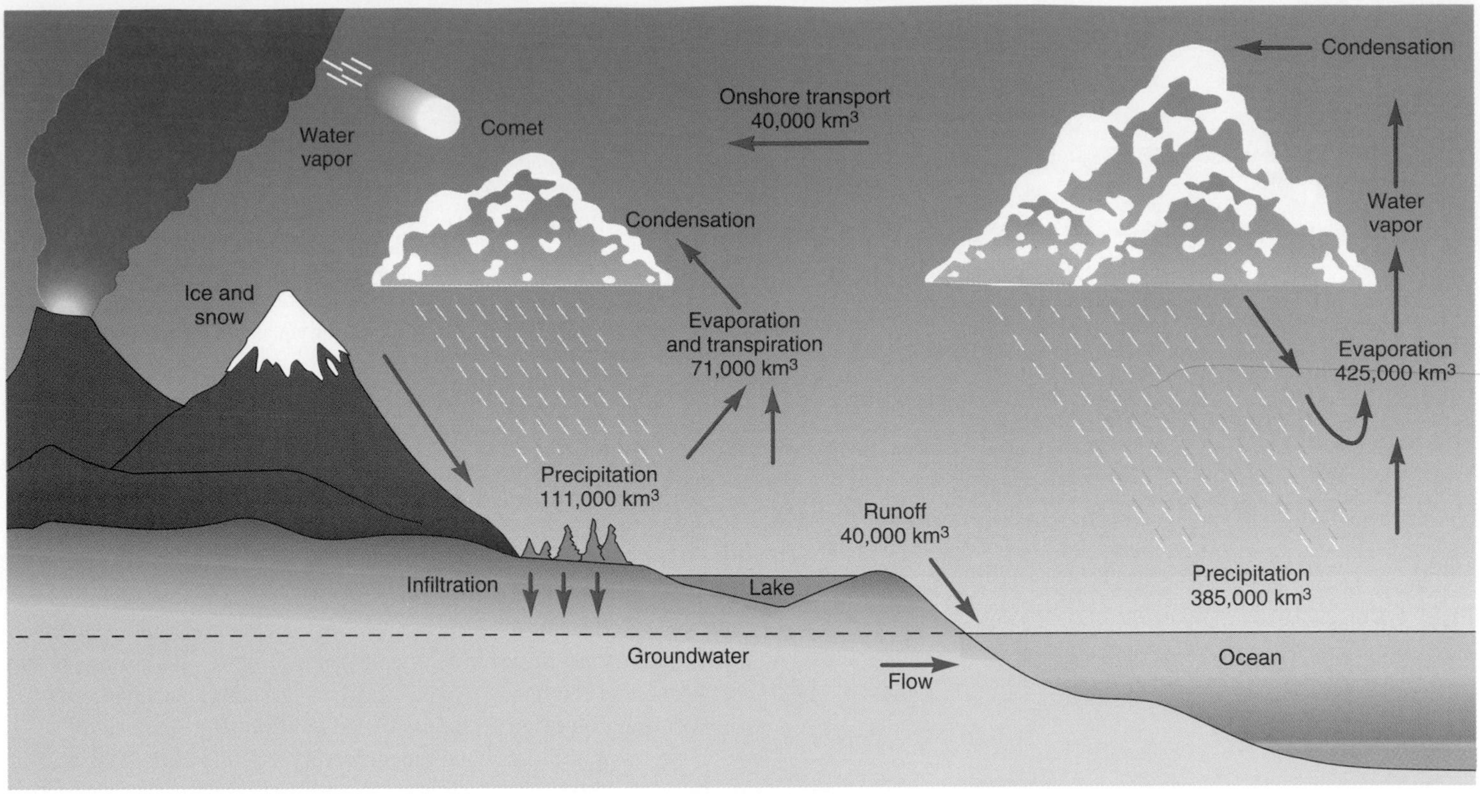

FIGURE 19.3 In the hydrologic cycle water moves constantly between aquatic, atmospheric, and terrestrial compartments driven by solar energy and gravity. The total annual runoff from land to the oceans is about 10.3×10^{15} gallons.

A cloud, then, is an accumulation of condensed water vapor in droplets or ice crystals. Normally, cloud particles are small enough to remain suspended in the air, but when cloud droplets and ice crystals become large enough, gravity overcomes uplifting air currents and precipitation occurs. Some precipitation never reaches the ground. Temperatures and humidities in the clouds where snow and ice form are ideal for their preservation, but as they fall through lower, warmer, and drier air layers, reevaporation occurs. Rising air currents lift this water vapor back into the clouds, where it condenses again; thus, liquid water and ice crystals may exist for only a few minutes in this short cycle between clouds and air (fig. 19.3).

Rainfall and Topography

Rain falls unevenly over the planet. In some places, it rains more or less constantly, while other areas get almost no precipitation of any kind. At Iquique, in the Chilean desert, for instance, no rain has fallen in recorded history. At the other end of the scale, 22 m (72 ft) of rain was recorded in a single year at Cherrapunji in India. Figure 19.4 shows broad patterns of precipitation around the world.

Very heavy rainfall is typical of tropical areas, especially where monsoon winds carry moisture-laden sea air onshore (chapter 17). Mountains act as both cloud formers and rain catchers. As air sweeps up the windward side of a mountain, pressure decreases and temperature falls, causing relative humidity to increase. Eventually, the air is supersaturated with moisture and condensation occurs. Further cooling of the air causes droplets of moisture to coalesce and become too heavy to remain suspended, so they fall as rain. The now cooler and drier air continues over the mountain to the leeward side. It descends and warms once again, reducing its relative humidity even further, not only preventing rainfall there, but absorbing moisture from other sources. As a result, a mountain range generally has two distinct climatic personalities. The windward side is usually cool, wet, and cloudy, while the leeward side is warm, dry, and sunny. We call the dry area on the downwind side of a mountain its **rain shadow.**

A striking example of this dichotomy is found in the Hawaiian Islands (fig. 19.5). The windward side of Mount Waialeale on the island of Kauai is one of the wettest places on earth, with an annual rainfall near 12 m (460 in). The leeward side, only a few kilometers away, is in the rain shadow of the mountain and has an average yearly rainfall of only 46 cm (18 in). On a broader scale, some mountain ranges cast rain shadows over vast areas. The Himalaya and Karakorum ranges of south Asia block moisture-laden monsoon winds from reaching central Asia. The Sierra Nevada of California and the coastal ranges of Oregon and Washington intercept moisture-laden Pacific winds, resulting in the arid intermountain Great Basin of the western United States.

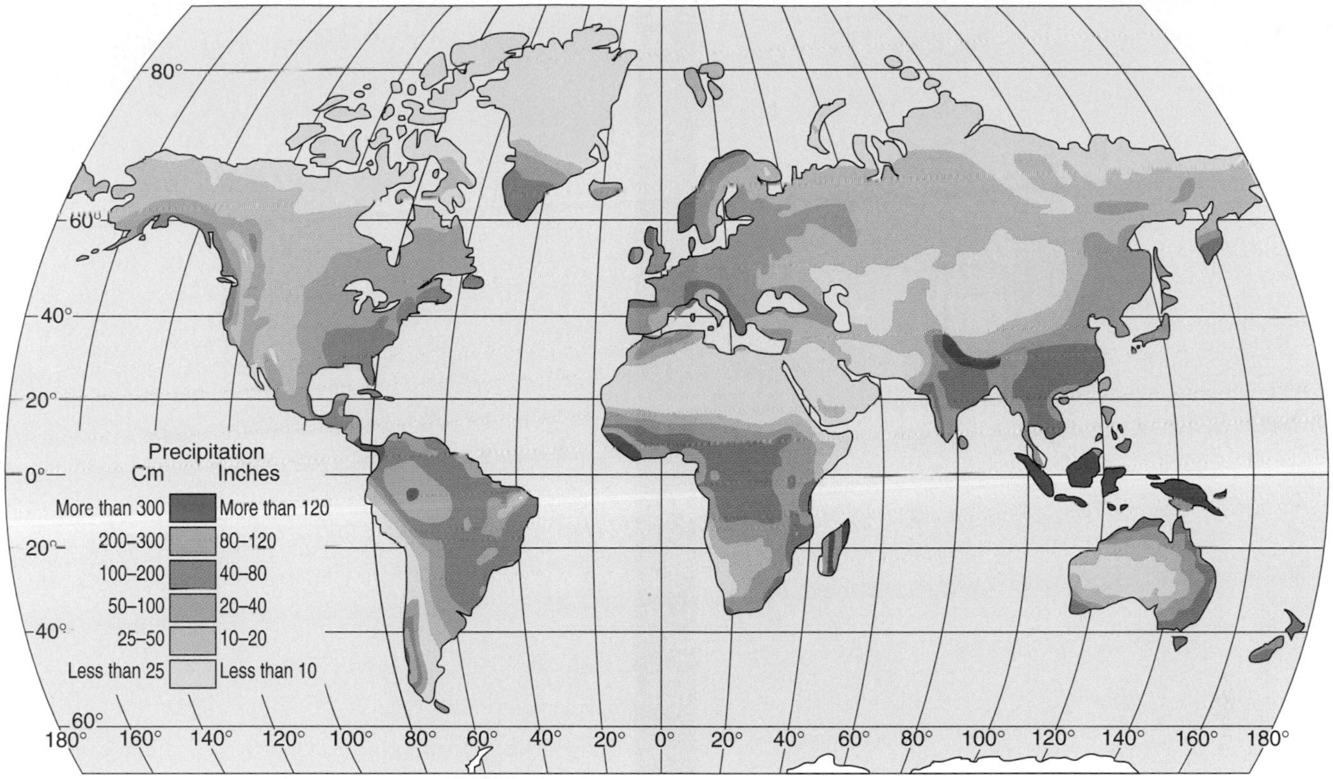

FIGURE 19.4 Mean annual precipitation. Note wet areas that support tropical rainforests occur along the equator while the major world deserts occur in zones of dry, descending air between 20 and 40 degrees North and South.

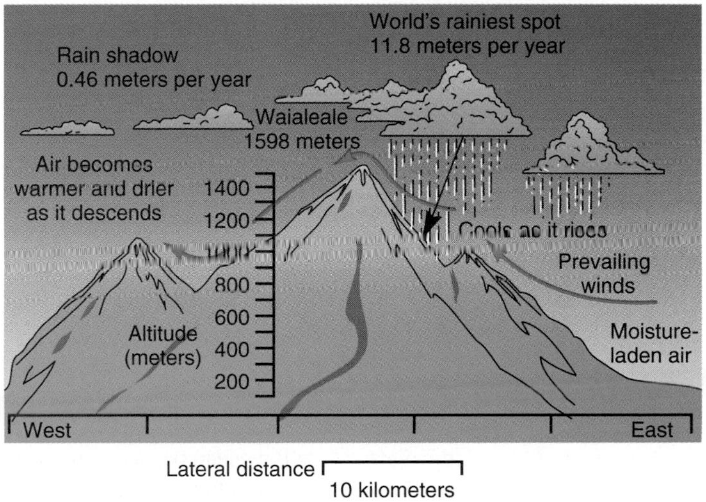

FIGURE 19.5 Rainfall on the east side of Mount Waialeale in Hawaii is more than twenty times as much as on the west side. Prevailing trade winds bring moisture-laden sea air onshore. The air cools as it rises up the flanks of the mountain and the water it carries precipitates as rain—11.8 m (38 ft) per year!

Desert Belts

Rising and falling air masses that result from global circulation patterns also help create deserts in two broad belts on either side of the equator around the world. Evaporation is highest near the equator where direct rays of the sun produce the greatest heat budgets. Hot air over the equator rises, cools, and drops its moisture as rain; thus, equatorial regions are areas of high precipitation. As this cooler, drier air moves toward the poles, it condenses and sinks earthward again along the Tropics of Cancer and Capricorn (23° north and south latitude, respectively), warming as it descends.

This hot, dry air causes high evaporative losses in these subtropical regions and creates great deserts on nearly every continent: the Sahara of North Africa, the Takla Makan and Gobi of China, the Sonoran and Chihuahuan of Mexico and the United States, the Kalahari and Namib of Southwest Africa, and Australia's Great Sandy Desert. Dry air cascading down the west side of the Andes creates some of the driest deserts in the world along the coasts of Chile and Peru.

Humans and domestic animals have expanded many of these deserts by destroying forests and stripping away protective vegetation from once fertile lands, exposing the bare soil to erosion. Local weather patterns and water supplies also are adversely affected by this process of desertification. Weather and climate are discussed in chapter 17.

TABLE 19.2 Earth's Water Compartments—Estimated Volume of Water in Storage, Percent of Total, and Average Residence Time

	VOLUME (THOUSANDS OF KM^3)	% TOTAL WATER	AVERAGE RESIDENCE TIME
Total	1,403,377	100	2800 years
Ocean	1,370,000	97.6	3000 years to 30,000 years*
Ice and snow	29,000	2.07	1 to 16,000 years*
Groundwater down to 1 km	4000	0.28	From days to thousands of years*
Lakes and reservoirs	125	0.009	1 to 100 years*
Saline lakes	104	0.007	10 to 1000 years*
Soil moisture	65	0.005	2 weeks to a year
Biological moisture in plants and animals	65	0.005	1 week
Atmosphere	13	0.001	8 to 10 days
Swamps and marshes	3.6	0.003	From months to years
Rivers and streams	1.7	0.0001	10 to 30 days

*Depends on depth and other factors.
Source: Data from U.S. Geological Survey.

Balancing the Water Budget

Everything about global hydrological processes is awesome in scale. Each year, the sun evaporates approximately 496,000 km^3 of water from the earth's surface. More water evaporates in the tropics than at higher latitudes, and more water evaporates over the oceans than over land. Although the oceans cover about 70 percent of the earth's surface, they account for 86 percent of total evaporation. Ninety percent of the water evaporated from the ocean falls back on the ocean as rain. The remaining 10 percent is carried by prevailing winds over the continents where it combines with water evaporated from soil, plant surfaces, lakes, streams, and wetlands to provide a total continental precipitation of about 111,000 km^3.

What happens to the surplus water on land—the difference between what falls as precipitation and what evaporates? Some of it is incorporated by plants and animals into biological tissues. A large share of what falls on land seeps into the ground to be stored for a while (from a few days to many thousands of years) as soil moisture or groundwater. Eventually, all the water makes its way back downhill to the oceans. The 40,000 km^3 carried back to the ocean each year by surface runoff or underground flow represents the renewable supply available for human uses and sustaining freshwater-dependent ecosystems.

The global water budget thus is balanced by circulation systems on land, in the atmosphere, and in the oceans that move water from areas of excess to areas of deficit. Rivers that carry water from the land to the sea are balanced by wind currents flowing in great swirls above the earth, moving moisture-laden air from one region to another. Ocean currents are equivalent to vast rivers, carrying warm water from the equator to higher latitudes on the surface and returning cold, nutrient-rich waters in deep currents. The Gulf Stream, which flows along the east coast of North America at a steady rate of 10 to 12 km per hour (6 to 7.5 mph), carries more than 100 times more water than all the rivers on land put together.

The redistribution of heat that results from the massive evaporation, precipitation, and transport of water is a major factor in keeping world temperatures relatively constant and making the world habitable. Without oceans to absorb and store heat, and wind currents to redistribute that heat in the latent energy of water vapor, the earth would probably undergo extreme temperature fluctuations like those of the moon, where it is 100°C (212°F) during the day and -130°C (-200°F) at night. Water is able to perform this vital function because of its unique properties in heat absorption and energy of vaporization (chapter 3).

MAJOR WATER COMPARTMENTS

The distribution of water often is described in terms of interacting compartments in which water resides for short or long times. Table 19.2 shows the major water compartments in the world.

Oceans

Together, the oceans contain more than 97 percent of all the *liquid* water in the world. (The water of crystallization in rocks is far larger than the amount of liquid water.) While the ocean basins really form a continuous reservoir, shallows and narrows between them reduce water exchange, so they have different compositions, climatic effects, and even different surface elevations. Oceans play a crucial role in moderating the earth's temperature (fig. 19.6), but they are generally too salty for most human uses. Nevertheless, over 90 percent of the world's living biomass is contained in the oceans.

FIGURE 19.6 The cold Japanese current keeps water temperatures low year round along the coast of Washington and Oregon but also delivers nutrients that make this a highly productive aquatic ecosystem.

In tropical seas, surface waters are warmed by the sun, diluted by rainwater and runoff from the land, and aerated by wave action. In higher latitudes, surface waters are cold and much more dense. This dense water subsides or sinks to the bottom of deep ocean basins and flows toward the equator. Warm surface water of the tropics stratifies or floats on top of this cold, dense water like cream on an unstirred cup of coffee. Sharp boundaries form between different water densities, different salinities, and different temperatures, retarding mixing between these layers.

While parts of the hydrologic cycle occur on a time scale of hours or days, other parts take centuries. The average **residence time** of water in the ocean (the length of time that an individual molecule spends circulating in the ocean before it evaporates and starts through the hydrologic cycle again) is about 3000 years. In the deepest ocean trenches, movement is almost nonexistent and water may remain undisturbed for tens of thousands of years.

Glaciers, Ice, and Snow

Of the 2.4 percent of all water that is fresh, nearly 90 percent is tied up in glaciers, ice caps, and snowfields (fig. 19.7). Glaciers are really rivers of ice flowing downhill very slowly (fig. 19.8). They now occur only at high altitudes or high latitudes, but as recently

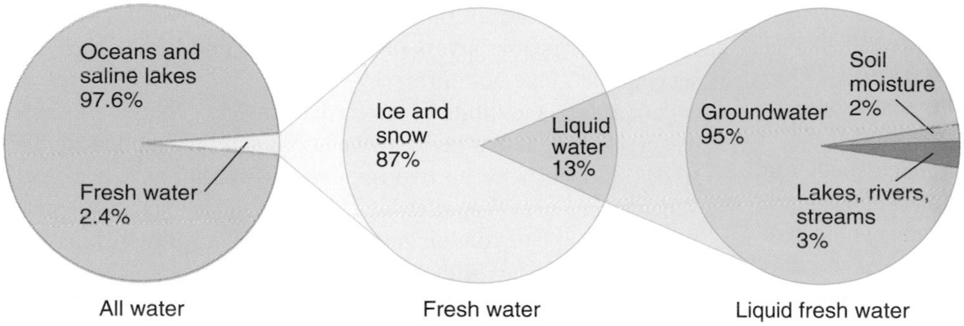

FIGURE 19.7 The easily accessible water in lakes, rivers, and streams represents only 3 percent of all liquid freshwater, which is 13 percent of all fresh water, which is 2.4 percent of all water.

FIGURE 19.8 Glaciers are rivers of ice sliding very slowly downhill. Together, polar ice sheets and alpine glaciers contain more than three times as much fresh water as all the lakes, ponds, streams, and rivers in the world. The dark streaks on the surface of this Alaskan glacier are dirt and rocks marking the edges of tributary glaciers that have combined to make this huge flow.

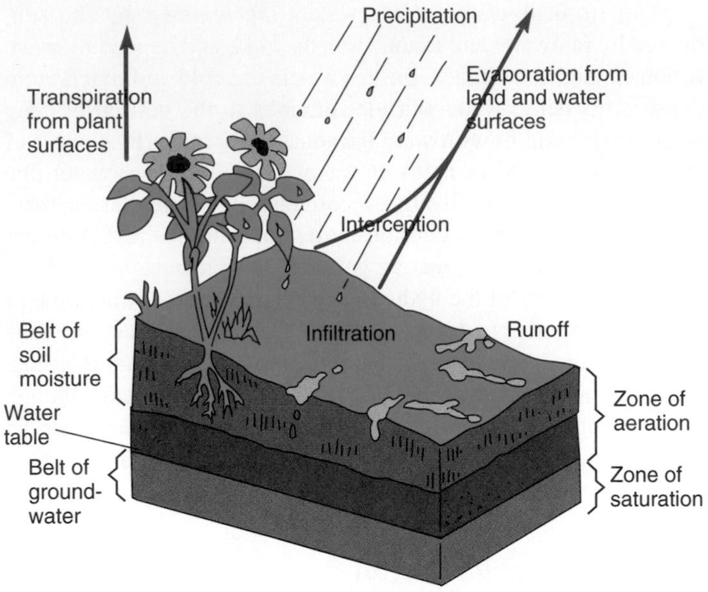

FIGURE 19.9 Precipitation that does not evaporate or run off over the surface percolates through the soil in a process called infiltration. The upper layers of soil hold droplets of moisture between air-filled spaces. Lower layers, where all spaces are filled with water, make up the zone of saturation or groundwater.

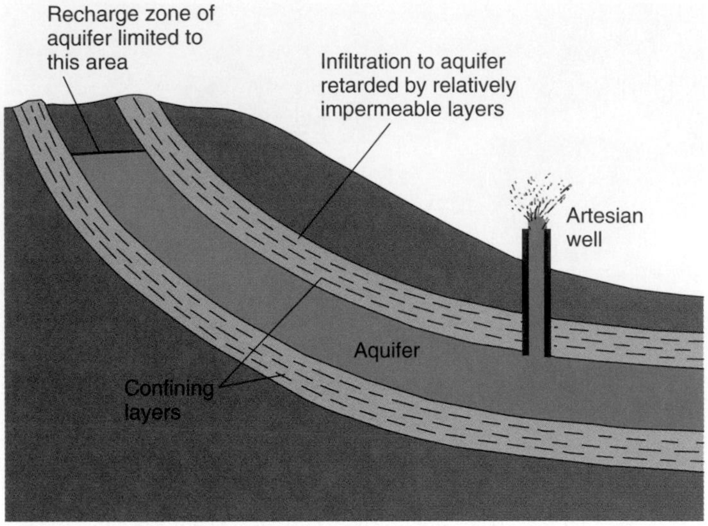

FIGURE 19.10 An aquifer is a porous, water-bearing layer of sand, gravel, or rock. This aquifer is confined between layers of rock or clay and bent by geological forces, creating hydrostatic pressure. A break in the overlying layer creates an artesian well or spring.

as 18,000 years ago about one-third of the continental landmass was covered by glacial ice sheets. Most of this ice has now melted and the largest remnant is in Antarctica. As much as 2 km (1.25 mi) thick, the Antarctic glaciers cover all but the highest mountain peaks and contain nearly 85 percent of all ice in the world.

An ice sheet that is similar in thickness but much smaller in volume covers most of Greenland. There is no landmass at the North Pole. A permanent ice pack made of floating sea ice covers much of the Arctic Ocean. Although sea ice comes from ocean water, salt is excluded in freezing so the ice is mostly fresh water. Together with the Greenland ice sheet, arctic ice makes up about 10 percent of the total ice volume. The remaining 5 percent of the world's permanent supply of ice and snow occurs mainly on high mountain peaks.

Groundwater

After glaciers, the next largest reservoir of fresh water is held in the ground as **groundwater.** Precipitation that does not evaporate back into the air or run off over the surface percolates through the soil and into fractures and spaces of permeable rocks in a process called **infiltration** (fig. 19.9). Upper soil layers that hold both air and water make up the **zone of aeration.** Moisture for plant growth comes primarily from these layers. Depending on rainfall amount, soil type, and surface topography, the zone of aeration may be very shallow or quite deep. Lower soil layers where all spaces are filled with water make up the **zone of saturation.** The top of this zone is the **water table.** The water table is not flat, but undulates according to the surface topography and subsurface structure. Nor is it stationary through the seasons, rising and falling according to precipitation and infiltration rates.

Porous layers of sand, gravel, or rock lying below the water table are called **aquifers.** Aquifers are always underlain by relatively impermeable layers of rock or clay that keep water from seeping out at the bottom.

Folding and tilting of the earth's crust by geologic processes can create shapes that generate water pressure in confined aquifers (those trapped between two impervious, confining rock layers). When a pressurized aquifer intersects the surface, or if it is penetrated by a pipe or conduit, an **artesian** well or spring results from which water gushes without being pumped.

Areas in which infiltration of water into an aquifer occurs are called **recharge zones** (fig. 19.10). The rate at which most aquifers are refilled is very slow, however, and groundwater presently is being removed faster than it can be replenished in many areas. Urbanization, road building, and other development often block recharge zones and prevent replenishment of important aquifers. Contamination of surface water in recharge zones and seepage of pollutants into abandoned wells have polluted aquifers in many places, making them unfit for most uses (chapter 20). Many cities protect aquifer recharge zones from pollution or development, both as a way to drain off rainwater and as a way to replenish the aquifer with pure water.

Some aquifers contain very large volumes of water. The groundwater within 1 km of the surface in the United States is more than 30 times the volume of all the freshwater lakes, rivers, and reservoirs on the surface. While water can flow through limestone caverns in underground rivers, most movement in aquifers is a dispersed and almost imperceptible trickle through tiny fractures and spaces. Depending on geology, it can take anywhere from a few hours to several years for contaminants to move a few hundred meters through an aquifer (chapter 10).

Rivers and Streams

Precipitation that does not evaporate or infiltrate into the ground runs off over the surface, drawn by the force of gravity back toward the sea. Rivulets accumulate to form streams, and streams join to form rivers. Although the total amount of water contained at any one time in rivers and streams is small compared to the other water reservoirs of the world (see table 19.2), these surface waters are vitally important to humans and most other organisms. Most rivers, if they were not constantly replenished by precipitation, meltwater from snow and ice, or seepage from groundwater, would begin to diminish in a few weeks.

The speed at which a river flows is not a very good measure of how much water it carries. Headwater streams are usually small and fast, often tumbling downhill in a continuous cascade. As the stream reaches more level terrain, it slows and generally becomes deeper and more quiet. The best measure of the volume carried by a river is its **discharge,** the amount of water that passes a fixed point in a given amount of time. This is usually expressed as liters or cubic feet of water per second. The 16 largest rivers in the world carry nearly half of all surface runoff on earth. The Amazon is by far the largest river in the world (table 19.3), carrying roughly ten times the volume of the Mississippi. Several Amazonian tributaries such as the Maderia, Rio Negro, and Ucayali would be among the world's top rivers in their own right.

TABLE 19.3 Major Rivers of the World

RIVER	COUNTRIES IN RIVER BASIN	AVERAGE ANNUAL DISCHARGE AT MOUTH (CUBIC METERS/SECOND)*
Amazon	Brazil, Peru	175,000
Orinoco	Venezula, Colombia	45,300
Congo	Congo	39,200
Yangtze	Tibet, China	28,000
Bramaputra	Tibet, India, Bangladesh	19,000
Mississippi	United States	18,400
Mekong	China, Laos, Burma, Thailand, Cambodia, Vietnam	18,300
Parana	Paraguay, Argentina	18,000
Yenisey	Russia	17,200
Lena	Russia	16,000
Irrawaddy	Burma	13,000
Ob	Russia	12,000
Ganges	Nepal, India, Bangladesh	11,600
Amur	China, Russia	11,000
St. Lawrence	Canada, United States	10,700
Mackenzie	Canada	9,600

*One cubic meter equals 264 gallons.
Source: Data from World Resources Institute.

Lakes and Ponds

Ponds are generally considered to be small temporary or permanent bodies of water shallow enough for rooted plants to grow over most of the bottom. Lakes are inland depressions that hold standing fresh water year-round. Maximum lake depths range from a few meters to over 1600 m (1 mi) in Lake Baikal in Siberia. Surface areas vary in size from less than one-half ha (one acre) to large inland seas, such as Lake Superior or the Caspian Sea, covering hundreds of thousands of square kilometers. Both ponds and lakes are relatively temporary features on the landscape because they eventually fill with silt or are emptied by cutting of an outlet stream through the barrier that creates them.

While lakes contain nearly 100 times as much water as all rivers and streams combined, they are still a minor component of total world water supply. Their water is much more accessible than groundwater or glaciers, however, and they are important in many ways for humans and other organisms.

Wetlands

Bogs, swamps, wet meadows, and marshes play a vital and often unappreciated role in the hydrological cycle. Their lush plant growth stabilizes soil and holds back surface runoff, allowing time for infiltration into aquifers and producing even, year-long stream flow. In the United States, about 20 percent of the 1 billion ha of land area was once wetland. In the past 200 years, more than one-half of those wetlands have been drained, filled, or degraded. Agricultural drainage accounts for the bulk of the losses.

When wetlands are disturbed, their natural water-absorbing capacity is reduced and surface waters run off quickly, resulting in floods and erosion during the rainy season and dry, or nearly dry, stream beds the rest of the year. This has a disastrous effect on biological diversity and productivity, as well as on human affairs (chapter 15).

The Atmosphere

The atmosphere is among the smallest of the major water reservoirs of the earth in terms of water volume, containing less than 0.001 percent of the total water supply. It also has the most rapid turnover rate. An individual water molecule resides in the atmosphere for about ten days, on average. While water vapor makes up only a small amount (4 percent maximum at normal temperatures) of the total volume of the air, movement of water through the atmosphere provides the mechanism for distributing fresh water over the landmasses and replenishing terrestrial reservoirs.

WATER AVAILABILITY AND USE

Clean, fresh water is essential for nearly every human endeavor. Perhaps more than any other environmental factor, the availability of water determines the location and activities of humans on earth (fig. 19.11). **Renewable water supplies** are made up, in general,

FIGURE 19.11 Water has always been the key to survival. Who has access to this precious resource and who doesn't has long been a source of tension and conflict.

of surface runoff plus the infiltration into accessible freshwater aquifers. About two-thirds of the water carried in rivers and streams every year occurs in seasonal floods that are too large or violent to be stored or trapped effectively for human uses. Stable runoff is the dependable, renewable, year-round supply of surface water. Much of this occurs, however, in sparsely inhabited regions or where technology, finances, or other factors make it difficult to use it productively. Still, the readily accessible, renewable water supplies are very large, amounting to some 1500 km^3 (about 400,000 gal) per person per year worldwide.

Water-Rich And Water-Poor Countries

As you can see in figure 19.4, South America, West Central Africa, and South and Southeast Asia all have areas of very high rainfall. Brazil and the Democratic Republic of Congo, because they have high precipitation levels and large land areas, are among the most water-rich countries on earth. Canada and Russia, which are both very large, also have some of the highest total annual water supplies of any country. The highest per capita water supplies generally occur in countries with moist climates and low population densities (table 19.4). Iceland, for example, has about 160 million gallons per person per year. In contrast, Kuwait, where temperatures are extremely high and rain almost never falls, has less than 3000 gallons per person per year from renewable natural sources (table 19.5). Almost all of Kuwait's water comes from imports and desalinized sea water. Egypt, in spite of the fact that the Nile River flows through it, has only about 11,000 gallons of water annually

TABLE 19.4 Water-Rich Countries

COUNTRY	RENEWABLE ANNUAL SUPPLY[1] (CUBIC METERS/CAPITA)[2]
Iceland	606,500
Surinam	452,000
Guyana	281,500
Papua New Guinea	174,000
Gabon	140,000
Solomon Islands	107,000
Canada	94,000
Norway	88,000
Panama	52,000
Brazil	31,000

[1]Renewable supplies include both surface runoff and groundwater replenishment.
[2]One cubic meter equals 264 gallons.
Source: Data from *World Resources, 1998–99.*

TABLE 19.5 Water-Poor Countries

COUNTRY	RENEWABLE ANNUAL SUPPLY[1] (CUBIC METERS/CAPITA)[2]
Kuwait	11
Egypt	43
United Arab Emirates	63
Malta	85
Jordan	114
Saudi Arabia	119
Singapore	172
Moldavia	225
Israel	289
Oman	393

[1]Renewable supplies include both surface runoff and groundwater replenishment.
[2]One cubic meter equals 264 gallons.
Source: Data from *World Resources, 1998–99.*

per capita, or about 15,000 times less than Iceland. Singapore is an unusual case. Although it has a humid climate, this small, island nation has very little land area for its 3.5 million people, and must depend on its neighbor, Malaysia, for almost all its water.

Another important consideration is rainfall interannual variability. In some areas, such as the African Sahel region, abundant rainfall occurs some years but not others. Unless steps are taken to even out water flows, the lowest levels encountered usually limit both ecosystem functions and human activities. Some of the world's earliest civilizations, such as the Sumerians and Babylonians of Mesopotamia, the Harappans of the Indus Valley, and the early Chinese cultures, were based on communal efforts to control water, to divert floods and drain marshes during wet seasons or wet years, and to store water in reservoirs or divert it from streams so that it would be available during the dry seasons or dry years.

Drought Cycles

Rainfall is never uniform in either geographical distribution or yearly amount. Every continent has regions where rainfall is scarce because of topographic effects or wind currents. In addition, cycles of wet and dry years create temporary droughts. Water shortages have their most severe effect in semiarid zones where moisture availability is the critical factor in determining plant and animal distribution. Undisturbed ecosystems often survive extended droughts with little damage, but introduction of domestic animals and agriculture disrupts native vegetation and undermines natural adaptations to low moisture levels.

In the United States, the cycle of drought seems to be about 30 years. There were severe dry years in the 1870s, 1900s, 1930s, 1950s, and 1970s. The worst of these in economic and social terms were the 1930s. Poor soil conservation practices and a series of dry years in the Great Plains combined to create the "dust bowl." Wind stripped topsoil from millions of hectares of land, and billowing dust clouds turned day into night. Thousands of families were forced to leave farms and migrate to cities. The El Niño, Southern Oscillation (ENSO) system plays an important role in droughts in North America and elsewhere. There now is a great worry that the greenhouse effect (see chapter 17) will bring about major climatic changes and make droughts both more frequent and more severe than in the past in some places.

Types of Water Use

In contrast to energy resources, which are consumed when used, water has the potential for being reused many times. In discussing water appropriations, we need to distinguish between different kinds of uses and how they will affect the water being appropriated.

Withdrawal is the total amount of water taken from a lake, river, or aquifer for any purpose. Much of this water is employed in nondestructive ways and is returned to circulation in a form that can be used again. **Consumption** is the fraction of withdrawn water that is lost in transmission, evaporation, absorption, chemical transformation, or otherwise made unavailable for other purposes as a result of human use. **Degradation** is a change in water quality due to contamination or pollution so that it is unsuitable for other desirable service. The total quantity available may remain constant after some uses, but the quality is degraded so the water is no longer as valuable as it was.

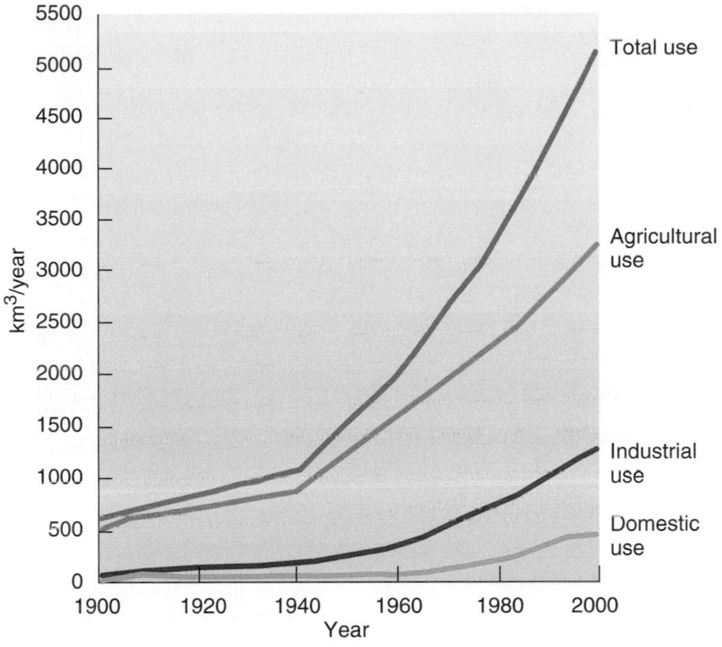

FIGURE 19.12 Growth of global water use 1900–2000. Amounts 1997–2000 are estimates.
Source: Data from L. A. Shiklomanov, "Global Water Resources" in *Nature and Resources,* vol. 26, p. 34–43, UNESCO, Paris.

Worldwide, humans withdraw about 10 percent of the total annual renewable supply. The remaining 90 percent is generally either uneconomical to tap (it would cost too much to store, ship, purify, or distribute), or there are ecological constraints on its use. Consumption and degradation together account for about half the water withdrawn in most industrial societies. The other half of the water we withdraw would still be valuable for further uses if we could protect it from contamination and make it available to potential consumers.

Many societies have always treated water as if there is an inexhaustible supply. It has been cheaper and more convenient for most people to dump all used water and get a new supply than to determine what is contaminated and what is not. The natural cleansing and renewing functions of the hydrologic cycle do replace the water we need if natural systems are not overloaded or damaged. Water is a renewable resource, but renewal takes time. The rate at which many of us are using water now may make it necessary to conscientiously protect, conserve, and replenish our water supply.

Quantities of Water Used

Human water use has been increasing about twice as fast as population growth over the past century (fig. 19.12). Water use is stabilizing in industrialized countries, but demand will increase in developing countries where supplies are available. The average

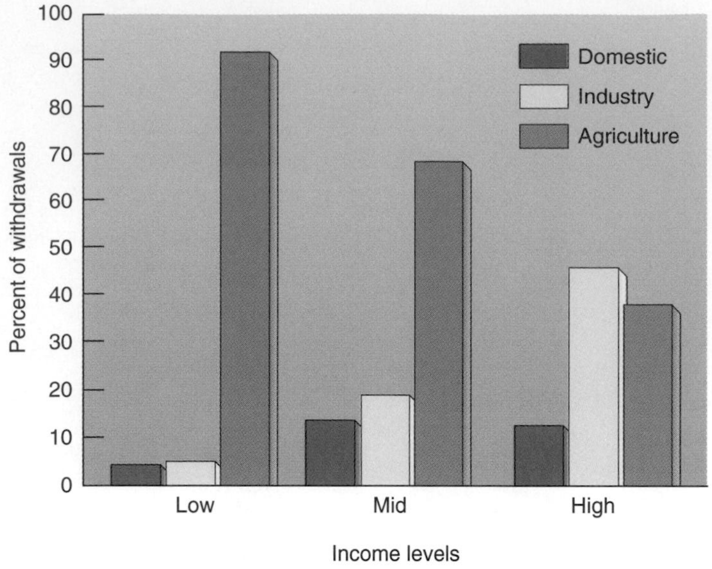

FIGURE 19.13 Water withdrawals by sector in low-, middle-, and high-income countries.
Source: Data from World Bank, 1992.

FIGURE 19.14 Rolling sprinklers allow farmers to irrigate crops on uneven terrain. Agriculture now uses 42 percent of all water withdrawn in high-income countries. In some areas, irrigation consumes 90 percent of all available water.

amount of water withdrawn worldwide is about 646 m^3 (170,544 gal) per person per year. This overall average hides great discrepancies in the proportion of annual runoff withdrawn in different areas. As you might expect, those countries with a plentiful water supply and a small population withdraw a very small percentage of the water available to them. Canada, Brazil, and the Congo, for instance, withdraw less than 1 percent of their annual renewable supply.

By contrast, in countries such as Libya and Israel, where water is one of the most crucial environmental resources, groundwater and surface water withdrawal together amount to more than 100 percent of their renewable supply. They are essentially "mining" water—extracting groundwater faster than it is being replenished. Obviously, this is not sustainable in the long run.

The total annual renewable water supply in the United States amounts to an average of about 9000 m^3 (nearly 2.4 million gal) per person per year. We now withdraw about one-fifth of that amount, or some 5000 l (1300 gal) per person per day. By comparison, the average water use in Haiti is less than 30 l (8 gal) per person per day.

Use by Sector

Water use can be analyzed by identifying three major kinds of use, or sectors: domestic, industry, and agriculture. Worldwide, agriculture claims about 69 percent of total water withdrawal, ranging from 93 percent of all water used in India to only 4 percent in Kuwait, which cannot afford to spend its limited water on crops. Canada, where the fields are well watered by natural precipitation, uses only 12 percent of its water for agriculture. As you can see in figure 19.13, water use by sector depends strongly on national wealth and degree of industrialization. Poorer countries with little industry and limited domestic supply systems use little of their water in these sectors.

In many developing countries, agricultural water use is notoriously inefficient and highly consumptive. Often, 60 to 70 percent of the water withdrawn for agriculture in these countries never reaches the crops for which it is intended. The most common type of irrigation is to simply flood the whole field or run water in rows between crops. As much as half is lost through evaporation or seepage from unlined irrigation canals bringing water to fields. Most of the rest runs off, evaporates, or infiltrates into the field before it can be used. While runoff can be channeled to other fields and infiltration replenishes underground aquifers, surplus water from fields is often contaminated with soil, fertilizer, pesticides, and crop residues, making it low quality. Sprinklers (fig. 19.14) are more efficient in distributing water evenly over the field than flooding and can be used on uneven terrain, but they can lose a great deal of water to evaporation. Water-efficient drip irrigation (see fig. 19.27) and other low-volume distribution systems can save significant amounts of water, but are used, currently, on only about 1 percent of the world's croplands.

Worldwide, industry accounts for about one-fourth of all water use, ranging from 70 percent of withdrawal in some European countries, such as Germany, to 5 percent in less industrialized countries, such as Egypt and India. Cooling water for power plants is by far the largest single industrial use of water, typically accounting for 50 to 100 percent of industrial withdrawal depending on the country. Although cooling water usually is not chemically contaminated, warm water can be thermal pollution if it is dumped directly into a stream or lake.

Industry often competes with agriculture for available water. In the case of the Yellow River that opened this chapter, China calculates that a given amount of water used in factories generates 60 times the value it would produce in agriculture, The future of farming looks bleak in such areas. Although Third World countries typically allocate only about 10 percent of their water withdrawal to industry, this could change rapidly as they industrialize. Water may be as important as energy in determining which countries develop and which remain underdeveloped.

FRESHWATER SHORTAGES

In 1977, the United Nations Water Conference declared that all people, regardless of their social or economic conditions, have a right to the clean drinking water and basic sanitation needed to prevent communicable diseases and provide for basic human dignity. More than two decades later, however, an estimated 1.5 billion people lack access to adequate quantity and quality of drinking water, while nearly 3 billion, or half the world's population, do not have acceptable sanitation. As populations grow, more people move into cities, and agriculture and industry compete for increasingly scarce water supplies, water shortages are expected to become even worse in the future. A country in which consumption exceeds more than 20 percent of the available, renewable supply is considered vulnerable to **water stress.**

Globally, water supplies are abundant but unevenly distributed, and the capital needed to collect, store, purify, and distribute water is unavailable in many developing countries. Worldwide, water consumption has increased sixfold over the past century, or about twice as fast as population growth. With easily accessible water already exploited in most places, the World Bank estimates that the financial and environmental costs of tapping new supplies will be two to three times more expensive than current water projects. If present trends continue, the U.N. cautions, some two-thirds of the world's population will live in countries experiencing water shortages by 2025. In countries with the greatest water stresses, it may be necessary to reallocate supplies to the most economically and socially beneficial uses.

FIGURE 19.15 Drawing water from a village well is a daily chore in many developing countries.

A Precious Resource

The World Health Organization considers 1000 m^3 (264,000 gal) of water per person per year to be the minimum level below which most countries are likely to experience chronic shortages on a scale that will impede development and harm human health. Currently, some 45 countries—most of them in Africa or the Middle East—are considered to have serious water stress, and cannot meet the minimum essential water needs of all their citizens. In some countries the problem is not so much total water supply as access to clean water. In Mali, for example, 88 percent of the population lacks clean water; in Ethiopia it is 94 percent. Rural people often have less access to clean water than do city dwellers. In the 33 worst affected countries, 60 percent of urban people can get clean water as opposed to only 20 percent of those in the country.

More than two-thirds of the world's households have to fetch water from outside the home (fig. 19.15). For many people, this is time-consuming and heavy work; it may take up to 2 hours per day to gather a meager supply. Women and children do most of this work. Improved public systems bring many benefits to poor families. In Mozambique, for example, the World Bank reports that the average time women spent carrying water decreased from 2 hours per day to only 25 minutes when village wells were installed. The time saved could be used to garden, tend livestock, trade in the market, care for children, or even rest!

The reasons for water shortages are many. In some cases, deficits are caused by natural forces: the rains fail; hot winds dry up reservoirs that normally would carry people through the dry season; rivers change their courses, leaving villages stranded. In other cases, shortages are human in origin: too many people compete for the resource; urbanization, overgrazing, and inappropriate agricultural practices allow water to run off before it can be captured; a lack of adequate sewage systems causes contamination of local supplies. Without money for wells, storage reservoirs, delivery pipes, and other infrastructure, people can't use the resources available to them.

CASE STUDY

RURAL WATER PROGRAMS IN MALAWI

Malawi is one of the poorest and least developed countries in the world. With an annual average per capita income of about $170 (US), Malawians can't afford many of the services that most of us take for granted. Having access to adequate supplies of clean, safe drinking water and basic sanitation are only a dream for many residents of Malawi. During the dry season, ponds usually turn green and smelly, or even dry up completely. Women—who do almost all domestic chores—often have to walk several kilometers to the nearest riverbed, dig down to the water table, scoop up what little water they can find, and carry it back home. It could take several hours just to get enough water for one day's drinking and cooking supply. There rarely is enough water in the dry season for bathing. Dysentery, parasitic worms, and skin diseases are considered to be normal.

But a program that provides simple tube wells and gravity-fed water systems is now bringing clean water to many people in Malawi. The gravity systems bring water to villages from year-round springs and streams in the mountains. Because the sources are located above the elevation where people live, the water is generally free of pollutants and human waste. Flowing downhill without pumps, the water passes through screens and filters to remove any particulate material, and then is distributed to public taps located in convenient spots around the villages. In villages that are too far from the mountains to be reached by piped water, shallow, hand-pumped tube wells provide access to groundwater that is much cleaner and more dependable than polluted rivers and ponds.

The water supply program was begun after the United Nations Water Conference in 1977. The first pilot project was in the Chigale region, where community members worked together to dig 10 km (16 mi) of trenches and lay pipes that brought fresh water to 25 taps serving 3000 people. Each person was provided with 381 (10 gal) of good quality water per day. because villagers provided most of the labor, the total cost of the project amounted to less than $3 (US) per person served. For the first time in their lives, many people have enough water to wash daily, even in the dry season. Women no longer need to spend so much time and energy gathering water. The community taps provide a social center where women can meet and talk to their neighbors and friends. The success of this initial undertaking convinced residents in other villages to construct their own water systems. As of 1999, more than a million people in Malawi, mostly in rural areas, now have access to safe community water supplies.

In addition to helping individuals who now have a reliable supply of safe water, the community-building aspects of these projects are among their greatest benefits. The key to success of the project is involvement of the villagers and shared responsibility between the village residents and the government. Local communities are responsible for constructing and maintaining the water supply system; technical advice and materials are provided by the government. Villagers generally are eager to volunteer their time and labor to ensure that their community will have a clean, lasting water supply. Many give up their ordinary wages to work on the 2 to 3 week project. When people have a sense of ownership, they are more likely to keep the system working. The Malawi rural water program shows what can be accomplished, quickly and at relatively low cost, when local communities are empowered and encouraged to organize and help themselves. How can those of us in the wealthy countries of the world help to spread this model elsewhere?

With the help of a loan from international development agencies, residents of this Malawian village have a simple hand-pumped tube well that provides convenient access to safe, clean water.

Safe, clean water is available in most countries—for those who can pay the price. Water from vendors often is the only source for many in the crowded slums and shanty towns around the major cities of the developing world. Although the quality is often questionable, this water generally costs about ten times more than a piped city supply. Naturally, sanitation levels decline when water is so expensive. A typical poor family in Lima, Peru, for instance, uses one-sixth as much water as a middle-class American family but pays three times as much for it. Following government recommendations that all water be boiled to prevent cholera would take up to one-third of the total income for such a poor family.

Investments in rural development have brought significant improvements over the past decade. Since 1990, nearly 800 million people—about 13 percent of the world population—have gained access to clean water (Case Study, above). The percentage of rural families with safe drinking water has risen from less than 10 percent to nearly 75 percent. Still, many people suffer from a lack of this essential resource.

Depleting Groundwater

Groundwater is the source of nearly 40 percent of the fresh water for agricultural and domestic use in the United States. Nearly half of all Americans and about 95 percent of the rural population depend on groundwater for drinking and other domestic purposes. Overuse of these supplies causes several kinds of problems, including drying of wells, natural springs, and disappearance of surface water features such as wetlands, rivers, and lakes.

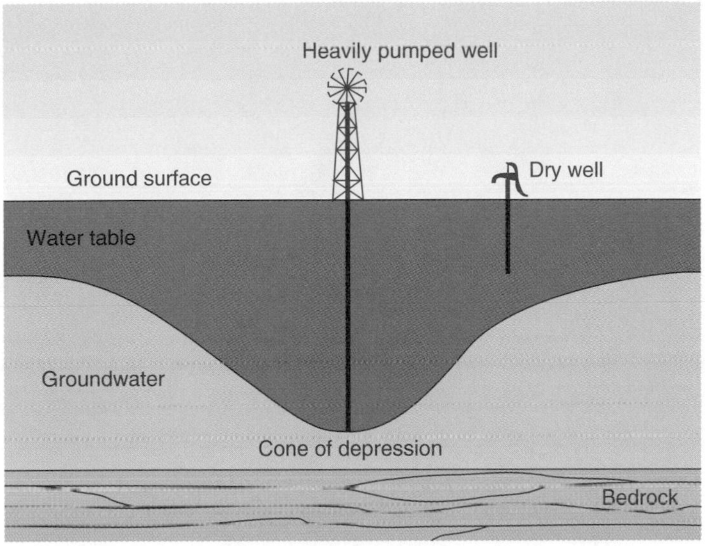

FIGURE 19.16 A cone of depression forms in the water table under a heavily pumped well. This may dry up nearby shallow wells or make pumping so expensive that it becomes impractical.

FIGURE 19.17 Sinkhole in Winter Park, Florida. Notice the cars and truck sliding into the hole.

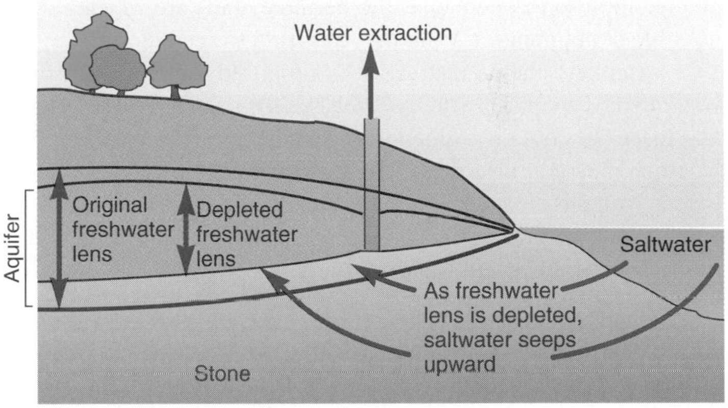

FIGURE 19.18 Saltwater intrusion into a coastal aquifer as the result of groundwater depletion. Many coastal regions of the United States are losing freshwater sources due to saltwater intrusion.

In many areas of the United States, groundwater is being withdrawn from aquifers faster than natural recharge can replace it. On a local level, this causes a cone of depression in the water table, as is shown in figure 19.16. A heavily pumped well can lower the local water table so that shallower wells go dry. On a broader scale, heavy pumping can deplete a whole aquifer. The Ogallala Aquifer, for example, underlies eight states in the arid high plains between Texas and North Dakota. As deep as 400 m (1200 ft) in its center, this porous bed of sand, gravel, and sandstone, once held more water than all the freshwater lakes, streams, and rivers on earth. Excessive pumping for irrigation and other uses has removed so much water that wells have dried up in many places and farms, ranches, even whole towns are being abandoned.

Many aquifers have slow recharge rates, so it will take thousands of years to refill them once they are emptied. Much of the groundwater we now are using probably was left there by the glaciers thousands of years ago. It is fossil water, in a sense. When we pump water out of a reservoir such as the Ogallala that cannot be refilled in our lifetime, we essentially are mining a nonrenewable resource. Covering aquifer recharge zones with urban development or diverting runoff that once replenished reservoirs ensures that they will not refill.

Withdrawal of large amounts of groundwater causes porous formations to collapse, resulting in **subsidence** or settling of the surface above. The U.S. Geological Survey estimates that the San Joaquin Valley in California has sunk more than 10 m in the last 50 years because of excessive groundwater pumping. Around the world, many cities are experiencing subsidence. Many are coastal cities, built on river deltas or other unconsolidated sediments. Flooding is frequently a problem as these coastal areas sink below sea level. Some inland areas also are affected by severe subsidence. Mexico City is one of the worst examples. Built on an old lake bed, it has probably been sinking since Aztec times. In recent years, rapid population growth and urbanization (chapter 24) have caused groundwater overdrafts. Some areas of the city have sunk as much as 8.5 m (25.5 ft). The Shrine of Guadalupe, the Cathedral, and many other historic monuments are sinking at odd and perilous angles.

Sinkholes form when the roof of an underground channel or cavern collapses, creating a large surface crater (fig. 19.17). Drawing water from caverns and aquifers accelerates the process of collapse. Sinkholes can form suddenly, dropping cars, houses, and trees without warning into a gaping crater hundreds of meters across. Subsidence and sinkhole formation generally represent permanent loss of an aquifer. When caverns collapse or the pores between rock particles are crushed as water is removed, it is usually impossible to reinflate these formations and refill them with water.

Another consequence of aquifer depletion is **saltwater intrusion.** Along coastlines and in areas where saltwater deposits are left from ancient oceans, overuse of freshwater reservoirs often allows saltwater to intrude into aquifers used for domestic and agricultural purposes (fig. 19.18).

INCREASING WATER SUPPLIES

Where do present and impending freshwater shortages leave us now? On a human time scale, the amount of water on the earth is fixed, for all practical purposes, and there is little we can do to make more water. There are, however, several ways to increase local supplies.

Seeding Clouds and Towing Icebergs

In the dry prairie states of the 1800s and early 1900s, desperate farmers paid self-proclaimed "rainmakers" in efforts to save their withering crops. Centuries earlier, Native Americans danced and prayed to rain gods. We still pursue ways to make rain. Seeding clouds with dry ice or potassium iodide particles has been tested for many years with mixed results. Recently, researchers have been having more success using hygroscopic salts that seem to significantly increase rainfall amounts. This technique is being tested in Mexico, South Africa, and the western United States. There is a concern, however, that rain induced to fall in one area decreases the precipitation somewhere else. Furthermore, there are worries about possible contamination from the salts used to seed clouds.

Another scheme that has been proposed for supplying fresh water to arid countries is to tow icebergs from the Arctic or Antarctic. Icebergs contain tremendous quantities of fresh water, but whether they can be towed without excessive melting to places where water is needed is not clear. Nor is it clear whether the energy costs to move an iceberg thousands of miles (if that were, in fact, possible) would be worthwhile.

Desalination

A technology that might have great potential for increasing freshwater supplies is **desalination** of ocean water or brackish saline lakes and lagoons. The most common methods of desalination are distillation (evaporation and recondensation) or reverse osmosis (forcing water under pressure through a semipermeable membrane whose tiny pores allow water to pass but exclude most salts and minerals). In 1990, the global capacity was estimated to be 13.3 million m^3 (3.4 billion gal) per day, or less than 0.1 percent of all freshwater withdrawals worldwide. Saudi Arabia, Kuwait, and the United Arab Emirates account for about 85 percent of all desalination, with South Africa and the United States making up most of the rest. Although desalination is still three to four times more expensive than most other sources of fresh water, it provides a welcome water supply in such places as Oman and Bahrain where there is no other access to fresh water. If a cheap, inexhaustible source of energy were available, however, the oceans could supply all the water we would ever need.

Dams, Reservoirs, Canals, and Aqueducts

People have been moving water around for thousands of years. Some of the great civilizations (Sumeria, Egypt, China, and the Inca culture of South America) were based on large-scale irrigation systems that brought river water to farm fields. In fact, some historians argue that organizing people to carry out large-scale water projects was the catalyst for the emergence of civilization. Roman aqueducts built 2000 years ago are still in use. Those early water engineers probably never even dreamed of moving water on a scale that is being proposed and, in some cases, being accomplished now.

FIGURE 19.19 Workers drive a team of 48 mules pulling a section of conduit for the Owens Valley aqueduct to Los Angeles in 1911. Water diversions such as this have allowed Los Angeles to grow to a metropolitan area of nearly 15 million people in what originally was a semiarid desert, but the ecological effects have been severe in the regions from which water has been diverted.

It is common to trap runoff with dams and storage reservoirs and transfer water from areas of excess to areas of deficit using canals, tunnels, and underground pipes. Some water transfer projects are truly titanic in scale. Los Angeles began importing water in 1913 through an aqueduct from the Owens Valley, 400 km (250 mi) to the north (fig. 19.19). This project led to a statewide program known as the California Water Plan in which a system of dams, reservoirs, aqueducts, and canals transfer water from the Colorado River on the eastern border and the Sacramento and Feather rivers in the north to Los Angeles and the San Joaquin Valley in the south.

There has been much controversy about this massive project. Some of the water now being delivered to southern California is claimed by other parts of the state, neighboring states, and even Mexico. Environmentalists claim that this water transfer upsets natural balances of streams, lakes, estuaries, and terrestrial ecosystems. Fishing enthusiasts, whitewater boaters, and others who enjoy the scenic beauty of free-running rivers mourn the loss of rivers drowned in reservoirs or dried up by diversion projects. These projects also have been criticized for using public funds to increase the value of privately held farmland and for encouraging agricultural development and urban growth in arid lands where other uses might be more appropriate.

FIGURE 19.20 Mono Lake in eastern California. Diversion of tributary rivers to provide water for Los Angeles has shrunk the lake's surface area by one-third, threatening migratory bird flocks that feed here. These formations were created underwater where calcium-rich springs entered the brine-laden lake. They show how far the lake surface has been drawn down.

Mono Lake, a salty desert lake in California just east of Yosemite National Park, is an example of environmental consequences of interbasin water transfers and is an important legal symbol as well (fig. 19.20). Diversion of tributary rivers has shrunk the surface area of this lake by one-third, threatening millions of resident and migratory wading birds, ducks, and gulls that seek shelter and food there. After years of legal wrangling, the California Water Resources Control Board ruled in 1994 that Los Angeles must restrict diversions and allow the lake's surface to rise 17 ft to 6391 ft above sea level. This is less than the 1941 level, but should improve conditions for birds and aquatic life.

In China, huge dams are being built in the Three Gorges section of the Chang Jiang (Yangtze) River. More than a million people are being forced to relocate as the huge reservoir (185 m, or 600 ft, deep and 600 km, or 375 mi, long) inundates some 150 cities and villages. Because the dam is being built on an active geological fault, there is a worry that a strong earthquake could rupture the dam and drown millions of people living downstream.

Many other huge dam and water diversion projects have generated controversy and protest around the world. In Turkey, construction of a network of 22 large dams and reservoirs for the Southeast Anatolia Project on the Tigris and Euphrates Rivers is regarded as a serious threat by Turkey's downstream neighbors, Syria and Iraq, which are now at the mercy of Turkey to allow the rivers to continue to flow. In India, the Sardar Sarovar Dam on the sacred Narmada River has been the focus of decades of protest. Many of the one million villagers and tribal people being displaced by this project have engaged in mass resistance and civil disobedience when police try to remove them forcibly. Some have vowed to drown rather than leave their homes. In Nepal, construction of the 240 m (850 ft) high Tehri Dam on the Bhagirathi River has stirred fears that a strong earthquake in this active seismic region might cause the dam to collapse and cause a catastrophic flood downstream. The Tehri Dam is only one of 17 high dams that Nepal and India plan for the Himalayan mountains.

Canada also has been the site of decades of protest by First Nations people over flooding of ancestral lands for hydroelectric projects. The James Bay project built by Hydro-Quebec has diverted three major rivers flowing west into Hudson Bay and created huge lakes that flooded more than 10,000 sq km (4000 sq mi) of forest and tundra, to generate 26,000 megawatts of electrical power. In 1984, shortly after Phase I of this project was completed, 10,000 caribou drowned when trying to follow migration routes across the newly flooded land. The loss of traditional hunting and fishing sites has been culturally devastating for native Cree people. In addition, mercury leeched out of rocks in newly submerged lands has entered the food chain and many residents show signs of mercury poisoning. In a similar, but less well-known case, Manitoba Hydro has diverted the Churchill River from the west shore of Hudson Bay so that it now flows into the Nelson River and then to Lake Winnipeg. More than 125,000 sq km (50,000 sq mi) has been inundated, much of it on permafrost, which is slowly melting and causing reservoir banks to expand continually. Here also, Cree people are being displaced against their will. As is the case in Quebec, most of the electricity generated by these dams is exported to the United States. In some places, however, dams are being removed (What Do You Think? p. 438).

Environmental Costs

Dams have been useful over the centuries for ensuring a year-round water supply, but they are far from perfect. In many cases, they reduce water availability and destroy both natural and human values. In this section, we will look at some of the disadvantages of dams.

Evaporation, Leakage, and Siltation

The main problem with dams is inefficiency. Some dams built in the western United States lose so much water through evaporation and seepage into porous rock beds that they waste more water than they make available. The evaporative loss from Lake Mead and Lake Powell on the Colorado River is about 1 km^3 (264 billion gal) per year. The salts left behind by evaporation and agricultural runoff nearly double the salinity of the river and make its water unusable when it reaches Mexico. To compensate, the United States has built a $350 million desalination plant at Yuma, Arizona, to improve water quality.

As the turbulent Colorado River slows in the reservoirs created by Glen Canyon and Boulder Dams, it drops its load of suspended material. More than 10 million metric tons of silt per year collect behind these dams. Imagine a line of twenty thousand dump trucks backed up to Lake Mead and Lake Powell every day, dumping dirt into the water. Within as little as one hundred years, these reservoirs could be full of silt and useless for either water storage or hydroelectric generation (fig. 19.21).

what do you think?

THE WORLD AROUND YOU

Should We Remove Dams?

For nearly a century, the U.S. Army Corps of Engineers and the Bureau of Reclamation have constructed dams to store water and to generate electricity. Building these dams provided cheap electricity, created jobs for workers, stimulated regional economic development, and allowed farming on lands that would otherwise be too dry. But not everyone agrees that big dam projects are an unmitigated benefit. Their storage reservoirs drown free-flowing rivers and often submerge towns, farms, cemeteries, and historic sites. They block fish migrations and change aquatic habitats essential for endemic species.

The tide may be turning, however, against dam building. In 1998, the Army Corps announced that it would no longer be building large dams and diversion projects. In the few remaining sites where dams might be built, public opposition is so great that getting approval for projects is unlikely. Instead, the new focus may be on removing existing dams and restoring natural habitats. In 1999, Interior Secretary, Bruce Babbitt, said, "Of the 75,000 large dams in the United States, most were built a long time ago and are now obsolete, expensive, and unsafe. They were built with no consideration of the environmental costs. As operating licenses come up for renewal, removal and restoration to original stream flows will be one of the options to be considered."

The first active hydroelectric dam in the United States to be breached against the wishes of its owners was the 162-year-old Edwards Dam, on the Kennebec River in Augusta, Maine. For many years, the U.S. Fish and Wildlife Service has advocated removal of this dam, which prevented migration of salmon, shad, sturgeon, and other species up the River. In a precedent-setting decision, the Federal Energy Regulatory Commission ordered the dam removed after concluding that the environmental and economic benefits of a free-flowing river outweigh the electricity generated by the dam. In July, 1999, the dam was breached and restoration work began on wetlands and stream banks long underwater.

The next dams likely to be taken down are the Elwha and Glines Dams on the Elwha River in Olympic National Park in Washington. Built nearly a century ago to provide power to lumber and paper mills in the town of Port Angeles, these dams blocked access to upstream spawning beds for six species of salmon on what once was one of the most productive salmon rivers in the world. Where 50 kg (110 lb) King salmon once fought their way up thundering waterfalls, there now are only concrete walls holding back still water and deep beds of silty sediment. Simply breaching the dams won't restore the salmon, however. Removing the sediment, uncovering gravel spawning beds, and finding suitable breeding stock to rebuild the population is a daunting task. And it won't be cheap. Congress will have to appropriate somewhere around $300 to $400 million to remove and rehabilitate these two relatively small dams.

Environmental groups, encouraged by these examples, have begun to talk about much more ambitious projects. Four giant dams on the Snake River in Washington State, for example, might be removed to restore salmon and steelhead runs to the headwaters of the Columbia River. The Hetch Hetchy Dam in Yosemite National Park might be taken down to reveal what John Muir called a valley "just as beautiful and worthy of preservation as the majestic Yosemite." Some groups have even suggested breaching the Glen Canyon Dam on the Colorado. In each of these cases, powerful interests stand in opposition. These dams generate low-cost electricity and store water that is needed for agriculture and industry. Local economies, domestic water supplies, and certain types of recreation all would be severely impacted by destruction of these dams.

What do you think? How would you go about weighing the many different economic, cultural, and aesthetic considerations for removing or not removing these dams? Do certain interests, such as the rights of native people, or the continued existence of endemic species of fish or wildlife take precedence over economic factors; or should this be a utilitarian calculation of the greatest good for the greatest number? Does that number include only humans or do other species count as well? If you were Secretary of the Interior, how many dams would you remove?

Protesters invoke the spirit of John Muir as they paint a symbolic crack on the Hetch Hetchy Dam in Yosemite National Park and call for restoration of the Tuolumne River.

FIGURE 19.21 This dam is now useless because its reservoir has filled with silt and sediment.

FIGURE 19.22 The recreational and aesthetic values of free-flowing wild rivers and wilderness lakes may be their greatest assets. Competition between *in situ* values and extractive uses can lead to bitter fights and difficult decisions.

The accumulating sediments that clog reservoirs and make dams useless also represent a loss of valuable nutrients. The Aswan High Dam in Egypt was built to supply irrigation water to make agriculture more productive. Although thousands of hectares are being irrigated, the water available is only about half that anticipated because of evaporation in Lake Nasser behind the dam and seepage losses in unlined canals which deliver the water. Controlling the annual floods of the Nile also has stopped the deposition of nutrient-rich silt on which farmers depended for fertilizing their fields. This silt is being replaced with commercial fertilizer costing more than $100 million each year. Furthermore, the nutrients carried by the river once supported a rich fishery in the Mediterranean that was a valuable food source for Egypt. After the dam was installed, sardine fishing declined 97 percent. To make matters worse, growth of snail populations in the shallow permanent canals that distribute water to fields has led to an epidemic of schistosomiasis. This debilitating disease is caused by blood flukes (parasitic flatworms) spread by snails living in permanent ponds and irrigation canals. In some areas, 80 percent of the residents are infected (chapter 9).

Loss of Free-Flowing Rivers

Among the environmental costs of many water projects are the loss of free-flowing rivers that are either drowned by reservoir impoundments or turned into linear, sterile irrigation canals. Conservation history records many battles between those who want to preserve wild rivers and those who would benefit from development (fig. 19.22).

One of the first and most divisive of these battles was over the flooding of the Hetch Hetchy Valley in Yosemite National Park. In the early 1900s, San Francisco wanted to dam the Tuolumne River to produce hydroelectric power and provide water for the city water system. This project was supported by many prominent San Francisco citizens because it represented an opportunity for both clean water and municipal power. Leader of the opposition was John Muir, founder of the Sierra Club and protector of Yosemite Park. Muir said that Hetch Hetchy Valley rivaled Yosemite itself in beauty and grandeur and should be protected. After a prolonged and bitter fight, the developers won and the dam was built. Hard feelings from this controversy persisted for many years. In 1987, Interior Secretary Donald Hodel suggested that the Hetch Hetchy Valley might be drained and restored to its former pristine condition. Supporters and opponents are lining up to fight the same battle again. Meanwhile, impoundment schemes for the remaining free stretches of the Tuolumne River are being proposed.

WATER MANAGEMENT AND CONSERVATION

Watershed management and conservation are often more economical and environmentally sound ways to prevent flood damage and store water for future use than building huge dams and reservoirs.

Watershed Management

A **watershed,** or catchment, is all the land drained by a stream or river. It has long been recognized that retaining vegetation and ground cover in a watershed helps hold back rainwater and lessens downstream floods. In 1998, Chinese officials acknowledged that unregulated timber cutting upstream on the Yangtze contributed to massive floods that killed 30,000 people.

Similarly, after disastrous floods in the upper Mississippi Valley in 1993, it was suggested that, rather than allowing residential, commercial, or industrial development on floodplains, these areas should be reserved for water storage, aquifer recharge, wildlife

what can you do?

Saving Water and Preventing Pollution

Each of us can conserve much of the water we use and avoid water pollution in many simple ways.

- Don't flush every time you use the toilet. Take shorter showers; don't wash your car so often.
- Don't let the faucet run while washing hands, dishes, food, or brushing your teeth. Draw a basin of water for washing and another for rinsing dishes. Don't run the dishwasher when half full.
- Dispose of used motor oil, household hazardous waste, batteries, etc., responsibly. Don't dump anything down a storm sewer that you wouldn't want to drink.
- Avoid using toxic or hazardous chemicals for simple cleaning or plumbing jobs. A plunger or plumber's snake will often unclog a drain just as well as caustic acids or lye. Hot water and soap will clean brushes more safely than organic solvents.
- If you have a lawn, or know someone who does, use water, fertilizer, and pesticides sparingly. Consider planting native plants, low-maintenance ground cover, a rock garden, or some other xeriphytic landscaping.
- Use water-conserving appliances: low-flow showers, low-flush toilets, and aerated faucets.
- Use recycled (gray) water for lawns, house plants, car washing.
- Check your toilet for leaks. A leaky toilet can waste 50 gallons per day. Add a few drops of dark food coloring to the tank and wait 15 minutes. If the tank is leaking, the water in the bowl will change color.

habitat, and agriculture (chapter 15). Sound farming and forestry practices can reduce runoff. Retaining crop residue on fields reduces flooding, and minimizing plowing and forest cutting on steep slopes protects watersheds. Wetlands conservation preserves natural water storage capacity and aquifer recharge zones. A river fed by marshes and wet meadows tends to run consistently clear and steady rather than in violent floods.

A series of small dams on tributary streams can hold back water before it becomes a great flood. Ponds formed by these dams provide useful wildlife habitat and stock-watering facilities. They also catch soil where it could be returned to the fields. Small dams can be built with simple equipment and local labor, eliminating the need for massive construction projects and huge dams.

In 1998, U.S. Forest Service chief, Mike Dombeck, announced a major shift in his agency's priorities. "Water," he said, "is the most valuable and least appreciated resource the National Forests provide. More than 60 million people in 33 states obtain their drinking water from national forest lands. Protecting watersheds is far more economically important than logging or mining, and will be given the highest priority in forest planning."

FIGURE 19.23 By using native plants in a natural setting, residents of Phoenix save water and fit into the surrounding landscape.

Domestic Conservation

We could probably save as much as half of the water we now use for domestic purposes without great sacrifice or serious changes in our lifestyles. Simple steps, such as taking shorter showers, stopping leaks, and washing cars, dishes, and clothes as efficiently as possible, can go a long way toward forestalling the water shortages that many authorities predict. Isn't it better to adapt to more conservative uses now when we have a choice than to be forced to do it by scarcity in the future?

The use of conserving appliances, such as low-volume shower heads and efficient dishwashers and washing machines, can reduce water consumption greatly (What Can You Do? above). If you live in an arid part of the country, you might consider whether you really need a lush green lawn that requires constant watering, feeding, and care. Planting native ground cover in a "natural lawn" or developing a rock garden or landscape in harmony with the surrounding ecosystem can be both ecologically sound and aesthetically pleasing (fig. 19.23). There are about 30 million ha (75 million acres) of cultivated lawns, golf courses, and parks in the United States. They receive more water, fertilizer, and pesticides per hectare than any other kind of land.

Our largest domestic water use is toilet flushing (fig. 19.24). We dispose of relatively small volumes of waste with very large volumes of water. In many cases it is much better to treat or dispose of waste at its origin before it is diluted or mixed with other materials. For instance, each person in the United States uses about 50,000 l (13,000 gal) of drinking-quality water annually to flush toilets. This is more than one-third of the amount supplied to homes each year. There are now several types of waterless or low-volume toilets. The Swedish-made Clivus Multrum (fig. 19.25) digests both human and kitchen wastes by aerobic bacterial action, producing a rich, nonoffensive compost that can be used as garden fertilizer. There are also low-volume toilets that use recirculating oil or aqueous chemicals to carry wastes to a holding tank, from which they

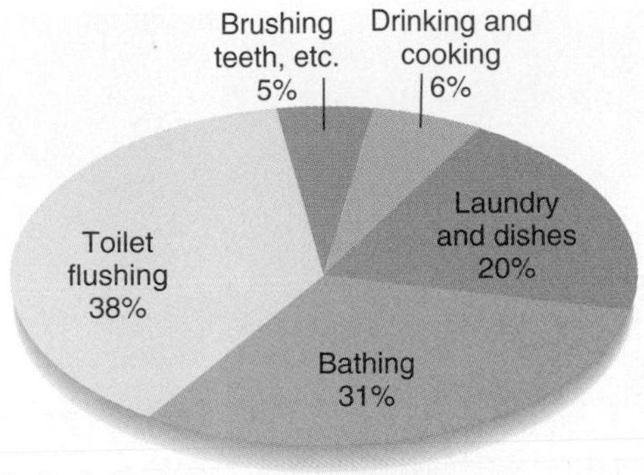

FIGURE 19.24 Typical household water use in the United States.

are periodically taken to a treatment plant. Anaerobic digesters use bacterial or chemical processes to produce usable methane gas from domestic wastes. These systems provide valuable energy and save water but are more difficult to operate than conventional toilets. Few cities are ready to mandate waterless toilets, but in 1988 a number of cities (including Los Angeles, California; Orlando, Florida; Austin, Texas; and Phoenix, Arizona) ordered that water-saving toilets, showers, and faucets be installed in all new buildings. The motivation was twofold: to relieve overburdened sewer systems and to conserve water.

Significant amounts of water can be reclaimed and recycled. In California, water recovered from treated sewage constitutes the fastest growing water supply, growing about 30 percent per year. Despite public squeamishness, purified sewage effluent is being used for everything from agricultural irrigation to flushing toilets (fig. 19.26). In a statewide first, San Diego is currently moving toward piping water from the local sewage plant directly into a drinking-water reservoir. Already, California uses more than 555 million m^3 450,000 acre-feet of recycled water annually. That's equivalent to about two-thirds of the water consumed by Los Angeles every year. Other places with growing populations and limited water supplies may soon find that it pays to follow California's example.

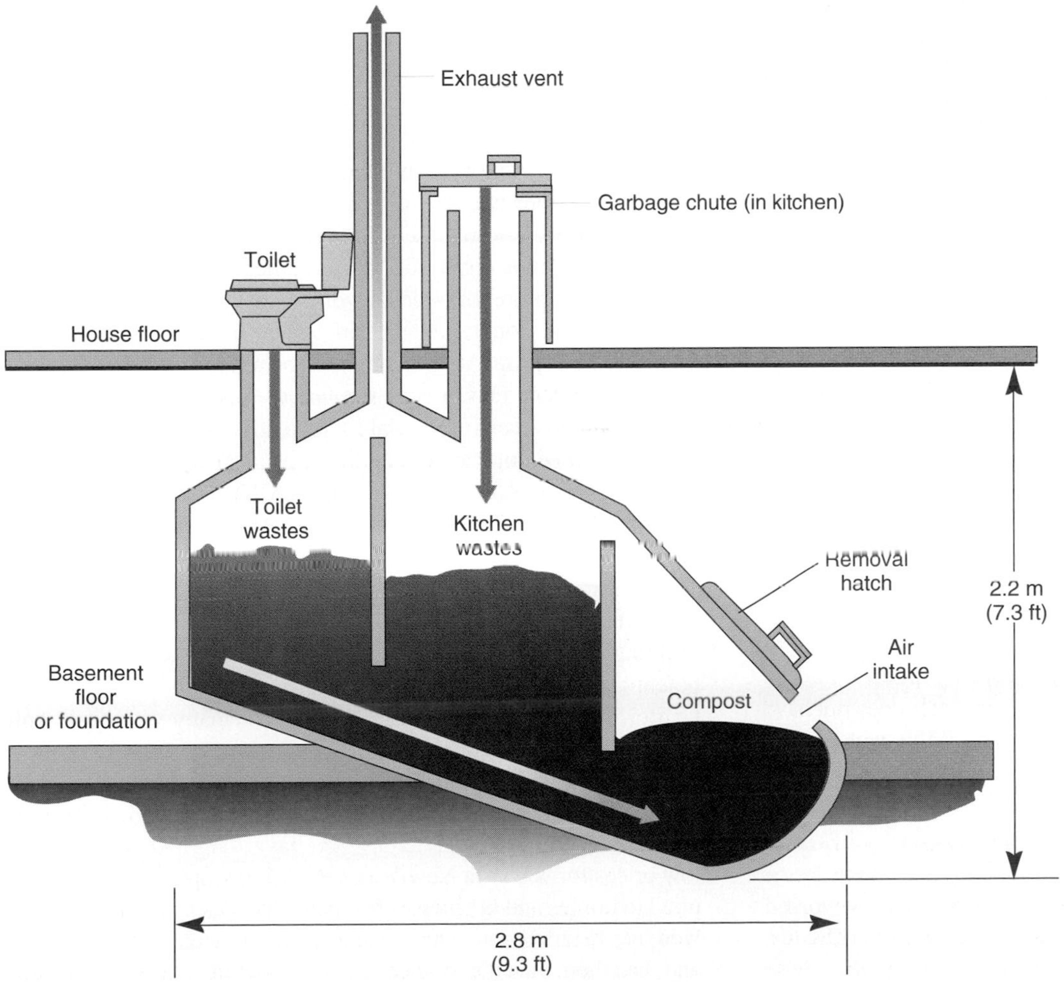

FIGURE 19.25 The Clivus Multrum waterless toilet. Wastes decompose and compost into an odorless, safe, rich fertilizer as they slowly slide down to the bottom compartment.

FIGURE 19.26 Recycled water is being used in California and Arizona for everything from agriculture, to landscaping, to industry. Some cities even use treated sewage effluent for human drinking-water supplies.

FIGURE 19.27 Drip irrigation delivers measured amounts of water exactly where the plants need and can use it. This technique can save up to 90 percent of irrigation water usage and reduces salt buildup. It is more expensive to install and operate, however, than simple flood irrigation.

Recycling and Water Conservation

In many developing countries as much as 70 percent of all the agricultural water used is lost to leaks in irrigation canals, application to areas where plants don't grow, runoff, and evaporation. Better farming techniques, such as leaving crop residue on fields and ground cover on drainage ways, intercropping, use of mulches, and low-volume irrigation could reduce these water losses dramatically (fig. 19.27).

Nearly half of all industrial water use is for cooling of electric power plants and other industrial facilities. Some of this water use could be avoided by installing dry cooling systems similar to the radiator of your car. In many cases, cooling water could be reused for irrigation or other purposes in which water does not have to be drinking quality. The waste heat carried by this water could be a valuable resource if techniques were developed for using it.

Price Mechansims and Water Policy

Through most of U.S. history, water policies have generally worked against conservation. In the well-watered eastern United States, water policy was based on riparian usufructuary (use) rights—those who lived along a river bank had the right to use as much water as they liked as long as they didn't interfere with its quality or availability to neighbors downstream. It was assumed that the supply would always be endless and that water had no value until it was used. In the drier western regions where water often is a limiting resource, water law is based primarily on the Spanish system of prior appropriation rights, or "first in time are first in right." Even if the prior appropriators are downstream, they can legally block upstream users from taking or using water flowing over their property. But the appropriated water had to be put to "beneficial" use by being consumed. This creates a policy of "use it or lose it." Water left in a stream, even if essential for recreation, aesthetic enjoyment, or to sustain ecological communities, is not being appropriated or put to "beneficial" (that is, economic) use. Under this system, water rights can be bought and sold, but water owners frequently are reluctant to conserve water for fear of losing their rights.

In most federal "reclamation" projects, customers were charged only for the immediate costs of water delivery. The costs of building dams and distribution systems was subsidized, and the potential value of competing uses was routinely ignored. Farmers in California's Central Valley, for instance, for many years paid only about one-tenth of what it cost the government to supply water to them. This didn't encourage conservation. Subsidies created by underpriced water amounted to as much as $500,000 per farm per year in some areas. Many cities also have long had unreasonably low prices for water. In New York City, for example, water was supplied to homes and businesses for many decades at a flat rate. There were no meters because it was considered more expensive to install and read them than the water was worth. With no incentive to restrict water use or repair leaks, it is estimated that 750,000 m^3 (200 million gal) of water was wasted each year from leaky faucets, toilets, and water pipes. The drought of 1988 convinced the city to begin a ten-year, $290 million program to install meters and reduce waste.

Growing recognition that water is a precious and finite resource has changed policies and encouraged conservation across the United States. Despite a growing population, the United States is now saving some 144 million l (38 million gal) per day—or enough water to fill Lake Erie in a decade—compared to per capita consumption rates of 20 years ago. With 37 million more people in the United Sates now than there were in 1980, we get by with 10 percent less water. New requirements for water-efficient fixtures and low-flush toilets in many cities help to conserve water on the home front. More efficient irrigation methods on farms also are a major reason for the downward trend. Instead of old systems that sprayed water up into the air, where it could evaporated or be blown away by the wind before reaching crops, new systems have small spray heads just a foot or so above the plant tops, and apply water much more directly. Even better is drip irrigation (see fig. 19.27), which applies water directly to plant roots. California and Florida farmers currently water about 500,000 ha (1.3 million acres) with this technique. Unfortunately, these efficient systems are too expensive for farms in most countries and are used on less than 1 percent of all farmland worldwide.

Charging a higher proportion of real costs to users of public water projects has helped encourage conservation, and so have water marketing policies that allow prospective users to bid on water rights. Both the United States and Australia have had effective water pricing and allocation policies that encourage the most socially beneficial uses and discourage wasteful water uses. Market mechanisms for water allotment can be sensitive, however, in developing countries where farmers and low-income urban residents could be outbid for irreplaceable water supplies.

Even in countries where water is abundant, national sovereignty and pride can be a barrier to water marketing. Canada, for instance, which has 20 percent of all the world's known freshwater resources, believes that it has none to spare. For years, Canada has resisted schemes to sell water to the United States. In 1999, the Canadian government passed federal legislation banning any bulk water export. At issue was a plan to ship supertankers of water from Grand Le Pierre on the southern coast of Newfoundland to thirsty Asian cities. Claiming that water is "too precious" to be treated like other commodities, the federal government is also urging provinces and territories to institute similar bans. Whether these policies will hold up under NAFTA and WTO trade policies remains to be seen.

It will be important, as water markets develop, to be sure that environmental, recreational, and wildlife values are not sacrificed to the lure of high-bidding industrial and domestic uses. Given prices based on real costs of using water and reasonable investments in public water supplies, pollution control, and sanitation, the World Bank estimates that everyone in the world could have an adequate supply of clean water by the year 2030 (fig. 19.28). We will discuss the causes, effects, and solutions for water pollution in chapter 20.

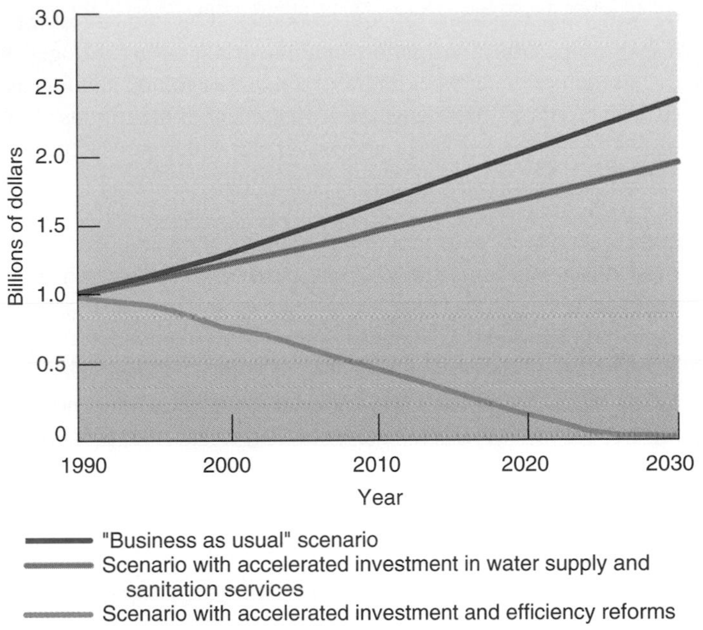

FIGURE 19.28 Three scenarios for government investments on clean water and sanitation services, 1990–2030.
Source: World Bank estimates based on research paper by Dennis Anderson and William Cavendish, "Efficiency and Substitution in Pollution Abatement: Simulation Studies in Three Sectors."

SUMMARY

The hydrologic cycle constantly purifies and redistributes fresh water, providing an endlessly renewable resource. The physical processes that make this possible—evaporation, condensation, and precipitation—depend upon the unusual properties of water, especially its ability to absorb and store solar energy. More than 97 percent of all water in the world is salty ocean water. Of the 33,400 km^3 (8.8 quintillion gal) that is fresh, 99 percent is locked up in ice or snow or buried in groundwater aquifers. Lakes, rivers, and other surface freshwater bodies make up only about 0.01 percent of all the water in the world, but they provide more than half of all water for human use and habitat and nourishment for aquatic ecosystems that play a vital role in the chain of life.

Water is essential for nearly every human endeavor. In the United States, direct personal use constitutes about one-tenth of the water we withdraw from our resources. Our two largest water uses are agricultural irrigation and industrial cooling. Altogether, the amount of water we withdraw from underground aquifers and surface reservoirs represents about 19 percent of the surplus runoff in the United States. Only about half the water we withdraw is consumed or degraded so that it is unsuitable for other purposes; much could be reused or recycled. Water conservation and recycling would have both economic and environmental benefits.

Water shortages in many parts of the world result from rising demand, unequal distribution, and increased contamination. Arid zones are especially vulnerable to the effects of natural droughts and land abuse by humans and domestic animals. Lakes, rivers, and groundwater reservoirs are being depleted at an alarming rate, leading not only to water shortages, but also to subsidence, sinkhole formation, saltwater intrusion, and permanent loss of aquifers.

Water storage and transfer projects are a response to flooding and water shortages. Giant dams and diversion projects can have environmental and social costs that are not justified by the benefits they provide. Among the problems they pose are evaporation and infiltration losses, siltation of reservoirs, and loss of recreation and wildlife habitat. Watershed management and small dams are preferred by many conservationists as means of flood control and water storage.

There is much we can do to save water. Charging users the true cost of water is a good start toward conservation. We can each use less water in our personal lives, and society can encourage development of water-saving appliances, natural yards, recycling, and efficient water use. Perhaps the most important change we can make is to treat wastes at their sources rather than use precious water resources for waste disposal. Not everyone can live upstream.

QUESTIONS FOR REVIEW

1. What is the difference between withdrawal, consumption, and degradation of water?
2. How does water use by sector differ between rich and poor countries?
3. What is subsidence? What are its results?
4. Describe some problems associated with dam building and water diversion projects.
5. Describe the path a molecule of water might follow through the hydrologic cycle from the ocean to land and back again.
6. Define *evaporation, sublimation, condensation, precipitation,* and *infiltration.* How do they work?
7. How do mountains affect rainfall distribution? Does this affect your part of the country?
8. What are the major water reservoirs of the world?
9. How much water is fresh (as opposed to saline) and where is it?
10. Define *aquifer.* How does water get into an aquifer?

QUESTIONS FOR CRITICAL THINKING

1. What changes might occur in the hydrologic cycle if our climate were to warm or cool significantly?
2. Why does it take so long for the deep ocean waters to circulate through the hydrologic cycle? What happens to substances that contaminate deep ocean water or deep aquifers in the ground?
3. Where would you most like to spend your vacations? Does availability of water play a role in your choice? Why?
4. Why do we use so much water? Do we need all that we use?
5. Are there ways you could use less water in your own personal life? Would that make any difference in the long run?
6. Should we use up underground water supplies now or save them for some future time?
7. How much should the United States invest to provide clean water to people in less-developed countries?
8. How should we compare the values of free-flowing rivers and natural ecosystems with the benefits of flood control, water diversion projects, hydroelectric power, and dammed reservoirs?
9. Would it be feasible to change from flush toilets and using water as a medium for waste disposal to some other system? What might be the best way to accomplish this?
10. How does water differ from other natural liquids? How do the properties of water make the hydrologic cycle, and life, possible? (You may need to review chapter 2 to answer this question.)

KEY TERMS

aquifer 428
artesian 428
condensation 423
condensation nuclei 423
consumption 431
degradation 431
desalination 436
dew point 423
discharge 429
evaporation 423

groundwater 428
infiltration 428
rain shadow 424
recharge zones 428
relative humidity 423
renewable water supplies 429
residence time 427
saltwater intrusion 435
saturation point 423
sinkholes 435
sublimation 423
subsidence 435
watershed 439
water stress 433
water table 428
withdrawal 431
zone of aeration 428
zone of saturation 428

ADDITIONAL INFORMATION ON THE INTERNET

Visit our website at http://www.mhhe.com/environmental science/ for specific information about each of the topics below:

Water Resources Information from the US Geological Survey
Water Use and Management: US Bureau of Reclamation
US Fish and Wildlife Service National Wetlands Inventory
Water Research Institute, Israel
Friends of the River: California River Conservation Organization
ESA Aquatic Ecology Section
Groundwater.com. An internet site with links to hundreds of other sites relating to groundwater
Massachusetts Water Resources Authority: is the water safe to drink?
10 Tips for Water Conservation
World Health Organization: Drinking Water Quality
European Rivers Network homepage
Israeli-Palestinian Water Dispute
Water Research Institute, Israel

CHAPTER 20

Water Pollution

Water, water, everywhere; nor any drop to drink.

—Samuel Taylor Coleridge—

OBJECTIVES

After studying this chapter, you should be able to:

- define *water pollution* and describe the sources and effects of some major types.
- appreciate why access to sewage treatment and clean water are important to people in developing countries.
- discuss the status of water quality in developed and developing countries.
- delve into groundwater problems and suggest ways to protect this precious resource.
- fathom the causes and consequences of ocean pollution.
- weigh the advantages and disadvantages of different human waste disposal techniques.
- judge the impact of water pollution legislation and differentiate between best available and best practical technology, and total maximum daily pollution loads.

ESSENTIAL STUDY PARTNER

The Essential Study Partner (ESP) CD-ROM that accompanies this textbook contains additional information on the following topics:

- waterborne disease: physical resources: water: quality
- eutrophication, sewage, and aquatic effects: physical resources: water: quality
- heavy metals, organic chemicals, pesticides, erosion, and thermal pollution: physical resources: water: quality
- biomagnification of dioxins: physical resources: water: pollution control
- groundwater contamination: physical resources: water: quality
- oil spills, and ocean pollution: physical resources: water: quality
- cleanup and source reduction: physical resources: water: pollution control
- Clean Water Act: physical resources: water: quality

Above: Water pollution is probably the single greatest cause of human disease worldwide.

A Flood of Pigs

When Hurricane Floyd roared ashore at Cape Fear in the early morning hours of September 16,1999, it brought much more than just strong winds to North Carolina. More than 50 cm (20 in) of rain poured down in less than 24 hours on ground already saturated from previous storms. The eastern part of North Carolina is a low, coastal plain drained by an intricate web of sluggish creeks and meandering rivers. As the deluge poured down, these myriad channels filled, spilled over, and merged to create an enormous lake more than 300 km (nearly 200 mi) across, covering towns, farms, factories, and forests. The Great Dismal Swamp in the northeast corner of the state became truly soggy and dismal. Worst of all were the thousands of hog and poultry operations and their open manure lagoons submerged by the raging water.

In the 1990s, North Carolina became the leading turkey-producing state in the United Sates, and second to Iowa in pork production, with more than 45 million turkeys and 10 million pigs (but only 7 million people), mostly raised in huge, factory operations. Manure is washed out of giant confinement barns and stored in huge ponds, each typically about the size of several football fields and holding around 40,000 m^3 (10 million gal) of liquid waste. The heart of corporate animal-farming is along the sluggish Neuse River as it winds across the flat coastal plain to its estuary on Pamlico Sound (fig. 20.1). Even now, no flood protection measures are required for large-scale animal operations. Environmentalists have warned for years that storage of such massive amounts of manure in a storm-prone floodplain is an accident waiting to happen. A 1996 editorial in the Raleigh *News and Observer* complained: "Are we so foolish as to think the winds won't blow, the rains won't fall, and the rivers won't rise again?"

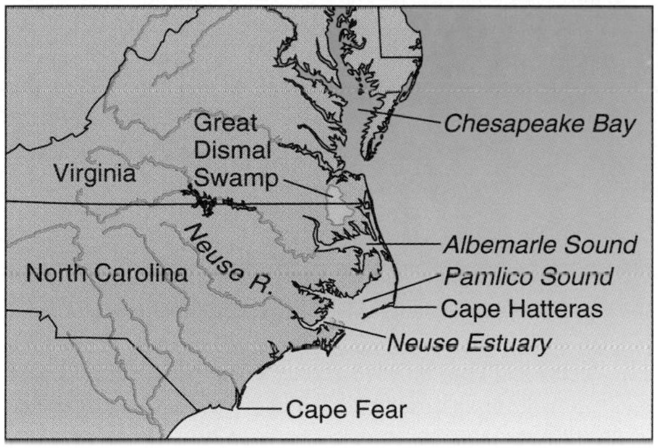

FIGURE 20.1 Heavy rains from Hurricane Floyd flooded a large area of eastern North Carolina in 1999, and washed millions of gallons of hog and turkey manure into Pamlico and Albemarle Sounds.

We'll probably never know the entire damage from Hurricane Floyd. Fifty-one people died in eastern North Carolina, about 8000 homes were damaged or destroyed, and a million ha (roughly 2.5 million acres) of crops were destroyed. Property losses were several billion dollars. Environmental damage is harder to measure. Millions of dead pigs, chickens, turkeys, and cattle littered the landscape or washed out to sea. Flooded junkyards, factories, and chemical plants oozed a toxic mix of gasoline, oil, and other chemicals. Twenty municipal wastewater treatment plants were swamped. Environmentalists claim that at least 250 overflowing manure ponds disgorged around 10 million m^3 (roughly 2.5 billion gal) of hog and poultry waste into the flood. The Carolina Pork Council, on the other hand, says that only three of 4000 waste lagoons overflowed. The fecal coliform bacteria now found in wells, industry claims, probably came from humans or wild Canadian geese. At the same time, North Carolina farmers are asking the federal government for $1 billion in flood relief to rebuild damaged barns and waste lagoons.

Whatever the source, decomposing sewage and dead organisms have created an anerobic "dead zone" in the Pamlico Sound, second largest estuary system in the U.S. Nutrients also are thought to contribute to blooms of toxic microorganisms such as the dinoflagellate, *Pfisteria,* which kills both finfish and shellfish, and has long-term adverse health effects on humans who eat seafood, get water on their skin, or even breathe toxic fumes from contaminated sources. When the floods finally receded, tens of thousands of dead animals were buried in shallow, watery graves, where they continue to contaminate groundwater.

Few environmental science issues are as widespread or emotionally charged as water pollution. According to the EPA, the vast majority of Americans—some 218 million—live within 16 km (10 mi) of a degraded water body. The dismal example of the flood of dead pigs introduces many of the topics we'll cover in this chapter: point and nonpoint water pollution, oxygen depletion and ocean dead zones, polluted groundwater, toxic and infectious aquatic organisms, waste disposal, and watershed management. We'll look at water quality in the United Sates and other countries, ocean pollution, stream and river health, and methods of water pollution control, as well as some important legislation and water policy issues that you, as a student of environmental science, should know something about. Let's get started.

WHAT IS WATER POLLUTION?

Any physical, biological, or chemical change in water quality that adversely affects living organisms or makes water unsuitable for desired uses can be considered pollution. Often, however, a change that adversely affects one organism may be advantageous to another. Nutrients that stimulate oxygen consumption by bacteria and other decomposers in a river or lake, for instance, may reduce some fish populations, but will stimulate a flourishing community of decomposers. Whether the quality of the water has suffered depends on your perspective. There are natural sources of water contamination, such as poison springs, oil seeps, and sedimentation from erosion, but in this chapter we will focus primarily on human-caused changes that affect water quality or usability.

FIGURE 20.2 Sewer outfalls, industrial effluent pipes, acid draining out of abandoned mines, and other point sources of pollution are generally easy to recognize.

Pollution control standards and regulations usually distinguish between point and nonpoint pollution sources. Factories, power plants, sewage treatment plants, underground coal mines, and oil wells are classified as **point sources** because they discharge pollution from specific locations, such as drain pipes, ditches, or sewer outfalls (fig. 20.2). These sources are discrete and identifiable, so they are relatively easy to monitor and regulate. It is generally possible to divert effluent from the waste streams of these sources and treat it before it enters the environment.

In contrast, **nonpoint sources** of water pollution are scattered or diffuse, having no specific location where they discharge into a particular body of water. Nonpoint sources include runoff from farm fields and feedlots (fig. 20.3), golf courses, lawns and gardens, construction sites, logging areas, roads, streets, and parking lots. Whereas point sources may be fairly uniform and predictable throughout the year, nonpoint sources are often highly episodic. The first heavy rainfall after a dry period may flush high concentrations of gasoline, lead, oil, and rubber residues off city streets, for instance, while subsequent runoff may have lower levels of these pollutants. Spring snowmelt carries high levels of atmospheric acid deposition into streams and lakes in some areas. The irregular timing of these events, as well as their multiple sources and scattered location, makes them much more difficult to monitor, regulate, and treat than point sources.

Perhaps the ultimate in diffuse, nonpoint pollution is **atmospheric deposition** of contaminants carried by air currents and precipitated into watersheds or directly onto surface waters as rain, snow, or dry particles. The Great Lakes, for example, have been found to be accumulating industrial chemicals such as PCBs and dioxins, as well as agricultural toxins such as the insecticide toxaphene that cannot be accounted for by local sources alone. The nearest sources for many of these chemicals are sometimes thousands of kilometers away.

FIGURE 20.3 This bucolic scene looks peaceful and idyllic, but allowing cows to trample stream banks is a major cause of bank erosion and water pollution. Nonpoint sources such as this have become the leading unresolved cause of stream and lake pollution in the United States.

Amounts of these pollutants can be quite large. It is estimated that there are 600,000 kg of the herbicide atrazine in the Great Lakes, most of which is thought to have been deposited from the atmosphere. Concentration of persistant chemicals up the food chain can produce high levels in top predators. Several studies have indicated health problems among people who regularly eat fish from the Great Lakes.

Ironically, lakes also can be pollution sources as well. In the past twelve years, about 26,000 metric tons of PCBs have "disappeared" from Lake Superior. Apparently, these compounds evaporate from the lake surface and are carried by air currents to other areas where they are redeposited.

TYPES AND EFFECTS OF WATER POLLUTION

Although the types, sources, and effects of water pollutants are often interrelated, it is convenient to divide them into major categories for discussion (table 20.1). Let's look more closely at some of the important sources and effects of each type of pollutant.

Infectious Agents

The most serious water pollutants in terms of human health worldwide are pathogenic organisms (chapter 9). Among the most important waterborne diseases are typhoid, cholera, bacterial and amoebic dysentery, enteritis, polio, infectious hepatitis, and schistosomiasis. Malaria, yellow fever, and filariasis are transmitted by insects that have aquatic larvae. Altogether, at least 25 million deaths each year are blamed on these water-related diseases. Nearly two-thirds of the mortalities of children under 5 years old are associated with waterborne diseases.

The main source of these pathogens is from untreated or improperly treated human wastes. Animal wastes from feedlots or fields near waterways and food processing factories with inadequate waste treatment facilities also are sources of disease-causing organisms.

TABLE 20.1 Major Categories of Water Pollutants

CATEGORY	EXAMPLES	SOURCES
A. CAUSES HEALTH PROBLEMS		
1. Infectious agents	Bacteria, viruses, parasites	Human and animal excreta
2. Organic chemicals	Pesticides, plastics, detergents, oil, and gasoline	Industrial, household, and farm use
3. Inorganic chemicals	Acids, caustics, salts, metals	Industrial effluents, household cleansers, surface runoff
4. Radioactive materials	Uranium, thorium, cesium, iodine, radon	Mining and processing of ores, power plants, weapons production, natural sources
B. CAUSES ECOSYSTEM DISRUPTION		
1. Sediment	Soil, silt	Land erosion
2. Plant nutrients	Nitrates, phosphates, ammonium	Agricultural and urban fertilizers, sewage, manure
3. Oxygen-demanding wastes	Animal manure and plant residues	Sewage, agricultural runoff, paper mills, food processing
4. Thermal	Heat	Power plants, industrial cooling

In developed countries, sewage treatment plants and other pollution-control techniques have reduced or eliminated most of the worst sources of pathogens in inland surface waters. Furthermore, drinking water is generally disinfected by chlorination so epidemics of waterborne diseases are rare in these countries. The United Nations estimates that 90 percent of the people in developed countries have adequate (safe) sewage disposal, and 95 percent have clean drinking water.

The situation is quite different in less-developed countries. The United Nations estimates that at least 2.5 billion people in these countries lack adequate sanitation, and that about half these people also lack access to clean drinking water. Conditions are especially bad in remote, rural areas where sewage treatment is usually primitive or nonexistent, and purified water is either unavailable or too expensive to obtain. The World Health Organization estimates that 80 percent of all sickness and disease in less-developed countries can be attributed to waterborne infectious agents and inadequate sanitation.

If everyone had pure water and satisfactory sanitation, the World Bank estimates that 200 million fewer episodes of diarrheal illness would occur each year, and 2 million childhood deaths would be avoided. Furthermore, 450 million people would be spared debilitating roundworm or fluke infections. Surely these are goals worth pursuing.

Detecting specific pathogens in water is difficult, time-consuming, and costly; thus, water quality control personnel usually analyze water for the presence of **coliform bacteria,** any of the many types that live in the colon or intestines of humans and other animals. The most common of these is *Eschericha coli* (or *E. coli*). Many strains of bacteria are normal symbionts in mammals, but some, such as *Shigella, Salmonella,* or *Lysteria* can cause fatal diseases. It is usually assumed that if any coliform bacteria are present in a water sample, infectious pathogens are present also.

To test for coliform bacteria, a water sample (or a filter through which a measured water sample has passed) is placed in a dish containing a nutrient medium that supports bacterial growth. After 24 hours at the appropriate temperature, living cells will have produced small colonies. If *any* colonies are found in drinking water samples, the U.S. Environmental Protection Agency considers the water unsafe and requiring disinfection. The EPA-recommended maximum coliform count for swimming water is 200 colonies per 100 ml, but some cities and states allow higher levels. If the limit is exceeded, the contaminated pool, river, or lake usually is closed to swimming (fig. 20.4).

Oxygen-Demanding Wastes

The amount of oxygen dissolved in water is a good indicator of water quality and of the kinds of life it will support. Water with an oxygen content above 6 parts per million (ppm) will support game fish and other desirable forms of aquatic life. Water with less than 2 ppm oxygen will support mainly worms, bacteria, fungi, and other detritus feeders and decomposers. Oxygen is added to water

FIGURE 20.4 Our national goal of making all surface waters in the United States "fishable and swimmable" has not been fully met, but scenes like this have been reduced by pollution control efforts.

by diffusion from the air, especially when turbulence and mixing rates are high, and by photosynthesis of green plants, algae, and cyanobacteria. Oxygen is removed from water by respiration and chemical processes that consume oxygen.

The addition of certain organic materials, such as sewage, paper pulp, or food-processing wastes, to water stimulates oxygen consumption by decomposers. The impact of these materials on water quality can be expressed in terms of **biochemical oxygen demand (BOD):** a standard test of the amount of dissolved oxygen consumed by aquatic microorganisms over a five-day period. An alternative method, called the chemical oxygen demand (COD), uses a strong oxidizing agent (dichromate ion in 50 percent sulfuric acid) to completely break down all organic matter in a water sample. This method is much faster than the BOD test, but normally gives much higher results because it oxidizes compounds not ordinarily metabolized by bacteria. A third method of assaying pollution levels is to measure **dissolved oxygen (DO) content** directly, using an oxygen electrode. The DO content of water depends on factors other than pollution (for example, temperature and aeration), but it is usually more directly related to whether aquatic organisms survive than is BOD.

The effects of oxygen-demanding wastes on rivers depends to a great extent on the volume, flow, and temperature of the river water. Aeration occurs readily in a turbulent, rapidly flowing river, which is, therefore, often able to recover quickly from oxygen-depleting processes. Downstream from a point source, such as a municipal sewage plant discharge, a characteristic decline and restoration of water quality can be detected either by measuring dissolved oxygen content or by observing the flora and fauna that live in successive sections of the river.

The oxygen decline downstream is called the **oxygen sag** (fig. 20.5). Upstream from the pollution source, oxygen levels support normal populations of clean-water organisms. Immediately below the source of pollution, oxygen levels begin to fall as decomposers metabolize waste materials. Rough fish, such as carp, bullheads, and gar, are able to survive in this oxygen-poor environment where they eat both decomposer organisms and the waste itself. Further downstream, the water may become so oxygen-depleted that only the most resistant microorganisms and invertebrates can survive. Eventually, most of the nutrients are used up, decomposer populations are smaller, and the water becomes oxygenated once again. Depending on the volumes and flow rates of the effluent plume and the river receiving it, normal communities may not appear for several miles downstream.

Plant Nutrients and Cultural Eutrophication

Water clarity (transparency) is affected by sediments, chemicals, and the abundance of plankton organisms, and is a useful measure of water quality and water pollution. Rivers and lakes that have clear water and low biological productivity are said to be **oligotrophic** (oligo = little + trophic = nutrition). By contrast, **eutrophic** (eu + trophic = truly nourished) waters are rich in organisms and organic materials. Eutrophication, an increase in nutrient levels and biological productivity, is a normal part of successional changes (chapter 5) in most lakes. Tributary streams bring in sediments and nutrients that stimulate plant growth. Over time, the pond or lake tends to fill in, eventually becoming a marsh and then a terrestrial biome. The rate of eutrophication and succession depends on water chemistry and depth, volume of inflow, mineral content of the surrounding watershed, and the biota of the lake itself.

Human activities can greatly accelerate eutrophication. An increase in biological productivity and ecosystem succession caused by human activities is called **cultural eutrophication.** Cultural eutrophication can be brought about by increased nutrient flows, higher temperatures, more sunlight reaching the water surface, or a number of other changes. Increased productivity in an aquatic system sometimes can be beneficial. Fish and other desirable species may grow faster, providing a welcome food source. Often, however, eutrophication has undesirable results. An oligotrophic lake or river usually has aesthetic qualities and species of organisms that we value.

The high biological productivity of eutrophic systems is often seen in "blooms" of algae or thick growths of aquatic plants stimulated by elevated phosphorus or nitrogen levels (fig. 20.6). Bacterial populations also increase, fed by larger amounts of organic matter. The water often becomes cloudy or turbid and has unpleasant tastes and odors. The deposition of silt and organic sediment caused by cultural eutrophication can accelerate the "aging" of a water body enormously over natural rates. Lakes and reservoirs that normally might exist for hundreds or thousands of years can be filled in a matter of decades.

Eutrophication also occurs in marine ecosystems, especially in near-shore waters and partially enclosed bays or estuaries. Partially enclosed seas such as the Black Sea, the Baltic, and the Mediterranean tend to be in especially critical condition. During the tourist season, the coastal population of the Mediterranean, for example, swells to 200 million people. Eighty-five percent of the effluents from large cities go untreated into the sea. Beach pollution, fish kills, and contaminated shellfish result. Extensive "dead zones" often form where rivers dump nutrients into estuaries and shallow seas. The

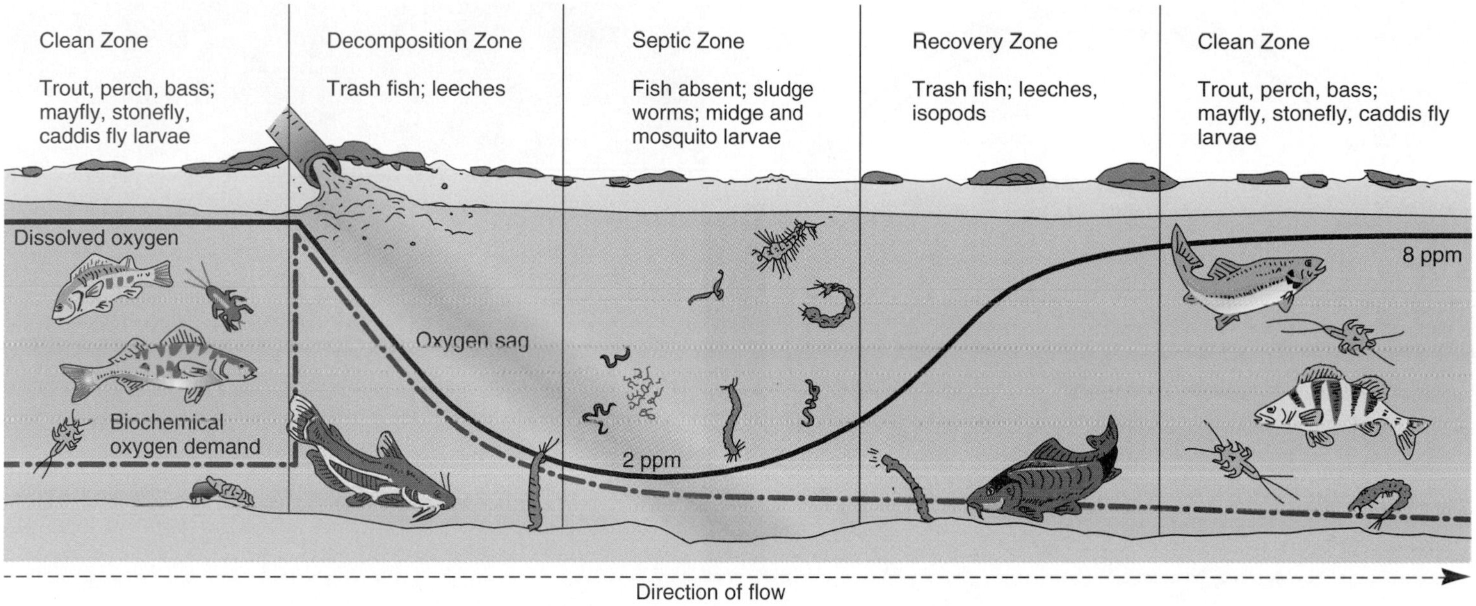

FIGURE 20.5 Oxygen sag downstream of an organic source. A great deal of time and distance may be required for the stream and its inhabitants to recover.

FIGURE 20.6 Eutrophic lake. Nutrients from agriculture and domestic sources have stimulated growth of algae and aquatic plants. This reduces water quality, alters species composition, and lowers recreational and aesthetic values of the lake.

largest in the world occurs during summer months in the Gulf of Mexico at the mouth of the Mississippi River. This hypoxic zone (less than 2 mg oxygen per liter) can cover 18,000 km^2 (7000 mi^2, or about as big as New Jersey). The hypoxic zone in Pamlico Sound following Hurricane Floyd was only about one-tenth this size.

Toxic Tides

According to the Bible, the first plague to afflict the Egyptians when they wouldn't free Moses and the Israelites was that the water in the Nile turned into blood. All the fish died and the people were unable to drink the water, a terrible calamity in a desert country. Some modern scientists believe this may be the first recorded history of a **red tide** or a bloom of deadly aquatic microorganisms called dinoflagellates. Red tides—and other colors, depending on the species involved—have become increasingly common in slow-moving rivers, brackish lagoons, estuaries, and bays, as well as near-shore ocean waters where nutrients and wastes wash down our rivers.

One of the most feared of these organisms is *Pfiesteria piscicida,* an extraordinarily poisonous dinoflagellate that only recently has been recognized as a killer of finfish and shellfish in polluted rivers and estuaries such as North Carolina's Pamlico Sound, where it recently has wiped out hundreds of thousands to millions of fish every year. Dinoflagellates are peculiar organisms with complex life cycles and many different shapes. They are unicellular, photosynthetic microorganisms distinct from both plants and animals. They typically swim with two slender, whiplike flagella, one of which vibrates in a shallow grove in the cell surface and causes the organism to rotate, while the other propels it through the water. *Pfiesteria* can change into at least two dozen distinct forms and sizes depending on water temperature, turbulence, and food supply available (fig. 20.7). In calm, warm, brackish water, rich in sewage or other nutrients, *Pfiesteria* assumes a spherical, nontoxic, but predatory, swimming form that preys on bacteria and plankton.

Under the right conditions, a population explosion can produce a dense bloom of these cells, which can reproduce either by binary fission or sexual fusion of small gametes. If fish blunder into this profuse swarm, *Pfiesteria* quickly turn to a toxic, swimming form that attacks with soluble poisons. These toxins paralyze fish, so they can't escape, and produce skin lesions that expose the fish to infections by pathogenic bacteria and fungi. The predatory zoospores feed on substances that leak from the sores, and also eat skin and flesh directly. When the fish die, zoospores change into star-shaped or lobose

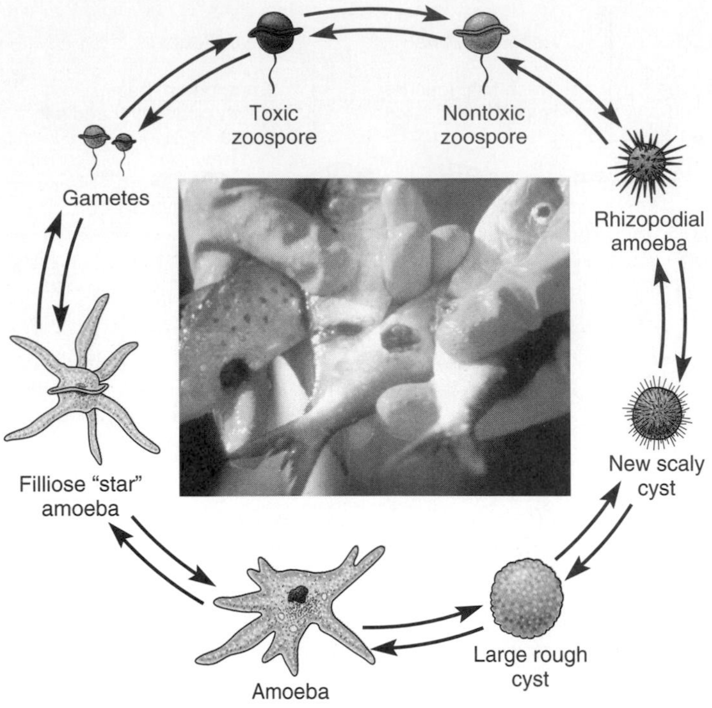

FIGURE 20.7 *Pfiesteria piscicida* life cycle. This toxic, predatory dinoflagellate has at least two dozen different sizes and cell shapes depending on nutrient supply and environmental conditions. They attack fish, causing large skin lesions that often are lethal.
Source: Adapted from J. M. Burkholder (1999), *Scientific American,* 281(2): pp. 42–49.

FIGURE 20.8 A mother from Minamata, Japan, bathes her daughter, who suffered permanent brain damage and birth defects from mercury-contaminated seafood the mother ate while pregnant. This kind of poisoning is now known as Minamata Disease.

(lobed) amoebae that crawl along the bottom and feed on dead carcasses. If the water is cold and calm, these amoebae can attack fish directly; or if it is very turbulent, they spend their time crawling on the bottom or burrowing into the mud looking for something to eat. If the environment becomes adverse, any of these forms can turn into resistant cysts that can withstand extremely harsh conditions.

Perhaps the scariest thing about *Pfiesteria* is the recently recognized fact that it can be extremely poisonous to humans, either through eating contaminated seafood, or by breathing aerosols (airborne mist or dust) containing *Pfiesteria* cells or their secretions. The symptoms of *Pfiesteria* poisoning include headaches, blurred vision, aching joints, burning muscles, difficulty breathing, disorientation, memory loss, and long-term damage to the brain, liver, pancreas, kidneys, and immune system. Just being on a boat for a few minutes in the midst of *Pfiesteria* bloom can result in debilitating symptoms and long-term injury. This does seem like a plague, indeed, and it appears to be provoked primarily by human actions. Runoff of agricultural wastes, fertilizer, human sewage, and other nutrients into shallow, near-shore water, seems to be the main cause of this terrible affliction.

Inorganic Pollutants

Some toxic inorganic chemicals are released from rocks by weathering, are carried by runoff into lakes or rivers, or percolate into groundwater aquifers. This pattern is part of natural mineral cycles (chapter 3). Humans often accelerate the transfer rates in these cycles thousands of times above natural background levels through the mining, processing, using, and discarding of minerals.

In many areas, toxic, inorganic chemicals introduced into water as a result of human activities have become the most serious form of water pollution. Among the chemicals of greatest concern are heavy metals, such as mercury, lead, tin, and cadmium. Supertoxic elements, such as selenium and arsenic, also have reached hazardous levels in some waters. Other inorganic materials, such as acids, salts, nitrates, and chlorine, that normally are not toxic at low concentrations may become concentrated enough to lower water quality or adversely affect biological communities.

Metals

Many metals such as mercury, lead, cadmium, and nickel are highly toxic. Levels in the parts per million range—so little that you cannot see or taste them—can be fatal. Because metals are highly persistent, they accumulate in food chains and have a cumulative effect in humans. A famous case of mercury poisoning in Japan in the 1950s was one of our first warnings of this danger (fig. 20.8).

Another mercury-poisoning disaster appears to be in process in South America. Since the mid-1980s, a gold rush has been under way in Brazil, Ecuador, and Bolivia. Forty thousand *garimperios* or prospectors have invaded the jungles along the Amazon River and its tributaries to pan for gold. They use mercury to trap the gold and separate it from sediments. Then, the mercury is boiled off with a blow torch. Miners and their families suffer nerve damage from breathing the toxic fumes. Estimates are that 130 tons of mercury per year are deposited in the Amazon, which will be impossible to clean up.

We have come to realize that other heavy metals released as a result of human activities also are concentrated by hydrological and biological processes so that they become hazardous to both natural ecosystems and human health. A condition known as Itai-Itai (literally ouch-ouch) disease that developed in Japanese living near the Jintsu River was traced to cadmium poisoning from mining and smelting waste-water discharges. Bacteria forming methylated tin have been found in sediments in Chesapeake Bay, leading to worries that this toxic metal also may be causing unsuspected health effects. The use of tin compounds as antifouling agents on ship bottoms has been banned because of its toxic effects.

Lead poisoning has been known since Roman times to be dangerous to human health. Lead pipes are a serious source of drinking water pollution, especially in older homes or in areas where water is acidic and, therefore, leaches more lead from pipes. Even lead solder in pipe joints and metal containers can be hazardous. In 1990, the EPA lowered the maximum limit for lead in public drinking water from 50 parts per billion (ppb) to 20 ppb. Some public health officials argue that lead is neurotoxic at any level, and the limits should be less than 10 ppb.

Mine drainage and leaching of mining wastes are serious sources of metal pollution in water. A survey of water quality in eastern Tennessee found that 43 percent of all surface streams and lakes and more than half of all groundwater used for drinking supplies was contaminated by acids and metals from mine drainage. In some cases, metal levels were 200 times higher than what is considered safe for drinking water.

Nonmetallic Salts

Desert soils often contain high concentrations of soluble salts, including toxic selenium and arsenic (In Depth, p. 454). You have probably heard of poison springs and seeps in the desert where these compounds are brought to the surface by percolating groundwater. Irrigation and drainage of desert soils mobilize these materials on a larger scale and can result in serious pollution problems, as in Kesterson Marsh in California where selenium poisoning killed thousands of migratory birds in the 1980s.

Such salts as sodium chloride (table salt) that are nontoxic at low concentrations also can be mobilized by irrigation and concentrated by evaporation, reaching levels that are toxic for plants and animals. Salt levels in the San Joaquin River in central California rose from 0.28 gm/l in 1930 to 0.45 gm/l in 1970 as a result of agricultural runoff. Salinity levels in the Colorado River and surrounding farm fields have become so high in recent years that millions of hectares of valuable croplands have had to be abandoned. The United States has built a huge desalinization plant at Yuma, Arizona, to reduce salinity in the river. In northern states, millions of tons of sodium chloride and calcium chloride are used to melt road ice in the winter. The corrosive damage to highways and automobiles and the toxic effects on vegetation are enormous. Leaching of road salts into surface waters may have a similarly devastating effect on aquatic ecosystems.

Acids and Bases

Acids are released as by-products of industrial processes, such as leather tanning, metal smelting and plating, petroleum distillation, and organic chemical synthesis. Coal mining is an especially important source of acid water pollution. Sulfur compounds in coal react with oxygen and water to make sulfuric acid. Thousands of kilometers of streams in the United States have been acidified by acid mine drainage, some so severely that they are essentially lifeless.

Coal and oil combustion also leads to formation of atmospheric sulfuric and nitric acids (chapter 18), which are disseminated by long-range transport processes and deposited via precipitation (acidic rain, snow, fog, or dry deposition) in surface waters. Where soils are rich in such alkaline material as limestone, these atmospheric acids have little effect because they are neutralized. In high mountain areas or recently glaciated regions where crystalline bedrock is close to the surface and lakes are oligotrophic, however, there is little buffering capacity (ability to neutralize acids) and aquatic ecosystems can be severely disrupted. These effects were first recognized in the mountains of northern England and Scandinavia about 30 years ago.

In recent years, aquatic damage due to acid precipitation has been reported in about 200 lakes in the Adirondack Mountains of New York State and in several thousand lakes in eastern Quebec, Canada. Game fish, amphibians, and sensitive aquatic insects are generally the first to be killed by increased acid levels in the water. If acidification is severe enough, aquatic life is limited to a few resistant species of mosses and fungi. Increased acidity may result in leaching of toxic metals, especially aluminum, from soil and rocks, making water unfit for drinking or irrigation, as well.

Organic Chemicals

Thousands of different natural and synthetic organic chemicals are used in the chemical industry to make pesticides, plastics, pharmaceuticals, pigments, and other products that we use in everyday life. Many of these chemicals are highly toxic (chapter 9). Exposure to very low concentrations (perhaps even parts per quadrillion in the case of dioxins) can cause birth defects, genetic disorders, and cancer. Some can persist in the environment because they are resistant to degradation and toxic to organisms that ingest them. Contamination of surface waters and groundwater by these chemicals is a serious threat to human health.

The two most important sources of toxic organic chemicals in water are improper disposal of industrial and household wastes and runoff of pesticides from farm fields, forests, roadsides, golf courses, and other places where they are used in large quantities. The U.S. EPA estimates that about 500,000 metric tons of pesticides are used in the United States each year. Much of this material washes into the nearest waterway, where it passes through ecosystems and may accumulate in high levels in certain nontarget organisms. The bioaccumulation of DDT in aquatic ecosystems was one of the first of these pathways to be understood. Dioxins, and other chlorinated hydrocarbons (hydrocarbon molecules that contain chlorine atoms) have been shown to accumulate to dangerous levels in the fat of salmon, fish-eating birds, and humans and to cause health problems similar to those resulting from toxic metal compounds (fig. 20.9).

Hundreds of millions of tons of hazardous organic wastes are thought to be stored in dumps, landfills, lagoons, and underground tanks in the United States (chapter 23). Many, perhaps most, of these sites have leaked toxic chemicals into surface waters or groundwater or both. The EPA estimates that about 26,000 hazardous waste sites will require cleanup because they pose an imminent threat to public health, mostly through water pollution.

IN DEPTH. . . . .

Arsenic in Drinking Water

When we think of water pollution, we usually visualize sewage or industrial effluents pouring out of a discharge pipe, but there are natural toxins that threaten us as well. One of these is arsenic, a common contaminate in drinking water that may be poisoning millions of people around the world. Arsenic has been known since the fourth century B.C. to be a potent poison. It has been used for centuries as a rodenticide, insecticide, and weed killer, as well as a way of assassinating enemies. Because it isn't metabolized or excreted from the body, arsenic accumulates in hair and fingernails, where it can be detected long after death. Napoleon Bonaparte was found recently to have high enough levels of arsenic in his body to suggest he was poisoned.

Perhaps the largest population to be threatened by naturally occurring groundwater contamination by arsenic is in West Bengal, India, and adjacent areas of Bangladesh. Arsenic, in the form of insoluble salts, occurs naturally in the bedrock that underlies much of this region. Under normal conditions, the groundwater stays relatively free of arsenic in a soluble form. Rapid population growth, industrialization, and intensification of agricultural irrigation, however, have put increasing stresses on the limited surface water supplies of this region (see Chapter 19). Groundwater has all but replaced other water sources for most people in West Bengal, especially in the dry season.

In the 1960s, thousands of deep tube wells were sunk throughout the region to improve water supplies. Much of this humanitarian effort was financed by loans from the world bank in the name of human development. At first, villagers were suspicious of well water, regarding it as unnatural and possibly evil. But as surface water supplies diminished, dusty Bengali villages became more and more dependent on this new source of supposedly fresh, clean water. By the late 1980s, health workers

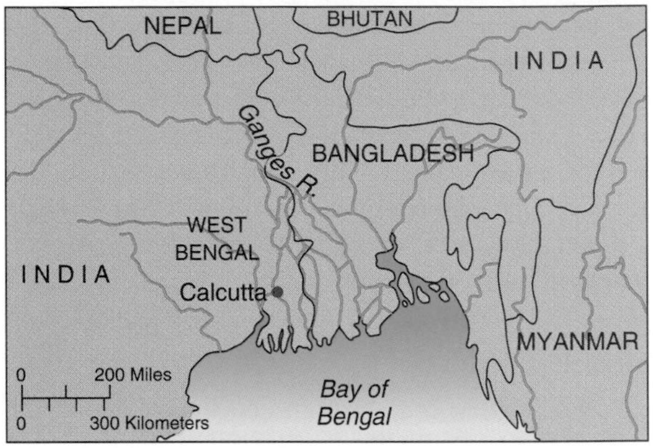

West Bengal and adjoining areas of Bangladesh have hundreds of millions of people who may be exposed to dangerous arsenic levels in well water.

Sediment

Rivers have always carried sediment to the oceans, but erosion rates in many areas have been greatly accelerated by human activities. As chapters 11 and 19 describe, some rivers carry astounding loads of sediment. Erosion and runoff from croplands contribute about 25 billion metric tons of soil, sediment, and suspended solids to world surface waters each year. Forests, grazing lands, urban construction sites, and other sources of erosion and runoff add at least 50 billion additional tons. This sediment fills lakes and reservoirs, obstructs shipping channels, clogs hydroelectric turbines, and makes purification of drinking water more costly. Sediments smother gravel beds in which insects take refuge and fish lay their eggs. Sunlight is blocked so that plants cannot carry out photosynthesis and oxygen levels decline. Murky, cloudy water also is less attractive for swimming, boating, fishing, and other recreational uses (fig. 20.10).

Sediment also can be beneficial. Mud carried by rivers nourishes floodplain farm fields. Sediment deposited in the ocean at river mouths creates valuable deltas and islands. The Ganges River, for instance, builds up islands in the Bay of Bengal that are eagerly colonized by land-hungry people of Bangladesh. In Louisiana, lack of sediment in the Mississippi River (it is being trapped by dams upstream) is causing biologically rich coastal wetlands to waste away. Sediment also can be harmful. Excess sediment deposits can fill estuaries and smother aquatic life on coral reefs and shoals near shore. As with many natural environmental processes, acceleration as a result of human intervention generally diminishes the benefits and accentuates the disadvantages of the process.

FIGURE 20.9 The deformed beak of this young robin is thought to be due to dioxins, DDT, and other toxins in its mother's diet.

(continued)

became aware of widespread signs of chronic arsenic poisoning among villagers in both India and Bangladesh. Symptoms of chronic arsenicosis include watery and inflamed eyes, gastrointestinal cramps, gradual loss of strength, scaly skin and skin tumors, anemia, confusion, and, eventually, death.

Why is arsenic poisoning appearing now? Part of the reason is increased dependence on well water, but some villages have had wells for centuries with no problem. One theory is that excessive withdrawals now lower the water table during the dry season, exposing arsenic-bearing rocks to air, which converts normally insoluble salts to soluble oxides. When aquifers are refilled during the next rainy season, dissolved arsenic can be pumped out. Health workers estimate that the total number of potential victims in India and Bangladesh may exceed 200 million people. But with no other source of easily accessible or affordable water, few of the poorest people have much choice. Although few places in North America have as high groundwater arsenic content as West Bengal, there are worries that millions of Americans also are exposed to dangerously high levels of this toxic element. In 1942, the U.S. Government set the acceptable level o arsenic in drinking water at 50 micrograms per liter (50 parts per billion or ppb). Although this standard was set before the connection between arsenic and cancer was understood, it has never been revised. Recent studies suggest that the risks of certain kinds of cancer from a lifetime of drinking water with 50 ppb of arsenic may be as high as 1 in 100, or 10,000 times the normally accepted threshold for acceptable risk. Repeated attempts to lower the standard to 10 ppb have been met with resistance from public officials and private water supply owners, who complain that it would cost too much to upgrade their systems. The government has no business, they maintain, in telling us what we can or cannot drink.

Ethical Considerations

What do you think? If you choose to drink water that gives you a 1 in 100 chance of cancer, is that your business or does the government have a right (or obligation) to stop you? Is it ethical to allow customers of water districts to be exposed to a poison that they know nothing about? Would it be enough to require public systems to inform their customers of the risks involved? Would your parents or friends understand the implications of 1 in 100 risk of cancer? Do they smoke? What could be done to help the hundreds of millions of people in India and Bangladesh who now drink contaminated water? Is it our responsibility? What if it was our money that installed the tube wells in the first place? Does that change the picture?

Thermal Pollution and Thermal Shocks

Raising or lowering water temperatures from normal levels can adversely affect water quality and aquatic life. Water temperatures are usually much more stable than air temperatures, so aquatic organisms tend to be poorly adapted to rapid temperature changes. Lowering the temperature of tropical oceans by even one degree can be lethal to some corals and other reef species. Raising water temperatures can have similar devastating effects on sensitive organisms. Oxygen solubility in water decreases as temperatures increase, so species requiring high oxygen levels are adversely affected by warming water.

Humans cause thermal pollution by altering vegetation cover and runoff patterns, as well as by discharging heated water directly into rivers and lakes. As chapter 19 shows, nearly half the water we withdraw is used for industrial cooling. Electric power plants, metal smelters, petroleum refineries, paper mills, food-processing factories, and chemical manufacturing plants all use and release large amounts of cooling water.

The cheapest way to remove heat from an industrial facility is to draw cool water from an ocean, river, lake, or aquifer, run it through a heat-exchanger to extract excess heat, and then dump the heated water back into the original source. A **thermal plume** of heated water is often discharged into rivers and lakes, where raised temperatures can disrupt many processes in natural ecosystems and drive out sensitive organisms. To minimize these effects, power plants frequently are required to construct artificial cooling ponds

FIGURE 20.10 Sediment and industrial waste flow from this drainage canal into Lake Erie.

or wet- or dry-cooling towers in which heat is released into the atmosphere and water is cooled before being released into natural water bodies. Wet cooling towers are cheaper to build and operate than dry systems, but lose large quantities of water to evaporation.

In some circumstances, introducing heated water into a water body is beneficial. Warming catfish-rearing ponds, for instance, can increase yields significantly. Warm water plumes from power plants often attract fish, birds, and marine mammals that find food and refuge there, especially in cold weather. This artificial environment can be a fatal trap, however. Organisms dependent on the warmth may die if they leave the plume or if the flow of warm water is interrupted by a plant shutdown. The manatee, for example, is an endangered marine mammal species that lives in Florida. Manatees are attracted to the abundant food supply and warm water in power plant thermal plumes and are enticed into spending the winter much farther north than they normally would. On several occasions, a midwinter power plant breakdown has exposed a dozen or more of these rare animals to a sudden thermal shock that they could not survive.

WATER QUALITY TODAY

Surface water pollution is often both highly visible and one of the most common threats to environmental quality. In more developed countries, reducing water pollution has been a high priority over the past few decades. Billions of dollars have been spent on control programs and considerable progress has been made. Still much remains to be done. In developed countries, poor water quality often remains a serious problem. In this section, we will look at progress as well as continuing obstacles in this important area.

Surface Waters in the United States and Canada

Like most developed countries, the United States and Canada have made encouraging progress in protecting and restoring water quality in rivers and lakes over the past 40 years. In 1948, only about one-third of Americans were served by municipal sewage systems, and most of those systems discharged sewage without any treatment or with only primary treatment (the bigger lumps of waste are removed). Most people depended on cesspools and septic systems to dispose of domestic wastes.

Areas of Progress

The 1972 Clean Water Act established a National Pollution Discharge Elimination System (NPDES), which requires an easily revoked permit for any industry, municipality or other entity dumping wastes in surface waters. The permit requires disclosure of what is being dumped and gives regulators valuable data and evidence for litigation. As a consequence, only about 10 percent of our water pollution now comes from industrial or municipal point sources. One of the biggest improvements has been in sewage treatment.

Since the Clean Water Act was passed in 1972, the United States has spent more than $180 billion in public funds and perhaps ten times as much in private investments on water pollution control. Most of that effort has been aimed at point sources, especially to build or upgrade thousands of municipal sewage treatment plants. As a result, nearly everyone in urban areas is now served by municipal sewage systems and no major city discharges raw sewage into a river or lake except as overflow during heavy rainstorms.

This campaign has led to significant improvements in surface water quality in many places. Fish and aquatic insects have returned to waters that formerly were depleted of life-giving oxygen. Swimming and other water-contact sports are again permitted in rivers, lakes, and at ocean beaches that once were closed by health officials.

The Clean Water Act goal of making all U.S. surface waters "fishable and swimmable" has not been fully met, but in 1999 the EPA reported that 91.4 percent of all monitored river miles and 87.5 percent of all assessed lake acres are suitable for their designated uses. This sounds good, but you have to remember that not all water bodies are monitored. Furthermore, the designated goal for some rivers and lakes is merely to be "boatable." Water quality doesn't have to be very high to be able to put a boat on it. Even in "fishable" rivers and lakes, there isn't a guarantee that you can catch anything other than rough fish like carp or bullheads, nor can you be sure that what you catch is safe to eat. Even with billions of dollars of investment in sewage treatment plants, elimination of much of the industrial dumping and other gross sources of pollutants, and a general improvement in water quality, the EPA reports that 21,000 water bodies still do not meet their designated uses. According to the EPA, an overwhelming majority of the American people—almost 218 million—live within 16 km (10 mi) of an impaired water body.

In 1998, a new regulatory approach to water quality assurance was instituted by the EPA. Rather than issue standards on a river by river approach or factory by factory permit discharge, the focus is being changed to watershed-level monitoring and protection. Some 4000 watersheds are monitored for water quality (fig. 20.11). You can find information about your watershed at *http://www.epa.gov/owow/tmdl/*. The intention of this program is to give the public more and better information about the health of their watersheds. In addition, states will have greater flexibility as they identify impaired water bodies and set priorities, and new tools will be used to achieve goals. States are required to identify waters not meeting water quality goals and to develop **total maximum daily loads (TMDL)** for each pollutant and each listed water body. A TMDL is the amount of a particular pollutant that a water body can receive from both point and nonpoint sources. It considers seasonal variation and includes a margin of safety.

By 1999, all 56 states and territories had submitted TMDL lists, and the EPA had approved most of them. Of the 3.5 million mi (5.6 million km) of rivers monitored, only 300,000 mi (480,000 km) fail to meet their clean water goals. Similarly, of 40 million lake acres (99 million ha), only 12.5 percent (in about 20,000 lakes) failed to meet their goal. To give states more flexibility in planning, the EPA

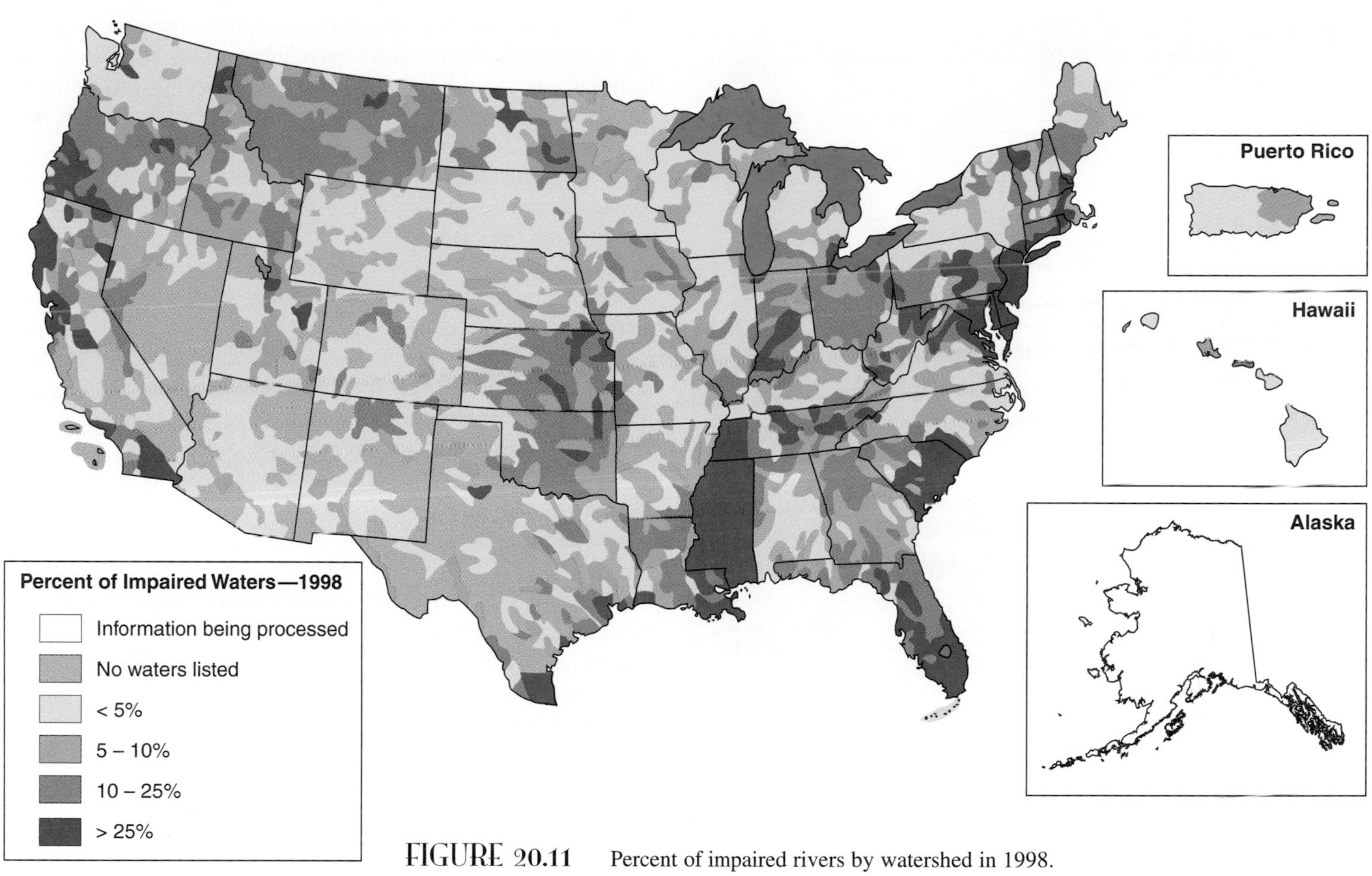

FIGURE 20.11 Percent of impaired rivers by watershed in 1998.

has proposed new rules that include allowances for reasonably foreseeable increases in pollutant loadings to encourage "Smart Growth." In the future, TMDLs also will include load allocations from all nonpoint sources, including air deposition and natural background levels.

An encouraging example of improved water quality is seen in Lake Erie. Although widely regarded as "dead" in the 1960s, the lake today is promoted as the "walleye capital of the world." Bacteria counts and algae blooms have decreased more than 90 percent since 1962. Water that once was murky brown is now clear. Interestingly, part of the improved water quality is due to immense numbers of exotic zebra mussels, which filter the lake water very efficiently. Swimming is now officially safe along 96 percent of the lake's shoreline. Nearly 40,000 nesting pairs of double-crested cormorants nest in the Great Lakes region, up from only about 100 in the 1970s. Anglers now complain that the cormorants eat too many fish. In 1998 wildlife agents found 800 cormorants shot to death in a rookery on Galloo Island at the east end of Lake Ontario.

Passage of the 1970 Water Act in Canada has produced comparable results. Seventy percent of all Canadians in towns over 1000 population are now served by some form of municipal sewage treatment. In Ontario, the vast majority of those systems include tertiary treatment. After ten years of controls, phosphorus levels in the Bay of Quinte in the northeast corner of Lake Ontario have dropped nearly by half, and algal blooms that once turned waters green are less frequent and less intense than they once were. Elimination of mercury discharges from a pulp and paper mill on the Wabigoon-English River system in western Ontario has resulted in a dramatic decrease in mercury contamination that produced Minamata-like symptoms in local native people 20 years ago. Extensive flooding associated with hydropower projects has raised mercury levels in fish to dangerous levels elsewhere, however.

Remaining Problems

The greatest impediments to achieving national goals in water quality in both the United States and Canada are sediment, nutrients, and pathogens, especially from nonpoint discharges of pollutants. These sources are harder to identify and to reduce or treat than are specific point sources. About three-fourths of the water pollution in the United States comes from soil erosion, fallout of air pollutants, and surface runoff from urban areas, farm fields, and feedlots. In the United States, as much as 25 percent of the 46,800,000 metric tons (52 million tons) of fertilizer spread on farmland each year is carried away by runoff (fig. 20.12).

Cattle in feedlots produce some 129,600,000 metric tons (144 million tons) of manure each year, and the runoff from these sites is rich in viruses, bacteria, nitrates, phosphates, and other contaminants. A single cow produces about 30 kg (66 lb) of manure per day, or about as much as that produced by ten people. Some feedlots have 100,000 animals with no provision for capturing or treating runoff water. Imagine

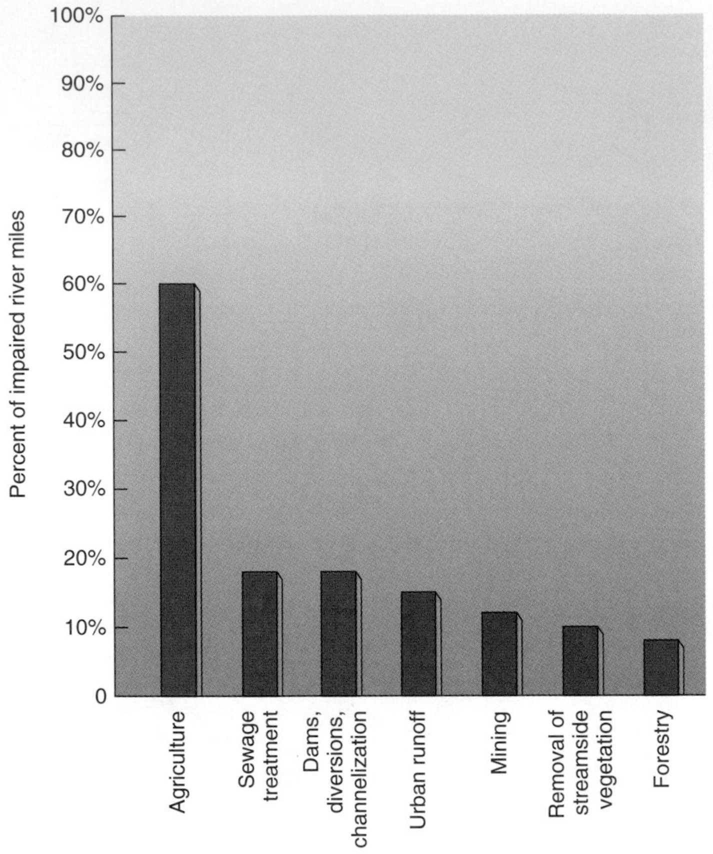

FIGURE 20.12 Percentage of impaired river miles in the United States by source of damage. Totals add up to more than 100 percent because one river can be affected by many sources.
Source: Data from USDA and Natural Resources Conservation Service, *America's Private Land: A Geography of Hope*, USDA, 1996.

drawing your drinking water downstream from such a facility. Pets also can be a problem. It is estimated that the wastes from about a half million dogs in New York City are disposed of primarily through storm sewers, and therefore do not go through sewage treatment.

Loading of both nitrates and phosphates in surface water have decreased from point sources but have increased about fourfold since 1972 from nonpoint sources. Fossil fuel combustion has become a major source of nitrates, sulfates, arsenic, cadmium, mercury, and other toxic pollutants that find their way into water. Carried to remote areas by atmospheric transport, these combustion products now are found nearly everywhere in the world. Toxic organic compounds, such as DDT, PCBs, and dioxins, also are transported long distances by wind currents.

Surface Waters in Other Countries

Japan, Australia, and most of Western Europe also have improved surface water quality in recent years. Sewage treatment in the wealthier countries of Europe generally equals or surpasses that in the United States. Sweden, for instance, serves 98 percent of its population with at least secondary sewage treatment (compared with 70 percent in the United States), and the other 2 percent have primary treatment. Poorer countries have much less to spend on sanitation. Spain serves only 18 percent of its population with even primary sewage treatment. In Ireland, it is only 11 percent, and in Greece, less than 1 percent of the people have even primary treatment. Most of the sewage, both domestic and industrial, is dumped directly into the ocean.

The fall of the "iron curtain" in 1989 revealed appalling environmental conditions in much of the former Soviet Union and its satellite states in eastern and central Europe. The countries closest geographically and socially to western Europe, the Czech Republic, Hungary, East Germany, and Poland have made massive investments and encouraging progress towards cleaning up environmental problems. Parts of Russia itself, however, along with former socialist states in the Balkans and Central Asia, remain some of the most polluted places on earth. In Russia, for example, only about half the tap water is fit to drink. In cities like St. Petersburg, even boiling and filtering isn't enough to make municipal water safe. About one-third of all Russians live in regions where air pollution levels are 10 times higher than World Health Organization safety standards. Life expectancies for Russian men have plummeted from about 72 years in 1980 to 59 years in 1999. Deaths now exceed births in Russia by about one million per year. Only about one-quarter of all Russian children are considered healthy. In heavily industrialized cities like Magnitogorsk, a steel-manufacturing center in the Ural Mountains, 9 out of 10 children suffer from pollution-related illnesses and birth defects.

High levels of radioactivity from the Chernobyl accident in 1986 remain in many former Eastern Bloc countries. The Danube River, which originates in Bavaria and Austria, and flows through Slovakia, Poland, Hungary, Serbia, Romania, and Bulgaria before emptying into the Black Sea, illustrates some of the problems besetting the area (fig. 20.13). Draining a landscape with a long history of unregulated mining and heavy industry, the Danube carries chrome, copper, mercury, lead, zinc, and oil to the Black Sea at 20 times the levels these materials flow into the North Sea. Just one city, Bratislava, the capital of Slovakia, dumps 73 million m^3 (about 2 billion gal) of industrial and municipal wastes into the river each year. With a population of about 100 million people in its catchment area, the beautiful blue Danube isn't blue anymore. Bombing in Serbia during the Kosovo War in 1999 further exacerbated the situation by releasing oil, industrial chemicals, agricultural fertilizers, pesticides, and other toxic materials into the Danube.

There are also some encouraging pollution control stories. In 1997, Minamata Bay in Japan, long synonymous with mercury poisoning, was declared officially clean again. Another important success is found in Europe, where one of its most important rivers has been cleaned up significantly through international cooperation. The Rhine, which starts in the rugged Swiss Alps and winds 1320 km through five countries before emptying through a Dutch delta into the North Sea, has long been a major commercial artery into the heart of Europe. More than 50 million people live in its catchment basin and nearly 20 million get their drinking water from the river or its tributaries. By the 1970s, the Rhine had become so polluted that dozens of fish species disappeared and swimming was discouraged along most of its length.

Efforts to clean up this historic and economically important waterway began in the 1950s, but a disastrous fire at a chemical

FIGURE 20.13 Two Gypsy children from Copsa Mică, Romania, play in a polluted river across from the carbon factory that turns everything in town black. Eventually this filthy water will be flushed down the Danube and into the Black Sea, which has lost nearly all its commercial fisheries in only 25 years.

warehouse near Basel, Switzerland, in 1986 provided the impetus for major changes. Through a long and sometimes painful series of international conventions and compromises, land-use practices, waste disposal, urban runoff, and industrial dumping have been changed and water quality has significantly improved. Oxygen concentrations have gone up fivefold since 1970 (from less than 2 mg/l to nearly 10 mg/l or about 90 percent of saturation) in long stretches of the river. Chemical oxygen demand has fallen fivefold during this same period, and organochlorine levels have decreased as much as tenfold. Many species of fish and aquatic invertebrates have returned to the river. In 1992, for the first time in decades, mature salmon were caught in the Rhine.

The less-developed countries of South America, Africa, and Asia have even worse water quality than do the poorer countries of Europe. Sewage treatment is usually either totally lacking or woefully inadequate. In urban areas, 95 percent of all sewage is discharged untreated into rivers, lakes, or the ocean. Low technological capabilities and little money for pollution control are made even worse by burgeoning populations, rapid urbanization, and the shift of much heavy industry (especially the dirtier ones) from developed countries where pollution laws are strict to less-developed countries where regulations are more lenient.

Appalling environmental conditions often result from these combined factors (fig. 20.14). Two-thirds of India's surface waters are contaminated sufficiently to be considered dangerous to human health. The Yamuna River in New Delhi has 7500 coliform bacteria per 100 ml (37 times the level considered safe for swimming in the United States) *before* entering the city. The coliform count increases to an incredible 24 *million* cells per 100 ml as the river leaves the city! At the same time, the river picks up some 20 million l of industrial effluents every day from New Delhi. It's no wonder that disease rates are high and life expectancy is low in this area. Only 1 percent of India's towns and cities have any sewage treatment, and only eight cities have anything beyond primary treatment.

FIGURE 20.14 Ditches in this Haitian slum serve as open sewers into which all manner of refuse and waste are dumped. The health risks of living under these conditions are severe.

In Malaysia, 42 of 50 major rivers are reported to be "ecological disasters." Residues from palm oil and rubber manufacturing, along with heavy erosion from logging of tropical rainforests, have destroyed all higher forms of life in most of these rivers. In the Philippines, domestic sewage makes up 60 to 70 percent of the total volume of Manila's Pasig River. Thousands of people use the river not only for bathing and washing clothes but also as their source of drinking and cooking water. China treats only 2 percent of its sewage. Of 78 monitored rivers in China, 54 are reported to be seriously polluted. Of 44 major cities in China, 41 use "contaminated" water supplies, and few do more than rudimentary treatment before it is delivered to the public.

Groundwater and Drinking Water Supplies

About half the people in the United States, including 95 percent of those in rural areas, depend on underground aquifers for their drinking water. This vital resource is threatened in many areas by overuse and pollution and by a wide variety of industrial, agricultural, and domestic contaminants. For decades it was widely assumed that groundwater was impervious to pollution because soil would bind chemicals and cleanse water as it percolated through. Springwater or artesian well water was considered to be the definitive standard of water purity, but that is no longer true in many areas.

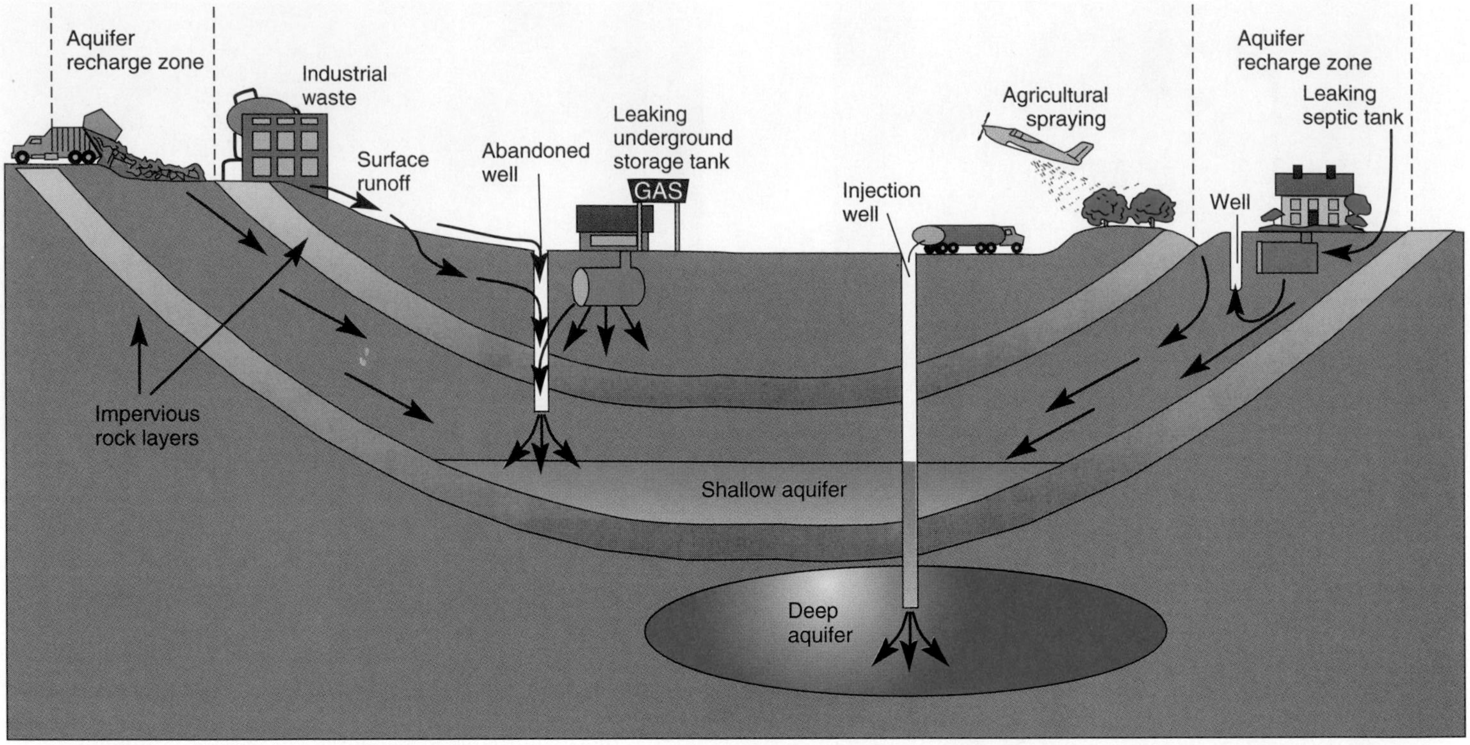

FIGURE 20.15 Sources of groundwater pollution. Septic systems, landfills, and industrial activities on aquifer recharge zones leach contaminants into aquifers. Wells provide a direct route for injection of pollutants into aquifers.

The U.S. EPA estimates that every day some 4.5 trillion l (1.2 trillion gal) of contaminated water seep into the ground in the United States from septic tanks, cesspools, municipal and industrial landfills and waste disposal sites, surface impoundments, agricultural fields, forests, and wells (fig. 20.15). The most important of these in terms of toxicity are probably waste disposal sites. Agricultural chemicals and wastes are responsible for the largest total volume of pollutants and area affected. Because deep underground aquifers often have residence times of thousands of years, many contaminants are extremely stable once underground. It is possible, but expensive, to pump water out of aquifers, clean it, and then pump it back. For very large aquifers, pollution may be essentially irreversible on a human timescale.

In farm country, especially in the Midwest's Corn Belt, fertilizers and pesticides commonly contaminate aquifers and wells. Herbicides such as atrazine and alachlor are widely used on corn and soybeans and show up in about half of all wells in Iowa, for example. Nitrates from fertilizers often exceed safety standards in rural drinking water. These high nitrate levels are dangerous to infants (nitrate combines with hemoglobin in the blood and results in "blue-baby" syndrome). They also are transformed into cancer-causing nitrosamines in the human gut. In Florida, one thousand drinking water wells were shut down by state authorities because of excessive levels of toxic chemicals, mostly ethylene dibromide (EDB), a pesticide used to kill nematodes (roundworms) that damage plant roots.

Although most of the leaky, single-walled underground storage tanks once common at filling stations and factories have now been removed and replaced by more modern ones, a great deal of soil in American cities remains contaminated by previous careless storage and disposal of petroleum products. Considering that a single gallon (3.8 l) of gasoline can make a million gallons of water undrinkable, soil contamination remains a serious problem.

Direct injection of liquid wastes such as oilfield brine, effluents from chemical plants, and treated sewage down wells into deep aquifers is much less common than in the past. It is still allowed in some circumstances, however, to the dismay of some critics.

Abandoned wells represent another major source of groundwater contamination. Some domestic wells lack grouting to prevent surface contaminants from leaking into aquifers that they penetrate. When these wells are no longer in use, they often are not capped adequately, and people forget where they are. They can become direct routes for drainage of surface contaminants into aquifers.

In addition to groundwater pollution problems, contaminated surface waters and inadequate treatment make drinking water unsafe in many areas. A 1996 survey concluded that nearly 20,000 public drinking water systems in the United States expose consumers to contaminants such as lead, pesticides, and pathogens at levels that violate EPA rules (fig. 20.16). A vast majority of these systems are small, serving fewer than 3000 people, but altogether some 50 million people are sometimes at risk. Problems often occur because small systems can't afford modern purification and distribution equipment, regular testing, and trained operators to bring water quality up to acceptable standards.

Every year epidemiologists estimate that around 1.5 million Americans fall ill from infections caused by fecal contamination. In 1993, for instance, a pathogen called cryptosporidium got into the Milwaukee public water system, making 400,000 people sick

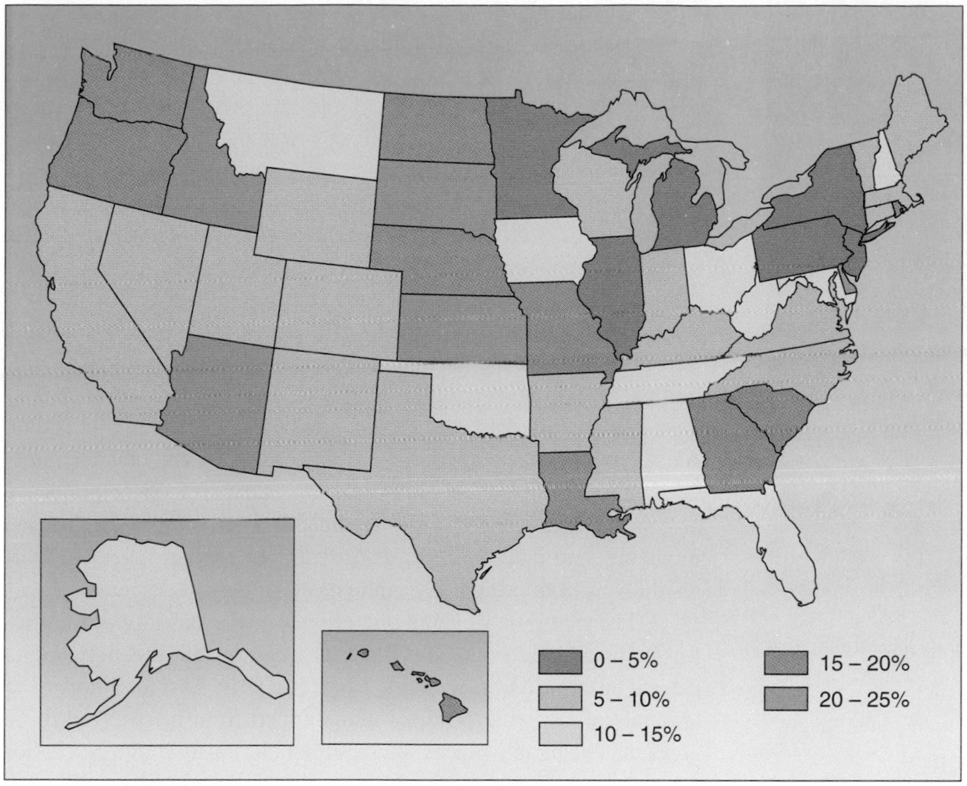

FIGURE 20.16 Percentage of drinking-water systems within states with violations of EPA health standards in 1996.
Source: Data from "Just Add Water," Environmental Working Group, 1997, table 2. Compiled from EPA Safe Drinking Water Information System 1994-1996.

and killing at least 100 people. The total costs of these diseases amount to billions of dollars per year. Preventive measures such as protecting water sources and aquifer recharge zones, providing basic treatment for all systems, installing modern technology and distribution networks, consolidating small systems, and strengthening the Clean Water Act and the Safe Drinking Water Act would cost far less. Unfortunately, in the present climate of budget-cutting and anti-regulation, these steps seem unlikely.

Ocean Pollution

As the opening story in this chapter illustrates, near-shore zones around the world, especially bays, estuaries, shoals, and reefs near large cities or the mouths of major rivers, often are overwhelmed by human-caused contamination. Suffocating and sometimes poisonous blooms of algae regularly deplete ocean waters of oxygen and kill enormous numbers of fish and other marine life. High levels of toxic chemicals, heavy metals, disease-causing organisms, oil, sediment, and plastic refuse are adversely affecting some of the most attractive and productive ocean regions. The potential losses caused by this pollution amount to billions of dollars each year. In terms of quality of life, the costs are incalculable.

Discarded plastic flotsam and jetsam are becoming an ubiquitous mark of human impact on the oceans (fig. 20.17). Since plastic is lightweight and nonbiodegradable, it is carried thousands of miles on ocean currents and lasts for years. Even the most remote beaches of distant islands are likely to have bits of polystyrene foam containers or polyethylene packing material that were discarded half a world away by some careless person. It has been estimated that some 6 million metric tons of plastic bottles, packaging material, and other litter are tossed from ships every year into the ocean where they ensnare and choke seabirds, mammals (fig. 20.18), and even fish. Sixteen states now require that six-pack yokes be made of biodegradable or photodegradable plastic, limiting their longevity as potential killers. The amount of municipal

FIGURE 20.17 Plastic litter covers this beach on the island of Niihau in the Hawaiian Islands.

FIGURE 20.18 A deadly necklace. Marine biologists estimate that castoff nets, plastic beverage yokes, and other packing residue kill hundreds of thousands of birds, mammals, and fish each year.

and industrial plastic that finds its way to the ocean is unknown but immense. In one day, volunteers in Texas gathered more than three hundred tons of plastic refuse from Gulf Coast beaches.

Few coastlines in the world remain uncontaminated by oil or oil products. Tar granules and sticky crude oil droplets stick to feet on beaches everywhere. Because oil floats on water, it is easily detected on the open ocean. Figure 20.19 shows locations where visible oil slicks have been reported. Oceanographers estimate that somewhere between 3 million and 6 million metric tons of oil are discharged into the world's oceans each year from both land- and sea-based operations. About half of this amount is due to maritime transport. Most of the 40 million liters of discharge (nearly 11 million gal) is not from dramatic, headline-making accidents, such as the 1989 oil spill from the *Exxon Valdez* in Prince William Sound, Alaska, but from routine open-sea bilge pumping and tank cleaning, which are illegal but, nonetheless, are carried out once ships are beyond sight of land. Much of the rest comes from land-based municipal and industrial runoff or from atmospheric deposition of residues from refining and combustion of fuels.

The transport of huge quantities of oil creates opportunities for major pollution episodes through a combination of human and natural hazards. Military conflict in the Middle East and increasing amounts of oil being pumped and shipped from off-shore drilling areas in inhospitable places, such as the notoriously rough North Sea and the Arctic Ocean, make it likely that more oil spills will occur. Plans to drill for oil along the seismically active California and Alaska coasts have been controversial because of the damage that oil spills could cause to these biologically rich coastal ecosystems.

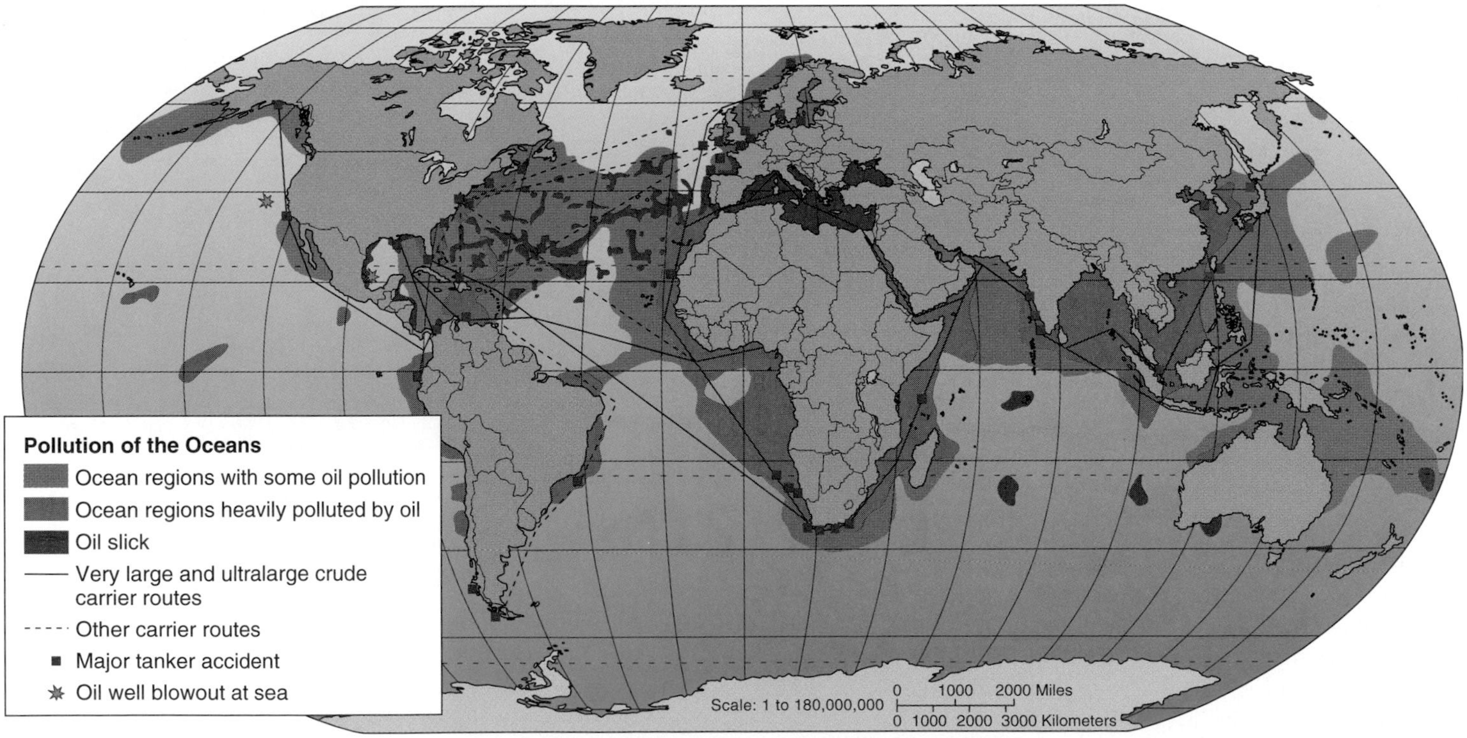

FIGURE 20.19 Location of oil pollution in the ocean. Major shipping routes of large oil tankers are very likely to be contaminated with oil.

CASE STUDY

WATERSHED PROTECTION IN THE CATSKILLS

New York City has long been proud of its excellent municipal drinking water. Drawn from the rugged Catskill Mountains 100 km (60 mi) north of the city, stored in hard-rock reservoirs, and transported through underground tunnels, the city water is outstanding for so large an urban area. Yielding 450,000 m^3 (1.2 billion gal) per day, and serving more than 9 million people, this is the largest surface water storage and supply complex in the world. As the metropolitan agglomeration has expended, however, people have moved into the area around the Catskill Forest Preserve, and water quality is not as high as it was a century ago.

When the 1986 U.S. Safe Drinking Water Act mandated filtration of all public surface water systems, the city was faced with building an $8 billion water treatment plant that would cost up to $500 million per year to operate. In 1989, however, the EPA ruled that the city could avoid filtration if it could meet certain minimum standards for microbial contaminants such as bacteria, viruses, and protozoan parasites. In an attempt to avoid the enormous cost of filtration, the city proposed land-use regulations for the five counties (Green, Ulster, Sullivan, Schoharie, and Delaware) in the Catskill/Delaware watershed from which it draws most of its water.

With a population of 50,000 people, the private land within the 520 km^2 (200 mi^2) watershed is mostly devoted to forestry and small dairy farms, neither of which are highly profitable. Among the changes the city called for was elimination of storm water runoff from barnyards, feedlots, or grazing areas into watersheds. In addition, farmers would be required to reduce erosion and surface runoff from crop fields and logging operations. Property owners objected strenuously to what they regarded as onerous burdens that would cost enough to put many of them out of business. They also bristled at having the huge megalopolis impose rules on them. It looked like a long and bitter battle would be fought through the courts and the state legislature.

To avoid confrontation, a joint urban/rural task force was set up to see if a compromise could be reached, and to propose alternative solutions to protect both the water supply and the long-term viability of agriculture in the region. The task force agreed that agriculture is the "preferred land use" on private land, and that agriculture has "significant present and future environmental benefits." In addition, the task force proposed a voluntary, locally developed and administered program of "whole farm planning and best management approaches" very similar to ecosystem-based, adaptive management (see chapter 10).

This grass-roots program, financed mainly by the city, but administered by local farmers themselves, attempts to educate land owners, and provides alternative marketing opportunities that help protect the watershed. Economic incentives are offered to encourage farmers and foresters to protect the water supply. Collecting feedlot and barnyard runoff in infiltration ponds together with solid conservation practices such as terracing, contour plowing, strip farming, leaving crop residue on fields, ground cover on waterways, and cultivation of perennial crops such as orchards and sugarbush have significantly improved watershed water quality. As of 1999, about 400 farmers—close to the 85 percent participation goal—have signed up for the program. The cost, so far, to the city has been about $50 million—or less than 1 percent of constructing a treatment plant.

In addition to saving billions of dollars, this innovative program has helped create good will between the city and its neighbors. It has shown that upstream cleanup, prevention, and protection are cheaper and more effective than treating water after it's dirty. Farmers have learned they can be part of the solution, not just part of the problem. And we have learned that watershed planning through cooperation is effective when local people are given a voice and encouraged to participate.

Investing in soil conservation and water-quality protection on small dairy farms and agroforestry programs within the Catskill/Delaware watershed has saved New York City billions of dollars in filtration costs and has also improved community relations.

WATER POLLUTION CONTROL

Appropriate land-use practices and careful disposal of industrial, domestic, and agricultural wastes are essential for control of water pollution.

Source Reduction

In many cases, the cheapest and most effective way to reduce pollution is to avoid producing it or releasing it to the environment in the first place. Elimination of lead from gasoline has resulted in a widespread and significant decrease in the amount of lead in surface waters in the United States. Studies have shown that as much as 90 percent less road deicing salt can be used in many areas without significantly affecting the safety of winter roads. Careful handling of oil and petroleum products can greatly reduce the amount of water pollution caused by these materials. Although we still have problems with persistent chlorinated hydrocarbons spread widely in the environment, the banning of DDT and PCBs in the 1970s has resulted in significant reductions in levels in wildlife (fig. 20.20).

Modifying agricultural practices in headwater streams in the Chesapeake Bay watershed and the Catskill Mountains of New York have had positive and cost-effective impacts on downstream water quality (Case Study, above).

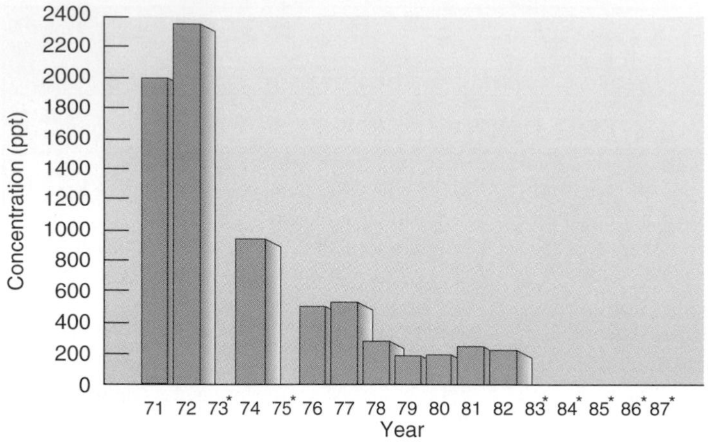

FIGURE 20.20 Dioxin concentrations in herring gull eggs on Scotch Bonnet Island in Lake Ontario have fallen sharply since 1971. Discontinued production and use of the herbicide 2,4,5-T are major reasons for this decline.
*Data not available.
Source: Data from *State of Canada's Environment,* Minister of the Environment.

FIGURE 20.21 Erosion on construction sites produces a great deal of sediment and is a major cause of nonpoint water pollution. Builders generally are required to install barriers to contain sediments, but these measures are often ineffectual.

Industry can modify manufacturing processes so fewer wastes are created. Recycling or reclaiming materials that otherwise might be discarded in the waste stream also reduces pollution. Both of these approaches usually have economic as well as environmental benefits. "Sewering" of heavy metals by industry has been outlawed. Producers are required instead to separate their wastes. It turns out that a variety of valuable metals can be recovered from wastes and reused or sold for other purposes. The company benefits by having a product to sell, and the municipal sewage treatment plant benefits by not having to deal with highly toxic materials mixed in with millions of gallons of other types of wastes.

Nonpoint Sources and Land Management

Among the greatest remaining challenges in water pollution control are diffuse, nonpoint pollution sources. Unlike point sources, such as sewer outfalls or industrial discharge pipes, which represent both specific locations and relatively continuous emissions, nonpoint sources have many origins and numerous routes by which contaminants enter ground and surface waters. It is difficult to identify—let alone monitor and control—all these sources and routes. Some main causes of nonpoint pollution are:

- *Agriculture:* The EPA estimates that 60 percent of all impaired or threatened surface waters are affected by sediment from eroded fields and overgrazed pastures; fertilizers, pesticides, and nutrients from croplands; and animal wastes from feedlots.
- *Urban runoff:* Pollutants carried by runoff from streets, parking lots, and industrial sites contain salts, oily residues, rubber, metals, and many industrial toxins. Yards, golf courses, parklands, and urban gardens often are treated with far more fertilizers and pesticides per unit area than farmlands. Excess chemicals are carried by storm runoff into waterways.
- *Construction sites:* New buildings and land development projects such as highway construction affect relatively small areas but produce vast amounts of sediment, typically ten to twenty times as much per unit area as farming (fig. 20.21).
- *Land disposal:* When done carefully, land disposal of certain kinds of industrial waste, sewage sludge, and biodegradable garbage can be a good way to dispose of unwanted materials. Some poorly run land disposal sites, abandoned dumps, and leaking septic systems, however, contaminate local waters.

Generally, soil conservation methods (see chapter 11) also help protect water quality. Applying precisely determined amounts of fertilizer, irrigation water, and pesticides saves money and reduces contaminants entering the water. Preserving wetlands that act as natural processing facilities for removing sediment and contaminants helps protect surface and groundwaters.

In urban areas, reducing materials carried away by storm runoff is helpful. Citizens can be encouraged to recycle waste oil and to minimize use of fertilizers and pesticides. Regular street sweeping greatly reduces contaminants. Runoff can be diverted away from streams and lakes. Many cities are separating storm sewers and municipal sewage lines to avoid overflow during storms.

A good example of watershed management is seen in Chesapeake Bay, America's largest estuary. Once fabled for its abundant oysters, crabs, shad, striped bass, and other valuable fisheries, the Bay had deteriorated seriously by the early 1970s. Citizens' groups, local communities, state legislatures, and the federal government together established an innovative pollution control program that made the Bay the first estuary in America targeted for protection and restoration.

Among the principal objectives of this plan is reducing nutrient loading through land-use regulations in the six watershed states to control agricultural and urban runoff. Pollution prevention measures such as banning phosphate detergents also are important, as are upgrading

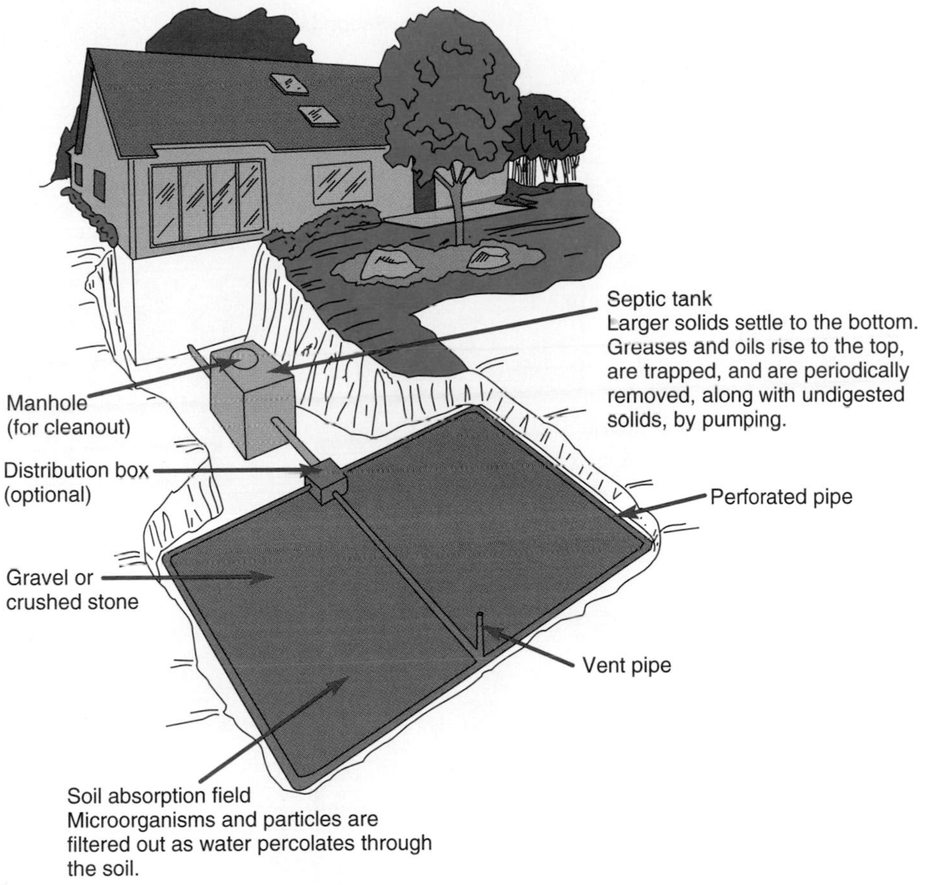

FIGURE 20.22 A domestic septic tank and drain field system for sewage and wastewater disposal. To work properly, a septic tank must have healthy microorganisms, which digest toilet paper and feces. For this reason, antimicrobial cleaners and chlorine bleach should never be allowed down the drain.

wastewater treatment plants and improving compliance with discharge and filling permits. Efforts are underway to replant thousands of hectares of seagrasses and to restore wetlands that filter out pollutants. Since the 1980s, annual phosphorous discharges into the Bay dropped 40 percent. Nitrogen levels, however, have remained constant or have even risen in some tributaries. Although progress has been made, the goals of reducing both nitrogen and phosphate levels by 40 percent and restoring viable fish and shellfish populations are still decades away. Still, as EPA Administrator Carol Browner says, it demonstrates the "power of cooperation" in environmental protection.

Human Waste Disposal

As we have already seen, human and animal wastes usually create the most serious health-related water pollution problems. More than 500 types of disease-causing (pathogenic) bacteria, viruses, and parasites can travel from human or animal excrement through water. In this section, we will look at how to prevent the spread of these diseases.

Natural Processes

In the poorer countries of the world, most rural people simply go out into the fields and forests to relieve themselves as they have always done. Where population densities are low, natural processes eliminate wastes quickly, making this an effective method of sanitation. The high population densities of cities make this practice unworkable, however. Even major cities of many less-developed countries are often littered with human waste which has been left for rains to wash away or for pigs, dogs, flies, beetles, or other scavengers to consume. This is a major cause of disease, as well as being extremely unpleasant. Studies have shown that a significant portion of the airborne dust in Mexico City is actually dried, pulverized human feces.

Where intensive agriculture is practiced—especially in wet rice paddy farming in Asia—it has long been customary to collect "night soil" (human and animal waste) to be spread on the fields as fertilizer. This waste is a valuable source of plant nutrients, but it is also a source of disease-causing pathogens in the food supply. It is the main reason that travelers in less developed countries must be careful to surface sterilize or cook any fruits and vegetables they eat. Collecting night soil for use on farm fields was common in Europe and America until about 100 years ago when the association between pathogens and disease was recognized.

Until about 50 years ago, most rural American families and quite a few residents of towns and small cities depended on a pit toilet or "outhouse" for waste disposal. Untreated wastes tended to seep into the ground, however, and pathogens sometimes contaminated drinking water supplies. The development of septic tanks and properly constructed drain fields represented a considerable improvement in public health (fig. 20.22). In a typical septic system, wastewater is first drained into a septic tank. Grease and oils rise to the top and solids settle to the bottom, where they

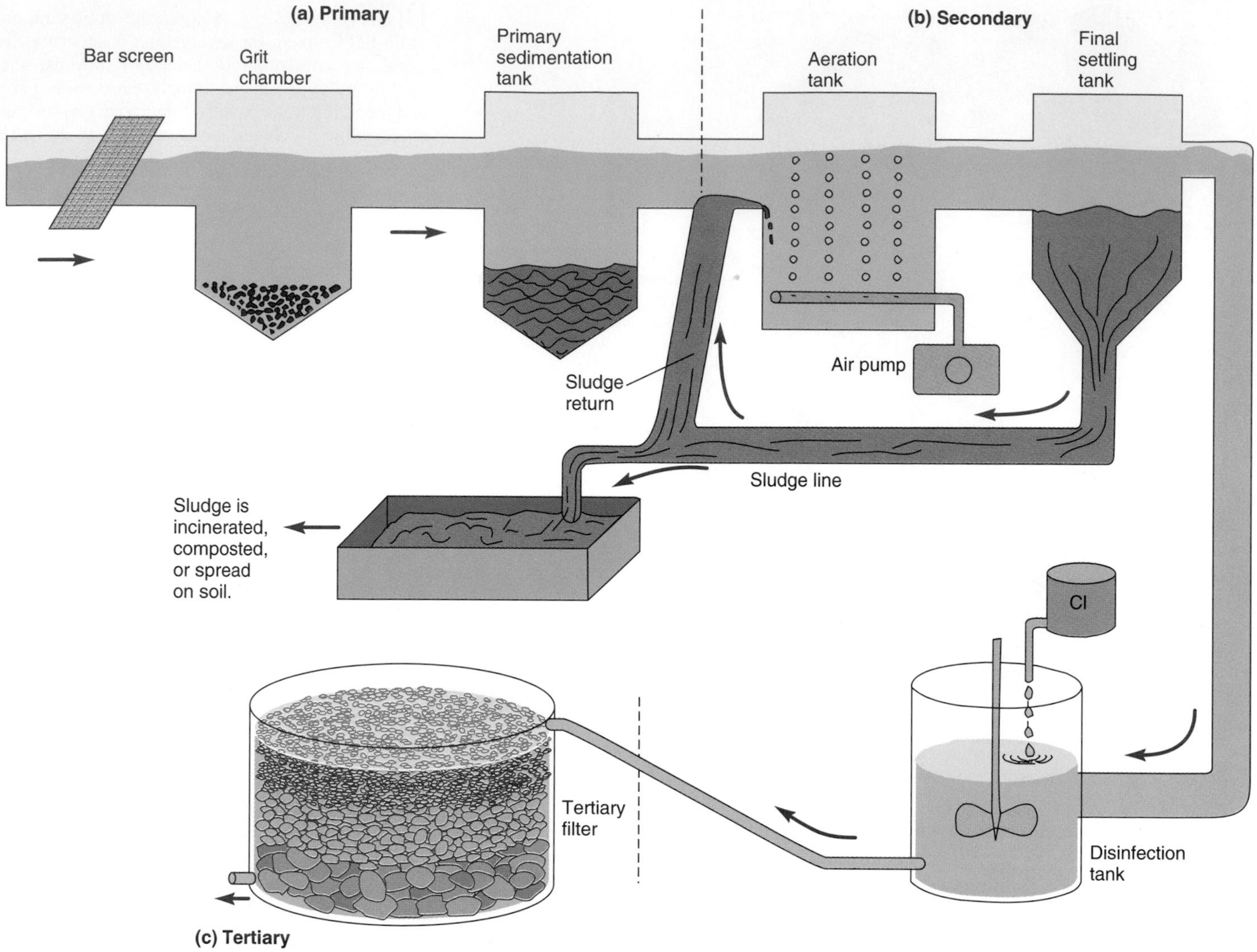

FIGURE 20.23 (*a*) Primary sewage treatment removes only solids and suspended sediment. (*b*) Secondary treatment, through aeration of activated sludge followed by sludge removal and chlorination of effluent, kills pathogens and removes most organic material. (*c*) During tertiary treatment passage through a trickling bed evaporator and/or a lagoon or marsh further removes inorganic nutrients, oxidizes any remaining organics, and reduces effluent volume.

are subject to bacterial decomposition. The clarified effluent from the septic tank is channeled out through a drainfield of small perforated pipes embedded in gravel just below the surface of the soil. The rate of aeration is high in this drainfield so that pathogens (most of which are anaerobic) will be killed, and soil microorganisms can metabolize any nutrients carried by the water. Excess water percolates up through the gravel and evaporates. Periodically, the solids in the septic tank are pumped out into a tank truck and taken to a treatment plant for disposal.

Where land is available and population densities are not too high, this can be an effective method of waste disposal. It is widely used in rural areas, but with urban sprawl, groundwater pollution often becomes a problem, indicating the need to shift to a municipal sewer system. It doesn't work well in cold, rainy climates where the drainfield may be too wet for proper evaporation, or where the water table is close to the surface.

Municipal Sewage Treatment

Over the past 100 years, sanitary engineers have developed ingenious and effective municipal wastewater treatment systems to protect human health, ecosystem stability, and water quality. This topic is an important part of pollution control, and is a central focus of every municipal government; therefore, let's look more closely at how a typical municipal sewage treatment facility works.

Primary treatment is the first step in municipal waste treatment. It physically separates large solids from the waste stream. As raw sewage enters the treatment plant, it passes through a metal grating that removes large debris (fig. 20.23*a*). A moving screen then filters out smaller items. Brief residence in a grit tank allows sand and gravel to settle. The waste stream then moves to the primary sedimentation tank where about half the suspended, organic solids settle to the bottom as sludge. Many pathogens remain in the effluent and it is not yet safe to discharge into waterways or onto the ground.

Secondary treatment consists of biological degradation of the dissolved organic compounds. The effluent from primary treatment flows into a trickling filter bed, an aeration tank, or a sewage lagoon. The trickling filter is simply a bed of stones or corrugated plastic sheets through which water drips from a system of perforated pipes or a sweeping overhead sprayer. Bacteria and other microorganisms in the bed catch organic material as it trickles past and aerobically decompose it.

Aeration tank digestion is also called the activated sludge process. Effluent from primary treatment is pumped into the tank and mixed with a bacteria-rich slurry (fig. 20.23*b*). Air pumped through the mixture encourages bacterial growth and decomposition of the organic material. Water flows from the top of the tank and sludge is removed from the bottom. Some of the sludge is used as an inoculum for incoming primary effluent. The remainder would be valuable fertilizer if it were not contaminated by metals, toxic chemicals, and pathogenic organisms. The toxic content of most sewer sludge necessitates disposal by burial in a landfill or incineration. Sludge disposal is a major cost in most municipal sewer budgets (fig. 20.24). In some communities this is accomplished by land farming, composting, or anaerobic digestion, but these methods don't inactivate metals and some other toxic materials.

Where space is available for sewage lagoons, the exposure to sunlight, algae, aquatic organisms, and air does the same job more slowly but with less energy costs. Effluent from secondary treatment processes is usually disinfected with chlorine, UV light, or ozone to kill harmful bacteria before it is released to a nearby waterway.

Tertiary treatment removes plant nutrients, especially nitrates and phosphates, from the secondary effluent. Although wastewater is usually free of pathogens and organic material after secondary treatment, it still contains high levels of inorganic nutrients, such as nitrates and phosphates. When discharged into surface waters, these nutrients stimulate algal blooms and eutrophication. To preserve water quality, these nutrients also must be removed. Passage through a wetland or lagoon can accomplish this. Alternatively, chemicals often are used to bind and precipitate nutrients. (see fig. 20.23*c*)

In many American cities, sanitary sewers are connected to storm sewers, which carry runoff from streets and parking lots. A large line called an interceptor delivers the combined stream of storm runoff and domestic and industrial waste to the municipal treatment plant. Storm sewers are routed to the treatment plant rather than discharged into surface waters because runoff from streets, yards, and industrial sites generally contains a variety of refuse, fertilizers, pesticides, oils, rubber, tars, lead (from gasoline), and other undesirable chemicals. During dry weather, this plan works well. Heavy storms often overload the system, however, causing bypass dumping of large volumes of raw sewage and toxic surface runoff directly into receiving waters. To prevent this overflow, cities are spending hundreds of millions of dollars to separate storm and sanitary sewers. These are huge, disruptive projects. When they are finished, surface runoff will be diverted into a river or lake and cause another pollution problem.

FIGURE 20.24 "Well, if *you* can't use it, do you know anyone who *can* use 3000 tons of sludge every day?"
(c) 1977 by Sidney Harris *What's So Funny About Science 7,* Wm. Kaufmann, Inc.

Low-Cost Waste Treatment

The municipal sewage systems used in developed countries are often too expensive to build and operate in the developing world where low-cost, low-tech alternatives for treating wastes are needed. One option is **effluent sewerage,** a hybrid between a traditional septic tank and a full sewer system. A tank near each dwelling collects and digests solid waste just like a septic system. Rather than using a drainfield, however, to dispose of liquids—an impossibility in crowded urban areas—effluents are pumped to a central treatment plant. The tank must be emptied once a year or so, but because only liquids are treated by the central facility, pipes, pumps, and treatment beds can be downsized and the whole system is much cheaper to build and run than a conventional operation.

Another alternative is to use natural or artificial wetlands to dispose of wastes. Arcata, California, for instance, needed an expensive sewer plant upgrade. By transforming a 65-hectare (160-acre) garbage dump into a series of ponds and marshes that serve as a simple, low-cost waste treatment facility, the city saved millions of dollars and improved the environment simultaneously. Sewage is piped to holding ponds where solids settle out and are digested by bacteria and fungi. Effluent flows through marshes where it is filtered and cleansed by aquatic plants and microorganisms. The marsh is a haven for wildlife and has become a prized recreation area for the city (see chapter 5). Eventually, the purified water flows into the bay where marine life flourishes.

Similar wetland waste treatment systems are now operating in many developing countries. Effluent from these operations can be used to irrigate crops or raise fish for human consumption if care is taken to first destroy pathogens. Usually 20 to 30 days of exposure to sun, air, and aquatic plants is enough to make the water safe. These systems make an important contribution to human food

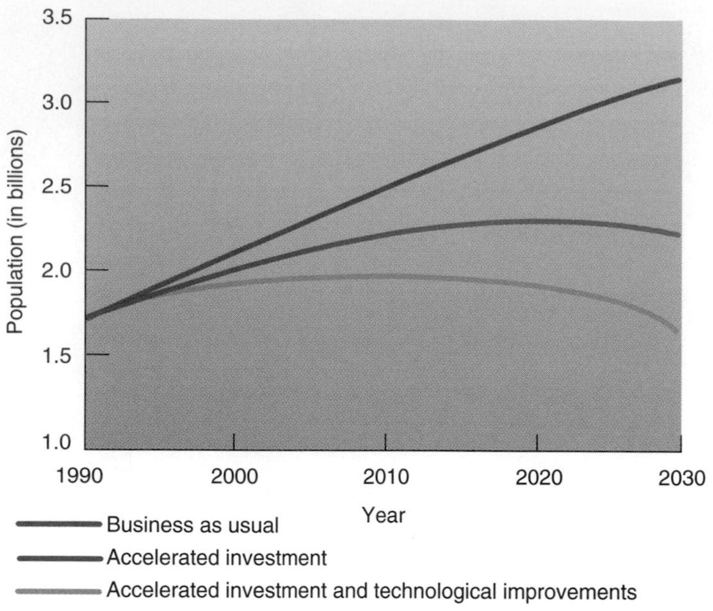

FIGURE 20.25 World population without adequate sanitation—three scenarios in the year 2030. If business as usual continues, more than 3 billion people will lack safe sanitation. Accelerated investment in sanitation services could lower this number. Higher investment, coupled with technological development, could keep the number of people without adequate sanitation from growing even though the total population increases.
Source: World Bank estimates based on research paper by Dennis Anderson and William Cavendish, "Efficiency and Substitution in Pollution Abatement: Simulation Studies in Three Sectors."

supplies. A 2500-hectare (6000-acre) waste-fed aquaculture facility in Calcutta, for example, supplies about 7000 metric tons of fish annually to local markets. The World Bank estimates that some 3 billion people will be without sanitation services by the middle of the next century under a business-as-usual scenario (fig. 20.25). With investments in innovative programs, however, sanitation could be provided to about half those people and a great deal of misery and suffering could be avoided.

Water Remediation

Remediation means finding remedies for problems. Just as there are many sources for water contamination, there are many ways to clean it up. New developments in environmental engineering are providing promising solutions to many water pollution problems.

Containment methods confine or restrain dirty water or liquid wastes *in situ* (in place) or cap the surface with an impermeable layer to divert surface water or groundwater away from the site and to prevent further pollution. Where pollutants are buried too deeply to be contained mechanically, materials sometimes can be injected to precipitate, immobilize, chelate, or solidify them. Bentonite slurries, for instance, can effectively stabilize liquids in porous substrates. Similarly, straw or other absorbent material is spread on surface spills to soak up contaminants.

Extraction techniques pump out polluted water so it can be treated. Many pollutants can be destroyed or detoxified by chemical reactions that oxidize, reduce, neutralize, hydrolyze, precipitate, or otherwise change their chemical composition. Where chemical techniques are ineffective, physical methods may work. Solvents and other volatile organic compounds, for instance, can be stripped from solution by aeration and then burned in an incinerator. Some contaminants can be removed by semipermeable membranes or resin filter beds that bind selectively to specific materials. Some of the same techniques used to stabilize liquids *in situ* can also be used *in vitro* (in a reaction vessel). Metals, for instance, can be chelated or precipitated in insoluble, inactive forms.

Often, living organisms can be used effectively and inexpensively to clean contaminated water. We call this bioremediation. Restored wetlands, for instance, along stream banks or lake margins can be very effective in filtering out sediment and removing pollutants. They generally cost far less than mechanical water treatment facilities and provide wildlife habitat as well.

Lowly duckweed (*Lemna* sp.), the green scum you often see covering the surface of eutrophic ponds, grows fast and can remove large amounts of organic nutrients from water. Under optimal conditions, a few square centimeters of these tiny plants can grow to cover nearly a hectare (about 2.5 acres) in four months. Large duckweed lagoons are being used as inexpensive, low-tech sewage treatment plants in developing countries. Where conventional wastewater purification typically costs $300 to $600 per person served, a duckweed system can cost one-tenth as much. The duckweed can be harvested and used as feed, fuel, or fertilizer. Up to 35 percent of its dry mass is protein—about twice as much as alfalfa, a popular animal feed.

Where space for open lagoons is unavailable, bioremediation can be carried out in reaction vessels. This has the advantage of controlling conditions more precisely and doesn't release organisms into the environment. Some of the most complex, holistic systems for water purification are designed by Ocean Arks International in Falmouth, Massachusetts. Their "living machines" combine living organisms—chosen to perform specific functions—in contained environments. In a typical living machine, water flows through a series of containers, each with a distinct ecological community designed for a particular function. Wastes generated by the inhabitants of one vessel become the food for inhabitants of another. Sunlight provides the primary source of energy.

OAI has created or is in the process of building water treatment plants in a dozen states and foreign countries. Designs range from remediating toxic wastes from superfund sites to simply treating domestic wastes (fig 20.26). Starting with microorganisms in aerobic and anaerobic environments where different kinds of wastes are metabolized or broken down, water moves through a series of containers containing hundreds of different kinds of plants and animals, including algae, rooted aquatic plants, clams, snails, and fish, each chosen to provide a particular service. Technically, the finished water is drinkable, although few people feel comfortable doing so. More often, the final effluent is used to flush toilets or for irrigation. Called ecological engineering, this novel approach can save resources and money as well as clean up our environment and serve as a valuable educational tool.

Solar sewage wall

FIGURE 20.26 A neighborhood sewage treatment facility designed by Jack Todd of Ocean Arks International. Water trickling through containers of aquatic plants is cleaned and purified by biological processes. Under optimal conditions, both odors and costs are minimal.

WATER LEGISLATION

Water pollution control has been among the most broadly popular and effective of all environmental legislation in the United States. It has not been without controversy, however. In this section, we will look at some of the major issues concerning water quality laws and their provisions (table 20.2).

The Clean Water Act

Passage of the Clean Water Act of 1972 was a bold, bipartisan step determined to "restore and maintain the chemical, physical, and biological integrity of the Nation's waters" that made clean water a national priority. Along with the Endangered Species Act and the Clean Air Act, this is one of the most significant and effective pieces of environmental legislation ever passed by the U.S. Congress. It also is an immense and complex law, with more than 500 sections regulating everything from urban runoff, industrial discharges, and municipal sewage treatment to land-use practices and wetland drainage.

The ambitious goal of the Clean Water Act was to return all U.S. surface waters to "fishable and swimmable" conditions. For specific "point" sources of pollution such as industrial discharge pipes or sewage outfalls, the act requires discharge permits and **best practicable control technology (BPT).** It sets national goals of **best available, economically achievable technology (BAT),** for toxic substances and zero discharge for 126 priority toxic pollutants. As we discussed earlier in this chapter, these regulations have had a positive effect on water quality. While not yet swimmable or fishable everywhere, surface water quality in the United States has significantly improved on average over the past quarter century. Perhaps the most important result of the act has been investment of $54 billion in federal funds and more than $128 billion in state and local funds for municipal sewage treatment facilities.

TABLE 20.2 Some Important U.S. Water Quality Legislation

1. *Federal Water Pollution Control Act* (1972). Established uniform nationwide controls for each category of major polluting industries.
2. *Marine Protection Research and Sanctuaries Act* (1972). Regulates ocean dumping and established sanctuaries for protection of endangered marine species.
3. *Ports and Waterways Safety Act* (1972). Regulates oil transport and the operation of oil handling facilities.
4. *Safe Drinking Water Act* (1974). Requires minimum safety standards for every community water supply. Among the contaminants regulated are bacteria, nitrates, arsenic, barium, cadmium, chromium, fluoride, lead, mercury, silver, pesticides; radioactivity and turbidity also regulated. This act also contains provisions to protect groundwater aquifers.
5. *Resource Conservation and Recovery Act* (RCRA) (1976). Regulates the storage, shipping, processing, and disposal of hazardous wastes and sets limits on the sewering of toxic chemicals.
6. *Toxic Substances Control Act* (TOSCA) (1976). Categorizes toxic and hazardous substances, establishes a research program, and regulates the use and disposal of poisonous chemicals.
7. *Comprehensive Environmental Response, Compensation, and Liability Act* (CERCLA) (1980) and *Superfund Amendments and Reauthorization Act* (SARA) (1984). Provide for sealing, excavation, or remediation of toxic and hazardous waste dumps.
8. *Clean Water Act* (1985) (amending the 1972 Water Pollution Control Act). Sets as a national goal the attainment of "fishable and swimmable" quality for all surface waters in the United States.
9. *London Dumping Convention 1990.* Calls for an end to all ocean dumping of industrial wastes, tank washing effluents, and plastic trash. The United States is a signatory to this international convention.

Not everyone, however, is completely happy with the Clean Water Act. Industries, state and local governments, farmers, land developers, and others who have been forced to change their operations or spend money on water protection often feel imposed upon. One of the most controversial provisions of the act has been Section 404, which regulates draining or filling of wetlands. Although the original bill only mentions wetlands briefly, this section has evolved through judicial interpretation and regulatory policy to become one of the principal federal tools for wetland protection. Environmentalists applaud the protection granted to these ecologically important areas that were being filled in or drained at a rate of about half a million hectares per year before the passage of the Clean Water Act. Farmers, land developers, and others who are prevented from converting wetlands to other uses often are outraged by what they consider "taking" of private lands.

Another sore point for opponents of the Clean Water Act are what are called "unfunded mandates," or requirements for state or local governments to spend money that is not repaid by Congress. You will notice that the $128 billion already spent by cities to install sewage treatment and stormwater diversion to meet federal standards far exceeds the $54 billion in congressional assistance for these projects. Estimates are that local units of government could be required to spend another $130 billion to finish the job without any further federal funding. Small cities that couldn't afford or chose not to participate in earlier programs in which the federal government paid up to 90 percent of water quality programs are especially hard hit by requirements that they upgrade municipal sewer and water systems. They now are faced with carrying out those same projects entirely on their own funds.

Clean Water Act Reauthorization

Opponents of federal regulation have tried repeatedly to weaken or eliminate the Clean Water Act. They regard restriction of their "right" to dump toxic chemicals and waste into wetlands and waterways to be an undue loss of freedom. They resent being forced to clean up municipal water supplies, and call for cost/benefit analysis that places greater weight on economic interests in all environmental planning. Most of all, they view any limitation on use of private property to be a "taking" for which they should be fully compensated.

Even those who support the Clean Water Act in principle would like to see it changed and strengthened. Among these proposals are a shift from "end-of-the-pipe" focus on removing specific pollutants from effluents to more attention to changing industrial processes so toxic substances won't be produced in the first place. Another important issue is nonpoint pollution from agricultural runoff and urban areas, which has become the largest source of surface water degradation in the United States. Regulating these sources remains a difficult problem.

Environmentalists also would like to see stricter enforcement of existing regulations, mandatory minimum penalties for violations, more effective community right-to-know provisions, and increased powers for citizen lawsuits against polluters. Under the current law, using data that polluters themselves are required to submit, groups such as the Natural Resources Defense Council and the Citizens for a Better Environment have won million-dollar settlements in civil lawsuits (the proceeds generally are applied to clean-up projects) and some transgressors have even been sent to jail. Not surprisingly, environmentalists want these powers expanded while polluters find them very disagreeable.

Other Important Water Legislation

In addition to the Clean Water Act, several other laws help to regulate water quality in the United States and abroad. Among these is the Safe Drinking Water Act, which regulates water quality in commercial and municipal systems. Critics complain that standards and enforcement policies are too lax, especially for rural water districts and small towns. Some researchers report pesticides, herbicides, and lead in drinking water at levels they say should be of concern (fig. 20.27). Atrazine, for instance, a widely used herbicide, was detected in 96 percent of all surface water samples in one study of 374 communities across twelve states. Remember, however, that simply detecting a toxic compound is not the same as showing dangerous levels.

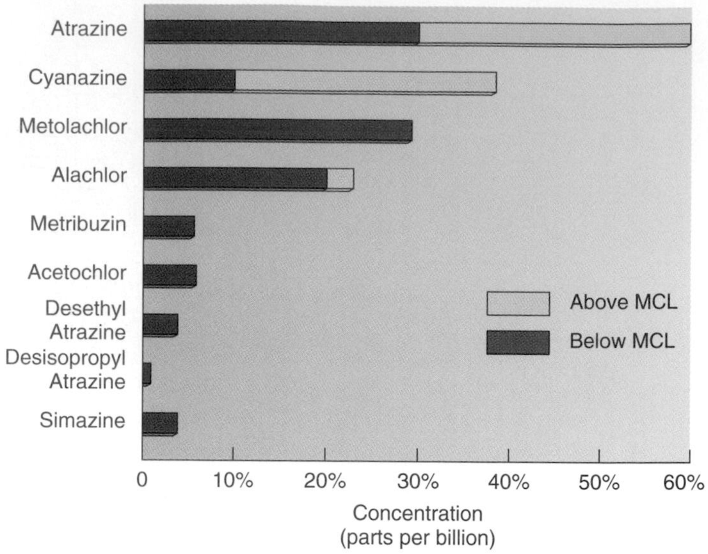

FIGURE 20.27 Nine herbicide active ingredients and metabolites found in drinking water samples in Fort Wayne, Indiana, Spring 1995 compared to Federal Maximum Concentration Levels.
Source: Data from B. Cohen and E. Bondoc, "Weed Killers by the Glass," Environmental Working Group 1995.

The Superfund program for remediation of toxic waste sites was created in 1980 by the Comprehensive Environmental Response, Compensation, and Liability Act (CERCLA) and was amended by the Superfund Amendments and Reauthorization Act (SARA) of 1984. This program is designed to provide immediate response to emergency situations and to provide permanent remedies for abandoned or inactive sites. These programs provide many jobs for environmental science majors in monitoring and removal of toxic wastes and landscape restoration. A variety of methods have been developed for remediation of problem sites.

Among the most important international agreements on water quality is the 1972 Great Lakes Water Quality Agreement between Canada and the United States. The agreement has produced encouraging progress in cleaning up the world's largest freshwater system. Another important international agreement is the 1990 London Dumping Convention, which calls for phasing out all ocean dumping of industrial waste, tank-washing effluent, and plastic trash by 1995. The 64 nations that have signed the Law of the Sea Treaty are bound by this agreement. The United States has already passed legislation to support its provisions. Whether this can be enforced remains to be seen, however.

Is it safe to assume that we are well on our way to solving our water pollution problems? We are better off in terms of legislation, policy, and practice than we were in 1960. Laws, however, are only as good as (1) the degree to which they are not weakened by subsequent amendments and exceptions and (2) the degree to which they are funded for research and enforcement. Economic interests cause continued pressure on both of these points, so that the importance of an overriding national attitude to maintain the intent of protective legislation must continually be stressed and retaught.

SUMMARY

Any physical, biological, or chemical change in water quality that adversely affects living organisms or makes water unsuitable for desired uses can be considered pollution. Worldwide, the most serious water pollutants, in terms of human health, are pathogenic organisms from human and animal wastes. We have traditionally taken advantage of the capacity of ecosystems to destroy these organisms, but as population density has grown, these systems have become overloaded and ineffective. Effective sewage treatment systems are needed that purify wastewater before it is released to the environment.

In industrialized nations, toxic chemical wastes have become an increasing problem. Agricultural and industrial chemicals have been released or spilled into surface waters and are seeping into groundwater supplies. The extent of this problem is probably not yet fully appreciated.

Ultimately, all water ends up in the ocean. The ocean is so large that it would seem impossible for human activities to have a significant impact on it, but pollution levels in the ocean are increasing. Major causes of ocean pollution are oil spills from tanker bilge pumping or accidents and oil well blowouts. Surface runoff and sewage outfalls discharge fertilizers, pesticides, organic nutrients, and toxic chemicals that have a variety of deleterious effects on marine ecosystems. We usually think of eutrophication (increased productivity due to nutrient addition) as a process of inland waterways, but this can occur in oceans as well.

The major water pollutants in terms of quantity are silt and sediments. Biomass production by aquatic organisms, land erosion, and refuse discharge all contribute to this problem. Addition of salts and metals from highway and farm runoff and industrial activities also damage water quality. In some areas, drainage from mines and tailings piles deliver sediment and toxic materials to rivers and lakes. Water pollution is a major source of human health problems. As much as 80 percent of all disease and some 25 million deaths each year may be attributable to water contamination.

Appropriate land-use practices and careful disposal of industrial, domestic, and agricultural wastes are essential for control of water pollution. Natural processes and living organisms have a high capacity to remove or destroy water pollutants, but these systems become overloaded and ineffective when pollution levels are too high. Municipal sewage treatment is effective in removing organic material from wastewater, but the sewage sludge is often contaminated with metals and other toxic industrial materials. Reducing the sources of these materials is often the best solution to our pollution problems.

QUESTIONS FOR REVIEW

1. Define *water pollution.*
2. List eight major categories of water pollutants and give an example for each category.
3. Describe eight major sources of water pollution in the United States. What pollution problems are associated with each source?
4. What is *Pfiesteria* and why is it dangerous?
5. What is eutrophication? What causes it?
6. What are the origins and effects of siltation?
7. Describe primary, secondary, and tertiary processes for sewage treatment. What is the quality of the effluent from each of these processes?
8. Why do combined storm and sanitary sewers cause water quality problems? Why does separating them also cause problems?
9. What pollutants are regulated by the Clean Water Act? What goals does this act set for abatement technology?
10. Describe remediation techniques and how they work.

QUESTIONS FOR CRITICAL THINKING

1. How precise is the estimate that 2 billion people lack access to clean water? Would it make a difference if the estimate is off by 10 percent or 50 percent?
2. How would you define *adequate* sanitation? Think of some situations in which people might have different definitions for this term.
3. Do you think that water pollution is worse now than it was in the past? What considerations go into a judgement like this? How do your personal experiences influence your opinion?
4. What additional information would you need to make a judgement about whether conditions are getting better or worse? How would you weigh different sources, types, and effects of water pollution?
5. Imagine yourself in a developing country with a severe shortage of clean water. What would you miss most if *your* water supply were suddenly cut by 90 percent?
6. Why has the EPA changed to total maximum daily pollution loads and watershed management? What are the major implications of this change?
7. Proponents of deep well injection of hazardous wastes argue that it will probably never be economically feasible to pump water out of aquifers more than 1 kilometer below the surface. Therefore, they say, we might as well use those aquifers for hazardous waste storage. Do you agree? Why or why not?
8. Under what conditions might sediment in water or cultural eutrophication be beneficial? How should we balance positive and negative effects?
9. Suppose that part of the silt in a river is natural and part is human-caused. Is one pollution but the other not?

10. Suppose that you own a lake but it is very polluted. An engineer offers options for various levels of cleanup. As you increase water quality, you also increase costs greatly. How clean would you want the water to be—fishable, swimmable, drinkable—and how much would you be willing to pay to achieve your goal? Make up your own numbers. The point is to examine your priorities and values.

KEY TERMS

atmospheric deposition 448
best available, economically achievable technology (BAT) 469
best practicable control technology (BPT) 469
biochemical oxygen demand (BOD) 450
coliform bacteria 449
cultural eutrophication 450
dissolved oxygen (DO) content 450
effluent sewerage 467
eutrophic 450
nonpoint sources 448
oligotrophic 450
oxygen sag 450
point sources 448
primary treatment 466
red tide 451
secondary treatment 467
tertiary treatment 467
thermal plume 455
total maximum daily loads (TMDL) 456

ADDITIONAL INFORMATION ON THE INTERNET

Visit our website at http://www.mhhe.com/environmentalscience/ for specific information about each of the topics below:

Hogwatch: Water Pollution and Industrial Farming
Pfiesteria, the Deadly Dinoflagellate
Office of Water, Environmental Protection Agency
Campaigns to Clean Up European rivers
International Rivers Network: linking human rights and environmental protection.
The EPA Office of Wetlands, Oceans, and Watersheds
"Red Tides" and Harmful Algal Blooms (Woods Hole Oceanographic Institute)
NOAA Office of Response and Restoration: oil spill remediation
Coral Reef Bleaching
Water Quality in the US, by State
EPA Drinking Water Quality Standards
Gound Water Quality
Water and Waste International

PROFILE

Environmental Manager

Kristin Pierskalla

Kristin Pierskalla is the environmental manager for SKB Environmental, a St. Paul, MN firm that operates construction/demolition debris landfill and composting facilities. Because her company is small, Kristin has a variety of duties, including research, marketing, compliance reporting, data management, and community relations. One of the aspects of her job that she enjoys most is the opportunity to serve on a number of boards and volunteer organizations. She considers volunteer activities to be an important part of her job. She serves on several boards including the Recycling Association of Minnesota, the Minnesota Environmental Education Association, and the America Recycles Day committee.

Kristin's path to her present position wasn't easy or straightforward. Her college major was in natural resources and environmental sciences, with an emphasis in resource planning and policy. Most students in her program went on to forestry or wildlife jobs, but neither area seemed right for her. She was interested in solid waste and recycling issues, but didn't know how to begin to look for work in this area. As an undergraduate, she did an unpaid summer internship at the Hennepin County Parks that gave her useful outdoor experience. After graduation, she worked for about two years in minimum-wage jobs in the forestry division of the Minnesota Department of Natural Resources. These assignments didn't lead to a professional path, but they provided some income, gave her practical skills, and helped her decide what she didn't want to do for a career. Her big break came from an internship she did with the Dakota County Solid Waste Unit. Her boss there knew that SKB had a position available and gave her a good recommendation. Having personal support from a professional in the field, who knew both Kristin's work and the folks at SKB, proved to be the boost she needed.

What are the best aspects of Kristin's job? She enjoys working with a wide variety of individuals in agencies, businesses, the public, and in her company. She also likes doing research on new technologies for composting and recycling, and studying alternatives to landfills. The creative aspects of writing marketing proposals, newspaper ads, agency reports, and educational materials adds variety to her day and gives her a chance to practice communication skills. The best part of her job, she reports, is the opportunity to be involved in volunteer work and to meet people in many different walks of life. She feels that her firm is helping to conserve resources and protect environmental quality—something that makes her feel like she's contributing to society.

There are both advantages and disadvantages in being in a small company. In a larger firm, she wouldn't have such a variety of opportunities, but in a smaller firm, there isn't much opportunity for advancement. She says that if you have the luxury of considering offers from several companies, think about your prospects for promotion. There's nowhere to go from being environmental manager at her firm. There also are relatively few chances for continuing education in her field. She attends a few seminars and workshops every year, but she'd like to do more to expand her horizons.

How could you prepare for a job like Kristin's? She recommends developing your computer and communications skills. Management proficiency also is important. You need to manage your own time and efforts even if you don't have responsibility for others. Extracurricular interests make you a more interesting person. Having been in the marching band in college turns out to be a good conversation starter for Kristin. Getting practical experience is extremely valuable. It's important to have a consistent work history. Being active in your chosen field—even if it's in volunteer or minimum-pay positions—gives you useful experience and shows prospective employers that you're serious. "Don't expect to find your dream job right out of college," she warns. "Be patient and persistent in your career preparation. Talking to as many people as possible will help open doors and keep you informed of opportunities. There may be careers out there that you don't even know about, as well as skills and interests you don't even know you have."

CHAPTER 21

Conventional Energy

Talk doesn't cook rice.

—Chinese proverb—

OBJECTIVES

After studying this chapter, you should be able to:

- summarize our current energy sources and explain briefly how energy use has changed through history.
- compare our energy consumption with that of other people in the world.
- itemize the ways we use energy.
- analyze the resources and reserves of fossil fuels in the world.
- evaluate the costs and benefits of using coal, oil, and natural gas.
- understand how nuclear reactors work, why they are dangerous, and how they might be made safer.
- discuss our options for radioactive waste storage and disposal.
- summarize public opinion about nuclear power and how you feel about this controversial energy source.

Above: Retreating Iraqi troops set more than 600 oil wells on fire as they left Kuwait in the 1990 Gulf War. At least 10 percent of Kuwait's oil was burned or spilled. Our insatiable appetite (some would say addiction) for oil plays a powerful role in international politics and our balance of payments deficit.

ESSENTIAL STUDY PARTNER

The Essential Study Partner (ESP) CD-ROM that accompanies this textbook contains additional information on the following topics:

- energy history and the Industrial Revolution, from wood to fossil fuels: physical resources: energy: overview
- coal, oil, and natural gas: physical resources: energy: fossil fuels
- types of coal, energy content, and impurities: physical resources: energy: fossil fuels
- sources, location and projected duration of proven reserves: physical resources: energy: fossil fuels
- nuclear power, fission and fusion reaction: physical resources: energy: nuclear
- pressurized-water reactors and breeder reactors: physical resources: energy: nuclear
- radioactive waste contamination, storage and melt down: physical resources: energy: nuclear
- world energy consumption: physical resources: energy: use patterns

Black Gold from the Caspian

Twice in four years, Russian troops pounded the rebellious province of Chechnya with bombs, rockets, and artillery in a war they can ill afford and probably never win conclusively. Why such a ferocious assault on a tiny, impoverished state at the fringes of their former empire? The answer is that Chechnya is the gateway, for Russia, to what may be one the world's richest reserves of oil and natural gas around—and under—the Caspian Sea (fig. 21.1).

It has long been known that Central Asia is rich in oil. Marco Polo wrote in the fourteenth century of "fountains from which oil springs in great abundance." An oil boom a century ago turned Baku, the capital of Azerbaijan, into a city of instant millionaires. But the Soviet Union never put much effort into developing Caspian oil, preferring to emphasize less politically challenging wells in Siberia. By the 1980s, a crumbling network of leaking pipelines, rusty drilling rigs, decaying cities, and patches of oil-soaked soil around the margins of the Caspian showed the effects of sloppy management and neglect. When the Soviet Union broke apart in 1991, a mad rush began as Western energy companies fought to be the first to exploit what may be the last really big, relatively accessible oil field in the world.

Oil deposits around the Caspian Sea are thought to hold up to 200 billion barrels, perhaps 25 percent of all the world's oil. If true, this resource would be worth about $4 trillion at today's prices, or about 30 times as much as Alaska's entire North Slope deposit. In addition, countries neighboring the Caspian are thought to have enormous reserves of natural gas. Turkmenistan alone is thought to sit on 9 trillion m^3 of gas, making it the world's fourth-largest holder of this valuable resource. In total, the Central Asian Republics may control a quarter of the world's natural gas supply.

The biggest difficulty is how to get these resources to market from their landlocked sources. It doesn't help that the area has some of the worst weather—temperatures ranging from –40°C to +50°C (–40°F to +122°F)—and most bellicose and unstable political climates in the world. Regional players include Russia, Chechnya, Dagestan, Abkhazia, Ingushetia, Ossetia, Kurdistan, Kazakhstan, Turkmenistan, Uzbekistan, Azerbaijan, Armenia, Georgia, Turkey, and Iran. There are more than 60 indigenous languages, and at least as many ethnic feuds in the region. Together in the 1990s, these groups had six major wars, two presidential assissination attempts, two coups, and countless guerrilla and bandit attacks.

The shortest—and probably cheapest—route for an oil or gas pipeline from the Caspian is across Iran to the Persian Gulf. The United States adamantly opposed that option, however, both to thwart Iran, and because of the risk of having additional oil passing through the vulnerable Persian Gulf. Russia has campaigned for a route following an existing pipeline northward through Chechnya—explaining why control of this area is so important—and then west to a Russian port on the Black Sea. This would require tankers to pass through the congested straits of Bosporus and the Dardanelles, and would make us dependent on Russia to keep the oil flowing. In 1999, a contract was signed to build a 2000 km (1250 mi) pipeline along the U.S.-preferred route from Baku, across Georgia, and then south across Turkey to the Mediterranean Sea. An extension nearly as long would cross under the Caspian and then run north to the Tengiz oil field in Kazakhstan. Costing at least $4 billion (US), the pipeline will take five years to build, and will carry about a million barrels of oil per day (42 million gal or about 6 percent of U.S. consumption) when finished.

Protecting this sprawling network of vulnerable pipelines in a rugged mountainous region of ancient but fierce ethnic, religious, and political hostilities is a daunting prospect. Does this story suggest to you that our dependence on oil creates odd bedfellows and difficult geopolitical problems? How far will we go to ensure our access to energy resources? In this chapter, we will focus on fossil fuels and nuclear power, which together supply about 97 percent of the world's commercial energy. Chapter 22 examines some energy conservation options as well as alternative, renewable energy sources.

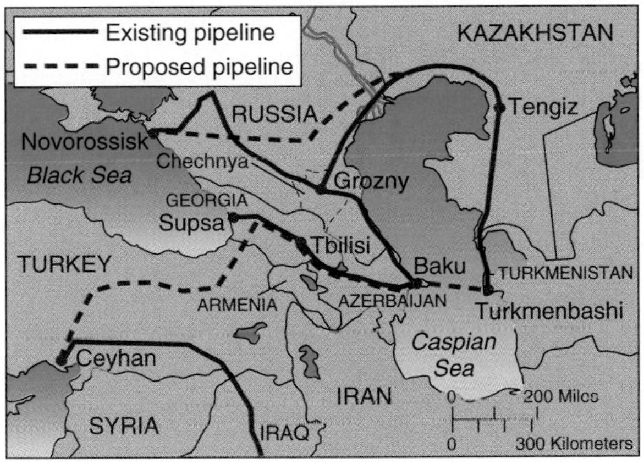

FIGURE 21.1 A U.S.-backed oil pipeline is proposed from Azerbaijan through Georgia, and then southwest across Turkey to the Mediterranean. Russia prefers a route across Chechnya to Novorossisk on the Black Sea.

WHAT IS ENERGY AND WHERE DO WE GET IT?

Work is the application of force through a distance. **Energy** is the capacity to do work. **Power** is the rate of flow of energy, or the rate at which work is done. The energy that we use to move our muscles, to think, and to carry out metabolic functions comes from stored chemical energy (or potential energy) in our food. Food energy is generally measured in calories (cal). One cal is the amount of energy to heat 1 gram of water 1°C. A kilocalorie (or food Calorie) is 1000 calories. In physics, the basic metric unit of force is a newton, which is the force necessary to accelerate one kilogram one meter per second. A **joule (J)** is the amount of work done when a force of 1 newton is exerted over 1 meter or one amp per second flows through 1 ohm. One J equals 0.238 cal. Some other common energy units are presented in table 21.1.

A Brief Energy History

Fire was probably the first human energy technology. Charcoal from fires has been found at sites occupied by our early ancestors 1 million years ago. Muscle power provided by domestic animals has been important at least since the dawn of agriculture some 10,000 years ago. Wind and water power have been used nearly as long. The invention of the steam engine, together with diminishing supplies of wood in industrializing countries, caused a switch to coal as our major energy source in the nineteenth century. Coal, in turn, has been replaced by oil in this century due to the ease of shipping, storing, and burning liquid fuels.

World oil use peaked in 1979, when daily production passed 66 million barrels per day. An Arab oil embargo in 1973 and the Iranian revolution in 1979 caused crude oil prices to rise nearly tenfold during the 1970s. These sudden price shocks were a major source of the crushing debt burdens that still hold back many developing countries. The results were less disruptive in richer countries, but Americans waiting in long lines at gas stations became aware for the first time that much of our luxurious lifestyle is dependent on a limited, unstable oil supply (fig. 21.2).

TABLE 21.1 Some Energy Units

1 joule (J) = the force exerted by a current of 1 amp per second flowing through a resistance of 1 ohm
1 watt (W) = 1 joule (J) per second
1 kilowatt (kwhr) = 1 thousand (10^3) watts exerted for 1 hour
1 megawatt (MW) = 1 million (10^6) watts
1 gigajwatt (GW) = 1 billion (10^9) watts
1 petajoule (PJ) = 1 quadrillion (10^{15}) joules
1 British thermal unit (BTU) = energy to heat 1 lb of water 1°F
1 PJ = 947 billion BTU, or 0.278 billion kwhr
1 standard barrel (bbl) of oil = 42 gal(160 l) or 5.8 million BTU
1 metric ton of standard coal = 27.8 million BTU or 4.8 bbl oil

FIGURE 21.2 Sudden price shocks in the 1970s caused by anticipated oil shortages showed Americans that our energy-intensive lifestyles may not continue forever.

The early 1980s saw an increased concern about conservation and development of renewable energy resources. This concern didn't last long, unfortunately, as reduced demand and increased production in the mid-1980s caused an oil glut that made prices fall almost as rapidly as they had risen a decade earlier. Through most of the 1990s, the price of a benchmark barrel of Texas light crude hovered around $15 when adjusted for inflation. In 2000, production curbs instituted by the Organization of Petroleum Exporting Countries (OPEC) pushed prices above $30 per barrel, still less, in real dollars, than prices 30 years earlier. The United States now imports more than half of its oil. These imports cost some $50 billion each year, more than one-half our total foreign trade deficit.

Current Energy Sources

Fossil fuels (petroleum, natural gas, and coal) now provide about 85 percent of all commercial energy in the world (fig. 21.3). Biomass fuels, such as wood, peat, charcoal, and manure, contribute about 6 percent of commercial energy. Other renewable sources—solar, wind, geothermal, and hydroelectricity—make up 4 to 5 percent of our commercial power (actually, hydro accounts for most of that). Although not included in these data, private systems (individual houses, businesses, industries) may capture as much renewable energy as that reported in commerical transactions. Similarly, this survey probably underrepresents the importance of biomass because these fuels often are gathered by the people using them, or they are sold in the informal economy and not reported internationally. In many poorer countries, such as Haiti, Bhutan, and Malawi, biomass supplies more than 90 percent of the energy used for heating and cooking. This is an important energy source for poor people, but can be a serious cause of forest destruction and soil loss (chapter 14).

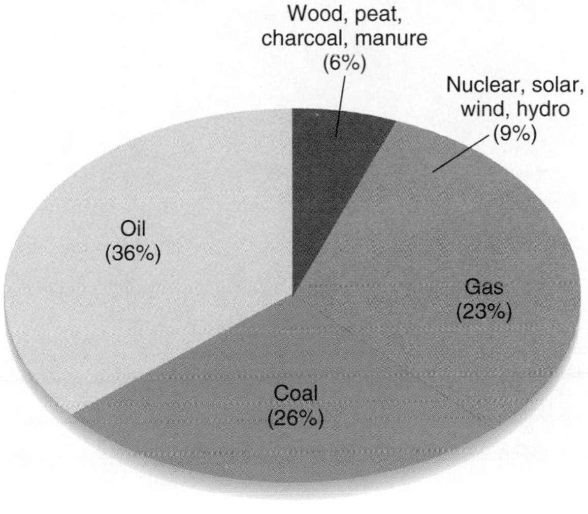

FIGURE 21.3 Worldwide commercial energy production. (Totals exceed 100 percent due to rounding.)
Source: Data from World Resources Institute, 1998–99.

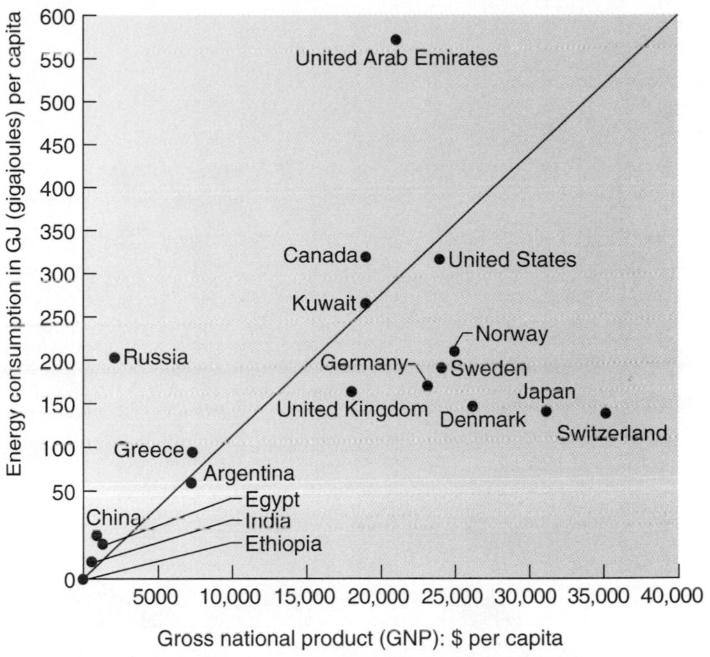

FIGURE 21.4 Per capita energy use and GNP. In general, higher energy use correlates with a higher standard of living. Denmark and Switzerland, however, use about half as much energy as we do and have a higher standard of living by most measures. Oil-rich states, such as the United Arab Emirates, use about twice as much energy as we do but are still relatively undeveloped in most areas.
Source: Data from World Resources Institute, 1996–97.

Nuclear power is roughly equal to hydroelectricity, worldwide (i.e., 4 to 5 percent of all commercial energy), but it makes up about 20 percent of all electric power in more developed countries such as Canada and the United States. We have enough nuclear fuel to produce power for a long time, and it has the benefit of not contributing to global warming (chapter 17), but as we discuss later in this chapter, safety concerns make this option unacceptable to most people.

Per Capita Consumption

Perhaps the most important facts about fossil fuel consumption are that we in the twenty richest countries consume nearly 80 percent of the natural gas, 65 percent of the oil, and 50 percent of the coal produced each year. Although we make up less than *one-fifth* of the world's population, we use more than *one half* of the commercial energy supply. The United States and Canada, for instance, constitute only 5 percent of the world's population but consume about one-quarter of the available energy. To be fair, however, some of that energy goes to grow crops or manufacture goods that are later shipped to developing countries.

How much energy do you use every year? Most of us don't think about it much, but maintaining the lifestyle we enjoy requires an enormous energy input. On average, each person in the United States and Canada uses more than 300 GJ (equivalent to about 60 barrels of oil) per year. By contrast, in the poorest countries of the world, such as Ethiopia, Kampuchea, Nepal, and Bhutan, each person generally consumes less than one GJ per year. This means that each of us consumes, on average, almost as much energy in a single day as a person in one of these countries consumes in a year.

Clearly, there is a link between energy consumption and the comfort and convenience of our lives (fig. 21.4). Those of us in the richer countries enjoy many amenities not available to most people in the world. The linkage is not absolute, however. Several European countries, including Sweden, Denmark, and Switzerland, have higher standards of living than does the United States by almost any measure but use about half as much energy as we do. These countries have had effective energy conservation programs for many years. Japan, also, has a far lower energy consumption rate than might be expected for its industrial base and income level. Because Japan has few energy resources of its own, it has developed very efficient energy conservation measures.

HOW ENERGY IS USED

The largest share (38 percent) of the energy used in the United States is consumed by industry (fig. 21.5). Mining, milling, smelting, and forging of primary metals consume about one-quarter of the industrial energy share. The chemical industry is the second largest industrial user of fossil fuels, but only half of its use is for energy generation. The remainder is raw material for plastics, fertilizers, solvents, lubricants, and hundreds of thousands of organic chemicals in commercial use. The manufacture of cement, glass, bricks, tile, paper, and processed foods also consumes large amounts of energy. Residential and commercial buildings use some 36 percent of the primary energy consumed in the United States, mostly for space heating, air conditioning, lighting, and water heating. Small motors and electronic equipment take an increasing

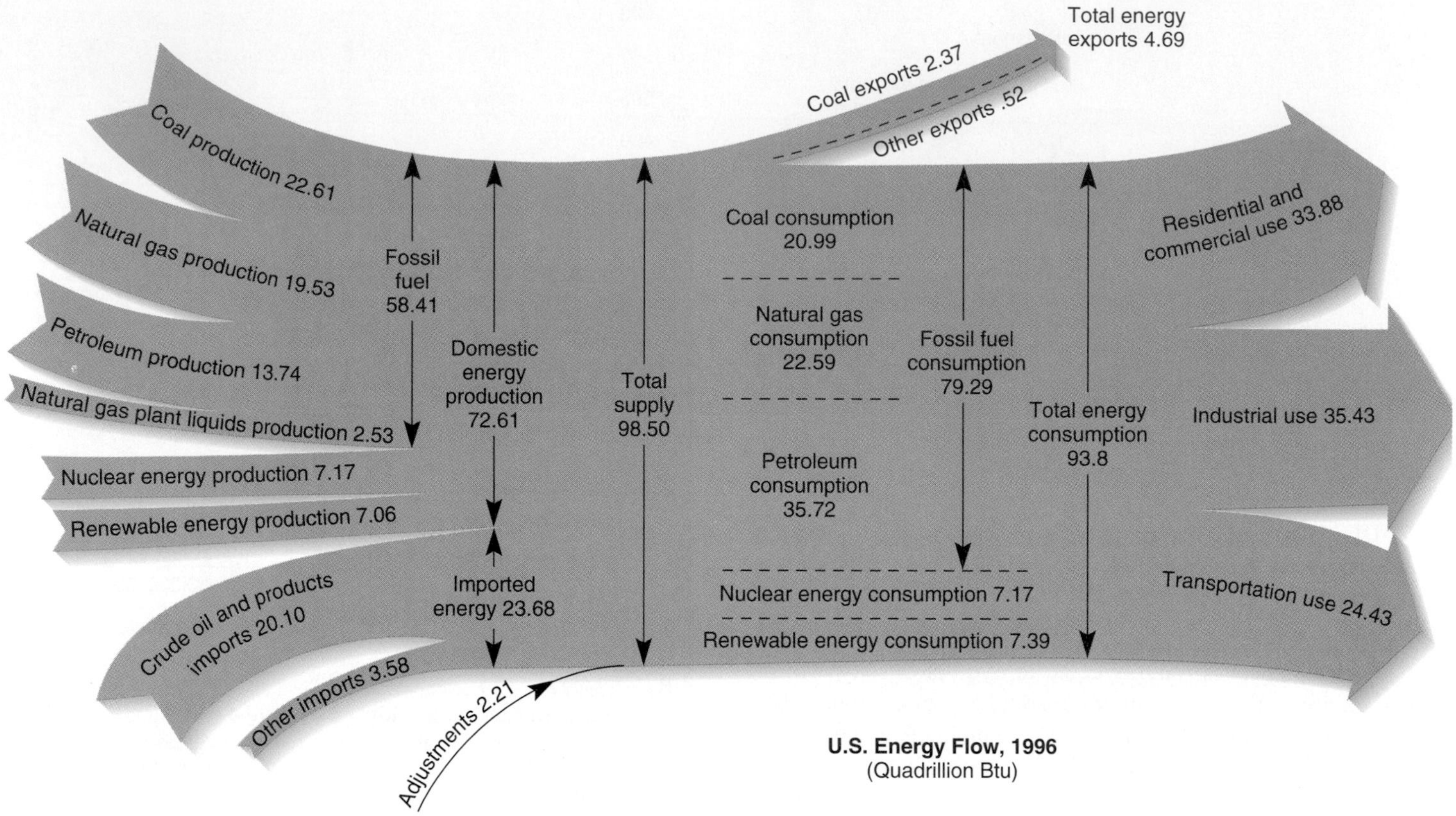

FIGURE 21.5 U.S. Energy Flow 1996. Quantities are in quadrillion BTUs (quads). In 1998, domestic energy production was 69.2 quads; imports were 26.8 quads, and total consumption was 90.0 quads, down about 3 percent from two years earlier.
Source: Data from U.S. Energy Flow, 1998. U.S. Department of Energy.

share of residential and commercial energy. In many office buildings, computers, copy machines, lights, and human bodies liberate enough waste heat so that a supplementary source is not required, even in the winter.

Transportation consumes about 26 percent of all energy used in the United States each year. About 98 percent of that energy comes from petroleum products refined into liquid fuels, and the remaining 2 percent is provided by natural gas and electricity. Almost three-quarters of all transport energy is used by motor vehicles. Nearly 3 trillion passenger miles and 600 billion ton miles of freight are carried annually by motor vehicles in the United States. About 75 percent of all freight traffic in the United States is carried by trains, barges, ships, and pipelines, but because they are very efficient, they use only 12 percent of all transportation fuel.

Finally, analysis of how energy is used has to take into account waste and loss of potential energy. About *half* of all the energy in primary fuels is lost during conversion to more useful forms, while it is being shipped to the site of end use, or during its use. Electricity, for instance, is generally promoted as a clean, efficient source of energy because when it is used to run a resistance heater or an electrical appliance almost 100 percent of its energy is converted to useful work and no pollution is given off. What happens before then, however? We often forget that huge amounts of pollution are released during mining and burning of the coal that fires power plants. Furthermore, nearly two-thirds of the energy in the coal that generated that electricity was lost in thermal conversion in the power plant. About 10 percent more is lost during transmission and stepping down to household voltages. Similarly, about 75 percent of the original energy in crude oil is lost during distillation into liquid fuels, transportation of that fuel to market, storage, marketing, and combustion in vehicles.

Natural gas is our most efficient fuel. Only 10 percent of its energy content is lost in shipping and processing since it moves by pipelines and usually needs very little refining. Ordinary gas-burning furnaces are about 75 percent efficient, and high-economy furnaces can be as much as 95 percent efficient. Because natural gas has more hydrogen per carbon atom than oil or coal, it produces about half as much carbon dioxide—and therefore half as much contribution to global warming—per unit of energy.

COAL

Coal is fossilized plant material preserved by burial in sediments and altered by geological forces that compact and condense it into a carbon-rich fuel. Coal is found in every geologic system since the Silurian Age 400 million years ago, but graphite deposits in very old rocks suggest that coal formation may date back to Precambrian times. Most coal was laid down during the Carboniferous period (286 million to 360 million years ago) when the earth's climate was warmer and wetter than it is now. Because coal takes so long to form, it is essentially a nonrenewable resource.

Coal Resources and Reserves

World coal deposits are vast, ten times greater than conventional oil and gas resources combined. Coal seams can be 100 m thick and can extend across tens of thousands of square kilometers that were vast swampy forests in prehistoric times. The total resource is estimated to be 10 trillion metric tons. If all this coal could be extracted, and if coal consumption continued at present levels, this would amount to several thousand years' supply. At present rates of consumption, these proven-in-place reserves—those explored and mapped but not necessarily economically recoverable—will last about 200 years.

Where are these coal deposits located? They are not evenly distributed throughout the world (fig. 21.6). North America has one-fourth of all proven reserves located in large deposits across the central and mountain regions (fig. 21.7). China and the former Soviet Union have nearly as much. Eastern and Western Europe have fairly large coal supplies in spite of their relatively small size. Africa and Latin America have very little coal despite their large size. Antarctica is thought to have large coal deposits, but they would be difficult, expensive, and ecologically damaging to mine.

It would seem that the abundance of coal deposits is a favorable situation. But do we really want to use all of the coal? What are the environmental and personal costs of coal extraction and use? In the next section, we will look at some of the disadvantages and dangers of mining and burning coal.

Mining

Coal mining is a hot, dirty, and dangerous business. Underground mines are subject to cave-ins, fires, accidents, and accumulation of poisonous or explosive gases (carbon monoxide, carbon dioxide, methane, hydrogen sulfide). Between 1870 and 1950, more than 30,000 coal miners died of accidents and injuries in Pennsylvania alone, equivalent to one man per day for 80 years. Untold thousands have died of respiratory diseases. In some mines, nearly every miner who did not die early from some other cause was eventually disabled by **black lung disease,** inflammation and fibrosis caused by accumulation of coal dust in the lungs or airways (chapter 9). Few of these miners or their families were compensated for their illnesses by the companies for which they worked. The U.S. Department of Labor now compensates miners and their dependents under the Black Lung Benefits Program, but workers often have difficulty proving that health problems are occupationally related. Even when compensation is provided, black lung remains a wretched occupational hazard of coal mining.

Strip mining or surface mining is cheaper than underground mining (fig. 21.8) but often makes the land unfit for any other use.

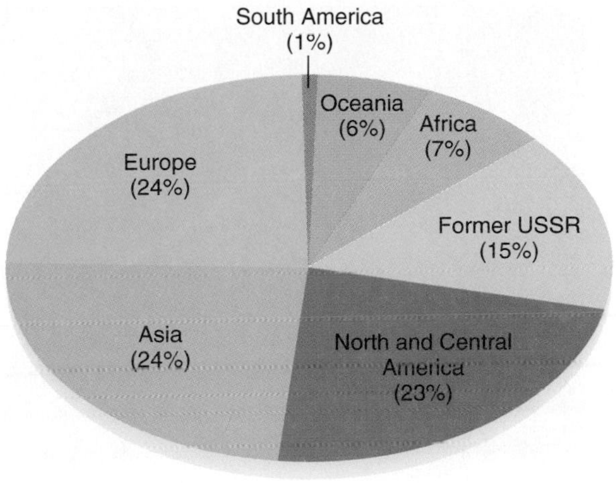

FIGURE 21.6 Proven-in-place coal reserves, by region as percent of a total 1.96 trillion metric tons.
Source: Data from World Resources Institute, 1994-95.

Mine reclamation is now mandated in the United States, but efforts often are shallow and ineffective. Coal mining also contributes to water pollution. Sulfur and other water soluble minerals make mine drainage and runoff from coal piles and mine tailings acidic and highly toxic. Thousands of miles of streams in the United States have been poisoned by coal-mining operations.

Air Pollution

Many people aren't aware that coal burning releases radioactivity and many toxic metals. Uranium, lead, cadmium, mercury, rubidium, thallium, and zinc—along with a number of other elements—are absorbed by plants and concentrated in the process of coal formation. These elements are not destroyed when the coal is burned, instead they are released as gases or concentrated in fly ash and bottom slag. You are likely to get a higher dose of radiation living next door to a coal-burning power plant than a nuclear plant under normal (nonaccident) conditions. Coal combustion is responsible for about 25 percent of all atmospheric mercury pollution in the United States.

Coal contains up to 10 percent sulfur (by weight). Unless this sulfur is removed by washing or flue-gas scrubbing, it is released during burning and oxidizes to sulfur dioxide (SO_2) or sulfate (SO_4) in the atmosphere. The high temperatures and rich air mixtures ordinarily used in coal-fired burners also oxidize nitrogen compounds (mostly from the air) to nitrogen monoxide, dioxide, and trioxide. Every year the 900 million tons of coal burned in the United States (83 percent for electric power generation) releases 18 million metric tons of SO_2, 5 million metric tons of nitrogen oxides (NO_x), 4 million metric tons of airborne particulates, 600,000 metric tons of hydrocarbons and carbon monoxide, and close to a trillion metric tons of CO_2. This is about three-quarters of the SO_2, one-third of the NO_x, and about half of the industrial CO_2 released in the United States each year. Coal burning is the largest single source of acid rain in many areas.

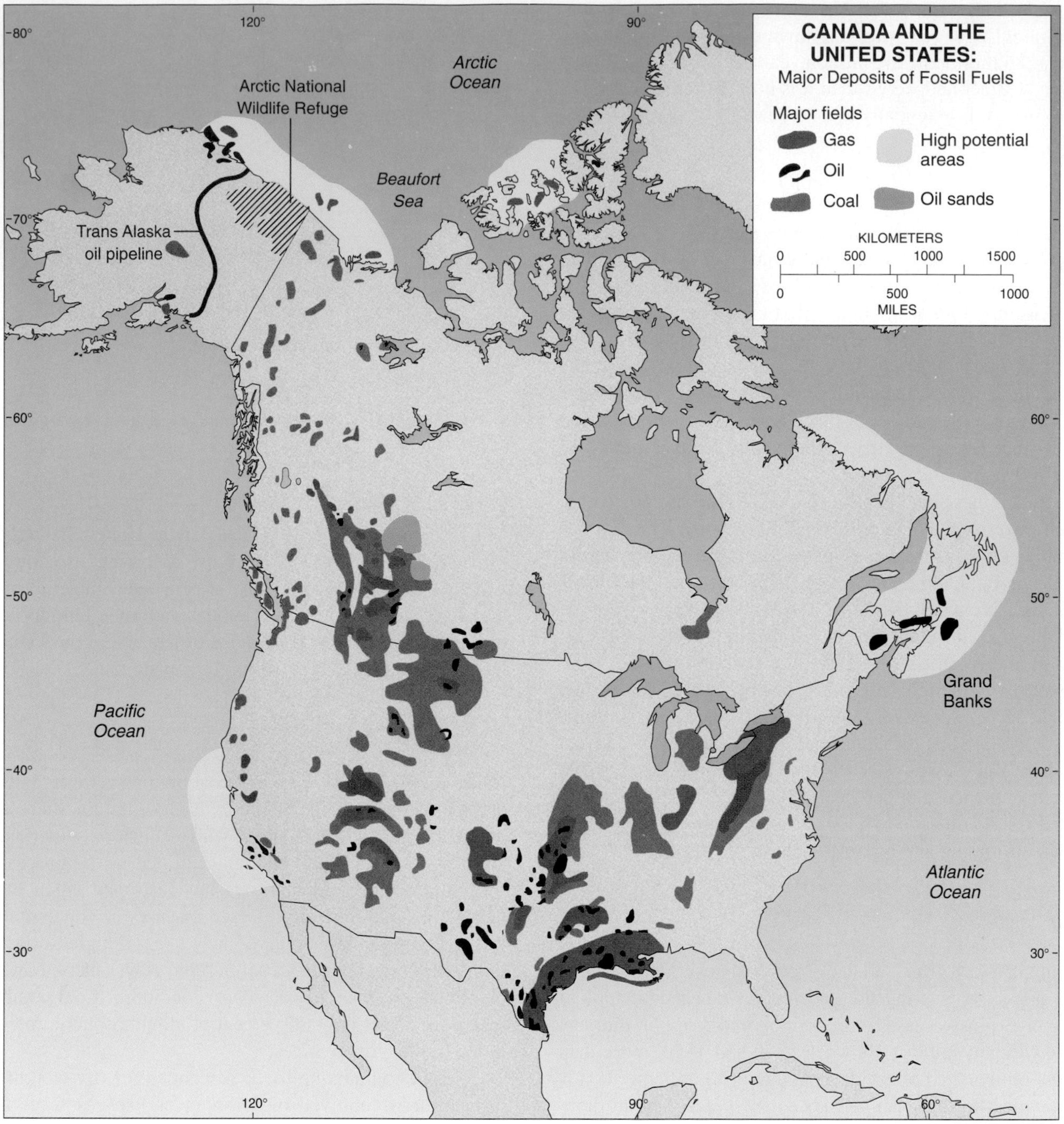

FIGURE 21.7 Canadian and U.S. fossil fuel deposits. Colored areas show off-shore sedimentary deposits with high potential for oil or gas. The ecological risks of drilling in these areas are high; many environmentalists argue that drilling should be prohibited in these places.
From *Geography: Regions and Concepts,* 6th ed., by H. J. DeBlij and P. O. Muller. Copyright © 1992 John Wiley & Sons, Inc. Redrawn by permission of John Wiley & Sons, Inc.

These air pollutants have many deleterious effects, including human health costs, injury to domestic and wild plants and animals, and damage to buildings and property (chapter 18). Total losses from air pollution are estimated to be between $5 billion and $10 billion per year in the United States alone. By some accounts, at least 5000 excess human deaths per year can be attributed to coal production and burning.

Sulfur can be removed from coal before it is burned, or sulfur compounds can be removed from the flue gas after combustion. Formation of nitrogen oxides during combustion also can be minimized. Perhaps the ultimate limit to our use of coal as a fuel will be the release of carbon dioxide into the atmosphere. As we discussed in chapter 17, carbon traps heat in the atmosphere and is a major contributor to global warming. It is difficult to imagine

FIGURE 21.8 A giant power shovel scoops up coal in an open pit mine in Montana. Careful recontouring and restoration can make this land productive once again, but costs are high and many old mines remain desolate and bare for decades, if not centuries.

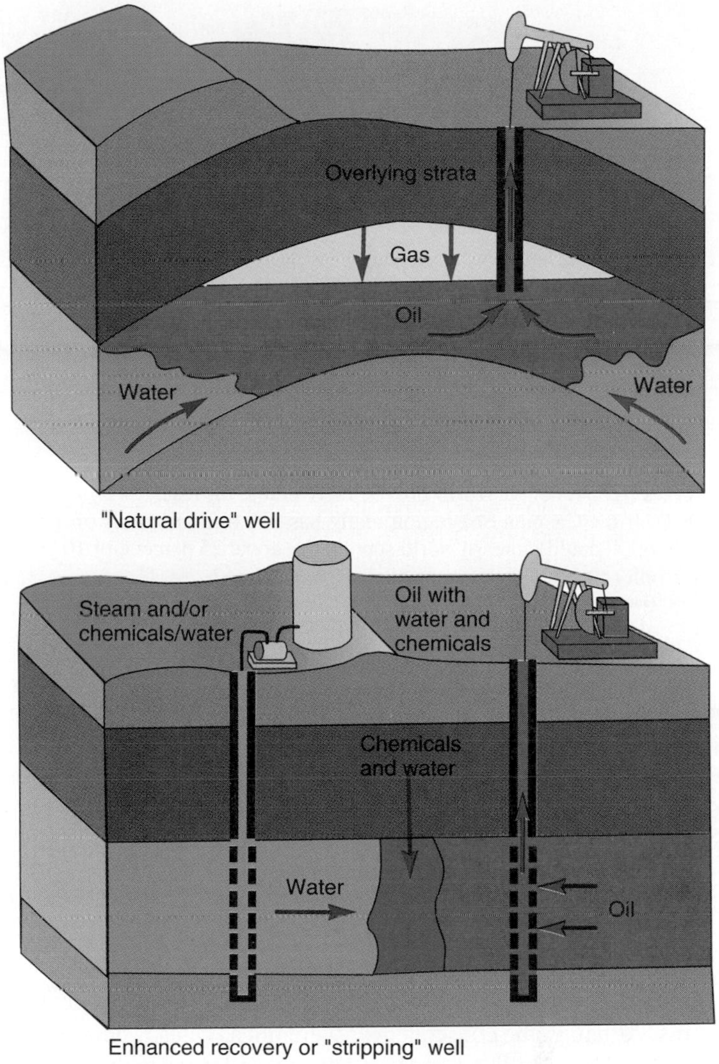

FIGURE 21.9 Recovery process for petroleum. In a "natural drive" well, water or gas pressure forces liquid petroleum into the well. In an enhanced recovery or "stripping" well, steam and/or water mixed with chemicals are pumped down a well behind the oil pool, forcing oil up the extraction well.

how we could trap and dispose of the 1 trillion tons of CO_2 generated each year in the United States by burning coal, not to mention the amount formed when we burn petroleum products. Alternative energy sources, such as solar and wind power (chapter 22), may be our only options.

OIL

Like coal, petroleum is derived from organic molecules created by living organisms millions of years ago and buried in sediments where high pressures and temperatures concentrated and transformed them into energy-rich compounds. Depending on its age and history, a petroleum deposit will have varying mixtures of oil, gas, and solid tarlike materials. Some very large deposits of heavy oils and tars are trapped in porous shales, sandstone, and sand deposits in the western areas of Canada and the United States.

Liquid and gaseous hydrocarbons can migrate out of the sediments in which they formed through cracks and pores in surrounding rock layers. Oil and gas deposits often accumulate under layers of shale or other impermeable sediments, especially where folding and deformation of systems create pockets that will trap upward-moving hydrocarbons (fig. 21.9). Contrary to the image implied by its name, an oil pool is not usually a reservoir of liquid in an open cavern but rather individual droplets or a thin film of liquid permeating spaces in a porous sandstone or limestone, much like water saturating a sponge.

Pumping oil out of a reservoir is much like sucking liquid out of a sponge. The first fraction comes out easily, but removing subsequent fractions requires increasing effort. We never recover all the oil in a formation; in fact, a 30 to 40 percent yield is about average. There are ways of forcing water or steam into the oil-bearing formations to "strip" out more of the oil, but at least half the total deposit usually remains in the ground at the point at which it is uneconomical to continue pumping. Methods for squeezing more oil from a reservoir are called **secondary recovery techniques** (fig. 21.9).

Oil Resources and Reserves

The total amount of oil in the world is estimated to be about 4 trillion barrels (600 billion metric tons), half of which is thought to be ultimately recoverable. Some 465 billion barrels of oil already have been consumed. In 1999, the proven reserves were roughly 1 trillion bbls, enough to last only 45 years at the current consumption rate of 22 billion barrels per year. It is estimated that another 800 billion barrels either remain to be discovered or are not recoverable at current prices with present technology. As oil resources become depleted and prices rise, it probably will become economical to

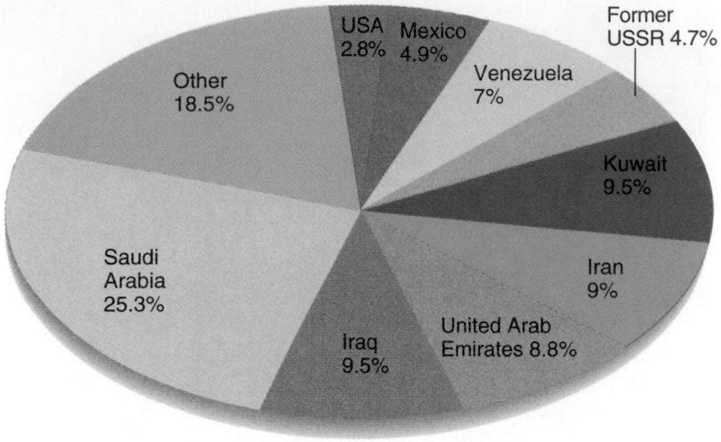

FIGURE 21.10 World proven recoverable oil reserves by country (1997). If the Caspian Sea region really has 200 billion barrels of recoverable oil, it could increase world supplies by about 25 percent, or 10 years at current rates of use.
Source: Data from World Resources Institute, 1998–99.

find and bring this oil to the market unless alternative energy sources are developed. This estimate of the resource does not take into account the very large potential from unconventional liquid hydrocarbon resources, such as shale oil and tar sands, which might double the total reserve if they can be mined with acceptable social, economic, and environmental impacts.

By far the largest supply of proven-in-place oil is in Saudi Arabia, which has 250 billion barrels, about one-fourth of the total proven world reserve (fig. 21.10). Kuwait had more than 10 percent of the proven world oil reserves before Iraq invaded in 1990. Some 600 wells were blown up and set on fire by the retreating Iraqis. Although the consequences were not as catastrophic as first feared, at least 5 billion barrels of Kuwait's oil was burned, spilled, or otherwise lost (see p. 474). Together, the Persian Gulf countries in the Middle East contain nearly two-thirds of the world's proven petroleum supplies. With our insatiable appetite for oil (some would say addiction), it is not difficult to see why this volatile region plays such an important role in world affairs.

Note that this discussion has been of *proven* reserves. If reservoirs around the Caspian Sea truly hold 200 billion barrels, they would by second only to Saudi Arabia in size.

Although oil consumption in Europe and North America has been relatively constant in recent years, Asian demand has been growing rapidly as economies there expand and people become more affluent. China, for instance, is expanding oil use 10 percent per year. By the year 2000, if current trends continue, about half of all oil produced by the Organization of Petroleum Exporting Countries (OPEC) will be going to Asia.

Oil Imports and Domestic Supplies

Originally, the United States is estimated to have had about 200 billion barrels of recoverable oil, or about as much as is thought to still be in Central Asia. This large supply of relatively accessible oil played an important part in development of America's economic and industrial power during the twentieth century. Until 1947, the United States was the leading oil export country in the world. By 1998, the United States was importing nearly 10 million barrels per day, more than half of total consumption. The major suppliers of this oil were Mexico and Venezuela in the 1970s but shifted to Saudi Arabia and other Middle Eastern countries in the 1990s, making the United States vulnerable to the unstable political situations in that volatile region.

Altogether, the United States has already used about 40 percent of its original recoverable petroleum resource. Of the 120 billion barrels thought to remain, about 58 billion barrels are proven-in-place. If we stopped importing oil and depended exclusively on indigenous supplies, our proven reserves would last only ten years at current rates of consumption.

The regions of North America with the greatest potential for substantial new oil discoveries are portions of the continental shelf along the California coast, the Beaufort Sea, and the Grand Banks (see fig. 21.7), all of which are prime wildlife areas. Proposals to do exploratory drilling in these areas have been strongly opposed by many people concerned about the potential for long-term environmental damage from drilling activities and oil spills like the *Exxon Valdez* in Prince William Sound. Other people argue that having an inexpensive, plentiful energy supply, even if only for a few decades, is worth the social and environmental costs. (Case Study, p. 483).

Oil Shales and Tar Sands

Estimates of our total oil supply usually do not reflect the very large potential from unconventional oil resources, such as shale oil and tar sands, which might double the total reserve if they can be extracted with reasonable social, economic, and environmental costs. Canada, for example, has an estimated 270 billion cubic meters of bituminous tar sands, mostly in northern Alberta (fig. 21.11). Liquid petroleum can be extracted from these sands with hot water, chemicals, or other stripping processes. If even part of this resource is recoverable, it would be the world's largest known oil deposit. There are severe environmental constraints, however. A typical plant producing 125,000 barrels of oil per day creates about 15 million cubic meters of toxic sludge and consumes or contaminates billions of liters of water each year.

Similarly, vast deposits of oil shale occur in the western United States. Actually, **oil shale** is neither oil nor shale but a fine-grained sedimentary rock rich in solid organic material called kerogen. When heated to about 480°C (900°F), the kerogen liquefies and can be extracted from the stone. Oil shale beds up to 600 m (1800 ft) thick occur in the Green River Formation in Colorado, Utah, and Wyoming, and lower-grade deposits are found over large areas of the eastern United States. If these deposits could be extracted at a reasonable price and with acceptable environmental impacts, they might yield the equivalent of several trillion barrels of oil.

Mining and extracting shale oil creates many problems, however. It is expensive; it uses vast quantities of water, a scarce resource in the arid west; it has a high potential for air and water

CASE STUDY

OIL VERSUS WILDLIFE

Beyond the jagged Brooks Range in Alaska's far northeastern corner lies one of the world's largest nature preserves, the 7-million ha (18-million acres) Arctic National Wildlife Refuge (ANWR). Geologists suspect that the narrow coastal plain bordering the Beaufort Sea contains oil and gas-bearing strata similar to the nearby fields of Prudhoe Bay. Debates have raged for years over whether to do exploratory drilling and seismic mapping of the area. A harsh and inhospitable wilderness in winter, this arctic tundra biome is a summer wetland of innumerable potholes and marshes that teem with wildlife. It is the central calving ground of the 180,000 caribou of the Porcupine herd, thought to be the world's second largest caribou herd. It also is an important habitat for thousands of snow geese and other migratory waterfowl, musk oxen, Alaskan brown bears, polar bears, and arctic wolves, along with unimaginable clouds of insects and many other species of rare plants and animals.

Energy companies are extremely interested in this area because any oil found there could be pumped out through the existing Trans-Alaska oil pipeline, thus extending the life of their multibillion dollar investment. The state of Alaska hopes that revenues from ANWR will replenish dwindling coffers as the oil supply from Prudhoe Bay dries up. Automobile companies, tire manufacturers, filling stations, and others who depend on our continued use of petroleum point out that domestic oil supplies are rapidly being depleted and that imported oil accounts for more than half of our balance of payments deficit.

Opponents to this project point out that scars left by mechanized equipment, blasting for seismic mapping, and drilling will persist in the tundra for centuries. A Department of the Interior environmental impact statement (EIS) predicted the loss of 40 percent of the caribou, 25 to 50 percent of the musk oxen, and 50 percent of the snow geese if drilling is allowed. Native Gwich'in people who depend on wildlife for subsistence have protested disruption of their traditional lifestyle.

Environmentalists argue that wildlife and wilderness values that the refuge was set up to protect should be paramount over short-term economic interests. Others feel that the potential for a valuable energy source overrides biological importance, and that, furthermore, wildlife will not be adversely affected if drilling is done carefully. They claim that caribou have actually increased at Prudhoe Bay since oil extraction began.

Few people ever visit this remote wilderness. Most of us, in fact, probably wouldn't do so even if we had a chance. Frigid winds howling off the Arctic Ocean make this bleak, barren landscape uninviting much of the year and swarming clouds of insects make it unbearable in the summer. The question becomes one of values: the rights of wildlife and ecosystems to exist undisturbed versus our desires to exploit resources. Ironically, we could easily do without this oil. The Department of the Interior estimates only a 19-percent chance of finding any oil, which at best would amount to only a 200-day supply if it were our only source. If we were to raise the average fuel efficiency of American automobiles to the levels now common in Europe, or if everyone in the United States who drives alone to school or work were to share the ride with one other person, we would save more oil in six months than we will ever pump out of ANWR even at the most optimistic estimates.

Alaska's Arctic National Wildlife Refuge, home to one of the world's largest caribou herds, has been called the crown jewel of the entire refuge system. Oil companies have proposed exploratory drilling here, but environmentalists fear irreparable damage to the fragile ecosystem.

Ethical Considerations

What do you think about the ethics of this situation? How would you weigh the potential for finding oil in the wildlife refuge against wilderness values and wildlife? Do human interests take precedence over nature? Would you feel differently about this issue if the oil there were the last available supply in the world? If we're going to have to wean ourselves from dependence on oil, why not do it sooner rather than later?

FIGURE 21.11 Canada has huge reserves of tar sands such as the one being mined here in Alberta. Mining and extraction of the petroleum products causes great environmental damage, however, and may never be safe or practical.

pollution; and it produces enormous quantities of waste. In the early 1980s, when the search for domestic oil supplies was at fever pitch, serious discussions occurred about filling whole canyons, rim to rim, with oil shale waste. One experimental mine used a nuclear explosion to break up the oil shale. All the oil shale projects dried up when oil prices fell in the mid-1980s, however. What do you think? Should we declare some western states sacrifice areas so we can have cheap gasoline?

NATURAL GAS

Natural gas is the world's third largest commercial fuel (after oil, and coal), making up 23 percent of global energy consumption. It is the most rapidly growing energy source because it is convenient, cheap, and clean burning. Because natural gas produces only half as much CO_2 as an equivalent amount of coal, substitution could help reduce global warming (chapter 17). Natural gas is difficult to ship across oceans or to store in large quantities. North America is fortunate to have an abundant, easily available supply of gas and a pipeline network to deliver it to market. Many developing countries cannot afford such a pipeline network, and much of the natural gas produced in conjunction with oil pumping is simply burned (flared off), a terrible waste of a valuable resource (fig. 21.12).

FIGURE 21.12 Fossil fuels provided the energy source for the Industrial Revolution and built the world in which we now live. We have used up much of the easily accessible supply of these traditional fuels, however, often through wasteful practices such as the flaring shown here.

Natural Gas Resources and Reserves

The republics of the former Soviet Union have 44 percent of known natural gas reserves (mostly in Siberia and the Central Asian republics) and account for 36.5 percent of all production. Both Eastern and Western Europe buy substantial quantities of gas from these wells. Figure 21.13 shows the distribution of proven natural gas reserves in the major producing countries of the world.

The total ultimately recoverable natural gas resources in the world are estimated to be 10,000 trillion cubic feet, corresponding to about 80 percent as much energy as the recoverable reserves of crude oil. The proven world reserves of natural gas are 3200 trillion cubic feet (73,000 million metric tons). Because gas consumption rates are only about half of those for oil, current gas reserves represent roughly a 60-year supply at present usage rates. Proven reserves in the United States are about 200 trillion cubic feet, or 6 percent of the world total. This is a ten-year supply at current rates of consumption. Known reserves are more than twice as large.

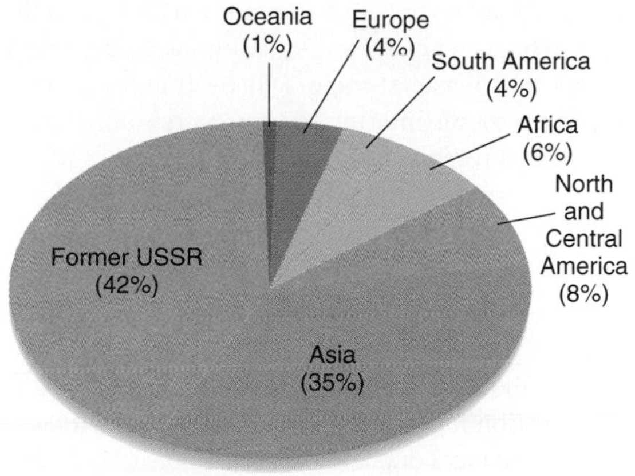

FIGURE 21.13 Percent of proven-in-place natural gas reserves, by region (1991).
Source: Data from World Resources Institute, 1994–95.

Unconventional Gas Sources

Natural gas resources have been less extensively investigated than petroleum reserves. There may be extensive "unconventional" sources of gas in unexpected places. Prime examples are recently discovered methane hydrate deposits in arctic permafrost and beneath deep ocean sediments. **Methane hydrate** is composed of small bubbles or individual molecules of natural gas trapped in a crystalline matrix of frozen water. At least fifty oceanic deposits and a dozen land deposits are known. Altogether, they are thought to hold some 10,000 gigatons (10^{13} tons) of carbon or twice as much as the combined amount of all coal, oil, and conventional natural gas. This could be a valuable energy source but would be difficult to extract, store, and ship. If climate change causes melting of these deposits, it could trigger a catastrophic spiral of global warming because methane is ten times as powerful a greenhouse gas as CO_2. Japan plans exploratory extraction of methane hydrate in the next few years, first on land near Prudhoe Bay, Alaska, and then in Japanese waters.

Methane also can be produced by digesting garbage or manure. Some U.S. cities collect methane from landfills and sewage sludge digestion. In developing countries, small-scale manure digesters provide a valuable, renewable source of gas for heating, lighting, and cooking (see chapter 22).

NUCLEAR POWER

In 1953, President Dwight Eisenhower presented his "Atoms for Peace" speech to the United Nations. He announced that the United States would build nuclear-powered electrical generators to provide clean, abundant energy. He predicted that nuclear energy would fill the deficit caused by predicted shortages of oil and natural gas. It would provide power "too cheap to meter" for continued industrial expansion of both the developed and the developing world. It would be a supreme example of "beating swords into plowshares." Technology and engineering would tame the evil genie of atomic energy and use its enormous power to do useful work.

FIGURE 21.14 The nuclear power plant at Three Mile Island near Harrisburg, PA, was the site of a partial meltdown of the reactor core in 1979. Most of the radioactive material was kept inside the containment buildings (short, cylindrical buildings in center). The four large open cylinders are cooling towers.

Glowing predictions about the future of nuclear energy continued into the early 1970s. Between 1970 and 1974, American utilities ordered 140 new reactors for power plants (fig. 21.14). Some advocates predicted that by the end of the century there would be 1500 reactors in the United States alone. The International Atomic Energy Agency (IAEA) projected worldwide nuclear power generation of at least 4.5 million megawatts (MW) by the year 2000, 18 times more than our current nuclear capacity and twice as much as present world electrical capacity from all sources.

Rapidly increasing construction costs, declining demand for electric power, and safety fears have made nuclear energy much less attractive than promoters expected. Electricity from nuclear plants was about half the price of coal in 1970 but twice as much by 1990. Wind energy is already cheaper than nuclear power in many areas and solar power or hydropower are becoming cheaper as well (chapter 22).

The United States led the world into the nuclear age, and it appears that we are leading the world out of it. After 1975, only 13 orders were placed for new nuclear reactors in the United States, and all of those orders subsequently were canceled (fig. 21.15). In fact, 100 of the 140 reactors on order in 1975 were canceled. It is beginning to look as if the much-acclaimed nuclear power industry may have been a very expensive wild goose chase that may never produce enough energy to compensate for the amount invested in research, development, mining, fuel preparation, and waste storage.

After the 1991 Persian Gulf war, the nuclear industry began advertising again, promoting reactors as clean and environmentally

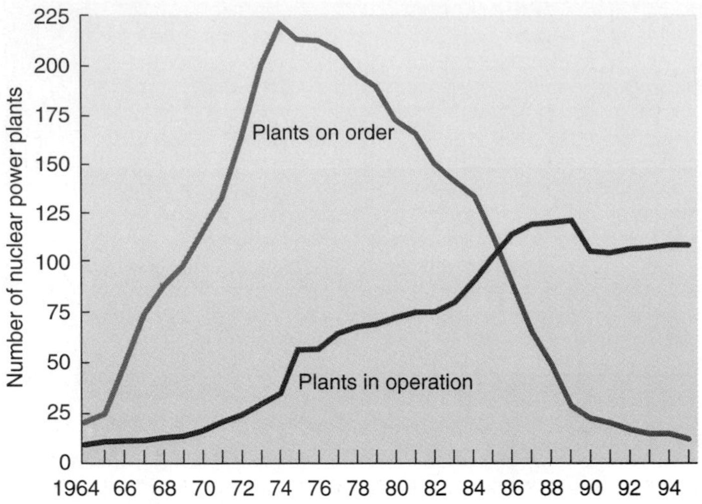

FIGURE 21.15 The changing fortunes of nuclear power in the United States are evident in this graph showing the number of nuclear plants on order and plants in operation. Since the early 1970s, few utilities have ordered new plants and many previous orders have been canceled.
Source: Data from World Resources Institute and *World Almanac,* 1997.

friendly because they do not contribute carbon dioxide to the greenhouse effect (chapter 17). Whether solutions to the problems of reactor safety and waste storage will be found so that nuclear power can serve as an interim energy source until we develop renewable sources remains to be seen.

How Do Nuclear Reactors Work?

The most commonly used fuel in nuclear power plants is U^{235}, a naturally occurring radioactive isotope of uranium. Ordinarily, U^{235} makes up only about 0.7 percent of uranium ore, too little to sustain a chain reaction in most reactors. It must be purified and concentrated by mechanical or chemical procedures (fig. 21.16). When the U^{235} concentration reaches about 3 percent, the uranium is formed into cylindrical pellets slightly thicker than a pencil and about 1.5 cm long. Although small, these pellets pack an amazing amount of energy. Each 8.5-gram pellet is equivalent to a ton of coal or four barrels of crude oil.

The pellets are stacked in hollow metal rods approximately 4 m long. About one hundred of these rods are bundled together to make a **fuel assembly.** Thousands of fuel assemblies containing

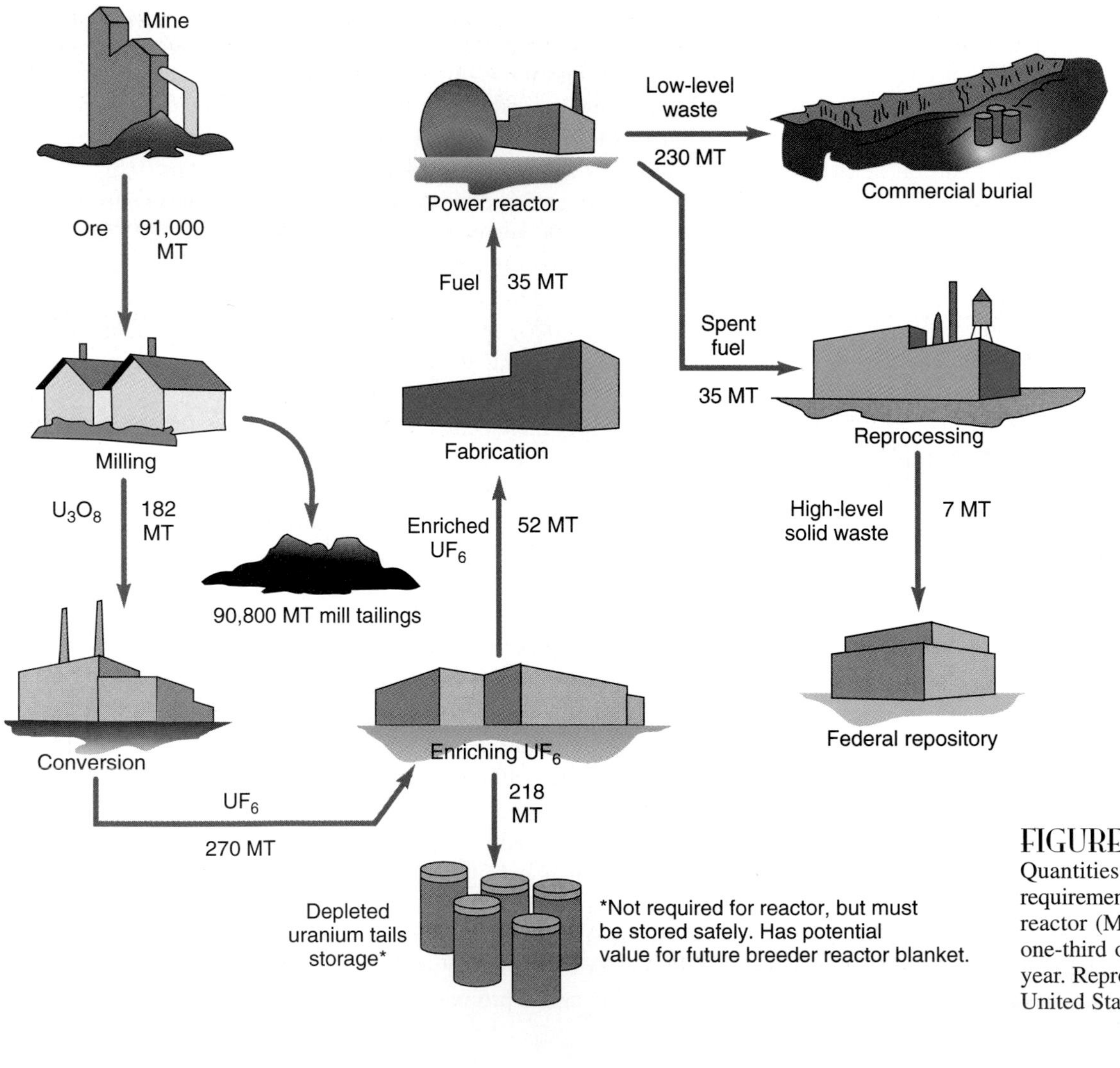

FIGURE 21.16 The nuclear fuel cycle. Quantities represent the average annual fuel requirements for a typical 1000 MW light water reactor (MT = metric tons). About 35 MT or one-third of the reactor fuel is replaced every year. Reprocessing is not currently done in the United States.

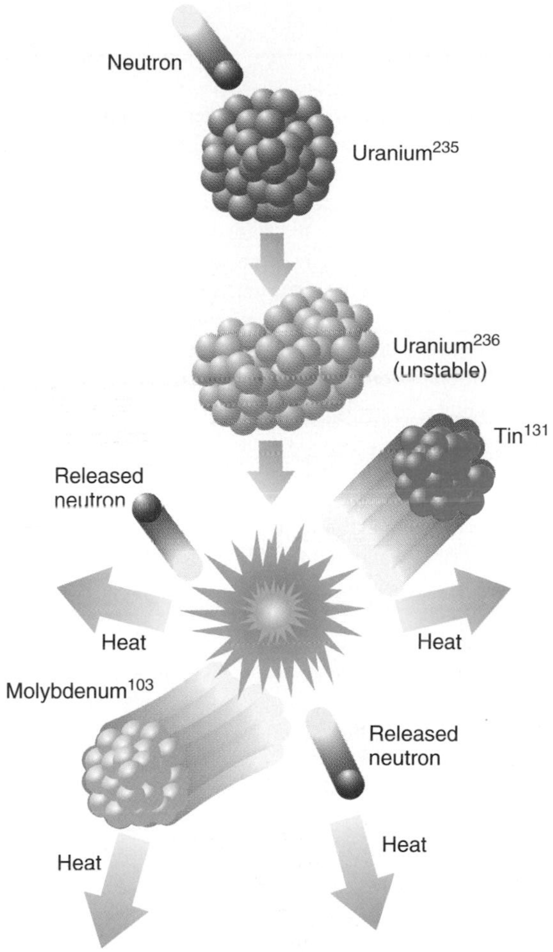

FIGURE 21.17 The process of nuclear fission is carried out in the core of a nuclear reactor. In the sequence shown here, the unstable isotope, uranium-235, absorbs a neutron and splits to form tin-131 and molybdenum-103. Two or three neutrons are released per fission event and continue the chain reaction. The total mass of the reaction product is slightly less than the starting material. The residual mass is converted to energy (mostly heat).

100 tons of uranium are bundled together in a heavy steel vessel called the reactor core. Radioactive uranium atoms are unstable—that is, when struck by a high-energy subatomic particle called a neutron, they undergo **nuclear fission** (splitting), releasing energy and more neutrons. When uranium is packed tightly in the reactor core, the neutrons released by one atom will trigger the fission of another uranium atom and the release of still more neutrons (fig. 21.17). Thus a self-sustaining **chain reaction** is set in motion and vast amounts of energy are released.

The chain reaction is moderated (slowed) in a power plant by a neutron-absorbing cooling solution that circulates between the fuel rods. In addition, **control rods** of neutron-absorbing material, such as cadmium or boron, are inserted into spaces between fuel assemblies to shut down the fission reaction or are withdrawn to allow it to proceed. Water or some other coolant is circulated between the fuel rods to remove excess heat.

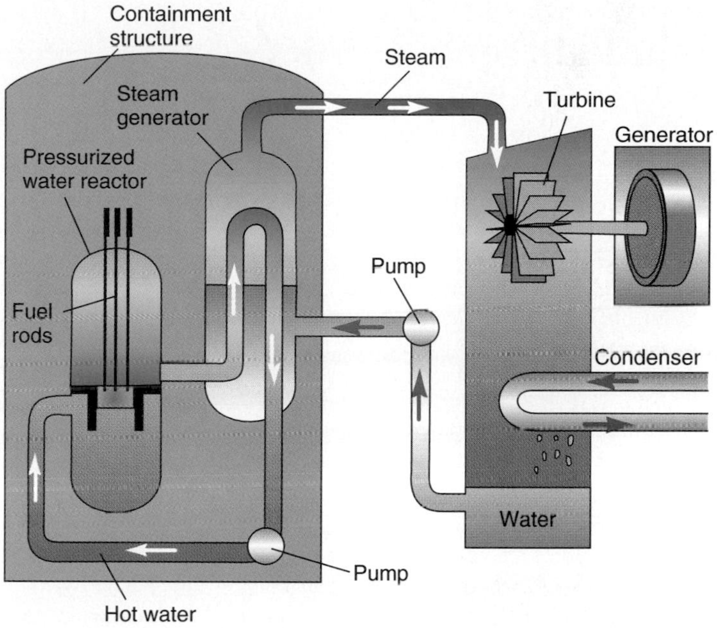

FIGURE 21.18 Pressurized water nuclear reactor. Water is superheated and pressurized as it flows through the reactor core. Heat is transferred to nonpressurized water in the steam generator. The steam drives the turbogenerator to produce electricity.

The greatest danger in one of these complex machines is a cooling system failure. If the pumps fail or pipes break during operation, the nuclear fuel quickly overheats and a "meltdown" can result that releases deadly radioactive material. (What Do You Think? p. 488). Although nuclear power plants cannot explode like a nuclear bomb, the radioactive releases from a worst-case disaster like Chernobyl are just as devastating as a bomb.

Kinds of Reactors in Use

Seventy percent of the nuclear plants in the United States and in the world are pressurized water reactors (PWR) (fig. 21.18). Water is circulated through the core, absorbing heat as it cools the fuel rods. This primary cooling water is heated to 317°C (600°F) and reaches a pressure of 2235 psi. It then is pumped to a steam generator where it heats a secondary water-cooling loop. Steam from the secondary loop drives a high-speed turbine generator that produces electricity. Both the reactor vessel and the steam generator are contained in a thick-walled concrete and steel containment building that prevents radiation from escaping and is designed to withstand high pressures and temperatures in case of accidents. Engineers operate the plant from a complex, sophisticated control room containing many gauges and meters to tell them how the plant is running.

Overlapping layers of safety mechanisms are designed to prevent accidents, but these fail-safe controls make reactors very expensive and very complex. A typical nuclear power plant has 40,000 valves compared to only 4000 in a fossil fuel-fired plant of similar size. In some cases, the controls are so complex that they what do you think?

THE WORLD AROUND YOU

Chernobyl: Could It Happen Here?

In the early morning hours of April 26, 1986, residents of the Ukrainian village of Pripyat saw a spectacular and terrifying sight. A glowing fountain of molten nuclear fuel and burning graphite was spewing into the dark sky through a gaping hole in the roof of the Chernobyl Nuclear Power Plant only a few kilometers away. Although officials assured them that there was nothing to worry about in this "rapid fuel relocation," the villagers knew that something was terribly wrong. They were witnessing the worst possible nuclear power accident, a "meltdown" of the nuclear fuel and rupture of the containment facilities, releasing enormous amounts of radioactivity into the environment.

The accident was a result of a risky experiment undertaken late at night by the plant engineers in violation of a number of safety rules and operational procedures. They were testing whether the residual energy of a spinning turbine could provide enough power to run the plant in an emergency shutdown if off-site power were lost. Reactor number four had been slowed down to only 6 percent of its normal operating level. To conserve the small amount of electricity being generated, they then disconnected the emergency core-cooling pumps and other safety devices, unaware that the reactor was dangerously unstable under these conditions.

The heat level in the core began to rise, slowly at first, and then faster and faster. The operators tried to push the control rods into the core to slow the reaction, but the graphite pile had been deformed by the heat so that the rods wouldn't go in. In 4.5 seconds, the power level rose 2000-fold, far above the rated capacity of the cooling system. Chemical explosions (probably hydrogen gas released from the expanding core) ripped open the fuel rods and cooling tubes. Cooling water flashed into steam and blew off the 1000-ton-concrete cap on top of the reactor. Molten uranium fuel puddled in the bottom of the reactor, creating a critical mass that accelerated the nuclear fission reactions. The metal superstructure of the containment building was ripped apart and a column of burning graphite, molten uranium, and radioactive ashes billowed 1000 m (3000 ft) into the air.

Panic and confusion ensued. Officials first denied that anything was wrong. The village of Pripyat was not evacuated for 36 hours. There was no public announcement for three days. The first international warning came, not from Soviet authorities, but from Swedish scientists 2000 km away who detected unusually high levels of radioactive fallout and traced airflows back to the southern Soviet Union.

There were many acts of heroism during this emergency. Firefighters climbed to the roof of the burning reactor building to pour water into the blazing inferno. Engineers dived into the suppression pool beneath the burning core to open a drain to prevent another steam explosion. Helicopter pilots hovered over the gaping maw of the ruined building to drop lead shot, sand, clay, limestone, and boron carbide onto the burning nuclear core to smother the fire and suppress the nuclear fission reactions. More than 600,000 workers participated in putting out the fire and cleaning up contamination. Thousands who were exposed to high radiation doses already have died and many more have dim prospects for the future.

The amount of radioactive fallout varied from area to area, depending on wind patterns and rainfall. Some places had heavy doses while neighboring regions had very little. One band of fallout spread across Yugoslavia, France, and Italy. Another crossed Germany and Scandinavia. Small amounts of radiation even reached North America. Altogether, about 7 tons of fuel containing 50 to 100 million curies were released, roughly 5 percent of the reactor fuel.

For several years after the accident, the Soviet government tried to suppress information and deny the consequences. The dissolution of the USSR has allowed many of these secrets to come to light. It's difficult to separate the effects of Chernobyl from other causes of ill-health, but at least a half million people live in areas where radiation levels were high enough to be of concern. One clear effect is

confuse operators and cause accidents rather than prevent them. Under normal operating conditions, a PWR releases very little radioactivity and is probably less dangerous for nearby residents than a coal-fired power plant.

A simpler, but dirtier and more dangerous reactor design is the boiling water reactor (BWR). In this model, water from the reactor core boils to make steam, which directly drives the turbine-generators. This means that highly radioactive water and steam leave the containment structure. Controlling leaks is difficult, and the chances of releasing radiation in an accident are very high.

Canadian nuclear reactors use heavy water containing deuterium (H^2 or ^2H, the heavy, stable isotope of hydrogen) as both a cooling agent and a moderator. These *Can*adian *deu*terium (CANDU) reactors operate with natural, unconcentrated uranium (0.7 percent U^{235}) for fuel, eliminating expensive enrichment processes. "Heavy water," deuterium-containing water (^2H_2O), is expensive, however, and these reactors are susceptible to overheating and meltdown if cooling pumps fail.

In Britain, France, and the former Soviet Union, a common reactor design uses graphite, both as a moderator and as the structural material for the reactor core. In the British MAGNOX design (named after the magnesium alloy used for its fuel rods), gaseous carbon dioxide is blown through the core to cool the fuel assemblies and carry heat to the steam generators. In the Soviet design, called RBMK (the Russian initials for a graphite-moderated, water-cooled reactor), low-pressure cooling water circulates through the core in thousands of small metal tubes.

These designs were originally thought to be very safe because graphite has high capacity for both capturing neutrons and dissipating heat. Designers claimed that these reactors could not possibly run out of control; unfortunately, they were proven wrong. The small cooling tubes are quickly blocked by steam if the cooling system fails and the graphite core burns when exposed to air. The two most disastrous reactor accidents in the world, so far, involved fires in graphite cores that allowed the nuclear fuel to melt and escape into the environment. In 1956, a fire at the Windscale seen in children in Belarus where thyroid cancers have increased a hundred-fold since 1986. Childhood leukemias and some autoimmune diseases also appear to be more prevalent in highly contaminated areas.

For the present, the damaged reactor has been entombed in a giant, steel-reinforced concrete "sarcophagus." Unfortunately, this containment structure was hastily built of inferior materials, and already has begun to deteriorate. Reconstruction started in 1999, but the Ukraine is demanding billions of dollars from other countries to finance this operation.

So far, more than 250,000 people have been relocated from the contaminated area. More than 70 villages have been destroyed and millions of hectares of the richest farmland in the Commonwealth of Independent States has been abandoned. The immediate direct costs were roughly $3 billion; total costs might be one hundred times that much.

Only one of the four Chernobyl reactors is still in operation. Many people would like to see all of them shut down but fuel shortages and a crippled economy make building replacement power plants prohibitively expensive. The United States, Canada, and the European Community are currently negotiating with Ukraine and Russia to close down all remaining Chernobyl-type reactors. What do you think the United States should contribute to help bring this about? How important is it to prevent further catastrophes?

The explosion and fire in Unit 4 (circled) at the Chernobyl nuclear power plant in the Soviet Ukraine in 1986 spread radioactive fallout across the entire Northern Hemisphere. Millions of people in Europe and Asia were exposed to dangerously high radiation levels.

Proponents of nuclear power point out that the Soviet-style reactors were much more dangerous than most of those now in the United States or Canada. Furthermore, they argue, the accident was caused by operator error and bureaucratic bungling. Something similar could never happen here, they claim. Opponents respond that we should learn from this tragedy and abandon this dangerous technology. What do you think? How likely is a similar accident in your neighborhood? Even if the chances are very low, does the potential harm make this an unacceptable risk? How would you weigh a low probability against a terrible potential outcome?

Plutonium Reactor in England released roughly 100 million curies of radionuclides and contaminated hundreds of square kilometers of countryside. Similarly, burning graphite in the Chernobyl nuclear plant in Ukraine made the fire much more difficult to control than it might have been in another reactor design. The worst nuclear accident in U.S. history, for instance, in 1979 at Three Mile Island near Harrisburg, Pennsylvania, released millions of times less radiation as did Chernobyl or Windscale (see fig. 21.14).

Alternative Reactor Designs

Several other reactor designs are inherently safer than the ones we now use. Among these are the modular High-Temperature, Gas-Cooled Reactor (HTGCR) and the Process-Inherent Ultimate-Safety (PIUS) reactor.

In HTGCR, uranium is encased in tiny ceramic-coated pellets; gaseous helium blown around these pellets is the coolant. If the reactor core is kept small, it cannot generate enough heat to melt the ceramic coating, even if all coolant is lost; thus, a meltdown is impossible and operators could walk away during an accident without risk of a fire or radioactive release. Fuel pellets are loaded into the reactor from the top, shuffle through the core as the uranium is consumed, and emerge from the bottom as spent fuel. This type of reactor can be reloaded during operation. Since the reactors are small, they can be added to a system a few at a time, avoiding the costs, construction time, and long-range commitment of large reactors. Only two of these reactors have been tried in the United States: the Brown's Ferry reactor in Alabama and the Fort St. Vrain reactor near Loveland, Colorado. Both were continually plagued with problems (including fires in control buildings and turbine-generators), and both were closed without producing much power.

A much more successful design has been built in Europe by General Atomic. In West German tests, a HTGCR was subjected to total coolant loss while running at full power. Temperatures remained well below the melting point of fuel pellets and

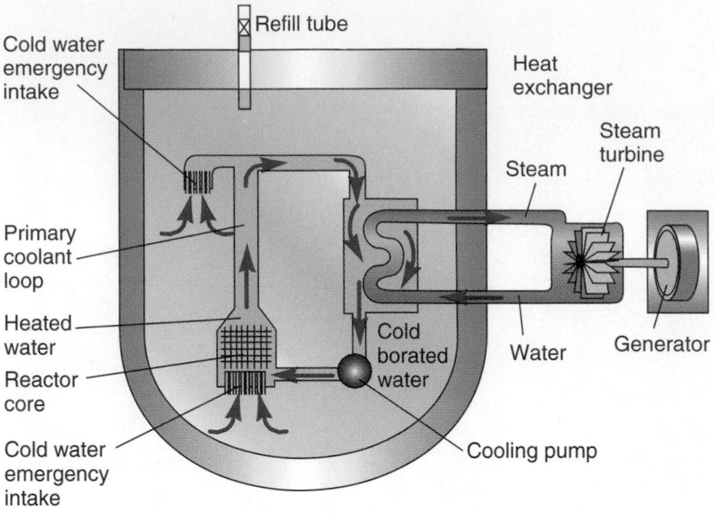

FIGURE 21.19 A PIUS reactor consists of a core and primary cooling system immersed in a very large pool of cold, borated water. As long as the reactor is operating, the cool water is excluded from the cooling circuit by the temperature and pressure differential. Any failure of the primary cooling system would allow cold water to flood the core and shut down the nuclear reaction. The large volume of the pool would cool the reactor for several days without any replenishment.

no damage or radiation releases occurred. These reactors might be built without expensive containment buildings, emergency cooling systems, or complex controls. They would be both cheaper and safer than current designs.

The PIUS design features a massive, 60-meters-high pressure vessel of concrete and steel, within which the reactor core is submerged in a very large pool of boron-containing water (fig. 21.19). As long as the primary cooling water is flowing, it keeps the borated water away from the core. If the primary coolant pressure is lost, however, the surrounding water floods the core, and the boron poisons the fission reaction. There is enough secondary water in the pool to keep the core cool for at least a week without any external power or cooling. This should be enough time to resolve the problem. If not, operators can add more water and evaluate conditions further.

The Canadian "slow-poke" is a small-scale version of the PIUS design that doesn't produce electricity but might be a useful substitute for coal, oil, or gas burners in district heating plants. The core of this "mini-nuke" is only about half the size of a kitchen stove and generates only 1/10 to 1/100 as much power as a conventional reactor. The fuel sits in a large pool of ordinary water, which it heats to just below boiling and sends to a heat exchanger. A secondary flow of hot water from the exchanger is pumped to nearby buildings for space heating. Promoters claim that a runaway reaction is impossible in this design and that it makes an attractive and cost-efficient alternative to fossil fuels. Despite a widespread aversion to anything nuclear, Switzerland and Germany are developing similar small nuclear heating plants.

If these reactors had been developed initially, the history of nuclear power might have been very different. Neither reactor type is suitable, however, for mobile power plants, such as nuclear submarines, which tells you something about the history of nuclear power and the motivation for its development. Aside from being inherently safer in case of coolant pressure loss, neither of these alternative reactor designs eliminates other problems that we will discuss later in this chapter. Still, one of these alternate forms of nuclear power might make it an attractive energy source some day.

Breeder Reactors

For more than 30 years, nuclear engineers have been proposing high-density, high-pressure, **breeder reactors** that produce fuel rather than consume it. These reactors create fissionable plutonium and thorium isotopes from the abundant, but stable, forms of uranium (fig. 21.20). The starting material for this reaction is plutonium reclaimed from spent fuel from conventional fission reactors. After about ten years of operation, a breeder reactor would produce enough plutonium to start another reactor. Sufficient uranium currently is stockpiled in the United States to produce electricity for 100 years at present rates of consumption, if breeder reactors can be made to work safely and dependably.

Several problems have held back the breeder reactor program in the United States. One problem is the concern about safety. The reactor core of the breeder must be at a very high density for the breeding reaction to occur. Water does not have enough heat capacity to carry away the high heat flux in the core, so liquid sodium is used as a coolant. Liquid sodium is very corrosive and difficult to handle. It burns with an intense flame if exposed to oxygen, and it explodes if it comes into contact with water. Because of its intense heat, a breeder reactor will melt down and self-destruct within a few seconds if the primary coolant is lost, as opposed to a few minutes for a normal fission reactor.

Another very serious concern about breeder reactors is that they produce excess plutonium that can be used for bombs. It is essential to have a spent-fuel reprocessing industry if breeders are used, but the existence of large amounts of weapons-grade plutonium in the world would surely be a dangerous and destabilizing development. The chances of some of that material falling into the hands of terrorists or other troublemakers are very high. Japan plans to purchase 30 tons of this dangerous material from France and ship it half way around the world through some of the most dangerous and congested shipping lanes on the planet to fuel a breeder program. In 1995, a serious accident at Japan's Moju breeder reactor caused reevaluation of the whole program.

A proposed $1.7 billion breeder-demonstration project in Clinch River, Tennessee, has been on and off for 15 years. At last estimate, it would cost up to five times the original price if it is ever completed. In 1986, France put into operation a full-sized commercial breeder reactor, the SuperPhenix, near Lyons. It cost three times the original estimate to build and produces electricity at twice the cost per kilowatt of conventional nuclear power. In 1987, a large crack was discovered in the inner containment vessel of the SuperPhenix, and in 1997 it was shut down permanently.

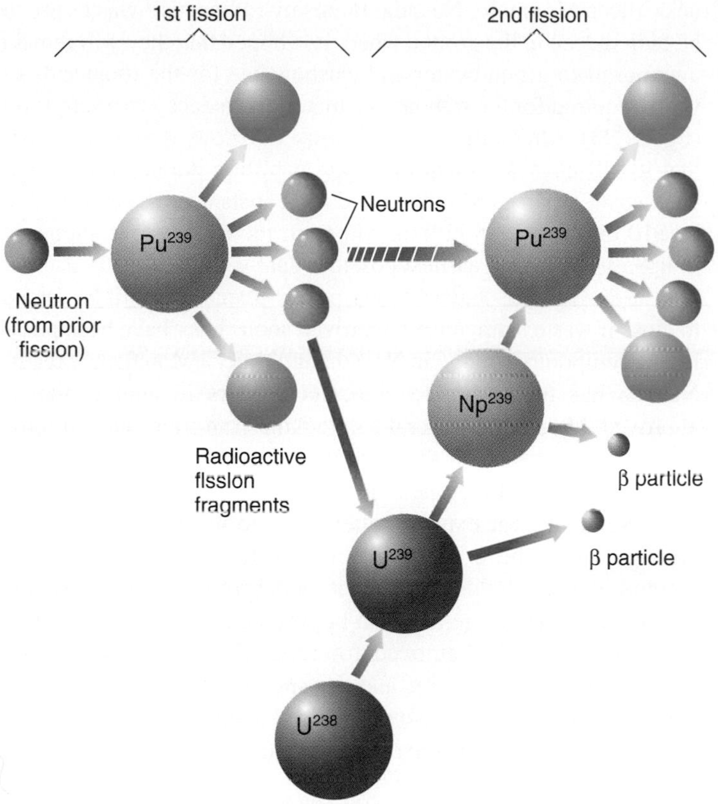

FIGURE 21.20 Reactions in a "breeder" fission process. Neutrons from a plutonium fission change U^{238} to U^{239} and then to Pu^{239} so that the reactor creates more fuel than it uses.

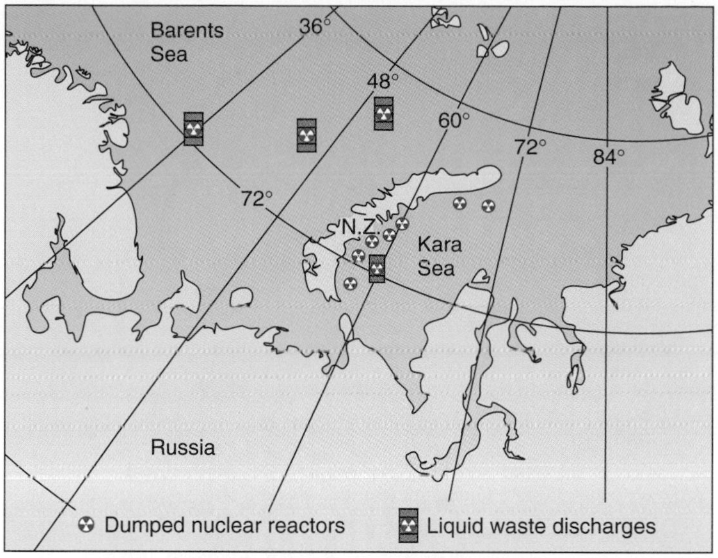

FIGURE 21.21 Sites of dumped nuclear reactors east of Novaya Zemlya. Dark blue areas indicate liquid waste discharges.

RADIOACTIVE WASTE MANAGEMENT

One of the most difficult problems associated with nuclear power is the disposal of wastes produced during mining, fuel production, and reactor operation. How these wastes are managed may ultimately be the overriding obstacle to nuclear power.

Ocean Dumping of Radioactive Wastes

Until 1970, the United States, Britain, France, and Japan disposed of radioactive wastes in the ocean. Dwarfing all these dumps, however, are those of the former Soviet Union, which has seriously—and some fear permanently—contaminated the Arctic ocean. Rumors of Soviet nuclear waste dumping had circulated for years, but it was not until after the collapse of the Soviet Union that the world learned the true extent of what happened. Starting in 1965, the Soviets disposed of eighteen nuclear reactors—seven loaded with nuclear fuel—in the Kara Sea off the eastern coast of Novaya Zemlya island, and millions of liters of liquid waste in the nearby Barents Sea (fig. 21.21). Two other reactors were sunk in the Sea of Japan.

Altogether, the former Soviet Union dumped 2.5 million curies of radioactive waste into the oceans, more than twice as much as the combined amounts that 12 other nuclear nations have reported dumping over the past 45 years. In 1993, despite protests from Japan, Russia dumped 900 tons of additional radioactive waste into the Sea of Japan. With aging storage tanks overflowing with millions of liters of highly radioactive wastes and no funds to maintain or monitor them, Russia is desperate for someplace to dispose of its nuclear wastes. This is an international crisis that threatens all of us.

Land Disposal of Nuclear Waste

Enormous piles of mine wastes and abandoned mill tailings in uranium-producing countries represent another serious waste disposal problem. Production of 1000 tons of uranium fuel typically generates 100,000 tons of tailings and 3.5 million liters of liquid waste. There now are approximately 200 million tons of radioactive waste in piles around mines and processing plants in the United States. This material is carried by the wind and washes into streams, contaminating areas far from its original source. Canada has even more radioactive mine waste on the surface than does the United States.

In addition to the leftovers from fuel production, there are about 100,000 tons of low-level waste (contaminated tools, clothing, building materials, etc.) and about 15,000 tons of high-level (very radioactive) wastes in the United States. The high-level wastes consist mainly of spent fuel rods from commercial nuclear power plants and assorted wastes from nuclear weapons production. For the past 20 years, spent fuel assemblies from commercial reactors have been stored in deep water-filled pools at the power plants. These pools were originally intended only as temporary storage until the wastes were shipped to reprocessing centers or permanent disposal sites.

With internal waste storage pools now full but neither reprocessing nor permanent storage available, a number of utility

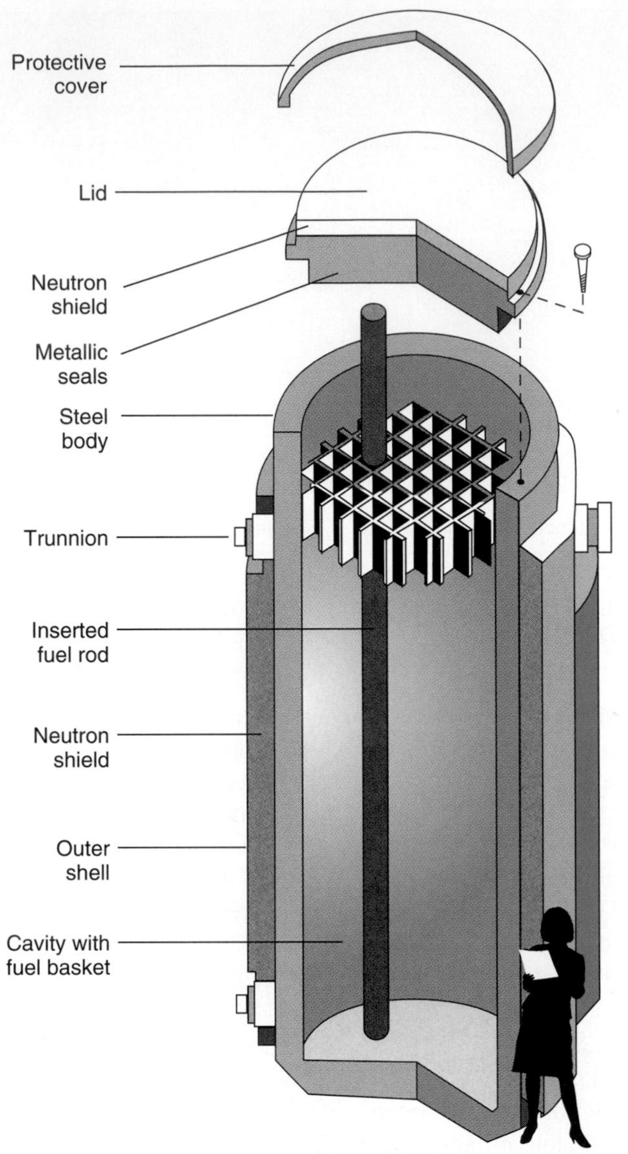

FIGURE 21.22 Dry cask storage for nuclear waste. Each cask is 17 ft tall, 8.5 ft in diameter, has steel walls 10 in. thick, and weighs 122 tons when fully loaded with 40 fuel assemblies. The casks are expected to last 40 years and cost $700,000 each. Critics are worried that the casks will leak, but proponents point out that they can be monitored easily and replaced if necessary.

companies are beginning to store nuclear waste in large metal dry casks placed outside power plants (fig. 21.22). These projects are meeting with fierce opposition from local residents who fear the casks will leak. Most nuclear power plants are built near rivers, lakes, or seacoasts. Extremely toxic radioactive materials could spread quickly over large areas if leaks occur. A hydrogen gas explosion and fire in 1997 in a dry storage cask at Wisconsin's Point Beach nuclear plant intensified opponents' suspicions about this form of waste storage.

In 1987, the U.S. Department of Energy announced plans to build the first **high-level waste repository** on a barren desert ridge near Yucca Mountain, Nevada. Intensely radioactive wastes are to be buried deep in the ground where it is hoped that they will remain unexposed to groundwater and earthquakes for the thousands of years required for the radioactive materials to decay to a safe level (fig. 21.23). Although the area is very dry now, we can't be sure that it will always remain that way. A billion dollars already has been spent at Yucca Mountain and total costs now are expected to be $10 billion to $35 billion. Although the facility was supposed to open in 1998, the earliest possible date is now 2010.

Several states have nuclear power plants that will be forced to close if waste storage isn't approved soon. They have been pushing for temporary storage in Nevada until Yucca Mountain is ready. Nevada has threatened to close its borders if such a plan is approved. Meanwhile several native American tribes have offered to store nuclear waste if the price is right. What do you think, is it ethical—or safe—to let them do so?

Some nuclear experts believe that **monitored, retrievable storage** would be a much better way to handle wastes. This method involves holding wastes in underground mines or secure surface facilities where they can be watched. If canisters begin to leak, they could be removed for repacking. Safeguarding the wastes would be expensive and the sites might be susceptible to wars or terrorist attacks. We might need a perpetual priesthood of nuclear guardians to ensure that the wastes are never released into the environment.

Decommissioning Old Nuclear Plants

Old power plants themselves eventually become waste when they have outlived their useful lives. Most plants are designed for a life of only 30 years. After that, pipes become brittle and untrustworthy because of the corrosive materials and high radioactivity to which they are subjected. Plants built in the 1950s and early 1960s already are reaching the ends of their lives. You don't just lock the door and walk away from a nuclear power plant; it is much too dangerous. It must be taken apart, and the most radioactive pieces have to be stored just like other wastes. This includes not only the reactor and pipes but also the meter-thick, steel-reinforced concrete containment building. The pieces have to be cut apart by remote-control robots because they are too dangerous to be worked on directly.

Only a few plants have been decommissioned so far, but it has generally cost two to ten times as much to tear them down as to build them in the first place. That means that the 100 reactors now in operation might cost somewhere between $200 billion and $1 trillion to decommission. No one knows how much it will cost to store the debris for thousands of years or how it will be done. However, we would face this problem, to some degree, even without nuclear electric power plants. Plutonium production plants and nuclear submarines also have to be decommissioned. Originally, the Navy proposed to just tow old submarines out to sea and sink them. The risk that the nuclear reactors would corrode away and release their radioactivity into the ocean makes this method of disposal unacceptable, however.

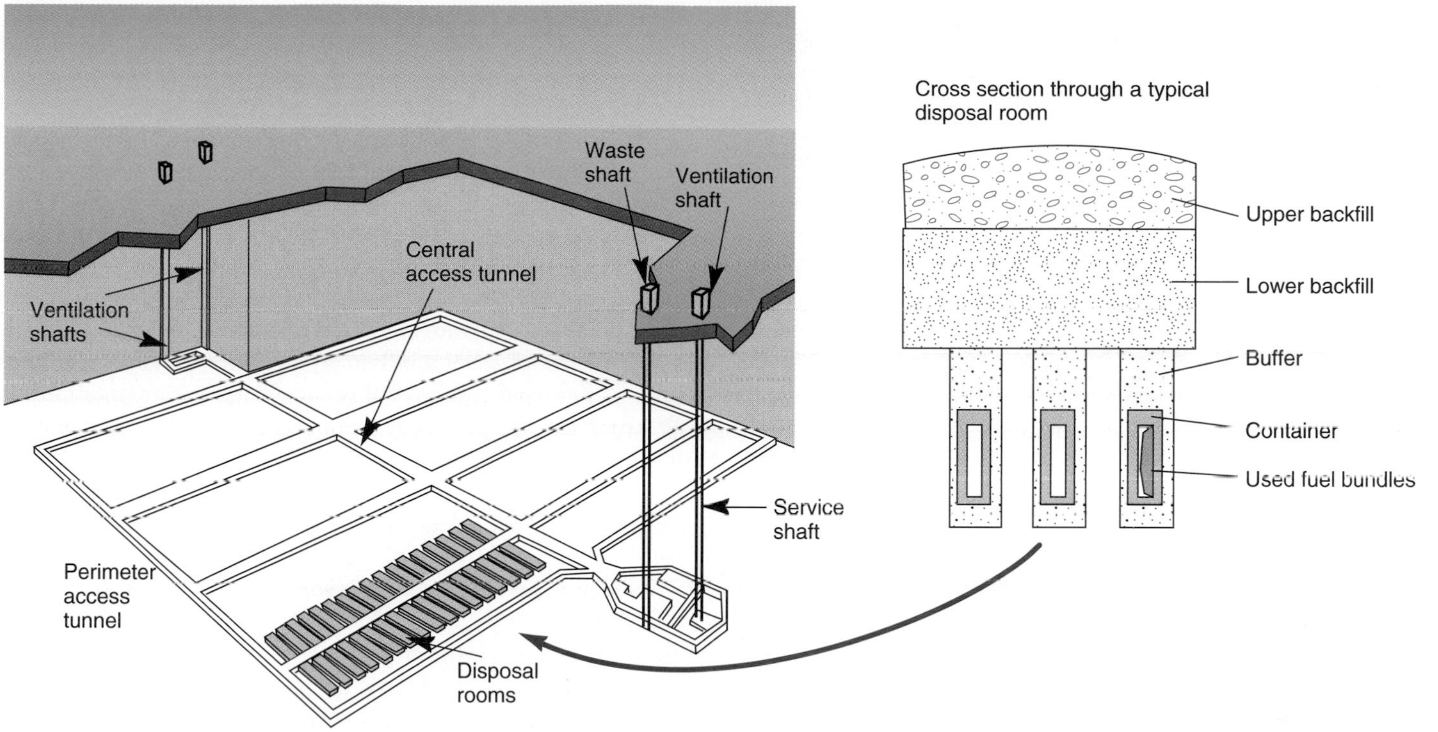

FIGURE 21.23 One proposal for the permanent disposal of used fuel is to bury the used fuel containers 500 to 1000 m deep in the stable rock formations. Spent fuel, in the form of ceramic fuel pellets, are sealed inside the corrosion-resistant containers. Glass beads are compacted around the used fuel bundles and into the spaces between the fuel and the container shell. The containers are then buried in specially built vaults that are packed and sealed with special materials to further retard the migration of radioactive material.

CHANGING FORTUNES OF NUCLEAR POWER

Although promoted originally as a new wonder of technology that could open the door to wealth and abundance, nuclear power has long been highly controversial. For many environmental organizations, opposition to nuclear power is a perennial priority. Antinuclear groups such as the Clamshell Alliance, Northern Sun, and Greenpeace have organized mass protest rallies featuring popular actors, musicians, and celebrities who help raise funds and attract attention to their cause. Protests and civil disobedience at some sites have gone on for decades, and plants that were planned—or in some cases, already built—have been abandoned or modified to burn fossil fuels because of public opposition.

On the other side, workers, consumers, utility officials, and others who stand to benefit from this new technology rally in support of nuclear power (fig. 21.24). They argue that abandoning this energy source is foolish since a great deal of money already has been invested and the risks may be lower than is commonly believed. Recently, engineering and utility companies have begun a new advertising campaign to promote nuclear power as a clean solution to air pollution and global warming.

Public opinion about nuclear power has fluctuated over the years but recently seems to have stabilized in the United States with

FIGURE 21.24 The Seabrook nuclear reactor in New Hampshire has been the site of protests and counter-protests for 20 years. Here pipefitters march in support of the plant and their jobs.

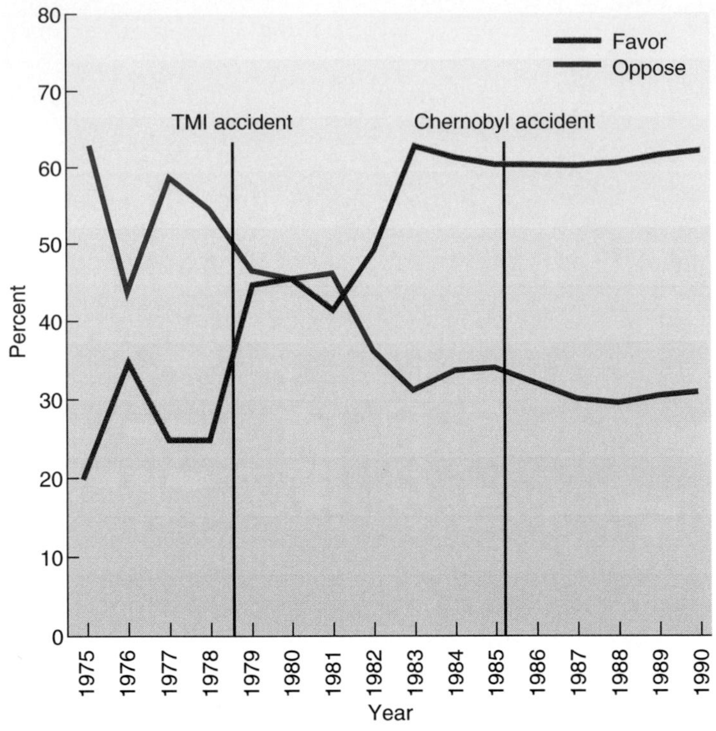

FIGURE 21.25 Public opinion on building more nuclear power plants in the United States 1975-1990. "Don't know" or "no opinion" make up the difference between these curves and 100%. Little change has occurred since 1990.

Source: Data from Louis Harris and Associates as appeared in E. Rosa and R. Dunlap, "Nuclear Power: 3 Decades of Public Opinion" in *Public Opinion Quarterly,* Vol. 58, 1994.

about one-third in favor and two-thirds opposed (fig. 21.25). Numerous revelations of sloppy management and careless operation at existing power plants fuels public suspicion about this dangerous technology. For example, an internal study in 1997 at Ontario Hydro, North America's largest electric utility, documented widespread safety violations, years of inadequate maintenance, and deteriorating conditions. Operators were filmed sleeping on the job and playing video games on control room computers. As a result, the company announced plans to shut down seven of its 19 nuclear reactors. These shut-downs will eliminate about one-fourth of the province's electric power and require increased use of fossil fuels, which will add to Canada's already high greenhouse emissions.

In the United States, 40,000 MW of nuclear power could face retirement as utility deregulation opens up competition with cheaper alternatives. Up to one-third of all existing nuclear plants could be priced out of the market by high operating costs. Altogether, about half of the nations with nuclear power have chosen to reduce or dismantle existing plants, and few new plants have been ordered in recent years. Opponents of nuclear power claim that the nuclear fission experiment may have been a very brief episode in our history, but one whose legacy will remain with us for a long time. Proponents of nuclear power claim that it may be our only near-term alternative to fossil fuels. Where do you stand on this issue; are the benefits of nuclear power worth its risks?

NUCLEAR FUSION

Fusion energy is an alternative to nuclear fission that could have virtually limitless potential. **Nuclear fusion** energy is released when two smaller atomic nuclei fuse into one larger nucleus. Nuclear fusion reactions, the energy source for the sun and for hydrogen bombs, have not yet been harnessed by humans to produce useful net energy. The fuels for these reactions are deuterium and tritium, two heavy isotopes of hydrogen.

It has been known for 40 years that if temperatures in an appropriate fuel mixture are raised to 100 million degrees Celsius and pressures of several billion atmospheres are obtained, fusion of deuterium and tritium will occur. Under these conditions, the electrons are stripped away from atoms and the forces that normally keep nuclei apart are overcome. As nuclei fuse, some of their mass is converted into energy, some of which is in the form of heat. There are two main schemes for creating these conditions: magnetic confinement and inertial confinement.

Magnetic confinement involves the containment and condensation of plasma, a hot, electrically neutral gas of ions and free electrons in a powerful magnetic field inside a vacuum chamber. Compression of the plasma by the magnetic field should raise temperatures and pressures enough for fusion to occur. The most promising example of this approach, so far, has been a Russian design called *tokomak* (after the Russian initials for "torodial magnetic chamber"), in which the vacuum chamber is shaped like a large donut (fig. 21.26).

Inertial confinement involves a small pellet (or a series of small pellets) bombarded from all sides at once with extremely high-intensity laser light. The sudden absorption of energy causes an implosion (an inward collapse of the material) that will increase densities by 1000 to 2000 times and raise temperatures above the critical minimum. So far, no lasers powerful enough to create fusion conditions have been built.

In both of these cases, high-energy neutrons escape from the reaction and are absorbed by molten lithium circulating in the walls of the reactor vessel. The lithium absorbs the neutrons and transfers heat to water via a heat exchanger, making steam that drives a turbine generator, as in any steam power plant. The advantages of fusion reactions, if they are ever feasible, include production of fewer radioactive wastes, the elimination of fissionable products that could be made into bombs, and a fuel supply that is much larger and less hazardous than uranium.

Despite 50 years of research and a $25 billion investment, fusion reactors never have reached the breakeven point at which they produce more energy than they consume. A major setback occurred in 1997, when Princeton University's Tokamak Fusion Test Reactor was shut down. Three years earlier, this reactor had set a world's record by generating 10.7 million watts for one second, but researchers conceded that the project was still decades away from self-sustaining power generation. Proponents of fusion power urge that research be continued, maintaining that success could be just around the corner. Opponents view this technology as just another expensive wild-goose chase and predict that it will never generate enough energy to pay back the fortune spent on its development.

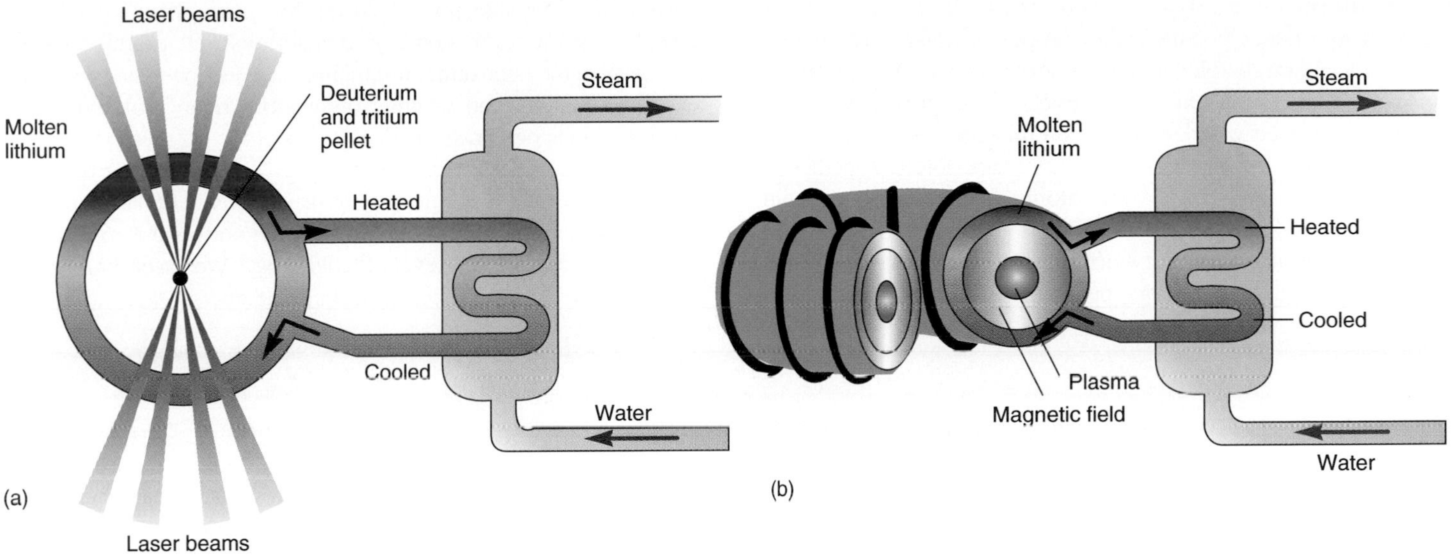

FIGURE 21.26 Nuclear fusion devices. (*a*) Inertial confinement is created by laser beams that bombard and ignite fuel pellets. Molten lithium absorbs energy from the fusion and transfers it as heat to a steam generator. (*b*) A powerful magnetic field confines the plasma and compresses it so that critical temperatures and pressures are reached. Again, molten lithium absorbs energy and transfers it to a steam generator.

SUMMARY

Energy is the capacity to do work. Power is the rate of doing work. Today, nearly 85 percent of all commercial energy is generated by fossil fuels, about 36 percent coming from petroleum. Next are coal, with 28 percent, and natural gas (methane), with 23 percent. Petroleum and natural gas were not used in large quantities until the beginning of this century, but supplies are already running low. Coal supplies will last several more centuries at present rates of usage, but it appears that the fossil fuel age will have been a rather short episode in the total history of humans. Nuclear power provides less than 5 percent of commercial energy worldwide.

Energy is essential for most activities of modern society. Its use generally correlates with standard of living, but there are striking differences between energy use per capita in countries with relatively equal standards of living. The United States, for instance, consumes nearly twice as much energy per person as does Switzerland, which is higher in many categories that measure quality of life. This difference is based partly on level of industrialization and partly on policies, attitudes, and traditions in the United States that encourage extravagant or wasteful energy use.

The environmental damage caused by mining, shipping, processing, and using fossil fuels may necessitate cutting back on our use of these energy sources. Coal is an especially dirty and dangerous fuel, at least as we currently obtain and use it. Some new coal treatment methods remove contaminants, reduce emissions, and make its use more efficient (so less will be used). Coal combustion is a major source of acid precipitation that is suspected of being a significant cause of environmental damage in many areas. We now recognize that CO_2 buildup in the atmosphere has the potential to trap heat and raise the earth's temperature to catastrophic levels.

Nuclear energy offers an alternative to many of the environmental and social costs of fossil fuels, but it introduces serious problems of its own. In the 1950s, there was great hope that these problems would be overcome and that nuclear power plants would provide energy "too cheap to meter." Recently, however, much of that optimism has been waning. No new reactors have been started in the United States since 1975. Many countries are closing down existing nuclear power plants, and a growing number have pledged to remain or become "nuclear-free."

The greatest worry about nuclear power is the danger of accidents that release hazardous radioactive materials into the environment. Several accidents, most notably the "meltdown" at the Chernobyl plant in the Soviet Ukraine in 1986, have convinced many people that this technology is too risky to pursue.

Other major worries about nuclear power include where to put the waste products of the nuclear fuel cycle and how to ensure that it will remain safely contained for the thousands of years required for "decay" of the radioisotopes to nonhazardous levels. Yucca Mountain, Nevada, was chosen for a high-level waste repository, but many experts believe that burying these toxic residues in nonretrievable storage is a mistake.

Nuclear fusion occurs when atoms are forced together in order to create new, more massive atoms. If fusion actions could be created under controlled conditions in a power plant, they might provide an essentially unlimited source of energy that would avoid many of the worst problems of both fossil fuels and fission-based nuclear power. So far, however, no one has been able to sustain controlled fusion reactions that produce more energy than they consume. It remains to be seen whether this will ever be possible, and whether other unforeseen problems will arise if it does become a reality.

None of our current major energy sources appear to offer security in terms of stable supply or environmental considerations. Neither coal nor nuclear power is a good long-term energy source with our present level of technology. We urgently need to develop alternative sources of sustainable energy.

QUESTIONS FOR REVIEW

1. What is energy? What is power?
2. What are the major sources of commercial energy worldwide and in the United States? Why are data usually presented in terms of commercial energy?
3. How is energy used in the United States?
4. How does our energy use compare with that of people in other countries?
5. How much coal, oil, and natural gas are in proven reserves worldwide? Where are those reserves located?
6. What are the most important health and environmental consequences of our use of fossil fuels?
7. Describe how a nuclear reactor works and why they are dangerous.
8. What are the four most common reactor designs? How do they differ from each other?
9. What are the advantages and disadvantages of the breeder reactor?
10. Describe methods proposed for storing and disposing of nuclear wastes.

QUESTIONS FOR CRITICAL THINKING

1. We have discussed a number of different energy sources and energy technologies in this chapter. Each has advantages and disadvantages. If you were an energy policy analyst, how would you compare such different problems as the risk of a nuclear accident versus air pollution effects from burning coal?
2. If your local utility company were going to build a new power plant in your community, what kind would you prefer?
3. The nuclear industry is placing ads in popular magazines and newspapers claiming that nuclear power is environmentally friendly since it doesn't contribute to the greenhouse effect. How do you respond to that claim?
4. Our energy policy effectively treats some strip-mine lands as national sacrifice areas since we know that they will never be restored to their original state when mining is finished. How do we decide who wins and who loses in this transaction?
5. Storing nuclear wastes in dry casks outside nuclear power plants is highly controversial. Opponents claim the casks will inevitably leak. Proponents claim they can be designed to be safe. What evidence would you consider adequate or necessary to choose between these two positions?
6. Since the environmental impact of a nuclear power plant accident cannot be limited to national boundaries (as in the case of Chernobyl), should there be international regulations for power plants? Who would define these regulations? How could they be enforced?
7. Although we have wasted vast amounts of energy resources in the process of industrialization and development, some would say that it was a necessary investment to get to a point at which we can use energy more efficiently and sustainably. Do you agree? Might we have followed a different path?
8. The policy of the United States has always been to make energy as cheap and freely available as possible. Most European countries charge three to four times as much for gasoline as we do. Think about the ramifications of our energy policy. Why is it so different from that in Europe? Who benefits and who or what loses in these different approaches? How have our policies shaped our lives? What does existing policy tell you about how our government works?

KEY TERMS

black lung disease 479
breeder reactor 490
chain reaction 487
control rods 487
energy 476
fossil fuels 476
fuel assembly 486
high-level waste repository 492
inertial confinement 494
joule (J) 476
magnetic confinement 494
methane hydrate 485
monitored, retrievable storage 492
nuclear fission 487
nuclear fusion 494
oil shale 482
power 476
secondary recovery techniques 481
work 476

ADDITIONAL INFORMATION ON THE INTERNET

Visit our website at http://www.mhhe.com/environmentalscience/ for specific information about each of the topics below:

Oil exploitation in the Caspian Sea area

Department of Energy—Home Page

Energy Analysis Department

Pressurized Water Reactor systems

U.S. Department of Energy

U.S. Nuclear Regulatory Commission

Energy Information Administration, Department of Energy: Information on sustainable and conventional energy resources

Oil spills: Exxon Valdez

Nuclear Waste Task Force Home Page-Sierra Club

Map of Nuclear Waste Shipment Routes

Oil Drilling in the Alaskan Arctic

CHAPTER 22

Sustainable Energy

Imagination is more important than knowledge.

—Albert Einstein—

OBJECTIVES

After studying this chapter, you should be able to:

- appreciate the opportunities for energy conservation available to us.
- understand how active and passive systems capture solar energy and how photovoltaic collectors generate electricity.
- comprehend why diminishing fuelwood supplies are a crisis in less-developed countries.
- evaluate the use of dung, crop residues, energy crops, and peat as potential energy sources.
- explain how hydropower, wind, and geothermal energy contribute to our power supply.
- describe how tidal and wave energy and ocean thermal gradients can be used to generate electrical energy.
- compare and contrast different options for storing electrical energy from intermittent or remote sources.

ESSENTIAL STUDY PARTNER

The Essential Study Partner (ESP) CD-ROM that accompanies this textbook contains additional information on the following topics:

- US energy consumption, fuel efficiency, pollution production, and toxic emissions: physical resources: energy: use patterns
- efficiency of energy conversion and machines: physical resources: energy: use patterns
- passive, active, and high-temperature solar energy and photovoltaic cells: physical resources: energy: solar
- environmental impacts of burning biomass and hydropower: physical resources: energy: other renewable
- crop residue, biogas, and ethanol fuels: physical resources: energy: other renewable
- wind and geothermal energy: physical resources: energy: other renewable
- renewable energy resource cost, technology, and research: physical resources: energy: other renewable

Above: Rows of 12-story-tall, 750 kv windmills stand on the prairie in western Minnesota.

Buffalo Ridge

A steady breeze flows most of the year across the Great Plains, rippling prairie grasses in the summer and sculpting snowdrifts through the winter. When it reaches western Minnesota, the wind rises over a broad upland known as Buffalo Ridge, where it encounters row after row of tall, white columns, each holding three long, black propellers that swoosh softly as they spin. The 200 wind turbines now standing in ordered ranks on the rolling farm and pasture land are the first stage in what promises to be the largest single wind farm in the world. By 2002, there will be 566 of these giant windmills generating a total of 425 MW of electrical power, or enough for about 42,000 average homes. Together, the wind turbines will be equivalent to one large coal-burning or nuclear electric power plant.

Each of the sleek, steel towers is 50 m (165 ft) tall, or about as high as a 12-story building. The hollow, fiberglass blades are each 23 m (65 ft) long and weigh about 2 metric tons apiece. From tip to tip, the blades cover 48 m or about the wingspread of a DC10 airplane. Every tower is topped by a 750 kilovolt generator about the size of a minivan. Although the entire project will eventually spread across 7000 ha (17,000 acres) of land, each tower has a footprint of only about 0.5 ha (roughly 0.25 acre), or 2 percent of the landscape.

Costing slightly less than $1 million each, the wind turbines will represent a total investment by Northern States Power—the principle public utility company in Minnesota—of about $500 million. This is slightly less than construction of a comparable-sized coal-fired power plant, and about one-fourth of the current cost to build a nuclear plant. Best of all, with wind power, the energy source is free and nonpolluting. Project managers estimate that burning enough coal to provide as much energy as the completed wind farm would release about 750,000 metric tons of carbon dioxide (the major gas in global warming) every year, together with about 2 metric tons each of sulfur dioxide (a major source of acid rain and respiratory diseases) and nitrogen oxides (a major source of smog) per year. Maintaining the wind turbines also provides far more jobs and a greater boost to the local economy than would fossil fuels or nuclear power. Although the residents of the small towns near Buffalo Ridge were skeptical of the project at first, they are now delighted with its economic benefits, and proud of their role in energy independence.

Perhaps the most important aspect of the Buffalo Ridge development is the promise it holds for the rest of the Great Plains. Millions of square kilometers, from Alberta and Saskatchewan to Texas, and from the Rocky Mountains to the tree line east of the 100th meridian, have the potential to supply considerably more energy than the total current world commercial production. As David Morris, from the Center for Local Self-Reliance says, "The Midwest could be the Saudi Arabia of safe, clean, sustainable wind power, and it would free us from dependence on foreign powers for our energy supply."

Chapter 21 details some of the disadvantages and limitations of fossil fuels and nuclear power; in this chapter, we will look at some options for energy conservation and renewable energy sources that might allow a real increase in comfort and standard of living for everyone, and yet also help preserve our global environment. Even with modest investment in projects like Buffalo Ridge, sustainable energy could meet more than half our energy needs 50 years from now (fig. 22.1).

CONSERVATION

One of the best ways to avoid energy shortages and to relieve environmental and health effects of our current energy technologies is simply to use less. Conservation offers many benefits both to society and to the environment.

Utilization Efficiencies

Much of the energy we consume is wasted. This statement is not a simple admonishment to turn off lights and turn down furnace thermostats in winter; it is a technological challenge. Our ways of using energy are so inefficient that most potential energy in fuel is lost as waste heat, becoming a form of environmental pollution. Of the energy we do extract from primary resources, however, much is used for frankly trivial or extravagant purposes. As chapter 21 shows, several European countries have higher standards of living than the United States, and yet use 30 to 50 percent less energy.

Many conservation techniques are relatively simple and highly cost effective. More efficient and less energy-intensive industry, transportation, and domestic practices could save large amounts of energy. Improved automobile efficiency, better mass transit, and

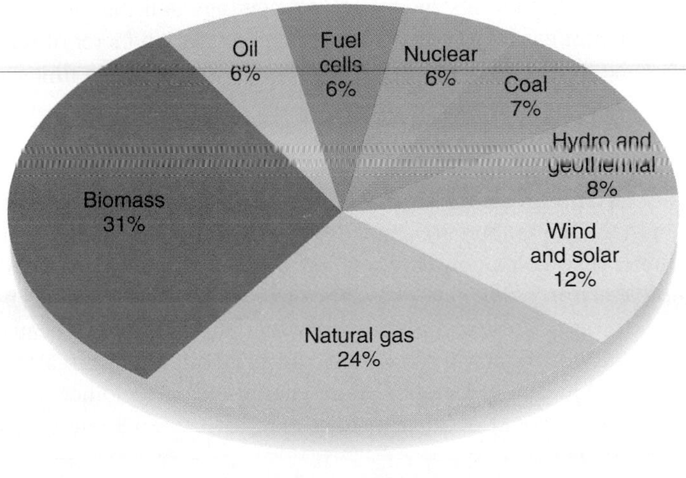

FIGURE 22.1 An alternative energy future for the United States in 2050. This scenario projects a 50 percent drop in total energy consumption and a 75 percent reduction in CO_2 production. Compare this to figure 21.3. *Source:* Environmental Protection Agency, 1992.

increased railroad use for passenger and freight traffic are simple and readily available means of conserving transportation energy. In response to the 1970s oil price shocks, automobile gas-mileage averages in the United States more than doubled from 13 mpg in 1975 to 28.8 mpg in 1988. Unfortunately, the oil glut and falling fuel prices of the late 1980s discouraged further conservation. Between 1990 and 1995, the average slipped to only 27.6 mpg.

Much more could be done. High-efficiency, low-emission automobiles are already available (What Do You Think? p. 501), and proposed rules that would make popular light trucks and sport utility vehicles (SUVs) meet the same mileage and emission standards as conventional cars will both clear our air and save energy. Amory B. Lovins of the Rocky Mountain Institute in Colorado estimated that raising the average fuel efficiency of the United States car and light truck fleet by one mile per gallon would cut oil consumption about 295,000 barrels per *day.* In one year, this would equal the total amount the Interior Department hopes to extract from the Arctic National Wildlife Refuge in Alaska (see chapter 21).

Many improvements in domestic energy efficiency have occurred in the past decade. Today's average new home uses one-half the fuel required in a house built in 1974, but much more can be done. Household energy losses can be reduced by one-half to three-fourths through better insulation, double or triple glazing of windows, thermally efficient curtains or window coverings, and by sealing cracks and loose joints. Reducing air infiltration is usually the cheapest, quickest, and most effective way of saving energy because it is the largest source of losses in a typical house. It doesn't take much skill or investment to caulk around doors, windows, foundation joints, electrical outlets, and other sources of air leakage.

For even greater savings, new houses can be built with extra thick superinsulated walls, air-to-air heat exchangers to warm incoming air, and even double-walled sections that create a "house within a house." The R-2000 program in Canada details how energy conservation can be built into homes. Special double-glazed windows that have internal reflective coatings and that are filled with an inert gas (argon or xenon) have an insulation factor of R11, the same as a standard 4-inch thick insulated wall or ten times as efficient as a single-pane window. Superinsulated houses now being built in Sweden require 90 percent less energy for heating and cooling than the average American home.

Orienting homes so that living spaces have passive solar gain in the winter and are shaded by trees or roof overhang in the summer also helps conserve energy. Earth-sheltered homes built into the south-facing side of a slope or protected on three sides by an earth berm are exceptionally efficient energy savers because they maintain relatively constant subsurface temperatures (fig. 22.2). In addition to building more energy-efficient homes, there are many personal actions we can take to conserve energy (In Depth, p. 502).

Industrial energy savings are another important part of our national energy budget. More efficient electric motors and pumps, new sensors and control devices, advanced heat-recovery systems, and material recycling have reduced industrial energy requirements significantly. In the early 1980s, U.S. businesses saved $160 billion per year through conservation. When oil prices collapsed, however, we returned to wasteful ways.

FIGURE 22.2 This earth-sheltered house made of rammed dirt-filled tires and empty aluminum cans saves energy and recycles waste. South-facing windows passively collect solar energy to provide most space heating.

Energy Conversion Efficiencies

Energy efficiency is a measure of energy produced compared to energy consumed. Table 22.1 shows the typical energy efficiencies of a variety of energy-conversion devices. Thermal-conversion machines, such as steam turbines in coal-fired or nuclear power plants, can turn no more than 40 percent of the energy in their primary fuel into electricity or mechanical power because of the need to reject waste heat. Does this mean that we can never increase the efficiency of fossil fuel use? No. Some waste heat can be recaptured and used for space heating, raising the net yield to 80 or 90 percent. In another kind of process, fuel cells convert the chemical energy of a fuel directly into electricity without an intermediate combustion cycle. Since this process is not limited by waste heat elimination, its efficiencies can approach 80 percent with such fuel as hydrogen gas or methane.

Another way to look at energy efficiency is to consider the total **net energy yield** from energy-conversion devices (table 22.2). Net energy yield is based on the total useful energy produced during the lifetime of an entire energy system minus the energy required to make useful energy available. To make comparisons between different energy systems easier, the net energy yield is often expressed as a *ratio* between the output of useful energy and the energy costs for construction, fuel extraction, energy conversion, transmission, waste disposal, etc.

Nuclear power is a good case for net energy yield studies. We get a large amount of electricity from a small amount of fuel in a nuclear plant, but it also takes a great deal of energy to extract

what do you think?

THE WORLD AROUND YOU

Hybrid Automobile Engines

It's not surprising that California was the first state to get tough on automobile emissions. Auto exhaust counts for 90 percent of the state's carbon monoxide, 77 percent of its nitrogen oxides, and 55 percent of its smog-producing hydrocarbons. In 1990, the California Air Resources Board shocked the automobile industry by ordering it to start producing emission-free vehicles or face stiff penalties. By 1998, 2 percent of all auto sales in California were supposed to be zero-emission vehicles, and 10 percent would have to be in this category by 2003. At the time this order was issued, only battery-powered electric vehicles were available. Although several manufacturers built all-electric autos—the General Motors EV1 was an example—the batteries were heavy, expensive, and required more frequent recharging than customers would accept. Even though 90 percent of all daily commutes are less than 80 km (50 mi), most people want the capability to take a long road trip of several hundred kilometers without needing to stop for fuel or recharging. Although some environmentally sensitive customers bought electric vehicles, for most people they were merely an expensive novelty.

An alternative now being promoted by Toyota and Honda that appears to have much more customer appeal is the hybrid gas-electric motor. There are two basic types of hybrid electric car: series and parallel motors. In a series hybrid, a small gas or diesel engine spins a generator that recharges a small battery pack and provides electric power to motors that drive the wheels. This is essentially the design of diesel-electric train engines. The internal combustion engine is a mobile charging station. Parallel hybrids have two discrete power systems, gasoline and electric. Either one can propel the vehicle, but they work together when more power is needed for acceleration or hill climbing. The electric system operates most of the time, and the gasoline motor only runs when the batteries need recharging. Alternatively, in some designs, the gasoline engine—which can be substantially smaller than a normal car—runs most of the time and the electrical system only kicks in to give an extra boost in acceleration. In either system, kinetic energy can be recaptured during "regenerative" braking and converted back to electricity to be stored in the battery.

The first of these hybrid cars to be marketed in the United States is the two-seat Honda Insight. Its 3-cylinder, 1.0 liter gas engine is the main power source. A 7 hp electric motor helps during acceleration and hill climbing. A "continuously variable" automatic transmission cuts down on energy lost during shifting, and regenerative braking increases efficiency. With a streamlined lightweight plastic and aluminum body, the Insight is reported to get 80 mpg (33.7 km/l) in highway driving and has low-enough emissions to satisfy California requirements. Cost is about $20,000. Top speed is 90 mph (145 kmph). In 1999, The Sierra Club awarded the Insight an "excellence in engineering" award, the first time this organization has ever endorsed a commercial product.

The other entrant in the hybrid vehicle race is the Toyota Prius. Larger than the Honda, the Prius seats five adult passengers and is about the same size as the popular Toyota Camry. During most city driving, it depends only on its quiet, emission-free, battery-powered, electric motor. The 1.5 liter gas engine kicks in to help accelerate or when the batteries need recharging. Getting about 60 mpg (25 km/l) on the highway, the Prius is still one of the most efficient cars on the road and can travel more than 625 mi (1000 km) without refueling. Some drivers are unnerved by the noiseless electric motor. Sitting at a stoplight, it makes no sound at all. You might think it was dead, but when the light changes, you glide off silently and smoothly. Introduced in Japan in 1997, the Prius also sells in the United States for about $20,000. The Sierra Club estimates that in 100,000 mi (160,000 km), a Prius will generate 27 tons of CO_2, a Ford Taurus will generate 64 tons, while the Ford Excursion SUV will produce 134 tons.

Most American automobile makers are skeptical of hybrid designs and are putting all their research into fuel cells, which have the promise of being truly zero-emission. What do you think? Would you buy a vehicle with a hybrid engine system? Would you sacrifice a little power, speed, and acceleration for a quiet, clean, efficient, environmentally friendly means of transportation? Do you think that automobile makers are wise to wait for fuel cells, or should they be producing hybrid vehicles as well?

The hybrid gas-electric Toyota Prius seats five adults and gets 60 mpg (25 km/l) while meeting the most strict emission standards.

and process nuclear fuel and to build, operate, and eventually dismantle power plants. As chapter 21 points out, we may never reach a break-even point where we get back more energy from nuclear plants than we put into them, especially considering the energy that may be required to decommission nuclear plants and guard their waste products in secure storage for thousands of years.

Net yields and overall conversion efficiencies are not the only considerations when we compare different energy sources. The yield/cost ratio and conversion-cycle efficiency is much higher for coal burning, for instance, than for photovoltaic electrical production, making coal appear to be a better source of energy than solar radiation. Solar energy, however, is free, renewable, and nonpolluting. If we can use solar energy to get electrical energy, it doesn't matter how *efficient* the process is, as long as we get more out of it than we put in.

Negawatt Programs

Utility companies are finding it much less expensive to finance conservation projects than to build new power plants. Rather than buy megawatts of new generating capacity, power companies are investing in "negawatts" of demand avoidance. Each of us can help

IN DEPTH. . . . . .

Personal Energy Efficiency

For some people, home energy efficiency means designing and building a new house that incorporates methods and materials that have been proven to conserve energy. Most of us, however, have to live in houses or apartments that already exist. How can we practice home energy conservation?

Think about the major ways we use energy. The biggest factor is space heating. Heat conservation is among the simplest, cheapest, and most effective ways to save energy in the home. Easy, surprisingly effective, and relatively inexpensive measures include weatherstripping, caulking, and adding layers of plastic to windows. Insulating walls, floors, and ceilings increases the energy conservation potential. Simply lowering your thermostat, especially at night, is a proven energy and money saver.

Water heating is the second major user of home energy, followed by large electric or gas appliances, such as stoves, refrigerators, washing machines, and dryers. Careful use of these appliances can save significant amounts of energy. Consider using less hot water, lowering the thermostat of your water heater, and buying an insulating blanket designed to wrap around it. Be sure that the refrigerator door doesn't stand open, and defrost it regularly (if it is not frost-free) to reduce the ice buildup that prevents the heat exchange system from working well. Make sure your oven door has a tight seal, and plan ahead when you use it; bake several things at once rather than reheating it repeatedly. Wash your clothes in cold or cool water rather than hot. Air dry your laundry, especially in the summer when sun-dried clothes are so pleasant to wear.

Some energy-efficient appliances can save substantial amounts of energy. Air conditioners and refrigerators with better condensers and heat exchangers use one-half to one-fourth as much energy as older models. New, improved furnaces can offer 95 percent efficiency, compared with conventional 70 percent efficiency in older models. The payback period may be as little as two to three years if you trade an old wasteful appliance for a newer, more efficient one. Some lightbulbs are made to save energy. High-efficiency fluorescent lights emit the same amount of light but use only one-fourth as much energy as conventional incandescent bulbs. They cost ten times more than an ordinary bulb, but last ten times as long. Total lifetime savings can be \$30 to \$50 per lamp. Almost all types of appliances, large and small, are available in efficient models and brands; we simply need to shop for them.

Comparison of an ordinary incandescent bulb with a new, compact fluorescent lamp. Both bulbs have standard screw-in bases, but the fluorescent ballast (arrows) is wider than the base and may require a socket "extender" to fit in some fixtures. A 15-watt compact fluorescent bulb puts out as much light as a 60-watt incandescent bulb and lasts about ten times longer.

The easiest way to save energy is to keep an eye on energy-use habits. Turning off lights, televisions, and other devices when they are not in use is elementary. Line drying clothes and hand washing dishes are not difficult. In some cases, the most beneficial result of living a frugal, conservative life is not so much the total amount of energy you save, as the effect that this lifestyle has on you and the people around you. When you live conscientiously, you set a good example that could have a multiplying effect as it spreads through society. Making conscious ethical decisions about this one area of your life may stimulate you to make other positive decisions. You will feel better about yourself and more optimistic about the future when you make even small, symbolic gestures toward living as a good environmental citizen.

TABLE 1 Energy Saving Potential

	MODEL AVERAGE	BEST ON MARKET	BEST PROTOTYPE	SAVING (PERCENT)
Automobile (mi/gal)	18	55	107	83
Home (1000 J/day)	190	68	11	94
Refrigerator (KWH/day)	4	2	1	75
Gas furnace (million J/day)	210	140	110	48
Air conditioner (KWH/day)	10	5	3	70

Source: U.S. Department of Energy.

save energy in everyday life (What Can You Do? p. 504). Pacific Gas and Electric in California and Potomac Power and Light in Washington, D.C., both have instituted large conservation programs. They have found that conservation costs about \$350 per kilowatt (kW) saved. By contrast, a new nuclear power plant costs between \$3000 and \$8000 per kW of installed capacity. New coal-burning plants with the latest air pollution-control equipment cost at least \$1000 per kW. By investing \$200 million to \$350 million in public education, home improvement loans, and other efficiency measures, a utility can avoid building a new power plant that would

TABLE 22.1 Typical Net Efficiencies of Energy-Conversion Devices

	YIELD (PERCENT)
ELECTRIC POWER PLANTS	
Hydroelectric (best case)	90
Combined-cycle steam	90
Fuel cell (hydrogen)	80
Coal-fired generator	38
Oil-burning generator	38
Nuclear generator	30
Photovoltaic generation	10
TRANSPORTATION	
Pipeline (gas)	90
Pipeline (liquid)	70
Waterway (no current)	65
Diesel-electric train	40
Diesel-engine automobile	35
Gas-engine automobile	30
Jet-engine airplane	10
SPACE HEATING	
Electric resistance	99*
High-efficiency gas furnace	90
Typical gas furnace	70
Efficient wood stove	65
Typical wood stove	40
Open fireplace	-10
LIGHTING	
Sodium vapor light	60*
Fluorescent bulb	25*
Incandescent bulb	5*
Gas flame	1

*Note that 60–70% of the energy in the original fuel is lost in electric power generation.
Source: U.S. Department of Energy.

TABLE 22.2 Typical Net Useful Energy Yields

ENERGY SOURCE	YIELD/COST RATIO
NONRENEWABLE SOURCES	
Coal (space or process heat)	20/1
Natural gas (as heat source)	10/1
Gasoline and fuel oil	7/1
Coal gasification (combined cycle)	5/1
Oil shale (as liquid fuel)	1/1
Nuclear (excluding waste disposal)	2/1*
RENEWABLE SOURCES	
Hydroelectric (best case)	20/1**
Wind (electric generation)	2/1
Biomass methane	2/1
Solar electric (10% efficient)	1/1
Solar electric (20% efficient)	2/1

*Decommissioning of old nuclear plants and perpetual storage of wastes may consume more energy than nuclear power has produced.
**Hydropower yield depends on availability of water and life expectancy of dam and reservoir. Most dams produce only about 40% of rated capacity due to lack of water. Some dams have failed or silted up without producing any net energy yield.

cost a billion dollars or more. Furthermore, conservation measures don't consume expensive fuel or produce pollutants.

Can application of this approach help alleviate energy shortages in other countries as well? Yes. For example, Brazil could cut its electricity consumption an estimated 30 percent with an investment of $10 billion (U.S.). It would cost $44 billion to build new power plants to produce that much electricity. South Korea has instituted a comprehensive conservation program with energy-saving building standards, efficiency labels on new household appliances, depreciation allowances, reduced tariffs on energy-conserving equipment, and loans and tax breaks for upgrading homes and businesses. It also forbids some unnecessary uses, such as air conditioning and elevators between the first and third floors.

Cogeneration

One of the fastest growing sources of new energy is **cogeneration,** the simultaneous production of both electricity and steam or hot water in the same plant. By producing two kinds of useful energy in the same facility, the net energy yield from the primary fuel is increased from 30-35 percent to 80-90 percent. In 1900, half the electricity generated in the United States came from plants that also provided industrial steam or district heating. As power plants became larger, dirtier, and less acceptable as neighbors, they were forced to move away from their customers. Waste heat from the turbine generators became an unwanted pollutant to be disposed of in the environment. Furthermore, long transmission lines, which are unsightly and lose up to 20 percent of the electricity they carry, became necessary.

By the 1970s, cogeneration had fallen to less than 5 percent of our power supplies, but interest in this technology is being renewed. The capacity for cogeneration more than doubled in the 1980s to about 30,000 megawatts (MW). District heating systems are being rejuvenated, and plants that burn municipal wastes are being studied. New combined-cycle coal-gasification plants or "mini-nukes" (chapter 21) offer high efficiency and clean operation that may be compatible with urban locations. Small neighborhood- or apartment building-sized power-generating units are being built that burn methane (from biomass digestion), natural gas, diesel fuel, or coal. The Fiat Motor Company makes a small generator for about $10,000 that produces enough electricity and heat for four or five houses.

what can you do?

Some Things You Can Do to Save Energy

1. Drive less: make fewer trips, use telecommunications and mail instead of going places in person.
2. Use public transportation, walk, or ride a bicycle.
3. Use stairs instead of elevators.
4. Join a car pool or drive a smaller, more efficient car; reduce speeds.
5. Insulate your house or add more insulation to the existing amount.
6. Turn thermostats down in the winter and up in the summer.
7. Weatherstrip and caulk around windows and doors.
8. Add storm windows or plastic sheets over windows.
9. Create a windbreak on the north side of your house; plant deciduous trees or vines on the south side.
10. During the winter, close windows and drapes at night; during summer days, close windows and drapes if using air conditioning.
11. Turn off lights, television sets, and computers when not in use.
12. Stop faucet leaks, especially hot water.
13. Take shorter, cooler showers; install water-saving faucets and showerheads.
14. Recycle glass, metals, and paper; compost organic wastes.
15. Eat locally grown food in season.
16. Buy locally made, long-lasting materials.

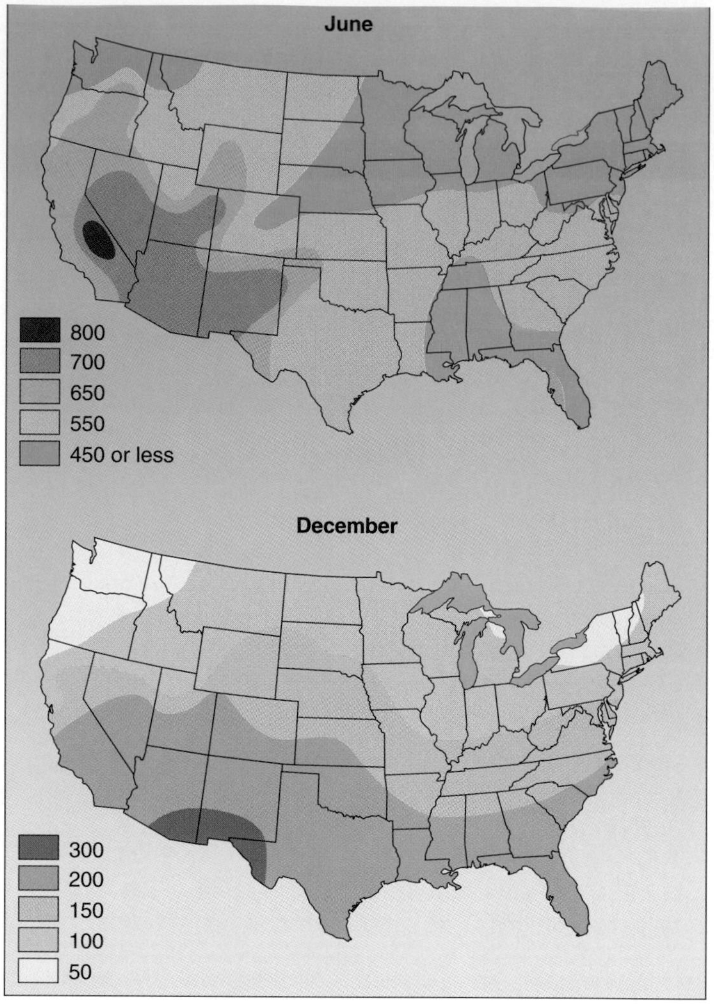

FIGURE 22.3 Average daily solar radiation in the United States in June and December. One langley, the unit for solar radiation, equals 1 calorie per square centimeter of earth surface (3.69 Btu/square foot).
Source: National Weather Bureau, U.S. Department of Commerce.

TAPPING SOLAR ENERGY

The sun serves as a giant nuclear furnace in space, constantly bathing our planet with a free energy supply. Solar heat drives winds and the hydrologic cycle. All biomass, as well as fossil fuels and our food (both of which are derived from biomass), results from conversion of light energy (photons) into chemical bond energy by photosynthetic bacteria, algae, and plants.

A Vast Resource

The average amount of solar energy arriving at the top of the atmosphere is 1330 watts per square meter. About half of this energy is absorbed or reflected by the atmosphere (more at high latitudes than at the equator), but the amount reaching the earth's surface is some 10,000 times all the commercial energy used each year. However, this tremendous infusion of energy comes in a form that, until this century, has been too diffuse and low in intensity to be used except for environmental heating and photosynthesis. Figure 22.3 shows solar energy levels over the United States for a typical summer and winter day.

Passive Solar Heat

Our simplest and oldest use of solar energy is **passive heat absorption,** using natural materials or absorptive structures with no moving parts to simply gather and hold heat. For thousands of years, people have built thick-walled stone and adobe dwellings that slowly collect heat during the day and gradually release that heat at night. After cooling at night, these massive building materials maintain a comfortable daytime temperature within the house, even as they absorb external warmth.

A modern adaptation of this principle is a glass-walled "sunspace" or greenhouse on the south side of a building. Incorporating massive energy-storing materials, such as brick walls, stone floors, or barrels of heat-absorbing water into buildings also collects heat to be released slowly at night (fig. 22.4). An interior, heat-absorbing wall called a Trombe wall is an effective passive heat collector. Some Trombe walls are built of glass blocks enclosing a water-filled space or water-filled circulation tubes so heat from solar rays can be absorbed and stored, while light passes through to inside rooms.

FIGURE 22.4 Stone floors or barrels of water are simple, heat-absorbing devices. Insulated shutters keep heat inside at night.

Active Solar Heat

Active solar systems generally pump a heat-absorbing, fluid medium (air, water, or an antifreeze solution) through a relatively small collector, rather than passively collecting heat in a stationary medium like masonry. Active collectors can be located adjacent to or on top of buildings rather than being built into the structure. Because they are relatively small and structurally independent, active systems can be retrofitted to existing buildings.

A flat black surface sealed with a double layer of glass makes a good solar collector. A fan circulates air over the hot surface and into the house through ductwork of the type used in standard forced-air heating. Alternatively, water can be pumped through the collector to pick up heat for space heating or to provide hot water (fig. 22.5). Water heating consumes 15 percent of the United States' domestic energy budget, so savings in this area alone can be significant. A simple flat panel with about 5 sq m of surface can reach 95°C (200°F) and can provide enough hot water for an average family of four almost anywhere in the United States. In California, 650,000 homes now heat water with solar collectors. In Greece, Italy, Israel, and other countries where fuels are more expensive, up to 70 percent of domestic hot water comes from solar collectors.

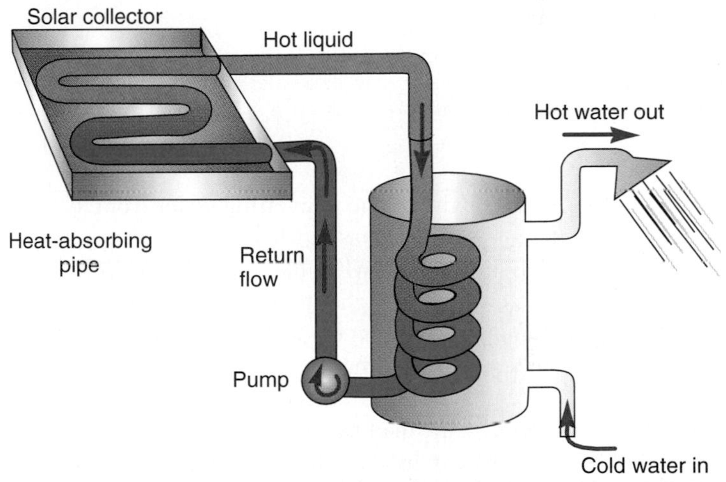

FIGURE 22.5 An active, closed-cycle solar water heating system. A cycling liquid, such as sodium or antifreeze, can reach hundreds of degrees Celsius in a parabolic reflector. Alternatively, a flat black panel enclosed with three layers of glass can absorb heat for the collecting pipes.

Sunshine doesn't reach us all the time, of course. How can solar energy be stored for times when it is needed? There are a number of options. In a climate where sunless days are rare and seasonal variations are small, a small, insulated water tank is a good solar energy storage system. For areas where clouds block the sun for days at a time or where energy must be stored for winter use, a large, insulated bin containing a heat-storing mass, such as stone, water, or clay, provides solar energy storage (fig. 22.6). During the summer months, a fan blows the heated air from the collector into the storage medium. In the winter, a similar fan at the opposite end of the bin blows the warm air into the house. During the summer, the storage mass is cooler than the outside air, and it helps cool the house by absorbing heat. During the winter, it is warmer and acts as a heat source by radiating stored heat. In many areas, six or seven months' worth of thermal energy can be stored in 10,000 gallons of water or 40 tons of gravel, about the amount of water in a very small swimming pool or the gravel in two average-sized dump trucks.

Another heat storage system uses **eutectic** (phase-changing) **chemicals** to store a large amount of energy in a small volume. Heating melts these chemicals and cooling returns them to a solid state. This is an important way the hydrologic cycle stores solar

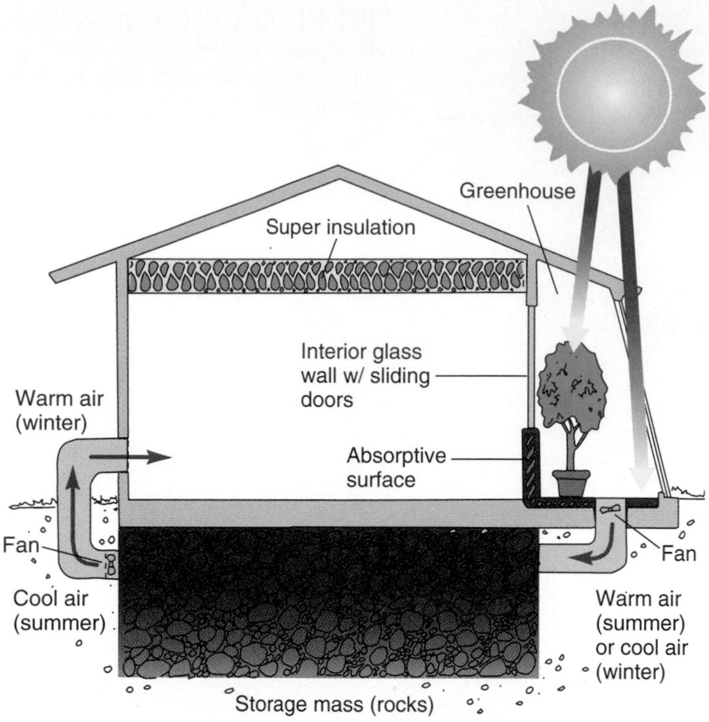

FIGURE 22.6 Underground massive heat storage unit. Heated air collected behind double- or triple-glazed windows is pumped down into a storage medium of rock, water, clay, or similar material, where it can be stored for a number of months.

energy in the natural environment, converting water from its solid phase to its liquid and gaseous phases. Most eutectic chemical systems (such as salts) do not swell when they solidify (as does water) and undergo phase changes at higher temperatures than water and ice, making them more convenient for storing heat. Since the latent heat of crystallization is usually much greater than the specific heat required to raise temperatures, a eutectic system can store more energy in a smaller volume than can a simple, one-phase system.

Heat also can be captured in chemical reactions in the storage material. Some clays, for instance, undergo reactions that release or store heat when they absorb or release water. An insulated bin filled with a clay, such as bentonite, can be baked dry (absorbing heat) by blowing warm air into it in the summer. Sprinkling water on the clay to rewet it will release heat in winter.

HIGH-TEMPERATURE SOLAR ENERGY

Parabolic mirrors are curved reflecting surfaces that collect light and focus it into a concentrated point. There are two ways to use mirrors to collect solar energy to generate high temperatures. One technique uses long curved mirrors focused on a central tube, containing a heat-absorbing fluid (fig. 22.7). Fluid flowing through the tubes reaches much higher temperatures than possible in a basic flat panel collector. Another high-temperature system uses thousands of smaller mirrors arranged in concentric rings around a tall central tower. The mirrors, driven by electric motors, track the sun and focus its light on a heat absorber at the top of the "power tower" where molten salt is heated to temperatures as high as 500°C (1000°F), which then drives a steam-turbine electric generator.

FIGURE 22.7 Parabolic mirrors focus sunlight on steam-generating tubes at this power plant in the California desert.

Under optimum conditions, a 50 ha (130 acres) mirror array should be able to generate 100 MW of clean, renewable power. The only power tower in the United States is Southern California Edison's Solar II plant in the Mojave Desert east of Los Angeles. Its 2000 mirrors focused on a 100 m (300 ft) tall tower generates 10 MW or enough electricity for 5000 homes at an operating cost far below that of nuclear power or oil. We haven't had enough experience with these facilities to know how reliable the mirrors, motors, heat absorbers, and other equipment will be over the long run.

If the entire U.S. electrical output came from such central tower solar steam generators, 60,000 km^2 of collectors would be needed. This is an area about half the size of South Dakota. It is less land, however, than would be strip-mined in a 30-year period if all our energy came from coal or uranium. In contrast with windmill farms, which can be used for grazing or farming while also producing energy, mirror arrays need to be carefully protected and are not compatible with other land uses.

FIGURE 22.8 A simple box of wood, plastic, and foil can help reduce tropical deforestation, improve women's lives, and avoid health risks from smoky fires in developing countries. These inexpensive solar cookers could revolutionize energy use in developing tropical countries.

Solar Cookers

Parabolic mirrors have been tested for home cooking in tropical countries where sunshine is plentiful and other fuels are scarce. They produce such high temperatures and intense light that they are dangerous, however. A much cheaper, simpler, and safer alternative is the solar box cooker (fig. 22.8). An insulated box costing only a few dollars with a black interior and a glass or clear plastic lid serves as a passive solar collector. Several pots can be placed inside at the same time. Temperatures only reach about 120°C (250°F) so cooking takes longer than an ordinary oven. Fuel is free, however, and the family saves hours each day usually spent hunting for firewood or dung. These solar ovens help reduce tropical forest destruction and reduce the adverse health effects of smoky cooking fires.

Promoting Renewable Energy

Utility restructuring currently being planned in the United States could include policies to encourage conservation and alternative energy sources. Among the proposed policies are: (1) "distributional surcharges" in which a small per KWH charge is levied on all utility customers to help renewable energy finance research and development, (2) "renewables portfolio" standards to require power suppliers to obtain a minimum percentage of their energy from sustainable sources, and (3) "green pricing" that allows utilities to profit from conservation programs and charge premium prices for energy from renewable sources.

Some individual states already are pursuing these policies. For example, Iowa has a Revolving Loan Fund supported by a surcharge on investor-owned gas and electric utilities. This fund provides low-interest loans for renewable energy and conservation. Several states have initiated **green pricing** programs as a way to encourage a transition to sustainable energy. One of the first is in Colorado, where 1000 customers have agreed to pay $2.50 per month above their regular electric rates to help finance a 10 MW wind farm being built on the Colorado-Wyoming border. Buying a 100 KW "block" of wind power provides the same environmental benefits as planting a half acre of trees or not driving an automobile 4000 km (2500 mi) per year. Not all green pricing plans are as straightforward as this, however. Some utilities collect the premium rates for facilities that already exist or for renewable energy bought from other utilities at much lower prices.

Interestingly, some nonutility companies are investing in sustainable energy. In 1997, British Petroleum PLC broke ranks with other fossil fuel corporations and put $20 million into a solar cell manufacturing facility to be located in Fairfield, CA. Although fossil fuel interests have generally tried to obstruct renewables, British Petroleum believes that the threat of global climate change requires us to search for new types of energy. Similarly, two European insurance companies, concerned about potential losses from storms and rising sea levels caused by global warming, are investing $5 million in Sunlight Power, a U.S. company that makes and services solar power systems for remote regions of developing countries where electric service is unavailable.

Photovoltaic Solar Energy

The photovoltaic cell offers an exciting potential for capturing solar energy in a way that will provide clean, versatile, renewable energy. This simple device has no moving parts, negligible maintenance costs, produces no pollution, and has a lifetime equal to that of a conventional fossil fuel or nuclear power plant.

Photovoltaic cells capture solar energy and convert it directly to electrical current by separating electrons from their parent atoms and accelerating them across a one-way electrostatic barrier formed by the junction between two different types of semiconductor material (fig. 22.9). The photovoltaic effect, which is the basis of these devices, was first observed in 1839 by French physicist Alexandre-Edmond Becquerel, who also discovered radioactivity. His discovery didn't lead to any useful applications until 1954, when researchers at Bell Laboratories in New Jersey learned how to carefully introduce impurities into single crystals of silicon.

These handcrafted single-crystal cells were much too expensive for any practical use until the advent of the U.S. space program. In 1958, when Vanguard I went into orbit, its radio was powered by six palm-sized photovoltaic cells that cost $2000 per peak watt of output, more than 2000 times as much as conventional energy at the time. Since then, prices have fallen dramatically. In 1970, they cost $100 per watt; in 1999 they were less than $10 per watt. This makes solar energy cost-competitive with other sources in some areas (fig. 22.10).

By 2020, photovoltaic cells could be less than $1 per watt of generating capacity. With 15 percent efficiency and 30-year life, they should be able to produce electricity for around 6¢ per kilowatt hour. At that time, coal-fired steam power will probably cost about half again as much and nuclear power will likely cost twice as much as photovoltaic energy.

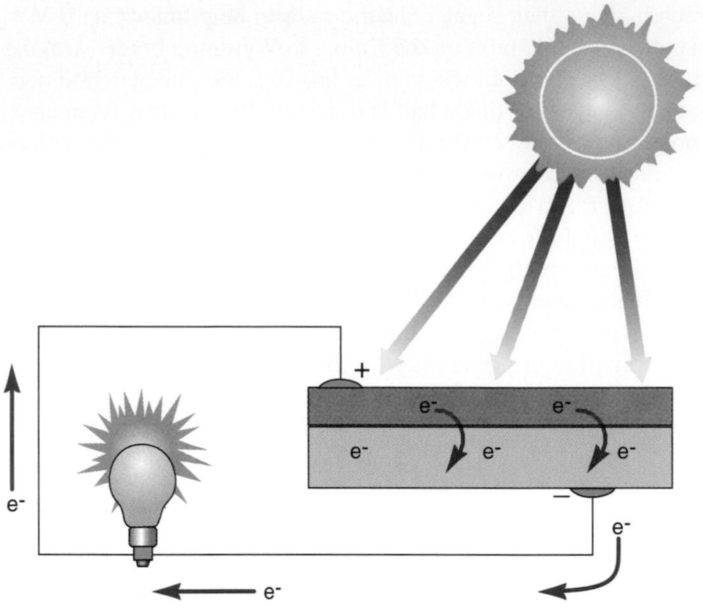

FIGURE 22.9 The operation of a photovoltaic cell. Boron impurities incorporated into the upper silicon crystal layers cause electrons (e-) to be released when solar radiation hits the cell. The released electrons move into the lower layer of the cell, thus creating a shortage of electrons, or a positive charge, in the upper layer and an oversupply of electrons, or negative charge, in the lower layer. The difference in charge creates an electric current in a wire connecting the two layers.

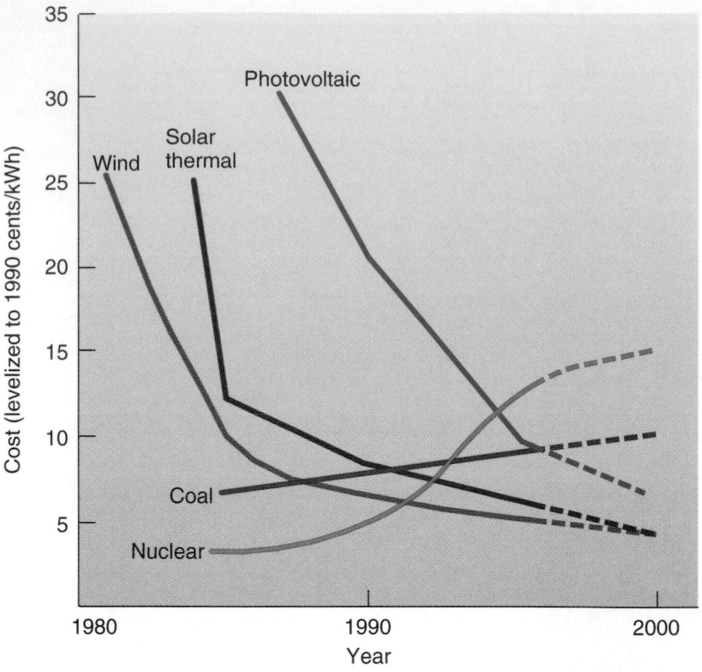

FIGURE 22.10 Costs for alternative and renewable energy sources have fallen dramatically in recent years. Wind power and solar-thermal electric power are already cheaper than coal in many places. Photovoltaic electricity should be less expensive than coal-fired electrical generation by the end of this decade.

Source: Data from *Wind Energy News*, 1997.

During the last 25 years, the efficiency of energy capture by photovoltaic cells has increased from less than 1 percent of incident light to more than 10 percent under field conditions and over 75 percent in the laboratory. Promising experiments are under way using exotic metal alloys, such as gallium arsenide, and semiconducting polymers of polyvinyl alcohol, which are more efficient in energy conversion than silicon crystals. Every year college students from all over North America race solar-powered cars across the country to test solar technology and publicize its potential (fig. 22.11). Perhaps you and your classmates would like to build a solar car and enter the race.

One of the most promising developments in photovoltaic cell technology in recent years is the invention of amorphous silicon collectors. First described in 1968 by Stanford Ovshinky, a self-taught inventor from Detroit, these noncrystalline silicon semiconductors can be made into lightweight, paper-thin sheets that require much less material than conventional photovoltaic cells. They also are vastly cheaper to manufacture and can be made in a variety of shapes and sizes, permitting ingenious applications. Roof tiles with photovoltaic collectors layered on their surface already are available. Even flexible films can be coated with amorphous silicon collectors. Silicon collectors already are providing power to places where conventional power is unavailable, such as lighthouses, mountaintop microwave repeater stations, villages on remote islands, and ranches in the Australian outback.

You probably already use amorphous silicon photovoltaic cells. They are being built into light-powered calculators, watches,

FIGURE 22.11 The solar race car built by students at the University of Minnesota finished fourth in the 1999 Sunrayce.

toys, photosensitive switches, and a variety of other consumer products. Japanese electronic companies presently lead in this field, having foreseen the opportunity for developing a market for photovoltaic cells. This market is already more than $100 million per year. Japanese companies now have home-roof arrays capable of providing all the electricity needed for a typical home at prices in some areas competitive with power purchased from a utility. By the end of this century, Japan plans to meet a considerable part of its energy requirements with solar power, an important goal for a country lacking fossil and nuclear fuel resources.

The world market for solar energy was about $700 million in 1999, and is expected to grow rapidly in the near future. Altogether, some 1.5 billion people around the world now have no access to electricity. Most would like to have a modern power source if it were affordable.

Think about how solar power could affect your future energy independence. Imagine the benefits of being able to build a house anywhere and having a cheap, reliable, clean, quiet source of energy with no moving parts to wear out, no fuel to purchase, and little equipment to maintain. You could have all the energy you need without commercial utility wires or monthly energy bills. Coupled with modern telecommunications and information technology, an independent energy source would make it possible to live in the countryside and yet have many of the employment and entertainment opportunities and modern conveniences available in a metropolitan area.

Storing Electrical Energy

Electrical energy is difficult and expensive to store. This is a problem for photovoltaic generation as well as other sources of electric power. Traditional lead-acid batteries are heavy and have low energy densities; that is, they can store only moderate amounts of energy per unit mass or volume. Acid from batteries is hazardous and lead from smelters or battery manufacturing is a serious health hazard for workers who handle these materials. A typical lead-acid battery array sufficient to store several days of electricity for an average home would cost about $5000 and weigh 3 or 4 tons. All the components for an electric car are readily available, but, as mentioned earlier, battery technology limits how far they can go between charges. Still, some communities are encouraging use of electric vehicles because they produce zero emissions (fig. 22.12).

Other types of batteries also have drawbacks. Metal-gas batteries, such as the zinc-chloride cell, use inexpensive materials and have relatively high-energy densities, but have shorter lives than other types. Sodium-sulfur batteries have considerable potential for large-scale storage. They store twice as much energy in half as much weight as lead-acid batteries. They require an operating temperature of about 300°C (572°F) and are expensive to manufacture. Alkali-metal batteries have a high storage capacity but are even more expensive. Lithium batteries have very long lives and store more energy than other types, but are the most expensive. Recent advances in thin-film technology may hold promise for future energy storage but often require rare, toxic elements that are both expensive and dangerous.

Another strategy is to store energy in a form that can be turned back into electricity when needed. Pumped-hydro storage involves pumping water to an elevated reservoir at times when excess electricity is available. The water is released to flow back down through turbine generators when extra energy is needed. Using a similar principle, pressurized air can be pumped into such reservoirs as natural caves, depleted oil and gas fields, abandoned mines, or special tanks. An Alabama power company currently uses off-peak electricity to pump air at night into a deep salt mine. By day, the air flows back to the surface through turbines, driving a generator that produces electricity. Cool night air is heated to 1,600°F by compression plus geothermal energy, increasing pressure and energy yield.

FIGURE 22.12 Leading the way toward a renewable-energy, zero-emissions future, the Sacramento Municipal Utility District has built the first solar-powered electric charging station in the western United States. Photovoltaic panels in the background charge a variety of electric vehicles.

An even better way to use surplus electricity is in electrolytic decomposition of water to H_2 and O_2. These gases can be liquefied (like natural gas) at –252°C (–423°F), making them easier to store and ship than most forms of energy. They are highly explosive, however, and must be handled with great care. They can be burned in internal combustion engines, producing mechanical energy, or they can be used as fuel in fuel cells to produce more electrical energy. Liquid hydrogen cars are already being produced. They could be very attractive in smoggy cities because they produce no carbon monoxide, smog-forming hydrocarbons, carcinogenic chemicals, or soot.

Flywheels are the subject of current experimentation for energy storage. Massive, high-speed flywheels, spinning in a nearly friction-free environment, store large amounts of mechanical energy in a small area. This energy is convertible to electrical energy. It is difficult, however, to find materials strong enough to hold together when spinning at high speed. Flywheels have a disconcerting tendency to fail explosively and unexpectedly, sending shrapnel flying in all directions.

FUEL CELLS

Rather than store and transport energy, another alternative would be to generate it locally, on demand. **Fuel cells** are devices that use ongoing electrochemical reactions to produce an electric current. They are very similar to batteries except that rather than recharging them with an electrical current, you add more fuel for the chemical reaction.

Fuel cells are not new; the basic concept was recognized in 1839 by William Grove, who was studying the electrolysis of water. He suggested that rather than use electricity to break apart water and produce hydrogen and oxygen gases, it should be possible to reverse the process by joining oxygen and hydrogen to produce water and electricity. The term "fuel cell' was coined in 1889 by Ludwig Mond and Charles Langer, who built the first practical device using a platinum catalyst to produce electricity from air and coal gas. The concept languished in obscurity until the 1950s when the U.S. National Aeronautics and Space Administration (NASA) was searching for a power source for spacecraft. Research funded by NASA eventually led to development of fuel cells that now provide both electricity and drinkable water on every space shuttle flight. The characteristics that make fuel cells ideal for space exploration—small size, high efficiency, low emissions, net water production, no moving parts, and high reliability—also make them attractive for a number of other applications.

All fuel cells consist of a positive electrode (the cathode) and a negative electrode (the anode) separated by an electrolyte, a material that allows the passage of charged atoms, called ions, but is impermeable to electrons (fig. 22.13). In the most common systems, hydrogen or a hydrogen-containing fuel is passed over the anode while oxygen is passed over the cathode. At the anode, a reactive catalyst, such as platinum, strips an electron from each hydrogen atom, creating a positively charged hydrogen ion (a proton). The hydrogen ion can migrate through the electrolyte to the cathode, but the electron is excluded. Electrons pass through an external circuit, and the electrical current generated by their passage can be used to do useful work. At the cathode, the electrons and protons are reunited and combined with oxygen to make water.

The fuel cell provides direct-current electricity as long as it is supplied with hydrogen and oxygen. For most uses, oxygen is provided by ambient air. Hydrogen can be supplied as a pure gas, but storing hydrogen gas is difficult and dangerous because of its volume and explosive nature. Liquid hydrogen takes far less space than the gas, but must be kept below –250°C (–400°F), not a trivial task for most mobile applications. The alternative is a device called a **reformer** or converter that strips hydrogen from fuels such as natural gas, methanol, ammonia, gasoline, ethanol, or even vegetable oil. Many of these fuels can be derived from sustainable biomass crops. Even methane effluents from landfills and wastewater treatment plants can be used as a fuel source. Where a fuel cell can be hooked permanently to a gas line, hydrogen can be provided by solar, wind, or geothermal facilities that use electricity to hydrolyze water.

A fuel cell run on pure oxygen and hydrogen produces no waste products except drinkable water and radiant heat. When a reformer is coupled to the fuel cell, some pollutants are released (most commonly carbon dioxide), but the levels are typically far less than conventional fossil fuel combustion in a power plant or automobile engine. Although the theoretical efficiency of electrical generation of a fuel cell can be as high as 70 percent, the actual yield is closer to 40 or 45 percent. This not much better than a very good fossil fuel power plant or a gas turbine electrical generator. On the other hand, the quiet, clean operation and variable size of fuel cells make them useful in buildings where waste heat can be

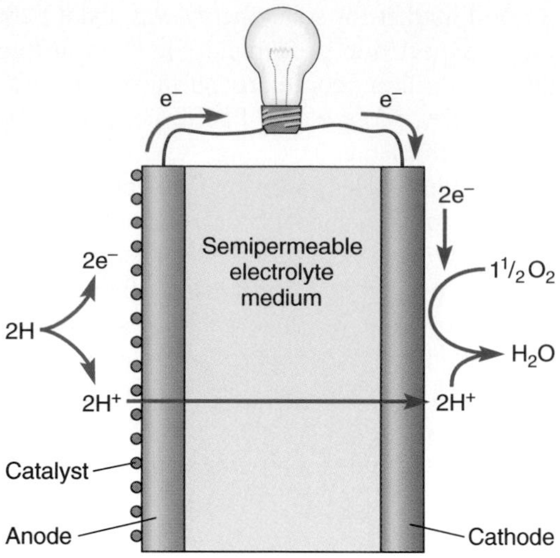

FIGURE 22.13 Fuel cell operation. Electrons are removed from hydrogen atoms at the anode to produce hydrogen ions (protons) that migrate through a semipermeable electrolyte medium to the cathode, where they reunite with electrons from an external circuit and oxygen atoms to make water. Electrons flowing through the circuit connecting the electrodes create useful electrical current.

captured for water heating or space heating. A new 45-story office building at 4 times square, for example, has two 200-kilowatt fuel cells on its fourth floor that provide both electricity and heat. This same building has photovoltaic panels on its façade, natural lighting, fresh air intakes to reduce air conditioning, and a number of other energy conservation features.

The current from a fuel cell is proportional to the size (area) of the electrodes, while the voltage is limited to about 1.23 volts per cell. A number of cells can be stacked together until the desired power level is reached. A fuel cell stack that provides almost all of the electricity needed by a typical home (along with hot water and space heating) would be about the size of a refrigerator. A 200 kilowatt unit fills a medium-size room and provides enough energy for 20 houses or a small factory. Tiny fuel cells running on methanol might soon be used in cell phones, pagers, toys, computers, videocameras, and other appliances now run by batteries. Rather than buy new batteries or spend hours recharging spent ones, you might just add an eyedropper of methanol every few weeks to keep your gadgets operating.

Fuel Cell Types

Several different electrolytes can be used in fuel cells, each with advantages and disadvantages (table 22.3). The design being developed for use in cars, buses, and trucks is called a proton exchange membrane (PEM). The membrane is a thin semipermeable layer of an organic polymer containing sulfonic acid groups that facilitate passage of hydrogen ions but block electrons and oxygen. The surface of the membrane is dusted with tiny particles of platinum catalyst. These cells have the advantage of being lightweight and operating at a relatively low temperature (80°C or 176°F). The fuel efficiency of PEM systems is typically less than 40 percent. Buses

TABLE 22.3 Fuel Cell Types

TYPE	PROTON ELECTROLYTE MEMBRANE	PHOSPHORIC ACID	MOLTEN CARBONATE	SOLID OXIDE
Electrolyte	Semipermeable organic polymer	Phosphoric acid	Liquid carbonate	Solid-oxide ceramic
Charge carrier	H^+	H^+	$CO_3^=$	$O^=$
Catalyst	Platinum	Platinum	Nickel	Perovskites (calcium titanate)
Operating temperature, °C	80	200	650	1000
Cell material	Carbon or metal-based	Graphite-based	Stainless steel	Ceramic
Efficiency (percent)	Less than 40	40 to 50	50 to 60	More than 60
Heat cogeneration	None	Low	High	High
Status	Demonstration systems	Commercially available	Demonstration systems	Under development

Source: Alan C. Lloyd, *Scientific American,* July 1999, vol. 281 (1) p. 83.

equipped with PEM stacks have been demonstrated in Chicago, Miami, and Vancouver, BC. In 1999, DaimlerChrysler unveiled their 5-passenger NECAR-4. A 500 kg (1100 lb) PEM stack that costs $30,000 (compared to about $3000 for a conventional gasoline engine) produces performance comparable to most passenger cars, but takes up considerable interior room.

For stationary electrical generation, the most common fuel cell design uses phosphoric acid immobilized in a porous ceramic matrix as the electrolyte. Because this system operates at higher temperatures than PEM cells, less platinum is needed for the catalyst. It has a higher efficiency, 40 to 50 percent, but is heavier and larger than PEM cells. It also is less sensitive to carbon dioxide contamination than other designs. Hundreds of 200 kilovolt phosphoric acid fuel cells have been installed around the world. Some have run for decades. They supply dependable electricity in remote locations without the spikes and sags and risk of interruption common in utility grids. The largest fuel cell ever built was an 11 MW unit in Japan, that provides enough electricity for a small town.

In 1999, the Central Park police station in New York City was equipped with a 200 kilowatt phosphoric acid fuel cell. The station is located in the middle of the park, so bringing in new electric lines would have cost $1.2 million and would have disrupted traffic and park use for months. A diesel generator was ruled out as too noisy and polluting. Solar photovoltaic panels were thought to be too obtrusive. A small, silent fuel cell provided just the right solution.

Carbonate fuel cells use an inexpensive nickel catalyst and operate at 650°C (1200°F). The electrode is a very hot (thus the name molten carbonate) solution trapped in a porous ceramic. The charge carrier is carbonate ion, which is formed at the cathode where oxygen and carbon dioxide react in the presence of a nickel oxide catalyst. Migrating through the electrolyte, the carbonate ion reacts at the anode with hydrogen and carbon monoxide to release electrons. The high operating temperature of this design means that it can reform fuels internally and ionize hydrogen without expensive catalysts. Heat cogeneration is very good, but the high temperature makes these units more difficult to operate. It takes hours for carbonate fuel cells to get up to operating temperature, so they aren't suitable for short-term, quick-response uses.

The least developed of the fuel cell design is called solid oxide, because it uses a coated zirconium ceramic as an electrolyte. Oxygen ions formed by the titanium catalyst carry the charge across the electrolyte. Operating temperatures are 1000°C (1800°F). They have the highest fuel efficiency of any current design, but mass production of components has not yet been mastered, and these cells are still in the experimental stage.

ENERGY FROM BIOMASS

Photosynthetic organisms have been collecting and storing the sun's energy for more than 2 billion years. Plants capture about 0.1 percent of all solar energy that reaches the earth's surface. That kinetic energy is transformed, via photosynthesis, into chemical bonds in organic molecules (chapter 3). A little more than half of the energy that plants collect is spent in such metabolic activities as pumping water and ions, mechanical movement, maintenance of cells and tissues, and reproduction; the rest is stored in biomass.

The magnitude of this resource is difficult to measure. Most experts estimate useful biomass production at 15 to 20 times the amount we currently get from all commercial energy sources. It would be ridiculous to consider consuming all green plants as fuel, but biomass has the potential to become a prime source of energy. It has many advantages over nuclear and fossil fuels because of its renewability and easy accessibility. Renewable energy resources account for about 18 percent of total world energy use, and biomass makes up three-quarters of that renewable energy supply. Biomass resources used as fuel include wood, wood chips, bark, branches, leaves, starchy roots, and other plant and animal materials.

Burning Biomass

Wood fires have been a primary source of heating and cooking for thousands of years. As recently as 1850, wood supplied 90 percent of the fuel used in the United States. Wood now provides less than 1 percent of the energy in the United States, but in many of the poorer countries of the world, wood and other biomass fuels provide up to 95 percent of all energy used. The 1500 million cubic meters of fuelwood collected in the world each year is about half of all wood harvested.

In northern industrialized countries, wood burning has increased since 1975 in an effort to avoid rising oil, coal, and gas prices. Most of these northern areas have adequate wood supplies to meet demands at current levels, but problems associated with wood burning may limit further expansion of this use. Inefficient and incomplete burning of wood in fireplaces and stoves produces smoke laden with fine ash and soot and hazardous amounts of carbon monoxide (CO) and hydrocarbons. In valleys where inversion layers trap air pollutants, the effluent from wood fires can present a major source of air quality degradation and health risk. Polycyclic aromatic compounds produced by burning are especially worrisome because they are carcinogenic (cancer-causing).

In Oregon's Willamette Valley or in the Colorado Rockies, where woodstoves are popular and topography concentrates contaminants, as much as 80 percent of air pollution on winter days is attributable to wood fires. Several resort towns, such as Vail, Aspen, and Telluride, Colorado, have banned installation of new woodstoves and fireplaces because of high pollution levels. Oregon, Colorado, and Vermont now have emission standards for new woodstoves. The Environmental Protection Agency ranks wood burners high on a list of health risks to the general population, and standards are being considered to regulate the use of woodstoves nationwide.

Highly efficient and clean-burning woodstoves are available but expensive. Running exhaust gases through heat exchangers recaptures heat that would escape through the chimney, but if flue temperatures drop too low, flammable, tarlike creosote deposits can build up, increasing fire risk. A better approach is to design combustion chambers that use fuel efficiently, capturing heat without producing waste products. Brick-lined fireboxes with afterburner chambers to burn gaseous hydrocarbons do an excellent job. Catalytic combusters, similar to the catalytic converters on cars, also are placed inside stovepipes. They burn carbon monoxide and hydrocarbons to clean emissions and to recapture heat that otherwise would have escaped out the chimney in the chemical bonds of these molecules.

Wood chips, sawdust, wood residue, and other plant materials (fig. 22.14) are being used in some places in the United States and Europe as a substitute for coal and oil in industrial boilers. In Vermont, for instance, where fossil fuels are expensive and 76 percent of the land is covered by forest, 250,000 cords of unmarketable cull wood are burned annually to fuel a 50-MW power plant in Burlington. Michigan's Public Service Commission estimated that the state's surplus forest growth could provide 1,300 MW of electricity each year. Pollution-control equipment is easier to install and maintain in a central power plant than in individual home units. Wood burning also contributes less to acid precipitation than does coal. Because wood has little sulfur, it produces few sulfur gases and burns at lower temperatures than coal; thus, it produces fewer nitrogen oxides. Burning wood as a renewable crop doesn't produce any net increase in atmospheric carbon dioxide (and, therefore, doesn't add to global warming) because all the carbon released by burning biomass was taken up from the atmosphere when the biomass was grown.

FIGURE 22.14 Harvesting marsh reeds (*Phragmites* sp.) in Sweden as a source of biomass fuel. In some places, biomass from wood chips, animal manure, food-processing wastes, peat, marsh plants, shrubs, and other kinds of organic material make a valuable contribution to energy supplies. Care must be taken, however, to avoid environmental damage in sensitive areas.

Fuelwood Crisis in Less-Developed Countries

Two billion people—about 40 percent of the total world population—depend on firewood and charcoal as their primary energy source. Of these people, three-quarters (1.5 billion) do not have an adequate, affordable supply. Most of these people are in the less-developed countries where they face a daily struggle to find enough fuel to warm their homes and cook their food. The problem is intensifying because rapidly growing populations in many developing countries create increasing demands for firewood and charcoal from a diminishing supply.

As firewood becomes increasingly scarce, women and children, who do most of the domestic labor in many cultures, spend

FIGURE 22.15 The firewood shortage in less-developed countries means that women and children must spend hours each day searching for fuel. Destruction of forests and removal of ground cover result in erosion and desertification.

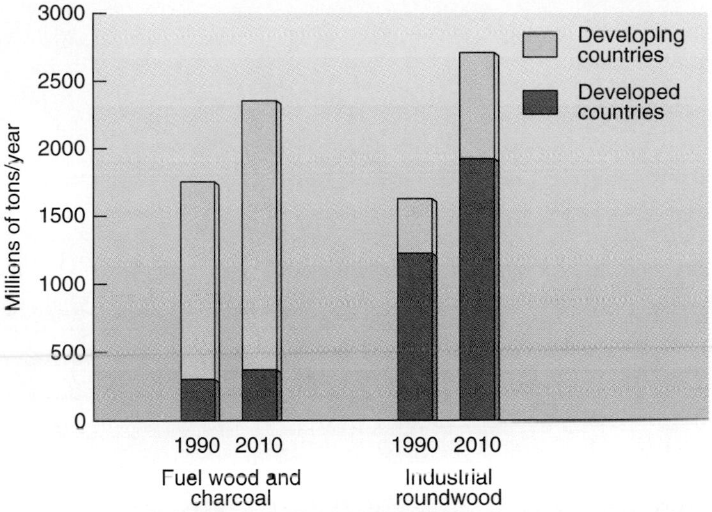

FIGURE 22.16 Worldwide, about half of all woody biomass harvested is used as fuel, mainly in developing countries. Industrial roundwood (lumber, timbers, beams) are primarily harvested and used in developed countries. A growing need for fuel in developing countries is expected to cause wood shortages in many places.
Source: Data from FAO, 1997.

more and more hours searching for fuel (fig. 22.15). In some places, it now takes eight hours, or more, just to walk to the nearest fuelwood supply and even longer to walk back with a load of sticks and branches that will only last a few days.

For people who live in cities, the opportunity to scavenge firewood is generally nonexistent and fuel must be bought from merchants. This can be ruinously expensive. In Addis Ababa, Ethiopia, 25 percent of household income is spent on wood for cooking fires. A circle of deforestation has spread more than 160 km (100 mi) around some major cities in India, and firewood costs up to ten times the price paid in smaller towns.

Currently, about half of all wood harvested each year worldwide is used as fuel (fig. 22.16). Eighty-five percent of that fuel is harvested in developing countries, whereas three-quarters of all industrial roundwood (lumber, poles, beams, and building materials) is harvested and consumed in developed countries. The poorest countries such as Ethiopia, Bhutan, Burundi, and Bangladesh depend on biomass for 90 percent of their energy. Often, the harvest is sustainable, consisting of deadwood, branches, trimmings, and shrubs. In Pakistan, for example, some 4.4 million tons of twigs and branches and 7.7 million tons of shrubs and crop residue are consumed as fuel each year with destruction of very few living trees.

In countries where fuel is scarce, however, desperate people often chop down anything that will burn. In Haiti, for instance, more than 90 percent of the once-forested land has been almost completely denuded and people cut down even valuable fruit trees to make charcoal they can sell in the marketplace. It is estimated that the 1700 million tons of fuelwood now harvested each year globally is at least 500 million tons less than is needed. By 2025 the worldwide demand for fuelwood is expected to be about twice current harvest rates while supplies will not have expanded much beyond current levels. Some places will be much worse than this average. In some African countries such as Mauritania, Rwanda, and the Sudan, firewood demand already is ten times the sustainable yield. Reforestation projects, agroforestry, community woodlots, and inexpensive, efficient, locally produced woodstoves could help alleviate expected fuelwood shortages in many places.

Dung and Methane as Fuels

Where wood and other fuels are in short supply, people often dry and burn animal manure. This may seem like a logical use of waste biomass, but it can intensify food shortages in poorer countries. Not putting this manure back on the land as fertilizer reduces crop production and food supplies. In India, for example, where fuelwood supplies have been chronically short for many years, a limited manure supply must fertilize crops and provide household fuel. Cows in India produce more than 800 million tons of dung per year, more than half of which is dried and burned in cooking fires (fig. 22.17). If that dung were applied to fields as fertilizer, it could boost crop production of edible grains by 20 million tons per year, enough to feed about 40 million people.

When cow dung is burned in open fires, more than 90 percent of the potential heat and most of the nutrients are lost. Compare that to the efficiency of using dung to produce methane gas, an excellent fuel. In the 1950s, simple, cheap methane digesters were designed for villages and homes, but they were not widely used. In China, 6 million households use biogas for cooking and lighting. Two large municipal facilities in Nanyang will soon provide fuel for more than 20,000 families. Perhaps other countries will follow China's lead.

Methane gas is the main component of natural gas. It is produced by anaerobic decomposition (digestion by anaerobic bacteria) of any moist organic material. Many people are familiar with

FIGURE 22.17 In many countries, animal dung is gathered by hand, dried, and used for fuel. This girl in Mali is carrying home fuel for cooking. Using dung as fuel deprives fields of nutrients and reduces crop production.

the fact that swamp gas is explosive. Swamps are simply large methane digesters, basins of wet plant and animal wastes sealed from the air by a layer of water. Under these conditions, organic materials are decomposed by anaerobic (oxygen-free) rather than aerobic (oxygen-using) bacteria, producing flammable gases instead of carbon dioxide. This same process may be reproduced artificially by placing organic wastes in a container and providing warmth and water (fig. 22.18). Bacteria are ubiquitous enough to start the culture spontaneously.

Burning methane produced from manure provides more heat than burning the dung itself, and the sludge left over from bacterial digestion is a rich fertilizer, containing healthy bacteria as well as most of the nutrients originally in the dung. Whether the manure is of livestock or human origin, airtight digestion also eliminates some health hazards associated with direct use of dung, such as exposure to fecal pathogens and parasites.

All sizes and forms of containment have been used for methane production, from coffee cans to oil drums to specially designed tanks. Like other bacteria and yeast cultures, such as those used for bread, yogurt, and beer, methane-producing cultures need heat to thrive. The

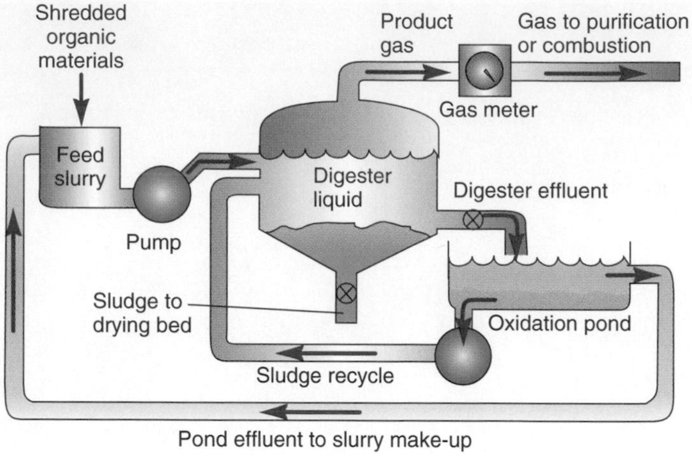

FIGURE 22.18 Continuous unit for converting organic material to methane by anaerobic fermentation. One kilogram of dry organic matter will produce 1-1.5 m^3 of methane, or 2500-3600 million calories per metric ton.

Source: Solar Energy as a National Energy Resource, NSF/NASA Solar Energy Panel, National Science Foundation, December 1972.

highest gas production is generally at 35°C (95°F), but once a culture is established, it produces enough of its own heat to continue. This warmth can be felt in all backyard compost heaps. It also accounts for early melting of snow over septic tanks.

How feasible is methane—from manure or from municipal sewage—as a fuel resource in developed countries? Methane is a clean fuel that burns efficiently. It is produced in a low-technology, low-capital process from any kind of organic waste material: livestock manure, kitchen and garden scraps, and even municipal garbage and sewage. In fact, municipal landfills are active sites of methane production, contributing as much as 20 percent of the annual output of methane to the atmosphere. This is a waste of a valuable resource and a threat to the environment because methane absorbs infrared radiation and contributes to the greenhouse effect (chapter 17). Some municipalities are drilling gas wells into landfills and garbage dumps. Cattle feedlots and chicken farms in the United States are a tremendous potential fuel source. Collectible crop residues and feedlot wastes each year contain 4.8 billion gigajoules (4.6 quadrillion BTUs) of energy, more than all the nation's farmers use. Municipal sewage treatment plants routinely use anaerobic digestion as a part of their treatment process, and many facilities collect the methane they produce and use it to generate heat or electricity for their operations. Although this technology is well-developed, its utilization could be much more widespread.

Alcohol from Biomass

Ethanol (grain alcohol) and methanol (wood alcohol) are produced by anaerobic digestion of plant materials with high sugar content, mainly grain and sugarcane. Ethanol can be burned directly in automobile engines adapted to use this fuel, or it can be mixed with gasoline (up to about 10 percent) to be used in any normal automobile engine. A mixture of gasoline and ethanol is often called

gasohol. Ethanol in gasohol raises octane ratings and is a good substitute for lead antiknock agents, the major cause of lead pollution. It also helps reduce carbon monoxide emissions in automobile exhaust. Gas stations in U.S. cities that exceed EPA air quality standards have been ordered to sell so-called oxygenated fuels containing ethanol or methanol.

Ethanol production could be a solution to grain surpluses and bring a higher price for grain crops than the food market offers. It also offers promise for reduced dependence on gasoline, which is refined from petroleum. Brazil has instituted an ambitious national program to substitute crop-based ethanol for imported petroleum. In 1995, the Brazilian sugar harvest produced 2.5 billion gal of ethanol, but falling oil prices have undercut plans to expand ethanol production to 4 billion gal per year.

Researchers at the Solar Energy Research Institute in Golden, Colorado, have developed a new biomass-to-gasoline process that yields a high-octane, high-energy product from wood scraps, crop residues, and paper or municipal wastes. In this one-step process, organic materials are vaporized and passed over a zeolite (silica alumina) catalyst, which converts hydrocarbons to long-chain, gasolinelike fluids, carbon dioxide, and water. A ton of biomass will produce about 225 l (60 gal) of liquid fuel suitable for direct use in engines, or to be blended with other fuels to raise octane levels.

Methanol might have many advantages as a motor vehicle fuel. Since it burns at a lower temperature than gasoline or diesel, the bulky, heavy radiator might be eliminated, making automobiles sleeker and more efficient. Combined with flywheel-energy storage, intermittent burning engines, turbochargers, and other high-tech devices, we might be able to have personal transportation that is both energy efficient and less polluting. Both methanol and ethanol make good fuels for fuel cells.

Crop Residues, Energy Crops, and Peat

Crop residues, such as cornstalks, corncobs, or wheat straw, can be used as a fuel source, but they are expensive to gather and often are better left on the ground as soil protection (chapter 11). The residue accumulated in food processing, however, can be a useful fuel. Hawaiian sugar growers burn bagasse, a fibrous sugarcane residue, to produce the state's second-largest energy source, surpassed only by petroleum.

Some crops are being raised specifically as an energy source. Fast-growing trees, such as *Leucaena,* and shrubs, such as alder and willow, are grown to provide wood for energy. Milkweeds, sedges, marsh grasses, cattails, and other biomass crops also might become useful energy sources, grown on land that is otherwise unsuitable for crops (see fig. 22.14). Unfortunately, there may be unacceptable environmental costs from disruption of wetlands and forests that are converted to energy crop plantations. Some plant species produce hydrocarbons that can be used directly as fuel in internal combustion engines without having to be fermented or transformed to hydrogen or methane gas. Sunflower oil, for instance, can be burned directly in diesel engines. In most cases, these oils and other high-molecular-weight hydrocarbons are more valuable for other purposes, such as food, plastics, and industrial chemicals; however, they might be an attractive energy source in some circumstances.

Several other forms of biomass are actual or proposed sources of energy. People in northern climates have been digging up peat (partially decomposed plant residues) from bogs and marshes for centuries. There has been a resurgence of interest in exploiting the vast Minnesota and Canadian peat bogs, which some people regard as wastelands. After mining some or all of the peat, the land could be used for energy crops, such as cattails. Critics of these schemes point out the damaging effects of bog draining on ecosystems, watersheds, and wildlife. Burning of municipal garbage ("waste to watts") is a popular proposal; however, many people are worried about air pollution from these facilities (chapter 23). Garbage may contain hazardous, volatile, or explosive materials that are difficult or dangerous to sort out. Effects on air quality from all forms of direct combustion of biomass are a serious concern.

ENERGY FROM THE EARTH'S FORCES

The winds, waves, tides, ocean thermal gradients, and geothermal areas are renewable energy sources. Although available only in selected locations, these sources could make valuable contributions to our total energy supply.

Hydropower

Falling water has been used as an energy source since ancient times. The invention of water turbines in the nineteenth century greatly increased the efficiency of hydropower dams. By 1925, falling water generated 40 percent of the world's electric power. Since then, hydroelectric production capacity has grown fifteenfold, but fossil fuel use has risen so rapidly that water power is now only one-quarter of total electrical generation. Still, many countries produce most of their electricity from falling water (fig. 22.19). Norway, for instance, depends on hydropower for 99 percent of its electricity; Brazil, New Zealand, and Switzerland all produce at least three-quarters of their electricity with water power. Canada is the world's leading producer of hydroelectricity, running 400 power stations with a combined capacity exceeding 60,000 MW.

The total world potential for hydropower is estimated to be about 3 million MW. If all of this capacity were put to use, the available water supply could provide between 8 and 10 terrawatt hours (10^{12} Whr) of electrical energy. Currently, we use only about 10 percent of the potential hydropower supply. The energy derived from this source in 1994 was equivalent to about 500 million tons of oil, or 8 percent of the total world commercial energy consumption.

Much of the hydropower development in recent years has been in enormous dams. There is a certain efficiency of scale in giant dams, and they bring pride and prestige to the countries that build them, but they can have unwanted social and environmental effects. The largest hydroelectric dam in the world at present is the Itaipu

FIGURE 22.19 Hydropower dams like this one in New South Wales, Australia, can prevent floods and provide useful energy, but often they destroy valuable farmland, forests, villages, and historic areas when valleys are flooded for water storage. In addition, dams can also lose large amounts of water to seepage and evaporation, or they could fail and thus be a threat to downstream areas.

Dam on the Parana River between Brazil and Paraguay. Designed to generate 12,600 MW of power, this dam should produce as much energy as 13 large nuclear power plants when completed. The lake that it is creating already has flooded about 1300 sq km (500 sq mi) of tropical rainforest and displaced many thousands of native people and millions of other creatures. An even larger dam that will produce twice as much electrical power is now under construction at the Three Gorges site on China's Yangtze River.

There are still other problems with big dams, besides human displacement, ecosystem destruction, and wildlife losses. Dam failure can cause catastrophic floods and thousands of deaths. Sedimentation often fills reservoirs rapidly and reduces the usefulness of the dam for either irrigation or hydropower. In China, the Sanmenxia Reservoir silted up in only four years, and the Laoying Reservoir filled with sediment before the dam was even finished. Rotting vegetation in artificial impoundments can have disastrous effects on water quality. When Lake Brokopondo in Suriname flooded a large region of uncut rainforest, underwater decomposition of the submerged vegetation produced hydrogen sulfide that killed fish and drove out villagers over a wide area. Acidified water from this reservoir ruined the turbine blades, making the dam useless for power generation.

Floating water hyacinths (rare on free-flowing rivers) have already spread over reservoir surfaces behind the Tucurui Dam on the Amazon River in Brazil, impeding navigation and fouling machinery. Herbicides sprayed to remove aquatic vegetation have contaminated water supplies. Herbicides used to remove forests before dam gates closed caused similar pollution problems. Schistosomiasis, caused by parasitic flatworms called blood flukes (chapter 9), is transmitted to humans by snails that thrive in slow-moving, weedy tropical waters behind these dams. It is thought that 14 million Brazilians suffer from this debilitating disease.

As mentioned before, dams displace indigenous people. The Narmada Valley project in India will drown 150,000 ha of tropical forest and displace 1.5 million people, mostly tribal minorities and low-caste hill people. China's Three Gorges project will displace one million people. The Akosombo Dam built on the Volta River in Ghana nearly 20 years ago displaced 78,000 people from 700 towns. Few of these people ever found another place to settle, and those still living remain in refugee camps and temporary shelters.

In tropical climates, large reservoirs often suffer enormous water losses. Lake Nasser, formed by the Aswan High Dam in Egypt, loses 15 billion cubic meters each year to evaporation and seepage. Unlined canals lose another 1.5 billion cubic meters. Together, these losses represent one-half of the Nile River flow, or enough water to irrigate 2 million ha of land. The silt trapped by the Aswan Dam formerly fertilized farmland during seasonal flooding and provided nutrients that supported a rich fishery in the Delta region. Farmers now must buy expensive chemical fertilizers, and the fish catch has dropped almost to zero. As in South America, schistosomiasis is an increasingly serious problem.

If big dams—our traditional approach to hydropower—have so many problems, how can we continue to exploit the great potential of hydropower? Fortunately, there is an alternative to gigantic dams and destructive impoundment reservoirs. Small-scale, **low-head hydropower** technology can extract energy from small headwater dams that cause much less damage than larger projects. Some modern, high-efficiency turbines can even operate on **run-of-the-river flow.** Submerged directly in the stream and small enough not to impede navigation in most cases, these turbines don't require a dam or diversion structure and can generate useful power with a current of only a few kilometers per hour. They also cause minimal environmental damage and don't interfere with fish movements, including spawning migration. **Micro-hydro generators** operate on similar principles but are small enough to provide economical power for a single home. If you live close to a small stream or river that runs year-round and you have sufficient water pressure and flow, hydropower is probably a cheaper source of electricity for you than solar or wind power (fig. 22.20).

Small-scale hydropower systems also can cause abuses of water resources. The Public Utility Regulatory Policies Act of 1978 included economic incentives to encourage small-scale energy projects. As a result, thousands of applications were made to dam or divert small streams in the United States. Many of these projects have little merit. All too often, fish populations, aquatic habitat, recreational opportunities, and the scenic beauty of free-flowing streams and rivers are destroyed primarily to provide tax benefits for wealthy investors.

FIGURE 22.20 Solar collectors capture power only when the sun shines, but hydropower is available 24 hours a day. Small turbines such as this one can generate enough power for a single-family house with only 15 m (50 ft) of head and 200 l (50 g) per minute flow. The turbine can have up to four nozzles to handle greater water flow and generate more power.

FIGURE 22.21 Millions of traditional windmills once pumped water, generated electricity, and provided power for farms and ranches across America. Most have been displaced by internal combustion engines and rural electric power systems, but there is still potential for capturing large amounts of renewable energy from the sun, wind, and other sustainable sources.

Wind Energy

The air surrounding the earth has been called a 20-billion-cubic-kilometer storage battery for solar energy. The World Meteorological Organization has estimated that 20 million MW of wind power could be commercially tapped worldwide, not including contributions from windmill clusters at sea. This is about 50 times the total present world nuclear generating capacity.

Wind power has advantages and disadvantages, as do other nontraditional technologies. Like solar power and hydropower, wind power taps a natural physical force. Like solar power (its ultimate source), wind power is a virtually limitless resource, is nonpolluting, and causes minimal environmental disruption. Like solar power, however, it requires expensive storage during peak production times to offset nonwindy periods.

Windmills played a crucial role in the settling of the American West. The Great Plains had abundant water in underground aquifers, but little surface water. The strong, steady winds that blew across the prairies provided the energy to pump water that allowed agriculture to move west across the prairies. By the end of the nineteenth century, nearly every farm or ranch west of the Mississippi River had at least one windmill. The manufacture, installation, and repair of windmills was a major industry in America. Even today, some 150,000 windmills still spin productively in the Great Plains (fig. 22.21).

Windmill technology was flourishing early in this century. Efficiencies increased ten times or more with the invention of aerodynamic propellers and affordable electric-wind generators in the 1920s. This promising technology was nipped in the bud, however, by passage of the Rural Electrification Act of 1935. This act brought many benefits to rural America, but it effectively killed wind-power development and encouraged the building of huge hydropower dams and coal-burning power plants. It is interesting to speculate what the course of history might have been if we had not spent trillions of dollars on fossil fuel and nuclear energy, but instead had invested that money on small-scale, renewable energy systems.

As the world's conventional fuel prices rise, interest in wind energy is resurging. In the 1980s, the United States was a world leader in wind technology and California hosted 90 percent of all existing windpower generators. Some 17,000 windmills marched across windy mountain ridges at Altamont, Tehachapi, and San

FIGURE 22.22 Together, these wind turbines in Altamont Pass, CA, produce as much electricity as a large coal-fired or nuclear power plant.

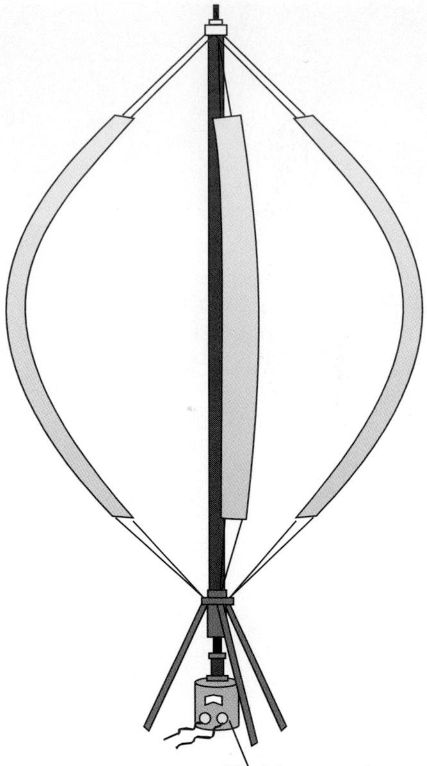

FIGURE 22.23 The Darrieus rotor. This "eggbeater" wind machine is a low-cost device capable of producing electricity at relatively low wind speeds. It has been in use since 1929 when it was patented by G. J. M. Darrieus in France.

Gorgino Passes (fig. 22.22). Poor management, technical flaws, and overdependence on subsidies, however, led to bankruptcy of major corporations including Kenetech, once the largest turbine producer in the United States. Now Danish, German, and Japanese wind machines are capturing the rapidly growing world market. Wind technology is now Denmark's second-largest export, employing 20,000 people and bringing in about $1 billion (U.S.) per year.

The 6200 MW of installed wind power currently in operation worldwide demonstrate the economy of wind turbines. Theoretically up to 60 percent efficient, windmills typically produce about 35 percent efficiency under field conditions. Where conditions are favorable, wind power is already cheaper than any other new energy source, with electric prices typically around 5 cents per KWH in places with steady winds averaging at least 24 km/hr (15 mph). Much of the Great Plains meets this requirement.

The World Energy Council predicts that wind could account for 200,000 to 500,000 MW of electricity by 2020, depending on how seriously politicians take global warming and how many uneconomical nuclear reactors go offline. One thousand MW meets the energy needs of about 50,000 typical U.S. households or is equivalent to about 6 million barrels of oil. By the middle of the next century, Shell Oil suggests that half of all the world's energy could be wind and solar generated.

The standard modern wind turbine uses only two or three propeller blades. More blades on a windmill provide more torque in low-speed winds, so that the traditional Midwestern windmill, with 20 or 30 blades, was most appropriate for small-scale use in less reliable wind fields. Fewer blades operate better in high-speed winds, providing more energy for less material cost at wind speeds of 25-40 km/hr (15-25 mph). A two-blade propeller can extract most of the available energy from a large vertical area and has less material to weaken and break in a storm. Three-bladed propellers often are preferred because they are easier to balance and spin more smoothly.

In 1929, the French engineer G. J. M. Darrieus proposed an alternative to the traditional horizontal-axis windmill. Developed in Canada in the 1960s, the Darrieus rotor, a vertical-axis "eggbeater" machine, is durable and inexpensive (fig. 22.23). It needs no adjustment to wind direction and operates in any wind stronger than 19 km/hr (12 mph), compared to conventional turbines that generally require at least 24 km/hr (15 mph). Quantity manufacturing of Darrieus generators has reduced the price rapidly, bringing the current cost to around 5.4 cents per KWH for a conservatively estimated 15-year life span.

Wind farms are large concentrations of wind generators producing commercial electricity. Construction of wind farms is not limited to mountain ridges. As this chapter's opening story shows, the Great Plains offer tremendous wind power potential. Seacoasts also provide excellent wind farm sites. Offshore wind farms have been proposed, with large (up to 8 MW each) generators to be anchored a short distance out at sea where competition for real estate development is low and winds are high year-round. According to such plans, collected energy would probably be stored in electrolyzed water (discussed earlier as storage for photovoltaic power) for transportation to shore.

Are there negative impacts of wind farms? They generally occupy places with wind and weather too severe for residential or other development. The wind speed at Altamont, for instance, averages 35 km/hr (22 mph) from March to September. Most wind farms are too far from residential areas to be heard or seen. But they do interrupt the view in remote isolated places and destroy the sense of isolation and natural beauty. Bird kills were a problem for some California wind farms. Careful placement outside of migration corridors and addition of warning devices can reduce mortality greatly.

Wouldn't wind power take up a huge land area if we were to depend on it for a major part of our energy supply? As table 22.4 shows, the actual space taken up by towers, roads, and other structures on a wind farm is only about one-third as much as would be consumed by a coal-fired power plant or solar thermal energy system to generate the same amount of energy over a 30-year period. Furthermore, the land under windmills is more easily used for grazing or farming than is a strip-mined coal field or land under solar panels.

When a home owner or community invests independently in wind generation, the same question arises as with solar energy: What should be done about energy storage when electricity production exceeds use? Besides the storage methods mentioned earlier, many private electricity producers believe the best use for excess electricity is to sell it to the public utility grid. When private generation is low, the public utility runs electricity through the meter and into the house or community. When the wind generator or photovoltaic systems overproduce, the electricity runs back into the grid and the meter runs backward. Ideally, the utility reimburses individuals for this electricity, for which other consumers pay the company. The 1978 Public Utilities Regulatory Policies Act required utilities to buy power generated by small hydro, wind, cogeneration, and other privately owned technologies at a fair price. Not all utilities yet comply, but some—notably in California, Oregon, Maine, and Vermont—are purchasing significant amounts of private energy.

TABLE 22.4 Jobs and Land Required for Alternative Energy Sources

TECHNOLOGY	LAND USE (m^2 PER GIGAWATT-HOUR FOR 30 YEARS)	JOBS (PER TERAWATT-HOUR PER YEAR)
Coal	3642	116
Photovoltaic	3237	175
Solar Thermal	3561	248
Wind	1335	542

Source: Data from Lester R. Brown, et al. *Saving the Planet*, 1991, W. W. Norton & Co., Inc.

Geothermal Energy

The earth's internal temperature can provide a useful source of energy in some places. High-pressure, high-temperature steam fields exist below the earth's surface. Around the edges of continental plates or where the earth's crust overlays magma (molten rock) pools close to the surface, this energy is expressed in the form of hot springs, geysers, and fumaroles (chapter 16). Yellowstone National Park is the largest geothermal region in the United States. Iceland, Japan, and New Zealand also have high concentrations of geothermal springs and vents. Depending on the shape, heat content, and access to groundwater, these sources produce wet steam, dry steam, or hot water.

Until recently, the main use of this energy source was in baths built at hot springs. More recently, **geothermal energy** has been used in electric power production, industrial processing, space heating, agriculture, and aquaculture. In Iceland, most buildings are heated by geothermal steam. A few communities in the United States, including Boise, Idaho, use geothermal home heating. Recently, geothermal steam also has been developed for electrical generation.

California's Geysers project is the world's largest geothermal electric-generating complex, with 200 steam wells that provide some 1,300 MW of power. This system functions much like other heat-driven generators. A shaft sunk into the subsurface steam reservoir brings pressurized steam up to a turbine at the surface. The steam spins the turbine to produce electricity and then is piped off to a condensing unit and then to a cooling pond. The remaining brine (water heavily laden with dissolved minerals) is pumped back down to the reservoir. When the steam source has high mineral concentrations that would corrode delicate turbine blades, a heat exchanger is used to generate clean steam as is done in a nuclear power plant.

Among the advantages of geothermal generators are a reasonably long life span (at least several decades), no mining or transporting of fuels, and little waste disposal. The main disadvantages are the potential danger of noxious gases in the steam and noise problems from steam-pressure relief valves.

While few places have geothermal steam, the entire earth is underlain by warm rocks that can be reached by drilling down far enough into the crust. Some heating systems pump groundwater from underground aquifers or inject surface water into deep strata and then withdraw it after it has been warmed, as a way of extracting geothermal energy (fig. 22.24). There have been experiments with pumping surplus heated water into underground reservoirs during the summer when the heat is not needed, and then withdrawing it in the winter as make-up water for steam-heating systems. A major concern with such systems, however, is to ensure that contaminants aren't injected into aquifers, which are important sources of drinking water.

Tidal and Wave Energy

Ocean tides and waves contain enormous amounts of energy that can be harnessed to do useful work. Tidal power exploitation is not new. The earliest recorded tide-powered mills were built in about

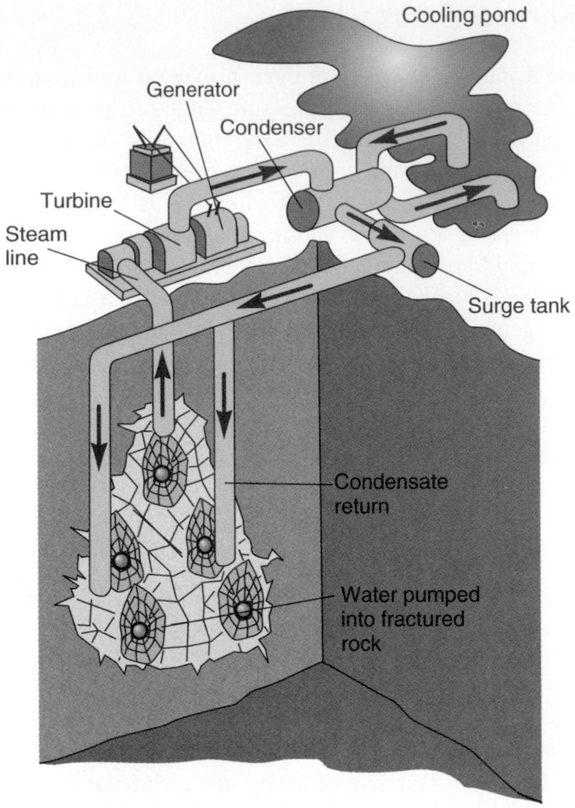

FIGURE 22.24 Geothermal steam extraction system for electricity production. Once the steam has spun the turbine, it is condensed, cooled, and pumped back into the underground steam field. The steam source for this relatively low-capital electric plant should last for several decades.
Sources: American Oil Shale Corporation and U.S. Atomic Energy Commission, *A Feasibility Study of a Plowshare of Geothermal Power Plant,* April 1971.

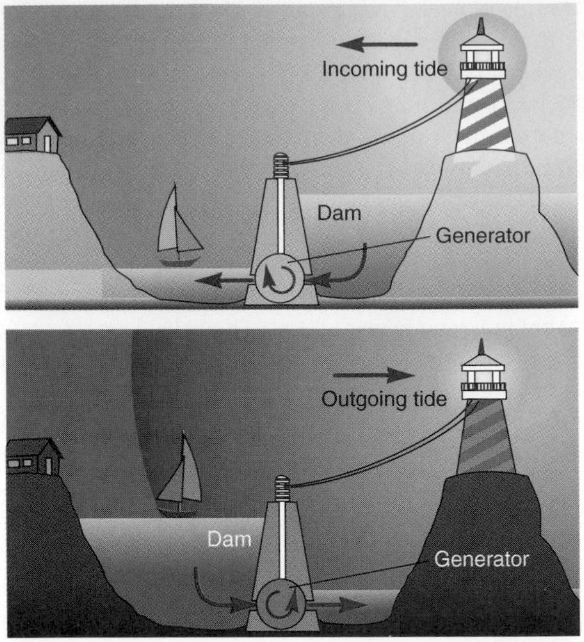

FIGURE 22.25 Tidal power station. Both incoming and outgoing tides are held back by a dam. The difference in water levels generates electricity in both directions as water runs through reversible turbogenerators.

1100 in England. One built in Woodbridge, England, in 1170 functioned productively for 800 years. Through the seventeenth century, similar mills were built by the Dutch in the Netherlands and in New York.

The Rance River Power Station in France, in operation since 1966, was the first large tidal electric generation plant, producing 160 MW. A **tidal station** works like a hydropower dam, with its turbines spinning as the tide flows through them (fig. 22.25). It requires a high-tide low-tide differential of several meters to spin the turbines. Unfortunately, the tidal period of $13^1/_2$ hours causes problems in integrating the plant into the electric utility grid, as it seldom coincides with peak-use hours. Nevertheless, demand has kept the plant running for more than two decades.

The first North American tidal generator, producing 20 MW, was completed in 1984 at Annapolis Royal, Nova Scotia. A much larger project has been proposed to dam the Bay of Fundy and produce 5000 MW of power on the Bay's 17-meter tides. The total flow at each tide through the Bay of Fundy theoretically could generate energy equivalent to the output of 250 large nuclear power plants. The environmental consequences of such a gargantuan project, however, may prevent its ever being built. The main worries are saltwater flooding of freshwater aquifers when seawater levels rise behind the dam and the flooding and destruction of rich shoals and salt flats, breeding grounds for aquatic species and a vital food source for millions of migrating shorebirds. There also would be heavy siltation, as well as scouring of the seafloor as water shoots through the dam. However, it appears that modest-sized plants like Annapolis Royal's will avoid most of these dangers. Three other medium-sized projects are under consideration for sites within the Bay of Fundy.

Ocean wave energy can easily be seen and felt on any seashore. The energy that waves expend as millions of tons of water are picked up and hurled against the land, over and over, day after day, can far exceed the combined energy budget for both insolation (solar energy) and wind power in localized areas. Captured and turned into useful forms, that energy could make a very substantial contribution to meeting local energy needs.

Numerous attempts have been made to use wave energy to drive electrical generators. Generally these take the form of a floating bar that moves up and down as the wave passes. When coupled to a dynamo, this mechanical energy can be converted to electricity in the same way that a waterwheel or steam turbine works. England, with a long coastline facing the stormy North Sea, plans to build an extensive system of wave-energy platforms. Unfortunately for developers of this energy source, the stormy coasts where waves are strongest are usually far from major population centers that need the power. In addition, the storms that bring this energy destroy the equipment intended to exploit it.

An intriguing proposal has been developed for small rotating cylinders called ducks that would sit just offshore, bobbing in the surf and generating electricity. There are claims that British energy officials deliberately suppressed this low-tech approach because it might compete with conventional sources.

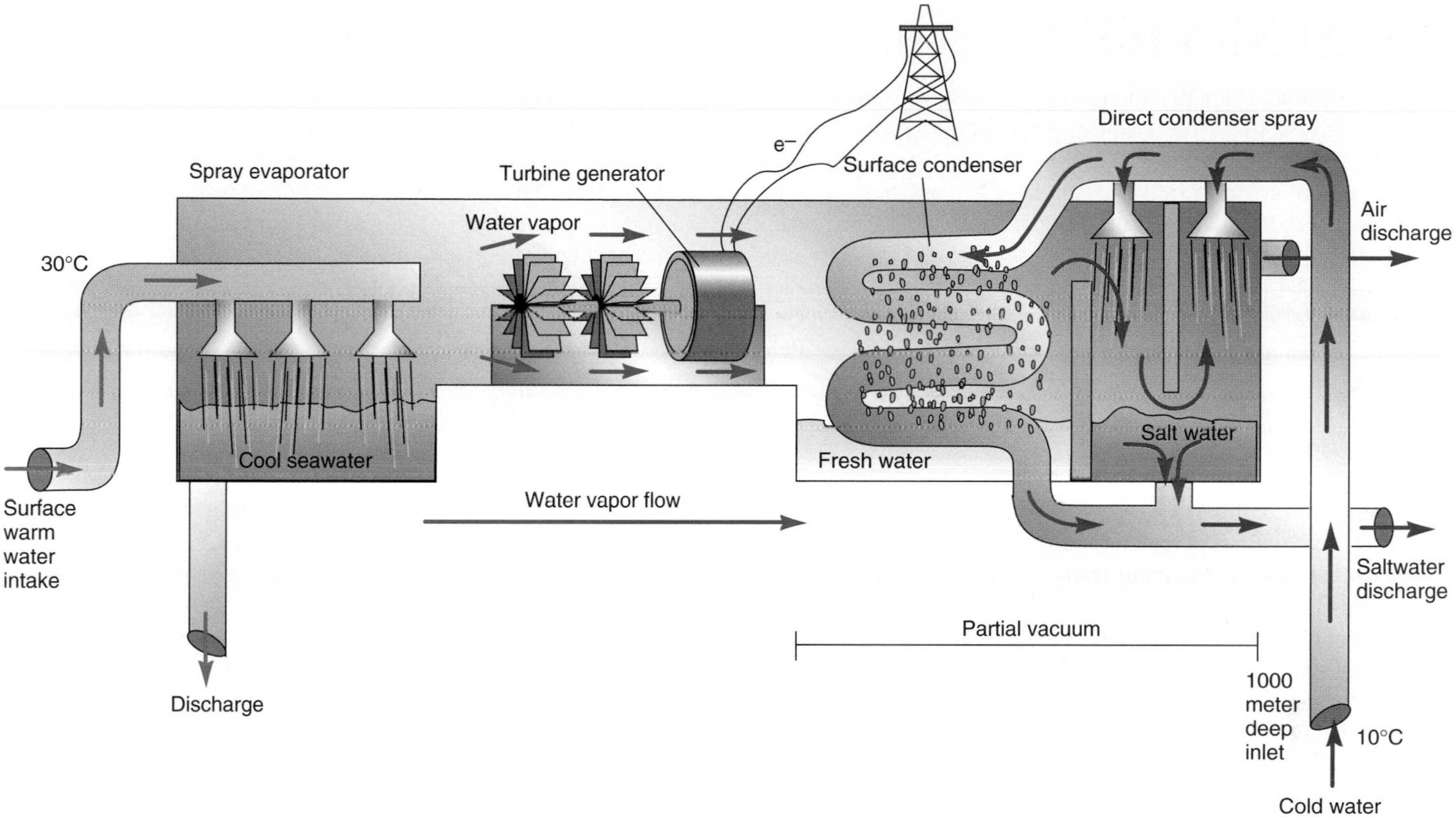

FIGURE 22.26 Open-cycle ocean thermal electric generator produces both fresh water and electricity as long as the water temperature differential is at least 20°C (36°F).

Ocean Thermal Electric Conversion

Temperature differentials between upper and lower layers of the ocean's water also are a potential source of renewable energy. Energy can be extracted in either a closed-cycle or open-cycle system. In a closed-cycle **ocean thermal electric conversion (OTEC)** system, heat from sun-warmed upper ocean layers is used to evaporate a working fluid, such as ammonia or Freon, which has a low boiling point. The pressure of the gas produced is high enough to spin turbines to generate electricity. Cold water then is pumped from the depths of the ocean to condense the gas.

An open-cycle system uses seawater itself as the working fluid (fig. 22.26). Warm seawater is sprayed into a chamber, where low air pressure causes some water to evaporate. The resulting water vapor drives a massive, low-pressure turbogenerator and then passes through two condensers cooled by deep ocean water. The first one, a closed surface condenser, produces distilled (desalinized) water. The other is a direct contact condenser in which cold water is sprayed directly into the steam flow, condensing the remaining water vapor and creating the vacuum that drives the system. The outflow from this condenser is discharged into the sea. This process produces fresh water, as well as electricity, but it requires very large turbines and inlet pipes, making it expensive to install.

As long as a temperature difference of about 20°C (36°F) exists between the warm upper layers and cooling water, useful amounts of net power can, in principle, be generated with one of these systems. This differential corresponds, generally, to a depth of about 1000 m in tropical seas. The places where this much temperature difference is likely to be found close to shore are islands that are the tops of volcanic seamounts, such as Hawaii, or the edges of continental plates along subduction zones (chapter 16) where deep trenches lie just offshore. The west coast of Africa, the south coast of Java, and a number of South Pacific islands, such as Tahiti, have usable temperature differentials for OTEC power.

Disadvantages of OTEC include the energy cost of pumping up deep waters (which could eliminate most of the energy advantage of the system), saltwater corrosion of pipes and equipment, vulnerability to storm damage, and ecological destabilization from the upwelling of deep, nutrient-rich water and local alterations of ocean temperatures.

Although their temperature differentials aren't as great as the ocean, deep lakes can have very cold bottom water. Ithaca, NY, has recently built a system to pump cold water out of Lake Cayuga to provide natural air conditioning during the summer. Cold water discharge from a Hawaiian OTEC system has been used to cool the soil used to grow cool-weather crops such as strawberries.

SUMMARY

Several sustainable energy sources could reduce or eliminate our dependence on fossil fuels and nuclear energy. Some of these sources have been used for centuries but have been neglected since fossil fuels came into widespread use. Passive solar heat, fuelwood, windmills, and waterwheels, for instance, once supplied a major part of the external energy for human activities. With increased concern about the dangers and costs associated with conventional commercial energy, these ancient energy sources are being reexamined as part of a more sustainable future for humankind.

Exciting new technologies have been invented to use renewable energy sources. Active solar air and water heating, for instance, require less material and function more quickly than passive solar collection. Wind is now the cheapest form of new energy in many places. It has potential to supply one-third or more of our energy requirements. Parabolic mirrors can produce temperatures high enough to be used as process heat in manufacturing.

Fuel cells use catalysts and semipermeable electrolytes to extract energy from fuels such as hydrogen or methanol at high efficiency and with very low emissions. Ocean thermal electric conversion, tidal and wave power stations, and geothermal steam sources can produce useful amounts of energy in some localities. One of the most promising technologies is direct electricity generation by photovoltaic cells. Since solar energy is available everywhere, photovoltaic collectors could provide clean, inexpensive, nonpolluting, renewable energy, independent of central power grids or fuel-supply systems.

Biomass also may have some modern applications. In addition to direct combustion, biomass can be converted into methane or ethanol, which are clean-burning, easily storable, and transportable fuels. These alternative uses of biomass also allow nutrients to be returned to the soil and help reduce our reliance on expensive, energy-consuming artificial fertilizers.

Many of these sustainable energy sources depend on technology that is still experimental and too expensive to compete well with established energy industries. If the economies of mass production and marketing were applied, these new technologies could be made available more cheaply and more dependably. It may take special funding and other governmental incentives to make sustainable energy competitive. The subsidies for renewable energy sources have been especially meager in comparison to the billions of dollars spent on nuclear energy, large-scale hydropower, and fossil fuel extraction and utilization.

Although conventional and alternative energy sources offer many attractive possibilities, conservation often is the least expensive and easiest solution to energy shortages. Even basic conservation efforts, such as turning off lights, can save large amounts of energy when practiced by many people. More major conservation methods, such as home insulation and energy-efficient appliances and transportation, can drastically reduce energy consumption and similarly reduce energy expenses. Our natural resources, our environment, and our pocketbooks all benefit from careful and efficient energy consumption.

QUESTIONS FOR REVIEW

1. What is cogeneration and how does it save energy?
2. Explain the principle of net energy yield. Give some examples.
3. What is the difference between active and passive solar energy?
4. How do photovoltaic cells work?
5. What is a fuel cell and how does it work?
6. Describe some problems with wood burning in both industrialized nations and developing nations.
7. How is methane made?
8. What are some advantages and disadvantages of large hydroelectric dams?
9. What are some examples of biomass fuel other than wood?
10. Describe how tidal power and ocean thermal conversion generate electricity.

QUESTIONS FOR CRITICAL THINKING

1. What alternative energy sources are most useful in your region and climate?
2. What can you do to conserve energy where you live? In personal habits? In your home, dormitory, or workplace?
3. What massive heat storage materials can you think of that could be attractively incorporated into a home?
4. Do you think building wind farms in remote places, parks, or scenic wilderness areas would be damaging or unsightly?
5. If you were the energy czar of your state, where would you invest your budget?
6. What are the advantages and disadvantages of being disconnected from central utility power?
7. Can you think of environmental consequences associated with tidal or geothermal energy? If so, how can they be mitigated?

8. You are offered a home solar energy system that costs $10,000 but saves you $1000 a year. Will you take it at this rate? If the cost were higher and the payoff time longer, what is the threshold at which you would not buy the system?

KEY TERMS

active solar systems 505
cogeneration 503
energy efficiency 500
eutectic chemicals 505
fuel cells 509
gasohol 515
geothermal energy 519
green pricing 507
low-head hydropower 516
micro-hydro generators 516
net energy yield 500
ocean thermal electric conversion (OTEC) 521
passive heat absorption 504
photovoltaic cell 507
reformer 510
run-of-the-river flow 516
tidal station 520
wind farms 518

ADDITIONAL INFORMATION ON THE INTERNET

Visit our website at http://www.mhhe.com/environmentalscience/ for specific information about each of the topics below:

Solstice, Sustainable Energy Info
American Wind Energy Association
Wind Energy Story: "Inherit the Wind"
Department of Energy—Alternative Fuels Data Center
Department of Energy—Efficiency and Renewable Energy Resources
American Solar Energy Society, Publications and Fact Sheets
National Renewable Energy Lab
Alternative Energy Course, the University of Oregon
American Wind Energy Association, information on the world's fastest growing energy source
Energy Information Administration from the Department of Energy
International Network for Sustainable Energy: links, information, resources
Solar Energy Industries Association: initiatives and FAQs
Hybrid Vehicles
Alternative Energy: automotive fuel cells
Hydrogen Energy
Biofuels and Bioenergy
Recycling Worldwide

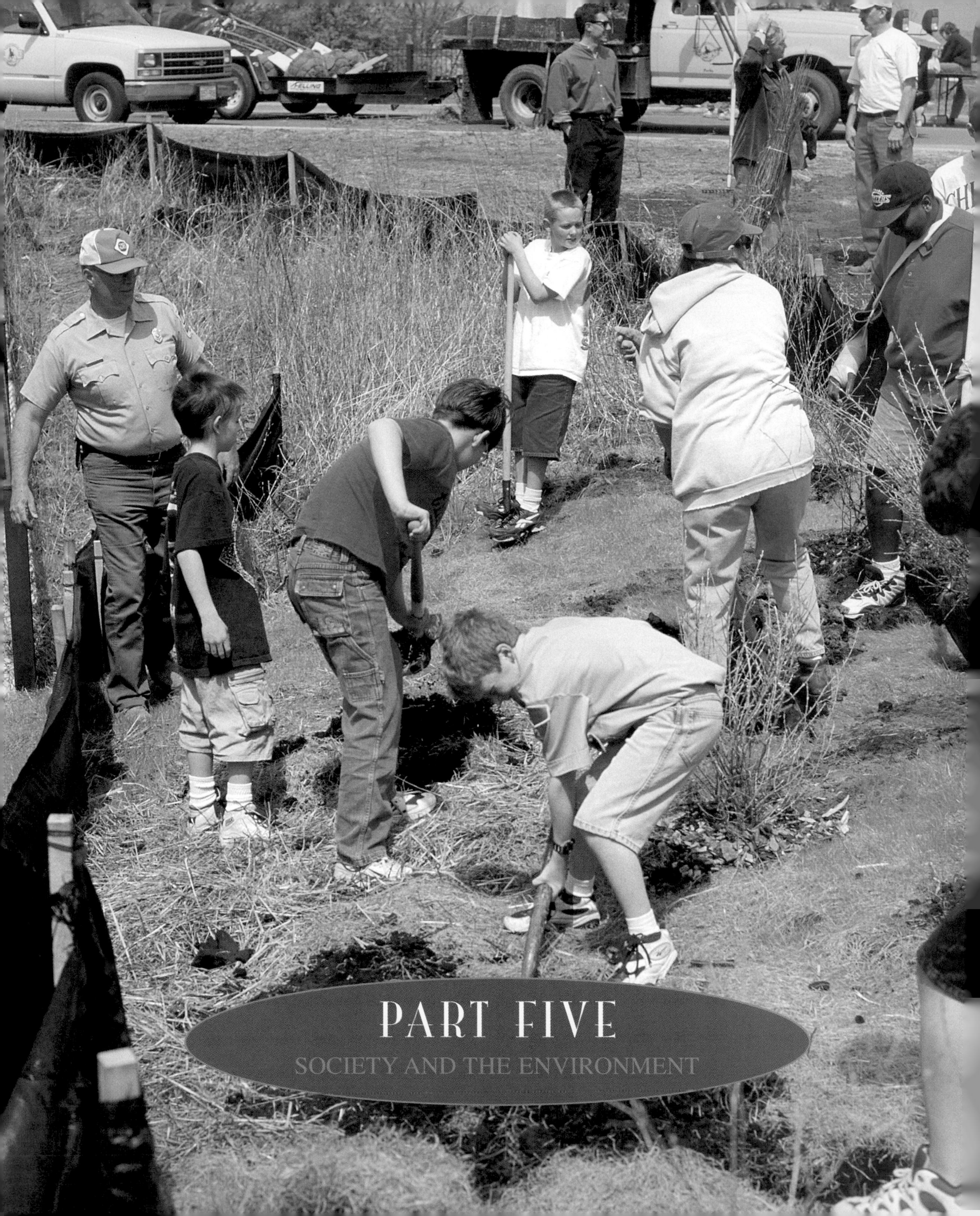

PART FIVE

SOCIETY AND THE ENVIRONMENT

CHAPTER 23

Solid, Toxic, and Hazardous Waste

We have no knowledge, so we have stuff;
but stuff without knowledge is never enough.

—Greg Brown—

OBJECTIVES

After studying this chapter, you should be able to:

- identify the major components of the waste stream and describe how wastes have been—and are being—disposed of in North America and around the world.
- explain the differences between dumps, sanitary landfills, and modern secure landfills.
- summarize the benefits, problems, and potential of recycling and reusing wastes.
- analyze some alternatives for reducing the waste we generate.
- understand what hazardous and toxic wastes are and how we dispose of them.
- evaluate the options for hazardous waste management.
- outline some ways we can destroy or permanently store hazardous wastes.

ESSENTIAL STUDY PARTNER

The Essential Study Partner (ESP) CD-ROM that accompanies this textbook contains additional information on the following topics:

- US and municipal solid waste generation: physical resources: waste: solid waste
- landfill, incinerator design, pollution and energy production: physical resources: waste: solid waste/physical resources: waste: hazardous waste
- reduce, reuse, recycle and the waste stream: physical resources: waste: the three r's
- glass to plastic, and plastic reduction: physical resources: waste: the three r's
- hazardous waste sources, treatment, and storage: physical resources: waste: hazardous waste
- bioremediation and phytoremediation: physical resources: waste: hazardous waste
- EPA Superfund cleanup sites: principles: overview: history/physical resources: waste: hazardous waste

Opposite: A scout troop plants native trees and shrubs in a watershed restoration project.

Above: Bulldozers pack down refuse at the Fresh Kills Landfill on Staten Island, NY.

What a Long, Strange Trip It Has Been

On August 31, 1986, the cargo ship *Khian Sea* loaded 14,000 tons (28 million pounds) of toxic incinerator ash from Philadelphia and set off on an odyssey that symbolizes a predicament we all share: what to do with our refuse. Starting in the 1970s, Philadelphia burned most of its municipal garbage and sent the resulting incinerator ash to a landfill in New Jersey. In 1984, when New Jersey learned that the ash contained enough arsenic, cadmium, lead, mercury, dioxin, and other toxins to be classified as hazardous waste, it refused to accept any more. When six other states also rejected incinerator ash shipments, Philadelphia was in a predicament. What would they do with 180,000 tons of the stuff every year? The answer was to send it offshore to countries with less stringent environmental standards. A local contractor offered to transport it to the Caribbean. The *Khian Sea* was to be the first of those shipments.

When the *Khian Sea* tried to unload its cargo in the Bahamas, however, it was turned away. Over the next 14 months, the ship also was refused entry by the Dominican Republic, Honduras, Panama, Bermuda, Guinea Bissau (in West Africa), and the Netherlands Antilles. Finally in late, 1987, the Haitian government issued a permit for "fertilizer" import, and the crew dumped 4000 tons of ash on the beach near the city of Gonaives. Alerted by the environmental group, Greenpeace, that the ash wasn't really fertilizer, Haitian officials canceled the permit and ordered everything returned to the ship, but the *Khian Sea* slipped away in the night, leaving behind a large pile of loose ash. Some of the waste has been moved inland and buried, but much of it remains on the beach, slowly being scattered by the wind and washed into the sea.

After it left Haiti, the *Khian Sea* visited Senegal, Morocco, Yugoslavia, Sri Lanka, and Singapore looking for a place to dump its toxic load. As it wandered the oceans looking for a port, the ship changed its name from *Khian Sea* to *Felicia* to *Pelacano*. Its registration was transferred from Liberia to the Bahamas to Honduras in an attempt to hide its true identity, but nobody wanted it or its contents. Like Coleridge's ancient mariner, it seemed cursed to roam the oceans forever. Two years, three names, four continents, and 11 countries later, the troublesome cargo was still on board (fig. 23.1). Then, somewhere in the Indian Ocean between Singapore and Sri Lanka all the ash disappeared. When questioned about this, the crew had no comment except that it was all gone. Everyone assumes, of course, that once out of sight of the land, it was just dumped overboard.

If this were just an isolated incident, perhaps it wouldn't matter much. However, some 3 million tons of hazardous and toxic waste goes to sea every year looking for a dumping site. A 1998 report by the United Nations Human Rights Commission listed the United States as a major exporter of toxic waste. In 1989—at least in part due to the misadventures of the *Khian Sea*—33 countries met in Basel, Switzerland, and agreed to limit international shipment of toxic waste, especially from the richer countries of the world to the poorer ones. Eventually 118 countries—not including the United States—ratified the Basel Convention. In 1995, the United States announced it would ratify the Convention but reserved the right to ship "recyclable" materials to whomever will take them. Since almost everything potentially can be recycled into something, that hardly puts any limits at all on what we send offshore.

The latest development in the saga of the *Khian Sea,* is that Haiti has asked Philadelphia to help pay for cleanup of the ash still sitting on the beach. Eastern Environmental Services, one of whose principal owners was responsible for dumping the load in Haiti 12 years ago, has agreed to retrieve what's left and bury it in a landfill in Pennsylvania. Philadelphia's share would be $200,000, or about one-third of the total cleanup cost. The city of brotherly love in spite of having a $130 million budget surplus last year, claims it can't afford to help out. Is this a case of environmental racism, or simply a matter of being fiscally responsible?

Although most of us don't have as big or world-famous a problem as Philadelphia, all of us contribute to some degree to related problems. We all generate vast amounts of unwanted stuff every year. Places to put our trash are becoming more and more scarce as the contents have become increasingly unpleasant and dangerous. We don't want it in our backyards, so it often ends up in those of the poorest and least powerful, both in this country and around the world. In this chapter, we will look at the kinds of waste we produce, who makes them, what problems their disposal cause, as well as how we might reduce our waste production and dispose of it in more environmentally friendly ways.

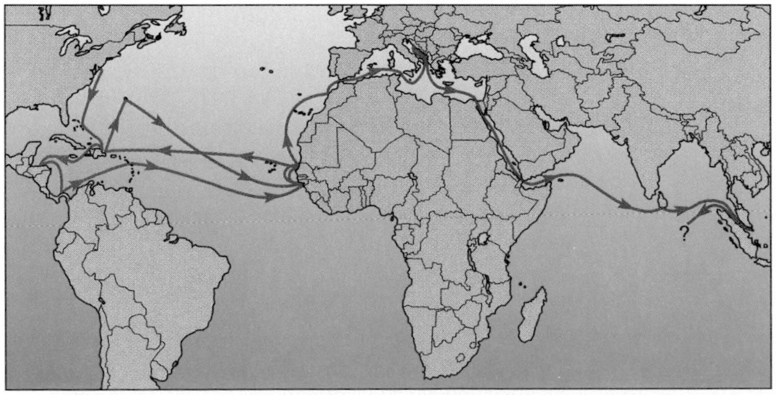

FIGURE 23.1 Route of the *Khian Sea.*

SOLID WASTE

Waste is everyone's business. We all produce wastes in nearly everything we do. According to the Environmental Protection Agency, the United States produces 11 billion tons of solid waste each year. Nearly half of that amount consists of agricultural waste, such as crop residues and animal manure, which are generally recycled into the soil on the farms where they are produced. They represent a valuable resource as ground cover to reduce erosion and fertilizer to nourish new crops, but they also constitute the single largest source of nonpoint air and water pollution in the country. About one-third of all solid wastes are mine tailings, overburden from strip mines, smelter slag, and other residues produced by mining and primary metal processing. Much of this material is stored in or near its source of production and isn't mixed with other kinds of wastes. Improper disposal practices, however, can result in serious and widespread pollution.

Industrial waste—other than mining and mineral production—amounts to some 400 million metric tons per year in the United States. Most of this material is recycled, converted to other forms, destroyed, or disposed of in private landfills or deep injection wells. About 60 million metric tons of industrial waste falls in a special category of hazardous and toxic waste, which we will discuss later in this chapter.

Municipal waste—a combination of household and commercial refuse—amounts to about 180 million metric tons per year in the United States. That's approximately two-thirds of a ton for each man, woman, and child every year—twice as much per capita as Europe or Japan, and five to ten times as much as most developing countries.

The Waste Stream

Does it surprise you to learn that you generate that much garbage? Think for a moment about how much we discard every year. There are organic materials, such as yard and garden wastes, food wastes, and sewage sludge from treatment plants; junked cars; worn out furniture; and consumer products of all types. Newspapers, magazines, advertisements, and office refuse make paper one of our major wastes (fig. 23.2). In spite of recent progress in recycling, many of the 200 *billion* metal, glass, and plastic food and beverage containers used every year in the United States end up in the trash. Wood, concrete, bricks, and glass come from construction and demolition sites, dust and rubble from landscaping and road building. All of this varied and voluminous waste has to arrive at a final resting place somewhere.

The **waste stream** is a term that describes the steady flow of varied wastes that we all produce, from domestic garbage and yard wastes to industrial, commercial, and construction refuse. Many of the materials in our waste stream would be valuable resources if they were not mixed with other garbage. Unfortunately, our collecting and dumping processes mix and crush everything together, making separation an expensive and sometimes impossible task. In a dump or incinerator, much of the value of recyclable materials is lost.

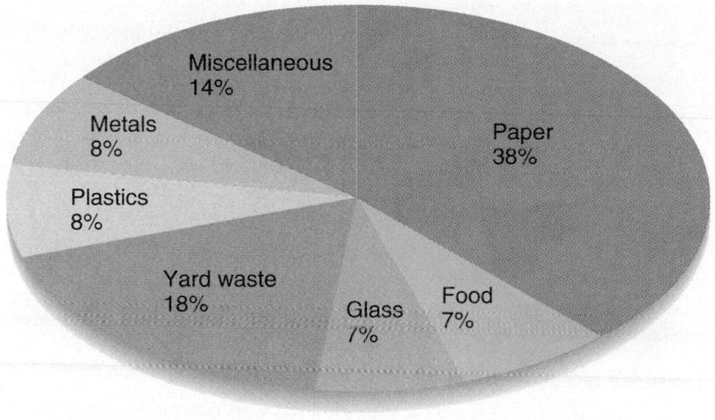

FIGURE 23.2 Composition of domestic waste in the United States by percent of total weight. Plastics make up only 8 percent by weight but up to 20 percent of total volume.
Source: Data from Environmental Protection Agency, 1997.

Another problem with refuse mixing is that hazardous materials in the waste stream get dispersed through thousands of tons of miscellaneous garbage. This mixing makes the disposal or burning of what might have been rather innocuous stuff a difficult, expensive, and risky business. Spray paint cans, pesticides, batteries (zinc, lead, or mercury), cleaning solvents, smoke detectors containing radioactive material, and plastics that produce dioxins and PCBs when burned are mixed willy-nilly with paper, table scraps, and other nontoxic materials. The best thing to do with household toxic and hazardous materials is to separate them for safe disposal or recycling, as we will see later in this chapter.

WASTE DISPOSAL METHODS

Where are our wastes going now? In this section, we will examine some historic methods of waste disposal as well as some future options. Notice that our presentation begins with the least desirable—but most commonly used—measures and proceeds to discuss some preferable options. Keep in mind as you read this that modern waste management reverses this order and stresses the "three R's" of reduction, reuse, and recycling before destruction or, finally, secure storage of wastes.

Open Dumps

For many people, the way to dispose of waste is to simply drop it someplace. Open, unregulated dumps are still the predominant method of waste disposal in most developing countries. The giant Third World megacities have enormous garbage problems. Mexico City, the largest city in the world, generates some 10,000 tons of trash *each day.* Until recently, most of this torrent of waste was left in giant piles, exposed to the wind and rain, as well as rats, flies, and other vermin. Manila, in the Philippines, has at least ten huge

FIGURE 23.3 Scavengers sort through the trash at "Smoky Mountain," one of the huge metropolitan dumps in Manila, Philippines. Some 20,000 people live and work on these enormous garbage dumps. The health effects are tragic.

open dumps. The most notorious is called Smoky Mountain because of its constant smoldering fires (fig. 23.3). Thousands of people live and work on this 30-meter-high heap of refuse. They spend their days sorting through the garbage for edible or recyclable materials. Health conditions are abysmal, but these people have nowhere else to go. The government would like to close these dumps, but how will the residents be housed and fed? Where else will the city put its garbage?

Most developed countries forbid open dumping, at least in metropolitan areas, but illegal dumping is still a problem. You have undoubtedly seen trash accumulating along roadsides and in vacant, weedy lots in the poorer sections of cities. Is this just a question of aesthetics? Consider the problem of waste oil and solvents. An estimated 200 million liters of waste motor oil are poured into the sewers or allowed to soak into the ground every year in the United States. This is about five times as much as was spilled by the *Exxon Valdez* in Alaska in 1989! No one knows the volume of solvents and other chemicals disposed of by similar methods.

Increasingly, these toxic chemicals are showing up in the groundwater supplies on which nearly half the people in America depend for drinking (chapter 20). An alarmingly small amount of oil or other solvents can pollute large quantities of drinking or irrigation water. One liter of gasoline, for instance, can make a million liters of water undrinkable. The problem of illegal dumping is likely to become worse as acceptable sites for waste disposal become more scarce and costs for legal dumping escalate. We clearly need better enforcement of antilittering laws as well as a change in our attitudes and behavior.

Ocean Dumping

The oceans are vast, but not so large that we can continue to treat them as carelessly as has been our habit. Every year some 25,000 metric tons (55 million lbs) of packaging, including half a million bottles, cans, and plastic containers, are dumped at sea. Beaches, even in remote regions, are littered with the nondegradable flotsam and jetsam of industrial society (see fig. 20.17). About 150,000 tons (330 million lbs) of fishing gear—including more than 1000 km (660 mi) of nets—are lost or discarded at sea each year. Environmental groups estimate that 50,000 northern fur seals are entangled in this refuse and drown or starve to death every year in the North Pacific alone.

Until recently, many cities in the United States dumped municipal refuse, industrial waste, sewage, and sewage sludge in the ocean. Federal legislation now prohibits this dumping. New York City, the last to stop offshore sewage sludge disposal, finally ended this practice in 1992. Still, 60 million to 80 million cubic meters of dredge spoil—much of it highly contaminated—are disposed of at sea. Some people claim that the deep abyssal plain is the most remote, stable, and innocuous place to dump our wastes. Others argue that we know too little about the values of these remote places or the rare species that live there to smother them with sludge and debris.

Landfills

Over the past 50 years most American and European cities have recognized the health and environmental hazards of open dumps. Increasingly, cities have turned to **sanitary landfills,** where solid waste disposal is regulated and controlled. To decrease smells and litter and to discourage insect and rodent populations, landfill operators are required to compact the refuse and cover it every day with a layer of dirt (fig. 23.4). This method helps control pollution, but the dirt fill also takes up as much as 20 percent of landfill space. Since 1994, all operating landfills in the United States have been required to control such hazardous substances as oil, chemical compounds, toxic metals, and contaminated rainwater that seeps through piles of waste. An impermeable clay and/or plastic lining underlies and encloses the storage area. Drainage systems are

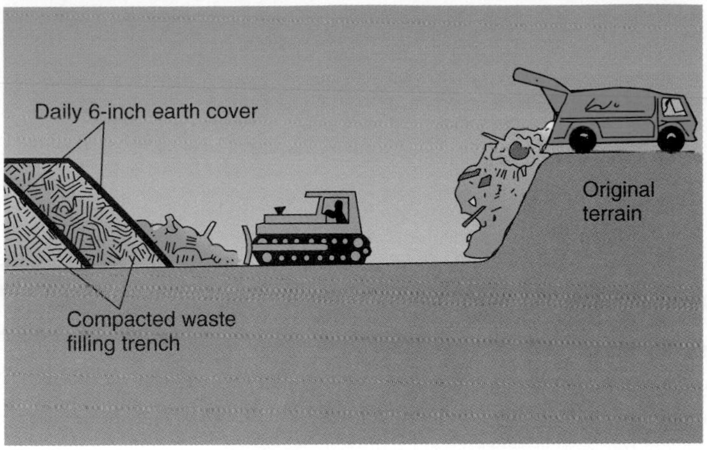

FIGURE 23.4 In a sanitary landfill, trash and garbage are crushed and covered each day to prevent accumulation of vermin and spread of disease. A waterproof lining is now required to prevent leaching of chemicals into underground aquifers.

installed in and around the liner to catch drainage and to help monitor chemicals that may be leaking. Modern municipal solid-waste landfills now have many of the safeguards of hazardous waste repositories described later in this chapter.

More careful attention is now paid to the siting of new landfills. Sites located on highly permeable or faulted rock formations are passed over in favor of sites with less leaky geologic foundations. Landfills are being built away from rivers, lakes, floodplains, and aquifer recharge zones rather than near them, as was often done in the past. More care is being given to a landfill's long-term effects so that costly cleanups and rehabilitation can be avoided.

Historically, landfills have been a convenient and relatively inexpensive waste-disposal option in most places, but this situation is changing rapidly. Rising land prices and shipping costs, as well as increasingly demanding landfill construction and maintenance requirements, are making this a more expensive disposal method. The cost of disposing a ton of solid waste in Philadelphia went from \$20 in 1980 to more than \$100 in 1990. Union County, New York, experienced an even steeper price rise. In 1987, it paid \$70 to get rid of a ton of waste; a year later, that same ton cost \$420, or about \$10 for a typical garbage bag. The United States now spends about \$10 billion per year to dispose of trash. A decade from now, it may cost us \$100 billion per year to dispose of our trash and garbage.

Suitable places for waste disposal are becoming scarce in many areas. Other uses compete for open space. Citizens have become more concerned and vocal about health hazards, as well as aesthetics. It is difficult to find a neighborhood or community willing to accept a new landfill. Since 1984, when stricter financial and environmental protection requirements for landfills took effect, more than 1200 of the 1500 existing landfills in the United States have closed. Many major cities are running out of local landfill space. They export their trash, at enormous expense, to neighboring communities and even other states. More than half the solid waste from New Jersey goes out of state, some of it up to 800 km (500 mi) away.

Exporting Waste

Although most industrialized nations in the world have agreed to stop shipping hazardous and toxic waste to less developed countries, the practice still continues. In 1999, for example, 3000 tons of incinerator waste from a plastics factory in Taiwan was unloaded from a ship in the middle of the night and dumped in a field near the small coastal Cambodian village of Bet Trang. The village residents thought they had been blessed with a windfall. They emptied out chunks of crumbling residue so they could use the white plastic shipping bags as bedding and roofing material. They rinsed out bags to use for rice storage, and they ripped them open with their teeth to get string to use as clotheslines and lashing for their oxcarts. Children played happily on the big pile of dusty, white material.

In the following weeks, unfortunately, the villagers discovered that rather than a treasure, they had a calamity. The first sign of trouble was when one of the dock workers who unloaded the waste died and five others were hospitalized with symptoms of nerve damage and respiratory distress. Villagers also began to complain of a variety of illnesses. The village was evacuated, and about 1000 residents of the nearby city of Sihanoukville fled in panic. Subsequent analysis found high levels of mercury and other toxic metals in the residue. The Formosa Plastics Corp., which shipped the waste, admitted paying a \$3 million bribe to Cambodian officials to permit its dumping. They said they couldn't dispose of it in Taiwan because of a threat of public protest. Following an international uproar, the plastics company agreed to go back and pick up the waste. But the villagers who handled the toxic wastes face an uncertain future. Is it safe to reinhabit their homes? Is it wise to have children? Will they suffer long-term health effects from exposure to this material?

As we will discuss later in this chapter, "garbage imperialism" also operates within richer countries as well. Poor neighborhoods and minority populations are much more likely than richer ones to be the recipients of dumps, waste incinerators, and other locally unwanted land uses (LULUs). In recent years, attention has turned, in the United States, to Indian reservations, which are exempt from some state and federal regulations concerning waste disposal. Virtually every tribe in America has been approached with schemes to store wastes on their reservation.

Another method of disposing of toxic wastes is to "recycle" them as asphalt or concrete filler for building highways. This is considered a beneficial use, but what happens to the toxins as the roadway is slowly worn away by traffic? Similar waste products are "land farmed" or sold as fertilizer and soil amendments. There are no safety standards for fertilizer composition because it's not intended for human consumption, but much of it is used on crops that humans will eat, or will be fed to livestock that are part of our food chain. For example, Florida has about a billion m^3 of phosphogypsum, a waste product of phosphate mining that producers want to market as a soil amendment. While it's true that phosphate is an essential plant nutrient, this particular product is also radioactive. In another case in Oregon, metal-rich dust and ash from steel mills is classified as hazardous waste when it leaves the mill. After being mixed with other minerals, however, it becomes fertilizer that will be spread on farm fields. Manufacturers are required to report the "active" ingredient—things like nitrogen, phosphorus, and phosphate—content of their product, but much can go unreported as "inert" matter.

Incineration and Resource Recovery

Landfilling is still the disposal method for the majority of municipal waste in the United States (fig. 23.5). Faced with growing piles of garbage and a lack of available landfills at any price, however, public officials are investigating other disposal methods. The method to which they frequently turn is burning. Another term commonly used for this technology is **energy recovery,** or waste-to-energy, because the heat derived from incinerated refuse is a useful resource. Burning garbage can produce steam used directly for heating buildings or generating electricity. Internationally, well over 1000 waste-to-energy plants in Brazil, Japan, and Western Europe generate much-needed energy while also reducing the amount that needs to be landfilled. In the United States, more than 110 waste incinerators burn 45,000 tons of garbage daily. Some of these are simple incinerators; others produce steam and/or electricity.

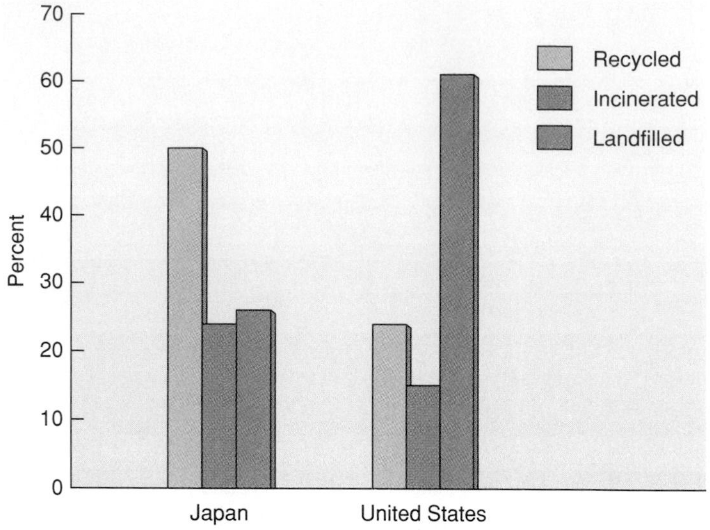

FIGURE 23.5 Percentages of municipal solid waste recycled, incinerated, and landfilled in Japan and the United States in 1995.

Types of Incinerators

Municipal incinerators are specially designed burning plants capable of burning thousands of tons of waste per day. In some plants, refuse is sorted as it comes in to remove unburnable or recyclable materials before combustion. This is called **refuse-derived fuel** because the enriched burnable fraction has a higher energy content than the raw trash. Another approach, called **mass burn,** is to dump everything smaller than sofas and refrigerators into a giant furnace and burn as much as possible (fig. 23.6). This technique avoids the expensive and unpleasant job of sorting through the garbage for nonburnable materials, but it often causes greater problems with air pollution and corrosion of burner grates and chimneys.

In either case, residual ash and unburnable residues representing 10 to 20 percent of the original volume are usually taken to a landfill for disposal. Because the volume of burned garbage is reduced by 80 to 90 percent, disposal is a smaller task. However, the residual ash usually contains a variety of toxic components that make it an environmental hazard if not disposed of properly. Ironically, one worry about incinerators is whether enough garbage will be available to feed them. Some communities in which recycling has been really successful have had to buy garbage from neighbors to meet contractual obligations to waste-to-energy facilities. In other places, fears that this might happen have discouraged recycling efforts.

Incinerator Cost and Safety

The cost-effectiveness of garbage incinerators is the subject of heated debates. Initial construction costs are high—usually between \$100 million and \$300 million for a typical municipal facility. Tipping fees at an incinerator, the fee charged to haulers for each ton of garbage dumped, are often much higher than those at a landfill. As landfill space near metropolitan areas becomes more scarce and more expensive, however, landfill rates are certain to rise. It may pay in the long run to incinerate refuse so that the lifetime of existing landfills will be extended.

Environmental safety of incinerators is another point of concern. The EPA has found alarmingly high levels of dioxins, furans, lead, and cadmium in incinerator ash. These toxic materials were more concentrated in the fly ash (lighter, airborne particles capable of penetrating deep into the lungs) than in heavy bottom ash. Dioxin levels can be as high as 780 parts per billion. One part per billion of TCDD, the most toxic dioxin, is considered a health concern. All of the incinerators studied exceeded cadmium standards, and 80 percent exceeded lead standards. Proponents of incineration argue that if they are run properly and equipped with appropriate pollution-control devices, incinerators are safe to the general public. Opponents counter that neither public officials nor pollution control equipment can be trusted to keep the air clean. They argue that recycling and source reduction efforts are better ways to deal with waste problems.

The EPA, which generally supports incineration, acknowledges the health threat of incinerator emissions but holds that the danger is very slight. The EPA estimates that dioxin emissions from a typical municipal incinerator may cause one death per million people in 70 years of operation. Critics of incineration claim that a more accurate estimate is 250 deaths per million in 70 years.

One way to reduce these dangerous emissions is to remove batteries containing heavy metals and plastics containing chlorine before wastes are burned. Bremen, West Germany, is one of several European cities now trying to control dioxin emissions by keeping all plastics out of incinerator waste. Bremen is requiring households to separate plastics from other garbage. This is expected to eliminate nearly all dioxins and other combustion by-products and prevent the expense of installing costly pollution-control equipment

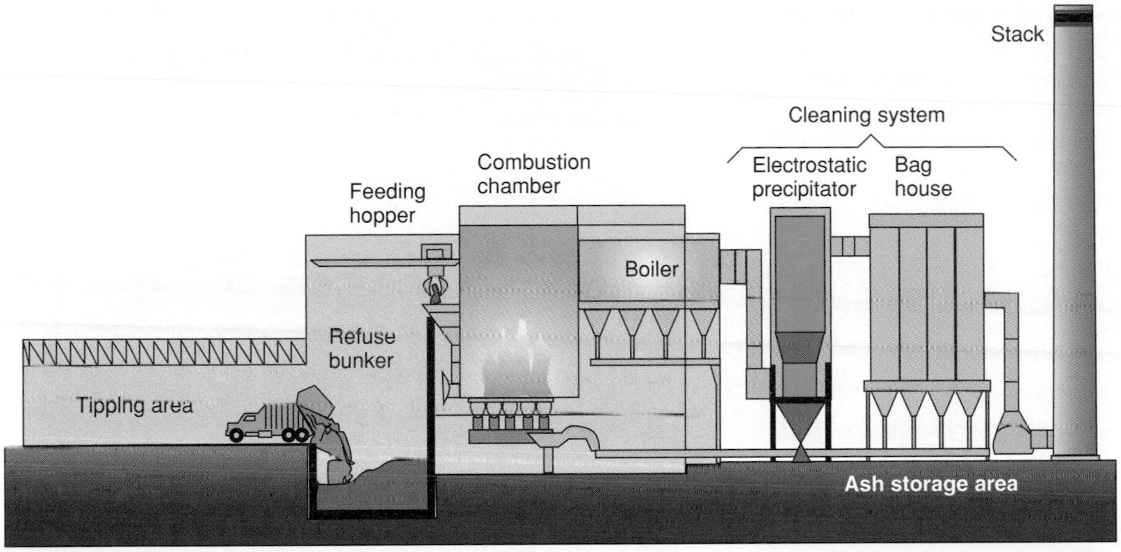

FIGURE 23.6 A diagram of a municipal "mass-burn" garbage incinerator. Steam produced in the boiler can be used to generate electricity or to heat nearby buildings.

that otherwise would be necessary to keep the burners operating. Minneapolis has initiated a recycling program for the small "button" batteries used in hearing aids, watches, and calculators in an attempt to lower mercury emissions from its incinerator.

SHRINKING THE WASTE STREAM

Having less waste to discard is obviously better than struggling with disposal methods, all of which have disadvantages and drawbacks. In this section we will explore some of our options for recycling, reuse, and reduction of the wastes we produce.

Recycling

The term *recycling* has two meanings in common usage. Sometimes we say we are *recycling* when we really are *reusing* something, such as refillable beverage containers. In terms of solid waste management, however, **recycling** is the reprocessing of discarded materials into new, useful products (fig. 23.7). Some recycling processes reuse materials for the same purposes; for instance, old aluminum cans and glass bottles are usually melted and recast into new cans and bottles. Other recycling processes turn old materials into entirely new products. Old tires, for instance, are shredded and turned into rubberized road surfacing. Newspapers become cellulose insulation, kitchen wastes become a valuable soil amendment, and steel cans become new automobiles and construction materials.

There have been some dramatic successes in recycling in recent years. Minnesota, for instance, now claims a 46-percent recycling rate, something thought unattainable a decade ago. The high value of aluminum scrap ($700 per ton in 1999) has spurred a large percentage of aluminum recycling nearly everywhere. About two-thirds of all aluminum cans are now recycled; up from only 15 percent 20 years ago. This recycling is so rapid that half of all the aluminum cans now on grocery shelves will be made into another can within two months. A big problem for recyclers is the wild fluctuation in market prices for commodities. Newsprint, for example, which peaked at $160 a ton in 1995, dropped to $42 per ton in 1999. One day, it's so valuable that people are stealing it off the curb; the next day it's literally down in the dumps. It's hard to build a recycling program when you can't count on a stable price for your product.

FIGURE 23.7 Trucks with multiple compartments pick up residential recyclables at curbside, greatly reducing the amount of waste that needs to be buried or burned. For many materials, however, collection costs are too high and markets are lacking for recycling to be profitable.

Another problem in recycling is contamination. Most of the 24 billion plastic soft drink bottles sold every year in the United States are made of PET (polyethylene terephthalate), which can be melted and remanufactured into carpet, fleece clothing, plastic-strapping, and nonfood packaging. However, even a smidgen of vinyl—a single PVC (polyvinyl chloride) bottle in a truckload, for example—can make PET useless. Although most bottles are now marked with a recycling number, it's hard for consumers to remember which is which. A looming worry is the prospect of single-use, plastic beer bottles. Already being test marketed, these bottles are made of PET but are amber colored to block sunlight and have a special chemical coating to keep out oxygen, which would ruin the beer. The special color, interior coating, and vinyl cap lining will make these bottles incompatible with regular PET, and it will probably cost more to remove them from the waste stream than the reclaimed plastic is worth. Plastic recycling already is down 50 percent from a decade ago because so many soft drink bottles are sold and consumed on the go, and never make it into recycling bins. Throw-away beer bottles may be the death knell for this industry.

Benefits of Recycling

Recycling is usually a better alternative to either dumping or burning wastes. It saves money, energy, raw materials, and land space, while also reducing pollution. Recycling also encourages individual awareness and responsibility for the refuse produced. Curbside pickup of recyclables costs around $35 per ton, as opposed to the $80 paid to dispose of them at an average metropolitan landfill. Many recycling programs cover their own expenses with materials sales and may even bring revenue to the community.

Another benefit of recycling is that it could cut our waste volumes drastically and reduce the pressure on disposal systems. Philadelphia is investing in neighborhood collection centers that will recycle 600 tons a day, enough to eliminate the need for a previously planned, high-priced incinerator. New York City, down to one available landfill but still producing 27,000 tons of garbage a day, has set a target of 50 percent waste reduction to be accomplished by recycling office paper and household and commercial waste. New York's curbside collection service, projected to be the nation's largest, should more than pay for itself simply in avoided disposal costs.

Japan probably has the most successful recycling program in the world (see fig. 23.5). Half of all household and commercial wastes in Japan are recycled while the rest is about equally incinerated or landfilled. By comparison, the United States landfills more than 60 percent of all solid waste. Japanese families diligently separate wastes into as many as seven categories, each picked up on a different day (fig. 23.8). Would we do the same? Some authors say that Americans are too lazy to recycle. North Stonington, Connecticut, however, faced with escalating disposal costs, reduced its waste volume by two-thirds in just two years. Overall, we already are making progress in removing material from the waste stream (fig. 23.9).

Recycling lowers our demands for raw resources. In the United States, we cut down 2 million trees every day to produce newsprint and paper products, a heavy drain on our forests. Recycling the print run of a single Sunday issue of the New York Times would spare 75,000 trees. Every piece of plastic we make reduces the reserves supply of petroleum and makes us more dependent on foreign oil. Recycling 1 ton of aluminum saves 4 tons of bauxite (aluminum ore) and 700 kg (1540 lb) of petroleum coke and pitch, as well as keeping 35 kg (77 lb) of aluminum fluoride out of the air.

Recycling also reduces energy consumption and air pollution. Plastic bottle recycling could save 50 to 60 percent of the energy needed to make new ones. Making new steel from old scrap offers up to 75 percent energy savings. Producing aluminum from scrap instead of bauxite ore cuts energy use by 95 percent, yet we still throw away more than a million tons of aluminum every year. If aluminum recovery were doubled worldwide, more than a million tons of air pollutants would be eliminated every year.

Reducing litter is an important benefit of recycling. Ever since disposable paper, glass, metal, foam, and plastic packaging began to accompany nearly everything we buy, these discarded wrappings have collected on our roadsides and in our lakes, rivers, and oceans. Without incentives to properly dispose of beverage

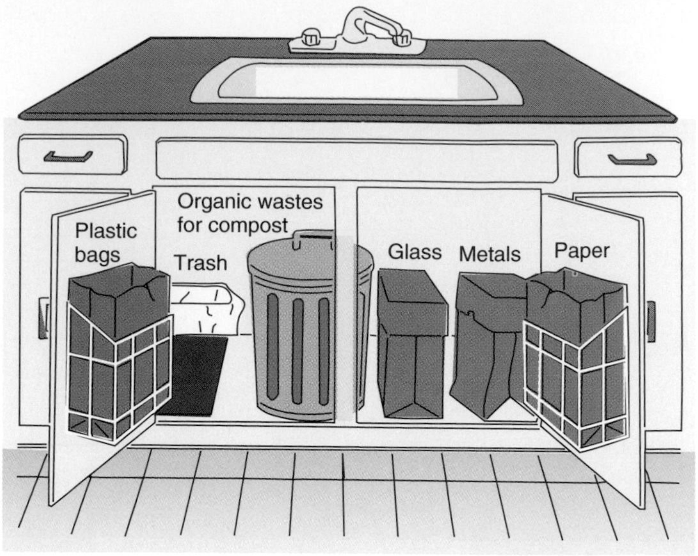

FIGURE 23.8 Source separation in the kitchen—the first step in a strong recycling program.

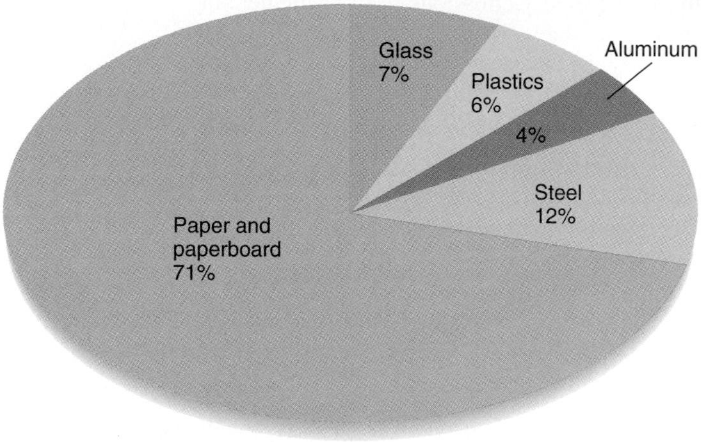

FIGURE 23.9 Percentage composition (by weight) of recycled materials in the United States in 1994. Some communities count composted yard waste—which accounts for about 18 percent of total waste stream—as recycled material.

Source: Data from Heather Hepler, "The Bulk of the Recycling Market" in *American City & County,* Vol. 109, July, 1994.

cans, bottles, and papers, it often seems easier to just toss them aside when we have finished using them. Litter is a costly as well as unsightly problem. We pay an estimated 32 cents for each piece of litter picked up by crews along state highways, which adds up to $500 million every year. "Bottle-bills" requiring deposits on bottles and cans have reduced littering in many states.

Creating Incentives for Recycling

In many communities, citizens have done such a good job of collecting recyclables that a glut has developed. Mountains of some waste materials accumulate in warehouses (fig. 23.10) because there are no markets for them. Too often wastes that we carefully separate for recycling end up being mixed together and dumped in a landfill or incinerator.

Our present public policies often tend to favor extraction of new raw materials. Energy, water, and raw materials are often sold to industries below their real cost to create jobs and stimulate the economy. For instance, in 1999, a pound of recycled clear PET, the material in most soft drink bottles, sold for about 40¢. By contrast, a pound of off-grade, virgin PET cost 25¢. Setting the prices of natural resources at their real cost would tend to encourage efficiency and recycling. State, local, and national statutes requiring government agencies to purchase a minimum amount of recycled material have helped create a market for used materials. Each of us can play a role in creating markets, as well. If we buy things made from recycled materials—or ask for them if they aren't available—we will help make it possible for recycling programs to succeed.

FIGURE 23.10 In some places, mountains of paper, plastic, and glass are accumulating in warehouses because collection programs have been more successful than material marketing efforts. Federal and state governments need to encourage use of recycled materials. Each of us can help by buying recycled products.

Composting

Pressed for landfill space, many cities have banned yard waste from municipal garbage. Rather than bury this valuable organic material, they are turning it into a useful product through **composting:** biological degradation or breakdown of organic matter under aerobic (oxygen-rich) conditions. The organic compost resulting from this process makes a nutrient-rich soil amendment that aids water retention, slows soil erosion, and improves crop yields. A home compost pile is an easy and inexpensive way to dispose of organic waste in an interesting and environmentally friendly way (In Depth, p. 534).

Large-scale municipal or industrial compost facilities have some of the same public relations and siting problems as incinerators and landfills. Neighbors complain of noise, odors, vermin, and increased traffic from poorly run facilities. These problems don't have to occur, but when you are dealing with huge amounts of garbage it's easy for the situation to get out of hand.

Energy from Waste

Every year we throw away the energy equivalent of 80 million barrels of oil in organic waste in the United States. In developing countries, up to 85 percent of the waste stream is food, textiles, vegetable matter, and other biodegradable materials. Worldwide, at least one-fifth of municipal waste is organic kitchen and garden refuse. In a landfill, much of this matter is decomposed by microorganisms generating billions of cubic meters of methane ("natural gas"), which contributes to global warming if allowed to escape into the atmosphere (chapter 17). Many cities are drilling methane wells in their landfills to capture this valuable resource. Fuel cells (chapter 22) are a good way to use this fuel.

This valuable organic material can be burned in an incinerator rather than being buried in landfills, but there are worries about air pollution from incineration. Organic wastes also can be decomposed in large, oxygen-free digesters to produce methane under more controlled conditions than in a landfill and with less air pollution than mass garbage burning.

Anaerobic digestion also can be done on a small scale. Millions of household methane generators provide fuel for cooking and lighting for homes in China and India (chapter 22). In the United States, some farmers produce all the fuel they need to run their farms—both for heating and to run trucks and tractors—by generating methane from animal manure.

Demanufacturing

Demanufacturing is the disassembly and recycling of obsolete consumer products, such as TV sets, personal computers, refrigerators, washing machines, and air conditioners. Together with deconstruction of houses, it is a good way to recover valuable materials. It also can be especially suited to inner cities where there is a large supply of materials to be demanufactured and a pool of skilled and unskilled laborers who need jobs. There are about 300 million TVs and personal computers in use in the Untied States. TVs often are discarded after only about five years, and computers, play-stations, and other electronics become obsolete even faster. Stoves, refrigerators, and other "white goods" have a much longer lifetime—typically about 12 years—but the EPA estimates that Americans dispose of 54 million of these household appliances every year. Many of these consumer products contain both valuable

IN DEPTH. . . . . .

Building a Home Compost Pile

Home composting is easy, beneficial, and educational. It takes very little effort or attention because millions of tiny microorganisms do almost all the work for you. All you have to do is supply them with a food supply, a little water, and oxygen.

To get started, you need a supply of organic material. Grass clippings, leaves, wood chips, vegetable peelings, sawdust, seaweed, manure, squash and tomato vines, watermelon rinds, coffee grounds, even hair, paper, and old rags are all good compost material. You can simply pile everything on the ground in an out-of-the-way place, or if you prefer a tidier operation, you can build a container or bin. Use boards, bricks, screen, an old barrel, or any other material that will hold the compost. Leave some spaces or holes in the side of your bin so that air can reach the lower layers—or turn the pile over frequently.

Layer material in the pile, alternating nitrogen-rich stuff such as grass, manure, and food scraps with carbon-rich substances such as cornstalks, dry leaves, straw, wood chips, and paper. A pile that is too high in carbon will take a long time to work. Too much nitrogen will produce an odor like urine or ammonia gas; it also will make the pile slimy and putrid. The ideal ratio of carbon to nitrogen is 30:1. Since you probably don't have the means to determine the composition of what you put in your pile, just use a variety of materials and don't overload with any one thing. If you have a very large amount of dry material—say leaves in the fall—you might want to add some extra manure or commercial fertilizer to give the compost a shot of nitrogen.

To function well, a compost pile shouldn't be either too big or too little. If the pile is too large, oxygen doesn't get into the middle. A pile that is too small doesn't retain enough heat for the microorganisms to grow optimally. One to two meters wide and a meter (3 ft) deep is about right. Given a good nutrient supply and plenty of air, bacteria and fungi growing in the compost pile will produce a temperature of about 70°C (160°F), enough to kill most pathogens and weed seeds. Turning the pile over frequently (every week or two) will mix the components and provide enough fresh air to keep the pile working well and will prevent the sour smell of anaerobic (oxygen-starved) fermentation.

The other essential ingredient for the microorganisms is water. If you live in a rainy climate or put lots of fresh vegetables and grass clippings in your compost pile, you won't need supplemental moisture, but if you use lots of dry leaves or live in a very dry place, you may need to add some water from time to time. The compost should be moist but not saturated. Too much water blocks oxygen penetration (that's why bogs are anaerobic and organic material doesn't decompose).

How long will your compost pile take to complete its work? That depends on temperature, moisture, texture, and contents. In the summer, a few weeks should produce a dark, soft, crumbly material that smells earthy and looks like rich potting soil. Composting will even work in the winter—but slowly if you live in a cold climate. Branches, stone fruit, and large chunks of material decay very slowly. Shredding, chipping, or chopping the starting material into small pieces will speed up the process.

There are some things that you shouldn't put into your compost pile. Don't add meat, fish, cheese, butter, milk, peanut butter, oil, grease, bones, skin or other animal products. They don't compost well, they tend to smell bad, and they attract pests and vermin. Don't add plants infected with disease or insects that could survive the compost pile's heat (such as scale insects, rusts, smuts, etc.). Avoid plants that take too long to break down such as pine needles, eucalyptus leaves or bark, or oak leaves. The tannins and acids they contain are natural bactericides and fungicides that inhibit composting. Cat and dog feces may contain harmful pathogens that can survive composting—especially if your pile doesn't reach optimal temperature. The compost produced could be infectious.

If you avoid these materials, keep pests and vermin out, and maintain a neat, well-aerated, odor-free compost pile, you shouldn't have any complaints from your neighbors. And after a few weeks, you will have a valuable soil amendment to spread on your yard or in your garden, or to use in potting house plants.

Composting is a good way to convert yard waste, vegetable scraps, and other organic materials into useful garden mulch. Mix everything together, keep it moist and well aerated, and in a few weeks, you will have a rich, odor-free mulch. This three-bin system allows you to have batches in different stages.

materials and toxins that must be kept out of the environment. Older refrigerators and air-conditioners, for example, have chloroflurocarbons (CFCs) that destroy stratospheric ozone and cannot be released into the air. Because new production has been banned, recycled CFCs are worth about $100 per pound. For both reasons, it pays to recycle them.

Similarly, computers and other electronic equipment contain both toxic metals (mercury, lead, gallium, germanium, nickel, palladium, beryllium, selenium, arsenic) as well as valuable ones such as gold, silver, and copper. A typical personal computer, for instance, has about $6 worth of gold, $5 of copper, and $1 of silver. It's estimated that 90 percent of the cadmium, lead, and mercury contamination in our solid waste stream comes from consumer electronics, batteries, mercury lamps, and switches. Small entrepreneurial firms are emerging in many urban centers to take advantage of this valuable resource. By forming alliances with other demanufacturing firms and large manufacturers, they can both develop the skills and experience to produce quality materials and also guarantee a market.

Reuse

Even better than recycling or composting is cleaning and reusing materials in their present form, thus saving the cost and energy of remaking them into something else. We do this already with some specialized items. Auto parts are regularly sold from junkyards, especially for older car models. In some areas, stained glass windows, brass fittings, fine woodwork, and bricks salvaged from old houses bring high prices. Some communities sort and reuse a variety of materials received in their dumps (fig. 23.11).

FIGURE 23.11 Reusing discarded products is a creative and efficient way to reduce wastes. This recycling center in Berkeley, CA, is a valuable source of used building supplies and a money saver for the whole community.

In many cities, glass and plastic bottles are routinely returned to beverage producers for washing and refilling. The reusable, refillable bottle is the most efficient beverage container we have. This is better for the environment than remelting and more profitable for local communities. A reusable glass container makes an average of 15 round-trips between factory and customer before it becomes so scratched and chipped that it has to be recycled. Reusable containers also favor local bottling companies and help preserve regional differences.

Since the advent of cheap, lightweight, disposable food and beverage containers, many small, local breweries, canneries, and bottling companies have been forced out of business by huge national conglomerates. These big companies can afford to ship food and beverages great distances as long as it is a one-way trip. If they had to collect their containers and reuse them, canning and bottling factories serving large regions would be uneconomical. Consequently, the national companies favor recycling rather than refilling because they prefer fewer, larger plants and don't want to be responsible for collecting and reusing containers. In some circumstances, life-cycle assessment shows that washing and decontaminating containers takes as much energy and produces as much air and water pollution as manufacturing new ones.

In less affluent nations, reuse of all sorts of manufactured goods is an established tradition. Where most manufactured products are expensive and labor is cheap, it pays to salvage, clean, and repair products. Cairo, Manila, Mexico City, and many other cities have large populations of poor people who make a living by scavenging. Entire ethnic populations may survive on scavenging, sorting, and reprocessing scraps from city dumps.

Producing Less Waste

What is even better than reusing materials? Generating less waste in the first place. The "What Can You Do?" box on page 536 describes some contributions you can make to reducing the volume of our waste stream. Industry also can play an important role in source reduction. In Minnesota, 3M Company saved over $500 million since 1975 by changing manufacturing processes, finding uses for waste products, and listening to employees' suggestions. What is waste to one division is a treasure to another.

Excess packaging of food and consumer products is one of our greatest sources of unnecessary waste. Paper, plastic, glass, and metal packaging material make up 50 percent of our domestic trash by volume. Much of that packaging is primarily for marketing and has little to do with product protection (fig. 23.12). Manufacturers and retailers might be persuaded to reduce these wasteful practices

what can you do?

Reducing Waste

1. Buy foods that come with less packaging; shop at farmers' markets or co-ops, using your own containers.
2. Take your own washable refillable beverage container to meetings or convenience stores.
3. When you have a choice at the grocery store between plastic, glass, or metal containers for the same food, buy the reusable or easier-to-recycle glass or metal.
4. When buying plastic products, pay a few cents extra for environmentally degradable varieties.
5. Separate your cans, bottles, papers, and plastics for recycling.
6. Wash and reuse bottles, aluminum foil, plastic bags, etc., for your personal use.
7. Compost yard and garden wastes, leaves, and grass clippings.
8. Write to your senators and representatives and urge them to vote for container deposits, recycling, and safe incinerators or landfills.

Source: Minnesota Pollution Control Agency.

if consumers ask for products without excess packaging. Canada's National Packaging Protocol (NPP) recommends that packaging minimize depletion of virgin resources and production of toxins in manufacturing. The preferred hierarchy is (1) no packaging, (2) minimal packaging, (3) reusable packaging, and (4) recyclable packaging. This plan sets a target of 50 percent reduction in excess packaging by December 31, 2000.

Where disposable packaging is necessary, we still can reduce the volume of waste in our landfills by using materials that are compostable or degradable. **Photodegradable plastics** break down when exposed to ultraviolet radiation. **Biodegradable plastics** incorporate such materials as cornstarch that can be decomposed by microorganisms. Several states have introduced legislation requiring biodegradable or photodegradable six-pack beverage yokes, fast-food packaging, and disposable diapers. These degradable plastics often don't decompose completely; they only break down to small particles that remain in the environment. In doing so, they can release toxic chemicals into the environment. And in modern, lined landfills they don't decompose at all. Furthermore, they make recycling less feasible and may lead people to believe that littering is OK.

Some environmental groups are beginning to think that we have put too much emphasis on recycling. Many people think that if they recycle aluminum cans and newspapers they are doing everything they can for the environment. In surveys conducted by the New Jersey-based Environmental Research Associates, only two in ten people surveyed could define waste reduction. Companies that make "throw away" or heavily packaged products generally favor recycling because it allows them to continue business-as-usual. While recycling is an important part of waste management, we have to remember that it is actually the third "R" in the waste hierarchy. The two preferred methods—reduction and reuse—get lost in our enthusiasm for recycling.

FIGURE 23.12 How much more do we need? Where will we put what we already have?
Reprinted with special permission of King Features Syndicate.

HAZARDOUS AND TOXIC WASTES

The most dangerous aspect of the waste stream we have described is that it often contains highly toxic and hazardous materials that are injurious to both human health and environmental quality. We now produce and use a vast array of flammable, explosive, caustic, acidic, and highly toxic chemical substances for industrial, agricultural, and domestic purposes (fig. 23.13). According to the EPA, industries in the United States generate about 265 million metric tons of *officially* classified hazardous wastes each year, slightly more than 1 ton for each person in the country. In addition, considerably more toxic and hazardous waste material is generated by industries or processes not regulated by the EPA. Shockingly, at least 40 million metric tons (22 billion lbs) of toxic and hazardous wastes are released into the air, water, and land in the United States each year. The biggest source of these toxins are the chemical and petroleum industries (fig. 23.14).

What Is Hazardous Waste?

Legally, a **hazardous waste** is any discarded material, liquid or solid, that contains substances known to be (1) fatal to humans or laboratory animals in low doses, (2) toxic, carcinogenic, mutagenic, or teratogenic to humans or other life-forms, (3) ignitable with a flash point less than 60°C, (4) corrosive, or (5) explosive or highly reactive (undergoes violent chemical reactions either by itself or when mixed with other materials). Notice that this definition

FIGURE 23.13 According to the U.S. Environmental Protection Agency, industries produce about one ton of hazardous waste per year for every person in the United States. Responsible handling and disposal is essential.

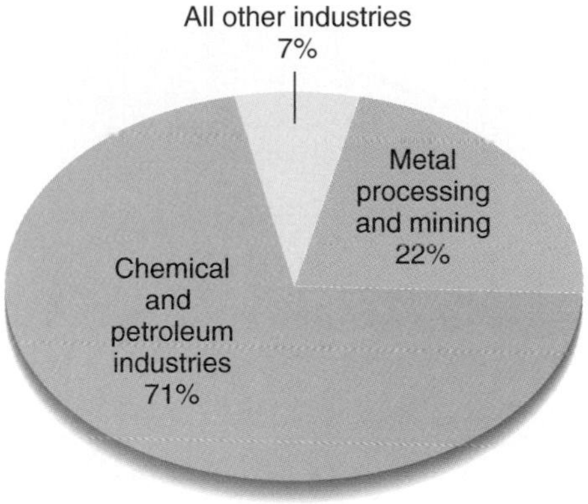

FIGURE 23.14 Producers of hazardous wastes in the United States.

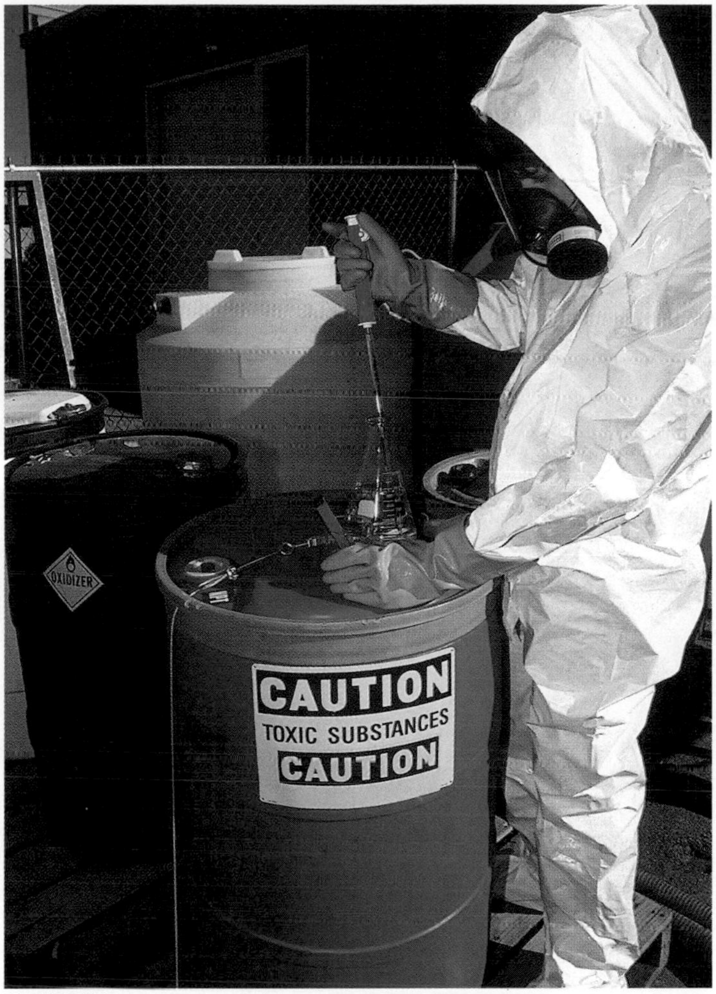

FIGURE 23.15 Careful handling, sampling, and record keeping for all toxic and hazardous wastes are now required by federal and state laws.

includes both toxic and hazardous materials as defined in chapter 9. Certain compounds are exempt from regulation as hazardous waste if they are accumulated in less than 1 kg (2.2 lb) of commercial chemicals or 100 kg of contaminated soil, water, or debris. Even larger amounts (up to 1,000 kg) are exempt when stored at an approved waste treatment facility for the purpose of being beneficially used, recycled, reclaimed, detoxified, or destroyed.

Hazardous Waste Disposal

Most hazardous waste is recycled, converted to nonhazardous forms, stored, or otherwise disposed of on-site by the generators—chemical companies, petroleum refiners, and other large industrial facilities—so that it doesn't become a public problem. Still, the hazardous waste that does enter the waste stream or the environment represents a serious environmental problem. And orphan wastes left behind by abandoned industries remain a serious threat to both environmental quality and human health. For years, little attention was paid to this material. Wastes stored on private property, buried, or allowed to soak into the ground were considered of little concern to the public. An estimated 5 billion metric tons of highly poisonous chemicals were improperly disposed of in the United States between 1950 and 1975 before regulatory controls became more stringent.

Federal Legislation

Two important federal laws regulate hazardous waste management and disposal in the United States. The Resource Conservation and Recovery Act (RCRA, pronounced "rickra") of 1976 is a comprehensive program that requires rigorous testing and management of toxic and hazardous substances (fig. 23.15). A complex set of rules require generators, shippers, users, and disposers of these materials to keep meticulous account of everything they handle and what happens to it from generation (cradle) to ultimate disposal (grave) (fig. 23.16).

The Comprehensive Environmental Response, Compensation and Liability Act (CERCLA or Superfund Act), passed in 1980 and modified in 1984 by the Superfund Amendments and Reauthorization Act (SARA), is aimed at rapid containment, cleanup, or remediation of abandoned toxic waste sites. This statute authorizes the Environmental Protection Agency to undertake emergency

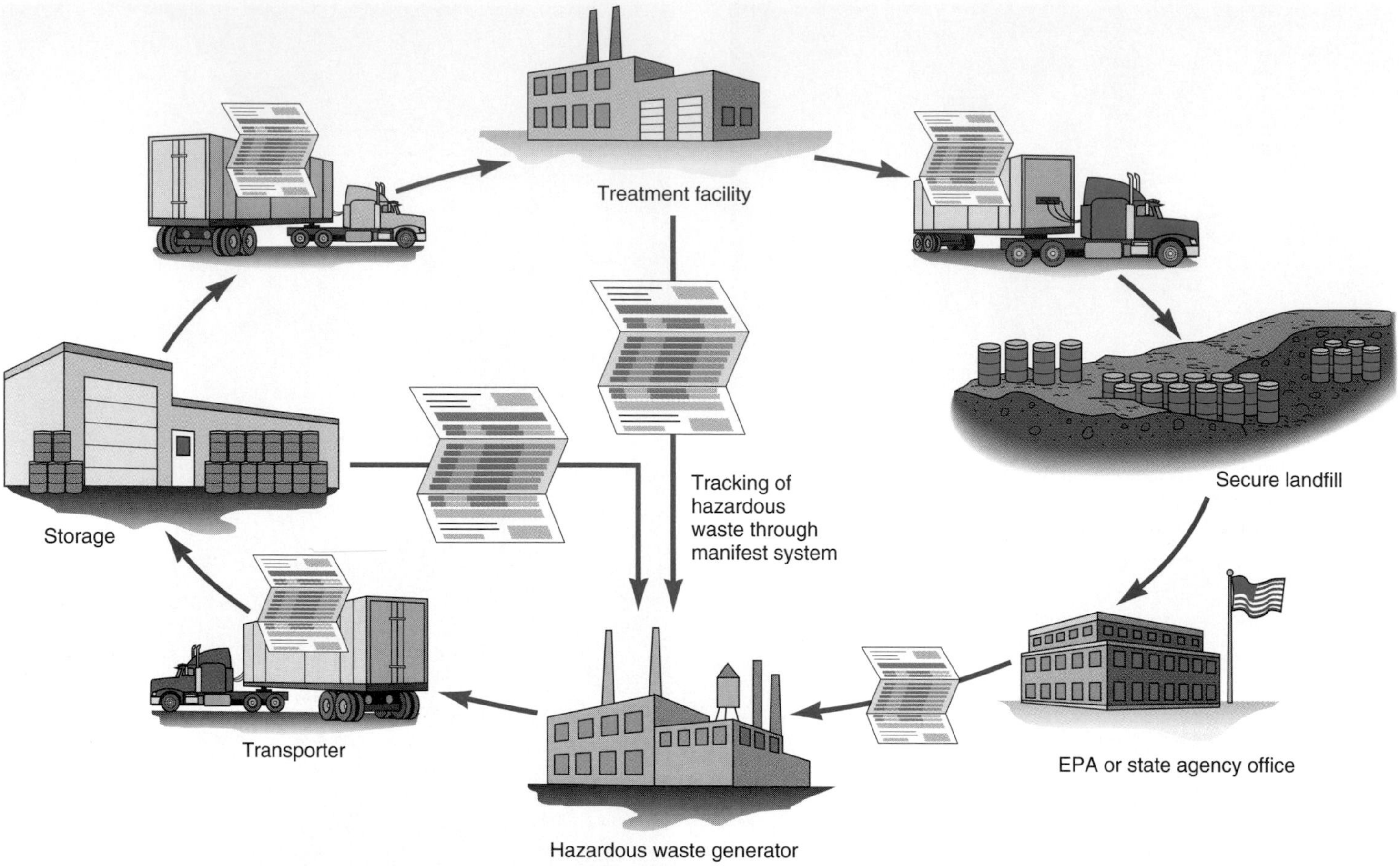

FIGURE 23.16 Toxic and hazardous wastes must be tracked from "cradle to grave" by detailed shipping manifests.

actions when a threat exists that toxic material will leak into the environment. The agency is empowered to bring suit for the recovery of its costs from potentially responsible parties such as site owners, operators, waste generators, or transporters.

SARA also established (under title III) community right to know and state emergency response plans that give citizens access to information about what is present in their communities. One of the most useful tools in this respect is the **Toxic Release Inventory,** which requires 20,000 manufacturing facilities to report annually on releases of more than 300 toxic materials. You can find specific information there about what is in your neighborhood.

The government does not have to prove that anyone violated a law or what role they played in a superfund site. Rather, liability under CERCLA is "strict, joint, and several," meaning that anyone associated with a site can be held responsible for the entire cost of cleaning it up no matter how much of the mess they made. In some cases, property owners have been assessed millions of dollars for removal of wastes left there years earlier by previous owners. This strict liability has been a headache for the real estate and insurance businesses.

CERCLA was amended in 1995 to make some of its provisions less onerous. In cases where treatment is unavailable or too costly and it is likely that a less-costly remedy will become available within a reasonable time, interim containment is now allowed. The EPA also now has the discretion to set site-specific cleanup levels rather than adhere to rigid national standards.

Superfund Sites

The EPA estimates that there are at least 36,000 seriously contaminated sites in the United States. The General Accounting Office (GAO) places the number much higher, perhaps more than 400,000 when all are identified. By 1997, some 1400 sites had been placed on the National Priority List (NPL) for cleanup with financing from the federal Superfund program. The **Superfund** is a revolving pool designed (1) to provide an immediate response to emergency situations that pose imminent hazards, and (2) to clean up or remediate abandoned or inactive sites. Originally a $1.6 billion pool, the superfund has since been enlarged tenfold. It is financed by fees assessed against hazardous or toxic waste producers and costs charged for remedial actions. By using the superfund to pay for immediate actions and then collecting later from responsible parties, we can take care of problems without waiting for litigation to run its course.

Total costs for hazardous waste cleanup in the United States are estimated to be between $370 billion and $1.7 trillion, depending on the remedial actions chosen and the degree to which sites are required to be cleaned. So far, about $38 billion in private and

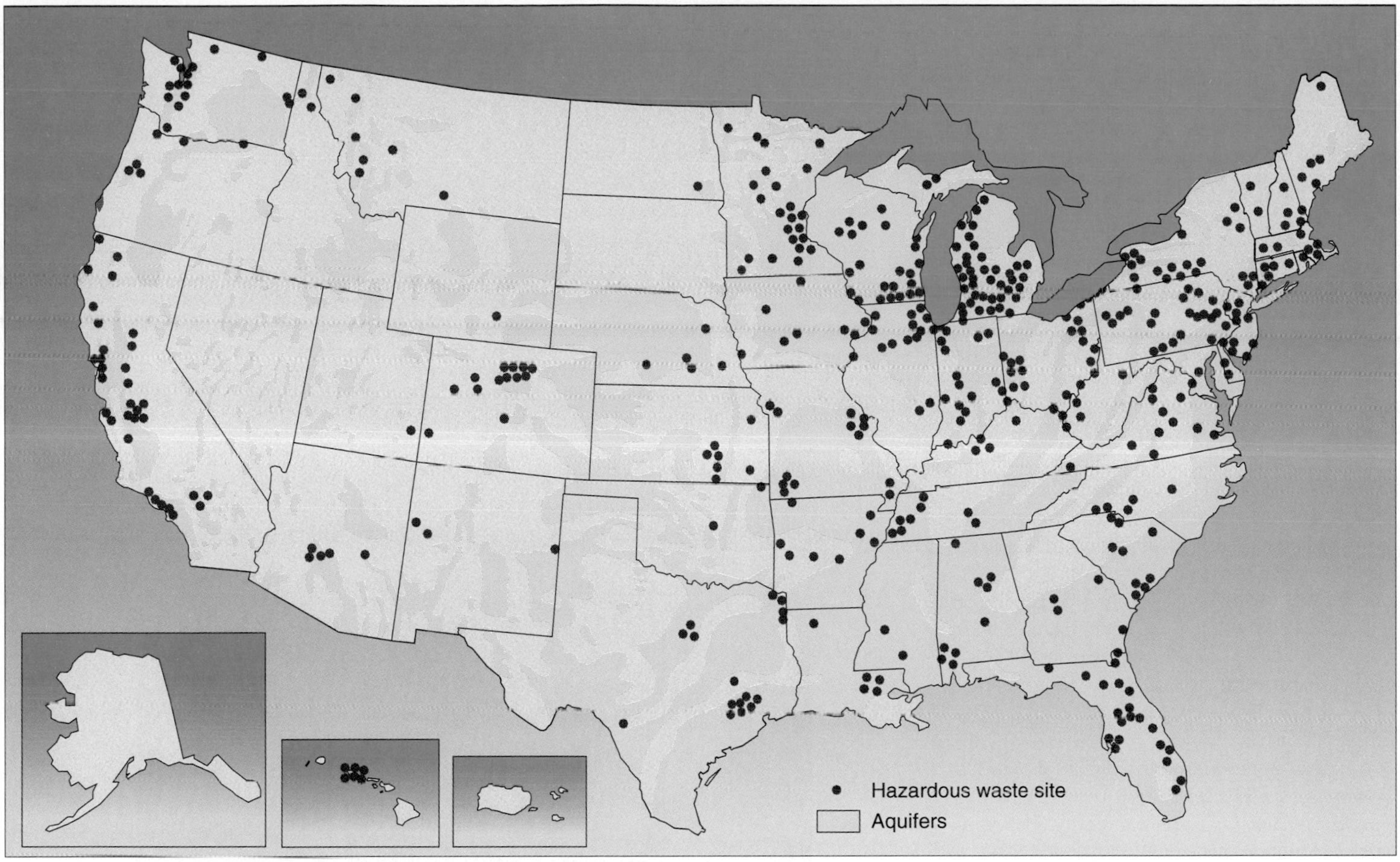

FIGURE 23.17 Some of the hazardous waste sites on the EPA priority cleanup list. Sites located on aquifer recharge zones represent an especially serious threat. Once groundwater is contaminated, cleanup is difficult and expensive. In some cases, it may not be possible.
Source: Data from the Environmental Protection Agency.

public funds has been spent on Superfund site remediation but much of this money has gone to consultants and lawyers. According to the GAO, only about 100 of the 1400 sites on the NPL have been completely rehabilitated. Many others are bogged down in disputes over who was liable or how best to carry out cleanup.

What qualifies a site for the NPL? These sites are considered to be especially hazardous to human health and environmental quality because they are known to be leaking or have a potential for leaking supertoxic, carcinogenic, teratogenic, or mutagenic materials (chapter 9). The ten substances of greatest concern or most commonly detected at Superfund sites are lead, trichloroethylene, toluene, benzene, PCBs, chloroform, phenol, arsenic, cadmium, and chromium. These and other hazardous or toxic materials are known to have contaminated groundwater at 75 percent of the sites now on the NPL. In addition, 56 percent of these sites have contaminated surface waters, and airborne materials are found at 20 percent of the sites.

Where are these thousands of hazardous waste sites, and how did they get contaminated? Old industrial facilities such as smelters, mills, petroleum refineries, and chemical manufacturing plants are highly likely to have been sources of toxic wastes. Regions of the country with high concentrations of aging factories such as the "rust belt" around the Great Lakes or the Gulf Coast petrochemical centers have large numbers of Superfund sites (fig. 23.17). Mining districts also are prime sources of toxic and hazardous waste. Within cities, factories and places such as railroad yards, bus repair barns, and filling stations where solvents, gasoline, oil, and other petrochemicals were spilled or dumped on the ground often are highly contaminated.

Some of the most infamous toxic waste sites were old dumps where many different materials were mixed together indiscriminately. For instance, Love Canal in Niagara Falls, New York, was an open dump used by both the city and nearby chemical factories as a disposal site. More than 20,000 tons of toxic chemical waste was buried under what later became a housing development. Another infamous example occurred in Hardeman County, Tennessee, where about a quarter of a million barrels of chemical waste were buried in shallow pits that subsequently leaked toxins into the groundwater. In other sites, liquid wastes were pumped into open lagoons or abandoned in warehouses.

Studies of who lives closest to Superfund and toxic release inventory sites reveal that minorities often are overrepresented in these neighborhoods. Charges of environmental racism have been made, but this is difficult to show conclusively (What Do You Think? p. 540).

what do you think?

THE WORLD AROUND YOU

Environmental Justice

Who do you suppose lives closest to toxic waste dumps, Superfund sites, or other polluted areas in your city or county? If you answered poor people and minorities, you are probably right. Everyday experiences tell us that minority neighborhoods are much more likely to have high pollution levels and unpopular industrial facilities such as toxic waste dumps, landfills, smelters, refineries, and incinerators than are middle- or upper-class, white neighborhoods.

One of the first systematic studies showing this inequitable distribution of environmental hazards based on race in the United States was conducted by Robert D. Bullard in 1978. Asked for help by a predominantly black community in Houston that was slated for a waste incinerator, Bullard discovered that all five of the city's existing landfills and six of eight incinerators were located in African-American neighborhoods. In a book entitled ***Dumping on Dixie,*** Bullard showed that this pattern of risk exposure in minority communities is common throughout the United States.

In 1987, the Commission for Racial Justice of the United Church of Christ published an extensive study of environmental racism. Its conclusion was that race is the most significant variable in determining the location of toxic waste sites in the United States. Among the findings of this study are:

- 3 of the 5 largest commercial hazardous waste landfills accounting for about 40 percent of all hazardous waste disposal in the United States are located in predominantly black or Hispanic communities.
- 60 percent of African-Americans and Latinos and nearly half of all Asians, Pacific Islanders, and Native Americans live in communities with uncontrolled toxic waste sites.
- The average percentage of the population made up by minorities in communities without a hazardous waste facility is 12 percent. By contrast, communities with one hazardous waste facility have, on average, twice as high (24 percent) a minority population, while those with two or more such facilities average three times as high a minority population (38 percent) as those without one.
- The "dirtiest" or most polluted zip codes in California are in riot-torn South Central Los Angeles where the population is predominantly African-American or Latino. Three-quarters of all blacks and half of all Hispanics in Los Angeles live in these polluted areas, while only one-third of all whites live there.

Race, not class or income, is claimed to be the strongest determinant of who is exposed to environmental hazards. While poor people in general are more likely to live in polluted

Native Americans march in protest of toxic waste dumping on tribal lands.

How Clean Is Clean?

Among the biggest problems in cleaning up hazardous waste sites are questions of liability and the degree of purity required. In many cities, these problems have created large areas of contaminated properties known as **brownfields** that have been abandoned or are not being used up to their potential because of real or suspected pollution. Up to one-third of all commercial and industrial sites in the urban core of many big cities fall in this category. In heavy industrial corridors the percentage typically is higher.

For years, no one was interested in redeveloping brownfields because of liability risks. Who would buy a property knowing that they might be forced to spend years in litigation and negotiations and be forced to pay millions of dollars for pollution they didn't create? Even if a site has been cleaned to current standards, there is a worry that additional pollution might be found in the future or that more stringent standards might be applied.

In many cases, property owners complain that unreasonably high levels of purity are demanded in remediation programs. Consider the case of Columbia, Mississippi. For many years a 35 ha (81 acre) site in Columbia was used for turpentine and pine tar manufacturing. Soil tests showed concentrations of phenols and other toxic organic compounds exceeding federal safety standards. The site was added to the Superfund NPL and remediation was ordered. Some experts recommended that the best solution was to simply cover the surface with clean soil and enclose the property with a fence to keep people out. The total costs would have been about $1 million.

Instead, the EPA ordered Reichhold Chemical, the last known property owner, to excavate more than 12,500 tons of soil neighborhoods than rich people, the discrepancy between the pollution exposure of middle class blacks and middle class whites is even greater than the difference between poorer whites and blacks. Where upper class whites can "vote with their feet" and move out of polluted and dangerous neighborhoods, minorities are restricted by color barriers and prejudice to less desirable locations.

Racial inequities also are revealed in the way the government cleans up toxic waste sites and punishes polluters. White communities see faster responses and get better results once toxic wastes are discovered than do minority communities. For instance, the federal Environmental Protection Agency takes 20 percent longer to place a hazardous waste site in a minority community on the Superfund National Priority List than it does for one in a white community. Penalties assessed against polluters of white communities average six times higher than those against polluters of minority communities. Cleanup is more thorough in white communities as well. Most toxic wastes in white communities are treated, that is, removed or destroyed. By contrast, waste sites in minority neighborhoods are generally only "contained" by putting a cap over them, leaving contaminants in place to potentially resurface or leak into groundwater at a later date. The growing environmental justice movement works to combine civil rights and social justice with environmental concerns to call for a decent, livable environment and equal environmental protection for everyone.

Ethical Considerations

What are the ethical considerations in waste disposal? Does everyone have a right to live in a clean environment or only a right to buy one if they can afford it? What would be a fair way to distribute the risks of toxic wastes? If you had to choose between an incinerator, a secure landfill, or a composting facility for your neighborhood, which would you take?

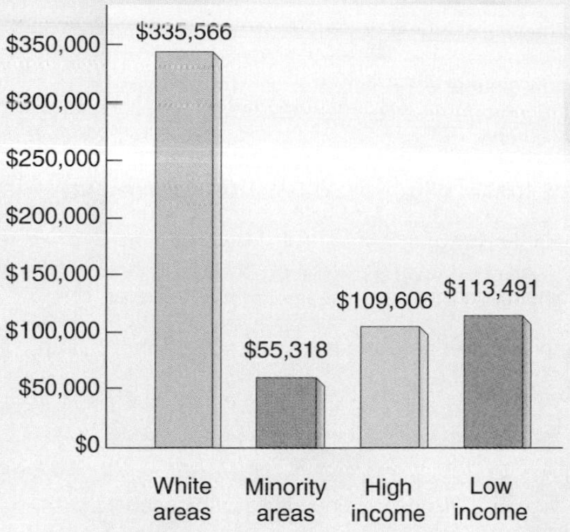

Hazardous waste law enforcement. The average fines or penalties imposed per site *for violation of the Resource Conservation and Recovery Act vary dramatically with racial composition—but not wealth—of the communities where waste was dumped.*

Source: M. LaVelle and M. Coyle, *The National Law Journal,* Vol. 15:52–56, No. 3, Sept. 21, 1992.

and haul it to a commercial hazardous waste dump in Louisiana at a cost of some $4 million. The intention is to make the site safe enough to be used for any purpose, including housing—even though no one has proposed building anything there. According to the EPA, the dirt must be clean enough for children to play in it—even eat it every day for 70 years—without risk.

Similarly, in places where contaminants have seeped into groundwater, the EPA generally demands that cleanup be carried to drinking water standards. Many critics believe that these pristine standards are unreasonable. Former Congressman, Jim Florio, a principal author of the original Superfund Act says, "It doesn't make any sense to clean up a rail yard in downtown Newark so it can be used as a drinking water reservoir." Depending on where the site is, what else is around it, and what its intended uses are, much less stringent standards may be perfectly acceptable.

Recognizing that reusing contaminated properties can play a significant role in rebuilding old cities, creating jobs, increasing the tax base, and preventing needless destruction of open space at urban margins, programs have been established at both federal and state levels to encourage brownfield recycling. Adjusting purity standards according to planned uses and providing liability protection for nonresponsible parties gives developers and future purchasers confidence that they won't be unpleasantly surprised in the future with further cleanup costs. In some communities, former brownfields are being turned into "eco-industrial parks" that feature environmentally friendly businesses and bring in much needed jobs to inner city neighborhoods.

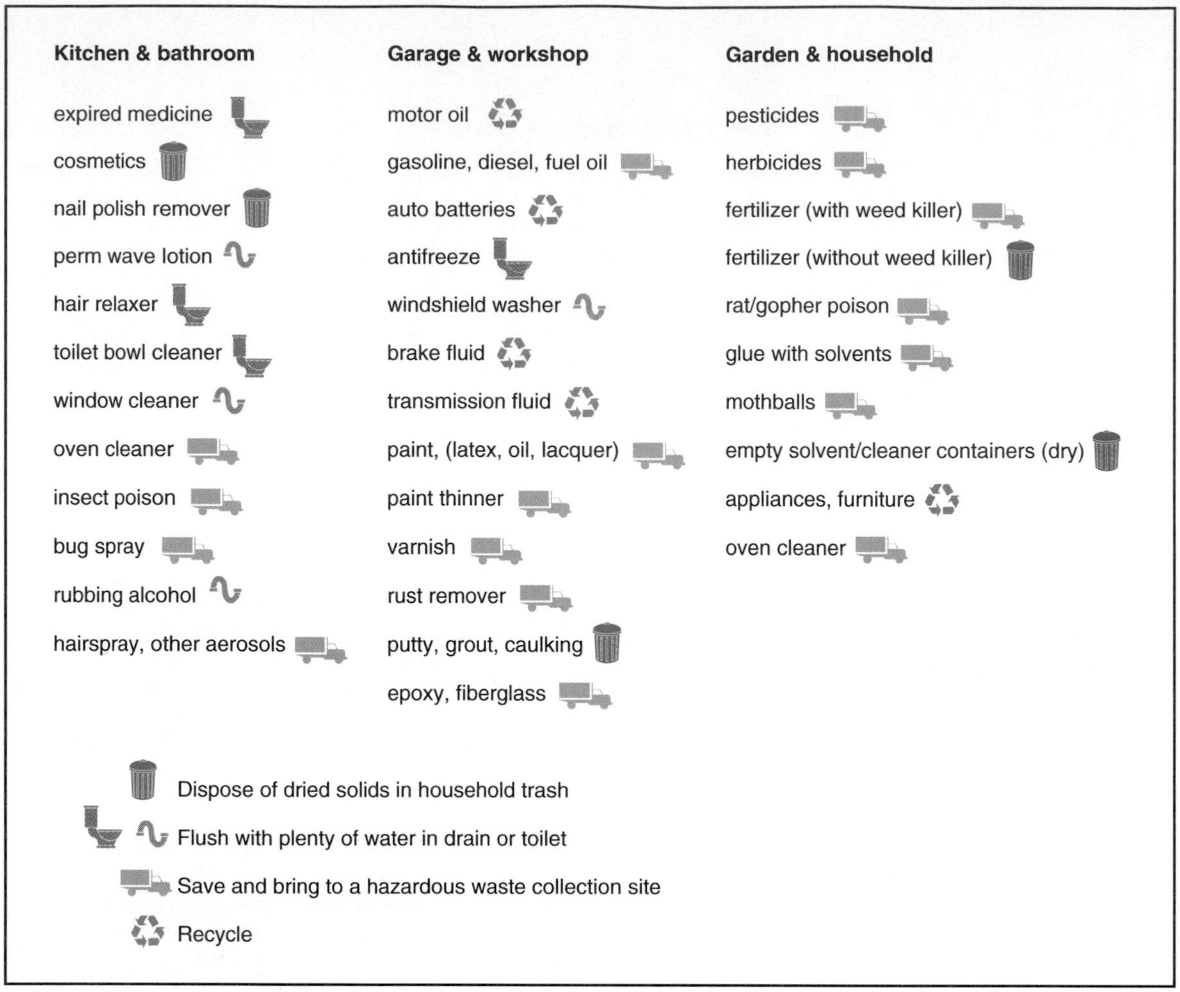

FIGURE 23.18 Household waste disposal guide. Solvent-containing products have the words "flammable," "combustible," or "petroleum distillates" on the labels. These materials should *not* be disposed of in a drain or toilet. Never mix products containing bleach with those containing ammonia. A toxic gas can form!

Options for Hazardous Waste Management

What shall we do with toxic and hazardous wastes? In our homes, we can reduce waste generation and choose less toxic materials. Buy only what you need for the job at hand. Use up the last little bit or share leftovers with a friend or neighbor. Many common materials that you probably already have make excellent alternatives to commercial products (What Can You Do?, p. 543). Dispose of unneeded materials responsibly (fig. 23.18).

Produce Less Waste

As with other wastes, the safest and least expensive way to avoid hazardous waste problems is to avoid creating the wastes in the first place. Manufacturing processes can be modified to reduce or eliminate waste production. In Minnesota, the 3M Company reformulated products and redesigned manufacturing processes to eliminate more than 140,000 metric tons of solid and hazardous wastes, 4 billion l (1 billion gal) of wastewater, and 80,000 metric tons of air pollution each year. They frequently found that these new processes not only spared the environment but also saved money by using less energy and fewer raw materials.

Recycling and reusing materials also eliminates hazardous wastes and pollution. Many waste products of one process or industry are valuable commodities in another. Already, about 10 percent of the wastes that would otherwise enter the waste stream in the United States are sent to surplus material exchanges where they are sold as raw materials for use by other industries. This figure could probably be raised substantially with better waste management. In Europe, at least one-third of all industrial wastes are exchanged through clearinghouses where beneficial uses are found. This represents a double savings: the generator doesn't have to pay for disposal, and the recipient pays little, if anything, for raw materials.

what can you do?

Alternatives to Hazardous Household Chemicals

Chrome cleaner: Use vinegar and nonmetallic scouring pad.

Copper cleaner: Rub with lemon juice and salt mixture.

Floor cleaner: Mop linoleum floors with 1 cup vinegar mixed with 2 gallons of water. Polish with club soda.

Brass polish: Use Worcestershire sauce.

Silver polish: Rub with toothpaste on a soft cloth.

Furniture polish: Rub in olive, almond, or lemon oil.

Ceramic tile cleaner: Mix 1/4 cup baking soda, 1/2 cup white vinegar, and 1 cup ammonia in 1 gallon warm water (good general purpose cleaner).

Drain opener: Use plunger or plumber's snake, pour boiling water down drain.

Upholstery cleaner: Clean stains with club soda.

Carpet shampoo: Mix 1/2 cup liquid detergent in 1 pint hot water. Whip into stiff foam with mixer. Apply to carpet with damp sponge. Rinse with 1 cup vinegar in 1 gal water. Don't soak carpet—it may mildew.

Window cleaner: Mix 1/3 cup ammonia, 1/4 cup white vinegar in 1 quart warm water. Spray on window. Wipe with soft cloth.

Spot remover: For butter, coffee, gravy, or chocolate stains: Sponge up or scrape off as much as possible immediately. Dab with cloth dampened with a solution of 1 teaspoon white vinegar in 1 quart cold water.

Toilet cleaner: Pour 1/2 cup liquid chlorine bleach into toilet bowl. Let stand for 30 minutes, scrub with brush, flush.

Pest control: Spray plants with soap-and-water solution (3 tablespoons soap per gallon water) for aphids, mealybugs, mites, and whiteflies. Interplant with pest repellent plants such as marigolds, coriander, thyme, yarrow, rue, and tansy. Introduce natural predators such as ladybugs or lacewings.

Indoor pests: Grind or blend 1 garlic clove and 1 onion. Add 1 tablespoon cayenne pepper and 1 quart water. Add 1 tablespoon liquid soap.

Moths: Use cedar chips or bay leaves.

Ants: Find where they are entering house, spread cream of tartar, cinnamon, red chili pepper, or perfume to block trail.

Fleas: Vacuum area, mix brewer's yeast with pet food.

Mosquitoes: Brewer's yeast tablets taken daily repel mosquitoes.

Note: test cleaners in small, inconspicuous area before using.

FIGURE 23.19 Actor Martin Sheen joins local activists in a protest in East Liverpool, OH, site of the largest hazardous waste incinerator in the United States. About 1000 people marched to the plant to pray, sing, and express their opposition. Involving celebrities draws attention to your cause. A peaceful, well-planned rally builds support and acceptance in the broader community.

Convert to Less Hazardous Substances

Several processes are available to make hazardous materials less toxic. *Physical treatments* tie up or isolate substances. Charcoal or resin filters absorb toxins. Distillation separates hazardous components from aqueous solutions. Precipitation and immobilization in ceramics, glass, or cement isolate toxins from the environment so that they become essentially nonhazardous. One of the few ways to dispose of metals and radioactive substances is to fuse them in silica at high temperatures to make a stable, impermeable glass that is suitable for long-term storage.

Incineration is applicable to mixtures of wastes. A permanent solution to many problems, it is quick and relatively easy but not necessarily cheap—nor always clean—unless it is done correctly. Wastes must be heated to over 1000°C (2000°F) for a sufficient period of time to complete destruction. The ash resulting from thorough incineration is reduced in volume up to 90 percent and often is safer to store in a landfill or other disposal site than the original wastes. Nevertheless, incineration remains a highly controversial topic (fig. 23.19).

Several sophisticated features of modern incinerators improve their effectiveness. Liquid injection nozzles atomize liquids and mix air into the wastes so they burn thoroughly. Fluidized bed burners pump air from the bottom up through burning solid waste as it travels on a metal chain grate through the furnace. The air velocity is sufficient to keep the burning waste partially suspended. Plenty of oxygen is available, and burning is quick and complete. Afterburners add to the completeness of burning by igniting gaseous hydrocarbons not consumed in the incinerator. Scrubbers and precipitators remove minerals, particulates, and other pollutants from the stack gases.

Chemical processing can transform materials so they become nontoxic. Included in this category are neutralization, removal of metals or halogens (chlorine, bromine, etc.), and oxidation. The Sunohio Corporation of Canton, Ohio, for instance, has developed a process called PCBx in which chlorine in such molecules as PCBs is replaced with other ions that render the compounds less toxic. A portable unit can be moved to the location of the hazardous waste, eliminating the need for shipping them.

Biological waste treatment or **bioremediation** taps the great capacity of microorganisms to absorb, accumulate, and detoxify a variety of toxic compounds. Bacteria in activated sludge basins, aquatic plants (such as water hyacinths or cattails), soil microorganisms, and other species remove toxic materials and purify effluents. Biotechnology offers exciting possibilities for finding or creating organisms to eliminate specific kinds of hazardous or toxic wastes. By using a combination of classic genetic selection techniques and high-technology gene-transfer techniques, for instance, scientists have recently been able to generate bacterial strains that are highly successful at metabolizing PCBs. There are concerns about releasing such exotic organisms into the environment, however (chapter 13). It may be better to keep these organisms contained in enclosed reaction vessels and feed contaminated material to them under controlled conditions.

Store Permanently

Inevitably, there will be some materials that we can't destroy, make into something else, or otherwise cause to vanish. We will have to store them out of harm's way. There are differing opinions about how best to do this.

Retrievable Storage. Dumping wastes in the ocean or burying them in the ground generally means that we have lost control of them. If we learn later that our disposal technique was a mistake, it is difficult, if not impossible, to go back and recover the wastes. For many supertoxic materials, the best way to store them may be in **permanent retrievable storage.** This means placing waste storage containers in a secure building, salt mine, or bedrock cavern where they can be inspected periodically and retrieved, if necessary, for repacking or for transfer if a better means of disposal is developed. This technique is more expensive than burial in a landfill because the storage area must be guarded and monitored continuously to prevent leakage, vandalism, or other dispersal of toxic materials. Remedial measures are much cheaper with this technique, however, and it may be the best system in the long run.

Secure Landfills. One of the most popular solutions for hazardous waste disposal has been landfilling. Although, as we saw earlier in this chapter, many such landfills have been environmental disasters, newer techniques make it possible to create safe, modern **secure landfills** that are acceptable for disposing of many hazardous wastes. The first line of defense in a secure landfill is a thick bottom cushion of compacted clay that surrounds the pit like a bathtub (fig. 23.20). Moist clay is flexible and resists cracking if the ground shifts. It is impermeable to groundwater and will safely contain wastes. A layer of gravel is spread over the clay liner and perforated drain pipes are laid in a grid to collect any seepage that escapes from the stored material. A thick polyethylene liner, protected from punctures by soft padding materials, covers the gravel bed. A layer of soil or absorbent sand cushions the inner liner and the wastes are packed in drums, which then are placed into the pit, separated into small units by thick berms of soil or packing material.

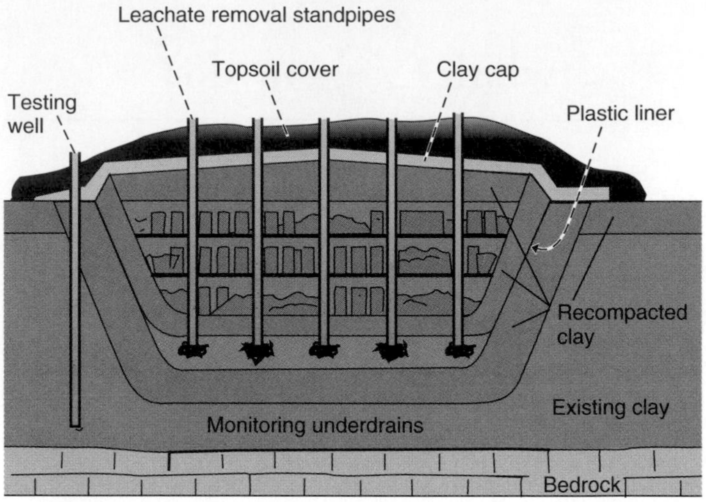

FIGURE 23.20 A secure landfill for toxic waste. Contents are enclosed by a thick plastic liner and two or more layers of impervious compacted clay. A gravel bed between the clay layers collects any leachate, which can then be pumped out and treated. Then, a test for escaping contaminants is done by testing a well sample.

When the landfill has reached its maximum capacity, a cover much like the bottom sandwich of clay, plastic, and soil—in that order—caps the site. Vegetation stabilizes the surface and improves its appearance. Sump pumps collect any liquids that filter through the landfill, either from rainwater or leaking drums. This leachate is treated and purified before being released. Monitoring wells check groundwater around the site to ensure that no toxins have escaped.

Most landfills are buried below ground level to be less conspicuous; however, in areas where the groundwater table is close to the surface, it is safer to build above-ground storage. The same protective construction techniques are used as in a buried pit. An advantage to such a facility is that leakage is easier to monitor because the bottom is at ground level.

Transportation of hazardous wastes to disposal sites is of concern because of the risk of accidents. Emergency preparedness officials conclude that the greatest risk in most urban areas is not nuclear war or natural disaster but crashes involving trucks or trains carrying hazardous chemicals through densely packed urban corridors. Another worry is who will bear financial responsibility for abandoned waste sites. The material remains toxic long after the businesses that created it are gone. As is the case with nuclear wastes (chapter 21), we may need new institutions for perpetual care of these wastes.

SUMMARY

We produce enormous volumes of solid waste in industrialized societies, and there is an increasing problem of how to dispose of those wastes in an environmentally safe manner. In this chapter, we have looked at the character of our solid and hazardous wastes. We have surveyed the ways we dispose of our wastes and the environmental problems associated with waste disposal.

Solid wastes are domestic, commercial, industrial, agricultural, and mining wastes that are primarily nontoxic. About 60 percent of our domestic and industrial wastes are deposited in landfills; most of the rest is incinerated or recycled. Old landfills were often messy and leaky, but new ones are required to have impermeable clay or plastic linings, drainage, and careful siting. Incineration can destroy organic compounds, but whether incinerators can or will be operated satisfactorily is a matter of debate. Recycling is growing nationwide, encouraged by the economic and environmental benefits it brings. City leaders tend to doubt the viability of recycling programs, but successful programs have been sustained in other countries and in some American cities.

Near major urban centers, land suitable for waste disposal is becoming increasingly scarce and expensive. Annual costs for solid waste disposal currently total nearly $10 billion in the United States and are expected to climb to $100 billion per year in a few decades. Some cities now ship their refuse to other states or even other countries, but worries about toxic and hazardous material in the waste are leading to increasing resistance to shipping or storing it.

Hazardous and toxic wastes, when released into the environment, cause such health problems as birth defects, neurological disorders, reduced resistance to infection, and cancer. Environmental losses include contamination of water supplies, poisoning of the soil, and destruction of habitat. The major categories of hazardous wastes are ignitable, corrosive, reactive, explosive, and toxic. Some materials of the greatest concern are heavy metals, solvents, and synthetic organic chemicals such as halogenated hydrocarbons, organophosphates, and phenoxy herbicides.

Disposal practices for solid and hazardous wastes have often been unsatisfactory. Thousands of abandoned, often unknown waste disposal sites still leak toxic materials into the environment. Some alternative techniques for treating or disposing of hazardous wastes include not making the material in the first place, incineration, secure landfill, and physical, chemical, or biological treatment to detoxify or immobilize wastes. People are often unwilling to have transfer facilities, storage sites, disposal operations, or transportation of hazardous or toxic materials in or through their cities. Questions of safety and liability remain unanswered in solid and hazardous waste disposal.

QUESTIONS FOR REVIEW

1. What are solid wastes and hazardous wastes? What is the difference between them?
2. Describe the difference between an open dump, a sanitary landfill, and a modern, secure, hazardous waste disposal site.
3. Why are landfill sites becoming limited around most major urban centers in the United States? What steps are being taken to solve this problem?
4. Describe some concerns about waste incineration.
5. List some benefits and drawbacks of recycling wastes. What are the major types of materials recycled from municipal waste and how are they used?
6. What is composting, and how does it fit into solid waste disposal?
7. Describe some ways that we can reduce the waste stream to avoid or reduce disposal problems.
8. List ten toxic substances in your home and how you would dispose of them.
9. What are brownfields and why do cities want to redevelop them?
10. What societal problems are associated with waste disposal? Why do people object to waste handling in their neighborhoods?

QUESTIONS FOR CRITICAL THINKING

1. A toxic waste disposal site has been proposed for the Pine Ridge Indian Reservation in South Dakota. Many tribal members oppose this plan, but some favor it because of the jobs and income it will bring to an area with 70 percent unemployment. If local people choose immediate survival over long-term health, should we object or intervene?
2. There is often a tension between getting your personal life in order and working for larger structural changes in society. Evaluate the trade-offs between spending time and energy sorting recyclables at home compared to working in the public arena on a bill to ban excess packaging.
3. Should industry officials be held responsible for dumping chemicals that were legal when they did it but are now known to be extremely dangerous? At what point can we argue that they *should* have known about the hazards involved?
4. Look at the discussion of recycling or incineration presented in this chapter. List the premises (implicit or explicit) that underlie the presentation as well as the conclusions (stated or not) that seem to be drawn from them. Do the conclusions necessarily follow from these premises?
5. Suppose that your brother or sister has decided to buy a house next to a toxic waste dump because it costs $20,000

less than a comparable house elsewhere. What do you say to him or her?

6. Is there an overall conceptual framework or point of view in this chapter? If you were presenting a discussion of solid or hazardous waste to your class, what would be your conceptual framework?
7. Is there a fundamental difference between incinerating municipal, medical, or toxic industrial waste? Would you oppose an incinerator for one type of waste in your neighborhood but not others? Why, or why not?
8. The Netherlands incinerates much of its toxic waste at sea by a shipborne incinerator. Would you support this as a way to dispose of our wastes as well? What are the critical considerations for or against this approach?
9. Do you believe that Philadelphia has an obligation to help remove its toxic ash from Haiti? Many African-Americans in innercity Philadelphia live in very poor conditions. Does the city have an obligation to help them before helping foreigners?

KEY TERMS

bioremediation 544
biodegradable plastics 536
brownfields 540
composting 533
demanufacturing 533
energy recovery 530
hazardous waste 536
mass burn 530
permanent retrievable storage 544
photodegradable plastics 536
recycling 531
refuse-derived fuel 530
sanitary landfills 528
secure landfills 544
Superfund 538
Toxic Release Inventory 538
waste stream 527

ADDITIONAL INFORMATION ON THE INTERNET

Visit our website at http://www.mhhe.com/environmentalscience/ for specific information about each of the topics below:

California Integrated Waste Management Board
EPA's Superfund Homepage
Agency for Toxic Substances and Disease Registry
Environmental Justice
Environmental Justice: Environmental Health Information Service
The Internet Consumer Recycling Guide
Environmental Systems of America, Inc: Facts on solid waste
Environmental Systems of America, Inc: Recycling Page
Oregon DEQ Solid Waste Program
Water and Waste International
Managing hazardous waste
Composting Resource Page

CHAPTER 24

Urbanization and Sustainable Cities

The problems that overwhelm us today are precisely those we failed to solve decades ago.

—Mostafa K. Tolba—

OBJECTIVES

After studying this chapter, you should be able to:

- distinguish between a rural village, a city, and a megacity.
- recognize the push and pull factors that lead to urban growth.
- appreciate the growth rate of giant metropolitan urban areas such as Mexico City, as well as the problems this growth engenders.
- visualize living conditions for ordinary citizens in the megacities of the developing world.
- understand the causes and consequences of crowding and pollution in cities.
- critique options for suburban design.
- see the connection between sustainable economic development, social justice, and the solution of urban problems.

ESSENTIAL STUDY PARTNER

The Essential Study Partner (ESP) CD-ROM that accompanies this textbook contains additional information on the following topics:

- rural-to-urban shifts and belts of habitation: populations: societal issues: urbanization
- food, water, energy needs and pollution protection of megacities: populations: societal issues: urbanization
- housing, slums, shantytowns, and squatter settlements in Third World Cities: populations: societal issues: urbanization
- urban sprawl, freeways, air and noise pollution: populations: societal issues: urbanization
- living spaces and preserving open spaces: populations: societal issues: urbanization

Above: The approach to downtown Miami, Florida, is dominated by a maze of freeways.

Chattanooga, A Model Sustainable City

Nestled in the Tennessee River valley, at the foot of the Cumberland Plateau, Chattanooga, Tennessee, is a medium-size industrial city that provides an inspiring example of participatory city planning that has produced a clean urban environment, a vibrant economy, and strong social capital. Twenty-five years ago, the U.S. Department of Health, Education, and Welfare named Chattanooga America's dirtiest city. Factory smoke trapped between the mountains gave city air the highest particulate count in the nation—dirtier, even, than Los Angeles. Per capita tuberculosis rates were three times the national average. Chattanooga Creek was so polluted by toxic waste from coke ovens and steel mills that the EPA designated 4 km (2.5 mi) of it a Superfund site.

Social conditions declined along with environmental quality. As industries moved out, and the upper class fled to the suburbs, the city was left with abandoned factories, boarded-up buildings, a declining tax base, unemployment, crime, and despair. Most of the worst pollution and highest unemployment were in predominantly African-American neighborhoods, where residents had the least resources to improve their situation. How could the city reinvent itself and turn these trends around?

In 1984, the city council, led by David Crockett—the great nephew of the famous frontiersman—brought together a variety of city residents to talk about their vision for the future. A series of town meetings drew more than 1700 people who brainstormed what their city might be like. Among the values mentioned most often were clean air and water, jobs, affordable housing, personal safety, a place to fish, and somewhere to walk along the river. Forty goals were identified, ranging from cleaning up the river to building low-cost housing. By giving everyone a voice, the process built community spirit and a broad base of support. Suddenly people with energy and good ideas from every part of the city got involved.

Public-private partnerships led to nearly $800 million in investments (two-thirds from the private sector) and created 8000 new permanent and temporary jobs. The Chattanooga Neighborhood Enterprise Corporation invested $60 million in affordable housing. More than 1000 first-time home buyers obtained low-interest loans, while other residents got home-improvement help. By 1990, Chattanooga was in compliance with national air quality standards. As part of urban redevelopment, the city encouraged use of public transport by building satellite parking lots at the city outskirts and providing free, convenient, rapid bus service to the city center. The electric buses designed for this system are now being sold to other cities and provide jobs and revenue for the city.

The riverfront has been cleaned up. Old warehouses that blocked river access have been removed and public fishing piers built. A 35-km (22 mi) long riverfront park with walking and bike trails has been built along both sides of the river through downtown (fig. 24.1). Fountains, flowers, shade trees, street musicians, and outdoor festivals now fill the park. A world-class freshwater aquarium attracts tourists. What had been an embarrassment of derelict buildings and weedy, vacant lots, has become the most desirable location in the city. Property values have shot up, and yuppies are moving back to the center city to take advantage of the recreational and cultural opportunities.

Chattanooga is now considered the "granddaddy" of the sustainable cities movement. Using principles of urban ecology, the city has created a livable environment, a strong sense of community, and a flourishing economy. As David Crockett says, it is "a city that is cleaner, greener, and safer; that values human and natural resources; and that provides an economic system that will keep our children here."

FIGURE 24.1 Chattanooga's waterfront has been cleaned up and has become an asset rather than a liability.

URBANIZATION

Since their earliest origins, cities have been centers of education, religion, commerce, record keeping, communication, and political power. As cradles of civilization, cities have influenced culture and society far beyond their proportion of the total population (fig. 24.2). Until recently, however, only a small percentage of the world's people lived permanently in urban areas and even the greatest cities of antiquity were small by modern standards. The vast majority of humanity has always lived in rural areas where farming, fishing, hunting, timber harvesting, animal herding, mining, or other natural resource-based occupations provided support.

Since the beginning of the Industrial Revolution some 300 years ago, however, cities have grown rapidly in both size and power. In every developing country, the transition from an agrarian society to an industrial one has been accompanied by **urbanization,** an increasing concentration of the population in cities and a transformation of land use and society to a metropolitan pattern of organization. Industrialization and urbanization bring many benefits—especially to the top members of society—but they also cause many problems, as we will detail in this chapter.

FIGURE 24.2 Since their earliest origins, cities have been centers of education, religion, commerce, politics, and culture. Unfortunately, they have also been sources of pollution, crowding, disease, and misery. The Parthenon, one of the most beautiful buildings in the world, is being eaten away by air pollution, vibrations, and the other urban ills of Athens, Greece.

FIGURE 24.3 A traditional hill-tribe village in northern Thailand is closely tied to the land through culture, economics, and family relationships. While the timeless pattern of life here gives a great sense of identity, it can also be stifling and repressive.

FIGURE 24.4 A city is a differentiated community with a large enough population and resource base to allow specialization in arts, crafts, services, and professions. Although there are many disadvantages in living so closely together, there are also many advantages. Cities are growing rapidly, and most of the world's population will live in urban areas if present trends continue.

Nearly half the people in the world now live in urban areas. Demographers predict that by the end of the twenty-first century, 80 or 90 percent of all humans will live in cities and that some giant interconnecting metropolitan areas could have hundreds of millions of residents. In this chapter, we will look at how cities came into existence, why people live there, and what the environmental conditions of cities have been, are now, and might be in the future. Some of the most severe urban problems in the world are found in the giant megacities of the developing countries. Far more lives may be threatened by the desperate environmental conditions in these cities than by any other issue we have studied. We also will look at a few of those problems and possible solutions that could improve the urban environment.

What Is a City?

Just what makes up an urban area or a city? Definitions differ. The U.S. Census Bureau considers any incorporated community to be a city, regardless of size, and defines any city with more than 2500 residents as urban. More meaningful definitions are based on *functions.* In a **rural area,** most residents depend on agriculture or other ways of harvesting natural resources for their livelihood. In an **urban area,** by contrast, a majority of the people are not directly dependent on natural resource-based occupations.

A **village** is a collection of rural households linked by culture, custom, family ties, and association with the land (fig. 24.3). A **city,** by contrast, is a differentiated community with a population and resource base large enough to allow residents to specialize in arts, crafts, services, or professions rather than natural resource-based occupations. While the rural village often has a sense of security and connection, it also can be stifling. A city offers more freedom to experiment, to be upwardly mobile, and to break from restrictive traditions, but it can be harsh and impersonal (fig. 24.4).

Beyond about 10 million inhabitants, an urban area is considered a supercity or **megacity.** Megacities in many parts of the world have grown to enormous size. Chongqing, China, having annexed a large part of Sichuan province and about 30 million

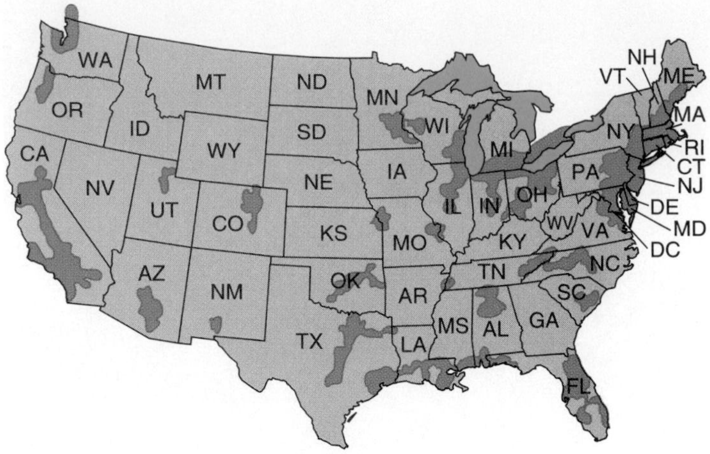

FIGURE 24.5 Urban core agglomerations (lavender areas) are forming megalopolises in many areas. While open space remains in these areas, the flow of information, capital, labor, goods, and services links each into an interacting system.
Source: Data from U.S. Census Bureau, 1998.

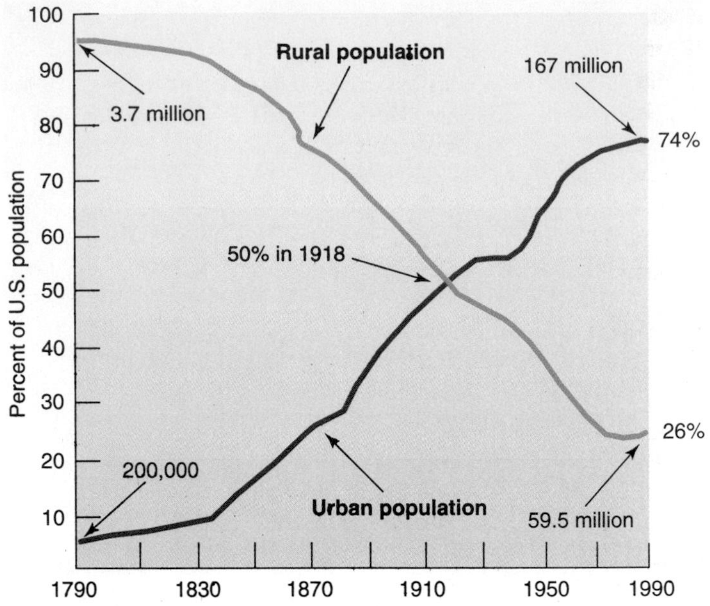

FIGURE 24.6 United States rural-to-urban shift. Percentages of rural and urban populations, 1790 to 1990.
Source: Data from U.S. Bureau of the Census.

people, claims to be the biggest city in the world. In the United States, urban areas between Boston and Washington, DC, have merged into a nearly continuous megacity (sometimes called Bos-Wash) containing about 35 million people. The Tokyo-Yokohama-Osaka-Kobe corridor contains nearly 50 million people. Because these agglomerations have expanded beyond what we normally think of as a city, some geographers prefer to think of urbanized **core regions** that dominate the social, political, and economic life of most countries (fig. 24.5). Architect and city planner C. A. Doxiadis predicts that as core regions of adjacent countries merge, the sea coasts and major river valleys of most continents will be covered with continuous strip cities, which he calls ecumenopolises, each containing billions of people.

World Urbanization

The United States underwent a dramatic rural-to-urban shift in the nineteenth and early twentieth centuries (fig. 24.6). Now many developing countries are experiencing a similar demographic movement. In 1850, only about 2 percent of the world's population lived in cities. By 2000, 47 percent of the people in the world were urban. Only Africa and South Asia remain predominantly rural, but people there are swarming into cities in ever-increasing numbers. About three-fourths of the people in Europe, North America, and Latin America already live in cities (table 24.1). Some urbanologists predict that by 2100, the whole world will be urbanized to the levels now seen in developed countries.

As figure 24.7 shows, 90 percent of the population growth over the next 25 years is expected to occur in the less-developed countries of the world. Most of that growth will be in the already overcrowded cities of the least affluent countries, such as India, China, Mexico, and Brazil. The combined population of these cities is expected to jump from its present 6 billion to more than 8 billion by the year 2025. Meanwhile, rural populations in these countries are expected to remain constant or even decline somewhat as rural people migrate into the cities.

TABLE 24.1 Urban Share of Total Population (percent)

REGION	1975	1995	2025*
Africa	25	34	54
Asia	25	35	55
Europe	67	74	83
Russian Federation	66	76	86
North and Central America	57	68	79
South America	64	78	88
Oceania	72	70	75
World	38	45	61

*Estimated.
Source: Data from *World Resources 1996–97.*

Recent urban growth has been particularly dramatic in the largest cities, especially those of the developing world. In 1900, 13 cities had populations over 1 million; all except Tokyo and Peking were in Europe or North America (table 24.2). By 1995, there were 235 metropolitan areas of more than one million people but only three of the largest cities were in developed countries.

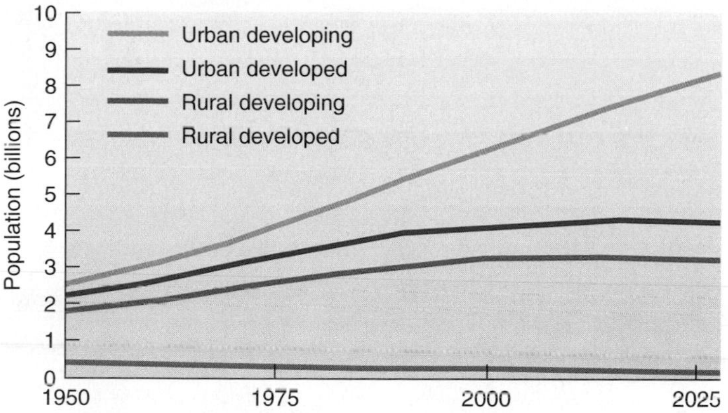

FIGURE 24.7 Urban and rural growth in developed and developing countries.
Source: Data from United Nations (U.N.) Population Division, *World Urbanization Prospects (The 1996 Revision),* on diskette (U.N. New York, 1996).

A century ago, London was the only city with more than 5 million people; now 19 cities have populations larger than that. Some futurists predict that by 2025 at least 400 cities will have populations of 1 million or more, and 93 supercities each will have 5 million or more residents. Three-fourths of those cities will be in the developing world (fig. 24.8). In just the next 25 years, Mumbay, India; Delhi, India; Karachi, Pakistan; Manila, Philippines; and Jakarta, Indonesia all are expected to grow by at least 50 percent.

The actual population of many of these cities is uncertain. Where to draw city boundaries and how to count everyone is controversial. If you live near a large metropolitan area, you probably have experienced how the city gradually merges into the countryside. Many people live beyond the official city boundaries but depend on the urban economy for their livelihoods. In megacities of the developing world, as much as half of urban populations are transient workers or residents of unplanned slums and shantytowns. These people are hard to count. By some estimates, Mexico City already has a population of 25 million and is adding another 750,000 people every year (fig. 24.9).

Can cities function with 20 or 25 million people? Can they supply the public services necessary to sustain a civilized life? Adding 750,000 new people annually to Mexico City amounts to building a new city the size of Baltimore or San Francisco every year. This growth is occurring, as is most urban growth in the world, in a country with a sagging economy, an unstable government, and a high foreign debt load.

CAUSES OF URBAN GROWTH

Urban populations grow in two ways: by natural increase (more births than deaths) and by immigration. Natural increase is fueled by improved food supplies, better sanitation, and advances in medical care that reduce death rates and cause populations to grow both within cities and in the rural areas around them (chapter 7). In Latin America and East Asia, natural increase is responsible for two-thirds of urban population growth. In Africa and West Asia, immigration is the largest source of urban growth. Immigration to cities can be caused both by **push factors** that force people out of the country and by **pull factors** that draw them into the city.

TABLE 24.2 The World's Largest Metropolitan Regions (populations in millions)

1900		1995		2015	
London, England	6.6	Tokyo, Japan	26.9	Tokyo, Japan	29.0
New York, USA	4.2	Mexico City, Mexico	16.6	Mumbay, India**	26.2
Paris, France	3.3	São Paulo, Brazil	16.5	Mexico City, Mexico	25.1
Berlin, Germany	2.4	New York, USA	16.3	Lagos, Nigeria	24.6
Chicago, USA	1.7	Mumbay, India**	15.1	São Paulo, Brazil	20.3
Vienna, Austria	1.6	Shanghai, China	13.6	Shanghai, China	18.0
Tokyo, Japan	1.5	Los Angeles, USA	12.4	New York, USA	17.6
St. Petersburg, Russia	1.4	Calcutta, India	11.9	Calcutta, India	17.3
Philadelphia, USA	1.4	Seoul, S. Korea	11.6	Beijing, China	16.0
Manchester, England	1.3	Beijing, China	11.3	Los Angeles, USA	14.2
Birmingham, England	1.2	Osaka, Japan	10.6	Buenos Aires, Argentina	13.9
Moscow, Russia	1.1	Lagos, Nigeria	10.3	Rio de Janeiro, Brazil	11.9
Peking, China*	1.1	Rio de Janeiro, Brazil	10.2	Osaka, Japan	10.6

*Now spelled Beijing.
**Formerly known as Bombay.
Source: Data from T. Chandler, *Three Thousand Years of Urban Growth,* 1974, Academic Press; and United Nations Populations Division, 1998.

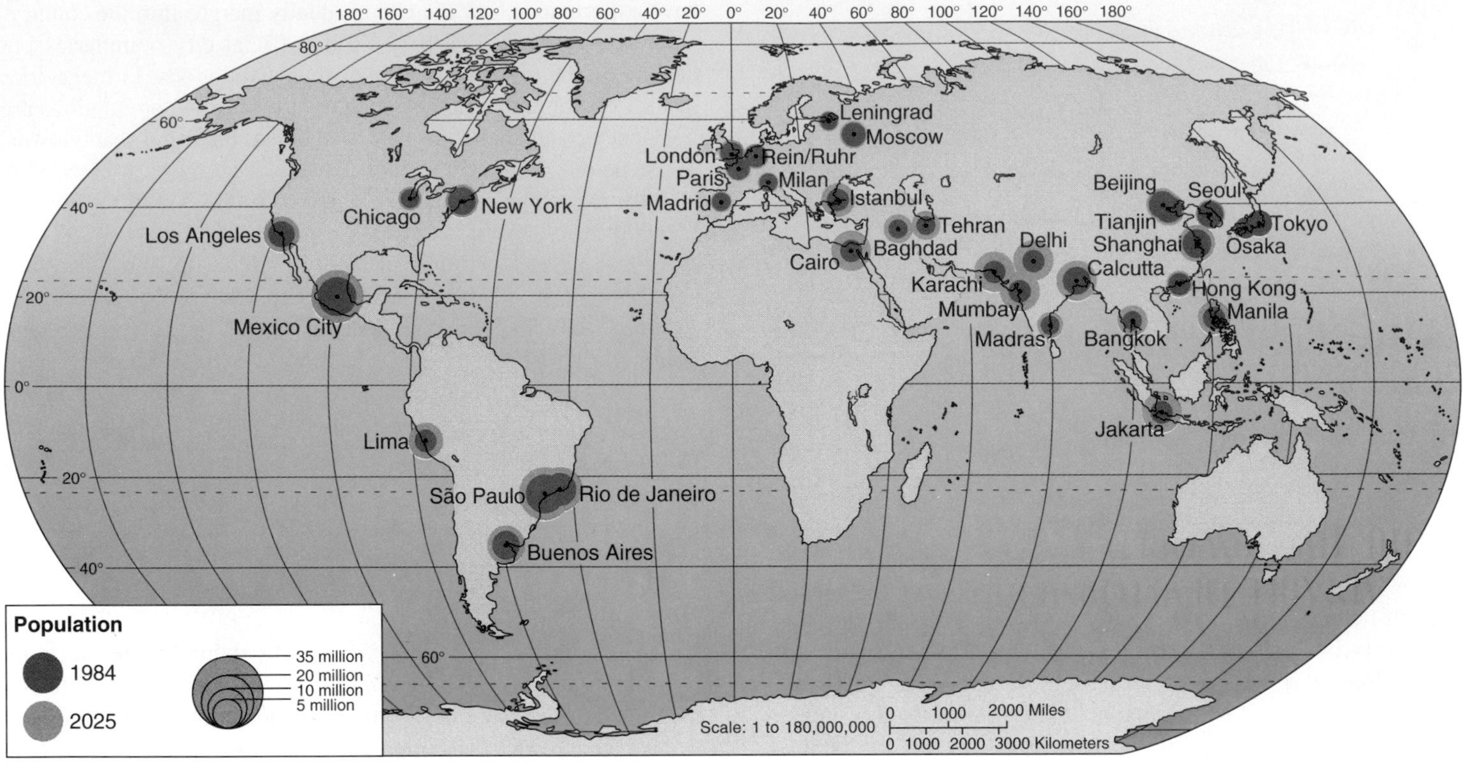

FIGURE 24.8 By 2025, at least 400 cities will have populations of 1 million or more, and 93 supercities will have populations above 5 million. Three-fourths of the world's largest cities will be in developing countries that already have trouble housing, feeding, and employing their people.

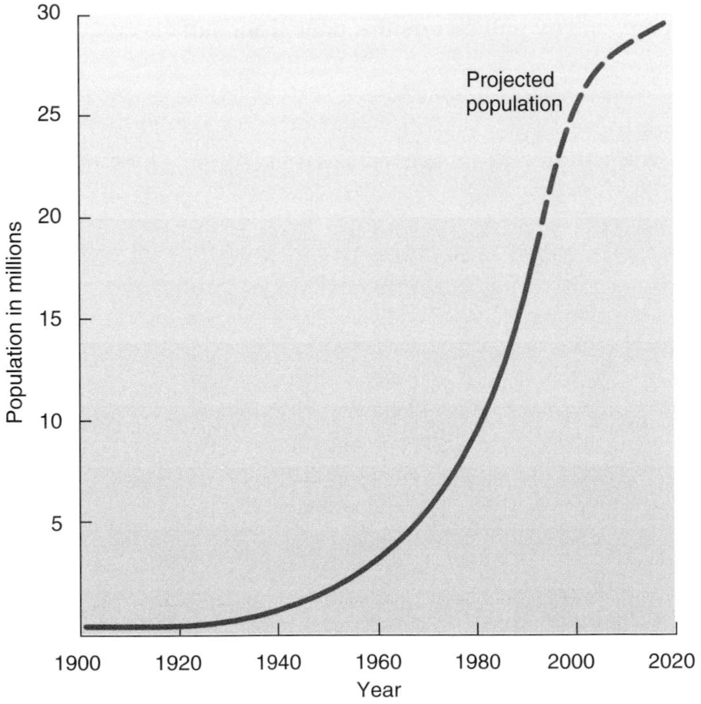

FIGURE 24.9 Estimated growth of Mexico City metropolitan area, 1900–2020.
Source: Data from U.N. Population Council.

Immigration Push Factors

People migrate to cities for many reasons. In some areas, the countryside is overpopulated and simply can't support more people. The "surplus" population is forced to migrate to cities in search of jobs, food, and housing. Not all rural-to-urban shifts are caused by overcrowding in the country, however. In some places, economic forces or political, racial, or religious conflicts drive people out of their homes. The countryside may actually be depopulated by such demographic shifts. The United Nations estimated that in 1992 at least 10 million people fled their native country and that another 30 or 40 million were internal refugees within their own country, displaced by political, economic, or social instability. Many of these refugees end up in the already overcrowded megacities of the developing world.

Land tenure patterns and changes in agriculture also play a role in pushing people into cities. The same pattern of agricultural mechanization that made farm labor largely obsolete in the United States early in this century is spreading now to developing countries. Furthermore, where land ownership is concentrated in the hands of a wealthy elite, subsistence farmers are often forced off the land so it can be converted to grazing lands or monoculture cash crops (chapter 10). Speculators and absentee landlords let good farmland sit idle that otherwise might house and feed rural families.

Immigration Pull Factors

Even in the largest and most hectic cities, many people are there by choice, attracted by the excitement, vitality, and opportunity to meet others like themselves. Cities offer jobs, housing, entertainment, and freedom from the constraints of village traditions. Possibilities exist in the city for upward social mobility, prestige, and power not ordinarily available in the country. Cities support specialization in arts, crafts, and professions for which markets don't exist elsewhere.

Modern communications also draw people to cities by broadcasting images of luxury and opportunity. An estimated 90 percent of the people in Egypt, for instance, have access to a television set. The immediacy of television makes city life seem more familiar and attainable than ever before. We generally assume that beggars and homeless people on the streets of teeming Third World cities have no other choice of where to live, but many of these people want to be in the city. In spite of what appears to be dismal conditions, living in the city may be preferable to what the country had to offer.

Government Policies

Government policies often favor urban over rural areas in ways that both push and pull people into the cities. Developing countries commonly spend most of their budgets on improving urban areas (especially around the capital city where leaders live), even though only a small percentage of the population lives there or benefits directly from the investment. This gives the major cities a virtual monopoly on new jobs, housing, education, and opportunities, all of which bring in rural people searching for a better life. In Peru, for example, Lima accounts for 20 percent of the country's population, but has 50 percent of the national wealth, 60 percent of the manufacturing, 65 percent of the retail trade, 73 percent of the industrial wages, and 90 percent of all banking in the country. Similar statistics pertain to São Paulo, Mexico City, Manila, Cairo, Lagos, Bogotá, and a host of other cities.

Governments often manipulate exchange rates and food prices for the benefit of more politically powerful urban populations but at the expense of rural people. Importing lower-priced food pleases city residents, but local farmers then find it uneconomical to grow crops. As a result, an increased number of people leave rural areas to become part of a large urban workforce, keeping wages down and industrial production high. Zambia, for instance, sets maize prices below the cost of local production to discourage farming and to maintain a large pool of workers for the mines. Keeping the currency exchange rate high stimulates export trade but makes it difficult for small farmers to buy the fuels, machinery, fertilizers, and seeds that they need. This depresses rural employment and rural income while stimulating the urban economy. The effect is to transfer wealth from the country to the city.

CURRENT URBAN PROBLEMS

Large cities in both developed and developing countries face similar challenges in accommodating the needs and by-products of dense populations. The problems are most intense, however, in rapidly growing cities of developing nations.

The Developing World

As figure 24.7 shows, 90 percent of the human population growth in the next century is expected to occur in the developing world, mainly in Africa, Asia, and South America. Almost all of that growth will occur in cities—especially the largest cities—which already have trouble supplying food, water, housing, jobs, and basic services for their residents. The unplanned and uncontrollable growth of those cities causes tragic urban environmental problems. You may have noticed that nearly every chapter in this book has a boxed reading presenting an ethical dilemma. This whole chapter is a dilemma of how people can survive under the conditions found in the megacities of the world. Consider as you read this what responsibilities we in the richer countries have to help others who are less fortunate and how we might do so.

Traffic and Congestion

A first-time visitor to a supercity—particularly in a less-developed country—is often overwhelmed by the immense crush of pedestrians and vehicles of all sorts that clog the streets. The noise, congestion, and confusion of traffic make it seem suicidal to venture onto the street. Jakarta, for instance, is one of the most densely populated cities in the world (fig. 24.10). Traffic is chaotic almost all the time. People commonly spend three or four hours each way commuting to work from outlying areas. Bangkok also has monumental traffic problems. The average resident spends the equivalent of 44 days a year sitting in traffic jams. About 20 percent of all fuel is consumed by vehicles standing still. Hours of work lost each year are worth at least $3 billion.

Air Pollution

The dense traffic (commonly old, poorly maintained vehicles), smoky factories, and use of wood or coal fires for cooking and heating often create a thick pall of air pollution in the world's supercities. Lenient pollution laws, corrupt officials, inadequate testing equipment, ignorance about the sources and effects of pollution, and lack of funds to correct dangerous situations usually exacerbate the problem. What is its human toll? An estimated 60 percent of Calcutta's residents are thought to suffer from respiratory diseases linked to air pollution. Lung cancer mortality in Shanghai is reported to be four to seven times higher than rates in the countryside. Mexico City, which sits in a high mountain bowl with abundant sunshine, little rain, high traffic levels, and frequent air stagnation, has one of the highest levels of photochemical smog (chapter 18) in the world.

FIGURE 24.10 Motorized rickshaws, motor scooters, bicycles, street vendors, and pedestrians all vie for space on the crowded streets of Jakarta. The heat, noise, smells, and sights are overpowering. In spite of the difficulties of living here, people work hard and have hope for the future.

Sewer Systems and Water Pollution

Few cities in developing countries can afford to build modern waste treatment systems for their rapidly growing populations. The World Bank estimates that only 35 percent of urban residents in developing countries have satisfactory sanitation services. The situation is especially desperate in Latin America, where only 2 percent of urban sewage receives any treatment. In Egypt, Cairo's sewer system was built about 50 years ago to serve a population of 2 million people. It is now being overwhelmed by more than 10 million people. Less than one-tenth of India's 3000 towns and cities have even partial sewage systems and water treatment facilities. Some 150 million of India's urban residents lack access to sanitary sewer systems. In Colombia, the Bogotá River, 200 km (125 mi) downstream from Bogotá's 5 million residents, still has an average fecal bacteria count of 7.3 million cells per liter, more than 700,000 times the safe drinking level and 3500 times higher than the limit for swimming.

Some 400 million people, or about one-third of the population in developing world cities do not have safe drinking water, according to the World Bank. Although city dwellers are somewhat more likely than rural people to have clean water, this still represents a large problem. Where people have to buy water from merchants, it often costs 100 times as much as piped city water and may not be any safer to drink. Many rivers and streams in Third World countries are little more than open sewers, and yet they are all that poor people have for washing clothes, bathing, cooking, and—in the worst cases—for drinking. Diarrhea, dysentery, typhoid, and cholera are widespread diseases in these countries, and infant mortality is tragically high (chapter 20).

FIGURE 24.11 In Mumbay, as many as half a million people sleep on the streets because they have no other place to live. Ten times as many live in crowded, dangerous slums and shantytowns throughout the city.

Housing

The United Nations estimates that at least 1 billion people—20 percent of the world's population—live in crowded, unsanitary slums of the central cities and in the vast shantytowns and squatter settlements that ring the outskirts of most Third World cities. Around 100 million people have no home at all. In Mumbay, India, for example, it is thought that half a million people sleep on the streets, sidewalks, and traffic circles because they can find no other place to live (fig. 24.11). In São Paulo, perhaps 1 million "street kids" who have run away from home or been abandoned by their parents live however and wherever they can. This is surely a symptom of a tragic failure of social systems.

Slums are generally legal but inadequate multifamily tenements or rooming houses, either custom built to rent to poor people or converted from some other use. The chals of Mumbay, India, for example, are high-rise tenements built in the 1950s to house immigrant workers. Never very safe or sturdy, these dingy, airless buildings are already crumbling and often collapse without warning. Eighty-four percent of the families in these tenements live in a single room; half of those families consist of six or more people.

Typically, they have less than 2 square meters of floor space per person and only one or two beds for the whole family. They may share kitchen and bathroom facilities down the hall with 50 to 75 other people. Even more crowded are the rooming houses for mill workers where up to 25 men sleep in a single room only 7 meters square. Because of this crowding, household accidents are a common cause of injuries and deaths in Third World cities, especially to children. Charcoal braziers or kerosene stoves used in crowded homes are a routine source of fires and injuries. With no place to store dangerous objects beyond the reach of children, accidental poisonings and other mishaps are a constant hazard.

Shantytowns are settlements created when people move onto undeveloped lands and build their own houses. Shacks are built of corrugated metal, discarded packing crates, brush, plastic sheets, or whatever building materials people can scavenge. Some shantytowns are simply illegal subdivisions where the landowner rents land without city approval. Others are spontaneous or popular settlements or **squatter towns** where people occupy land without the owner's permission. Sometimes this occupation involves thousands of people who move onto unused land in a highly organized, overnight land invasion, building huts and laying out streets, markets, and schools before authorities can root them out. In other cases, shantytowns just gradually "happen."

Called *barriads, barrios, favelas,* or *turgios* in Latin America, *bidonvillas* in Africa, or *bustees* in India, shantytowns surround every megacity in the developing world. They are not an exclusive feature of poor countries, however. Some 250,000 immigrants and impoverished citizens live in the *colonias* along the southern Rio Grande in Texas. Only 2 percent have access to adequate sanitation. Many live in conditions as awful as you would see in any Third World city.

About three-quarters of the residents of Addis Ababa, Ethiopia, or Luanda, Angola, live in squalid refugee camps. Two-thirds of the population of Calcutta live in unplanned squatter settlements and nearly half of the 20 million people in Mexico City live in uncontrolled, unauthorized shantytowns. Many governments try to clean out illegal settlements by bulldozing the huts and sending riot police to drive out the settlers, but the people either move back in or relocate to another shantytown.

These popular but unauthorized settlements usually lack sewers, clean water supplies, electricity, and roads. Often the land on which they are built was not previously used because it is unsafe or unsuitable for habitation. In Bhopal, India, and Mexico City, for example, squatter settlements were built next to deadly industrial sites. In Rio de Janeiro, La Paz (Bolivia), Guatemala City, and Caracas (Venezuela), they are perched on landslide-prone hills. In Jakarta, Indonesia, thousands of people live in shacks built over fetid tidal canals that serve as open sewers (fig. 24.12). In Lima (Peru), Khartoum (Sudan), and Nouakchott, (Mauritania), shantytowns have spread onto sandy deserts. In Manila, 20,000 people live in huts built on towering mounds of garbage and burning industrial waste in city dumps (see fig. 23.3).

As desperate and inhumane as conditions are in these slums and shantytowns, many people do more than merely survive there. They keep themselves clean, raise families, educate their children, find jobs, and save a little money to send home to their parents. They learn to live in a dangerous, confusing, and rapidly changing world and have hope for the future. The people have parties; they sing and laugh and cry. They are amazingly adaptable and resilient. In many ways, their lives are no worse than those in the early industrial cities of Europe and America a century ago. Perhaps continuing development will bring better conditions to cities of the Third World as it has for many in the First World.

FIGURE 24.12 More than half of all city dwellers in Indonesia live in fetid slums and shantytowns like this one in Jakarta. Appalling conditions like this now affect more than one billion people worldwide.

The Developed World

For the most part, the rapid growth of central cities that accompanied industrialization in nineteenth- and early twentieth-century Europe and North America has now slowed or even reversed. London, for instance, once the most populous city in the world, has lost nearly 2 million people, dropping from its high of 8.6 million in 1939 to about 6.7 million now. While the greater metropolitan area surrounding London has been expanding to about 10 million inhabitants, it is now only the twelfth largest city in the world.

Urban Problems

Many of the worst urban environmental problems of industrialized countries have been substantially reduced in recent years. Air and water quality have improved greatly. Working conditions and

FIGURE 24.13 A homeless man sleeps on a park bench in Santa Monica, California, with all his belongings in a shopping cart nearby. Welfare agencies estimate that 3.5 million people in the United States are homeless and 10 to 20 times that number live in substandard housing.

housing are better for most people. Improved sanitation and medical care have reduced or totally eliminated many of the worst communicable diseases that once were the scourge of cities. Dispersal of the people and businesses to the suburbs has significantly reduced crowding and congestion. Still, city life is generally regarded as being more stressful than life in the country or suburbs.

The major problems now facing cities of the United States tend to be associated with decay and blight. Ironically, the United States Agency for International Development (USAID) has recently begun turning its attention from the developing world to inner cities in this country, where many people live in more desperate and degrading conditions than do residents of Third World countries.

Many of the social problems of central cities are a result of a concentration of racial minorities, new immigrants, and the poor, elderly, handicapped, or others who were left behind when the middle and upper classes moved to the suburbs. Unemployment in some inner-city areas is routinely 50 percent or higher, and jobs that are available generally are menial, minimum-wage work. Homelessness is an increasing problem of cities (fig. 24.13). Welfare agencies estimate that 3.5 million people are homeless in the United States and 10 to 20 times that number live in substandard housing. About one-third of the homeless are now families. Of the single females living on the street, perhaps three-fourths are mentally disturbed and would formerly have been cared for in hospitals or institutions. Of the single, homeless men, two-thirds are elderly, poor, disabled, drug addicts, or alcoholics.

Although freeways and cheap gasoline have allowed the upper classes to escape from center-city problems, not all is well in the American suburbs. Urban sprawl consumes productive farmland and wildlife habitat. As new office buildings, roads, and homes go up farther and farther out, commutes lengthen, air quality deteriorates, and we become even more dependent on fossil fuels. Malls and "big-box" retailers surrounded by a sea of parking present a forbidding barrier to pedestrians or those without cars. Newly formed suburbs have no sense of history, and no center where people can gather. In wealthier housing developments, "starter castles" occupy vast expanses of land. Houses set back from streets require long driveways and are too far apart for a sense of neighborhood. Too thinly scattered to be served by city water and sewer systems, each has its own well and septic system. Mothers with young children are far from schools, shopping, and playmates; a large part of their day is spent in an automobile running errands and taking children to events they might walk to in a traditional city. Old people often are stuck alone in their homes. If you even know your neighbors, at most you wave to each other as you drive past.

Urban Renewal

In the 1950s, the United States began a massive program of **urban renewal.** Billions of dollars were spent to revitalize blighted sections of inner cities. Old neighborhoods and decaying industrial areas were cleared or renovated, but often to create luxury housing, high-rise office towers, freeways, sports arenas, concert halls, or other projects that benefit the upper classes but do little for the people who once lived there. Many cohesive neighborhoods with strong traditions, cultural identity, and affordable housing were torn down to make way for "festival centers." People displaced from these neighborhoods can't afford to move to the suburbs and don't have transportation to commute to work in distant sites.

Much of the public housing built in the 1950s and 1960s consisted of massive blocks of high-rise apartment buildings that tend to concentrate and aggravate the problems that brought people there in the first place. Bleak, industrial architecture tends to isolate residents rather than promote community identity. There are few jobs nearby and few opportunities to escape from the culture of poverty and violence. Police often refuse to respond to calls from these projects because they are so dangerous. Among some infamous examples are the Cabrini-Green project in Chicago and the Pruitt-Igoe project in St. Louis, Missouri. Built in 1958 to house about 10,000 people, Pruitt-Igoe won architectural awards for modern design. It was poorly built and badly maintained, however, and lacked space in which residents could feel secure. By 1970, it was so run down and dangerous that most residents had left. The city concluded that all 33 buildings were beyond repair and demolished the entire project in 1976 (fig. 24.14).

Not all public housing has been so disastrous. In some projects, residents have formed associations to clean up and manage the buildings. They hire their own security guards, set standards for cleanliness and orderliness, and evict incorrigible troublemakers. Some associations also mediate disputes, organize day care, and help residents find jobs. By taking charge of their own lives and building a sense of community, people can create a better place to live (What Do You Think? p. 557).

what do you think?

THE WORLD AROUND YOU

People for Community Recovery

The Lake Calumet Industrial District on Chicago's far South Side is an environmental disaster area. A heavily industrialized center of steel mills, oil refineries, railroad yards, coke ovens, factories, and waste disposal facilities, much of the site is now a marshy wasteland of landfills, toxic waste lagoons, and slag dumps, around a system of artificial ship channels.

At the southwest corner of this degraded district sits Altgeld Gardens, a low-income public housing project built in the late 1940s by the Chicago Housing Authority. The 2000 units of "The Gardens" or "The Projects," as they are called by the largely minority residents, are low-rise rowhouses, many of which are vacant or in poor repair. But residents of Altgeld Gardens are doing something about their neighborhood. People for Community Recovery (PCR) is a grassroots citizen's group organized to work for a clean environment, better schools, decent housing, and job opportunities for the Lake Calumet neighborhood.

PCR was founded in 1982 by Mrs. Hazel Johnson, an Altgeld Gardens resident whose husband died from cancer that may have been pollution-related. PCR has worked to clean up more than two dozen waste sites and contaminated properties in their immediate vicinity. Often this means challenging authorities to follow established rules and enforce existing statutes. Public protests, leafleting, and community meetings have been effective in public education about the dangers of toxic wastes and have helped gain public support for cleanup projects.

PCR's efforts successfully blocked construction of new garbage and hazardous waste landfills, transfer stations, and incinerators in the Lake Calumet district. Pollution prevention programs have been established at plants still in operation. And PCR helped set up a community monitoring program to stop illegal dumping and to review toxic inventory data from local companies.

Education is an important priority for PCR. An environmental education center administered by community members organizes workshops, seminars, fact sheets, and outreach for citizens and local businesses. A public health education and screening program has been set up to improve community health. Partnerships have been established with nearby Chicago State University to provide technical assistance and training in environmental issues.

PCR also works on economic development. Environmentally responsible products and services are now available to residents. Jobs that are being created as "green" businesses are brought into the community. Wherever possible local people and minority contractors from the area are hired to clean up waste sites and restore abandoned buildings. Job training for youth and adults as well as retraining for displaced workers is a high priority. Funding for these projects has come from fines levied on companies for illegal dumping.

PCR and Mrs. Johnson have received many awards for their fight against environmental racism and despair. In 1992, PCR was the recipient of the President's Environmental and Conservation Challenge Award. PCR is the only African-American grassroots organization in the country to receive this prestigious award.

Although Altgeld Gardens is far from clean, much progress has been made. Perhaps the most important accomplishment is community education and empowerment. Residents have learned how and why they need to work together to improve their living conditions. Could these same lessons be useful in your city or community? What could you do to help improve urban environments where you live?

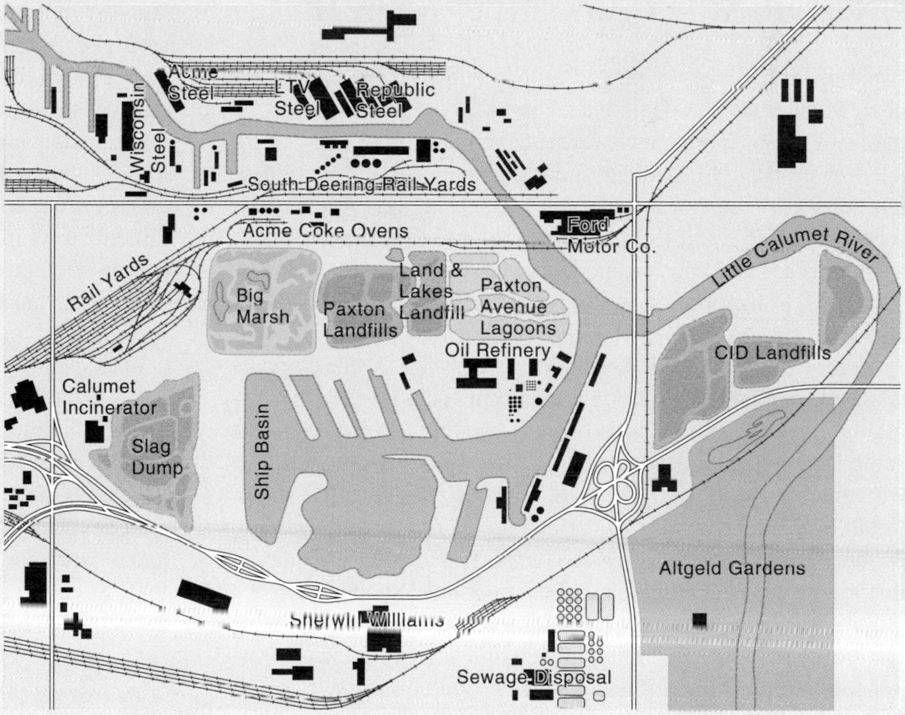

The Calumet Industrial district in South Chicago.

Ethical Considerations

Is the proximity of low-income housing and industrial waste dumps evidence for environmental injustice or merely a question of economics? Most residents of Altgeld Gardens are racial minorities; does that make this an example of environmental racism? What evidence would you need to decide these questions? Is a clean, safe environment a basic human right or something that needs to be earned?

FIGURE 24.14 The Pruitt-Igoe public housing project in St. Louis was a dismal failure in urban planning. All 33 high-rise buildings were demolished in 1976.

TRANSPORTATION AND CITY GROWTH

Transportation has always played an essential role in city development. Most cities are located at crossroads, river fords, seaports, junctions of major rivers, agricultural centers, or other sites where travelers were brought together and merchandise could be traded, bought, or sold. Transportation is needed to bring building materials, energy sources, and food into the city and to remove wastes. Most older cities have been remodeled and reshaped repeatedly by the changing transportation systems that provide their life-giving circulation.

Before the Industrial Revolution, most cities were compact and densely populated (fig. 24.15*a*). Except for a few broad ceremonial boulevards, streets were narrow, twisting, and often barely passable for wheeled traffic. Transportation was by sailboat, horse-drawn vehicles, or on foot. The wealthiest citizens usually lived in the middle of the city close to the centers of power and prestige, while the poorer people lived on the outskirts. Housing was mainly multistoried apartment complexes over street-level shops. Railroads and streetcars provided mass transportation, allowing people to live farther from the center of industrialized cities (fig. 24.15b). Housing tended to cluster along these transportation corridors.

Automobiles brought many changes to cities. The middle class moved to single-family housing developments that filled in open spaces between transit lines (fig. 24.15*c*). The wealthy moved out of town to rural estates and satellite cities. Business offices remained downtown, but shopping centers sprang up at the nodes where major streets intersected transit lines. In the country, small towns and crossroads stores that had been built at about 5-mile intervals to accommodate horse-and-buggy travelers dwindled when people were able to travel to the county seat or the nearest big city to do their shopping. The development of chain stores and brand-name merchandise was largely due to this new mobility of shoppers.

The decision to build freeways has been called the most important land-use decision ever made in the United States. Freeways have profoundly reshaped where we live, work, and shop, and how we get from place to place. Shopping malls located at freeway interchanges have largely replaced downtown department stores and have become the urban centers of expanding rings of suburbs around major cities (fig. 24.15*d*). Freeways allow us to travel with greater privacy, freedom, convenience, and speed (usually) than a mass transit system.

But freeways also bring many problems. They have torn through neighborhoods, choked cities with traffic, enormously increased energy consumption, and caused pollution, noise, and urban sprawl (fig. 24.16). Some of the most contentious U.S. environmental battles of the last two decades concerned freeways planned to cut across residential neighborhoods, parks, scenic and historic areas, or farmlands. Many people were initiated into environmental activism through their opposition to a particular stretch of freeway that threatened an area they cared about.

Los Angeles epitomizes the modern postindustrial city. Its multiple suburban centers are linked by a 600-mile-long network of freeways. It has been estimated that two-thirds of downtown Los Angeles and one-third of the total metropolitan area is devoted to roads, parking lots, service stations, and other automobile-related uses. Five million vehicles crowd onto the roads and highways each day, causing about 85 percent of both the air pollution and urban noise in the metropolitan area. Most Los Angeles freeways no longer have morning rush in one direction and evening rush in the other; traffic now comes to a standstill morning and evening in both directions on many sections, while busy intersections have slowdowns throughout the day and into the night. It can take three hours to drive 20 miles at peak traffic periods. Between commuting to work and driving to beaches, the mountains, or for shopping, many people spend a substantial portion of their lives on the freeways.

What's next now that we seem to have saturated our urban areas with automobiles? Some cities are using modern mass transit systems to redesign where people live and work. Shopping centers are developing into urban villages with a complete set of services and facilities (see fig. 24.15*e*). The old downtown, freed of the congestion and pollution of heavy industry, is undergoing revitalization as a cultural center. In the next section we will look at some ideas for how cities might be reconceived and rebuilt as better places to live.

CITY PLANNING

City planning has a long history. Many of the earliest cities in Mesopotamia, India, and Egypt were built in an orderly and civilized fashion. The Greek city-states of the first millennium B.C. reached a level of architectural beauty and rational city planning that has rarely been surpassed (see fig. 24.2). Miletus and Pergamum on the Ionian coast of Turkey, for instance, and Alexandria on the Nile delta in Egypt, were splendid examples of Greek city planning. They were beautifully situated, with an orderly, unified design, magnificent buildings, and a pleasant mix of public and private spaces.

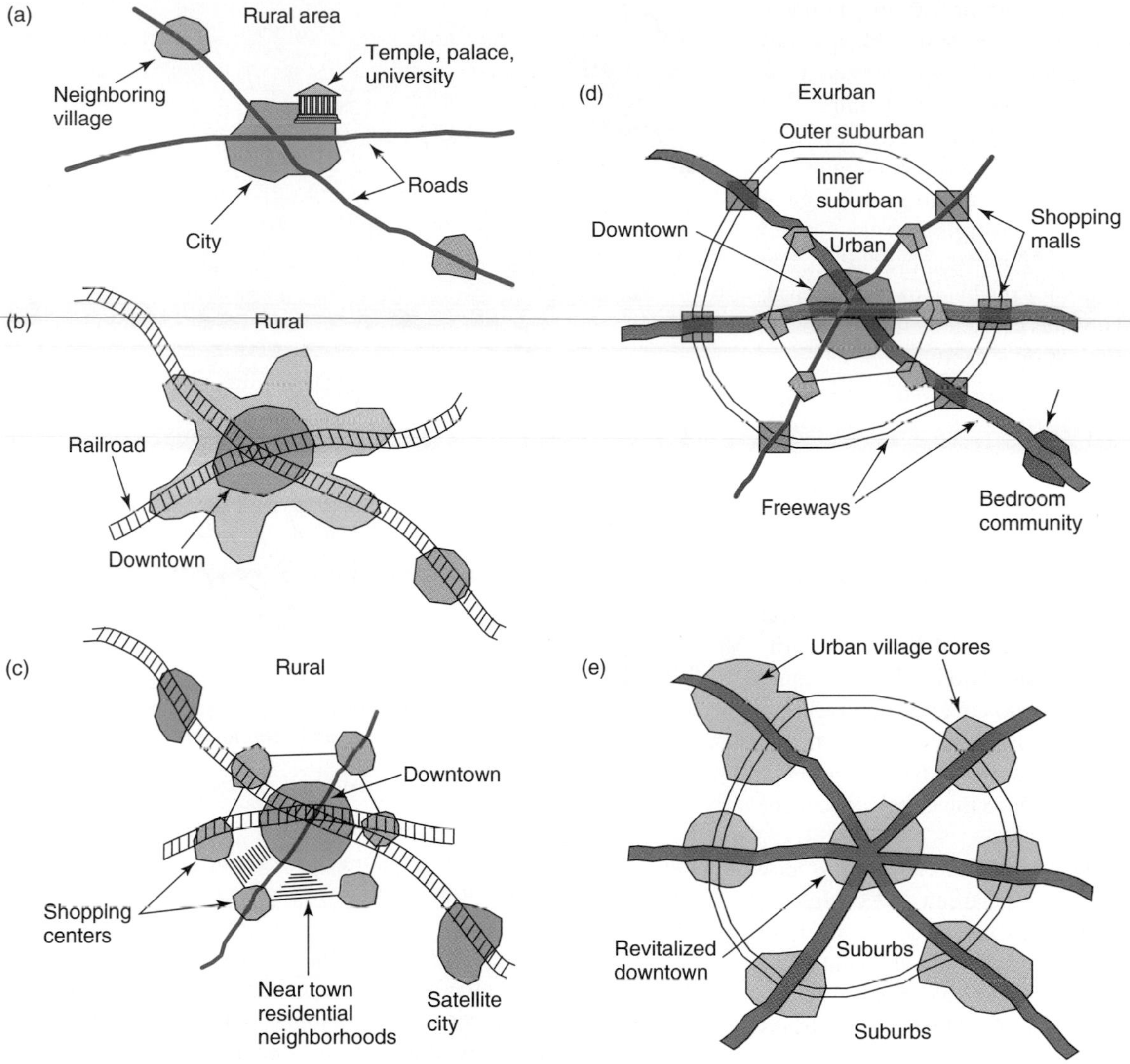

FIGURE 24.15 Patterns of city development from preindustrial (*a*) to polycentric postindustrial (*e*). Note how housing areas follow rail lines in early industrial stage (*b*). Introduction of the automobile allows people to move away from the rail lines (*c*). Freeways (*d*) divert business to shopping malls and encourage growth of suburban rings beyond the central city. Finally, (*e*) these urban village cores develop a full range of services and amenities that rival the central core. Removal of heavy manufacturing from downtown allows gentrification and revitalization.

FIGURE 24.16 The freeway makes it easy for suburban residents to get into and out of the city, but it consumes land, brings pollution into the city, creates noise and visual blight, and destroys neighborhoods.

Residents of these cities regarded them as oases of comfort, convenience, safety, and delight in an otherwise cruel and dangerous world (see fig. 2.12). Cities were seen as elysian refuges protected by a sacred wall from the chaos of evil spirits, wild animals, and dangerous people in the profane world outside. City temples were regarded as homes for the gods; their gardens were sanctuaries of beauty and sacred order. As centers of education, religion, political power, commerce, communication, and technology, cities offered many advantages over life in the country.

The Greeks considered optimum city size to be 30,000 to 50,000 people. This size was large enough to support a theater, university, library, temple, lively marketplace, and an active political and intellectual life, yet small enough to draw food from the surrounding region and compact enough so that a person could walk out into the country within a few minutes from anywhere in the city. These planned cities resemble efficient ecosystems in which inhabitants participate in exchanges of matter and energy with their environment.

In the sixteenth and seventeenth centuries, Rome, London, Paris, Berlin, Vienna, and Brussels all were rebuilt with spacious boulevards, parks, and imposing public buildings following the rediscovery of Greek and Roman principles of city planning. The design of these European capitals is reflected in Pierre L'Enfant's plan for Washington, DC, which is one of the most notable city plans in the United States (fig. 24.17). Washington's rectangular street grid is cut by diagonal avenues converging on the Capitol and Executive Mansion, forming numerous squares, circles, and triangles that make interesting spaces for pocket parks. The broad, tree-shaded streets and impressive vistas created by this plan make Washington a grand setting for the national capital.

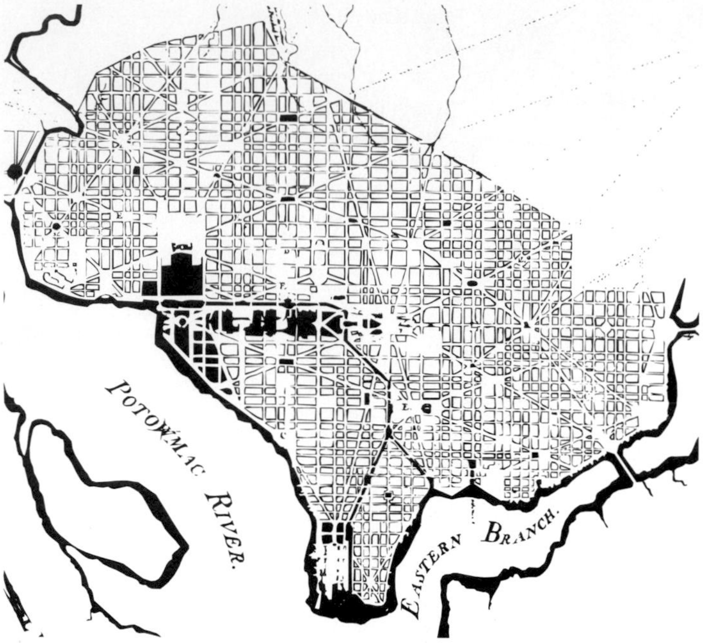

FIGURE 24.17 Pierre L'Enfant's plan for Washington, DC, is considered one of the best urban designs in North America. Spacious boulevards, gracious traffic circles, and grand ceremonial malls create open space and interesting patterns.
Reproduced by the U.S. Geological Survey for the Library of Congress.

Garden Cities and New Towns

The twentieth century has seen numerous experiments in building **new towns** for society at large that try to combine the best features of the rural village and the modern city. One of the most influential of all urban planners was Ebenezer Howard (1850–1929), who not only wrote about ideal urban environments but also built real cities to test his theories. In his *Garden Cities of Tomorrow,* written in 1898, Howard proposed that the congestion of London could be relieved by moving whole neighborhoods to **garden cities** separated from the central city by a greenbelt of forests and fields. Howard was the first modern urban planner to advocate comprehensive land-use planning and to reintroduce the Greek concept of organic growth and human scale to the city.

In the early 1900s, Howard worked with architect Raymond Unwin to build Letchworth and Welwyn Garden just outside of London. Interurban rail transportation provided access to these cities. Houses were clustered in "superblocks" surrounded by parks, gardens, and sports grounds. Streets were curved. Safe and convenient walking paths and overpasses protected pedestrians from traffic. Businesses and industries were screened from housing areas by vegetation. Each city was limited to about 30,000 people to facilitate social interaction. Housing and jobs were designed to create a mix of different kinds of people and to integrate work, social activities, and civic life. Trees and natural amenities were carefully preserved and the towns were laid out to maximize social interactions and healthful living. Care was taken to meet residents' psychological needs for security, identity, and stimulation.

Letchworth and Welwyn Garden each have 70 to 100 people per acre. This is a true urban density, about the same as New York City in the early 1800s and five times as many people as most suburbs today. By planning the ultimate size in advance and choosing the optimum locations for housing, shopping centers, industry, transportation, and recreation, Howard believed he could create a hospitable and satisfying urban setting while protecting open space and the natural environment. He intended to create parklike surroundings that would preserve small-town values and encourage community spirit in neighborhoods.

Letchworth and Welwyn Garden were the first of 32 new towns established in Great Britain, which now house about 1 million people. The Scandinavian countries have been especially successful in building garden cities. The former Soviet Union also built about 2000 new towns. Some are satellites of existing cities, and others are entirely new communities far removed from existing urban areas.

Planned communities also have been built in the United States following the theories of Ebenezer Howard, but most plans have been based on personal automobiles rather than public transit. In the 1920s, Lewis Mumford, Clarence Stein, and Henry Wright drew up plans that led to the establishment of Radburn, New Jersey, and Chatham Village (near Pittsburgh). Reston, Virginia, and Columbia, Maryland, both were founded in the early 1960s and are widely regarded as the most successful attempts to build new towns of their era (fig. 24.18). Another movement to build new towns according to Howard's principles has sprung up in the 1990s. Towns such as Seaside in northern Florida and Laguna West near Sacramento cluster houses to save open space and create a sense of community. Commercial centers are located within a few minutes walk of most houses, and streets are designed to encourage pedestrians and to provide places to gather and visit.

Cities of the Future

Is the expansion of cities—especially onto agricultural land—a waste of land and resources? Some people feel that growing cities decrease local agricultural resources and corrupt once pristine countryside and quaint rural towns. Others see this as a healthy opportunity to decongest central cities and build smaller, more pleasant, more livable communities that have the benefits of both technological cities and rural villages. Perhaps the demographic shifts of the

FIGURE 24.18 By clustering buildings together, Reston, VA, has maintained a high percentage of open space as public commons. Pedestrian walkways are separated from roads to provide a safe, pleasant way to get anywhere in the village.

last 40 years have merely been a transition between nineteenth-century industrial cities and garden cities of tomorrow. The polycentric network of urban villages in surrounding metropolitan areas may be a step toward more human-scale communities.

An alternative to spreading the population across a wide area of the countryside is to build upward. This model, which depends strongly on technology, has been called the **technopolis,** vertical city, or city of the future. The central hub of most big American cities now is dominated by skyscrapers and a highly technological environment (fig. 24.19). The emerging supercities of the Third World are also moving toward this style, in part because of its association with wealth, power, and progress.

There have been many proposals for gigantic, vertical supercities based on skyscrapers and completely artificial environments—what the Swiss architect Le Corbusier called "a machine for living." An example was Buckminster Fuller's proposed mile-high geodesic dome to cover most of Manhattan. Fuller also designed a hollow, tetrahedronal building 200 stories high that would float on the ocean and could house 1 million people. It could be built in stages, he suggested, like a beehive, with trailerlike modules inserted in successive layers into a mountainous, open-truss framework. Some architects have talked seriously about the eventual possibility of erecting 500-story, mile-high buildings! The psychological and physiological effects of living 5000 feet above street level can only be imagined.

Japan is now building 18 new high-technology cities intended to be centers for economic and scientific growth in the next century. With names like Teletopoia, Agripolis, and New Media City, these regional research, education, and marketing centers will have innovative housing, enclosed shopping malls, and high-technology communication and transportation systems. They will concentrate on leading-edge research and industries, such as fifth-generation supercomputers, biotechnology, lasers, ceramics, and bioelectronics. Some such high-tech cities may be giant, floating structures similar to Buckminster Fuller's visionary proposals. Others may consist of a maze of tunnels and chambers entirely underground, not unlike a giant ant nest, opening onto huge 20-story-deep air wells. This plan would conserve energy and preserve scarce land surface; however, many technical and psychological problems will need to be overcome.

FIGURE 24.19 For many people, the gleaming, climate-controlled, artificial environment of the modern high-rise city is the crowning achievement of modern civilization. To others, it seems brutal and dehumanizing. What do you think?

Urban Redesign

Most people in the United States now live in suburbs and will probably continue to do so in the foreseeable future. What can be done to make them more humane and environmentally sustainable?

Although suburbs have many advantages, they also have disadvantages. Most suburbs are too spread out for people to meet and interact, except with their immediate neighbors. Residential streets are empty during much of the day. Suburbs often have no sense of community and can be places of alienation and loneliness, just as cities often are. Although many urban problems have been eliminated, so have many of the finer aspects of city living. Suburbs, for instance, lack the activities, energy, and diversity that make cities exciting and dynamic. They have very limited artistic, cultural, and educational opportunities, compared to cities.

The low population density of the suburb makes public transport prohibitively expensive and private automobiles essential. Car pools for work, school, and extracurricular activities help fill group transportation needs. The uniform, single-family, detached houses of the suburbs offer few options to those who don't fit into the traditional, middle-class, nuclear family. As children grow up and leave home, parents often find the house bigger than they need. Single-parent families, households of single adults, the elderly, and the poor tend to have difficulty finding affordable suburban housing that fits their needs.

What can or should be done to create ideal suburban environments? Obviously not everyone in America will want to leave established suburban communities for new garden communities or wilderness utopias. Nor should they. Abandoning existing buildings and civic infrastructures would be a terrible waste, and if everyone moves to the country, natural areas will be obliterated. We need, instead, to find ways to remodel and revitalize existing suburban cities, reduce their problems, and adapt to the changing needs of their residents.

How can cities be redesigned to make them more diverse, flexible, and energy efficient? The following proposals have been suggested to give them some of the better aspects of both the rural village and the big city. Many of these ideas are already in use in European cities such as Stockholm, Sweden; Helsinki, Finland; Leichester, England; and Neerlands, the Netherlands. Architects Bill McDonough and Sym Van Der Ryn have been U.S. leaders in this movement.

- Limit city size or organize them in modules of 30,000 to 50,000 people, large enough to be a complete city but small enough to be a community. A greenbelt of agricultural and recreational land around the city limits growth while promoting efficient land use. By careful planning and cooperation with neighboring regions, a city of 50,000 people can have real urban amenities such as museums, performing arts centers, schools, hospitals, etc.
- Determine in advance where development will take place. This protects property values and prevents chaotic development in which the lowest uses drive out the better ones. It also recognizes historical and cultural values, agricultural resources, and such ecological factors as impact on wetlands, soil types, groundwater replenishment and protection, and preservation of aesthetically and ecologically valuable sites.
- Turn shopping malls into real city centers that invite people to stroll, meet friends, or listen to a debate or a street musician (fig. 24.20). If there aren't 100 places for an impromptu celebration, a place isn't a real city. Another test of a city is a vital nightlife. Design city spaces with sidewalk cafes, pocket parks, courtyards, balconies, and porticoes that shelter pedestrians, bring people together, and add life and security to the street. Restaurants, theaters, shopping areas, and public entertainment that draw people to the streets generate a sense of spontaneity, excitement, energy, and fun.
- Locate everyday shopping and services so people can meet daily needs with greater convenience, less stress, less automobile dependency, and less use of time and energy. This might be accomplished by encouraging small-scale commercial development in or close to residential areas. Perhaps we should once again have "mom and pop" stores on street corners or in homes.
- Increase jobs in the community by locating offices, light industry, and commercial centers in or near suburbs, or by enabling work at home via computer terminals. These alternatives save commuting time and energy and provide local jobs. There are also concerns, however, about work-at-home employees being exploited in low-paying "sweatshop" conditions by unscrupulous employers. Some safeguards may be needed.

FIGURE 24.20 Many cities have redesigned core shopping areas to be more "user friendly." Pedestrian shopping malls, such as this one in Boulder, Colorado, create space for entertainment, dining, and chance encounters and provide opportunities for building a sense of community.

FIGURE 24.21 Bicycles provide low cost, nonpolluting, energy-efficient urban transportation. Guangzhou (Canton), China, a city of 4 million people, is said to have 3 million bicycles.

- Encourage walking or the use of small, low-speed, energy-efficient vehicles (microcars, motorized tricycles, bicycles, etc.) for many local trips now performed by full-size automobiles. Creating special traffic lanes, reducing the number or size of parking spaces, or closing shopping streets to big cars might encourage such alternatives (fig. 24.21).
- Promote more diverse, flexible housing as alternatives to conventional, detached single-family houses. "In-fill" building between existing houses saves energy, reduces land costs, and might help provide a variety of living arrangements. Allowing owners to turn unused rooms into rental units provides space for those who can't afford a house and brings income to retired people who don't need a whole house themselves. Allowing single-parent families or groups of unrelated adults to share housing and to use facilities cooperatively also provides alternatives to those not living in a traditional nuclear family. One of the great "discoveries" of urban planning is that mixing various types of housing—individual homes, townhouses, and high-rise apartments—can be attractive if buildings are aesthetically arranged in relation to one another.
- Create housing "superblocks" that use space more efficiently and foster a sense of security and community. Widen peripheral arterial streets and provide pedestrian overpasses so traffic flows smoothly around residential areas; then reduce interior streets within blocks to narrow access lanes with speed bumps and barriers to through traffic so children can play more safely. The land released from streets can be used for gardens, linear parks, playgrounds, and other public areas that will foster community spirit and encourage people to get out and walk. Cars can be parked in remote lots or parking ramps, especially where people have access to public transit and can walk to work or shopping.
- Make cities more self-sustainable by growing food locally, recycling wastes and water, using renewable energy sources, reducing noise and pollution, and creating a cleaner, safer environment (fig. 24.22). Reclaimed inner-city space or a greenbelt of agricultural and forestland around the city provides food and open space as well as such valuable ecological services as purifying air, supplying clean water, and protecting wildlife habitat and recreation land.

FIGURE 24.22 Communal gardens in Seattle, Washington, use once-vacant land to grow food and give local residents a chance to experience nature in a highly urban setting.

- Rooftop gardens and natural plantings can absorb up to 70 percent of rain water. They provide habitat for birds and insects, and ameliorate climate. They save energy and provide contact with nature for building residents (see fig. 8.21).
- Invite public participation in decision making as was done in Chattanooga. Emphasize local history, culture, and environment to create a sense of community and identity. Create local networks in which residents take responsibility for crime prevention, fire protection, and home care of children, the elderly, sick, and disabled. Coordinate regional planning through metropolitan boards that cooperate with but do not supplant local governments.

Ecologists and conservation biologists have, in the past, tended to study pristine nature, untouched—as much as possible—by human influences. In recent years, however, we have come to recognize that cities are ecological systems too. If ecology is the relationship among organisms and their environment, what could be more ecological than studying how humans impact other species and our environment? And for a vast majority of us, that means studying urban ecosystems where we are the keystone species. Rather than merely seeing humans as disturbing factors in these ecosystems, some scientists are welcoming people as participants in research and project-based service learning.

Two of the most important examples of urban ecology research are Long-Term Ecological Research (LTER) sites in Phoenix and Baltimore. Funded by the National Science Foundation for an initial period of six years at nearly a million dollars per year, but with an understanding that the research will go on much beyond that, these projects use modern technology to study every aspect of urban ecology. How does the city shape its own weather? What plants and animals live in the city, and where, and how? What ecological processes cycle materials and energy? How do human activities influence those processes? One of the first things that scientists are discovering is that cities are not homogenous; like other ecosystems, they have patches at varying scales that change over time. Using computer models and sophisticated mapping techniques, ecologists can explore how patches shift around in space and time.

Environmental justice concerns are at the forefront of many urban ecology research projects. Where are toxic and hazardous materials generated, stored, and released in the city? How do they move around and where do they accumulate? In Detroit, for example, a group of students worked with experts to map data on more than 5000 children with elevated blood lead levels. Not surprisingly, they found a correlation between low income, old housing, incidence of poisoning, and concentration of special-education students. Sometimes public awareness of a problem is one of the best outcomes of this research. Another large group of students in the Detroit area, used geographic information systems together with chemical and biological analysis to prepare a detailed study of water quality and ecosystem health in the Rouge River. Participatory planning coupled with citizen science enable some cities to identify indicators of urban sustainability (table 24.3).

TABLE 24.3 Urban Sustainability Indicators

Children in poverty
Violent crime
Access to health care
Air and water quality, litter
Vacant or deteriorating housing
Participation in neighborhood organizations
Money earned and spent in neighborhood
Access to public transportation
Shopping and services within walking distance
Quality of schools
Cultural and recreational opportunities

Source: MN Citizen's Environment Action Committee, 1999.

Designing for Open Space

Traditional suburban development typically divides land into a checkerboard layout of nearly identical 1 to 5 ha parcels with no designated open space (fig. 24.23, *left*). The result is a sterile landscape consisting entirely of house lots and streets. This style of development, which is permitted—or even required—by local zoning and ordinances, consumes agricultural land and fragments wildlife habitat. Many of the characteristics that people move to the country to find—space, opportunities for outdoor recreation, access to wild nature, a rural ambience—are destroyed by dividing every acre into lots that are "too large to mow but too small to plow."

An interesting alternative known as **conservation development**, cluster housing, or open space zoning preserves at least half of a subdivision as natural areas, farmland, or other forms of open space. Among the leaders in this design movement are landscape architects Ian McHarg, Frederick Steiner, and Randall Arendt. They have shown that people who move to the country don't necessarily want to own a vast acreage or to live miles from the nearest neighbor; what they most desire is long views across an interesting landscape, an opportunity to see wildlife, and access to walking paths through woods or across wildflower meadows.

By carefully clustering houses on smaller lots, a conservation subdivision can provide the same number of buildable lots as a conventional subdivision and still preserve 50 to 70 percent of the land as open space (fig. 24.23, *right*). This not only reduces development costs (less distance to build roads, lay telephone lines, sewers, power cables, etc.) but also helps foster a greater sense of community among new residents. Walking paths and recreation areas get people out of their houses to meet their neighbors. Home owners have smaller lots to care for and yet everyone has an attractive vista and a feeling of spaciousness.

Some good examples of this approach are Farmview near Yardley, Pennsylvania, and Hawksnest in Delafield Township, Wisconsin. In Farmview, 332 homes are clustered in six small villages set in a 160 ha (414 acre) rural landscape, more than half of which is dedicated as permanent farmland. House lots and villages were strategically placed to maximize views, helping the development to lead its county in sales for upscale developments. Hawksnest is situated in dairy-farming country outside of Waukesha, Wisconsin. Seventy homes are situated amid 70 ha (180 acres) of meadows, ponds, and woodlands. Restored prairies, neighborhood recreational facilities, and connections to a national scenic trail have proved to be valuable marketing assets for this subdivision.

And urban habitat can make a significant contribution toward saving biodiversity. In a ground-breaking series of habitat conservation plans triggered by the need to protect the endangered California gnatcatcher, some 85,000 ha (210,000 acres) of coastal scrub near San Diego was protected as open space within the rapidly expanding urban area. This is an area larger than Yosemite Valley, and will benefit many other species as well as humans.

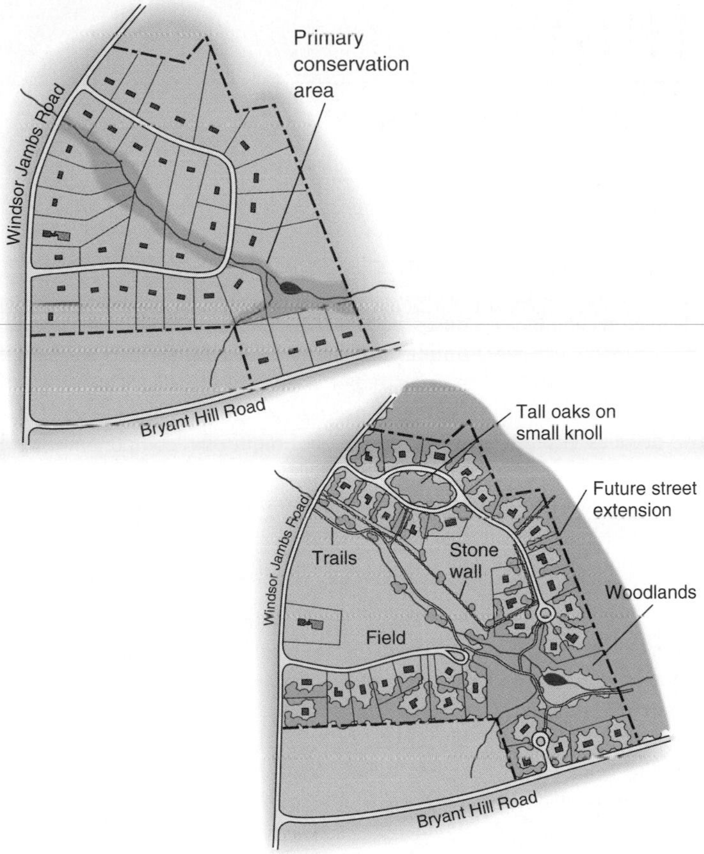

FIGURE 24.23 Conventional subdivision (*left*) and an open space plan (*right*). Although both plans provide 36 home sites, the conventional development allows for no open space. Cluster housing on smaller lots in the open space design preserves at least half the area as woods, prairie, wetlands, farms, or other conservation lands, while providing residents with more attractive vistas and recreational opportunities than a checkerboard development.

SUSTAINABLE DEVELOPMENT IN THE THIRD WORLD

What can be done to improve conditions in Third World cities? Curitiba, Brazil, is an outstanding example of what can be done, even in relatively undeveloped countries to improve transportation, protect central cities, and create a sense of civic pride (see Case Study, p. 566). Other cities have far to go, however, before they reach this standard. Among the immediate needs are housing, clean water, sanitation, food, education, health care, and basic transportation for their residents. The World Bank estimates that interventions to improve living conditions in urban households in the developing world could average the annual loss of almost 80 million "disability-free" years of life. This is about twice the feasible benefit estimated from all other environmental programs studied by the World Bank.

Some countries, recognizing the need to use vacant urban land, are redistributing unproductive land or closing their eyes to illegal land invasions. Indonesia, Peru, Tanzania, Zambia, and Pakistan have learned that squatter settlements make a valuable contribution to meeting national housing needs. Squatters' rights are being upheld in some cases, and such services as water, sewers, schools, and electricity are being provided to the settlements (fig. 24.24). Some countries intervene directly in land distribution and land prices. Tunisia, for instance, has a "rolling land bank" to buy and sell land. This strong and effective program controls urban land prices and reduces speculation and unproductive land ownership.

FIGURE 24.24 In this *colonia* on the outskirts of Mexico City, residents work with the government to bring in electricity, water, and sewers to shantytowns and squatter settlements. Like many Third World countries, Mexico recognizes that helping people help themselves is the best way to improve urban living.

Many planners argue that social justice and sustainable economic development are answers to the urban problems we have discussed in this chapter. If people have the opportunity and money to buy better housing, adequate food, clean water, sanitation, and other things they need for a decent life they will do so. Democracy, security, and improved economic conditions help in slowing population growth and reducing rural-to-city movement. An even more important measure of progress may be institution of a social welfare safety net guaranteeing that old or sick people will not be abandoned and alone.

Some countries have accomplished these goals even without industrialization and high incomes. Sri Lanka, for instance, has lessened the disparity between the core and periphery of the country. Giving all people equal access to food, shelter, education, and health care eliminates many incentives for interregional migration. Both population growth and city growth have been stabilized, even though the per capita income is only $800 per year. China has done something similar on a per capita income of around $300 per year.

Whether sustained, environmentally sound economic development is possible for a majority of the world's population remains one of the most important and most difficult questions in environmental science. The unequal relationship between the richer "Northern" countries and their impoverished "Southern" neighbors is a major part of this dilemma. Some people argue that the best hope for developing countries may be to "delink" themselves from the established international economic systems and develop direct south-south trade based on local self-sufficiency, regional cooperation, barter, and other forms of nontraditional exchange that are not biased in favor of the richer countries.

CASE STUDY

CURITIBA: AN ENVIRONMENTAL SHOWCASE

Curitiba, a Brazilian city of about 2 million people located on the Atlantic coast about 650 km (400 mi) southwest of Rio de Janeiro, has acquired a worldwide reputation for its innovative urban planning and environmental protection policies. Tree-lined streets, clean air, smoothly flowing traffic, and absence of litter and garbage make this one of Latin America's most liveable cities.

The architect of this remarkable program is mayor Jaime Lerner, who began in 1962 as a student leader protesting the proposed destruction of Curitiba's historic downtown center. Elected mayor nine years later, he has worked for more than two decades to preserve the city and make it a more beautiful and habitable place. Now Governor of the State of Parana, Lerner's success has thrust the city into the international spotlight as a shining example of what conservation and good citizenship can do to improve urban environments, even in Third World cities.

The heart of Curitiba's environmental plan is education for both children and adults. Signs posted along roadways proclaim "50 kg of paper equals one tree" and "recycle; it pays." School children study ecology along with Portuguese and math. With the assistance of children who encourage their parents, the city has successfully instituted a complex recycling plan that requires careful separation of different kinds of materials. The city calculates that 1200 trees per day are saved by paper recycled in this program. "Imagine if the whole of Brazil did this with an urban population 60 times greater than Curitiba," Lerner exclaims. "We could save 26 million trees per year!" More than 70 percent of the residents now recycle. Food and bus coupons are given out in exchange for recyclables, providing an alternative to welfare while also protecting the environment.

Another area in which Curitiba is setting an example is transportation. Faced with a population that tripled in two decades, bringing increasing levels of traffic congestion and air pollution, Curitiba had a transportation dilemma. The choices were either to bulldoze freeways through the historic heart of the city or to institute mass transit. The city chose mass transit.

Special articulated buses each carry up to 300 passengers on dedicated transit lanes. You pay your fare before entering the platform waiting area. When the bus pulls in, multiple doors open flush with the platform so that many people can enter or exit at once (even in wheelchairs) and each stop takes only seconds. A feeder network of smaller buses and vans collects riders from neighborhoods, so that everyone has convenient, frequent, and inexpensive transit service. Now more than three-quarters of the city's population leave their automobiles at home every day and take special express buses to work. The system is so successful that ridership has increased from 25,000 passengers a day 20 years ago to more than 1.5 million per day now, or 70 percent of all trips in the city. The result is not only less congestion and pollution but major energy savings.

Other measures to clear the air and reduce congestion include a limit on building height and construction of an industrial park outside city boundaries. Maximum use is made of all materials and buildings. Worn-out buses become city training centers, an old military fort is a cultural center, and a gunpowder depot is now a theater. Water and energy conservation are practiced widely. Even litter—an ubiquitous component of most Brazilian cities—is absent in Curitiba. More than one million trees have been planted and the green area ratio of 52 square meters of park per capita is higher than most American cities.

People are so imbued with city pride that they keep their surroundings spotless. To further beautify their city, civic volunteers have planted 1.5 million trees—more than any other place in Brazil.

Although many residents initially were skeptical of this environmental plan, now a remarkable 99 percent of the city's inhabitants would not want to live anywhere else. The World Bank uses Curitiba as an example of what can be done through civic leadership and public participation to clean up the urban environment. Some people claim that Curitiba, with its cool climate and high percentage of European immigrants, may be a special case among Brazilian cities. Lerner claims that Curitiba has no special features except concern, creativity, and communal efforts to care for its environment. Could you start a similar program in your hometown?

SUMMARY

A rural area is one in which a majority of residents are supported by methods of harvesting natural resources. An urban area is one in which a majority of residents are supported by manufacturing, commerce, or services. A village is a rural community. A city is an urban community with sufficient size and complexity to support economic specialization and to require a higher level of organization and opportunity than is found in a village.

Urbanization in the United States over the past 200 years has caused a dramatic demographic change. A similar shift is occurring in most parts of the world. Only Africa and South Asia remain predominantly rural, and cities are growing rapidly there as well. By the end of this century, we expect that more than half the world's people will live in urban areas. Most of that urban growth will be in the supercities of the Third World. A century ago only 13 cities had populations above 1 million; now there are 235 such cities. In the next century, that number will probably double again, and three-fourths of those cities will be in the Third World.

Cities grow by natural increase (births) and migration. People move into the city because they are "pushed" out of rural areas or because they are "pulled" in by the advantages and opportunities of the city. Huge, rapidly growing cities in the developing world often have appalling environmental conditions. Among the worst problems faced in these cities are traffic congestion, air pollution, inadequate or nonexistent sewers and waste disposal systems, water pollution, and housing shortages. Millions of people live in slums and shantytowns where conditions would crush any but the strongest spirit, yet these people raise families, educate their children, learn new jobs and new ways of living, and have hope for the future.

The problems of developed world cities tend to be associated with urban sprawl around the outskirts and decay and blight in the core. Over the past 50 years, many in the urban middle and upper classes moved to the suburbs, leaving elderly, very poor, handicapped, or economically marginal people in the inner city. Increasing joblessness and poverty in the inner city create a cycle of poverty from which it is difficult to escape. Still, there are ways that we can improve cities in both the developed and the developing world to make them healthier, safer, and more environmentally sound, socially just, and culturally fulfilling than they are now.

QUESTIONS FOR REVIEW

1. What is the difference between a city and a village and between rural and urban?
2. How many people now live in cities, and how many live in rural areas worldwide?
3. What changes in urbanization are predicted to occur in the next 50 years, and where will that change occur?
4. Identify the 13 largest cities in the world. Has the list changed in the past 50 years? Why?
5. When did the United States pass the point at which more people live in the city than the country? When will the rest of the world reach this point?
6. Describe the current conditions in a typical megacity of the developing world. What forces contribute to its growth?
7. Describe the difference between slums and shantytowns.
8. Why are urban areas in U.S. cities decaying?
9. How has transportation affected the development of cities? What have been the benefits and disadvantages of freeways?
10. Describe some ways that American cities and suburbs could be redesigned to be more ecologically sound, socially just, and culturally amenable.

QUESTIONS FOR CRITICAL THINKING

1. Picture yourself living in a rural village or a Third World city. What aspects of life there would you enjoy? What would be the most difficult for you to accept?
2. Are there fundamental differences between the lives of homeless people in First World and Third World cities? Where would you rather be?
3. A city could be considered an ecosystem. Using what you learned in chapters 2 and 3, describe the structure and function of a city in ecological terms.
4. Look at the major urban area(s) in your state. Why were they built where they are? Are those features now a benefit or drawback?
5. Weigh the costs and benefits of automobiles in modern American life. Would we have been better off if the internal combustion engine had never been invented?
6. Boulder, Colorado, has been a leader in controlling urban growth. One consequence is that housing costs have skyrocketed and poor people have been driven out. If you lived in Boulder, would you vote for additional population limits? What do you think is an optimum city size?
7. Eleven proposals are presented in this chapter for suburban redesign. Which of them would be appropriate or useful for your community? Try drawing up a plan for the ideal design of your neighborhood.
8. How much do you think the richer countries are responsible for conditions in the developing countries? How much have people there brought on themselves? What role should, or could, we play in remedying their problems?

KEY TERMS

city 549
conservation development 564
core regions 550
garden cities 560
megacity 549
new towns 560
pull factors 551
push factors 551
rural area 549
shantytowns 555
slums 554

squatter towns 555
technopolis 561
urban area 549
urbanization 548
urban renewal 556
village 549

ADDITIONAL INFORMATION ON THE INTERNET

Visit our website at http://www.mhhe.com/environmentalscience/ for specific information about each of the topics below:

Chattanooga Environmental Information

Sustainable San Francisco

Environmental Urban and Regional Planning

Homes and Communities, HUD: links concerning urban development.

Ecological "Footprints" of Different Countries and Cities, from IGC—Food, Agriculture & Trade

Sprawl Campaign Index-Sierra Club

NATURE: The Wild Side of New York

The Effect of City Living on Urban Wildlife

Phoenix Long-Term Ecological Research—an urban ecological research program

Ecologists Go to Town Investigations in Baltimore: urban LTER sites.

Urban Development Issues

Planning and Sustainable Cities

The Center for Urban Ecology

Sustainable Community Indicators

New Town Locator: find new towns in your state

Designs for New Towns (commercial site)

Celebration, Florida: opinions and information

CHAPTER 25

What Then Shall We Do?

All history consists of successive excursions from a single starting point to which man returns again and again to organize yet another search for a durable set of values.

—Aldo Leopold—

OBJECTIVES

After studying this chapter, you should be able to:

- be aware of the goals and opportunities in environmental education and environmental careers.
- recognize opportunities for making a difference through the goods and services we choose as well as the limits of green consumerism.
- compare the differences between radical and mainline environmental groups and the tactics they employ to bring about social change.
- summarize some practical suggestions for how to reduce consumption, organize an environmental campaign, communicate effectively with the media, and write to elected officials.
- appreciate the need for sustainable development and explain how nongovernmental groups work toward this goal.
- evaluate how green politics and environmental citizenship can help protect the earth.
- formulate your own philosophy and action plan for what you can and should do to create a better world and a sustainable environment.

Above: How can we ensure a liveable environment and a fair share of resources for everyone?

ESSENTIAL STUDY PARTNER

The Essential Study Partner (ESP) CD-ROM that accompanies this textbook contains additional information on the following topics:

- sustainable development: populations: economics: sustainability
- environmental practices and businesses, and green consumerism: populations: societal issues: public action
- environmental advocacy groups and individual participation: populations: societal issues: public action
- public action, radical groups and anti-environmental backlash: populations: societal issues: public action
- poor farming, over-exploitation of wildlife, and increasing desires for material things in developing countries: populations: societal issues: public action
- green political parties: populations: societal issues: public action
- international environmental issues and the Earth Summit and Conference on Global Warming: populations: societal issues: public action

The Battle in Seattle

The streets of Seattle, Washington, filled with billowing clouds of tear gas and pepper spray during the first week in December, 1999, as squadrons of police in full riot gear skirmished with surging masses of protestors in the most confrontational political demonstrations in the United States since the end of the Vietnam War (fig. 25.1). What was all this sound and fury about? The angry crowds were there to confront delegates from 135 member nations of the **World Trade Organization (WTO)** who were meeting to hammer out an agenda for the next round of negotiations to regulate international trade.

More than 50,000 people had come to Seattle from all over the world for this historic event. French farmers, Zapatista rebels from Mexico, Tibetan refugees, German anarchists, First Nations people from Canada, labor unionists, environmentalists, animal rights activists, and organic farmers were among those who filled the streets. They held teach-ins, workshops, and peaceful marches with thousands of participants, but what got headline coverage was civil disobedience including blocking streets, unfurling banners from giant construction cranes, and, for a few of the most radical anarchists, breaking windows, setting fires in dumpsters, and looting stores. These destructive acts were condemned by mainstream groups but got most of the press attention anyway.

The police, unprepared for such a massive protest movement, reacted erratically. Ordered to avoid confrontation on the first day of the protests, the police stood by while a small contingent of demonstrators smashed windows and vandalized property. The next day, stung by criticism of being too soft, the police used excessive force to clear the streets, firing rubber bullets and tear gas indiscriminately, spraying innocent bystanders with pepper spray, and clubbing nonviolent groups engaged in passive sit-ins. The mayor declared a civil emergency and a 24-hour curfew in the area around the Civic Center. The national guard was called in to assist the thousands of city police.

Few people in America had ever heard of the WTO before the historic protest in Seattle, and yet, this exclusive body has power that affects us all. Created in 1995 by an international treaty, the WTO is the successor to the General Agreement on Tariffs and Trade (GATT), established at the end of the Second World War to eliminate tariffs and trade barriers. But where GATT was limited to considering economic issues, the scope of the WTO has been expanded to "noneconomic trade barriers" such as food safety laws, quality standards, product labeling, workers rights, and environmental protection standards. With legal standing equivalent to the United Nations, the WTO operates largely in secret. When considering trade disputes, it meets at undisclosed times and places, making decisions behind closed doors based on confidential evidence.

WTO judges are trade bureaucrats, usually corporate lawyers with ties to the industries being regulated. There are no rules against conflicts of interest, nor are there requirements that judges know anything about the culture or circumstances of the countries they judge. No appeal of WTO rulings is allowed. A country that loses a trade dispute has three options: (1) amend laws to comply with WTO rules, (2) pay annual compensation—often millions of dollars—to the complainants, or (3) face nonnegotiable trade sanctions. Critics claim that the WTO always serves the interest of transnational corporations and the world's richest countries.

FIGURE 25.1 Police arrest protesters at the World Trade Organization meetings on December 1, 1999.

Most of the people protesting in Seattle agreed that the current WTO represents a threat to democracy, quality of life, environmental health, environmental justice, labor rights, and national sovereignty. Underneath these complaints is a broader unease about trends towards globalization and the power of transnational corporations. Interestingly, the protestors argued bitterly about the best solutions to these problems and how to achieve them. While many claimed they wanted to shut down the WTO, others actually want a stronger trade organization that can enforce rules to protect workers, environmental quality, and endangered species.

Among the most controversial issues brought up in this round of WTO negotiations are agricultural subsidies, child labor laws, occupational health and safety standards, protection of intellectual property, and environmental standards. Environmentalists, for example, were outraged by a 1998 WTO ruling that U.S. laws prohibiting the import of shrimp caught in nets that can entrap sea turtles are a barrier to trade. The United States must either accept shrimp regardless of how they are caught, or face large fines. Some other WTO rulings that overturn environmental or consumer safety laws require Europeans to allow importation of U.S. hormone-treated beef, Americans must accept tuna from Mexico that endangers dolphins, and the U.S. Environmental Protection Agency cannot bar import of low-quality gasoline that causes excessive air pollution. In some pending cases, Denmark wants to ban 200 lead compounds in consumer products; France wants to prohibit asbestos; and several countries want to eliminate electronic devices containing lead, mercury, and cadmium. Under current WTO rules, all of these cases probably will be ruled illegal.

In the end, the delegates adjourned without agreement on an agenda for the "Millennium Round" of the WTO. Developing countries, such as Malaysia, Brazil, Egypt, and India, refused to allow labor conditions into the debate. Major agricultural exporters, such as the United States, Canada, Brazil, and Australia, continued to demand an end to agricultural subsidies and protective policies. Japan and the European Union, on the other hand, maintain that they have a right to preserve small, family farms, rural lifestyles, and traditional methods of food production against foreign competition. Developing countries insist that protection of their environment and wildlife is no one's business but their own.

This historic event connects many of the topics in this book, and it introduces a number of issues we will consider in this final chapter. It shows how economics and social concerns together with seemingly obscure and remote groups such as the WTO can have important environmental consequences. It also raises questions about the best ways to bring about social change. Are direct action and civil disobedience appropriate and effective, or should we focus on teach-ins and peaceful marches? Are confrontations useful, or is it better to try to work within the generally accepted social paradigms and established institutions? Of course the answers to these questions depend on both the situations and individuals involved, but in this chapter, we will look at some things we can do collectively and individually to help preserve our global environment.

ENVIRONMENTAL EDUCATION

In 1990 Congress passed the National Environmental Education Act establishing two broad goals: (1) to improve understanding among the general public of the natural and built environment and the relationships between humans and their environment, including global aspects of environmental problems, and (2) to encourage postsecondary students to pursue careers related to the environment. Specific objectives proposed to meet these goals include developing an awareness and appreciation of our natural and social/cultural environment, knowledge of basic ecological concepts, acquaintance with a broad range of current environmental issues, and experience in using investigative, critical-thinking, and problem-solving skills in solving environmental problems (fig. 25.2). Several states, including Arizona, Florida, Maryland, Minnesota, Pennsylvania, and Wisconsin, have successfully incorporated these goals and objectives into their curricula.

Environmental Literacy

Speaking to the first of these broad goals, former Environmental Protection Agency administrator William K. Reilly called for broad **environmental literacy** in which every citizen is fluent in the principles of ecology and has a "working knowledge of the basic grammar and underlying syntax of environmental wisdom." Environmental literacy, according to Reilly can help create a stewardship ethic—a sense of duty to care for and manage wisely our natural endowment and our productive resources for the long haul. "Environmental education," he says, "boils down to one profoundly important imperative: preparing ourselves for life in the next century. When the twenty-first century rolls around, it will not be enough for a few specialists to know what is going on while the rest of us wander about in ignorance."

While you've made a great start toward learning about your environment by reading this book and taking a class in environmental science, I hope you will continue to read and expand your horizons after your class is finished. A large body of environmental literature can help you in this quest. Some of the most influential environmental books of all time are listed in table 25.1. To this list I'd add some personal favorites: *The Singing Wilderness* by Sigurd F. Olson, *My First Summer in the Sierra* by John Muir, and *Remaking Society: Pathways to a Green Future* by Murray Bookchin.

Citizen Science

While university classes often tend to be theoretical and abstract, many students are discovering they can make authentic contributions to scientific knowledge through active learning and undergraduate

FIGURE 25.2 Outdoor activities are an essential part of environmental education, which strives to introduce environmental literacy to the general population and to encourage students to prepare for environmental careers.

research programs. Internships in agencies or environmental organizations are one way of doing this. Another is to get involved in organized **citizen science** projects in which ordinary people join with established scientists to answer real scientific questions. Community-based research was pioneered in the Netherlands, where several dozen research centers now study environmental issues ranging from water quality in the Rhine River, cancer rates by geographic area, and substitutes for harmful organic solvents. In each project, students and neighborhood groups team with scientists and university personnel to collect data. Their results have been incorporated into official government policies.

Similar research opportunities exist in the United States and Canada. Among these are programs at the Cornell Lab of Ornithology, which recruits citizens to gather bird information in their own backyards. The Audubon Society conducts an annual Christmas bird count and a summer-breeding bird survey that represent the world's largest and oldest database on bird distribution and frequency. Earthwatch offers a much smaller but more intense opportunity to take part in research. Every year hundreds of Earthwatch projects each field a team of a dozen or so volunteers who spend a week or two working on issues ranging from loon nesting behavior to archaeological digs. The American River Watch organizes teams of students to measure water quality. You might be able to get academic credit as well as helpful practical experience in one of these research experiences.

TABLE 25.1 The Environmentalist's Bookshelf

What are some of the most influential and popular environmental books? A survey of environmental experts and leaders around the world listed the following among the 500 best books on nature and the environment[1]:

A Sand County Almanac by Aldo Leopold (100)[2]
Silent Spring by Rachel Carson (81)
State of the World by Lester Brown and the Worldwatch Institute (31)
The Population Bomb by Paul Ehrlich (28)
Walden by Henry David Thoreau (28)
Wilderness and the American Mind by Roderick Nash (21)
Small is Beautiful: Economics as if People Mattered by E. F. Schumacher (21)
Desert Solitaire: A Season in the Wilderness by Edward Abbey (20)
The Closing Circle: Nature, Man, and Technology by Barry Commoner (18)
The Limits to Growth: A Report for the Club of Rome's Project on the Predicament of Mankind by Donella H. Meadows, et al. (17)
The Unsettling of America: Culture and Agriculture by Wendell Berry (16)
Man and Nature by George Perkins Marsh (16)
Encounters with the Archdruid by John McPhee (16)
Man's Role in Changing the Face of the Earth by William Thomas (ed.), with Carl Sauer, Marston Bates, and Lewis Mumford (16)
The Monkey Wrench Gang by Edward Abbey (14)
World Resources by the World Resources Institute (14)
Gaia: An Atlas of Planet Management by Norman Meyers (13)
Soft Energy Paths: Toward a Durable Peace by Amory Lovins (12)
Our Common Future by the World Commission on Environment and Development (12)
The Immense Journey by Loren Eisley (11)
Only One Earth: The Care and Maintenance of a Small Planet by Barbara Ward and Rene Dubos (11)
For the Common Good: Redirecting the Economy Toward Community, the Environment, and a Sustainable Future by Herman Daly and John Cobb (10)
Deep Ecology: Living as if Nature Mattered by Bill Devall and George Sessions (10)
Arctic Dreams: Imagination and Desire in a Northern Landscape by Barry Lopez (9)
Pilgrim at Tinker Creek by Annie Dillard (8)

[1]Robert Merideth, 1992, G. K. Hall/Macmillan Inc.
[2] Indicates number of votes for each book. Because the preponderance of respondents were from the United States (82 percent), American books are probably overrepresented.

Environmental Careers

The need for both environmental educators and environmental professionals opens up many job opportunities in environmental fields. The EPA estimates, for example, that 100,000 new professionals will be required in the United States in the next five years alone to deal with hazardous waste problems. Scientists are needed to understand the natural world and the effects of human activity on the environment. Lawyers and other specialists are needed to develop government and industry policy, laws, and regulations to protect the environment. Engineers are needed to develop technologies and products to clean up pollution and to prevent its production in the first place. Economists, geographers, and social scientists are needed to evaluate the costs of pollution and resource depletion and to develop solutions that are socially, culturally, politically, and economically appropriate for different parts of the world. In addition, business will be looking for a new class of environmentally literate and responsible leaders who appreciate how products sold and services rendered affect our environment.

Trained people are essential in these professions at every level, from technical and clerical support staff to top managers. Perhaps the biggest national demand over the next few years will be for environmental educators to help train an environmentally literate populace. We urgently need many more teachers at every level who are trained in environmental education. Outdoor activities and natural sciences are important components of this mission, but environmental topics such as responsible consumerism, waste disposal, and respect for nature can and should be incorporated into reading, writing, arithmetic, and every other part of education. Table 25.2 presents some guidelines for an environmental education program.

Can environmental protection and resource conservation—a so-called green perspective—be a strategic advantage in business? Many companies think so. An increasing number are jumping on the environmental bandwagon, and most large corporations now have an environmental department. A few are beginning to explore integrated programs to design products and manufacturing processes to minimize environmental impacts. Often called "design for the environment," this approach is intended to avoid problems at the beginning rather than deal with them later on a case-by-case basis. In the long run, executives believe this will save money and make their businesses more competitive in future markets. The alternative is to face increasing pollution control and waste disposal costs—now estimated to be more than $100 billion per year for all American businesses—as well as to be tied up in expensive litigation and administrative proceedings.

The market for pollution-control technology and know-how is also expected to be huge. Cleaning up the former East Germany alone is expected to cost some $200 billion. Many companies are positioning themselves to cash in on this enormous market. Germany and Japan appear to be ahead of America in the pollution control field because they have had more stringent laws for many years, giving them more experience in reducing effluents.

The rush to "green up" business is good news for those looking for jobs in environmentally related fields, which are predicted to be among the fastest growing areas of employment during the next few years. The federal government alone projects a need to hire some 10,000 people per year in a variety of environmental disciplines (fig. 25.3). How can you prepare yourself to enter this market? The best bet is to get some technical training: environmental engineering, analytical chemistry, microbiology, ecology, limnology, groundwater hydrology, or computer science all have great potential. Currently, a chemical engineer with a graduate degree and some experience in an environmental field can practically name his or her salary. Some other very good possibilities are environmental law and business administration, both rapidly expanding fields.

FIGURE 25.3 Many interesting, well-paid jobs are opening up in environmental fields. Here an environmental technician uses a video probe to inspect a residential sewer line.

TABLE 25.2 Outcomes from Environmental Education

The natural context: An environmentally educated person understands the scientific concepts and facts that underlie environmental issues and the interrelationships that shape nature.

The social context: An environmentally educated person understands how human society is influencing the environment, as well as the economic, legal, and political mechanisms that provide avenues for addressing issues and situations.

The valuing context: An environmentally educated person explores his or her values in relation to environmental issues; from an understanding of the natural and social contexts, the person decides whether to keep or change those values.

The action context: An environmentally educated person becomes involved in activities to improve, maintain, or restore natural resources and environmental quality for all.

Source: A Greenprint for Minnesota, Minnesota Office of Environmental Education, 1993.

FIGURE 25.4 Is this our highest purpose?

For those who aren't inclined toward technical fields there are still opportunities for environmental careers. A good liberal arts education will help you develop skills such as communication, critical thinking, balance, vision, flexibility, and caring that should serve you well. Large companies need a wide variety of people; small companies need a few people who can do many things well.

I hope you have noticed the profiles throughout this book describing a variety of young people and the environmental careers they are pursuing. The tips they offer for how you might prepare yourself for similar jobs should help in your job search. There are many opportunities if you think creatively and are persistent.

INDIVIDUAL ACCOUNTABILITY

Some prime reasons for our destructive impacts on the earth are our consumption of resources and disposal of wastes. Technology has made consumer goods and services cheap and readily available in the richer countries of the world. As you already know, we in the industrialized world use resources at a rate out of proportion to our percentage of the population. If everyone in the world were to attempt to live at our level of consumption, given current methods of production, the results would surely be disastrous. In this section we will look at some options for consuming less and reducing our environmental impacts.

How Much Is Enough?

A century ago, economist and social critic, Thorstein Veblen, in his book, *The Theory of the Leisure Class,* coined the term **conspicuous consumption** to describe buying things we don't want or need just to impress others. How much more shocked he would be to see current trends. The average American now consumes twice as many goods and services as in 1950. The average house is now more than twice as big as it was 50 years ago, even though the typical family contains half as many people. We need more space to hold all the stuff we buy. Shopping has become the way many people define themselves (fig. 25.4). As Marx predicted, everything has become commodified; getting and spending have eclipsed family, ethnicity, even religion as the defining matrix of our lives. But the waste and tastelessness of much American Consumerism leaves a psychological void. Once we possess things, we find they don't make us young, beautiful, smart, and interesting as they promised. So we buy even more stuff to try to fill the empty feeling that we don't have time to have real friends, to cook real food, to have creative hobbies, or to do work that makes us feel we have accomplished something with our lives. Some social critics call this drive to possess stuff "affluenza."

A growing number of people find themselves stuck in a vicious circle: they work frantically at a job they hate, to buy things they don't need, so they can save time to work even longer hours. Seeking a measure of balance in their lives, they opt out of the rat race and voluntarily abandon stressful jobs for simpler, less complex ones, usually at lower pay. They give up big houses and choose smaller, less expensive ones, perhaps moving from busy metro areas to small towns where the pace is slower and the cost of living lower. They strip away the clutter from their lives and deliberately live with less stuff. As Thoreau wrote in *Walden,* "Our life is frittered away by detail . . . simplify, simplify."

Some more recent authors of **voluntary simplicity** and creative downscaling include Elaine St. James, whose 1994 book, *Simplify Your Life,* points out how even with innumerable labor-saving devices, our lives are much busier and more hectic than our grandparents. Similarly, Joe Dominguez and Vicky Robin write in *Your Money or Your Life,* "before you buy something, calculate how long you'll have to work to pay for it. Ask yourself, is it worth giving up that much of my life?" Some other books in this genre include Duane Elgin's *Voluntary Simplicity: Toward a Way of Life That is Outwardly Simple, Inwardly Rich,* and Sara Ban Brethnach's *Simple Abundance,* which suggest that simplicity is less about what to do than how to be. Many people see simplicity as frugality: make less and live on less. Some hope it will reduce stress, while others tie simplicity to environmental protection and social justice.

Perhaps you're at a point where you're considering careers and lifestyles. What kind of a career do you want? Would you trade more time and less stress for a lower salary? If you have a

what can you do?

Reducing Consumption

Purchase Less

Ask yourself whether you really need more stuff.
Avoid buying things you don't need or won't use.
Use items as long as possible (and don't replace them just because a new product becomes available).
Use the library instead of purchasing books you read.
Make gifts from materials already on hand, or give nonmaterial gifts.

Reduce Excess Packaging

Carry reusable bags when shopping and refuse bags for small purchases.
Buy items in bulk or with minimal packaging; avoid single-serving foods.
Choose packaging that can be recycled or reused.

Avoid Disposable Items

Use cloth napkins, handkerchiefs, and towels.
Bring a washable cup to meetings; use washable plates and utensils rather than single-use items.
Buy pens, razors, flashlights, and cameras with replaceable parts.
Choose items built to last and have them repaired; you will save materials and energy while providing jobs in your community.

Conserve Energy

Walk, bicycle, or use public transportation.
Turn off (or avoid turning on) lights, water, heat, and air conditioning when possible.
Put up clotheslines or racks in the backyard, carport, or basement to avoid using a clothes dryer.
Carpool and combine trips to reduce car mileage.

Save Water

Water lawns and gardens only when necessary.
Use water-saving devices and fewer flushes with toilets.
Don't leave water running when washing hands, food, dishes, and teeth.

Based on material by Karen Oberhauser, Bell Museum Imprint, University of Minnesota, 1992. Used by permission.

family, is it more important to earn a lot of money to send your children to the best schools, or to have time to spend with them when they're growing up? The ultimate irony of this movement is that merchandizers are now beginning to capitalize on it. Expensive clothes, household appliances, even luxury automobiles claim they can "simplify your life" and make big profits along the way. The "What Can You Do?" box above offers some suggestions for reducing our consumption levels by buying only what we need.

Shopping for Green Products

Obviously, we can never reduce our consumption levels to zero. We can, however, make sound, informed decisions about the products we buy to select those that are least environmentally damaging in production, use, and disposal. As consumers demand environmentally friendly products, manufacturers, food producers, and merchants are moving to safer, more humane, and more sustainable consumer items. Although each of our individual choices makes a small impact, collectively they can be important.

With surveys showing nine out of ten of us willing to pay more for products and packaging that don't harm the environment, many merchandisers are moving quickly to cash in on green consumerism. Green sales and services are projected to nearly quintuple, from $1.8 billion in 1990 to $8.8 billion in 1995. Some claims made by green marketers are of questionable validity, however. Consumers must look closely to avoid "green scams." Many terms used in advertising are vague and have little meaning. For example:

- "Nontoxic" suggests that a product has no harmful effects on humans. Since there is no legal definition of the term, however, it can have many meanings. How nontoxic is the product? And to whom? Substances not poisonous to humans can be harmful to other organisms.
- "Biodegradable," "recyclable," "reusable," or "compostable" may be technically correct but not signify much. Almost everything will biodegrade *eventually,* but it may take thousands of years. Similarly, almost anything is potentially recyclable or reusable; the real question is whether there are programs to do so in your community. If the only recycling or composting program for a particular material is half a continent away, this claim has little value.
- "Natural" is another vague and often misused term. Many natural ingredients—lead or arsenic, for instance—are highly toxic. Synthetic materials are not necessarily more dangerous or environmentally damaging than those created by nature.
- "Organic" can connote different things in different places. Some states have standards for organic food, but others do not. On items such as shampoos and skin-care products, "organic" may have no significance at all. Most detergents and oils are organic chemicals, whether they are synthesized in a laboratory or found in nature. Few of these products are likely to have pesticide residues anyway.
- "Environmentally friendly," "environmentally safe," and "won't harm the ozone layer" are often empty claims. Since there are no standards to define these terms, anyone can use them. How much energy and nonrenewable material are used in manufacture, shipping, or use of the product? How much waste is generated, and how will the item be disposed of when it is no longer functional? One product may well be more environmentally benign than another, but be careful who makes this claim.

FIGURE 25.5 American (*a*) and Canadian (*b*) symbols that will be used to indicate products that are "environmentally superior."
(*a*) Reproduced with permission, Green Seal, Inc., Washington, DC. (*b*) Official mark of Environment Canada.

Blue Angels and Green Seals

Products that claim to be environmentally friendly are being introduced at twenty times the normal rate for consumer goods. To help consumers make informed choices, several national programs have been set up to independently and scientifically analyze environmental impacts of major products. Germany's Blue Angel, begun in 1978, is the oldest of these programs. Endorsement is highly sought after by producers since environmentally conscious shoppers have shown that they are willing to pay more for products they know have minimum environmental impacts. To date, more than 2000 products display the Blue Angel symbol. They range from recycled paper products, energy-efficient appliances, and phosphate-free detergents to refillable dispensers.

Similar programs are being proposed in every Western European country as well as in Japan and North America. Some are autonomous, nongovernmental efforts like the United States' new Green Seal program (managed by the Alliance for Social Responsibility in New York). Others are quasigovernmental institutions such as the Canadian Environmental Choice programs (fig. 25.5).

The best of these organizations attempt "cradle-to-grave" **life-cycle analysis** (fig. 25.6) that evaluates material and energy inputs and outputs at each stage of manufacture, use, and disposal of the product. While you need to consider your own situation in making choices, the information supplied by these independent agencies is generally more reliable than self-made claims from merchandisers.

Limits of Green Consumerism

To quote Kermit the Frog, "It's not easy being green." Even with the help of endorsement programs, doing the right thing from an environmental perspective may not be obvious. Often we are faced with complicated choices. Do the social benefits of buying rainforest nuts justify the energy expended in transporting them here, or would it be better to eat only locally grown products? In switching from Freon propellants to hydrocarbons, we spare the stratospheric ozone but increase hydrocarbon-caused smog. By choosing

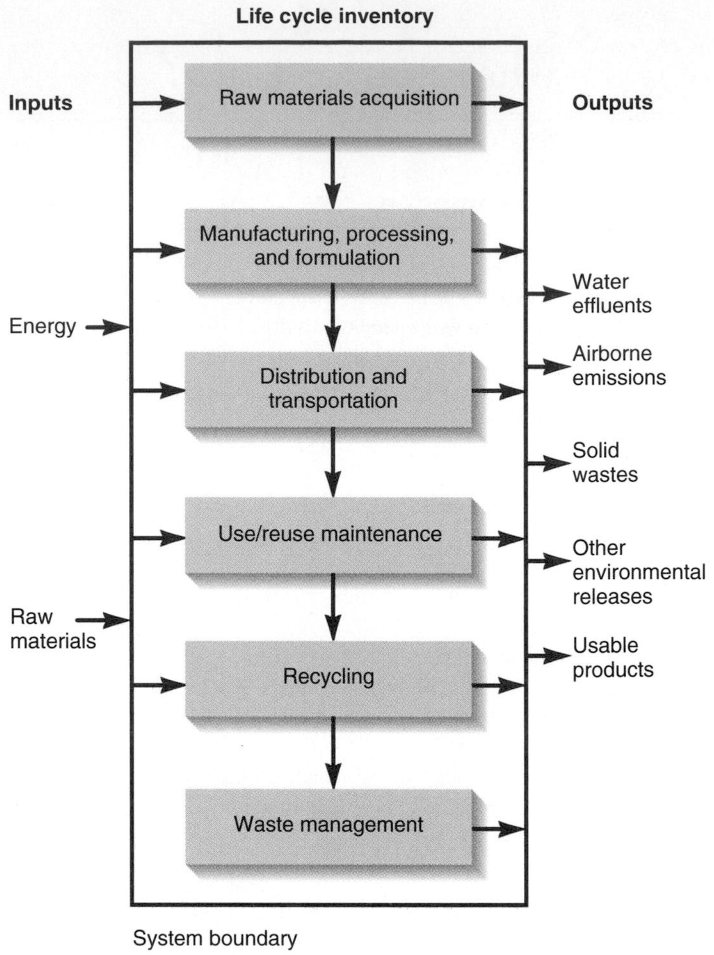

FIGURE 25.6 At each stage in its life cycle, a product receives inputs of materials and energy and produces outputs of materials or energy that move to subsequent phases and wastes that are released into the environment.
Source: After Society of Environmental Toxicology and Chemistry, 1991.

reusable diapers over disposable ones, we decrease the amount of material going to the landfill, but we also increase water pollution, energy consumption, and pesticide use (cotton is one of the most pesticide-intensive crops grown in the United States).

When the grocery store clerk asks you, "Paper or plastic?" you probably choose paper and feel environmentally virtuous, right? Everyone knows that plastic is made by synthetic chemical processes from nonrenewable petroleum or natural gas. Paper from naturally growing trees is a better environmental choice, isn't it? Well, not necessarily. In the first place, paper making consumes water and causes much more water pollution than does plastic manufacturing. Paper mills also release air pollutants, including foul-smelling sulfides and captans as well as highly toxic dioxins.

Furthermore, the brown paper bags used in most supermarkets are made primarily from virgin paper. Recycled fibers aren't strong enough for the weight they must carry. Growing, harvesting, and transporting logs from agroforestry plantations can be as environmentally disruptive as oil production. It takes a great deal of energy to pulp wood and dry newly made paper. Paper is also heavier and bulkier to ship than plastic. Although the polyethylene used to make a plastic bag contains many calories, in the end, paper bags are generally more energy-intensive to produce and market than plastic ones.

If both paper and plastic go to a landfill in your community, the plastic bag takes up less space. It doesn't decompose in the landfill, but neither does the paper in an air-tight, water-tight landfill. If paper is recycled but plastic is not, then the paper bag may be the better choice. If you are lucky enough to have both paper and plastic recycling, the plastic bag is probably a better choice since it recycles more easily and produces less pollution in the process. The best choice of all is to bring your own reusable cloth bag.

Complicated, isn't it? We often must make decisions without complete information, but it's important to make the best choices we can. Don't assume that your neighbors are wrong if they reach conclusions different from yours. They may have valid considerations of which you are unaware. The truth is that simple black and white answers often don't exist.

Paying Attention to What's Important

Taking personal responsibility for our environmental impact can have many benefits. Recycling, buying "green" products, and other environmental good works not only set good examples for your friends and neighbors, they also strengthen your sense of involvement and commitment in valuable ways. There are limits, however, to how much we can do individually through our buying habits and personal actions to bring about the fundamental changes needed to save the earth. Green consumerism generally can do little about larger issues of global equity, chronic poverty, oppression, and the suffering of millions of people in the Third World. There is a danger that exclusive focus on such problems as whether to choose paper or plastic bags will divert our attention from the greater need to change basic institutions.

During the 1970s, many people joined the back-to-the-land movement. Abandoning mass society, they moved to places like the mountains of North Carolina or remote valleys in Vermont where they experimented with self-sufficient lifestyles, living in converted school buses, teepees, or geodesic domes; growing their own food; cutting firewood; carrying water; and performing other aspects of alternative, utopian living. This lifestyle can be satisfying and morally uplifting but may contribute little toward helping other people live better lives.

Remaining a part of mass society, living an ordinary life, and devoting time and energy to bringing about important structural changes may ultimately be more valuable than withdrawing to "do your own thing." Rather than sort recyclables for which there is no market, for instance, your time might be better spent working for legislation to prevent excess packaging and to create markets for recycled materials. As in many other cases, there is no simple choice between these alternatives. Some of us will focus primarily on living sustainably, others will work to bring about change in the system, and still others will find some balance between these two options.

FIGURE 25.7 Student volunteers plant native trees and bushes in a watershed protection project. Working together on a practical problem can be enjoyable and productive.

COLLECTIVE ACTIONS

While a few exceptional individuals can be effective working alone to bring about change, most of us find it more productive and more satisfying to work with others.

Collective action multiplies individual power (fig. 25.7). You get encouragement and useful information from meeting regularly with others who share your interests. It's easy to get discouraged by the slow pace of change; having a support group helps maintain your enthusiasm. You should realize, however, that there is a broad spectrum of environmental and social action groups. Some will suit your particular interests, preferences, or beliefs more than others. In this section, we will look at some environmental organizations as well as options for getting involved.

Student Environmental Groups

A number of organizations have been established to teach ecology and environmental ethics to elementary and secondary school students, as well as to get them involved in active projects to clean up their local community. Groups such as Kids Saving the Earth or Eco-Kids Corps are an important way to reach this vital audience. Family education results from these efforts as well. In a World Wildlife Fund survey, 63 percent of young people said they "lobby" their parents about recycling and buying environmentally responsible products.

Organizations for secondary and college students often are among our most active and effective groups for environmental change. The largest student environmental group in North America is the Student Environmental Action Coalition (SEAC). Formed in 1988 by students at the University of North Carolina at Chapel Hill, SEAC has grown rapidly to more than 30,000 members in some 500 campus environmental groups. SEAC is both an

FIGURE 25.8 Making a ruckus on behalf of environmental protection can attract attention to your cause.

umbrella organization and grassroots network that functions as an information clearinghouse and a training center for student leaders. Member groups undertake a diverse spectrum of activities ranging from politically neutral recycling promotion to confrontational protests of government or industrial projects (fig. 25.8). National conferences bring together thousands of activists who share tactics and inspiration while also having fun. If there isn't a group on your campus, why not look into organizing one?

Another important student organizing group is the network of Public Interest Research Groups active on most campuses in the United States. While not focused exclusively on the environment, the PIRGs usually include environmental issues in their priorities for research. By becoming active, you could probably introduce environmental concerns to your local group if they are not already working on problems of importance to you.

One of the most important skills that you are likely to learn in SEAC or other groups committed to social change is how to organize. Organizing is a dynamic process in which you must constantly adapt to changing conditions. Some basic principles apply in most situations, however (table 25.3). Remember that you are not alone. Others share your concerns and want to work with you to bring about change; you just have to find them. There is power in numbers. As Margaret Mead once said, "Never doubt that a small, highly committed group of individuals can change the world; indeed, it is the only thing that ever has."

Using the communication media to get your message out is an important part of the modern environmental movement. Table 25.4 suggests some important considerations in planning a media campaign.

TABLE 25.3 Organizing an Environmental Campaign

1. What do you want to change? Are your goals realistic, given the time and resources you have available?
2. What and who will be needed to get the job done? What resources do you have now and how can you get more?
3. Who are the stakeholders in this issue? Who are your allies and constituents? How can you make contact with them?
4. How will your group make decisions and set priorities? Will you operate by consensus, majority vote, or informal agreement?
5. Have others already worked on this issue? What successes or failures did they have? Can you learn from their experience?
6. Who has the power to give you what you want or to solve the problem? Which individuals, organizations, corporations, or elected officials should be targeted by your campaign?
7. What tactics will be effective? Using the wrong tactics can alienate people and be worse than taking no action at all.
8. Are there social, cultural, or economic factors that should be recognized in this situation? Will the way you dress, talk, or behave offend or alienate your intended audience? Is it important to change your appearance or tactics to gain support?
9. How will you know when you have succeeded? How will you evaluate the possible outcomes?
10. What will you do when the battle is over? Is yours a single-issue organization, or will you want to maintain the interest, momentum, and network you have established?

Source: Based on material from "Grassroots Organizing for Everyone" by Claire Greensfelder and Mike Roselle from *Call to Action,* 1990 Sierra Book Club Books.

Mainline Environmental Organizations

Among the oldest, largest, and most influential environmental groups in the United States are the National Wildlife Federation, the World Wildlife Fund, the Audubon Society, the Sierra Club, the Izaak Walton League, Friends of the Earth, Greenpeace, Ducks Unlimited, the Natural Resources Defense Council, and The Wilderness Society. Sometimes known as the "group of 10," these organizations are criticized by radical environmentalists for their tendency to compromise and cooperate with the establishment. Although many of these groups were militant—even extremist—in their formative stages, they now tend to be more staid and conservative. Members are mostly passive and know little about the inner workings of the organization, joining as much for publications or social aspects as for their stands on environmental issues.

Still, these groups are powerful and important forces in environmental protection. Their mass membership, large professional staffs, and long history give them a degree of respectability and influence not found in newer, smaller groups. The Sierra Club, for instance, with about half a million members and chapters in almost every state, has a national staff of about 400, an annual budget over $20 million, and 20 full-time professional lobbyists in Washington. These national groups have become a potent force in Congress, especially when they band together to pass specific legislation such as the Alaska National Interest Lands Act or the Clean Air Act.

In a survey that asked congressional staff and officials of government agencies to rate the effectiveness of groups that

TABLE 25.4 Using the Media to Influence Public Opinion

Shaping opinion, reaching consensus, electing public officials, and mobilizing action are accomplished primarily through the use of the communications media. To have an impact in public affairs, it is essential to know how to use these resources. Here are some suggestions:

1. *Assemble a press list.* Learn to write a good press release by studying books from your public library on press relations techniques. Get to know reporters from your local newspaper and TV stations.
2. *Appear on local radio and TV talk shows.* Get experts from local universities and organizations to appear.
3. *Write letters to the editor, feature stories, and news releases.* You may include black and white photographs. Submit them to local newspapers and magazines. Don't overlook weekly community shoppers and other "freebie" newspapers, which usually are looking for newsworthy material.
4. *Try to get editorial support from local newspapers, radio, and TV stations.* Ask them to take a stand supporting your viewpoint. If you are successful, send a copy to your legislator and to other media managers.
5. *Put together a public service announcement and ask local radio and TV stations to run it* (preferably not at 2 A.M.). Your library or community college may well have audiovisual equipment that you can use. Cable TV stations usually have a public access channel and will help with production.
6. *If there are public figures in your area who have useful expertise, ask them to give a speech or make a statement.* A press conference, especially in a dramatic setting, often is a very effective way of attracting attention.
7. *Find celebrities or media personalities to support your position.* Ask them to give a concert or performance, both to raise money for your organization and to attract attention to the issue. They might like to be associated with your cause (fig. 25.9).
8. *Hold a media event that is photogenic and newsworthy.* Clean up your local river and invite photographers to accompany you. Picket the corporate offices of a polluter, wearing eye-catching costumes and carrying humorous signs. Don't be violent, abusive, or obnoxious; it will backfire on you. Good humor usually will go farther than threats.
9. *If you hear negative remarks about your issue on TV or radio, ask for free time under the Fairness Doctrine to respond.* Stations have to do a certain amount of public service to justify relicensing and may be happy to accommodate you.
10. *Ask your local TV or newspaper to do a documentary or feature story about your issue or about your organization and what it is trying to do.* You will not only get valuable free publicity, but you may inspire others to follow your example.

attempt to influence federal policy on pollution control, the top five were national environmental organizations. In spite of their large budgets and important connections, the American Petroleum Institute, the Chemical Manufacturers Association, and the Edison Electric Institute ranked far behind these environmental groups in terms of influence.

FIGURE 25.9 Celebrities can help raise money and bring visibility to your cause. Here Whitney Houston, Sting, Luciano Pavarotti, and Elton John perform at a Rainforest Foundation Benefit at Carnegie Hall.

Although much of the focus of the big environmental groups is in Washington, Audubon, Sierra Club, and Izaak Walton have local chapters, outings, and conservation projects. This can be a good way to get involved. Go to some meetings, volunteer, offer to help. You may have to start out stuffing envelopes or some other unglamourous job, but if you persevere, you may have a chance to do something important and fun. It's a good way to learn and meet people.

Some environmental groups such as the Environmental Defense Fund (EDF), The Nature Conservancy (TNC), National Resources Defense Council (NRDC), and the Wilderness Society (WS) have limited contact with ordinary members except through their publications. They depend on a professional staff to carry out the goals of the organization through litigation (EDF and NRDC), land acquisition (TNC), or lobbying (WS). Although not often in the public eye, these groups can be very effective because of their unique focus. The Nature Conservancy buys land of high ecological value that is threatened by development. With more than $50 million in cash purchases and $44 million in donated lands, TNC now owns some nine hundred parcels of land, the largest privately owned system of nature sanctuaries in the world. Altogether, it has saved more than 1 million ha (2.47 million acres) of land in its 40-year history.

Still, there often is bickering between groups. Members of Greenpeace and the Sierra Club angrily criticized The Nature Conservancy for taking "tainted money" from corporations such as Procter and Gamble (which tests products on animals), the Royal Caribbean Cruise Line (which was convicted twice in a year of dumping garbage in the ocean), and Gulf Oil, among others. The Nature Conservancy responds that they are trying to be part of the solution rather than just criticizing these companies. When is confrontation called for (What Do You Think? p. 580)? When is cooperation the best strategy (What Do You Think? p. 581)?

what do you think?

THE WORLD AROUND YOU

When Is Confrontation Appropriate?

As the opening story for this chapter shows, some people see civil disobedience and publicity-generating stunts to be both ethical and effective in winning the hearts and minds of the larger public. Others disagree, regarding the kinds of protests and vandalism that occurred in Seattle to be counterproductive: nothing more than immature, self-indulgent behavior. What do you think? Would you ever break the law in support of an environmental goal? If so, under what circumstances?

The issue of civil disobedience was discussed in an internet debate following the "Battle in Seattle." The following sentiments were among some of the messages posted. With which would you agree? Although these are only excerpts, what underlying worldviews or ethical perspectives do you think they represent? Do you find unspoken assumptions or stereotypes here?

- "*Non-violent* civil disobedience has a long tradition—Henry Thoreau, Mohandas Gandhi, Martin Luther King, and Nelson Mandela are the inspiration for movements around the world that brought down corrupt governments and secured civil and religious liberty."
- "The ends never justify the means. This is a country based on law, not emotionally charged mobs and vigilante justice. They do more harm by offending potential supporters than they can ever offset with their small victories."
- "The women's suffrage movement in Britain did a fair amount of civil disobedience, and breaking windows was part of that. When people feel they have no other way to make their voice heard, they resort to desperate acts."
- "When we accept the premise that we can break the law to suit our purposes, we give up the moral high ground. If you claim you are entitled to destroy property or endanger lives, then your opponents will feel justified in doing the same things to you."
- "There are times when a person is not only justified in violating the law, but morally obligated to do so. Civil disobedience and monkey wrenching have been called radical; I consider the damage done to our environment in the name of world trade to be much more radical and destructive."

What would you add to this debate? Is civil disobedience an effective tool? Are there conditions under which breaking the law is justified, perhaps even demanded for an ethical person?

Broadening the Environmental Agenda

The environmental movement in the United States tends to be overwhelmingly white, middle-class, and suburban. Few blue-collar workers and even fewer minorities are involved, especially in leadership or professional positions. This is unfortunate in several ways. The movement will probably never be truly successful until it has a broad-based coalition of support. Furthermore, as we saw in chapter 2, poor people are often most seriously affected by toxic waste dumps, noise, urban blight, and air and water pollution (fig. 25.10). They should be included in discussions of how to remedy these problems. Unfortunately, most environmental groups have acquired a reputation for caring about plants, animals, and wilderness areas more than about people. Recently, a number of grassroots citizen's groups have begun to demand cleaner local environments. Often organized around tenant councils or other social justice issues, these groups could be a powerful force for general environmental protection.

FIGURE 25.10 Mainline environmental organizations tend to concentrate on wildlife and wilderness, paying scant attention to environmental justice and inner-city issues.

Deep or Shallow Environmentalism?

Norwegian philosopher Arne Naess criticizes the superficial commitment of environmentalists who claim to be green but are quick to compromise and who do little to bring about fundamental change. He characterizes this as **shallow ecology** and contrasts it with what he calls **deep ecology,** a profoundly biocentric worldview that calls for a fundamental shift in our attitudes and behavior. An equally radical but more humanist philosophy called **social ecology** is proposed by Murray Bookchin, who draws on the communitarian anarchism of the Russian geographer Peter Kropotkin.

what do you think? THE WORLD AROUND YOU

When Is Cooperation Helpful?

Rather than use confrontation and denunciation as methods for solving disputes, Dan Dagget, author of *Beyond Rangeland Conflict*, calls for grassroots cooperation and compromise. "Name-calling" Dagget says, "won't solve our environmental problems. Nor does it help to paint the world in shades of guilt or look for someone to blame. I've decided to join the growing number of people who have made a commitment to work together in spite of diverse politics, religious beliefs, and culture to try to improve our common environment . . . This is something each of us uses every day when we work to achieve our goals—to plant a garden, to paint a house, to do our job. Frequently, in these situations, we work with people we don't agree with politically, but we still get the job done. And we get better results than the politicians do when they try to solve our problems for us."

Dagget describes how a diverse group of ranchers, government land managers, wise-users, radical environmentalists, and just plain citizens in central Nevada formed a group called the Toiyabe Wetlands and Watersheds Management Team to work on reversing the effects of more than a century of overgrazing. Although the federal government had plowed loads of money into trying to protect and restore rangelands in this area, they made little progress until local residents began a program of collaborative stewardship. Similarly, on a northern Arizona ranch belonging to the family of Secretary of the Interior Bruce Babbitt, a team of local environmentalists, ranchers, students, and National Park Service staffers is studying better ways to sustainably manage grasslands of the arid Southwest. And in northeastern California, a group of townsfolk in Quincy got together to try to save their regional economy and to hammer out a management plan for the surrounding fire-prone forests.

In working together, people come to understand perspectives different from their own. They begin to see their opponents as decent individuals rather than faceless members of some vast, evil conspiracy. And when a group has some small, tangible successes, they are more likely to tackle bigger issues. Critics of these efforts, on the other hand, say that working on mundane problems distracts your attention from structural relations that cause those problems in the first place. In the jargon of the 1960s, you become co-opted—too busy with meaningless details to pay attention to what's really important. While personal friendships may facilitate dialogue, they also mute disagreements.

What do you think? Is it better to remain independent and critical, or to join together to try to find workable solutions? How would you decide when to compromise and when to stand firm?

Among the tenets of both these "dark green" philosophies are voluntary simplicity; rejection of anthropocentric attitudes; intimate contact with nature; decentralization of power; support for cultural and biological diversity; a belief in the sacredness of nature; and direct personal action to protect nature, improve the environment, and bring about essential societal change. They differ mainly on the necessity for personal freedom and communal interaction.

Both Naess and Bookchin criticize mainstream environmental groups that favor pragmatic work within the broadly agreed-upon social agenda and the established political system to bring about incremental, progressive reform rather than radical revolution. Most progressives naturally object to being called shallow, light, or superficial; they prefer instead to characterize their position as moderate rather than radical, or reformist rather than revolutionary. They tend to view the deep ecologists and anarchists as unrealistic zealots who would rather be righteous than effective.

Aside from name calling, there are some important philosophical questions in this debate. One of the questions is whether we face an environmental crisis that requires radical action or only "problems" that can best be overcome by working within established channels. Another important question is whether it is better to overturn society to explore new ways of living or to work for progressive change within existing political, economic, and social systems. Finally, is it more important to work for personal perfection or collective improvement? There may be no single answer to any of these questions; it's good to have people working in many different ways to find solutions.

Is environmentalism a religion? A group of Minnesota loggers thinks so. In 1999 they sued the U.S. Forest Service for violating the separation of church and state by granting timber permit appeals filed by two environmental groups, Forest Guardians and the Superior Wilderness Action Network. The loggers claim that these environmental groups are actually espousing the "religion of deep ecology" in trying to prevent logging. According to the loggers, any philosophy that venerates the earth is a pagan religion similar to American Indian spirituality that should not be the basis for Forest Service policy. Deep ecology, their attorney charges, is the belief system behind efforts to preserve biodiversity. As evidence of the pantheism inherent in deep ecology, he points to a statement on a Forest Guardian web page that says, "If we see ourselves as a part of the Tree of Life—the interconnected web of beings we call Gaia—the Deep Ecological approach to spirituality might emphasize our relationship with this larger whole." What do you think? Is there anything in this statement incompatible with your religious beliefs?

While most environmentalists dismiss these claims as paranoid hysteria, they do raise some interesting questions about the differences between worldviews and religions. Both are concerned with the ultimate principles that give meaning and value to the world and explain how we should live. Both tend to have symbols, rituals, and practices carried out by their adherents (fig. 25.11). Both have revered founders or historic figures and a canon of cherished books regarded as a source of authority. But are beliefs in free-market capitalism or libertarian property rights any less an explanation of how things work and how we should live than deep ecology? How would you distinguish between a worldview and a religion?

FIGURE 25.11 Gaia or Earth Mother bestows her largess to the little creatures in a May Day pageant. When do worldviews become a religion?

FIGURE 25.12 Street theater can be a humorous, yet effective, way to convey a point in a nonthreatening way. Confrontational tactics get attention, but they may alienate those who might be your allies and harden your opposition.

Radical Environmental Groups

A striking contrast to the mainline conservation organizations are the **direct action** groups, such as Earth First!, Sea Shepherd, and a few other groups that form either the "cutting edge" or the "radical fringe" of the environmental movement, depending on your outlook. Often associated with the deep ecology philosophy and bioregional ecological perspective, the strongest concerns of these militant environmentalists tend to be animal rights and protection of wild nature. Their main tactics are civil disobedience and attention-grabbing actions, such as guerrilla street theater, picketing, protest marches, road blockades, and other demonstrations. Many of these techniques are borrowed from the civil rights movement and Mahatma Gandhi's nonviolent civil disobedience. While often more innovative than the mainstream organizations, pioneering new issues and new approaches, the tactics of these groups can be controversial.

Members of Earth First! chained themselves to a giant tree-smashing bulldozer to prevent forest clearing in Texas, and blockaded roads being built into wilderness areas. In one remarkable case, a 25-year-old activist named Julia "Butterfly" Hill spent more than two years sitting on a small, tarp-covered platform 55 m (180 ft) up in a redwood tree in northern California. She was protesting the clear-cutting of the Headwaters forest by the Maxxam/Pacific Lumber Company. She came down from her perch in 1999 when her supporters raised $50,000 to buy the tree from Maxxam. Some protests are humorous and lighthearted, such as skits or street theater that get a point across in a nonthreatening way (fig. 25.12).

A more problematic tactic is **monkey wrenching,** or environmental sabotage, a concept made popular by author Edward Abbey's book *The Monkey Wrench Gang.* Among the actions advocated by some Earth First!ers are driving large spikes in trees to protect them from loggers, vandalizing construction equipment, pulling up survey stakes for unwanted developments, and destroying billboards.

This raises some difficult ethical questions. While each of us has a duty to resist immoral authority and destruction of nature, what kinds of action are justified? Suppose that in spiking trees you endanger loggers or mill workers; what moral responsibility do you bear? On a pragmatic level, does violent action engender or validate counterviolence?

Sea Shepherd, a radical offshoot from Greenpeace, is determined to stop the killing of marine mammals. Convinced that peaceful protests were not working, the members took direct action by sinking whaling vessels and destroying machinery at whale processing stations in Iceland. Greenpeace criticized this destructive approach, perhaps because of memories of the sinking of its ship *Rainbow Warrior* by the French secret service in New Zealand in 1985. They and other critics of ecotage (ecological sabotage) see these acts as environmental terrorism that give all environmentalists a bad name. Sea Shepherd's Paul Watson says, "Our aggressive nonviolence is just doing what must be done, according to the dictates of our own conscience." Earth First!'s Dave Foreman (once described as a kamikaze environmentalist) said, "Extremism in defense of Mother Earth is no vice." What do you think? Do ends justify means?

Antienvironmental Backlash

As the world becomes increasingly urbanized and interconnected, social controls become necessary for us to live together safely and peacefully. Some people, however, find these restrictions confining. Especially in the western United States and Alaska, where people cling to traditions of independence, self-reliance, and personal freedom, a widespread feeling prevails that outsiders and the government represent a threat to local customs and values. Those whose livelihood is based on mining, logging, ranching, fishing, farming, and other methods of resource-extraction see jobs, communities, and ways of life coming to an end (fig. 25.13).

To many who live in sparsely populated areas—like much of the West—big government seems to be alien yet all-invasive, interfering constantly and unnecessarily in their lives. Environmental rules and regulations are an important part of this cultural clash. Restrictions on resource access, recreational opportunities, development rights, and other ways that people have become accustomed to using the land seem an infringement. Often the rhetoric on both sides of this argument rises to shrill extremes (What Do You Think?, p. 584).

Drawing on the anger, suspicion, and resentment felt by many rural residents, a number of antienvironmental groups have sprung up in recent years. Grouped together under the banner of the **wise use movement,** these organizations generally stand for little or no government, unlimited access to resources, less pollution control, and reduced protection for wild nature and endangered species. The Center for the Defense of Free Enterprise, founded by conservative fund raiser Alan Gottleib and former Sierra Clubber Ron Arnold, leads the way in setting an agenda for the wise use movement. Sending out 30 million pieces of direct mail each year, Gottleib and Arnold raise $5 million annually for themselves and for antienvironmental causes.

The National Cattlemen's Association and the National Farm Bureau, with hundreds of thousands of members and annual budgets approaching $10 million, compare in size and influence to some of the largest environmental groups. Their main concerns are with farming and ranching issues such as water projects and grazing on federal lands. People for the West, with 15,000 members in more than 200 local chapters, claims to be a grassroots, populist organization. However, 12 of its 13 original board members and almost all of its $1 million budget come from big timber or mining corporations. The Mountain States Legal Foundation, founded by former Interior Secretary James Watt and funded by donations from Coors Beer and other corporations, provides legal services to antienvironmental groups such as ranchers who tried to stop wolf reintroduction in Yellowstone National Park.

The California Desert Coalition is an unlikely association of off-road bikers, miners, ranchers, and oilmen who resent passage of the 1995 California Desert Protection Act and the new national parks it established. An offshoot of this group is the "Sahara Club," which advocates intimidation and violence—including assassination—to terrorize opponents. As James Watt said in 1993, "If the troubles from environmentalists cannot be solved in the jury box or at the ballot box, perhaps the cartridge box should be used." The link between some of these groups and armed survivalists, tax protesters, and antigovernment militias is a stark contrast to radical environmental groups, whose activism generally is directed against property and ideas, not people.

FIGURE 25.13 To families dependent on jobs in logging, mining, ranching, or other resource-based occupations, environmental regulations may seem threatening. How do we balance human needs against those of nature?

Antienvironmental groups gained significant congressional power in 1994 when a large group of new conservative legislators was elected. Drastically reducing the power of governmental agencies, easing restrictions on access to public lands and natural resources, and weakening many environmental regulations were high on the list of priorities for this conservative revolution. Hundreds of bills were introduced in the 104th and 105th Congresses to accomplish these goals. Some inroads were made but these threats also reenergized environmental protection groups.

Still, some powerful foes of environmental protection—especially representatives and senators from Western states—remain in charge of key congressional committees. The swarm of antienvironmental riders attached to recent spending bills attests to their power.

Among the issues that remain high priorities for this group are: (1) weakening clean water regulations—especially wetland protection; (2) easing restrictions on air pollution; (3) reducing powers of the endangered species act; (4) opening up more federal lands to motorized recreational use and to resource extraction; (5) forcing government, under so-called "takings bills" to pay landowners

what do you think?

THE WORLD AROUND YOU

Evaluating Extremist Claims

Attaining a sustainable future requires us to change some of the ways we impact the earth. Some problems will likely turn out less serious and others more so, than we currently believe. Undoubtedly new concerns have yet to be uncovered. The issues will continue to stimulate strong feelings on both sides for several reasons. The problems are potentially very serious. Uncertainty is likely to continue to surround many of the issues. And, most importantly, needed changes often will challenge our world views.

Sometimes advocates on both sides get carried away and attempt to persuade through the use of biased language instead of relying on objective evidence to influence the thought of others. Critical thinking requires that conclusions be based on the objective evaluation of objective evidence. Bias, a person's inclination to favor or oppose a particular conclusion, is a formidable barrier to critical thinking.

Being able to detect bias enables a person to determine whether a claim or position is being presented objectively. There are a number of strategies to help identify bias. One clue as to whether or not a claim is being presented fairly is the person's choice of language. Many words have positive or negative connotations that can reflect emotion rather than objectivity.

The viewpoints presented below are positions that have appeared in print. Notice that bias can range from obvious to subtle.

Writer A. *Despite the hysterics of a few pseudo-scientists, there is no reason to believe in global warming.*

This writer seeks to persuade by using language to ridicule the view that the global warming threat is real. Would you believe a *"pseudo-scientist,"* especially one using *"hysterics?"* Such emotionally-charged language attacks the credibility of the opposing view. The implication is that only kooks think global warming likely. Would the sentence have lost its emotional punch if it had said: "Despite the claims of researchers, there is no reason to believe in global warming"?

Writer B. *With the collapse of Marxism, environmentalism has become the new refuge of socialist thinking. The environment is a great way to advance a political agenda that favors central planning and an intrusive government. What better way to control someone's property than to subordinate one's private property rights to environmental concerns?*

This writer attempts to discredit environmental concerns not with evidence, but by linking them to an unpopular political ideology. *Marxism* and *socialist* are highly negatively charged terms in the minds of most Americans. The writer seeks to influence the reader's opinion by claiming environmental concern is akin to accepting a distasteful political belief.

Writer C. *The battle to feed humanity is over. In the 1970s and 1980s hundreds of millions of people will starve to death in spite of any crash programs embarked upon now. At this late date nothing can prevent a substantial increase in the world death rate, although many lives could be saved through dramatic programs to 'stretch' the carrying capacity of the earth. . . . But these programs will only provide a stay of execution unless they are accompanied by determined and successful efforts at population control.*

Words and phrases conveying bias include: *"The battle . . . is over."* "Hundreds of millions *will* starve to death. . . . *Crash programs*. . . . will only provide a *stay of execution*. . . ." The writer hopes to persuade the reader through an emotional appeal rather than convince the reader through reason.

All people have biases. They are a part of our conceptual frameworks, which in turn, are rooted in our level of knowledge, past experiences, purposes, and vested interests. Identifying bias in ourselves and others is a necessary skill in the critical thinking process. What are some other strategies one could use to detect bias in addition to looking at language?

(Writer of A and B: Rush Limbaugh; C: Paul Ehrlich)

whose rights to develop property commercially is limited by environmental regulations; and (6) cutting requirements for business to keep track of and report pollution releases and use of toxic materials. Groups like the League of Conservation Voters can help keep you up-to-date on what is happening in the legislative arena.

GLOBAL ISSUES

At the first United Nations Conference on the Environment in 1972 in Stockholm, Prime Minister Indira Gandhi of India declared that "poverty is the greatest danger to the environment." Twenty years of research and subsequent international conferences have shown the truth of that statement. In 1992, the United Nations Conference on Environment and Development focused on the plight of the poor and the need for sustainable development to alleviate poverty and provide ways to avoid environmental destruction.

The needs are tremendous. The United Nations estimates that more than 1.3 billion people worldwide live in acute poverty and lack access to secure food supplies, safe drinking water, education, health services, infrastructure (roads, markets, etc.), land, credit, and jobs (fig. 25.14). Even more are exposed to dangerous air pollution, especially in urban areas and in homes where smoky fires provide the only means of cooking. At least a fifth of the world's population lacks access to adequate sanitation, leading to the spread of infectious diseases that debilitate the poor and prevent them from working effectively to improve their conditions.

The connection between poverty and the environment has been a major focus of this book. Having to meet urgent short-term needs for survival, the poor are forced to "mine" their natural capital through excessive tree cutting, poor farming and grazing practices that lead to erosion and nutrient depletion, and overharvesting of fisheries and wildlife. The consequent degradation of the land results in further impoverishment of the people who depend on that land for survival. Often the poorest people live in areas of high biodiversity and also high ecological vulnerability (fig. 25.15). As the World Bank *World Development Report* says, "Without adequate environmental protection, human development is undermined; without human development, environmental protection will fail."

FIGURE 25.14 More than 1 billion people—like this Indonesian family—live in acute poverty, lacking access to secure food supplies, safe drinking water, housing, education, health services, and decent jobs.

Public Opinions and Environmental Protection

In spite of the recent gains in political power by antienvironmental forces in the United States mentioned earlier in this chapter, people around the world are becoming increasingly aware of the damaging effects of pollution on their own health and prosperity. A 1999 *Vancouver Sun* poll found that environmental pollution is Canadians' top concern, having replaced economic hardship as the "greatest threat to future generations." Depletion of natural resources was third ranked in this poll. In a World Values survey carried out in 43 countries representing 70 percent of the world population, almost everyone polled—96 percent of all respondents—said they are in favor of environmental protection. Perhaps even more important, 65 percent said that they would pay more taxes if they were assured that the money would go toward pollution control.

In general, public concern about environmental protection tends to be highest where pollution is most severe. More than 75 percent of urban residents of China, South Korea, and Chile, for instance, would pay higher taxes for clean air. In places where urban air reaches hazardous levels more than 200 days per year, it

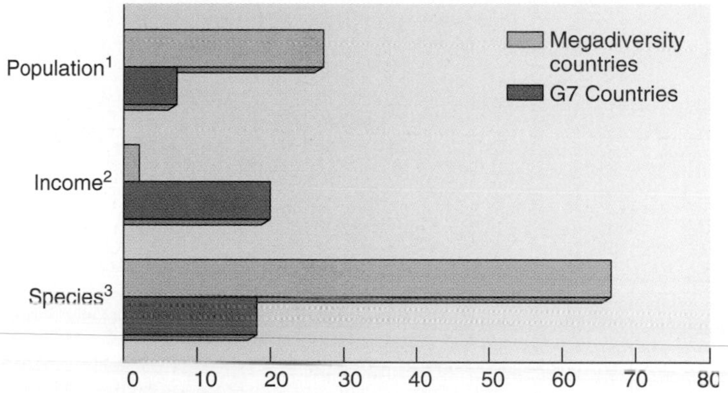

[1]total population × 100 million
[2]average per capita income × \$1000 U.S.
[3]total endemic plant species × 1000

FIGURE 25.15 Countries with the highest biodiversity also tend to have large human populations and high poverty rates. Preserving nature often requires improving options for poor people living in regions of high ecological vulnerability. The G7 countries are the United States, Britain, France, Germany, Japan, Italy, and Canada.
Source: Data from *World Resources 1996–97.*

is not surprising that pollution control is a popular topic. But our attitudes about our environment also are shaped by cultural values. Interestingly, some of the highest levels of public support for environmental protection are observed in countries such as Norway, Sweden, Denmark, and the Netherlands where environments are among the world's cleanest.

These countries also have high levels of what are called **postmaterialist values** that emphasize quality of life over acquisition of material goods. Postmaterialists place democracy, freedom, creativity, and humane society above making money or accumulating power. Countries with high levels of postmaterialist values are significantly more likely to take action to protect their environment than are those where the public is preoccupied with materialism (fig. 25.16). An encouraging trend is that young people in many societies tend to be much less materialistic than their parents or grandparents. In a survey of West European countries, people over 65 years were ten times more likely to agree with materialistic than postmaterialistic goals. Among people under 25 years in those same countries, on the other hand, postmaterialistic values predominate.

A majority of United States residents also strongly support environmental protection. A 1994 Times-Mirror poll of a random sample of Americans found the following:

- 82 percent wanted stricter laws to protect the environment.
- 76 percent said that water pollution laws have not gone far enough.
- 67 percent would pay higher prices for environmental defense.
- 66 percent believe that a clean environment and economic growth go hand in hand.
- 66 percent said that air pollution laws should be stronger.

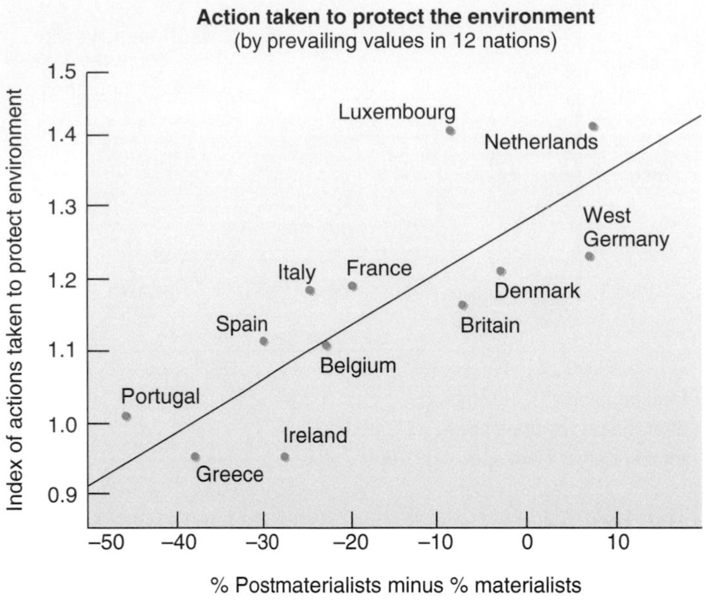

FIGURE 25.16 Countries where postmaterialist (quality of life) values outweigh materialistic values tend to be significantly more likely to act in favor of environmental protection than are countries where people are primarily concerned with acquiring money and material goods.
Source: R. Inglehart, "Public Support for Environmental Protection," *Political Science and Politics,* March 1995.

- 63 percent support more federal spending for the environment.
- 60 percent would choose a clean environment over economic growth if compromise is impossible.
- 54 percent think we need more protection for wild and natural areas.

These results show that politicians who claim a mandate from voters to eliminate environmental regulations are not really representing majority views. Perhaps as a new generation takes control, postmaterialist values will infuse the decision-making process in this country as it already has in many Nordic countries.

Sustainable Development

How can we accomplish the twin goals of improving human well-being and protecting our common environment? The developing countries of the world need access to more efficient, less-polluting technologies and to learn from the successes and failures of the developed countries (table 25.5). Technology transfer is essential. Environmental and resource protection benefits both the richer countries of the world and the poorer ones. It is in our best interests to help finance protection of our common future in some equitable way. Maurice Strong, chair of the Earth Charter Council, estimates that development aid from the richer countries should be some $150 billion per year, while internal investments in environmental protection by developing countries will need to be about twice that amount. Many scholars and social activists believe that poverty is at the core of many of the world's most serious human problems: hunger, child deaths, migrations, insurrections, and environmental degradation. One way to alleviate poverty is to foster economic growth so there can be a bigger share for everyone.

TABLE 25.5 Goals for Sustainable Development

1. Ensure every person the benefits of a healthy environment.
2. Sustain a healthy economy that affords the opportunity for a high quality of life.
3. Ensure equity and opportunity for economic, social, and environmental well-being.
4. Protect and restore natural resources for current and future generations.
5. Encourage stewardship.
6. Encourage people to work together to create healthy communities.
7. Create full opportunity for citizens, businesses, and communities to participate in and influence the natural resource, environmental, and economic decisions that affect them.
8. Move toward stabilization of the U.S. population.
9. Lead in developing and carrying out sustainable development policies globally.
10. Ensure access to formal education and lifelong learning that will prepare citizens for meaningful work and a high quality of life, and give them an understanding of concepts involved in sustainable development.

Source: President's Council on Sustainable Development, 1997.

Strong economic growth already is occurring in many places. The World Bank projects that if current trends continue, economic output in developing countries will rise by 4 to 5 percent per year in the next 40 years. Economies of industrialized countries are expected to grow more slowly but could still triple over that period. Altogether, the total world output could be quadruple what it is today.

That growth could provide funds to clean up environmental damage caused by earlier, wasteful technologies and misguided environmental policies. It is estimated to cost $350 billion per year to control population growth, develop renewable energy sources, stop soil erosion, protect ecosystems, and provide a decent standard of living for the world's poor. This is a great deal of money, but it is small compared to the $1 trillion per year spent on wars and military equipment.

While growth simply implies an increase in size, number, or rate of something, development, in economic terms, means a real increase in average welfare or well-being. **Sustainable development** based on the use of renewable resources in harmony with ecological systems is an attractive compromise to the extremes of no growth versus unlimited growth. Perhaps the best definition of this goal is that of the World Commission on Environment and Development, which defined sustainable development in *Our Common Future* as

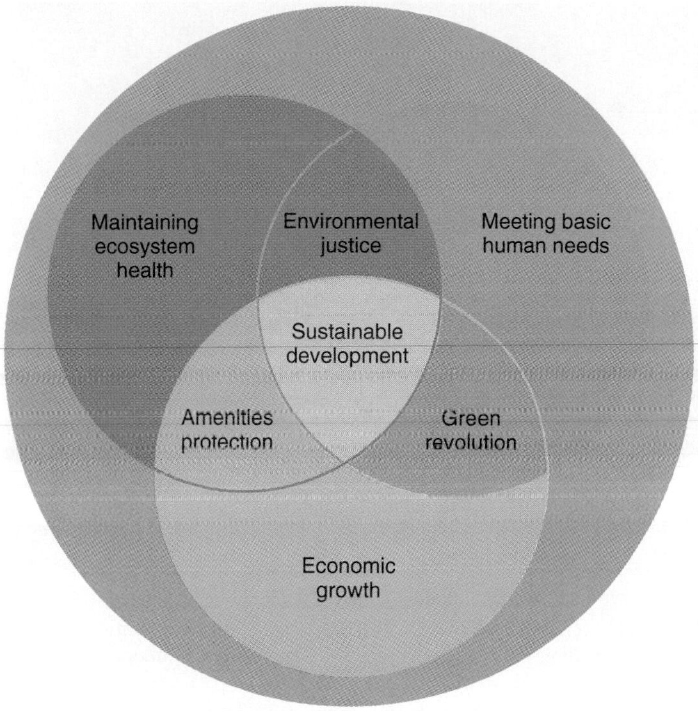

FIGURE 25.17 A model for integrating ecosystem health, human needs, and sustainable economic growth.
(Modified from Raymond Grizzle and Christopher Barrett, personal communication.)

"meeting the needs of the present without compromising the ability of future generations to meet their own needs." Some goals of sustainable development include:

- A demographic transition to a stable world population of low birth and death rates.
- An energy transition to high efficiency in production and use, coupled with increasing reliance on renewable resources.
- A resource transition to reliance on nature's "income" without depleting its "capital."
- An economic transition to sustainable development and a broader sharing of its benefits.
- A political transition to global negotiation grounded in complementary interests between North and South, East and West.
- An ethical or spiritual transition to attitudes that do not separate us from nature or each other.

In the United States, the President's Council on Sustainable Development has recommended 145 specific actions to accomplish three dozen policy goals. At the heart of the Council's findings is a conviction that economic, environmental, and social equity issues are inextricably linked and must be considered together (fig. 25.17). Another core belief is that to achieve sustainability, some things—jobs, productivity, wages, capital, savings, profits, information, knowledge, education, social welfare, and justice—must grow, while others—pollution, waste, and poverty—must not. Ten core goals and indicators of progress suggested by the Council are listed in table 25.5.

FIGURE 25.18 As whale blood pours from a Japanese whaling factory ship, Greenpeace protesters in a rubber dinghy draw attention to the killing of endangered species.

International Nongovernmental Organizations

The rise in international environmental organizations in recent years has been phenomenal. At the Stockholm Conference in 1972, only a handful of environmental groups attended, almost all from First World countries. Twenty years later, at the Rio Earth Summit, more than 30,000 individuals representing several thousand environmental groups, many from the Third World, held a global Ecoforum to debate issues and form alliances for a better world. We call these groups working for social change **nongovernmental organizations (NGOs).** They have become a powerful aspect of environmental protection.

Some NGOs are located primarily in the more highly developed countries of the North and work mainly on local issues. Others are headquartered in the North but focus their attention on the problems of developing countries in the South. Still others are truly global, with active groups in many different countries. A few are highly professional, combining private individuals with representatives of government agencies on quasi governmental boards or standing committees with considerable power. Others are at the fringes of society, sometimes literally voices crying in the wilderness. Many work for political change, more specialize in gathering and disseminating information, and some undertake direct action to protect a specific resource.

Public education and consciousness-raising using protest marches, demonstrations, civil disobedience, and other participatory public actions and media events are generally important tactics for these groups. Greenpeace, for instance, carries out well-publicized confrontations with whalers, seal hunters, toxic-waste dumpers, and others who threaten very specific and visible resources (fig 25.18). Greenpeace may well be the largest environmental organization in the world, with some 2.5 million contributing members.

In contrast to these highly visible groups, others choose to work behind the scenes, but their impact may be equally important. Conservation International has been a leader in debt-for-nature swaps to protect areas particularly rich in biodiversity. It also has some interesting initiatives in economic development, seeking products made by local people that will provide income along with environmental protection.

GREEN GOVERNMENT AND ENVIRONMENTAL CITIZENSHIP

What would it mean to be a global environmental citizen? Can the principles of environmental protection and sustainable development be incorporated into local and national government? In this section, we'll look at some ways that governments are responding to environmental problems as well as how you might get involved.

Environmental Citizenship

Have you ever thought about what the term citizenship really means? It's enlightening to look at its origins. In Latin, *civitas,* or citizen, forms the root for civil, city, civic, civilian, and civilization. A citizen is more than a mere resident; she or he must be an active participant to enjoy the privileges and benefits due members of society. A city is more than just a place to live; it is a community in which civic (community) duties and civil (polite) behavior contribute to civilization (an advanced state of development and culture). Civil defense, civil law, and civil rights all imply an organized effort for mutual protection, recognition of fundamental worth, and self-restraint. A citizen, then, is a member of a specific community who enjoys certain rights and duties, has certain attributes or virtues, and plays a meaningful role in maintaining the harmony and balance of both the place to live and the way of life. Within this broad definition, however, different perspectives give unequal weight to the various dimensions of citizenship. Liberals, for instance, emphasize rights, communitarians value community, populists highlight skills, immigration officers focus on formal membership.

How might this apply to environmental citizenship? As you've seen throughout this book, there are many different perspectives on nature and our role in it. Still, there are some principles with which many people now agree. It is clear that we depend on other organisms and ecological processes to sustain the environmental conditions on which our life depends. Where some people once regarded nature as external—and perhaps irrelevant—to humans, the lessons of modern ecology show us that we remain dependent on the natural world. As biologist E. O. Wilson says, "The little things run the world. Without them, we couldn't exist." Or as Aldo Leopold wrote in his essay on the land ethic, "We abuse land because we regard it as a commodity belonging to us. When we see land as a community to which we belong, we may begin to use it with love and respect."

FIGURE 25.19 Delegates to the Green Party convention in Germany debate their political, social, economic, and environmental platform.

Similarly, the spread of information technology and global travel reveals that environmental and social conditions in far away places have repercussions in our lives. We can no longer remain indifferent to the lives of those people with whom we are interlinked in a global economy and with whom we share a global environment.

Green Politics

In many countries with a parliamentary form of government, **"green" political parties** based on environmental interests have become a political force in recent years. The largest and most powerful green party is Germany's *Die Grünen.* It is a grassroots, egalitarian, council-style movement, committed to participatory democracy, environmental protection, and a fundamental transformation of society (fig. 25.19). An essential premise is that ceaseless industrial growth destroys both people and the environment and must, therefore, be stopped. The Greens have formed coalitions with antinuclear weapons protesters, feminists, human rights advocates, and other public interest groups, but have struggled over whether to work with or oppose the ruling centrist party. Political realists argue that they could be more effective within a larger coalition; idealists vow never to compromise their principles.

In 1998 federal elections, the Greens won 6.7 percent of the total vote, giving them 47 seats in the Bundestag, Germany's lower house. Forming a historic "red-green" coalition with the Social Democratic Party, the Greens, for the first time in world history, became a part of the ruling government in an industrialized country. In spite of the sudden access to power for environmentalists, it isn't clear that the coalition will hold. Some of the far left in the Greens continue to take a hard line on things that most Germans hold as cherished rights, such as the ability to drive at unlimited speeds on the Autoban, and to take fuel-guzzling vacations via jet airplane to far-away places.

The majority-rule political system in the United States makes it extremely difficult to start a new political party at the national level. Green candidates have been successful at the local and state levels, however. In 1990, Alaska became the first state to give a Green Party official standing. Jim Sykes, the Green gubernatorial candidate, received 3.2 percent of the total vote, enough to guarantee the party a spot on future ballots. National groups such as the League of Conservation Voters, and "green committees of correspondents," work to introduce environmental issues into party politics and support candidates who share their environmental concerns. Campus Greens are strong at many colleges and universities. The four key values they espouse are (1) ecological wisdom, (2) peace and social justice, (3) grassroots democracy, and (4) freedom from violence.

What Can Individuals do?

Although there is little likelihood of a green party gaining power at the United States federal level, there are many opportunities for environmentally concerned citizens to have an influence on government policies. Get to know the positions of your state legislators, congressional representatives, and senators. Be active in party politics. Going to your district, county, and state political conventions gives you a voice in choosing candidates and establishing party platforms. Individuals can also contact legislators directly. Table 25.6 gives some suggestions for how to write effectively. You would probably be surprised at how little mail most legislators get from their constituents. Even on controversial issues, a representative might receive less than 100 personal letters or telephone calls. Since each congressional district includes about a half a million people, the legislator tends to assume that each letter represents the views of 5000 to 10,000 people. Your voice can have great impact. A single, well-written letter can make a difference in whether important legislation gets passed.

All legislators now have e-mail access, but it isn't clear how many of them read internet messages or how much attention they pay to them. They may fall in the category of form letters and petitions. Individually they don't count for much, but if a decision maker gets a million of them, they probably will notice. If you don't have time to write, a telephone call to the local or Washington office of your senator or representative can be effective. Follow the same steps in formulating your argument that you would use in a letter. You probably won't talk directly to the senator or representative, but your opinion will be registered by an aide. Often congressional staff are the ones who are really knowledgeable about a specific topic. Their bosses depend on them for advice about how they should vote. A well-organized, factual argument can be very persuasive and may change the position of your legislator. Chapter 10 has more on environmental legislation, planning, policy, and law.

Don't forget that many important decisions also are made at the local level. Your city planning and zoning board, your county commissioners, state departments of natural resources or environmental protection, and a host of other agencies have power to establish many significant policies. These bodies are much more accessible for ordinary citizens. You might even run for a seat on a local board. What better way to ensure that local actions are sustainable than to become a policy maker yourself! Participating in practical environmental projects like litter cleanup or restoration projects can help build a sense of community, be educational, and also do good (fig. 25.20).

FIGURE 25.20 Schoolchildren collect litter and trash along the roadside in Queensland, Australia. Each of us can do something to improve our shared environment.

Finally, not everyone agrees with all the protest tactics used at the Seattle WTO meeting described at the beginning of this chapter, but public demonstrations have played an important role in getting issues on the public agenda and influencing public opinion and official policies. You might think about what kinds of protest you believe to be effective and acceptable, and whether you would be willing to participate the next time there is an important issue at stake.

The Earth Charter

The U.N. Conference on Environment and Development in Rio de Janiero in 1992 was a landmark in global conservation history. More than 30,000 people—including 178 heads of state and representatives from more than 1000 NGOs—gathered in Rio de Janeiro for this historic occasion. They considered a wide range of international environmental issues including protecting forests and biodiversity, prevention of global climate change, rights of indigenous and tribal peoples, the role of poverty in environmental problems, sustainable development, and aid for environmental protection programs in poor countries. The final document of the Earth Summit, called Agenda 21, was long on promise but lacking in specific detail about how to accomplish its lofty goals. Still, many important issues were discussed for the first time and the contacts made in this process, both government-to-government and person-to-person networking, may turn out to have been an important contribution.

Following the Rio Earth Summit, a U.N. Council was established in Costa Rica to draft an **Earth Charter** of principles for environmental protection and sustainable development. Intended to be a

TABLE 25.6 How to Write to Your Elected Officials

1. Address your letter properly.
   a. *Your representative:*
      The Honorable ________,
      House Office Building
      Washington, DC 20515
      Dear Representative ________,
   b. *Your senators:*
      The Honorable ________,
      Senate Office Building
      Washington, DC 20510
      Dear Senator ________,
   c. *The president:*
      The President
      The White House
      1600 Pennsylvania Avenue N.W.
      Washington, DC 20500
      Dear Mr. President,
2. Tell who you are and why you are interested in this subject. Be sure to give your return address.
3. Always be courteous and reasonable. You can disagree with a particular position, but be respectful in doing so. You will gain little by being shrill, hostile, or abusive.
4. Be brief. Keep letters to one page or less. Cover only one subject, and come to the point quickly. Trying to cover several issues confuses the subject and dilutes your impact.
5. Write in your own words. It is more important to be authentic than polished. Don't use form letters or stock phrases provided by others. Speak or write from your own personal experiences and interests. Try to show how the issue affects the legislator's own district and constituents.
6. If you are writing about a specific bill, identify it by number (for instance, H.R. 321 or S. 123). You can get a free copy of any bill or committee report by writing to the House Document Room, U.S. House of Representatives, Washington, DC 20515 or the Senate Document Room, U.S. Senate, Washington, DC 20510.
7. Ask your legislator to vote a specific way, support a specific amendment, or take a specific action. Otherwise you will get a form response that says: "Thank you for your concern. Of course I support clean air, pure water, apple pie, and motherhood."
8. If you have expert knowledge or specifically relevant experience, share it. But don't try to intimidate, threaten, or dazzle your representative. Don't pretend to have vast political influence or power. Legislators quickly see through artifice and posturing; they are professionals in this field.
9. If possible, include some reference to the legislator's past action on this or related issues. Show that you are aware of his or her past record and are following the issue closely.
10. Follow up with a short note of thanks after a vote on an issue that you support. Show your appreciation by making campaign contributions or working for candidates who support issues important to you.
11. Try to meet your senators and representatives when they come home to campaign, or visit their office in Washington if you are able. If they know who you are personally, you will have more influence when you call or write.
12. Join with others to exert your combined influence. An organization is usually more effective than isolated individuals.

statement of fundamental principles of enduring significance widely shared by people of all races, cultures, and religions, this statement draws on traditional wisdom, scientific insights, and a growing world literature on environmental ethics. It also draws on the experience of indigenous and tribal people and others whose cultural practices and belief systems most effectively promote environmental protection and sustainable livelihoods. The Benchmark II draft prepared for the "Millennium" meeting in January 2000 contained 49 principles drawn from hundreds of consultations over a three-year period in many countries and with people from all sectors of society. A summary of some key principles is presented in table 25.7.

In the preamble to the Earth Charter, the Earth Council delegates made the following poignant statement that sums up what we have discussed in this book: "In an increasingly interdependent world, it is imperative that we, the citizens of the Earth, declare our responsibility to one another, the greater community of life, and future generations. The Earth community stands at a defining moment. Environmental degradation, biodiversity loss, and depletion of natural resources threaten to destroy the ecological systems that sustain life. Injustice, poverty, and armed conflict deepen the world's suffering. Fundamental changes in our attitudes, values, and ways of living are necessary. A shared vision of basic values is urgently needed to provide an ethical foundation for the emerging world community. Therefore, in a spirit of human solidarity and kinship with all life, we affirm the following principles for sustainable development, which are interrelated and indivisible. We commit ourselves to implement these principles and to create a global partnership in support of their fulfillment.

As never before in human history, common destiny beckons us to redefine our priorities and to seek a new beginning. Such renewal is the promise of these Earth Charter principles. Fulfillment of this promise requires an inner change—a change of mind and heart. It requires that we take decisive action to adopt, apply, and develop the vision of the Earth Charter. Every individual, family, organization, business enterprise, and government has a critical role to play. Youth are fundamental actors for change. We can, if we will, take advantage of the creative possibilities before us and inaugurate an era of fresh hope."

In the end, although a global perspective is essential for both environmental protection and social justice, we need to, as E. F. Schumacher once said, "think globally but act locally." We hope this book will have given you the inspiration and the tools to do both.

TABLE 25.7 Earth Charter Principles

Together in hope, we pledge to:

1. Resect Earth and all life, recognizing the interdependence and intrinsic value of all beings.
2. Care for the community of life in all its diversity as a responsibility shared by everyone.
3. Strive to build free, just, participatory, and sustainable communities.
4. Secure peace and Earth's abundance and beauty for present and future generations.

In pursuit of these goals, we will:

5. Protect and restore the integrity of Earth's ecological systems with special concern for biological diversity and the natural processes that sustain and renew life.
6. Prevent harm to the environment as the best method of ecological protection, and when knowledge is limited, take the path of caution.
7. Treat all living beings with compassion, and protect them from cruelty and wanton destruction.
8. Adopt patterns of consumption, production, and reproduction that respect and safeguard Earth's regenerative capacities, human rights, and community well-being.
9. Ensure that economic activities support and promote human development in an equitable and sustainable manner.
10. Eradicate poverty as an ethical, social, economic, and ecological imperative.
11. Honor and defend the right of all persons, without discrimination, to an environment supportive of their dignity, bodily health, and spiritual well-being.
12. Advance worldwide the cooperative study of ecological systems, the dissemination and application of knowledge, and the development, adoption, and transfer of clean technologies.
13. Establish access to information, inclusive democratic participation in decision making, and transparency, truthfulness, and accountability in governance.
14. Affirm and promote gender equality as a prerequisite to sustainable development.
15. Make the knowledge, values, and skills needed to build just and sustainable communities an integral part of formal education and lifelong learning for all.
16. Create a culture of peace and cooperation.

Source: Earth Council

SUMMARY

Current trends are not predictions of what must be, but rather warnings of what may be, depending on the course we follow. Protests surrounding the meeting of the World Trade Organization in 1999 show how many cultural and institutional factors affect our global environment as well as how we are all interconnected. Knowledge about, and concern for, our environment is at an all-time worldwide high. Perhaps it is not unreasonable to hope that we will find ways to achieve social goals and environmental protection. The main goals of environmental education are to expand environmental literacy among the general public and to prepare students for environmental careers. Environmental literacy is a basic understanding of the natural and built environments and the relationships between humans and their environment. There are many opportunities for careers at a variety of different levels in the environmental field.

Businesses are rushing to invent and market environmental technology and "green" consumer goods. We have to look carefully at the claims made for "environmentally friendly" products. Some are true but others are not. We also must ask ourselves how much we need or have a right to consume. There is much we can do through our individual choices of goods and services, but we must be careful to avoid green scams and fraudulent marketing. Independent, scientific, lifetime analysis can help us make informed decisions. Some people are opting for voluntary simplicity to reduce stress, create time, and lessen their consumption levels.

There are many opportunities for collective action to bring about social change. Student environmental groups offer chances to network with others, learn useful organizing techniques, do good work, and have fun. The large national "mainline" environmental organizations have a degree of respectability, power, and influence unmatched by smaller, independent groups. Monkey wrenching, tree-sitting, and other forms of direct action are highly controversial tactics espoused by some radical environmentalists. A majority of the world's population supports environmental safeguards. Many people are willing to pay more for cleaner air and water, but antienvironmental forces also are becoming more vocal.

Sustainable development promises to alleviate acute poverty while also protecting the environment. Many nongovernmental organizations (NGOs) as well as intergovernmental treaties and conventions work toward these twin goals. In many countries, green politics espouses four key values of ecological wisdom, peace and social justice, grassroots democracy, and freedom from violence. While it is difficult to introduce new national parties in the United States, there are opportunities to work within the legislative, judicial, and administrative agencies to bring about change.

The Earth Charter affirms many key values of environmental science and sustainable development including respect for all life, ecosystem protection and restoration, human rights, equitable sharing of resources, eradication of poverty, democratic decision making, and a sense of shared responsibility for the well-being of the Earth community.

QUESTIONS FOR REVIEW

1. Summarize the goals of environmental education.
2. What are deep and shallow ecology? Why do progressives object to being called shallow ecologists?
3. List some things you can do to reduce resource consumption.
4. Discuss some of the limits to green consumerism and how you can spot fraudulent or meaningless green marketing claims.
5. Review the trade-offs between paper and plastic bags. Which is better in your view?
6. List ten key issues in organizing an environmental campaign or using media to influence public opinion.
7. What are the ten largest, oldest mainline environmental organizations in the United States?
8. Describe and evaluate the tactics used by radical environmental groups. Why is direct action sometimes called monkey wrenching?
9. Describe the goals and tactics of the antienvironmental wise use groups. In what way are their claims extremist?
10. List the four key values of the Green party.
11. Define sustainable development and identify its six goals.
12. In a few words, summarize the main concept in each of the four main goals of the Earth Charter.

QUESTIONS FOR CRITICAL THINKING

1. How would it change your life if all the principles of the Earth Charter were really taken seriously?
2. Do you agree with the principles presented in Table 25.7? Why or why not?
3. Put yourself in the place of someone who might object to one or more of these principles. What would they find challenging or objectionable?
4. The wise use movement appeals to people who fear that their livelihood and way of life are threatened by those who want to "lock up" resources in parks and nature preserves. What would you say to convince them that their loss is worthwhile?
5. People in many cultures traditionally regard wild nature as disagreeable and dangerous. How would you approach environmental education if you were assigned to such a country?
6. Do you support the conservative, mainstream approach to conservation followed by the mainline environmental groups or do you prefer the more challenging, innovative approaches of more radical groups? What are the advantages and disadvantages of each?
7. Extremists on both sides of the environmental debate sometimes use violence against property—and sometimes against people—to advance their agendas. Would you ever condone taking the law into your own hands for an environmental cause? Under what circumstances and with what limits might you do so?
8. Do you agree that sustainable development has the potential to simultaneously reduce poverty *and* protect the environment? What responsibility do those of us in rich nations bear toward the rest of the world?
9. Choose one of the ambiguous dilemmas presented in this chapter such as whether it is better to use paper or plastic bags. What additional information would you need to choose between alternatives? How should competing claims or considerations be weighed in an issue such as this?
10. The Talmud says, "If not now, when?" How might this apply to environmental science?

KEY TERMS

citizen science 572
conspicuous consumption 574
deep ecology 580
direct action 582
Earth Charter 589
environmental literacy 571

green political parties 588
life-cycle analysis 576
monkey wrenching 582
nongovernmental organizations (NGOs) 587
postmaterialist values 585
shallow ecology 580
social ecology 580
sustainable development 586
voluntary simplicity 574
wise use movement 583
World Trade Organization (WTO) 570

ADDITIONAL INFORMATION ON THE INTERNET

Visit our website at http://www.mhhe.com/environmentalscience/ for specific information about each of the topics below:

Handbook for a Better Future
Earthwatch: citizen participation in research and monitoring
League of Conservation Voters
Committee for the National Institute for the Environment
Natural Resource Defense Fund: Legislative Watch
Earth Justice
Union of Concerned Scientists: activist scientists on public policy
The Global Citizen: commentaries from Donella Meadows
Student Environmental Action Coalition
Current News Sources:
New York Times
Christian Science Monitor
National Public Radio
BBC News (British Broadcasting Corporation)
E: the Environmental Magazine
EE-Link: environmental publications
CNN science news
Living on Earth, National Public Radio's environmental news program
Information on international environmental agreements and enforcement (INECE)
Greenlines: news from Defenders of Wildlife
Natural Resources Defense Council news page
Institute for Global Communications

UNITS OF MEASUREMENT METRIC/ENGLISH CONVERSIONS

Length

1 meter = 39.4 inches = 3.28 feet = 1.09 yard
1 foot = 0.305 meters = 12 inches = 0.33 yard
1 inch = 2.54 centimeters
1 centimeter = 10 millimeters = 0.394 inch
1 millimeter = 0.001 meter = 0.01 centimeter = 0.039 inch
1 fathom = 6 feet = 1.83 meters
1 rod = 16.5 feet = 5 meters
1 chain = 4 rods = 66 feet = 20 meters
1 furlong = 10 chains = 40 rods = 660 feet = 200 meters
1 kilometer = 1,000 meters = 0.621 miles = 0.54 nautical miles
1 mile = 5,280 feet = 8 furlongs = 1.61 kilometers
1 nautical mile = 1.15 mile

Area

1 square centimeter = 0.155 square inch
1 square foot = 144 square inches = 929 square centimeters
1 square yard = 9 square feet = 0.836 square meters
1 square meter = 10.76 square feet = 1.196 square yards = 1 million square millimeters
1 hectare = 10,000 square meters = 0.01 square kilometers = 2.47 acres
1 acre = 43,560 square feet = 0.405 hectares
1 square kilometer = 100 hectares = 1 million square meters = 0.386 square miles = 247 acres
1 square mile = 640 acres = 2.59 square kilometers

Volume

1 cubic centimeter = 1 milliliter = 0.001 liter
1 cubic meter = 1 million cubic centimeters = 1,000 liters
1 cubic meter = 35.3 cubic feet = 1.307 cubic yards = 264 US gallons
1 cubic yard = 27 cubic feet = 0.765 cubic meters = 202 US gallons
1 cubic kilometer = 1 million cubic meters = 0.24 cubic mile = 264 billion gallons
1 cubic mile = 4.166 cubic kilometers
1 liter = 1,000 milliliters = 1.06 quarts = 0.265 US gallons = 0.035 cubic feet
1 US gallon = 4 quarts = 3.79 liters = 231 cubic inches = 0.83 imperial (British) gallons
1 quart = 2 pints = 4 cups = 0.94 liters
1 acre foot = 325,851 US gallons = 1,234,975 liters = 1,234 cubic meters
1 barrel (of oil) = 42 US gallons = 159 liters

Mass

1 microgram = 0.001 milligram = 0.000001 gram
1 gram = 1,000 milligrams = 0.035 ounce
1 kilogram = 1,000 grams = 2.205 pounds
1 pound = 16 ounces = 454 grams
1 short ton = 2,000 pounds = 909 kilograms
1 metric ton = 1,000 kilograms = 2,200 pounds

Temperature

Celsius to Fahrenheit ° F = (°C × 1.8) + 32
Fahrenheit to Celsius ° C = (° F – 32) ÷ 1.8

Energy and Power

1 erg = 1 dyne per square centimeter
1 joule = 10 million ergs
1 calorie = 4.184 joules
1 kilojoule = 1,000 joules = 0.949 British Thermal Units (BTU)
1 megajoule = MJ = 1,000,000 joules
1 kilocalorie = 1,000 calories = 3.97 BTU = 0.00116 kilowatt-hour
1 BTU = 0.293 watt-hour
1 kilowatt-hour = 1,000 watt-hours = 860 kilocalories = 3,400 BTU
1 horsepower = 640 kilocalories
1 quad = 1 quadrillion kilojoules = 2.93 trillion kilowatt-hours

Quantitative Prefixes

Large Numbers	Description	Small Numbers
exa 10^{18}	quintillion	alto 10^{-18}
peta 10^{15}	quadrillion	femto 10^{-15}
tera 10^{12}	trillion	pico 10^{-12}
giga 10^{9}	billion	nano 10^{-9}
mega 10^{6}	million	micro 10^{-6}
kilo 10^{3}	thousand	milli 10^{-3}

(e.g., a kilogram = 1,000 gm; a milligram = one-thousandth of a gram)

Glossary

A

abundance The number or amount of something.

acid precipitation Acidic rain, snow, or dry particles deposited from the air due to increased acids released by anthropogenic or natural resources.

active solar system A mechanical system that actively collects, concentrates, and stores solar energy.

acute poverty Insufficient income or access to resources needed to provide the basic necessities for life such as food, shelter, sanitation, clean water, medical care, and education.

adaptive management A management plan designed from the outset to "learn by doing," and to actively test hypotheses and adjust treatments as new information becomes available.

administrative courts Courts that hear enforcement cases for agencies or consider appeals to agency rules.

administrative law Executive orders, administrative rules and regulations, and enforcement decisions by administrative agencies and special administrative courts.

aerosols Minute particles or liquid droplets suspended in the air.

aesthetic degradation Changes in envionmental quality that offend our aesthetic senses.

albedo A description of a surface's reflective properties.

allergens Substances that activate the immune system and cause an allergic response; may not be directly antigenic themselves but may make other materials antigenic.

alpine The high, treeless biogeographic zone of mountains that consists of slopes above the timberline.

ambient air The air immediately around us.

amino acid An organic compound containing an amino group and a carboxyl group; amino acids are the units or building blocks that make peptide and protein molecules.

anaerobic respiration The incomplete intracellular breakdown of sugar or other organic compounds in the absence of oxygen that releases some energy and produces organic acids and/or alcohol.

anemia Low levels of hemoglobin due to iron deficiency or lack of red blood cells.

annual A plant that lives for a single growing season.

anthropocentric The belief that humans hold a special place in nature; being centered primarily on humans and human affairs.

antigens Substances that stimulate the production of, and react with, specific antibodies.

appropriate technology Technology that can be made at an affordable price by ordinary people using local materials to do useful work in ways that do the least possible harm to both human society and the environment.

aquifers Porous, water bearing layers of sand, gravel, and rock below the earth's surface; reservoirs for groundwater.

arbitration A formal process of dispute resolution in which there are strigent rules of evidence, cross-examination of witnesses, and a legally-binding decision made by the arbitrator that all parties must obey.

arithmetic growth A pattern of growth that increases at a constant amount per unit time, such as 1, 2, 3, 4 or 1, 3, 5, 7.

artesian well The result of a pressurized aquifer intersecting the surface or being penetrated by a pipe or conduit, from which water gushes without being pumped; also called a spring.

asthma A distressing disease characterized by shortness of breath, wheezing, and bronchial muscle spasms.

atmospheric deposition Sedimentation of solids, liquids, or gaseous materials from the air.

atom The smallest unit of matter that has the characteristics of an element; consists of three main types of subatomic particles: protons, neutrons, and electrons.

atomic number The characteristic number of protons per atom of an element. Used as an identifying attribute.

autotroph An organism that synthesizes food molecules from inorganic molecules by using an external energy source, such as light energy.

B

barrier islands Low, narrow, sandy islands that form offshore from a coastline.

BAT *See* best available, economically achievable technology.

Batesian mimicry Evolution by one species to resemble the coloration, body shape, or behavior of another species that is protected from predators by a venomous stinger, bad taste, or some other defensive adaptation.

benthos The bottom of a sea or lake.

best available, economically achievable technology (BAT) The best pollution control available; the Clean Water Act effectively negates this category by stipulating that equipment must be economically feasible.

best practical control technology (BPT) The best technology for pollution control available at reasonable cost and operable under normal conditions.

beta particles High-energy electrons released by radioactive decay.

bill A piece of legislation introduced in Congress and intended to become law.

bioaccumulation The selective absorption and concentration of molecules by cells.

biocentric preservation A philosophy that emphasizes the fundamental right of living organisms to exist and to pursue their own goods.

biocentrism The belief that all creatures have rights and values; being centered on nature rather than humans.

biocide A broad-spectrum poison that kills a wide range of organisms.

biodegradable plastics Plastics that can be decomposed by microorganisms.

biodiversity The genetic, species, and ecological diversity of the organisms in a given area.

biogeochemical cycles Movement of matter within or between ecosystems; caused by living organisms, geological forces, or chemical reactions. The cycling of nitrogen, carbon, sulfur, oxygen, phosphorus, and water are examples.

biogeographical area An entire self-contained natural ecosystem and its associated land, water, air, and wildlife resources.

biological community The populations of plants, animals, and microorganisms living and interacting in a certain area at a given time.

biological controls Use of natural predators, pathogens, or competitors to regulate pest populations.

biological or biotic factors Organisms and products of organisms that are part of the environment and potentially affect the life of other organisms.

biological oxygen demand (BOD) A standard test for measuring the amount of dissolved oxygen utilized by aquatic microorganisms.

biological pests Organisms that reduce the availability, quality, or value of resources useful to humans.

biological resources The earth's organisms.

biomagnification Increase in concentration of certain stable chemicals (for example, heavy metals or fat-soluble pesticides) in successively higher trophic levels of a food chain or web.

biomass The total mass or weight of all the living organisms in a given population or area.

biomass fuel Organic material produced by plants, animals, or microorganisms that can be burned directly as a heat source or converted into a gaseous or liquid fuel.

biomass pyramid A metaphor or diagram that explains the relationship between the amounts of biomass at different trophic levels.

biome A broad, regional type of ecosystem characterized by distinctive climate and soil conditions and a distinctive kind of biological community adapted to those conditions.

bioregionalism Organization of human activities according to natural geographic or ecological boundaries and associations. This philosophy emphasizes a sense of place and living within the resources of one's local ecosystem.

bioremediation Use of biological organisms to remove or detoxify pollutants from a contaminated area.

biosphere The zone of air, land, and water at the surface of the earth that is occupied by organisms.

biosphere reserves World heritage sites identified by the IUCN as worthy for national park or wildlife refuge status because of high biological diversity or unique ecological features.

biota All organisms in a given area.

biotic Pertaining to life; environmental factors created by living organisms.

biotic potential The maximum reproductive rate of an organism, given unlimited resources and ideal environmental conditions. Compare with environmental resistance.

birth control Any method used to reduce births, including celibacy, delayed marriage, contraception; devices or medication that prevent implantation of fertilized zygotes, and induced abortions.

black lung disease Inflammation and fibrosis caused by accumulation of coal dust in the lungs or airways. *See* respiratory fibrotic agents.

blue revolution New techniques of fish farming that may contribute as much to human nutrition as miracle cereal grains but also may create social and environmental problems.

bog An area of waterlogged soil that tends to be peaty; fed mainly by precipitation; low productivity; some bogs are acidic.

boreal forest A broad band of mixed coniferous and deciduous trees that stretches across northern North America (and also Europe and Asia); its northernmost edge, the **taiga,** intergrades with the arctic **tundra.**

BPT *See* best practical control technology.

breeder reactor A nuclear reactor that produces fuel by bombarding isotopes of uranium and thorium with high-energy neutrons that convert inert atoms to fissionable ones.

bronchitis An inflammation of bronchial linings that causes persistent cough, copious production of sputum, and involuntary muscle spasms that constrict airways.

brownfields Abandoned or underused urban areas in which redevelopment is blocked by liablilty or financing issues related to toxic contamination.

C

cancer Invasive, out-of-control cell growth that results in malignant tumors.

capital Any form of wealth, resources, or knowledge available for use in the production of more wealth.

captive breeding Raising plants or animals in zoos or other controlled conditions to produce stock for subsequent release into the wild.

carbohydrate An organic compound consisting of a ring or chain of carbon atoms with hydrogen and oxygen attached; examples are sugars, starches, cellulose, and glycogen.

carbon cycle The circulation and reutilization of carbon atoms, especially via the processes of photosynthesis and respiration.

carbon monoxide (CO) Colorless, odorless, nonirritating but highly toxic gas produced by incomplete combustion of fuel, incineration of biomass or solid waste, or partially anaerobic decomposition of organic material.

carbon sink Places of carbon accumulation, such as in large forests (organic compounds) or ocean sediments (calcium carbonate); carbon is thus removed from the carbon cycle for moderately long to very long periods of time.

carbon source Originating point of carbon that reenters the carbon cycle; cellular respiration and combustion.

carcinogens Substances that cause cancer.

carnivores Organisms that mainly prey upon animals.

carrying capacity The maximum number of individuals of any species that can be supported by a particular ecosystem on a long-term basis.

case law Precedents from both civil and criminal court cases.

cash crops Crops that are sold rather than consumed or bartered.

catastrophic systems Dynamic systems that jump abruptly from one seemingly steady state to another without any intermediate stages.

cell Minute biological compartments within which the processes of life are carried out.

cellular respiration The process in which a cell breaks down sugar or other organic compounds to release energy used for cellular work; may be **anaerobic** or **aerobic,** depending on the availability of oxygen.

chain reaction A self-sustaining reaction in which the fission of nuclei produces subatomic particles that cause the fission of other nuclei.

chaotic systems Systems that exhibit variability, which may not be necessarily random, yet whose complex patterns are not discernible over a normal human timescale.

chemical bond The force that holds molecules together.

chemical energy Potential energy stored in chemical bonds of molecules.

chemosynthesis Autotrophic synthesis of organic compounds by certain bacteria; uses energy from inorganic compounds.

chlorofluorocarbons Chemical compounds with a carbon skeleton and one or more attached chlorine and fluorine atoms. Commonly used as refrigerants, solvents, fire retardants, and blowing agents.

chloroplasts Chlorophyll-containing organelles in eukaryotic organisms; sites of photosynthesis.

chronic effects Long-lasting results of exposure to a toxin; can be a permanent change caused by a single, acute exposure or a continuous, low-level exposure.

chronic food shortages Long-term undernutrition and malnutrition; usually caused by people's lack of money to buy food or lack of opportunity to grow it themselves.

citizen science Projects in which trained volunteers work with scientific researchers to answer real-world questions.

city A differentiated community with a sufficient population and resource base to allow residents to specialize in arts, crafts, services, and professional occupations.

civil law A body of laws regulating relations between individuals or between individuals and corporations concerning property rights, personal dignity and freedom, and personal injury.

classical economics Modern, western economic theories of the effects of resource scarcity, monetary policy, and competition on supply and demand of goods and services in the marketplace. This is the basis for the capitalist market system.

clear-cut Cutting every tree in a given area, regardless of species or size; an appropriate harvest method for some species; can be destructive if not carefully controlled.

climate A description of the long-term pattern of weather in a particular area.

climax community A relatively stable, long-lasting community reached in a successional series; usually determined by climate and soil type.

closed canopy A forest where tree crowns spread over 20 percent of the ground; has the potential for commercial timber harvests.

cloud forests High mountain forests where temperatures are uniformly cool and fog or mist keeps vegetation wet all the time.

coal gasification The heating and partial combustion of coal to release volatile gases, such as methane and carbon monoxide; after pollutants are washed out, these gases become efficient, clean-burning fuel.

coal washing Coal technology that involves crushing coal and washing out soluble sulfur compounds with water or other solvents.

Coastal Zone Management Act Legislation of 1972 that gave federal money to 30 seacoast and Great Lakes states for development and restoration projects.

co-composting Microbial decomposition of organic materials in solid waste into useful soil additives and fertilizer; often, extra organic material in the form of sewer sludge, animal manure, leaves, and grass clippings are added to solid waste to speed the process and make the product more useful.

coevolution The process in which species exert selective pressure on each other and gradually evolve new features or behaviors as a result of those pressures.

cofactor Nonprotein components needed by enzymes in order to function; often minerals or vitamins.

cogeneration The simultaneous production of electricity and steam or hot water in the same plant.

cold front A moving boundary of cooler air displacing warmer air.

coliform bacteria Bacteria that live in the intestines (including the colon) of humans and other animals; used as a measure of the presence of feces in water or soil.

commensalism A symbiotic relationship in which one member is benefited and the second is neither harmed nor benefited.

common law the body of court decisions that constitute a working definition of individual rights and responsibilities where no formal statutes define these issues.

communal resource management systems Resources managed by a community for long-term sustainability.

community ecology The study of interactions of all populations living in the ecosystem of a given area.

competitive exclusion A theory that no two populations of different species will occupy the same niche and compete for exactly the same resources in the same habitat for very long, disputed by some ecologists who see biological communities as highly individualistic and variable.

complexity (ecological) The number of species at each trophic level and the number of trophic levels in a community.

composting The biological degradation of organic material under aerobic (oxygen-rich) conditions to produce compost, a nutrient-rich soil amendment and conditioner.

compound A molecule made up of two or more kinds of atoms held together by chemical bonds.

condensation The aggregation of water molecules from vapor to liquid or solid when the saturation concentration is exceeded.

condensation nuclei Tiny particles that float in the air and facilitate the condensation process.

conifers Needle-bearing trees that produce seeds in cones.

conservation development Consideration of landscape history, human culture, topography, and ecological values in subdivision design. Using cluster housing, zoning, covenants, and other design features, at least half of a subdivision can be preserved as open space, farmland, or natural areas.

conservation of matter In any chemical reaction, matter changes form; it is neither created nor destroyed.

Consolidated Metropolitan Statistical Area (CMSA) An area with more than one million people formed by the merger of two or more population centers meeting other specified requirements of the Census Bureau.

conspicuous consumption A term coined by economist and social critic Thorstein Veblen to decribe buying things we don't want or need to impress others.

consumer An organism that obtains energy and nutrients by feeding on other organisms or their remains. *See also* heterotroph.

consumption The fraction of withdrawn water that is lost in transmission or that is evaporated, absorbed, chemically transformed, or otherwise made unavailable for other purposes as a result of human use.

contour plowing Plowing along hill contours; reduces erosion.

control rods Neutron-absorbing material inserted into spaces between fuel assemblies in nuclear reactors to regulate fission reaction.

convection currents Rising or sinking air currents that stir the atmosphere and transport heat from one area to another. Convection currents also occur in water; *see* spring overturn.

conventional pollutants The seven substances (sulfur dioxide, carbon monoxide, particulates, hydrocarbons, nitrogen oxides, photochemical oxidants, and lead) that make up the largest volume of air quality degradation; identified by the Clean Air Act as the most serious threat of all pollutants to human health and welfare; also called criteria pollutants.

cool deserts Deserts such as the American Great Basin characterized by cold winters and sagebrush.

coral reefs Prominent oceanic features composed of hard, limy skeletons produced by coral animals; usually formed along edges of shallow, submerged ocean banks or along shelves in warm, shallow, tropical seas.

core The dense, intensely hot mass of molten metal, mostly iron and nickel, thousands of kilometers in diameter at the earth's center.

core region The primary industrial region of a country; usually located around the capital or largest port; has both the greatest population density and the greatest economic activity of the country.

Coriolis effect The influence of friction and drag on air layers near the earth; deflects air currents to the direction of the earth's rotation.

cornucopian fallacy The belief that nature is limitless in its abundance and that perpetual growth is not only possible but essential.

corridor A strip of natural habitat that connects two adjacent nature preserves to allow migration of organisms from one place to another.

cost-benefit analysis An evaluation of large-scale public projects by comparing the costs and benefits that accrue from them.

cover crops Plants, such as rye, alfalfa, or clover, that can be planted immediately after harvest to hold and protect the soil.

criminal law A body of court decisions based on federal and state statutes concerning wrongs against persons or society.

criteria pollutants *See* conventional pollutants.

critical factor The single environmental factor closest to a tolerance limit for a given species at a given time. *See* limiting factors.

critical thinking An ability to evaluate information and opinions in a systematic, purposeful, efficient manner.

croplands Lands used to grow crops.

crude birthrate The number of births in a year divided by the midyear population.

crude death rate The number of deaths per thousand persons in a given year; also called crude mortality rate.

crust The cool, lightweight, outermost layer of the earth's surface that floats on the soft, pliable underlying layers; similar to the "skin" on a bowl of warm pudding.

cultural eutrophication An increase in biological productivity and ecosystem succession caused by human activities.

D

debt-for-nature swap Forgiveness of international debt in exchange for nature protection in developing countries. Environmental groups and nongovernmental organizations often pay banks to write off uncollectable debts of developing countries at a steep discount in exchange for a promise by the debtor country to establish nature preserves.

deciduous Trees and shrubs that shed their leaves at the end of the growing season.

decline spiral A catastrophic deterioration of a species, community, or whole ecosystem; accelerates as functions are disrupted or lost in a downward cascade.

decomposers Fungi and bacteria that break complex organic material into smaller molecules.

deep ecology A philosophy that calls for a profound shift in our attitudes and behavior based on voluntary simplicity; rejection of anthropocentric attitudes; intimate contact with nature; decentralization of power; support for cultural and biological diversity; a belief in the sacredness of nature; and direct personal action to protect nature, improve the environment, and bring about fundamental societal change.

degradation (of water resource) Deterioration in water quality due to contamination or pollution; makes water unsuitable for other desirable purposes.

Delaney Clause A controversial amendment to the Federal Food, Drug, and Cosmetic Act, added in 1958, prohibiting the addition of any known cancer-causing agent to processed foods, drugs, or cosmetics.

delta Fan-shaped sediment deposit found at the mouth of a river.

demand The amount of a product that consumers are willing and able to buy at various possible prices, assuming they are free to express their preferences.

demanufacturing Disassembly of products so components can be reused or recycled.

demographic transition A pattern of falling death rates and birthrates in response to improved living conditions; could be reversed in deteriorating conditions.

demography Vital statistics about people: births, marriages, deaths, etc.; the statistical study of human populations relating to growth rate, age structure, geographic distribution, etc., and their effects on social, economic, and environmental conditions.

denitrifying bacteria Free-living soil bacteria that converts nitrates to gaseous nitrogen and nitrous oxide.

dependency ratio The number of nonworking members compared to working members for a given population.

desalinization (or desalination) Removal of salt from water by distillation, freezing, or ultrafiltration.

desert A type of biome characterized by low moisture levels and infrequent and unpredictable precipitation. Daily and seasonal temperatures fluctuate widely.

desertification Denuding and degrading a once-fertile land, initiating a desert-producing cycle that feeds on itself and causes long-term changes in soil, climate, and biota of an area.

detritivore Organisms that consume organic litter, debris, and dung.

dew point The temperature at which condensation occurs for a given concentration of water vapor in the air.

dieback A sudden population decline; also called a population crash.

diminishing returns A condition in which unrestrained population growth causes the standard of living to decrease to a subsistence level where poverty, misery, vice, and starvation makes life permanently drab and miserable. This dreary prophecy has led economics to be called "the dismal science."

direct action Civil disobedience, guerrilla street theater, picketing, protest marches, road blockades, demonstrations, and other techniques borrowed from the civil rights movement and applied to environmental protection.

disability-adjusted life years (DALY) A measure of premature deaths and losses due to illnesses and disabilities in a population.

discharge The amount of water that passes a fixed point in a given amount of time; usually expressed as liters or cubic feet of water per second.

discount rate The amount we discount or reduce the value of a future payment. When you borrow money from the bank at 10 percent annual interest, you are in effect saying that having the money now is worth 10 percent more to you than having the same amount one year from now.

disease A deleterious change in the body's condition in response to destabilizing factors, such as nutrition, chemicals, or biological agents.

dissolved oxygen (DO) content Amount of oxygen dissolved in a given volume of water at a given temperature and atmospheric pressure; usually expressed in parts per million (ppm).

diversity (species diversity, biological diversity) The number of species present in a community (species *richness*), as well as the relative *abundance* of each species.

DNA Deoxyribonucleic acid; the long, double-helix molecule in the nucleus of cells that contains the genetic code and directs the development and functioning of all cells.

dominant plants Those plant species in a community that provide a food base for most of the community; they usually take up the most space and have the largest biomass.

domination To rule or dominate. Some religions and moral philosophies regard humans as superior to nature and hold that we are justified in controlling and using other organisms in any way we choose.

downbursts Sudden, very strong, downdrafts of cold air associated with an advancing storm front.

drip irrigation Uses pipe or tubing perforated with very small holes to deliver water one drop at a time directly to the soil around each plant.

dry alkali injection Spraying dry sodium bicarbonate into flue gas to absorb and neutralize acidic sulfur compounds.

E

earth charter A set of principles for sustainable development, environmental protection, and social justice developed by a council appointed by the United Nations.

earthquakes Sudden, violent movement of the earth's crust.

ecocentric (ecologically centered) A philosophy that claims moral values and rights for both organisms and ecological systems and processes.

ecofeminism A pluralistic, nonhierarchical, relationship-oriented philosophy that suggests how humans could reconceive themselves and their relationships to nature in nondominating ways as an alternative to patriarchal systems of domination.

ecojustice Justice in the social order and integrity in the natural order.

ecological development A gradual process of environmental modification by organisms.

ecological economics Application of ecological insights to economic analysis in a holistic, contextual, value-sensitive, ecocentric manner.

ecological equivalents Different species that occupy similar ecological niches in similar ecosystems in different parts of the world.

ecological niche The functional role and position of a species (population) within a community or ecosystem, including what resources it uses, how and when it uses the resources, and how it interacts with other populations.

ecological succession The process by which organisms occupy a site and gradually change environmental conditions so that other species can replace the original inhabitants.

ecology The scientific study of relationships between organisms and their environment. It is concerned with the life histories, distribution, and behavior of individual species as well as the structure and function of natural systems at the level of populations, communities, and ecosystems.

economic development A rise in real income *per person;* usually associated with new technology that increases productivity or resources.

economic growth An increase in the total wealth of a nation; if population grows faster than the economy, there may be real economic growth, but the share per person may decline.

economic thresholds In pest management, the point at which the cost of pest damage exceeds the costs of pest control.

ecosystem A specific biological community and its physical environment interacting in an exchange of matter and energy.

ecosystem management An integration of ecological, economic, and social goals in a unified systems approach to resource management.

ecosystem restoration To reinstate an entire community of organisms to as near its natural condition as possible.

ecotage Direct action (guerrilla warfare) or sabotage in defense of nature. *See* monkey wrenching.

ecotone A boundary between two types of ecological communities.

ecotourism A combination of adventure travel, cultural exploration, and nature appreciation in wild settings.

edge effects A change in species composition, physical conditions, or other ecological factors at the boundary between two ecosystems.

effluent sewerage A low-cost alternative sewage treatment for cities in poor countries that combines some features of septic systems and centralized municipal treatment systems.

electron A negatively charged subatomic particle that orbits around the nucleus of an atom.

electrostatic precipitators The most common particulate controls in power plants; fly ash particles pick up an electrostatic surface charge as they pass between large electrodes in the effluent stream, causing particles to migrate to the oppositely charged plate.

element A molecule composed of one kind of atom; cannot be broken into simpler units by chemical reactions.

El Niño A climatic change marked by shifting of a large warm water pool from the western Pacific Ocean towards the east. Wind direction and precipitation patterns are changed over much of the Pacific and perhaps around the world.

emergent diseases A new disease or one that has been absent for at least 20 years.

emigration The movement of members from a population.

emission standards Regulations for restricting the amounts of air pollutants that can be released from specific point sources.

emphysema An irreversible, obstructive lung disease in which airways become permanently constricted and alveoli are damaged or destroyed.

endangered species A species considered to be in imminent danger of extinction.

endemism A state in which species are restricted to a single region.

energy The capacity to do work (that is, to change the physical state or motion of an object).

energy efficiency A measure of energy produced compared to energy consumed.

energy pyramid A representation of the loss of useful energy at each step in a food chain.

energy recovery Incineration of solid waste to produce useful energy.

environment The circumstances or conditions that surround an organism or group of organisms as well as the complex of social or cultural conditions that affect an individual or community.

environmental ethics A search for moral values and ethical principles in human relations with the natural world.

environmental hormones Chemical pollutants that substitute for, or interfere with, naturally-occurring hormones in our bodies; these chemicals may trigger reproductive failure, developmental abnormalities, or tumor promotion.

environmental impact statement (EIS) An analysis, required by provisions in the National Environmental Policy Act of 1970, of the effects of any major program a federal agency plans to undertake.

environmental indicators Organisms or physical factors that serve as a gauge for environmental changes. More specifically, organisms with these characteristics are called bioindicators.

environmentalism Active participation in attempts to solve environmental pollution and resource problems.

environmental justice A recognition that access to a clean, healthy environment is a fundamental right of all human beings. Some biocentric or ecocentric philosophers would extend this concept to include nonhuman organisms or ecological entities.

environmental law The special body of official rules, decisions, and actions concerning environmental quality, natural resources, and ecological sustainability.

environmental literacy Fluency in the principles of ecology that gives us a working knowledge of the basic grammar and underlying syntax of environmental wisdom.

environmental policy The official rules or regulations concerning the environment adopted, implemented, and enforced by some governmental agency.

environmental racism Decisions that restrict certain people or groups of people to polluted or degraded environments on the basis of race.

environmental resistance All the limiting factors that tend to reduce population growth rates and set the maximum allowable population size or carrying capacity of an ecosystem.

environmental resources Anything an organism needs that can be taken from the environment.

environmental science The systematic, scientific study of our environment as well as our role in it.

enzymes Molecules, usually proteins or nucleic acids, that act as catalysts in biochemical reactions.

epidemiology The study of the distribution and causes of disease and injuries in human populations.

epiphyte A plant that grows on a substrate other than the soil, such as the surface of another organism.

equilibrium community Also called a **disclimax community;** a community subject to periodic disruptions, usually by fire, that prevent it from reaching a climax stage.

estuary A bay or drowned valley where a river empties into the sea.

eukaryotic cell A cell containing a membrane-bounded nucleus and membrane-bounded organelles.

eutectic chemicals Phase-changing chemicals used in heat storage systems to store a large amount of energy in a small volume.

eutrophic Rivers and lakes rich in organisms and organic material (*eu* = truly; *trophic* = nutritious).

evaporation The process in which a liquid is changed to vapor (gas phase).

evergreens Coniferous trees and broad-leaved plants that retain their leaves year-round.

evolution A theory that explains how random changes in genetic material and competition for scarce resources cause species to change gradually.

exhaustible resources Generally considered the earth's geologic endowment: minerals, nonmineral resources, fossil fuels, and other materials present in fixed amounts in the environment.

existence value The importance we place on just knowing that a particular species or a specific organism exists.

exotic organisms Alien species introduced by human agency into biological sommunities where they would not naturally occur.

exponential growth Growth at a constant rate of increase per unit of time; can be expressed as a constant fraction or exponent. *See* geometric growth.

external costs Expenses, monetary or otherwise, borne by someone other than the individuals or groups who use a resource.

extinction The irrevocable elimination of species; can be a normal process of the natural world as species out-compete or kill off others or as environmental conditions change.

extirpate To destroy totally; extinction caused by direct human action, such as hunting, trapping, etc.

F

family planning Controlling reproduction; planning the timing of birth and having as many babies as are wanted and can be supported.

famines Acute food shortages characterized by large-scale loss of life, social disruption, and economic chaos.

fauna All of the animals present in a given region.

fecundity The physical ability to reproduce.

fen An area of waterlogged soil that tends to be peaty; fed mainly by upwelling water; low productivity.

feral A domestic animal that has taken up a wild existence.

fermentation (alcoholic) A type of anaerobic respiration that yields carbon dioxide and alcohol; used in commercial fermentation processes, including production of raised bakery dough products and alcoholic beverages.

fertility Measurement of actual number of offspring produced through sexual reproduction; usually described in terms of number of offspring of females, since paternity can be difficult to determine.

fetal alcohol syndrome A tragic set of permanent physical and mental and behavioral birth defects that result when mothers drink alcohol during pregnancy.

fibrosis The general name for accumulation of scar tissue in the lung.

fidelity A principle that forbids misleading or deceiving any creature capable of being mislead or deceived. We are to be truthful in our dealings with others.

filters A porous mesh of cotton cloth, spun glass fibers, or asbestos-cellulose that allows air or liquid to pass through but holds back solid particles.

fire-climax community An equilibrium community maintained by periodic fires; examples include grasslands, chaparral shrubland, and some pine forests.

first law of thermodynamics States that energy is *conserved;* that is, it is neither created nor destroyed under normal conditions.

First World The industrialized capitalist or market-economy countries of Western Europe, North America, Japan, Australia, and New Zealand.

flood An overflow of water onto land that normally is dry.

floodplains Low lands along riverbanks, lakes, and coastlines subjected to periodic inundation.

flora All of the plants present in a given region.

flue-gas scrubbing Treating combustion exhaust gases with chemical agents to remove pollutants. Spraying crushed limestone and water into the exhaust gas stream to remove sulfur is a common scrubbing technique.

fluidized bed combustion High pressure air is forced through a mixture of crushed coal and limestone particles, lifting the burning fuel and causing it to move like a boiling fluid.

food aid Financial assistance intended to boost less-developed countries' standards of living.

food chain A linked feeding series; in an ecosystem, the sequence of organisms through which energy and materials are transferred, in the form of food, from one trophic level to another.

food security The ability of individuals to obtain sufficient food on a day-to-day basis.

food surpluses Excess food supplies.

food web A complex, interlocking series of individual food chains in an ecosystem.

forest management Scientific planning and administration of forest resources for sustainable harvest, multiple use, regeneration, and maintenance of a healthy biological community.

fossil fuels Petroleum, natural gas, and coal created by geological forces from organic wastes and dead bodies of formerly living biological organisms.

Fourth World A political/economic category describing very poor nations that have neither market economies nor central planning and are either not developing or are developing very slowly. Also used to describe indigenous communities within wealthier nations.

freezing condensation A process that occurs in the clouds when ice crystals trap water vapor. As the ice crystals become larger and heavier, they begin to fall as rain or snow.

fresh water Water other than seawater; covers only about 2 percent of earth's surface, including streams, rivers, lakes, ponds, and water associated with several kinds of wetlands.

freshwater ecosystems Ecosystems in which the fresh (nonsalty) water of streams, rivers, ponds, or lakes plays a defining role.

front The boundary between two air masses of different temperature and density.

frontier An unexploited natural area at the leading edge of human settlement.

frontier mentality The idea that the world has an unlimited supply of resources for human use regardless of the consequences to natural ecosystems and the biosphere.

fuel assembly A bundle of hollow metal rods containing uranium oxide pellets; used to fuel a nuclear reactor.

fuel cells Mechanical devices that use hydrogen or hydrogen-containing fuel such as methane to produce an electric current. Fuel cells are clean, quiet and highly efficient sources of electricity.

fuel-switching A change from one fuel to another.

fuelwood Branches, twigs, logs, wood chips, and other wood products harvested for use as fuel.

fugitive emissions Substances that enter the air without going through a smokestack, such as dust from soil erosion, strip mining, rock crushing, construction, and building demolition.

fungi One of the five kingdom classifications; consists of nonphotosynthetic, eukaryotic organisms with cell walls, filamentous bodies, and absorptive nutrition.

fungicide A chemical that kills fungi.

G

Gaia hypothesis A theory that the living organisms of the biosphere form a single, complex interacting system that creates and maintains a habitable Earth; named after Gaia, the Greek Earth mother goddess.

gamma rays Very short wavelength forms of the electromagnetic spectrum.

gap analysis A biogeographical technique of mapping biological diversity and endemic species to find gaps between protected areas that leave endangered habitats vulnerable to disruption.

garden city A new town with special emphasis on landscaping and rural ambience.

gasohol A mixture of gasoline and ethanol.

gene A unit of heredity; a segment of DNA nucleus of the cell that contains information for the synthesis of a specific protein, such as an enzyme.

gene banks Storage for seed varieties for future breeding experiments.

general fertility rate Crude birthrate multiplied by the percentage of reproductive age women.

genetic assimilation The disappearance of a species as its genes are diluted through crossbreeding with a closely related species.

genetic engineering Laboratory manipulation of genetic material using molecular biology techniques to create desired characteristics in organisms.

geometric growth Growth that follows a geometric pattern of increase, such as 2, 4, 8, 16, etc. *See* exponential growth.

geothermal energy Energy drawn from the internal heat of the earth, either through geysers, fumaroles, hot springs, or other natural geothermal features, or through deep wells that pump heated groundwater.

germ plasm Genetic material that may be preserved for future agricultural, commercial, and ecological values (plant seeds or parts or animal eggs, sperm, and embryos).

global environmentalism A concern for, and action to help solve, global environmental problems.

grasslands A biome dominated by grasses and associated herbaceous plants.

greenhouse gas Gasses added to the atmosphere by human actions that trap heat and cause global warming.

green plans Integrated national environmental plans for reducing pollution and resource consumption while achieving sustainable development and environmental restoration.

green political parties Political organizations based on environmental protection, participatory democracy, grassroots organization, and sustainable development.

green revolution Dramatically increased agricultural production brought about by "miracle" strains of grain; usually requires high inputs of water, plant nutrients, and pesticides.

gross domestic product (GDP) The total economic activity within national boundaries.

gross national product (GNP) The sum total of all goods and services produced in a national economy. Gross domestic product (GDP) is used to distinguish economic activity within a country from that of offshore corporations.

groundwater Water held in gravel deposits or porous rock below the earth's surface; does not include water or crystallization held by chemical bonds in rocks or moisture in upper soil layers.

gully erosion Removal of layers of soil, creating channels or ravines too large to be removed by normal tillage operations.

H

habitat The place or set of environmental conditions in which a particular organism lives.

habitat conservation plans Agreements under which property owners are allowed to harvest resources or develop land as long as habitat is conserved or replaced in ways that benefit resident endangered or threatened species in the long run. Some incidental "taking" or loss of endangered species is generally allowed in such plans.

Hadley cells Circulation patterns of atmospheric convection currents as they sink and rise in several intermediate bands.

half-life The time required for one-half of a sample to decay or change into some other form.

hazardous Describes chemicals that are dangerous, including flammables, explosives, irritants, sensitizers, acids, and caustics; may be relatively harmless in diluted concentrations.

hazardous waste Any discarded material containing substances known to be toxic, mutagenic, carcinogenic, or teratogenic to humans or other life-forms; ignitable, corrosive, explosive, or highly reactive alone or with other materials.

health A state of physical and emotional well-being; the absence of disease or ailment.

heap-leach extraction A technique for separating gold from extremely low-grade ores. Crushed ore is piled in huge heaps and sprayed with a dilute alkaline-cyanide solution, which percolates through the pile to extract the gold, which is separated from the effluent in a processing plant. This process has a high potential for water pollution.

heat A form of energy transferred from one body to another because of a difference in temperatures.

heat capacity The amount of heat energy that must be added or subtracted to change the temperature of a body; water has a high heat capacity.

heat of vaporization The amount of heat energy required to convert water from a liquid to a gas.

herbicide A chemical that kills plants.

herbivore An organism that eats only plants.

heterotroph An organism that is incapable of synthesizing its own food and, therefore, must feed upon organic compounds produced by other organisms.

high-level waste repository A place where intensely radioactive wastes can be buried and remain unexposed to groundwater and earthquakes for tens of thousands of years.

high-quality energy Intense, concentrated, and high-temperature energy that is considered high-quality because of its usefulness in carrying out work.

holistic science the study of entire, integrated systems rather than isolated parts. Often takes a descriptive or interpretive approach.

homeostasis Maintaining a dynamic, steady state in a living system through opposing, compensating adjustments.

Homestead Act Legislation passed in 1862 allowing any citizen or applicant for citizenship over 21 years old and head of a family to acquire 160 acres of public land by living on it and cultivating it for five years.

host organism An organism that provides lodging for a parasite.

hot desert Deserts of the American Southwest and Mexico; characterized by extreme summer heat and cacti.

human ecology The study of the interactions of humans with the environment.

human resources Human wisdom, experience, skill, labor, and enterprise.

humus Sticky, brown, insoluble residue from the bodies of dead plants and animals; gives soil its structure, coating mineral particles and holding them together; serves as a major source of plant nutrients.

hurricanes Large cyclonic oceanic storms with heavy rain and winds exceeding 119 km/hr (74 mph).

hypothyroidism Listlessness and other metabolic symptoms caused by low thyroid hormone levels.

I

igneous rocks Crystalline minerals solidified from molten magma from deep in the earth's interior; basalt, rhyolite, andesite, lava, and granite are examples.

inbreeding depression In a small population, an accumulation of harmful genetic traits (through random mutations and natural selection) that lowers viability and reproductive success of enough individuals to affect the whole population.

industrial revolution Advances in science and technology that have given us power to understand and change our world.

industrial timber Trees used for lumber, plywood, veneer, particleboard, chipboard, and paper; also called roundwood.

inertial confinement A nuclear fusion process in which a small pellet of nuclear fuel is bombarded with extremely high-intensity laser light.

infiltration The process of water percolation into the soil and pores and hollows of permeable rocks.

inflammatory response A complex series of interactions between fragments of damaged cells, surrounding tissues, circulating blood cells, and specific antibodies; typical of infections.

informal economy Small-scale family businesses in temporary locations outside the control of normal regulatory agencies.

inherent value Ethical values or rights that exist as an intrinsic or essential characteristic of a particular thing or class of things simply by the fact of their existence.

inholdings Private lands within public parks, forests, or wildlife refuges.

insecticide A chemical that kills insects.

insolation Incoming solar radiation.

instrumental value Value or worth of objects that satisfy the needs and wants of moral agents. Objects that can be used as a means to some desirable end.

intangible resources Factors such as open space, beauty, serenity, wisdom, diversity, and satisfaction that cannot be grasped or contained. Ironically, these resources can be both infinite and exhaustible.

integrated pest management (IPM) IPM is an ecologically based pest-control strategy that relies on natural mortality factors, such as natural enemies, weather, cultural control methods, and carefully applied doses of pesticides.

internal costs The expenses, monetary or otherwise, borne by those who use a resource.

internalizing costs Planning so that those who reap the benefits of resource use also bear all the external costs.

interplanting The system of planting two or more crops, either mixed together or in alternating rows, in the same field; protects the soil and makes more efficient use of the land.

interpretive science Explanation based on observation and description of entire objects or systems rather than isolated parts.

interspecific competition In a community, competition for resources between members of *different* species.

intraspecific competition In a community, competition for resources among members of the *same* species.

ionizing radiation High-energy electromagnetic radiation or energetic subatomic particles released by nuclear decay.

ionosphere The lower part of the thermosphere.

ions Electrically charged atoms that have gained or lost electrons.

irruptive growth *See* Malthusian growth.

island biogeography The study of rates of colonization and extinction of species on islands or other isolated areas based on size, shape, and distance from other inhabited regions.

island effects Reductions in species diversity caused by reduction in ecosystem area.

isotopes Forms of a single element that differ in atomic mass due to a different number of neutrons in the nucleus.

J

J curve A growth curve that depicts exponential growth; called a J curve because of its shape.

jet streams Powerful winds or currents of air that circulate in shifting flows; similar to oceanic currents in extent and effect on climate.

joule A unit of energy. One joule is the energy expended in 1 second by a current of 1 amp flowing through a resistance of 1 ohm.

K

keystone species A species whose impacts on its community or ecosystem are much larger and more influential than would be expected from mere abundance. This could be a top predator, a plant that shelters or feeds other organisms, or an organism that plays a critical ecological role.

kinetic energy Energy contained in moving objects such as a rock rolling down a hill, the wind blowing through the trees, or water flowing over a dam.

known resources Those that have been located but are not completely mapped but, nevertheless, are likely to become economical in the foreseeable future.

kwashiorkor A widespread human protein deficiency disease resulting from a starchy diet low in protein and essential amino acids.

L

landfills Land disposal sites for solid waste; operators compact refuse and cover it with a layer of dirt to minimize rodent and insect infestation, wind-blown debris, and leaching by rain.

land reform Democratic redistribution of landownership to recognize the rights of those who actually work the land to a fair share of the products of their labor.

landscape ecology The study of the reciprocal effects of spatial pattern on ecological processes. A study of the ways in which landscape history shapes the features of the land and the organisms that inhabit it as well as our reaction to, and interpretation of, the land.

landslide The sudden fall of rock and earth from a hill or cliff. Often triggered by an earth quake or heavy rain.

LD50 A chemical dose lethal to 50 percent of a test population.

less-developed countries (LDC) Nonindustrialized nations characterized by low per capita income, high birthrates and death rates, high population growth rates, and low levels of technological development.

life-cycle analysis Evaluation of material and energy inputs and outputs at each stage of manufacture, use, and disposal of a product.

life-expectancy The average age that a newborn infant can expect to attain in a particular time and place.

life span The longest period of life reached by a type of organism.

lignite Soft, brown to black coal of low BTU value with original plant components still discernible.

limiting factors Chemical or physical factors that limit the existence, growth, abundance, or distribution of an organism.

lipid A nonpolar organic compound that is insoluble in water but soluble in solvents, such as alcohol and ether; includes fats, oils, steroids, phospholipids, and carotenoids.

liquid metal fast breeder A nuclear power plant that converts uranium 238 to plutonium 239; thus, it creates more nuclear fuel than it consumes. Because of the extreme heat and density of its core, the breeder uses liquid sodium as a coolant.

lobbying Using personal contacts, public pressure, or political action to persuade legislators to vote in a particular manner.

logistic growth Growth rates regulated by internal and external factors that establish an equilibrium with environmental resources. *See* S curve.

longevity The length or duration of life; compare to survivorship.

low-head hydropower Small-scale hydro technology that can extract energy from small headwater dams; causes much less ecological damage.

low-quality energy Diffuse, dispersed energy at a low temperature that is difficult to gather and use for productive purposes.

LULUs *L*ocally *U*nwanted *L*and *U*ses such as toxic waste dumps, incinerators, smelters, airports, freeways, and other sources of environmental, economic, or social degradation.

M

magma Molten rock from deep in the earth's interior; called lava when it spews from volcanic vents.

magnetic confinement A technique for enclosing a nuclear fusion reaction in a powerful magnetic field inside a vacuum chamber.

malignant tumor A mass of cancerous cells that have left their site of origin, migrated through the body, invaded normal tissues, and are growing out of control.

malnourishment A nutritional imbalance caused by lack of specific dietary components or inability to absorb or utilize essential nutrients.

Malthusian growth A population explosion followed by a population crash; also called irruptive growth.

Man and Biosphere (MAB) program A design for nature preserves that divides protected areas into zones with different purposes. A highly protected core is surrounded by a buffer zone and peripheral regions in which multiple-use resource harvesting is permitted.

mantle A hot, pliable layer of rock that surrounds the earth's core and underlies the cool, outer crust.

marasmus A widespread human protein deficiency disease caused by a diet low in calories and protein or imbalanced in essential amino acids.

marginal costs The cost to produce one additional unit of a good or service.

marine Living in or pertaining to the sea.

market equilibrium The dynamic balance between supply and demand under a given set of conditions in a "free" market (one with no monopolies or government interventions).

marsh Wetland without trees; in North America, this type of land is characterized by cattails and rushes.

mass burn Incineration of unsorted solid waste.

matter Anything that takes up space and has mass.

mediation An informal dispute resolution process in which parties are encouraged to discuss issues openly but in which all decisions are reached by consensus and any participant can withdraw at any time.

Mediterranean climate areas Specialized landscapes with warm, dry summers; cool, wet winters; many unique plant and animal adaptations; and many levels of endemism.

megacity *See* megalopolis.

megalopolis Also known as a megacity or supercity; megalopolis indicates an urban area with more than 10 million inhabitants.

megawatt (MW) Unit of electrical power equal to one thousand kilowatts or one million watts.

mesosphere The atmospheric layer above the stratosphere and below the thermosphere; the middle layer; temperatures are usually very low.

metabolism All the energy and matter exchanges that occur within a living cell or organism; collectively, the life processes.

metamorphic rock Igneous and sedimentary rocks modified by heat, pressure, and chemical reactions.

methane hydrate Small bubbles or individual molecules of methane (natural gas) trapped in a crystalline matrix of frozen water.

micro-hydro generators Small power generators that can be used in low-level rivers to provide economical power for four to six homes, freeing them from dependence on large utilities and foreign energy supplies.

Milankovitch cycles Periodic variations in tilt, eccentricity, and wobble in the earth's orbit; Milutin Milankovitch suggested that it is responsible for cyclic weather changes.

milpa agriculture An ancient farming system in which small patches of tropical forests are cleared and perennial polyculture agriculture practiced and is then followed by many years of fallow to restore the soil; also called swidden agriculture.

mineral A naturally occurring, inorganic, crystalline solid with definite chemical composition and characteristic physical properties.

mitigation Repairing or rehabilitating a damaged ecosystem or compensating for damage by providing a substitute or replacement area.

mixed perennial polyculture Growing a mixture of different perennial crop species (where the same plant persists for more than one year) together in the same plot; imitates the diversity of a natural system and is often more stable and more suitable for sustainable agriculture than monoculture of annual plants.

molecule A combination of two or more atoms.

monitored, retrievable storage Holding wastes in underground mines or secure surface facilities such as dry casks where they can be watched and repackaged, if necessary.

monkey wrenching Environmental sabotage such as driving large spikes in trees to protect them from loggers, vandalizing construction equipment, pulling up survey stakes for unwanted developments, and destroying billboards.

monoculture agroforestry Intensive planting of a single species; an efficient wood production approach, but one that encourages pests and disease infestations and conflicts with wildlife habitat or recreation uses.

monsoon A seasonal reversal of wind patterns caused by the different heating and cooling rates of the oceans and continents.

montane coniferous forests Coniferous forests of the mountains consisting of belts of different forest communities along an altitudinal gradient.

moral agents Beings capable of making distinctions between right and wrong and acting accordingly. Those whom we hold responsible for their actions.

moral extensionism Expansion of our understanding of inherent value or rights to persons, organisms, or things that might not be considered worthy of value or rights under some ethical philosophies.

morals A set of ethical principles that guide our actions and relationships.

moral subjects Beings that are not capable of distinguishing between right or wrong or that are not able to act on moral principles and yet are capable of being wronged by others. This category assumes some rights or inherent values in moral subjects that gives us duties or obligations towards them.

morbidity Illness or disease.

more-developed countries (MDC) Industrialized nations characterized by high per capita incomes, low birth and death rates, low population growth rates, and high levels of industrialization and urbanization.

mortality Death rate in a population; the probability of dying.

Müellerian mimicry Evolution of two species, both of which are unpalatable and, have poisonous stingers or some other defense mechanism, to resemble each other.

mulch Protective ground cover, including manure, wood chips, straw, seaweed, leaves, and other natural products, or synthetic materials, such as heavy paper or plastic, that protect the soil, save water, and prevent weed growth.

multiple use Many uses that occur simultaneously; used in forest management; limited to mutually compatible uses.

mutagens Agents, such as chemicals or radiation, that damage or alter genetic material (DNA) in cells.

mutation A change, either spontaneous or by external factors, in the genetic material of a cell; mutations in the gametes (sex cells) can be inherited by future generations of organisms.

mutualism A symbiotic relationship between individuals of two different species in which both species benefit from the association.

N

NAAQS *N*ational *A*mbient *A*ir *Q*uality *S*tandard; federal standards specifying the maximum allowable levels (averaged over specific time periods) for regulated pollutants in ambient (outdoor) air.

natality The production of new individuals by birth, hatching, germination, or cloning.

natural history The study of where and how organisms carry out their life cycles.

natural increase Crude death rate subtracted from crude birthrate.

natural landscaping A landscaping design that uses carefully contrived meadows, forests, and natural-looking ponds and marshes to create an illusion of wild nature.

natural resources Goods and services supplied by the environment.

natural selection The mechanism for evolutionary change in which environmental pressures cause certain genetic combinations in a population to become more abundant; genetic combinations best adapted for present environmental conditions tend to become predominant.

neo-classical economics A branch of economics that attempts to apply the principles of modern science to economic analysis in a mathematically rigorous, non- contextual, abstract, predictive manner.

neo-Luddities People who reject technology as the cause of environmental degradation and social disruption. Named after the followers of Ned Ludd who tried to turn back the Industrial Revolution in England.

neo-Malthusian A belief that the world is characterized by scarcity and competition in which too many people fight for too few resources. Named for Thomas Malthus, who predicted a dismal cycle of misery, vice, and starvation as a result of human overpopulation.

net energy yield Total useful energy produced during the lifetime of an entire energy system minus the energy used, lost, or wasted in making useful energy available.

neurotoxins Toxic substances, such as lead or mercury, that specifically poison nerve cells.

neutron A subatomic particle, found in the nucleus of the atom, that has no electromagnetic charge.

new towns Experimental urban environments that seek to combine the best features of the rural village and the modern city.

nihilists Those who believe the world has no meaning or purpose other than a dark, cruel, unceasing struggle for power and existence.

NIMBY *N*ot *I*n *M*y *B*ack*Y*ard: the rallying cry of those opposed to LULUs.

nitrate-forming bacteria Bacteria that convert nitrites into compounds that can be used by green plants to build proteins.

nitrite-forming bacteria Bacteria that combine ammonia with oxygen to form nitrites.

nitrogen cycle The circulation and reutilization of nitrogen in both inorganic and organic phases.

nitrogen-fixing bacteria Bacteria that convert nitrogen from the atmosphere or soil solution into ammonia that can then be converted to plant nutrients by nitrite- and nitrate-forming bacteria.

nitrogen oxides Highly reactive gases formed when nitrogen in fuel or combustion air is heated to over 650° C (1200° F) in the presence of oxygen or when bacteria in soil or water oxidize nitrogen-containing compounds.

noncriteria pollutants *See* unconventional air pollutants.

nongovernmental organizations (NGOs) A term referring collectively to pressure and research groups, advisory agencies, political parties, professional societies, and other groups concerned about environmental quality, resource use, and many other issues.

noninterference A principle that requires us to refrain from interfering with the lives or destinies of other moral agents or moral subjects except to satisfy essential needs.

nonmalfeasance Doing no harm to any being that has goods (values, interests, moral values) of their own.

nonpoint sources Scattered, diffuse sources of pollutants, such as runoff from farm fields, golf courses, construction sites, etc.

nonrenewable resources Minerals, fossil fuels, and other materials present in essentially fixed amounts (within human time scales) in our environment.

North/South division A description of the fact that most of the world's wealthier countries tend to be in North America, Europe, and Japan while the poorer countries tend to be located closer to the equator.

nuclear fission The radioactive decay process in which isotopes split apart to create two smaller atoms.

nuclear fusion A process in which two smaller atomic nuclei fuse into one larger nucleus and release energy; the source of power in a hydrogen bomb.

nucleic acids Large organic molecules made of nucleotides that function in the transmission of hereditary traits, in protein synthesis, and in control of cellular activities.

nucleus The center of the atom; occupied by protons and neutrons. In cells, the organelle that contains the chromosomes (DNA).

nuées ardentes Deadly, denser-than-air mixtures of hot gases and ash ejected from volcanoes.

numbers pyramid A diagram showing the relative population sizes at each trophic level in an ecosystem; usually corresponds to the biomass pyramid.

O

oceanic islands Islands in the ocean; formed by breaking away from a continental landmass, volcanic action, coral formation, or a combination of sources; support distinctive communities.

ocean shorelines Rocky coasts and sandy beaches along the oceans; support rich, stratified communities.

ocean thermal electric conversion (OTEC) Energy derived from temperature differentials between warm ocean surface waters and cold deep waters. This differential can be used to drive turbines attached to electric generators.

offset allowances A controversial component of air quality regulations that allows a polluter to avoid installation of control equipment on one source with an "offsetting" pollution reduction at another source.

oil shale A fine-grained sedimentary rock rich in solid organic material called kerogen. When heated, the kerogen liquefies to produce a fluid petroleum fuel.

old-growth forests Forests free from disturbance for long enough (generally 150 to 200 years) to have nature trees, physical conditions, species diversity, and other characteristics of equilibrium ecosystems.

oligotrophic Condition of rivers and lakes that have clear water and low biological productivity (*oligo* = little; *trophic* = nutrition); are usually clear, cold, infertile headwater lakes and streams.

omnivore An organism that eats both plants and animals.

open access system A commonly held resource for which there are no management rules.

open canopy A forest where tree crowns cover less than 20 percent of the ground; also called woodland.

open range Unfenced, natural grazing lands; includes woodland as well as grassland.

open system A system that exchanges energy and matter with its environment.

optimum The most favorable condition in regard to an environmental factor.

orbital The space or path in which an electron orbits the nucleus of an atom.

organic compounds Complex molecules organized around skeletons of carbon atoms arranged in rings or chains; includes biomolecules, molecules synthesized by living organisms.

overburden Overlying layers of noncommercial sediments that must be removed to reach a mineral or coal deposit.

overnutrition Receiving too many calories.

overshoot The extent to which a population exceeds the carrying capacity of its environment.

oxygen cycle The circulation and reutilization of oxygen in the biosphere.

oxygen sag Oxygen decline downstream from a pollution source that introduces materials with high biological oxygen demands.

ozone A highly reactive molecule containing three oxygen atoms; a dangerous pollutant in ambient air. In the stratosphere, however, ozone forms an ultraviolet absorbing shield that protects us from mutagenic radiation.

P

parabolic mirrors Curved mirrors that focus light from a large area onto a single, central point, thereby concentrating solar energy and producing high temperatures.

parasite An organism that lives in or on another organism, deriving nourishment at the expense of its host, usually without killing it.

particulate material Atmospheric aerosols, such as dust, ash, soot, lint, smoke, pollen, spores, algal cells, and other suspended materials; originally applied only to solid particles but now extended to droplets of liquid.

parts per billion (ppb) Number of parts of a chemical found in one billion parts of a particular gas, liquid, or solid mixture.

parts per million (ppm) Number of parts of a chemical found in one million parts of a particular gas, liquid, or solid mixture.

parts per trillion (ppt) Number of parts of a chemical found in one trillion (10^{12}) parts of a particular gas, liquid, or solid mixture.

passive heat absorption The use of natural materials or absorptive structures without moving parts to gather and hold heat; the simplest and oldest use of solar energy.

pasture Enclosed domestic meadows or managed grazing lands.

patchiness Within a larger ecosystem, the presence of smaller areas that differ in some physical conditions and thus support somewhat different communities; a diversity-promoting phenomenon.

pathogen An organism that produces disease in a host organism, disease being an alteration of one or more metabolic functions in response to the presence of the organism.

peat Deposits of moist, acidic, semidecayed organic matter.

pellagra Lassitude, torpor, dermatitis, diarrhea, dementia, and death brought about by a diet deficient in tryptophan and niacin.

peptides Two or more amino acids linked by a peptide bond.

perennial species Plants that grow for more than two years.

permafrost A permanently frozen layer of soil that underlies the arctic tundra.

permanent retrievable storage Placing waste storage containers in a secure building, salt mine, or bedrock cavern where they can be inspected periodically and retrieved, if necessary, for repacking or for transfer if a better means of disposal or reuse is developed.

persistent organic polutants (POPs) Chemical compounds that persist in the environment and retain biological activity for long times.

pest Any organism that reduces the availability, quality, or value of a useful resource.

pesticide Any chemical that kills, controls, drives away, or modifies the behavior of a pest.

pesticide treadmill A need for constantly increasing doses or new pesticides to prevent pest resurgence.

pest resurgence Rebound of pest populations due to acquired resistance to chemicals and nonspecific destruction of natural predators and competitors by broadscale pesticides.

pH A value that indicates the acidity or alkalinity of a solution on a scale of 0 to 14, based on the proportion of H^+ ions present.

phosphorus cycle The movement of phosphorus atoms from rocks through the biosphere and hydrosphere and back to rocks.

photochemical oxidants Products of secondary atmospheric reactions. *See* smog.

photodegradable plastics Plastics that break down when exposed to sunlight or to a specific wavelength of light.

photosynthesis The biochemical process by which green plants and some bacteria capture light energy and use it to produce chemical bonds. Carbon dioxide and water are consumed while oxygen and simple sugars are produced.

photosynthetic efficiency The percentage of available light captured by plants and used to make useful products.

photovoltaic cell An energy-conversion device that captures solar energy and directly converts it to electrical current.

physical or abiotic factors Nonliving factors, such as temperature, light, water, minerals, and climate, that influence an organism.

phytoplankton Microscopic, free-floating, autotrophic organisms that function as producers in aquatic ecosystems.

pioneer species In primary succession on a terrestrial site, the plants, lichens, and microbes that first colonize the site.

plankton Primarily microscopic organisms that occupy the upper water layers in both freshwater and marine ecosystems.

plasma A hot, electrically neutral gas of ions and free electrons.

poachers Those who hunt wildlife illegally.

point sources Specific locations of highly concentrated pollution discharge, such as factories, power plants, sewage treatment plants, underground coal mines, and oil wells.

policy A societal plan to statement of intentions intended to accomplish some social good.

policy cycle The process by which problems are identified and acted upon in the public arena.

political economy the branch of economics concerned with modes of production, distribution of benefits, social institutions, and class relationships.

pollution To make foul, unclean, dirty; any physical, chemical, or biological change that adversely affects the health, survival, or activities of living organisms or that alters the environment in undesirable ways.

pollution charges Fees assessed per unit of pollution based on the "polluter pays" principle.

polycentric complex Cities with several urban cores surrounding a once dominant central core.

population A group of individuals of the same species occupying a given area.

population crash A sudden population decline caused by predation, waste accumulation, or resource depletion; also called a dieback.

population explosion Growth of a population at exponential rates to a size that exceeds environmental carrying capacity; usually followed by a population crash.

population hurdle The need for investment in infrastructure and social services in a rapidly growing population that prevents the capital investment necessary for real economic development.

population momentum A potential for increased population growth as young members reach reproductive age.

postmaterialist values A philosophy that emphasizes quality of life over acquisition of material goods.

post-modernism A philosophy that rejects the optimism and universal claims of modern positivism. Post-modernists' claims admit to no grand narrative of history and no universal philosophical truths. Nature—or at least our perception of it—is an arbitrary, social construction. No single viewpoint or set of preferences—according to this point of view—is inherently better than any other.

potential energy Stored energy that is latent but available for use. A rock poised at the top of a hill or water stored behind a dam are examples of potential energy.

power The rate of energy delivery; measured in horsepower or watts.

precautionary principle The decision to leave a margin of safety for unexpected developments. This principle says we should take precautionary measures to prevent harm to human health and the environment even if some cause and effect relationships are not fully understood.

precedent An act or decision that can be used as an example in dealing with subsequent similar situations.

precycling Making environmentally sound decisions at the store and reducing waste before we buy.

predation The act of feeding by a predator.

predator An organism that feeds directly on other organisms in order to survive; live-feeders, such as herbivores and carnivores.

prevention of significant deterioration A clause of the Clean Air Act that prevents degradation of existing clean air; opposed by industry as an unnecessary barrier to development.

price elasticity A situation in which supply and demand of a commodity respond to price.

primary pollutants Chemicals released directly into the air in a harmful form.

primary productivity Synthesis of organic materials (biomass) by green plants using the energy captured in photosynthesis.

primary standards Regulations of the 1970 Clean Air Act; intended to protect human health.

primary succession An ecological succession that begins in an area where no biotic community previously existed.

primary treatment A process that removes solids from sewage before it is discharged or treated further.

principle of competitive exclusion A result of natural selection whereby two similar species in a community occupy different ecological niches, thereby reducing competition for food.

producer An organism that synthesizes food molecules from inorganic compounds by using an external energy source; most producers are photosynthetic.

production frontier The maximum output of two competing commodities at different levels of production.

prokaryotic Cells that do not have a membrane-bounded nucleus or membrane-bounded organelles.

Promethean environmentalism A form of technological optimism that predicts that human ingenuity and enterprise will find cures for all our problems.

promoters Agents that are not carcinogenic but that assist in the progression and spread of tumors; sometimes called cocarcinogens.

pronatalist pressures Influences that encourage people to have children.

proteins Chains of amino acids linked by peptide bonds.

proton A positively charged subatomic particle found in the nucleus of an atom.

proven resources Those that have been thoroughly mapped and are economical to recover at current prices with available technology.

public trust A doctrine obligating the government to maintain public lands in a natural state as guardians of the public interest.

pull factors (in urbanization) Conditions that draw people from the country into the city.

push factors (in urbanization) Conditions that force people out of the country and into the city.

R

radioactive An unstable isotope that decays spontaneously and releases subatomic particles or units of energy.

radioactive decay A change in the nuclei of radioactive isotopes that spontaneously emit high-energy electromagnetic radiation and/or subatomic particles while gradually changing into another isotope or different element.

radioactive isotope An unstable form of an element that spontaneously emits either high-energy electromagnetic radiation or subatomic particles (or both). The decay rate describes the time necessary for a radioactive isotope to change to a stable element.

radionucleides Isotopes that exhibit radioactive decay.

rainforest A forest with high humidity, constant temperature, and abundant rainfall (generally over 380 cm [150 in] per year); can be tropical or temperate.

rain shadow Dry area on the downwind side of a mountain.

rangeland Grasslands and open woodlands suitable for livestock grazing.

raprenox A process to reduce nitrogen oxide emissions by mixing combustion gases with isocyanic acid, which converts nitrogen oxides to dimolecular nitrogen gas.

rational choice Public decision making based on reason, logic, and science-based management.

recharge zone Area where water infiltrates into an aquifer.

reclamation Chemical, biological, or physical clean-up and reconstruction of severely contaminated or degraded sites to return them to something like their original topography and vegetation.

recoverable resources Those accessible with current technology but not economical under current conditions.

re-creation Construction of an entirely new biological community to replace one that has been destroyed on that or another site.

recycling Reprocessing of discarded materials into new, useful products; not the same as reuse of materials for their original purpose, but the terms are often used interchangeably.

red tide A population explosion or bloom of minute, single-celled marine organisms called dinoflagellates. Billions of these cells can accumulate in protected bays where the toxins they contain can poison other marine life.

reduced tillage systems Systems, such as minimum till, conserve-till, and no-till, that preserve soil, save energy and water, and increase crop yields.

reformer A device that strips hydrogen from fuels such as natural gas, methanol, ammonia, gasoline, or vegetable oil so they can be used in a fuel cell.

refuse-derived fuel Processing of solid waste to remove metal, glass, and other unburnable materials; organic residue is shredded, formed into pellets, and dried to make fuel for power plants.

regenerative farming Farming techniques and land stewardship that restore the health and productivity of the soil by rotating crops, planting ground cover, protecting the surface with crop residue, and reducing synthetic chemical inputs and mechanical compaction.

regulations Rules established by administrative agencies; regulations can be more important than statutory law in the day-to-day management of resources.

rehabilitate land A utilitarian program to make an area useful to humans.

rehabilitation To rebuild elements of structure or function in an ecological system without necessarily achieving complete restoration to its original condition.

relative humidity At any given temperature, a comparison of the actual water content of the air with the amount of water that could be held at saturation.

relativists Those who believe moral principles are always dependent on the particular situation.

remediation Cleaning up chemical contaminants from a polluted area.

renewable resources Resources normally replaced or replenished by natural processes; resources not depleted by moderate use; examples include solar energy, biological resources such as forests and fisheries, biological organisms, and some biogeochemical cycles.

renewable water supplies Annual freshwater surface runoff plus annual infiltration into underground freshwater aquifers that are accessible for human use.

residence time The length of time a component, such as an individual water molecule, spends in a particular compartment or location before it moves on through a particular process or cycle.

resilience The ability of a community or ecosystem to recover from disturbances.

resistance (inertia) The ability of a community to resist being changed by potentially disruptive events.

resource In economic terms, anything with potential use in creating wealth or giving satisfaction.

resource partitioning In a biological community, various populations sharing environmental resources through specialization, thereby reducing direct competition. *See also* ecological niche.

resource scarcity A shortage or deficit in some resource.

respiratory fibrotic agents Special class of irritants, including chemical reagents and particulate materials, that damages the lungs, causing scar tissue formation that lowers respiratory capacity.

restoration To bring something back to a former condition. Ecological restoration involves active manipulation of nature to re-create conditions that existed before human disturbance.

restoration ecology Seeks to repair or reconstruct ecosystems damaged by human actions.

restore ecosystems To reinstate an entire community of organisms to naturally occurring association.

riders Amendments attached to bills in conference committee, often completely unrelated to the bill to which they are added.

rill erosion The removing of thin layers of soil as little rivulets of running water gather and cut small channels in the soil.

risk Probability that something undesirable will happen as a consequence of exposure to a hazard.

risk assessment Evaluation of the short-term and long-term risks associated with a particular activity or hazard; usually compared to benefits in a cost-benefit analysis.

RNA Ribonucleic acid; nucleic acid used for transcription and translation of the genetic code found on DNA molecules.

rock a solid, cohesive, aggregate of one or more crystalline minerals.

rock cycle The process whereby rocks are broken down by chemical and physical forces; sediments are moved by wind, water, and gravity, sedimented and reformed into rock, and then crushed, folded, melted, and recrystallized into new forms.

ruminant animals Cud-chewing animals, such as cattle, sheep, goats, and buffalo, with multichambered stomachs in which cellulose is digested with the aid of bacteria.

runoff The excess of precipitation over evaporation; the main source of surface water and, in broad terms, the water available for human use.

run-of-the-river flow Ordinary river flow not accelerated by dams, flumes, etc. Some small, modern, high-efficiency turbines can generate useful power with run-of-the-river flow or with a current of only a few kilometers per hour.

rural area An area in which most residents depend on agriculture or the harvesting of natural resources for their livelihood.

S

S curve A curve that depicts logistic growth; called an S curve because of its shape.

salinity Amount of dissolved salts (especially sodium chloride) in a given volume of water.

salinization A process in which mineral salts accumulate in the soil, killing plants; occurs when soils in dry climates are irrigated profusely.

saltwater intrusion Movement of saltwater into freshwater aquifers in coastal areas where groundwater is withdrawn faster than it is replenished.

sanitary landfills A landfill in which garbage and municipal waste is buried every day under enough soil or fill to eliminate odors, vermin, and litter.

saturation point The maximum concentration of water vapor the air can hold at a given temperature.

scavenger An organism that feeds on the dead bodies of other organisms.

scientific method A systematic, precise, objective study of a problem. Generally this requires observation, hypothesis development and testing, data gathering, and interpretation.

secondary pollutants Chemicals modified to a hazardous form after entering the air or that are formed by chemical reactions as components of the air mix and interact.

secondary recovery technique Pumping pressurized gas, steam, or chemical-containing water into a well to squeeze more oil from a reservoir.

secondary standards Regulations of the 1972 Clean Air Act intended to protect materials, crops, visibility, climate, and personal comfort.

secondary succession Succession on a site where an existing community has been disrupted.

secondary treatment Bacterial decomposition of suspended particulates and dissolved organic compounds that remain after primary sewage treatment.

second law of thermodynamics States that, with each successive energy transfer or transformation in a system, less energy is available to do work.

Second World The industrialized, socialist, centrally planned economy nations of Eastern Europe and the former Soviet Union and its allies.

secure landfill A solid waste disposal site lined and capped with an impermeable barrier to prevent leakage or leaching. Drain tiles, sampling wells, and vent systems provide monitoring and pollution control.

sedimentary rock Deposited material that remains in place long enough or is covered with enough material to compact into stone; examples include shale, sandstone, breccia, and conglomerates.

sedimentation The deposition of organic materials or minerals by chemical, physical, or biological processes.

selective cutting Harvesting only mature trees of certain species and size; usually more expensive than clear-cutting, but it is less disruptive for wildlife and often better for forest regeneration.

seriously undernourished Those who receive less than 80 percent of their minimum daily caloric requirements; are likely to suffer permanently stunted growth, mental retardation, and other social and developmental disorders.

shallow ecology A critical term applied to superficial environmentalists who claim to be green but are quick to compromise and who do little to bring about fundamental change. Many of those accused of being shallow ecologists counter that they are merely pragmatic, progressive reformers who prefer to work within the established social contract rather than try to overthrow it.

shantytowns Settlements created when people move onto undeveloped lands and build their own shelter with cheap or discarded materials; some are simply illegal subdivisions where a landowner rents land without city approval; others are land invasions.

sheet erosion Peeling off thin layers of soil from the land surface; accomplished primarily by wind and water.

sinkholes A large surface crater caused by the collapse of an underground channel or cavern; often triggered by groundwater withdrawal.

sludge Semisolid mixture of organic and inorganic materials that settles out of wastewater at a sewage treatment plant.

slums Legal but inadequate multifamily tenements or rooming houses; some are custom built for rent to poor people, others are converted from some other use.

smog The term used to describe the combination of smoke and fog in the stagnant air of London; now often applied to photochemical pollution products or urban air pollution of any kind.

social ecology A socialist/humanist philosophy based on the communitarian anarchism of the Russian geographer Peter Kropotkin. It shares much with deep ecology except that it is more humanist in its outlook.

social justice Equitable access to resources and the benefits derived from them; a system that recognizes inalienable rights and adheres to what is fair, honest, and moral.

soil A complex mixture of weathered mineral materials from rocks, partially decomposed organic molecules, and a host of living organisms.

soil horizons Horizontal layers that reveal a soil's history, characteristics, and usefulness.

southern pine forest United States coniferous forest ecosystem characterized by a warm, moist climate.

species A population of morphologically similar organisms that can reproduce sexually among themselves but that cannot produce fertile offspring when mated with other organisms.

species diversity The number and relative abundance of species present in a community.

species recovery plan A plan for restoration of an endangered species through protection, habitat management, captive breeding, disease control, or other techniques that increase populations and encourage survival.

spring overturn Springtime lake phenomenon that occurs when the surface ice melts and the surface water temperature warms to its greatest density at 4° C and then sinks, creating a convection current that displaces nutrient-rich bottom waters.

squatter towns Shantytowns that occupy land without owner's permission; some are highly organized movements in defiance of authorities; others grow gradually.

stability In ecological terms, a dynamic equilibrium among the physical and biological factors in an ecosystem or a community; relative homeostasis.

stable runoff The fraction of water available year-round; usually more important than total runoff when determining human uses.

Standard Metropolitan Statistical Area (SMSA) An urbanized region with at least 100,000 inhabitants with strong economic and social ties to a central city of at least 50,000 people.

standing The right to take part in legal proceedings.

statute law Formal documents or decrees enacted by the legislative branch of government.

statutory law Rules passed by a state or national legislature.

steady-state economy Characterized by low birth and death rates, use of renewable energy sources, recycling of materials, and emphasis on durability, efficiency, and stability.

stewardship A philosophy that holds that humans have a unique responsibility to manage, care for, and improve nature.

stomates The small openings in leaves, herbaceous stems, and fruits through which gases and water vapor pass.

strategic lawsuits against public participation (SLAPP) Lawsuits that have no merit but are brought merely to intimidate and harass private citizens who act in the public interest.

strategic metals and minerals Materials a country cannot produce itself but that it uses for essential materials or processes.

stratosphere The zone in the atmosphere extending from the tropopause to about 50 km (30 mi) above the earth's surface; temperatures are stable or rise slightly with altitude; has very little water vapor but is rich in ozone.

stress Physical, chemical, or emotional factors that place a strain on an animal. Plants also experience physiological stress under adverse environmental conditions.

stress-related disease *See* stress shock.

stress shock A loose set of physical, psychological, and/or behavioral changes thought to result from the stress of excess competition and extreme closeness to other members of the same species.

strip-cutting Harvesting trees in strips narrow enough to minimize edge effects and to allow natural regeneration of the forest.

strip-farming Planting different kinds of crops in alternating strips along land contours; when one crop is harvested, the other crop remains to protect the soil and prevent water from running straight down a hill.

strip-mining Removing surface layers over coal seams using giant, earth-moving equipment; creates a huge open-pit from which coal is scooped by enormous surface-operated machines and transported by trucks; an alternative to deep mines.

structure (in ecological terms) Patterns of organization, both spatial and functional, in a community.

sublimation The process by which water can move between solid and gaseous states without ever becoming liquid.

subsidence A settling of the ground surface caused by the collapse of porous formations that result from withdrawal of large amounts of groundwater, oil, or other underground materials.

subsoil A layer of soil beneath the topsoil that has lower organic content and higher concentrations of fine mineral particles; often contains soluble compounds and clay particles carried down by percolating water.

sulfur cycle The chemical and physical reactions by which sulfur moves into or out of storage and through the environment.

sulfur dioxide A colorless, corrosive gas directly damaging to both plants and animals.

Superfund A fund established by Congress to pay for containment, cleanup, or remediation of abandoned toxic waste sites. The fund is financed by fees paid by toxic waste generators and by cost-recovery from cleanup projects.

supply The quantity of that product being offered for sale at various prices, other things being equal.

surface mining Some minerals are also mined from surface pits. *See* strip-mining.

surface tension A condition in which the water surface meets the air and acts like an elastic skin.

survivorship The percentage of a population reaching a given age or the proportion of the maximum life span of the species reached by any individual.

sustainable agriculture An ecologically sound, economically viable, socially just, and humane agricultural system. Stewardship, soil conservation, and integrated pest management are essential for sustainability.

sustainable development A real increase in well being and standard of life for the average person that can be maintained over the long-term without degrading the environment or compromising the ability of future generations to meet their own needs.

sustained yield Utilization of a renewable resource at a rate that does not impair or damage its ability to be fully renewed on a long-term basis.

swamp Wetland with trees, such as the extensive swamp forests of the southern United States.

swidden agriculture *See* milpa agriculture.

symbiosis The intimate living together of members of two different species; includes **mutualism, commensalism,** and, in some classifications, **parasitism.**

synergistic effects When an injury caused by exposure to two environmental factors together is greater than the sum of exposure to each factor individually.

systemic A condition or process that affects the whole body; many metabolic poisons are systemic.

T

taiga The northernmost edge of the boreal forest, including species-poor woodland and peat deposits; intergrading with the arctic tundra.

tailings Mining waste left after mechanical or chemical separation of minerals from crushed ore.

taking Unconstitutional confiscation of private property.

technological optimists Those who believe that technology and human enterprise will find cures for all our problems. Also called Promethean environmentalism.

technopolis Also called a vertical city; this model of city development proposes that cities grow vertically instead of horizontally.

tectonic plates Huge blocks of the earth's crust that slide around slowly, pulling apart to open new ocean basins or crashing ponderously into each other to create new, larger landmasses.

temperate rainforest The cool, dense, rainy forest of the northern Pacific coast; enshrouded in fog much of the time; dominated by large conifers.

temperature A measure of the speed of motion of a typical atom or molecule in a substance.

teratogens Chemicals or other factors that specifically cause abnormalities during embryonic growth and development.

terracing Shaping the land to create level shelves of earth to hold water and soil; requires extensive hand labor or expensive machinery, but it enables farmers to farm very steep hillsides.

territoriality An intense form of intraspecific competition in which organisms define an area surrounding their home site or nesting site and defend it, primarily against other members of their own species.

tertiary treatment The removal of inorganic minerals and plant nutrients after primary and secondary treatment of sewage.

thermal plume A plume of hot water discharged into a stream or lake by a heat source, such as a power plant.

thermocline In water, a distinctive temperature transition zone that separates an upper layer that is mixed by the wind (the epilimnion) and a colder, deep layer that is not mixed (the hypolimnion).

thermodynamics A branch of physics that deals with transfers and conversions of energy.

thermodynamics, first law Energy can be transformed and transferred, but cannot be destroyed or created.

thermodynamics, second law With each successive energy transfer or transformation, less energy is available to do work.

thermoplastics Soft plastics composed of single-chain, unlinked polymers, such as polyethylene, polypropylene, polyvinylchloride, polystyrene, and polyester, that can be remelted and reformed to make useful products.

thermoset polymers Hard plastics composed of cross-linked molecular networks, such as acrylic, phenolic, or epoxy resins, that cannot be remelted or recycled.

thermosphere The highest atmospheric zone; a region of hot, dilute gases above the mesosphere extending out to about 1,600 km (1,000 mi) from the earth's surface.

Third World Less-developed countries that are not capitalistic and industrialized (First World) or centrally-planned socialist economies (Second World); not intended to be derogatory.

threatened species While still abundant in parts of its territorial range, this species has declined significantly in total numbers and may be on the verge of extinction in certain regions or localities.

tidal station A dam built across a narrow bay or estuary traps tide water flowing both in and out of the bay. Water flowing through the dam spins turbines attached to electric generators.

timberline In mountains, the highest-altitude edge of forest that marks the beginning of the treeless alpine tundra.

tolerance limits *See* limiting factors.

tool-making revolution The invention of tools—probably first of stone, wood, or bone—sometime in the early Paleolithic period. Knowledge of tool-making and tool-using technologies may have predated modern humans, as evidenced by tool utilization by other primates.

topsoil The first true layer of soil; layer in which organic material is mixed with mineral particles; thickness ranges from a meter or more under virgin prairie to zero in some deserts.

tornado A violent storm characterized by strong swirling winds and updrafts; tornadoes form when a strong cold front pushes under a warm, moist air mass over the land.

tort law Court cases that seek compensation for damages.

total fertility rate The number of children born to an average woman in a population during her entire reproductive life.

total growth rate The net rate of population growth resulting from births, deaths, immigration, and emigration.

total maximum daily loads (TMDL) The amount of particular pollutant that a water body can receive from both point and nonpoint sources and still meet water quality standards.

toxic colonialism Shipping toxic wastes to a weaker or poorer nation.

toxic release inventory A program created by the Superfund Amendments and Reauthorization Act of 1984 that requires manufacturing facilities and waste handling and disposal sites to report annually on releases of more than 300 toxic materials. You can find out from the EPA whether any of these sites are in your neighborhood and what toxics they release.

toxins Poisonous chemicals that react with specific cellular components to kill cells or to alter growth or development in undesirable ways; often harmful, even in dilute concentrations.

tradable permits Pollution quotas or variances that can be bought or sold.

tragedy of the commons An inexorable process of degradation of communal resources due to selfish self-interest of "free riders" who use or destroy more than their fair share of common property. *See* open access systems.

transitional zone A zone in which populations from two or more adjacent communities meet and overlap.

transpiration The evaporation of water from plant surfaces, especially through stomates.

trauma Injury caused by accident or violence.

Trombe wall An interior, heat-absorbing wall; may be water-filled glass tubes that absorb heat rays and let light into interior rooms.

trophic level Step in the movement of energy through an ecosystem; an organism's feeding status in an ecosystem.

tropical rainforests Forests in which rainfall is abundant—more than 200 cm (80 in) per year—and temperatures are warm to hot year-round.

tropical seasonal forest Semievergreen or partly deciduous forests tending toward open woodlands and grassy savannas dotted with scattered, drought-resistant tree species; distinct wet and dry seasons, hot year-round.

tropopause The boundary between the troposphere and the stratosphere.

troposphere The layer of air nearest to the earth's surface; both temperature and pressure usually decrease with increasing altitude.

tsunami Giant seismic sea swells that move rapidly from the center of an earthquake; they can be 10 to 20 meters high when they reach shorelines hundreds or even thousands of kilometers from their source.

tundra Treeless arctic or alpine biome characterized by cold, harsh winters, a short growing season, and potential for frost any month of the year; vegetation includes low-growing perennial plants, mosses, and lichens.

U

unconventional air pollutants Toxic or hazardous substances, such as asbestos, benzene, beryllium, mercury, polychlorinated biphenyls, and vinyl chloride, not listed in the original Clean Air Act because they were not released in large quantities; also called noncriteria pollutants.

unconventional oil Resources such as shale oil and tar sands that can be liquefied and used like oil.

undernourished Those who receive less than 90 percent of the minimum dietary intake over a long-term time period; they lack energy for an active, productive life and are more susceptible to infectious diseases.

undiscovered resources Speculative or inferred resources or those that we haven't even thought about.

universalists Those who believe that some fundamental ethical principles are universal and unchanging. In this vision, these principles are valid regardless of the context or situation.

upwelling Convection currents within a body of water that carry nutrients from bottom sediments toward the surface.

urban area An area in which a majority of the people are not directly dependent on natural resource-based occupations.

urbanization An increasing concentration of the population in cities and a transformation of land use to an urban pattern of organization.

urban renewal Programs to revitalize old, blighted sections of inner cities.

utilitarian conservation A philosophy that resources should be used for the greatest good for the greatest number for the longest time.

utilitarianism *See* utilitarian conservation.

V

values An estimation of the worth of things; a set of ethical beliefs and preferences that determine our sense of right and wrong.

vertical city *See* technopolis.

vertical stratification The vertical distribution of specific subcommunities within a community.

village A collection of rural households linked by culture, custom, and association with the land.

visible light A portion of the electromagnetic spectrum that includes the wavelengths used for photosynthesis.

vitamins Organic molecules essential for life that we cannot make for ourselves; we must get them from our diet; they act as enzyme cofactors.

volatile organic compounds Organic chemicals that evaporate readily and exist as gases in the air.

voluntary simplicity Deliberately choosing to live at a lower level of consumption as a matter of personal and environmental health.

vulnerable species Naturally rare organisms or species whose numbers have been so reduced by human activities that they are susceptible to actions that could push them into threatened or endangered status.

W

warm front A long, wedge-shaped boundary caused when a warmer advancing air mass slides over neighboring cooler air parcels.

waste stream The steady flow of varied wastes, from domestic garbage and yard wastes to industrial, commercial, and construction refuse.

water cycle The recycling and reutilization of water on Earth, including atmospheric, surface, and underground phases and biological and nonbiological components.

water droplet coalescence A mechanism of condensation that occurs in clouds too warm for ice crystal formation.

waterlogging Water saturation of soil that fills all air spaces and causes plant roots to die from lack of oxygen; a result of overirrigation.

watershed The land surface and groundwater aquifers drained by a particular river system.

water stress A situation when residents of a country don't have enough accessible, high-quality water to meet their everyday needs.

water table The top layer of the zone of saturation; undulates according to the surface topography and subsurface structure.

weather Description of the physical conditions of the atmosphere (moisture, temperature, pressure, and wind).

weathering Changes in rocks brought about by exposure to air, water, changing temperatures, and reactive chemical agents.

wetlands Ecosystems of several types in which rooted vegetation is surrounded by standing water during part of the year. *See also* swamp, marsh, bog, fen.

wicked problems Problems with no simple right or wrong answer where there is no single, generally agreed on definition of or solution for the particular issue.

wilderness An area of undeveloped land affected primarily by the forces of nature; an area where humans are visitors who do not remain.

Wilderness Act Legislation of 1964 recognizing that leaving land in its natural state may be the highest and best use of some areas.

wildlife Plants, animals, and microbes that live independently of humans; plants, animals, and microbes that are not domesticated.

wildlife refuges Areas set aside to shelter, feed, and protect wildlife; due to political and economic pressures, refuges often allow hunting, trapping, mineral exploitation, and other activities that threaten wildlife.

windbreak Rows of trees or shrubs planted to block wind flow, reduce soil erosion, and protect sensitive crops from high winds.

wind farms Large numbers of windmills concentrated in a single area; usually owned by a utility or large-scale energy producer.

Wise Use Movement A coalition of ranchers, loggers, miners, industrialists, hunters, off-road vehicle users, land developers, and others who call for unrestricted access to natural resources and public lands.

withdrawal A description of the total amount of water taken from a lake, river, or aquifer.

woodland A forest where tree crowns cover less than 20 percent of the ground; also called open canopy.

work The application of force through a distance; requires energy input.

world conservation strategy A proposal for maintaining essential ecological processes, preserving genetic diversity, and ensuring that utilization of species and ecosystems is sustainable.

World Ocean The interconnected world seas and oceans.

World Trade Organization An association of 135 nations that meet to regulate international trade.

X

X ray Very short wavelength in the electromagnetic spectrum; can penetrate soft tissue; although it is useful in medical diagnosis, it also damages tissue and causes mutations.

Y

yellowcake The concentrate of 70 to 90 percent uranium oxide extracted from crushed ore.

Z

zero population growth (ZPG) The number of births at which people are just replacing themselves; also called the replacement level of fertility.

zone of aeration Upper soil layers that hold both air and water.

zone of leaching The layer of soil just beneath the topsoil where water percolates, removing soluble nutrients that accumulate in the subsoil; may be very different in appearance and composition from the layers above and below it.

zone of saturation Lower soil layers where all spaces are filled with water.

Figure Credits

Introduction

What do I need to think critically and reflectively?: Courtesy of Karen Warren, Philosophy Department, Macalester College, St. Paul, MN.

Chapter 1

Fig. 1.11: From *Student Atlas of Environmental Issues*, First Edition, edited by John L. Allen, copyright © 1997 by The McGraw-Hill Companies, Inc. Reprinted by permission of Dushkin/McGraw-Hill; Fig. 1.14: Data from World Bank and United Nations as appeared in Jerome Fellmann, et al., *Human Geography*, 4th edition. Copyright © McGraw-Hill Company, Inc. All Rights Reserved. Reprinted by permission; Fig. 1.15: From United Nations (U.N.) Critical Trends: Global Change and Sustainable Development, 1997, United Nations, New York. The United Nations is the author of the original material.

Chapter 3

Fig. 3.18: From Michael D. Moran, et al., *Environmental Science*. Copyright © 1991 McGraw-Hill Company, Inc. All Rights Reserved. Reprinted by permission.

Chapter 4

Fig. 4.5: From Peter H. Raven and George B. Johnson, *Biology,* 4th edition. Copyright © 1996 McGraw-Hill Company, Inc. All Rights Reserved. Reprinted by permission; Fig. 4.23: From Verner, Jared. *Wildlife 2000*. Copyright © 1986. Reprinted by permission of University of Wisconsin Press.

Chapter 7

Fig. 7.2: Figure redrawn with permission from *Population Bulletin*, vol. 18, no. 1, 1985. Population Reference Bureau; Fig. 7.5: From Thomas Merrick, with PRB staff, "World Population in Transition," *Population Bulletin*, vol. 41, no. 2 (reprinted 1991); Fig. 7.6: From World Resources Institute, *World Resources 1998–99*. Data source from CIESIN, Columbia University; Fig. 7.13: Reprinted from *Beyond the Limits*, copyright © 1992 by Meadows, Meadows and Randers. With permission from Chelsea Green Publishing Co., White River Junction, Vermont; Fig. 7.15: From *World Resources 1998–99*. Reprinted by permission of World Resources Institute.

Chapter 8

Fig. 8.15 & 8.16: Reprinted from *Beyond the Limits*, copyright © 1992 by Meadows, Meadows and Randers. With permission from Chelsea Green Publishing Co., White River Junction, Vermont.

Chapter 9

Fig. 9.7: From Dr. Steven Morse, Rockefeller University in Time, September 12, 1994. Copyright © Time, Inc.

Chapter 10

Fig. 10.14: From K. Sexton, et al., *Better Environment Decisions*, 1999. Copyright © 1999 Island Press. Reprinted by permission of Alexander Hoyt & Associates; Fig. 10.15: From Buzz Holling, *Barriers & Bridges*, edited by Gunderson, Holling, and Light, 1995, Columbia University Press.

Chapter 11

Fig. 11.2: From *World Resources 1998–99*, World Resource Institute. Reprinted by permission; Fig. 11.6: Data from USDA as appeared in Peter H. Raven and George B. Johnson, *Understanding Biology*, 3rd edition. Copyright © 1995 McGraw-Hill Company, Inc. All Rights Reserved. Reprinted by permission; Fig. 11.13: From Jerome Fellmann, et al., *Human Geography*, 4th ed., from data in *World Resources 1992–1993*, International Source and Information Center. Copyright © 1995 McGraw-Hill Company, Inc. All Rights Reserved. Reprinted by permission.

Chapter 13

Fig. 13.10: Reprinted with permission from J. Curtis in William L. Thomas (ed.), *Man's Role in Changing the Face of the Earth*. Copyright © 1956 by The University of Chicago Press. All rights reserved.

Chapter 14

Fig. 14.24: Data from H.E. Dregne, *Desertification of Arid Lands*, Harwood Academic Press, 1982; and *A World Map of Desertification*, UNESCO/FAQ; as appeared in Jerome Fellmann, et al., *Human Geography*, 4th edition. Copyright © 1995 McGraw-Hill Company, Inc. All Rights Reserved. Reprinted by permission.

Chapter 15

Fig. 15.20: Copyright © Rodger Doyle 1999.

Chapter 16

Fig. 16.2: From Carla Montgomery, *Physical Geology*, 3rd edition. Copyright © 1993 McGraw-Hill Company, Inc. All Rights Reserved. Reprinted by permission.

Chapter 17

Fig. 17.14 A–C: From the Intergovernmental Panel on Climate Change (1990) and Minister of Interior, Government of Canada, 1991. *The State of Canada's Environment,* Canada. Reproduced with the permission of the Minister of Public Works and Government Services Canada, 1999; Fig. 17.21: © Laurie Grace.

Chapter 18

Fig. 18.13: From Government of Canada, 1991. *The State of Canada's Environment.* Reproduced with the permission of the Minister of Public Works and Government Services Canada, 1999; Fig. 18.25: Used with permission of General Motors Corporation.

Chapter 19

Fig. 19.4: From Jerome Fellmann, et al., *Human Geography*, 4th edition. Copyright © 1995 McGraw-Hill Co. All Rights Reserved. Reprinted by permission.

Chapter 20

Fig. 20.19. From *Student Atlas of Environmental Issues*. First Edition, edited by John L. Allen, copyright © 1997 by The McGraw-Hill Companies, Inc. Reprinted by permission of Dushkin/McGraw-Hill; Fig. 20.26: Drawing is based on an original drawing by John Todd. Reprinted by permission.

Chapter 21

Fig. 21.18: Courtesy of Northern States Power Co., Minneapolis, MN; Fig. 21.23: Redrawn with permission of Atomic Energy of Canada Limited.

Chapter 24

Fig. 24.23: From Randall Arenct, "Creating Open Space Networks" in *Environment & Development*, May/June 1996, American Planning Association, Chicago, IL. Reprinted by permission.

Chapter 25

Fig. 25.5 (a): Reproduced with permission, Green Seal, Inc., Washington, DC. (b): M-Official mark of Environment Canada; Fig. 25.17: Modified from Raymond Grizzle and Christopher Barrett, personal communication.

All page heads, In Depth boxes, Case Study box heads: © 1999 Artville, Ken Shearer (scenics); © 1999 Digital Vision (Globe)

All Globe usages: © 1999 Digital Vision

What Can You Do?: #1 PhotoDisc 2; #2 PhotoDisc 2; #3 PhotoDisc 2; #4 Digital Stock

Photo Credits

Inside cover: © RMC, R.L. 00-S-18, www.randmcnally.com.

Introduction

Opener: © PhotoDisc/Jeff Maloney/Education 2; I.1: PhotoDisk, I.2: PhotoDisk.

Part Opening photos by William P. Cunningham.

Chapter One

Opener 1: © The McGraw-Hill Companies, Inc./Barry W. Barker, photographer; 1.1: Courtesy of the Minnesota Pollution Control Agency; 1.3: © Kurgan-Lisnet/Gamma Liaison International; 1.4: © Bettmann/Corbis; 1.5: Courtesy David Swanlund, Save-The-Redwoods League; 1.6: © AP/Wide World Photos; 1.7: © Earth Images/Tony Stone Images; 1.8: © Craig Tuttle/The Stock Market; 1.9: © Norbert Schiller/The Image Works; p. 23: © AP Photo/Elaine Thompson; 1.12: © Gary Milburn/Tom Stack & Associates; 1.13: © Kit King/Gamma Liaison International; 1.18: © The McGraw-Hill Companies, Inc./Barry W. Barker, photographer; 1.19: © Tony Stone Images; 1.20: © The McGraw-Hill Companies, Inc./Barry W. Barker, photographer; p. 32: © John Chiasson/Gamma Liaison International; 1.22: "The Slaughter of the Jacques on the Bridge at Meaux," From Froissart's Chronicles, Louis de Bruges copy, c. 1460. Courtesy of Bibliotheque Nationale.

Chapter Two

Opener: © J. Lotter/Tom Stack & Associates; 2.1: © Hollyman/Gamma Liaison International; 2.2: Courtesy Monumenti Musei E Gallerie Pontificie; 2.3: © B. Hemphill/Photo Researchers, Inc.; 2.4: © Grant Heilman; 2.5: © Carr Clifton; 2.6: © Inga Spence/Tom Stack & Associates; 2.7, 2.10, 2.13, 2.14: William P. Cunningham; 2.8: "Nataraja: Shiva as King of Dance." Bronze, h. 111.5 cm. South India, Chola Period, 11th c. © The Cleveland Museum of Art, Purchase from the J. H. Wade Fund, 1930.331; 2.9: © Sam Kittner; 2.11: © Sam Kittner/GREENPEACE; 2.16: © Bettman/Corbis; 2.17: © Martha Cooper/Peter Arnold, Inc.; p. 53: Courtesy of Micaela Martinez.

Chapter Three

Opener: © Tom Tietz/Tony Stone Images; 3.4: © E. H. Newcomb/W. P. Wergin/Biological Photo Service; 3.5: © Carl Purcell/Photo Researchers, Inc.; 3.17: © David Dennis/Tom Stack & Associates; p. 71: © G. I. Bernard/Animals Animals/Earth Scenes.

Chapter Four

Opener: © Larry Ulrich/Tony Stone Images; 4.1: © Leonard L. Rue/Visuals Unlimited; 4.3: © S. J. Krasemann/Photo Researchers, Inc.; 4.4 a: © Breck Kent/Animals Animals/Earth Scenes; 4.4 b: © Breck Kent/Animals Animals/Earth Scenes; 4.6: © Robert C. Simpson/Valen Photos; 4.9: © Ray Coleman/Photo Researchers, Inc.; 4.10: © D. P. Wilson/Photo Researchers, Inc.; 4.11: © Randy Morse/Tom Stack & Associates; p. 86: © John Gerlach/Tony Stone Images; 4.12: © Bill Beatty/Visuals Unlimited; p. 88: © Milton Tierny/Visuals Unlimited; 4.13, 4.14, 4.22, 4.25, 4.26: William P. Cunningham; 4.15: © Michael Fogden/Animals Animals/Earth Scenes; 4.16 A: © Edward S. Ross; 4.16 B: © Edward S. Ross; 4.17: © Brian Parker/Tom Stack & Associates; 4.21: © Fred Bavendam/Peter Arnold, Inc., 4.27: © Gerard Lacz/Peter Arnold, Inc.; p. 120: Courtesy Marion Taylor.

Chapter Five

Opener: © Larry Kolvoord/The Image Works; 5.1, 5.4, 5.5, 5.6, 5.7, 5.8, 5.10: William P. Cunningham; 5.9: © The McGraw-Hill Companies, Inc./Barry W. Barker, photographer; 5.13: © John D. Cunningham/Visuals Unlimited; 5.14: © Andrew Martinez/Photo Researchers, Inc.; 5.16: © Mike Bacon/Tom Stack & Associates; 5.20: © LW Richardson/Photo Researchers, Inc.; 5.21: Courtesy South Florida Water Management District; 5.22: Courtesy University of Wisconsin-Madison Archives; 5.23: Courtesy University of Wisconsin Arboretum; 5.24: Courtesy State of Minnesota, Department of Natural Resources. Reprinted with permission; p. 120: William P. Cunningham; 5.25: © Peter Menzel Photography; 5.26: © RJ Goldstein/Visuals Unlimited.

Chapter Six

Opener: © Mark Peterson/Tony Stone Images; 6.4: Courtesy National Museum of Natural History, Smithsonian Institution; p. 140: Courtesy James L. Bartlett.

Chapter Seven

Opener: James L. Amos/Peter Arnold, Inc.; 7.1: Courtesy John Cunningham; 7.16a-b,d: © The McGraw-Hill Companies, Inc./Bob Coyle, photographer; 7.16e: © The McGraw-Hill Companies, Inc./Vincent Ho, photographer; 7.16f: © The Mcgraw-Hill Companies, Inc./Bob Coyle, photographer.

Chapter Eight

Opener: © John Miller/AP Wide World; 8.1, 8.9: William P. Cunningham; 8.7: © Lowell Georgia/Photo Researchers, Inc.; 8.8: © Audrey Lang/Valan Photos; 8.20: © The Mcgraw-Hill Companies, Inc./Barry W. Barker, photographer; p. 182: Courtesy Ray Anderson; 8.21: © Mark Luthringer; p. 187: Courtesy Jeremy Sterk.

Chapter Nine

Opener: © Klaus Reisinger/Blackstar; 9.2: Courtesy Dr. Frederick A. Murphy, Centers for Disease Control and Prevention; 9.3: Courtesy of Stanley Erlandsen, University of Minnesota; 9.5: Courtesy WHO/World Bank/R. Witlin; p. 193: Courtesy Donald R. Hopkins; 9.8: © John Maier, Jr./The Image Works.

Chapter Ten

Opener: © Bob McKeever/Tom Stack & Associates; 10.4: © Doug Sokell/Tom Stack & Associates; 10.5: © Esther Henderson/Photo Researchers, Inc.; 10.7: © Bob Daemmrich/The Image Works; 10.8, 10.12: William P. Cunningham; 10.9: © Sygma; 10.16: © Jon Riley/Tony Stone Images; 10.17: © Phil Degginger/Animals Animals/Earth Scenes; 10.18: © David L. Brown/Tom Stack & Associates; p. 234: Courtesy Kristin Groce.

Chapter Eleven

Opener: © Tony Carddock/Tony Stone Images; 11.3: © Paul Souders Photography; 11.4: © Scott Daniel Peterson/Gamma Liaison International; 11.5: © Lester Bergman/Corbis Media; 11.7: © Norbert Schiller/The Image Works; 11.8: Courtesy Thierry A. Brun; 11.12: © Jacques Jangoux/Tony Stone Images; 11.15: © Larry Miller/Photo Researchers, Inc.; 11.16, 11.18, 11.19, 11.24: William P. Cunningham; 11.21: © Michael Rosenfeld/Tony Stone Images; 11.23: © Joe Munroe/Photo Researchers, Inc.; p. 256: © Still Pictures/Peter Arnold, Inc.; 11.25: Courtesy of Dave Hanson, College of Agriculture, University of Minnesota; 11.26: © Tom Sweeny/Minneapolis Star Tribune.

Chapter Twelve

Opener: © Inga Spence/Tom Stack & Associates; 12.1: © Tom J. Ulrich/Visuals Unlimited; 12.2: William P. Cunningham; 12.3; © Steve Wilson/Entheos; 12.4. © Bettmann/Corbis; 12.5: © Yoram Lehmann/Peter Arnold, Inc.; 12.6: © The McGraw-Hill Companies, Inc./Barry W. Barker, photographer; 12.7: © Stephen Dalton/Photo Researchers, Inc.; 12.8: © SPL/Photo Researchers, Inc.; 12.9: © Joe Munroe/Photo Researchers, Inc.; p. 268: © Howard Suzuki, Aquatic Life Sculptures/Discover Magazine; 12.12: © Norm Thomas/Photo Researchers, Inc.; 12.13: © AP/Wide World Photos; 12.16: © Cary Wolinsky/Stock Boston; 12.17: Courtesy Florida Department of Agriculture and Consumer Services; p.273: © Terry Gips/International Alliance for Sustainable Agriculture; 12.18: Photo by the late J. K. Holloway, from "Biological Control," C. B. Huffaker, ed. © Plenum Press, NY 1971; 12.20: © W. C. Wood/Tanimura & Antle, Inc.; p. 277: William P. Cunningham; Profile 12: Courtesy Christy Corse and Arnold Phifer.

Chapter Thirteen

Opener: © The McGraw-Hill Companies, Inc./Barry W. Barker, photographer; 13.1: © Mark Newman/Tom Stack & Associates; 13.3: © The McGraw-Hill Companies, Inc./Barry W. Barker, photographer; 13.5, 13.8, 13.13: William P. Cunningham; 13.7: Courtesy INBIO, San Jose, Costa Rica; 13.11: © Bill Beatty/Visuals Unlimited; 13.14: © The McGraw-Hill Companies, Inc./Barry W. Barker, photographer; 13.15: © Lynn Funkhouser/Peter Arnold, Inc.; 13.16: Courtesy U.S. Department of Agriculture Animals & Plant Health Inspection Service; (Kudzu vine): © David M. Dennis/Tom Stack & Associates; (Purple loosestrife): William P. Cunningham; (Leafy spurge): William P. Cunningham; (Zebra mussels): © Ken Cole/Animals Animals/Earth Scenes; (Round Goby): Courtesy Dr. David Jude; (Beetle): © Taina Litwak; 13.18: © Tom McHugh/Photo Researchers, Inc.; 13.19: © The McGraw-Hill Companies, Inc./Barry W. Barker, photographer; 13.24: William P. Cunningham; 13.25: © Tom Myers Photography; 13.26: Courtesy Dr. Ronald Tilson, Minnesota Zoo.

Chapter Fourteen

Opener: Courtesy Karen Oberhauser; 14.2, 14.11, 14.14, 14.20, 14.25: William P. Cunningham; 14.7: Courtesy Dr. Daniel Kammen, Princeton University; 14.6, 14.9: © The McGraw-Hill Companies, Inc./Barry W. Barker, photographer; 14.12: © The McGraw-Hill Companies, Inc./Barry W. Barker, photographer; 14.13: Courtesy Andy White, University of Minnesota, College of Agriculture; 14.15: © Greg Vaughn/Tom Stack & Associates; 14.16: © Shahn Kermani/Gamma Liaison International; 14.17: © Gary Braasch/Tony Stone Images; 14.18: © Earl Dotter; 14.19: © Seattle Times/Gamma Liaison International; 14.23: © Eastcott/Momatiuk/The Image Works; 14.26: © Francois Gohier/Photo Researchers, Inc.; 14.27: © The McGraw-Hill Companies, Inc./Barry W. Barker, photographer; 14.28: © AP/Wide World Photos; 14.29: © G. Prance/Visuals Unlimited; p. 329: Courtesy Andi Moffatt.

Chapter Fifteen

Opener: © Greg Probst/Tony Stone Images; 15.1: © Galen Rowell, 1987 Mountain Light Photography; 15.2, 15.3, 15.8, 15.9, 15.15: William P. Cunningham; 15.4: © John Gerlach/Visuals Unlimited; 15.5: © Lee Battaglia, 1973/Photo Researchers, Inc.; 15.6: © American Heritage Center, University of Wyoming; p. 336: © Gary Milburn/Tom Stack & Associates; 15.12: Courtesy R. O. Bierregaard; 15.13: © The McGraw Hill Companies, Inc./Barry W. Barker, photographer; p. 344: © Bruce Paton/Panos Pictures; 15.16: © Francois Gohier/Photo Researchers, Inc.; 15.17: © Jeff Foott Productions; 15.18: © Steve Raymer/National Geographic Society; 15.19: © Fred Whitehead/Animals Animals/Earth Scenes; 15.21: © Chris Harris/Gamma Liaison International; 15.22: © James Shaffer; 15.23: © The McGraw-Hill Companies, Inc., Bob Coyle, photographer; 15.24: © Stephen Rose/Gamma Liaison International.

Chapter Sixteen

Opener: © The McGraw-Hill Companies, Inc./Barry W. Barker, photographer; 16.1: © Yannis Kontos/Corbis Sygma; 16.7, 16.11: William P. Cunningham; p. 361: © Bryan F. Peterson; 16.10: © James P. Blair/National Geographic Image Collection; p. 363: © Greenpeace/Dorreboom; 16.12: Courtesy Library of Congress; 16.14: © Joseph Nettis/Photo Researchers, Inc.; 16.16: © Reuter/Sankei Shimbun; 16.17: Courtesy K. Jackson, USAF; 16.18: © Al Seib/LA Times, p. 372: Coutesy Robyn Angliss.

Chapter Seventeen

Opener: © Wm. L. Wantland/Tom Stack & Associates; 17.1: © Reuters/Jeff Mitchell/Archive Photos; 17.2: William P. Cunningham; 17.9: Courtesy NOAA Coastal Services Center; 17.10: © Science VU/Visuals Unlimited; 17.13: © Vic Englebert/Photo Researchers, Inc.; 17.16: Courtesy Peter Gilday; p. 394: Courtesy Tom Moffatt.

Chapter Eighteen

Opener: © Peter May/Peter Arnold, Inc.; 18.2a: © Leonard L. T. Rhodes/Animals Animals/Earth Scenes; 18.2b: © Joseph Sohm/Photo Researchers, Inc.; 18.3, 18.7,

18.18, 18.19, 18.21: William P. Cunningham; 18.11: © 1988 Ted Spiegel/Black Star; 18.14: Courtesy Ronald Nagatani/National Centers for Environmental Prediction; 18.22: © John D. Cunningham/Visuals Unlimited; 18.27: © James Nachtwey/Magnum Photos Inc.

Chapter Nineteen

Opener; 19.2, 19.6, 19.8, 19.14, 19.23, 19.27: William P. Cunningham; 19.11: © Ray Ellis/Photo Researchers, Inc.; 19.15: © The McGraw-Hill Companies, Inc./Barry W. Barker, photographer; p. 434: © Penny Tweedie/Tony Stone Images; 19.17: © M. Timothy O'Keefe/Tom Stack & Associates; 19.19: Courtesy Los Angeles Department of Water and Power; 19.20: © John Gerlach/Visuals Unlimited; p. 438: Courtesy Patagonia/David Cross, photographer; 19.21: Courtesy Tim McCabe, Soil Conservation Service, USDA; 19.22: © Josh Baker/Colorado Whitewater Photography; 19.26: © Sylvan H. Wittwer/Visuals Unlimited.

Chapter Twenty

Opener: © Tom Stack/Tom Stack & Associates; 20.2: © Joe McDonald/Animals Animals/Earth Scenes; 20.3: © Simon Fraser/Photo Researchers, Inc.; 20.4: © Roger A. Clark, Jr./Photo Researchers, Inc.; 20.6: William P. Cunningham; 20.7b: JoAnn M. Burkholder/"The Lurking Perils of Pfisteria" Scientific American Vol 281(2) Aug. 1999; pg. 42–49; 20.8: © The Heirs of W. Eugene Smith/Black Star; 20.9: © Rob & Melissa Simpson/Valan Photos; 20.10: © Lawrence Lowry/Photo Researchers, Inc.; 20.13: © James Nachtwey/Magnum Photos; 20.14: © Les Stone/Sygma; 20.17: © Ken Sakamoto/Black Star; 20.18: © Frans Lanting/Photo Researchers, Inc.; p. 463: William P. Cunningham; 20.21: © The McGraw-Hill Companies, Inc./Bob Coyle, photographer; p. 473: Courtesy Kristin Pierskalla.

Chapter Twenty One

Opener: © Noel Quidu/Gamma Liaison International; 21.2: © Owen Franken/Stock Boston; 21.8: © Charles E. Rotkin/Corbis Media; p. 483: © Mary Cunningham; 21.11: Courtesy American Petroleum Institute; 21.12: © W. Ormerod/Visuals Unlimited; 21.14: © Jerry Irwin/Photo Researchers, Inc.; p. 489: © TASS/Sovfoto/Eastfoto; 21.24: © Arthur Grace/Sygma.

Chapter Twenty Two

Opener: © 2000 David Brewster/Star Tribune/Minneapolis-St. Paul; 22.2: © The McGraw-Hill Companies, Inc./Barry W. Barker, photographer; p. 501: Courtesy Toyota; 22.7: © The McGraw-Hill Companies, Inc./Doug Sherman, photographer; 22.8: William P. Cunningham; 22.11: Charles Haberman; 22.12: © G. Turner/Sacramento Municipal Utility District; 22.14: Courtesy Dr. Douglas Pratt; 22.15: © Robert E. Ford/BPS; 22.17: © Wolfgang Kaehler; 22.19: © Robert Frerck/Odyssey Productions; 22.20: Courtesy Burkhardt Turbines; 22.21: © W. Shields/Valan Photos; 22.22: © George Steinmetz Photography.

Chapter Twenty Three

Opener: © Ray Pfortner/Peter Arnold, Inc.; 23.3: © Fred McConnaughey/Photo Researchers, Inc.; 23.7: William P. Cunningham; 23.10: © Mike Brisson; 23.11: Courtesy Urban Ore, Inc., Berkeley, CA; 23.13: © Michael Greenlar/The Images Works; 23.15: © J. Cancalosi/Valan Photos; p. 540: © Barbara Gauntt/The Clarion-Ledger, Jackson, MS; 23.19: © Piet Van Lier/Impact Visuals.

Chapter Twenty Four

Opener: © Malcolm S. Kirk/Peter Arnold, Inc.; 24.1: © Bob Boyer; 24.2: © Robert Frerck/Odyssey Productions; 24.3, 24.10, 24.12, 24.18, 24.19, 24.21, 24.24: William P. Cunningham; 24.4: © The McGraw-Hill Companies, Inc./Barry W. Barker, photographer; 24.11: © Louis Psihoyos/Matrix International, Inc.; 24.13: © Joseph Sohm/Photo Researchers, Inc.; 24.14: © AP/Wide World Photo; 24.16: © Tom Myers Photography; 24.20: © Tom Myers/Tom Stack & Associates; 24.22: © Kevin R. Morris/Tony Stone Images.

Chapter Twenty Five

Opener: © Jeremy Horner/Tony Stone Images; 25.1: © Tim Matsui/Liaison Agency; 25.2: © Robert Queen, Wisconsin Department of Natural Resources; 25.3: © Frank Pedrick/The Image Works; 25.7, 25.8, 25.11, 25.12, 25.14: William P. Cunningham; 25.9: © Mark Greenberg/Gamma Liaison International; 25.10: © Sam Kittner/Greenpeace Photo; 25.13: © Bryan F. Peterson; 25.18: © Cully/Greenpeace Photo; 25.19: © Photoreporters, Inc./International Press On Line; 25.20: © Bill Bachman/Photo Researchers, Inc.

A

- Abbey, Edward, 342, 582
- abiotic factors, 136
- abortion, 159, 162
- Absaroka-Beartooth Wilderness (Montana), 343
- abstinence, 159, 160
- abundance in biological communities, 91–92
- acacias, 89–90
- Acadia National Park (Maine), air pollution at, 334
- acaricides, 263
- acclimation, 80
- accuracy, in critical thinking, 8
- acid precipitation (acid rain), 410–12, 453
  - aquatic effects, 410–11
  - buildings and monuments, 21–22, 24, 412, 419
  - coal burning and, 479
  - damage in Poland, Czech Republic and Slovakia, 419
  - ecosystems, damage to, 21–22, 24
  - forest damage, 411–12
  - legislation, 416
  - national parks and, 334
  - pH and atmospheric acidity, 410
  - sulfur dioxide and, 398, 400
  - visibility reduction, 412
- acids
  - deposition, 21–22, 24, 410–12
  - as water pollutants, 453
- acquired immunodeficiency syndrome (AIDS), 190–91
- actinomycetes, 244
- action context, from environmental education, 573
- active learners, 4
- active solar heat, 505–6
- acute effects, 204
- acute poverty, 24
- adaptation, 80–82
- adaptive management, 122, 226
- additive, materials, 202
- Adirondack Mountains (New York)
  - acid precipitation damage, 410
  - water pollution of lakes, 453
- administrative courts, 224
- administrative law, 215
- adversarial approaches, in American legal system, 222
- adzuki beans, nitrogen-fixing root nodules on, 69
- aeration, zone of, 428
- aerobic digestion, 514
- aerosols, 388, 401
  - in atmosphere, 376
  - defined, 376
  - long-range transport, 405–6
  - *Pfisteria* in, 451
- aesthetic degradation, 402
- aesthetic values, 176
- Afghanistan
  - fertility rates, 149
  - population growth rates, 150
- Africa
  - AIDS, 190–91
  - cropland, distribution, 245–46
  - crude death rates, 150
  - deforestation, 312
  - desertification, 322
  - deserts, 425
  - ecotourism, 344
  - elephant conservation, 290
  - erosion rates, 248, 254
  - family planning in, 162, 163
  - famine in, 241
  - fertility, 149, 159
  - fertilizer usage, 250
  - food production in, 237
  - food shortages in, 238
  - guinea worm, eradication programs, 193
  - human development index, 28, 175
  - human disturbance of ecosystems, 113
  - indoor air pollution in, 403
  - integrated pest management in, 275
  - land degradation in, 247
  - landless poor in, 324
  - land reform in, 325
  - life expectancy, 151
  - monsoons, 383
  - onchocerciasis (river blindness) in, 192
  - poaching in, 345–46
  - poverty in, 22, 25, 26
  - protected areas in, 338
  - swidden agriculture in, 255
  - urban population growth, 551
  - water quality, 459
  - water-rich and water-poor countries, 430
  - water stress, 433
  - wild animals, harvesting, 322–23
  - wildlife and wildlife products, trade in, 290
  - wildlife preserves, 345–46
- African-Americans, life expectancy, 151
- Agassiz, Louis, 353
- agriculture
  - in Africa, 345
  - biomass accumulation and, 91, 92
  - biotechnology (*see* biotechnology)
  - cropland, distribution, 245–46
  - crop rotation, 272, 273
  - energy and, 250
  - energy crops, 515
  - erosion, intensive farming as cause of, 248–49
  - fertilizer, 249–50
  - food production, gains in, 237
  - genetic engineering, 251, 253
  - green revolution, 251, 252
  - mechanization as factor in urbanization, 552
  - methane generators, use of, 533
  - monoculture, 248
  - new crops, 250, 251
  - polyculture, mixed perennial, 313
  - regenerative, 273
  - resources, 245–46, 249–50
  - soil (*see* soil)
  - stewardship and, 41–42
  - sustainable (*see* sustainable agriculture)
  - swidden, 255, 312–13
  - transgenic crops, 191
  - water use, 249, 432, 442, 443
  - world land use for crops, 306
- Agriculture, Department of (USDA)
  - balanced diet recommendations, 240
  - erosion, data on, 248
  - mandate and organization, 223
  - pesticide regulation, 275
- AIDS (acquired immunodeficiency syndrome), 190–91
- Air Pollutant Standards Index, 416
- air pollution, 19–20, 22–24. *See also* emissions
  - acid rain (*see* acid precipitation)
  - aquatic effects, 410–11
  - buildings and monuments, damage to, 412
  - carbon oxides, 400–401
  - in Chattanooga (Tennessee), 548
  - coal and, 479–81
  - contamination of Lake Labarge (Canada), 55
  - control, 412–16
  - conventional or criteria pollutants, 398–402
  - dust domes, 405
  - effects of, 188, 408–12
  - from fires, 396
  - flue gas desulfurization, 414
  - fluidized bed combustion, 413–14
  - forest damage, 411–12
  - fuel switching and fuel cleaning, 413
  - halogens, 401
  - heat islands, 405
  - human-caused, 398–404
  - human health, effects on, 408–9
  - hydrocarbon controls, 415–16
  - indoor, 403–4
  - from large mechanized farms, 257
  - legislation, clean air, 416–17
  - limestone injection, 413–14
  - long-range transport, 405–6
  - marketing pollution rights, 417
  - metals, 401
  - moving, to remote areas, 412–13
  - at national parks, 334, 412
  - national plan to reduce, in Netherlands, 231
  - natural sources of, 397–98

air pollution—*Cont.*
nitrogen compounds, 400
nitrogen oxide control, 414–15
noncriteria pollutants, 402
ozone (*see* ozone)
particulate material, 401, 413
photochemical oxidants, 402, 403
plant pathology and, 409–10
primary pollutants, 398
secondary pollutants, 398
from smelting, 364, 409
from solid waste, 527
sulfur compounds, 398–400, 409
sulfur recovery process, 414
sulfur removal, 413
temperature inversions, 404–5
unconventional pollutants, 402
in urban areas of developing countries, 553
visibility reduction, 412
volatile organic compounds, 401–2, 403
from woodstoves and fireplaces, 512
Alabama
air pollution in Birmingham, 419
Brown's Ferry reactor, 489
Alar, 207
Alaska
ancient forests, 315
glaciers, 427
Green Party, 589
Alaska Lands Act (1980), 333
Alaska National Interest Land Act (1980), 344
albedo, 377
albendazole, 192
Alberta Research Council (Canada), 391
alcohol
from biomass, 514–15
fetal alcohol syndrome, 197
as teratogen, 197
alder, as energy crop, 515
alfalfa
as cover crop, 255
feed for domestic animals, 322
algae
blooms, 450, 457, 461
in soil, 244
allergens, 196
Alliance for Social Responsibility (New York), 576
Alliance of Small Island States, 389
alligators
hormone disrupting chemicals in, 268
protection and recovery, 296
alochlor, 270
alpine tundra, 106
Alps (Europe), acid rain damage, 411
Altamont Pass (California), 518, 519
aluminum
acid rain damage and, 412
downward price trend, 173–74
in earth, 355
as economic resource, 358, 359
extraction, 365
recycling, 173, 365, 531, 532
amaranth, 250
Amazonian rainforest (South America), 115
Amazon River (South America), 429
mercury poisoning, 452
water hyacinths blocking Tucurui Dam, 516
ambient air, 398
American Arbitration Association, 228
American bison, 170
American Cancer Society, 197, 198
American Farm Bureau, 336
American Petroleum Institute, 579
American redstarts, declining populations of, 86
American River Watch program, 572
America Recycles Day, 473
Ames, Bruce, 198, 277
Amigos de Sian Ka'an, 341
amino acids, in nitrogen cycle, 67, 68, 69
ammonia, 231
fuel cells and, 510
in nitrogen cycle, 68, 69, 70
odor from large mechanized farms, 257
amoebae, lobose, 451–52
amphibians
developmental abnormalities in, 16
hormone disturbances in, 268
Amu Dar'ya River (Kazakhstan), diversion of water from, 422
anaerobic digestion, 514
analytical thinking, 7
Anderson, Anne, 212
Anderson, Jimmy, 212
Anderson, Ray, 181, 182
andesite, 357
androgens, 268
anemia, 239
Angliss, Robyn, 372
Angola
fertility rates, 149
food production in, 237
refugee camps in Luanda, 555
Animal Damage Control Program, 291–92
animals. *See also* wildlife; *individual species*
animal rights, 39, 40, 42–43
captive breeding programs, 299, 301–2
climate change and, 389
drugs and medicines from, 284–86
environmental stress in, 198
exotic species, introduction of, 292–94
logging with draft, 318
overgrazing, 320–22
species survival plans, 301–2
toxins in, 197
value and, 40
wild, edible, 322–23
wild, saving rare species in the, 302
animal testing, 203, 204
Annapurna Conservation Area Project (ACAP) (Nepal), 331
Annapurna Range (Asia), 331
anodes, 510
Anopheles mosquitoes, 265–66
antagonistic reactions, 202
Antarctica
air pollution to, long-range transport of, 406
climate change, effects of, 389
coal reserves, 479
glaciers, 428
marine food web, 92, 93
ozone depletion over, 376, 406–7
Antarctic Atmospheric Survey (Great Britain) (1985), 406
anthropocentrism, 41
antibiotics, 263
resistance to, 194–95
antienvironmental backlash, 583–84
antigens, 196
antimony, recycling, 173
ants
in soil, 244
symbiotic relationships and, 89–90
Appleseed Fundraising, 30
appliances, energy efficient, 502
appropriate technology, 50–51
aquaculture
shrimp, 236
waste-fed, 467
aquatic ecosystems, 109–13
acid precipitation, effects of, 410–11
barrier islands, 112
bogs, 111
coral reefs, 112–13
estuaries, 110–11, 113
eutrophication, 450–51
fens, 111
freshwater ecosystems, 109–10
human disturbance of, 113–15
marshes, 111, 112, 113
saline ecosystems, 110, 427
shorelines, 111–13
swamps, 111
wetlands, 111, 113
aqueducts, 436–37
aquifers, 428
contamination at superfund sites, 539
depletion of, 435
groundwater pollution and, 460–61
replenishment, 111
Arabian Sea, long-range transport of air pollution to, 405
Aral Sea, 422
arbitration, 228–29
Arctic
air pollution to, long-range transport of, 406
climate change, effects of, 389
Arctic hares, response to stress shock, 137
Arctic lemmings, response to stress shock, 137
Arctic National Wildlife Refuge (ANWR) (Alaska), 344, 345, 483
Arctic Ocean
nuclear wastes, contaminated by, 491
sea ice at, 428
arctic tundra, 106
Arendt, Randall, 564
Argentina
genetically engineered crops in, 251
grasslands, 246
argon, atmospheric, 375
arguments, clues for unpacking, 9
Aristotle, 38
Arizona
desalination plant at Yuma, 453
Mt. Graham, telescope versus red squirrel debate, 48
water conservation efforts in Phoenix, 440
Arkansas, lawsuit over drunk driver checkpoints, 317
Army Corps of Engineers, U.S., 117, 118
dam construction, 438
flood control efforts on Mississippi River, 348
Arnold, Ron, 583
Arntzen, Charles, 191
arsenic, 198
as air pollutant, 401
in electronic equipment, 535
in inorganic pesticides, 264
at superfund sites, 539
as water pollutant, 454–55, 458
artesian wells, 428
asbestos, as air pollutant, 402
Asia. *See also individual countries*
AIDS in, 191
air pollution, 405
cropland, distribution, 245–46
endangered mammals, threats to, 288
family planning in, 162, 163
fertility, 149, 155
food production, gains in, 237
food shortages in, 238
human disturbance of ecosystems, 113
indoor air pollution in, 403
iron deficiency in, 239
land degradation in, 247
landless poor in, 324
land reform in, 325
malaria, return of, 195
oil consumption, 482
political economies in, 27–28

B

C

D

F

G

groundwater
arsenic in, 454
contamination at superfund sites, 539
depleting, 434–35
as freshwater reservoir, 428
pollution and, 212, 459–61
groupers, overharvesting of, 291
Group of Seven, 180
Grove, William, 510
Grover Archbold Park (Washington, D.C.), 86
Grumbine, R.E., 122, 300
Guanacaste Conservation Area (Costa Rica), 118
Guanacaste National Park (Costa Rica), 313
guano, mining, 363
Guatemala
artisans in Appleseed Fundraising cooperative, 30
Mayan Indians, 46, 47, 143
shantytowns in Guatemala City, 555
soil degradation, rangeland, 320
guinea worm, 192, 193
Gulf Stream, 426
gully erosion, 248
Gunderson, Lance, 226
Gwich'in people, 483
gypsum, 360

H

habitat, 82
destruction, 288
protection, 299–300
habitat conservation plans (HCP), 297–98
Haiti
biomass energy in, 476
colonialism, exhausted resources from, 157
degraded soil, 249
forests denuded for fuelwood, 513
human development index ranking, 175
human disturbance of ecosystems, 113
landless poor, 324
poverty, 25
reforestation projects, 314
water use, 432
Hales, Stephen, 18
half-life, 57
halite, 358, 360
Hall-Martin, Anthony, 37
halogens
as air pollutants, 401
chemical processing of, 543
halons, 407
Hanta fever, 189, 194
Harappans (Indus Valley), 431
Hardin, Garret, 157, 171
hares, population oscillations, 136–37
Harr, Jonathan, 212
Harrowsmith, 278
Hassayampa River Preserve (Arizona), 121
Haugenbeck, Karl, 301
Hawaiian Islands
air chemistry study at Mauna Loa Observatory, 387
air pollution to, long-range transport of, 405
bagasse, energy from, 515
biodiversity on, 299
feral pig, elimination, 118
litter covering beach on Niihau, 461
mongooses introduced to control rats, 98
rainfall around Mount Waialeale, 424, 425
tsunami (1960), 367
Hawken, Paul, 181, 182
hazardous, defined, 195
hazardous wastes, 536–44
bioremediation, 544
brownfields, 540–41
defined, 536–37
disposal, 537–38, 543
environmental racism and, 540–41
federal legislation, 537–38
in groundwater, 460, 528
incineration of, 543
in incinerator ash, 530
international transport of, 526
less hazardous substances, converting to, 543–44
less waste, producing, 542
management options, 542
Native Americans' storage of, 45
open dumping of, 528
permanent storage of, 544
recycling, 529, 542
reusing contaminated properties, debate over, 540–41
superfund sites, 538–41
toxic colonialism, 45
Headwaters Redwood Forest (California), 41
Healing the Wounds, 44
health, defined, 190
Health, Education and Welfare, Department of, Chattanooga (Tennessee) as dirtiest city (1975), 548
Health Care Without Harm (HCWH), 200
health hazards, 190–99
air pollution, 408–9
antibiotic and pesticide resistance, 194–95
diet, 198–99
emergent diseases, 192–94
infectious organisms, 190–91
morbidity and quality of life, 191–92
from pesticides, 270–71
physical agents, 198
public policy, establishing, 207–8
risk assessment and acceptance, 205–7
stress, 198
toxic chemicals, 195–97, 198 (*see also* toxins)
trauma, 198
heap-leach extraction, 364, 365
heart attacks, 198, 408
heart disease, 198
heat, 58
heat islands, 405
heavy metals, 452–53
heavy water, 488
Heliothis worms, 269
helium
atmospheric, 375
in nuclear reactors, 490
hemlock trees, 315
hemorrhagic fevers, 189, 192, 194
hepatitis, 449
herbicides, 448
biotechnology and, 253
Canada, use of, 264
cancer and, 271
defined, 262, 263
deformed amphibians and, 16
herbicide-filled tanker car accident in California, 267
persistence of, 201
United States, use of, 264
use in U.S., 270
water pollution by, 516
herbivores, 64, 65, 84, 92
heterotrophs, 375
hierarchy, in ecosystem management, 122
high-level nuclear waste repository, 492
High-Temperature, Gas-Cooled Reactors (HTGCR), 489–90
high-temperature solar energy, 506–7
Hill, Julia "Butterfly," 582
Himalaya Mountains (Asia), 331, 355
dam construction projects, 437
effect on precipitation, 424
monsoons, 383
historic areas, 339
Historic Roots of Our Ecological Crisis, The, 41
Hodel, Donald, 439
Hodgkin's disease, periwinkles used to treat, 285, 286
holistic science, 49
Holling, C.S., 226–27
homelessness, 554–55, 556
homeostasis, 63
Honduras, poverty in, 176
honeybees, pesticides and, 265, 266–67
Hong Kong, importer of wildlife and wildlife products, 290
Hooke, Roger, 20, 358
hormones
birth control and, 159–60
hormone-disrupting chemicals in environment, 268, 277
insect control and, 274
horse latitudes, 379
horseradish, 117
horses, logging with, 318
Houghton, Richard A., 400
housefly, biotic potential of, 130, 131
household chemicals, alternatives to, 543
house mouse, fecundity, 138
housing
cluster housing, 564–65
energy efficient, 499–500, 502
passive heat absorption, 504–5
"superblocks" use of, 563
in urban areas of developing countries, 554–55
Howard, Ebenezer, 560
Huang He River (China)
diversion of water from, 422, 432
sediment in, 249
Huascaran National Park (Ecuador), 338
Hubbard Brook Experimental Forest (New Hampshire), acid rain damage, 411
Hubbert, Stanley, 172
Hubbert Curves, 172
human capital, 169
human development, 28–31
index, 28
indigenous peoples, 30–31
overview, 28
sustainable development, 28–30
20:20 Compact for Human Development, 30, 31
human development index (HDI), 28, 175–76
human population, 142–63
birth dearth, 155–56
birthrates (*see* birthrates)
birth reduction pressures, 155
carrying capacity of earth, 133
death rates, 150
demographics (*see* demographics)
demographic transition (*see* demographic transition)
developing versus developed countries, 24–28
and development, 156
emigration, 151–52
family planning, 153, 158–60
fertility, 148–50, 154
future of, 160, 162–63
growth, 21, 23, 142–45, 147, 150, 550–51, 552
history of, 143–44
immigration, 151–52
larger populations, debate over, 147
life expectancy, 134, 150, 190

I

Interior, Department of, 483
mandate and organization, 223
internal costs, 178
internal parasites, incidence and mortality, 191–92
International Atomic Energy Agency (IAEA), 485
international development, 179–80
International Geophysical Year (1957), 387
international nongovernmental organizations (NGOs), 225, 587–88
International Rice Institute (Philippines), 251, 301
International Soil Reference and Information Centre (Netherlands), 246, 322
International Species Information System, 302
international trade, 179
international treaties and conventions, 224–25
International Union for the Conservation of Nature and Natural Resources (IUCN), 335, 338
world conservation strategy, 338–39
International Whaling Ban, 23
International Whaling Commission, 289
international wildlife preserves, 345–46
international wildlife treaties, 300
Internet, 10
interpretive science, 49
interspecific competition, 86, 88
interspecific interactions, 136–37
intraspecific competition, 86–88
intraspecific interactions, 137
intrauterine devices (IUDs), 159, 160, 161, 162
intrinsic factors, 136
intrinsic value, 40
introduced species, 97–98
Inuit people (Canada), 326, 327
PCBs in blood of, 406
pesticides in breast milk, 270
inversions, temperature, 404–5
inviolable preserves, 339
iodine
as air pollutant, 401
deficiency, 239–40
half-life, 57
ionosphere, 376
ions, 56–57
in fuel cells, 510
Iowa
diversified crop and livestock farm near Boone, 273
nitrates in rural drinking water, 460
Iran, population, 147, 148
Iraq
burning of Kuwait's oil wells, 474, 482
soil degradation, rangeland, 320
Ireland, Republic of, sewage treatment, 458
iron
dietary, 239
in earth, 355
as economic resource, 358, 359
minimills, 366
recycling, 173, 365–66
substituting new materials for old, 366
irrigation
drip, 441, 443
efficiency, 249, 432, 442, 443
land degradation by, 249
moving water for, 436–37
with recovered water, 441, 442
water use for, 432
irruptive population growth, 130–31
island biogeography, 298
isotopes, 56–57
Israel
plastic mulch used in Negev Desert, 255
reforestation projects, 311
water use, 432
wealth in, 25
Itaipu Dam (Brazil and Paraguay), 515–16
Italy
air pollution damage to Colosseum in Rome, 412
birthrate below replacement rate, 155
fertility rates, 149
Group of Seven, member of, 180
life expectancy, 151
pesticide use, 264
ivermectin, 192
ivory trade, ban on, 290, 291
Izaak Walton League, 578, 579

J

jack pine
harvesting, 316
population growth of, 136
jaguar, 281
Jamaica, population control, 157
James Bay hydropower project (Canada), 29, 437
Janzen, Dan, 313
Japan
air pollution, 397
cadmium poisoning of Jintsu River, 452
demographic transition, 156
fertility rates, 155
geothermal energy, 519
green business in, 184
Group of Seven, member of, 180
high-technology cities, 561
human development index ranking, 175
incentives for child bearing, 155
life expectancy, 151
megacity, Tokyo-Yokohama-Osaka-Kobe corridor, 550
metals, consumption of, 358–59
methane hydrate extraction, proposed, 485
Minamata Disease in, 452, 458
Moju breeder reactor accident, 490
ocean dumping of nuclear wastes, 491
per capita energy consumption, 477
photovoltaic cells, marketing of devices with, 508
recycling programs, national, 532
reforestation, 309
resource consumption in, 27
steel production, 366
trade, positions on, 570
waste-to-energy plants, 530
water quality, improvements in, 458
wealth in, 25, 26
whale hunting, support of limited, 289
wildlife and wildlife products, importer of, 290
wind technology in, 518
wood products, importer of, 308, 309
Japan, Sea of, nuclear wastes dumped in, 491
Jasper National Park (Canada), 334
Java, human disturbance of ecosystems, 113, 115
Javanese rhinos, program to save, 302
jays, Florida scrub, 300
J curve, 129, 132
jetsam, as ocean pollution, 461
jet streams, 379–80
Jiko stove, 309
Jintsu River (Japan), cadmium poisoning of, 452
jobs, and the environment, 184
Johnson, Hazel, 557
Jordan, demographic transition, 156
Joshua Tree National Park (California), 105
Joshua trees, 105
joules, 58, 476
Journal of Public Health, 25
judicial branch
administrative courts, 224
case law and the, 219–22
justice, environmental, 44–47

K

Kalahari Desert (Africa), 425
Kampuchea
deforestation, 312
per capita energy consumption, 477
Kansas
Flint Hills restoration project, 118
tornado (1999), 374
Kant, Immanuel, 38
Karakorum Mountains (Asia), precipitation and, 424
Kara Sea, nuclear wastes dumped in, 491
Karban, Richard, 88
Keillor, Garrison, 115
kelp, 78, 85
Kennebec River (Maine), removal of dam, 438
Kennedy, John F. (president, U.S.), 29
Kennet, Lord, 395
Kenya
demographic transition, 156
guinea worm, eradication programs, 193
Jiko stoves in, 309
population growth rate, 152
wildebeest crossing Mara River, 127
wildlife in Serengeti ecosystem, 345–46
Keolado National Park (India), 342
kerogen, 482
Kesterson Wildlife Refuge (California), selenium poisoning of, 345, 453
ketones, excretion of, 202
Kew Gardens (England), 301
keystone species, 85
Khian Sea, odyssey of, 526
Kids Saving the Earth, 577
Kimfumu, 189
kinetic energy, 58
King, Martin Luther, 580
King, Ynestra, 44
King's Canyon National Park (California), 19
Kirtland's warbler
nest parasitism and, 87
population growth of, 136
Kissimmee River (Florida), restoration, 118
Klamath weed, biological control of, 274
Kluane National Park (Canada), 333
KM Minnesota, 302
knowledge, approaches to, 7–8
known resources, 171
Komondor, 292
Koop, C. Everett, 200
Kootenay National Park (Canada), 334
Krakatoa volcano (Indonesia) (1883), 367
Kropotkin, Peter, 580
krypton, atmospheric, 375
kudu, harvesting, 322
kudzu vine, 292, 293
Kuna Indians (Panama), 31, 325
Kung! people (Africa), fertility control, 159
Kuwait
burning of oil wells by Iraq, 474, 482
desalination, 436
oil reserves, 482
as water-poor country, 430
water use, 432
kwashiorkor, 238–39
Kyoto Protocol (U.N.), 391

L

Labor, Department of, 223
- Black Lung Benefits Program, 479

Labrador, indigenous rights in, 326
ladybird beetles, as biological controls, 272, 273
ladybugs, as biological controls, 272, 273
laissez faire market systems, 167
Lake Apopka (Florida), alligator infertility in, 268
Lake Manyara National Park (Tanzania), 345
lakes, 109–10, 429. *See also individual lakes*
- acid precipitation damage, 410
- eutrophication, 450, 451
- sediment and, 454
- water budget and, 426
- water pollution, 410, 456–59

land degradation, 246–47
land deregulation, 320–22
land disposal, water pollution and, 464
"Land Ethic, The," 103
landfills, 528–29, 530
- groundwater pollution and, 460
- methane wells in landfills, 533
- sanitary landfills, development of, 528–29
- secure, for hazardous wastes, 544
- superfund sites and, 539
- water pollution from, 460, 527
- wetlands, dumping of "landfill" into, 346

land ownership, 324–26, 327
land reform, 325
land resources, 245–46
landscape, defined, 115
landscape dynamics, 115–16
landscape ecology, 115–16
landscape heterogeneity, 115
landslides, 369
land use
- agriculture (*see* agriculture)
- forests (*see* forests)
- rangelands (*see* rangelands)
- world land uses, 306

Langer, Charles, 510
languages, dying, 30–31
La Niña, climate and, 386
Laos, fertility rates, 149
Lassa fever, 189
Lassen National Forest (California) management plan, 230
latent heat, 378, 379
laterization, 247
Latin America. *See also* Central America; South America; *individual countries*
- aquaculture in, 236
- cropland, distribution, 245–46
- dry tropical forests, destruction of, 25
- family planning in, 162, 163
- food shortages in, 238
- indoor air pollution in, 403
- landless poor in, 324
- land reform in, 325
- pesticide exposure by farmworkers, 270, 271
- poverty in, 25
- sanitation in, 554
- urban population growth, 551
- wildlife and wildlife products, trade in, 290

Latvia
- fertility rates, 149
- population growth rates, 150

law
- administrative law, 222–24
- adversarial approaches, 222
- arbitration and mediation, 228–29
- case law, 215, 219–22
- civil law, 221–22
- court system, 220
- criminal law, 220–21
- dispute resolution and planning, 225–31
- environmental (*see* environmental law)
- international treaties and conventions, 224–25
- legal thresholds, 220
- legislative riders, 218–19
- lobbying, 219
- path of a bill from inception to law, 216–18
- regulatory agencies, 222–24
- statute law, 215, 216–19
- types of, 215

lawsuits
- events in a civil, 221
- by private property owners, 317

LD50, 203, 204
lead, 406
- as air pollutant, 401, 479, 530
- from coal burning, 479
- in electronic equipment, 535
- in incinerator ash, 530
- in inorganic pesticides, 264
- poisoning, 44–45
- recycling, 173, 365–66
- shortage, 360
- as water pollutant, 452, 453, 460

lead poisoning, 294
leafy spurge, 292, 293
League of Conservation Voters, The, 218, 589
leakage, dams and, 437
learning styles, 4
Le Corbusier, 561
Lee, Kai, 226
legal thresholds, 220
legislation
- clean air, 396, 416–17
- hazardous wastes, federal legislation, 537–38
- water pollution, 469–70

legislative branch, statutory law and the, 215, 216–19
legislative riders, 218–19
legumes, nitrogen-fixing bacteria on, 250
Leidy's comb jelly, 294
lemmings, response to stress shock, 137
L'Enfant, Pierre, 560
leopards
- in Serengeti ecosystem, 345
- smuggling of skins, 290

Leopold, Aldo, 36, 569
- on benefits of nature, 342
- on conservation, 33, 103, 310
- ecosystem management and, 121–22, 123
- on the intrinsic rights and value of all organisms, 42
- on land being treated as commodity, 235
- on land ethic, 588
- predator studies, 85
- *Sand County Almanac, A,* 103, 310
- Sand County farm (Wisconsin), 103, 121
- on saving species, 281

Lerner, Jaime, 566
less developed countries. *See* developing countries
lesser snow geese, growth of populations along Gulf Coast, 282, 283
lethal dose (LD), 203
Leucaena, 311
- as energy crop, 515

leukemias
- Chernobyl nuclear accident (Soviet Union, former) and rates of, 489
- periwinkles used to treat, 285, 286

Levi-Strauss, Claude, 77
Lewis, Martin, 33
Liberia
- deforestation, 312
- fertility rates, 149

"Liberty Link" crops, 253
Libya
- demographic transition, 156
- water use, 432

lichens, 80
- mutualism and, 89
- as pioneer species, 97

Liebig, Justus von, 79
lifeboat ethics, 157
life-cycle analysis, 576
life expectancy, 134, 150, 190, 408
life span, 150
lifestyle, as health hazard, 198
Light, Steven, 226
light bulbs, energy efficiency and, 502
light pollution, as atmospheric degradation, 402
limestone, 67
- acid precipitation damage, 412
- cave formation in, 358
- as economic resources, 359–60
- injection, 413–14
- oil and, 481

Limits to Growth, 174
Lincoln, Abraham (president, U.S.), 332
lions, in Serengeti ecosystem, 345
lipids, as bioorganic compounds, 57
liquids, 56
literacy, environmental, 571
lithium, in fusion reactions, 494, 495
litter
- collection efforts, 589
- lack of, in Curitiba (Brazil), 566
- as ocean pollution, 461, 528
- open dumping, problems with, 528
- recycling as way to reduce litter, 532–33

"little ice age," 385
Lixophaga diatraeae, 256
lizard, Coachella Valley fringe-toed, 296
llamas, forage conversion, 322
lobbying, 219
loblolly pine, harvesting, 316
locoweeds, 80, 117
lodgepole pine, 98
- harvesting, 316

loess, 249, 358
logging
- below-cost and salvage sales, 318–19
- economic benefits from, in U.S., 315–16
- harvest methods, 316–18
- "takings" litigation, 317
- in temperate forests, 297, 315–16
- in tropical forests, 313, 314

logical errors, avoiding, 9
logical learners, 4
logistic growth, 131
logistic population strategies, 133
London Dumping Convention (1990), 469, 470
longhorn beetle, mimicry in, 90
Long-Term Ecological Research (LTER), 564
Lorenz, Edward, 131
Louisiana
- Louisiana Toxics March, 46
- sediment, lack of, 454
- wetland destruction, 347

Lovelock, James, 63, 375
Lovins, Amory, 50, 181, 499
low-head hydropower, 516
low-input sustainable agriculture, 257–58
Ludd, Ned, 50
Luddites, 50
LULUs (locally unwanted land uses), 44, 46, 529
lynx, population oscillations, 136–37
Lyotard, Jean-Francois, 39
Lysteria, 449

M

MacArthur, R.H., 298
macaws, hyacinth, 290
"machine for living," 561
Maclean, Norman, 361
Madagascar
debt-for-nature swap, 314
deforestation, 312
human disturbance of ecosystems, 113
Madagascar periwinkle, 285, 286
magma, 355, 356, 357
magnesium
in earth, 355
plants and, 249
magnetic confinement, in nuclear fusion, 494, 495
MAGNOX nuclear reactor, 488
Maine, private energy purchased in, 519
maize
Bacillus thuringiensis used with, 252
cornstalks, energy from, 515
as domestic animal feed, 322
engineered sterility in, 254
herbicides applied to, 270
interplanting of, 255
as major world crop, 241
in mixed perennial polyculture, 313
transgenic varieties, 251, 253
yields, 251
Makah tribe (Washington), 23
malaria, 194–95, 449
climate change, effects of, 389
incidence and mortality, 191, 265–66
Malawi
biomass energy in, 476
fertility rates, 149
poverty in, 25
rural water program, 434
Malaysia
deforestation, 312, 313
plant species, 284
trade, positions on, 570
water quality, 459
Maldive Islands, 389
male sterilization, 159, 160, 161, 162
Mali
clean water shortages, 433
dung used as fuel, 514
mallards, genetic assimilation and, 295
malls, shopping, 558
malnourishment, 191–92, 238
Malthus, Thomas, 33
diminishing returns, theory of, 167, 172, 174
on population, 144–45, 157, 167
Malthusian population strategies, 130, 132
Man and Biosphere (MAB) program, 341
Man and Nature (Marsh), 18
manatees, thermal pollution and, 456
Manchurian cranes, 119
Mandela, Nelson, 580
manganese, 406
as economic resource, 358, 359
MMT gasoline additive, 225
as strategic mineral, 360
mangosteens, 285
mangrove forests, aquaculture damaging, 236
mantle of the earth, 355
manufactured capital, 169
manure, 250, 257, 273, 457–58
commercial energy derived from, 476, 477
maps, concept, 9, 10–12
maquiladoras, 45
marasmus, 239
marble, 358
acid precipitation damage, 412
marbled murrelet, 315
Marburg fever, 189, 192
marginal costs, 167
Marine Protection Research and Sanctuaries Act (1972), 469
market equilibrium, 167
Marsh, George Perkins, 18
marshes, 111, 112, 113, 429
artificial, for sewage treatment, 121
preservation of, 346–48
Martinez, Micaela, 53
Martinique, nuèes ardentes (1902), 368
Marx, Karl, 144–45, 168, 574
Maryland
air pollution in Baltimore, 418
Columbia, design and building of, 560
Gwynns Falls watershed, 116
Masa Mara National Reserve (Kenya), 345
Masri Mara National Park (Kenya), 127
Massachusetts, contaminated well water in Woburn, 212
mass burn incinerators, 530, 531
mass extinctions, 287–88
Matamek Ecological Reserve (proposed) (Canada), 337
material cycles, 67–74
carbon cycles, 67, 68
nitrogen cycle, 67–70
phosphorus cycle, 71–72
sulfur cycle, 72–74
Mather, Stephen, 19
matter, 56–57
conservation of matter, 58
Mauritania
fuelwood demand above sustainable yield, 513
shantytowns in Nouakchott, 555
Mauritius, 18
Mayan Indians (Central America), 30, 46, 47, 143
Mayflower, 46
McArthur, Robert, 93
McClellan Air Force Base (California), 140
McDonough, William, 181, 182–83
McHarg, Ian, 564
McLuhan, Marshall, 20
Mead, Lake (Nevada), 437
Mead, Margaret, 330, 578
Meadows, Donella, 174
Meadows group, 174
meadow spittlebug, 88
measurement, English/metric conversions, 594
meat, in human diets, 242
mechanical cultivation, 273
mechanical weathering, 357
media, influencing public opinion through, 579
mediation, 228–29
medical supplies, PVCs in, 200
medicines, as benefit of biodiversity, 284–86
Mediterranean climates, 108
biodiversity and, 285
Mediterranean Sea
dead dolphins in, 196
eutrophication, 450
exotic species introduction, 294
megacity, defined, 549–50
megadiversity countries, 31
"megarat" experiments, 204
meltdown, 487
Menominee Indian Nation (Michigan, Wisconsin), sustainable forestry, 310
Merced River (California), 15
Merchant, Carolyn, 44
mercury
as air pollutant, 401, 402
from coal burning, 479
in electronic equipment, 535
in inorganic pesticides, 264
as water pollutant, 452, 457, 458
Mesa Verde National Park (Colorado), 332
mesocyclone, 381
mesolimnion, 110
Mesopotamia, city planning in, 558
mesosphere, 376
metabolic degradation, 202
metabolism, 57
metals
as air pollutants, 401
chemical processing of toxic, 543
heavy, 452–53
as resources, 358–59, 360
strategic metals, 360
substituting new materials for old, 366
as water pollutant, 452–53
metamorphic rocks, 358
methane, 485
atmospheric, 375, 397
emissions, attempts to cut, 391
energy, used for generating, 533
as fuel, 513–14
fuel cells and, 510
global warming and, 387
from natural wetlands and bacteria, 402
methane hydrate, 389, 485
methanesulfonic acid, 204
methanol
from biomass, 514–15
fuel cells and, 510
methotrexate, 159
methyl chloroform, 407
methyl parathion, 277
metric/English conversions, 594
Mexico
air pollution in Mexico City, 418, 553
artisans in Appleseed fundraising cooperative, 30
colonias, 45, 565
debt-for-nature swap, 314
deserts, 425
fertility rates, 149
land degradation, 247, 322
land reform in, 325
maquiladoras, 45
monarch butterflies overwintering in, 304, 305–6
parks, diversity in, 333
pesticide exposure, study of, 271
photochemical smog in Mexico City, 553
population, 151, 157, 550, 551, 552
scavenging in Mexico City, 535
sewage treatment in Mexico City, 465
shantytowns in Mexico City, 555
shrimp aquaculture, 236
subsidence in Mexico City, 435
waste disposal problems in Mexico City, 527
Mexico, Gulf of, eutrophication, 451
Meyers, Norman, 284
mice, Hanta fever spread by deer mice, 194
Michigan, air pollution in Muskegon, 418
microbial agents, as pesticides, 265
micro-hydro generators, 516
microlending programs, 180
Mid-Course Correction: Toward a Sustainable Enterprise, 182
Middle East
oil reserves, 482
water stress, 433
mifegyne, 159
mifepristone, 159
migration, 340
monarch butterflies, 305
Milankovich, Milutin, 385
Milankovich cycles, 385

milk
growth-hormone-treated, 253
in human diets, 242
Mill, John Stuart, 30, 38, 167–68, 169
Mille Lacs Refuge (Minnesota), 344
millet, 241
milpa agriculture, 312–13
mimicry, 90
Minamata Disease, 452
Minar, Dave, 258
Mineral Policy Center (Washington, D.C.), 362
minerals
conservation of, 364–66
defined, 356
extraction, environmental effects of, 360, 362–64
metal resources, 358–59, 360
nonmetal mineral resources, 359–60
as nonrenewable resources, 169
recoverability of, 171
recycling, 365–66
strategic, 360
substituting new materials for old, 366
water pollution and, 452–53
Minimata Bay (Japan), mercury poisoning, 452, 458
minimills, iron and steel, 366
minimum tillage system, 257
minimum viable populations, 298–99
mining, 360, 362–64. *See also* economic mineralogy
coal, 362, 364, 479, 481
copper, 362
energy usage, 477, 478
environmental effects, 360, 362–64
national parks and, 334
open-pit, 362
pollution from, 361, 362, 364
processing, environmental effects of, 364
on public lands, controversy over, 358, 361
radioactive mine waste, disposal of, 491
reclamation, 364
shale oil, 482, 484
strip mining, 362, 364
superfund sites, 539
"takings" litigation, 317
water, 432
water pollution and, 452–53
Ministry of Natural Resources (Canada), 101
minke whales, 289
Minnesota
Boundary Waters Canoe Area Wilderness, mediation of dispute at, 228
environmentalism as religion, court case over, 581
hazardous waste reduction by 3M Company, 542
peat bogs, as source of energy, 515
recycling programs, 531
subsidies for conventional versus low-input, study of, 258
sustainable farm of Minar family, 257, 258
tornado in Boundary Waters Canoe Area (1999), 382
windmills on Buffalo Ridge, 498, 499
Wood Lake Nature Center, 394
Minnesota Environmental Education Association, 473
Minnesota Zoo
International Species Information System, 302
save Javanese rhinos in the wild, program to, 302
minorities
environmental health risks and, 44–47
racism, environmental, 44–45, 539–41
misoprostol, 159
Mississippi, brownfield remediation in Columbia, 540–41
Mississippi River (United States)
dead zone at mouth of, 451
flood control efforts, 348
flooding, 347–48
Missouri
earthquake at New Madrid (1812), 367
Pruitt-Igoe project in St. Louis, 556
Missouri River (United States), flooding of, 347–48
Misty Fjords National Monument (Alaska), mining at, 334
Mitchell, Mount (North Carolina), acid rain damage, 411
mites, 244
mitigation, 117
mixed perennial polyculture, 313
MMT (manganese-based gasoline additive), 225
modernism, 38–39
Moffat, Tom, 394
Moffatt, Andi, 329
moldboard plow, 257
mold spores, as indoor air pollution, 404
molecules, 57
moles, in soil, 244
Molina, Mario, 407
mollusks, decline in, 242
molybdenum-103, 487
monarch butterflies, overwintering in Mexico, 304
Monarch Butterfly Sanctuary Foundation, 305
Mond, Ludwig, 510
mongooses, 98
monitoring, in ecosystem management, 122
Monkey Wrench Gang, The, 582
monkey wrenching, 582
monoculture
agriculture, 248, 272, 273
forestry, 310–11
Mono Lake (California), 437
Monserrat, island biogeographical effects and, 298
monsoons, 382–83
Montana, cloud seeding debate with North Dakota, 384
Montreal Protocol (1987), 225
monuments, acid rain damage, 412
moral agents, 39
moral extensionism, 39
morals, 37, 39
moral subjects, 39
morbidity, 190, 191–92
Morocco, population control, 157
Morris, David, 499
Morrisette, Bill, 300
mortality, 134
infant, 157
rates, 150
mosquitoes, 265–66
biological control of, 274
dengue fever spread by, 194
malaria-causing mosquitoes resistant to insecticides, 195
mosses, as pioneer species, 97
moths, competition and, 88
mountain gorillas, 299
mountains. *See also individual mountains*
earthquakes and formation of, 355–56
precipitation and, 424, 425
Mountain States Legal Foundation, 583
mouse, fecundity, 138
Mozambique
colonialism, exhausted resources from, 157
food production in, 237
foreign debt, accumulated, 180
poverty in, 25
water, time spent fetching, 433
Mt. Graham red squirrel, protection of, 48
mudslides, 368
Mueller, Fritz, 90
Muellerian mimicry, 90
Muir, John, 18, 19, 42, 333, 439, 571
mulch, use of, 255
mules, logging with, 318
Müller, Paul, 263
Multilateral Agreement on Investments (MAI), 225
Mumford, Lewis, 50, 560
murrelets, marbled, 315
musk oxen, 483
mussels, orange-footed pimple-back, 296
mutagens, 196–97, 204
mutation, 287
mutualism, 89
mycorrhizae, 85, 412
My First Summer in the Sierra, 571
Myxobolus cerebralis, 294

N

Nabhan, Gary, 250
Naess, Arne, 580, 581
Namib Desert (Africa), 425
Narmada River (India), dam construction project, 437
Nassar, Lake (Egypt), 516
natality, 134
National Academy of Science, 205, 323
Indonesian fruits study, 284
National Aeronautics and Space Administration (NASA), fuel cells, development of, 510
National Air Toxics Program, 402
National Cattlemen's Association, 583
National Elk Refuge (Wyoming), 343
National Environmental Education Act (1990), 571
National Environmental Education Advancement Project (Wisconsin), 17
National Environmental Policy Act (NEPA) (1970), 214–15
National Farm Bureau, 583
National Flood Insurance Program, 348
national forests, logging in, 318–19
National Institute for Occupational Safety and Health (NIOSH), 223
National Marine Fisheries Service (NMFS), 372
National Museum of Natural History, 130
National Packaging Protocol (NPP) (Canada), 536
national parks. *See* parks; *individual parks*
National Park Service, 223
ecosystem management, 122
establishment, 19
founding, 332
grasslands project, sustainable, 581
National People of Color Environmental Leadership Summit (1991), 46
National Pollution Discharge Elimination System (NPDES), 456
National Priority List (NPL), 538–39
National Religious Partnership for the Environment, 41
National Science Foundation, 564
National Space Research Institute (INPE) (Brazil), 311
National Wildlife Federation, 578
Native Americans
approached with reservation dumping schemes, 529
environmental racism and, 540
forced privatization of reservations, 172
hazardous waste disposal on reservations, 45
interplanting of two different crops by, 255
life expectancy, 151
native peoples, 30–31
"natural," product claims of being, 575
natural capital, 169
natural context, from environmental education, 573

O

P

palladium
in electronic equipment, 535
as strategic mineral, 360
Pamlico Sound (North Carolina)
"dead zone" in, 447
hypoxic zone in, 451
Pfisteria in, 451
Pan, 46
Panama, San Blas Islands, 15
pandas, 82, 83
Pangaea, 356, 357
Pangkalanbuun Conservation Reserve (Indonesia), 396
panther, Florida, protection and recovery, 296
papayas, 312
Papua New Guinea
indigenous peoples in, 31, 326
old-growth forests, 308
parabolic mirrors, for solar energy collection, 506, 507
Paracelsus, 203
paradise, 46, 47
Paramillo National Park (Colombia), 338
Parana River (South America), damming of, 516
parasites, 84
parasitic diseases, 190–92, 194–95
parasitism, 89, 137
parataxonomists, 286
parks, 332–42. *See also individual parks*
fire policy in national parks, 319
historical overview, 332–33
new, 337
North American, 333–37
problems, 333–34
world, 338–42
particulate materials
as air pollutants, 401
removal of, 413, 418
Partners for Wildlife, 347
Pasig River (Philippines), 459
passenger pigeons, 170, 288–89
Passeo Pantera (Central America), 341
passive heat absorption, 504–5
passive solar heat, 504–5
pastoralism, sustainable, 319–20
pasture, defined, 319
patchiness in spatial distribution, 94–95
pathogenic organisms, 190–91
pathogens, 84
as biological controls, 272
as water pollution, 449, 450, 457–58, 460–61
PCBs (polychlorinated biphenyls)
as air pollutant, 402
banning, 463
biodiversity and, 294
bioremediation of, 544
in blood of Inuit people, 406
contamination of Lake Labarge (Canada), 55
in Great Lakes, 448
at superfund sites, 539
in water pollution, 458
wildlife mortality and, 196
peat, 107
energy from, 476, 477, 515
Pelacano, odyssey of, 526
Pelee, Mount, volcano (Martinique) (1902), 368
pelicans, brown, DDT and, 262
Pelto, Mauri S., 389
penguins, 94
Pennsylvania
air pollution in Philadelphia, 418
Chatham Village, design and building of, 560
coal mining fatalities, 479
Farmview conservation subdivision, 564
nuclear accident at Three Mile Island, 485, 489
People for Community Recovery (PCR) group in Lake Calumet (Illinois), 557
People for the West, 583
peppered moths, 81
peregrine falcons
DDT and, 262
protection and recovery, 297
permanent retrievable storage of hazardous wastes, 544
Permian period, mass extinction at end of, 287, 288
peroxyacetyl nitrate (PAN), 402
persistant organic pollutants (POPs), 270
persistence
of chemical compounds, 194
toxin, 201–2, 269–70
persistence, in critical thinking, 8
personal energy efficiency, 502, 504
Peru
cotton yields and insecticide usage in Canete Valley, 269
debt-for-nature swap, 314
deserts, 425
land reform in, 325
population control, 157
shantytowns in Lima, 555
squatter settlements, 565
urban areas, government policies benefiting, 553
Pest Control Products Act and Regulations (Canada), 275
pesticides, 262–78
alternatives, 271–75
behavioral changes by farmers, 272
benefits of, 265–66
biological controls, 256, 264, 265, 272–74
carbamates, 265
chlorinated hydrocarbons, 264–65
crop protection, 266
defined, 262–63
deformed amphibians and, 16
detection limits, 205
disease control from, 265–66
endogenous insecticides, 251–52
exposure, reducing, 275–78
fumigants, 264
fungicides, 263, 264
herbicides (*see* herbicides)
historical overview, 263–64
human health problems, 270–71
inorganic, 264
insecticides (*see* insecticides)
integrated pest management, 274
microbial agents, 265
national plan to reduce usage, in Netherlands, 231
natural, in diets, 198
new pests, creation of, 269
nontarget species, effects on, 266–77
organic pesticides, natural, 264
organophosphates, 265
persistant organic pollutants, 270
personal plan for reducing pesticide exposure, 277–78
problems, 19, 20, 266–71
public policy, establishing, 207–8
regulating, 275–77
resistance to, 194–95, 267, 269
synthetic chemical pesticides, 263–64
types, 264–65
uses, 264
as water pollutant, 453–54, 460
pesticide treadmill, 267, 269
pests
behavioral changes in agriculture as means of control, 272
defined, 262
endogenous insecticides, 251–52
new pests, creation of, 269
pest control, as threat to biodiversity, 291–92
resurgence, pest, 267, 269
wild species, pest suppression by, 286
Peto, Richard, 277
PET (polyethylene terephthalate), 531–32, 533
petroleum
Arctic National Wildlife Refuge, proposal to explore in, 483
characteristics, 481
recovery process for, 481
from tar sands, 482
U.S. production, 172
pets, waste materials as water pollution, 458
Pfisteria, 447, 451–52
pH, 410
pheasants, as boundary zone inhabitants, 95
Pheidole megacephala, 256
phenols
in air pollution, 402
as organic pesticide, 264
at superfund sites, 539
Phifer, Arnold, 280
Philippines
human disturbance of ecosystems, 115
indigenous peoples, 31, 325, 326
Pinatubo, Mt., volcano (1991), 368, 385, 388
population growth, 551
reforestation projects, 311
rice terraces at Chico River Valley, 254
scavenging in Manila, 535
shantytowns in Manila, 555
Smoky Mountain dump in Manila, 527–28
water quality, 459
phocomelia, 197
phosphates
mining, 363
removal of, in sewage treatment systems, 467
in water pollution, 250, 458
phosphogypsum, 529
phosphoric acid, in fuel cells, 511
phosphorus
plants and, 249
water pollution, 457, 465
phosphorus cycle, 71–72
photochemical oxidants, 402, 403
photochemical smog, 404, 553
photodegradable plastics, 536
photosynthesis, 59–60
atmosphere, in formation of the, 375
biomass production and, 91, 92, 511–12
in carbon cycle, 67, 68
energy capture in, 60–61, 511
photovoltaic cells, 507–9
photovoltaic solar energy, 507–9
pH scale, 70, 71
phthalates, 200
physical agents as health hazards, 198
physical treatment of hazardous wastes, 543
physiological modifications, 80
phytoplankton, in biomagnification process, 200
Pierskalla, Kirstin, 473
pigeon, extermination of American passenger, 288–89
pigs, as introduced species, 98
Pimentel, David, 145
Pinatubo, Mt., volcano (Philippines) (1991), 368
Pinchot, Gifford, 18, 19, 38
Pindos National Park (Greece), 338
pineapples, mulch used with, 255
pine trees
bristlecone, lifespan of, 134
harvesting, 316, 317, 318
jack pines, population growth of, 136

pioneer species, 96
placer mining, 360, 362
plankton
eutrophication, 450
predation by, 84–85
in standing water ecosystems, 110
planning
community-based planning, collaborative approaches to, 229–30
dispute resolution and, 225–31
plant nutrients, as water pollution, 449, 450–51, 457–58
plants. *See also individual species*
air pollution and plant pathology, 409–10
biomass production (*see* biomass)
biotechnology and (*see* biotechnology)
botanical gardens, 301
climate change and, 389
clustering in, 94
competition and, 86–89
drugs and medicines from, 284–86
elements required by, 249
endangered, 295, 300
environmental stress in, 198
eutrophication, 450–51
exotic species, introduction of, 292–94
extinction (*see* extinction)
forage conversion by domestic animals, 322
in hydrologic cycle, 423
as organic pesticides, 264, 265
overharvesting, 290
perennial species, managing topography with, 255
photosynthesis (*see* photosynthesis)
rangeland, 320, 321
seed dispersal mechanisms, 86–87
as source of volatile organic compounds, 401–2
species, numbers of, 283–84, 285
threatened species, 295, 300, 320, 321
toxins in, 197–98
transgenic plants, 251–53
water conservation and, 440
wild, edible, 322–23
plasmids, 195
Plasmodium protozoans, 265
plastics
biodegradable, 536
photodegradable, 536
recycling, 531–32
as replacement for metals, 366
waste reduction and, 535–36
plate boundaries, 355–56
platinum
mining, 360
recycling, 365
shortage, 360
as strategic mineral, 360
Plato, 18, 38
Plumas National Forest (California) management plan, 230
plutonium
as air pollutant, 401
in breeder reactors, 490, 491
half-life, 57
plutonium-239, 491
poaching, 345–46
pocket veto, 218
point sources, of water pollution, 448
poison arrow frogs, 90
Poivre, Pierre, 18
Poland
acid rain damage, 411
air pollution in, 178, 188, 408, 419
air pollution-related disease in, 188
cost/benefit of removing air pollution in Tarnobrzeg, 178
sewage treatment, 458
polar bears, in Arctic National Wildlife Refuge, 483
policy
defined, 213
environmental (*see* environmental policy)
public policy, establishing, 207–8
policy cycle, 214–15
polio, 449
eradication efforts, 191
political decision making, 213–14
political economies, 27–28, 168
politics
green, 588–89
as power, 213
pollutants, detection limits, 205
pollution, 21–22, 24, 29
air (*see* air pollution)
defined, 397
historical view of, 19–20
from mining, 361, 362, 364
persistant organic pollutants (pesticides), 270
pollution-control technology, market for, 573
public opinion and, 585–86
public policy, establishing, 207–8
recycling and, 542
rights, marketing, 417
as threat to biodiversity, 294, 295
toxic metal, 345
water (*see* water pollution)
Polo, Marco, 475
polyculture, farming, 272, 273
polycyclic aromatic compounds, 512
polymers, as replacement for steel, 366
polyvinyl chloride (PVCs), 200
ponderosa pine, harvesting, 318
ponds, 109–10, 429
acid precipitation damage, 410
eutrophication, 450, 451
water pollution, 410, 456–59
Population and Community Development Association (PCDA) (Thailand), 153
population dynamics, 128–38
age structure, 135
biotic potential, 129–30, 131
carrying capacity, 133
catastrophic, 132
chaotic systems, 131–32
crowding, 137–38
density-dependent factors, 136–38
density-independent factors, 136
doubling times, 129, 130, 143
emigration, 136
exponential growth, 129, 130
fecundity, 134
fertility, 134
immigration, 134
interspecific interactions, 136–37
intraspecific interactions, 137
irruptive growth, 130–31
logistic strategies, 133
Malthusian population strategies, 130, 132
mortality, 134
natality, 134
oscillations, 130–31
stable population, growth to a, 131, 132
strategies of population growth, 132–34
stress, 137–38
survivorship, 134–35
population momentum, 135, 151
populations
defined, 62
human (*see* human population)
Ports and Waterways Safety Act (1972), 469
Portugal, human disturbance of wetlands, 113
positive crankcase ventilation (PCV), 415
postmaterialist values, 585, 586
postmodernism, 38–39
poststructuralism, 38–39
potash, 360
potassium
in earth, 355
plants and, 249
potatoes, 241
Bacillus thuringiensis used with, 252
transgenic varieties, opposition to, 253
potential energy, 58
poverty
absolute poverty, World Bank study of, 324, 584
acute, 24, 584
developing versus developed countries, 24–28
effects of, 24–28
environment, connection between poverty and the, 584–85
food security and, 238
population growth and, 144–45
reduction, 28, 176
Powell, Lake, 437
power
defined, 476
politics as, 213
prairie dogs, control of, 291
prairies, 105–6
restoration projects, 118, 119
praying mantis, as biological control, 272, 273
precautionary principle, 227–28
precedent, 220
precipitation. *See also* rainfall
biomes and, 103
changes in, this century, 388
contamination of Lake Labarge (Canada) through, 55
countries, water-rich and water-poor, 430–31
drought cycles, 431
El Niño, 386
frontal weather and, 380–81
in hydrologic cycle, 423–24
infiltration of, 428
mean annual, map of, 425
water budget and, 426
weather modification, attempts at, 384, 436
predation, 83–85, 86–87, 136–37
predators
clustering behavior, 94
control, 291–92
diseases as, 294
premises, 9
preservation or restoration, debate over, 117–18
President's Council on Sustainable Development, 182
President's Environmental and Conservation Challenge Award, 557
pressurized water reactors, 487–88
price elasticity, 167
primary air pollutants, 398
primary consumer organisms, 65
primary productivity, 62, 91
primary sewage treatment, 466–67
primary standards for air quality, 416
primary succession, 96, 97
Prince William Sound (Alaska), 462, 482
Principles of Political Economy, 168
pristine research areas, 339
private property
and endangered species, 297–98
takings, controversy over, 317
Process-Inherent Ultimate-Safety (PIUS) reactor, 489–90
producer organisms, 64, 65

production, as ecological resource, 170
productivity, 62, 91, 92
progesterone analogs, 159, 160
Promethean environmentalism, 33
pronatalist pressures, 143, 153–54
proteins, as bioorganic compounds, 57
proton exchange membrane (PEM), 510–11
protozoans, 244
proven reserves, 171
providing, as ecological resource, 170
pseudoscorpions, 244
public activism, 215–16
public housing, 556–57
Public Interest Research Groups (PIRGs), 578
public opinions, environmental protection and, 585
public policy, establishing, 207–8
Public Utilities Regulatory Policies Act (1978), 516, 519
puffins, 54
pull factors, 551, 553
pumpkins, interplanting of, 255
purple loosestrife, 292–93
push factors, 551–52
pyramids, ecological, 64–67
pyrethrum, as organic pesticide, 264, 265
pythons, albino, 290

Q

quality of life, 25, 26, 191–92
quartzite, 358
Quinault Rain Forest (Washington), 315
Quincy Library Plan (California), 230
Quinte, Bay of (Canada), 457

R

race
 racial make-up and agendas of green organizations, 45–46
 and socioeconomic status, 28
racism, environmental, 44–45, 539–41
radiation, as health hazard, 198
radical environmental groups, 582–83
radioactive isotopes, 57
radioactive materials, physical treatment of, 543
radioactive wastes. *See* nuclear wastes
radioactivity
 from coal burning, 479
 discovery, 507
radon, as indoor air pollution, 404
Rainbow Warrior, 582
rainfall. *See also* precipitation
 cloud seeding, attempts at, 384, 436
 drought cycles, 431
 rainfall and topography, 424–25
rainforests
 biomass accumulation and, 91, 92
 species in, 284
 temperate, 107, 113, 114, 315
 tropical rainforests, 22, 24, 109, 113, 114
Rainier National Park, Mount (Washington), 330, 332
rain shadow, 424, 425
rangelands, 319–26
 forage conversion by domestic livestock, 322
 grazing fees, 323–24
 land deregulation, 320–22
 management, 320
 overgrazing, 320–22, 323
 rotational grazing, 324
 in U.S., 323–24
 wild animals, harvesting, 322–23
 world land use, as percentage of, 306
Raphael, 38
raprenox, 415
rational choice, 213–14, 225–26
rats, as introduced species, 98
RBMK nuclear reactor, 488
Reagan, Ronald (president, U.S.), 177, 222–23
Reality (female condom), 159
rebound, pest, 267
recharge zones, 428
reclamation, defined, 117
recoverable resources, 171
recovery plans for endangered species, 296–97
re-creation, defined, 117
recreation areas, 339
"recyclable," product claims of being, 575
recycling, 173, 184, 536
 benefits, 532–33
 of "button" batteries, 531
 defined, 531
 in ecological economics, 169
 hazardous wastes, 529, 542
 incentives for, creating, 533
 metals, 365–66
 minerals, 365–66
 plastics, 531–32
 program in Curitiba (Brazil), 566
 solid waste, 531–33
 water, 441, 442
Recycling Association of Minnesota, 473
red oak, harvesting, 317
red snapper, overharvesting, 291
red tide, 451
reduced tillage systems, 257
redwood, 315
Redwood National Park (California), 19
redwood trees
 existence value, 287
 harvesting, 318
reefs
 coral (*see* coral reefs)
 pollution of, 461–62
reflective thinking, 7
reformer, in fuel cells, 510
refuse-derived fuel incinerators, 530
regeneration, forest, 306, 309–11, 318
regenerative agriculture, 273. *See also* sustainable agriculture
regulation, as ecological resource, 170
regulatory agencies, 222–24
rehabilitation, defined, 116–17
Reilly, William K., 571
relative humidity, 423
relativism, 38, 39
relevance, in critical thinking, 8
Remaking Society: Pathways to a Green Future, 571
remediation
 defined, 117
 water, 468–69
renewable resources, 170, 511–12
renewables portfolio standards, 507
renewable water supplies, 429–30
renewal, 93
Repeto, Robert, 175
reptiles, hormone disturbances in, 268
Reserva de la Biosfera Mariposa Monarca (Mexico), 305
reserves, proven, 171
reservoirs, 436–37, 439
 eutrophication, 450–51
 sediment and, 454, 516
 water losses, 516
residence time, 427
resiliency
 in biological communities, 92–94
 in institutions and ecosystems, 226–27
 of nature, 46
resistance, pesticide, 194–95, 267, 269
Resource Conservation and Recovery Act (RCRA) (1976), 469, 537, 541
resource partitioning, 83, 84
resources
 categories of economic, 170–71
 communal property resources, 171–72
 conservation, debate over, 174–75
 defined, 169
 depletion, economic models of, 174, 175
 environmental carrying capacity, increasing, 173–74
 mineral (*see* economic mineralogy)
 types, 169–70
respiration, cellular, 61
respiratory disease, coal mining and, 479
restoration ecology, 116–21
 artificial ecosystems, creating, 121
 authenticity, 119–21
 goals, debate over, 117–18, 121
 letting nature heal itself, 118–19
 reclamation, defined, 117
 rehabilitation, defined, 116–17
 remediation, defined, 117
 restoration, defined, 116
 restoration or preservation, debate over, 117–18
 tools of, 118
Restoring the Earth (California, 1998), 116, 117
retrievable storage of hazardous wastes, 544
reuse
 of hazardous wastes, 535
 "reusable," product claims of being, 575
 waste reduction through, 535
reverse osmosis, 436
Rhine River (Europe), water quality, 458–59, 572
rhinoceros
 black, commercial value of horns, 290
 Javanese rhinos, program to save, 302
 poaching, 345, 346
rhyolite, 357
ribonucleic acid (RNA) viruses, 189
Ricardo, David, 167
rice
 continuous farming of, 254, 255
 as major world crop, 241
 "miracle," 251
 in mixed perennial polyculture, 313
 snow geese habitat and, 282
ricin, 196, 203
rickettsia, 266
riders, 218–19
ridge-tilling, 257, 273
Rift Valley fever, 189
right whales, hunting of, 289
rill erosion, 248
"ring of fire," 356
risk, defined, 205
risk assessment and acceptance, 205–7
river blindness, 192, 266
River Runs Through It, A, 361
rivers. *See* streams
Rivers and Harbors Act (1899), 215
RNA (ribonucleic acid) viruses, 189
Robert, Dr., 181
Robin, Vicky, 574
robins
 killed by Azodrin insecticide, 267
 territoriality and, 89
Robinson, Frances, 4
Rock Creek Park (Washington, D.C.), 86
rock cycle, 356, 357
rocks
 defined, 357
 igneous rocks, 357–58

shoals, pollution of, 461–62
shopping malls, 558, 562–63
redesigning, 562
shorelines, 111–13
short-leaf pine, harvesting, 317
Should Trees Have Standing?, 40
shrikes, DDT and, 262
shrimp
aquaculture, 236
overfishing, 291
shrubland, biomass accumulation, 91, 92
Sian Ka'an Reserve (Mexico), as MAB reserve, 341
"sick building" syndrome, 196, 404
Sierra Club, 336, 578, 579
carbon dioxide emissions estimates for automobiles, 501
Sierra Club *versus* Disney Corporation (1969), 40, 220
Sierra Leone
deforestation, 312
human development index, 28
poverty in, 25
Sierra Nevada Mountains (California)
acid precipitation damage, 410
effect on precipitation, 424
Silent Spring, 19, 20, 215
silicate minerals, as economic resources, 359–60
silicon
amorphous silicon collectors, development of, 508
in earth, 355
silt, 358
siltation, dams and, 437
silver
recycling, 365
shortage, 360
silver iodide, 384
Simon, Julian, 147, 173
Simple Abundance, 574
Simplify Your Life, 574
Singapore, Republic of
child bearing, incentives for, 155
fertility rates, 155
as water-poor country, 430–31
wealth in, 25, 27
Singing Wilderness, The, 571
sinkholes, 435
Sitka spruce, 315
skepticism, in critical thinking, 8
slag, 413
slash and burn agriculture, 312–13
slate, 358
sleeping sickness, 266
SLOSS debate, 339
Slovakia, air pollution in, 408, 419
Slovenia, fertility rates, 149
sludge disposal, 467
slugs, 244
slums, 554–55
Small Is Beautiful, 50
smallpox, eradication of, 191
smells, as atmospheric degradation, 402
smelting, 364, 409
energy usage, 477, 478
Smith, Adam, 166, 324
Smith, Robert Angus, 410
Smithsonian Institute, 339
National Museum of Natural History, 130
smog
Arctic, accumulation over, 406
California pollution control efforts, 417
in developing countries, 553
eastern U.S. in summer, 412
legislation, 416
photochemical, 404, 553
temperature inversions and, 404–5
Smokey the Bear, 319
smoking
illness and death from, 408–9
and indoor air pollution, 403, 404
snail darter, 297
snails, 244
Snake River (United States), possible dam removal, 438
snow geese, 282, 283, 483
snowshoe hares, population oscillations, 136–37
snowy egrets, 295
social capital, 169
social considerations, in environmental policies, 213
social context, from environmental education, 573
social ecology, 580–81, 582
social justice view of demographic transition, 157, 158
sockeye salmon, 128
Socrates, 37, 38
sodium
in breeder reactors, 490–91
in earth, 355
sodium bicarbonate, 414
sodium chloride, 453
soil
as an ecosystem, 243
composition, 243
conservation, 253–57, 464
as economic resource, 359–60
erosion (*see* erosion)
infiltration, 428
land degradation, 246–47
land resources, 245–46
organisms in, 243–44
profiles, 244–45
rangeland degradation, 320–21, 323
as renewable resource, 242–45
topsoil, 243, 244–45
types, 243, 245–46
use and abuse of, 249
Soil Conservation Service, U.S., 292
soil fungus, 244
solanine, 198
solar cookers, 507
solar energy, 59–60, 500, 504–9
active solar heat, 505–6
atmospheric heating by, 376–78
cars powered by, 508
commercial energy derived from, 476, 477
cookers, solar, 507
earth heated by, 59–60
as essential to life, 59–60
high-temperature, 506–7
in hydrologic cycle, 423–24
passive solar heat, 504–5
photochemical oxidants and, 402, 403
photovoltaic, 507–9
reflection of, 377
as renewable resource, 170, 504
switching to, 413
unequal striking of earth, 378
world market for, 509
Solar Energy Research Institute (Colorado), 515
solid oxide fuel cells, 511
solids, 56
solid waste, 515
solubility of toxins, 199–200
solvents, disposal of, 528
Somalia
food production, 237
hunger, 240
land degradation in, 246
soil degradation, rangeland, 320
songbirds, declining populations of, 86–87
Sonoran Desert (Mexico), 425
Sonoran panicgrass, 250
sophism, 38
sorghum, 241
South Africa
desalination, 436
human needs versus wildlife protection, 37
metals, production of, 359
South America. *See also individual countries*
cholera, 194
cropland, distribution, 245–46
dry tropical forests, destruction of, 25
erosion rates in, 248
family planning in, 162
mercury poisoning, 452
metals, production of, 359
old-growth forests in, 307
swidden agriculture in, 255
water quality, 459
water-rich and water-poor countries, 430
South Carolina, "takings" case (1994), 317
South Coast Air Quality Management District (California), 417
Southeast Chicago Development Commission (SCDC), 234
Southern Baptist Convention, 287
Southern Oscillations, 128, 386
South Korea
ecosystems in DMZ, reestablishing, 118–19
energy conservation programs, 503
land reform in, 325
public opinion and environmental protection, 585
reforestation campaign, 309
South Pole, ozone depletion over, 376, 406–7
Soviet Union (former)
acid rain damage, 411
air pollution in, 418
Chernobyl nuclear accident, 458, 488–89
coal reserves, 479
food production in, 237
land reform in, 325
metals, production of, 359
natural gas reserves, 484
new towns, 560
nuclear reactors, 488–89
ocean dumping of nuclear wastes, 491
population, 147, 148
sewage treatment, 458
wood products, producer of, 308
sow bugs, 244
soybean farming, snow geese habitat and, 282
soybeans
air pollution and, 409
genetically engineered, 251
herbicides applied to, 270
transgenic plants, 251, 253
Spain
birthrate below replacement rate, 155
fertility rates, 150
population growth rates, 150
sewage treatment, 458
species
abundance, 91–92
adaptation, 80–82
biological communities (*see* biological communities)
in Biosphere 2, losses, 166
captive breeding plans, 301–2
competition, 86–89
complexity, 92, 93
connectedness, 92, 93
critical factors, 79–80
defensive mechanisms, 90, 91
defined, 61, 283
diversity, 91–92, 283
ecological niche, 82–83, 84

T

U

V

W